U0592116

魏江春
科学论文选集

魏江春　主编

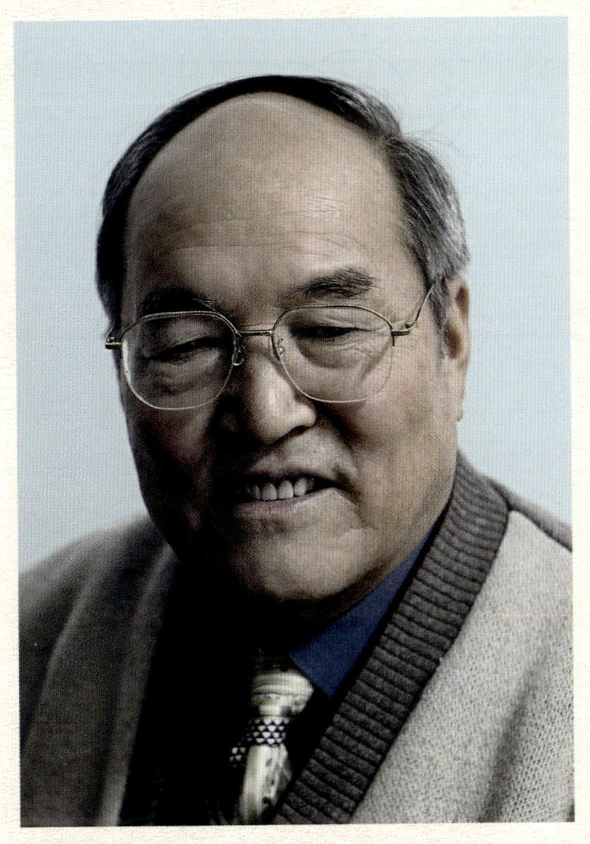

科学出版社
北京

内容简介

本书是魏江春院士六十年以来学术活动的概括性总汇，包括作者简介、前言、生日贺词、展示往事轮廓的影集、往事回眸、科学论文选集、著作目录及培养的研究生名录和后记。科学论文选集是由魏江春院士发表的一百多篇论文中选出八十九篇予以编辑出版。它们分别隶属于地衣学概论，地衣物种多样性及其系统演化与资源生物学，荒漠地衣生理生态学，地衣型真菌基因组学，部分药用真菌，以及真菌学科展望六大类。在地衣物种多样性研究中提出地衣分类学与三大存取系统相结合的新概念。此外，除发表一系列新种和新记录属种以外，还有新属、新科和新目，以及石耳目的建立。在基因组研究中发表了世界第一个地衣型真菌全基因组测序和生物学分析。

本书可供地衣型真菌学相关领域的科研工作者和高等学校相关专业师生参考。

图书在版编目（CIP）数据

魏江春科学论文选集 / 魏江春主编. —北京：科学出版社，2016. 3

ISBN 978-7-03-047184-0

Ⅰ. ①魏…　Ⅱ. ①魏…　Ⅲ. ①菌类植物－文集　Ⅳ. ①Q949.3-53

中国版本图书馆CIP数据核字（2016）第013648号

责任编辑：王　静　王　好 / 责任校对：李　影

责任印制：肖　兴 / 书籍设计：北京美光设计制版有限公司

科学出版社 出版

北京东黄城根北街16号

邮政编码：100717

http://www.sciencep.com

中国科学院印刷厂 印刷

科学出版社发行　各地新华书店经销

*

2016年3月第　一　版　　开本：890×1240 A4

2016年3月第一次印刷　　印张：55 1/4　插页：28

字数：1 850 000

定价：420.00元

（如有印装质量问题，我社负责调换）

WEI JIANG-CHUN'S
COLLECTION OF SCIENTIFIC PAPERS

Wei Jiang-chun

Science Press
Beijing

总 目 录

魏江春简介

魏江春，字青川，地衣型真菌学家，生于 1931 年 11 月 1 日，陕西省咸阳市秦都区古都镇魏家泉人。中国科学院微生物研究所研究员，生物科学副博士 (PhD) 及生物科学博士，中国科学院院士。

兼任中国科学院微生物研究所顾问委员会主任，中国科学院中国孢子植物志编辑委员会主编，中国菌物学会名誉理事长，中华人民共和国濒危物种科学委员会委员，国际生物多样性中国委员会顾问委员，创新中国智库首席科学家《中国生物物种名录》编委会副主任（副主编）。

学历

西北农学院（今西北农林科技大学）(1951—1955)

苏联科学院研究生院（考玛诺夫植物研究所孢子植物研究室），获生物科学副博士 (=PhD) 学位 (1958—1962)

俄罗斯科学院（考玛诺夫植物研究所），获生物科学博士 (DBioSci.) 学位 (1995)

主要经历

中国科学院微生物研究所学术委员会副主任 (1973—1977)

中国科学院中国孢子植物志编辑委员会常务委员 (1973—1976)、副主编 (1976—1986)、常务副主编 (1986—2006)、主编 (2007—)

中国科学院微生物研究所副所长 (1980—1984)

瑞典乌普萨拉大学和芬兰赫尔辛基大学等高级访问学者 (1981—1982)

《真菌学报》副主编 (1982—1997)

中华人民共和国濒危物种科学委员会委员 (1982—)

南极乔治王岛进行科学考察 (1983—1984)

中国科学院真菌地衣系统学开放研究实验室主任 (1985—1993)

中国科学院真菌地衣系统学开放研究实验室学术委员会主任 (1985—1997)

中国科学院真菌地衣系统学开放研究实验室 *MYCOSYSTEMA* 年报主编 (1986—1993)

中国科学院微生物研究所学位委员会主任 (1987—1993)

英国自然历史博物馆、美国哈佛大学及田纳西大学等高级访问学者 (1989—1990)

国际孢子植物学报编辑委员会委员 (1989—)

中国植物学会真菌学分会副理事长 (1990—1993)

中国菌物学会理事长 (1993—2003) 和名誉理事长 (2003—)

亚洲菌物学会执委会委员 (1993—2003)

美国芝加哥菲尔德自然历史博物馆，以及密执安州立大学和密尼苏达大学进行合作研究 (1995—1997)

美国芝加哥菲尔德自然历史博物馆植物系客座研究员 (1997—2007)

《菌物学报》顾问 (1997—)

中国科学院微生物研究所微生物资源前期开发国家重点实验室学术委员会主任 (1998—2006)

美国哈佛大学高级访问学者 (1999.9—1999.10)

中国科学院微生物研究所微生物资源中心顾问委员会主任 (2009—)

国际生物多样性中国委员会顾问委员 (2010—)

创新中国智库首席科学家 (2014—)

主要科研进展

带着填补国家空白学科之任务于 1958 年被派往苏联科学院研究生院留学。在完成学习任务之后于 1962 年回国。在进行多年的全国地衣真菌普查采集和研究基础上，建立了我国第一个地衣标本室：中国科学院微生物研究所国家菌物标本馆地衣标本室 (HMAS-L)。

在"青藏高原及其对自然环境和人类活动影响的综合研究"中，作为《西藏地衣》主编，《珠穆朗玛峰地区科学考察报告，生物与高山生理》的作者，于 1986 年获中国科学院科技进步奖特等奖。在《中国南极考察科学研究》中于 1997 年获国家海洋局科技进步奖特等奖。在主持并进行《南极菲尔德斯半岛及其附近地区生态系统研究》中于 1997 年获国家海洋局科技进步奖二等奖。

通过中国科学院研究生培养系统，以及有关大学培养并代理培养了一批以地衣学博士为核心的我国年轻的地衣学科技团队；发起并筹建了包括编研《中国地衣志》在内的中国科学院孢子植物志编辑委员会，启动了《中国地衣志》的编研工作。

基于上述一系列工作，基本实现了填补我国空白学科的任务。

在此基础上，通过地衣系统分类研究，发现和发表新目、新科、新属及一系列新种和中国新纪录种属。在表型与基因型相结合的综合研究中，确立了新属——岛蕊属 (*Cetradoria* J. C. Wei & Ahti) 的系统学地位。

在石耳科系统分类研究中，基于综合分析，对当时流行的五属系统修订为新二属系统 (J. C. Wei 1966)；在表型与基因型相结合的综合研究中建立了，包括盾衣新科 (Rhizoplacopsidaceae Wei et al. 2006) 在内的新目，石耳目 Umbilicariales J. C. Wei et al. (2007. 2. 22.)，被国际权威工具书 *Dictionary of Fungi* 所收录。近来又为石耳科属的共祖起源提供了表型和基因型证据，使基因型特征与表型特征可以同时展现在科属等类群描述中。当前正在进行石耳科在内的《中国地衣志》的编前研究和在研究基础上的编写以及世界范围石耳科的系统演化生物学研究。

对荒漠地衣石果衣进行的光合固碳功能实验研究结果表明，如果给腾格里 30 万平方公里的沙漠铺上以石果衣为优势物种的微型结皮生物，便能将腾格里沙漠变为年固碳量为 43.8 万吨的碳汇，不仅能够固沙防沙尘暴，而且能够为缓解温室效应和全球变暖有所贡献。

在干旱饥饿双胁迫实验中，荒漠地衣型真菌石果衣 (*Endocarpon pusillum*) 展现出 7 个月的耐受力。

为了探明石果衣真菌的抗旱基因功能，为构建抗干旱转基因植物提供基因资源，完成了石果衣真菌全基因组的测序和分析；这也是世界第一个地衣型真菌全基因组的正式发表 (2014)。

当前正在领导其科研团队进行石果衣真菌抗旱基因在植物中表达的研究，抗旱地衣在沙漠治理中重建的研究，以及地衣菌藻资源研发新途径的探索。

前　　言

自从大学毕业参加工作以来已经整整 60 年，其中除去“十年动乱”外，仅 47 年是为填补我国空白学科地衣学，亦即地衣型真菌学而学习和工作。本选集正是我近半个世纪以来在我国地衣学发展中所做的工作的概括总结。

本书中生日贺词是有关领导、专家和朋友对我八十岁生日的祝福。往事回眸后的影集是对八十多年来往事的点滴图注。正文部分则是四十七年来在地衣学研究中的部分论文选编。

论文选编中，从填补我国地衣学空白学科开始，遵循面向国家经济建设需求，面向世界科学前沿的方针，以地衣物种多样性及其系统演化与资源生物学为基础，在中国科学院微生物研究所菌物标本馆建立了地衣标本室 (HMAS-L)。在表型与基因型相结合的综合研究中发现并发表新种、新属、新科和新目，石耳目 (Umbilicariales J. C. Wei et al.) 的同时，又提出了地衣分类学与三大存取系统相结合的新概念。在国家研究生教育体系中结合科学研究实践培养了一批年轻的，具有博士学位的地衣学家。

结合干旱沙漠治理，以“沙漠生物地毯工程”为目标，开展了荒漠地衣生理生态学研究。基于荒漠地衣石果衣真菌所展示出极高的抗干旱性能，构建了世界地衣型真菌中的第一个全长序列的 cDNA 文库及其转录组分析；完成了石果衣真菌全基因组的测序与生物学分析，为以石果衣真菌抗旱基因构建抗旱转基因植物的研究提供了研究平台，也使石果衣真菌基因组成为世界第一个被测序和分析的地衣型真菌全基因组。

本书内容原则上为原文照登，只是根据不同内容加以归类编辑。对于个别文章有局部调整。

限于水平，不足之处，在所难免，有待读者指正。

魏江春
2015 年 12 月 12 日
于北京中关村

生日贺词

BIRTHDAY CONGRATULATIONS

贺魏江春院士八十华诞

孢子志书理资源

沙漠地毯谱新篇

二〇一一年十月　白春礼

物小天地大
志远路正长

祝贺魏江春院士八十寿诞

陈宜瑜
二〇一一年十月十五日

青春常在
夕陽更紅

賀江春同志八十華誕

李振聲

当岩石从原始的海洋中露出，
地衣就是陆地植被的先行者；
当地极的冻原从冰川下解脱，
地衣即是覆被它裸体的外衣；
最初的叶绿素是由它而萌生，
成为维系着地球生命的源泉；
为高山裸岩绘上斑斓的地图，
给流动沙漠生成固结的外壳；
在砾石戈壁长出甜蜜的吗哪[1)]，
在盐碱滩地层叠的地衣发糕；
地衣坚韧而低调的高风亮节，
也蕴藏在它的研究者的身上。

——贺魏江春院士八十华诞

敬撰

1 1）对地衣的最早文献记载可能是4000年前的《出埃及记》，当时以色列人脱离埃及奴役出行，在西奈荒漠中徘徊40年，靠食用一种白霜状的地衣“吗哪”为生。

八十随心欲
百岁再高歌

贺
汪李学长八十华诞

许绍雄
2011-10-15

祝贺

江春同志八十华诞

相识五十年
相处属偶然
互抱真诚心
祝君永向前

山仓

2011年10月15日

贺　词

值此魏江春院士八十华诞之际，谨向魏江春院士致以崇高的敬意和诚挚的祝福，衷心祝愿魏院士福如东海长流水，寿比南山不老松。

春秋迭易，岁月轮回，金秋十月，天清气和，吉祥喜庆，好事多多。先生虽至耄耄，但自信人生二百一、一百二三不稀奇，七十八十小弟弟！先生岂不正当时宜！

先生年高德众，学贯中西，“苦渡学海不知苦，乐登书山自有乐”，兢兢业业，孜孜以求。无论是在颠沛流离，危机四伏的岁月，还是在太平盛世，功成名就之时，先生始终坚守在科学第一线，以忘我的大无畏精神，积极创造条件，努力开拓，为我国的菌物事业奠定了坚实的基础。先生的身影不仅留在了世界三极，留在了浩渺的林海草原，留在了茫茫大野荒漠，也留在了丰富的史料文献里，更留在了中国菌物学人的心中！先生不仅在科研上业绩赫然、著作等身；更在教书育人上诲人不倦、硕果累累，为国家培养出众多优秀的科技人才。吾与先生神交数十载，经历了菌物学界在中国成长的日日月月、点点滴滴，是我敬重的良师更是益友，先生永远是学界更是人生的楷模。

“壮志凤飞逸情云上，灵芝献瑞仙鹤同年”。恭祝老寿星：笑口常开，身体康健，生活之树常绿，生命之水长流！

同时祝福所有来祝福先生寿辰的至爱亲朋，身体康健、工作顺利、合家欢乐、万事如意！更望在魏院士的米寿之年，望九之年我们重聚，颂日月之昌明，贺松鹤之长春！

先生之春辉永绽！

中国工程院院士

吉林农业大学教授　李玉

辛卯年深秋

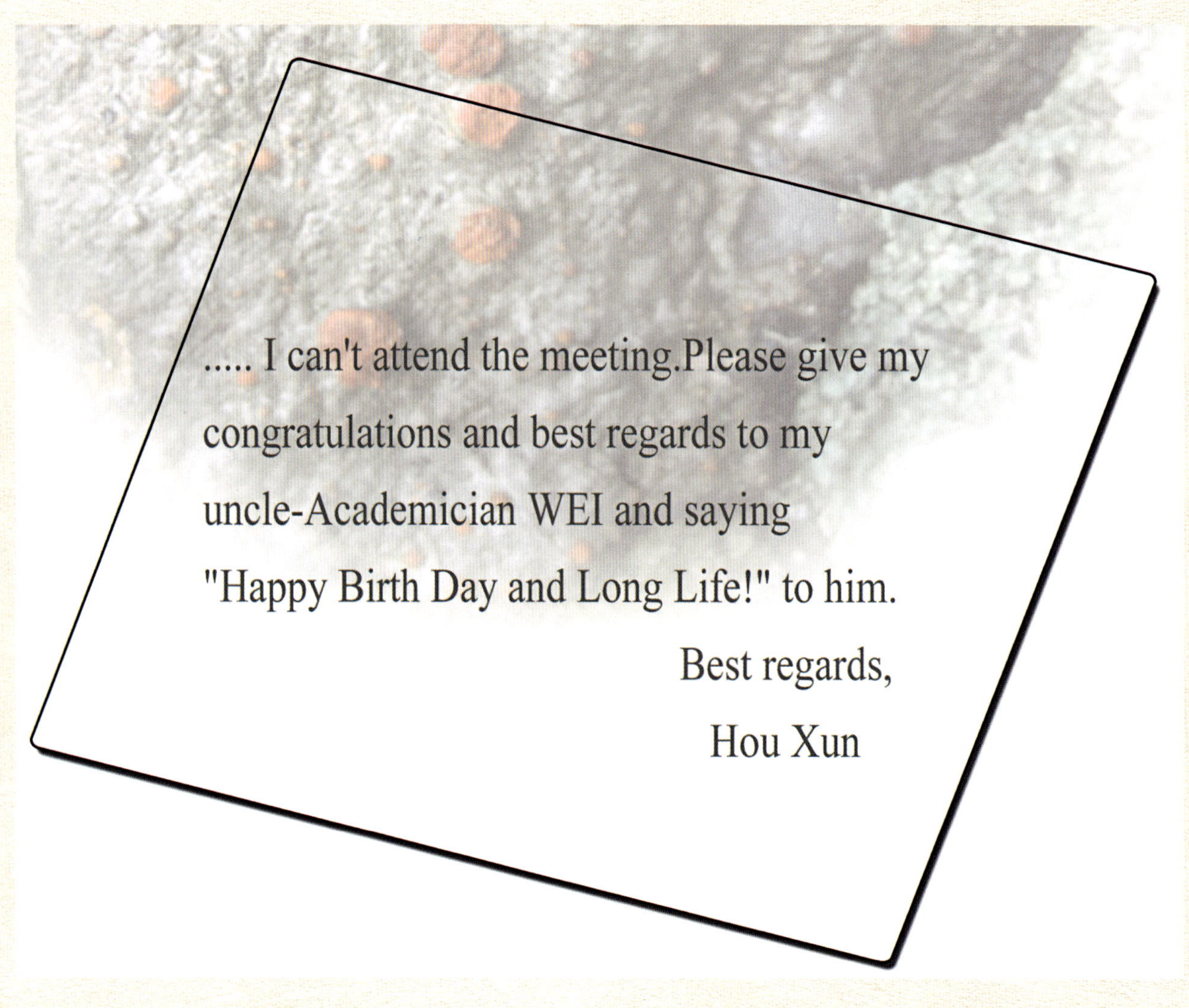

..... I can't attend the meeting.Please give my congratulations and best regards to my uncle-Academician WEI and saying "Happy Birth Day and Long Life!" to him.

Best regards,

Hou Xun

中国科学院

贺　信

尊敬的魏江春院士:

欣悉先生即将迎来八十华诞，我局向您致以热烈的祝贺和最诚挚的敬意！

您是我国著名的菌物学家，也是我国地衣学主要奠基者和学术带头人，长期致力于地衣型真菌的系统与演化生物学的研究，取得了一系列重要研究成果。在世界范围石耳科地衣系统生物学研究中，以表型与基因型相结合的整体系统学方法论述了石耳科地衣新系统，得到世界同行科学家的承认。在地衣型真菌的系统发育地理学研究中，为中国和北美间断分布的美味石耳和大叶石耳可能由同一祖先分化而来提供了分子生物学证据。您发起并主持中国孢子植物志编研、提出“干旱沙漠生物地毯工程”，为推动我国孢子植物志编研和真菌学科的发展做出了重要贡献。

几十年来，您一直活跃在菌物研究前沿领域，治学严谨，造诣精深；您十分关心青年人才成长，在科研和教学中以身作则，诲人不倦，为国家培养了大批优秀人才，桃李满天下。

衷心感谢您在研究和教育事业做出的突出贡献，祝您生日快乐、生活幸福、健康长寿！

中国科学院生命科学与生物技术局

2011年10月12日

贺 信

春秋迭易，岁月轮回，值此辛卯金秋之际，我们迎来了尊敬的魏江春院士八十大寿。

魏院士是著名的地衣真菌学家，我国地衣学研究的主要奠基人，长期从事地衣型真菌生物多样性及其系统与演化生物学研究，取得了卓越的成就，为我国地衣学研究在国际上赢得了盛誉；魏院士广种桃李，弟子遍天下，为我国培养了大批地衣真菌学人才。

至此魏院士八十大寿之际，谨送上最真诚、最温馨的祝福，祝魏院士如月之恒，如日之升，如南山之寿，不骞不崩！

河北省微生物多样性研究与应用重点实验室

2011 年 11 月 1 日

贺 信

金秋桂香，硕果盈枝。值此魏江春院士八十华诞暨“魏江春与中国地衣学术研讨会”召开之际，山东农业大学生命科学学院及其代表山东农业大学谨向魏先生致以亲切问候和温馨的祝福。

魏先生是享誉海内外的真菌学家，是中国地衣生物学的奠基人；魏先生是中国孢子植物志编研的首席科学家，为中国的生物多样性研究做出了重要贡献；魏先生厚德正气，治学严谨，具有大家风范，为我国培养了大批优秀的地衣真菌学研究人才。

辛苦追梦人生路，终得正果效祖国！再次，表达我们美好祝愿，祝魏院士及其家人身体健康、幸福美满！

山东农业大学生命科学学院

2011年11月1日

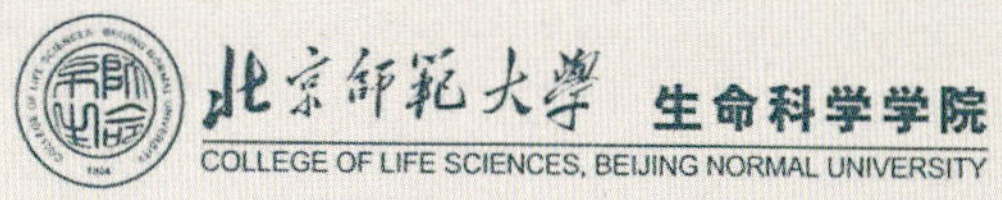

北京師範大學 生命科学学院
COLLEGE OF LIFE SCIENCES, BEIJING NORMAL UNIVERSITY

贺信

中国菌物学会

转致魏江春先生:

喜逢魏江春院士八十华诞，北京师范大学生命科学学院全体师生致以热烈祝贺!

先生德高身正，一生不懈追求真理、领导学术群伦、胸襟博大豁达令后辈敬仰，堪称我国菌物学界的一代大师；先生严谨治学、成就非凡，深邃的学术思想、丰硕的研究成果令世人赞叹，可谓后人崇仰的学术泰斗；先生虚怀若谷，育人有道，桃李满天下，奠基了我国真菌地衣整合生物学的宏基伟业，丰富与发展了地衣学的理论与实践，实为学行的楷模。

您的健康是我国菌物学科学研究和人才培养事业的福祉。在此先生八十华诞之际，诚挚恭祝魏老幸福安康!

北京师范大学生命科学学院

二〇一一年十月二十日

地址：北京市新街口外大街19号 19 Xinjiekouwai Avenue, Beijing 100875 P.R. CHINA 网址(Web)：http://cls.bnu.edu.cn 电话/传真(Tel/Fax)：010-58807720

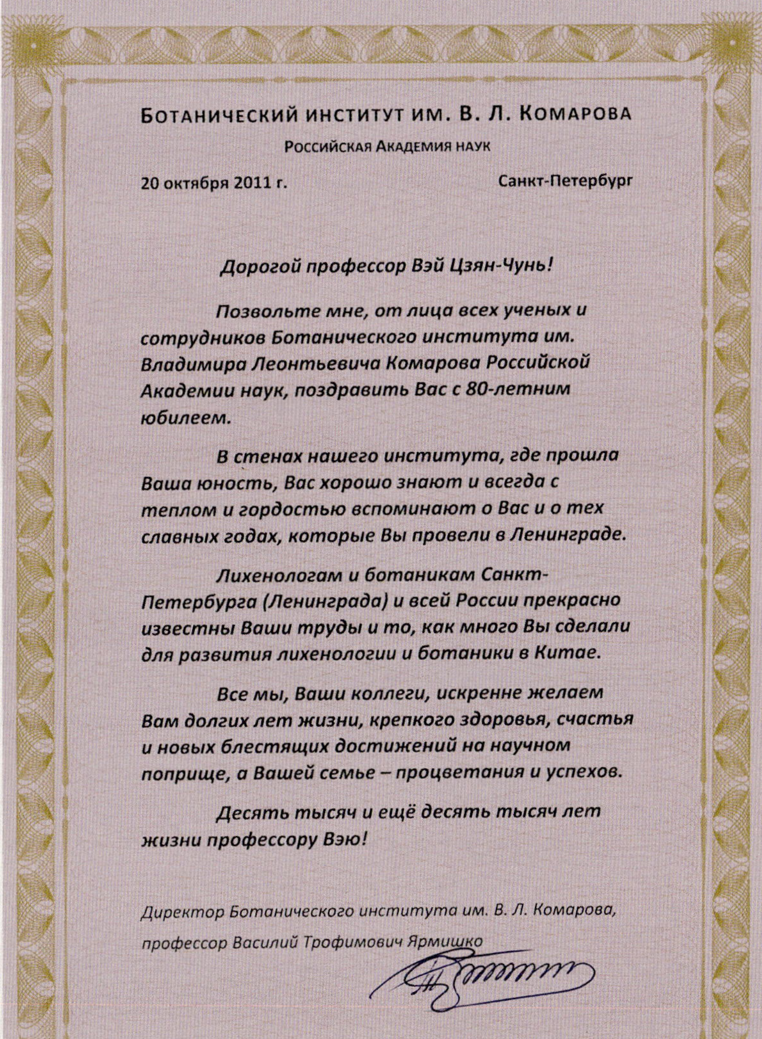

Ботанический институт им. В. Л. Комарова

Российская Академия наук

20 октября 2011 г. Санкт-Петербург

Дорогой профессор Вэй Цзян-Чунь!

Позвольте мне, от лица всех ученых и сотрудников Ботанического института им. Владимира Леонтьевича Комарова Российской Академии наук, поздравить Вас с 80-летним юбилеем.

В стенах нашего института, где прошла Ваша юность, Вас хорошо знают и всегда с теплом и гордостью вспоминают о Вас и о тех славных годах, которые Вы провели в Ленинграде.

Лихенологам и ботаникам Санкт-Петербурга (Ленинграда) и всей России прекрасно известны Ваши труды и то, как много Вы сделали для развития лихенологии и ботаники в Китае.

Все мы, Ваши коллеги, искренне желаем Вам долгих лет жизни, крепкого здоровья, счастья и новых блестящих достижений на научном поприще, а Вашей семье – процветания и успехов.

Десять тысяч и ещё десять тысяч лет жизни профессору Вэю!

Директор Ботанического института им. В. Л. Комарова,

профессор Василий Трофимович Ярмишко

Komarov Botanical Institute

Russian Academy of Sciences

20 October 2011 St. Petersburg

Dear Professor Wei Jiangchun!

On behalf of all scientists and collaborators of the Komarov Botanical Institute of the Russian Academy of Sciences let me express you my compliments in the day of yours 80 anniversaries.

Within the precincts of our institute, where you worked as young scientist, everybody still knows you very well and always remember you and nice years which you have spent in Leningrad cordial and with pride.

Lichenologists and botanists of St. Petersburg (Leningrad) and of the whole Russia know very well your works and great contribution you have made for the development of botany in China.

We, your colleagues, sincerely wish you many years of life, good health, happiness, and new major breakthrough in sciences, and to your family – prosperity and success.

Ten thousand and another ten thousand years of life to Professor Wei!

Director of the Komarov Botanical Institute,

Professor Vasilij Trofimovich Yarmishko

国际友人

UNIVERSITY OF CALIFORNIA, BERKELEY

BERKELEY • DAVIS • IRVINE • LOS ANGELES • MERCED • RIVERSIDE • SAN DIEGO • SAN FRANCISCO  SANTA BARBARA SANTA CRUZ

PROFESSOR JOHN W. TAYLOR
DEPARTMENT OF PLANT AND MICROBIAL BIOLOGY
111 KOSHLAND HALL
TELEPHONE: (510) 642-5366 TELEFAX: (510) 642-4995
INTERNET: jtaylor@berkeley.edu

BERKELEY, CALIFORNIA 94720-3102

October 11, 2011

Prof. Jiangchun Wei
Key Laboratory of Systematic Mycology and Lichenology
Institute of Microbiology, Chinese Academy of Sciences
1-3 West Beichen Rd., Chaoyang District, Beijing 100101, China

Dear Professor Wei:

Congratulations from the University of California at Berkeley and from the International Mycological Association on the occasion of the eightieth anniversary of your birth. Or, as we would say to each other in California,

"Happy 80th birthday."

The field of lichenology, including endolichenic fungi, and that of cordycepology are the richer for your efforts and I wish you many more years of inspired and inspiring mycology.

Sincerely,

John Taylor

John W. Taylor

国际友人

UNIVERSIDAD COMPLUTENSE
FACULTAD DE FARMACIA
DEPARTAMENTO DE BIOLOGIA VEGETAL II

16 October 2011

¡My dear Professor Wei!

Congratulations on becoming an octogenarian.
I count it as a great privilege and honour
to have known and corresponded with you for
so many years, and to follow your achievements.
That is specially so as you have had to experience
such hardships — and come through smiling.

You have been an inspiration to so many, and
had the vision of bringing lichenologists together
with other mycologists so early on — and of
launching MYCOSYSTEMA (I smile at the
discussions over the changing red tone of
the early covers....).

With best wishes for a healthy and fulfilling
future to you and your family,

David Hawksworth

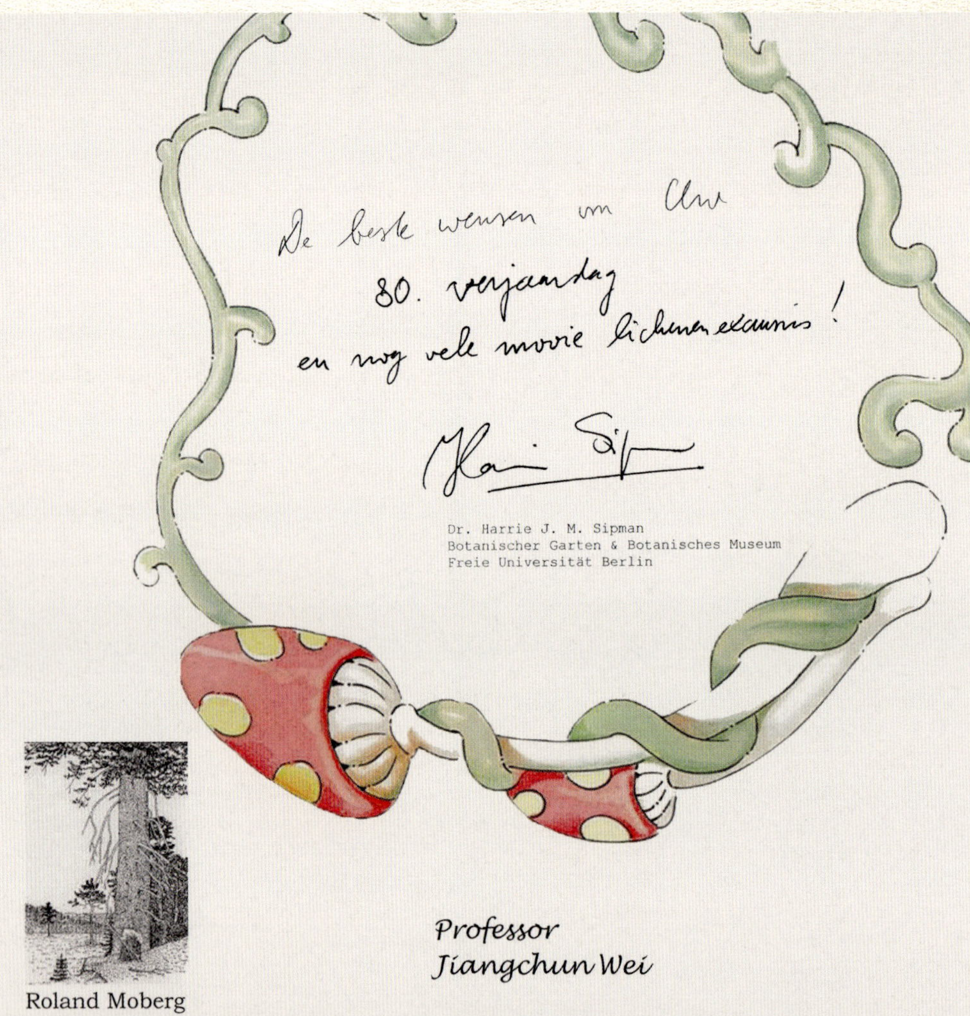
De beste wensen voor Uw
80. verjaardag
en nog vele mooie lichenexcursies!
Harrie Sipman
Dr. Harrie J. M. Sipman
Botanischer Garten & Botanisches Museum
Freie Universität Berlin

Roland Moberg

Professor
Jiangchun Wei

Hjärtliga gratualtioner på 80-årsdagen från oss alla i Uppsala.

Vi minns med stor glädje de besök vi gjort i Kina och blivit varmt mottagna av Dig och Din stab.

Roland

Professor
Jiangchun Wei
Warmest congratulations on Your 80th birthday from us all in Uppsala
We remember with joy the visits we have made in China
and been cordially received by You and Your staff.
Roland

THE UNIVERSITY of TENNESSEE
College of Arts and Sciences
Department of Ecology & Evolutionary Biology
569 Dabney Hall
Knoxville, TN 37996-1610
Phone: (865) 974-3065
Fax: (865) 974-3067

Oct. 12, 2011

Dear Jiang-Chun –

Many years ago you were very kind to me. You arranged a fine fieldtrip to Jilin Province, with an introduction to the mycological diversity of northern China. In return, it was an honor to host your visit to Tennessee and to introduce you to the biodiversity of our Smoky Mountains.

Because of my very early visits to China, I have several friends at the Institute of Microbiology in Beijing, as well as in other parts of the country.

It was a pleasure to eat Chinese dumplings at your home and to learn about your family. Since those days, China has been transformed, and I hope that someday, we can eat Chinese dumplings together again.

I thank you for your important role in my career, and I wish you a very happy 80th birthday.

Ron Petersen

国际友人

Soest, oct. 2011

Hooggeleerde Heer Wei,

Van Harte gefeliciteerd met uw verjaardag!

Hoogachtend, Dr. André Aptroot

[Highly esteemed sir Wei,
cordial greetings for your birthday!
Sincerely, André Aptroot]

寿

影集

PHOTOGRAPH ALBUM

01 中学

西安一中高一甲班 1948. 10

看望西安一中初三语文老师 2000. 11

父母女母校百年校庆留影 2007. 10

西安中学（原西安一中）百年校庆与王兰君校长留影 2007. 10

母校百年校庆同学聚会 2007. 10

02 大学

西北农学院四年级 1955. 08

师（李振岐）生同年分别当选工程院和科学院院士聚会 1997. 11

同班同学 45 年后回母校聚会留影 2000. 11

同班同学回母校师生聚会留影 2000. 11

在母校 80 华诞庆典上致辞2014. 09

在母校 80 华诞暨合校 15 周年植保学院庆典上 2014. 09

携小孙访母校在周尧教授蜡像前留影

03 留学

苏联科学院研究生院考玛诺夫植物研究所列宁格勒（圣彼得堡）1962. 05

在孢子植物研究室进行显微镜观察 1959. 09

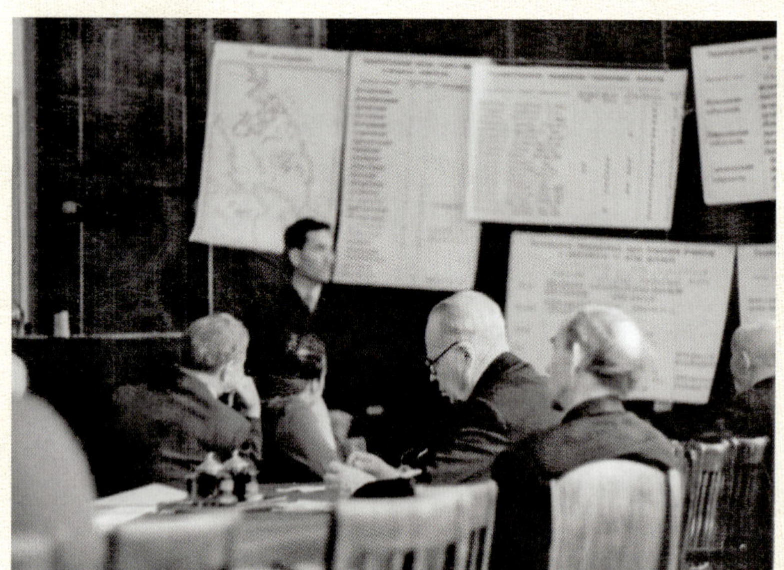

进行毕业论文报告 1962. 05

在科学导师 V. P. Savicz 教授森林别墅外驱车兜风 1962. 06

毕业论文答辩后师生合影 1962. 05

莫斯科留影 1959. 01

04 科研

I: 国内外科研工作

和陈建斌进行野外考察及标本采集

与瑞典地衣学家 Roland Moberg 等在云南考察 1987. 07

在斯密斯研究院工作期间与同事留影　Washengtong D. C. 1990. 04

在北美 Mt. Appalachian 考察采集时留影 1996. 07

在中国科学院微生物研究所真菌地衣开放实验室工作

在美国哈佛大学工作期间留影　Boston,Cambridge 1999. 09

II: 博士论文答辨

做博士论文报告前的准备，圣彼得堡 1995. 05. 23

做博士论文报告 1995. 05. 24

Golubkova 祝贺

俄罗斯农业科学院院士祝贺

地植物学家 Nina 祝贺

和 Nina Dombrovska 等进行学术交流

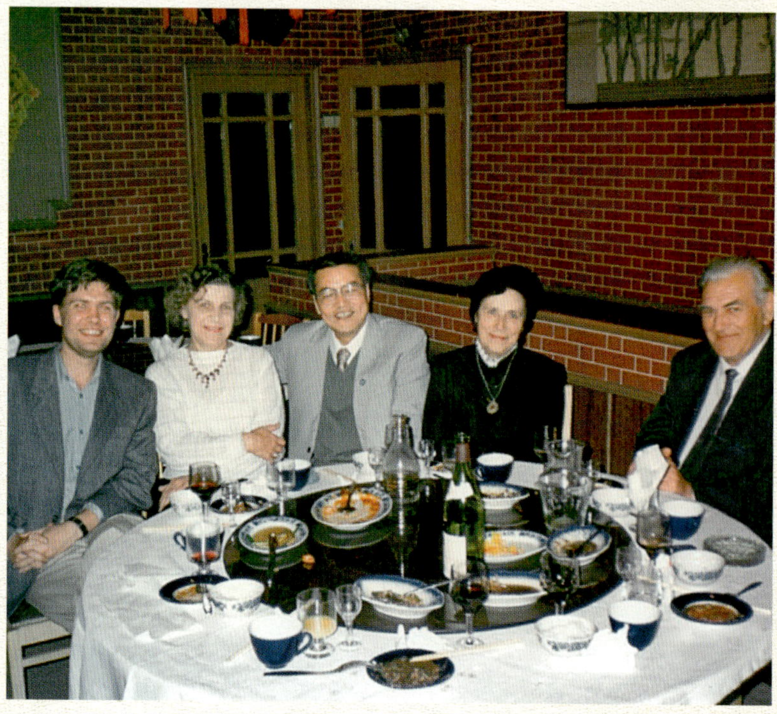

和考玛诺夫植物研究所所长及同事聚餐

III: 国内科研工作

进行荒漠地衣摄像 2007. 07

在长白山进行地衣摄像 2000. 07

在滇南老君山进行野外考察采集期间野餐 1999. 08

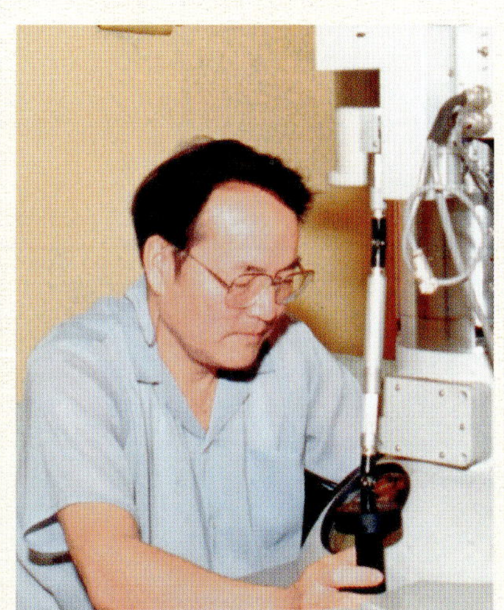

进行透射电镜观察

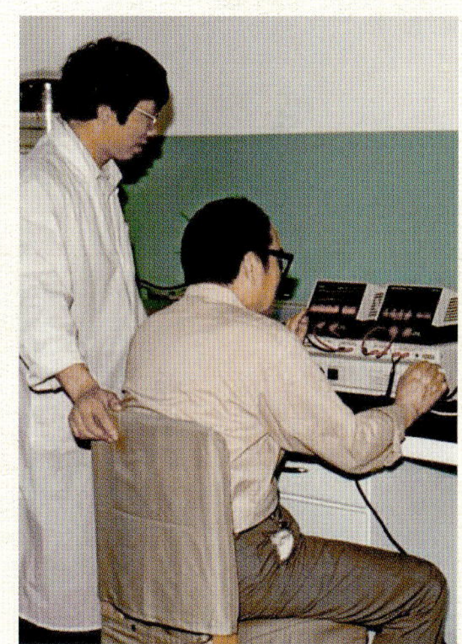

在分子生物学实验室工作

在昆明出席学术讨论会 2014. 05

05 编志

中国孢子植物志编委会（简称“孢编会”）第四次扩大会议 武汉 1986. 11

孢编会主席台

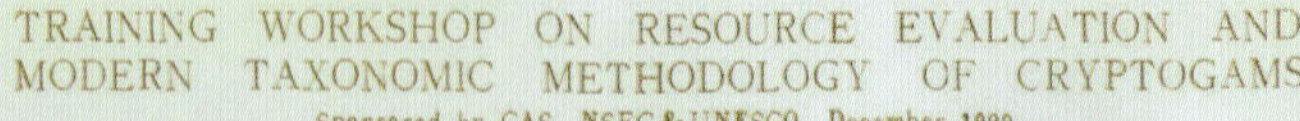

孢编会主办编志讲习班 1999. 12

和曾呈奎主编讨论编目

讨论地衣志编研工作

第五届孢编会主副编留影 2007

真菌地衣系统学开放实验室学术委员会 1986

真菌地衣系统学开放实验室学术委员会 1987

和开放实验室学术委员会委员 Korf 教授

开放实验室学术委员会国内委员

真菌地衣系统学开放实验室学术委员会 1997

07 学会

在中国菌物学会成立大会上 1993. 05

和“菌物”术语建议人裘维番院士合影

和中国台北真菌学代表陈瑞清商谈两岸学术交流事宜

中国菌物学会第二届理事会留影 1998

第一届地衣学研讨会，青岛

首届中国菌物产业与科技研讨会暨展览会，北京 2005. 10. 27

全国药用真菌学术研讨会，江西 2008. 10

全国第六届地衣生物学研讨会，乌鲁木齐

银婚照

金婚照 1997

祝贺四姐九十大寿

庐山留影 2011. 08. 29

母女照

姐弟照

父母子女大孙女照

家庭首次聚会照 2005

马年春节家庭二次聚会照 2014

马年春节和女儿一家聚会照

马年春节和大儿一家聚会照

和小儿夫妇及小孙女合照

和大孙女冰合照

奶奶和大孙女冰合照

小孙女佳佳

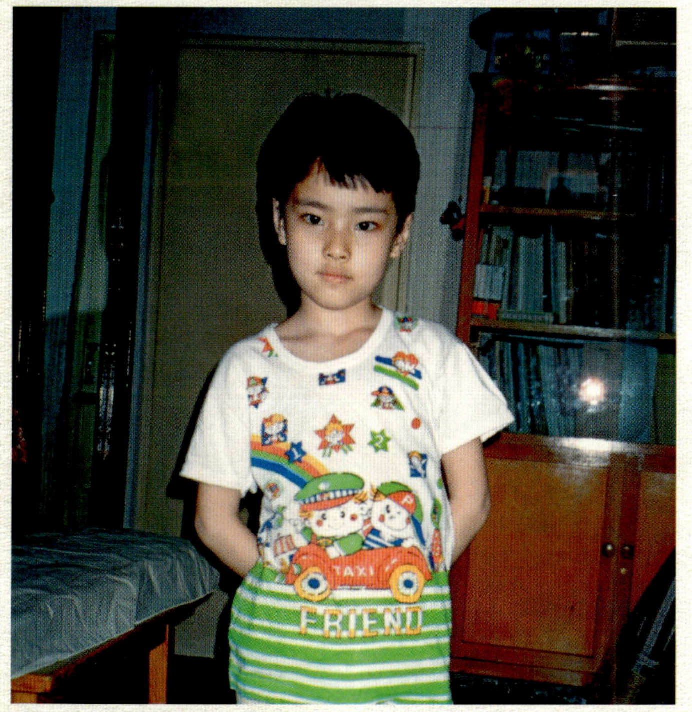

小孙女佳佳

和小孙女佳佳合影 2011. 09. 03

三代机场留影 1999

奶奶和小孙子牛牛 1998

爷孙家里留影

重孙子子雄

重孙女子棋

重孙女若桐

重孙女玉洁

往事回眸

GAMCEBACK

魏江春

中国科学院微生物研究所真菌学国家重点实验室

摘要：本文是作者对他八十多年来的往事所做的简要回顾。该回顾分为求学；十年动乱中编志工作的启动；与科研唱反调的闹剧；家庭；改革开放后学科发展中的跨越；展现科学道德的学风，以及畅想未来的展望七部分分别予以叙述。在学风方面，作者回顾了他自己与之交往过的两位不诚实的地衣学家所为的两件不愉快的往事。作者希望他的点滴回顾能对后来者有所借鉴和警戒。

人生的路是自己走出来的，走什么路以及如何走是由自己选择的，主动权在于自己。然而，客观条件和机会是自己无法选择的，主动权不在自己。不过，一切机会永远属于有准备的人，而准备的主动权却在于自己。

儿时总觉时间过得太慢，期盼欢度新年，快快长大。人到老年则深感日月如梭，欲做之事太多，时不我待。自从我来到这个世界，搭载地球已经围绕太阳周转了八十多圈！我对这八十多年来的往事的点滴回忆和对未来的畅想，如能对后来者起到某些借鉴，我将感到欣慰。

求学

我出生在国耻日“九一八”事变后的第四十九天。我三岁丧母，是在姐姐的关照下直到八岁入学。那年正是 1939 年秋季，陕西咸阳“学道门小学”为躲避日寇轰炸，由城中心迁至被扩大了的我们魏家泉村小学，以能起防空作用的窑洞为教室。于是我就在该校完成了四年的初级小学学业。当我刚学会写字的时候，我便用粉笔在我家二门楼门楣上写下了“打倒日本帝国主义”八个大字。从那时起，我的脑海里就一直存在着一个百思不解的问号：为什么具有五千年历史文化和九百多万平方公里国土的泱泱大国的国土和人民竟能长期遭受到一个小小邻居的践踏和蹂躏！？

由于太平洋战争的爆发，日寇对我大后方咸阳的狂轰滥炸也随之停息。于是，在 1943 年秋季，“学道门小学”又迁回咸阳城中心原校址，我也随校进城继续就读五年级。我虽身处抗日大后方的陕西咸阳，然而，小学时光却是在日寇轰炸机的隆隆响声中度过。

我于 1945 年小学毕业。就在那年暑假的一天，我正在门外玩耍，一位叫陈彪的村民从县城回来，在村前路上边走边不断地高声呼喊：“日本投降了！”这是多么令人兴奋的消息！更是人们期盼已久的时刻！这消息迅速传遍全村的家家户户，人人兴高采烈，全村欢腾。抗战终于胜利了！这一情景至今还回荡在我的脑海里，犹如昨日事，历历在目。

我怀着抗战胜利后的和平喜悦走进了初中课堂，并希望像诸兄长那样继续走进高中和大学的校园。由于我的家是由父辈兄弟四人及其后代所组成的大家庭，诸兄长均在外从业，家父为了让我继承家业，决定在我初中毕业后便终止学业。这一决定重重地打击了我继续求学的愿望。为此我曾向我舅父写过一封长信，希望他能说服家父满足我继续求学的愿望。然而，1948 年的暑假，高中升学考试日益逼近，家父却仍未放行。当时我在想，如果母亲健在，我的处境会这样吗？无望之下，我便去母亲

本文系在“八旬回眸展未来”[1] 基础上增补改写而成，并更名为“往事回眸”。

坟地大哭一场！此情此景被我儿时以来的邻居朋友王希钧看见并告诉了家父。于是家父便勉强改变注意，同意我继续升学高中。由初中到高中一年级的四年学习生活，实际上是在内战中度过的。至于在 1951 年背着家父参加高考已是解放后的事了。国家对全体大学生发放助学金是我顺利完成大学学业的最大保证，这是后话。

1951 年的高考考场设在西安西南门外历史悠久的西北大学。化学系的讲解员是我高中的高年级同学。他在用魔术般的实验向考生们讲解化学专业的神奇，希望考生们报考化学系。然而，当我征求他对我报考志愿的建议时，他却毫不犹豫地建议我报考西北农学院，即今西北农林科技大学之前身。理由是那里有很多名教授。因此，我的全部报考志愿都填在西北农学院。其中第一志愿便是植物病虫害系。

入学后，又进一步选修了著名教授周尧所在的昆虫学专业。经过一学期的学习体验，我发现自己对于真菌植物病理学专业更有兴趣，因而中途改学真菌植物病理学专业。在大三期间，我曾为探讨小麦抗条锈病与易感染条锈病的小麦品种之间在生理生化方面的差异及其研究方法向授课老师，毕业于英国剑桥大学的著名植物生理生化学家石声汉教授请教。在暑假教学实习期间，我曾根据番茄白星病的病原真菌 Septoria lycopersici Speg. 的生活史及其传播途径向当时带领我们实习的李振岐老师提出下列建议：由于番茄白星病的病原真菌随同番茄植株病叶落地入土越冬；次年越冬后的病原真菌又从最接近地面的第一层叶子逐层向上侵染。因而建议采用不受该病原真菌感染的豆科植物三叶草密植于番茄地段，从而切断病原真菌的第一侵染通道。该设想既可达到防病目的，又可通过三叶草植物固氮以肥沃土壤，一箭双雕。

1955 年大学毕业后，我被统一分配至中国科学院西北农业生物研究所。所长是著名胶体化学家，中国科学院学部委员（院士），西北农学院虞宏正教授。在这期间，我曾在著名植物病理学家，西北农学院李建义教授指导下从事小米黑穗病的研究。1956 年，我被调入中国科学院应用真菌学研究所，在著名真菌学家王云章教授指导下进行蔷薇科锈菌分类研究。

随后，遵照国家关于填补空白学科的政策，根据我国真菌学的主要奠基人，中国科学院应用真菌学研究所所长，中国科学院学部委员（院士）戴芳澜教授的建议，为了填补我国真菌学中地衣空白学科，我于 1958 年被派往苏联科学院研究生院攻读地衣学专业，师从著名地衣学家 V. P. Savicz 教授。在完成学位论文之后，于 1962 年毕业并获苏联生物科学副博士学位 (=PhD)。近四年的研究生学习机会是我一生中最为珍惜的。因为在当时的情况下，出国留学不是由个人选择的，而是由国家选派的。既然我有幸被选派赴苏留学，我当然非常珍惜这一机会，为更好地完成填补空白学科任务而努力学习。在学习期间，我基本上是按照由研究生宿舍、研究生院课堂及实验室这三个点所连成的线作为行动路线完成了四年的留学生活。

按照苏联的学位制度，在获得了相当于 PhD 的科学副博士学位之后，在生物学领域通常要经过 10 年左右的独立科研工作，发表与博士学位论文有关的研究论文 20 篇以上，才能有资格提交博士学位论文并经过答辩，以获得生物科学博士学位。我本来打算回国后经过 8 年左右的努力以完成上述计划。然而，十年动乱使我的计划推迟了 30 多年，直到 1995 年才以《东亚石耳科地衣分类学与地理学分析》的博士学位论文，在俄罗斯科学院考玛诺夫植物研究所通过答辩而获得了生物科学博士学位 (Dr. Sci.)。

编志

1958 年，我带着填补空白学科的任务和一张地衣学专业的白纸进入了苏联科学院研究生院；又带着一颗落实学科任务的心和一本学位论文于 1962 年回国。我如饥似渴地希望了解国内自然界的地衣区系，因而，自 1963 年初夏直至十年动乱之初，从陕西秦岭太白山和贵州梵净山开始，沿着西南至东北的对角线进行了地衣区系的实地考察。正当我全力投身于我国地衣区系的野外考察和采集活动时，1966 年 7 月开始的十年政治动乱中断了我们的珠穆朗玛峰地区科学考察活动。

十年动乱期间，全国科学研究与文化教育基本停止。在这种情况下，有关科研的事，也只能一有空，遇有机会，便见缝插针似地作为插曲予以推进。

在周恩来总理关于恢复科教工作指示的背景下，中国科学院于1972年夏季在北京友谊宾馆召开了计划工作会议。我是出席计划工作会议的微生物所代表。开会时，我的座位和出席会议的植物所代表林鎔先生相邻。当我看到林先生正在审阅植物所拟向会议提出恢复《中国科学院中国植物志编辑委员会》工作的报告时，我便产生了一个想法：借此机会向会议提出筹备成立包括仍游离于动植物志规划之外的藻类、真菌、地衣和苔藓在内的《中国科学院中国孢子植物志编辑委员会》的建议，以启动我国孢子植物志的编前研究和在研究基础上的编写工作。会间休息时，我便向出席会议的海洋所代表，藻类学专家曾呈奎先生说了我的上述想法，并得到他的积极支持，随后又向出席会议的动物所代表朱弘复先生通报了上述情况，并试探关于恢复动物志编委会的可能性时，也获得了朱先生的积极回应。在此基础上，我又与出席会议的其他有关研究所代表进行了沟通并得到了积极响应。于是，我们出席会议的各有关生物研究所的代表就恢复《中国科学院中国植物志编辑委员会》、《中国科学院中国动物志编辑委员会》工作并筹备成立新的《中国科学院中国孢子植物志编辑委员会》(简称“三志”)达成了共识；推选海洋所代表曾呈奎代表大家向大会正式提出尽快召开“三志”工作会议的倡议。该倡议受到在那次会议上刚恢复工作的原生物局局长过兴先的大力支持；会后又得到当时主持中国科学院工作的武衡的批准。这就是正当全国科学研究基本陷于停滞时期中国科学院于1973年召开的影响广泛的广州“三志”工作会议鲜为人知的发起过程。

《中国科学院中国孢子植物志编辑委员会》的成立，对于由《中国海藻志》、《中国淡水藻志》、《中国真菌志》、《中国地衣志》和《中国苔藓志》五志组成的中国孢子植物志的编前研究和在研究基础上的编写工作的启动和推进起到了重要的关键作用，从而有机会在十年动乱的夹缝中开展部分科研工作。然而，在随后的所谓批林批孔运动中，微生物所出现了将《中国科学院中国孢子植物志编辑委员会》的成立及其志的编研当作复旧靶子予以批判的大字报。对此，我在《中国科学院中国孢子植物志编辑委员会》工作通讯中明确指出：“《中国科学院中国孢子植物志编辑委员会》的成立及其运行，在我国是史无前例的，因而是新生事物！”以稳定全国编志人员的情绪，鼓励大家的编志积极性。在此期间，我曾于1974年只身前往黑龙江小兴安岭的带岭凉水林场进行地衣区系考察与标本采集。

闹剧

在1973年夏天，一位名叫乔林的中国科学院造反派领导成员之一，带领了一群随员来到微生物所810会议室；主持了一个事后才知道为“抓辫子”的座谈会。我在座谈会上为了说明科研人员务必搞好科研而强调：“科学院是搞科学实验的”。参加座谈会的本所人员，包括乔林及其随员在内的所有出席者均未提出异议。

然而不久后，乔林便在中关村大礼堂作为主讲人，对于科学院的所谓“修正主义科研路线”展开了公开大批判。当她为科学院的所谓“修正主义科研路线”提出证据时便说：“有人说：‘科学院是搞科学实验的’，我说不对，科学院首先是搞阶级斗争的”。如果否认科学院是搞科学实验的，那么就等于否认学校是搞教学的，农场是搞农业生产的，工厂是搞工业生产的，军队是保卫国家的一样荒谬绝伦！

我于1974年春节期间回陕西老家，突然接到所领导要我马上返回研究所参加批林批孔运动的电报。当我回所后，才知道是所领导专门组织全所大会，不点名地针对我曾说过的“科学院是搞科学实验的”所谓“修正主义科研路线”展开大批判。

将“科学院是搞科学实验的”这句话当作所谓的“修正主义科研路线”在大小会上的批判而并未点出说这句话的人名，可能是极左路线与正确路线较量的折中产物吧！

在1974年底至1975年初，小平同志复出后，根据周恩来总理所做的第四次全国人民代表大会第一次会议开幕式上的《政府工作报告》精神，从1975年3月份开始了一系列的调整工作。当时主持科学院工作的耀邦同志来到微生物所进行视察。他在听取汇报的会上和我谈起我母校西北农学院(今西北农林科技大学)的辛树帜院长和虞宏正教授。后者是该校两位中国科学院学部委员(今院士)之一。会后临走时，耀邦同志特意绕过会议桌一圈，舍近求远，走近我并和我握手告别。当时我很受感动！我在想，耀邦同志的言行举止，传达给我们一个很重要的信息，即教授，科学家，科学实验应该受到尊重，这是国家四个现代化的支柱；同时也可能是对于我曾受到不点名批判的安抚吧！

当“四人帮”被粉碎后的 1976 年冬季，中国科学院在首都体育馆召开了全院京区人员庆祝大会。大会主席的头一句话就说：“现在看来，科学院就是搞科学实验的嘛！”在十年动乱期间把“科学院是搞科学实验的”这句朴素的大实话当作“修正主义科研路线”的荒唐批判随着“四人帮”的倒台而被彻底抛入历史的垃圾堆！

家庭

我的妻子儿女生活在陕西农村，原本计划在完成留学任务之后与家人来京安居落业。然而，不幸的是 1959 年风云突变，当从异国他乡的广播中传出了与我国宪法有关公民享有自由迁居权利相悖的冻结农村户口的消息时，犹如一瓢冷水从我头顶浇下！在此政策下，我与家人长期分居两地。

回国后，在我如饥似渴地投入地衣研究的那几年，有一次，当我听到别人的孩子喊“爸爸”时，我的心如刀绞，想到了自己远在他乡的儿女！在心烦意乱之下，便搭车前往动物园公交车枢纽站，在人来车往中独自坐在天文馆前台阶上，陷入了关于儿女家人的深思中久久不能自拔。

在这期间，生长在红旗下的我三个儿女由于受到老家成分的株连，在受教育和谋出路方面均遭受了政治歧视。悔当初，未能在出国留学之前办理家属来京落户之事！否则我和家人不会长期分居两地；也不会因此而从学科任务中分心而浪费精力和时间！

为了完成填补空白学科的任务，曾多次通过组织和朋友关系向有关领导提出解决家人来京落户的申请而毫无结果。无奈之下，便决定放弃学科任务而离京赴外地另谋出路。通过各种关系，先后联系过云南、黑龙江、山西、青海及西藏等地，其结果仍以失败告终。

正在走投无路之际，小平同志于 1975 年复出。当时，耀邦同志来我院主持工作。为了调动科技人员积极性而着手解决包括部分科技人员两地分居在内的实际困难采取措施。然而，这些措施却随着小平同志在所谓“黑猫白猫”的“罪名”下被再次废黜而付诸东流。直到“四人帮”被粉碎后，小平同志再次复出，有关解决科技人员实际困难的措施才得以落实。我们夫妇在 50 岁时终于得以团聚。当地的一切政治歧视政策也烟消云散，小儿也有机会参加十年动乱结束后的第一批高考；女儿在自己努力下，也在当地被作为第一批国家公务员予以录取；大儿也当上了当地牛奶场的职工。

我那时的心情才真正感受到拨开乌云见晴天，轻装上阵无后顾之忧的愉快。

跨越

“四人帮”被粉碎后，小平同志再次出山，制定了改革开放的治国方略，为我中华民族的复兴指明了航向。全国人民的积极性和生产力获得了解放，各行各业展开了探索快速发展的途径。全国科技大会的召开，进一步调动了全国科教界知识分子为发展和繁荣我国科教事业的积极性。“中国科学院真菌地衣系统学开放实验室”作为全院，也是全国第一批 17 个对外开放的实验室（所）于 1985 年正式成立。开放实验室本着面向全国，面向世界，面向未来的方针，积极开展了真菌地衣的系统生物学研究。

在地衣真菌演化系统生物学的研究方面，我自 1962 年回国后，一直在思考并寻求如何使基于表型性状的地衣真菌系统分类学与其基因型相结合的途径和方法。我于 1975 年，曾使用常规方法从石耳标本中提取 DNA 的实验失败后，直到改革开放后的 1989 年，当我在美国杜克大学工作时得知 PCR 技术用于地衣真菌分子系统学分析后，才迅速地在我主持的中国科学院真菌地衣系统学开放实验室推开。先是在开放实验室及当时的真菌学会报告了该技术在生物演化系统学中的应用，接着在真菌地衣系统学开放实验室成立了真菌地衣分子系统学实验室，随即于 1991 年请美国杜克大学的 Ryttas Vilgalis 博士来华主讲，举办了一个月的真菌地衣分子系统学训练班，向国内外开放，从而启动了我国真菌地衣分子系统学的研究。

为了配合《中国地衣志》的编前研究与在研究基础上的编写，我编撰并出版了《中国地衣综览》[2]。在野外考察方面，沿着从西南至东北，从华东至西北的两个对角线，以及从华北至华南直至海南岛地区进行了广泛的野外考察，标本采集和室内分析研究。在此基础上，连同前人和同事们的标本积累，建立了我国第一个地衣标本室。该标本室由创建时的约一千号地衣标本已经发展至今天的 12 万份地衣标本的中国科学院微生物研究所国家菌物标本馆地衣标本室 (HMASL)。它已成为我国最大的，现代化管理的，供地衣系统与演化生物学研究和资源开发利用参考的地衣原型标本保藏基地。

为了强调包括地衣在内的菌物演化系统生物学与菌物原型标本在科学研究与资源开发中的重要性，曾明确提出并强调菌物多样性及其资源研发与三大存取系统密切结合的必要，亦即原型标本存取系统（标本馆）、地衣菌、藻培养物存取系统（地衣菌、藻物种培养物保藏库）和种系综合信息存取系统（分类学著作和数据库等）之间的密切结合[3]。在这期间，我们先后与瑞典、芬兰、美国、英国、德国等许多国家的真菌、地衣学家建立了人员交往、学术交流与标本交换关系。这对于我国地衣学的发展和繁荣产生了重要影响，是我国菌物科学事业空前繁荣的时代。

中国植物学会真菌学分会于 1980 年在北京成立。在学会的主持下，中国第一届地衣学研讨会在青岛召开。随着科学技术的飞速发展，Whittaker(1969) 的生物五界系统已被修订。其中真菌界 (Fungi) 的黏菌和假菌已被分别划归原生动物界 (Protozoa) 和管毛生物界 (Chromista)，留在真菌界的只包括狭义的真菌。由于习惯和传统，包括黏菌和假菌在内的 Whittaker 生物五界系统中的所谓“真菌界”，实际上可被理解为广义的“真菌”，又被称为“由真菌学家进行研究的生物”或“真菌联合体”等；在我国则被称之为“菌物”。中国菌物学会于 1993 年正式成立。在我主持菌物学会工作期间，本着民主、公平、公开的原则，为将学会办成全国菌物学家之家而和同事们努力工作。在《中国科学院中国孢子植物志编辑委员会》的主持下，我国第二届和第三届地衣学研讨会在北京召开；第四、五、六届地衣学研讨会在中国菌物学会地衣学专业委员会的主办以及在齐齐哈大学、山东农业大学和新疆大学的承办下，于 2011 年、2012 年和 2013 年分别在黑龙江齐齐哈尔、山东泰安和新疆乌鲁木齐召开。

我国地衣学科起步于二十世纪 60 年代之初，基本上是围绕地衣分类学进行研究。十年浩劫中包括刚起步的地衣分类学在内的全国生物分类学以及其他基础科学研究与教育事业被完全停止。在周总理的关心下全国科教事业有所恢复。中国科学院于 1973 年召开的《三志》工作会议重新启动了包括地衣分类学在内的全国生物分类研究，从而使我国生物分类学家们有可能继续为我国生物资源的开发提供信息存取系统。为了保护和发展有基础的科学领域，中国科学院于 1985 年率先在全国创立了一批向国内外开放的重点实验室。真菌地衣系统学开放研究实验室有幸被选入。这些事件的发生都为我国地衣分类学的发展和年轻科技人才的成长起到了关键性作用。

在地衣分类研究中，除发现并发表一系列新种和新记录种属以及新属、新科外，在表型与基因型相结合的综合研究中，以基于新二属系统的石耳科为模式的石耳目 (Umbilicariales J. C. Wei et al.) 新目于 2007 年 2 月发表后，便被新版 (2008) 的世界“菌物辞典”(Dictionary of the Fungi) 所收录。

二十一世纪之初，当我们在调查研究中发现地衣在干旱荒漠中所起的结皮固沙作用以后，便以“沙漠生物地毯工程”为目标，开展了荒漠地衣生理生态学研究。在对干旱荒漠优势地衣物种之一的石果衣 (Endocarpon pusillum Hedwig) 的光合速率测定及其固碳量的估算中发现，如果利用石果衣在腾格里沙漠实施生物地毯工程，便可将腾格里沙漠变为年固碳量为 34 万吨的碳汇[4]。在对石果衣菌藻的抗旱研究中发现，其地衣型真菌在经受 7 个月的干旱饥饿双胁迫之后仍能正常生活，而其中的共生藻柯氏复球藻 (Diplosphaera chodatii Bialosuknia) 则只能坚持两个月的干旱胁迫[5]。在石果衣中，共生藻柯氏复球藻被其共生菌组织包裹在内，犹如酷暑天人们借助房屋避暑一样。于是，我们对于帮助柯氏复球藻避暑的“房屋”，即其共生菌所具有的抗旱基因产生了兴趣。在对石果衣真菌的抗旱基因研究中，首先，构建了石果衣真菌全长序列的 cDNA 文库[6]；随后，又完成了石果衣真菌的全基因组测序及其基本生物学分析[7]。这既是构建的世界第一个地衣型真菌的全长序列 cDNA 文库，又是世界第一个正式发表的地衣型真菌全基因组测序及其基本生物学分析。

这正是以地衣生物多样性及其系统演化与资源生物学为基础，在面向国家经济建设需求，面向世界科学前沿的方针指引下，以“沙漠生物地毯工程”为目标的地衣生物学科研进展和现状，也是我国地衣学科在发展中的跨越式进展。

学风

所谓学风，其词意为学术界之风气，即学术界流行的爱好和习惯。学风正与不正取决于学者的诚信、人格和道德。自然科学领域的学风取决于科学家的诚信、人格和道德，即科学道德。在科学活动中被称为三项主罪 (cardinalsins) 的伪造 (fabrication)、弄虚作假 (falsification) 和剽窃 (plagiarism)，简称“FFP”的科研不端行为[8]是与科学家的诚信、人格和道德背道而驰的。实际上，把别人的工作当成自己的工作，把别人的发现当作自己的发现并私下赠予第三者均构成剽窃。

苗德岁在美国科学三院撰写的《科研道德——倡导负责行为》一书的中译版的译者序中指出：“在人际交往方面，诚实是信任的基础；在科研工作中，同样如此。因而，科研人员之间的相互信任，是以科研人员本身的诚实为前提的。事实上，科学的进展首先建立在科研人员之间相互信任的基础之上，其次才是社会公众对科学事业和科学人员的信任。换言之，如果说科学研究旨在探求真理的话，那么科研人员的道德底线应该是诚实。”[9]赫胥黎在致朋友的信中说：“你对科学圣殿内发生的令人惊讶的事情一无所知。我担心科学并不比人类活动的其他任何领域更为纯洁，尽管它理应如此。仅诉诸道德是无济于事的，还得让公众的知情和了解起点作用。”[10,11]爱因斯坦有句名言：“大多数人说，是才智造就了伟大的科学家。他们错了：是人格。”[12]

在与国外科学家的交流中，多数科学家在做学问和做人方面是值得我们敬重的。然而，作为对年轻人的借鉴，有必要在此回顾一下我自己与犯有科学不端行为的科学家的交往经历。

通过中国科学院与英国皇家学会之间的人员交换计划，我于 1989 年 9 月前往英国有关科研机构进行世界范围石耳科地衣的研究。当我在林奈学会标本馆*里发现林奈关于两种石耳地衣的原始标本被后人混淆并张冠李戴，尤其令我深感惊奇的是，从瑞士的 Frey，挪威的 Scholander, 到美国的 Llano 这些专门研究石耳科地衣的大师们，竟无一人发现这一错误！于是我便围绕这一问题撰写了一篇文章。我将该文的初稿于 1989 年 11 月离英赴美前夕，寄往北国一位著名的地衣学家，一向受我敬重的长者 R. Santesson，希望能得到他的意见和建议。此后，我一方面在等待对方的回音，另方面在访美期间，根据新的资料继续修改并完善我的文稿。然而，出我意料的是，等了整整三年有余，却未能得到曾受我一向敬重的长者的任何信息！！直到 1993 年，当我的《亚洲石耳科》地衣专著即将问世之时，我不得不主动写了一封催问受我敬重的长者的信件。更加令我吃惊的是，曾受我一向敬重的长者的回信居然毫无愧疚地告诉我，某人正在伦敦林奈学会标本馆里进行这件事，并要我等待他即将发表的文章！！看来，如果我不催问，这位长者根本就没有打算给我回信！！这正是把别人的发现当作自己的发现并私下赠予第三者的剽窃行为！！使我百思不解的是，在新千年即将来临的 19 世纪末，竟然还会有如此不诚实的科学家，而且是我一直敬重的科学家，实在令我失望！！！这使我想起了爱因斯坦关于造就科学家的是人格的名言。实际上，自然科学领域的学风取决于科学家的诚信、人格和道德，即科学道德，这是科学家的底线。

当我和我的芬兰朋友 T. Ahti 教授在针对以 *Cladonia linearis* Evans 为基原异名的 *Gymnoderma lineare* (Evans) Yaoshimura 建立新属岛蕊属 (*Cetradonia*，2002) 时，由于参考了 1998 年发表于美国植物学杂志上基于 S. Stenroos 的分子数据而将 *Gymnoderma lineare* 置于远离石蕊科系统地位的文章，我们又以岛蕊属为模式，进而建立了新科岛蕊科 (Cetradoniaceae，2002)。虽然文章发表了，但是，我对于岛蕊属如此远离石蕊科的那篇文章仍有质疑。于是，我和我的学生便在核查 GenBank 中 1998 年文章的原作者提供的 *Cetradonia linearis*

* 瑞典作为生物科学的开创者 Linnaeushe 和地衣学之父 Acharius 的故乡是我改革开放后出国交流的第一站。然而，他们的珍贵标本却不在瑞典的标本馆保藏。前者的标本保藏在英国伦敦的林奈学会，后者的标本则保存在芬兰的赫尔辛基大学。

(=*Gymnoderma lineare*) 的 SSU rDNA 碱基序列数据时发现，在 1645 个碱基序列中，竟有 185 个碱基未被测出，亦即仅有 1460 个碱基序列参与了系统发育的分析。基于如此残缺不全的 SSU rDNA 碱基序列数据所进行的系统发育分析，其系统发育树的可靠性究竟有多高？岛蕊属能如此远离石蕊科吗？我带着这些质疑，由我的学生开始对岛蕊属的模式种 Cetradonia linearis 的 SSU rDNA 碱基序列进行了尽可能完整的测序工作。当我们通过大量实验将包括那 185 个未知碱基序列完全测出并进行了系统发育分析之后发现，它从远离石蕊科的位置回归至石蕊科内。我和我的学生基于这一新的结果撰写了新的文稿并发给了我的芬兰朋友 T. Ahti 教授，2002 年文章的合作者，以便共同发表。我的芬兰朋友 T. Ahti 教授告诉我，他的学生 S. Stenroos，即 1998 年文章的主作者，希望加入本文作者行列的意向；同时也考虑到涉及推翻她 1998 年文章的有关结论，为了不使原作者被动，我不仅同意 S. Stenroos 参加，而且还向 1998 年文章的另一原作者 P. Depriet 发去了加入我们新文稿的邀请。

不久后，S. Stenroos 令其助手重复了我们的实验结果之后告诉我，我们的实验结果可能因污染而数据有误，并希望能同意其助手也被作为作者之一，自己能作为通讯作者之一。鉴于我朋友学生的要求我也同意了。

但是，为了进一步讨论双方数据的可靠性，我要求对方提供其所重复的所谓正确的碱基序列。经过我们仔细分析比对后发现，S. Stenroos 提供的 SSU rDNA 碱基序列根本就不是该地衣型真菌的序列，而是其共生藻 Trebouxia erici 的碱基序列！因此，不是我们搞错了，而是 S. Stenroos 将该种地衣的共生藻碱基序列误认为是共生菌的序列！当对方得知这一结果后，便不好意思地回信说：“我从未错过，怎么这次就错了！”

后来，S. Stenroos 在文稿修改和资料补充方面做了一些贡献，至于其助手，除了在实验中错将共生藻序列误为共生菌序列之外，毫无贡献可言。最令我不可接受的是，在文章发表前的校样中，我和 S. Stenroos 同是合作通讯作者，然而，当文章正式发表后，通讯作者却只剩下 S. Stenroos 一人，并以刊物只允许一位通讯作者为由作为对我的回答！！！即使真如此，也应听取我的意见，无论如何也绝不能自作主张地保留她一人为通讯作者！！更令我百思不解的是，在新千年之初，竟然还会有这样缺乏科学道德的年轻人！！实在令人鄙视！！这又使我想起了两句中国成语：“得寸进尺”和“喧宾夺主”。

这些点滴往事使我想起了我的朋友，著名真菌学家，美国田纳西大学的 Ronald Petersen 教授曾向我提醒过：“要当心别人侵占你的成果哟！”我希望我国年轻地衣学家，一方面，能从上述二人的不端行为中吸取教训并与之划清界限；另方面，更重要的是，我们要堂堂正正做人，踏踏实实做学问，在科学征途上，严谨求实，开拓创新。

展望

半个多世纪以来，我为我国地衣科学的贡献仅在于遵照我国真菌学的奠基人戴芳澜的夙愿，填补了学科空白；为后来者铺了路，期盼着后来者的更大作为。我现在能做的仅限于提出自己的看法，供后来者参考。

所谓生物多样性，实际上是指存在于地球生物圈内多种多样生态系统中的，含有多种多样基因的物种多样性。一个物种便是一个基因库，物种是基因的载体。基因离开了生物体本身便毫无意义。有了物种多样性才有基因多样性；没有物种多样性便没有基因多样性。因此，生物多样性的保护，归根结底是物种多样性的保护。物种多样性的保护又在于生态系统多样性的保护。

对于地衣物种多样性的研究和保护在于分类学中三大存取系统的密切结合。首先，是从自然界地衣物种多样性中采集地衣标本，将其中的一部分作为地衣原型保存于标本馆，即所谓地衣原型证据存取系统。其次，将其中的另一部分，进行地衣型菌、藻的分离培养和保藏，作为地衣型菌、藻物种资源，同时也是基因资源的活体培养物存取系统，即菌种和藻种保藏库。同时，对于地衣原型及其活体生物学进行的研究成果，包括论文著作及其数据库作为地衣物种和基因综合信息的存取系统。

地衣分类学家，或称地衣演化系统生物学家，在对地球生物圈内生态系统多样性中的地衣物种多样性进行采集、分析、识别、归类、命名、描述等一系列研究的基础上，使其研究结果不断丰富上述三大存取系统，是新世纪生命科学研究和生物资源研究开发和利用不可缺少的上游环节和重要支撑系统与后盾。这三大存取系统越丰富，则资源筛选的基数就越大，筛选效率就越高。如果能将我国地衣分类学或地衣系统演化生物学及其三大存取系统作为一个整体予以重视并给予强力资助，必将在生物经济时代的生物资源创新与人类可持续发展中起到重要作用。

在进行《中国地衣志》的编前研究和编写工作及《真菌地衣系统学开放实验室》运行期间，我高兴地看到一批年轻的地衣学博士陆续成长起来。在《中国科学院中国孢子植物志编辑委员会》主持下，我积极建议启用一批年轻的地衣学博士作为实习主编，通过承担十二五国家基金委重大项目中的《中国地衣志》任务，使我国这批年轻的地衣学博士得到锻炼和成长，成为国家地衣学科的栋梁。

我国地衣学的发展应在分类学的基础上，围绕国家经济建设需求，沿着世界科学前沿，面向环境与健康，进行尚未开垦的“处女地”——地衣资源的开发，是我国地衣学发展的重要途径。对于地衣及其菌、藻的初生与次生代谢产物在人类健康上的潜力以及在优化人类环境方面的功能生物学进行研究与开发是我国地衣学家不可推卸的责任。在解决国家需求中不断克服遇到的科学技术难题，在攻克科学技术难题中完成国家需求任务，以任务带学科，学科促任务的方式促进我国地衣科学的发展与繁荣。这一光荣任务历史性地落在我国年轻地衣学家的肩上。

在进行地衣生物学，包括地衣演化系统生物学、地衣及其菌藻功能生物学等的研究中，表型组学、基因组学与环境组学相结合是我们的努力方向；尤其是对于作为抗逆境基因资源宝库的地衣型真菌的研发应该受到重视。在主客观条件许可的前提下将尽力向前推进。

在研究生培养中，我所遵循的理念是希望培养出具有重视人格和科学道德，具有求实、勤奋、实干、谦虚和创新精神的年轻人。在这一过程中，导师和学生之间的关系是很重要的。《科研道德——倡导负责行为》一书告诉我们：“成功的师生关系是建立在相互尊重、相互信任、相互理解和相互同情的基础之上的。师生关系能够持续到科学生涯的各个阶段，正因为如此，这种关系有时又被称为师徒关系而不只是一般的师生关系。”“尊重导师、忠诚于研究群体、热衷于科学事业、对科研课题全力以赴、仔细做实验、实验记录完整准确、实验结果汇报精确、对口头及书面报告和论文的发表要尽心尽力。”[13,14]这些忠告，无论是对于我国年轻的地衣学家，还是有意投身于我国地衣学家行列的青年学生，都是极为有益的。

我衷心期待我国地衣学不断发展与繁荣。我国地衣学的发展与繁荣，寄希望于我国年轻地衣学家之间在相互信任和团结基础上的勤奋、严谨、开拓与创新。愿我国涌现出更多世界一流的地衣学家。

参考文献

[1] 刘华杰，贾泽峰，任强，周启明(主编). 2011. 中国地衣学现状与潜力：祝贺魏江春院士八旬华诞地衣学文集. 北京：科学出版社 .(Liu Huajie et al. Ed. 2011. The present stayus and potentialities of the lichenology in china: a collection of lichenological papers in congratulation of the eightieth birthday of cas member, dr. sc. prof. Wei Jiangchun. Beijing: Science Press.)

[2] Wei Jiang-Chun. 1991. An Enumeration of Lichens in China. Beijing: International Academic Publishers: 1-278.

[3] 魏江春 . 2010. 菌物生物多样性与人类可持续发展 . 中国科学院院刊 , 25(6):645-650。

[4] Ding LiPing, Zhou QiMing, WEI Jiang Chun. 2013. Estimation of *Endocarpon pusillum* Hedwig carbon budget in the Tengger Desert based on its photosynthetic rate. SCIENCE CHINA Life Sciences 56(9): 848–855.

[5] Zhang Tao, WEI JiangChun. 2011. Survival analyses of symbionts isolated from Endocarpon pusillum Hedwig to desiccation and starvation stress. SCIENCE CHINA Life Sciences, 54(5):480-489.

[6] Yan-Yan WANG, Tao ZHANG, Qi-Ming ZHOU and Jiang-Chun WEI. 2011. Construction and characterization of a full-length cDNA library from mycobiont of Endocarpon pusillum (lichen-forming Ascomycota). World Journal of Microbiology and Biotechnology, 27(12): 2879-2884.

[7] Yan-Yan Wang, Bin Liu, Xin-Yu Zhang, Qi-Ming Zhou, Tao Zhang, Hui Li, Yu-Fei Yu, Xiao-Ling Zhang, Xi-Yan Hao, Meng Wang, Lei Wang, Jiang-Chun Wei. 2014. Genome characteristics reveal the impact of lichenization on lichen-forming fungus Endocarpon pusillum Hedwig (Verrucariales, Ascomycota). BMC genomics, 15:34

[8] 美国医学科学院，美国科学三院，国家科研委员会(撰), 苗德岁(译). 2007. 科研道德——倡导负责行为 . 北京：北京大学出版社：7-8.

[9] 同上 , pp.1-2.

[10] 同上 , pp.4.

[11] Huxley L. 1900. Life and Letters of Thomas Henry Huxley. London: Macmillan.

[12] 美国医学科学院，美国科学三院，国家科研委员会(撰)， 苗德岁(译). 2007. 科研道德——倡导负责行为 . 北京：北京大学出版社：18.

[13] 同上 , pp.50-51.

[14] NAS. 1997. Advisor, techer, role model, friend: on being a mentor to students in science and engineering. Washington, DC: National Academy Press.

A glance back over the past

Wei Jiang-Chun

The State Key Laboratory of Mycology, Inst. of Microbiology, Chinese Academy of Sciences

Abstract: Author of the present paper briefly looks back on his past more than eighty years. The past events are divided into the following seven parts: studies, compilation of cryptogamic flora, farce, family, leap forward, style of study, and prospect of the lichenology in China is given by the author.

As to the style of study, the author reviewed the unpleasant past events made by the two dishonest lichenologists he met.

Keywords: studies, cryptogamic flora, farce, family, leap forward, style of study, prospect

科学论文选集

目录

第一篇　地衣学概论
Introduction to Lichenology

第二篇　地衣物种多样性及系统演化与资源生物学
Lichen Species Diversity, their Systematic Evolution and Resources Biology

第三章　新分类群 Taxa New to Science ……404

第四章　中国新纪录 Taxa New to China ……542

第五篇　部分药用真菌
Some Medicinal Fungi

第六篇　真菌学科发展
Development of the Fungal Sciences

第一篇　地衣学概论 Introduction to Lichenology

第一章　地衣形态学
Morphology of Lichens

地衣形态是以地衣型真菌的菌丝体组织内包含着相应的共生藻或蓝细菌而形成的具有各种外形的地衣体。地衣体在形态上的多样性不仅表现在生长型上的壳状、鳞状、叶状、枝状和丝状，而且还表现在多种多样的营养体、营养繁殖体，及奇特的内部结构上。至于地衣子实体的形态与发育则基本上与非地衣型真菌类似。

一、地衣体的形态及内部结构

用徒手切片法将叶状地衣体切成用于显微镜下观察的薄片，一幅奇妙的图案便展现在显微镜视野中：由上向下分成上皮层、绿色的藻细胞层（简称藻层）、蛛网状的无色菌

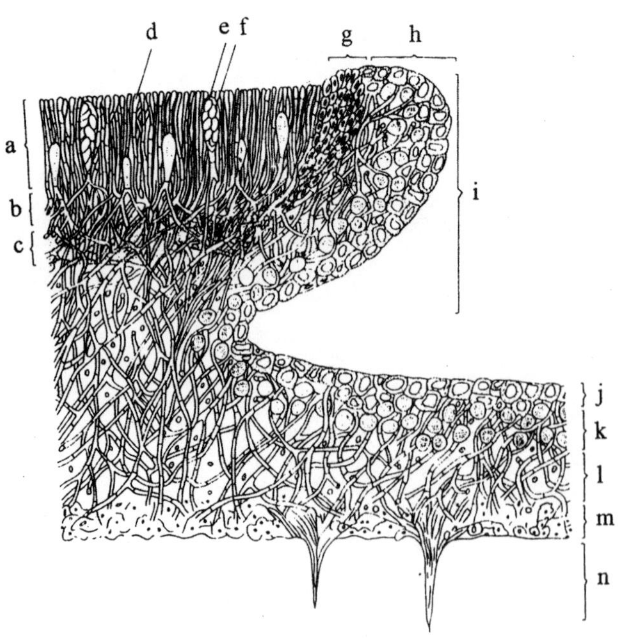

图 6-1-1 地衣子囊盘及叶状体的结构

a. 子囊层；b. 囊层基；c. 子实下层；d. 侧丝；e. 子囊；f. 子囊孢子；g. 果壳边缘；h. 果托边缘；i. 果托；j. 上皮层（假薄壁组织）；k. 藻层；l. 髓层；m. 下皮层；n. 假根。

本文原载于《菌物学大全》（裘维蕃主编）: 445-548, 1998.

丝区，亦即髓层，下皮层以及由菌丝束形成的假根从下皮层向外伸出（图6-1-1）。这种层次分明的地衣体被称为异层型地衣体。具有这种地衣体的地衣为异层型地衣（heteromerous lichen, heterolichen）。枝状地衣体的解剖特征类似于将叶状地衣体的下表面向内卷成一个圆筒状的结构。这种圆筒状结构的最外层为假皮层组织；紧贴皮层内侧的是藻层；再向内则是由菌丝组成的髓层。圆筒状结构的内侧分空心型和实心型两类。空心型包括由稀疏菌丝组成蛛网状髓层所形成，如小孢发属（*Bryoria*）和树发属（*Alectoria*）；或在髓层的内侧由坚韧的硬组织或称纤维骨架组织（stereome）构成的管状内壁，使髓层处于藻层和管状结构外壁之间，如石蕊属（*Cladonia*）；实心型包括有密集的髓层所形成，如扁枝衣（*Evernia*）和球粉衣属（*Sphaerophorus*）；或由坚韧的菌索作为中轴位于管状结构的中央，使髓层处于藻层和中轴之间，如松萝属（*Usnea,* 图6-1-2）；此外，由坚韧的菌索构成的枝状体外围或多或少地散生小鳞芽，如珊瑚枝属（*Stereocaulon*）。

另一种类型的地衣体在上、下皮层之间并无明显的分层现象。因此共生的蓝细菌往往呈念珠状成串地或多或少均匀地散布于上、下皮层之间疏松蛛网状菌丝区的整个厚度之内。这种类型的地衣体被称为同层型地衣体（图6-1-3）。具有这种地衣体的地衣为同层型地衣（homoeomerous lichen, homogeneous lichen）。

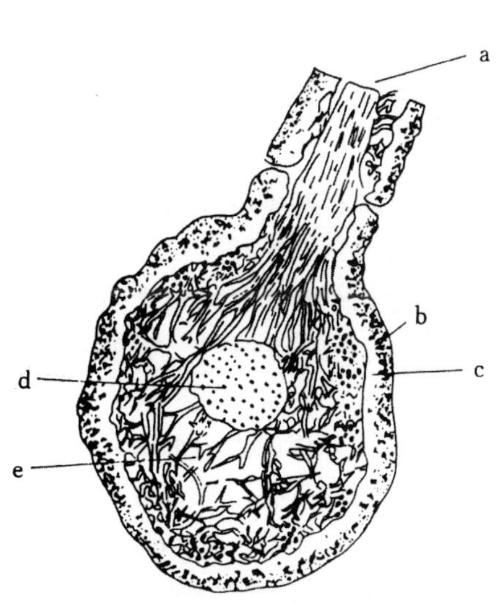

图 6-1-2 光松萝（*Usnea glabrata*）
枝状体的横断面
（仿 Ozenda,1963）
a. 纤毛；b. 皮层；c. 藻层；
d. 中轴；e. 髓层。

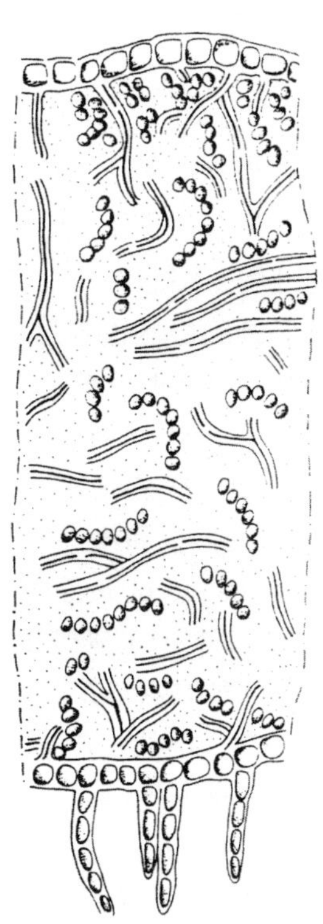

图 6-1-3 裸果猫耳衣
（*Leptogium hildenbrandii*）
叶状体横断面
（仿Ozenaa,1963）

上、下皮层均为单层细胞组成的假薄壁组织。下皮层伸出的为单个菌丝构成的茸毛型假根。

上、下皮层之间为念珠蓝细菌和菌丝借助于粘胶物而交织在一起.

（一）地衣体的结构

1. 皮层及各种菌丝组织

地衣的皮层是由一面分裂的菌丝交织在一起所形成的菌丝组织，因而不同于高等植物所特有的真组织，故称假组织。地衣的皮层实际上也是假皮层，但是通常则简称为“皮层”。真组织的发育在地衣中是极为稀少的。这样的组织是由三面分裂的细胞形成的。高等植物所特有的细胞分裂法仅被发现于若干腔囊菌的子座中和某些地衣，如多囊衣属(*Polyblastia*)及矮疣衣属(*Staurothele*)的砖壁型子囊孢子中。除了这些特别情况之外，几乎所有的地衣组织都是假组织，即菌丝组织。在不同菌丝之间的交织联合很紧密，以至于难以分辨出其中的单个菌丝。菌丝组织的断面在显微镜下观察时，和高等植物的真组织相类似，因而，就起用了能反映这种相似性的相应名称。类似于高等植物薄壁组织的菌丝组织，称假薄壁组织(pseudoparenchyma 或 paraplectenchyma)（图 6-1-1，6-1-3）。由于菌丝细胞壁强烈胶化而形成的类似于高等植物长轴组织(prosenchyma)或厚角组织(collenchyma) 的菌丝组织称假长轴组织(prosoplectenchyma)。

(1) 假薄壁组织(paraplectenchyma)：假薄壁组织是由短而无一定方向的薄壁细胞菌丝交织在一起而构成的。可以由挤压在一起的不同菌丝的短而圆的薄壁细胞而产生，最终形成一个有棱角的等径细胞的假组织。起着支撑和保护作用的皮层常常是由这一组织构成的。在较多地衣中，上、下皮层都是由多层次的假薄壁组织所构成。在有些地衣，如猫耳属（*Leptogium*）中，上皮层或上、下皮层仅由一层假薄壁组织所构成（图 6-1-3）。

(2) 假长轴组织(prosoplectenchyma)：绝大多数假长轴组织是具有多棱角的或细胞形状不规则的网状结构发育而来的。细胞壁强烈胶化而形成一个同质型的菌丝团块，在这个菌丝团块中已不可能区分出单个的菌丝。在假组织的生长期间细胞腔的形状常常发生变化，因而假组织的外貌在细微结构方面能有相当大的变化。不仅是短细胞的菌丝，而且甚至呈平行方向的长细胞菌丝也能形成假长轴组织。

和假薄壁组织一样，假长轴组织常常是皮层中存在的保护组织和支撑组织。在有些地衣中假薄壁组织在上皮层中，而假长轴组织则在下皮层中。

假长轴组织也发现在一些枝状地衣中，如松萝属(*Usnea*) 中的髓部形成中轴型菌索(图 6-1-2)。此外，地衣的假根和缘毛看来也都是假长轴组织。

(3) 假珊栏组织(palisade plectenchyma)：该组织是由或多或少直立走向或垂直的短菌丝构成。他们在外貌上类似于维管束植物珊栏状排列的叶内组织。在一些地衣，如染料衣属(*Roccella*)及球针黄叶属(*Relicina*)的皮层中存在有假珊栏组织。

绝大多数地衣是具有皮层的。皮层往往是有颜色的，并覆盖在地衣体的上表面，称上皮层。有许多地衣也具有下皮层。所有的假组织类型都能构成地衣的皮层。有时两种不同的组织构成一个皮层，因而该皮层便呈现为双层构造。在长角树花（*Ramalina siliquosa*）中，外皮层是由稍平行的至网状排列的菌丝所构成；而其他厚皮层则由帚状排列而具有胶化壁的菌丝所构成。这两种假组织的菌丝彼此成直角排列着。在犬地卷(*Peltigera canina*)中，上皮层的末端细胞呈交织状无定向的无色透明茸毛，使皮层的表面呈现毡状外貌。

有些地衣特有的次生代谢产物，如一些色素［石黄酮（parietin），松萝酸（usnic-acid）］和无色物质［黑茶渍素（atranorin），地衣氧杂蒽酮（lichexanthone）］存在于皮层或外皮层中，使地衣呈现为五彩缤纷的外貌。在蜈蚣衣科（Physciaceae）及鳞网衣属（*Psora*）的一些种类的皮层表面还分泌一些草酸盐或碳酸盐或相关晶体，甚至是其他无定形的物质或地衣最外层细胞的坏死层的积累，而使皮层表面呈现明显的白色粉霜层。

上、下皮层表面微形态的多样性因地衣种类的不同而各异。有些叶状地衣上皮层表面呈现不同程度的网状裂纹，如网大叶梅（*Parmotrema reticulatum*）或更加明显的网状棱脊如网脊平盘石耳（*Umbilicaria lyngei*）；有些叶状地衣下皮层的表面呈现精致的乳头状微突，如硬石耳（*Umbilicaria rigida*）或更加明显而粗大的乳突，如疱脐衣属（*Lasallia*）。

2. 藻层

藻层位于上皮层与髓层之间菌丝疏松交织的部位（图 6-1-1）。这个部位在深度上既能满足共生藻细胞所需要的非直射光照强度，也具有足够的空间以储存空气和水分，这是共生藻进行光合作用的良好环境。

地衣的菌、藻互惠共生关系正是发生在藻层内的菌、藻之间的相互接触中。菌丝和藻细胞之间的联系方式在不同地衣中有很大的变化。菌丝与藻细胞之间要莫缺乏直接的接触，要莫菌丝在不同程度上缠绕和包围着藻细胞。有些地衣的菌丝细胞以吸盘紧紧地压住藻细胞。另一些地衣的菌丝吸器穿入藻细胞膜之内。穿入活的藻细胞内的菌丝吸器可以杀死藻细胞。有些地衣中的藻细胞及正在侵入的菌丝吸器同时分裂。藻的两个子细胞被菌丝吸器的两个分枝所缠绕。

在藻层中藻细胞借助于有丝分裂和静孢子进行繁殖。在共球藻属（*Trebouxia*）中，一个藻细胞的原生质体分裂成若干个原生质体。每一个原生质体随后便分泌出细胞壁而形成新的静孢子。这些静孢子借助于母细胞壁的破裂而被释放出来。这一繁殖过程的各个阶段在地衣体的徒手切片中能够很容易地被观察到。借助于游动孢子形成的有性繁殖在地衣体中从未见过。游动孢子阶段只在共生藻的人工纯培养中才能见到。藻层的厚度及其连续性程度因地衣种属的不同而有所变化。

3. 髓层

髓层是由蛛网状疏松的交织菌丝体所构成，呈疏松交织的纤维状或棉絮状外貌（图 6-1-1，图 6-1-2）。由于它具有大量空隙而比任何其他假组织具有较大的持水量以及对于甘露（糖）醇及绝大多数地衣产物具有较大的储存空间。这些地衣产物固结在髓层菌丝细胞外表面。在缺乏下皮层的地衣中，如地卷属（*Peltigera*）的下表面经常具有海绵状与毡状外貌的脉纹。在这一属中脉纹是由在特定部位增殖的髓层菌丝所组成。

（二）地衣体的生长型

地衣体的生长型是地衣生长中形成的外部形状。这些外部形状是地衣给人们的直观留下的第一印象。因此，在传统上地衣一直被划分为壳状、叶状、枝状 3 大生长类型。尽管它并不反映地衣在自然系统中的演化关系，但是，作为地衣的诸多特征之一，对于人们认识、描述和掌握地衣综合特征来说是不可缺少而重要的直观特征之一。

1. 壳状地衣体

壳状地衣体通常是指在形体上微小的那些地衣。它们可分为 3 个亚型：

（1）亚型一通常是既无上、下皮层结构，更无内部分层现象，只是由菌丝缠绕的藻细胞群排列成不定形至圆形的黄绿色或灰绿色至灰白色粉斑，如癞屑衣（*Lepraria*）等属。

（2）亚型二的地衣体完全生长在它的基物内部，仅以其微型子实体外露于基物表面或因含有色素的地衣体使基物变色而标志着地衣的存在。生长在岩石内部的地衣称石内生的(endolithic)，如黑瘤衣属(*Buellia*）和网衣属(*Lecidea*)中的少数种。它们生长在岩石表面以下，但是能在岩石晶体之间形成分层的，多少有些连续性的组织：菌丝缠着共球藻(*Trebouxia*）的黑色藻层约 1mm 厚，以及可达 10mm 厚的白色髓层(Hale，1983)。而穿透于树木内部的种类称树皮内生的(*endophloeodic*)，如针孢衣(*Leptorhaphis epidermidis*)。该种通常生长在桦树皮内，仅以微小黑点状子囊壳孔外露于白色的树皮外表。

（3）亚型三的地衣呈痂状，即类似于伤口表面上凝结而成的块状物。它们通常生长在基物表面而形成一层痂状物。它的外形、颜色及厚度因种属不同而各异。它们有的呈现为黄绿色近圆形的，如地图衣(*Rhizocarpon geographicum*)，米黄色不定形的，如赤星衣(*Haematomma*)斑块或圆形而周缘呈明显小裂瓣型深裂的黄绿色的，如墙茶渍(*Lecanora muralis*)或桔黄色的，如莲座粉橙衣(*Caloplaca decipiens*)小鳞片。这一类型虽然都是有上皮层，但是，它们都缺乏下皮层，借助于髓层而直接紧固于基物表面。

绝大多数壳状地衣体是由称做小网块(areoles）的小鳞片所构成的。在这一地衣体的下面菌丝通常比地衣体的主要部分生长得较快，并在地衣体周围形成一个稍大于地衣体周缘的微薄菌丝层。这个微薄的菌丝层通常为暗色至黑色，有时为白色，称地衣原体(prothallus)。

2. 叶状地衣体

叶状地衣体通常是具有背腹构造的叶片状地衣。它们借助于腹面的脐，如石耳科(Umbilicariaceae）的种类，或假根或假根状菌丝而紧固于基物表面。它们的直径 1～50cm。在我国西藏的皱梅衣(*Flavoparmelia caperata*)直径可达 34cm 以上（魏、姜，1986)。叶状地衣体又可分为 3 个亚型：

（1）亚型一为脐叶型地衣体。该亚型通常以其腹面具有作为固着器的脐状突起为特征。该亚型包括较小的鳞状，如脐鳞属(*Rhizoplaca*）及较大的叶状，如大叶石耳(*Umbilicaria mammulata*)。有些石耳属种类的腹面往往也有不少假根，但是，它们并不起固着作用，而在水分关系中具有重要意义(Larson，1981)。它们的背面（上表面）和腹面（下表面）都具有良好的皮层。最典型的代表为盘果类石耳科（Umbilicariaceae)的全部种类及属于核果类的皮果衣属(*Dermatocarpon*)。此外还有聚盘衣属(*Glypholecia*)，脐叶属（*Omphalodium*)，皮盘衣属（*Dermatiscum*)，脐鳞属（*Rhizoplaca*)，黄盾衣属(*Xanthopeltis*)以及粉芽盾衣(*Peltula euploca*)等都为这一亚型。它们的直径由 1 cm 左右，如小石耳(*Umbilicaria kisovana*)及粉芽盾衣(*Peltula euploca*)，可达 30cm，如大叶石耳(*Umbilicaria mammulata*)。

(2) 亚型二为多型性叶状体。它们的直径通常较大，从若干厘米到几十个厘米。皱梅衣的直径可达 34cm(魏、姜，1986)。它们的多型性不仅表现在直径大小上，而且还表现在与基物的固着方式上，以及叶缘的裂形和叶状体的解剖构造上。至于背腹叶面微形态的多样性则更加复杂。

绝大多数叶状地衣的背腹两面都具有良好的皮层，如梅衣科(Parmeliaceae)，肺衣科(Lobariaceae) 等。有的叶状体则只有上皮层，如地卷属(*Peltigera*)。它们的腹面则缺乏下皮层，而代之以海绵型或棉絮状脉纹和绒毛。这些脉纹和扫帚状绒毛都是直接来自髓层菌丝的衍生结构。而梅衣科种类的叶状体整个腹面或部分腹面是借助于假根固着于基物上。叶状体周缘的浅裂和深裂的多种外形因种类不同而各异。所有这些多型性特征在叶状地衣的分类鉴定中都具有重要的参考意义。

(3) 亚型三为胶质状地衣体。它们在干燥时呈软骨质状；在潮湿生境中则立刻吸水膨胀而呈胶质状。它们中的大多数种类都具有叶状的外形，如胶衣科(Collemataceae)。但是，也有些胶质地衣体近似于壳状或枝状的类型。胶衣属(*Collema*) 的绝大多数种类都缺乏皮层。猫耳属(*Leptogium*) 的皮层是由一层薄壁的等径细胞，即假薄壁组织所构成(图 6-1-3)。

3. 枝状地衣体

一般呈现为圆柱形、管形、灌木状、带状、线状至丝状外形。它们的内部结构多为异层型地衣体的枝状变体。枝状体的单枝可以细到用微米计量直径，如绒衣属(*Coenogonium*)，也可以粗壮到用毫米计量直径，如石蕊属(*Cladonia*)、珊瑚枝属(*Stereocaulon*)、松萝属(*Usnea*) 等。其长度可由数毫米，如绒衣属，到数米，如长松萝(*Usnea longissima*)。根据枝状地衣体的特点可分为下列 6 个亚型：

(1) 亚型一为直立的管状体。它们在外形上类似于柱状体，但是并非实心的，而是空心的管状体，如地茶(*Thamnolia vermicularis*)，喇叭石蕊(*Cladonia pyxidata*)、北极地指衣(*Dactylina arctica*)等。石蕊属直立的管状体是从围绕着性器官的生殖组织而形成的，是有性子实体的一部分，其顶端产生子囊盘。这种管状体称之为果柄(podetia)(图版 I-d)。

(2) 亚型二为直立的柱状体。它们的柱状体呈现为实心的圆柱形或近圆柱形，无分枝或只有简单的分枝，如珊瑚枝属(*Stereocaulon*)，柱衣属(*Pilophorus*)。这种柱状体是由一部分鳞片状或颗粒状营养体向上突起并伸长，因而不是有性子实体的一部分；有性子实体的原基只是在业已形成的这种柱体顶端才出现的。因而称为拟果柄(pseudopodetia)。

(3) 亚型三为灌木状地衣体。该亚型的主要特征是呈现为直立而具有繁茂分枝的微小灌木型外貌。它们包括多数的石蕊(*Cladonia*)、鹿蕊(*Cladina*) 及少数松萝(*Usnea* spp.)和槽枝衣(*Sulcaria sulcata*)等。

(4) 亚型四为带状体。它们的代表有扁枝衣属(*Evernia*)及树花衣属(*Ramalina*)的种类。

(5) 亚型五为线状体。它们通常是以向下悬垂的方式生长在树木上。这种地衣体看起来类似于挂在树上的“挂面”，因而在我国民间往往称它们为“树挂面”或“树挂子”，如长松萝（*Usnea longissima*）等（图6-0-c）；有的则类似于长发，故当地人称它们为“树头发”，或“头发七”，如小孢发属（*Bryoria*） 及树发属（*Alectoria*）的某些种类。

（6）亚型六为丝状体。这种地衣体通常是由单个菌丝缠绕着单行排列的丝状绿藻体（*Cladophora*）而形成丛生的细微丝状体，如林氏绒衣（*Coenogonium linkii*）（图6-1-4：a）；或皮层发育良好的微型丝状体，如柔毛毡衣（*Ephebe pubescens*）（图6-1-4：b）。

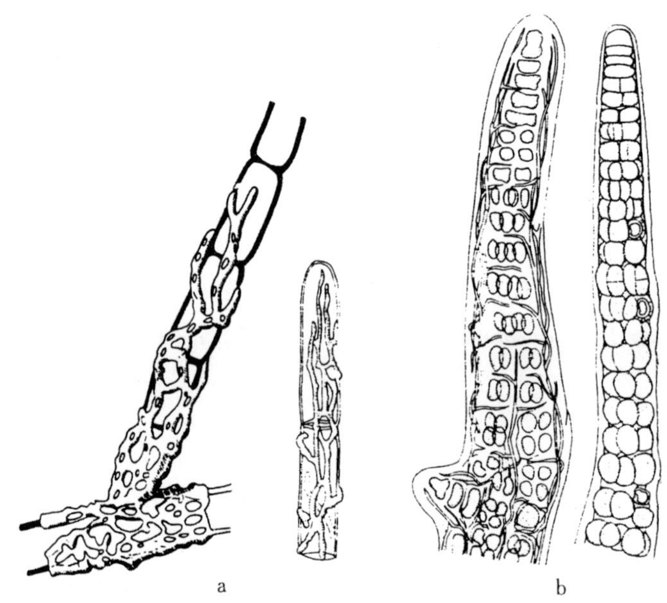

图6-1-4 丝状地衣体

a. 林氏绒衣（*Coenogonium linkii*）；b. 柔毛毡衣（*Ephebe pubescens*）.

根据壳状、叶状和枝状地衣3大生长型的不同特征又细分为上述的12个亚型。在这些亚型之间还存在着一些难以归属的中间型。

（三）地衣体的营养结构及营养性繁殖体

地衣具有独特的营养结构与营养性繁殖体。有些营养结构，如假根、茸毛及缘毛等在非地衣型真菌中也有存在。然而，像杯点（cyphellae）、假杯点（pseudocyphellae）、粉芽（soredia）、裂芽（isidia）、小裂片（lobules）及衣瘿（cephalodia）则只存在于地衣中。它们是地衣的独特结构。有的结构有传播与繁殖作用，有的则可能有某些生理功能。所有这些结构在地衣分类鉴定中都具有重要意义。

1. 杯点

杯点是位于牛皮衣属（*Sticta*）地衣腹面的圆形凹穴。杯点周围的下皮层延伸而围成稍微凸起的边缘，包围着一个小小的火山口状的凹穴。这个凹穴类似于不同深度的小碗。小碗底部及内侧组织的质地较疏松，由一层球形细胞构成平滑的内侧（图6-1-5）。

2．假杯点

假杯点是位于若干叶状地衣体的背面或腹面以及某些枝状地衣体的表面。假杯点往往较小，直径多在1ram左右，而且很少分化，只不过是一个不规则的开口而已。髓层菌丝通过这个开口而外露（图6-1-6）。在叶状的假杯点属（*Pseudocyphellaria*）中假杯点呈现为不规则的小疣。其开孔处被呈网状交错的短菌丝细胞所充满。在老熟而不规则的大型假杯点中，藻细胞往往混杂在疏松的菌丝之间，呈现为类似于粉芽堆的假杯点，它可能还具有传播与营养性繁殖功能。在斑叶属（*&trelia*）的种类中，如领斑叶（*C. collata*）及星点梅属（*Punctelia*）的种类中，如粉斑星点梅（*P. borreri*）等以及一些枝状地衣，如散藓茎衣（*Cornicularia divergens*）及砖孢发属的种类（*Oropogon* spp.）中，表皮上呈现为外形不规则的凹穴使髓层外露；或者它们形成一些穿透至髓的不规则形小孔。假杯点的大小在绝大多数种类中是比较稳定的，因而在分类学上具有一定的意义。它们在橄榄斑叶（*Cetrelia olivetorum*）中的直径通常小于0. 5 mm，在奇氏斑叶（*C. chicitae*）中则小于1 mm，而在领斑叶（*C. collata*）中则大于lmm。

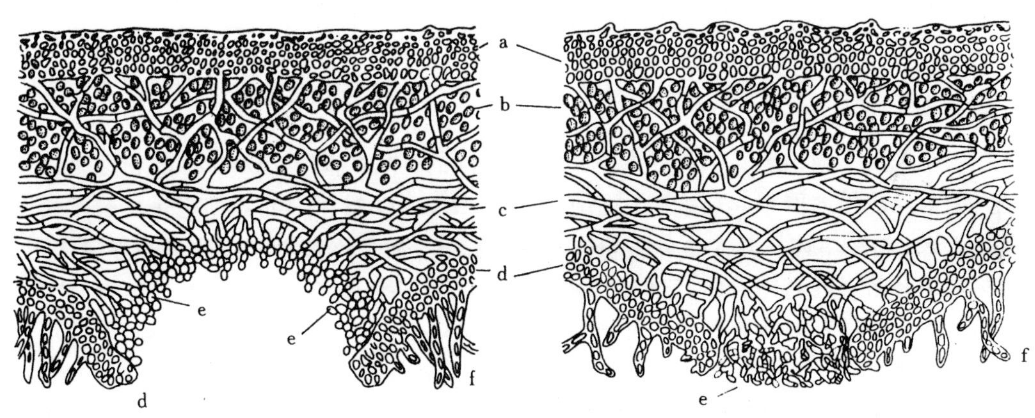

图6-1-5　牛皮叶属（*Sticta damaecornis*）杯点的横断面（仿Schneider）

a. 上皮层；b. 藻层；c. 髓层；d. 下皮层；e. 构成杯点内侧表面的球形细胞；f. 茸毛型假根.

图 6-1-6　假杯点属（*Pseudocyphellaria* sp.）叶状体下表面假杯点横断面（仿 Schneider）

a. 上皮层；b. 藻层；c. 髓层；d. 下皮层；e. 假杯点开口处网状交织的髓层菌丝；f 茸毛型假根.

3. 假根（rhizines）

它是位于叶状地衣体腹面无色的或变黑了的菌丝束或菌索。它们是由髓层与下皮层组织共同构成的坚韧的菌索（图6-1-1）。最简单的假根是不分枝的，如石梅衣[*Parmelia saxatilis*（L.）Ach.]；有的则具有羽状分枝，如羽根梅衣（*Parmelia squarrosa* Hale）等。石耳属　　的一些种的假根虽然不起固着器的作用，但是有实验证明在水分关系中起重要作用（Larson，1981）。而绝大多数起固着器作用的假根也可能同时在水分关系中也起重要作用。

4. 茸毛(tomentum)

茸毛是由疏松的菌丝束组成的毡状的、毛发状或棉絮状的团块，通常存在于缺乏下皮层或下皮层发育不良的地衣体腹面，如胶衣科(Collemataceae)、地卷科(Peltigeraceae)和肺衣科(Lobariaceae)中。茸毛有时还存在于某些地衣的上表面，即背面，如犬地卷(*Peltigera canina*)、地卷(*P. rufescens*)、毛面衣属(*Erioderma*)。生长在茸褐梅[*Melanelia glabra* (Schaer.) Essl.]上表面的细微茸毛是散生的单个短小菌丝。

5. 缘毛(cilia)

缘毛是睫毛型的菌丝结构，是由紧密结合的具有一定坚韧的菌丝束构成。缘毛通常存在于在结构上较高级的叶状地衣中，如梅衣科(Parmeliaceae)种类的叶缘或子囊盘的盘托上。在地卷科、肺衣科和胶质地衣中迄今未见缘毛存在。

6. 小裂片

小裂片是地衣体上生长的不定芽。这些不定芽往往是沿着叶状地衣体边缘，如梅衣属及肾盘衣属(*Nephroma*)中以及缝隙边缘，如地卷属及肾盘衣属等种类。小裂片与裂芽的区别在于前者具有明显的背腹之分，即通常为扁平型，上表面和下表面有显著的差异；而裂芽则无背腹之分，即多呈圆柱状或小球状等。有实验证明，地卷属叶面沿着缝隙边缘生长的小裂片是由于皮层创伤而受到刺激所产生的(Hale，1983)。

7. 粉芽及粉芽堆(soredia and soralia)

粉芽是被地衣共生菌丝缠绕着的少数藻细胞丛并位于地衣体皮层之外，而癞屑衣属(*Lepraria*)地衣基本上是由一层连续不断的粉芽所形成的近圆形或不定形的斑块。单个粉芽的直径约在25—100μm之间。

在亲缘关系不同的地衣中粉芽的构造也是不同的。梅衣属(*Parmelia*)的粉芽是由地衣型真菌菌丝疏松地网络起藻细胞，而地卷属(*Peltigera*)粉芽中的藻细胞是由菌丝紧密的包裹起来。粉芽单体按其大小分为粉质状和颗粒状。粉芽不被限制在明确的界限之内，而是弥散地分布于地衣体表面。它们的形状、构造及其在地衣体上分布位置在地衣分类上具有重要意义。具有明确界限的粉芽群称为粉芽堆(soralia)。粉芽堆按其不同形状可分为下列7个类型：

(1) 斑点状粉芽堆呈圆形、近圆形至长方形或凹穴。有时许多斑点状粉芽堆彼此相连而汇合成一大片。它们往往分布于地衣体的上表面。

(2) 球形粉芽堆呈球形或半球形。它们通常分布于叶状或壳状地衣体表面；若分布在裂片顶部或枝状地衣体末稍时，则往往被称为头状粉芽堆。

(3) 袖口状粉芽堆是因为分布在突起的环状裂片边缘而呈现为具有中央穿孔的外形。实际上这种袖口状主要是由突起的环状裂片形成的，而并非由分布在其上的粉芽形成的。这种现象多见于袋衣属(*Hypogymnia*)及孔衣属(*Menegazzia*)中。

(4) 唇形粉芽堆也和袖口状粉芽堆相似，是由于粉芽分布于貌似上、下两张“咀唇”般开裂的小裂片顶端边缘及其内侧而得名。有时唇形裂片之一呈拱形增大而形成类似于钢盔帽状时则称为盔形粉芽堆。

（5）镶边粉芽堆是沿着叶状地衣体的边缘发育成一个连贯的粉芽堆，类似于镶了花边的衣服，如粉芽斑叶衣（*Cetrelia cetrarioides*）及粉芽盾衣（*Peltula euploca*）等。

（6）裂缝状粉芽堆是沿着叶状地衣体皮层裂缝分布的粉芽。它的形状取决于地衣体皮层裂缝的形状，往往呈长椭圆形、长方形以及若干个这样的粉芽堆相连而汇合成较长而有分枝的粉芽堆，如槽梅衣（*Parmelia sulcata*）的粉芽堆。

（7）疱状粉芽堆位于地衣体内而被膨胀的疱状物所包围。疱状物破裂之后则粉芽才被释放出来，如钝形树花（*Ramalina obtusata*）的粉芽堆。

粉芽及粉芽堆的形成及形状正同地衣的其它表型特征一样都是由相应地衣的遗传性决定的。在自然界，具有粉芽或粉芽堆的地衣种类往往是缺乏子囊盘的；而具有子囊盘的地衣种类则往往缺乏粉芽或粉芽堆。看来粉芽及粉芽堆在地衣的传播与繁殖中起着重要作用。至于在形态上相似而只由于有无粉芽或粉芽堆的几对地衣种类之间究竟是同一种类或不同种类之争只能在获得足够证据之后才能得出结论。

8．裂芽

裂芽是由地衣体表皮层的局部突起而形成的球形、棒状或珊瑚状的小瘤。其直径范围在0.01～0.03mm之间，高度在0.5～3mm之间。它和小裂片之间的主要区别在于前者无背腹之分，而后者有明显的背腹之别，而且后者具有次生皮层。裂芽的皮层并不是次生的，而是由地衣体的皮层向上生长或突起，从一开始就包围着带有藻细胞的髓层组织，如疱脐衣（*Lasallia pustulata*）中的裂芽（图6-1-7）。裂芽在无皮层的胶衣中是由共生的蓝细菌在地衣体边缘开始分裂而形成的小疣。这个疣状构造后来才被菌丝所网络。在具有皮层的地衣体上则裂芽形成之初就被皮层所包围，从而同髓层菌丝及光合共生物同步形成。

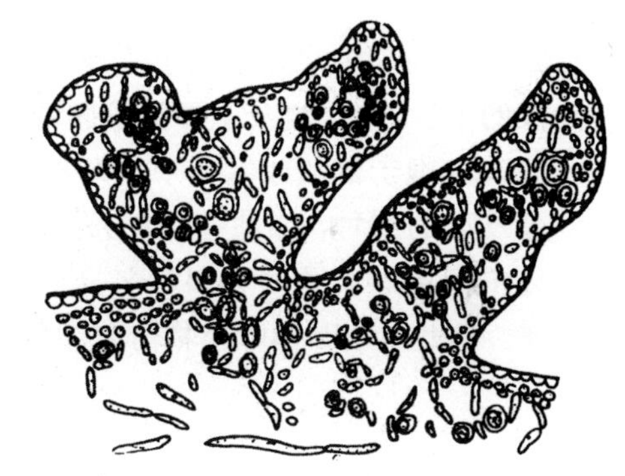

图6-1-7　疱脐衣（Lasallia pustulata）的幼小裂芽纵剖面（仿Moreau）

在壳状、叶状、枝状以及胶质状地衣上都能形成裂芽。由于裂芽的存在而增大了地衣体的上表面积，因而，在增强光合作用面积和提高同化率方面可能具有一定意义。由于裂芽的断裂而在地衣传播与繁殖方面所起的作用在许多壳状地衣中已被证实（Jahns，1973）。

9．衣瘿

衣瘿是位于地衣体上的小瘤，类似于高等植物的虫瘿。不过衣瘿小瘤包含的并非小虫，而是第二光合共生物，即不同于地衣体内的光合共生物。在绿皮地卷（*Peltigera aphthosa*）叶状体表面的饼状衣瘿看来是由外来的念珠蓝细菌（*Nostoc* sp.）群体降落在地衣叶状体表面后被茸毛状共生菌丝网络而形成的。由于鳞地卷（*Peltigera lepidophora*）上表面的圆饼状或圆盘形小裂片的形成途径之一也类似于绿皮地卷上的衣瘿，因此，也有称这些小裂片为衣瘿的。

衣瘿根据其位于地衣体内或地衣体外而分为内衣瘿和外衣瘿两类。内衣瘿是由位于地衣体髓层的光合共生物的细胞丛所构成。它们变得较大，以致于使地衣体局部膨大而形成可见的小瘤。它们多见于肺衣科（*Lobariaceae*）中的肺衣属（*Lobaria*）及牛皮叶属（*Sticta*）中。这些内衣瘿的光合共生物通常是念珠蓝细菌。外衣瘿的形状比内衣瘿要更加多样化。绿皮地卷上的饼状暗色衣瘿正是外衣瘿中最常见的类型。有的外衣瘿呈盘状，石竹红色，往往比寄主的子囊盘还大，如瘿茶渍衣属（*Placopsis*）的种类。在珊瑚枝属（*Stereocaulon*）中外衣瘿往往被分为无皮层的和有皮层的两类。后者往往是呈现为外表具有不同程度凹穴的囊袋状。

在一般情况下，具有衣瘿的地衣体内光合共生物为绿藻，而它的衣瘿内的光合共生物为蓝细菌。叶上枝属（*Dendriscocaulon*）地衣的共生生物为蓝细菌中的念珠蓝细菌（*Nostoc*）或伪枝蓝细菌（*Scytonema*）由于它往往生长在绿藻为共生物的蕨牛皮叶（*Sticta filix*）、宽牛皮叶（*S. latifrons*）， 亚皱牛皮叶（*S. subcaperata*）, 及大叶肺衣（*Lobaria amplissima*）上而的曾被 误作为牛皮叶和肺衣上的小裂片型衣瘿，实为地衣型真菌肺衣科的成员。

二、分生孢子器、分生孢子梗及分生孢子

在地衣中，分生孢子器及分生孢子作为一个物种的无性阶段的结构而与该物种有性阶段的子囊器、子囊及子囊孢子通常是存在于同一个地衣体上的。因而与非地衣型子囊菌相反，其分生孢子器及分生孢子从来不被当作独立的物种而另起属名和种名，除非有性阶段尚未被发现的属种，如地茶（*Thamnolia*）等。虽然有个别真菌学家也曾为具有有性阶段地衣的分生孢子发表了属名和种名，如将叶上衣属（*Strigula*）的分生孢子结构命名为*Discosiella*属。但是，它是国际植物命名法规所不允许的，因而是不合法的。

（一）分生孢子器

分生孢子器（conidiomata）通常是球形或瓶状（分生孢子壳pycnidia）、盘状（分生孢子盘acervula）、垫状（分生孢子座sporodochia）、盔状或盾状（分生孢子盔：Campylidia）或直立的菌丝束（联丝体：Synnemata）或盾梗（hyphophores）。在地衣型真菌的人工培养中，偶然也有分生孢子直接产生于单个菌丝上，如孔鸡皮衣（*Pertusaria pertusa*）以及麸屑口果粉衣（*Chaenotheca furfuracea*）等的无性型。

1．分生孢子壳

地衣分生孢子壳通常呈球形或瓶状，位于地衣体内而仅以其褐色至黑色，偶然也为白色的小孔微微突起于地衣体表面，如石耳科（Umbilicariaceae）等。在另一些地衣中分生孢子壳往往位于叶状地衣体边缘，如岛衣型地衣（Cetrarioides）。迄今已知的地衣分生孢子壳有4个类型（图6-1-8：a～d）：

（1）染料衣型（*Roccella*-type），见于染料衣科（Roccellaceae）中（图6-1-8：a）。

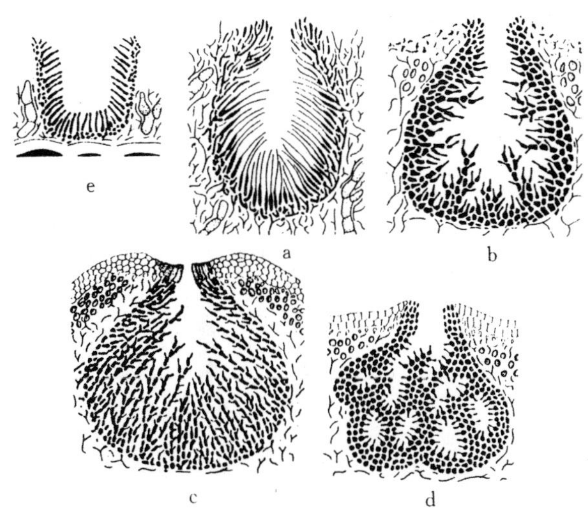

图6-1-8　分生孢子器中的分生孢子壳（a～d）及分生孢子盘（e）
（仿Hawksworth,1988）

a. 染料衣型（*Roccella*-type）分生孢子壳；b. 石耳型（*Umbilicaria*-type）分生孢子壳；
c. 肺衣型（*Lobaria*-type）分生孢子壳；d. 石黄衣型（*Xanthoria*-type）分生孢子壳；
e. 碗衣型（*Lecanactis*-type）分生孢子盘。

（2）石耳型（*Umbilicaria*-type），除在石耳科中之外，还存在于顶杯衣属（*Acroscyphus*）、岛衣属（*Cetraria*）、袋衣属（*Hypogymnia*）、及梅衣属（*Parmelia*）中（图 6-1-8：b）。

（3）肺衣型（*Lobaria*-type），见于肺衣科中（图 6-1-8：c）。

（4）石黄衣型（*Xanthoria*-type），除在石黄衣中之外，也见于皮果衣属（*Dermatocarpon*）及石果衣属（*Endocarpon*）中（图 6-1-8：d）。

此外，在胶囊衣属（*Physma*，Stahl，1877）及毡衣科（Ephebeaceae）中（Henssen，1963）地衣中的产囊丝在分生孢子壳下面形成。随后，子囊便在此处的分生孢子梗之间发育而成，甚至生出真正的侧丝，从而出现了分生孢子壳也能起子囊壳作用的奇特现象。这种结构通常被称为“分孢器子囊果”（Pycnoascocarps, Letrouit-Galinou，M.-A. 1973）。

2．分生孢子盘

分生孢子盘为盘状分生孢子器（disk-like 或 acervular conidiomata）。这一结构在地衣中比较少见。分生孢子盘往往被称作碗衣型（*Lecanactis*-type），存在于碗衣属（*Lecanactis*）、斑衣属（*Arthonia*）及旋衣属（*Byssoloma*）中（图 6-1-8：e）。

3．分生孢子盔

分生孢子盔是由子囊器状的原体随后扩大变为鳞状或帽盖状或盔分生孢子着生在它的内侧表面（Serusl‘aux，1986）。分生孢子盔主要发生在火焰锈衣

（*Loflammia flammea*, syn. *Lopadium flameum*），以及孢足衣属（*Sporopodium*）及褐盘衣属（*Badimia*）等一些叶生地衣中。而这些叶生地衣曾被许多作者当做生长在地衣上的真菌而被划归腔孢纲（*Coenomycetes*）的核盔菌属（*Pyrenotrichum*，Santesson，p.1952），（图 6-1-9）。

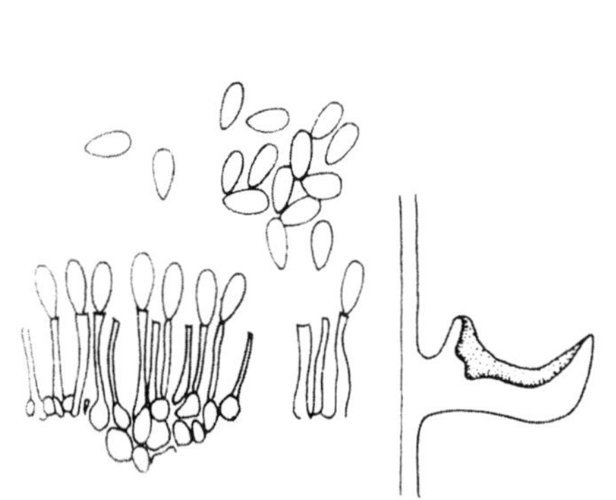

图 6-1-9 黄锈疣衣（*Lopadium gilvum*）
的分生孢子盔及其分生孢子梗（仿 Vězda,1984）

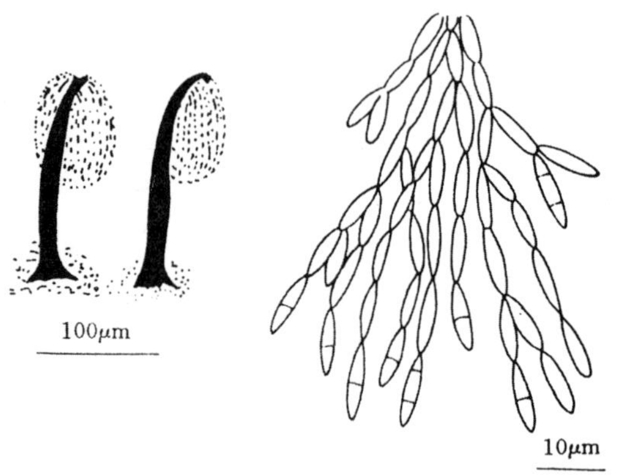

图 6-1-10 叶面瘤盘衣（*Aulaxina epiphiylla*）
的分生孢子柱及其分生孢子

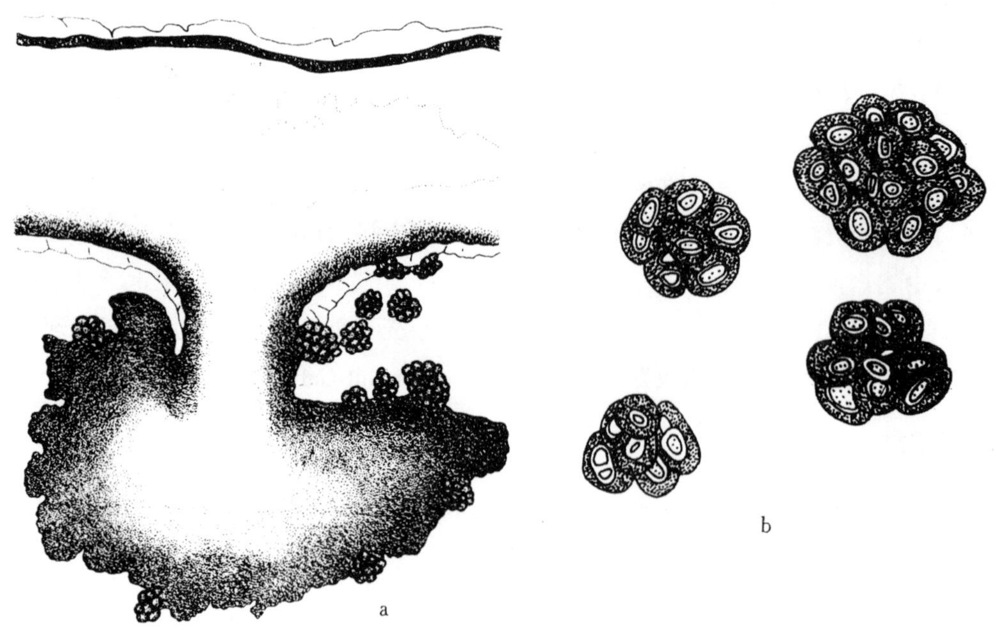

图 6-1-11 肉根石耳（*Umbilicaria tylorrhiza*）下表面及假根上的体裂分生孢子（thalloconidia）
a.假根及其上的体裂分生孢子；b.地衣体下表面的体裂分生孢子（仿 J. C. Wei & Y. M. Jiang，1993）

4. 分生孢子柱(hyphophores)

分生孢子柱是由延长了的菌丝所产生的束状菌丝结构。分生孢子链着生在这个束状菌丝的顶端(图 6-1-10)。这个专化型的菌丝束被称为分生孢子柱。这一类型的分生孢子器只见于文字衣目(Graphidales) 的叶面瘤盘衣(*Aulaxina epiphylla*)中(图 6-1-10)。

5. 分生孢子座(porodochia)

分生孢子座是分生孢子梗产生于一个垫状菌丝簇。分生孢子座见于茶渍目(Lecanorales)的贴生亚网衣(*Micarea adnata*) 中。

6. 体裂分生孢子结构

此外，在石耳属(*Umbilicaria*) 的许多种的叶状体下表面及假根上，偶然也在上表面的凹穴处或子囊盘的盘托附近直接产生大量厚壁的分生孢子。因为这种分生孢子是直接产生于地衣体的表面，而不是产生于专化的结构上，因此，它们并无专化的分生孢子器，或者，在某种意义上，地衣体本身就是它们的分生孢子器 (图 6-1-11)。体裂分生孢子(thalloconidia) 即因此而得名。

(二)分生孢子梗

分生孢子梗(conidiophores)是一根载有分生孢子及产孢细胞的分枝或不分枝的专化型菌丝。在过去的地衣文献中分生孢子梗按照推测的功能被称做担子(basidia) 、支柱(fulcra)或精子梗(spermatiophores)。分生孢子梗及其产孢细胞的形状与分枝方式因地

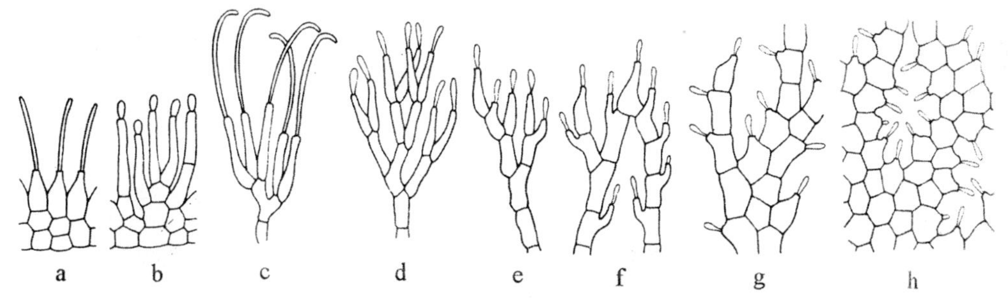

图 6-1-12 分生孢子梗及产孢细胞的排列方式

(仿 Hawksworth，1988)

a. 斑衣属(*Arthonia*)；b. 旋衣属(*Byssoloma*)、碗衣属(*Lecanactis*)、地卷属(*Peltigera*) 及温衣属(*Thermutis*)；c. 梳衣属(*Combea*)、碳质衣属(*Dirina*)、长果衣属(*Dolichocarpus*)、染料衣属(*Roccella*)及小染衣属(*Roccellina*)；d. 石蕊属(*Cladonia*)及树花属(*Ramalina*)；e. 树发属(*Alectoria*)、岛衣属(*Cetraria*)、杯褐梅(*Melanelia acetabulum*)及脐叶属(*Omphalodium*)；f. 顶杯衣属(*Acroscyphus*)、袋衣属(*Hypogymnia*)、梅衣属(*Parmelia* 部分种)、黑蜈蚣衣属(*Phaeophyscia*)、大孢蜈蚣衣属(*Physconia*) 及石耳属(*Umbilicaria*)；g. 雪花衣属(*Anaptychia*)、肺衣属(*Lobaria*)、肾盘衣属(*Nephroma*)及鳞藓衣属(*Psoroma*)；h. 皮果衣属(*Dermatocarpon*)、黄枝衣属(*Teloschistes*)及石黄衣属(*Xanthoria*)。

衣属的不同而各异。分生孢子及产孢细胞顶生的分生孢子梗(*exobasidia fulcra*) 或除顶生之外也侧生的分生孢子梗(*endobasidial fulcra*) 在地衣分类中曾广泛应用。虽然这种概括大体上也是符合实际情况的，但是，过于简单化了，不宜在地衣分类中使用。因为它将分生孢子及产孢细胞的形状，分生孢子梗的形状及其是否分枝以及分枝方式等一系列特征均排除在外。根据这些特征，分生孢子及产孢细胞在分生孢子梗上的着生方式及其形状与分枝方式迄今已知者至少有下列 8 种(图 6-1-12)：

(三) 分生孢子

地衣的分生孢子产生于分生孢子器内，如分生孢子壳，分生孢子盘、分生孢子盔、分生孢子柱和分生孢子座内或直接产生于地衣表面菌丝上；在热带一些地衣如孢足衣属(*Sporopodium*) 及毛蜡属(*Tricharia*) 中，分生孢子直接产生于子囊内的子囊孢子上的现象也偶见发生(Santesson，1952)。

分生孢子的形状通常呈亚球形、短杆状、椭圆形、镰刀形、S 形、线状，有分枝的、不分枝的以及有色的或无色的等(图 6-1-13)。分生孢子有大、小两种类型。小型分生孢子的大小约在 1μm×0.5μm 范围内，而大型分生孢子则在 100μm×2μm 范围之内，如大孢斑衣(*Arthonia macrosperma*)(Santesson，1952)。小型分生孢子肯定在减数分裂繁殖 (有性繁殖) 中起精子作用，而大型分生孢子则只能在有丝分裂繁殖 (无性繁殖) 中发挥作用而不在减数分裂繁殖中起精子作用。

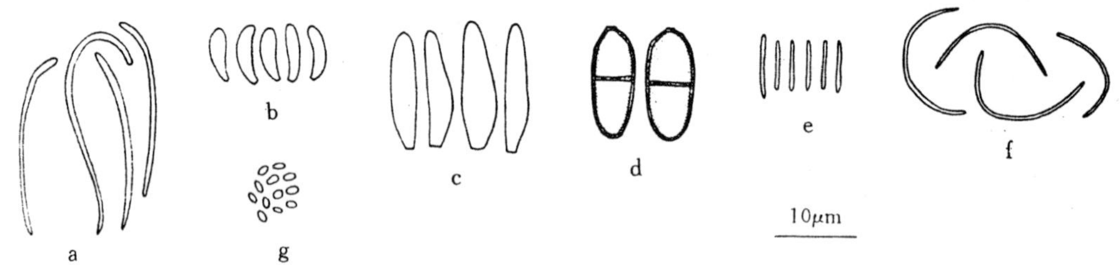

图 6-1-13 分生孢子

(仿 Hawksworth，1988)

a. 丽盘茶渍(*Lecanora pulicaris*)；b. 柳茶渍(*L. saligna*)；c. 杉茶渍(*L. abietina*)；
d. 黑孢菌疣核(*Mycomicrothelia melanospora*)；e. 褐杯梅(*Melanelia acetabulum*)；
f. 普孔文衣(*Opegrapha vulgata*)；g. 双异形衣 (*Anisomeridium biforme*)。

三、子囊器、子囊及子囊孢子

(一) 子 囊 器

子囊器(ascomata)是包含子囊的结构。子囊器通常按其形状与结构而分为三大类型：子囊盘、子囊壳和子囊腔。根据子囊在子囊器中的排列方式则可分为层囊型与腔囊型两类。前者是子囊在子囊器内基本上整齐地排列成一层，即子囊层或子实层；而后者则缺乏真正的子囊层或子实层，却代之以零乱的子囊散布于子囊腔内。有些地衣则缺乏真正的子囊器，而代之以柔软疏松而呈薄垫状的菌丝组织位于地衣体表面，其外形类似于一片片发育不良的粉芽。

1．层囊型子囊器（hymenial ascomata）

层囊型子囊器包括子囊盘及子囊壳，通常也统称为子囊果（*ascocarp*）。

（1）子囊盘（*apothecia*）：子囊盘通常呈圆盘形，具有一个圆盘形的子实层表面，直径约为 lmm，有时较大或较小。在地卷科（*Peltigeraceae*）及梅衣科（*Parmeliaceae*）中，其直径可达 10～20mm。子囊盘位于地衣叶状体上表面或边缘，凸出于或凹陷于地衣体表面。

子囊盘包括子实上层、子实层（子囊、侧丝）及子实下层、果壳及盘被（图 6-1-1）。子囊盘按其结构及质地可分为下列若干类型（图 6-1-14）：

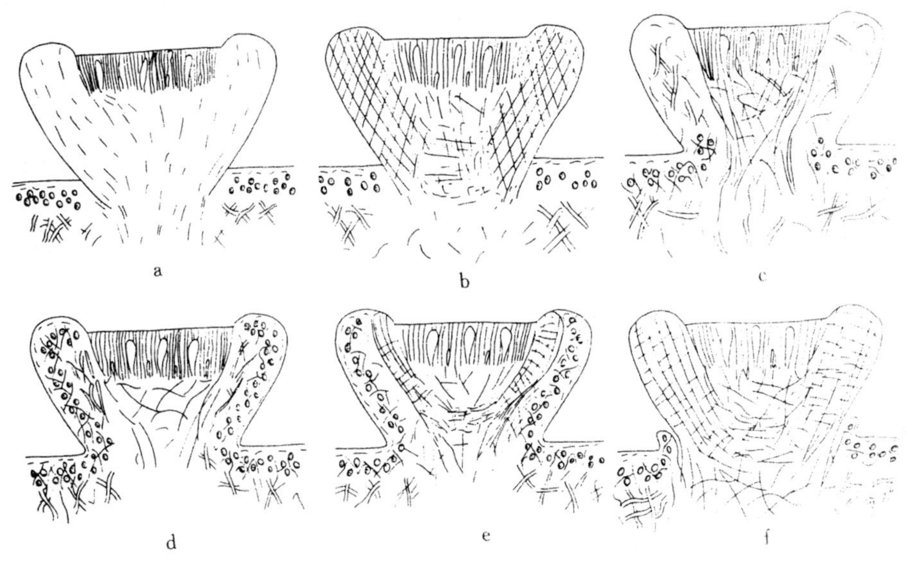

图 6-1-14　地衣型子囊盘的类型

a．蜡盘型；b．网衣型；c．亚茶渍型；

d．茶渍型；e．双缘型；f．中间型。

1）子囊盘缺乏由地衣体上皮层形成的盘托。

A．网衣型（*Lecideine*）：网衣型子囊盘的盘被是由地衣共生菌体单独构成。盘被及盘缘质地坚硬，颜色通常为暗色，往往与盘面颜色相同而与地衣体颜色相异（图 6-1-14:b）。

网衣型子囊盘周围带有向上翘起的叶状地衣体，貌似原始的或退化了的茶渍型盘托（图 6-1-14:f），可被看作为中间型（*Intermediate*）。

B．蜡盘型（*Biatorine*）：蜡盘型子囊盘基本与网衣型者相同，只是盘被质地呈蜡拉质型柔软，颜色通常淡色或较鲜艳（图 6-1-14:a）。

2）子囊盘具有由地衣体上皮层形成的盘托。

A. 亚茶渍型（Sublecanorine）：亚茶渍型子囊盘的盘托由地衣体上皮层形成，但是盘托内并无藻细胞层。这一类型常见于石耳科的某些种类以及肺衣属（*Lobaria*）的个个别种类中（图 6-1-14:c）。

B. 茶渍型（Lecanorine）：茶渍型子囊盘的盘托由地衣体上皮层形成，而且盘托内有可见的藻层。但是由纯菌丝形成的固有盘被往往缺乏。这一类型普遍存在于茶渍科（Lecanoraceae）的种类中（图 6-1-14:d）。

C. 双缘型（Zeorine）：双缘型子囊盘既具有清晰可辨的固有盘被，也具有内含藻层的盘托结构。该类型的子囊盘可见于双缘衣属（*Diploschistes*）的种类中（图 6-1-14:e）。

（2）子囊壳（perithecia）：子囊壳通常呈亚球形或烧瓶状结构而埋置于地衣体内，仅以其褐色至黑色乳头状小点而外露于地衣体上皮层。乳头状小黑点中央可见细微小孔。在一定湿度和温度条件下，成熟的子囊及子囊孢子便通过这一小孔喷射而出。子囊壳通常单生于地衣体上，有时则群生于子座中。

子囊壳在解剖显微镜下难以和分生孢子壳相区别。在这种情况下，只能通过切片或压片法在透射光学显微镜下检查，以辨认是否含有子囊而与分生孢子壳相区别。子囊壳内含有子囊及子囊孢子，而分生孢子壳内则含有分生孢子，而且当分生孢子壳在水中破裂时，分生孢子便迅速地从分生孢子壳内成群地涌流而出。

2. 腔囊型子囊器（locular ascomata）

腔囊型子囊器通常是具有分枝型的假侧丝和易分离的双层子囊壁。子囊不是整齐成层地分布于子囊腔内或柔软的菌丝体垫内。

可分离的双层子囊壁在成熟的子囊释放时最容易被观察到。因为子囊的外壁较薄而易破裂。当孢子成熟后，在一定的温、湿度条件下，子囊外壁破裂，带有子囊孢子的子囊内壁连同孢子一起呈圆柱体状部分地通过外壁破裂处向外伸出。子囊孢子便通过子囊内壁顶端有弹性的小孔而被释放出来，如冬青斑衣（*Arthonia illicina*）及芝麻粒衣（*Arthothelium spectabile*）等。

（1）子囊腔或假囊壳（locules or pseudothecia）：子囊腔型地衣的子囊器变化较大，其外形有时像圆形子囊盘，如染料衣科，或呈不规则星状或线条状子囊盘，如斑衣科（Arthoniaceae）、孔文字衣科（Opegraphaceae），或貌似子囊壳的小核衣科（Pyrenulaceae）和瓶口衣科（Verrucariaceae）。乳嘴衣科（Trypetheliaceae）的子囊器则类似于若干个子囊壳聚集于子座状群体上，而其他科的类似子囊器则为单个的子囊腔或假囊壳（Hale，1983）。

（2）子囊垫（patsches）：有些地衣则缺乏真正的子囊器，而代之以柔软菌丝体形成的薄垫，其外形看起来类似于一片发育不良的粉芽，多为白色或浅灰色，如隐果衣科（Cryptotheciaceae）（Santesson，1952；Eriksson，1981）。它们与子囊腔型地衣的主要区别除了缺乏真正的子囊器之外，就是子囊间组织很疏松，甚至即使在潮湿情况下也并不黏着在一起（Santesson，1952）。

（二）子　　囊

子囊（asci）是一个圆柱形的，如口果粉衣属（*Chaenotheca*），棒状、长棒状或梨形，如斑衣科（Arthoniaceae）以及长颈瓶状，如乳头衣属（*Thelocarpon*）的大细胞，其中经过核配和减数分裂以及随后的有丝分裂而形成一至若干个子囊孢子。

地衣的子囊通常较小，长约50μm；有些则较大，其长度超过150μm，如大孢网盘属（*Stenhammarela*）以及甚至其长度有超过300μm的，如鸡皮衣属（*Pertusaria*）。较小的子囊通常呈棒状。有的子囊则呈亚球形，如斑衣科；或瓶状，如乳头衣科（Thelocarpaceae）；
或圆柱形，如粉衣目（Caliciales）。

1. 子囊壁的层次

几乎所有的子囊在幼小时都具有较薄的壁，而在成熟时便逐渐变厚。在绝大多数核果地衣和许多盘果地衣中都具有内外两层壁（bitunicate）并带有外壁开裂型子囊。所谓的双壁子囊及单壁（unitunicate）子囊并不是结构上的层次，而是和子囊孢子释放方式有密切相关的。双壁子囊的内壁与外壁，如果要用形象的说法加以比喻的话，那么，子囊的内壁和外壁犹如子囊孢子的“睡衣”和“睡袋”。当孢子释放时，首先作为“睡袋”的外壁破裂，接着作为“睡衣”的内壁裹着孢子伸出“睡袋”并通过其“睡衣”的顶孔而释放孢子。至于单壁子囊则只有“睡袋”而缺乏“睡衣”。关于子囊壁在结构上的层次，则远不止双层，即使单壁子囊也会有四层之多。以海橙衣（*Caloplaca marina*）为例，其子囊壁虽薄，但却由清晰可辨的四层子囊壁构成（图 6-1-15）：

A 层薄而遇碘呈强烈的正反应而显示出多糖和胶质化的外套。

B 层很薄而无多糖，遇碘呈负反应，因而难以辨认。

C 层占子囊壁的大部分（厚约0.5μm），其下部遇碘呈轻微正反应。

D 层比c层薄（即0.1～0.2μm），遇碘呈正反应，而且除顶部以外呈波浪状分层现象（Hawksworth，1988）。

实际上类似的四层结构存在于全部地衣子囊壁中。

2. 子囊开裂方式

地衣型与非地衣型子囊顶器都是子囊壁的加厚及其进一步分化的结果。它们的多样性并结合开裂方式在地衣型与非地衣型子囊菌的系统分类中具有重要意义。

地衣子囊的开裂方式有下列5种（图6-1-16）：

（1）黄枝衣型（Teloschistes type）：子囊壁的各层次同时延伸至子实层表面。子囊壁破裂后孢子被释放（图6-1-16:a）。这一类型可见于黄枝衣科（Teloschistaceae）及相近的属，如综网属（*Fuscidea*）；甚至在多极孢衣属（*Letrouitia*）及鸡皮衣属（*Pertusaria*）中也可见到。

（2）鸟喙型（Rostrum type）：鸟喙开裂方式是a,b,c外三层先在顶部破裂。然后顶部变厚的d层延伸成一个鸟喙状而抵达子囊层顶部并开裂。伴随而来的便是中轴体也延伸并胶质化，这有利于孢子向外喷射（图 6-1-16:b）。

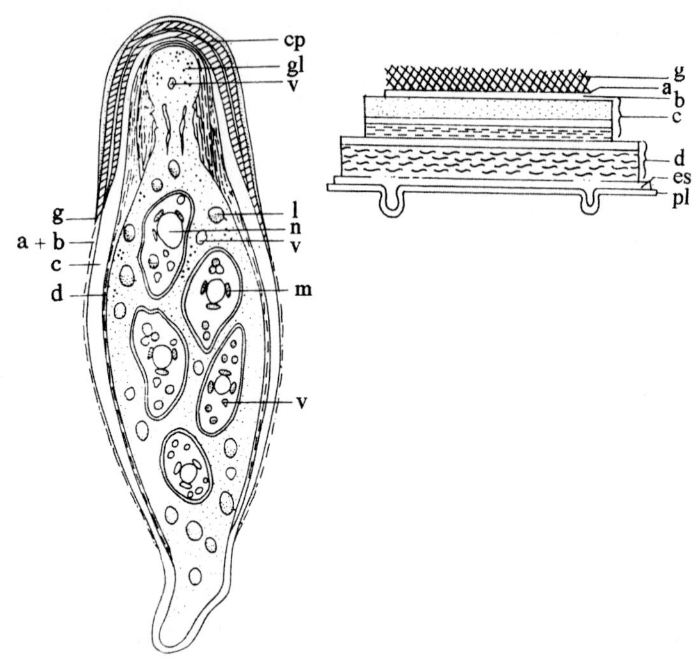

图 6-1-15 子囊壁的层次
(仿 Hawksworth，1988)
a ＝ a 层；b ＝ b 层；c ＝ c 层；cp ＝ 顶帽；es ＝ 周质空间；g ＝ 细毛型胶质外层覆盖物；
gl ＝ 糖原粒；l ＝ 脂质小球；m ＝ 线粒体；n ＝ 核；pl ＝ 质膜；v ＝ 液泡。

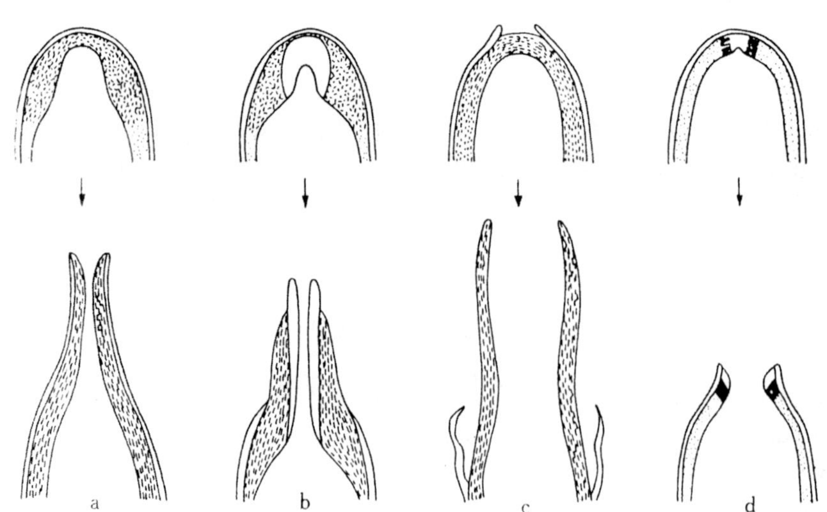

图 6-1-16 子囊开裂方式
(Hawksworth，1988)
a ＝ 黄枝衣型；b ＝ 鸟喙型；
c ＝ 缝裂型；d ＝ 外翻型。

鸟喙型开裂方式在地衣中十分普遍，如茶渍衣科(Lecanoraceae)、梅衣科

(Parmeliaceae)、蜈蚣衣科(Physciaceae) 及石蕊科(Cladoniaceae)等。

(3) 缝沟型(Fissitunicate)：缝沟型又称玩偶匣型或 JB 型(jack-in-the-box) 的第一步类似鸟啄型，首先是 a、b、c 外三层在顶部开裂，然后是含有孢子的 d 层大部分由顶端开裂处延伸至子囊顶外。随后，延伸部分伸长至子实层表面并常常冒出而外露于表面。在子囊开裂期间,d 层则发生了强烈的改变。

缝沟型常见于具有双壁子囊的核果地衣中，如小核衣目(Pyrenulales) 及瓶口衣目(Verrucariales) 中；而且在子囊腔型地衣，如斑衣科及孔文字衣目(Opegraphales)中也很普遍。地卷衣的子囊也属于缝沟型开裂方式(图 6-1-16:c)。

(4) 外翻型(Eversion type)：外翻型开裂方式借助于子囊顶孔地轻微外翻而释放孢子。该类型在地衣型子囊菌中很少见，但是在非地衣型子囊菌中却十分普遍(图 6-1-16:d)。

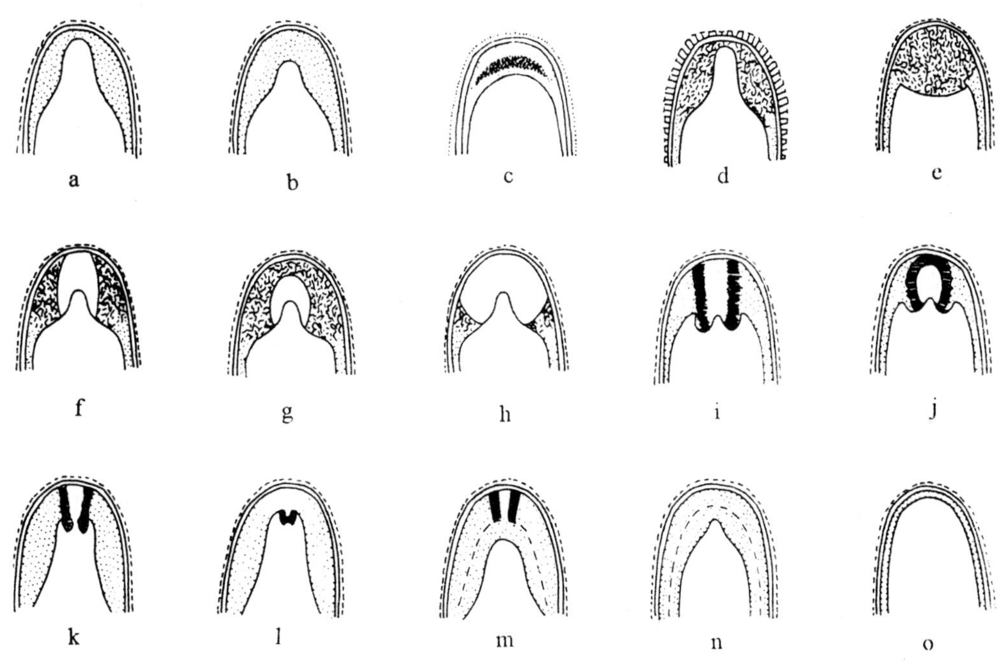

图 6-1-17　子囊顶器结构

(a，b，d～o，仿 Hawksworth，1988)。

a ＝ 脐鳞衣型(*Rhizoplaca* type)幼小期；b ＝ 蜈蚣衣型(*Physcia* type)，幼小期；
c ＝ 石耳型(*Umbilicaria* type)；　d ＝ 黄枝衣型(*Teloschistes* type)；
e ＝ 腊肠衣型(*Catillaria* type)；　f ＝ 茶渍型(*Lecanora* type)；
g ＝ 小网衣型(*Lecidella* type)；　h ＝ 黄茶渍型(*Candelariella* type)；
i ＝ 鳞网衣型(*Psora* type)；　j ＝ 网衣型(*Lecidea* type)；
k ＝ 胶衣型(*Collema* type)；　l ＝ 地卷衣型(*Peltigera* type)；
m ＝ 地图衣型(*Rhizocarpon* type)；　n ＝ 孔文衣型(*Opegrapha* type)；o ＝ 疱衣型(*Phlyctis* type)。

3. 子囊顶器结构

当子囊发育到足够长度并进行减数分裂时子囊顶器便开始发育。子囊顶器的不同结构主要表现在d层子囊壁的各种变化上（图 6-1-17）。在地衣中常见的子囊顶器结构如下：

（1）脐鳞衣型（*Rhizoplaca* type）：在未成熟的子囊中d 层只在两侧加厚。在蜈蚣衣属（*Physcia*）中有轻微的变异。

（2）黄枝衣型（*Teloschistes* type）：在脐鳞型的基础上有进一步的分化而加厚的d层遇碘成正反应。

（3）腊肠衣型（*Catillaria* type）：d层在顶部强烈加厚并遇碘成正反应而缺乏眼室。

（4）茶渍型（*Lecanora* type）：类似于黄枝衣型而不同之处在于强烈加厚的顶部d层处出现一个明显的遇碘呈负反应的轴体（axial body）；眼室有所退化。在小网衣属（*Lecidella*）及黄茶渍属（*Canderiella*）中有轻微的变异。

（5）鳞网衣型（*Psora* type）：眼室强烈退化而轴体周围遇碘呈强烈的正反应带。在网衣属（*Lecidea*）及胶衣属（*Collema*）中有轻微的变异。

（6）石耳型（*Umbilicaria* type）：在d 层加厚的顶部出现一个圆屋顶状或帽状遇碘呈强烈正反应的部分。

（7）孔文衣型（*Opegrapha* type）：d 层不仅在子囊顶部，而且在子囊周围均有明显的加厚并分化为内外两个亚层。

（8）地图衣型（*Rhizocarpon* type）：在孔文衣型基础上在d层外亚层顶部出现了一个遇碘呈强烈正反应的管形体。

（9）地卷衣型（*Peltigera* type）：为界于双壁子囊型与胶衣型之间的过渡类型。

（10）疱衣型（*Phlyctis* type）：c, d 两层明显可见而并不加厚。

（三）子囊孢子

子囊孢子是产囊体中双核融合后进行两次减数分裂和一次有丝分裂而形成的八孢子囊。多数地衣的子囊含有8个孢子，而有些地衣的子囊只含有1～2个孢子，如疱脐衣属（*Lasallia*）、黑红衣属（*Mycoblastus*）及鸡皮衣属（*Pertusaria*）等；有些地衣子囊则含有16～64或更多的孢子，如微孢衣属（*Acarospora*），多孢衣属（*Polysporina*）、网盘衣属（*Sarcogyne*）、黄烛衣属（*Candelaria*）及绵腹衣属（*Anzia*）等。成熟子囊孢子的大小、形状、有无隔膜、隔膜的纵与横以及孢子的颜色等随着科、属、种的不同而各异。地衣中最小的子囊孢子直径约1μm，如微孢衣属，而最大的孢子长约510μm，如缘杆孢衣（*Bacidia marginalis*）。虽然地衣子囊孢子按其特征可细分到17种之多，然而，基本上则可归纳为4大类型：

1. 单细胞孢子

单细胞孢子通常较小而壁薄，呈圆球形（如球粉衣：*Sphaerophorus*）、椭圆形（如灰衣属：*Tephromela*，网衣属：*Lecidea*，茶渍属：*Lecanora*，梅衣属：*Parmelia* 及松萝属：*Usnea*等）。孢子较大而壁厚者往往少见（如鸡皮衣属）。此外，还有S型孢子，如萨拉属的一种*Sarrameana paradoxa* （Hawksworth，1988）。

2．横隔多胞孢子

横隔膜的多胞孢子通常具有1～40个彼此平行的横隔膜。孢子的形状与长度随着横隔膜的多少而变化。具有一个横隔膜的孢子通常为椭圆形，如斑衣属（*Arthonia*），鳞瘤衣属（*Diploicia*）及鳞管杯属（*Tholurna*）等。此外还有肾形弯曲状的，如白蜡树花（*Ramalina fraxinea*）。具有2～40个横隔膜的孢子则呈现为纺锤形至线形孢子[如孔文衣属：*Opegrapha*（纺锤形）、线孢属：*Conotrema*（线形）]。界于纺锤形与线形之间的还有多种形状。

3．砖壁型孢子

砖壁型孢子是由纵横排列的细胞构成的多细胞孢子，形似砖石所砌的墙壁，故而得名。此类孢子成熟时往往呈褐色至暗褐色，如疱衣属（*Phlyctis*）、疱脐衣属（*Lasallia*）、双缘衣属（*Diploschistes*）、锈疣衣属（*Lopadium*）及地图衣属（*Rhizocarpon*）等。

4．对极型孢子

对极型孢子为双细胞孢子，是有较厚的中央隔膜和一个较细的峡道，如黄枝衣科（Teloschistaceae）。此类孢子是地衣型子囊菌所特有的类型。

第二章　地衣共生藻及蓝细菌
Lichen Phycobiont & Cyanobiont

地衣是一群专化型真菌，它们经常与一些低等光合共生物，如藻类及蓝细菌紧密结合成体内胞外互惠共生型生态系统。

参与这一生态系统的光合生物近 200 种，分隶于绿藻门（Chlorophyta）中轮藻纲（Charophyceae）的 1 属，绿藻纲（Chlorophyceae）的 1 目 10 属，石莼纲（Ulvophyceae）的 5 目 11 属；黄藻门（Xanthophyta）的 1 目 1 种；褐藻门（Phaeophyta）的 1 目 1 种；以及蓝细菌（Cyanobacteria）的 4 目 12 属。在这一类共生型生态系统中最常见的光合生物为共球藻属（*Trebouxia*）、桔色藻属（*Trentepohlia*）及念珠蓝细菌属（*Nostoc*）中之种类，约占参于共生的低等光合共生物的 90%。

对于地衣体内胞外状态下的共生藻及蓝细菌的鉴定是十分困难的，而且往往是不可能的。因为它们在形态上受菌丝缠绕的影响而发生了畸变。只是在某些含有丝状光合生物的地衣中，位于地衣体裂片幼嫩末梢的光合生物由于尚未受到菌丝的过多影响而可能还保持着丝状体原形。在这种情况下，对于光合生物的鉴定则是可能的（Henssen，1963；Ahmadjian，1967a）。

因此，为了鉴定地衣共生光合生物，通常需要对共生光合生物进行分离和培养。现将最常见的地衣共生光合生物在培养条件下形态特征及其分离培养方法介绍如下：

（一）绿藻纲（Chlorophyceae）常见属的形态特征

1. 桔色藻属（*Trentepohlia*）

丝状体，细胞圆柱形至条带状，其长度大于宽度的 1～3 倍而带有一些桔红色色素滴（β-胡萝卜素）因而常常使周壁的载色体变暗。游动孢子具 4 根鞭毛。配子具两根鞭毛。有静孢子。在地衣体内，丝状藻体可能很小并且不规则的被分隔成单个细胞。为热带及亚热带地衣中普遍的共生光合生物（图 6-2-1）。

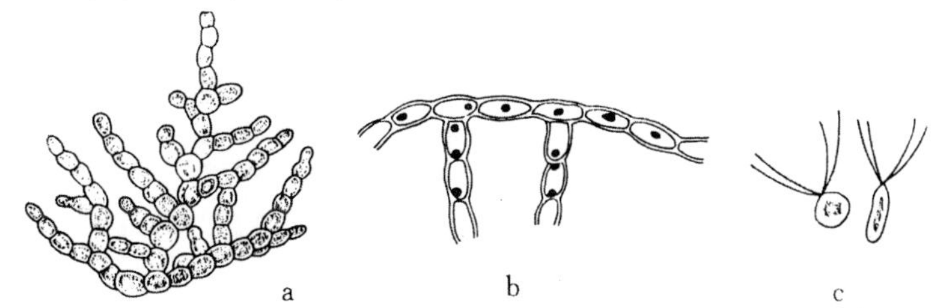

图 6-2-1　桔色藻属（*Trentepohlia*）的丝状体及游动孢子

a. *Trentepohlia umbrina* 的丝状体（来自 *Chaenotheca phaeocephala* var. *hispidula*，仿 Raths，1938）；

b. *T. annulata* 的丝状体（来自 *Graphis scripta*，仿 Verseghy，1961）；

c. *T. annulata* 的游动孢子（来自自由生活者，仿 Printz，1940）。

2．共球藻属（*Trebouxia*）

细胞呈球形至椭圆形或梨形；营养细胞直径可达34μm；在地衣体内的每一个共球藻细胞比在人工纯培养下具有更明显可见的中央载色体和淀粉核。细胞核位于近细胞壁处和载色体外缘褶页之间；以双鞭毛而裸露的（即缺乏坚硬的细胞壁）游动孢子（每囊可含256个游动孢子）和静孢子（每囊可含32个静孢子）繁殖（图 6-2-2）。共球藻在培养基中呈现多种多样的形状。许多共球藻呈现为颗粒状并隆起于培养基表面之上。该属可分为两个类群：

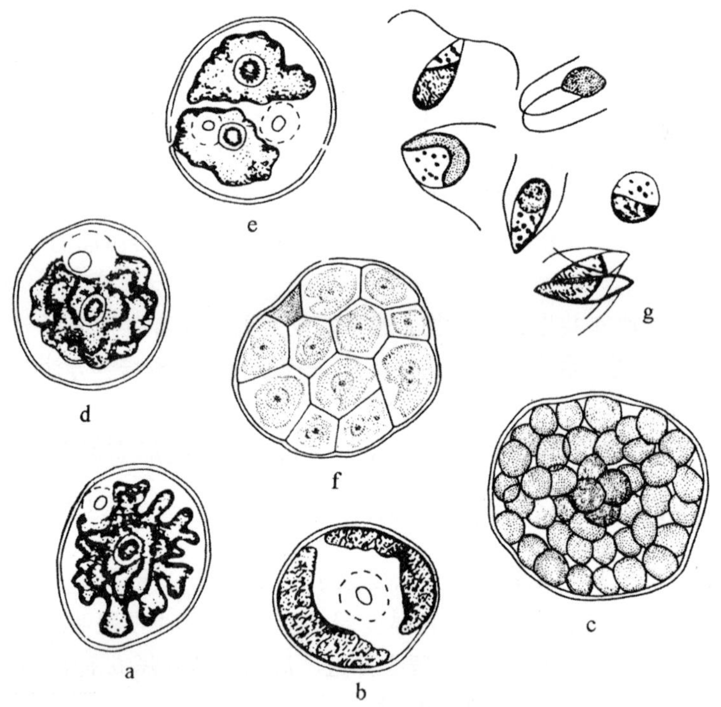

图6-2-2 共球藻属（*Trebouxia*）的形态

类群1.a～c:a.营养细胞载色体边缘分化为狭细分枝；b.载色体在细胞分裂期间变为带状而位于细胞周壁处；c.由滞留的游动孢子形成的静孢子。类群2.d～f:d.营养细胞载色体的边缘光滑而不分化为狭细分枝；e.载色体在细胞分裂期间不变为带状，也不扩展至细胞壁，而仍位于细胞中央；f.静孢子既由一个多边形的细胞分裂而形成，也由滞留的游动孢子所形成；g.双鞭毛而裸露的游动孢子。（仿Waren, 1920 和 Ahmadjian, 1960）

类群1．载色体的边缘分化为狭细裂片。这些细裂片可延伸至细胞壁。载色体在细胞分裂期间变为带状而位于细胞周壁处。静孢子是由滞留的游动孢子所形成。这一类群共球藻通常是枝状地衣的共生藻（图 6-2-2:a～c；表 6-2-1）。

类群2．载色体的边缘光滑而不分化为细裂片。载色体也不扩展至细胞壁。载色体在细胞分裂期间并不变为带状而仍位于细胞中央位置。静孢子除了借助于一个“多边形的”细胞分裂而形成之外，也来自滞留的游动孢子。这一类群的共球藻通常是叶状地衣和壳状地衣的共生藻（图 6-2-2：d～f；表 6-2-2）。

表6-2-1 类群1的共球藻及其共生真菌

共 球 藻	地衣型真菌
绵共球藻（*Trebouxia erici*） 团聚共球藻（*T. glomerata*）	鸡冠石蕊（*Cladonia cristatella*）（Ahmad., 1960） 红石蕊（*C. coccifera*）（Waren, 1920） 角石蕊（*C. cornuta*）（Waren, 1920） 正硫石蕊（*C. deformis*）（Waren, 1920） 细石蕊（*C. gracilis* v.*chordalis*）（Waren, 1926） 瘦柄红石蕊（*C. macilenta*）（Waren, 1920） 鹿石蕊（*C. rangiferina*）（Waren, 1920） 散珊瑚枝（*Stereocaulon evolutoides*）（Ahmad., 1960） 粉帽珊瑚枝（*S. pileatum*）（Ahmad., 1960）

表6-2-2 类群的共球藻及其共生真菌

共球藻	地衣型真菌
退色共球藻（*Trebouxia decorans*） 凹共球藻（*T. impressa*）	点黑瘤衣（*Buellia punctata*）（Ahmad., 1960） 石黄衣（*Xanthoria parietina*）（Ahmad., 1960） 蜈蚣衣（*Physcia stellaris*）（Ahmad., 1960）

3.小球藻属（*Chlorella*）

细胞呈圆球形至椭圆形并含有薄而盘状载色体及淀粉核；借助于静孢子繁殖。每一细胞中最多可形成16个静孢子（图 6-2-3；表 6-2-3）。

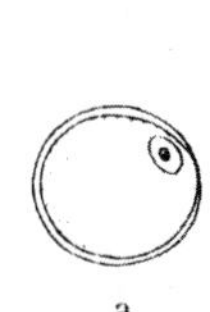

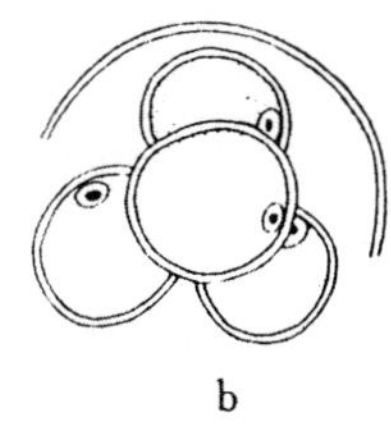

a　　b

图6-2-3 小球藻属（*Chlorella*）（仿Runnthaler,1915）
a. 营养细胞；b. 四个静孢子连同破裂了的孢囊壁（自由生活类型）。

表6-2-3 小球藻及其共生真菌

小球藻属	地衣型真菌
异极小球藻（*Chlorella lichina*）	粉衣属种类（*Calicium* sp.） 癞屑衣（*Lepraria chlorina*）（Raths, 1938）
类椭圆小球藻（*C. ellipsoidea*）	粗网衣（*Lecidea coarctata*）（T.Woess,1948）

4.胶球藻属（*Coccomyxa*）

细胞长椭圆形，椭圆形或几乎圆球形；有时有胶鞘；细胞平均为 7～12μm 长和 4～10μm 宽；每一个细胞中含有一个盘状的周壁载色体，通常缺乏淀粉核；借助于静孢子

（每孢子囊内含有静孢子2～4个）繁殖。胶球藻属的种类有时也普遍附生在地衣体表面（图6-2-4；表6-2-4）。

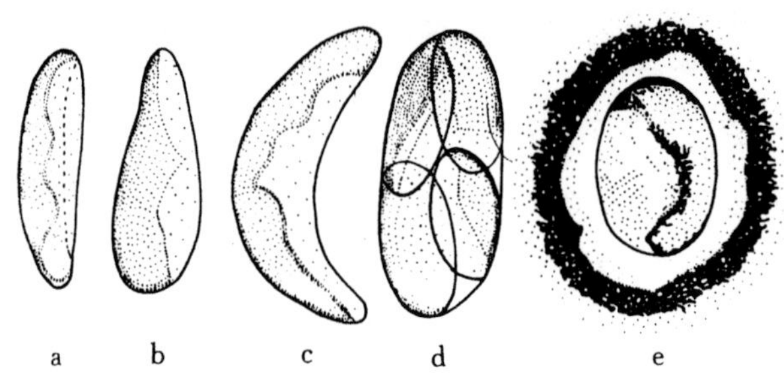

图6-2-4 胶球藻属（*Coccomyxa*）

a～c. 不同形状的营养细胞；d. 位于一个孢子囊内的四个静孢子；

e. 用印度墨汁染色后显示的细胞外胶质鞘。

表6-2-4 胶球藻及其共生真菌

胶球藻属	地衣型真菌
地卷胶球藻（*Coccomyxa peltigerae*）	绿皮地卷（*Peltigera aphthosa*）
黏胶球藻（*C. mucigena*）	绿皮地卷（*P. aphthosa*）
小地卷胶球藻（*C. peltigerae-venosae*）	小地卷（*P. venosa*）
瘤地卷胶球藻（*C. peltigerae-variolosae*）	绿皮地卷一变种（*P. aphthosa* var. *variolosa*）
橙散盘胶球藻（*C. solorinae-croceae*）	橙散盘衣（*Solorina crocea*）
梯诺胶球藻（*C. tiroliensis*）	橙散盘衣（*S. crocea*）
凹散盘胶球藻（*C. soloriae-saccatae*）	凹散盘衣（*S. saccata*）
小石胶球藻（*C. glaronensis*）	凹散盘衣（*S. saccata*）
椭圆胶球藻（*C. ovalis*）	凹散盘衣（*S. saccata*）
双孢散盘胶球藻（*C. solorinae-bisporae*）	双孢散盘衣（*S. bispora*）
霜降胶球藻（*C. icmatophylae*）	八孢散盘衣（*S. octospora*）
	霜降衣（*Icmadophila ericetorum*）

（二）蓝细菌（*Cyanobacteria*）常见属的形态特征

1．念珠蓝细菌属（*Nostoc*）

念珠状，由一个结实的胶质所包裹的扭曲的丝状体；藻殖段普遍。这些藻殖段通常是由于在异形胞与营养细胞之间断裂而形成的（图6-2-5；表6-2-5）。

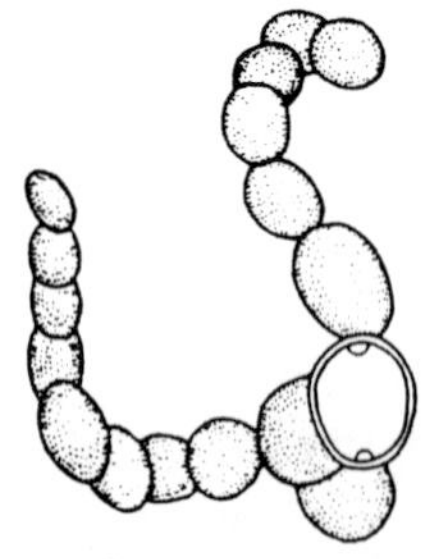

图 6-2-5 念珠蓝细菌属（*Nostoc*）具有一个异形胞（仿Bornet, 1873）

表6-2-5　念珠蓝细菌及其共生真菌

念璇蓝细菌	地衣型共生真菌
念珠蓝细菌（*Nostoc commune*）	胶衣属的种类（*Collema* spp.）（Degelius, 1954）
藓念珠蓝细菌（*N. muscorum*）	胶衣属的种类（*Collema* spp.）（Degelius, 1954）
点念珠蓝细菌（*N. punctiforme*）	犬地卷（*Peltigera canina*）
球念珠蓝细菌（*N. sphaericum*）	平盘地卷（*P. horizontalis*） 软地卷（*P. mazacea*） 多指地卷（*P. polydactyla*）（Linkola, 1923） 胶衣属的种类（*Collema* spp.）（Degelius, 1954） 鳞叶衣属种类（*Pannaria* sp.）（Schiman, 1957） 伊萨氏猫耳衣（*Leptogium issatschenkii*）（Danilov, 1927）

2．伪枝蓝细菌属（*Scytonema*）

丝状体由单行细胞排列成的假分枝（图6-2-6；表6-2-6）。

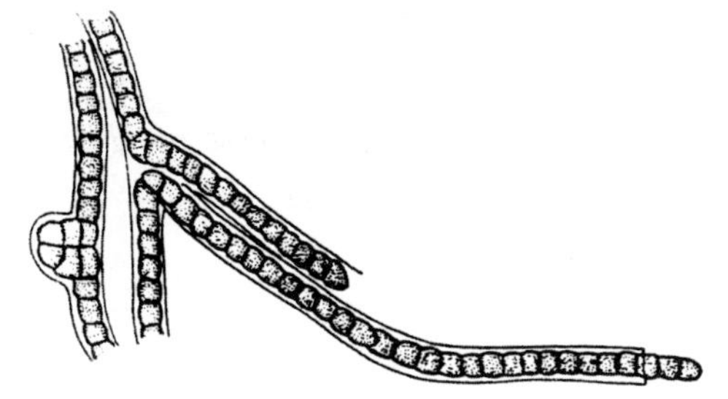

图 6-2-6　伪伎蓝细菌属（*Scytonema* sp.）
（仿Bornet, 1873）

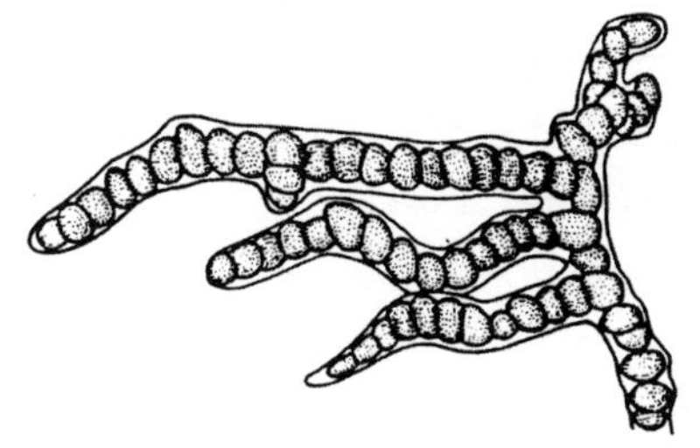

图 6-2-7　真枝蓝细菌属
（*Stigonema　hormoides* var. *africana*）
（仿Geitler, 1932）

表6-2-6　伪枝蓝细菌及其共生真菌

伪枝蓝细菌属	地衣型真菌
伪枝蓝细菌属的种类 *Scytonema* sp.	鳞叶衣属（*Pannaria*） 瓦衣属（*Coccocarpia*） 毛面衣属（*Erioderma*）（Bornet, 1873） 瓦衣属（*Coccocarpia*）（Sant., 1952） 扎氏衣属（*Zahlbrucknerella*） 多枝衣属（*Polychidium* spp.）（Hensserl, 1963） 锈疣衣属（*Lopadium*） 珊瑚枝属（*Stegeocaulon*）（Bornet, 1873; Lamb, 1951; Sant., 1952; Duvigneaud, 1955）

3．真枝蓝细菌属（*Stigonema*）

丝状体由单行排列的细胞构成；较老熟的丝状体则由2～4行排列的细胞组成；具有真正的分枝（图 6-2-7；表 6-2-7）。

表6-2-7 真枝蓝细菌及其共生真菌

真枝蓝细菌	地衣型真菌
真枝蓝细菌属的种类（*Stigonema* spp.）	毡衣属（*Ephebe*）（Bornet, 1873; Henssen, 1963） 珊瑚枝属（*Stereocaulon*）（Bornet, 1873; Lamb, 1951; Duvigneaud, 1955）（作为

第三章　地 衣 共 生

Lichen Symbiosis

地衣共生真菌、藻类及蓝细菌的分离、培养，以及菌、藻或蓝细菌共生的人工重建不仅是菌、藻共生生态系统理论研究中的重要内容，而且在地衣资源的开发利用中具有现实意义。

一、培 养 基

给每一升溶液中添加15～20g洋菜（琼脂）制成固体培养基。

（一）共生藻及蓝细菌的培养基

1．Bold’s Basal Medium （BBM）（Deason and Bold, 1960; Lawrey 1984; Ahmadjian, 1993）

该培养基适合于地衣共生藻和共生蓝细菌。

原液Ⅰ：

将下列6种化合物分别各溶于400ml的蒸馏水中而制备成6种原液：

硝酸钠（$NaNO_3$）	10 g	磷酸氢二钾 （K_2HPO_4）	3 g
氯化钙（CaCl2·$2H_2O$）	1 g	磷酸二氢钾（KH_2PO_4）	7 g
硫酸镁（$MgSO_4·7H_2O$）	3g	氯化钠（NaCl）	1 g

（从上述6种原液的每一种原液中取10ml加到940ml的蒸馏水中）

原液Ⅱ:

① 硼酸 （H_3BO_3）	11.42 g / L	③三氧化钼（MoO_3）	0.71 g / L
② 酸亚铁（绿矾：$FeSO_4·7H_2O$）	4.98 g / L	硫酸铜（$CuSO_4·5H_2O$）	1.57 g / L
硫酸锌（皓矾$ZnSO_4·7H_2O$）	8.82 g / L	硝酸钴[Co（NO_3）$_3·6H_2O$]	0.49 g / L
氯化亚锰即氯化锰（$MnCl_2·4H_2O$）	1.44 g / L	④乙二胺四醋酸（EDTA）（即乙底酸:editic acid）	50.0 g / L
		KOH	31.0 g / L

从上述每一种原液中各取1ml 加入上述原液Ⅰ中。BBM培养基可以采取下列途径予以改变：A. 增加更多的氮源（BBM^3N）；（$NaNO_3$） = 30g/l，它将可增加共生生物的生长对数期，或者 B. 不含氮源（BBMON）可用于蓝细菌的培养。如果要制备成固体培养基，则可加添15g的洋菜（琼脂）。

2．BBM3N + Glucose Medium

Bold’s无机盐培养基BBM3N（990ml），葡萄糖（10g）。为制备固体培养基在每一升培养基中加15g洋菜（琼脂）。

3．共球藻培养基 （*Trebouxia* Medium: TM）

Bold’s无机盐培养基BBM（970ml），蛋白胨（10g），葡萄糖（20g）。为制备固体培养基在每一升培养基中加15g洋菜（琼脂）。

（二）地衣型真菌常用培养基

1．麦芽汁-酵母膏培养基（Malt-extract yeast-extract medium:MEYE, Ahmadjian，1993）

麦芽汁　（Malt-extract broth）	20g
酵母膏　（Yeast-extract）	2g
蒸馏水　（Water）	978ml

2．土壤浸汁-琼脂培养基（Soil-exrac medium:SM, Ahmadjian，1993）

Bold’s矿物盐溶液（Bold’s Basal Medium）	960ml
土壤-水（Soil-water）	40ml
琼脂（Agar）	15g

3．Lilly and Barnett（1951）培养基（Medium:LB, Ahmadjian，1993 ）

葡萄糖（Glucose）	10.0g	硫酸锌　（$ZnSO_4 \cdot 7H_2O$）	0.2mg
天冬酰胺（Asparagine）	2.0g	硫酸锰（$MnSO_4 \cdot 4H_2O$）	0.1mg
磷酸二氢钾（KH_2PO_4）	1.0g	硫胺素（Thiamine）	100μg
硫酸镁（$MgSO_4 \cdot 4H_2O$）	0.5g	生物素（Biotin）	5.0μg
氯化铁（$FeCl_4 \cdot 6H_2O$）	0.2mg	蒸馏水	986 ml

通常是在20℃下进行6～8周培养以后，地衣型真菌的菌落才能出现。如果菌落过早地出现，那就可能是被污染的杂菌，而不是要培养的地衣型真菌。

然而，根据近年来的研究进展表明，虽然在固体培养基上培养的地衣型真菌生长速度较慢，但是在液体培养中的生长速度则较快。此外，在进行鸡冠石蕊（*Cladonia cristatella*）的真菌培养中用核糖醇（ribitol）则比用葡萄糖作为碳源时生长速度要快。因为鸡冠石蕊中的共生光合生物是共球藻（*Trebouxia*）。在地衣菌、藻共生中共球藻向真菌所提供的碳源正是核糖醇，而不是葡萄糖。

二、共生真菌的分离与培养

地衣共生真菌主要是子囊菌。为了对于菌、藻共生中的真菌进行单独研究或者用于菌、藻共生的人工重建，地衣共生真菌的分离及培养是不可缺少的。

（一）子囊孢子释放法

分离地衣真菌的最佳途径是通过子囊孢子的释放。首先，将子囊器用水浸湿后用吸水纸除去其表面的多余水层，用凡士林将子囊器固定在盛有琼脂的无菌皮氏培养皿盖顶内侧，使子囊上层面对盛有一薄层琼脂的皮氏培养皿底部。将皮氏培养皿盖好之后约几分钟或几小时之内，子囊孢子便可被释放在琼脂表面上。一个子囊器通常可以连续释放好几天。为了记录孢子释放期限，可以将子囊器固定在皮氏培养皿盖顶内侧近边缘的位置，并定期转动培养皿盖。对于子囊孢子被释放在培养皿底部琼脂层上的撞击区可以用彩笔在培养皿底部外侧予以标记。观察孢子时无需打开皮氏培养皿盖，而将培养皿底部朝上置于倒置显微镜下检查即可。

当孢子已经萌发后，应将含有子囊孢子的一小块琼脂切下并移接到盛有培养基的无菌试管、皮氏培养皿或三角瓶内。为了有利于单孢子分离培养，可以用减少孢子释放时间以控制孢子数量；也可以用加大孢子释放距离以便在释放后的孢子漂落中彼此拉开距离，便于单孢子分离的进行。以上操作都应在无菌条件下进行。

（二）组织培养法

首先，在自然界很多地衣并不形成子囊器；其次，并非所有地衣子囊孢子都能萌发；因此，用地衣子囊孢子进行菌、藻共生的人工重建就难以在所有地衣中实现。在这种情况下，组织培养法就具有更重要的现实意义。

根据日本Yamamoto（1991）的组织培养法，首先将地衣体切成碎片并置于漏斗滤纸上用自来水冲洗。然后将冲洗后的地衣体碎片在研钵中磨碎并加水混匀成悬浮液。经过滤后的滤液再经过第二次过滤。将残渣用水洗3次。从第二次过滤的滤纸上取出少部分接种至斜面培养基上。

Yamamoto通过地衣组织培养法已成功地分离出地衣共生菌和共生藻。迄今在Yamamoto实验室已保存有400多株不同地衣培养物及其共生菌和共生藻。它们分别隶属于16个科，34个属和124个种。

三、共生藻及蓝细菌的分离与培养

（一）地衣体碎片法

首先，取一小块新鲜地衣体，用自来水将其表面的杂物及附生苔藓清洗干净后，再用无菌蒸馏水冲洗之。用小解剖镊子将备好的干净地衣体撕下一小片，大小约1 mm，并移入有光照条件的盛有液体培养基的三角瓶中。若用较多的地衣体时，则可用研钵或置于两张载玻片之间轻轻磨碎后移入液体培养基中。对于叶状和枝状地衣体还可在解剖镜下去其皮层而取其藻层或蓝细菌细胞备用。

其次，将培养物置于18～20℃条件下进行培养。但是有些种、属的生长温度范围为1～27℃值得注意的是切勿将培养物置于直射光下，而应置于弥散光下，如背光方向的窗台上，或者置于配备连续不断的人工光照条件下。光照强度范围一般为100～250 ft-c （ft-c=foot candle:英尺烛光），但是，有些共生光合生物如共球藻的某些品系若生长在高于100个英尺烛光时它们的色素便会受到破坏。

两周之后，可除去瓶中的地衣体碎片，在显微镜下观察。然后，每隔一周进行镜检一次，直到藻细胞生长成自由生活状态下的形状，便可用于属的鉴定。

（二）微量吸移法

这是保证从单细胞获得纯净培养的最好方法。用自制的微量吸移管（玻璃管直径3～4mm，长约10cm，逐渐变尖的末端开口直径约50～75μm，而另一端开口则装上一定长度的橡皮管以便用口吹吸空气以协助操作）进行显微操作。

将小块冲洗干净的地衣体置于两张载玻片之间轻轻研磨，使地衣体被磨成绿色羹汤状悬胶体为止。这个悬胶体是由藻细胞和破碎了的菌丝所组成。将它倒入一专门小玻璃瓶中备用。

取一滴上述悬胶体置于载玻片上，并连续通过4～5滴水进行稀释，直至水滴中的藻细胞纯化为止。这一过程必须在显微镜下进行操作。最后将纯化了的藻细胞接种在洋菜斜面培养基上。

（三）离　心　法

将地衣体用匀浆器制备悬胶体后进行离心。对于由藻细胞组成的沉淀物进行稀释纯化后用吸管移至无机盐洋菜培养基平板上。

（四）其他条件

1. 对温度的适应范围：表6-3-1

南极地衣共生藻对温度的适应范围：10～20℃；最适温度：15℃

其他地衣共生藻对温度的适应范围：（1～18℃）～（20～27℃）；最适温度：18～20℃

2. 对酸碱度的最适范围

pH 4.0～7.4

3. 对光照强度的适应范围

150～250ft-c

有些共球藻在超过100ft-c时因色素受到破坏而退色。

四、地衣菌、藻共生的人工重建

自从Schwendner发现并提出地衣是由菌、藻共生的观点以后，关于地衣体的人工重建问题一直吸引着科学家的兴趣。

最早由Möller（1887）对于地衣真菌进行培养以来，几乎所有这一类的培养实验以及地衣的人工重建实验都是从孢子开始的。尽管许多人曾试图实现地衣菌、藻共生的人工重建，然而，除了Stahl等人的实验之外，很少有人能实现这一目标。在本世纪70年代，Ahmadjian及Heikkila（1970）曾重复了Stahl关于壳状地衣石果衣（*Endocarpon pusillum*）的人工重建实验，从而进一步证实了由该地衣的无藻子囊孢子和无菌共生藻细胞重建了共生联合之后能够形成具有子囊壳的地衣体。Ahmadjian等人（1980）在鸡冠石蕊（*Cladonia cristatella*）及红脐鳞（*Rhizoplaca chrysoleuca*）等枝、叶状地衣的人工重建中取得了成功。在鸡冠石蕊的人工重建实验中，首先是将经过纯培养的子囊孢子与团聚共球藻（*Trebouxia glomerata*）细胞混合在盛有琼脂的皮氏培养皿中。然后很快出现菌藻之间的连接；一个月之后便有粉芽状共生体形成，三个月后将它转接一次，在18℃下培养两个月以后便有子囊盘及分生孢子器形成。整个果柄的形成需要一年的时间。

此外，Ahmadjian（1980）还发现，鸡冠石蕊子囊孢子除了和它自己体内的共球藻（*Trebouxia erici*）形成共生联合以外，还能和来自其他12个不同属、种的共球藻（*T.*spp.）形成共生联合，却不能和来自另外10个不同属、种的拟共球藻的种类（*Pseudotrebouxia* spp.）实行共生联合。这说明地衣型真菌在与藻的共生联合中具有高度的专化性。

在分子生物学技术和地衣细胞工程飞速发展的今天，日本的Kinoshita等（1991）已经从石蕊科地衣的共生菌培养物中成功的分离出原生质体。据估计，在不久之后，地衣的人工培养物，包括地衣型真菌的纯菌种与纯藻种和其它微生物菌种一样被保藏起来。它们的DNA将被分离和鉴定。由人工培养的地衣菌藻所产生的地衣化学物质将能提供给医药、化妆品、染料等方面的需要。

第四章　生　态　学

Ecology

地衣是一群专化型或地衣型真菌。它们经常和藻类或蓝细菌形成稳定的共生生态系统，亦即适应不利于生物的外界环境的生命支撑系统。这种系统本身又因地理和生境的不同而表现为各种各样的群聚和分布类型。

一、共生生态系统

“生态系统就是包括特定地段中的全部生物和物理环境相互作用的任何统一体，并且在系统内部，能量的流动导致形成一定的营养结构、生物多样性和物质循环”。“只要有主要成分，并能互相作用，得到某种机能上的特定性，那怕是短暂的，这个整体就可视为生态系统”[奥德姆（Odum），1981]。根据这一概念，生物的共生现象，从广义上讲，无疑也属于生态系统，即共生生态系统。因为所谓共生现象是指两种以上不同生物生活在一起所形成的稳定的生态系统或群落。地衣菌藻共生这一生态系统中的生产者为藻类或蓝细菌，消费者为真菌，其内部能量流动导致形成稳定的营养结构及其与环境之间的物质交换。不过其净生产力在总的初级产量中所占的比率是很低的，其营养流也是很短的（Farrar，1976）。由于这种系统内部的物种多样性同样也受时空关系，亦即特定地段或特定地理范围的制约，因此无论是物种或群落，包括共生生态群落都是生物地理学的研究对象。

体内胞外或胞内共生现象则是共生生态系统演化中最高级的类型。而地衣正是一些专化型真菌与藻类或蓝细菌生活在一起所形成的体内胞外的共生生态系统。

这种共生生态系统一般是由一种地衣型真菌和一种藻类，或一种地衣型真菌和一种蓝细菌所形成；在少数这类系统中也有一种地衣型真菌同时和一种藻类及一种甚至两种蓝细菌所形成。在后一情况下的蓝细菌往往在地衣体局部形成小瘤或使局部肿大，这就是所谓的“衣瘿”。衣瘿有外衣瘿与内衣瘿之分。外衣瘿通常是以盘状、瘤状或囊状体而附着于地衣体表面。衣瘿内的光合生物通常都是蓝细菌，也称第二光合生物。而第一光合生物为地衣体内的主要共生光合生物，其中以绿藻为主，偶然也有黄藻和褐藻，如在瓶口衣属（*Verrucaria*）的种类中；而在另一些类群中则是蓝细菌。具有外衣瘿的常见地衣有绿皮地卷（*Peltigera aphthosa*）。此外还有瘿茶渍属（*Placopsis*）、柱衣属（*Pilophorus*）及珊瑚枝属（*Stereocaulon*）的种类都具有外衣瘿。而在网衣属的多共体网衣（*Lecidea pelobotrya*）的地衣体表面毗邻的两个衣瘿中各自却含有不同的光合生物，即第二光合生物和第三光合生物，它们分别为粘球蓝细菌（*Goleocapsa*）和真枝蓝细菌（*Stigonema*）。内衣瘿则是由位于地衣体髓层的蓝细菌簇所形成。由于内衣瘿的发育成长而使地衣体外表面相应部位呈现局部肿胀外貌，如肺衣属（*Lobaria*）及牛皮叶属（Sticta）的一些种类。

在极少数地衣中的第二光合生物并不是以衣瘿方式存在，而是和第一光合生物一样作为两个光合生物层而并存于地衣体内。在番红散盘衣（*Solorina crocea*）中，位于皮层以下的第一光合生物层是由绿藻中的胶球藻(*Coccomyxa* sp.)所组成;而第二光合生物则为念珠蓝细菌(*Nostoc* sp.),位于绿藻层之下。在复合枝属(*Compsocladium*)、珊瑚枝科(Stereocaulaceae)中,位于皮层下的第一光合生物为绿藻中的小球藻属(*Chlorella*)的种类;而第二光合生物为丝状蓝细菌,位于枝状地衣体髓层中。

在这一共生生态系统中,除了一种真菌分别和一种光合生物或同时和两种甚至三种光合生物共生之外,往往还有第二种非地衣型真菌与地衣型真菌共享共生光合生物所提供的碳源。这种生长在地衣或地衣型真菌上的第二种真菌虽然并非地衣型真菌,然而也可能是已经退化了的地衣型真菌。

地衣型真菌与藻类或蓝细菌以及非地衣型真菌所形成的共生生态系统多样性还远不止这些。因为自然界还存在着一系列程度不同的菌、藻共生现象,即松散而稳定的共生和以藻为主或以藻体包围菌体的紧密而长期的共生生态系统。在一种叫做沟鹿角菜[*Pelvitia canaliculata* (L.) Dene & Thur.]的滨海褐藻上经常生长着一种球腔菌,即鹿角菜球腔菌(*Mycosphaerella pelvetiae* Suth.),以及生长在另一种叫做泡叶藻[*Ascophyllum nodosum* (L.) Le Jolis]的褐藻上的泡叶藻球腔菌(*Mycosphaerella ascophylli* Cotton)的菌、藻共生联合,以及广泛分布于西南极海岸岩石上一种叫做南极乳壳衣[*Mastodia antarctica* (Kaetzing)Dodge]和东南极海岸的短孢乳壳衣(*M. maesoni* Dodge)的地衣,实际上也是生长在溪菜属(*Prasiola* sp.)绿藻上的核菌褶柯氏壳[*Kohlmeyera complicatula* (Nyl.) Schatz]。至于这些以藻为主的或以藻体包围菌体的菌、藻共生联合体被叫做菌藻生物(Mycophycobioses)(Kohlmeyer 和 Kohlmeyer,1979)。

此外,在热带地区往往有伪枝蓝细菌(*Scytonema* sp.)和桔色藻(*Trentepohlia* sp.)的种群被一些未定名的真菌菌丝所包围。球壳菌目(Sphaeriales)的子囊菌[珊瑚查氏壳[*Chadefaudia corallinarum* (Crouan et Crouan) Muller et Von Arx]和至少七个属的海藻经常处于相当松散而稳定的共生联合中。它们被称为"原始地衣"(Kohlmeyer 和 Kohlmeyer,1979)。

还有一些形态上密切相关的子囊菌,由于是否与藻类或蓝细菌形成共生生态系统而在分类上曾经被划归为真菌或地衣,如粉衣科(Caliciaceae)的粉盘菌属(*Roesleria* Thum & Pers.)与粉头衣属(*Coniocybe* Ach.);网衣科(Lecideaceae)的隔孢黑盘菌属(*Karschia* Koerb)与黑瘤衣属(*Buellia* DNot.),线孢软盘菌(*Mycobilimbia* Rehm.)与线孢蜡盘衣属(*Bilimbia* DNot.),黑斑菌属(*Mycomelaspila Reinke*)与黑斑衣属(*Melaspilea* Nyl.),盾状菌属(*Scutula* Tul.)与腊肠衣属[*Catillaria* (Ach.) Th. Fr.];文字衣科(Graphidaceae)的盾文菌属(*Placographa* Th. Fr.)与黑果文衣属(*Lithographa* Nyl.);斑衣科(Arthoniaceae)的星斑菌属(*Celidium* Tul.)与斑衣属(*Arthonia* Ach.);格孢腔菌科(Pleosporaceae)的亚隔孢壳属(*Didymella* Sacc.)与星核衣属(*Arthopyrenia* Massal. em. Mull. Arg.)等等。还有一些在地衣与真菌之间界限不清而曾被多数地衣学家和真菌学家所忽略的属,如狭头衣属(*Stenocybe* Nyl.)、托马氏属(*Tomasellia* Massal.)、小疣核衣属(*Microthelia* Koerb.)、针孢属(*Leptorhaphis* Koerb.)、类蜡盘属(*Dimerella* Trev.)以及类多囊衣属(*Polyblastiopsis* A. Zahlbr.)等。此外,在形态上与非地衣型子囊菌密切相关的地衣型子囊菌中还存在着明显的脱藻趋势。

上述形形色色的真菌与藻类或蓝细菌的共生联合现象不仅是地衣型子囊菌共生生态系统多样性的表现，而且还可能是子囊菌从水生向陆生过渡以及地衣起源和演化历程的印迹。在这一漫长的演化历程中，真菌与藻类或蓝细菌的共生联合并形成地衣作为适应不利环境的生命支撑系统可能在子囊菌从水生向陆生过渡中起到了重要的桥梁作用，从而导致了“共生登陆说”（symbiosis landing theory）的提出（魏江春，1983；Wei J.C. and Jiang Y.M. 1993）。

二、地衣群落学

地衣学的发展和植物学的发展一直是密切相关的。地衣学所遵循的理论和方法，基本上是来自植物学的理论和方法。形态上相似的一群地衣个体可被归类为居群或群体（population），若干相似的居群或群体归为物种（species）；形态上不同而生活习性上相似的一群地衣物种则被归为群落。前者涉及到分类学，其研究的基本单位是物种；而后者则涉及群落学，其研究的基本单位是群落中的群丛（association）。地衣群落学所遵循的理论和方法也是植物群落学的理论和方法。地衣群落学在欧洲的研究积累较多，水平也高。

在植物群落分类系统中存在着两大学派，即由Du Rietz创立的乌普萨拉（Uppsala）学派或称斯堪的那维亚（Scandinavia）学派和由Braun—Blanquet创立的苏黎世-蒙彼利埃（Zurich—Montpellier）学派或称法瑞学派。这两个学派在划分群落的标准以及对于具体群落的命名方法都很相似，只是使用的等级名称有所不同。乌普萨拉学派将群落的基本单位称做“联体”（union），而法瑞学派则称为“群丛”（association）。每一种“联体”或“群丛”都是以各该“联体”或“群丛”中的优势种或特有种的属名来命名，其词尾均为-etum，如斜漏斗石蕊群丛为Cladonietum cenoteae（来自*Cladonia cenotea*）。相近的联体”或“群丛”命名为高一级的“大联体”（federation）（乌普萨拉学派）或“群团”（alliance，也可称为群落属，法瑞学派）。每一个“大联体”或“群团”（群落属）都是以各该“大联体”或“群团”中的优势种或特有种的属名来命名，词尾均为-ion,如枪石蕊群团为Cladonion coniocraeae（来自*Cladonia coniocraea*）。此外，法瑞学派在“群丛”之下还使用“亚群丛”（subassociation）或“变体”（variant）；“群团”之上还可使用“目”（order），其每一目名的拉丁词尾为 -etalia，如癞屑衣群落目为 Leprarietalia（来自*Lepraria*）（表6-4-1）。这两个学

表6-4-1 两大学派在群落分类系统命名方面之异同

	群落 亚变体	群落 变体	群落 亚种	群落 种	群落 亚属	群落 属	群落 目
乌普 萨拉 学派				联体 Union -etum’		大联体 federat. -ion	
法瑞 学派	亚变体 Subvar. -osum	变体 variant -osum	亚群丛 Subass. -etum	群丛 Ass. -etum	亚群团 Suball. -ion	群团 Alliance -ion	群目 order -etalia

*指各群落中特有种或优势种的属名或种加词的词干。

派的另一重要区别还在于乌普萨拉学派强调优势种在每一群落中的经常性而法瑞学派则强调确限度(fidelity)，即它的存在只局限于这一群落中却并不强调他的经常性(Hawksworth，1974)。乌普萨拉学派的观点在北欧、英国和北美比较流行，而法瑞学派则在其它多数欧洲国家和其它地区较为流行。而近年来法瑞学派的观点日益得到更多研究者的采纳和使用。

以下列地衣群落与生态描述为例，简介法瑞学派的群落分类系统(Barkman，1958)：

翘叶蜈蚣衣群目

Order：**Physcietalia ascendentis** Hattick (1951) emand. Barkm. (1958).

模式群团：石黄衣群团

type-alliance：*Xanthorion parietinae* Ochsn. (1928).

组成(composition)：该群目除了石黄衣群团(*Xanthorion*)之外，Mattick 还包括了亚热带的黄枝衣-雪花衣群团(*Teloschisto-Anaptychion leucomelaenae* Mattick n. n.)。在群团之下，他并未描述过群丛。至于金黄枝衣群丛(*Teloschistetum chrysophtalmae* Ochsn. 1934.)则为裸名。因此，关于热带及亚热带黄枝衣科(*Teloschistaceae*)及蜈蚣衣科(Physeiaceae)的附生群落尚缺乏可用的资料。

在翘叶蜈蚣衣群目中，包括有两个群团，即石黄衣群团(*Xanthorion parietinae*)及灰鳞瘤衣群团(*Buellion canescentis*=*Diploicion canescentis*)。前者以枝叶状地衣为主，而后者为树皮表生鳞状地衣以及某些枝状地衣。

该群目的确限种(faithful species)：点状黑瘤衣(*Buellia punctiformis*)、灰鳞瘤衣[*Diploicia canescens*(=*Buellia canescens*)]、黄茶渍(*Candelariella vitellina*)、石黄衣(*Xanthoria parietina*)、长缘毛蜈蚣衣(*Physcia tenella*)、灰色大孢蜈蚣衣(*Physconia grisea*)及翘叶蜈蚣衣(*Physcia ascendens*)。其中有些种为群团的确限种。

生态学：喜光型，多为阳生型；通常为喜雨的(ombrophytic)，旱生的(aeraxerophytic)；中性至喜碱性的(neutrophent)，强喜氮的(nitrophytic)或耐氮的，中度耐毒素的地衣群目，通常存在于粗糙的树皮上。

灰鳞瘤衣群团 Buellion canescentis Barkm. 1958

(*Diploicion canescentis* Barkm. 1958)(Wei：按改变后的地衣名称群落名称应改为 *Diploicion*)。

模式群丛(type-association)：杜氏树花群丛[*Ramalinetum duriaei* (Duvign.) Barkm.](1958)。

组成(composition)：确限种：灰鳞瘤衣(*Diploicia canescens*)及点状黑瘤衣(*B. punctiformis*)。在该群目中的识别种：黑孔文衣(*Opegrapha atra*)，灰孔文衣(*O. cinerea*)及土赭桔色藻(*Trentepohlia umbrina*)。粉衣群团(*Calicion hyperelli*)在区系方面的界限模糊，但是在群落外貌以及畸茶渍群团(*Lecanorion variae*)方面是清晰可辨的。它们在各方面具有相似性：石黄衣群团(*Xanthorion*)，树皮茶渍群团(*Lecanorion carpineae*)，粉衣群团以及畸茶渍群团。灰鳞瘤衣群团主要为石黄衣群团与树皮茶渍群团之间的过渡型。

生态学(Ecology)：灰鳞瘤衣群团(*Diploicion canescentis*)存在于粗糙树皮上，喜生于榆属(*Ulmus*)，椴属(*Tilia*)，及柳属(*Salix*)树上，有时也生长在杨属(*Populus*)树皮上，很少生于白蜡树皮属(*Fraxinus*)，胡桃属(*Juglans*)，及栎属(*Quercus*)树皮上，从不生于赤杨属(*Alnus*)，桦木属(*Betula*)，山毛榉属(*Fagus*)以及针叶树皮上。它在光照强度及透风方面具有很宽广的生境范围。

该群团是相对耐盐的，主要存在于近沿海地带富有氯化钠的榆属(*Ulmus*)树皮上。灰鳞瘤衣甚至还生长在海岸岩堤上。

该群团下有 5 个群丛：星核衣群丛(*Arthopyrenietum gemmatae* Barkm.) 1958；灰孔文衣群丛(*Opegraphetum cinereodubiae* Barkm.) 1958；点状黑瘤衣群丛(*Buellietum punctiformis* Barkm.) 1958；杜氏树花群丛[*Ramalinetum duriei* (Duvign.) Barkm.] 1958；粉粒树花群丛[*Ramalinetum pollinariae*(Almb.) Barkm.] 1958。

石黄衣群团 Alliance：Xanthorion parietinae Ochsn. 1928. emend.Barkm. 1958.

翘叶蜈蚣衣亚群团 Suball. A. Physcion ascendenti Barkm. 1958.

模式群丛：同于群团模式群丛。

确限种：翘叶蜈蚣衣(*Physcia ascendens*)，圆叶黑蜈蚣衣(*Phaeophyscia orbicularis*)，及蜈蚣衣(*Physcia stellaris*)；在荷兰还有灰色大孢蜈蚣衣(*Physconia grisea*)及石黄衣(*Xanthoria parietina*)。它不同于杯褐梅群团(*Parmelion acetabulae* = *Melaneion acetabulae*)者在于生态学方面更为中性，喜氮，喜光和旱生的。

油蜈蚣衣群丛 Ass. Physcietum elaeinae Barkm. 1958.

模式记录(Type-record)：Barkm. 1958，Tab. XLII. no. 4.

组成(Composition)：

确限类群：油蜈蚣衣(*Physcia elaeina*)，赛氏蜈蚣衣(*P. cernohorskyi* var. *cernohorskyi*)，珊瑚芽蜈蚣衣(*P. clementi*)、反应蜈蚣衣(*P. reagensis*)及狭叶石黄衣(*Xanthoria stenophylla*)。

油蜈蚣衣群丛(*Physcietum elaeinae*)为一地中海群丛，沿着大西洋海岸向北延伸至中欧的干热盐地。在那里有两个群落变体，一个是地中海型的，另一个是大西洋型的。

a. 油蜈蚣衣群丛原变体 Physcietum elaeinae typicum

除了油蜈蚣衣(*Physcia elaeina*)之外，全部确限种只局限于该变体内。识别种：金黄枝衣(*Teloschistes chrysophtalmus*)，光茶渍(*Lecanora laevis*)及锈橙衣(*Caloplaca ferruginea*)。种数的存在度等级为：V 级 4 种，IV 级无，III 级 4 种，II 级 7 种，I 级 15 种。

壳状地衣层的平均覆盖度为 65%，叶状地衣 13%。该变体是石黄衣群团(*Xanthorion*)中以壳状地衣层占优势的唯一群丛，而实际结果却并不引人注目，而占优势的油蜈蚣衣(*Physcia elaeina*)却强烈地紧贴于树皮上，因而一看即发现该属占优势。

该变体的群落外貌是很不引人注目的，它是以多种深浅不同的灰色为主，唯一鲜艳的色彩是狭叶石黄衣(*Xanthoria stenophylla*)及石黄衣(*X. parietina*)的黄色。

生态学：主要生长在紧密而无深沟的、稍呈鳞状的冬青栎(*Quercus ilex*)树皮上；有时也生长于七叶树属(*Aesculus*)的树皮上。虽然冬青栎的树皮正常情况下为酸性的，pH 为 4.9～5.7，但是，当树皮被石黄衣属及蜈蚣衣属的种类覆盖时，其 pH 则为 6.0～7.5。当这些地衣覆盖物获得最适发育时则 pH 为 8.4(Ochsner，1934)。这种现象可能是由于强烈地受到含钙尘埃浸渍的结果。因此它被认为是中性的甚或是喜碱的。

该变体生长在空旷的树林内以及远离住宅和农村的小树丛中，在中等隐蔽至外露而又多荫的条件下。

分布：该变体可见于法国东南部的蒙彼利埃与尼姆之间，海拔为0—100m。它主要分布于从葡萄牙至南斯拉夫的地中海地区，以及也许还会向东扩展至维也纳盆地，摩拉维亚和波希米亚。

b. 油螟蚣衣群丛灰鳞瘤衣变体 Physcietum elaeinae buelliosum canescentis Barkm. 1958.

（=*Physcietum elaeinae diploiciosum canescentis*）

识别种：灰鳞瘤衣(*Diploicia canescens*)、灰鳞大孢螟蚣衣(*Physconia grisea*)、圆叶黑螟蚣衣(*Phaeophyscia orbicularis*)、黄茶渍(*Candelariella vitellina*)、及细片石黄衣(*Xanthoria candelaria*)。有些种和原变体中的一些种为替代种，而另一些则不是。

油螟蚣衣(*Physcia elaeina*)是唯一存在的确限种。呈阳生性，光照型，只是比荫生者更需光照，而不是完全无荫，故在路北侧。

该变体的恒有种(constant sp.)为油螟蚣衣(*Physcia elaeina*)、翘叶螟蚣衣(*P. ascendens*)、灰色大孢螟蚣衣(*Physconia grisea*)、圆叶黑螟蚣衣(*Phaeophyscia orbicularis*)以及石黄衣(*Xanthoria parietina*)。它们在该群丛中可能全都是有子囊盘的。在荷兰翘叶螟蚣衣(*Physcia ascendens*)及圆叶黑螟蚣衣(*Phaeophyscia orbicularis*)的子囊盘并不普遍存在，而灰色大孢螟蚣衣(*Physconia grisea*)的子囊盘则非常少见。

生态学：很明显，灰鳞瘤衣变体(*Physcetum elaeinae buelliosum canescentis*)对生境的要求正好与原变体相反。该变体是中等光照型和阳生型(heliophytic)，喜生在孤立的或路边树的基部南侧，尤其是位于东-西向公路的北侧。灰鳞瘤衣变体(buelliosum = diploiciosum)多见于富氮地方：位于农村或农场内或附近的行道树基部。这种情况只限于中小型农村和小镇附近，而不包括中等城镇和工业区附近的行道树。

细片石黄衣群丛 Ass. *Xanthorietum candelariae* Barkm. 1958.

翘叶螟蚣衣群丛 Ass. *Physcietum ascendentis* Frey & Ochsner 1926.

(1)茸褐梅变体 *variant*：*Parmeliosum glabrae* Barkm. 1958.

（=*Melaneliosum glabrae*）

模式记录(Type-record)：Ochsner，1928，Tab. V，no. 3.

确限种：螟蚣衣(*Physcia stellaris*)

识别种：微糙褐梅(*Melanelia exasperatula*)及茸褐梅(*M. glabra*)，而毛边雪花衣(*Anaptychia ciliaris*)则通常缺乏。

分布于中欧和东欧地带。

(2)亚星石黄衣变体 var. *Xanthoriosum substellaris* (Steiner) Barkm. 1958.

(3)黑胶衣变体 var. *Collemosum nigrescentis* Ochsner 1928.

(4)灰色大孢螟蚣衣变体 var. *Physciosum griseae* Barkm. 1958.

（=*Physconiosum griseae*）

(5)半羽螟蚣衣变体 var. *Physciosum leptaleae* Klament 1948.

（=*Physciosum semipinatae*）

杯褐梅亚群团 Suball. Parmelion acetabulae Barkm. 1958.

(*Melanelion acetabulae*)

模式群丛(Type-association)：杯褐梅群丛[*Parmelietum acetabulae* Ochoner. 1928)

(=*Melanelietum acetabulae*)].

确限类群(faithful)：杯褐梅(*Melanelia acetabulum*)、亚粗星点梅(*Punctelia subrudecta*)、袋衣(*Hypogymnia physodes* var. *subcrustacea*)及微白鸡皮衣[*Pertusaria globulifera* var. *henrici*. (*Pertusaria albescens*)].

识别种：槽梅衣(*Parmelia sulcata*)、金杯褐梅(*Melanelia subaurifera*)、皱梅衣(*Flavoparmelia caperata*)、栎扁枝衣(*Evernia prunastri*)及银疱衣 (*Phlyctis argena*)。在荷兰还有绿色原球藻(*Protococcus viridis*)。

蜈蚣衣科(Physciaceae)的种、属在翘叶蜈蚣衣亚群团中(20 种) 占优势，而梅衣科(19 种)的属、种则在杯褐梅亚群团中占优势。因此，这两个亚群团的群落外貌也各不相同。杯褐梅亚群团的群落外貌由更加茁壮的叶状地衣所支配，从而形成较大的莲座丛状，更疏松地固着于树皮上。因而其最小面积较大。

杯褐梅亚群团比翅叶蜈蚣衣亚群团更加喜酸性，并且存在于栎属(*Quercus*)树上，偶见于单独的山毛榉属(*Fagus*)及赤杨属(*Alnus*)树上。它是弱喜氮型至中性的。存在于树干上部及树冠中，常常位于具有充足弥散光的荫面。

丛树花群丛 Ass. Ramalinetum fastigiatae Duvigneaud 1942.

①**蜈蚣衣亚群丛 subass. Physcietosum** Barkm. 1958.

模式记录(Type-record)：Tab. XLII no. 10.

识别种：白腊树花(*Ramalina fraxinea*)，灰鳞瘤衣(*Diploicia canescens*)，长缘毛蜈蚣衣(*Physcia tenella*)，灰鳞大孢蜈蚣衣(*Physconia grisea*)，翘叶蜈蚣衣(*Physcia ascendens*)，圆叶黑蜈蚣衣(*Phaeophyscia orbicalaris*)，细片石黄衣(*Xanthoria candelaria*)，石黄衣(*X. parietina*)，及多果石黄衣(*X. polycarpa*)。

很明显，该亚群丛和翘叶蜈蚣衣亚群团，尤其是细片石黄衣群丛有密切的关系。该亚群丛是个亚中性类型，主要存在阳光充足的榆属(*Ulmus*)树上，遍布于荷兰群丛地区。

②**袋衣亚群丛 subass. Parmelietosum physodis Barkm. 1958.**

(=*Hypogymnietosum physodis*)

模式记录同于群丛的模式记录。

识别种：栎扁枝衣(*Evernia prunastri*)，袋衣(*Hypogymnia physodes*)，及管袋衣(*H. tubulosa*)。

喜酸性，在荷兰主要存在于西部海岸地带和北部的栎属(*Quercus*)和杨属(*Populus*)微隐蔽处和荫处的树干上。

第五章　地衣与大气污染

Lichen & Air Pollution

环境污染对人类的威胁并不亚于一次世界大战。但是，并非人们都能从这样的高度和紧迫性去认识和对待这一问题。近年来，我国在大气污染监测方面虽已开展了一些工作，但是，在所采用的监测手段方面只侧重于现代化进口和国产仪器设备和理化检验技术的使用，而未能考虑对大气污染比较敏感的地衣监测法。地衣监测法在云南（李天庆，1979）、杭州（Hawksworth 和翁月霞，1990）和北京（魏江春等，1991）的使用结果证明是行之有效的；此外，它不仅经济和灵敏，而且所监测出的污染后果不可能在短期内予以消除，从而杜绝了排污单位为逃避责任而采取临时对策的可能性。

一、大气污染影响地衣生长的研究历史

由于地衣的结构特点和生理特性，使得它们对环境污染相当敏感。除了地衣学工作者外，越来越多的应用生物学家和环境生态学家对地衣作为大气污染指示生物发生了浓厚的兴趣。这不仅进一步认清了地衣的应用价值，而且也为大气污染的监测工作提供了既经济又有效的手段。

（一）早期研究

地衣的生长对大气污染相当敏感。早在一百多年前，Nylander（1866）就发现巴黎市区地衣的种类和数量在不断减少；他把这归咎于密集的建筑物导致黑色烟雾的产生和有毒气体的释放，使得空气不适宜地衣的生长。在他发表的文章中，还列举了一些地衣种，如杯褐梅（*Melanelia acetabulum*）、石黄衣（*Xanthoria parietina*）以及蜈蚣衣属（*Physcia*）的一些种，包括它们的生态习性及出现频度（frequency）。

在 Nylander（1866）之后和 Sernander（1912）之前的这段时间内，除了 Arnold（1891～1901）对德国慕尼黑地区及其周围的地衣进行了详细的报道外，没有其他人进行过类似的尝试。因此，直到本世纪初，大气污染对地衣影响的工作没有取得多大进展。

（二）地衣的区带分布现象

地衣与大气污染的研究工作首先是在欧洲开展的，特别是在斯堪的纳维亚地区。Sernander（1912，1926）被认为是第一个认真研究城区周围地衣并进行分带处理的科学家；他选择斯德哥尔摩地区作为研究对象。根据地衣分布及生长情况，把该地区分为三个区带：中央地衣荒漠区（lichen desert），很少几种壳状地衣或根本没有地衣生长；在这区域外面是一个地衣挣扎生长区（struggle zone），有某些种类的地衣出现在该区域内，但盖度(coverage)不大；最外面的是地衣正常生长区（normal zone）。Haugsja(1930）依据 Sernander 的三区方法，画出了奥斯陆的地衣分布图。Vaarna(1934)研究了赫尔辛基的地衣分布后，将 Sernander 认为的地衣挣扎生长区再划分为：内部发育不良叶状地衣生长区和外部发育不良灌木状地衣生长区，即四区方法。这两个区带分别是以名为袋衣(*Hypogymnia physodes*）的叶状地衣和称做栎扁枝衣(*Evernia prunastri*）的灌木状地衣作为内限边缘。

（三）等级做图方法

第一篇阐明地衣分布与二氧化硫之间关系的文章是由 Mrose(1941)所发表。他在调查了德国一个工业污染区后发现：雨水中硫酸的含量与松萝(*Usnea* sp.)的消失有相互密切的关系。此后，由于许多城市或工业污染区空气中二氧化硫浓度的测定，使得地衣学家有可能把该地区大气污染程度与地衣生长及分布情况相联系，分为几个等级，然后制作成图表。Jones(1952)首先进行了这种工作，他注意到：在没有地衣生长的区域之外，随着污染程度降低（即二氧化硫浓度减小），从污染源由近到远可分为三个带：第一个带里只有耐污茶渍(*Lecanora conizaeoides*）出现；第二个带里有袋衣和梅衣(*Parmelia*)的一些种；第三个带里有扁枝衣(*Evernia*）和树花(*Ramalina*)以及其它属的一些种。在 Jones 之后，更为复杂的分等级做图文章陆续发表，其中最有代表性的是 Barkman (1963)，Hawksworth 和 Rose（1970)以及 Gilbert（1970）的工作。

Barkman（1963)在调查地区选择了 15 种地衣进行鉴定和做图，并按照地衣丰度(richness）的大小依次排列，但没有直接跟二氧化硫水平相联系。他根据等级作图的结果，将所调查的地区分为三类：第一类地区大气污染不严重，可在该地区内建造工厂。第三类地区严重污染，现有的工厂和矿山应关闭，或采取其他更为严格的措施来减轻污染程度。第二类地区的污染程度界于前二者之间，但不允许盖新的工厂。Barkman 的调查结果和建议被当地政府采纳，这说明了地衣监测大气污染的应用价值。这种方法也可用于规划住宅和工厂的建造地点。

Hawksworth 和 Rose 调查了一些常见地衣种的分布，依据大气中二氧化硫的浓度，把这些种耐污染的程度分为 11 个等级，并列表与二氧化硫的浓度一一对应。但他们只限于某些种的调查，地衣生长的基物也只是树干。而 Gilbert 则与此不同，他调查的不是单个种而是地衣群落。其基物除了树干以外，还有岩石以及石头建筑物等。

(四) 干旱假说

大多数的地衣学家都认为：大气污染是导致地衣消失的主要原因。但 Rydzak(1958，1968)，Ryazak & Krysiak(190) 报道了他在波兰几个城市和工业污染区的研究结果，坚持地衣荒漠区的形成是由于有害的微气候环境，特别是湿度的降低，气温的升高，以及干湿交替频率的加快。他称这个理论为"干旱假说"(drought hypothesis)。这种观点正好和 Steiner 和 Schulze-Horn (1955)的发现相一致，即波恩地区的地衣分布区带正好与相对湿度区带相吻合。同时，也得到了 Beshel(1958)的支持。不过，他们谁也没有报道过有关地区的二氧化硫浓度。而且 LeBlanc (1961) 和 Tallis (1964)分别调查了湿度大的城区，同样发现了地衣荒漠区，因此，"干旱假说" 理论还欠说服力。而 Skye (1958)，Fenton (1960)，Rao & LeBlanc(1966) 以及 Gilbert(1965)有关空气污染导致地衣消失的报道有充分的二氧化硫数据，令人信服。

(五) 其 他 研 究

除了以上提到的几个方面，在研究大气污染影响地衣生长和分布过程中，地衣移植也是一个经常使用的方法。Brodo (1961，1967) 移植地衣后，通过照像记录颜色变化来研究地衣损伤程度与离城区中心距离的关系。Schonbeck(1969) 在没有树木的地方用木板固定移植地衣，通过比较损伤程度来评价不同地区的污染程度。

此外，DeSloover 和 LeBlanc (1968)，LeBlanc 和 DeSloover (1970)在研究了加拿大两个地区的地衣分布后，通过计算 "大气纯净指数"(Index of Atmospheric Purity，缩写为 IAP；以下统称为"IAP"值)的方法来表示该地区的大气污染程度。

在有些地区，空气中除了二氧化硫，还含有氟化物，它对地衣的生长和分布也起一定的作用。Gilbert(1971)，LeBlanc，Rao 和 Comeau (1972b) 在氟化物污染的工业区调查了地衣的生长和分布；Nash (1971)，LeBlanc，Comeau 和 Rao (1971) 在氟化物污染区进行了地衣移植；Nash (1971)，Comeau 和 LeBlanc(1972)还在实验室进行了不同氟化物水平影响地衣生长的研究。

有关大气污染影响地衣生长机制的研究起步较晚，直到 1966 年，Rao 和 LeBlanc (1966)才开始在实验室研究二氧化硫对地衣的影响，紧接着，Coker(1967)，Gilbert (1968)，Hill(1971)，Puckett，Nieber，Flora 和 Richardson(1973)也进行了有关地衣损伤机理的研究。尽管他们的实验结果能成功地解释一些野外观察到的现象，但更为精确的机制仍在探索之中。

二、大气污染影响地衣生长的因素

(一) 二 氧 化 硫

是什么因素导致地衣荒漠区的形成？从 50 年代至今，一直存在着两种观点。一种就是 Rydzak 提出的“干旱假说”;持这种观点的人认为，干旱是导致地衣消失的真正原因。另外一种称之为“毒气理论”(toxic gas theory);赞成这种观点的人则坚持空气污染（主要为二氧化硫污染）是导致地衣荒漠区形成的主要因素。目前，后一种观点为大多数地衣学家所接受，而且，一些野外观察和实验结果都与这种观点相符。

Skye(1958)调查瑞典一家乡村炼油厂附近的地衣后发现，在工厂的盛行风下方一定距离内，地衣损伤严重，而上方的地衣生长良好。当时，他测定了空气中的主要污染成分是二氧化硫。紧接着，他又进行了如下实验：将一个盛装二氧化硫的圆柱桶放在乡村一棵长满着地衣的橡树附近，打开圆柱桶的阀门，使二氧化硫慢慢渗漏出去。一个月后，地衣形态变化明显：树上丰度最大的地衣槽梅衣(*Parmelia sulcata*)呈紫红色，袋衣、栎扁枝衣和粉树花（*Ramalina farinacea*)地衣体的裂片或顶端变为黑色。6 个月后，树上靠近圆桶那边的地衣脱落。一年后，整个树上的地衣全部消失。

从另外的城区或工业污染区的地衣分布图也可以看出：绝大多数地衣荒漠区的形状呈椭圆形，并朝着盛行风方向拉长，经常延伸离污染源数公里或几十公里。这一结果清楚地表明空气中有某种成分影响地衣的生长和分布。LeBlanc，Rao 和 Comeau（1972a）在研究了加拿大一地区的 IAP 值后，根据 IAP 值的大小作出了一个五区图；而这个图的区带与二氧化硫浓度的梯度区带非常相似。从二氧化硫测量的结果也可以知道：二氧化硫浓度低的地方，地衣的种类多，盖度大；二氧化硫浓度高的地方，地衣的种类少，盖度小。

Skye 和 Halberg（1969）在瑞典上述一家炼油厂停办两年后，调查了周围地衣的生长情况，发现同一种地衣与 10 年前比，地衣体裂片增厚。Hafellner 和 Grill（1981）在奥地利发现，一家纤维厂停办 8 年后，周围曾严重污染的地区，又有新的地衣种出现。还有，Showman（1981）在华盛顿，Rose 和 Hawksworth（1981）在伦敦，也分别注意到，由于大气污染程度减轻，新的地衣种又不断在城内出现。

Rao 和 LeBlanc(1966)将地衣菌丝体放在二氧化硫浓度为 5×10^{-6}的容器中，于不同的湿度下培养 24 小时，菌丝表面出现褐色斑点，藻细胞发生质壁分离现象。Hill（1971）也进行了类似的实验。

以上野外观察和实验结果都证明了：二氧化硫对地衣具有损伤作用，是导致地衣荒漠区形成的主要因素。

（二）氟　化　物

氟化物主要是炼铝厂、炼锌厂、陶瓷厂、磷酸盐工厂以及砖厂等的附产品，有时烧煤火力发电站也有少量氟化物释放于空气尘埃中。显然，氟化物对大气的污染不如二氧化硫那样普遍，但又由于这些工厂常常在大都市里，要区别氟化物和其它污染物对地衣的影响是很困难的。到目前为止，氟化物影响地衣生长的研究还只限于以下 3 个方面：

首先，野外观察证明氟化物污染源附近的地衣非常贫乏。LeBlanc，Rao 和 Comeau（1972b）在加拿大一家炼铝厂周围调查了 42 个点，他们发现：靠近工厂 1～2km 的地方没有地衣，随着离工厂距离的增加，开始出现地衣，随后则盖度增加，种类增多。他们根据调查的结果计算出各点的 IAP 值，最后作出反映不同污染水平的六区图。同时，还分析了地衣体中的含氟量，他发现，随着离炼铝厂距离的增加，含氟量逐渐减少，并作出了 IAP 值和氟含量与离污染距离的关系曲线图。Gilbert (1971)也调查了苏格兰一家炼铝厂附近地区的地衣，可以清楚地辨认出地衣的三个分区，而且受影响的地区朝当地盛行的西南风方向延长达 4km，区带呈狭窄的椭圆形。

其次，通过地衣移植，Nash (1971) 发现，移植到一家释放氟化物的化学工厂附近 100m 处的正常生长地衣，不久就遭到损伤，同时分析该处放置的钙化滤纸，发现有氟的富集。而 6km 外的对照却安然无恙，也没有氟的富集。LeBlanc，Comeau 和 Rao (1971) 在炼铝厂周围 10km 的距离内设置几个移植点，经过 4 个月和 12 个月后分别照像记录颜色变化和分析地衣体的含氟量，结果表明：距离工厂越近，氟含量越高，地衣体损伤越严重。

另外，Comeau 和 LeBlanc (1972) 的实验结果表明：在含氟化氢为 13×10^{-6}的容器中，袋衣培养 8 小时没有变化，培养 36 小时、72 小时或 108 小时，地衣体出现斑点，边缘折卷。而地衣体在含氟化氢 65×10^{-6}的容器中培养 12 小时，就会发生类似的情况。

(三) 其 他 因 素

除了二氧化硫和氟化物外，空气中还有一些污染成分对地衣的生长和分布产生一定的影响。

1. 重金属元素

有些重金属，特别是铜、钙、铁、铅、镁、镍和锌，它们在地衣体内的含量高出周围环境许多倍。这些元素主要来源于石油和煤的不完全燃烧所产生的烟雾而形成的颗粒，也与冶炼厂及汽车所排出的废气有关。含这些重金属元素的尘埃随雨水到达地衣体表面影响其生长。

2. 放射性物质

放射性元素，如铯-137、锶-90，它们都是长衰期的，短时期内不会丧失其对地衣的影响作用。这些物质主要来自核电站和实验原子反应堆。

此外，电离辐射、人工肥料以及杀虫剂也会对地衣生长产生影响。至于臭氧、氧化氮、硫化氢、氨和硝酸钾是否对地衣生长产生影响，至目前为止，还没有积累足够的资料和数据。

三、大气污染与地衣生理

与其它高等植物相比，地衣体缺乏具有保护作用的真皮层和蜡质层，因此，周围环境的物质非常容易进入地衣体内，并在体内富集，这样，就会导致一些有害成分往往高出周围环境许多倍，因此，地衣对空气的污染相当敏感。另外，90% 以上的地衣体中所含藻类是共球藻 (*Trebouxia* sp.)，它的结构非常脆弱，容易被损坏，这也是地衣对空气污染非常敏感的原因之一。

空气中的二氧化硫浓度低至 $30\mu g/m^3$ 时，就能对某些地衣产生显著的影响。二氧化硫等有害成分在地衣体内积累后，就会导致细胞膜的渗透性改变，叶绿素转变为脱镁叶绿素或是其基本结构遭到破坏。有人认为，当二氧化硫浓度低时，它竞争性地抑制了磷酸核酮糖碳化酶（ribulous diphosphate carboxylase），二氧化碳吸收受阻；二氧化硫浓度高时，细胞膜和叶绿素受到不可逆损伤，使得物质交换和光合作用受到影响。许多实验证明：光合作用是受大气污染影响最敏感的过程。此外，空气中的二氧化硫除了直接进入地衣体，还通过形成酸雨降低环境 pH 值的方式间接影响地衣生长。

Smith（1960）最早提出大气污染损伤地衣的可能性机制：他认为部分原因是由于地衣具有从周围环境富集物质的能力；部分原因是由于地衣的生长几乎全部依赖于雨水中的物质。

Skye(1958)分析了受到大气污染损伤的地衣槽梅衣的叶绿素，与正常生长的地衣相比，其叶绿素含量减少。他由此提出了一个假设，即由于地衣的损伤，叶绿素含量减少，光合作用也减弱。为了证实该假设成立，Pearson 和 Skye(1965)从 4 个不同的地点采集地衣槽梅衣，在实验室培养，观察其生理行为。与乡村采集的地衣相比，从城内采集的地衣其光合作用明显减弱。

Rao 和 LeBanc（1966）在二氧化硫浓度为 5ppm 时于不同湿度下培养地衣体，观察到质壁分离、叶绿素渗漏和藻层颜色消失等现象，并能在菌丝抽提液中检测到亚硫酸、镁离子和脱镁叶绿素。他们推测：可能是从亚硫酸离解下来的氢离子取代了叶绿素中镁的位置而转变为脱镁叶绿素。由于 HSO、SO 和 H 浓度的增加，细胞内压力增大，导致质壁分离，叶绿素渗漏。藻细胞颜色的消失可能是二氧化硫与之反应生成了无色物质，也可能是游离 H 还原有色物质为无色成分。紧接着，Coker(1967）也进行了类似的实验，相同的结果进一步证实了 Rao 和 LeBlanc 的解释。

$$H_2O + SO_2 \rightarrow H_2SO_3$$
$$H_2SO_3 \rightarrow HSO_3^- + H^+$$
$$HSO_3^- \rightarrow SO_3^{2-} + H^+$$
$$2H^+ + \text{chlorophyll} \rightarrow Mg^{2+} + \text{phaeophytin}$$

Hill(1971)就有关二氧化硫损伤地衣的机理进行了更为深入的研究。他采集不同种类的正常生长地衣，放在盛亚硫酸水溶液和硫酸水溶液的烧杯内培养相同的时间，然后在光照时测定其固定二氧化碳的速率。结果表明：①不同种类的地衣对亚硫酸水溶液的敏感度不同，而且正好与野外调察的结果一致。②同一种地衣对不同 pH 值的相同浓度的亚硫酸水溶液敏感度不同，pH 值越低，损伤越严重。③硫酸水溶液对地衣的损伤作用不如亚硫酸水溶液明显。但 Hill 对实验结果没有进行任何理论上的解释。后来，Puckett（1973）等人也进行了类似的实验，结果一样。所不同的是：他们直接从城里或乡村采来地衣样品进行二氧化碳固定速率测定。他们对其实验结果解释如下：二氧化硫水溶液（即亚硫酸溶液）既具有氧化性又具有还原性，pH 值的高低影响溶液中的离子存在状态。当 pH2～5 时，HSO 为溶液的主要存在状态，它的氧化性强于还原性；pH 值增加，氧化性减弱，还原性增强。所以，低的 pH 值有利于二氧化硫水溶液氧化叶绿素，使其结构遭到破坏，光合作用就会减弱。

由于光合作用对空气污染极为敏感，相应就会导致地衣体一些结构物质的合成以及次生代谢产物受到影响。这可以通过 Pearson(1967) 及 Pyatt(1967)的实验得到证实：

Pearson 在野外人工设置二氧化硫污染源影响垫脐鳞(*Rhizoplaca melanophthalma*)的生长，然后提取制备蛋白质的氨基酸溶液，通过纸层析观察到胱氨酸和半胱氨酸的 Rf 值都减小，而且与污染程度直接成反比关系，这种现象两天后就可以观察到，10 天后更为明显。同样，甲硫氨酸的 Rf 值也稍稍减小。但天冬氨酸和其它 3 种未知氨基酸都没有受到影响。

许多地衣能产生抗生素。Pyatt(1967) 注意到：正常生长的犬地卷(*Peltigera canina*)菌丝体浸泡液能抑制一些野草种子的萌发，而在钢铁厂附近被二氧化硫严重污染的该地衣就没有这种功能(Pyatt，1968)。然后，他将正常生长的犬地卷拿到实验室，分别用 2×10^{-6}和 10×10^{-6}的二氧化硫培养一段时间，其菌丝体浸泡液用来进行萌发实验，结果表明：10×10^{-6}二氧化硫培养的菌丝体浸泡液完全丧失了抑制能力，而 2×10^{-6}二氧化硫培养的菌丝体的抑制功能只是部分减弱。

到目前为止，大部分有关大气污染影响地衣生理的实验都是在非自然污染条件下进行的，其污染成分都高出自然环境许多倍。因此，有人认为这种实验结果不能完全可信。但是，这些实验还的确非常重要.因为通过这些实验可以弄清楚污染成分从生理上作用于地衣的方式。

四、大气污染影响地衣分布的研究方法

地衣种的分布及其频度与空气污染程度的关系是讨论单个种对污染敏感度和作为其指示生物的基础。在过去的几十年中，许多地衣工作者在这个领域进行了全面而深入的研究，运用几种不同的方法，将野外调查得到的有关地衣分布、频度及盖度的数据与空气污染成分浓度相联系，并将二者划分为相应的等级或制成图表，再运用这些等级或图表去估计其它地区的污染程度。

(一) 种的分布

就某个污染地区内单个种的分布进行做图被认为是研究空气污染影响地衣分布最经典的方法。Haugsja(1930) 最先作出了奥斯陆受大气污染的地区 20 个种的分布图。到目前为止，已有十几个大气污染地区的地衣分布图陆续发表，其中，Rose(1973)对英格兰东南地区的地衣种分布图较为详细。后来，经过 Gilbert(1970) 和 Rose(1970，1973) 改进的方法是做图时指出该种在这个地区内某基物上的内限边缘。不过，这种做图方法只表明了某种地衣在一个地区内的存在与否，大部分工作者主要是调查或是仅调查树生地衣。而且，没有从地衣体的角度去考虑，如频度、盖度等。但是，这种方法对于更为深入的研究是必要的，因为只有通过单个种的调查，才能了解它对大气污染的敏感性，从而找到耐污染的种和那些最适合作为指示生物的地衣种。

（二）种的数目

在相同的某些点(如相同环境下的同种同龄树)出现的地衣种类的数目也可以反映出大气污染的程度。许多早期工作者都很清楚地明白这一点，但有关这类资料却很少以图表的方式积累，只有很少几人进行了这方面的工作。Jones(1952)将 Haugsja(1930)的有关奥斯陆地衣分布的资料制作成图表，结果表明，越接近城内，地衣种类数目越少。Gilbert(1965) 调查了从纽卡斯尔到一家工厂约 16 000m 长的一段距离，发现所有基物上的地衣种类从 56 种减少到 5 种。Pyatt (1970)将这种方法进一步完善，他基于每 $1km^2$ 内地衣种类的数目，作出了南威尔士纽波特港沿海 1.6km 长的横断面内地衣种类数目分布曲线。

（三）盖度和频度

Jones(1952) 首先注意到：当污染地区的中心接近时，某种地衣在基物上的盖度发生变化。当所有的地衣种都考虑时，盖度变化非常复杂，用它来指示污染程度也许没有多大价值。相反，单个种的盖度变化却相当重要，可以与其它数据一起跟空气污染相联系制作曲线图或划分为等级。

盖度曲线图也可以用于某一生长类型的地衣(Fenton，1960)。因为在大气污染较严重的地带，以壳状地衣为主；而在没有遭受多大污染的地带，灌木状地衣和叶状地衣占优势。

此外，Brodo(1967) 在调查纽约长岛的地衣分布时，调查一些点，每个点选择 50 棵相同的树，观察某种地衣在这些树上出现的百分比-频度，再以该种地衣在各点频度为纵坐标，各点到市中心的距离为横坐标，作出一条曲线。这种依据频度做图的方法也可以反映该地区的污染程度。

（四）分 区 做 图

自从 Sernander 后，许多地衣学家就某个地区单个地衣种的分布和频度进行了仔细的研究，并根据该地区一系列不同地衣种的频度和分布做出分区图。这些地区小到一个工厂区或一个城镇，大到一个都市，一个省份，甚至于一个国家。这些分区图中最常见的为三区或四区系统(Haugsja，1930；Vaarna，1934)。

但 Barkman(1958)所采用的做图方法却与一般的不同，他不是依据各种地衣的分布和频度，而是基于一些常见地衣的群落，将它们划分为 4 个等级，再根据这些等级作出地衣分布区带图来估价大气污染程度。

最适合实际应用而又令人满意的分区图莫过于利用那些能在不同污染程度下出现而又容易辨认的地衣作指示种。Hawksworth 和 Rose(1970) 的 0～10 分级系统就是利用这些地衣种，同时还考虑它们所附生的树龄，树干上的高度以及周围的环境差异。他们利用这种方法作出了英格兰和威尔士的地衣分布图，从这个图也可以估计出大气中二氧化硫的浓度。

（五）IAP 做图

大气纯净指数 IAP 是由 DeSloover (1964)提出的。IAP 是依据地衣的种类数目、频度和对污染的抵抗力来对大气污染的程度进行数量评价。IAP 是按如下公式定义的：

$$IAP = \frac{n}{100}\sum_{i=1}^{n} Qi \times Fi$$

公式中 n = 某调查点的地衣种类数目；

F = 某个种在调查点的频度；

Q = 某个种的生态指数或污染敏感指数。

最早，式中的 Q 是采纳 Barkman (1958) 在荷兰建立的 12 点地衣敏感指数，即不同的地衣对大气污染的敏感度不同，通过野外调查，将它们分为 12 个等级，Q 值为 12 的地衣对污染最敏感，而 Q 值为 1 的地衣最耐污染。但这个指数太粗略，而且主观的因素较大。所以，DeSloover 建议利用 CMS 指数——在调查地区内所有各点与某种地衣同时出现的地衣种类数的平均值，即为该地衣的 CMS 值。这种指数较客观地反映了调查地区内某种地衣对污染的敏感度。

后来，LeBlanc 和 DeSloover(1970)在用该方法调查蒙特利尔地区的地衣分布时，把以上公式稍加修改如下：

$$IAP = \sum_{i=1}^{n}(Qi \times Fi)/10$$

LeBlanc，Rao 和 Comeau (1972a，1972b)又利用该公式评价了加拿大两个工业区的大气污染程度，并依照 IAP 值做出五区和六区图，并且这两个区带图与污染成分的浓度梯度非常一致。

（六）地 衣 移 植

地衣的移植实验首先是由 Arnold (1891～1901) 进行，他将乡村生长的地衣移植到城内，不久即死掉，但此举并未引起人们的注意。直到 Brodo(1961)改进了地衣移植方法，并应用于纽约长岛的地衣研究。他借用一种装置，从树上挖下一块带地衣的树皮，再将移植点的树上挖下同样大小的一块树皮，并将移植地衣固定在上面，定期观察照像，并与对照比较，Brodo 发现，移植地衣死亡与其距布鲁克林中心的距离有关，距离越近，地衣体死得越早。LeBlanc 和 Rao (1966) 在加拿大进行了同样的实验，结果也与 Brodo 的一致。

由于在建筑物密集的地方没有合适的树木进行移植，Schonbeck (1969)利用同样的方法，将移植的带地衣的树皮固定在一个离地面有一定高度的木板上。将观察结果与对照木板进行比较，通过比较移植地衣的损伤程度，可以估计不同地区的污染程度。这种方法也被 LeBlanc，Comeau 和 Rao (1971)，LeBlanc 和 Rao (1972) 采纳。

此外，Gilbert（1968），Hawksworth（1969，1971b）等人在英国，Nash（1971）等在美国进行了一系列与大气污染有关的移植实验。

为了排除因地衣移植所造成的地衣生理损伤，魏江春等(1991)在进行北京地区大气质量监测研究中，首先使用了地衣移植法，即将固着于石块表面的地衣连同石块一起移植至试验区的方法。

第六章　地衣共生生理学
Lichen Symbiotic Physiology

我们所说的地衣生理特征，是指菌、藻或蓝细菌共生状态下的地衣生理特征，即共生生理特征，而不包括地衣共生菌及共生藻或蓝细菌分别在人工分离培养中的生理特征。

一、共生的基本概念

既然我们要讨论的课题是共生生理特征，那么，澄清有关“共生”的基本概念理应首当其冲。

“共生”(symbiosis）一词是 de Bary（1879）最先使用的。其原意是指不同的生物生活在一起，它们也许是互惠的、偏惠的或拮抗的。只是后来在真菌学中，“共生”一词才渐渐被局限于互惠关系的共生概念中。现在看来，“共生”的概念正在出现一个恢复其原意的趋势，即“共生”是指不同生物之间的一切“联合”。在这种联合中的各方要莫不同程度地从联合中受益（互惠共生：mutualistic symbiosis），要莫一方从中受害（相克共生：antagonistic symbiosis），要莫各方从中既不受益，也不受害，处于中性状态，即中性共生(neutral symbiosis)，或者更确切地说，是处于偏惠状态，即一方受益而另一方并不受害，即偏惠共生(commensalism)。恢复“共生”原意这一趋势的倡导者是 Starr(1975)，并得到 Cooke（1977），Ahmadjian 和 Paracer（1987），Smith 和 Douglas（1987）以及 Hawksworth(1988)的赞同和支持。

（一）相克共生

相克共生(antagonistic symbioses)是一方受益，另一方受害的共生联合，即一般所理解的寄生。相克共生真菌即寄生真菌，或病原真菌。相克共生包括那些在联合状态下才能在自然界生存的寄生菌及其寄主。此外需要指出的是，相克共生与抗生素工作中的拮抗现象是不同的。前者是在共生联合中的拮抗现象，而后者则只有拮抗，毫无共生。

(二) 中 性 共 生

中性共生(neutral symbioses)或偏惠共生(commensalistic symbioses)是介于相克共生和互惠共生之间的一种共生现象。共生的各方从联合中既不受益,也不受害,而处于中性状态。在真菌的中性共生现象中,有许多真菌与动物,主要是无脊椎动物处于密切联合之中而不能离开这些动物单独生活在自然界,同时各方既不从中受益,也不从中受害。这类真菌中还包括那些在其生命活动中的某个阶段需要在动物体内(通常是在肠道中)生活以完成其生活史的一部分,而大部分时间则单独地生活在自然界。这类真菌可称为"继代真菌"(passage fungi)。所谓中性共生状态,实际上是有利于其中的一方而无害于也无利于另一方。因此称它偏惠共生(commensalistic symbioses)也许更为确切(表 6-6-1)。

表 6-6-1 真菌与其他生物的中性共生或偏惠共生

真菌		动物	
门类	种类	种类	部位
毛菌纲 (Trichomycetes) Zygomycota	肠藓毛菌属一种 *Enterobryus elegans*	千足虫 *Spirobolus americanus*	肠道中
虫囊菌目 (Laboulbeniales) Ascomycota	二托虫囊菌属 *Herpomyces stylopygae*	东方蟑螂 *Blatta orientalis*	触角上
	虫囊菌属一些种 *Laboulbenia* spp.	甲虫 *Bembidion picipes*	爪、背部、甲壳、头部

(三) 互惠共生(mutualistic symbioses)

在两种以上不同生物联合中各方不同程度地从中受益。

1. 虫菌共生

(1) 与食木甲虫共生的真菌:

1) 丝孢菌类: Hyphomycetes: *Ambrosiella*, *Raffaelea*, *Monacroporium* 及 *Phialophoropsis*

2) 其他:镰刀菌属(*Fusarium*)、头孢霉菌属(*Cephalosporium*)、假丝酵母属(*Candida*)、粘囊孢菌属(*Graphium*)、浆霉菌属(*Ascoidea*)及拟内孢霉菌属(*Endomycopsis*)等。

3) 引起木材变蓝色者:长喙壳菌属(*Ceratocystis*)、球二孢菌属(*Botryodiplodia*)及 *Leptographium* 等。

(2) 与食木黄蜂共生的真菌:

1) 担子菌门(Basidiomycota):

非褶菌目(=多孔菌目)(Aphyllophorales=Polyporales)

a. 韧革菌科(Stereaceae)：韧革菌属(*Stersam*)及淀粉韧革菌属(*Amylostereum*)。

b. 多孔菌科(Polyporaceae)：迷孢菌属(*Daedalea*)。

2）核菌纲的种(Pyrenomycetes)：

球壳菌目(Sphaeriales)：黑轮层炭壳(*Daldinia concentrica*)。

(3）与蚁类共生的真菌：

1）担子菌门(Basidiomycota)：木耳属(*Aricularia*)、伞菌属(*Agaricus*)、及环柄菇属(*Lepiota*)等。

2）核菌纲(Pyrenomycetes)：轮层炭壳属(*Daldinia*)及炭角菌属(*Xylaria*)等。

(4）与介壳虫共生的真菌：担子菌门(Basidiomycota)：隔担耳属(*Septobasidium*)。

(5）与其他许多昆虫营内共生的真菌：球拟酵母属(*Torulopsis*)、假丝酵母属(*Candida*)及外囊菌属(*Taphrina*)等。

2. 菌根共生

(1）外生菌根(Ectomycorrhiza)

1）担子菌门(Basidiomycota)

A. 层菌纲(Hymenomycetes)

伞菌目(Agaricales)：鹅膏属(*Amanita*)、牛肝菌属(*Boletus*)及口蘑属(*Tricholoma*)等。

B. 腹菌纲(Gasteromycetes)

2）子囊菌门(Ascomycota)

(2）内生菌根(Endomycorrhiza)

3. 菌与藻或蓝细菌共生

真菌与藻类或蓝细菌之间的共生是自然界中比较普遍的现象。根据现有知识，有半数以上的陆生真菌和1/3左右的海洋真菌参与这种共生联合。这些真菌中的绝大多数为子囊菌。依据它们的特点可归纳为3种类型：

(1）子囊菌与微观海藻形成松散的不稳定而又无新的形态学与生理学特征的共生联合。这一类型被称为原始海洋地衣(Kohlm 和 Kohlm.，1979)。

1）珊瑚查氏壳(*Chadefaudia corallinarum*)为核菌纲(Pyrenomycetes)、球壳目(Sphaeriales)、海壳科(Halosphaeriaceae)的子囊菌。它和一些微观海藻共生。这些海藻附生在海底大型海藻如珊瑚藻(*Corallina officinalis* L.)、仙掌藻[*Halimeda opuntia*(L.) Lann.]、标准仙掌藻[*H. tuna*(Ellis & Solander) Lamx.]、叉珊藻[*Jania rubens* (L.) Lamx.]、宽角叉珊藻(*J. adhaerns* Lamx.)及海带属(*Laminaria*)、石叶藻属(*Lithophyllum*)、马尾藻属(*Sargassum*)、钙扇藻属(*Udotea*)的种类上。在这种共生联合中的子囊菌生长在共生藻的细胞间而并不穿入共生藻的细胞内，也不穿入其附生的大型海藻植物即载附植物(phorophyte)的细胞内。它们分布于大西洋、印度洋、太平洋以及黑海与地中海。

2）鹿角光皮腔(*Leiophloea pelvetiae*)为腔菌壳纲(Loculoascomycetes)、座囊目(Dothideales)、圆孔腔菌科(Mycoporaceae)的子囊菌。它和一种附生在沟鹿角菜(*Pelvetia canaliculata*)上的蓝细菌形成共生联合体。该种分布于法国和英国以及西班牙大西洋海岸。

3) 巴兰皱壳[*Pharcidia balani* (Winter)Bauch]为核菌纲(Pyrenomycetes),球壳菌目(Sphaeriales)、瓶口衣科(Verrucariaceae)中皱壳属(*Pharcidia*)的子囊菌巴兰皱壳与生长在潮涧带软体动物甲壳或蔓脚动物甲壳里或石灰质岩上的各种微观蓝细菌形成兼性共生联合体。共生菌的子囊果和分生孢子器虽然埋生于钙质基物内,但是菌丝体并不穿入甲壳深部,因而动物并不受害于它。它广泛分布于热带、亚热带、温带及北极海潮涧带的60多种动物的甲壳上。

4) 海带皱壳(*Pharcidia laminariicola*)为瓶口衣科中皱壳属的海底子囊菌海带皱壳与一种附生在带状海带(*Laminaria digidata* Lamx.)柄上的叫做束枝水云(*Ectocarpus fasciculatus*)的小型丝状褐藻形成地衣状共生联合(Kohlm.,1973)。在这种共生联合中菌藻之间紧密的缠绕着,但是菌丝并不穿透载附植物海带细胞。该种只在挪威海岸采到过一次。

5) 岩溪皱壳(*Pharcidia rhachiana*)与带状海带柄表面的附生蓝细菌或附生褐藻形成共生联合并形成微薄的壳状地衣体。该种初看类似于海带皱壳,但不同之处在于子囊孢子和子实体的大小以及子囊的形状。该种分布于挪威海岸。

(2) 真菌与宏观海藻形成宏观海藻形态为主的专性共生联合。这种共生被称为菌藻生物(mycophycobioses)。

1) 念珠布氏菌(*Blodgettia bornetii* Wright)为有丝分裂型真菌。它的菌丝体在共生藻聚枝刚毛藻(*Cladophora fuliginosa* Kutz.)等的体内与轴相平行地位于内外细胞壁之间。厚壁孢子往往在菌丝顶端形成2~8个细胞的念珠串,不过菌丝从不穿透至共生藻的细胞内。它主要分布于大西洋的巴哈马群岛、百慕大群岛、瓜德罗普岛、波多黎各岛及美国的佛罗里达及维尔京群岛。太平洋的澳大利亚(昆士兰)、日本、中国台湾、美国夏威夷。

2) 泡叶球腔菌(*Mycosphaerella ascophylli*)经常与叫做泡叶藻(*Ascophyllum nodosum*)及沟鹿角菜(*Pelvetia canaliculata*)的褐藻处于共生状态。菌丝位于共生藻皮层及髓层的细胞间。子囊果和分生孢子器埋生于这两种褐藻孢子托的三层细胞深处。Kohlm.和Kohlm.(1979)认为这是菌藻生物的典型例子。该菌和这两种褐藻在自然界从不分开单独存在。

3) 溪菜肿孔壳(*Turgidosculum complicatulum*)和两种绿藻北方溪菜(*Prasiola borealis*)及棋盘溪菜(*P. tessellata*)处于共生状态。菌丝在共生藻体内将藻细胞分割成四个成团或成排的藻细胞群。大而暗色的子囊果及微小的分生孢子器埋生于这样的菌、藻联合体内并呈现为小瘤状外形。这些共生藻也能够自由生活在同一岩石上(Kohlm.和Kohlm.,1979)。

分布于南北极水域及英属哥伦比亚。

4) 石莼肿孔壳(*Turgidosculum ulvae*)的菌丝体同样也生长在共生藻的细胞间,而共生藻石莼(*Ulva vexata*)的形态起主导作用。该种分布于太平洋的加拿大(英属哥伦比亚),美国加利福尼亚、华盛顿及俄勒冈。

(3) 真菌与微观藻及蓝细菌形成以真菌形态、生理为主体的共生联合,亦即狭义的地衣。

在复杂而长期的共生联合中以藻为主的现象并非地衣，因为整个生物体是以藻体为主，其形态特征无异于一般藻类，而真菌的菌丝只位于藻体组织内部细胞之间，并不形成独特的构造。这种现象在海洋中分布较为广泛。这种共生联合的生物实际上既非真菌，亦非地衣型真菌或地衣，而是菌藻生物。而以真菌为主的现象由于其子实体实际上为真菌的子实体，其形态特征几乎完全由参与共生的真菌所决定。而藻细胞或蓝细菌位于菌体组织内部细胞之间。所不同的是它们共同形成了具有独特结构的地衣体及其一系列独特的生物学特征。这就是我们将要讨论的地衣。只有那些与一定的藻类或蓝细菌处于共生联合状态下度过了长期的生存斗争与演化过程而生存下来的真菌才能与相应的藻类或蓝细菌共生而形成地衣。这类真菌也叫做地衣型真菌。

关于地衣二重性被发现以来经历了一百多年的研究和争论，我们仍然还不完全了解地衣共生的本质。虽然大量实验结果表明地衣中菌与藻或蓝细菌的共生关系基本上是属于互惠性质的。但是，共生现象的多样性在地衣物种多样性中的存在也是毋庸置疑的。因此，绝大多数地衣中的菌、藻或蓝细菌的共生可能是互惠型的，而在有些地衣中则可能是寄生型或腐生型或二者或三者同时存在。下面是地衣中关于菌与藻或蓝细菌共生的几种观点。

1)寄生(paratism)或拮抗共生(antagonistic symbioses):Schwendener (1867) 把地衣中的真菌和藻类或蓝细菌比做主与奴的关系，故称为"主奴共生"(helotism)，尤如人类与家禽之间的关系。人类在喂养鸡、鸭、鹅，而鸡、鸭、鹅为人类提供蛋品和肉食。

2) 互惠共生(mutulism):地衣中的菌与藻或菌与蓝细菌之间的共生联合是彼此有利的。地衣体内的藻或蓝细菌借助光合作用向真菌提供碳源，而真菌菌丝形成的地衣假皮层所起的过滤作用为藻类或蓝细菌提供了适合于藻类或蓝细菌光合作用所需要的光照强度。地衣中的菌、藻互惠共生关系最先是由 Reinke 在 1873 年所强调的。

3) 内腐生(endosaprophytism):俄国地衣学家 Elenkin(1902，1904) 根据地衣体内藻层中含有少量坏死的藻细胞而认为地衣型真菌具有内腐生的营养方式。

4) 准性共生(parasymbiosis):准性共生是专门生长在菌与藻或蓝细菌共生的地衣体上的非地衣型真菌与地衣型真菌和藻或蓝细菌形成的多种生物共生联合体。它和地衣型真菌与藻类或蓝细菌之间的生理学相互关系尚待进一步研究。近年来，这一术语有被淘汰的趋势(Hawksworth，1988)。

二、地衣的生长发育

当地衣共生菌的孢子成熟以后，在一定的温、湿度条件下从子囊中释放出来并落在一定的基物(岩石、土地或树木等)上，于适宜条件下便开始萌发和生长，形成菌丝体。这些菌丝体与相应的地衣共生藻相遇而开始共生生活，这就是前地衣体阶段。当共生藻以静孢子繁殖并被共生菌丝所分开时，前地衣体便开始逐渐增大，并形成假皮层。藻细胞在假皮层下排列成层，这便是初生地衣体阶段。接着便是髓层、下皮层以及假根的出现。在进行营养生长之后，地衣的子实体便开始形成。但是，自然界的许多地衣并不形成子实体。它们的生活起点并不是孢子，而是菌与藻已经处于共生状态的粉芽、裂芽及地衣体碎片等。

地衣为多年生菌物，以鹿蕊(*Cladina rangiferina*) 为例，其年龄可由 10～55 年或更长(Prince，1974)。而高山地图衣(*Rhizocarpon alpicola*)的年龄可达 9000 年左右(Denton 和 Karlen，1973)。不过，它的生长速度却是相当缓慢的。有实验证明，地衣的生长速度在其生长发育的不同阶段是不尽相同的。在鹿蕊地衣体的生长发育初期通常表现为稳定生长和分枝。在成熟期地衣体基部开始腐烂而上部则继续生长，而且基部腐烂的速度与上部生长速度大致相等。在衰老期则基部腐烂速度大于上部生长速度直至最后坏死，其节间平均生长速度为每年 4～7mm(Prince，1974)。

在对巴尔底摩拟梅衣(*Pseudoparmelia baltimonensis*)的幼小叶状地衣体(直径为 0.1～7.0mm)的定期观察中表明：在生长初期其半径为每年生长 2mm；在成熟期的半径为每年生长 5mm；而衰老期则为每年 3mm，而且直径为 14cm 的叶状体中央已开始坏死。该种的整个年龄估计为 25～35 年(Lawrey 和 Hale，1977)。

正因为地衣具有生长慢而生命长的特点，因而，它在测算未知年龄的基质方面发挥着有益的作用，此即地衣测年法(lichenometry)。

冰川学家在推算冰川退缩年代时往往使用黄绿地图衣(*Rhizocarpon geographicum*)作为推算根据。黄绿地图衣是北极和高山岩石上常见的一种壳状地衣。该种的年龄在阿尔卑斯山区可以从 600～1300 年，而在西格陵兰的年龄则可从 1000～4500 年(Beschel，1961)。它的生长速度，经研究证明，在北极和高山地区，该种地衣体直径为 10～25mm 的年龄为 100 年。因此，对于地衣体直径为 150mm 的黄绿地图衣的年龄便可推知为 1000 年左右(Beschel，1973)。这就是说，生长黄绿地图衣的冰碛岩石是在距今约 1000 年以前由于冰川退缩而形成。Denton 和 Karlen(1973)以类似的方法用高山地图衣(*Rhizocarpon alpicola*)将瑞典拉普兰冰川退缩时间推算为距今约 9000 年。

此外，地衣测年法也被用于地震活动年代(Porter 和 Orombelli，1981)及古迹文物年代(McCartney，1978)的推算。

三、真菌与藻或蓝细菌在共生中的相互关系

关于地衣共生的本质确实是个极为复杂的问题，迄今的实验证据还不足以全面回答这个难题。但是，根据现有资料，地衣的共生菌是依赖于共生光合生物即共生藻或共生蓝细菌的光合作用提供碳源(葡萄糖及多元醇)而生活的。如果地衣的共生藻为念珠蓝细菌，它还能从大气中固定氮素。地衣是通过共生菌伸出的吸器从地衣体内胞外共生藻或共生蓝细菌的生活细胞中获得有机营养(寄生现象)。有时吸器将部分藻或蓝细菌细胞致死而继续从中吸收残余养分(腐生现象)。同时，藻细胞膜的透性也被共生菌所改变，从而加强了藻或蓝细菌细胞内养分的外渗，有利于共生菌的吸收利用。由于藻或蓝细菌遭受真菌吸器的侵袭而本能地产生一定程度的拮抗作用。另一方面藻或蓝细菌细胞由于被交织的菌丝组织所包围而使藻或蓝细菌细胞免遭有害元素的影响与机械作用的损伤，使光照强度适当减弱，从而有利于依赖弱光照生活的共生藻或蓝细菌的生命活动；有利于改善共生光合生物的水分状况，提高抗旱能力；通过菌丝组织的吸水与失水作用，菌丝体内积累着高浓度的可溶性矿物盐以利于共生光合生物对矿物盐的需要等等。此外，共生菌的某种分泌物还可增进光合生物细胞的光合作用，因而既有利于共生光合生物的生存，也有利于共生

菌所需碳源的积累。但是，当环境条件只有利于共生菌一方并使其徒长时，或者相反，只有利于共生光合生物一方并使其徒长时，地衣的菌、藻或菌与蓝细菌之间的共生状态便发生不同程度的解体。

四、碳、氮营养

（一）碳素营养

地衣生活中所需要的碳源是来自其体内胞外共生藻或共生蓝细菌的光合作用。它的光合速率远比高等植物为低，而呼吸速率却与高等植物相近。这样，其净同化作用也就非常低，因而能提供给共生菌的碳化物也是少量的。这也许正是地衣生长缓慢的原因之一。实验证明，参与地衣共生的藻类或蓝细菌在光合作用下所释放的碳化物的类型由于共生藻或蓝细菌的类群不同而各异，如多指地卷（*Peltigera polydactyla*）中的念珠蓝细菌及瓦衣属（*Coccocarpia*）中的伪枝蓝细菌（*Scytonema*）则向共生菌释放葡萄糖（I-①）；染料衣（*Roccella fuciformis*）中的桔色藻则释放赤藓醇（I-②）；皮果衣（*Dermatocarpon miniatum*）中的明球藻属（*Hyalococcus*）则释放山梨糖醇（I-③），而绿皮地卷（*Peltigara aphthosa*）中的胶球藻属（*Coccomyxa*）及石梅衣（*Parmelia saxatilis*）中的共球藻则释放核糖醇（I-④）。这些碳化物虽然各不相同，但是，它们在分子结构上的共性则表现为最后3个碳原子的一致性。根据Smith及Drew用多指地卷所做的实验证明，念珠蓝细菌在光合作用下所固定的14C-葡萄糖在半小时之内便释放出来并从光合生物层向髓层移动，而且立刻转化为14C-甘露糖醇，随后以较慢的速度再转化为不溶性地衣酸而在髓层等菌丝组织内储存下来。这是地衣菌、藻或蓝细菌共生之间碳化物移动速度最快的类型。桔色藻释放的赤藓醇的移动速度最慢，而共球藻等所释放的核糖醇及山梨糖醇的移动速度则界于上述二者之间（Richardson，1973）。

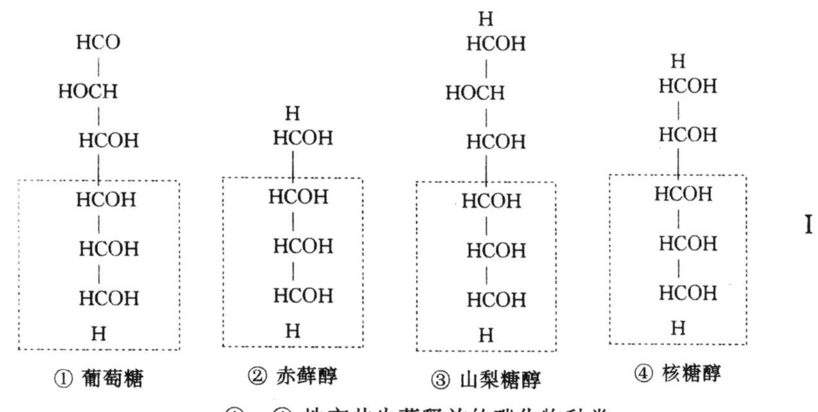

①～④ 地衣共生藻释放的碳化物种类

（二）氮素营养

氮素是地衣蛋白质合成中不可缺少的重要元素。然而，与碳素的快速转化相反，无论是在氮源存在时地衣蛋白质的合成速度，还是在缺乏氮源时蛋白质的损坏速度都是很缓慢的。自然界存在着广泛而丰富的氮源。地衣除了从其周围生境中吸收利用氮素之外，作为地衣共生物的念珠蓝细菌还可以直接固定空气中的氮以供地衣利用。

五、水 分 关 系

水是一切生物新陈代谢的介质，所以说，没有水就没有生命。但是，地衣水分关系的特点是吸水快，失水也快，这是其它生物所没有的现象。干燥的地衣体在降雨时只需1～2分钟就被水饱和。而当雨过天晴、空气干燥时，只要几个钟头就会失水变干。在严酷的干旱条件下地衣的最低含水量为2.0%～14.5%。可靠资料证实，与任何生物相比，各种地衣能够度过最长期限的干旱条件。当地衣的旱情解除之后，其呼吸速率便快速上升并远远超过正常的呼吸速率，然后才缓慢的降至正常速率。与此相反，光合作用却缓慢地恢复并逐步上升至正常的光合速率。因此，在旱情解除之后的这段时期内，呼吸速度大大超过光合速率，因而，在地衣内可能有一个二氧化碳的净损。即使在正常情况下，地衣每个单位面积的光合速率也远比高等植物为低，而呼吸速率却与高等植物相近。所以地衣的净化作用就比高等植物低得多。这也是地衣生长缓慢的原因之一。

六、矿物元素及其他

(一) 矿 物 元 素

关于矿物元素在地衣生活中的作用问题，人们的知识还比较缺乏。但是，地衣从它所生长的基物中吸取矿物元素的能力却是十分惊人的。例如，有一种貌似龟甲的双缘衣(*Diploschistes scruposus*)的土壤地衣，其含锌(Zn)量为93.4×10^{-6}，占该种地衣体干重的9.34%，而生长这种地衣的土壤的含锌量仅为10.9×10^{-6}，占土壤干重的1.09%。地衣所积累的锌、镉(Cd)、铅(Pb)以及锡(Sn)的量远比高等植物为高。石生地衣往往含有大量的锶(Sr)、钛(Ti)、钒(V)、钇(Yt)、铁(Fe)、铜(Cu)。附生在树皮上的地衣往往积累有硅(Si)、磷(P)、镁(Mg)、铝(Al)等。

(二) 抗 辐 射

一个很值得注意的现象是，在紫外线照射较强的高山上，仍生长有繁茂的地衣以及地衣对于核爆炸后的散落物所具有的惊人抗性，为人类寻找生物抗幅射机理提供了应该探讨的线索。

(三) 岩 石 风 化

由于地衣能够直接生长在裸露岩石上，所以地衣在土壤形成的岩石风化过程中起着先锋生物的作用。地衣在其生命活动中分泌有多种地衣酸，而地衣正是通过地衣酸，如茶渍酸(lecanoric acid)的螯合作用而对岩石起风化作用的(II)。

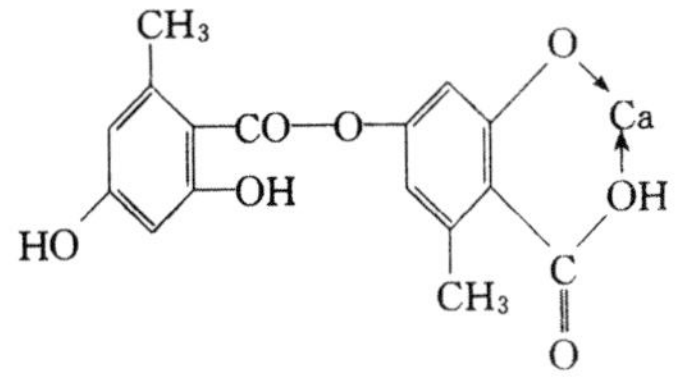

II

茶渍酸对钙的螯合作用

此外，对于地衣共生体分别进行生理特性的研究也是很重要的。对于地衣共生藻或共生蓝细菌与共生菌进行分离是研究其各自生理特性的首要条件。其分离与纯培养的方法与一般藻类、蓝细菌及真菌所用的方法是一致的。所不同的是共生菌的缓慢生长为研究者带来了许多困难。尽管如此，在这一领域中的研究报告日益增多(Ahmadjian，Yoshimura 和 Yamada，1991；Crittenden 和 Porter，1991；Yamamoto，1991 等)，其研究成果正在丰富着我们的知识宝库，也为地衣资源的开发利用开辟着可喜的前景。

第七章　地 衣 化 学
Lichen Chemistry

一、地衣化学物质及其生物合成途径

（一）乙酸酯-多聚丙二酸酯途径

（二）甲羟戊酸途径

（三）莽草酸途径

二、地衣化学物质在体内的分布及其含量

（一）在体内的分布

（二）含量

地衣体内所含的多种独特的化学物质引起了有机化学家与药物学家的极大兴趣。这些地衣化学物质是以次生的脂（肪）族的和（苯）酚的代谢产物为主。它们几乎都是胞外的，聚积在菌丝表面的水不溶性晶体物质（Culberson. C. F.，1969）。

1. 初生代谢产物

地衣初生代谢产物除了核酸、蛋白质以及类胡萝卜素、多元醇、多种游离氨基酸以及由地衣藻或蓝细菌细胞所合成的B、C族维生素等之外，地衣多糖（Lichenan＝Lichenin）（III-①）及异地衣多糖（isolichenan＝isolichenin）（III-②）乃是地衣所含的主要化学成分

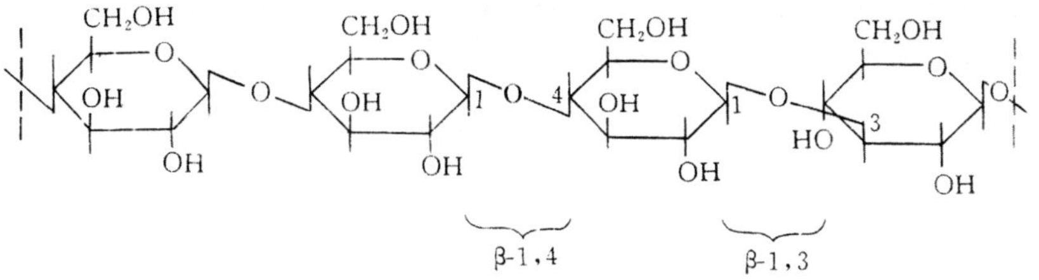
β-1，4
β-1，3

Ⅲ-①地衣多糖

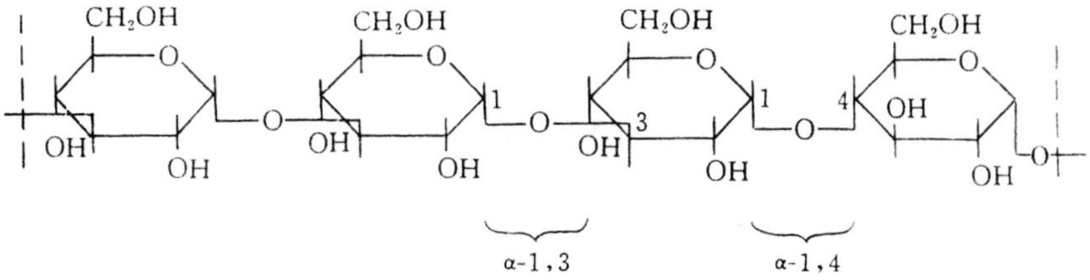
α-1，3
α-1，4

III

Ⅲ-②异地衣多糖

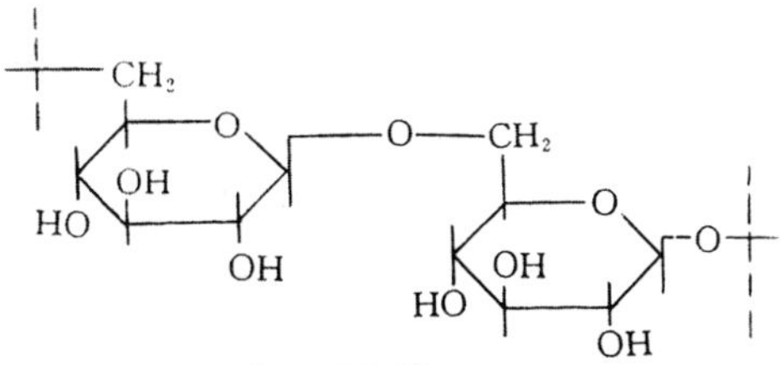

Ⅲ-③石耳多糖

Ⅲ①～③

之一。它们是地衣菌丝细胞壁的主要成分，而石耳多糖(pustulan)(III-③)以及其它多糖与几丁质则是地衣细胞的次要成分。地衣多糖是一个由β-D-葡萄糖以1，3-及1，4键合所构成的线形聚合物，其分子量为20 000至40 000。而异地衣多糖则是由D-葡萄糖残留物以α-1,3及α-1,4葡萄糖苷键合所构成的线形聚合物。

虽然人们很久以来就以为地衣多糖也可以从燕麦种子中获得，但是，动物实验证明，地衣多糖在抗癌(用Sarcoma180)方面具有极高的活性，而从燕麦种子中获得的所谓地衣多糖(lichenin)却毫无抗癌活性。这一实验结果可以说明二者在分子结构上，至少是在空间结构上可能是有差别的。据知大约80%以上的地衣都含有地衣多糖。

上述地衣初生代谢产物主要是包含在细胞壁及原生质内的，即胞内产物，为水溶性成分。而储存于地衣菌丝细胞表面的化学组分往往为水不溶性的胞外产物，它们都是地衣的次生代谢产物。这些地衣次生代谢产物开始是在菌丝细胞内合成，而最后则以结晶形式增大为不定形物而聚积于菌丝表面。

2. 次生代谢产物

在地衣的次生代谢产物中，草酸盐及碳酸盐也广泛的存在于高等植物中。通常称为地衣酸的地衣次生代谢产物主要是指缩酚(羧)酸类化合物，其中绝大部分，尤其是地衣二酚(IV-①)及β-地衣二酚(IV-②)衍生物几乎只存在于地衣体内。

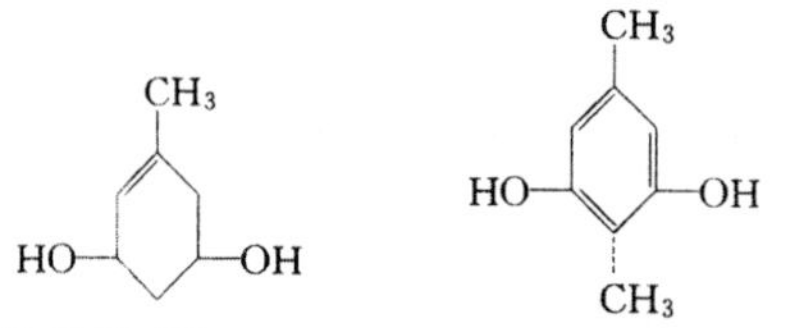

CH_3 HO OH ① 地衣二酚

CH_3 HO OH CH_3 ② β-地衣二酚

IV

一、地衣化学物质及其生物合成途径

迄今已知的地衣化学物质为300多种。地衣化学物质的生物合成有3大途径，即乙酸酯多聚丙二酸酯(acetate-pohymalonate)途径，甲羟戊酸(mevalonic acid)途径及莽草酸(shikimic acid)途径(Culberson C.，Culberson W，和 Johnson，1977；Hale，1983：119)(图6-7-4)。

乙酸酯-多聚丙二酸酯途径导致绝大多数地衣化学物质的形成。它们是缩酚(羧)酸类及其衍生物，松萝酸及其相关产物，蒽醌类以及高级酯(肪)酸及酯类。甲羟戊酸途径主要形成萜类、类胡萝卜素及甾醇类。而许多黄色素则是通过莽草酸途径产生的。

按照地衣化学物质的三大生物合成途径对于地衣的主要产物作一简要介绍。

(一)乙酸酯-多聚丙二酸酯途径

大约有25种地衣脂肪酸。由于它们是无色物质，因而在色谱分析中往往易被忽略。现将常见几种脂肪酸的结构式展示如下：

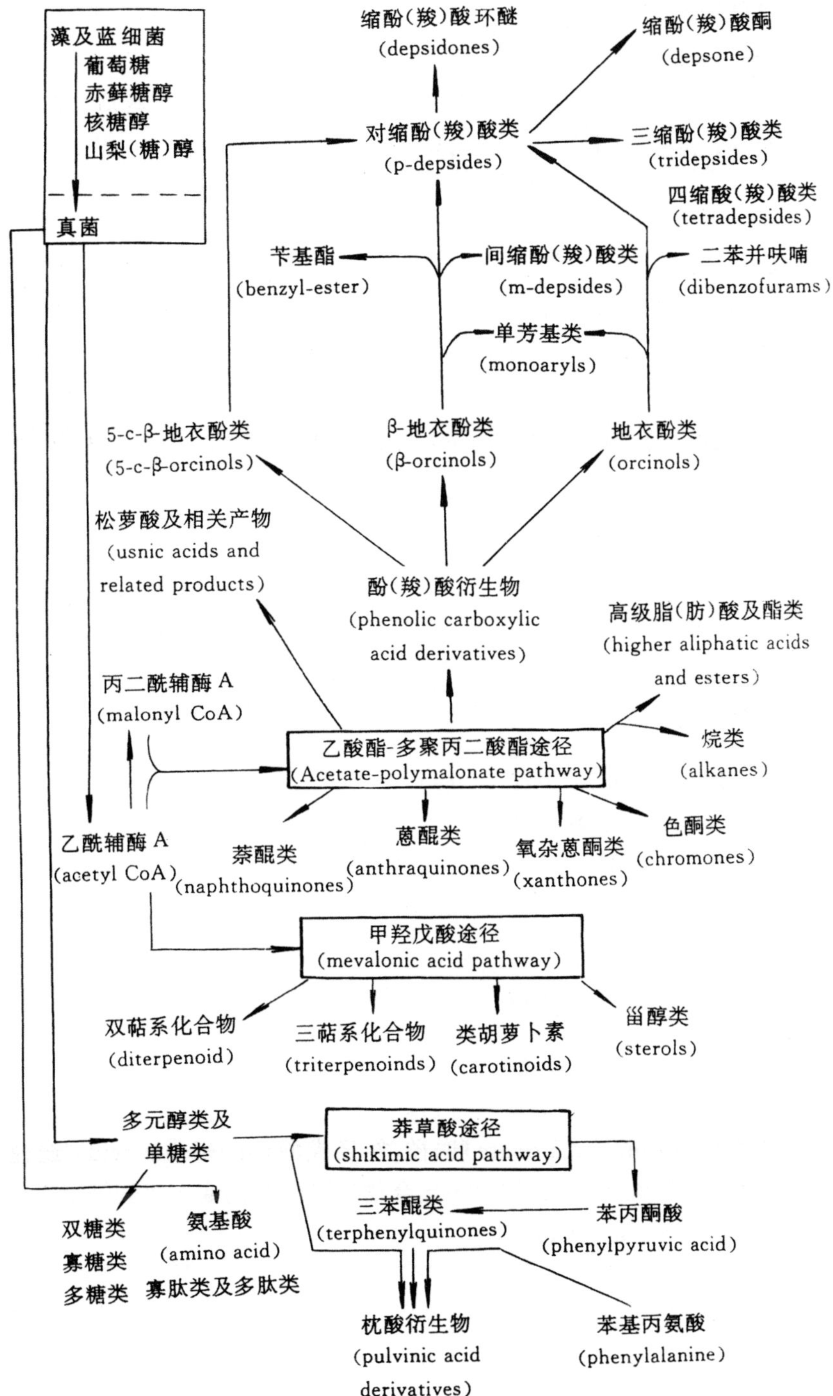

图 6-7-1 地衣化学物质的生物合成途径

(根据 Culbersons 和 Johnson，1977；Hale，1983)

1. **高级脂(肪)酸及酯类**

(1) 皱梅衣酸(caperatic acid)(V)

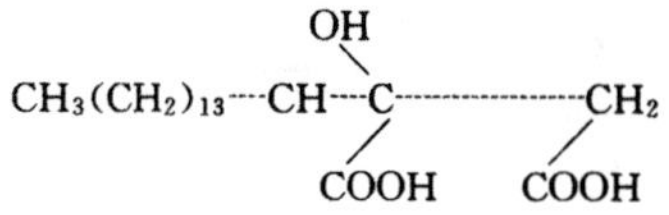

OH

$CH_3(CH_2)_{13}$—CH—C—CH_2 V

COOH COOH

(2) 地衣硬酸(lichesterinic acid)(VI)

HOOC—C═C—CH

$CH_3(CH_2)_{12}$—CH CO

O

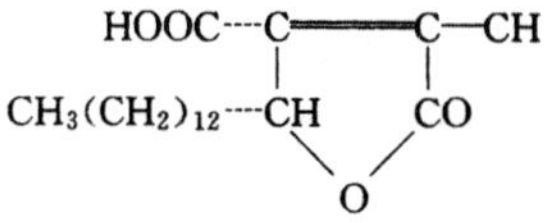

VI

(3) 原地衣硬酸[(+)-Protolichesterinic acid](VII)

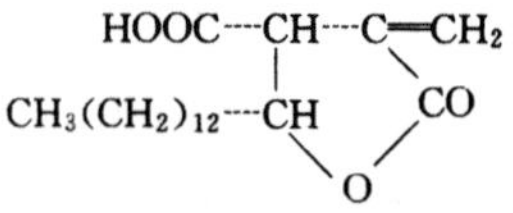

HOOC—CH—C═CH_2

$CH_3(CH_2)_{12}$—CH CO VII

O

2. **烷类**

3. **酚(羧)酸衍生物**

(1) 单环衍生物(monocyclic derivatives)

苔色酸(orsellinic acid)(VIII)

4,6-二羟-2-甲苯甲酸

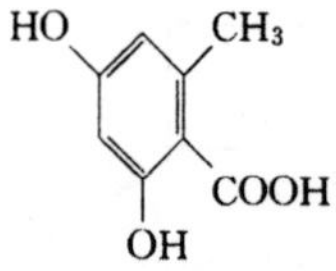

HO CH_3

COOH VIII

OH

(2) 二芳基,三芳基,及四芳基酯类

1) 地衣酚系

A. 对(位)--缩酚(羧)酸类,三缩酚(羧)酸类,四缩酚(羧)酸类迄今已知的该类化合物约 60 多种,主要存在于地衣中,其中很多化合物遇次氯酸钙或次氯酸钠呈正反应。

a. 绵腹衣酸(anziaic acid)(IX)

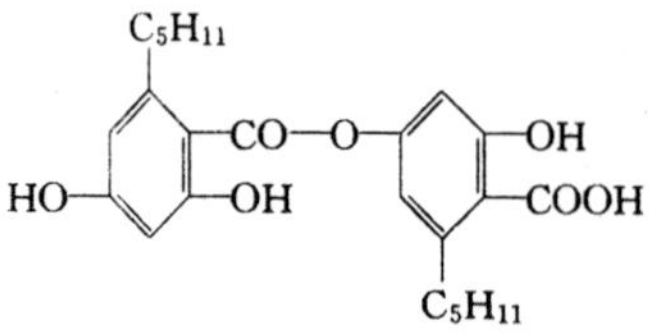

C_5H_{11}

—CO—O— —OH

HO— —OH —COOH IX

C_5H_{11}

b. 柔扁枝衣酸(divaricatic acid)(X)

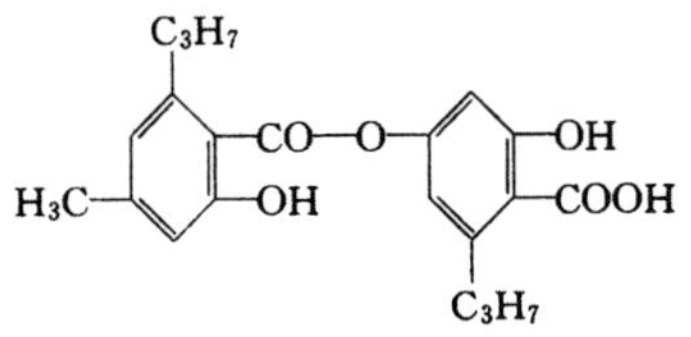

X

c. 去甲环萝酸(evernic acid)(XI)

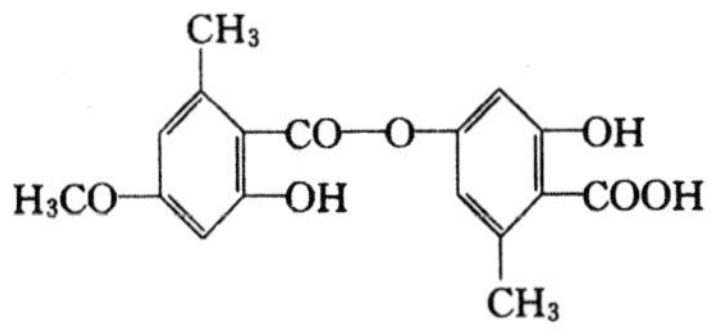

XI

d. 茶渍酸(lecanoric acid)(XII)

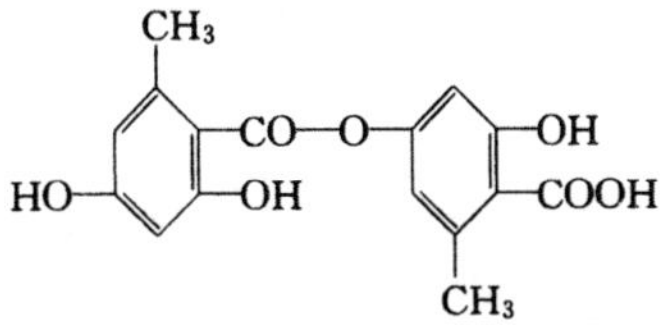

XII

e. 三苔色酸(gyrophoric acid)(XIII)

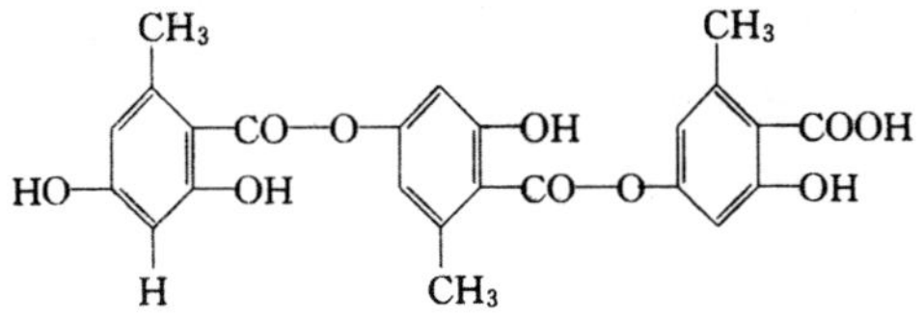

XIII

f. 石花酸(hiascic acid)(XIV)

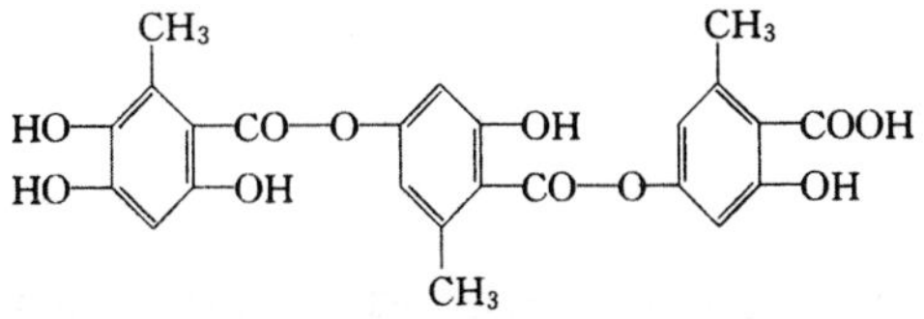

XIV

g. 橄榄陶酸(olivetoric acid)(XV)

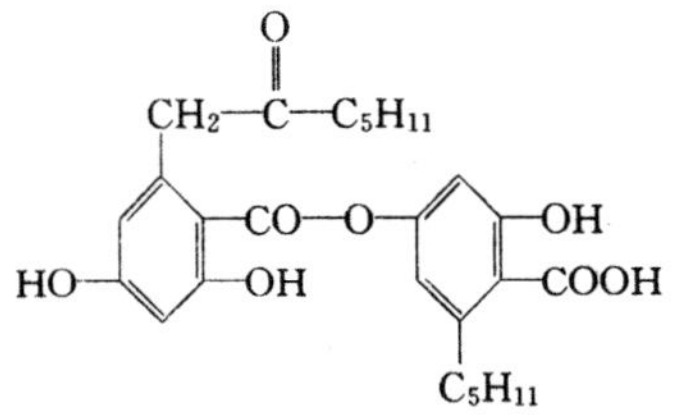

XV

h. 石耳酸(umbilicaric acid)(XVI)

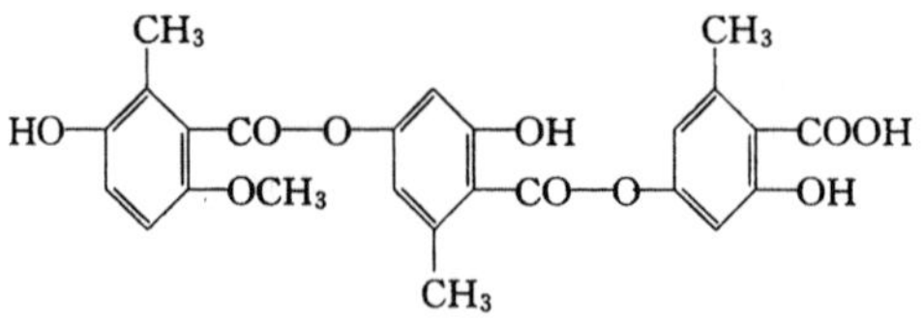

XVI

B. 间(位)缩酚(羧)酸类有 15 种之多。

a. 包宁酸(boninic acid)(XVII)

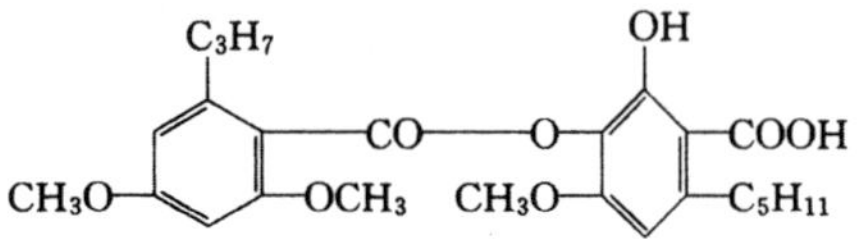

XVII

b. 隐氯地衣酸(crygtochlorophaic acid)(XVIII)

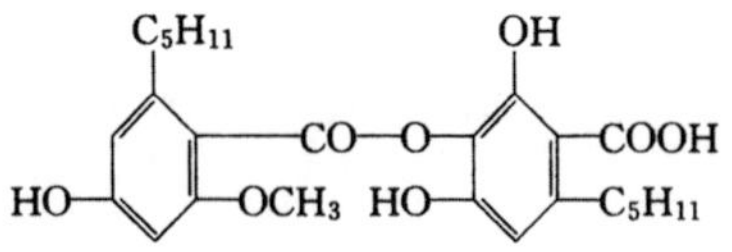

XVIII

C. 缩酚(羧)酸环醚

缩酚(羧)酸环醚为地衣酚类化合物，在生物发生学上可能是由缩酚(羧)酸衍生物而来的 。它们是地衣成分中第二个最大的化合物类群。

a. 树发衣酸(alectoronic acid)(XIX)

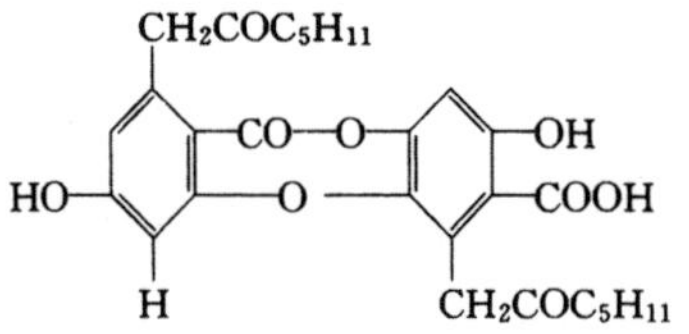

XIX

b. 肺衣酸(lobaric acid)(XX)

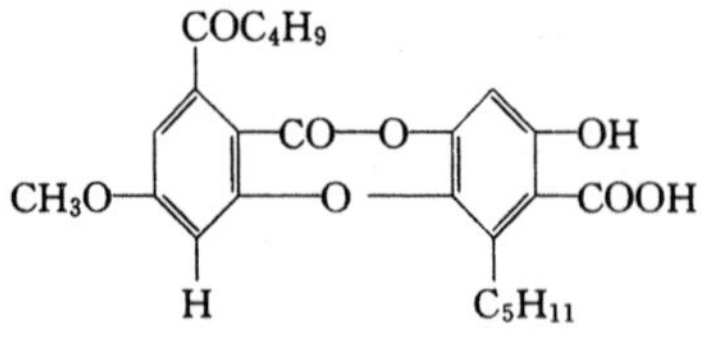

XX

c. 袋衣酸(physodic acid)(XXI)

对于革兰氏阳性细菌具有高度抗菌活性，对于结核杆菌及真菌(trichophyton，candida 等)的抗菌活性较弱。

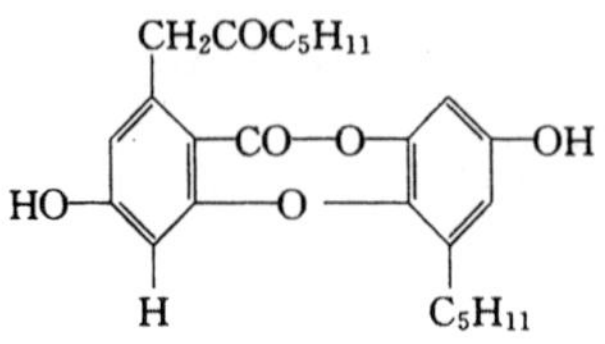

XXI

d. 颗瘤衣酸(variolaric acid)(XXII)

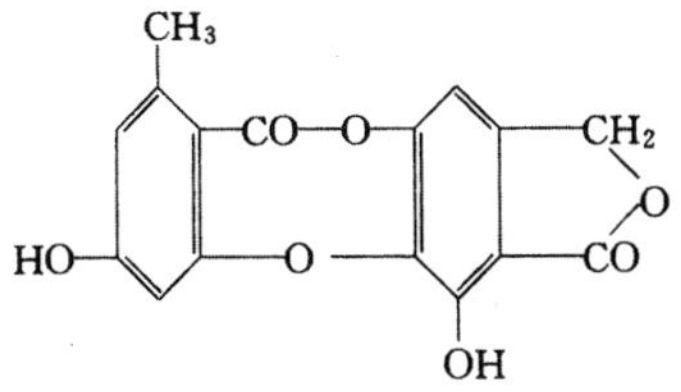

XXII

D. 缩酚(羧)酸酮

苦苔地衣酸(picrolichenic acid)(XXIII)

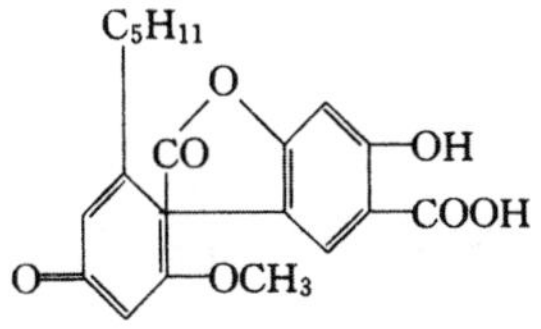

XXIII

2) β-地衣酚系

A. 对(位)-缩酚(羧)酸类

a. 黑茶渍素(atranorin)(XXIV)

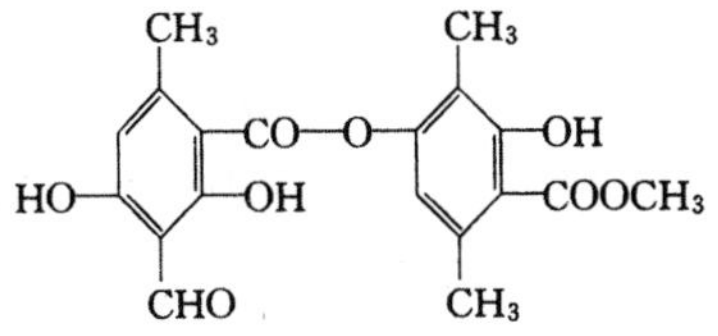

XXIV

b. 羊角衣酸(baeomycesic acid)(XXV)

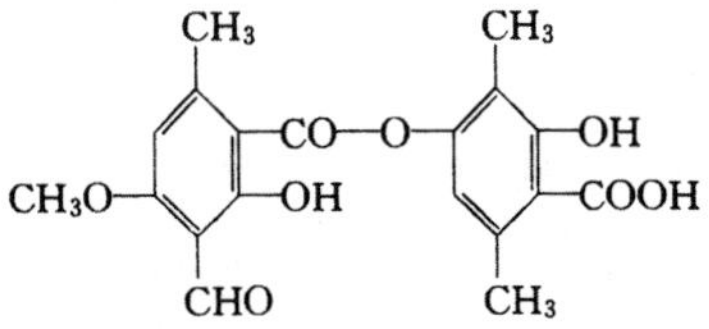

XXV

c. 巴巴酸(barbatic acid)(XXVI)

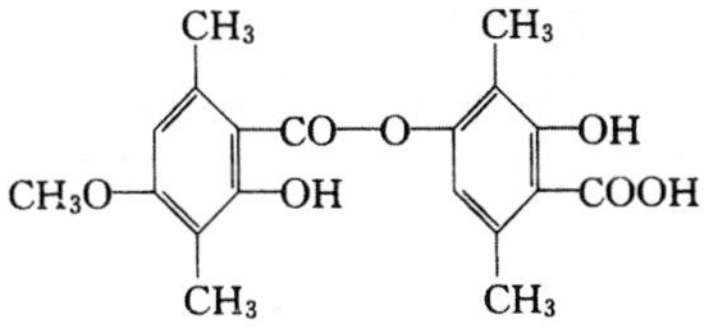

XXVI

d. 环萝酸(diffractaic acid)(XXVII)

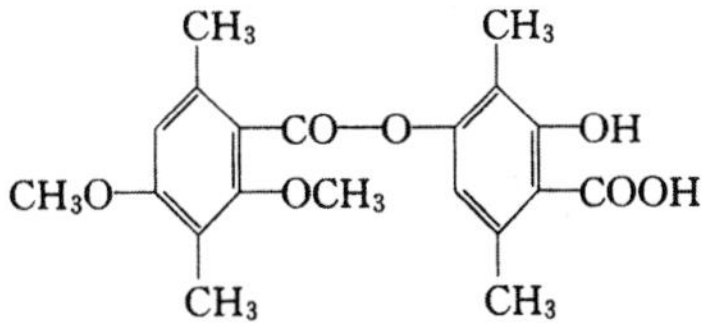

XXVII

e. 鳞片衣酸(squamatic acid)(XXVIII)

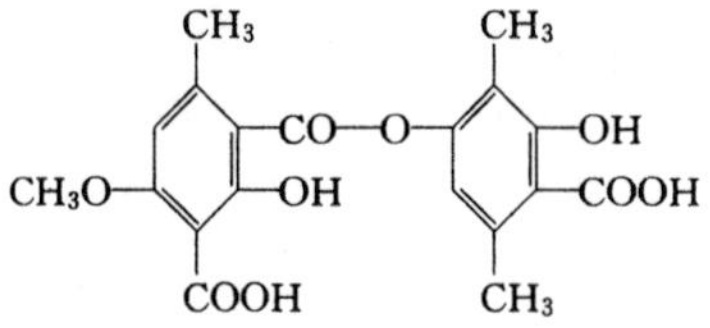

XXVIII

B. 间(位)-缩酚(羧)酸

地茶酸(thamaolic acid) (XXIX)

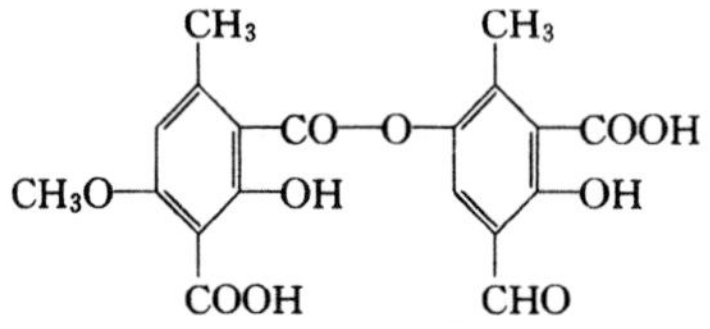

XXIX

C. 苄基酯(benzyl-ester)

坝巴醇酸(barbatolic acid)(XXX)

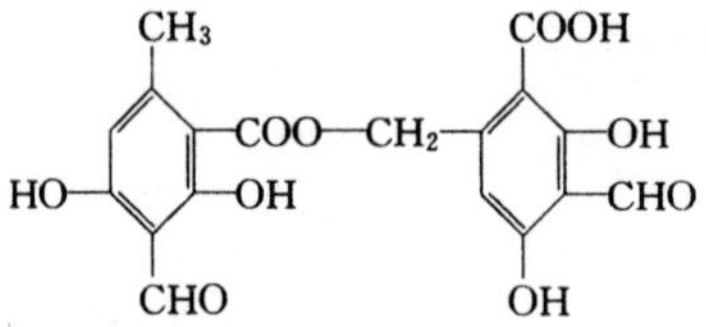

XXX

D. 缩酚(羧)酸环醚

该类化合物为β-地衣酚类化合物，迄今已知约40种，其中绝大多数遇氢氧化钾或对苯二胺呈正反应。

a. 降斑点酸(norstictic acid)(XXXI)

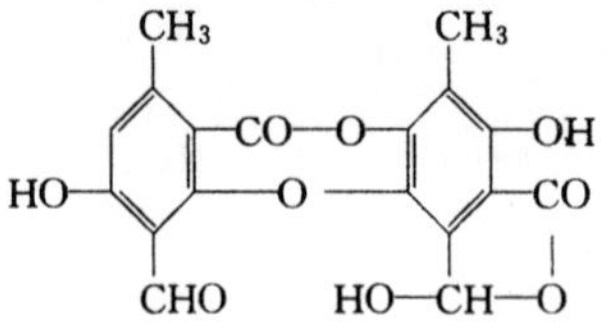

XXXI

b. 牛皮衣酸(stictic acid)(XXXII)

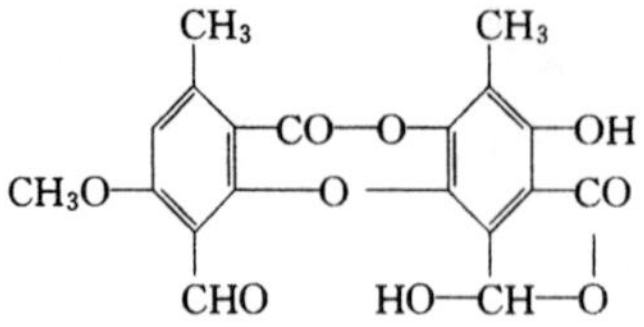

XXXII

c. 水杨嗪酸(salazinic acid)(XXXIII)

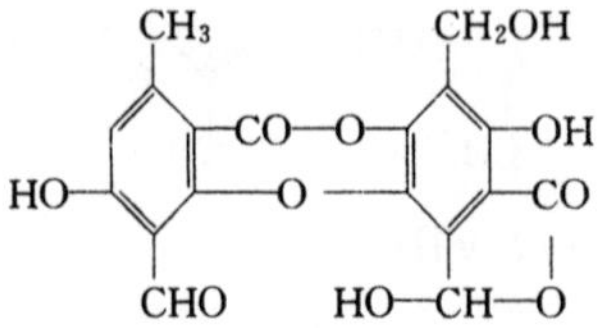

XXXIII

d. 富马原岛衣酸(fumarprotocetraric acid)(XXXIV)

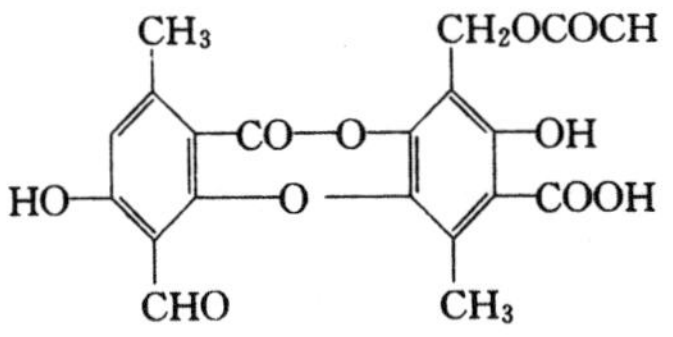

XXXIV

e. 原岛衣酸(protocetraric acid)(XXXV)

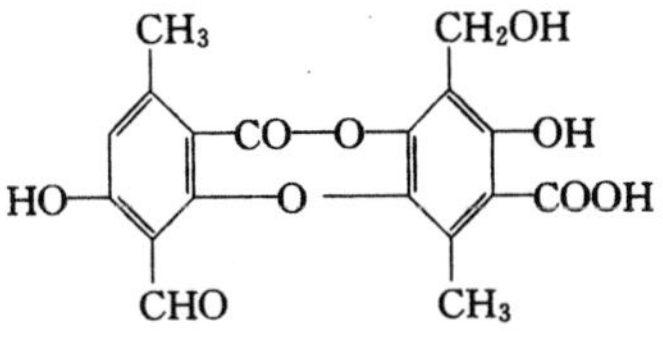

XXXV

f. 袋衣甾酸(physodalic acid)(XXXVI)

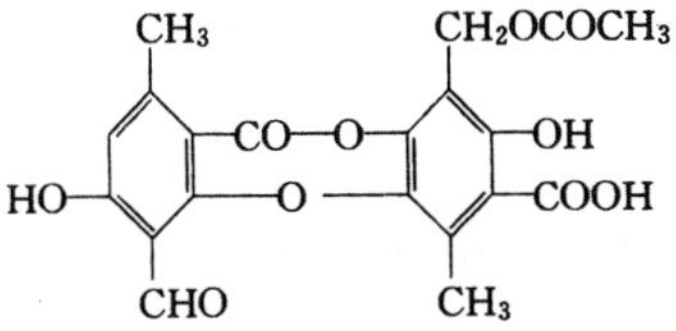

XXXVI

g. 茶痂衣酸(psoromic acid)(XXXVII)

存在于沟槽枝衣(sulcaria sulcata)中,具有一定的抗癌活性。

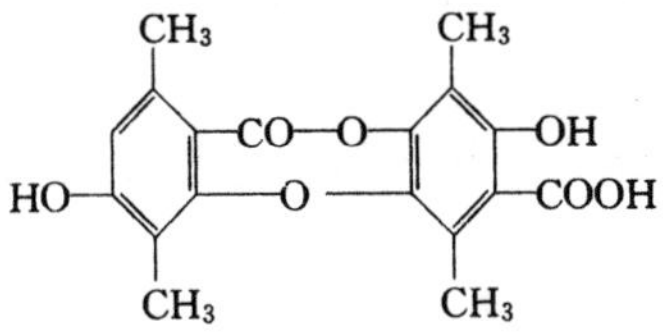

XXXVII

(3) 二酚基类,5. 5′-=-C1-取代了 β-地衣酚对(位)-缩酚(羧)酸

(4) 二苯并呋喃:下列两种化合物为石蕊属地衣所特有。此类化合物往往具有很强的抗癌活性。

1) 小红石蕊酸(didymic acid)(XXXVIII)

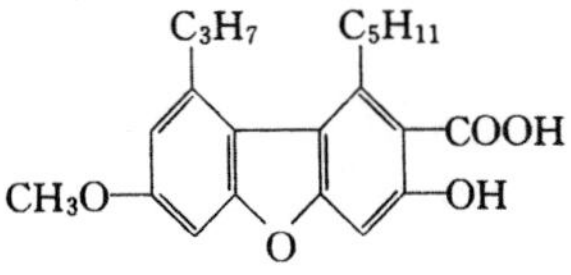

XXXVIII

2)链丝素(strepsilin)(XXXIX)

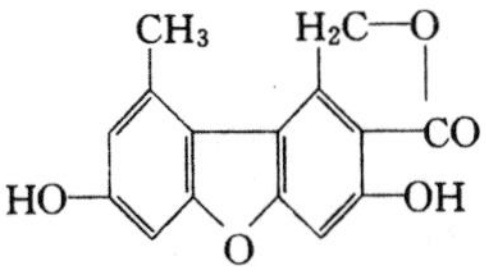

XXXIX

(5) 松萝酸及其有关产物:松萝酸是广布于地衣皮层中的淡黄色化合物。它对于革兰氏阳性细菌,尤其是结核杆菌具有较强的抗菌活性,是著名的地衣抗生素。

松萝酸[(+)-usnic acid](XL)

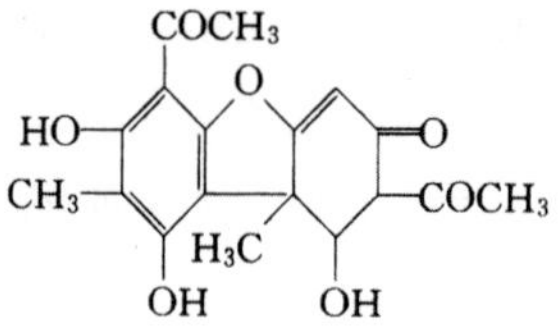

XL

(6) 色酮类

色酮类(siphulin)(XLI)

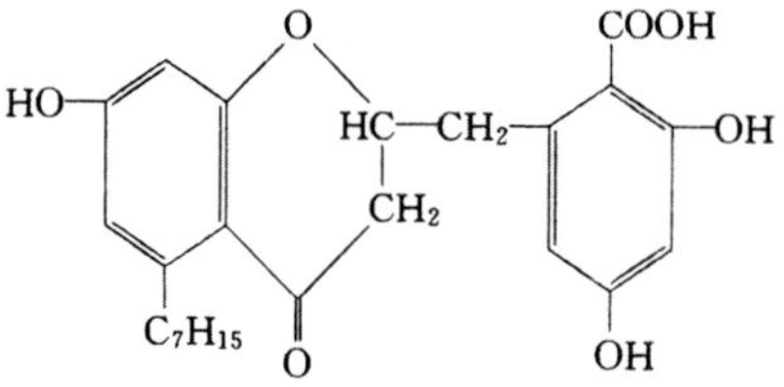

XLI

(7) 氧杂蒽酮类

地衣黄(lichexanthone)(XLII)

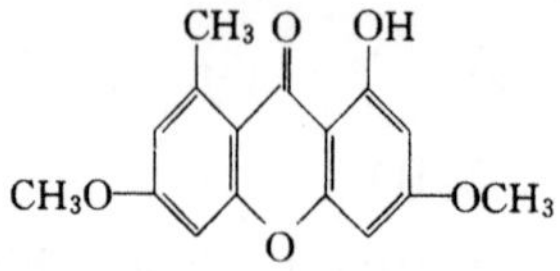

XLII

该化合物是较常见的地衣皮层中的物质。它在长波紫外光下呈鲜艳橙黄色。

(8) 蒽醌类

在地衣中已有约40种蒽醌类化合物被提取。它们都是浅黄色至深橙红色色素。

1) 石黄酮(parietin)(XLIII)

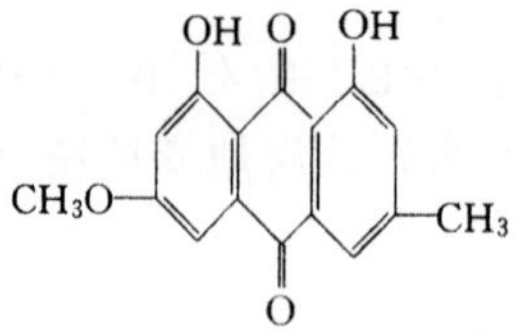

XLIII

2) 石黄酸(parietic acid)(XLIV)

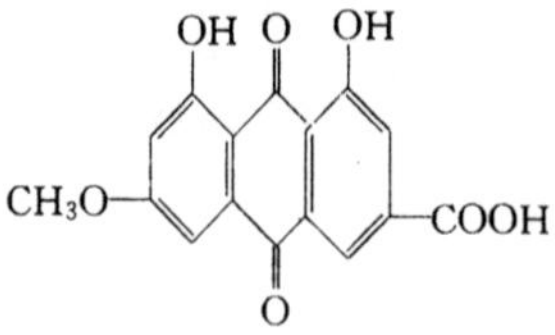

XLIV

(9) 萘醌类

座盘衣酸(chiodectonic acid)(XLV) (=pyxiferin)

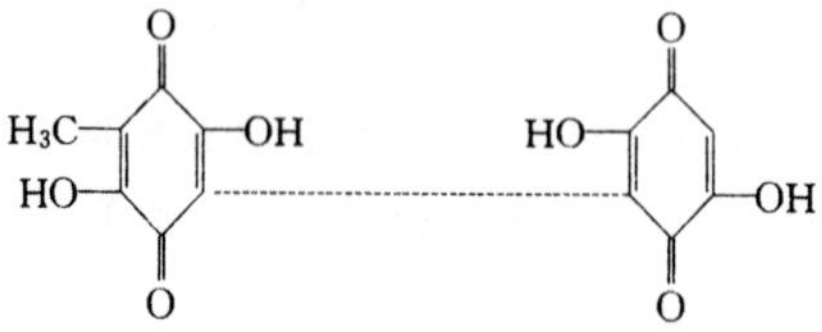

XLV

（二）甲羟戊酸途径

大约 60 种不同的五环三萜系化合物在地衣中被发现。它们主要存在于蜈蚣衣科(Physciaceae)及肺衣科(Lobariaceae)的地衣中。

三萜系化合物

泽屋萜(zeorin)(XLVI)

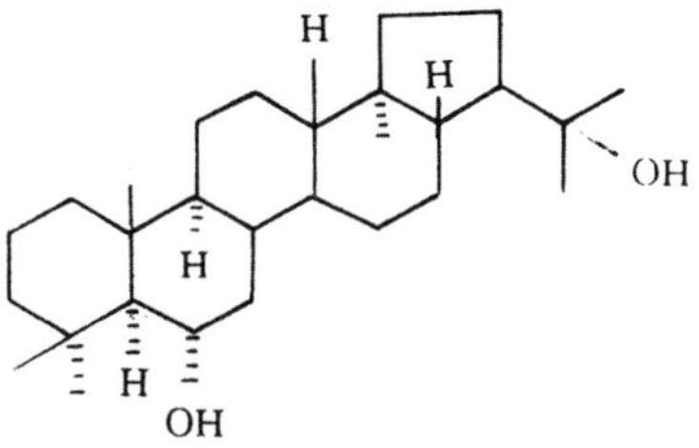

XLVI

（三）莽草酸途径

有三苯醌类及枕酸衍生物

(1) 黄花衣酸(pinastric acid)(XLVII)

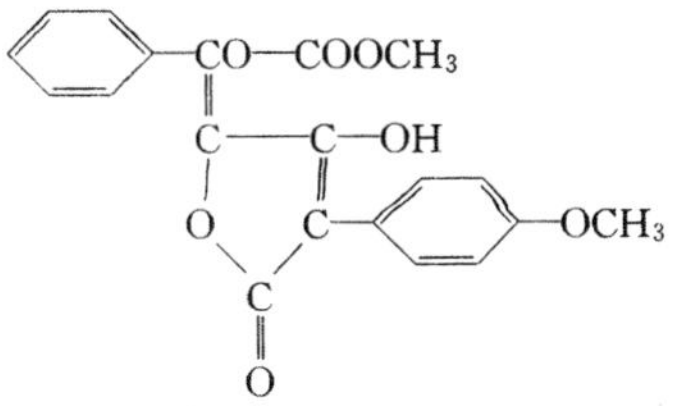

XLVII

(2) 狐衣酸(吴尔品酸:Vulpinic acid)(XLVIII)

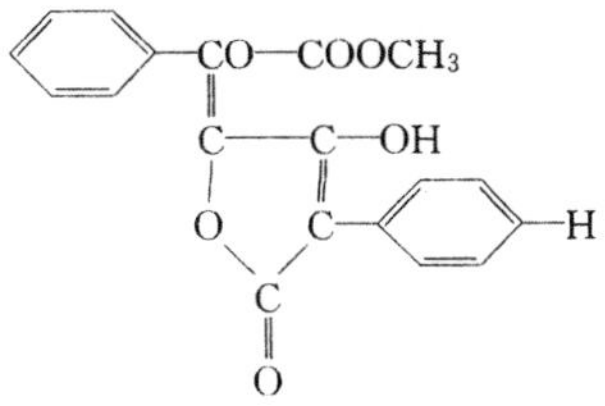

XLVIII

二、地衣化学物质在体内的分布及其含量

（一）在体内的分布

地衣次生代谢产物存在于地衣体内的部位是具有高度专化性的。具有各种颜色的绝大部分蒽醌(anthraquinones) 的衍生物，淡黄色的松萝酸以及无色的黑茶渍素(atranorin) 等则储存于地衣的假皮层组织内，因而使地衣呈现为各种不同的色彩；而绝大部分缩酚(羧)酸及缩酚(羧)环醚则储存于地衣的髓层内。

（二）含 量

地衣酸在地衣体内的含量是相当高的。作为地衣抗生素的松萝酸在松萝科（Usneaceae）、梅衣科（Parmeliaceae）及石蕊科（Cladonianceae）地衣中的含量一般占地衣体干重的 0.2%～4%，但在金黄树发［*Alectoria ochroleuca*（Hoffm）Mass］中则其含量可达 8%。茶渍酸在梅衣（*Parmotrema tinctorum*=*Parmelia tinctorum*）中的含量竟高达 23.5%。

第八章　地衣化学成分微量综合检验法
Microchemical Techniques for Identification of Lichen Substances

所谓“化学分类学”（chemotaxonomy）无论是对植物分类学或菌物分类学来说都是不确切的。作为菌物学之一的地衣分类学或者更确切地说是地衣型真菌分类学，实际上是对它们的种系从生物学的各个角度（形态学、解剖学、生理学、发育学、生物化学、生态学以及分布学等）和不同层次（个体、群体、组织、细胞以及大分子等）进行综合性的分类研究。而有关地衣的化学知识只是作为地衣分类或地衣型真菌分类中生物学的许多角度和层次之一而已。对于地衣分类学来说，上述每一个角度和层次只是地衣分类学这门科学中的一个组成部分。离开地衣分类学，上述每一个角度或层次，作为单独的分类学，如化学分类学、细胞分类学等似无独立存在的可能和必要。因此，所谓化学分类学实际上是有关地衣或地衣型真菌化学知识在地衣或地衣型真菌分类研究中的应用问题。

地衣能够产生许多其它菌物及植物不产生的独特的次生代谢产物。迄今已知的这类产物有 300 多种。地衣化学知识在现代地衣分类研究中，除个别类群之外，已是不可缺少的重要组成部分。任何不考虑化学知识的地衣分类研究在现代地衣学家看来都是不完全的，过时的，因而也是不能接受的。

地衣化学知识在地衣分类研究中的应用和其它菌物及植物相比是最有成效的，历史也是最悠久的。早在 19 世纪，芬兰地衣学家 Nylander（1865）就首先发现了碘反应在地衣分类鉴定中的作用。后来，他又分别发现了漂白粉（$CaCl_2O_2$，次氯酸钙，简写为 C）及氢氧化钾（KOH，简写为 K）的显色反应的分类学价值（Nylander，1866a，b）。英国的 Leighton（1867）曾发现了氢氧化钾（K）及漂白粉（C）的联合显色反应（简写为 KC）的分类学意义。对苯二胺（Paraphenylene diamine，简写为 P 或 PD）的显色反应是日本朝比奈泰彦（1934）首先报道的。这是地衣化学知识在地衣分类研究应用方面的经典时期。

本世纪 30 年代朝比奈泰彦（1936～1940）报道了微量结晶试验法（MCT）在地衣分类研究中的应用。这个简而易行的测定法使地衣化学知识在地衣分类研究中从经典时期跃入真正的科学时期。50 年代初期（1952）Wachmeister 使用纸层色谱法进行地衣化学测定，而 60 年代末期（1967）Santesson 使用薄层色谱法使地衣化学知识在地衣分类研究中的应用成为现代地衣分类研究中不可缺少的重要组成部分。Culberson 和 Kristinsson（1970）使薄层色谱法在地衣化学成分测定方法标准化以及有关资料积累方面做出了巨大贡献。标准化的地衣化学薄层色谱法现已普及到每一个地衣分类实验室，成为现代地衣分类研究中不可缺少的常规手段之一。至于高压液相色谱、气相色谱与质谱等分析手段由于仪器设备昂贵而无法在地衣分类实验室推广使用。

一、化学显色反应法

此法虽较原始，但仍可检出某些地衣次生代谢产物。因而迄今仍在地衣分类鉴定中延用不衰。

（一）试　　剂

1. C. 次氯酸钙

[calcium hypochlorite：$Ca(OCl)_2$]或称漂白粉(bleachpowder)。也可以家用漂白液作为代用品，如在北美可用 Clorox 或 Javex(均为商品名，下同)，在英国为 Parazone。实际上它们都是次氯酸钠(sodium hypochlorite：NaOCl) 的水溶液。因此，我们可用北京西中化工厂的次氯酸钠溶液以代替次氯酸钙。

无论采用次氯酸钙的饱和水溶液或次氯酸钠溶液，使用时都应检验它们是否仍具有活性。为了检验它们的活性，可以选用一些含有遇 C 起正反应的地衣酸的地衣种类进行，如大叶梅(*Parmotrema tinctorum* = *Parmotrema tinctorum*)，细裂岛衣(*Cetraria delisei*)，石耳(*Umbilicaria esculenta*)，红腹石耳(*U. hypococcinea*)等均可使用。若上述地衣髓部遇 C+红色，说明试液尚有活性，若不起红色反应，说明试液已经失活，不能使用，需换用有活性的新鲜试液。一般使用的次氯酸钠溶液最好存放在干燥阴凉处，可以保持较长时间而不易失活。

2. K. 10%～25%的氢氧化钾(potassium hydroxide：KOH)水溶液

或将 70g 氢氧化钾溶于 200ml 水中形成的标准试液。

3. KC. 氢氧化钾与漂白粉(KC)

联合反应时，先使用 K，紧接着补加 C。

4. P. 对苯二胺(p-phenylenediamine)

必须在暗处保存，使用时配成 5%的乙醇溶液。配好的溶液一天之后便会丧失活性。为了克服上述缺点，可以采用下述两种方法之一：一是随用随配，即在使用时，可取出 1～3 颗对苯二胺晶体置于比色瓷盘中，当即加一两滴乙醇，使之溶解，便可使用。检验时，用小画笔或毛细管将对苯二胺的乙醇溶液涂于待检标本上即可。二是取 1 g 对苯二胺晶体，加 10g 亚硫酸钠，再加 0.5ml 中性洗涤液，最后加 100ml 水，即可形成混合剂。这种混合剂最先是由 Steiner(1955) 首先配制的，故名为 Steiner 稳定 PD 溶液。这种稳定 PD 试液至少可以保持活性达 1 个月或 3 个月。不过，在使用这种混合试液时其显色反应的速度比新鲜的乙醇溶液缓慢一些。

5. I. 碘用 70%乙醇作为溶剂，配成 1% 碘溶液

（二）用　　法

将上述试液分别滴入地衣体皮层或髓层部分并观察有无显色反应。由于显色试验中的正负反应结果，直接关系到该种地衣是否含有某种地衣酸；而不同种的地衣则含有不同的地衣酸，因而也就直接关系到地衣种类的鉴定结果是否正确。因此在进行显色试验时必须十分仔细，切勿粗枝大叶。因此，显色试验最好是在立体显微镜下进行。在用 K 进行地衣皮层显色试验时，可用一根纤细的玻璃针将氢氧化钾水溶液轻轻滴在地衣体皮层表面，观察其是否显色。为了精确辨认反应结果，还可将一片滤纸条浸于地衣体表面的氢氧化钾小滴中，从洁白的滤纸条上便可精确的辨认出反应结果。对于髓层的显色试验与上述方法基本相同，只是为使髓层露出而用小解剖刀或小刀片轻轻剥去地衣体皮层时更须仔细进行，一定不能使皮层碎屑粘留在待试的髓部。在滴试 K 或 C 时，应将试液滴于露出的髓层中央，绝不能使试液小滴接触周围的皮层。有时在髓层上往往将氢氧化钾与漂白粉联合使用，即先滴氢氧化钾，后加漂白粉。至于对苯二胺的使用方法，并不像使用氢氧化钾及漂白粉那样严格要求，只是用浸有对苯二胺试液的小画笔在地衣体表面轻轻涂抹即可。不过值得注意的是越靠近地衣的生长点部位，其显色反应越灵敏。因此，石蕊果柄末梢的反应比果柄基部更为灵敏。显色试验结果的记录方法，习惯写法是：地衣体 K＋黄→红（即地衣体遇 KOH 变黄，随后变红）；髓层 C＋红（即髓层遇 $CaCl_2O_2$）变红）；髓层 KC＋红（即髓层遇 KOH 及 $CaCl_2O_2$）联合反应则变红）；髓层 P＋桔红（即髓层遇对苯二胺变桔红）。若为负反应时：则为 K-；C-；KC-；P-。此外，还有其它几种试液。不过，上述 3 种试液是地衣显色试验中最常用的。这一试验法，可以证明某种地衣酸是否存在，因而长期以来一直是地衣分类鉴定工作中简而易行和不可缺少的。但是，它仍有不足之处。因为这些试验的灵敏度较低。而且对于某些地衣酸，如茶渍酸（lecanoric acid）及橄榄陶酸（olivetoric acid）来说，它们虽然都具有遇 C 显红色的反应基——间位羟基，但是漂白粉并不能精确证明显红色的物质究竟是茶渍酸还是橄榄陶酸。尽管有人证明茶渍酸遇 C 起红色反应，而橄榄陶酸遇 C 起桔红色反应（Hale，1961），但是，实际上这两种颜色反应在检验中往往难以区分。

（三）常见地衣次生代谢产物显色反应检索表

1. 地衣皮层内所含地衣物质
    2. 无色化学物质
        3. 皮层 P＋深黄（进而变红色）；K＋深黄色（进而变红色）……………… **地茶酸（thamnolic acid）**
        3. 皮层 P-；K＋黄 ……………………………………………………………… **黑茶渍素（atranorin）**
    2. 色素物质
        4. 皮层橙色，K＋深紫红色 ………………………………………………… **石黄酮（parietin）**
        4. 皮层橙黄色或黄绿色，或髓层柠檬黄色；K-
            5. 皮层 KC＋黄 …………………………………………………………… **松萝酸（usnic acid）**
            5. 皮层 KC- …………………………………………………… **枕酸（pulvic acid 或 pulvinic acid）**

1. 地衣髓层内所含地衣物质
  6. 髓层 K+
    7. 髓层. K+ 红或黄；C-；P+橙色或黄色
      8. 髓层. K+黄；P+橙红色 ………… **斑点酸(stictic acid)**
      8. 髓层. K+血红；P+深黄 ………… **水杨嗪酸(salazinic acid)；降斑点酸(norstitic acid)**
    7. 髓层. K+ 浅褐色，或 K-；C-；KC-；P+砖红色
  6. 髓层. K- ………… **富马原冰岛衣酸(fumarprotocetraric acid)**
    9. 髓层. C+；P-
      10. 髓层. C+蔷薇红色 ………… **三苔色酸(gyrophoric acid)**
      10. 髓层. C+血红色 ………… **茶渍酸(lecanoric acid)**
      10. 髓层. C+橙色 ………… **橄揽陶酸(olivetoric acid)**
    9. 髓层. C-；
      11. 髓层. KC+
        12. 髓层. P+砖红色；KC+蔷薇红色 ………… **原冰岛衣酸(protocetraric acid)**
        12. 髓层. P-；KC+红或橙色
          13. 地衣体 UV-… **袋衣酸(physodic acid)；隐氯地衣酸(cryptochlorophaeic acid)**
          13. 地衣体 UV+
            14. 髓层. UV+++ ………… **树发酸(alectoronic acid)**
            14. 髓层. UV+ ………… **巴巴酸(barbatic acid)**
      11. 髓层. KC-
        15. 髓层. P+
          16. 髓层. P+砖红色 ………… **富马原冰岛衣酸(fumarprotocetraric acid)**
          16. 髓层. P+黄色 ………… **茶痂衣酸(psoromic acid)；羊角衣酸(baeomysic acid)**
        15. 髓层. P-
          17. 地衣体 UV-… **原地衣硬酸(protolichesteric acid) 及皱梅衣酸(caperatic acid)**
          17. 地衣体 UV+ ………… **珠光酸(perlatolic acid) 及柔扁枝衣酸(divaricatic acid)**

二、微量结晶检验法

地衣微量化学结晶检验法(microcrystal test，MCT)是将下列试液中的一种滴入载玻片上干涸了的地衣化学提取物上，使其溶解后重新结晶为盐类。不同的地衣酸所形成的盐类重结晶形体不同。对于这些不同形体的重结晶必须在显微镜下观察检验。此法在灵敏度和精确性上远比显色反应法为高，因而作为显色检验法的补充而成为地衣现代分类研究中不可缺少的常规方法之一。

(一) 试液及提取器

(1) G.E 液：甘油：冰醋酸=1：3；

(2) G.A.W 液：甘油：乙醇：水=1：1：1；

(3) G.A.An 液：甘油：乙醇：苯胺=2：2：1；

(4) G.A.O-T 液：甘油：乙醇：邻-甲苯胺=2：2：1；

(5) G. A. Q 液：甘油：乙醇：喹林＝2：2：1。

以上试液及丙酮、氯仿等溶剂分别装入带有磨口的小滴管的小瓶内备用。

微量提取器是将直径约 8mm 的玻璃管拉制成下图所示的形状(管身长 10cm，细管尾长约 2.5cm) 备用。

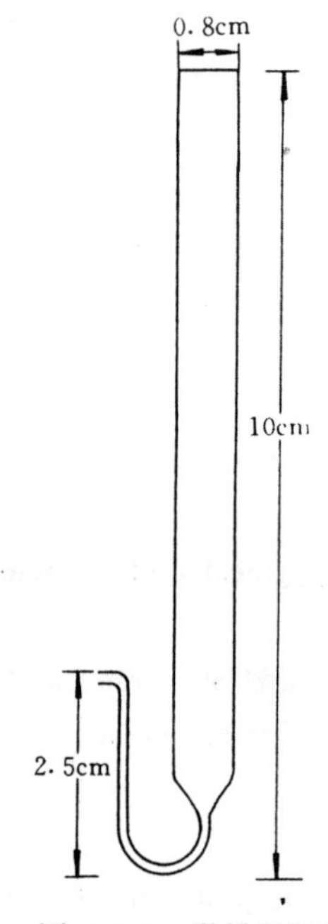

图 6-8-1 微量提取器
(仿朝比奈泰彦)

(二) 操 作 程 序

(1) 取 $1cm^2$ 大小的叶状地衣体碎片或 2 cm 长的枝状地衣体一段冲洗干净之后剪碎置于微量提取器内。

(2) 给装有地衣体碎片的微量提取器内加入几滴丙酮或氯仿。

(3) 将盛有样品及溶剂的提取器置于酒精灯火苗上方文热(为了不使溶剂因加热而突然喷出，应采取间歇加热法，即加热数秒钟后将提取器从火苗上方移开，过一两秒钟后再行加热)。照此法不断进行直至提取液稍微浓缩后，通过毛细管尾滴注于备用的载玻片上，待提取液挥发变干后在显微镜下进行观察。

(4) 将上述一种试液滴加于载玻片上干涸后的提取物上，加盖玻片后，在酒精灯火苗上方文热数秒钟使提取物溶于试液。根据需要，静置冷却一定时间，提取物便在上述不同的试液中重新结晶成不同的盐类。

(5) 在显微镜下观察上述结晶，于有关地衣书刊中的地衣化学成分的盐类结晶照片相查对，以确定其为何种物质。

以上(1)至(3)的提取程序也可以简化为将地衣体碎片直接置于载玻片上，然后加几滴丙酮或氯仿，以溶解地衣体内之地衣酸。

虽然微量化学结晶法比显色法远为精确，但是，和常量法相比，其精确度仍然是不足的。而且其含量在 0.9μg 以下的物质也无法得出可见的结晶，因而无法达到测定的目的。

当前在地衣工作中广泛采用的薄层色谱法则是更为灵敏的方法。一般来说，只要地衣酸的含量在 0.12μg 以上时均可检出。

三、薄层色谱法

薄层色谱法(thin-layer chromatography，简写 TLC)的优点在于简便易行，灵敏快速，因而已成为现代地衣分类实验室的常规技术。

(一) 设 备

(1) 硅胶板：① 成品：青岛有两种规格可用：即 15cm×15cm 玻璃硅胶板和 20cm×10cm 玻璃硅胶板。以塑料薄片为支撑物的硅胶板不宜用于本实验。② 自制：将硅胶

GF254 粉剂用水调制成糊状，用金属或有机玻璃涂板器在 15cm×15cm 或 20cm×10cm 的玻璃板上自制硅胶板。

(2) 层析缸：3 个；

(3) 毛细管：大量备用；

(4) 小试管：直径 1 cm，长度 5cm，可以用直径为 1cm 以内的玻璃管在实验室内即可自制；

(5) 三用紫外线分析仪（上海顾村电光仪器厂）；

(6) 烘箱；

(7) 实验室喷雾器或喷枪；

(8) 木制小试管台；

(9) 丙酮；

(10) 10% 硫酸溶液；

(11) 硅润滑油。

(二) 溶 剂 系 统

(1) 基本溶剂系统：全部待检样都应通过下列三个溶剂系统进行分离：

A 系统.(T-D-A)：toluene(=methhylbenzene)：dioxan (=dioxane)

：acetic acid.

=甲苯：二氧杂环己烷：乙酸

=180：60：8

B 系统.(H-E-F)：hexane：diethylether：formic acid

=(正)己烷：乙醚：甲酸

=130：100：20ml

C 系统.(T-A)：toluene：acetic acid

=甲苯：乙酸

=200：30ml

(2) 补充溶剂系统，专门用于茶渍酸(lecanoric) 和三苔色酸(gyrophoric acids)分离的：

E. A. 溶剂系统：diethylether：acetic acid

=乙醚：乙酸

=200：2ml

本溶剂系统应在使用时随配随用。不过，实验证明上述B 溶剂系统对于石耳酸、三苔色酸及茶渍酸的检出则更加灵敏。

(三) 基 本 方 法

(1) 溶剂的配制与准备：三种基本溶剂系统应置于通风橱内保存。所用玻璃器皿应预先洗刷干净备用。B 溶剂易于挥发，致使溶剂成分比例失调，影响分离效果。为了克服此缺陷，可用补加乙醚的办法予以解决。

(2) 硅胶板：本实验室通常使用青岛海洋化工厂制造的硅胶玻板成品，有 15cm×15cm 及 20cm×10cm 两种规格。以塑料为支撑物的硅胶薄板不能用于本实验。但是以铝板为支撑物的硅胶板可以使用。

在使用时，首先用软铅笔(2B)在距玻璃硅胶板底部边缘 1.5～2cm 处轻轻画一直线。在直线上每隔 1cm 处标记一点。这样在 15cm×15cm 规格的玻璃硅胶板上即可标出 13 个点，除第 1、7、13 三个点不予编号、以便于用分区标准样品外，其余 10 个样点应依次编号为 1～10 以便待检样品点样之用。以同样方法，在 20cm×10cm 玻璃硅胶板上可依次编号为 1～15。或者将玻璃硅胶板置于三用紫外线分析仪点样台上直接编号点样。

玻璃硅胶板的优点除了有国产成品可供订购之外，还在于必要时可以自制，以利于节约。第三个优点还在于能检验脂肪酸之有无。在喷雾 10%硫酸之后，趁湿将玻璃硅胶板面对光源透视，凡脂肪酸类化合物均可显出白色半透明斑点。而铝制硅胶板则无此优点，但可随需要而随意裁剪不同大小之板面，以利于节约使用。

(3) 分区标准样品的选用：按照 C. F. Culberson 及 H- D. Kristinsson (1970)的标准方法，选用同时含有黑茶渍素(atranorin)及降斑点酸(norstictic acid)的地衣杯褐梅(*Melanelia acetabulum*)(欧洲)或穿孔大叶梅(*Parmotrema perforatum*)(北美)作为分区标准样品。我们实验室选用国产特有种金刷耙(*Letharielia cladonioides*)作为分区标准样品。此外，也可选用分别含有黑茶渍素及降斑点酸的两种地衣混合使用。含有黑茶渍素的国产地衣种类比较普遍，而含有降斑点酸的国产地衣种类较少。后者可以选用针芽肺衣(*Lobaria isidiophora*)，因为此种地衣在国内分布较广，容易获得。

(4) 样品的准备与提取：首先，将木制小试管台上的孔洞依次编号，然后，依编号次序取少量(1 cm 叶片或 1～2cm 长枝状地衣体或相当此量的壳状地衣体)待检地衣样品剪碎后逐个放入小试管。每一小试管装入样品后随即对号插入试管台孔洞中。分区标准样品按同样方法插入另一空洞内备用。全部待检样品分别按号装入各试管之后，给每一试管内加入少量丙酮，以使丙酮埋没样品为止。过 5 分钟后，即可按号依次点样。

(5) 展层

1) 检查层析缸是否置于水平面上。

2) 将点样的层析板小心置于层析缸内液面以上 10 分钟，使之饱和后，再置入溶剂液面以下约 1cm，使样品原点于溶剂液面相距约 0.5cm，并使原点线与液面相平行。

3) 待溶剂前缘从原点走至 8cm (用 20cm×10cm 玻璃硅胶板时)或 9cm(用 15cm×15cm 玻璃硅胶板时)处时即可取出层析板，并用吹风机吹干板面溶剂。

4) 在三用紫外线分析仪下观察层谱斑点位置及其有无荧光现象，用软铅笔随及标出斑点位置及有无荧光现象。

5) 对于已有标记的斑点进行 P、K、C 及 KC 等显色反应试验，并做记录。

6) 喷雾 10% 硫酸后，趁湿透视有无脂肪斑，并做记录。随后置入预先调好的烘箱内(110℃)约 10～15 分钟以便色谱显示良好为止。

7) 分区：在显色后的色谱原点上、下缘各取一点分别画一切线。上、下切线之间的区域为第一区。以同样方法在分区标准样品的黑茶渍素及降斑点酸斑点分别画出第四区及第七区。然后，在第一区与第四区之间画一中线，使之等分为第二区及第三区；以同样方法在第四区与第七区之间画出第五区与第六区。

8）求 Rf 值：必要时应求出各斑点的 Rf 值。

9）成分鉴定：按照 Culberson 与 Kristinsson（1970）及 Culberson（1972）所提供的地衣次生代谢产物层析资料进行检索。将上述资料制成带孔卡片，以供检索。在进行层析时，或直接使用纯化样品，或使用确知含有某一成分的地衣种类作为对照以代替纯化样品进行层析（表 6-8-1）。

表 6-8-1　常见地衣化学成分在 TLC 中的分区资料

溶剂系统			化学成分	斑点颜色	含有关成分的地衣种类
A.	B.	C.			
1	3	2	fumarproto-cetraric acid	暗灰;P+	*Cladina rangiferina* *Cladina arbuscula* *Cladonia furcata* *Cladonia cornuta* *Cetraria laevigata*
1	2～3	2	protocetraric acid	暗灰;P+	*Flavoparmelia caperata*
1	3	2	thamnolic acid	淡褐;P+	*Thamnolia vermicularis* *Platismatia glauca*
2	2	2	salazinic acid	黄;K+,P+	*Everniastrum cirrhatum* *E. nepalensis* *Xanthoparmelia camtschadalis* *X. tinctina* *X. mexicana* *X. taractica*
2	3	3	squamatic acid	黄;UV+++	*Cladonia uncialis* *Thamnolia subuliformis*
2	4～5	3～4	physodalic acid	暗灰;P+	*Hypogymnia hypotrypa* *H. hypotrypella*
3	2	3	stictic acid	橘黄;P+	*Menegazzia terebrata* *M. asahinae*
3	4	3	alectoronic acid	灰白;KC+,UV+++	*Cetrelia nuda*
3	4～5	3	physodic acid	淡绿;KC+	*Hypogymnia austerodes*
3	5	3	gyrophoric acid	黄或灰;C+	*Cetraria delisei*
3	5	3	lecanoric acid	黄;C+	*Parmotrema tinctorum* *Melanelia glabra*
3	5	4～5	lobaric acid	浅灰绿;KC+	*Protoparmelia badia*
3	5	5	psoromic acid	暗黄或褐;C+	*Sulcaria sulcata* *Cladina aberrans*

续表

溶剂系统			化学成分	斑点颜色	含有关成分的地衣种类
A.	B.	C.			
					Rhizoplaca melanophthalma
3～4	4～5	3	olivetoric acid	黄;C+	*Cetrelia olivetorum*
3～4	5	5	baeomycesic acid	黄;P+	*Thamnolia subulifformis*
4	4	4	norstictic acid	黄;P+,K+	*Lethariella cladonioides*
					Lobaria isidiophora
					Xanthoparmelia conspersa
4	5	5	α-collatolic acid	灰白;KC+	*Cetrelia nuda*
4	6	5～6	barbatic acid	黄	*Cladia aggregata*
					Cladonia floerkeana
					Cladonia amaurocraea
4	6	5～6	divaricatic acid	黄	*Evernia divaricata*
4	6	5～6	diffractaic acid	黄	*Alectoria ochroleuca*
					Usnea diffracta
4～5	6	5～6	homosekikaic acid	黄褐	*Cladonia rei*
4～5	7	5～6	perlatolic acid	黄	*Cladina pseudoevansii*
					Cladina stellaris
4～5	7	5～6	sphaerophorin	黄	*Sphaerophorus* sp.
5	5	5	zeorin	紫或浅灰褐;UV+	*Heterodermia boryi*
6	5	6	rhizocarpic acid	黄色素;UV+	*Acarospora gobiensis*
6	6	6	usnic acid	淡绿	*Usnea longissima*
7	6	7	pinastric acid	黄色素	*Cetraria juniperina*
					C. pinastri
7	6	7	tenuiorin	黄	*Peltigera horizontalis*
7	7	7	atranorin	黄	*Platismatia glauca*
					Sulcaria sulcata
7	6	7～8	vulpinic acid	黄色素	*Sulcaria sulcata* var. *vulpinoides*
					S. virens
					Cetraria juniperina
					C. pinastri
7	7	7～8	parietin	黄色素;K+,UV+++	*Xanthoria mandschurica*

四、综合检验法

有些化合物在薄层色谱法中难以辨认，如茶渍酸(lecanoric acid)与三苔色酸

(gyrophoric acid)遇 C 分别变红色及蔷薇红色，在 TLC A、C 溶剂系统的色谱上都处于同一 Rf 区，而且其 Rf 值也极接近，因而实难区分。但是这两种地衣酸除了在 TLC 的 B 溶剂系统中易于辨认之外,在微量结晶检验法(MCT) 中也极易区分。

此外，还可直接将地衣标本置于三用紫外线分析仪下检验其是否有荧光现象，以帮助鉴定某些含有荧光化合物的种类，如含有鳞片衣酸(squamatic acid)的地衣种类等。

总之，在确定某些疑难成分时，应该实行显色检验法、微量结晶检验法，薄层色谱法(A. B. C. 3 个溶剂系统同时使用)以及荧光检验法等的综合使用。

第九章　地衣分类学与系统学
Lichen Taxonomy & Systematics

“系统学”是研究和识别客观事物“顺序”的科学。这种“顺序”即“系统”。“系统”(system)一词来自希腊文“systema”，后来被引入拉丁文中为“systema”。该词的含义是指同类事物按一定关系组成的整体，也就是说一个整体由几个部分按一定顺序排列而成。“系统学”是一门具有普遍意义的科学。在人类活动的各个领域，系统学的研究对象虽然不尽相同，但是，它的最终目标是研究和识别客观事物的顺序，它们之间的内在联系和各种自然关系，并用语言和图表予以表达，以进入使用领域。这种系统往往是很有用的，因为它可能只是基于一种或几种关系而建立的，但是，它却可以解释其他多种关系。

在生物学中，系统学(systematics)指的是将自然界形形色色的生物归成若干类群；把这些类群按其亲缘关系和顺序加以排列，并对这一顺序中各类群的多样性和分化，以及它们之间的演化关系所进行的科学研究。

虽然“系统学”和“分类学”(taxonomy)这两个术语在生物学中往往互换使用，但是，实际上，二者还是有所区别的。“分类学”指的是对于形形色色的生物进行归类时的有关根据、原理、步骤和方法所进行的科学研究；而“系统学”却是在归类的基础上对于类群的多样性、类群的分化、类群之间的亲缘关系及其自然顺序所进行的科学研究。而“分类”(classification)则是指对于自然界形形色色的生物按一定关系进行归类并排列的过程。“分类”这一术语往往被误解为“鉴定”(identification，determination)。“鉴定”是按照一个业已存在的分类安排将每一种生物进行命名或“对号入座”。

综上所述，系统学是在分类及分类学的基础上对于自然界形形色色的生物多样性和分化与亲缘关系及其自然顺序所进行的科学研究。因此分类和分类学是系统学的组成部分和基础工作。

“系统学”和“生物系统学”(biosystematics)这两个术语在概念上容易产生混淆。“系统学”正如以上所阐述的含义，而“生物系统学”则是系统学中关于生物种内群体(population)之间及群体内生物个体变异、分化与演化关系的研究。这是有关物种生态学或群体遗传生态学范畴的，即物种在其生长环境中的基因型变异与表型变异以及物种分化与物种形成生物学的研究。有人将“biosystematics”一词按其含义译为“物种生物学”以区别于“系统学”(systematics)。

系统学对于生物资源开发利用等应用生物学具有重要意义。如果说自然界形形色色的生物看起来好像是一堆杂乱无章的“乱麻”，那么，系统学的任务就在于对于这一堆“乱麻”进行清理或称“梳辫子”，按其内在联系将它们清理得有条有理，以便生物资源学家按照系统学所提供的生物种系之间的各种关系的信息和规律去“顺藤摸瓜”，这就是系统学为人们所提供的所谓具有预期价值的“信息存取系统”。

既然现代科学已在生物演化系统中为地衣找到了它的位置，那么，对于作为真菌的地衣，就不能停留在分类和分类学研究阶段，而必须深入到真菌的系统学领域。对于地衣系统学的深入研究不仅在菌物科学上，尤其是真菌科学上具有理论价值，而且在生物资源开发利用中也具有重要的现实意义。

一、地衣分类学的发展简史

（一）表型分类阶段

地衣分类学与系统学的研究从历史角度可以划分为 6 个时期，它们是宏观形态学时期、微观形态学时期、经典化学时期及现代化学时期、形态地理化学时期以及表型、基因型相结合的生物学物种时期。

1. 宏观形态学时期

这一时期从 18 世纪中叶(1753)开始一直到 19 世纪中叶，这一时期的特点是研究者主要借助于手持扩大镜将地衣体及子囊盘的外部宏观形态特征作为分类依据进行分类。这里有地衣学的奠基者，瑞典医生 Acharius(1803-)的巨大贡献。这一阶段一直延至 Schaerer(1850)的工作。

2. 微观形态学时期

在地衣分类研究中 Fee(1824) 和 De Notarius(1846) 先后借助于显微镜将孢子作为分类依据，使地衣分类研究进入了马-柯-尼时期。在这一时期内由于围绕孢子特征作为分属依据的激烈争论而形成了以意大利的马萨隆哥(Massalongo，1852) 和西里西亚的柯尔柏(Koerber，1854～1855) 为创始人的意大利- 西里西亚① (Italian-Silesian School)学派与芬兰的尼兰德(Nylander) 学派之争。意大利- 西里西亚学派的创始人之一马萨隆哥在地衣属的概念中按重要性顺序将特征列为孢子 (颜色、隔膜及数量)、子囊及侧丝、子实下层、果壳及地衣体特征。从这一顺序可以看出，该学派的主导思想是将与有性繁殖有关的器官特征置于首位。这在当时应该是一大进步。然而，这一学派却遭到了芬兰的尼兰德(Nylander，1854～1855)学派的强烈反对，从而构成了马-柯-尼时期的主要特色。

3. 化学显色法时期

19 世纪 60 年代，尼兰德开创了化学显色法在地衣分类鉴定中的应用。他首先发现了碘遇地衣体时的显色反应在分类鉴定中的应用(Nylander，1865)，随后又发现了漂白粉($CaCl_2O_2$，简写为 C)及氢氧化钾(KOH，简写为 K)的显色反应在分类鉴定中的应用(Nylander，1866)。在此基础上英国的 Leighton(1866，1867)又发现了氢氧化钾与漂白粉联合反应法(简写为 KC)在地衣分类鉴定中的应用价值。虽然当时并不清楚不同的颜

① 西里西亚中欧东部地区位于今波兰西南部。

色反应与所含的何种地衣酸有关，因而具有一定的盲目性，但是，这并没有妨碍化学显色法在地衣分类鉴定中的有效使用。实际上，地衣体遇 C 起紫色反应表明含有三苔色酸(gyrophoric acid)，起红色反应表明含有茶渍酸(lecanoric acid)，以及起桔红色反应表明含有橄榄陶酸(olivetoric acid)，而起绿色反应则表明含有链丝素(strepsilin)；遇 K 起黄色反应表明含有黑茶渍素(atranorin) 或斑点酸(stictic acid)，起深黄色反应表明含有地茶酸(thamnolic acid)，起桔黄色反应表明含有降斑点酸(norstictic acid)，起红色反应表明含有水杨嗪酸(salazinic acid)；遇 KC 起红色反应表明含有树发酸(alectoronic acid)或领斑叶酸(a-collatolic acid)或肺衣酸(lobaric acid) 或袋衣酸(physodic acid) (Hale, 1961)。化学显色反应法迄今仍在地衣分类鉴定中延用不衰。

除了化学显色法以外，地衣体解剖学特征也受到极大重视。

4. 化学重结晶法时期

20 世纪上半叶在 Zahlbruckner 编著的“世界地衣总汇”(Catalogus Lichenum Universalis，1922～1940) 这一巨著所表达的分类系统影响下，地衣分类研究虽然缺乏重大进展，然而，开始于 19 世纪中叶的地衣化学分类还是获得了长足的进步。在多年研究的基础上，德国的 Zopf(1907)发表了地衣化学名著，其中描述了当时几乎全部已知的和许多新发现的地衣化学物质共 150 多种。朝比奈泰彦(Asahina，1936～1940)还开创了地衣微量化学重结晶法 (简称 MCT)在地衣分类鉴定中的有效应用，从而使地衣化学分类在精确性和灵敏度上跃入现代科学水平。在这一时期，朝比奈泰彦还为化学显色法增加了对苯二胺(paraphenylene diamine，简写为 P 或 PD) 这一重要试剂。

5. 形态-地理-化学综合时期

20 世纪下半叶以来，地衣分类研究领域进入了一个突飞猛进的时期。这一时期的特点是在化学分类方法中成功地引进了更加灵敏而又简便易行的薄层色谱法(TLC)。此外，扫描电子显微镜(SEM) 的出现以及生物地理学方法在地衣分类研究中的应用使地衣分类学进入了形态-地理-化学综合时期。虽然地衣化学分类法从上一世纪 60 年代以来所取得的成就是其它任何生物分类学无法比拟，然而，地理学方法在地衣分类中的应用却比高等植物落后了近一个世纪。在形态与化学相结合中有 J. Poelt(1965)对于广义的蜈蚣衣属(*Physcia*) 所进行的细分在地衣分类学领域引起了链锁反应，从而使本世纪 60 年代以前被普遍接受的 500 多个地衣属在短短的几年之内就增加到 600 多个。以广义的梅衣属(*Parmelia*)为例，迄今已被再分为 20 多个属(表 6-9-1)。

由此看来，对于广义属实行进一步细分的趋势在本世纪结束之前仍将处于高潮之中。在这一细分高潮中，强调属的概念应体现形态- 地理- 化学综合特征已迫在眉睫。W. & C. Culbersons(1968)关于斑叶属(*Cetrelia*)和宽叶衣属(*Platismatia*)的建立以及 Hale (1974，1976)关于球针叶属(*Bulbothrix*)和球针黄叶属(*Relicina*)的建立正是体现了这一综合特征的典范(表 6-9-2)。

实际上，并非所有因细分而新建立的属都能体现这一综合特征。如果按照有差异就要进一步细分，就有可能误入模式概念的危险。形态-地理-化学综合法为使地衣表型分类免于误入模式概念提供了有效手段。

表 6-9-1 广义与狭义属名对照表

广义属名	狭义属名
蜈蚣衣属 (*Physcia*)	大孢蜈蚣衣属(*Physconia* Poelt) 蜈蚣衣属[*Physcia* (Ach.) Vain.] 外蜈蚣衣属(*Hyperphyscia* Muell. Arg.) 黑蜈蚣衣属(*Phaeophyscia* Moberg)
雪花衣属 (*Anaptychia*)	雪花衣属(*Anaptychia* Koerb. em. Poelt) 亚铃孢属(*Heterodermia* Trev.)
树发属 (*Alectoria*)	树发属(*Alectoria* Ach.) 小孢发属(*Bryoria* Brodo & Hawks.) 拟毡衣属(*Pseudephebe* Choisy) 槽枝属[*Sulcaria* (Motyka) Bystr.]
岛衣属 (*Cetraria*)	裸腹叶(*Asahinea* Culb.) 岛衣属(*Cetraria* Ach.) 类岛衣属(*Cetrariopsis* Kurok). 斑叶属(*Cetrelia* Cule. & Culb.) 艾斯林属(*Esslingeria* Lai) 麦松氏属(*Masonhalea* Karnef.) 宽叶衣属(*Platismatia* Culb. & Culb.) 土克曼衣属(*Tuckermannopsis* Gyel.)
网衣属 (*Lecidea*)	扁桃盘衣属(*Amygdalaria* Norm.) 星网属(*Astroplaca* Bagl.) 蜡肠衣属[*Catillaria* (Ach.) Th. Fr.] 下网属(*Hypocenomyce* Choisy) 网衣属(*Lecidea* Ach.) 小网衣属(*Lecidella* Koerb.) 圆顶衣属(*Lecidoma* G. Schneid. & Hertel.) 亚网衣属(*Micarea* Fr.) 准网衣属(*Paraporpidia* Rambold & Pietschmann.) 假网衣属(*Porpidia* Koerb.) 鳞网衣属(*Psora* Hoffm.) 小鳞网属(*Psorinia* G. Schneid.) 小鳞衣属(*Psorula* G. Schneid.) 谢氏衣属(*Schaereria* Th. Fr.) 类褐边衣属(*Trapeliopsis* Hertel & G. Schneid.) 节体衣属(*Tylothallia* Jamesw & Kilias) 黄鳞网属(*Xanthopsora* G. Schneid. & W. A. Weber)
梅衣属 (*Parmelia*)	实袋属[*Allantoparmelia* (Vain.) Essl.] 北极梅属(*Arctoparmelia* Hale) 球针叶属(*Bulbothrix* Hale)

续表

广义属名	狭义属名
	腹穴衣属(*Cavernularia* Degel.)
	小条属(*Cetrariastrum* Sipm.)
	斑叶属(*Cetrelia* Culb. & Culb.)
	条衣属(*Everniastrum* Hale ex Sipman)
	蜡带属(*Concamerella* Culb.)
	皱梅属(*Flavoparmelia* Hale)
	点黄梅属[*Flavopunctelia* (Krog) Hale]
	袋衣属[*Hypogymnia* (Nyl.) Nyl.]
	双歧根属[*Hypotrachyna* (Vain.) Hale]
	褐梅属(*Melanelia* Essl.)
	孔叶衣属(*Menegazzia* Mass.)
	新棕梅衣属(*Neofuscelia* Essl.)
	黄髓梅属(*Parmelina* Hale.)
	大叶梅属(*Parmotrema* Mass.)
	宽叶衣属(*Platismatia* Culb. & Culb.)
	拟扁枝衣属(*Pseudevernia* Zopf.)
	拟梅衣属(*Pseudoparmelia* Lynge)
	星点梅属(*Punctelia* Krog)
	球针黄叶属[*Relicina* (Hale & Kurok.) Hale]
	黄梅属[*Xanthoparmelia* (Vain.) Hale]
	梅衣属(*Parmelia* Ach. sensu str.)
树花属	双扁枝属(*Dievernia* Choisy)
(*Ramalina*)	空心花属(*Fistulariella* Bowler & Rundel)
	树花属(*Ramalina* Ach.)
	类树花属(*Ramalinopsis* Follm.)

表 6-9-2 形态-地理-化学综合特征举例

属名	形态特征	地理分布	化学特征
Cetrelia	分生孢子双纺锤形	东亚分布中心	地衣酚衍生物
Platismatia	分生孢子圆柱形	北方太平洋两岸	β-地衣酚衍生物或脂肪酸
Bulbothrix	分生孢子圆柱形	干旱或温带-亚热带异地分布	黑茶渍素
Relicina	分生孢子双纺锤形	热带东南亚	松萝酸

实际上,并非所有因细分而新建立的属都能体现这一综合特征。如果按照有差异就要进一步细分,就有可能误入模式概念的危险。形态-地理-化学综合法为使地衣表型分类免于误入模式概念提供了有效手段。

（二）表型与基因型相结合阶段

肺衣科（Lobariaceae）牛皮叶属（*Sticta*）的两个形态近似种杜氏牛皮叶（*Sticta dufourii*）和金丝雀牛皮叶（*S. canariensis*）之间的主要区别在于前者的共生光合生物为蓝细菌而后者为绿藻。同样的情况也存在于假杯点属（*Pseudocyphellaria*）的两个形态近似种穆氏假杯点衣（*P. murrayi*）和红绿假杯点衣（*P. rufovirescens*）中。基于DNA的限制性多态分析结果表明，上述形态近似而共生光合生物不同的两对种的共生真菌之间不存在种级差异（Armaleo and Clere，1991）。这说明仅仅根据不同的共生光合生物作为不同种的依据是难以成立的。从上述分子系统学证据所得到的启示是某种次生代谢产物之有无和异同，分生孢子形状之异同，以及某个营养体结构上之异同究竟能在多大程度上反映出各个地衣或地衣型真菌物种的基因型特征，有待于进一步深入探索。尽管现代地衣物种概念和属的划分注意到了形态-地理-化学综合特征的运用，但是，从18世纪中叶以来，在地衣分类研究领域的进步就在于随着研究工具的不断更新而识别了越来越多的地衣特征，而且其中大多数特征是无性型的或营养型的，何况所有这些特征归根到底都是表型特征。尽管这些表型特征仍是基因型的末端产物，但是，它们究竟还不是基因型本身。值得庆幸的是，由于分子生物学技术在许多方面所取得的突破性进展，使地衣分子系统学便应运而生。在以表型性状分析为基础的地衣分类学中着重掌握形态-地理-化学综合法并结合地衣分子系统学方法的运用，为地衣系统学在表型与基因型相结合的生物学物种水平上进行深入探索开辟了广阔的前景。

二、分子系统学的兴起

生物物种多样性的物质基础是物种各自所固有的基因库。每一个物种的基因库负载着各自物种生物学特性的全部遗传信息。这些信息则包含在它们各自DNA的碱基序列之中。每一种生物都具有各自不同的DNA碱基序列。这些DNA碱基序列通过精确的复制而传递给子代，从而保持了各个生物物种的稳定性。通过对于DNA碱基序列的测定分析以探索生物种系的亲缘关系，多年来一直是生物分类学家和系统学家的夙愿。近年来，由于分子生物学技术所取得的突破性进展，尤其是PCR技术的出现，对于从极少量生物体或单个孢子以及少量干标本中提取的DNA粗制样品进行体外扩增，以获得足够用于系统学分析的纯DNA样品成为现实，从而为生物分类学家和系统学家夙愿得偿开辟了有效途径。

从事分子系统演化研究的生物学家通常是使用各种生物的核糖体DNA（rDNA）作为分析研究材料。这是因为核糖体普遍存在于所有原核与真核细胞中。它不仅存在于细胞核中，而且真核生物的线粒体和叶绿体也都含有各自独立的核糖体。从生物学功能角度上看，核糖体是细胞中蛋白质的合成场所，与细胞的生长、代谢和繁殖有密切关系。核糖体DNA碱基序列包含着足够的遗传信息以利于生物演化系统学分析。而且，不同的生物细胞中核糖体的组分也被人们阐明得较多；用于PCR扩增引物碱基顺序的资料也比较丰富，便于选用。此外，文献中报道的研究结果也比较多，便于进行各种生物之间的比较分析。

核糖体 DNA 的重复单位通常是由小亚基和大亚基及二者之间的非编码区所构成。为了简便起见，人们将真核生物核糖体小亚基的rDNA称作16s类rDNA（16s like rDNA）或者 Ss-rDNA（Small subunit ribosomal DNA）；将大亚基的 rDNA 作为 25s 类 rNDA（25s like rDNA）或 Ls-rDNA（Large subunit ribosomal DNA）。在小亚基与大亚基之间的还有转录间区或非编码区。

研究结果表明，rDNA 小亚基，即 16s 类 rDNA 的序列演化速度较慢，对于研究远缘生物之间的亲缘关系颇为有用。而转录间区的演化速度最快，对于研究属内种间或种内群体（population）间的亲缘关系有用。而线粒体 rDNA 的演化速度界于上述二者之间，因而对于研究科、目一级亲缘关系有用（White et al.，1990）。不过，核的 rDNA 大亚基，即 25s 类 rDNA 也常被用于种、属间的比较分析（Armaleo and Clere，1991）。

地衣型真菌经常与藻类或蓝细菌组成共生联合体。地衣的科学名称实际上是这一共生联合体中真菌的名称。因此，地衣分类学和系统学的基础乃是地衣型真菌。地衣分子系统学的研究对象当然也是地衣型真菌的 DNA 及其碱基序列。在一般情况下，一种地衣是由一种真菌和一种藻类，或者一种真菌和一种蓝细菌形成共生联合体。然而有时侯，一种真菌同时和一种藻类及一种蓝细菌形成共生联合体，如在绿皮地卷（*Peltigera aphthosa*）中那样。该种地衣叶状体中的藻层是由绿藻组成，而叶状体表面的衣瘿内却含有大量蓝细菌。如果用绿皮地卷作为提取 DNA 的材料，那么，所得到的总 DNA 中便含有 6 种 rDNA，即该种地衣型真菌核的 rDNA，地衣型真菌线粒体的 rDNA，绿藻核的 rDNA，绿藻中线粒体的 rDNA，绿藻中叶绿体的 rDNA 以及蓝细菌的 rDNA。如果一种地衣含有 3 种藻类和蓝细菌，那么，所得到的总 DNA 中便会含有更多种 rDNA。

为了获得地衣型真菌核的核糖体小亚基 DNA 序列（16s 类 rDNA）进行系统学分析，*有下列 5 种方法可供选用*：

（1）用于地衣型真菌核 16s 类 rDNA PCR 扩增的非藻引物对（Gargas and Taylor，1992）可供选用。这些引物只适于扩增地衣型真菌核的 16s 类 rDNA，而不扩增绿藻核的 16s 类 rDNA。但是，它对真菌核 16s 类 rDNA 也并无专化性，因为它也能扩增其它高等植物细胞核 16s 类 rDNA。

（2）在提取 DNA 过程中，使用含有几丁质酶的缓冲液，以最大限度的破坏地衣型真菌细胞壁以达到释放 DNA 的目的，而藻细胞的细胞壁完好。这样，所提取的便只能是真菌的 DNA（Armaleo and Clerc，1991）。不过这一方法的缺点在于缺乏绝对纯净而不含纤维素酶的几丁质酶，而纤维素酶则能破坏藻细胞壁。

（3）地衣型真菌的 DNA 可以从共生菌的纯培养物中提取（Armaleo and Clere，1991）。地衣型真菌培养物既可从孢子萌发开始，也可从地衣型真菌的菌丝细胞培养开始（Yamamoto et al.，1985）。

（4）不含藻细胞或蓝细菌的网衣型或亚茶渍型子囊盘或假根均可用作提取地衣型真菌 DNA 的材料（Niu and Wei，1993）。

（5）从单个孢子中提取用于 PCR 扩增的 DNA 时，孢子数不能少于 10 个（Lee and Taylor，1990）。这一方法也可用于地衣型真菌孢子 DNA 的提取。

从少量生物材料中提取的 DNA，通过 PCR 技术进行体外扩增所获得的 PCR 产物，可以用于DNA序列的系统学比较分析。用于分子系统学分析的方法很多，但是，以DNA为材料的分析方法除了DNA-DNA杂交以外，还有DNA指纹法(DNA fingerprinting)，限制性酶切片段多态(restriction fragment length polymorphism，简写RFLPs)分析，随机扩增法(RADP-PCR)以及直接测序法等。上述方法中最直接而精确的方法要算是直接测序法。对于DNA特定片段的直接测序可以通过DNA序列自动测定仪，也可通过人工进行测定，甚至还可通过PCR技术进行序列测定。

三、现代地衣分类所依据的特征及主要分类类群

现代地衣分类所依据的特征概括为形态学、化学、地理学及生态学特征。形态学特征则侧重于由子囊所衍生而来的特征，主要是子囊顶部结构及其开裂方式等。此外，还有从子囊器及其孢子的个体发育中衍生而来的特征。无性阶段的分生孢子器的特征及营养体的结构也受到相应的重视。

(一) 子 囊 菌 门

属于子囊菌门的地衣类群主要有下列各目：

1. 子囊缝沟型

(1) 斑衣目(Arthoniales)：子囊器为不规则形至近圆形的假囊壳；子囊缝沟型，通常宽棒状；孢子有隔膜；地衣体壳状，棉絮状；分布于北方至热带，如斑衣科(Arthoniaceae)及金絮衣科(Chrysotrichaceae)。

(2) 孔文衣目(Opegraphales)：子囊器为近圆形或线形的假囊壳，单生或陷于子座内；子囊缝沟型；分布于温带至热带地区。如座盘衣科(Chiodectonaceae)、孔文衣科(Opegraphaceae)、染料衣科(Roccellaceae)及碗衣科(Lecanactidaceae)。

(3) 缝囊盘目(Gomphillales)：子囊器为子囊盘，具有清晰可辨的果壳；子囊缝沟型；非淀粉质；孢子具有孢梗，地衣体壳状，主要分布于热带地区。代表科为缝囊盘科(Gomphillaceae)。

(4) 胶皿菌目(Lecanidiales = Patellariales)：子囊器为盘状假囊壳，通常具有碳化了的果壳；子囊缝沟型；非淀粉质；至少部分地具有侧丝；只有很少的壳状地衣型种类，如节针科(Arthrorhaphidaceae)。

(5) 瓶口衣目(Verrucariales)：子囊器为子囊壳状假囊壳，往往碳质化；子囊非淀粉质并且顶壁较厚；开裂方式不肯定；假侧丝往往胶化；缘丝存在；孢子单胞至砖壁型，无色透明至有色；地衣体壳状至叶状，不含地衣物质；分布于北极至亚热带地区；代表科为瓶口衣科(Verrucariaceae)。

(6) 小核衣目(Pyrenulales)：子囊器为子囊壳状假囊壳，单个地或联合地生于子座中；子囊缝沟型，非淀粉质；通常具有假侧丝；孢子具隔膜至砖壁型，往往有色，具有加厚的隔膜；地衣体壳状；主要分布于热带，但也可延伸至温带。代表科为小核衣科(Pyrenulaceae)及乳嘴衣科(Trypetheliaceae)。

(7) 座囊菌目(Dothideales):子囊器为类似于子囊壳的假囊壳;单个地或联合地生于子座内;子囊缝沟型,非淀粉质,具有假侧丝或无侧丝;孢子绝大多数具隔膜,往往一端较厚;只有少数壳状的地衣型成员,如星核衣科(Arthopyreniacae)及圆孔腔菌科(Mycoporaceae)等。

(8) 地卷衣目(Peltigerales):子囊器为半被果型;子囊缝沟型并具有一个明显的淀粉质环;地衣体为叶状,常与蓝细菌共生。分布于北方至热带的潮湿生境中。它们有肺衣科(Lobariaceae)、肾盘衣科(Nephromataceae)、地卷衣科(Peltigeraceae)及散盘衣科(Solorinaceae)。

该目以及鸡皮衣目(Pertusariales)和黄枝衣目(Teloschistales)是近几年来根据其子囊顶部结构特征而从茶渍目中划出而成3个独立的目。

2. 子囊为茶渍型

(9) 茶渍目(Lecanorales):该目是子囊菌中最大的类群。它包括几乎全部盘果类地衣型或生长在地衣上的非地衣型真菌,约69个科,常见的有微孢衣科(Acarosporaceae)、树发衣科(Alectoriaceae)、绵腹衣科(Anziaceae)、石蕊科(Cladoniaceae)、胶衣科(Collemataceae)、赤星衣科(Haematommataceae)、袋衣科(Hypogymniaceae)、茶渍衣科(Lecanoraceae)、网衣科(Lecideaceae)、梅衣科(Parmeliaceae)、鳞网衣科(Psoraceae)、树花衣科(Ramalinaceae)、地图衣科(Rhizocarpaceae)、珊瑚枝科(Stereocaulaceae)以及石耳科(Umbilicariaceae)等。这是一个相当异源的大目,需要进行大量的深入研究。

(10) 鸡皮衣目(Pertusariales):它的子囊器为典型的盘果型或类似于子囊壳而陷入瘤状地衣体中;较厚的子囊壁具有一个淀粉质的内壁层并在顶部加厚;开裂方式不是缝沟型的;侧丝交织型;孢子无色透明往往大型;地衣体壳状;常与绿藻共生;分布于北极至赤道地带。包括鸡皮衣科(Pertusariaceae)。

(11) 黄枝衣目(Teloschistales):子囊器为典型的盘果类;子囊顶部具有一个较厚的淀粉质外壁层,圆顶存在或退化;开裂方式不是缝沟型的;孢子对极型或多极型;地衣体呈壳状、叶状及枝状,含有蒽醌类色素。世界分布型。包括棕网衣科(Fuscideaceae)、多极孢衣科(Letrouitiaceae)及黄枝衣科(Teloschistaceae)。

3. 子囊为功能单壁型

(12) 凹盘衣目(Gyalectales):子囊器由半被果型子囊盘衍生而来,固有盘缘为假薄壁组织;子囊壁较薄而无圆顶,而无淀粉质或子囊层胶质部分具有微弱的淀粉质;以顶孔开裂;地衣体壳状;多与桔色藻共生,分布于温带至热带的潮湿生境中。包括凹盘衣科(Gyalectaceae)。

(13) 厚顶盘目(Ostropales):子囊器为开放型至点状子囊盘,往往陷入于地衣体内;盘壁为假薄壁组织;类绿丝存在;子囊非淀粉质,壁厚而非缝沟型;地衣体壳状;许多非地衣型附生种类也包括在此目;主要分布于热带,也可延伸至温带。如点盘菌科(Stictidaceae)及齿孔菌科(Odontotremataceae)。

(14) 粉衣目(Caliciales):子囊器往往具有短柄,类似于微型大头针;子囊为假原囊壁子囊类,子实层往往分解为孢丝粉(mazaedium);子囊具有加厚了的顶部,孢子有纹饰,并通过顶孔而被释放；地衣体壳状；绝大部分成员为地衣型的，而少部分为自由生活菌类，包括粉衣科(Caliciaceae)、粉头衣科(Coniocybaceae)、微粉衣科(Microcaliciaceae)、菌粉衣科(Mycocaliciaceae)、硬梗科(Sclerophoraceae)、球粉衣科(Sphaerophoraceae)及粉苔衣科(Sphintrinaceae)。

(15) 柔膜菌目(Helotiales)：子囊器多为盘果类子囊盘，有柄或无柄贴生；子囊棍棒状至圆柱状，有顶孔，淀粉环存在或缺如；只有羊角衣科(Baeomycetaceae)为地衣型的。

(16) 盘菌目(Pezizales)：子囊器通常为肉质子囊盘，稀为革质或碳化了的；子囊圆柱形至棍棒形，顶部具有囊盖或稀为裂口；子囊孢子经常为单胞，其中少数为壳状地衣型种类，如谢氏菌科(Schaereriaceae)。

(17) 文字衣目(Graphidales)：子囊器线条形或龙舟形至近圆形子囊盘或有时类似于子囊壳；子囊顶壁加厚，非淀粉质，非缝沟型，以顶孔释放孢子；孢子具有横隔膜或砖壁型，具有加厚了的淀粉隔膜，地衣体壳状，分布于温带至热带，如文字衣科(Graphidaceae)及疣孔衣科(Thelotremataceae)。

(18) 球壳目(Sphaeriales)：子囊器为子囊壳；子囊非缝沟型，子囊顶部结构多样，淀粉质环常见；具有侧丝或缺如；孢子多形型；只有少数壳状种类为地衣型成员；这是一个高度异源性的目。具有地衣型成员的科有叶上衣科(Strigulaceae)等。

(二) 担 子 菌 门

属于担子菌的地衣型真菌约 50 种，它们分别隶属于白蘑目(Tricholomatales)的白蘑科(Tricholomataceae)，如荷菇属(*Botrydina*)或初生的小荷叶状共生体被称做小荷叶属(*Coriscium*)；鸡油菌目(Cantharellales)中的珊瑚菌科(Clavariaceae)；展齿革菌目(Phanerochaetales)中的云片衣科(Dictyonemataceae)等。

由于地衣型担子菌与藻类多处于共生的初级阶段，因而，这些地衣型真菌，实际上是由真菌学家进行研究的。

(三) 其　　他

至于常见的无子实体地衣或地衣型真菌，如白角衣属(*Siphula*)、地茶属(*Thamnolia*)及癞屑衣属(*Lepraria*)等的归属问题，尚待今后研究结果而定。

第二篇　地衣物种多样性及系统演化与资源生物学 Lichen Species Diversity, their Systematic Evolution and Resources Biology

第一章　地衣区系与资源 Lichen Flora and Resouces

第二章　专科属研究 Studies of Special Lichen Groups

第三章　新分类群 Taxa New to Science

第四章　中国新纪录 Taxa New to China

第五章　系统演化生物学 Systematic & Evolutionary Biology

第六章　命名法 Nomenclature

第七章　地衣化学 Lichen Chemistry

第一章　地衣区系与资源
Lichen Flora and Resouces

Вэй Цзян-чунь

Лихенофлора северо-восточной части Карельского Перешейка (Ленинградская Область)

(Получено 12 I 1962)

Первые сведения по лихенофлоре Ленинградской области непосредственно связаны с работами Гортера (Gorter, 1761), Георги (Georgi, 1790, 1800) и Соболевского (Sobolevskj, 1799). Сейчас эти работы имеют лишь историческое значение. Затем появились большие работы Вейнманна (Weinmann, 1837) и Вайнио (Vainio, 1878). Первый автор приводит для территории нынешней Ленинградской области 114 видов и много разновидностей лишайников без точного указания их местонахождения, в более ранней его работе (1822) Вейнманн приводит для окрестностей Павловска 5 видов лишайников. Вайнио дает очень полный список, заключающий в себе 421 лишайник, собранный в окрестностях Выборга. В первой половине XX в. появляется ряд работ, уже специально посвященных лихенофлоре Ленинградской области. Первую работу опубликовал К. С. Мережковский (1904—1906), в ней приводится 29 лишайников для окрестностей Репино (б. Куоккала).

Большой интерес имеет работа В. П. Савича «Из жизни лишайников юго-западной части Петербургской губ. и прилегающей части Эстляндской» (1909). В этой работе приведено 106 видов лишайников и впервые для Ленинградской области дается анализ лишайниковых ассоциаций, указываются их происхождение и изменения под влиянием естественного развития растительности во времени и под влиянием изменений среды. В другой работе, основанной на материале из Репино (б. Куоккала), В. П. Савич (1910) приводит 36 видов лишайников, из которых большой редкостью для Ленинградской области является арктический реликтовый здесь лишайник *Cetraria nivalis* (L.) Ach.

Вслед за работой Савича (1909), посвященной не только систематике, но и лихеноассоциациям, появляется ряд статей А. А. Еленкина (Еленкин и Бекетов, 1919, Еленкин и Петров, 1919, Еленкин 1921), К. А. Рассадиной (1930а, 1930б) и Н. А. Миняева (1936, 1940), в которых большое внимание уделено анализу лихеноассоциаций, особенно в работе Рассадиной «О лишайниках б. Петергофского уезда Ленинградской губ.» (1930а), содержащей список 132 видов и хороший анализ лишайниковых синузий на стволах деревьев.

В другой своей работе Рассадина (1930б) приводит 70 видов лишайников с двумя новыми для науки разновидностями. Что касается работ Миняева, то в них указывается несколько интересных и новых для Ленинградской области лишайников и осуществляется первая интересная попытка флористического анализа лишайниковой флоры Ленинградской области.

Необходимо отметить, что имеется очень немного работ, посвященных лихенофлоре северо-восточной части Карельского перешейка. Хотя в работе Рязянена (Räsänen, 1939) и содержится полный список, насчитывающий 785 видов лишайников с 383 разновидностями и формами, однако и в ней для исследуемого нами района мы находим всего лишь 29 видов с 4 разновидностями и формами. Рязянен в своей работе не упоминает

Работа была опублиkована в. *БОТ. ЖУРН. АН СССР*, 47（6）: 830-837, 1962.

лишайниковых ассоциаций, и не дает общей картины лишайниковой растительности.

Кроме вышеупомянутых, мы имеем ряд работ различных авторов (Norrlin, 1878 г.; Vainio, 1940 г. и т. д.), посвященных лихенофлоре районов, расположенных к северу от исследуемого нами района. В геоботанических работах также упоминаются более или менее широкораспространенные лишайники.

В результате изучения выяснилось, что лишайниковая флора северо-восточной части Карельского перешейка насчитывает 239 видов с многочисленными разновидностями и формами.[1] Из этих лишайников новыми для Ленинградской области являются: *Peltigera scabrosa* Th. Fr., *Nephroma arcticum* (L.) Torss., *Lecidea farinosa* H. Magn., *Lecidea silacea* Ach., *Rhizocarpon postumum* (Nyl.) Arn., *Lecanora bicincta* Ram., *L. subfuscata* H. Magn., *L. chlorotera* Nyl., *L. campestris* (Schaer.) Hue, *Parmelia bitteriana* A. Z., *Cetraria fahlunensis* (L.) Vain., *Cornicularia odontella* (Ach.) Röhl., *Rinodina arenaria* (Hepp.) Th. Fr. и *Physcia caesiella* (B. de Lesd.) Suza.

Северо-восточная часть Карельского перешейка простирается между 60°40′ и 61°10′ с. ш. и 29°30′ 30°30′ и в. д. Территория исследуемого нами района находится в пределах Вуоксинской низины. На севере территория граничит с Карельской АССР, с востока омывается Ладожским озером, на западе ограничена поселком Мельниково. Южной границей данного района являются озера Вуокса и Суходольское. Северная часть[2] района расположена на окраине балтийского кристаллического щита, являющегося массивом древнейших кристаллических пород, непосредственно выходящих на поверхность, с образованием скалистых возвышенностей (сельги). Рельеф местности сильно пересеченный. Характерно чередование вытянутых скалистых гряд с понижениями. Лишайниковая растительность довольно интересна и разнообразна. В противоположность северной, южная часть исследуемого района представляет собой беспорядочное нагромождение высоких песчаных холмов. Лишайниковая растительность более однообразна и характерна для лесной зоны. Что касается горных, арктических и редких видов лишайников, таких, как *Nephroma arcticum*, *Peltigera scabrosa*, *P. horizontalis* (Huds.) Baumg., *Lecidea silacea*, *Gyrophora proboscidea* Ach., *G. erosa* Ach., *G. hirsuta* Ach., *Cetraria hepatizon* (Ach.) Vain., *C. juniperina* (L.) Ach., *Cornicularia odontella*, *Lecanora bicincta*, *Rinodina arenaria* и др., то они очень часто встречаются в северной части района и совершенно не найдены в южной. Лишь на скалистом берегу оз. Вуокса нам удалось обнаружить образцы лишайника *Gyrophora hirsuta*, однако эти образцы обладают особыми морфологическими признаками и мы вынуждены были описать новую разновидность[3] данного вида, которая в северной части района совершенно не встречается.

Резкая разница между лихенофлорами северной и южной частей исследуемого района объясняется не только своеобразием рельефа местности, но и тем, что климатические условия северной части района очень отличны от условий южной. Так, например, осенние заморозки на севере района наблюдаются более рано, весна и лето прохладнее, а зима, особенно в приладожской полосе, холоднее, с характерными, более сильными ветрами. Интересно, что на северных склонах скал в северной части района часто встречается *Cladonia gracilis* var. *chordalis* f. *abortiva* Schaer.; подеции этого лишайника всегда несут бесформенные или серповидные галлы коричневого или темно-бурого цвета. Как полагают некоторые лихенологи, эти галлы появляются вследствие повреждения лишайников морозами. Таким образом, климатические условия в исследуемом нами неболь-

[1] 18 видов приведено нами из работ Рязянена (Räsänen, 1939), остальные 221 вид собраны нами во время исследований в 1959—1961 гг.

[2] От северной границы района до Приозерска.

[3] *Gyrophora hirsuta* var. *vuoxaensis* var. nova.

шом районе, по-видимому, также играют известную роль как в распространении лишайников, так и в их морфологических особенностях.

На основании сказанного можно заключить, что в тех местах перешейка, где климат холоднее и дольше в течение года сохраняются снег и лед, в условиях, несколько приближающихся к условиям ледникового периода, до сих пор удерживались высокогорные и арктические виды лишайников, можно встретить реликты лишайников ледникового периода. Это особенно подтверждается местообитаниями арктических видов, как например *Nephroma arcticum* и др., найденных нами только на северных склонах скал по берегам озер в прибрежной полосе Ладожского озера в северной части района.

Лихенофлора на скалах

Гребни и вершины скалистых кряжей в изученных нами местах в значительной мере свободны от покрова высших растений. На них растут преимущественно лишайники, особенно накипные. Лишь местами встречаются отдельные сосны, чахлые березы и рассеянные кусты можжевельника. Низины скал покрыты вереском, черникой, брусникой и багульником. Флора лишайников насчитывает здесь более 87 видов. Из них накипные лишайники большей частью селятся на вертикальных склонах и на сильно освещенных скалистых холмах, тогда как лишайники, характерные для земляного субстрата, встречаются на более отлогих или плоских склонах скал, где легко накопляются мелкозем и растительные остатки. Эти виды лишайников почти всегда встречаются вместе с зелеными мхами и сплошь покрывают скалу на большом пространстве или же на небольших участках и придают ей заметную уже издали пеструю окраску. На вертикальных склонах скал господствующими видами часто являются *Rhizocarpon geographicum* (L.) DC. с его формами, которые придают скале ярко-зеленый узорчатый вид, и *Aspicillia cinerea* (L.) Körb. Наряду с этими господствующими видами в большом количестве найдены нами также *Lecanora rupicola* (L.) A. Z., *L. polytropa* (Ehrh.) Rabh., *L. cenisea* Ach. и *L. badia* (Hoffm.) Ach. В небольшом количестве встречаются *L. campestris*, *Diploschistes scruposus* (Schreb.) Norrl., *Rhizocarpon grande* (Flk.) Arn., *Rinodina arenaria* и другие. Среди многочисленных накипных лишайников здесь же в большом количестве встречаются также и листоватые, из которых господствующими видами являются *Gyrophora erosa*, местами *Parmelia stygia* (L.) Ach. В небольшом количестве встречаются *Parmelia centrifuga* (L.) Ach., *P. conspersa* (Ehrh.) Ach., *P. incurva* (Pers.) Fr., *P. sorediata* (Ach.) Röhl., *Cetraria hepatizon* и особенно характерные для скал виды *Gyrophora*.

На южных, освещенных солнцем вертикальных склонах скал, по которым часто течет дождевая или талая вода, всегда можно найти *Umbilicaria pustulata* (L.) Hoffm. На северных, затененных и увлажненных вертикальных склонах встречаются преимущественно зеленые мхи и некоторые влаголюбивые лишайники, большей частью: *Cladonia pleurota* (Flk.) Schaer., *C. squamosa* (Scop.) Hoffm., *Nephroma arcticum*, встречаются также *Menegazzia pertusa* (Schrank) Stein. и накипные лишайники *Rhizocarpon eupetreum* (Nyl.) Arn. и *Haematomma ventosum* (L.) Mass.

На холмистых, хорошо освещенных кряжах господствуют преимущественно *Rhizocarpon geographicum* (с его формами) и *Aspicilia cinerea*. Среди них разбросанно встречаются *Rhizocarpon grande*, *Gyrophora deusta* (L.) Ach., *G. polyphylla* (L.) Funck. (последняя часто с сильно курчавыми лопастями), *Cetraria hepatizon* и *Cornicularia odontella*. В противоположность сторонам хорошо освещенным, на слабо освещенной стороне скал господствует *Parmelia centrifuga*, которая придает скалистым холмам светло-зеленоватую окраску и узорчатый вид.

Нам приходилось наблюдать ряд интересных случаев, когда растущая здесь же *Gyrophora polyphylla* пробивалась сквозь слоевища *Parmelia incurva*. В результате часть слоевища этой *Parmelia* под слоевищем *Gyrophora* постепенно разрушилась; в то же время при соприкосновении краями слоевище *Gyrophora* часто заходит на слоевище *Parmelia*. Однако когда сталкиваются своими краями *Gyrophora erosa* и *Parmelia incurva* или слоевище *G. erosa* пробивается сквозь слоевище *P. incurva*, то картина меняется, а именно, зачастую слоевище *Parmelia incurva* заходит на слоевище *Gyrophora erosa*.

Это объясняется несомненно тем, что слоевище у *Gyrophora erosa* обычно очень плотно прикрепляется краями к субстрату.

Такая морфологическая особенность слоевища *Gyrophora erosa* создает благоприятные условия для прорастания слоевища *Parmelia incurva* на слоевище *Gyrophora erosa*. Наоборот, та особенность, что слоевище *Gyrophora polyphylla* обычно очень приподнято по краям, способствует тому, что слоевище *G. polyphylla* часто заходит на слоевище *P. incurva*.

Из сказанного следует, что существующая в природе борьба между различными видами лишайников заключает в себе много своеобразного и характер ее часто зависит от различных морфологических и биологических особенностей лишайников.

Что же касается напочвенных лишайников, то они обычно встречаются на плоских кряжах скал, где образуют сплошные или разрозненные ковры. Преимущественное положение здесь занимает *Cladonia sylvatica* (L.) Hoffm. С ней вместе в большом количестве всегда наблюдаются *Cladonia rangiferina* (L.) Web., а также *C. uncialis* (L.) Web., *C. amauroc raea* (Flk.) Schaer., *C. gracilis* (L.) Willd. и другие виды *Cladonia*. Что касается *C. alpestris* (L.) Rabh., то она обычно сплошь покрывает более или менее затененные склоны скал или понижения. В местах сильного освещения *C. alpestris* встречается обычно единично.

В затененных и увлажненных местах среди мхов часто встречаются *Cetraria islandica* (L.) Ach., *Cladonia turgida* (Ehrh.) Hoffm., *C. squamosa* и *C. alpicola* (Flot.) Vain.

У оснований скал и в более или менее затененных местах чаще всего поселяются *Nephroma parile* Ach. и виды *Peltigera*, а также *Parmelia fuliginosa* (Wib.) Nyl. К этой же экологической группе следует отнести лишайники, поселяющиеся на валунах в открытых местах и в сухих сосновых лесах. Здесь нужно отметить, что господствующими видами в открытых местах чаще всего являются или *Aspicilia cinerea*, или *Physcia tribacia* (Ach.) Nyl., а также *Lecanora cenisea* или *Parmelia sorediata* и в сосновых лесах *Cladonia sylvatica* и *C. amaurocraea*.

Среди лишайникового покрова всегда встречаются мозаичные пятна зеленых мхов, которые иногда образуют кайму вокруг лишайников. Нередко также встречаются отмершие стебельки мхов под лишайниковым ковром. Можно констатировать, что вслед за накипными и листоватыми лишайниками на скалах и валунах чаще всего поселяются мхи, чему особенно способствуют скальные трещины. В свою очередь в моховом покрове, особенно способствующем накоплению минеральных и органических остатков, начинают поселяться кустистые лишайники: *Cladonia sylvatica*, *C. rangiferina*, иногда и *C. coccifera* (L.) Willd., для которых мхи сохраняют достаточно влаги на обнаженных и сильно освещенных скалах. Таким образом, на скалах при сравнительно постоянном содержании воды и при хорошей инсоляции лишайники постепенно начинают господствовать. Однако в более затененных и влажных местах преимущество остается за мхами, хотя и здесь нередко можно встретить такие тенелюбивые лишайники, как *Cladonia squamosa*, *C. turgida*, *C. alpicola* и *Cetraria islandica*.

Итак, между скальными мхами и лишайниками, поселяющимися на растительных остатках и тонком слое мелкозема на скалах, также существует борьба за площадь, за существование. Наряду с этим со временем среди мохово-лишайниковых пятен начинают вклиниваться цветковые растения, особенно вереск, далее молодые березы, рябинки, можжевельник, сосенки, а местами даже группы сосен. Небольшие группы сосен очень медленно, но неизбежно будут изменять в захваченном ими месте окружающую их растительную среду, что постепенно приводит к облесению всей местности.

Лихенофлора сосновых лесов

В сухих сосновых лесах Карельского перешейка земля покрыта сплошными сероватыми коврами, состоящими в основном из лишайников *Cladonia sylvatica*, *C. rangiferina* и местами *C. alpestris*. При детальном рассмотрении можно заметить, что среди господствующих видов, особенно на кочках, по краям старых ям или там, где мало *C. sylvatica* и *C. rangiferina*, в большом количестве встречаются *C. cornuta* (L.) Schaer., *C. gracilis*, *C. crispata* (Ach.) Flot. и многочисленные представители из подрода *Cenomyce* Th. Fr. (*Cladonia* [Hill.] Vain.). Среди вереска на земле часто растет *Peltigera aphthosa* (L.) Willd. По опушкам или в глубине леса можно найти широко распространенный лишайник *Peltigera canina* (L.) Willd., а на старых гарях *P. spuria* (Ach.) DC. Под тенью сосен земля большей частью обильно покрыта широколопастной формой лишайника *Cetraria islandica*. На сильно прогреваемой солнцем почве у стволов сосен или на кочках иногда встречается *Biatora uliginosa* (Schrad.) Fr. Часто заходят непосредственно с почвы на основания стволов лишайники рода *Cladonia*, такие, как *C. deformis* (L.) Hoffm. и др.

Обычными видами на нижней части стволов сосен являются *Parmeliopsis ambigua* (Wulf) Nyl., *P. hyperopta* (Ach.) Arn., *Cetraria caperata* (L.) Vain., но вообще господствующими видами на стволах сосен являются *Parmelia physodes* (L.) Ach., а иногда *Evernia furfuracea* (L.) Mann. Постоянными элементами стволовой лихенофлоры являются также *Usnea hirta* (L.) Wig. em. Mot., *U. comosa* (Ach.) Roehl., *U. dasypoga* (Ach.) Roehl., *Alectoria implexa* (Hoffm.) Nyl., *Evernia prunastri* (L.) Ach., *Cetraria glauca* (L.) Ach. Часто встречаются *Cetraria chlorophylla* (Humb.) Vain., *Parmelia tubulosa* (Schaer.) Bitt., *Calicium abietinum* Pers., *Buellia punctiformis* Mass., *Lecanora coilocarpa* (Ach.) Nyl., *Biatora symmicta* Fr., *Opegrapha pulicaris* (Hoffm.) Schrad. На сухих ветвях часто встречаются *Cetraria saepincola* Ach., *C. islandica* f. *ramulicola* Savicz и др.

Сильно изменяется видовой состав лишайников в сосновых лесах после рубок. На вырубках сосновых лесов господствующими видами лишайников обычно являются: *Biatora uliginosa*, поселяющаяся на сильно прогреваемой почве, *Cladonia floerkeana* (Fr.) Somf. и *C. botrytes* (Hag.) Willd.; последний вид особенно обильно растет на верхней горизонтальной поверхности пней. По бокам пней на коре селятся те виды, которые характерны для оснований стволов сосен. На пнях после отпада с них коры, кроме *C. botrytes*, часто поселяются виды из сем. *Caliciaceae*.

На полусгнивших пнях поселяются в большом количестве виды *Cladonia*, обычно встречающиеся на земле. Нужно отметить, что вырубки быстро заселяются молодыми соснами. В обычных условиях они образуют и новые сосновые леса, так что видовой состав лишайников снова изменяется. Отметим, что на молодых соснах встречаются обычно лишь молодые слоевища *Parmelia physodes* (L.) Ach., на основаниях стволов — *Cetraria caperata*, *Parmeliopsis ambigua*. На верхней же части стволов молодых сосен лишайники почти совсем отсутствуют. Слоевища *P. physodes* вслед за ростом сосны постепенно распространяются по всему стволу, начинают поселяться и другие лишайники: виды *Usnea*, *Alectoria*, *Cetraria*, *Parmelia*, *Lecanora* и др.

Лихенофлора еловых лесов

При смене соснового леса еловым соответственно изменяется и лихенофлора. Однако в еловых лесах флора лишайников очень бедна по сравнению с сосновыми лесами. Земля здесь почти сплошь покрыта черникой и другими, травянистыми растениями; лишь в более освещенных местах или на кочках иногда встречаются *Cladonia carneola* Fr., *C. deformis* и некоторые другие виды. На гнилых пнях обильно встречается *Cladonia digitata* (Ach.) Schaer.

На стволах ели обычно встречаются в небольшом количестве *Parmelia physodes*, иногда *Calicium viride* Pres. Внизу стволов — *Parmeliopsis ambigua* и *Cetraria caperata*. Ветки ели обычно почти сплошь покрыты лишайниками, преимущественно *Cetraria chlorophylla*, *C. glauca*, *Usnea dasypoga*, *U. glabrescens* (Nyl.) Vain., *Alectoria implexa*, *Parmelia physodes*, *Biatora symmicta* и др.

Лихенофлора смешанных лесов

В исследуемом нами районе наблюдается явная тенденция смены соснового леса смешанным. В сосновых лесах под пологом сосны часто попадается молодняк березы, ольхи, рябины, осины, можжевельника и др.

Вслед за этой сменой несомненно наступит изменение и лишайников, обитавших в сосновых лесах. Смешанные леса обычно затененные, увлажненные; земля сплошь покрыта травянистыми растениями. Здесь, обычно только на валунах, попадаются *Peltigera aphthosa*, *P. canina*, *Baeomyces rufus* (Huds.) Rebent., *Cladonia cornuta*, *C. turgida*, *C. gracilis*, *C. furcata* (Huds.) Schrad. и др. Стволы лиственных пород, как правило, покрыты накипными лишайниками, в открытых же местах — большей частью кустистыми и листоватыми лишайниками.

Характерными для стволов осины лишайниками являются *Xanthoria parietina* (L.) Th. Fr. на гладкой коре и *Lecanora allophana* (Ach.) Röhl на шероховатой коре. Среди них часто можно встретить *Lecidea glomerulosa* (DC.) Steud., *Caloplaca pyracea* (Ach.) Th. Fr., *C. cerina* (Ehrh.) Th. Fr., *Ramalina populina* (Ehrh.) Vain., *Physcia tenella* (Scop.) DC., *Ph. ascendens* (Fr.) Oliv., *Ph. ciliata* (Hoffm.) D. R., *Ph. aipolia* (Ehrh.) Hampe, *Ph. stellaris* (L.) Nyl., *Anaptychia ciliaris* (L.) Körb. и виды *Collemaceae*. У основания стволов нередко растут *Peltigera canina*, *Nephroma resupinatum* (L.) Ach., *N. helveticum*, *N. laevigatum* Ach., *Parmelia sulcata* Tayl. и *P. aspidota* (Ach.) Poetsch.

На стволах березы большей частью поселяются *Parmelia olivacea* Nyl., *P. aspidota*, *P. sulcata*, *P. physodes*; реже встречается *P. tubulosa*. Среди них обычно в небольшом количестве встречаются *Cetraria chlorophylla*, *C. glauca*, *Evernia prunastri*, *E. thamnodes* (Flot.) Arn., *E. furfuracea*, *Usnea comosa* и др., а из накипных — *Lecanora varia* (Ehrh.) Ach., вместе с которой часто поселяются *Biatora symmicta*, *Lecanora coilocarpa*, *L. chlarona* и *Leptorhaphis epidermidis* (Ach.) Th. Fr. У основания стволов наблюдаются *Cetraria caperata*, *Parmeliopsis ambigua*, нередко также встречаются *Cladonia cenotea* (Ach.) Schaer., *C. fimbriata* (L.) Fr., *C. deformis* и близкие виды, заходящие с почвы.

Флора лишайников на стволах ольхи представлена главным образом *Buellia disciformis* (Fr.) Mudd. и *Lecidea glomerulosa*. Среди них встречаются *Lecanora carpinea* (L.) Ach., *Biatora symmicta* и только местами *Parmelia physodes*, *Evernia prunastri*, *Usnea sublaxa* Vain, *Usnea comosa* и *Alectoria implexa*. В открытых местообитаниях стволы ольхи обычно обильно покрыты видами из родов *Parmelia*, *Cetraria*, *Evernia*, *Usnea* и *Alectoria*. В нижней части стволов почти всегда находимы *Graphis scripta* (L.) Ach. и *Arthonia radiata* Ach. с разбросанными розетками *Cetraria caperata*. Интересно, что в окрестностях с. Кузнечного на стволах ольхи *Graphis scripta* встречается очень редко.

В смешанных лесах довольно обильно произрастает можжевельник, на котором обычно встречаются такие лишайники, как *Parmelia physodes*, *Cetraria saepincola*, *C. caperata* и *Bilimbia subsphaeroides* (Nyl.) Galloe. В окрестностях с. Кузнечного на нижних ветвях можжевельника была найдена в большом количестве хорошо развитая *Cetraria juniperina*.

Среди указанных видов нередко можно было обнаружить также *Calicium trabinellum* Ach., *Chaenotheca chrysocephala* (Turn.) Th. Fr., *Ch. stemonea* (Ach.) Zwackh., а иногда и *Evernia furfuracea* и *E. thamnodes*, которые сплошь покрывают ветки можжевельника.

Лихенофлора на крышах строений

Старые деревянные крыши обычно обильно покрыты лишайниками, характерными для старых пней, заборов и для земли. Из них преимущественными видами обычно являются: *Cladonia sylvatica* на северных склонах и *C. floerkeana*, *C. gracilis* на южных склонах. Южные склоны крыш обычно более богаты видами, чем северные, но зато на северных склонах лишайники вообще лучше развиты, чем на южных. Среди указанных основных видов часто встречаются лишайники из подрода *Cenomyce* Th. Fr. (*Cladonia* [Hill.] Vain.), а также и широко распространенные представители *Parmeliaceae*, *Usneaceae*, *Lecanoraceae* и др. К этой группе следует отнести и лишайники, поселяющиеся на старых заборах, на которых обычно преимущественное положение занимает *Lecanora varia*.

На основании приведенного материала следует отметить, что после наших исследований количество видов лишайников для лихенофлоры северо-восточной части Карельского перешейка возросло до 239 видов, из которых ряд видов найден нами впервые для Ленинградской области. Большой интерес во флористическом отношении имеет нахождение *Nephroma arcticum*, *Peltigera scabrosa* и других горно-арктических видов, а также и таких довольно интересных и редких для СССР видов, как например *Lecidea farinosa*, *Parmelia bitteriana* и др. Необходимо отметить, что почти все эти виды лишайников уже известны для северного побережья Ладожского озера и для Финляндии.

Заслуживает внимания резкая разница между лихенофлорами северной и южной части исследуемого района. Вообще окрестности с. Кузнечного по побережью Ладожского озера являются наиболее интересной местностью для изучения лишайниковой флоры, принимающей здесь горноарктический характер.

Напротив, типичные горноарктические виды, столь характерные для окрестностей с. Кузнечного по побережью Ладожского озера, в южной части района большей частью совершенно отсутствуют или представлены очень скудно. Здесь же, особенно в смешанных лесах, лишайник *Graphis scripta* с его спутниками из родов *Arthonia* и *Buellia* представлен в изобилии самыми разнообразными формами на коре различных лиственных пород, как например ольхи, рябины и др., покрывая иногда почти сплошь кору стволов деревьев своими хорошо развитыми узорчатыми апотециями. В то же время в окрестностях с. Кузнечного по побережью Ладожского озера *Graphis scripta* представляет большую редкость. Все это, по-видимому, обусловлено не только наличием на севере района многочисленных скал и другими топографическими факторами, но и более суровым климатом.

Что касается распределения лишайников на скалах, то оно в значительной степени зависит от крутизны и экспозиции склонов скал. Наши исследования показывают, что наскальная лихенофлора района более богата и разнообразна по сравнению с лихенофлорой лесов.

Сосновые леса обладают сравнительно богатой лихенофлорой; кроме многочисленных эпифитов, сильно развит лишайниковый покров на почве, образованный главным образом видами *Cladonia*, а также *Cetraria*, *Stereocaulon* и *Peltigera*. В старых еловых лесах лишайники встречаются преимущественно на ветвях. Преобладают виды сем. *Parmeliaceae* и *Usneaceae*. В противоположность тому, что наблюдается в еловых лесах, в смешанных лесах они чаще встречаются на стволах. Господствуют виды *Arthonia*, *Graphis*, *Lecanora*, *Lecidea* или *Buellia*. В открытых местах на деревьях преимущественное положение занимают *Parmeliaceae* и *Usneaceae*. Однако накипные лишайники встречаются здесь же не менее часто, чем в лесах.

Таким образом, лишайниковая флора не является постоянной и неизменной. В лесах она изменяется в зависимости от смены типа леса, на скалах — от выветривания поверхности скал и влияния деятельности человека. В процессе изменения состава флоры между видами лишайников, между лишайниками и мхами происходит борьба за существование.

ЛИТЕРАТУРА

Биркенгоф А. Л. и др. (1958). Ленинградская область, природа и хозяйство. — Еленкин А. А. (1906—1911). Флора лишайников Средней России, I—IV. — Еленкин А. А. (1921, 1933). Лишайники как объект педагогики и научного исследования. Экскурсионное дело, 2 и 3, 4—6. — Еленкин А. А. и И. А. Бекетов. (1919). Четырнадцать спорологических экскурсий в окрестностях Приморской ж. д. от ст. Лахта до ст. Дюны, предпринятых в 1918 году. Изв. бот. сада РСФСР, 19, 1. — Еленкин А. А. и В. А. Петров. (1919). О некоторых лишайниках для Петроградской губ. Изв. Главн. бот. сада РСФСР, 19, 1. — Ладыженская К. И. (1927). Материалы к экологии мхов окрестностей Петергофа. Журн. русск. бот. общ., XII, 4. — Мережковский К. С. (1904—1906). К познанию лишайников севера России. Приложение к протоколам засед. общ. естествоиспыт. при Имп. Казанск. унив., № 234. — Миняев Н. А. (1936). Новые лишайники для флоры окрестностей Ленинграда. Тр. Бот. инст. АН СССР, сер. 2. Споровые растения, 2, 3. — Миняев Н. А. (1940). Реликтовые элементы в современной флоре лишайников восточной Прибалтики. Бот. журн. СССР, 4—5. — Миняев Н. А. (1949). Основные проблемы географии лишайников. Тр. Второго Всесоюзн. географ. съезда, III. — Ниценко А. А. (1951). О процессах развития растительности на обнаженных скалах. Уч. зап. ЛГУ, 143, сер. биолог. наук, 30. — Ниценко А. А. (1959). Геоботанический очерк территории Ладожской станции Ленинградского университета. Очерки растительности Ленинградской области. — Рассадина К. А. (1930а). О лишайниках б. Петергофского уезда Ленинградской губернии. Тр. Бот. музея АН СССР, 22. — Рассадина К. А. (1930б). Лишайники, собранные С. С. Ганешиным в Лужском уезде и в окрестностях Новосиверской Ленинградской губернии. Тр. Бот. муз. АН СССР, 22. — Савич В. П. (1909). Из жизни лишайников югозападной части Петербургской губ. и прилегающей части Эстляндской. Тр. СПб. общ. естеств., 50, 2. — Савич В. П. (1910). Лишайники, собранные в окрестностях Куоккала (Выборгской губ.) И. И. Воронихиным в 1907 г. Тр. студенч. науч. кружк. физ.-мат. фак. СПб. унив., I, 2. — Савич В. П. (1921). Новые лишайники для Петроградской губ. Изв. Главн. бот. сада РСФСР, 12, 1. — Смирнова З. Н. (1928). Лесные ассоциации северо-западной части Ленинградской области. Тр. Петергофск. естеств. научн. инст., 5. — Gorter David. (1761). Flora Ingrica ex schedis Stephani Krascheninnikow confecta et propriis observationibus aucta. — Räsänen V. (1925). Lichenes novi vel rariores, e taeniis Ladogensibus. Medd. Soc. F. Fl. Fenn. — Räsänen V. (1939). Die Flechtenflora der nördlichen Küstengegend am Laatokka-See. Ann. Bot. Soc. Vanami, 12, 1. — Sobolevskj Greg. (1799). Flora Petropolitana. — Vainio E. (Wainio). (1878). Lichens in viciniis Viburgi observati. Medd. af Soc. pro Fauna et Flora Fennica, 2. — Weinmann A. (1822). «Correspondenz». Flora oder Bot. Zeitung. — Weinmann A. (1837). Enumeratio stirpium in Agro Petropolitano sponte crescentium.

Ботанический институт
им. В. Л. Комарова
Академии наук СССР,
Ленинград.

LICHENOFLORA OF THE NORTH-EASTERN PART OF THE KARELIAN ISTHMUS (LENINGRAD REGION).

By Vej Tzjan-czunj

SUMMARY

As the result of this investigation the number of species known for the north-eastern part of the Karelian Isthmus reached 239. The distribution of lichens among the ecological stations is described and a number of florogenetic considerations are stated. A brief outline of history of the investigations of lichens in the Leningrad Region is given.

珠穆朗玛峰地区地衣区系资料

魏江春 陈健斌*

（中国科学院微生物研究所）

珠穆朗玛峰（以下简称珠峰）地区地衣区系的研究是中国科学院西藏科学考察队在1966—1968 年对珠峰地区进行多学科综合性科学考察的组成部分。由于我们的工作正在进行中，因此，本文的任务不是要反映珠峰地区地衣区系的全貌，也不涉及珠峰地区地衣区系成份的综合分析，而只是工作当中的阶段性初步报告，仅就掌握的部分资料对地衣在珠峰地区自然带谱中的基本情况作一概述。

由于所谓"第三极"的珠峰在自然地理方面所具有的独特性，这一地区的地衣区系组成越来越引起区系学家和地衣学家们的极大注意。尽管对喜马拉雅山脉地衣区系的研究工作日益增多（Asahina、Awasthi、Bhatia、Chopra 等），但是，对山脉的主峰——珠峰地区地衣区系的研究工作却极为少见（Paulson）。本文涉及的珠峰地区包括珠峰南翼的樟木、曲乡、聂拉木等地和珠峰北翼的定日、绒布寺、中绒布冰川和西绒布冰川地区。

珠峰地区位于喜马拉雅山脉中段(即主脉地段)，是"世界屋脊"——青藏高原的南缘。由于喜马拉雅山脉的走向位置和巨大高程所形成的天然屏障，使山脉南北两翼自然特征产生显著差异。主脉南翼山势陡峻，比降极大；下接恒河平原，濒临湿润的印度洋，为海洋性季风地区，呈现出以热带季雨林为基带的垂直分带系统。自然分带的垂直带谱，随着海拔高度的增加而变化。在我国境内，从常绿阔叶林带到终年积雪的高山冰雪带共七个带谱。主脉北翼高原因受喜马拉雅山高大山体的屏障作用，西南季风受阻，北与干燥的亚洲大陆相连，呈现为大陆半干旱高原地区的自然分带系统，自然分带的垂直带谱比较单纯。

地衣在自然分带中的分布特点不仅受喜马拉雅山高大山体及其南北两翼自然特征差异的影响，而且也受山脉中错综复杂的小气候和林型差异的影响。因此，地衣在本区自然带谱中的分布情况也和其他自然特征一样，呈现出极为明显的南北分异和垂直变化。

分布在珠峰地区北翼自然带谱中的地衣主要是壳状种类。在高原草原带中由于适于干旱条件的草原植被发育较好，因而地衣的种类和数量在本带并不占居优势；而在高山草甸、地衣带及高山寒冻地衣带，地衣的种类和盖度却占居明显的优势地位。

高山草甸、地衣带(海拔 5,000—5,600 米)位于高原草原带与高山寒冻地衣带之间。这里气候比较寒冷，冰川冷风较大，年平均气温在 −2℃—+5℃ 之间。由于空气稀薄，太阳直接辐射很强，地面辐射也大，因而地表昼夜温差悬殊，以七月份为例，温差超过 30℃。此带地貌是以冰川堆积地形为特征，地面多为冰碛石块和寒冻风化岩屑所覆盖，仅在较为细碎的岩屑地段和冰碛石块间隙分布有发育比较原始的高山原始草甸土。这里的植被不仅有冰川苔草(*Carex atrata* var. *glacialis*)、小嵩草(*Kobresia pygmaea*)以及一些垫状植物，

---

* 参加部分室内工作的尚有本所徐连旺、孙增美、郑儒永、李惠中、魏淑霞和韩者芳等。

本文原载于《珠峰地区科考报告》(生物与高山生理)，科学出版社: 173-118, 1974.

如垫状蚤缀（*Arenaria pulvinata*）等有花植物，而且还有其盖度不小于上述有花植物的地衣。由于生长在本带岩隙原始草甸土上的这些地衣不大引人注目，因而往往被人们所忽视。它们以伴藓蜈蚣衣（*Physcia muscigena*）、雪地茶小枝变型（*Thamnolia subvermicularis* f. *minor*）和红鳞网衣（*Psora decipiens*）等较为常见。此外，几乎布满地表的冰碛石块和寒冻风化岩屑上则固着有大量鲜艳夺目的各种地衣，呈现出一幅幅美丽的花色图案。这些地衣有糙聚盘衣（*Glypholecia scabra*）、红橙衣（*Caloplaca elegans*）、岩表地图衣（*Rhizocarpon superficiale*）、红盘鳞茶渍（*Squamaria rubina*）、石茶渍（*Lecanora polytropa*）、耳盘网衣（*Lecidea auriculata*）和菊叶梅衣（*Parmelia conspersa*）等种类。因此，相当于上述有花植物盖度的岩隙土壤地衣以及岩石地衣一起所构成的地衣盖度在此带便居于明显的优势地位，约占全部植被（包括上述有花植物）总盖度的 70% 左右。

高山寒冻地衣带（海拔 5,600—6,100 米）位于高山草甸地衣带与高山冰雪带之间。此带气候条件比高山草甸地衣带更为严酷，年平均气温在－4℃——8℃之间。冰碛石块和寒冻风化岩屑几乎布满地面。土壤发育极差。上述有花植物只在局部有零星分布，而地衣则相应地居于绝对优势。但主要地衣的种类组成大致类同于高山草甸地衣带；只是岩隙土壤地衣明显减少，石生地衣相应地处于明显优势。

位于高山寒冻地衣带之上的为高山冰雪带。此带年降雪量大于年融雪量，因而终年积雪。但是本带极为陡峻的山峰则往往并不积雪，呈现为常年裸露的岩壁。在这些背风面的陡峻岩壁上，很可能仍有地衣存在。

珠峰地区南翼，因面对西南季风，属于湿润的海洋性气候，因而分布着各种茂密的森林。随着海拔的升高和气温的下降，自然带呈现出明显的垂直变化，由常绿阔叶林带过渡到针阔混交林带，进而呈现出针叶林带和灌丛草甸带等等。在这里地衣的种类组成同样也随之表现出有规律的变化。

在山地常绿阔叶林带（海拔 1,600—2,500 米）常见的地衣有梅衣（*Parmelia tinctorum*）、小珊芽梅衣（*P. nimandairana*）、珊瑚缘梅衣（*P. pseudolivetorum*）和亚灌松萝（*Usnea arborea*）等种类。在山地针阔混交林带（海拔 2,500—3,100 米）的地衣则以黄果梅衣（*Parmelia xanthocarpa*）、黄髓梅衣（*P. aurulenta*）、漂红梅衣（*P. olivetorum*）、类粉缘梅衣（*P. neglecta*）、皱梅衣（*P. caperata*）、条梅衣（*P. cirrhata*）、黄粉袋梅衣（*P. hypotrypella*）、粉缘梅衣（*P. cetrarioides*）以及皮革岛衣（*Cetraria pallescens*）、中国树花（*Ramalina sinensis*）和筛石蕊（*Cladonia aggregata*）等为主。

至于山地针叶林带（海拔 3,100—3,700 米）的地衣种类就更加丰富，其中主要种类有刺树发（*Alectoria acanthodes*）、珊粉树发（*A. smithii*）、长松萝（*Usnea longissima*）、拟长松萝（*U. pectinata*）、灌松萝（*U. thomsonii*）、粉斑梅衣（*Parmelia pseudoborreri*）、广布梅衣（*P. arnoldii*）、东方梅衣（*P. simodensis*）、条袋梅衣（*P. vittata*）、指袋梅衣（*P. enteromorpha*）、孔梅衣（*Menegazzia pertusa*）、麸石蕊（*Cladonia pityrea*）、细石蕊（*Cl. gracilis*）、黑穗石蕊（*Cl. amaurocraea*）、分枝石蕊（*Cl. furcata*）、角石蕊（*Cl. cornuta*）、鳞片石蕊（*Cl. squamosa*）、小瘤地卷（*Peltigera scabrosa*）、多根脐衣（*Umbilicaria polyrrhiza*）和石耳（*Umb. esculenta*）等。

分布于针叶林带上限与亚高山灌丛、草甸带下限的地衣种类主要为黄条梅衣（*Parmelia sinuosa*）以及狭杯红石蕊（*Cladonia transcendens*）、石蕊（*Cl. rangiferina*）、软石蕊（*Cl.

mitis)、喇叭石蕊（*Cl. pyxidata*)、黑穗石蕊（*Cl. amaurocraea* f. *oxycers*)、黑岛衣(*Cetraria nigricans*)、栗卷岛衣(*C. crispa*)以及瓦衣(*Coccocarpia pellita*)和孔疱脐衣(*Lasallia pertusa*)、稀根石耳(*Umbilicaria grisea*)、粗根石耳(*Umb. hirsuta*)、多盘石耳(*Umb. proboscidea*)与红石耳(*Umb. hypococcinea*)等。

分布在亚高山灌丛、草甸带的地衣主要有扁枝衣(*Evernia thamnodes*)、石梅衣(*Parmelia saxatilis*)、沟槽梅衣(*P. sulcata*)、白粉石蕊(*Cladonia glauca*)、喇叭粉石蕊(*Cl. chlorophaea*)、黄粉石蕊(*Cl. cyanipes*)、黄卷岛衣(*Cetraria cucullata*)以及地茶(*Thamnodes vermicularis*)、雪地茶(*Th. subvermicularis*)和皮果衣(*Dermatocarpon miniatum*)、短绒皮果衣(*D. vellereum*)等。

综上所述，从樟木常绿阔叶林带到聂拉木以上亚高山灌丛、草甸带的各垂直带谱中地衣种类组成的基本情况和从华南、西南地区向北到秦岭之颠的分布特点有很多相似性。至于珠峰绒布冰川地带地衣区系的特点由于自然条件极度严酷而表现出相当的独特性。原先只知分布于地中海沿岸山地的雪地茶小枝变型（*Thamnolia subvermicularis* f. *minor*）在珠峰冰川地带的大量存在，为进一步进行本区地衣区系成分的综合分析提供了值得注意的资料。关于本区地衣区系成分问题，我们将在今后的报告中进行讨论。

珠峰地区地衣名录如下：

1. ***Dermatocarpon miniatum*** **(Lin.) Mann.** **皮果衣**

var. ***imbricatum*** **(Mass.) Vain.**

聂拉木：波曲河谷古侧碛南坡岩石北垂面，1528 号，海拔 4,050 米。

var. ***complicatum*** **(Lighft) Hellb.**

聂拉木：波曲河谷古侧碛南坡岩石缝隙处，1529—① 号，海拔 4,070 米。

2. ***Dermatocarpon vellereum*** **Zschacke** **短绒皮果衣**

聂拉木：波曲河谷古侧碛南坡岩石缝隙处密生，1529 号，海拔 4,070 米及 1537 号，海拔 3,700 米。

3. ***Coccocarpia pellita*** **(Ach.) Müll. Arg.** **瓦衣**

曲乡：波曲河谷岩石上，976 号，海拔 3,580 米及 1078 号，海拔 3,350 米。

聂拉木：桦林岩石藓层上，1878 号，海拔 3,860 米。

4. ***Peltigera scabrosa*** **Th. Fr.** **小瘤地卷**

樟木：岩石藓层上，557 号，海拔 3,400 米。

5. ***Lecidea auriculata*** **Th. Fr.** **耳盘网衣**

珠峰：中绒布冰川第三古侧碛石上，1281 号，海拔 5,450 米。

6. ***Psora decipiens*** **(Ehrh.) Hoffm.** **红鳞网衣**

珠峰：绒布寺附近绒布河谷岸旁土壤上，1302 号，海拔 4,905 米。

7. ***Rhizocarpon superficiale*** **(Schaer.) Vain.** **岩表地图衣**

珠峰：西绒布冰川第二古侧碛大岩石上，1241 号，海拔 5,580 米；绒布寺附近岩石上，1310 号，海拔 5,000 米。

8. ***Cladonia rangiferina*** **(Lin.) Weber** **石蕊**

曲乡：岩石细土层上，758 号，海拔 3,795 米、984 号，海拔 3,950 米及 1005 号，海拔 3,780 米；岩石藓丛间，981 号，海拔 3,950 米。

9. ***Cladonia mitis*** **Sandst.** **软石蕊**

曲乡：岩石细土层上，942 号及 1041 号，海拔 3,650 米、1047 号，海拔 3,450 米及 1093 号，海拔

3,680 米;片麻岩带岩石细土层上, 754 号及 756 号, 海拔 3,760 米; 岩石藓丛间, 1011 号, 海拔 3,750 米;德青塘附近岩石藓丛间, 951 号,海拔 3,550 米。

10. *Cladonia aggregata* (Sw.) Ach. 筛石蕊

樟木: 林间空地石上, 574 号及 583 号,海拔 2,600 米。

曲乡: 德青塘附近岩石细土层上, 1088 号,海拔 3,730 米。

11. *Cladonia coccifera* (Lin.) Willd. 红石蕊

樟木: 林中石上, 703 号,海拔 3,560 米。

曲乡: 岩石细土层上, 977 号,海拔 3,560 米;德青塘附近岩石细土层上, 1089 号,海拔 3,730 米。

12. *Cladonia pleurota* (Floerk.) Schaer. 粉芽红石蕊

樟木: 林边岩石细土层上, 416 号,海拔 2,600 米;康巴吉巴附近岩石细土层上,195 号,海拔 2,700 米。

曲乡: 德青塘附近岩石细土层上, 956 号,海拔 3,550 米及 1113 号,海拔 3,350 米。

聂拉木: 波曲河谷沿岸古侧碛南坡岩石北垂面, 1524 号,海拔 3,910 米。

13. *Cladonia transcendens* Vain. 狭杯红石蕊

var. yunnana Vain.

曲乡: 德青塘附近岩石细土层上, 1094 号,海拔 3,680 米及岩石藓丛间, 1121 号,海拔 3,650 米。

14. *Cladonia amaurocraea* (Floerk.) Schaer. 黑穗石蕊

f. *amaurocraea*

聂拉木: 波曲河沿岸山坡竹林及杜鹃丛中岩石细土层上, 1555 号,海拔 3,650 米。

f. *oxyceras* (Ach.) Vain.

曲乡: 岩石细土层上, 1014 号,海拔 3,760 米。

15. *Cladonia furcata* (Huds.) Schrad. 分枝石蕊

樟木: 林中石上, 624 号,海拔 3,550 米。

曲乡: 阳坡土壤上, 971 号, 海拔 3,600 米及 972 号, 海拔 3,530 米以及藓丛间, 1018 号, 海拔 3,670 米;德青塘附近岩石细土层上, 1101 号,海拔 3,610 米。

16. *Cladonia squamosa* (Scop.) Hoffm. 鳞片石蕊

曲乡: 德青塘以西小河边岩石藓丛间, 933 号,海拔 3,550 米。

17. *Cladonia glauca* Floerk. 白粉石蕊

f. *subacuta* Asahina

聂拉木: 北坡桦林内朽树桩上, 1788—③ 号,海拔 3,990 米。

18. *Cladonia gracilis* (Lin.) Willd. 细石蕊

var. *elongata* Floerk. ex Vain.

f. *laontera* (Del.) Arn.

樟木: 朽木上, 466 号,海拔 3,600 米。

曲乡: 岩石细土层上, 1024 号,海拔 3,720 米。

19. *Cladonia cornuta* (Lin.) Schaer. 角石蕊

f. *phyllotoca* (Floerk.) Vain.

樟木: 岩石藓丛间, 559 号,海拔 3,400 米。

曲乡: 岩石细土层上, 1029 号及 1030 号,海拔 3,580 米。

20. *Cladonia pyxidata* (Lin.) Hoffm. 喇叭石蕊

曲乡: 岩石细土层上, 1021 号, 海拔 3,700 米; 德青塘附近道旁岩石藓丛间, 890 号, 海拔 3,500 米。

聂拉木：波曲河沿岸山坡竹林及小杜鹃丛林下岩石藓丛间，1566 号，海拔 3,670 米；北坡桦林内腐朽树桩上，1789 号，海拔 3,990 米。

21. ***Cladonia chlorophaea*** **(Floerk.) Spreng. 喇叭粉石蕊**

聂拉木：北坡桦林内朽树桩上，1790 号，海拔 3,990 米及 1793 号，海拔 3,950 米。

22. ***Cladonia pityrea*** **(Floerk.) Fr. 麸石蕊**

樟木：杉树杆上，493 号，海拔 3,300 米。

23. ***Cladonia cyanipes*** **(Sommerf.) Vain. 黄粉石蕊**

聂拉木：北坡桦林内朽树桩上，1788 号，海拔 3,990 米。

24. ***Lasallia pertusa*** **(Rassad.) Llano. 孔疱脐衣**

聂拉木：县机关食堂附近大岩石上，92号，海拔 3,700 米、河谷岩石北垂面，1648号，海拔 3,720 米。

25. ***Umbilicaria grisea*** **(Swartz) Ach. 稀根石耳**

曲乡：德青塘附近大岩石南垂面，887 号，海拔 3,500 米。

聂拉木：波曲河谷沿岸大岩石上，1610 号及 1618 号，海拔3，900 米；县机关食堂附近大岩石上，90 号，海拔 3,700 米。

26. ***Umbilicaria hirsuta*** **(Swartz) Ach. 粗根石耳**

聂拉木：波曲河岸岩石上，1632 号，海拔 3,720 米；县机关食堂附近大岩石上，90—① 号，海拔 3,700 米；波曲河谷沿岸桦林边缘大岩石西垂面，1540 号，海拔 3,900 米。

27. ***Umbilicaria polyrrhiza*** **(Lin.) Ach. 多根石耳**

樟木：岩石上，547 号，海拔 3,610 米。

曲乡：林中岩石上，502 号，海拔 3,500 米。

28. ***Umbilicaria esculenta*** **(Miyoshi) Minks 石耳**

樟木：小河旁岩石上，289 号，海拔 3,375 米。

曲乡：波曲河西岸岩石上，1073 号，海拔 3,470 米；大岩石上，1036 号及 1070 号，海拔 3,580 米及 880 号，海拔 3,250 米。

29. ***Umbilicaria proboscidea*** **(Lin.) Schrad. 多盘石耳**

f. *proboscidea*

聂拉木：波曲河沿岸古侧碛岩石上，1440 号，海拔 3,800 米；波曲河沿岸阶地岩石上，1535 号，海拔3,750 米。

f. *rhizophora* Vain.

聂拉木：岩石上，1816 号，海拔 3,870 米。

30. ***Umbilicaria hypococcinea*** **(Jatta) Llano 红石耳**

聂拉木：波曲河沿岸阶地岩石上，1534 号，海拔 3,750 米；大岩石上，1617 号，海拔 3,900 米。

31. ***Glypholecia scabra*** **(Pers.) Th. Fr. 糙聚盘衣**

珠峰：中绒布冰川附近第三古侧碛岩石上，1169 号，海拔 5,470 米；中绒布冰川第五古侧碛岩石上，1194 号，海拔 5,650 米。

32. ***Lecanora polytropa*** **(Ehrh.) Rabh. 石茶渍**

f. *polytropa*

珠峰：绒布寺附近大岩石上，1321 号，海拔 5,000 米；绒布河谷旁山上岩石表面，1418 号，海拔 5,700 米。

f. *illusoria* **(Ach.) Leight.**

珠峰：绒布寺附近阶地岩石上，1298 号，海拔 4,905 米及 1421 号，海拔 5,900 米；中绒布冰川附近第五古侧碛北坡岩石上，1148 号及 1201 号，海拔 5,650 米。

33. ***Squamaria rubina*** **(Vill.) Elenk. 红盘鳞茶渍**

珠峰：绒布寺绒布河沿岸岩石上，1308 号，海拔 5,000 米；绒布寺北岩石上，1374号，海拔4,820米。

34. ***Menegazzia pertusa*** **(Schrank.) Stein. 孔梅衣**

曲乡：林中枯枝上，1053 号，海拔 3,560 米；朽树桩上，1075—① 号，海拔 3,460 米；岩石上，876 号，海拔 3,450 米及 1079 号，海拔 3,500 米；德青塘附近——花楸树杆上，1124 号，海拔3，530 米、桦树杆上，920 号，海拔 3,550 米，冷杉树杆上，1097 号，3,450 米。

35. ***Parmelia obscurata*** **Bitt. 暗粉袋梅衣**

聂拉木：岩石藓丛上，1818 号，海拔 3,870 米及 1849 号，海拔3，880 米；桦林岩石上，1863—②号，海拔 3,890 米及 1882—① 号，海拔 3,870 米。

36. ***Parmelia enteromorpha*** **Ach. 指袋梅衣**

曲乡：德青塘附近桦树杆上，1110 号，海拔 3,660 米及杜鹃树杆上，1117 号，海拔 3,660 米。

37. ***Parmelia hypotrypella*** **Asahina 黄粉袋梅衣**

樟木：林中杜鹃树上，488 号，海拔 3,300 米；桦树枝上，668—③ 号，海拔 2,650 米。

曲乡：林中枯枝上，105—① 号，海拔 3,560 米；岩石表面藓层上，783号，海拔 3,800 米及 1016号，海拔 3,700 米；阳坡冷杉树上，860 号，海拔 3,680 米；杜鹃花科植物上，1104 号，海拔3，620 米；德青塘附近岩石藓层上，889 号，海拔 3,500 米及桦树杆上，1111 号，海拔 3,660 米。

38. ***Parmelia physodes*** **(Lin.) Ach. 袋梅衣**

聂拉木：桦林内岩石上，1863 号，海拔 3,890 米。

39. ***Parmelia vittata*** **(Ach.) Nyl. 条袋梅衣**

樟木：林中杜鹃树上，489 号，海拔 3,300 米、杉树杆上，491 号，海拔 3,300 米及 634—① 号，海拔 3,350 米。

曲乡：杜鹃花科植物上，1103 号，海拔 3,620 米；朽树桩上，1074 号及 1075 号，海拔 3,460 米；德青塘附近——杜鹃树杆上，1117—① 号，海拔 3,660 米、冷杉树杆上，1098 号，海拔 3,450 米、桦树杆上，1116 号，海拔 3,700 米、枯树杆上，1106 号，海拔 3,100 米、岩石表面藓层上，1095 号，海拔 3,680 米及1119 号，海拔 3,710 米。

40. ***Parmelia cirrhata*** **Fr. 条梅衣**

f. *cirrhata*

樟木：高山栎树枝上，374 号，海拔 2,600 米、385—① 号，海拔 2,600 米及 381—①号，海拔 3,500 米；杜鹃树上，439 号，海拔 2,700 米；康巴吉巴附近荚蒾树上，118 号，海拔 2,740 米。

曲乡：阳坡杉树杆上，861 号，海拔 3,680 米；小灌木上，762 号，海拔 3,740 米；林场杉树杆上，218 号，海拔 2,500 米。

f. *vermicularis* **(Vain.) Asahina**

樟木：灌丛枝上，461 号，海拔 3,620 米，灌丛枝上，675 号，海拔 2,450 米；桦树枝上，688 号，海拔 2,650 米，林中水旁枯枝上，306 号，海拔3,360米、杜鹃树上，527 号，海拔3,250 米及 264 号，海拔 2,620 米；林中杉树枝上，492 号，海拔 3,300 米；林中枯枝上，400 号，海拔 2,400 米、樟科植物上，473 号，海拔 3,620 米、荚蒾灌丛枝上，483 号，海拔 3,600 米及 571 号，海拔 2,850 米、冬青树上，570 号，海拔 2,820 米；林中空地岩石藓层上，531 号，海拔 2,900 米；雪布岗附近——林中枯枝上，133 号及 145 号，海拔 3,210 米、灌丛枝上，167 号，海拔 3,000 米、蔷薇科植物上，149 号，海拔 3,000 米；康巴吉巴附近——林中枯枝上，181 号，海拔2，620 米。

曲乡：岩石藓层上，838号，海拔 3,200 米及 1077号，海拔3,350米；杜鹃枝上，806号，海拔3,500 米。

41. ***Parmelia sinuosa*** **(Sm.) Ach. 黄条梅衣**

曲乡：小灌丛带灌木上，762—② 号，海拔 3,740 米；德青塘附近岩石上，1129 号，海拔 3,360 米及枯枝上，1109 号，海拔 3,500 米。

聂拉木：河谷岸旁岩石上，1082 号，海拔 3,600 米。

42. *Parmelia conspersa* (Ehrh.) Ach. 菊叶梅衣

var. *latior* Schaer.

珠峰：绒布寺附近岩石上，1358 号，海拔 5,000 米。

var. *isidiata* (Anzi) Stzbg.

聂拉木：波曲河谷沿岸大岩石上，1966.6. 采(无号)，海拔 3,700 米。

珠峰：中绒布冰川附近第三古侧碛西坡石块与地面接触处，原植物体一半长在石上，另一半长在地表，1219 号，海拔 5,500 米及 1157 号，海拔 5,490 米。

var. *isidiosula* Hillm.

珠峰：绒布寺附近大岩石上，1349 号，海拔 5,000 米。

var. *hypoclysta* Nyl.

聂拉木：波曲河谷沿岸大岩石上，1966.6. 采(无号)，海拔 3,700 米。

var. *imbricata* Mass.

f. *microphylla* Hillm.

珠峰：绒布寺附近岩隙土壤上，1328 号，海拔 5,000 米。

43. *Parmelia vagans* Nyl. 旱梅衣

聂拉木：波曲河谷沿岸岩石上，1539 号，海拔 3,700 米。

44. *Parmelia aurulenta* Tuck. 黄髓梅衣

樟木：康巴吉巴附近林中树皮上，190—① 号，海拔 2,670 米。

45. *Parmelia xanthocarpa* Hue 黄果梅衣

樟木：林中岩石上，566 号，海拔 2,700 米、树杆上，739 号，海拔 2,720 米、花楸树上，744—① 号，海拔 2,720 米。

46. *Parmelia pseudoborreri* Asahina 粉斑梅衣

曲乡：枯枝上，814—① 号，海拔 3,500 米。

47. *Parmelia saxatilis* Ach. 石梅衣

聂拉木：桦林中岩石上，1893 号，海拔 3,870 米及 1863—① 号，海拔 3,890 米。

48. *Parmelia sulcata* Tayl. 沟槽梅衣

f. *rubescens* Bouly De Lesd.

聂拉木：桦林中岩石上，1862 号，海拔 3,845 米及 1865 号，海拔 3,850 米。

f. *ulophylla* Bouly De Lesd.

聂拉木：大岩石藓土层上，1966.6. 采（无号），海拔 3,700 米、桦林内岩石上，1863 号，海拔 3,890 米、桦树皮上，1966.6.22. 采(无号)，海拔 3,840 米。

49. *Parmelia caperata* (Lin.) Ach. 皱梅衣

樟木：岩石表面，691 号，海拔 2,530 米。

曲乡：波曲河谷大岩石上，870 号，海拔 3,220 米；德青塘附近岩石上，1128 号，海拔 3,250 米。

50. *Parmelia ulophyllodes* (Vain.) Savics 卷叶梅衣

聂拉木：大岩石垂面，1631 号，海拔 3,800 米、桦树皮上，1885 号，海拔 3,900 米。

51. *Parmelia tinctorum* Despr. 梅衣

樟木：达来玛桥附近河边岩石上，225 号，海拔 1,680 米及 344 号，海拔 1,700 米、树杆上，609 号及 619 号，海拔 1,720 米。

52. *Parmelia cetrarioides* Del. 粉缘梅衣

樟木：波曲河谷沿岸西南坡向林中杜鹃枝上，367 号，海拔 2,650 米、437 号及 440—① 号，海拔 2,700 米；雪布岗附近枯枝上，131 号，海拔 3,210 米。

曲乡：岩石藓层上，985 号，海拔 3,910 米、枯枝上，804 号，海拔 3,480 米及 814 号，海拔 3,500 米。

聂拉木：桦林中岩石上，1838 号，海拔 3,850 米。

53. *Parmelia olivetorum* Nyl. 漂红梅衣

樟木：茜草科灌木枝上，477 号，海拔 2,650 米。

54. *Parmelia pseudolivetorum* Asahina 珊瑚缘梅衣

樟木：林中岩石上，566—① 号，海拔 2,700 米；枯枝上，383 号，海拔 2,900 米、高山栎树枝上，382—② 号，海拔 2,500 米、林中灌木枝上，426 号，海拔 2,700 米；杜鹃枝上，368 号，海拔 2,650 米、436 号，海拔 2,700 米和 724 号，海拔 2,610 米、冬青枝上，663 号，海拔 2,830 米；二号桥附近山坡枯树杆上，357 号，海拔 2,290 米；康巴吉巴附近铁杉树皮上，176 号，海拔 2,610 米。

曲乡：枯枝上，814—② 号，海拔 3,500 米。

55. *Parmelia arnoldii* Du Rietz. 广布梅衣

曲乡：德青塘附近阴坡桦林中桦树杆上，893 号，海拔 3,570 米。

56. *Parmelia nimandairana* A. Zahlbr. 小珊芽梅衣

樟木：达来玛桥附近岩石上，344—① 号，海拔 1,680 米、岩石上，733 号，海拔 2,700 米、枯树杆上，730 号、735—① 号及 736 号，海拔 2,680 米以及 568—① 号，海拔 2,550 米、高山栎上，711 号，海拔 2,580 米及 382—① 号，海拔 2,500 米、杜鹃枝上，442 号，海拔 2,700 米及 717 号，海拔 2,650 米、乌醉木上，649 号，海拔 2,600 米；康巴吉巴附近片麻岩面藓层上，102 号，海拔 2,480 米及高山栎上，190 号，海拔 2,670 米以及荚蒾上，391—① 号，海拔 2,580 米。

57. *Parmelia neglecta* Asahina 类粉缘梅衣

樟木：枯树杆上，728 号，海拔 2,680 米、杜鹃枝上，153 号，海拔 3,200 米。

58. *Parmelia simodensis* Asahina 东方梅衣

曲乡：波曲河谷岩石上，822 号，海拔 3,230 米及 835 号，海拔 3,200 米；德青塘附近桦树皮上，893—①号，海拔 3,570 米。

59. *Cetraria caperata* (Lin.) Vain. 黄花岛衣

聂拉木：北坡桦林内桦树杆上，1785—① 号、1787 号，海拔 3,930 米及 1860 号，海拔 3,900 米、朽树桩上，1788—① 号及 1790 号，海拔 3,990 米。

60. *Cetraria laureri* (Krplh.) Koerb. 麦黄岛衣

樟木：林中杜鹃枝上，490 号，海拔 3,300 米及 526 号，海拔 3,250 米、枯枝上，934 号，海拔 3,500 米。

聂拉木：桦林中岩石藓层上，1863—① 号，海拔 3,890 米、桦树杆上，1898 号，海拔 3,900 米。

61. *Cetraria pallescens* Schaer. 皮革岛衣

樟木：西南坡林中杜鹃枝上，369 号，海拔 2,650 米。

62. *Cetraria ambigua* Bab. 黄条岛衣

聂拉木：波曲河谷沿岸古侧碛阶地草丛中，1439—① 号及 1498 号，海拔 3,950 米；古侧碛南坡土壤上，1521 号，海拔 4,060 米。

63. *Cetraria cucullata* (Bell.) Ach. 黄卷岛衣

聂拉木：波曲河谷沿岸古侧碛北坡土壤上，1491 号，海拔 4,290 米。

64. *Cetraria nigricans* (Retz.) Nyl. 黑岛衣

曲乡：岩石细土层上，1022—② 号，海拔 3,730 米。

65. *Cetraria crispa* Nyl. 栗卷岛衣

var. *japonica* Asahina ex Sato

曲乡：岩石细土层上，1022—⑤ 号，海拔 3,730 米。

66. *Evernia thamnodes* (Fw.) Arn. 扁枝衣

曲乡：小杜鹃丛上，763 号，海拔 3,720 米及 755 号，海拔 3,760 米、小灌丛基部及岩石上，753 号，海拔 3,770 米、片麻岩上，774 号，海拔 3,800 米、小灌木上，761 号，海拔 3,734 米。

聂拉木：波曲河沿岸古侧碛山上岩隙处，1548 号，海拔 4,200 米及桦林内岩石上，1884 号，海拔 3,870 米、小杜鹃枝上，1843 号，海拔 3,910 米及灌木上，1435 号，海拔 3,840 米、小杜鹃上，1784 号，海拔 4,000 米、岩石藓层上，1859 号，海拔 3,890 米。

67. *Alectoria acanthodes* Hue 刺树发

樟木：岩石上，469 号，海拔 3,660 米、蔷薇科灌木上，463 号，海拔 3,620 米、杜鹃枝上，541 号，海拔 3,650 米。

68. *Alectoria smithii* Du Rietz. 珊粉树发

樟木：蔷薇科灌木上，463—① 号，海拔 3,620 米、杜鹃枝上，541—① 号，海拔 3,650 米、林中枯枝上，306—① 号，海拔 3,360 米。

69. *Ramalina sinensis* Jatta 中国树花

樟木：林中枯枝上，394 号，海拔 2,520 米、林中空地岩石上，579—① 号。

70. *Usnea longissima* Ach. 长松萝

樟木：杜鹃枝上，528 号，海拔 3,250 米及 655 号，海拔 3,560 米；桦树枝上，512 号，海拔 3,620 米及 464 号，海拔 3,580 米；蔷薇科植物上，522 号，海拔 3,600 米；灌木上，456 号，海拔 3,350 米；冷杉树上，633 号，海拔 3,100 米。

曲乡：桦树上，802 号，海拔 3,470 米。

71. *Usnea pectinata* Tayl. 拟长松萝

樟木：蔷薇科植物上，551—① 号，海拔 3,400 米。

72. *Usnea thomsonii* Stirt. 灌松萝

樟木：桦树上，637 号，海拔 3,660 米。

曲乡：小灌木上，766 号，海拔 3,610 米。

73. *Usnea arborea* Stirt. 亚灌松萝

樟木：枯枝上，392 号、393 号，海拔 2,520 米及 402 号，海拔 2,400 米。

74. *Thamnolia vermicularis* (Sw.) Schaer. emend. Asahina 地茶

曲乡：岩石细土层上，1022—① 号，海拔 3,730 米及 1966.5.20. 所采(无号)，海拔 3,760 米。

聂拉木：桦林岩石土层上，1864 号，海拔 3,860 米。

75. *Thamnolia subvermicularis* Asahina 雪地茶

f. *subvermicularis*

聂拉木：波曲河沿岸古侧碛阶地草丛中，1439 号，海拔 3,950 米。

f. *minor* (Lamy) Mot.

珠峰：中绒布冰川附近第四古侧碛北坡岩石背风面下侧土壤上，1149 号及 1210 号，海拔 5,510 米，同上，第二古侧碛平台岩隙土壤上，1240 号，海拔 5,580 米；西绒布冰川附近第二古侧碛阶地石块背风处土壤上，1262 号，海拔 5,580 米。

76. *Caloplaca elegans* (Link.) Th. Fr. 红橙衣

珠峰：中绒布冰川附近古侧碛岩石上，1966.5.28. 采(无号)，海拔 5,500 米。

77. *Anptychia leucomelaena* (Lin.) Mass. 白腹雪花衣

曲乡：冷杉上，867 号，海拔 3,650 米、桦树上，1035 号，海拔 3,630 米及 1039 号，海拔 3,650 米。

78. ***Anaptychia hypoleuca*** **(Mühl.) Mass.**

曲乡：波曲河谷岩石上，823—① 号，海拔 3,200 米。

79. ***Physcia muscigena*** **(Ach.) Nyl. 伴藓蜈蚣衣**

珠峰：绒布寺附近山上岩隙土层上，1396 号，海拔 5,570 米、1344 号，海拔 5,000 米、1301 号，海拔 4,905 米、1425 号，海拔 5,630 米；中绒布冰川附近第三古侧碛西坡石块下背风处藓土层上，1173 号，海拔 5,460 米、1181 号，海拔 5,490 米及第四古侧碛北坡岩隙藓土层上，1225 号，海拔 5,510 米；西绒布第二古侧碛阶地石块下背风处藓土层上，1252 号及 1254 号，海拔 5,580 米。

参考文献

Asahina, Y. 1952—53. "Lichens" in Kihara; Fauna and Flora of Nepal Himalaya, scientific results of Japanese Expedition to Nepal Himalaya. 1:44—63.

Awasthi, D. D. 1960. On a collection of macrolichens by the Indian expedition to Cho-Oyu, East Nepal in Proceeding of the Indian Academy of Sciences. **51B**:169—180.

————, 1961. Some foliose and fructicose lichens from Assam and North-East Frontier Agency(NEFA) of India. *Ibid.* **54B**:24—44 (1961).

————, 1965. Catalogue of the Lichens From India, Nepal, Pakistan and Ceylon.

Bhatia, K. K. 1957. Some observations on the lichen communities of the Western Himalaya. in Bull. *Bot. Soc. Univ. Saugor.* **9**(1, 2):36—39.

Chopra, G. L. 1934. Lichens of the Himalayas, pt. I. Lichens of Darjeeling and Sikkim Himalayas.

Paulson, R. 1925. Lichens of Mount Everest. in Jourm. *Bot.* **63**:189—193.

A biogeographical analysis of the lichen flora of Mt. Qomolangma region in Xizang

Wei Jiang-chun*　　Jiang Yu-mei
(Institute of Microbiology, Academia Sinica)

The lichens of Mt. Qomolangma were first reported by R. Paulson in a paper entitled "Lichens of Mount Everest" (1925) which included 31 species. Two of them were described as new species. But the author discussed little on the elements of the lichen flora of this region. Much later. Awasthi (1965) discussed and touched only on the geographical elements of the lichen flora in the Himalayas. Recently, Poelt (1976) also touched slightly on the geographical elements of the lichen genus *Umbilicaria* from the Himalayas and East Asia.

The present paper deals with the relationships between the upheaveal of the Himalayas and the origin and development of lichen flora of Mt. Qomolangma region.

Regions Investiged

Mt. Qomolangma is situated in the south rim of Qinghai-Xizang Plateau. It is one of the lofty Himalan peaks and the highest one in the world. Mt. Qomolangma region extends from 27°29' N. 1atitude and from 85-89°E. longitude.

The string of the lofty Himalayan peaks blocks the moist air currents which blow up from Indian Ocean. As a result, the southern slopes of this region have abundant precipitation and a lush covering Vegetation while the northern slopes, in sharp contrast, have an extremely dry climate and with sparse vegetation. As the mountains increase in height. the natural landscape keeps changing in a series of belts of different vegetations. Naturally, lots of lichens are present in both the forest and shrub belts. The constitution of lichen species also changes with increasing in height and with the changing in different forest belts.

Materials and Methods

The biogeographical analysis of the lichen flora of the region studied is based on the lichen specimens collected from the Mt. Qomolangma region at 1,600-6,100 m above sea level in 1966 by the senior author and Chen J.B. as member of the Comprehensive Scientific Expedition of Xizang, Academia Sinica, such as Lang K.Y., Li W.H., Zang M., Zhang J.W., Zhao K.Y., Zheng D., Zong Y. C. etc. The authors with to express their sincere appreciation for the specimens they proyided.

The analysis is on the basis of 113 species belonging to 36 genera collected from Mt. Qomolangma region. These species are classified according to the bioclimatic and chorological principle (A. S. Lazarenko. 1956: M. F. Makarevicz. 1963 and A. N. Oxner. 1974) into 9 geographical elements. (A. S. Lazarenko. 1956: M. F. Makarevicz. 1963 and A. N. Oxner. 1974) into 9 geographical elements.

---

*Wei Jiang-chun = Vej Tzjan-czunj
This paper was originally published in *Environment and ecology of Qinghai-Xizang plateau*, Proceedings of Symposium on Qinghai-Xizang (Tibet) Plateau (Beijing, China) Vol. 2: 1145-1151, 1981.

Results

Analysis of the Biogeographical Elements

1. Arctalpine element. The following 5 species found in Mt. Qomolangma region belong to this element: *Peltigera scabrosa* Th.Fr., *Lecidea auriculata* Th. Fr.,*Cetraria cucullata* (Bell.) Ach., *C. hepatizong* (Ach.) Vain., *Sporastatia testudinea* (Ach.) Massal.

2. Montane subarctic element. Species belonging to the element are *Lichenomphalia hudsoniana* (H.S. Jenn.) Redhead, Lutzoni, Moncalvo & Vilgalys (*Coriscium viride*, has been used for the anamorph (lichenized squamules) of the lichenized species of *Omphalina* and is now listed nom. utique rej. (Redhead & Kuyper: Arctic alpine Mycology 2:319, 1987), *Peltigera aphthosa* (L.) Willd., *Cladonia amaurocrea* (Flk.) Schaer., *Cl. cyanipes* (Sommerf.) Nyl., *Umbilicaria hirsuta* (Sw.) Ach., and *Physcia muscigena* (Ach.) Nyl.

3. Boreal element. This element can be divided into 3 areal types: Amphipacific areal which contains *Cetraria laevigata* Räsänen, and *Oropogon formosanus* Asahina; Eurasian-Northern-American areal type which contains *Parmelia soredica* Nyl., *Hypogymnia bitteri* (Lynge.) Ahti, *Cornicularia odontella* (Ach.) Röhl., and *Ramalina sinensis* Jatta; and Panboreal areal type which contains *Cladonia glauca* Flk., *Cl. scabriuscula*(Del. ex Duby)Nyl., *Cetraria pinastri* (Scop.) Röhl., *Evernia divaricata* (L.) Ach., and *Usnea longissima* Ach.

4. Montane element. This element can also be divided into areal types as follows: Eurasian areal type which contains only one species (*Cetraria laureri* Krempehl.), Asian-African areal type (*Lasallia pertusa* (Rassad.) Llano), and Eastern Asian areal type which contains 36 species. The latter can be also divided into Pan-eastern subtype and Sino-Himalayan areal subtype.

Among the 36 species Eastern Asian areal type, the following is species are endemic to Sino-Himalayan region: *Umbilicaria yunnana* (Nyl.) Hue, *U. nanella* Frey et Poelt, *U. hypococcinea* (Jatta) Llano, *U. thamnodes* Hue, *U. trabeculata* Frey et Poelt, *Usnea thomsonii* Stirt., *U. dentritica* Stirt., *U. arborea* Stirt., *Hypogymnia laccata* J. C. Wei et Y. M. Jiang, *H. sinica* J. C. Wei et Y. M. Jiang, *H. pruinosa* J. C. wei et Y. M. Jiang, *Parmelia subverruculifera* J. C. Wei et Y. M. Jiang, *Cetraria ambigua* Bab., *C. xizangensis* J. C. Wei et Y. M. Jiang, *Lethariella cashimeriana* Krog, *L. cladonioides* (Nyl.) Krog emend J. C. Wei, *L. flexuosa* (Nyl.) J. C. Wei et Y. M. Jiang, and *Sulcaria sulcata* f. *vulpinoides* (A.Z.) Hawksw. They cover 16% of the total elements of the lichen flora in this region. However, all species of the Sino-Himalayan areal subtype are distributed not only in Himalayas but also in northwestern Yunnan, western Sichuan, and southern Gansu or Mt. Qinling of Shanxi.

The following 18 species belong to Pan-eastern Asian areal subtype:*Lobaria kurokawae* Yoshim., *L. isidiophora* Yoshim., *L. meridionalis* Vain., *L. orientalis* (Asahina) Yoshim., *Cladonia yunnana* (Vain.) Abb., *Lasallia asiae-orientalis* Asahina, *Hypogymnia hypotrypella* (Asahina) Rassad., *Parmelia crassata* (Hale) J. C. Wei et Y. M. Jiang, *P. entotheiochroa* Hue, *P.irrugana* Nyl., *Cetraria asahinae* Sato, *C. pallescens* Schaer. ex Moritzi, *C. wallichiana* Müll. Arg., *C. everniella* (Nyl.) Krempelh., *Nephromopsis strachcheyi* f. *ectocarpisma* Hue, *Cetrelia pseudolivetorum* (Asahina) C.Culb. et W. Culb., *C. braunsiana* (Müll. Arg.) C. Culb. et W.Culb., and *Platismatia erosa* W.Culb. et C.Culb.

5. Noto-boreal element. The following species belong to this element: *Peltigera erumpens* (Tayl.) Vain., *Stereocaulon tomentosum* Fr., *Lecanora polytropa* (Ehrh.) Rabh., *Hypogymina vittata* (Ach.) Gat. and *Usnea comosa* (Ach.) Röhl.

6. Nemoral elment. There are only 2 species which belong to this element, i.e. *Parmelia glabra* (Schaer.) Nyl., and *Cetrelia cetrarioides* (Del. Et Duby) C. Culb. et W. Culb.

7. Arid element. The species *Diploschistes bryophilus* (Ehrh.) A. Z., *Psora decipiens* Hoffm., *Parmelia camtschadalis* (Ach.) Eschw. in Martius. belong to this element.

8. Pantropical element. The following species belong to this element: *Sticta gracilis* A. Z., *Cladia aggregata* (Sw.) Nyl., *Parmelia nilgherrensis* Nyl., *P. tinctorum* Despr. apud Nyl., *P. wallichiana* Tayl., *P. santiangelii* Lynge, *P. perisidians* Nyl., *P. subaurulenta* Nyl., *Everniastrum cirrhatum* (Fr.) Hale, *E. nepalense* (Tayl.) Hale, *E. sorocheilum* (Vain.) Hale, and *E. rhizodendroideum* J. C. Wei et Y. M. Jiang, *Anaptychia comosa* (Eschw.) Massal., and *A. fragillissima* Kurok.

9. Multirange's element. This element can also be classified into 6 areal types as follows: (1) Arctalpine-holarctic areal type: *Cladonia elongata* (Jaqcg.) Hoffm., *Umbilicaria proboscidea* (L.) Schrad., *Thamnolia vermicularis* (Sw.) Ach. ex Schaer, and *Th. subuliformis* (Ehrh.) W. Culb.; (2) Subarctic-montane holarctic areal type: *Icmadophila ericetorum* (L.) A. Z. and *Parmelia saxatilis* (L.) Ach.; (3) Boreal areal type: *Peltigera spuria* (Ach.) DC., *Cladina rangiferina* (L.) Harm., *Cladonia bacillaris* (Ach.) Nyl., *Cl. pleurota* (Flk.) Schaer., *Cl. verticillata* (Hoffm.) Schaer., *Cl. ochrochlora* Flk., *Cl. pityrea* (Flk.) Fr. and *Parmelia caperata* (L.) Ach.; (4) Nemoral areal type: *Normandina pulchella* (Borr.) Nyl., *Candelaria concolor* (Dicks.) Stein. and (5) Montane-holartic areal type: *Parmelia sinuosa*(Sm.)Ach., and *Menegazzia pertusa*(Schrank.)Stein.;(6)Holarctic areal type: *Peltigera canina* (L.) Willd., *P. rufescens* (Weis.) Humb., *Rhizocarpon tinei* (Tornadb.) Run., *Rh. superficiale* (Schaer.) Vain., *Cladonia fulcata* (Huds.) Schrad., *Cl. pyxidata* (L.) Hoffm., *Cl. chlorophaea* (Flk.) Spreng., *Parmelia sulcata* Tayl., *Xanthoria elegans* (Link.) Th. Fr. and *X. fallax* (Hepp) Arn.

Disjunctive Distributions

The subgenus *Chlorea* of the lichen genus *Lethariella* consists of 5 species, namely *Lethariella zahlbruckneri*, *L. cashimeriana*, *L. flexuosa*, *L. cladonioides* and *L. canariensis*. One of them, i.e. *L. canariensis* is distributed only in Canaries while the other 4 species mentioned above are distributed in Sino-Himalayan region (Fig.1).

In addition, *Thamnolia subuliformis* f. *minor* which is distributed in the Mt. Dore of France is also found in the Mt. Qomolangma region (Fig.2). Another disjunction, i. e. the pantropical species *Parmelia wallichiana* is distributed not only in the southern rim of the continent of Asia including the region studied, but also in islands of Asia and even southern Africa (Fig. 2). The Other species of Montane element *Lasallia pertusa* is distributed in Siberia, Himalayas and Africa (Krog 1973).

Conclusions

To sum up,we shall come to the following conclusions:

l. The position of the Mt. Qomolangma region is corresponding to the northern rim of the subtropics in geographical latitude. However,most of the elemerlts of the flora are not basically constituted by pantroptcal elements which cover 11%, but rather by alpine, arctic, holarctic and noto-boreai elements which cover 89% of the total elements of the lichen flora in this region.

In addition, neither' endemic genus nor endemic subgenus is found either in the Mt. Qomolangma region or in all over the Himalayas.

The above facts can be explained by the hypothesis that the climate of the Mt. Qomolangma region became cooler and cooler after the Tertiary.

As a result, the influence of the factors of latitutdinal zonality was replaced by that of a cool climate which was caused by upheaving of the Hinlalayas. Therefore, most of the species here can be explanined by the southward migration of the northern species.

2. Many taxonomically well-defined lichen species,have extremely broad range. This is often referred to as evidence that evolution in the lichens is very slow (Culberson, 1972). Therefore, a lot of above mentioned species have their main distributions in the Himalayas but some of them are also known in Northwestern Yunnan, western Sichuan,southern Gansu or Mt. Qinling of Shaanxi and is named "Sino-Himalayan areal subtype". These species cover 16% of the total elements of the lichen flora in this region.

3. In the light of the above analysis of the Himalayan-Canarian distribution and Himalayan-Mt. Dore disjunction, it can be seen that the subgenus *Chlorea* of *Lethariella* and *Thamnolia subuliformis* f. *minor* historically had a continual distribution. The present disjunction seems to be resulted from the extinctions of some of the species caused by the vicissitude of the ancient climate in Palaeomed terranean region during the late Tertiary, particularly after the upheaval of the Himalayas.

On the other hand, the southern rim of Asiatic continent-Islands of Asia and South Africa disjunction seems to be resulted from contimental drift.

Acknowledgments

The authors wish to express their sincere appreciation of the kind help given by Dr. T. Ahti and Mr. O. Vitikainen (H). Dr. S. Kurokawa and Dr. H. Kashiwadani (TNS) and Dr. L. Tibell and Dr. Roland Moberg (UPS) for the loan of type specimens from their herbaria.

References

(1)Awasthi, D. D. Catalogue of the Lichens from India, Nepal. Pakistan and Ceylong. Beih. Nowa Hedwigia. 17 (1965).

(2) Culberson, W. L. Disjunctive distributions in the lichenforming fungi·Ann. Mo. Bot. Gard. 59 (2). 165 (1972).

(3) Hale, Jr. M. E. , A Monograph of the Lichen Genus *Parmelina* Hale (Parmeliaceae). Smithsonian Contributions to Botany. No. 33. Washinton (1976).

(4) Krog, H. On *Umbilicaria pertusa* Rassad. and some related lichen species· Bryologist. 76 (4). 550 (1973).

(5) Krog, H. *Lethariella and Protousnea*, two new lichen genera in Parmeliaceae. Norw J. Bot. 23:83 (1976).

(6) Lazarenko, A. S. The principle situation of areas of mosses in the Soviet Far East. Bot. J. of Ukratinian S. S. R. 13 (1) :31-38 (1956). (in Ukrainian).

(7) Makarevicz, H. F. Analiz likhenoflori Ukrayinskikh Karpat. Akademii Nauk Ukrainskoi RSR. Kiev. (Analysis of Lichenoflora of Ukrainian Karpathen l-263 (1963) (in Ukrainian).

(8) Oxner, A. N. Handbook of the Lichens of the U. S. S. R. 2. Morphology. Systematic and Geographical Distribution. Leningrad (1974)

(9) Paulson, R. Lichens of Mount Everest. Journal of Botany. 63: 189 (1925).

(10) Poelt, J. Die Gattung *Umbilicaria* Hofrm.Flechten des Hirnalaya Khumbu Himal. 6 (3) : 397 (1976).

(11) Roderick, W. R. Lichens of hot arid and semi-arid Lands. in M. R. D. Seaward: Lichen Ecology. London. New York. Francisco p.211 (1977).

(12) Wei, J. C., Chen, J. B. Materials for a lichen flora of the Mount Qomolangma region in southern Xizang, China. In Report on the Scientific Investigations (1966-1968) in Mt. Qomolangma district, Science Press. Beijing. 173-182 (1974) (in Chinese).

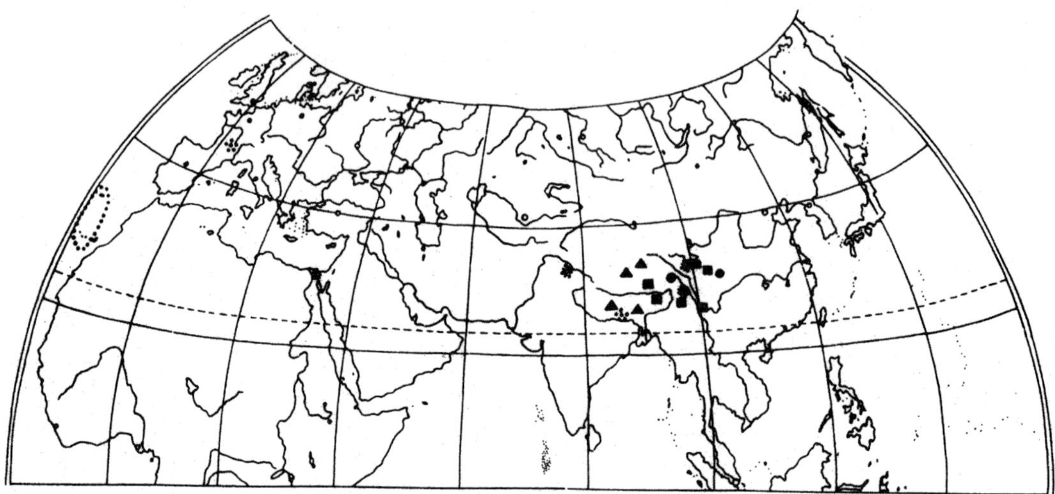

Fig. 1 Distribution map of species of *Lethariella* subg. *Chlorea*

- Lethariella canariensis (Ach.) Krog
- Lethariella cashmeriana Krog
- Lethariella cladonioides (Nyl.) Krog ement. Vej
- Lethariella flexuosa (Nyl.) Vej et Jiang
- Lethariella zahlbruckneri (DR.) Krog
- Thamnolia subuliformis f. minor (Lamy) Vej et Jiang

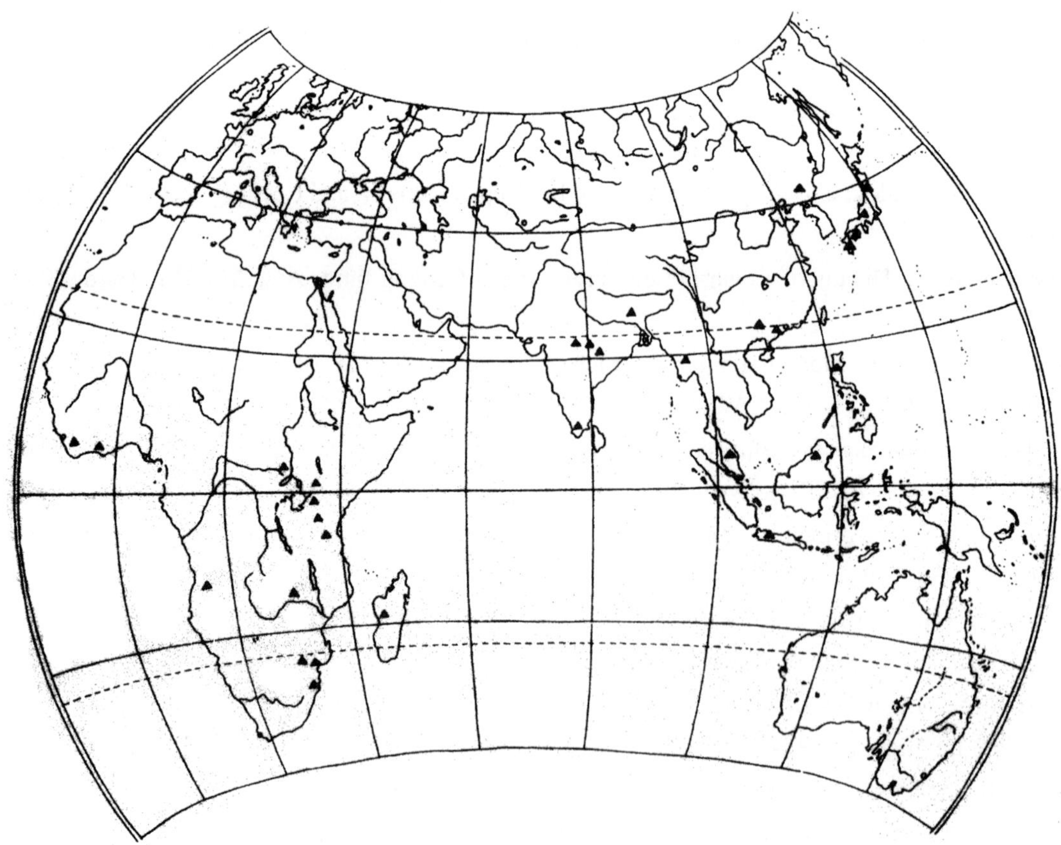

Fig. 2 Distribution map of *Parmelia wallichiana* Taylor

（Taken from M. E. Hale 1976 and supplemented）

Some foliicolous lichens in Xishuangbanna, China

J.C. WEI and Y.M. JIANG
Systematic Mycology and Lichenology Laboratory, Institute of Microbiology, Academia Sinica, Beijing, China

Abstract

Twenty species of foliicolous lichens from 12 genera in 10 families, collected in October 1980 from Xishuangbanna are reported in the present paper. Amongst them seven foliicolous genera, *Asterothyrium*, *Byssolecania*, *Dimerella*, *Fellhanera*, *Gyalectidium*, *Opegrapha*, and *Trichothelium*, and 15 species including one variety are new to China, and *Byssolecania deplanata* is new to Asia.

Introduction

Xishuangbanna, an autonomous prefecture of the Dai nationality in southern Yunnan of China, 21°10′–22°40′ N, 99°55′–101°50′ E, includes Jinghong, Mengla, and Menghai counties (Fig. 14.1, 14.2). The area of the whole prefecture is about 19 220 square kilometres with an elevation of 420–2400 m. Most of the land comprises terraces, hills, and mountains at an elevation of 540–1200 m. The area of forest is about 640 000 hectares which covers *c*. 34 per cent of the whole area.

The flora of Xishuangbanna belongs to Yunnan, Burma, and Thailand and is within the Malaysian subkingdom of the Palaeotropic Kingdom according to Wu's regionalization of the Chinese flora (Wu 1979). The flora of this area is identical to the tropical flora of Burmese Shan State, of northern Thailand, and of northern Laos. It is very rich in palaeotropical floral elements. About 5000 species of phanerogams and ferns in Xishuangbanna are equivalent to one-sixth of the plant species in the whole of China, although the area covers only one-five

This paper was originally published in *Tropical lichens: their systematics, conservation and ecology* （1990, D. J. Galloway ed.）. Systematics Association Special Volume No. 43. Oxford: Clsrendon Press: 201-216, 1991.

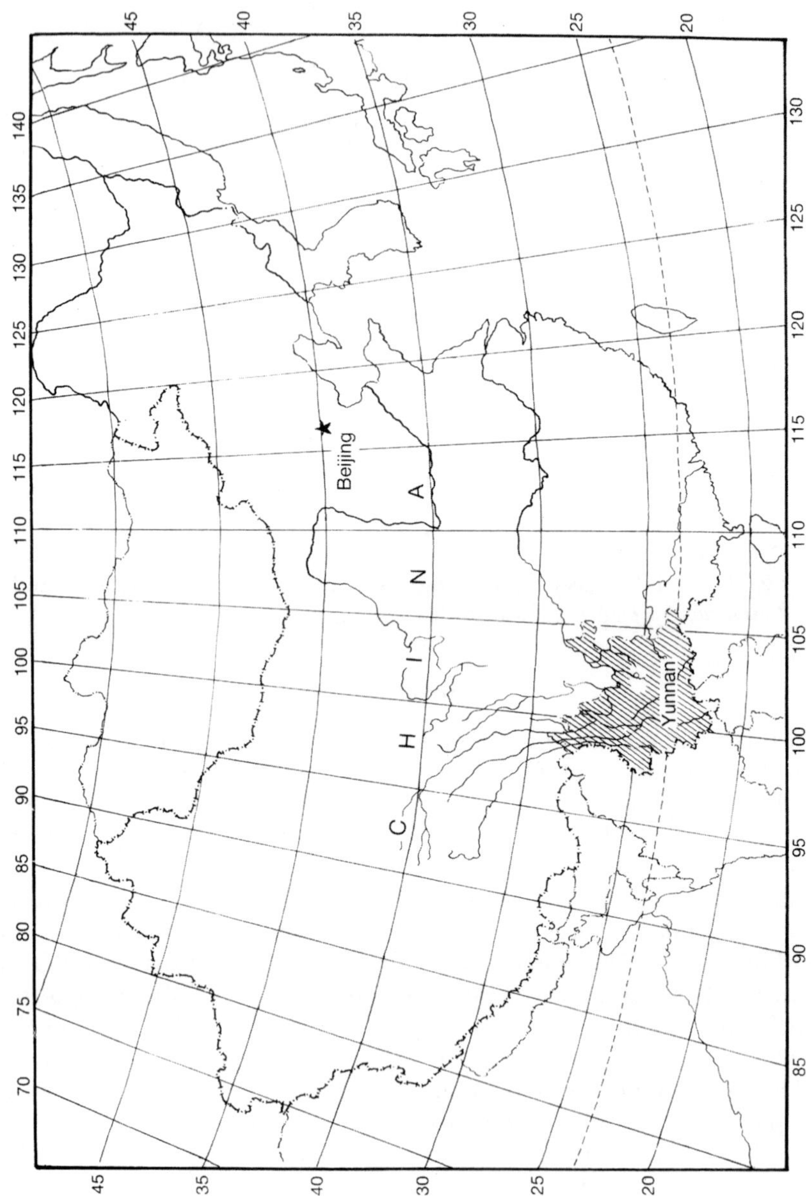

Fig. 14.1. Map showing the position of Yunnan province in China.

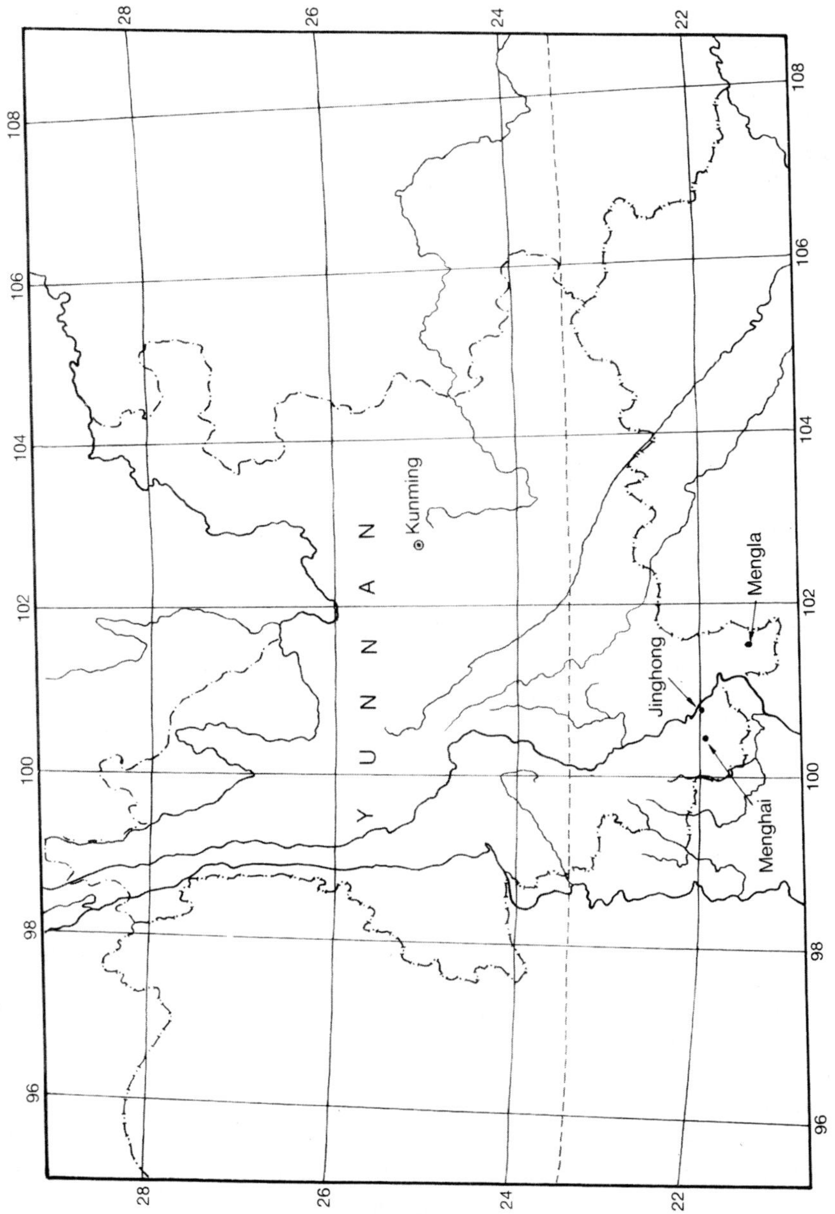

Fig. 14.2. Map showing the position of Xishuangbanna prefecture in Yunnan.

hundredth of the whole territory of China. Therefore, the foliicolous lichen flora must also be rich in this area.

Studies on the foliicolous lichens in this area are fragmentary. The following foliicolous genera have previously been reported from China: *Bacidia*, *Byssloma*, *Calopadia*, *Echinoplaca*, *Mazosia*, *Porina*, *Sporopodium*, and *Strigula*. In recent years, Hertel [1980], Eriksson and Wei [1980], Santesson *et al.* [1987] visited some of the reservations in Jinghong and collected a large number of lichens there. A series of papers concerning foliicolous lichens from China will soon be published and will bring together the current knowledge of foliicolous lichen flora in China (Wei 1990).

Materials and methods

This paper is based in part on materials collected by Wei, mainly from the reservation near Menglun in Jinghong in October 1980. The materials of the genus *Mazosia* examined were not only from Xishuangbanna, but also from tropical Guangdong, the Mt. Dinghu shan. Observations on anatomical features of these lichens were made by light microscopy (LM) using OPTON II, and those of morphological features through a dissecting microscope (DM) using OPTON, and sometimes with scanning electron microscopy (SEM) using an HITACHI instrument. The chemical data for the testing of lichen products were obtained by standardized thin-layer chromatographic methods (TLC) (White and James 1985).

Results and discussion

Results show that, of the 20 taxa studied in this paper, seven genera and 15 species including one variety are new to China and *Byssolecania deplanata* is new to Asia.

It is obvious that our knowledge of the Chinese foliicolous lichens is very poor, which makes it imperative that studies should proceed with some urgency, particularly when faced with the serious denudation of tropical forest today.

Only two species of the genus *Mazosia* were previously reported from China; *M. rotula* by Zahlbruckner (1930) and *M. melanophthalma* by Thrower (1988). However, the former identification seems less certain, as pointed out by Santesson (1952). After a careful re-examination of specimens collected by Handel-Mazzetti from Tonkin (no. 5 in HMAS-L and BM) in 1914, and from Yunnan (no. 1035 in BM) in 1915, and which were identified by Zahlbruckner as *M. rotula*, the following results were obtained; the radiate ridges typically found in

M. rotula are not seen on the whitened and felted surface of the thallus (Handel-Mazzetti, no. 5 in HMAS-L). The description given by Zahlbruckner (1930, p. 64) also indicates that the thallus is smooth ('thallus . . . laevigatus'). Another specimen from Tonkin (Handel-Mazzetti no. 5 in BM) bears not only *M. phyllosema*, but also *M. melanophthalma*, and the specimen from Yunnan (Handel-Mazzetti no. 1035 in BM) bears *M. melanophthalma* alone.

Numerous verrucose specimens of a species of *Mazosia* were collected by us from both Xishuangbanna in Yunnan and Dinghu shan in Guangdong. This lichen may be identical with *M. melanophthalma*. However, the verrucae are concolorous with the thallus. A few thalli with white or whitish verrucae were found in some collections during this study, but neither asci nor spores were seen. They seem to be dead thalli with whitened verrucae. In addition, the spores of all Chinese specimens of the species in *Mazosia* are longer then those collected from other countries (Santesson 1952). Chemical data for a few species were obtained but the compounds were not identified.

THE SPECIES

Asterothyriaceae

1. Asterothyrium pittieri Müll.Arg. YUNNAN, Xishuangbanna: near Menglun on the way to the reservation, on leaves of *Thunbergia grandiflora* (Roxb. ex Rottl.) Roxb. (Acanthaceae), 18 October, 1980, *Wei* 3588 (HMAS-L).

Thallus white, smooth, nitid, round, *c.* 1–4 mm diam., bearing usually one or sometimes several apothecia. *Apothecia* up to 1 mm diam., immersed in thallus at first, then erumpent, and later somewhat prominent, round; *disc* badious, nitid; margins formed mainly by the ragged, irregular lobes of the ruptured tissue originally covering the apothecia like the top of a pomegranate. *Paraphyses* simple, non-septate. *Asci* oval to ovoid, 2-spored, 90–99 × 27–36 μm. *Spores* hyaline, 1-septate, ellipsoid with round ends, 50–61 × 22.5–24 μm.

New to China. Pantropical.

Bacidiaceae

2. Bacidia apiahica (Müll.Arg.) A.Zahlbr. YUNNAN, Xishuangbanna: on the way to the reservation near Menglun, on leaves of *Thunbergia grandiflora* (Roxb. ex Rottl.) Roxb., 18 October, 1980, *Wei* 3588-(1).

Most similar to *Dimerella epiphylla* (Müll.Arg.) Malme but with 3–4-septate, filiform spores.

New to China. Pantropical.

3. *Bacidia sp.* YUNNAN, Xishuangbanna: in the reservation near Menglun, on leaves of *Capparis caudata* B. S. Sun (Capparidaceae), 18 October, 1980, *Wei* 3658.

Thallus crustose, sordid white, partially slightly rust-coloured. *Apothecia* up to 542 μm diam., *disc* red-brown to dark brown. *Asci* 8-spored 38–50 × 8–11 μm. *Spores* needle-shaped, hyaline, 5–7-septate, 18–30.5 × 2 μm.

The species may be related to or identical with *Bacidia ziamensis* Vězda, but the spores are shorter.

New to China.

4. *Bacidia palmularis (Müll.Arg.) A.Zahlbr.* YUNNAN, Xishuangbanna: in the reservation near Menglun, on leaves of *Knema furfuraceum* (Hook. f. et Thoms.) Warb. (Myristicaceae), 18 October, 1980, *Wei* 3691-(1).

Thallus sordid whitish-grey, felty. *Apothecia c.* 325 μm diam., *c.* 300 μm high, only slightly constricted at base; margins almost invisible, often with aerial root-like, white mycelium threads at base, *disc* orange yellowish to yellowish brown. *Asci* 8-spored, 81–99 × 16–18 μm. *Spores* hyaline, 15–25-septate, 61–74 × 2.5–3 μm.

It may possibly be a young stage of its development.

New to China. Pantropical.

Ectolechiaceae

5. *Sporopodium leprieurii Mont. var. leprieurii* YUNNAN, Xishuangbanna in the reservation nearly Menglun, on leaves of *Knema furfuracea*, 18 October, 1980, *Wei* 3697.

Thallus thin, felty, verrucose, sordid whitish, irregular, *c.* 10–20 × 6–10 mm; verrucae concolorous with thallus. *Hypothallus* faintly visible. *Apothecia* sublecanorine, without algae in the thalline margin, *c.* 1 mm diam. *Asci* single-spored, 75.5–90 × 20–21.5 μm. *Spores* muriform, 47–83 × 12.5–18 μm. Algae in epithecium *c.* 3.5–4.5 × 3–3.5 μm.

New to China. Pantropical.

Gomphillaceae

6. *Gyalectidium filicinum Müll.Arg.* YUNNAN, Xishuangbanna: in the forest of the Limestone Hill near the Tropical Botanical Garden of the Kunming Botanical Institute of Academia Sinica, on leaves of *Pseuduvaria indochinensis* Merr., 19 October, 1980, *Wei* 3715-(3).

Thallus crustose, wax-white, strongly nitid, densely verrucose, suborbicular, *c.* 4–5 × 3 mm. *Apothecia* round, dish-shaped, *c.* 380 μm diam.; margins concolorous with thallus, crenulate; *disc* wax-yellowish to wax-brownish, nitid. *Asci* single-spored. *Spores* hyaline, muriform, curved within asci, straight after discharge, 41–52 × 16–23 μm.

Chemistry: One unidentified substance R_F class 6 (bottom) in C, giving grey-purple colour after acid and heat treatment.

New to China. Pantropical.

7. Tricharia vainioi R.Sant. YUNNAN: Xishuangbanna, on leaves of *Polyalthia viridis* Craib (Anonaceae) in the Limestone Hill near the Tropical Botanical Garden of the Kunming Botanical Institute of Academia Sinica, 19 October, 1980, *Wei* 3713 and on leaves of a broadleaved tree *Wei* 3702, 3707-(1).

Thallus crustose, thin, membraneous, nitid, subcircular to irregular, *c.* 5 mm diam., or 12 × 4 mm, sordid whitish, furnished with *c.* 1 mm long, black and tapering hairs. *Apothecia c.* 280 μm diam., almost not constricted at base, wax-gold with brownish colour to dark brown, biatorine type; disc plane to faintly concave. *Asci* single-spored. *Spores* hyaline, muriform, *c.* 34–47 × 16–22 μm.

New to China.

Gyalectaceae

8. Dimerella epiphylla (Müll.Arg.) Malme YUNNAN, Xishuangbanna: in forest of Limestone Hill near the Tropical Botanical Garden of the Kunming Botanical Institute of Academia Sinica, on leaves of *Pseuduvaria indochinensis* Merr. (Anonaceae), 18 October, 1980, *Wei* 3715.

Thallus membraneous, grey green, *c.* 1.5 cm diam. *Apothecia* round, strongly constricted at base, *c.* 0.5–1.2 mm diam.; *disc* orange yellow to pale buff brown, margins yellowish whitish, usually growing along the veins and margin of leaves. *Asci* single-spored. *Spores* fusiform, 1-septate, 5–11 × 2 μm.

New to China. Pantropical.

Literature records: This lichen was previously reported by Zahlbruckner (1930, p. 70) from Yunnan. However, Santesson (1952) pointed out that this record is still unverified. So, it seems that *Dimerella* as a genus is new to China.

Opegraphaceae

9. Opegrapha filicina Mont. YUNNAN, Xishuangbanna: in the reservation near Menglun, on leaves of *Knema furfuracea* (Hook. f. et Thoms.) Warb. (Myristicaceae), 18 October, 1980, *Wei.*

Asci 8-spored, 34–40 × 14–16 μm. *Spores* 4- (rarely 5) septate, 12.5–30.5 × 4.5–6 μm.

Spores of this species are shorter than those of the specimens collected from south-east Asia, Africa, and South America, usually 4-septate, and rarely 5-septate; one end is larger than the other.

New to China. Pantropical.

Phragmopelthecaceae

10. Mazosia melanophthalma (Müll.Arg.) R.Sant. (Fig. 14.3, 14.4, 14.5) YUNNAN: Xishuangbanna, in the reservation near Menglun, on leaves of *Knema furfuracea*, 18 October, 1980, *Wei* 3565, 3572, 3593, 3686, 3705, and on leaves of *Celtis giganticarpa* Hu, in the forest of Limestone Hill near the Tropical Botanical Garden of the Kunming Botanical Institute of Academia Sinica, 19 October, 1980, *Wei* 3714. GUANGDONG: Dinghu shan, on leaves of *Tectaria sp.* in the forest near Qingyunsi, 9 November, 1980, *Wei* 3784.

Thallus membraneous, thin, yellow-green or grey-green to olive-green, slightly nitid, up to 5 mm diam., verrucose; verrucae concolorous with thallus. *Ascomata* scutate, *disc* plane, dark brown to the naked eye or through a DM, but hyaline and colourless through a LM. *Asci*

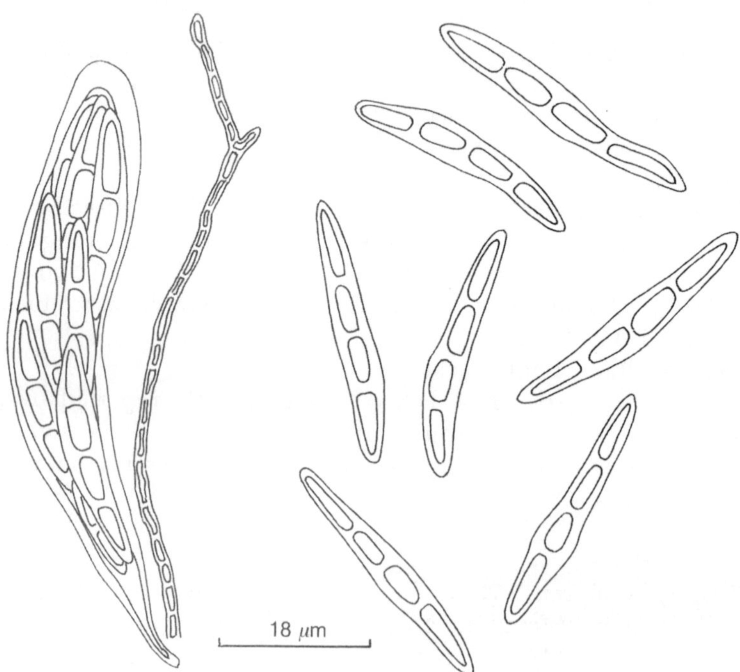

Fig. 14.3. *Mazosia melanophthalma* (*Wei* 3572), ascus, ascospores and a paraphysis.

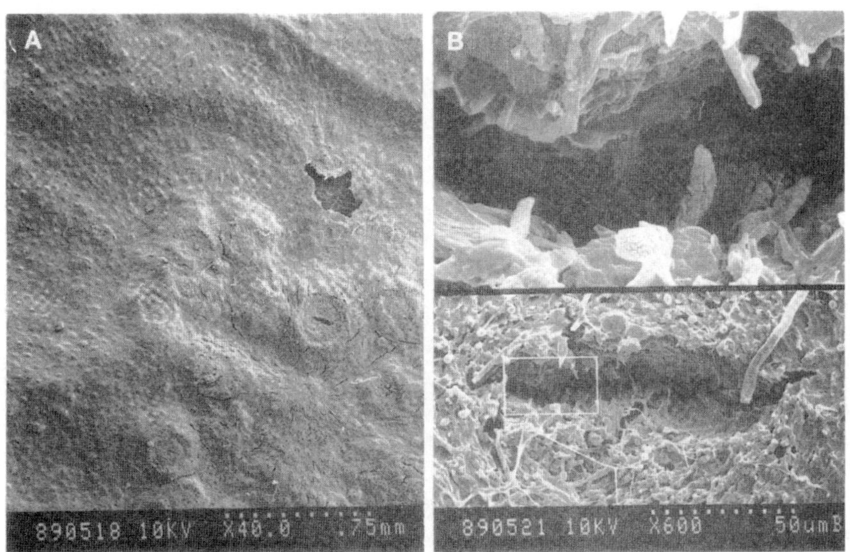

Fig. 14.4. (A) *Mazosia melanophthalma* (*Wei* 9074), verrucose thallus with ascomata. (B) *Mazosia melanophthalma* (*Wei* 9074), a fissure in one of the ascomata.

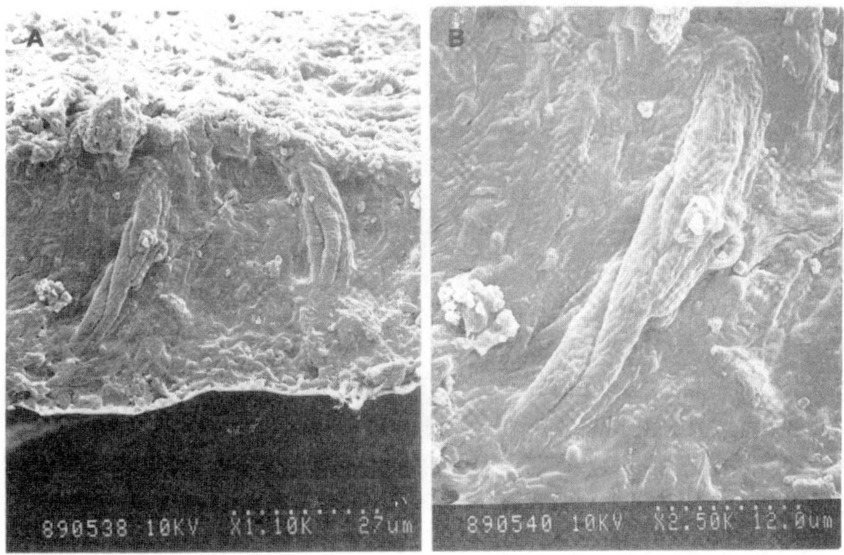

Fig. 14.5. (A,B) *Mazosia melanophthalma* (*Wei* 9074), asci in the hymenium.

clavate, 8-spored, *c*. 54.5 × 10.5 μm. *Spores* fusiform, hyaline and colourless, 3-septate; the second cell from the obtuse end often larger than others, *c*. 21–36 × 3.5–4.5 μm.

Specimens examined: YUNNAN: near Manhao, *c*. 650 m, 6 March, 1915, *Handel-Mazzetti* no. 1035 (as *Mazosia rotula*, determined by Zahlbruckner (BM!). TONKIN: 2 November, 1914, *Handel-Mazzetti* no. 5 (as *M. rotula*, determined by Zahlbruckner, in BM!).

Chemistry: Three unidentified substances R_F class 2–3 in C, giving yellow colour, R_F class 5 (bottom) in C, giving yellowish-green, and R_F class 6 (bottom) in C, giving yellowish-green colour after acid and heat treatment.

Pantropical.

Literature record: Xianggang (Hong Kong: Thrower, 1988, p. 119).

11. Mazosia paupercula (Müll.Arg.) R.Sant. (Fig. 14.6) YUNNAN, Xishuangbanna: in the reservation near Menglun, on leaves of *Knema furfuracea*, 18 October, 1980, *Wei* 3641.

Thallus crustose, thin, smooth, grey-green, usually *c*. 2.5–3 mm, sometimes up to 20 mm diam. *Apothecia* up to 1 mm diam., prominent. *Asci* 8-spored, *c*. 63–88 × 18–21.5 μm. *Spores* hyaline, 5–7-septate, 25–44 × 3.5–7 (–9) μm.

New to China. Pantropical.

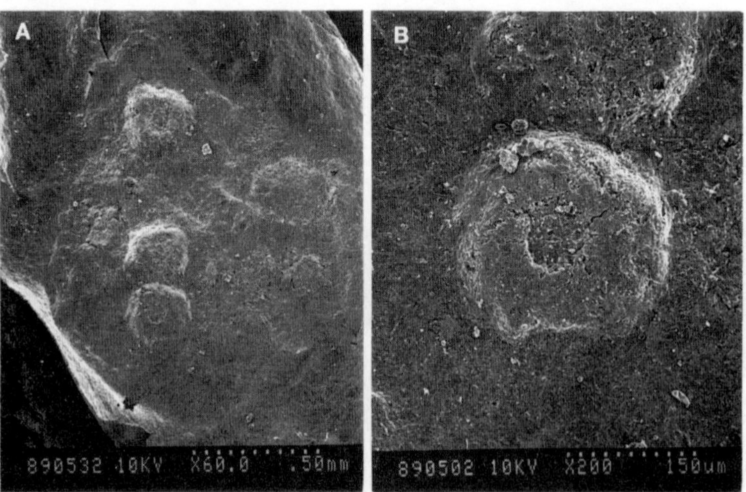

Fig. 14.6. (A) *Mazosia paupercula* (*Wei* 3641), smooth thallus with ascomata. (B) *Mazosia paupercula* (*Wei* 3641), two ascomata.

12. Mazosia phyllosema (Nyl.) A.Zahlbr. (Fig. 14.7) YUNNAN, Xishuangbanna: in the reservation near Menglun, on leaves of *Knema furfuracea*, 18 October, 1980, *Wei* 3568, 3571, 3697.

Thallus thin, membraneous, grey-green, smooth. *Spores* 3-septate, 16–27 × 3.5–5 μm.

Specimens examined: *Tonkin*: 2 November, 1914, *Handel-Mazzetti* no. 5 (as *Mazosia rotula*, determined by Zahlbruckner, in HMAS-L and BM, respectively). The radiate ridges are invisible on the whitened and felted surface of the thallus of the specimen, which is preserved in HMAS-L. The surface of the specimen thallus preserved in BM, is also completely smooth, in addition to having *Mazosia melanophthalma* on the same collection. It seems that both specimens of the same number labelled as *Mazosia rotula* are probably *Mazosia phyllosema*, as considered by Santesson (1952, p. 115).

New to China. Pantropical.

13. Mazosia rotula (Mont.) Massal. (Fig. 14.8) YUNNAN, Xishuangbanna: in the reservation near Menglun, on leaves of *Knema furfuracea*, 18 October, 1980, *Wei* 3570, 3572-(1), 3686; on leaves of *Prosartema stellaria*, *Wei* 3598-(1).

Thallus yellowish-green with olive green tint, *c.* 1–3.5 mm diam., sometimes several different small individuals confluent, with numerous

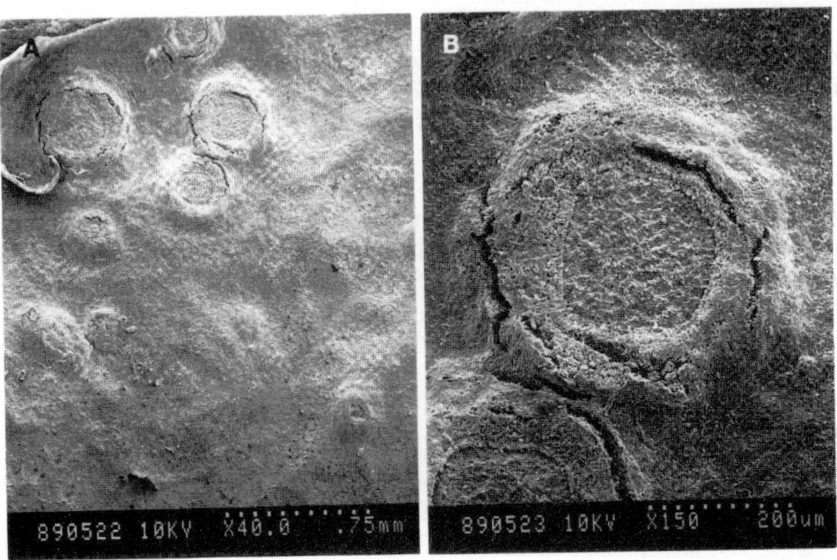

Fig. 14.7. (A) *Mazosia phyllosema* (*Wei* 3571), smooth thallus with ascomata. (B) *Mazosia phyllosema* (*Wei* 3571), two ascomata.

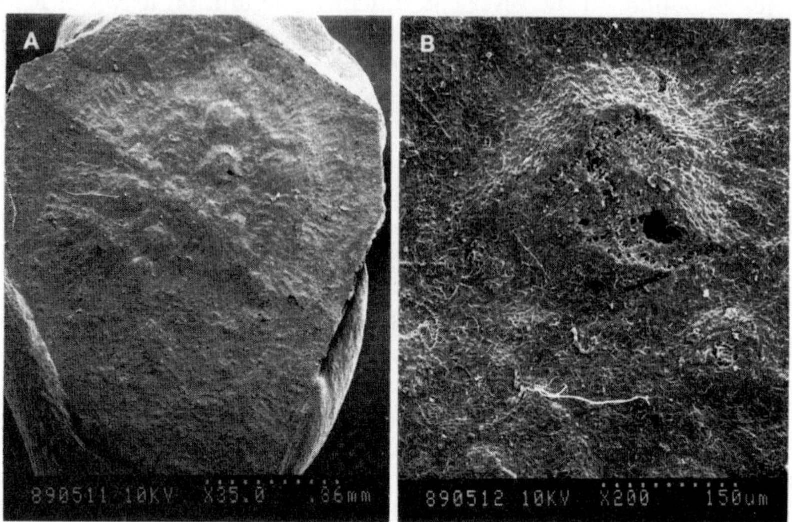

Fig. 14.8. (A) *Mazosia rotula* (without a number on leaves of *Knema furfuracea*), thallus with radiate ridges and ascomata. (B) *Mazosia rotula* (as in Fig. 14.8 (A)), two ascomata.

gentle and thin radiate ridges, simple or rarely slightly branched, lighter in colour than thallus. *Ascomata* scutiform; *disc* plane, non-pruinose, leaden black to the naked eye or through a DM, but hyaline and colourless through an LM, *c.* 300 μm in diam. *Asci* clavate, 8-spored, *c.* 38–54 × 10.5–14.5 μm. *Spores* usually 3-septate, fusiform, ends usually rather obtuse, one of the median cells often larger than the others, *c.* 21–26 × 4 μm, the larger cell among them about 5 μm wide. *Paraphyses* thin and branched.

Chemistry: One unidentified substance R_F class 5 (bottom) in C, giving yellowish-green colour after acid and heat treatment.

New to China.

Pilocarpaceae

14. Byssolecania deplanata (Müll.Arg.) R.Sant. YUNNAN, Xishuangbanna: in the reservation near Menglun on leaves of *Cleidion brevipetiolatum* Pax. et Hoffm., 18 October, 1980, *Wei* 3645.

Thallus thin, membraneous, dimly seen, grey-green, 10–20 mm diam. *Apothecia* oblate, totally adnate, *c.* 448–471 μm diam., very regularly circular, white, byssoid hyphae of the thallus around the apothecia; *disc* dark brown, margins not seen. *Asci* 8-spored, *c.* 45–65 × 14–20 μm. *Spores*, 5–7-septate, 23–40 × 3.5–4.5 μm.

The spores are longer than those of this species collected from tropical Africa and America (Santesson 1952).

New to Asia. Pantropical.

15. Fellhanera semecarpi (Vainio) Vězda YUNNAN, Xishuangbanna: *Jiang* 960-(3).

Thallus sordid whitish. *Apothecia* biatorine, numerous, small, *c.* 240 μm diam., *disc* brownish to yellowish-brown. *Asci* 8-spored 23–32 × 7–11 μm. *Spores* hyaline, 1-septate, 9 × 3.5 μm.

New to China. Pantropical.

Strigulaceae

16. Strigula elegans (Fée) Müll.Arg. YUNNAN: Xishuangbanna, in the reservation near Menglun on leaves of *Phoebe lanceolata*, 18 October, 1980, *Wei* 3587, and on leaves of *Celtis giganticarpa* (Ulmaceae) in the Limestone Hill near the Tropical Botanical Garden of the Kunming Botanical Institute of Academia Sinica, 19 October, 1980, *Wei* 3714.

Pantropical.

Literature records: Yunnan (Zahlbruckner 1930, p. 31; Santesson 1952, p. 160), Guizhou, Hubei, Hunan (Santesson 1952, p. 160), Fujian (Zahlbruckner 1930, p. 31; Santesson 1952, p. 160), Taiwan (Zahlbruckner 1933, p. 205; Wang and Lai 1973, p. 97; Santesson 1952, p. 160).

17. Strigula macrocarpa Vainio YUNNAN: Xishuangbanna, in the reservation near Menglun, 18 October, 1980: on leaves of *Amaplelopris sp.* (*Wei* 3592), on leaves of *Prosartema stellaria* (*Wei* 3598), and on leaves of *Tetrastigma sp.* (Vitaceae) (*Wei* 3626).

The species is similar to *Strigula elegans* but differs by its large ascomata.

New to China. Tropical species of the old world.

18. Strigula melanobapha (Krempelh.) R.Sant. YUNNAN: Xishuangbanna, in the reservation near Menglun on leaves of *Phoebe lanceolata* (*Wei* 3584) and of *Drypetes indica* (Müll.Arg.) Pax. et Hoffm. (Euphorbiaceae) (*Wei* 3615), 18 October, 1980 and on leaves of a broadleaved tree in Limestone Hill near the Tropical Botanical Garden of the Kunming Botanical Institute of Academia Sinica, 19 October, 1980, *Wei* 3712-(1).

Pycnidia black, *c.* 144 μm diam., containing numerous bacillar, hyaline conidia, 4.5 × 1.5 μm.

Pantropical.

Literature records: Fujian (Zahlbruckner 1930, p. 31, as *Strigula fibrilosa*, *Chung* 488—lectotype, Chung 319, 487b—syntypes; Santesson 1952, p. 188).

19. Strigula subelegans Vainio YUNNAN: Xishuangbanna, on leaves of a broad-leaved tree in Limestone Hill near the Tropical Botanical Garden of the Kunming Botanical Institute of Academia Sinica, 19 October, 1980, *Wei* 3706.

New to China. Tropical and subtropical species in Asia and America.

Trichotheliaceae

20. Trichothelium annulatum (Karsten) R.Sant. (Fig. 14.9, 14.10)

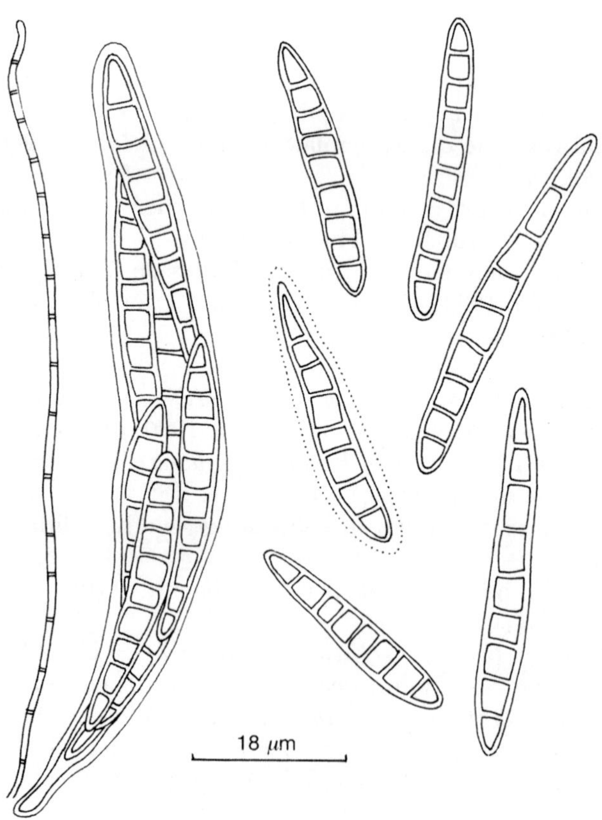

Fig. 14.9. *Trichothelium annulatum* (*Wei* 3714), ascus, ascospores, and a paraphysis.

YUNNAN: Xishuangbanna, on leaves of *Celtis giganticarpa* (Ulmaceae) in Limestone Hill near the Tropical Botanical Garden of the Kunming Botanical Institute of Academia Sinica, 19 October, 1980, *Wei* 3714.

Thallus crustose, thin and smooth, faintly nitid, olive green to badious, subcircular to irregular, *c*. 5 mm or sometimes up to 10–12 mm diam., very similar to *Trichothelium epiphyllum* Müll.Arg. but differing in the olive-greenish to badious thallus. *Spores* with 7 to 9 transverse septa and the apical setae of perithecia are no more than 12 'rays'. *Ascomata c*. 283 μm diam.; *Asci* 8-spored, *c*. 81–99 × 11.5–18 μm. *Spores* hyaline, *c*. 21.5–43 × 3.5–5 μm.

This species usually has 8–15-septate spores, but the specimens studied from Yunnan seem to be young.

Chemistry: Two unidentified substances R_F class 5 (bottom) and 6 (bottom) in C, both of which give a yellowish-green colour after acid and heat treatment.

New to China. Pantropical.

Acknowledgements

We are deeply grateful to Professor R. Santesson (Uppsala) for checking the identification of some collections and for his very helpful

Fig. 14.10. (A) *Trichothelium annulatum* (*Wei* 3714), apical setae of an ascoma. (B) *Trichothelium annulatum* (*Wei* 3714), top of one seta.

revisions, and to Dr Zhuang W.Y. for revising the English language of the manuscript. Our thanks are also due to Mr Dong G.J. for technical assistance in making scanning electron microscopic investigations; to Ms Yuan L.C. for printing the photographs, and to Ms Han Z.F. for inking the line drawings.

References

*Hawksworth, D.L. (1972). A new species of *Tricharia* Fée em. R.Sant. from Hong Kong. *The Lichenologist*, **5**, 321.

Santesson, R. (1952). Foliicolous lichens. I. A revision of the taxonomy of the obligately foliicolous, lichenized fungi. *Symbolae Botanicae Upsalienses*, **12**, 1–590.

Thrower, S.L. (1988). *Hong Kong lichens*. An urban council publication.

*Vězda, A. (1977). Beitrag Zur Kenntnis foliikoler Flechten. *Vietnams Casopsis slezskeho Musea Serie A*, **26**, 21.

*Vězda, A. (1983). Foliicole Flechten aus der Kolchis (West-Transkaukasien, UdSSR). *Folia Geobotanica et Phytotaxonomica, Praha*, **18**, 45.

*Vězda, A. (1984). Foliicole Flechten der Insel Kuba. *Folia Geobotanica et Phytotaxonomica, Praha*, **19**, 177.

Wang-Jang, J.-R. and Lai, M.-J. (1973). A checklist of the lichens of Taiwan. *Taiwania*, **18**, 83–104.

Wei, J.C. (1990). *An enumeration of the lichens in China* (to be published).

Wu, C.Y. (1979). The regionalization of Chinese Flora. *Acta Botanica Yunnanica*, **1**(1), 1–22.

White, F.J. and James, P.W. (1985). A new guide to microchemical techniques for the identification of lichen substances. *Bulletin of the British Lichen Society*, **57** (supplement), 1–41.

Zahlbruckner, A. (1930). Lichens. In *Symbolae Sinicae* III. Handel-Mazzetti. pp. 254.

List of Plants in Xishuangbanna (ed. Yunnan Institute of Tropical Botany). Academia Sinica.

* Used in specimen identification.

Some disjunctions and vicariisms in the Umbilicariaceae (Ascomycotina)

WEI Jiang-chun[1] Lev G. BIAZROV[2]

ABSTRACT: Two species of *Umbilicaria*, *U. angulata* Tuck. and *U. phaea* Tuck., new to Asia and extra America, and one species of *Lasallia*, *L. papulosa* (Ach.) Llano, very rare to Asia are reported. *U. haplocarpa* Nyl. has been confirmed as present in the southern part of Africa in addition to South America. Disjunction and vicariism of these and some other species in the Umbilicariaceae are analyzed and discussed.

KEY WORDS: Disjunction; vicariism; Asia; America; Umbilicariaceae; *Umbilicaria angulata*; *U. phaea*; *U. haplocarpa*

INTRODUCTION

The disjunction and vicariism of species with regard to their geographical distributions are common phenomena in various plants, and also lichens. Owing to their close relations to the history of speciation, the discovery and analysis of these phenomena has been attracting the attention of systematists and of flora writers.

In the course of a systematic study of the Umbilicariaceae we were surprised to discover new localities for *Umbilicaria angulata* Tuck. and *U. phaea* Tuck. in the specimens collected by one of the authors (Biazrov) from Mongolia. So, the two species are new not only to Asia, but also outside America. In addition, though Frey (1949, p.450, Pl.3, right above) reported that *U. haplocarpa* Nyl. was found in Africa, but Llano (1950, p.66) considered that Frey's report might be based on erroneous identification. However, in materials (US!) collected by Hale from Lesotho, and in those (HMAS–L!) by L. Kofler from Basutoland in southern Africa we found *U. haplocarpa* Nyl. indeed. So, the species in question has been confirmed to be a species distributed in both South America and southern Africa.

This paper attempts to analyze and discuss the disjunctions and vicariisms, and also other species in the Umbilicariaceae based on new information.

①Systematic Mycology & Lichenology Laboratory, Institute of Microbiology, Academia Sinica, Beijing 100080.

②Institute of Evolutionary Morphology and Ecology of Animals, Academy of Sciences of U.S.S.R., Moscow 117071.

This paper was originally published in *Mycosystema*, 4: 65-72, 1991.

MATERIALS AND METHODS

The specimens examined are preserved in the Herbarium of the Smithsonian Institution in Washington, D.C. (US), the herbarium of the Institute of Evolutionary Morphology and Ecology of Animals in Moscow, and HMAS–L in Beijing. The analysis of the materials was carried out according to biogeographical principles and methods.

RESULTS AND DISCUSSIONS

Disjunctions

1. Northern-Pacific disjunction

   Under the disjunction between Eastern Asia and North America, two patterns are reported:

   A. Disjunction "Eastern Asia – western North America"

      a. **Umbilicaria angulata** Tuck. Syn. Lich. New Engl. 74 (1848).

         Type: California, Monterey, 1875, coll. Menzies in 1842. It is preserved in Tuck.-Herb., sheet 994 in FH (!).

         = *Umbilicaria semitensis* Tuck. Gen. Lich. 31 (1872).

         Type: California, Yosemite Valley, 1876, coll. H.N.Bolander (265), alt. 7–8000 ft. It is preserved in Tuck.-Herb. in FH (!).

         Mongolia: Ara-Khangai, on rocks, alt. 2000m, July 17, 1970, L. G. Biazrov. no. 6591 (Herb. of the Inst. of Evolutionary Morphology & Ecology of Animals, Academy of Sciences of U. S. S. R., Moscow).

         On careful examination by the first author the collection from Mongolia (Biazrov no. 6591) proved to be typical *Umbilicaria angulata* Tuck. So, the species is new not only to Asia, but also outside America. It had been known as a species endemic to western North America and is restricted within narrow limits to the western coastal range of California and Washington State in the United States through Yukon and British Columbia in Canada, and north to southern Alaska from the Andreanof Island in the U.S.A. (Fig. 1).

      b. **Umbilicaria phaea** Tuck. Lich. Calif. 15 (1866).

         Lectotype: California, coll. in 1864 by H.N.Bolander No.11 on the right, preserved in Tuck.- Herb. in FH (!), and isolectotype in Exs. Reliquiae Tuckermanianae No.94 (FH!), and selected by J.C. Wei here.

         Mongolia: Ara-Khangai, on rocks, July 22, 1970, L. G. Biazrov, no. 6591 (Herb. of the Inst. of Evolutionary Morphology & Ecology of Animals, Academy of Sciences of U. S. S. R., Moscow).

         A careful examination, made by the first author, shows the collection from Mongolia (Byazrov no. 6590) to be typical *Umbilicaria phaea* Tuck.

         This species is new to Asia and outside America. It had previously been considered as endemic to America, with a range mainly extending through the western United States, such as California, Washington, Oregon, Montana, Colorado, Nebraska, and south to Mexico: Lower California, and north to Alberta, British Columbia and the

Northwest Territories of Mackenzie, and Great Slave Lake in Canada, and rarely in the South America (Fig. 2).

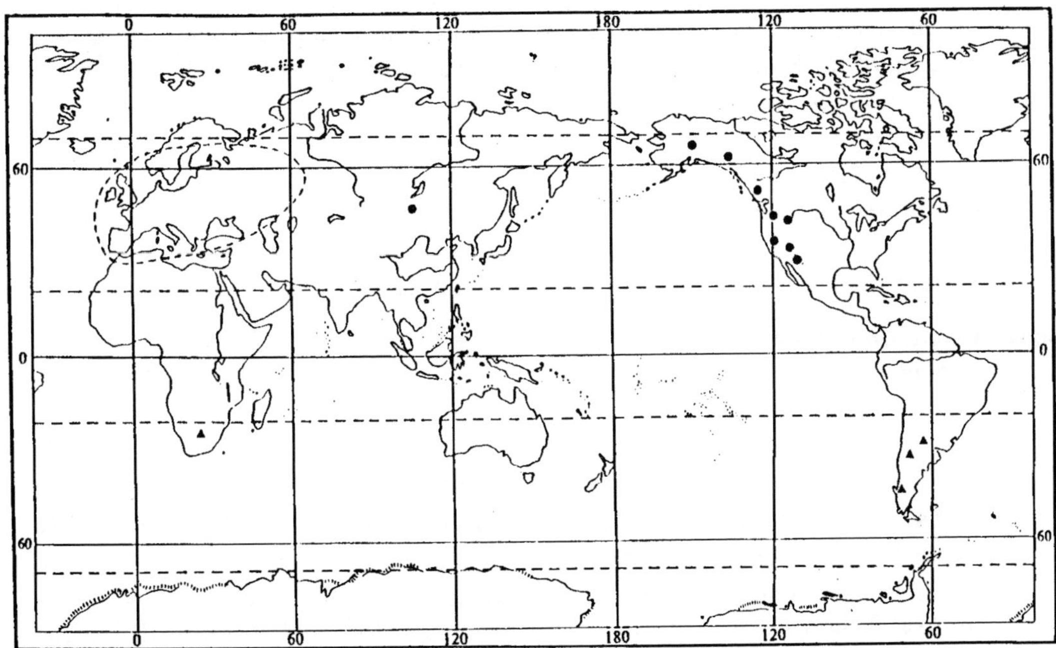

Fig. 1. ● Disjunctive pattern "Eastern Asia – Western North America" of *Umbilicaria angulata* Tuck. ▲ Disjunctive pattern "South America – Southern Africa" of *Umbilicaria haplocarpa* Nyl. (after Llano, 1950, with additional recoards).

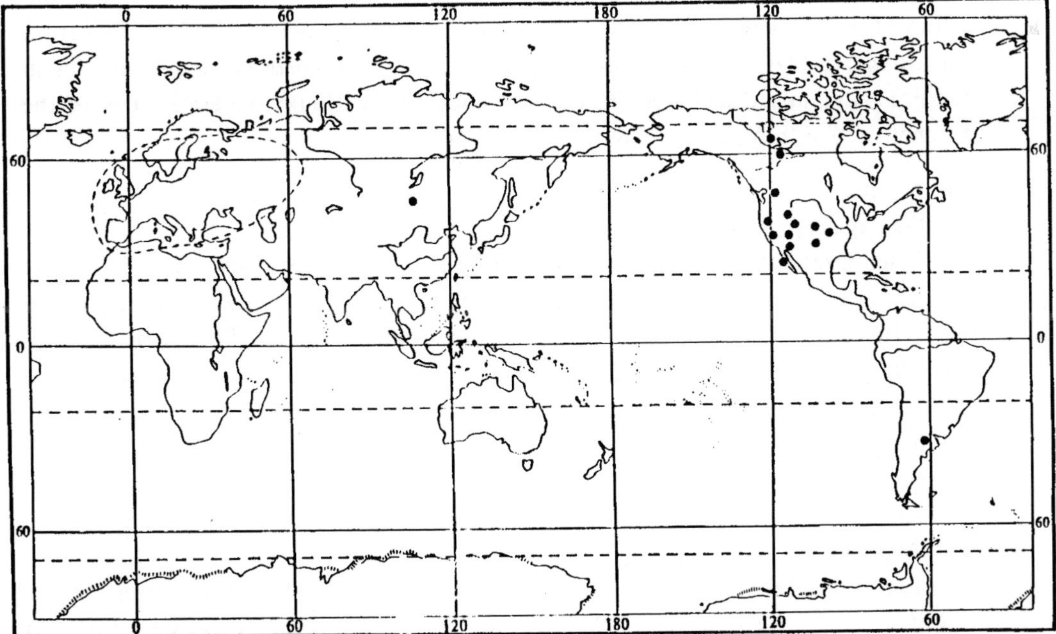

Fig. 2. Disjunctive pattern "Eastern Asia – Western North (and South) America" of *Umbilicaria phaea* Tuck. (after Llano, 1950, with additions).

B. Disjunction "Eastern Asia – Eastern North America"

a. **Umbilicaria muehlenbergii** (Ach.) Tuck. is distributed mainly in the eastern part of the United States along the Appalachian Mountains from South Carolina in the United States north to Fort Chimo of Quebec in Canada; northwest through Montana to Alaska near the point Barrow in U.S.A. The species in Asia is distributed from Krasnojarsk in Siberia, U.S.S.R., and south east to Da Hinggan Ling in China and Honshu in Japan (Llano, 1950; Yoshimura, 1974; Wei and Jiang, 1988).

b. **Umbilicaria caroliniana** Tuck. has the disjunction " North Carolina – Alaska in U.S.A. – east Siberia in U.S.S.R. – Da Hinggan Ling in China and several localities in Japan (Llano, 1950; Yoshimura, 1974; Wei and Jiang, 1988).

2. South Atlantic disjunction

Umbilicaria haplocarpa Nyl.

South Africa: Lichens of Lesotho, Thaba Tseka District, Low dolerite Ledges facing river valley, 7.5 km NW of Sani Pass on road to Mokhotlong Alt. 2900m, [GRID: 2929CB]. 6 May 1988, M. E. Hale, no. 81524 (US!); Basutoland, L. Kafler (HMAS–L!). Frey (1949, p. 450, P1.3, right above) reported this species from Africa (South Africa: Capland, Natal, New Hut, Mont-aux-sources, on basalt boulder. E-exposed, locally Common, 10,000ft., leg. Schelpe, Herb. Bolusianum. Cape Town, on.1284). However, Llano (1950, p.66) pointed out: "Frey (1949) reports this species from Africa. The specimen which he referred to has not been examined. Previous specimens from Africa referred to *U. haplocarpa* have all turned out to be otherwise. For the purpose of this study this species is held to be endemic to South America." Although the collection cited by Frey (1949) has not been examined, the careful examination of the specimens collected by Hale from Lesotho (US!) and by L. Kafler from Basutoland (HMAS–L!) show them to be typical *U. haplocarpa*. So the species in question is mainly distributed in, but not endemic to, South America. It is of the disjunctive "South America – South Africa" pattern (Fig. 1).

3. Multirange disjunction

Lasallia papulosa (Ach.) Llano

Mongolia: Dzabhan, on wood (branches), July 1, 1976, L.G.Biazrov, nos. 6575, 7468; Ara-Khangai, 17 km from southern Somon, on the left bank of Caicairlaig River, on rocks, July 21, 1979, L. G. Biazrov, no. 5736 & September 7, 1979, L.G.Biazrov no. 6190; Bulgan, Dashinchilen residential area, Mt. Caicairlat-Ula, southern residential area, on Betula sp., June 29, 1972, L.G.Biazrov nos. 5109, 6689; Ara-Khangai, on the left bank of Chulata River, on tree bark of *Larix sibirica*, August 22, 1970, L.G.Biazrov, no. 6570. Above cited collections are deposited in Herb. of the Inst. of Evolutionary Morphology and Ecology of Animals of Academy of Sciences of U. S. S. R. Moscow.

This species which is mainly distributed in America and Africa, and rarely in the Yakut Autonomous Republic of the S.U. and Mongolia in Asia. It is interesting that the species almost always are growing on rocks and only occasionally on wood (Llano, 1950, p.34) in America, but are often on wood in both Asia and Africa in addition to rocks.

Vicariism

With our present knowledge, *Lasallia pustulata* (L.) Mérat is a ubiquitous species in Europe, and its distribution is rather spotty in southern Africa, central America (Frey, 1936, p. 416, Fig. 1), and North America (Llano, 1950, p. 31). However, the records of this species from China (Yunnan, Taiwan: Zahlbr., 1930, p. 138 & 1933, p. 49; Asahina, 1931, p. 102) were based on erroneous identifications (Sato, 1950, p. 168; Wei & Jiang, 1988, p. 93). The first author checked Biazrov's collection of this species from Mongolia (Biazrov, no. 5923, in the same Herb. in Moscow) and confirmed his report (Biazrov, 1986, p. 99). Before him several other authors had also reported this species from Mongolia (Schubert et al., 1969, p. 406; Golubkova, 1981, p. 134; Olech, 1984, pp. 215, 216). It seems that Mongolia is the southern limit of its distributional area in Asia. So far it has never been found in China, where the vicarious species *Lasallia sinorieatalis* Wei occurs instead of *L. pustulata*. The latter one is ubiquitous in Europe (Fig. 3). In addition, the other vicarious species, *U. mammulata* (Ach.) Tuck. and *U. esculenta* (Miyoshi) Minks, mentioned by Culberson (1972), are distributed respectively in the eastern Asia and the eastern North America (Fig. 4). It is uncertain whether all these vicarious species result from speciation or convergence. Therefore it is necessary to study them further in the future.

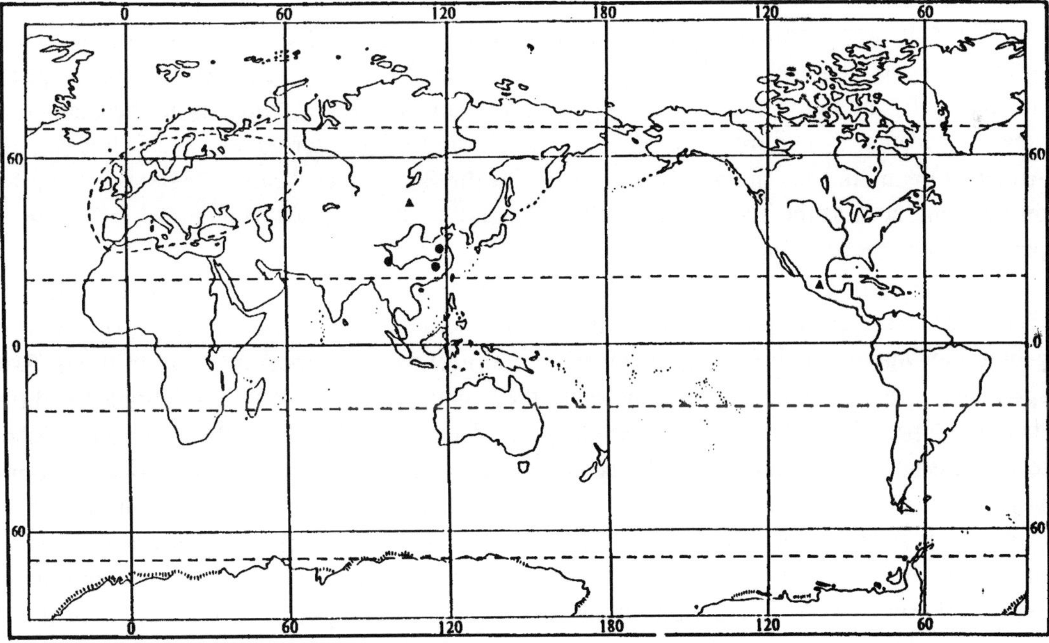

Fig. 3. Distribution pattern of ▲ *Lasallia pustulata* (L.) Mérat and ● *Lasallia sinorientalis* Wei. (after Frey, 1936, with additional records).

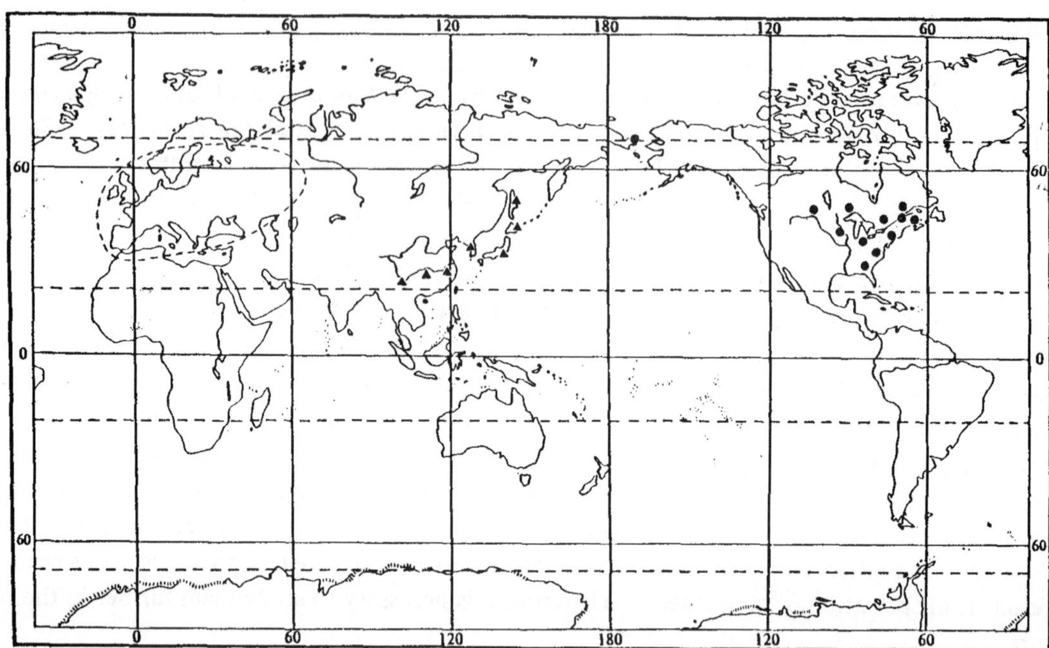

Fig. 4. Distribution pattern of ▲ *Umbilicaria esculenta* (Miyoshi) Minks and ● *Umbilicaria mammulata* (Ach.) Tuck. (after Llano, 1950, with additional records).

The Northern-Pacific disjunction of plants can be explained by three theories, i.e. landbridges, a polar origin of the floras, and continental drift, because all three theories accept the role played by the Bering land in the migration of plants. However, there has not been any convincing evidence to confirm the existence of the landbridges across the Pacific Ocean, the Atlantic Ocean, and the Indian Ocean, except for the Bering Landbridge. Therefore, the other disjunctions, such as the Tropical, South Pacific, and South Atlantic disjunctions can not be explained by the theory of landbridges.

The theory of the polar origin of the floras also seems unacceptable because it is based on theory of the permanence of continents and oceans. According to the principles of geothermodynamics and palaeogeomagnetism and new data evidences a theory of ocean floor spreading – plate tectonics was established in the beginning of the 1960s, and inherited and further developed the theory of continental drift. The theory of ocean floor spreading-plate tectonics is now well established and accepted by most scientists in the fields of geology, geography and biogeography, and is playing a more and more important role in the explanation of the disjunctions between continents. According to the present knowledge of palaeogeography, in the Miocene, about 150 million years ago, the Bering land was replaced by the Bering Strait, and in the early Cretaceous period, about 135 million years ago, the southern part of South America and the southern part of Africa began isolating. Thus, the origin of the disjunction between Mongolia and western North America of *U. angulata* and *U. phaea*, and that between north-eastern Asia and eastern North America of *U. caroliniana* and *U. muehlenbergii* in the North-pacific zone, and the disjunction between South America and southern Africa of *U. haplocarpa* in South Atlantic zone can be explained by the theory of ocean floor spreading-plate tectonics.

ACKNOWLEDGEMENTS

The information contained in this paper was obtained during the first author's stay in the Herbarium of the Smithsonian Institution in Washington, D.C., in the Herbarium of the Institute of Evolutionary Morphology and Ecology of Animals in Moscow, and in HMAS–L in Beijing. A fellowship from the Smithsonian Institution made it possible to study in the US herbarium. Dr. L. Skog, director of the Botany Department of the Smithsonian Institution, Ms S. Pittam (US), Dr. T. Topeisheva of the University of Moscow gave considerable assistance to the first author during his stay in the above mentioned herbaria, respectively. For all the above mentioned the first author wishes to express his sincere thanks to them. We also wish to thank Dr. D. L. Hawksworth (CMI) and Ms S. Pittam (US) for critically reading the manuscript and improving the English language in it.

REFERENCES

Biazrov LG (1986) Additions to the lichen flora in Khangai (Mongolian People's republic). 1. The family Umbilicariacene Fee [*BUL. Mosc. O–va Ispeitatelei Prirodei. Otd. Biol.*] **91**(6):99 (in Russian)

Culberson WL (1972) Disjunctive distrivbutions in the lichen forming fungi. *Annales of the Missouri Botanical Garden* **59**(2):165

Frey E (1936) Dieglographische Verberitung der Umbilicariaceen und einiger alpiner Flechten. *Berichte der Schweizerischen Botanischen Gesellschaft* **46**:412

Frey E (1949) Neue Beitrage zu einer Monographie des Genus *Umbilicaria* Hoffm., Nyl. Ibid. **59**:427

Llano G (1950) A Monograph of the Lichen Family Umbilicariaceae in the Western Hemisphere

Olech M (1984) Materials to the lichen flora of the Khentei monutains (Mongolia). *Fragm. Florist. Geobot.* vol.28, N.2.

Schubert R, Klement O (1971) Beitrag zur Flechtenflora der Mongolischen Volksrepublik. *Feddes Repertorium* **82**:187

Wei JC (1982) Some new species and materials of *Lasallia* Mérat em. Wei. *Acta Mycol. Sin.* **1**(1):19

Wei JC, Jiang YM (1988) A conspectus of the ascomycetes Umbilicariaceae in China. *Mycosystema* **1**:73

Wulff EV (1932) Introduction to the Historical Geography of Plants. Leningrad

Yoshimura I (1974) Lichen Flora of Japan

腾格里沙漠沙坡头地区地衣物种多样性研究

刘萌[1,2]　　魏江春[1,2*]

[1]山东农业大学生命科学学院 山东 泰安 271000

[2]中国科学院微生物研究所真菌学国家重点实验室 北京 100101

摘　要：报道了腾格里沙漠东南缘沙坡头地区地衣 22 种，分隶于 16 属，13 科，5 目，包括中国新记录 4 种：异色杆孢衣 *Bacidia heterochroa*，安奈污核衣 *Porina aenea*，白黑瘤衣 *Buellia alboatra* 及美丽黑瘤衣 *Buellia venusta*。该研究丰富了关于该地区荒漠地衣物种多样性的知识，并将为沙漠生物地毯工程的前期研究提供更多的地衣种质资源。

关键词：沙坡头，地衣分类，中国新记录

Lichen diversity in Shapotou region of Tengger Desert, China

LIU Meng[1,2]　　WEI Jiang-Chun[1,2*]

[1]College of Life Sciences, Shandong Agriculture University, Tai'an, Shandong 271000, China

[2]State Key Laboratory of Mycology, Institute of Microbiology, Chinese Academy of Sciences, Beijing 100101, China

Abstract: Twenty-two lichens belonging to 16 genera, 13 families, and 5 orders from Shapotou region of Tengger Desert are reported in this paper. *Bacidia heterochroa*, *Porina aenea*, *Buellia alboatra* and *Buellia venusta* are new to China. The research has enriched the knowledge of lichen diversity in the studied region, and will provide more lichen resources for the basic research of desert biocarpet engineering.

Key words: Shapotou, lichen taxonomy, China new record

腾格里沙漠东南缘沙坡头（37°32′N，105°02′E）位于宁夏回族自治区中卫市境内，濒临黄河；海拔 1,300–1,700m；年均降水量为 186.2mm；年均气温 9.7℃；年均风速为 2.8m/s（李守中等 2008）；气候干旱而多风，属温带草原化荒漠。1956 年，为保护包兰铁路沙坡头地段而在铁路两侧铺置草方格，种植沙生植物，营造防风固沙人工植被。在水分平衡规律作用

基金项目：国家自然科学基金（No. 31070018，No. 30670004）；国家科技支撑项目（No. 2011BAC07B03）

*Corresponding author. E-mail: weijc2004@126.com

收稿日期: 2012-02-11，接受日期: 2012-03-15

本文原载于《菌物学报》，32（1）：42-50, 2013.

下，人工植被逐年衰退，虽然未能实现规划中的人工植被防护体系，却为微型生物结皮逐年发育创造了条件，从而在生物演替中形成了以微型生物结皮为主体，以稀疏灌木和短命草本植物为次的结皮、灌、草固沙的新防护体系，引起了国际同行科学家的普遍关注，从而于1995年被联合国评为环球保护“全球500佳”之一，使沙坡头地区成为沙漠治理的典范。

有作者指出，在沙坡头地区微型生物结皮中未能发现地衣（Li 2002）。然而，初步研究表明，沙坡头地区微型生物结皮中存在有大量地衣，而且往往占据优势地位。

本实验室对我国荒漠微型生物结皮中的地衣多样性及其生态学意义有长期的研究和科学积累。其中沙坡头地区荒漠微型生物结皮中发现地衣多种，其中1属6种为中国新记录，1种1亚种为新种。前人对我国荒漠地区地衣物种多样性虽多有报道（Magnusson 1933，1940），然而并未涉及到沙坡头地区。本文作为沙坡头地区沙漠生物地毯工程（魏江春2005）的前期研究之一，对于该地区微型生物结皮中的地衣物种多样性做了进一步的补充研究。

1 材料与方法

本研究所依据的100余份地衣物种标本采自腾格里沙漠东南缘沙坡头地区，保存在中国科学院微生物研究所国家菌物标本馆地衣标本室（HMAS-L）。体视显微镜Motic SMZ-168及显微镜Zeiss Axioscope 2 plus分别用于外部形态和内部结构研究。在地衣化学检测中，10% KOH（K），饱和CaOCl水溶液（C），及对苯二胺（P）用于显色反应；薄板层析法（TLC）用于地衣化学组分分析（Culberson & Kristinsson 1970；Culberson 1972）。

2 结果与讨论

在进行沙坡头地区荒漠地衣物种多样性补充研究中进一步发现了地衣22种，分隶于16属，13科，5目。其中4种为中国新记录并以方块号“♦”标记：异色杆孢衣 *Bacidia heterochroa* (Müll. Arg.) Zahlbr.，安奈污核衣 *Porina aenea* (Wallr.) Zahlbr.，黑白黑瘤衣 *Buellia alboatra* (Hoffm.) Th. Fr.，及美丽黑瘤衣 *Buellia venusta* (Körber) Lettau（Wei 1991），并附有外部形态及其内部结构图片（图1，2）。红鳞网衣 *Psora decipiens* (Hedw.) Hoffm.已有报道（Yang & Wei 2012），但由于该种在研究地区比较普遍，因而仍被列入本文。本实验室在荒漠地衣石果衣 *Endocarpon pusillum* Hedw.共生菌藻耐干旱饥饿实验中获得了共生菌能存活7个月的最长耐干旱耐饥饿数据（Zhang & Wei 2011）。从该菌的全长cDNA文库中鉴定了111个基因，其中11个基因在已知真菌中从未发现（Wang *et al.* 2011）。这说明，对于作为基因载体的荒漠地衣物种多样性研究，在丰富地衣物种多样性知识，为沙漠生物地毯工程提供所需要的地衣种质资源基因资源方面具有重要意义。

下列地衣名录按照目、科系统（Kirk *et al.* 2008）归类排列。在目、科范围内按属、种第一字母顺序排列。

微孢衣目 Acarosporales

微孢衣科 Acarosporaceae

1 糙聚盘衣

Glypholecia scabra (Pers.) Müll. Arg., *Hedwigia* 31: 156, 1892.

石生：翠柳沟，1,600m，2009.5.29，张涛

NX09004，NX09030（2010.5.28），1,669m，2010.7.3，曹叔楠和刘文婧 SPT10010-2。

黄茶渍目 Candelariales

黄茶渍科 Candelariaceae

2 灰体黄茶渍

Candelariella antennaria Räsänen, Anales Soc. Ci. Argent. 128: 137, 1939.

树枝：红卫，1,598m，2011.4.13，刘萌 SPT11008，SPT11004-2。

该种在沙坡头的主要特征是地衣体灰色，不明显，子囊盘大量，黄绿色。

3 莲座黄茶渍

Candelariella rosulans (Müll.Arg.) Zahlbr., Cat. Lich. Univ. 5: 802, 1928.

土生：翠柳沟，1,684m，2008.8.10，张涛和曹叔楠 SPT0-3，1,619m，2011.4.13，刘萌 SPT11032；香山，1,377m，2011.4.17，刘萌 SPT11045，SPT11046，SPT11047；沙坡头至青山途中，1,410m，2011.4.19，周启明和曹叔楠 ZW11016。

该种地衣体黄色，K+淡红色，表面光滑，稍有光泽，子囊盘较少，盘面黄棕色。

茶渍目 Lecanorales

茶渍科 Lecanoraceae

4 加州茶渍衣

Lecanora californica Brodo, Beih. Nova Hedwigia 79: 107, 1984.

石上土层：翠柳沟，1,600m，2010.4.9，张涛 SPT10087。

5 碎茶渍衣

Lecanora argopholis (Ach.) Ach., Cat. Lich. Univers. 5: 456, 1810.

土生：翠柳沟，1,669m，2010.7.3，曹叔楠和刘文婧 SPT10011。

6 红鳞茶渍衣

Rhizoplaca chrysoleuca (Sm.) Zopf, Justus Liebigs Annln Chem. 340: 291, 1905.

石生：翠柳沟，1,639m，2011.4.13，刘萌 SPT11028。

梅衣科 Parmeliaceae

7 旱黄梅

Xanthoparmelia camtschadalis (Ach.) Hale, Phytologia 28: 486, 1974.

土生：中卫至青山根途中，1,410m，2011.4.19，周启明和曹叔楠 ZW11016。

鳞网衣科 Psoraceae

8 红鳞网衣

Psora decipiens (Hedw.) Hoffm., Descr. Adumb. Plant. Lich. 2: 681, 1794.

土生：翠柳沟，1,600–1,669m，2010.7.3，曹叔楠和刘文婧 SPT10007，SPT10016，SPT10033，2010.4.9，张涛 SPT10071，SPT10104（2010.4.16）；青山根，1,535–1,569m，2011.4.19，周启明和曹叔楠 ZW11005，ZW11023，ZW11023，ZW11030，ZW11044，ZW11071；青山，1,549m，2011.4.19，周启明和曹叔楠 ZW11058。

沙坡头常见种，主要特征是地衣体鳞片状，表面橘红色，边缘多覆粉霜，子囊盘黑色，多在地衣体边缘。

树花衣科 Ramalinaceae

9 异色杆孢衣♦

Bacidia heterochroa (Müll. Arg.) Zahlbr., Cat. Lich. Univ. 4: 204, 1926.

地衣体不明显。子囊盘蜡盘型，突生于基物上；盘面黑色，表面平坦或稍微上凸，

0.4–1mm diam.，少数子囊盘盘面覆少量粉霜（图 1-C）。果壳深蓝色至黑色，厚 30–45μm，囊层被深蓝色至褐色，厚 15–22.5μm，子实层无色，厚 105–120μm，囊层基灰白色至淡黄褐色，厚 15–30μm。子囊大多 8 孢，少数 4–6 孢，长棍型，105–120×19.5–22.5μm（图 1-B），子囊孢子无色，横隔多室，7–10 室，长条形，笔直或稍有弯曲，一头较粗，一头较细，52.5–67.5×7.5–9μm（图 1-A）。

化学：K-，C-，KC-，P-。无化学物质。

基物：树枝。

研究标本：红卫，1,598m，2011.4.13，刘萌 SPT11007。

世界分布：世界广布种（Nash III *et al.* 2004）。中国新记录。

讨论：该种的主要特征是树皮生，地衣体不明显，黑色子囊盘，子囊孢子无色，7–10 室。该种与绿囊被杆孢衣 *Bacidia salazarensis* 较为相近，但后者囊层被为绿色，子囊孢子 6–8 室，且含有少量黑茶渍素。

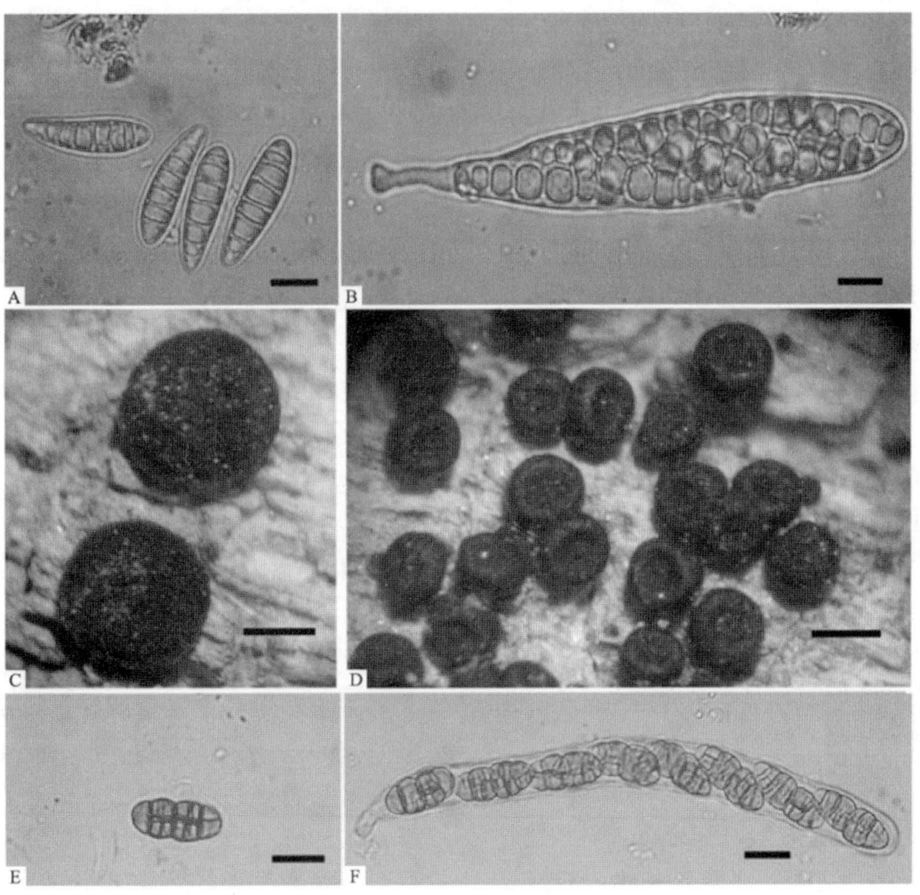

图 1 A: *Bacidia heterochroa* 子囊孢子（标尺=10μm）; B: *Bacidia heterochroa* 子囊（标尺=10μm）; C: *Bacidia heterochroa* 子囊盘（标尺=0.5mm）; D: *Buellia alboatra* 子囊盘（标尺=0.5mm）; E: *Buellia alboatra* 子囊孢子（标尺=10μm）; F: *Buellia alboatra* 子囊及子囊孢子（标尺=10μm）.

Fig. 1 A: *Bacidia heterochroa* ascospores (bar=10μm); B: *Bacidia heterochroa* ascus (bar=10μm); C: *Bacidia heterochroa* apothecia (bar=0.5mm); D: *Buellia alboatra* apothecia (bar=0.5mm); E: *Buellia alboatra* ascospore (bar=10μm); F: *Buellia alboatra* ascus and ascospores (bar=10μm).

珊瑚枝科 Stereocaulaceae

10 条斑鳞茶渍

Squamarina lentigera (G.H. Weber) Poelt, Mitt. bot. StSamml., Münch. 19-20: 536, 1958.

土生：翠柳沟，1,684m，2008.8.29，张涛和曹叔楠 SPT1-17，SPT6-2（1,720m，2008.4），张涛 Z08062，Z08063，Z08064（1,600m，2008.8.28），NX09011，NX09013，NX09023，NX09024（2009.5.28），SPT10096（2010.4.9）；青山根，1,360–1,569m，2011.4.19，刘萌 SPT11062，11080，周启明和曹叔楠 ZW11017，11031，11057，11077；沙坡头路北，1,400m，2010.4.10，张涛 SPT10037，杨军和张涛 SPT231（2007.4.17）；香山，1,500m，2007.4.13，杨军和张涛 SPT255。

沙坡头常见种，地衣体表面龟裂，灰绿色，具裂片，子囊盘茶渍型，盘面棕黄色，大多具子囊盘，少数标本子囊盘大量。少数标本具黑色点状分生孢子器。

厚顶盘目 Ostropales

污核衣科 Porinaceae

11 安奈污核衣♦

Porina aenea (Wallr.) Zahlbr., Cat. Lich. Univ. 1: 363, 1922.

地衣体灰色，不明显。子囊壳半球形，半埋生于地衣体表面，约 0.2mm diam.，黑色，开口不明显（图 2-C）。子囊壳 400–500×400–500μm，果壳黄褐色，厚 15–20μm，具黑色包顶组织，子实层无色。子囊长条状，8 孢，112.5–150×9–12μm（图 2-F），子囊孢子单列排列，无色，梭形，大多 4 室，两端小室较小，中间小室较大，少数未成熟孢子 2–3 室，22.5–27×（4.5）6.5–9μm（图 2-E）。

化学：K-，C-，KC-，P-。无化学物质。

基物：枯树枝。

研究标本：红卫，1598，2011.4.13，刘萌 SPT11010。

世界分布：欧洲，北美（Nash III *et al.* 2002），澳大利亚（McCarthy 2001）。中国新记录。

讨论：该种的主要特征是树枝生或树皮生，地衣体表面灰色，不明显，子囊壳黑色，半埋生，子囊孢子无色 4 室。该种与 *Porina bryophila* 相近，但后者多生于树皮上苔藓，孢子 8 室，26–38×3–5.5μm。中国标本孢子比澳大利亚和索诺拉沙漠（16–23×3–5.5μm）的较大。

茱孔衣科 Thelotremataceae

12 藓生双缘衣

Diploschistes muscorum (Scop.) R. Sant., Lichenologist 12(1): 106, 1980.

土生：翠柳沟，1,604m，2011.4.13，刘萌 SPT11038。

黄枝衣目 Teloschistales

粉衣科 Caliciaceae

13 黑白黑瘤衣♦

Buellia alboatra (Hoffm.) Th. Fr., Gen. Heterolich. Eur.: 91, 1861.

地衣体不明显。子囊盘网衣型，大量，圆盘状，直径 0.2–0.4mm diam.，突生；盘面黑色，大多中央凹下，无粉霜或零星粉霜（图 1-D SPT11009）。果壳黑褐色，厚 15–30μm，囊层被黑褐色/黄褐色，厚 12–22.5μm，子实层无色，厚 60–75μm，囊层基黄褐色，厚 30–45μm，侧丝厚 1.5–3μm，顶端有黄褐色的帽子（4–6μm）。子囊棍棒状，顶端稍微膨大，8 孢，Bacidia-type，75–90×12–15μm（图 1-F SPT11009），子囊孢子长椭圆型，未成熟无色，成熟褐色，亚砖壁型至砖壁型，横向 4 室，中间小室多被分成 4 个小胞，少数两头小室被分成 2 个小胞，

16.5–21×6–9μm（图 1-E SPT11009）。

化学：K-，C-，KC-，P-。无化学物质。

基物：树枝。

研究标本：红卫，1,598m，2011.4.13，刘萌 SPT11004-1，SPT11009。

世界分布：欧洲，北美，北非，亚洲（Nash III *et al.* 2007）。中国新记录。

讨论：该种的主要特征是树枝生，子囊孢子褐色，亚砖壁型至砖壁型。在形态上和美丽黑瘤衣和 *Buellia morsina* 较为相近，但前者多为石生，且孢子为横隔 4 室，后者地衣体 K+ 黄至红色，含有降斑点酸，孢子较大，20.4–26.4×8.5–10.1μm。

14 美丽黑瘤衣♦

Buellia venusta (Körber) Lettau, Hedwiggia 52: 244, 1912.

地衣体壳状，龟裂，较厚，表面石灰白色，无粉芽。子囊盘网衣型，盘面黑色，较小时盘面平坦，成熟时表面轻微隆起，少数未成熟子囊盘边缘轻微被地衣体包被，呈假茶渍型；子囊盘大量，大多圆盘状，少数不规则，0.5–1（2）mm diam.，大部分突生，少数深埋于地衣体中；盘面黑色，覆少量粉霜，表面轻微上凸（图 2-D SPT11015）。果壳黄褐色 20–35μm，囊层被黑褐色/黄褐色，22.5–37.5μm，I+蓝色，子实层无色，75–90μm，囊层基黄褐色，75–90μm，侧丝 2–4μm，顶端有黄褐色的帽子（4–6μm）（图 2-B ZW11078）。子囊棍棒状，顶端稍微膨大，8 孢，Bacidia-type，45–60×15–18μm（图 2-B），子囊孢子长椭圆型，未成熟无色，成熟褐色/黑褐色，横隔 4 室，16.5–21×6–7.5μm（图 2-A ZW11078）。

化学：皮层和髓层 K-，C-，KC-，P-。无化学物质。

基物：石生。

研究标本：翠柳沟，1,601m，2011.4.13，刘萌 SPT11013，SPT11015，SPT11021；香山，1,338m，2011.4.13，刘萌 SPT11043；沙坡头至青山根途中，1,410m，2011.4.19，周启明和曹叔楠 ZW11014；中卫青山根，1,451m，2011.4.19，周启明和曹叔楠 ZW11078。

世界分布：欧洲，北美，北非，亚洲（Nash III *et al.* 2007）。中国新记录。

讨论：此种的主要特征是地衣体较厚，石灰白色，突生于地衣体表面的黑色网衣型子囊盘，孢子褐色，4 室，且不含有任何化学物质。黑白黑瘤衣和美丽黑瘤衣常被归入多孢瘤属 *Diplotomma*，但未被广泛认可，本文仍将其归入黑瘤衣属。

15 椰子黑盘衣

Pyxine cocoës (Sw.) Nyl. Mém. Soc. Imp. Sci. Nat. Cherbourg 5: 108, 1857.

土生：路北实验区，1,400m，2007.8.25，杨军和张涛 SPT368，SPT230；青山，1,420m，2011.4.19，刘萌 SPT11079；翠柳沟，1,684m，2008.8.29，张涛和曹叔楠 SPT1-4，张涛 Z08071，NX09027（1,600m，2009.5.29），SPT10069（2010.4.9）；青山根，1,535m，2011.4.19，周启明和曹叔楠 ZW11037。

蜈蚣衣科 Physciaceae

16 陆生饼干衣

Rinodina terrestris Tomin, Cat. Lich. Univ. 7: 557, 1928.

土生：路北实验区，1,400m，2004.8.13，杨军 SPT140，2006.8.26，杨军和张涛 SPT285。

该种主要特征是地衣体覆大量粉霜，呈亮白色，子囊盘茶渍型，大量。盘面黑色，时有少量粉霜，子囊孢子蜈蚣衣型（physcia-type），黄褐色，2 室（McCune & Rosentrete 2007）。该种曾在新疆报道（Obermayer 2004），生境同为干燥土生，海拔 3,600m。

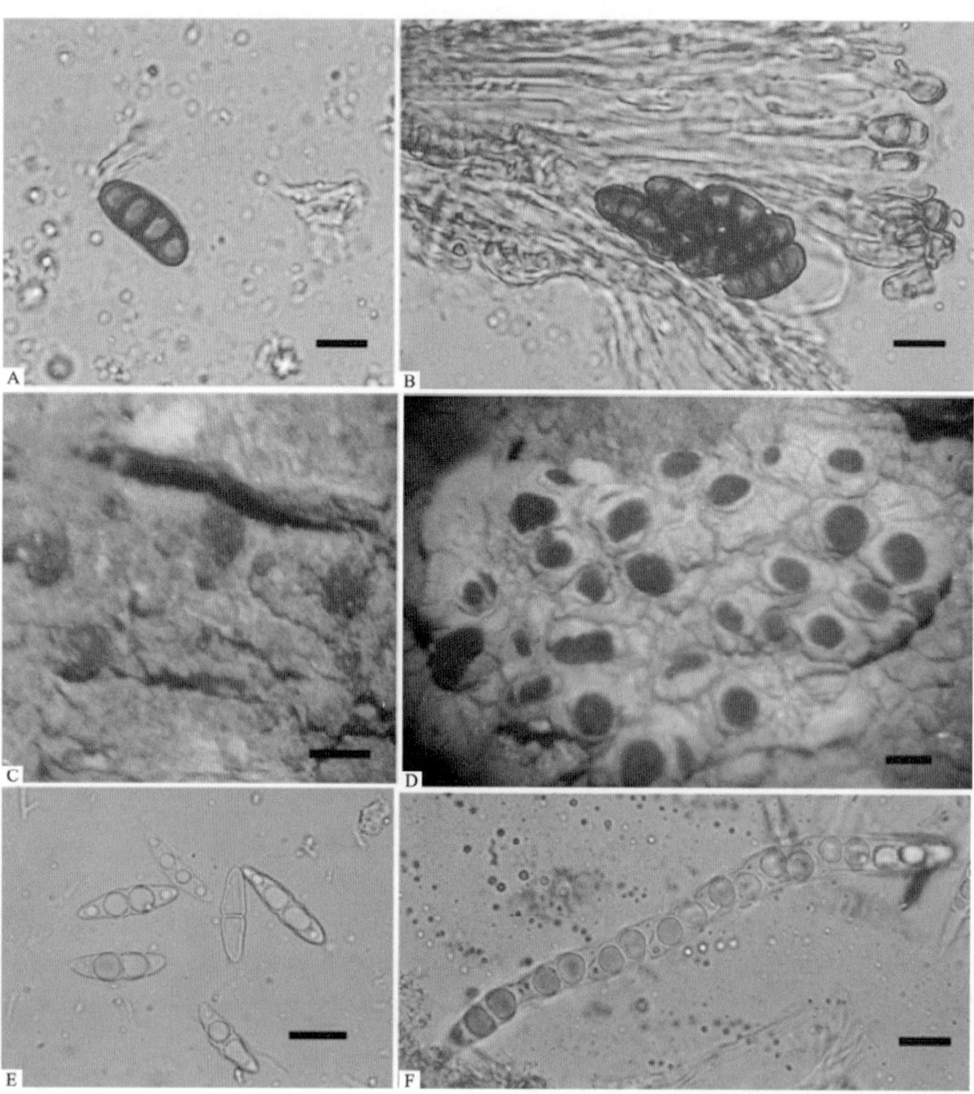

图 2 A：*Buellia venusta* 子囊孢子（标尺=10μm）；B：*Buellia venusta* 子囊（标尺=10μm）；C：*Porina aenea* 子囊壳（标尺=0.5mm）；D：*Buellia venusta* 子囊盘（标尺=0.5mm）；E：*Porina aenea* 子囊孢子（标尺=10μm）；F：*Porina aenea* 子囊（标尺=10μm）.

Fig. 2 A: *Buellia venusta* ascospore (bar=10μm); B: *Buellia venusta* ascus (bar=10μm); C: *Porina aenea* perithecia (bar=0.5mm); D: *Buellia venusta* apothecia (bar=1mm); E: *Porina aenea* ascospores (bar=10μm); F: *Porina aenea* ascus (bar=10μm).

黄枝衣科 Teloschistaceae

17 南方拟橙衣

Fulgensia australis (Arnold) Poelt, Mitt. Bot. München 5: 594, 1965.

石生：翠柳沟，1,469m，2010.7.5，曹叔楠和刘文婧 SPT10042，1,451m，2011.4.19，周启明和曹叔楠 ZW11070；香山，1,359m，2011.4.17，刘萌 SPT11053。

18 拟橙衣

Fulgensia bracteata (Hoffm.) Räsänen, Cat. Lich.

Univers. 8: 585, 1931.

土生：翠柳沟，1,600m，2010.4.9，张涛 SPT10088。

19 荒漠拟橙衣

Fulgensia desertorum (Tomin) Poelt, Mitt. Bot. München 5: 600, 1965.

土生：翠柳沟，1,600–1,669m，2010.4.9–16，张涛 SPT10033，SPT10076，SPT10094，SPT10095，曹叔楠和刘文婧 SPT10015，SPT10027（2010.7.5）；青山，1,360m，2011.4.19，刘萌 SPT11058，SPT11060，1,451m，周启明和曹叔楠 ZW11062，ZW11075。

讨论：荒漠拟橙衣系沙坡头常见种，其主要特征是地衣体表面覆粉霜，边缘有明显的长条状裂片，子囊孢子为 2 室；而拟橙衣地衣体表面覆粉霜，边缘无裂片，子囊孢子单胞；南方拟橙衣石生，地衣体表面黄色，龟裂，无粉霜。

20 凹面茸枝衣

Seirophora lacunosa (Rupr.) Frödén, Lichenologist 36(5): 297, 2004.

土生：翠柳沟，1,600m，2010.4.17，张涛 SPT10110。

21 柔毛茸枝衣

Seirophora villosa (Ach.) Frödén, Lichenologist 36(5): 297, 2004.

=*Teloschistes brevior* (Nyl.) Vain. Magnusson 1940.

土层：翠柳沟，1,669m，2010.7.3，曹叔楠和刘文婧 SPT10019。

22 丽黄石衣

Xanthoria elegans (Link) Th. Fr., Lich. Arct.: 69, 1860.

石生：翠柳沟，1,600m，2009.5.29，张涛 NX09017，曹叔楠和刘文婧 SPT11019，SPT10051（1,469m，2010.7.5），刘萌 SPT11021-1，SPT11022（1,628m，2011.4.13）；青山根，1,549m，2011.4.19，周启明和曹叔楠 ZW11059，ZW11036；香山，1,377m，2011.4.17，刘萌 SPT11048。

土生：翠柳沟，1,700m，2008.9.4，张涛和曹叔楠 SPT6-4，1,600m，2010.5.28，张涛 NX09029；青山，1,508m，2011.4.19，刘萌 SPT11076。

[REFERENCES]

Culberson CF, 1972. Improved conditions and new data for identification of lichen products by standardized thin-layer chromatographic method. *Journal of Chromatography A*, 72(1): 113-125

Culberson CF, Kristinsson HD, 1970. A standardized method for the identification of lichen products. *Journal of Chromatography A*, 46: 85-93

Kirk PM, Cannon PF, Minter DW, Stalpers JA (eds.), 2008. Dictionary of the fungi. 10th edition. CABI Europe, UK. 1-784

Li SZ, Zheng HZ, Li SL, Shen BC, 2008. Development characteristics of biotic crusts on Shapotou vegetated sand dunes. *Chinese Journal of Ecology*, 27(10): 1675-1679 (in Chinese)

Li XR, Wang XP, Li T, Zhang JG, 2002. Microbiotic soil crust and its effect on vegetation and habitat on artificially stabilized desert dunes in Tengger Desert. North China. *Biology and Fertility of Soils*, 35: 147-154

Magnusson AH, 1940. Lichens from Central Asia I. Rep.Sci. Exped. N.W.China S.Hedin – The Sino-Swedish expedition – (Publ.13). XI. Bot.I: 1-168. Pl.1-12, 1 folded map.f.1-3

Magnusson AH, 1944. Lichens from Central Asia II. Rep.Sci. Exped. N.W.China S.Hedin – The Sino-Swedish expedition – (Publ.13). XI. Bot.II: 1-68. Pl.1-8, 1 text map

McCarthy PM, 2001. Trichotheliaceae, Fl. *Australia*, 58A:

105-157

McCune B, Rosentreter R (eds.), 2007. Biotic soil crust lichens of the Columbia Basin. Northwest Lichenologists, Corvallis. 1-105

Nash III TH, Gries C, Bungartz F (eds.), 2007. Lichen flora of the Greater Sonoran Desert Region. Vol. 3. Lichens Unlimited, Arizona State University, Arizona. 1-567

Nash III TH, Ryan BD, Diederich P, Gries C, Bungartz F (eds.), 2004. Lichen flora of the Greater Sonoran Desert Region, Vol.2. Lichens Unlimited, Arizona State University, Arizona. 1-742

Nash III TH, Ryan BD, Gries C, Bungartz F (eds.), 2002. Lichen flora of the Greater Sonoran Desert Region, Vol. 1. Lichens Unlimited, Arizona State University, Arizona. 1-532

Obermayer W, 2004. Additions to the lichen flora of the Tibetan region. *Bibliotheca Lichenologica*, 88: 476-526

Wang YY, Zhang T, Zhou QM, Wei JC, 2011. Construction and characterization of a full-length cDNA library from mycobiont of *Endocarpon pusillum* (lichen-forming Ascomycota). *World Journal of Microbiology and Biotechnology*, 27(12): 2879-2884

Wei JC, 1991. An enumeration of lichens in China. International Academic Publishers, Beijing. 1-278

Wei JC, 2005. Desert biological carpet engineering—new way of arid desert control. *Arid Zone Research*, 22(3): 287-288 (in Chinese)

Zhang T, Wei JC, 2011. Survival analyses of symbionts isolated from *Endocarpon pusillum* Hedwig to desiccation and starvation stress. *Science China Life Sciences*, 54(5): 480-489

[附中文参考文献]

李守中，郑怀舟，李守丽，沈宝成，2008. 沙坡头植被固沙区生物结皮的发育特征，生态学杂志，27(10): 1675-1679

魏江春，2005. 沙漠生物地毯工程——干旱沙漠治理的新途径. 干旱区研究，22(3): 287-288

海南地衣多样性考察及其资源研发前景

魏江春[1*]，贾泽峰[2]，吴兴亮[3]

（1. 中国科学院微生物研究所真菌学国家重点实验室，北京 100101；2. 聊城大学生命科学学院，聊城 252059；3. 海南大学，海口 570228）

摘　要：文中对中国海南岛地区地衣多样性进行了考察并对其资源研究与开发提出了建议。笔者对生物多样性的概念做了阐明；对于考察中的有关地衣资源做了简报，对地衣多样性及其资源的研究与开发及其三大存取系统进行了论述。报道了海南地衣 138 种，其中中国新记录种 14 种。

关键词：物种多样性；中国新记录种；存取系统；地衣型真菌

中图分类号：Q949.34　　**文献标识码**：A　　**文章编号**：1672-3538(2013)04-0224-15

引文格式：魏江春，贾泽峰，吴兴亮. 海南地衣多样性考察及其资源研发前景[J]. 菌物研究，2013，11(4)：224-238.

An Investigation of Lichen Diversity from Hainan Island of China and Prospect of the R &D of Their Resources

—Dedicated to the famous mycologist Prof. Wang Yun-chang

WEI Jiang-chun[1], JIA Ze-feng[2], WU Xing-liang[3]

(1. *State Key Laboratory of Mycology*, *Institute of Microbiology*, *CAS*, *Beijing* 100101, *China*; 2. *College of Life Sciences*, *Liaocheng University*, *Liaocheng* 252059, *China*; 3. *Hainan University*, *Haikou* 570228, *China*)

Abstract: This study investigated the lichen diversity of Hainan island of China. A review of the lichen resources from the island and prospect of R &D of them are also given. A concept and definition of the biodiversity and three storage &retrieval systems are also discussed. One hundred and thirty eight lichen species, including 14 species new to China, from the island are reported.

Key words: species diversity; species new to China; storage &retrieval system; lichen-forming fungi

1　概　述

地衣多样性是生物多样性的组成部分。生物多样性通常被理解为物种多样性、基因多样性和生态系统多样性 3 个层次。物种是由表型与基因型彼此相似的生物居群组成的。而居群是由表型与基因型彼此相似的生物个体组成的。基因本身在生物体之外是没有生存价值的。因此，物种是基因的载体，1 个物种 1 个基因库。没有物种便没有基因。没有物种多样性便没有基因多样性。没有物种，就无法进行结构基因组学和功能基因组学的研究、开发和利用。所以，生物多样性是指生存于地球上多种多样生态系统中的，含有多种多样基因的物种多样性；简言之，在多样性的生态系统中生存着丰富的物种多样性[1]。

菌物物种多样性在人类可持续发展中，尤其

* **作者简介**：魏江春，男，院士，研究员，研究方向：地衣真菌学。

收稿日期：2013-10-13

通讯作者：魏江春，E-mail：weijc2004@126.com

本文原载于《菌物研究》，11（4）：224-238, 2013.

在人类健康与环境保障中具有极为重要的意义，是取之不尽、用之不竭的可再生资源宝库。保护生态系统多样性是物种多样性，也是基因多样性保护的关键。在此基础上，对物种进行采集、分析、识别、归类、命名、描述等一系列研究；将其研究结果储备于信息、原型标本和培养物3个存取系统中，是菌物物种资源和基因资源保护、研究、开发和利用的上游关键环节[1]。

地衣在生物学概念上是一群专化型真菌和藻类及/或蓝细菌共生的生物学类群，是稳定的菌藻共生群落的生态系统。这一稳定系统中的每一种成员各自具有各自的科学名称及其演化系统地位。所谓地衣的科学名称，正是参与共生的地衣型真菌的科学名称，当然，这里并不包括一切内生真菌。因此，多种多样的地衣物种在系统演化和分类学上隶属于真菌界的各级有关阶元。

据专家对地球生物圈中生存的菌物物种多样性的定量估计，以菌物中的真菌为例，按全世界有25万种维管束植物为基数，每种维管束植物按6种真菌估计，应有真菌150万种[2]；每种维管束植物按4种内生真菌估计，应有100万种[3-4]。以此估计，地球生物圈中生存的真菌至少有250万种。然而，已被人类所认识和命名的真菌仅9.7861万种[5]，占估计种数的3.9%，尚有96.1%的真菌有待人类去发现、认识、命名、描述、研究和开发利用。

我国已知维管束植物为3万种，按照上述方法估计，我国真菌至少有30万种。不过，已被命名的真菌仅1.47万种[6]，占估计种数的4.9%；尚有95.1%的真菌物种有待我国真菌学家去发现、认识、命名、描述、研究和开发利用[1]。

除内生菌外，在全世界150万种真菌[2]中，地衣型真菌按20%估计[5]应为30万种。我国已知维管束植物为3万种，按照上述方法估计，我国不包括内生真菌在内的真菌约18万种，其中地衣型真菌按20%[5]计应为3.6万种左右。然而，已有报道的我国地衣型真菌近2 000种，占估计种数的5.5%，尚有94.5%的地衣型真菌有待我国地衣学家去发现、认识、命名、描述、研究和开发利用。

2 海南地衣资源考察与初步研究

海南岛地处热带地区，生物多样性极为丰富，其中地衣型真菌的物种多样性，尤其是附生在植物茎叶上的物种多样性更为丰富。

在考察期间，我们对海南岛的尖峰岭、霸王岭、五指山、黎母山、吊罗山、七仙岭、甘什岭、铜鼓岭等自然保护区与东郊椰林等沿海地带以及蜈支洲岛等生态条件下的地衣物种多样性进行了考察和标本采集。除尚未定名的大量地衣标本外，已经定名和入柜的地衣标本为3 762号。

2.1 地衣型真菌次生代谢产物及其生物学活性初报

从采自海南岛霸王岭树干上的星果衣（*Astrothelium* sp.）地衣（图1）的共生菌中发现了3种具有抗菌活性的次生代谢产物，它们是Trypethelone和trypethelone methyl ether，另一种为新结构的萘醌类化合物，被命名为7-hydroxyl-8-methoxyl-trypethelone。它和已知的化合物trypethelone methyl ether对于*Enterococcus faecalis* 850E（CGMCC 1.2135）均具有极强的抗菌活性；3种化合物对金黄色葡萄球菌[*Staphylococcus aureus* col（MRSA）（CGMCC 1.2465）]均具有极强的抗菌活性。而其中的Trypethelone则对枯草芽孢杆菌[*Bacillus subtilis*（ATCC6633）]具有显著的抗菌活性[7]。

图1 星果衣（*Astrothelium* sp.）原型

Fig. 1. Prototype of *Astrothelium* sp.

2.2 地衣型真菌内生菌的次生代谢产物及其生物学活性初报

从采自海南岛霸王岭的中华藻珊瑚（*Multiclavula sinensis* R. Petersen et M. Zaang.，图2）中分离出地衣内生真菌类盘多毛菌（*Pestalotiopsis* sp.，图3）中发现了23种次生代谢产物，其中21种为首次发现的新结构化合物，这些化合物中有的具有明显的抗金黄色葡萄球菌活性[8]。

2.3 潜在的石蕊资源地衣种类的发现

众所周知的酸碱度指示剂石蕊（litmus，pH 4.5为红色，pH 8.3为蓝色）是色素的复杂混合物，产于染料衣属（*Roccella* spp.）的不同种类中，如 *R*. *phycopsis*（Ach.）Ach. 和 *R*. *tinctoria* DC. 等[9-10]。

产于沿海岩石上的中华染料衣（*R*. *sinensis* Nyl.，图 4）的模式产地为我国港澳地区[11-12]。蔓氏染料衣（*R*. *montagnei* Bél.，图 5）曾有报道在我国浙江树皮上有分布[12]。上述 2 种地衣在海南考察中均有所发现，也是近 23 年来（前者）和 83 年来（后者）在国内的首次发现。它们分别分布于铜鼓岭沿海岩石（图 4）和沿海椰林树干上（图 5）。

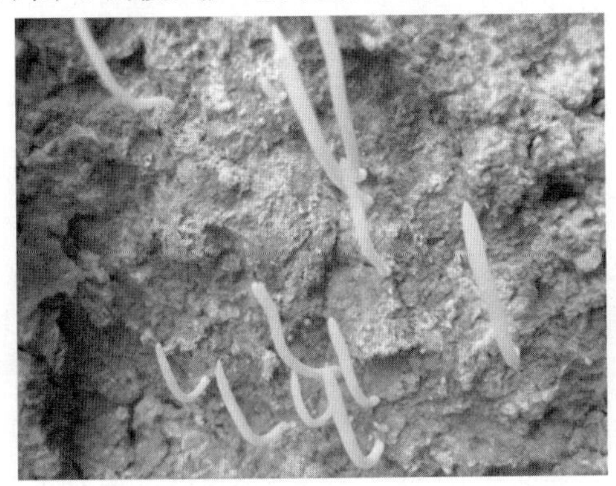

图 2 中华藻珊瑚（*Multiclavula sinensis*）原型

Fig. 2. Prototype of *Multiclavula sinensis*

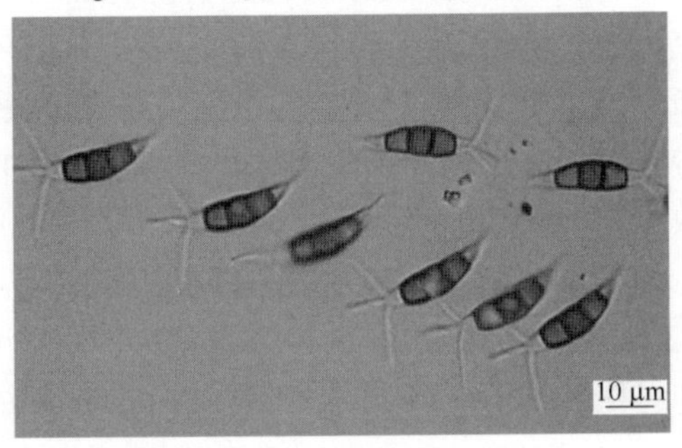

图 3 内生真菌（*Pestalotiopsis* sp.）的分生孢子

Fig. 3. Conidia of *Pestalotiopsis* sp.

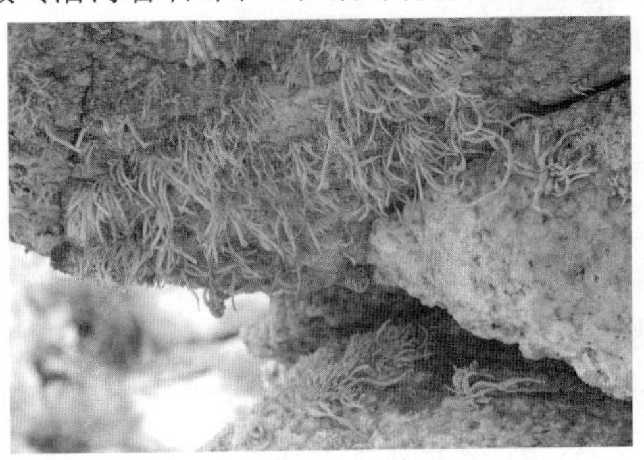

图 4 中华染料衣（*Roccella sinensis*）原型

Fig. 4. Prototype of *Roccella sinensis*

图 5 曼氏染料衣（*Roccella montagnei*）原型

Fig. 5. Prototype of *Roccella montagnei*

3 地衣对三华李果园丰产的危害

三华李属于蔷薇科李属中的一种。三华李果肉含有丰富营养物质，肉色深红，气味芳香，肉质松脆爽口，无涩味，果味清甜。成熟期在农历芒种到夏至之间，史有“岭南夏令果王”之称。

在考察期间，根据果农的反映，由于地衣的危害，导致三华李果粒无收。我们在对海南岛尖峰岭地区的三华李果园的考察中发现，有的三华李果园中果树竟然成片干枯。果树茎秆和枝条上均长满了地衣，其中以叶状地衣和壳状地衣为主，也有少量枝状地衣。

4 海南地衣资源评估与开发前景

4.1 地衣抗癌多糖等的筛选与开发

绝大多数地衣均含有抗癌多糖，但不一定每种地衣均含有抗癌活性最高的乙酰基-β-1，6-葡聚糖。此外，地衣还富含抗辐射、抗细菌和抗真菌的活性物质。对于上述有利于人类健康的海南地衣产物的筛选与开发具有相当看好的前景。

4.2 地衣型菌藻工业化生产的前景

地衣型菌藻的生长速度通常要比非地衣型真菌慢。但是，试验证实，地衣型真菌利用液体深层培养要比固体培养的速度快。因此，通过培养基以及培养条件的优化试验，提高地衣型真菌的培养速度，使之符合大规模工业化生产水平的可能性是存在的。

4.3 三华李果树地衣病害问题

因为地衣是菌藻共生生态系统，地衣型真菌是借助于其共生藻或蓝细菌的光合作用获得有机营养。因此，人们通常认为生长在树皮上的地衣应该不会给树木的生命活动造成严重危害。然而，在海南岛，长满地衣的三华李果树成片干枯，果农果粒无收的现实，给人们提出了需要对上述观念进行认真检验的必要。因此，进一步通过试验观察和分析研究以揭示内幕，探明真相，以便采取对症措施是必要的。

4.4 地衣物种多样性研究、开发与三大存取系统的建立

含有多样性基因的多样性地衣物种，存在于地球生物圈中多样性的生态系统中。地衣生物多样性的保护在于地衣物种多样性和基因多样性的保护，归根结底是物种多样性的保护。地衣物种多样性的保护又在于生态系统多样性的保护。地衣物种多样性的另一种保护途径还在于迁地保护。对于地衣物种多样性的迁地保护就在于从自然界地衣物种多样性中采集并分离培养出尽可能多的地衣型菌藻的物种多样性并保存在菌藻物种库里。因此，地衣资源研究的上游工作，进行地衣物种多样性研究、开发与三大存取系统，即信息存取系统菌藻培养物存取系统以及地衣生物原型存取系统的建立是必不可少的。

(1)对于它们在表型与基因型相结合中进行研究，按照其各自的演化关系，以物种为基本单位，将它们排列成不同等级的演化系统，从而形成地衣型真菌的物种综合信息存取系统，即演化系统生物学或分类学著作以及菌物学报（Mycosystema）等，为生命科学研究和生物资源开发利用提供地衣生物参考。

(2)将多样性物种的活体菌种藻种403份保藏于中国科学院微生物研究所作为物种资源和基因资源存取系统的菌藻种库，为生命科学研究和生物资源开发利用提供包括基因多样性的地衣菌种和藻种的物种资源存取系统。

(3)从地衣中分离培养并保存备用的活体菌种和藻种，与之生长在自然界多种生态系统中的地衣生物原型在表型特征上截然不同。为了辨认和对接保存在菌藻种库中的活体菌种藻种和自然界的地衣生物原型，将采自自然界的地衣生物的多样性物种原型标本12万号保存于中国科学院微生物研究所国家菌物标本馆地衣标本室（HMAS－L），作为地衣物种原型证据存取系统，对生命科学研究和地衣物种资源及其基因资源的研究与开发具有重要意义。

3个存取系统，是新世纪生命科学研究和生物资源研究开发和利用不可缺的上游环节和重要支撑系统与后盾。这3个存取系统越丰富，则资源筛选的基数就越大，筛选效率就越高。如果能将海南，乃至全国的地衣以及菌物物种多样性及其演化系统生物学与资源生物学的研究与三大存取系统作为一个整体予以重视并给予强力资助，必将在可再生资源创新与人类可持续发展中起到重要作用。

4.5 建议

建议在该专项结题的基础上，似应针对具有资源开发前景的生物类群进行进一步试验研究与开发给予资助，使物种多样性及其资源考察与评估专项与进一步资源开发有所衔接。

5 海南岛地衣型真菌名录

本名录中的物种名称仅系海南岛菌物考察中所采集的大量标本中已被定名的部分。其中个别种类虽未正式定名，但因进行过资源方面的初步研究，也一并列入名录。至于其余标本，包括可能是新种的标本，将在后续研究中逐步研究和报道，为进行海南岛地衣资源研究与开发提供必要的上游数据和科学依据。

该名录包括138种地衣，分别隶属于真菌界的43属，17科，10目，4纲和2门。其中中国新记录种14个：厚粉霜盘衣［*Diorygma hololeucum* (Mont.) Kalb, Staiger & Elix.］、鸽色裂痕衣［*Fissurina columbina* (Tuck.) Staiger］、灰绿裂痕衣［*Fissurina glauca* (Müll. Arg.) Staiger］、仲马裂痕衣（*Fissurina dumastii* Fée）、小孢裂痕衣［*Fissurina nitidescens* (Nyl.) Nyl.］、密集文字衣（*Graphis conferta* Zenker）、线丝文字衣（*Graphis filiformis* Adaw. & Makhija）、骨针文字衣（*Graphis oxyclada* Müll. Arg.）、短盘文字衣（*Graphis plagiocarpa* Fée）、根生文字衣［*Graphis rhizocola* (Fée) Lücking & Chaves］、黑厚基衣（*Leiorreuma melanostalazans* (Leight.) A. W. Archer）、米勒凸唇衣（*Platygramme mülleri* (A. W. Archer) Staiger）、平滑平盘衣［*Platythecium leiogramma* (Nyl.) Staiger］及蛇形平盘衣［*Platythecium serpentinellum* (Nyl.) Staiger］。名录中各门、纲、目、科中的属、种名称按字母顺序排列。

5.1 子囊菌门（Ascomycota）

座囊菌纲（Dothideomycetes）

乳嘴衣目（Trypetheliales）

乳嘴衣科（Trypetheliaceae）

(1)星果衣 *Astrothelium* sp.

基物：树皮。

标本信息：海南霸王岭，900 m，2007-05-19，孟庆峰 M256（HMAS-L-120613，图 1）

茶渍衣纲（Lecanoromycetes）

茶渍衣亚纲（Lecanoromycetidae）

茶渍衣目（Lecanorales）

梅衣科（Parmeliaceae）

（2）裂芽球针叶 *Bulbothrix isidiza*（Nyl.）Hale

基物：树皮。

标本信息：海南乐东县，尖峰岭，1993-04-02，姜玉梅、郭守玉，H-0727（HMAS-L-021998）；五指山，1993-04-20，姜玉梅、郭守玉，H-1107（HMAS-L-021993）；陵水县，吊罗山，2001-06-26，陈建斌、胡光荣、徐蕾，20552（HMAS-L-021997）。

（3）烟草球针叶 *Bulbothrix tabacina*（Mont. & Bosch）Hale

基物：树皮。

标本信息：海南昌江县，霸王岭，1993-03-28，姜玉梅、郭守玉，H0278（HMAS-L-021987）；五指山，2001-06-28，陈建斌、胡光荣、徐蕾，20603（HMAS-L-021971）；乐东县，尖峰岭，2001-06-29，陈建斌、胡光荣、徐蕾，20770（HMAS-L-021986）。

（4）大叶斑衣 *Canomaculina subtinctoria*（A. Zahlbr.）Elix

基物：树皮。

标本信息：海南昌江县，霸王岭雅加林场，1993-03-29，姜玉梅、郭守玉，0429（HMAS-L-018606）；五指山，2001-06-28，陈建斌、胡光荣、徐蕾，20645（HMAS-L-019891）。

（5）针芽条衣 *Everniastrum vexans*（A. Zahlbr.）Hale ex Sipman

基物：树皮。

标本信息：海南，1957-04-24，巴良斯基，HMAS-L-008520。

（6）义笃双岐根 *Hypotrachyna ikomae*（Asahina）Hale

基物：树皮。

标本信息：海南五指山，2001-06-28，陈建斌、胡光荣、徐蕾，20635（HMAS-L-018771）。

（7）覆瓦双崎根 *Hypotrachyna imbricatula*（Zahlbr.）Hale

基物：朽木。

标本信息：海南五指山，2001-06-28，陈建斌、胡光荣、徐蕾，20641（HMAS-L-018798）。

（8）双歧根 *Hypotrachyna koyaensis*（Asahina）Hale

基物：树皮。

标本信息：海南昌江县，霸王岭雅加林场，1993-03-29，姜玉梅、郭守玉，H0414-1（HMAS-L-018819）。

（9）骨白双歧根 *Hypotrachyna osseoalba*（Vain.）Y. S. Park & Hale

基物：岩石。

标本信息：海南昌江县，霸王岭雅加林场，1993-03-29，姜玉梅、郭守玉，H0385（HMAS-L-018676）；乐东县，尖峰岭，1993-04-01，姜玉梅、郭守玉，H-0550（HMAS-L-018678）；五指山，1993-04-20，姜玉梅、郭守玉，H-1074（HMAS-L-018701）。

（10）灰条双歧根 *Hypotrachyna pseudosinuosa*（Asahina）Hale

基物：岩石。

标本信息：海南乐东县，尖峰岭，2001-06-30，陈建斌、胡光荣、徐蕾，20779（HMAS-L-018713）。

（11）金髓缘毛衣 *Myelochroa aurulenta*（Tuck.）Elix & Hale

基物：藓层。

标本信息：海南五指山，1993-04-20，姜玉梅、郭守玉，H-1088（HMAS-L-016357）。

（12）裂芽大缘毛衣 *Parmelinella wallichiana*（Tayl.）Elix & Hale

基物：岩石。

标本信息：海南昌江县，霸王岭雅加林场，1993-03-29，姜玉梅、郭守玉，H0392（HMAS-L-015918）。

（13）白髓缘毛衣 *Parmelinopsis horrescens*（Taylor）Elix & Hale

基物：树皮。

标本信息：海南五指山，2001-06-28，陈建斌、胡光荣、徐蕾，20614-1（HMAS-L-115020）。

（14）原岛白髓缘毛衣 *Parmelinopsis protocetrarica* Elix

基物：树皮。

标本信息：海南五指山，2001-06-28，陈建斌、胡光荣、徐蕾，20614（HMAS-L-018269）。

（15）泡沫白髓缘毛衣 *Parmelinopsis spumosa*（Asah.）Elix & Hale

基物：树皮。

标本信息：海南五指山，2001-06-28，陈建斌、胡光荣、徐蕾，20635-2(HMAS-L-018254)。

(16)亚消白髓缘毛衣 *Parmelinopsis subfatiscens* (Kurok.) Elix &Hale

基物：树皮。

标本信息：海南五指山，2001-06-28，陈建斌等，20635-3(HMAS-L-018259)。

(17)佩缘毛大叶梅 *Parmotrema arnoldii* (Du Rietz) Hale

基物：树皮。

标本信息：海南昌江县，保护区，1993-03-30，姜玉梅、郭守玉，H-0461(HMAS-L-016072)；乐东县，尖峰岭，1993-04-01，姜玉梅、郭守玉，H-0557(HMAS-L-016700)。

(18)毛大叶梅 *Parmotrema crinitum* (Ach.) Choisy

基物：树干。

标本信息：海南陵水县，吊罗山，2000-12-14，黄满荣，288(HMAS-L-085744)。

(19)鸡冠大叶梅 *Parmotrema cristiferum* (Tayl.) Hale

基物：岩石，树皮。

标本信息：海南五指山，1993-04-20，姜玉梅、郭守玉，H-1108(HMAS-L-016128)；陵水县，吊罗山，2001-06-26，陈建斌、胡光荣、徐蕾，20429(HMAS-L-085742)。

(20)宽大叶梅 *Parmotrema dilatatum* (Vain.) Hale

基物：树皮，岩石。

标本信息：海南乐东县，尖峰岭，1993-04-02，姜玉梅、郭守玉，H-0729(HMAS-L-020044)；五指山，2001-06-27，陈建斌、胡光荣、徐蕾，20555(HMAS-L-018846).

(21)罗微大叶梅 *Parmotrema lovisianae* (Hale) Hale

基物：树枝。

标本信息：海南乐东县，尖峰岭，2001-06-29，陈建斌、胡光荣、徐蕾，20802(HMAS-L-018850)。

(22)玛尔嘎大叶梅 *Parmotrema margaritatum* (Hue) Hale

基物：树皮。

标本信息：海南昌江县，霸王岭，1993-03-28，姜玉梅、郭守玉，H-0312(HMAS-L-020048)。

(23)麦氏大叶梅 *Parmotrema mellissii* (Dodge) Hale

基物：岩石，树枝。

标本信息：海南昌江县，霸王岭雅加林场，1993-03-29，姜玉梅、郭守玉，H-0399(HMAS-L-016171)；五指山，1993-04-20，姜玉梅、郭守玉，H-1064(HMAS-L-016172)。

(24)裂芽大叶梅 *Parmotrema praeisidiosum* Fleig.,

基物：树皮。

标本信息：海南昌江县，昌江保护区，1993-03-30，H0483(HMAS-L-018856)。

(25)粉芽大叶梅 *Parmotrema rampoddense* (Nyl.) Hale

基物：树皮，树枝。

标本信息：海南陵水县，吊罗山，2001-06-26，陈建斌、胡光荣、徐蕾，20525(HMAS-L-018859)；五指山，2001-06-28，陈建斌、胡光荣、徐蕾，20628(HMAS-L-018860)。

(26)囊瓣大叶梅 *Parmotrema saccatilobum* (Tayl.) Hale

基物：树皮。

标本信息：海南乐东县，1993-04-03，尖峰岭，姜玉梅、郭守玉，H-0713(HMAS-L-020088)；乐东县，尖峰岭，1993-04-03，姜玉梅、郭守玉，H-0717(HMAS-L-016843)；万宁市，兴隆农场15队，1993-04-06，郭守玉，H-0750(HMAS-L-019954)；陵水县，2001-06-26，陈建斌、胡光荣、徐蕾，20537(HMAS-L-018862)；五指山，2001-06-28，陈建斌、胡光荣、徐蕾，20601(018863)。

(27)拟佩缘毛大叶梅 *Parmotrema subarnoldii* (des Abb.) Hale

基物：树皮。

标本信息：海南乐东县，尖峰岭，1993-04-01，姜玉梅、郭守玉，H-0548(HMAS-L-018865)。

(28)亚珊瑚大叶梅 *Parmotrema subcorallinum* (Hale) Hale

基物：树皮。

标本信息：海南昌江县，雅加林场，1993-03-29，姜玉梅、郭守玉，H0422(HMAS-L-020059)；乐东县，尖峰岭，1993-04-02，姜玉梅、郭守玉，H-0721(HMAS-L-020062)。

(29)亚赭大叶梅 *Parmotrema subochraceum* Hale

基物:树皮。

标本信息:海南昌江县,霸王岭,1993-03-28,姜玉梅、郭守玉,320-1(HMAS-L-020063)。

(30)大叶梅 *Parmotrema tinctorum* (Dilese ex Nyl.) Hale

基物:树皮。

标本信息:海南昌江县,霸王岭,1993-03-28,姜玉梅、郭守玉,H0286(HMAS-L-110682);乐东县,尖峰岭,1993-04-01,姜玉梅、郭守玉,H-0571(HMAS-L-016026);陵水县,吊罗山,2001-06-26,陈建斌、胡光荣,徐蕾,20494(HMAS-L-018868)。

(31)卓氏大叶梅 *Parmotrema zollingeri* (Hepp) Hale

基物:树枝。

标本信息:海南五指山,1993-04-20,姜玉梅、郭守玉,H-1001-1(HMAS-L-016271)。

(32)玄球针黄叶 *Relicina abstrusa* (Vain.) Hale

基物:树皮,朽木。

标本信息:海南昌江县,霸王岭,1993-03-28,姜玉梅、郭守玉,H0288(HMAS-L-022019);五指山,2001-06-28,陈建斌、胡光荣、徐蕾,20622(HMAS-L-110473)。

(33)悉尼球针黄叶 *Relicina sydneyensis* (Gyeln.) Hale

基物:树皮。

标本信息:海南昌江县,霸王岭,1993-03-28,姜玉梅、郭守玉、H-0310(HMAS-L-022009);五指山,1993-04-20,姜玉梅、郭守玉,H-1052(HMAS-L-022011)。

(34)睫毛网裂梅 *Rimelia cetrata* (Ach.) Hale & Fletcher

基物:树枝。

标本信息:海南昌江县,霸王岭雅加林场,1993-03-29,姜玉梅、郭守玉,H0439(HMAS-L-110864)。

(35)网裂梅 *Rimelia reticulata* (Tayl.) Hale & Fletcher

基物:树皮,岩石,枯枝。

标本信息:海南昌江县,霸王岭,1993-03-28,姜玉梅、郭守玉,H0287(HMAS-L-020108);乐东县,尖峰岭,1993-04-01,姜玉梅、郭守玉,H-0552(HMAS-L-015675);五指山,1993-04-20,姜玉梅、郭守玉,H-1060(HMAS-L-015680)。

(36)亚裂芽网裂衣 *Rimelia subisidiosa* (Muell. Arg.) Hale & Fletcher

基物:岩石。

标本信息:海南五指山,1993-04-20,姜玉梅、郭守玉,H-1072(HMAS-L-016713)。

(37)棒黄梅 *Xanthoparmelia claviculata* Kurok.

基物:岩石。

标本信息:海南五指山,2001-06-28,陈建斌、胡光荣、徐蕾,20637(HMAS-L-104292)。

粉果衣科(Sphaerophoraceae)

(38)扁枝粉果衣 *Bunodophoron melanocarpum* (Sw.) Wedin

基物:树皮。

标本信息:海南五指山一峰,1 850 m,2000-12-16,黄满荣,362(HMAS-L-X003795);霸王岭,1 170 m,2000-12-07,黄满荣,123(HMAS-L-003810)。

石蕊科(Cladoniaceae)

(39)聚筛蕊 *Cladia aggregata* (Sw.) Nyl.

基物:地上。

标本信息:海南乐东县,尖峰岭,2001-06-30,陈建斌、胡光荣、徐蕾,20784(HMAS-L-110901)。

(40)粉杆石蕊 *Cladonia bacillaris* Nyl.

基物:朽木。

标本信息:海南琼中县,黎母山,1993-03-23,姜玉梅、郭守玉,H0052(HMAS-L-020794)。

(41)亚麸石蕊 *Cladonia fruticulosa* Kremp.

基物:岩石。

标本信息:海南琼中县,1993-03-23,姜玉梅、郭守玉,H0050(HMAS-L-018896);昌江县,霸王岭,1993-03-28,姜玉梅、郭守玉,H0266(HMAS-L-018901);昌江县,霸王岭雅加林场,1993-03-29,姜玉梅、郭守玉,H0334(HMAS-L-018902);乐东县,尖峰岭,1993-04-01,姜玉梅、郭守玉,H-0574(HMAS-L-018895)。

(42)柄红石蕊 *Cladonia macilenta* Hoffm.

基物:地上。

标本信息:海南昌江县,保护区,1993-03-30,姜玉梅、郭守玉,琼494(HMAS-L-084022)。

(43)蒙古石蕊 *Cladonia mongolica* Ahti

基物：朽木。

标本信息：海南昌江县，保护区，1993-03-30，姜玉梅、郭守玉，H-0508（HMAS-L-029728）。

（44）麸皮石蕊 *Cladonia ramulosa* （With.） Laundon

基物：地上，石上，朽木上。

标本信息：海南琼中县，黎母山，1993-03-24，姜玉梅、郭守玉，H0060（HMAS-L-020592）；昌江县，霸王岭，1993-03-28，姜玉梅、郭守玉，H0267（HMAS-L-020591）；昌江县，霸王岭，1993-03-29，姜玉梅、郭守玉，H0343（HMAS-L-018239）；昌江县，霸王岭雅加林场，1993-03-29，姜玉梅、郭守玉，H0340（HMAS-L-029745）；乐东县，尖峰岭，1993-04-01，姜玉梅、郭守玉，H-0580（HMAS-L-020442）。

（45）平杯石蕊 *Cladonia rappii* A. Evans

基物：地上。

标本信息：海南昌江县，霸王岭，1993-03-28，姜玉梅、郭守玉，H0269（HMAS-L-018446）；昌江县，霸王岭，1993-03-28，姜玉梅、郭守玉，H-0296（HMAS-L-018447）。

（46）裂杯石蕊 *Cladonia rei* Schaer.

基物：朽木。

标本信息：海南陵水县，吊罗山，1993-04-11，郭守玉，H-0816-1（HMAS-L-029593）。

（47）拟枪石蕊 *Cladonia subradiata* （Vain.） Sandst.

基物：地上。

标本信息：海南昌江县，霸王岭，1993-03-29，姜玉梅、郭守玉，H0350（HMAS-L-023992）；乐东县，尖峰岭，1993-04-04，姜玉梅、郭守玉，H-711（HMAS-L-023995）；乐东县，尖峰岭，1993-04-04，姜玉梅、郭守玉，H-711（HMAS-L-023995）。

（48）缝裂扇盘衣 *Thysanothecium scutellatum* （Fr.） D. Galloway

基物：朽木。

标本信息：海南昌江县，霸王岭，1993-03-28，姜玉梅、郭守玉，H-304、H-305、H-306（HMAS-L-012880，012878，012879）。

珊瑚枝科（Stereocaulaceae）

（49）小珊瑚枝 *Stereocaulon octomerellum* Müll. Arg.

基物：岩石。

标本信息：海南乐东县，尖峰岭，2001-06-30，陈建斌、胡光荣、徐蕾，20796-1（HMAS-L-028823）。

（50）大珊瑚枝 *Stereocaulon sorediiferum* Hue

基物：岩石。

标本信息：海南陵水县，吊罗山，2000-12-13，黄满荣，275（HMAS-L-085393）。

（51）有珊瑚枝 *Stereocaulon verruculigerum* Hue

基物：岩石。

标本信息：海南乐东县，尖峰岭，2001-06-30，陈建斌、胡光荣、徐蕾，20796（HMAS-L-028930）。

赤星衣科（Haematommataceae）

（52）赤星衣 *Haematomma* sp.

基物：三华李果树皮。

标本信息：海南尖峰岭三华李果园，果树皮上。

树花衣科（Ramalinaceae）

（53）瘤枝树花 *Ramalina intermediella* Wain.

基物：树干。

标本信息：海南乐东县，尖峰岭，1974-05-20，郭英兰，HMAS-L-029515；2011-06-28，陈建斌、胡光荣、徐蕾，20618（HMAS-L-110490）。

地卷目（Peltigerales）

瓦衣科（Coccocarpiaceae）

（54）粗瓦衣 *Coccocarpia palmicola* （Spreng.） Arvidsson & D. Galloway

基物：树皮。

标本信息：海南，1934-02-18，彦圣，T. L. Tso，2726（HMAS-L-011581）；Mt. Qinlin，1957-02-22，巴良斯基，HMAS-L-011577。

胶衣科（Collemataceae）

（55）扁胶衣 *Collema complanatum* Hue

基物：树皮，苔藓上。

标本信息：海南昌江县，保护区，1993-03-30，姜玉梅、郭守玉，H-0499（HMAS-L-021361）；五指山，2001-06-27，陈建斌、胡光荣、徐蕾，20574（HMAS-L-021360）。

（56）亚石胶衣 *Collema subflaccidum* Degel.

基物：岩石，树皮。

标本信息：海南陵水县，吊罗山，1993-04-11，郭守玉，H-0841（HMAS-L-021242）；五指山，2001-06-28，陈建斌、胡光荣、徐蕾，20662（HMAS-L-021241）。

肺衣科（Lobariaceae）

(57)针芽肺衣 *Lobaria isidiophora* Yoshim.

基物:土层。

标本信息:海南,1957-04-24,巴良斯基,1023(HMAS-L-006954)。

(58)银白假杯点衣 *Pseudocyphellaria argyracea* (Del.) Vain.

基物:岩石。

标本信息:海南五指山,2001-06-28,陈建斌、胡光荣、徐蕾,20660(HMAS-L-019964)。

(59)黄褐假杯点衣 *Pseudocyphellaria cinnamomea* (Rich.) Vain.

基物:树皮。

标本信息:海南陵水县,吊罗山,2001-06-26,陈建斌、胡光荣、徐蕾,20491(HMAS-L-110334)。

(60)金缘假杯点衣 *Pseudocyphellaria crocata* (L.) Vain.

基物:树皮。

标本信息:海南五指山,2001-06-28,陈建斌、胡光荣、徐蕾,20612(HMAS-L-019966)。

(61)垫状牛皮叶 *Sticta pulvinata* (Mey. et Flot.) Vain.

基物:松树,岩石。

标本信息:海南陵水县,吊罗山,2001-06-26,陈建斌、胡光荣、徐蕾,20532(HMAS-L-110489)。

黄枝衣目(Teloschistales)

粉果衣科(Caliciaceae)

(62)扁平黑囊基衣 *Dirinaria applanata* (Fee) Awasthi

基物:树皮。

标本信息:海南陵水县,吊罗山,1993-04-10,郭守玉,H-0813(HMAS-L-023775);通什市,南圣镇公路边,2001-06-27,陈建斌、胡光荣、徐蕾,20556(HMAS-L-025877);西沙群岛珊瑚岛上,1992-05-29,Xing Fu-wu,HMAS-L-016936。

(63)汇黑囊基衣 *Dirinaria confluens* (Fr.) Awasthi

基物:岩石。

标本信息:海南乐东县,尖峰岭热带作物研究所,1993-04-03,姜玉梅、郭守玉,H-0716(HMAS-L-023761)。

(64)蜜果黑囊基衣 *Dirinaria confuse* Awasthi=*D. aegialita* (Afz. in Ach.) Moore

基物:岩石。

标本信息:海南五指山,2001-06-27,陈建斌、胡光荣、徐蕾,20559(HMAS-L-024610)。

(65)光面黑盘衣 *Pyxine berteriana* (Fée) Imsh.

基物:树皮。

标本信息:海南乐东县,尖峰岭,1993-04-03,姜玉梅、郭守玉,H-0716-1(HMAS-L-023692)。

(66)椰子黑盘衣 *Pyxine cocoes* (Sw.) Nyl.

基物:柳树。

标本信息:海南西沙群岛,霜花树,1992-05-29,Xing Fu-wu,HMAS-L-023695;陵水县,吊罗山,2001-06-25,陈建斌、胡光荣、徐蕾,20485(HMAS-L-023694);乐东县,尖峰岭,2001-06-30,陈建斌、胡光荣、徐蕾,20800(HMAS-L-111182)。

(67)黑盘衣 *Pyxine consocians* Vain.

基物:树皮。

标本信息:海南乐东县,尖峰岭,1993-04-03,姜玉梅、郭守玉,H-0716-2(HMAS-L-023698)。

(68)柯普兰氏黑盘衣 *Pyxine copelandii* Vain.

基物:树皮。

标本信息:海南陵水县,吊罗山,2001-06-26,陈建斌、胡光荣、徐蕾,20531-2(HMAS-L-111217)。

蜈蚣衣科(Physciaceae)

(69)白哑铃孢 *Heterodermia leucomela* (L.) Poelt

基物:树皮。

标本信息:海南五指山,1981-04-29,赵从福,2033(HMAS-L-017327)。

(70)黑纹蜈蚣衣 *Physcia atrostriata* Moberg

基物:柳树皮。

标本信息:海南陵水县,吊罗山,陈建斌、胡光荣、徐蕾,20485-1(HMAS-L-110956)。

厚顶盘亚纲(Ostropomycetidae)

羊角衣目(Baeomycetales)

羊角衣科 Baeomycetaceae

(71)小羊角衣 *Baeomyces absolutus* Tuck.

基物:岩石。

标本信息:海南乐东县,尖峰岭,2000-12-10,黄满荣,259(HMAS-L-113675)。

厚顶盘目(Ostropales)

文字衣科(Graphidaceae)

(72)象形霜盘衣 *Diorygma hieroglyphicum* (Pers.) Staiger & Kalb

基物:树皮。

标本信息：海南琼山市红树林，0 m，2007-05-24，孟庆峰，M401（HMAS-L 109811）、M402、M405；保亭县七仙岭路边，110 m，2009-07-25，刘萌，HN09386、HN09392、HN09399（HMAS-L 115519、115518、115550）；乐东县尖峰岭鸣凤谷，1 030 m，2007-05-11，孟庆峰 M6、M7、M10；乐东县尖峰岭热带植物园，130 m，2007-05-17，孟庆峰，M303-1；文昌市铜鼓岭，320 m，2007-05-23，孟庆峰，M421、M422；琼中县黎母山国家森林公园，630 m，2008-09-24，贾泽峰，HN003（LHS）。

（73）厚粉霜盘衣 *Diorygma hololeucum*（Mont.）Kalb，Staiger &Elix.（中国新记录种）

基物：树皮。

标本信息：海南乐东县尖峰岭核心区，950 m，2008-10-02，李静 HN081449（HMAS-L-117001）。

（74）容氏霜盘衣 *Diorygma junghuhnii*（Mont. &Bosch）Kalb，Staiger &Elix

基物：树皮。

标本信息：海南乐东县尖峰岭热带植物园，130 m，2007-05-17，孟庆峰，M301（HMAS-L-109813）、M304（HMAS-L-109817）、M305（HMAS-L-109815）、M308（HMAS-L-109814）、M309（HMAS-L-109816）、M292、M297、M302、M303、M306、M312、M503；乐东县尖峰岭瞭望塔，980 m，2007-05-11～12，孟庆峰，M22、M41、M193；乐东县尖峰岭天池，830～1000 m，2007-05-18，孟庆峰，M490；尖峰岭雨林谷，630 m，2007-05-16，孟庆峰，M225；琼中县黎母山路边，600 m，2008-09-26，高斌，HN080935；儋州热带植物园，50 m，2008-10-05，贾泽峰，HN08079；五指山路边，680 m，2008-09-28，李静，HN081259；霸王岭雅加瀑布，490 m，2008-10-05，李静，HN08147；佳西南黎山，400 m，2009-07-22，李静，HN09117（HMAS-L-115526）。

（75）马基高霜盘衣 *Diorygma macgregorii*（Vain.）Kalb，Staiger &Elix

基物：树皮。

标本信息：海南昌江霸王岭，980 m，2007-05-18，孟庆峰，M488；琼中县黎母山路边，480 m，2008-09-26，李静，HN142。乐东县尖峰岭，180 m，2008-10-01，李静，HN081360。

（76）厚唇霜盘衣 *Diorygma pachygraphum*（Nyl.）Kalb，Staiger &Elix

基物：树皮。

标本信息：海南乐东县尖峰岭瞭望塔，855 m，2007-05-11，张涛，Z1014（HMAS-L）；860 m，2007-05-12，孟庆峰，M49；乐东县尖峰岭天池，810 m，2007-05-12，孟庆峰，M475；810 m，2008-10-01，贾泽峰，HN080721（HMAS-L-117011）；五指山路边，680 m，2008-09-28，李静，HN081274（HMAS-L-117012）；霸王岭雅加瀑布，800 m，2008-10-05，李静，HN081495（HMAS-L-117013）。

（77）粉霜霜盘衣 *Diorygma pruinosum*（Eschw.）Kalb，Staiger &Elix

基物：树皮。

标本信息：海南乐东县尖峰岭雨林谷，670 m，花李木树皮，2007-05-16，孟庆峰，M231、M333、M362；尖峰岭鸣凤谷，1 030 m，2007-05-11，孟庆峰，M8；尖峰岭天池，830 m，花李木树皮，2007-05-11，孟庆峰，M28；琼州市红树林，0 m，红树皮，2007-05-24，孟庆峰 M390（HMAS-L-109820）、M392（HMAS-L-109823）、M393（HMAS-L-109821）、M395（HMAS-L-109822）、M404（HMAS-L-109819）。

（78）白果白唇衣 *Dyplolabia afzelii*（Ach.）A. Massal.

基物：树皮。

标本信息：海南乐东县尖峰岭路边，200 m，2008-10-01，李静，HN081433（HMAS-L-117015）；霸王岭雅加瀑布，800～840 m，2008-10-05，李静，HN081497（HMAS-L-117017）、HN081481（HMAS-L-117016）；五指山路边，670 m，2008-09-27，贾泽峰，HN028；佳西南黎山，400 m，2009-07-22，李静，HN237；保亭县七仙岭路边，200 m，2009-07-23，李静，HN240。

（79）鸽色裂痕衣 *Fissurina columbina*（Tuck.）Staiger（中国新记录种）

基物：树皮。

标本信息：海南乐东县尖峰岭鸣凤谷，1 080 m，2007-05-14，孟庆峰，M131。霸王岭雅加瀑布，800 m，2008-10-05，李静，HN081494（HMAS-L-117018）；五指山路边，730 m，2009-07-21，刘萌，HN09339（HMAS-L-115557）。

（80）仲马裂痕衣 Fissurina dumastii Fée（中国新记录种）

基物：树皮。

标本信息：海南乐东县尖峰岭鸣凤谷，1 020 m，2007-05-13，孟庆峰，M103，琼中县黎母山路边，650 m，2008-09-25，李静，HN081238（HMAS-L-117019）。

（81）灰绿裂痕衣 *Fissurina glauca*（Müll. Arg.）Staiger（中国新记录种）

基物：树皮。

标本信息：海南乐东县尖峰岭鸣凤谷，1 000 m，2007-05-14，孟庆峰，M151。

（82）小孢裂痕衣 *Fissurina nitidescens*（Nyl.）Nyl.（中国新记录种）

基物：树皮，石头。

标本信息：海南乐东县尖峰岭雨林谷，650 m，石生，2007-05-16，孟庆峰，M352；五指山，780 m，树皮，2001-06-28，陈建斌、胡光荣、徐蕾，20657（HMAS-L-076854）；乐东县尖峰岭，1 150 m，树皮，2007-05-15，孟庆峰，M170；琼中县黎母山路边，630 m，2008-09-25，李静，HN013、HN170。

（83）黑脉文字衣 *Graphis assimilis* Nyl.

基物：树皮。

标本信息：海南尖峰岭自然保护区，830 m，2007-10-01，李静，HN081387（HMAS-L-117024）；尖峰岭核心区，980 m，2009-07-28，贾泽峰，HNJ57（HMAS-L-117077）；尖峰岭植物园，650 m，2008-09-30，贾泽峰，HN080679、HN080687、HN080688（HMAS-L 117021、117022、117023）；霸王岭雅加，840 m，2008-10-05，李静，HN081480（HMAS-L-117020）。

（84）淡兰文字衣 *Graphis caesiella* Vain.

基物：树皮，石生。

标本信息：海南乐东县，尖峰岭，180 m，树皮，2008-10-01，李静，HN081333、HN081359（HMAS-L-116705、117025）；昌江县，霸王岭雅加林场，贾泽峰，HN080780（HMAS-L-116706）；文昌市东郊镇椰树湾，0 m，2007-05-23，孟庆峰，M438（HMAS-L-117079）[cf.]。

（85）环带文字衣 *Graphis cincta*（Pers.）Aptroot

基物：树皮。

标本信息：海南尖峰岭自然保护区，740 m，2008-10-01，李静，N081378（HMAS-L-117026）。

（86）毛文字衣 *Graphis cleistoblephara* Nyl.

基物：树皮。

标本信息：海南尖峰岭上山路，750 m，2009-07-29，李静，HN252-1。

（87）密集文字衣 *Graphis conferta* Zenker（中国新记录种）

基物：树皮。

标本信息：海南昌江县雅加林场，1 500 m，1993-03-29，姜玉梅、郭守玉，H0448（HMAS-L-067969）；五指山，680 m，2008-09-28，李静，HN081251（HMAS-L-117027）。

（88）裂出文字衣 Graphis descissa Müll. Arg.

基物：树皮。

标本信息：海南五指山自然保护区，树皮，670 m，2008-09-27，贾泽峰，HN080638（HMAS-L-117029）；尖峰岭自然保护区，树皮，200 m，2008-10-01，李静，HN081425（HMAS-L-117028）。

（89）无鳞文字衣 *Graphis desquamescens*（Fée）Zahlbr.

基物：树皮。

标本信息：海南尖峰岭，950 m，树皮，2008-10-02，李静，HN081442（HMAS-L-117005）。

（90）曲盘文字衣 *Graphis dupaxana* Vain.

基物：树皮。

标本信息：海南五指山 680 m，树皮，2009-07-21，刘萌，HN09334（HMAS-L-115515），李静，HN09062-2（HMAS-L-115563）；乐东县尖峰岭核心区，900 m，树皮，2009-07-27，李静，HN09103（HMAS-L-115500）。

（91）线丝文字衣 *Graphis filiformis* Adaw. & Makhija（中国新记录种）

基物：树皮。

标本信息：海南黎母山，630 m，2008-09-25，李静，HN081201（HMAS-L-117100）。

（92）黑白文字衣 *Graphis glauconigra* Vain.

基物：树皮。

标本信息：海南琼中县黎母山上山路（Lat. 19.27，Long. 109.79），490 m，2008-09-26，贾泽峰，HN080624（HMAS-L-117031）、HN021（HMAS-L-117030）。

（93）汉氏文字衣 *Graphis handelii* Zahlbr.

基物：树皮。

标本信息：海南乐东县，尖峰岭鸣凤谷，980 m，2010-11-27，贾泽峰，10-535（HMAS-L-117841）。

（94）裂文字衣 *Graphis hiascens*（Fée）Nyl.

基物：树皮。

标本信息：海南尖峰岭天池，750 m，2009-07-

29, 贾泽峰, HNJ62、HNJ68-3 (HMAS-L-117035、117036)。

(95)泰北文字衣 *Graphis hossei* Vain.

基物:树皮。

标本信息:海南昌江县,霸王岭雅加,800 m, 2007-10-05, 李静, HN081479 (HMAS-L-117037); 五指山市,五指山,2009-07-20,李静,09011 (HMAS-L-115484)。

(96)沉文字衣 *Graphis immersicans* A. W. Archer

基物:树皮。

标本信息:海南乐东县,尖峰岭天池,2008-10-01,贾泽峰,HN080740 (HMAS-L-116720); 陵水县,吊罗山,970 m,松树,2010-12-01,贾泽峰,Z. F. Jia 10-657。

(97)缠结文字衣 *Graphis intricata* Fée

基物:树皮。

标本信息:海南乐东县,尖峰岭天池,2009-07-27,李静,HN09198(HMAS-L-115499)。

(98)日本文字衣 *Graphis japonica* (Müll.Arg.) A. W. Archer & Lücking

基物:树皮,石生。

标本信息:海南尖峰岭自然保护区,980 m, 2008-10-01,李静,HN081357 (HMAS-L-117038); 尖峰岭,700 m,2009-07-27,李静,HN243-2 (HMAS-L-117088); 尖峰岭核心区,980 m,2009-07-28,李静,HN248(HMAS-L-117089); 五指山上山路边,640~800 m,2009-07-20,李静,HN09314、HN201、HN211、HN220-2、HN229 (HMAS-L-115558、117090、117091、117092、117093); 五指山上山路边,720 m,2009-07-21,李静,HN09084 (HMAS-L-115488); 琼中自治县黎母山,630 m,2008-09-25,李静,HN081212 (HMAS-L-117039)。

(99)基隆文字衣 *Graphis kelungana* Zahlbr.

基物:树皮。

标本信息:海南尖峰岭上山路,200 m,树皮,2008-10-01,李静,HN081419(HMAS-L-117004)。

(100)岩生文字衣 *Graphis lapidicola* Fée

基物:树皮。

标本信息:海南五指山,390 m,2009-07-20,李静,HN215 (HMAS-L-117094)。

(101)梭盘文字衣 *Graphis librata* C. Knight

基物:树皮。

标本信息:海南黎母山,650 m,2008-09-25,李静,HN081214 (HMAS-L-117040); 昌江县,霸王岭雅加山庄,2008-10-05,高斌,HN081043 (HMAS-L-116725); 尖峰岭,2008-10-01,李静,HN081356 (HMAS-L-116726)以及900 m,2009-07-29,贾泽峰,HNJ68-4(HMAS-L-117036)。

(102)长枝文字衣 *Graphis longiramea* Müll. Arg.

基物:树皮。

标本信息:海南通什文化市,海拔未记录,1934-02-20,采集者未记录,2769 (HMAS-L-047830); 琼中县,黎母山,2008-08-25,李静,HN081206(HMAS-L-116727); 乐东县,尖峰岭天池,750m,2009-07-29,贾泽峰,HNJ58 (HMAS-L-117080); 尖峰岭森林公园,2008-10-01,高斌,NH080996(HMAS-L-116728)。

(103)骨针文字衣 *Graphis oxyclada* Müll. Arg. (中国新记录)

基物:树皮,树枝。

标本信息:海南乐东县,尖峰岭鸣凤谷,900 m,2009-07-26,刘萌,HN09417 (HMAS-L-115446); 尖峰岭鸣凤谷,980 m,树皮,2010-11-27,贾泽峰,10-531; 尖峰岭鸣凤谷,980 m,树皮,2010-11-27,贾泽峰,10-532; 尖峰岭鸣凤谷 900 m,树皮,2009-07-26,刘萌,HN09417 (HMAS-L-115446)。江县霸王岭东四区,700 m,树皮,2010-11-25,贾泽峰,10-453; 霸王岭雅加霸道,820 m,树皮,2010-11-26,贾泽峰,10-462。

(104)近杜氏文字衣 *Graphis paradussii* Z. F. Jia

基物:树皮。

标本信息:海南乐东县尖峰岭,910 m,2008-10-01,高斌,HN081009-b (HMAS-L-117098)。

(105)松皮文字衣 *Graphis pinicola* A. Zahlbr.

基物:树皮,树枝。

标本信息:海南乐东县,尖峰岭上山路,200~750 m,树皮,2008-10-01,李静,HN081421、HN081403 (HMAS-L-117003、117002); 尖峰岭避暑山庄,960 m,2008-10-01,高斌 HN081009。

(106)短盘文字衣 *Graphis plagiocarpa* Fée (中国新记录种)

基物:树皮。

标本信息:海南尖峰岭自然保护区,960 m,

2008-10-01，高斌，HN081009（HMAS-L-117041）。

（107）多层文字衣 *Graphis proserpens* Vain.

基物：树皮。

标本信息：海南乐东县，尖峰岭雨林谷，650 m，2007-05-16，孟庆峰，M351（HMAS-L-117081）；尖峰岭核心区，900 m，2009-07-27，李静，HN09191。

（108）伦施文字衣 *Graphis renschiana* （Müll. Arg.）Stizenb.

基物：树皮。

标本信息：海南五指山保护区，670 m，树皮，2008-09-27，贾泽峰，HN080641（HMAS-L-117043）。

（109）根生文字衣 *Graphis rhizocola* （Fée）Lücking & Chaves（中国新记录种）

基物：树皮。

标本信息：海南尖峰岭鸣凤谷，1 030 m，2007-05-11，孟庆峰，M1、M5（HMAS-L-117082、117083）。

（110）小隙文字衣 *Graphis rimulosa* （Mont.）Trevis.

基物：树皮。

标本信息：海南尖峰岭核心区，900 m，2009-07-28，贾泽峰，HNJ53、HNJ54（HMAS-L-117084、117085）。

（111）皱沟文字衣 *Graphis striatula* （Ach.）Spreng.

基物：树皮。

标本信息：海南五指山，680 m，树皮，2008-09-28，李静，HN081262、HN081264（HMAS-L-117047、117046）。

（112）亚蛇皮文字衣 *Graphis subserpentina* Nyl.

基物：树皮。

标本信息：海南文昌市，东郊椰林湾，3 m，椰树树皮，2010-12-03，贾泽峰，10-663、10-664、10-667、10-668、10-670、10-671、10-672、10-673、10-674、10-677、10-683（HMAS-L 117970、117971、117974、117975、117977、117978、117979、117980、117981、117984、117990）。

（113）细柔文字衣 *Graphis tenella* Ach.

基物：树皮。

标本信息：海南霸王岭雅加瀑布，880 m，2008-10-05，高斌，HN081038。

（114）条纹文字衣 *Graphis vittata* Müll. Arg.

基物：树皮。

标本信息：海南琼中县黎母山，650 m，2008-09-25，李静，HN081213（HMAS-L-117049），贾泽峰，HN080604（HMAS-L-117050）；五指山市，五指山保护区，2010-11-30，贾泽峰，10-600（HMAS-L-117908）。

（115）巴氏半实衣 *Hemithecium balbisii* （Fée）Trevis.

基物：树皮。

标本信息：海南保亭县七仙岭路旁，200 m，2009-07-23，李静，HN240-1；乐东县佳西南黎山，180 m，2009-07-22，刘萌，HN09357（HMAS-L-115556）。

（116）双砖孢半实衣 *Hemithecium duomurisporum* Z.F. Jia

基物：树皮。

标本信息：海南五指山市，五指山，640 m，2009-07-20，刘萌，HN09318（HMAS-L-115536）。

（117）交织半实衣 *Hemithecium implicatum* （Fée）Staiger

基物：树皮。

标本信息：海南乐东县尖峰岭，950 m，2009-07-28，李静，HN09218（HMAS-L-115502）；霸王岭，880 m，2008-09-25，李静，HN081456、HN081466（HMAS-L）。

（118）白壳半实衣 *Hemithecium oshioi* （M. Nakan.）M. Nakan. & Kashiw.

基物：树皮。

标本信息：海南五指山市五指山，700 m，2009-07-21，李静，HN09092（HMAS-L-115471）。

（119）黑厚基衣 *Leiorreuma melanostalazans* （Leight.）A. W. Archer（中国新记录种）

基物：树皮。

标本信息：海南乐东县尖峰岭植物园，630 m，2008-09-30，贾泽峰，HN080683（HMAS-L）；乐东县尖峰岭，700 m，2008-10-01，李静，HN081422（HMAS-L）；佳西南黎山路边，400 m，2009-07-22，李静，HN235-1、HN236（LHS）；五指山路边，680 m，2009-07-21，李静，HN226-3（LHS）。

（120）台地灰线衣 *Pallidogramme chapadana* （Redinger）Staiger，Kalb & Lücking

基物：树皮。

标本信息：海南五指山，1 340 m，1980-11-28，姜玉梅、郭守玉，H-1029（HMAS-L-305420-1）。

(121)绿果灰线衣 *Pallidogramme chlorocarpoides* (*Nyl.*) *Staiger*, *Kalb* &*Lücking*

基物:树皮。

标本信息:海南五指山,无海拔记录,1993-04-20,姜玉梅、郭守玉,H-1029(HMAS-L-X005420-2);吊罗山,900 m,2001-06-16 日,陈健斌、胡光荣、徐蕾,20496((HMAS-L-076911)。

(122)乳黄半实衣 *Pallidogramme chrysenteron* (Mont.) Staiger, Kalb &Lücking

基物:树皮。

标本信息:海南五指山市五指山,700 m,2009-07-21,李静,HN09078、HN09093(HMAS-L-115435、115463)。

(123)错综黑文衣 *Phaeographis intricans* (Nyl.) Staiger

基物:树皮。

标本信息:海南陵水县,吊罗山,海拔未记录,2006-06-06,魏江春,无采集号;乐东县,尖峰岭鸣凤谷,950 m,孟庆峰,M167。

(124)细枝黑文衣 *Phaeographis subdividens* (Leight.) Müll. Arg.

基物:树皮。

标本信息:海南乐东县,尖峰岭,740 m,2008-10-01,李静,HN081379。

(125)米勒凸唇衣 *Platygramme muelleri* (A. W. Archer) Staiger(中国新记录种)

基物:树皮。

标本信息:海南乐东县尖峰岭鸣凤谷,2007-05-16,670 m,孟庆峰,M371-1(LHS);尖峰岭核心区,930 m,2007-05-15,孟庆峰,M380;尖峰岭核心区,960 m,2008-10-01,李静,HN081330(HMAS-L-117063);尖峰岭核心区,770 m,2008-10-01,李静,HN081400(HMAS-L-117064)。

(126)圆丘平盘衣 *Platythecium colliculosum* (Mont.) Staiger

基物:树皮。

标本信息:海南乐东县,1 100 m,2001-06-29,陈健斌,等,20773(HMAS-L-076856)。

(127)双型平盘衣 *Platythecium dimorphodes* (Nyl.) Staiger

基物:树皮。

标本信息:海南乐东县尖峰岭,1 300 m,2001-06-30,陈健斌,等,20786(HMAS-L-076855)。

(128)平滑平盘衣 *Platythecium leiogramma* (Nyl.) Staiger(中国新记录种)

基物:树皮。

标本信息:海南乐东县尖峰岭鸣凤谷,880 m,2009-07-26,刘萌,HN09404(HMAS-L-115453)。

(129)蛇形平盘衣 *Platythecium serpentinellum* Staiger(中国新记录种)

基物:岩石。

标本信息:海南乐东县尖峰岭核心区,850 m,2008-10-01,李静,HN081324(HMAS-L)。

(130)蔓氏板文衣 *Thecaria montagnei* (Bosch) Staiger

基物:树皮。

标本信息:海南霸王岭雅加瀑布,820 m,2008-10-05,李静,HN081、HN082;乐东县尖峰岭核心区,900 m,2009-07-27,李静,HN243-1;霸王岭三分区,850 m,2009-7-23,李静,HN253。

(131)苦木板文衣 *Thecaria quassiicola* Fée

基物:树皮。

标本信息:海南乐东县,尖峰岭鸣凤谷,1 030 m,2007-05-11,孟庆峰,M4(HMAS-L-117076);琼中县,黎母山路边,630 m,2008-09-25,李静,HN081205(HMAS-L-116740);黎母山路边,480 m,2008-09-26,李静,HN139;五指山市,五指山路边,680 m,2008-09-29,李静,HN081257(HMAS-L-116741);五指山路边,640 m,2009-07-20,李静,HN209。

鸡皮衣目(Pertusariales)

鸡皮衣科(Pertusariaceae)

(132)苦味鸡皮衣 *Pertusaria amara* (Ach.) Nyl.

基物:树皮。

标本信息:海南五指山,1993-04-20,姜玉梅,郭守玉,H-1030-2(HMAS-019305);琼中县,黎母山,1977-06-06,韩树金等,HMAS-L-019566。

(133)中国鸡皮衣 *Pertusaria chinensis* Müll. Arg.

基物:树皮。

标本信息:海南热带植物所,刘藹堂,810014(HMAS-L-027813)。

(134)包被鸡皮衣 *Pertusaria velata* (Turner) Nyl.

基物:树皮。

标本信息：海南五指山，1993-04-20，姜玉梅、郭守玉，H-1030-1(HMAS-L-019802)。

无座盘菌目(Agyriales)

无座盘菌科(Agyriaceae)

(135)海南类褐边衣 *Trapeliopsis hainanensis* Hertel

基物：树皮。

标本信息：海南乐东县，尖峰岭，1980-05-23，H.Hertel，HMAS-L-003351。

斑衣纲 Arthoniomycetes

斑衣目(Arthoniales)

染料衣科(Roccellaceae)

(136)曼氏染料衣 *Roccella montagnei* Bél.

基物：椰树皮。

标本信息：海南文昌东郊椰林，18 m，2011-06-25，魏江春，张颖，王延延，HN1106-112(HMAS-L-120614)

(137)中华染料衣 *Roccella sinensis* Nyl.

基物：岩石。

标本信息：海南文昌铜鼓岭附近石头公园，5 m，魏江春、张颖、王延延，HN1106-100(HMAS-L120615)。

5.2 担子菌门(Basidiomycota)

伞菌纲 Agaricomycetes

鸡油菌目(Cantharellales)

锁瑚菌科(Clavulinaceae)

(138)中华藻珊瑚 *Multiclavula sinensis* R. Petersen et M.Zang.

基物：地上。

标本信息：海南霸王岭，2007，杨军，HMAS-L，2008，魏江春，HMAS-L；尖峰岭，2008，魏江春，HMAS-L。

致谢：参加本项目野外考察及室内研究的有中国科学院微生物研究所真菌学国家重点实验室、海南大学、山东农业大学生命科学学院、聊城大学生命科学学院以及河北大学生命科学学院的下列人员(按拼音字母顺序排列)：曹淑楠[1,4]、邓红[1]、高斌[1,3]、郭威[1]、韩乐琳[1,4]、贾泽峰[3]、李静[3]、刘萌[1,3]、孟庆峰[1,3]、王海英[1]、王延延[1]、魏江春[1]、吴兴亮[2]、杨军[1]、张涛[1]和张颖[1,3](1. 中国科学院微生物研究所真菌学国家重点实验室；2. 海南大学；3. 山东农业大学生命科学学院；4. 聊城大学生命科学学院；5. 河北大学生命科学学院)，特此致谢！

参考文献：

[1] 魏江春. 菌物生物多样性与人类可持续发展[J]. 中国科学院院刊, 2010, 25(6): 645-650.

[2] Hawksworth D L. The fungal demention of biodiversity: magnitude, significance, and conservation[J]. Mycol Res, 1991, 95(6): 641-655.

[3] Petrini O, Sieber T N, Toti L, Viret O. Ecology, metabolit production, and substrate utilization in endophytic fungi[J]. Natural Toxins, 1992, 1: 185-196.

[4] Strobel G, Daisy B. Bioprospecting for microbial endophytes and their natural products[J]. Microbiology and Molecular Biology Reviews, 2003, 67(4): 491-502.

[5] Kirk P M, Cannon P F, Minter D W, Stalpers J A. Dictionary of the Fungi[M]. 10th ed. UK: CABI, 2008.

[6] Dai Yucheng, Zhuang Jianyun. Numbers of fungal species hitherto known in China[J]. Mycosystema, 2010, 29(5): 625-628.

[7] Sun L Y, Liu Z L, Zhang T, et al. Three antibacterial naphthoquinone analogues from cultured mycobiont of lichen *Astrothelium* sp.[J]. Chinese Chemical Letters, 2010, 21: 842-845.

[8] Ding G, Li Y, Fu S, Liu S, Wei J C, Che Y S. Ambuic acid and torreyanic acid derivatives from the endolichenic fungus *Pestalotiopsis* sp.[J]. J Nat Prod, 2009, 72(1): 182-186[DOI: 10.1021/np800733y Publication Date (Web): 31 December 2008].

[9] Kok A. A short history of the orchil dyes[J]. Lichenologist, 1966, 3: 248-272.

[10] Huneck S, Yoshimura. Identification of Lichen Substances[M]. [s. l.]: Springer, 1996.

[11] Nylander W. Synops[J]. Lich, 1860, 1: 261.

[12] Wei J C. An Enumeration of Lichens in China[M]. Beijing: International Academic Publishers, 1991: 1-278.

第二章　专科属研究

Studies of Special Lichen Groups

中国黄梅属地衣的初步订正

魏 江 春

（中国科学院微生物研究所，北京）

摘 要 本文是作者对中国黄梅属地衣从形态与化学两方面进行比较研究之后的初步订正。研究结果表明，文献中所记载的中国 *Parmelia conspersa* Ach.实际上为 *Xanthoparmelia tinctina*（Mah. et Gillet）Hale 及 *X. mexicana*（Gyeln.）Hale 的误用名称；长白山的 *Parmelia subramigera* Gyeln.则为 X. *mexicana* 的误用名；*Parmelia stenophylla*（Ach.）Heug.及 *Parmelia subconspersa* Nyl 为 *X. taractica*（Kremph.）Hale 的误用名；*Parmelia vagans*（Nyl.）Nyl. 为 *X. camtschadalis*（Ach.）Hale 的误用名。此外，文中还报道了新组合一种，新变种一个，中国新记录一种。

关键词 黄梅属；暗腹黄梅；暗腹黄梅，西藏变种；淡腹黄梅；拟菊叶黄梅；旱黄梅；齿裂黄梅；曲黄梅

由于黄梅属地衣均含有松萝酸，为地衣抗菌素的重要原料，因此，它是药用地衣的类群之一。

全世界已描述的黄梅属地衣计 94 种（HALE，1974）。中国的黄梅属尚缺乏专门研究，而且文献记载也较混乱。以 *Parmelia* 为属名，作为黄梅组（sect. *Xanthoparmelia*）地衣曾报道过 13 种及 17 个种下单位，其中作为 *Parmelia conspersa* Ach. 及其种下单位者有 15 个（KREMPH，1873，1874； RABENHORST，1873；BARONI，1894；ZAHLBRUCKNER，1930，1933；TCHOU，1935；MAGNUSSON，1944；MOREAU et MOREAU，1951；SATO，1952，1981；ASAHINA，1952；赵，1964；WANG & LAI，1973；魏、陈，1974；赵等，1982）。此外，还以 *P. mougeotii* Schaer.（KREMPH，1873，1874；RABENHORST，1873；ZAHLBRUCKNER，1930），*P. molliuscula* Ach.（ZAHLBRUCKNER, 1930），*P. conspersula* Nyl. apud Cromb.（ZAHLBRUCKNER, 1933； WANG & LAI, 1973），*P. squamaraeformis* Gyeln.（1934）， *P. vagans* （Nyl.）Nyl.（MAGNUSSON, 1944；赵，1964；魏、陈，1974； WEI & JIANG, 1981；赵等，1982），*P. subconspersa* Nyl.（赵，1964；赵等，1982；陈等，1981）*P. stenophylla*（Ach.）Heug.（以“*dentata*”为加词，描述一新变型），（赵，1964； 赵等，1982；陈等，1981），*P. subramigera* Gyeln.（陈等，1981），*P. tinctina* Mah. et Gillet（陈等，1981），*P. mexicana* Gyeln.（魏等，1982）以及 *P. taractica* Kremph.）陈等，1981；魏等，1982）为种 名作过报道。至于 *P. sinuosa*（Sm.）Ach.（赵，1964；WANG & LAI，1973；魏、陈，1974；魏等，1982；赵等，1982），因其具二叉分枝型假根等特征而划归双歧根属：*Hypotrachyna*（Vain.）Hale.

根据形态及化学两方面的研究结果表明，在作者研究过的标本中，除 *X. dentata*（Zhao）

本文原载于《真菌学报》, 2（4）: 221-227, 1983.

Wei 一种仅含松萝酸而未伴有其它副成分以外，其它各种均伴有水杨嗪酸，因而，文献中记载的中国 *Parmelia conspersa* Ach.实际上为 *Xanthoparmelia tinctina*（Mah. et Gillet）Hale 及 *X. mexicana*（Gyeln.）Hale 两种的误用名称；长白山的 *P. subramigera* Gyeln.（陈等，1981）为 *X. mexicana* 的误用名；*P. subconspersa* Nyl.及 *P. stenophylla*（Ach.）Heug.为 *X. taractica*（Kaemph.）Hale 的误用名；中国以及亚洲的 *P. vagans*（Nyl.）Nyl.则为 *X. camtschadalis*（Ach.）Hale 的误用名。其余各种，因未能见到标本，留待以后订正。

黄 梅 属

Xanthoparmelia（Vain.）Hale

Phytologia, **28**（5）: 485（1974）.

= *Parmelia* sect. *Xanthoparmelia* Vain. Acta Soc. Faun. Fl. Fenn. **7**（7）:60（1980）.

Type species: *Lichen conspersus* Ach. Prod. Lich. Suec. 1798, p.118.

地衣体叶状，边缘深裂，叶片宽度多为 3-5mm 左右，但其幅度范围在 0.5-10mm 之间，无缘毛；上表面黄绿色，上皮层为珊瑚状假组织，并具外皮小孔，无假杯点，有时具裂芽或粉芽；下表面淡色（白色至淡褐色）或暗色（暗褐色，暗紫褐色至黑色）；假根散生，不分叉，有时只具零星假根或缺如。子囊盘茶渍型；子襄内含 8 孢；孢子椭圆形，无色，单胞。

衣体内含松萝酸，往往分别伴有斑点酸（stictic acid），降斑点酸（norstictic acid），水杨嗪酸（salazinic acid），富马原岛衣酸（fumarprotocetraric acid），树发酸（alectoronic acid）或黑茶渍素（atranorin）等副成分。中国黄梅属地衣多数种类均伴有水杨嗪酸，有些种除伴有水杨嗪酸外，还伴有痕量降斑点酸，或只含有松萝酸而无任何副成分。

分 种 检 索 表

1.地衣体内伴生水杨嗪酸

  2.衣体上表面具裂芽

    3. 衣体下表面暗色至黑色…………………………………… 暗腹黄梅：*X. tinctina*

    3. 衣体下表面淡色至白色…………………………………… 淡腹黄梅：*X. mexicana*

  2.衣体上表面无裂芽

    4.衣体较紧密地固着于基物上，至少部分裂片较宽而平展,不卷成半管状………………

    ……………………………………………………………… 拟菊叶黄梅：*X. taractica*

    4.衣体脱离基物，往往卷曲成团；裂片狭长，卷成半管状 ·· 旱黄梅：*X. camtschadalis*

1.地衣体内不伴生水杨嗪酸

    5.地衣体内仅含松萝酸，上表面散生少量裂芽……………… 齿裂黄梅：*X. dentata*

    5.地衣体内伴生树发酸，上表面具粉芽堆…………………… 曲黄梅：*X. incurva*

1.暗腹黄梅

Xanthoparmelia tinctina（Mah. et Gillet）Hale, Phytologia, **28**（5）:489（1974）.

=*Parmelia tinctina* Mah. et Gillet, Bull. Soc. Bot. Fr.**72**:860（1925）.

Parmelia conspersa auct.non Ach.: Magnusson, H. Lich. Cent. Asia,1944, p. 46;Zhao, Acta Phytotaxon. Sin.9（2）:147（1964）; Zhao et al Prod. Lich.Sin.1982, p.20.

Parmelia conspersa var. *isidiosula* auct. non Hillm.: Zhao, Acta Phytotaxon. Sin. 9（2）:147（1964）（pr. p.）; Zhao et al. Prod. Lich. Sin. 1982, p. 21（pr. p.）.

原变种

var. **tinctina**

本变种的主要特征是衣体裂片较宽，上表面具亚球形或倒卵形及圆柱形裂芽，下表面暗色。

TLC: usnic and salazinic acids.（± norstictic acid）

Specimens Examined：**北京**：西山，陈玉本，nos. 45，46。**河北**：雾灵山，赵，no.818。**内蒙**：Bohlin，no. 169，（S）. **浙江**：浦阳镇，陈超英，no. 5088，杭州，赵、徐，no. 5007（此外还伴有微量降斑点酸）；天目山，赵、徐，no. 6433。**新疆**：天山，卯等，no. 1291。**江苏**：南 京紫金山，S. N. Lei. 1923. 3. 20（H-Vain. no. 2817， TUR）。**四川**：H. Smith, no. 5176（UPS）。（此外，还伴有微量降斑点酸）**西藏**：拉萨市郊，魏、陈，nos. 24，26； 聂拉木，江宁，魏、陈，no. 1676；珠峰，中绒布，5430 m，1966. 5.28. 魏、陈（无号）；泽当，宗，no. 12；昌都，向达乡，宗、廖，no. 218。**安徽**：狮子山，旧矿坑，赵等，no. 116。

西藏变种　新变种　图 1

var. **xizangensis** Wei, var. nov. Fig. 1

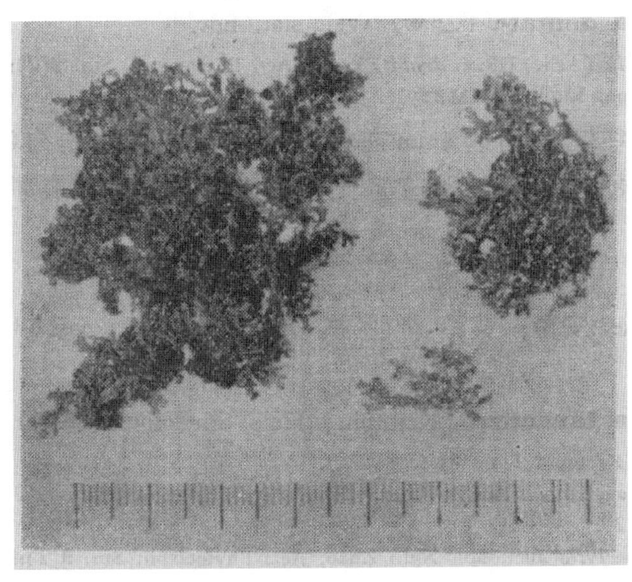

图 1　暗腹黄梅,西藏变种

Fig. 1 *Xanthoparmellia tinctina* var. *xizangensis* Wei

Varietas haec a var. *tinctina* differt lacinulis perangustis et isidii bacilliformibus gracilibus ramossimis.

Thallus acidum usneicum et acidum salazinicum cotinens.（TLC）

Typus: Xizang（Tibet）: Changdu, Xiangdaxiang, 1 VI 1976, Zong Yuchen et al. no. 218-（1）, in HMAS conservatur.

本变种不同于原变种者在于裂片狭长，裂芽细棒状而多分枝。

模式标本：**西藏**：昌都，向达乡，1976. 6. 1. 宗毓臣等，no. 218-（1），保存于中国科学院微生物研究所真菌标本室。

2.淡腹黄梅

Xanthoparmelia mexicana（Gyeln.）Hale, Phytologia **28**（5）:488（1974）.

=*Parmelia mexicana* Gyeln. Feddes Repert. **29**:281（1931）.

Parmelia conspersa auct. non Ach.: Magnusson, H. Lich. Cent. Asia, 1944, p. 46. *Parmelia conspersa* var. *isidiosula* auct. non Hillm.: Zhao, Acta Phytotaxon. Sin. **9**（2）: 147（1964）(pr.p.)；Zhao et al. Prod. Lich. Sin. 1982, p. 21（pr. p.）.

P. conspersa var. *latior* auct. non Schaer.: Zhao, Acta Phytotaxon. Sin. **9**（2）: 147（1964）；Zhao et al. Prod. Lich. Sin. 1982, p. 21.

Parmelia subramigera auct. non Gyeln.: Chen et al. Journ. N. E. Forestry Inst. **4**:154（1981）.

本种下表面淡褐色至白色，并具同色假根。

基物：岩石及树皮。

TLC: usnic and salazinic acids.

Specimens Examined：**北京**：潭拓寺，陈玉本，nos. 7，7a，19，73。**河北**：雾灵山，季克恭，no. 143。**山西**：太原，桑志华，nos. 95。**内蒙**：Bohlin, no. 107，(S)；百灵庙，刘慎谔，no. 2052。**吉林**：长白山，侯家龙，no. 78089；长白山冰场附近，1963.9. 3.（无采集人姓 名及编号）。**安徽**：黄山温泉，赵、徐，nos. 5974，5991。**浙江**：杭州，赵、徐，no. 5042；天 目山，赵、徐，no. 6077。**新疆**：天山天池，王，no. 1409。

3.齿裂黄梅 改级新组合

Xanthoparmelia dentata（Zhao）Wei, stat. nov.

= *Parmelia stenophylla*（Ach.）DR. f. *dentata* Zhao, Acta Phytotaxon. Sin. **9**（2）:148（1964）.（Holotypus: Huang Shan, Zhao et al. no. 5860, in HMAS!）

衣体边缘裂片往往较宽，可达 4mm 左右，中央部分裂片较狭，约 0.3-1 mm 左右，裂片边缘有黑色镶边；上表面具蜡样光泽，基本无裂芽，偶见零星球状裂芽。

基物：岩石。

TLC: usnic acid.

Specimens Examined：**安徽**：黄山，光明顶，1800m，赵、徐，nos. 5860，5861。

4.拟菊叶黄梅

Xanthoparmelia taractica（Kremph.）Hale, Phytologia **28**（5）:489（1974）.

=*Parmelia taractica* Kremph. Flora 61:439（1878）.

Parmelia stenophylla auct. non（Ach.）Heug.:Zhao, Acta Phytotaxon. Sin. 9（2）: 148（1964）; Zhao et al. Prod. Lich. Sin. 1982, p. 21.

P. conspersa var. *hypoclysta* auct. non Nyl.:Zhao, Acta Phytotaxon. Sin. **9**（2）: 147（1964）；Wei et Chen, Report, on the Scientific Investigations（1966- 1968）in Mt. Qomclangma District（Sect. Biology）, Science Press, Beijing, 1974, p.179; Zhao et al. Prod. Lich. Sin. 1982, p.22.

Parmelia subconspersa auct. non Nyl.:Zhao, Acta Phytotaxon. Sin. **9**（2）: 147（1964）: Zhao et al. Prod. Lich. Sin. 1982, p.22.

本种衣体较紧密地固着于基物表面，其外部形态类似于 *Xanthoparmelia stenophylla*（Ach.）Hale 及 *X. camtschadalis*（Ach.）Hale，不同于前者之处在于衣体内伴有水杨嗪酸，有时除水杨嗪酸外，同时还伴有微量降斑点酸（Liou K. M. no. 4565；韩等，no.2025），下表面淡色；不同于后者之处在于衣体较紧密地固着于基物上，至少部分裂片较宽而平展，不卷成半管状。

基物：岩石，岩隙土及草茎。

TLC: usnic and salazinic acids.（± norstictic acid）

Specimens Examined: **北京**：西山，陈玉本，nos. 98, 500；八大处，陈超英、陈玉本，

nos. 4927, 4935。**内蒙**：大兴安岭，古莲，魏，no. 3283； 集宁西郊，1981.8. 30.李伯堂（无号）。**宁夏**：贺兰山，韩，nos.2016a，2025（此外还伴有微量降斑点酸）及韩等，no. 2010。**四川**:H.Smith，nos. 2534，2564， 2565，5322（UPS）。**西藏**：聂拉木，魏、陈，no. 1460。河南：刘继孟，no. 4565（此外还伴有微量降斑点酸）。**新疆**：乌鲁木齐南山，王，no. 139。

5.旱黄梅

Xanthoparmelia camtschadalis（Ach.）Hale, Phytologia **28**（5）:486（1974）.

=*Borrera camtschadalis* Ach. Synops. Lich. 1814, p. 223（Holotypus: Kamchatka, leg. W. G. Tilesius, H-Ach. no. 1450, in H!）.

= *Parmelia camtschadalis*（Ach.）Eschw. apud Martius, Flora Brasil.1: 202（1833）（not seen）.

Parmelia conspersa ssp. *molliuscula* var. *vagans* auct. non Nyl.: Elenkin, A. Acta Horti. Petropolitani, **19**（1）:20（1901）.

Parmelia molliuscula var. *vagans* auct. non Ny.: Elenkin, A. Bull. Jard. Imp. Bot. St. -Petersbourg, 4:55（not seen）et 136（1901）.

Parmelia vagans auct. non Nyl,: Elenkin, A. Lich. Fl. Ross. Med. 1: 142（1906）; Magnusson, H. Lich, Cent. Asia, 1944, p. 47; Savicz, V. P. Notulae Syst. e sect. Crypt. Inst. Bot. Nom. V. L. Komarovii A. S. URSS, 5（10-12）:131（1945）; Zhao, Acta Phytotaxon. Sin. 9（2）: 148（1964）; Vezda, Acta Mus. Siles. ser. A, 14: l89（1965）（not seen）; Klem. Feddes Repert. 72（2-3）: 117（1966）; Schub. & Klem, Feddes Repert , **82**（3-4）242（1971）; Golubkova, Bot. Rhurn.（in Russian）（Journ. of Bot.）, **56**（6）: 784（1971）; Wei & Chen, Report on the Scientific Investigations（1966-1968）in Mt. Qomolangma District（Biological Sect.）, Science Press, Beijing, 1974, p.179; Cogt, Feddes Repert. 90（7-8）:436（1979）; Golubkova, Konspekt Fl. Lishaykov MNR（in Russian）（Conspect of Lichen Flora from PRM）, 1981, p. 97; Zhao et al. Prod. Lich. Sin. 1982, p.22.

在地衣学文献中本种长期以来被误作 *Parmelia vagans*（Nyl）Nyl. 当作者在赫尔辛基大学工作期间，在 T. Ahti 教授建议下，对有关模式标本进行了对比研究之后证实，所谓的亚洲 *Parmelia vagans*（Nyl）Nyl. 实则为 *Parmelia camtschadalis*（Ach.）Eschw.。其主要特征为裂片上表面具明显的黄白色花斑，体内伴有水杨嗪酸，模式标本（H-Ach. no. 1450, in H!）产于亚洲堪察加半岛；而 *Parmelia vagans*（Nyl.）Nyl.的裂片上表面无任何淡色花斑，体内伴有斑点酸及降斑点酸，模式标本（H-Nyl. no. 1730, in H!）产于南美洲厄爪多尔。本种与 *X. taractica*（Kremph.）Hale 之区别在于裂片卷成半管状，与基物脱离，衣体往往卷曲成团；而后者衣体不同程度地紧贴于基物表面，至少部分裂片宽而平展，不卷成半管状。

TLC: usnic and salazinic acids.

Specimens Examined：**山西**：关帝山长立沟，马赛，no. 15361。**新疆**：吐鲁番，徐，1958。6. 15（无号）；塔城，刘恒英、刘荣，no. 531；夏塔，王，no. 899。**西藏**：聂拉木，魏、陈，nos. 1439-（2），1524-（1）, 1539。

6.曲黄梅

Xanthoparmelia incurva（Pers.）Hale, Phytologia, **28**（5）:488（1974）.

=*Lichen incurvus* Pers. Neue Annal. der Botan. 1. stück, 1794, p. 24.

= *Parmelia incurva*（Pers.）Fr. Nov. Sched. Critic. 1826, p. 31. A.Z. Cat. Lich. Univ. 6:138（1930）.

本种叶状体近圆形，紧密贴生于岩石表面，裂片极细而上表面鼓起，散生头状粉芽堆。

这是本种在中国的首次报道。

TLC: Alectoronic and usnic acids.

Specimens Examined:

吉林：长白山南坡，长白县，红头山山顶岩石表面，海拔 1900m，1983.7.28.魏、陈 no.6257.

参 考 文 献

[1] Asahina, Y. 1952. Lichens of Japan II. p. 1-162.

[2] Baroni, E.1894, Sopra alcuni licheni della China raccolti nella provincia dello Schen-si settentrionale. *Bull. Soc. Bot. Ital.* 46-59.

[3] 陈锡龄、赵从福、罗光裕。1981。东北地衣名录。东北林学院学报 4：154。

[4] Gyelnik, V. 1934. Additamenta ad cognionem Parmeliarum V. XIX. *Xanthoparmeliae. Fedde Repertorium* **36**: 163.

[5] Hale, M. 1955. *Xanthoparmelia* in North America, I. The *Parmelia conspersa-stenophylla* group. *Bull. Torrey Bot, Club* **82**: X-Y.

[6] Hale, M. E. Jr. 1964. The *Parmelia conspersa* group in North. America and Europe. *Bryologist* **67**: 462-473.

[7] _____________. 1973. Fine structure of the cortex in the lichen family Parmeliaceae viewed with the scanningelectron microscope. *Smithsonian Contr. Bot.* **10**: 1-92.

[8] _____________, and Kurokawa, S.1964. Studies on *Parmelia* subgenus *Parmelia*.Contr. *U. S. Nat, Herb.* **36**: 121-191.

[9] Huneck, S., Poelt,J., Ahti, T., Vitikainen, O.und Cogt, U.1983. Zur Verbbreitung und Chemie von Flechten der Mongolischen Volksrepublik（in press）,von Flechten der Monogolischen.

[10] Krempelhuber, A. 1873. Chineslache Flechten. *Flora* **56**: 465-471.

[11] _____________. 1874. Chinesische Fiechten. H **13**（4）: 59-no. 16. **13**（5）: 65-no. 32, 33.

[12] Moreau, F. et Moreau. Mme. F. 1951. Lichens de *Chine. Rev. Bryol. et Lich.* **20**: 183.

[13] Rabenhorst, L. 1873. Chinesische Flechtea in der Umgegend von Saison, Hongkong, Wampoa, Shanghay U.S.W.gesammelt im J.1871-72 von Rudolph Rabenhorst fil. bestimmt von Dr. A. V. Krempelhuber in Munchen. *Flora* **56**: 286-287.

[14] Rassadina, K. A. 1968. De Speciebus *Xanthoparmeliae* pro URSS novis of Curiosis notula. Novitates Syst. plant, non Vascularium p. 245-248.

[15] Sato, M. 1952. Lichens Khinganenses: or a list of lichens collected by Prof. T. Kira in the Great Khingan Range. Manchuria. *Bot. Mag. Tokyo* **65**（769-770）: 172.

[16] Tchou, Yen-tch'eng. 1935. Note preliminaire sur les lichens de Chine *Contr. Inst. Bot. Nat. Acad.* Peiping **3**: 299-322.

[17] Waug-yang, Jen-rong and Lai Ming-jou. 1973. A checklist of the lichens of Taiwania **18**（1）: 83- 104.

[18] 魏江春、陈健斌。1974。珠穆朗玛峰地区地衣区系资料。珠穆朗玛峰地区科学考察报告。(生物与高山生理)。179 页。科学出版社。北京。

[19] 魏江春等著。1982。《中国药用地衣》。科学出版社。第 1-65 页。

[20] Zahlbruckner, A. 1930. Lichenes, in Handel-Mazzetti, *Symbolae Sinicae* 3: 1-254.

[21] Zahlbruckner, A. 1933. Flechten der Insel Formosa. *Feddes Repertorium* 33: 54-55.

[22] 赵继鼎。1964。中国梅花衣属的初步研究。植物分类学报 9（2）: 139-166。

[23] 赵继鼎、徐连旺、孙增美。1982。《中国地衣初编》。科学出版社。北京。

A taxonomic revision of the lichen genus *Xanthop armelia*（vain.）hale from China

Wei Jiang-chun

（*Institute of Microbiology, Academia Sinica, Beijing*）

ABSTRACT The lichen species of *Xanthaparmelia* from China have been confused in the literature. In the present paper a taxonomic revision of this genus is made on the basis of Chinese materials preserved in HMAS, S, UPS, H, and TUR. Six species, including a new status *X. dentata*（Zhao）Wei, a new variety *X. tinctina* var. *xizangensis* Wei, and a species *X. incurva*（Pers.）Hale new to China, are reportel.

The results of morphological and chemical reexamination of Chinese materials, and the type specimens of some species of this group show that *Parmelia conspersa* Ach. from China in the literature should belong to *X. tinctina*（Mah. et Gillet）Hale and *X. mexicana*（Gyeln.）Hale, and *P. subramigera* Gyeln. to *X. mexicana* as well. Both *P. stenophylla*（Ach.）Heug. and *P. subconspersa* Nyl, from China belong to *X. taractica*（ Kremph.）Hale. Moreover, some Chinese specimens of *X. tinctina* and *X. taractica* contain a trace of norstictic acid in addition to usnic and salazinic acids. *P. camtschadalis*（Ach.）Eschw. has been mistaken for *P. vagans*（Nyl.）Nyl. by most authors. Actually, *P. camtschadalis*（=*X. camtschadalis*）is quite different from *P. vagans*（=*X. vagans*）in morphological and chemical characters and geographical distributions as well. The former is always marked with yellowish or white lines and spots on the upper surface of lobes, and contains salazinic acid besides usnic acid. The type specimen（H-Ach. no. 1450, in H）was collected from Kamchatka in Asia. But the latter lacks yellowish or white lines and spots on the upper surface of lobes, and contains stictic and norstictic acids in addition to usnic acid. The type specimen（H-Nyl. no. 1730, in H）was collected from Ecuador in South America. The rest of this group from China in the literature remain to be revised in future.

KEY WORDS *Xanthoparmelia*; *X. tinctina*; *X. tinctina* var. *xizangensis*; *X. mexicana*; *X. dentata*; *X. taractica*; *X. camtschadalis*; *X. incurva*

I wish to thank Prof. T. Ahti who kindly reminded me that the asiatic *P. camtschadalis* has been mistaken for *P. vagans* by the most authors, and gave me access to unpublished data. I am also indebted to the herbarium directors and curators Prof. N. Lundqvist（Stockholm）, Dr. R. Moberg（Uppsala）, O.Vitikainen（Helsinki）, and Dr. Y. Mäkinen（Turku）who have supported my work in every possible way and given me access to all the facilities during my stay in their herbaria, and sent on loan the specimens. For the kind help I am also grateful to Prof. R. Santesson, Dr. L. Tibell, Mr. G. Thol, Dr. R. Alava, Mr. U. Laine, and Mrs. Eila Kastari. My visit to Sweden and Finland was a part of exchange programme between Academia Sinica and Royal Swedish Academy of Sciences, and the Academy of Finland respectively. I am particularly grateful to Dr. O. Eriksson and Prof. T. Ahti who have done everything to make my visit to Sweden and Finland.

中国脐鳞属地衣的初步研究

魏 江 春

（中国科学院微生物研究所，北京）

摘要 在泛北极地区迄今已知的脐鳞属地衣共有三种；中国有记录的两种。本文报道了四种，其中之一为新种，另一种为中国新记录。此外，有一改级新组合。

关键词 脐鳞属；红脐鳞；盾脐鳞；垫脐鳞；华脐鳞

脐鳞属（*Rhizoplaca*）是 Zopf 于 1905 年以 *Lecanora chrysoleuca opaca* Ach.（1810）为基础而建立的。但是，它却是 *Lichen melanophthalmus* Ram,（1805）的异名（C. Leuckert, J. Poelt und G. Hähnel, 1977）。在泛北极地区迄今已经描述而又被作者所接受的本属地衣计有红脐鳞 [*Rhizoplaca chrysoleuca*（Smith）Zopf]，盾脐鳞[*Rh. peltata*（Ram.）Leuckert et Poelt]及垫脐鳞[*Rh.melanophthalma*（Ram,）Leuckert et Poelt]等三种。有文献记载的中国脐鳞属地衣有红脐鳞及盾脐鳞两种。H. Magnusson（1940）所描述的新种 *Lecanora regalis*，根据形态特征与化学成分应为盾脐鳞内一变种[*Rh. peltata* var. *regalis*（H. Magn.）Wei]。采自西藏的垫脐鳞为本种在中国的首次报道，而华脐鳞则为本属一新种。

至于文献有记载的中国脐鳞属地衣，包括红脐鳞一变种（=*Lecanora chrysoleuca* var. *subdiscrepans* Nyl.），由于它们既无化学成分的报道，又无原始标本可资利用，因而未能列入本文相应的分类单位之内。

脐 鳞 属

RHIZOPLACA Zopf

Zopf, *Ann. Chem.* **340**: 291（1905）. C. Leuckert, J Poelt und G. Hähnel. *Nova Hedwigia* **28**（1）: 72（1977）.

地衣体盾片型小叶状或大而坚硬厚实的鳞片状，小叶片状单叶型至垫状的复叶型，直径 1-3cm；上表面黄绿色至枯草黄色，表面光滑，平坦，或具细微皱纹及光泽；下表面淡褐色至蓝黑色，光滑或粗糙，以粗壮的脐状物紧固于岩石表面。

子囊盘茶渍型，盘面淡绿色，蓝绿色，淡褐色至黑色，或橙黄色至橙红色。子囊内含 8 孢，孢子椭圆形单胞，无色，透明。

衣体内含松萝酸以及其他副成分。

模式种为 *Lecanora chrysoleuca opaca* Ach. [= *Lichen melanophthalmus* Ram.（1805）= *Rhizoplaca melanophtalma*（Ram）Leuckert et Poelt]。

本文于 1983 年 12 月 22 日收到。

文中线条图由简荔同志清绘；照片由苑兰翠同志洗印，在此表示感谢。

本文原载于《真菌学报》，3（4）：207-213，1984.

分 种 检 索 表

（1）子囊盘淡黄绿色，淡褐色，暗褐色至黑色，但从不呈橙红色。
（2）子囊盘淡褐色，但从不呈黑色。地衣体下表面粗糙，内含泽渥萜（zeorin）…… 2.盾脐鳞 *Rh. peltata*
（2）子囊盘淡黄绿色，淡褐色，暗褐色至黑色。地衣体下表面光滑，不含泽渥萜。
（3）地衣体复叶型，垫状,下表面淡色……4 .垫脐蛾 *Rh. melanophthalma*
（3）地衣体单叶型，盾片状,下表面黑色…… 1.华脐鳞 *Rh.huashanensis*
（1）子囊盘橙红色或橙黄色 ……3.红脐鳞 *Rh.chrysoleuca*

1. 华脐鳞 新种 图版 I 插图 I

Rhizoplaca huashanensis Wei, sp. nov P1. I, Fig. 1

Species colore disci similis *Rhizoplacae melanophthlmae* est, sed a qua thallis monophyllis peltatis incrassis et subtus atris differt. Species affinis *Rhizoplacae peltatae* est, sed a qua disco atro ct absque zeorine differt.

Thallus peltatus, rotundatus, monophyllus, praccrassus, ca. 0.7 mm margine et 2.5mm centro crassus, durus, usque ad 15mm diametro, margine obtuse lobatus, ca. 4-8 mm latus, supra viridi-flavus vel stramineus, impolitus vel nitidiusculus, laevigatus, sine rugis, sed sulcatus inter duos lobos, concavus in centro thalli vel lobi, elevatus in ambitu, margine devexus; subtus ater, impolitus, margine incrassatus crista lineolata, umbilico ca. 5-10 mm diametro, brunneus in peripheria umbilici.

Cortex superus pseudoprosenchymaticus, 8-16μm crassus. Stratum algarum 24-32μm crassum. Medulla 628-2447μm crassa. Cortex inferus pseudoprosenchymaticus 20.8-24μm crassus, hyphis reticularis atris.

Apothecia lecanorina usque ad 2.8mm diametro, disco juventute viridi-flavo et demum brunneo et atro pruinis albis tecto.

Epithecium 47.7-63.6μm crassum. Thecium 63.6-79.5μm crassum. Hypothecium 63.6-79.5μm crassum.Asci clavati, 61 × 16μm. Sporae 8: nae, decolores, ovoideo-oblongae, eseptatae, 11.55-12.32 × 6.93-7.7μm.

Cortex superus K + flavens.

Medulla K + aurantiaca, C-, KC-, P-.

Acidum usneicum.（TLC）

Typus: CHINA, Shaanxi, Mt. Huashan，Bei-Feng, in rupe, 15/VI 1964, Wei J. C. no.59（Holotypus in HMAS-L）.

地衣体盾片状，近圆形，单叶型，厚而坚硬，边缘厚约 0.7mm，中央厚约 2.5mm，直径可达 1.5cm，周缘浅裂为钝圆形裂片，裂片宽为 4-8mm，沿浅裂线方向往往具延伸的表面型沟线；上表面黄绿色，光滑而无褶皱，中央凹下，周围鼓起，周缘微平展或微向下倾斜，无蜡样光泽，或稍具微弱光泽；下表面黑色，具微弱的凹凸表面，无光泽，周缘具棱状钝圆边缘，以脐状物紧固于基物表面，脐直径为 0.5-1cm，近脐处周围局部褐色；由于黑色的下表面而使裂片边缘往往呈现黑色镶边。

上皮层假厚壁组织厚约 8-16μm。藻层厚为 24-32 μm。髓层厚为 628-2447μm。下皮层假厚壁组织，厚为 20.80-24μm，具有黑色网状菌丝。

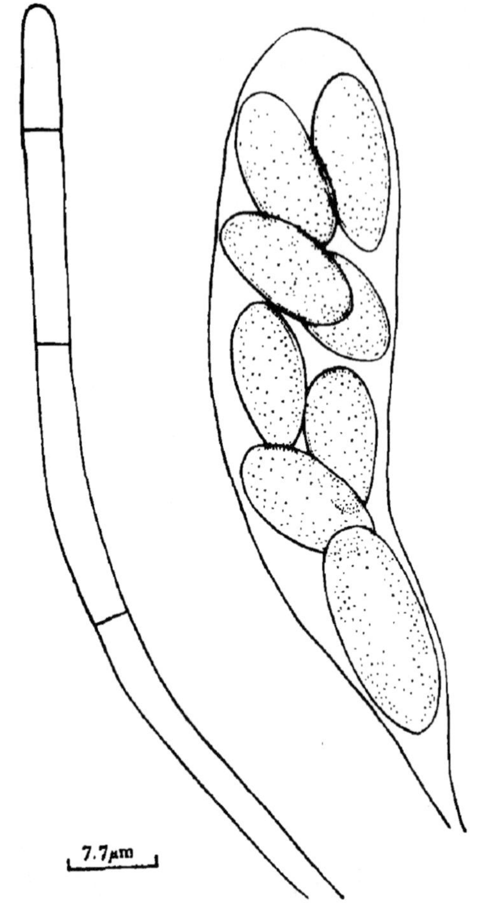

图 1 华脐鳞及其含有 8 个子囊孢子的子囊及侧系
Fig. 1 Ascus with 8 ascospores and paraphysis of *Rhizoplaca huashanensis* Wei

子囊盘茶渍型，幼小时盘面与衣体上表面同色，即黄绿色，随后则呈淡褐色，进而暗褐色至黑色，直径可达 2.8mm 左右，往往覆盖以白色粉霜层。

子实上层厚为 47.70-63.6μm。子实层厚为 63.60-79.50μm。子实下层厚为 63.60-79.50μm。子囊棒状，61× 16μm，内含 8 个孢子。孢子广椭圆形，单胞，无色，11.55 -12.32 × 6.93-7.70μm。

上皮层遇 K+黄色。髓层遇 K +橙黄色，C-，KC-，P-。上皮层含松萝酸。

模式标本：中国，陕西：华山，北峰，岩石上，1964 VI 15，魏江春，no. 59（主模式保存于中国科学院微生物研究所真菌标本室地衣部）。

本种按其子囊盘色泽类似于垫脐鳞，但不同之处在于衣体单叶型而较厚，下表面黑色。本种近似于盾脐鳞，但不同之处在于子囊盘黑色以及不含泽渥萜。

2. 盾脐鳞

Rhizoplaca peltata（Ram.）Leuckert et Poelt, *Nova Hedwigia*, **28**（1）：73（1977）.

Basion. *Lichen peltatus* Ram in Lam. & DC.《 Flore Franc.》ed. 3, **2**:377（1805）.

=*Lecanora peltata*（Ram.）Steuel, «Nomencl. bot. » 237（1824）. H. Magn.《Lichens from Central Asia》，121（1940）,Moreau et Moreau, *Rev. Bryol. et Lichenol*, **20**: 189（1951）.

原变种

Var. ***peltata***

地衣体鳞片状，厚而坚硬，直径 2.4 - 3cm，叶缘不深裂为细长裂片；上表面黄绿色，枯草黄色，表面光滑而具蜡样光泽；下表面粗糙而无光泽，土色至污灰色，淡污灰褐色，有裂纹，边缘有时呈蓝黑色，以中央脐固着于基物上。

子囊盘褐色至黄褐色，无粉霜层，几乎密布于全叶表面。

髓层：P +橙红色。

Usnic & lecanoric caids, atranorin & zeorin,（TLC）

Specimens Examined:

新疆：博格多山，1928 IX 4，D. Hummel alt.2550 m（S）. 天山，1978 VIII 10，王先业 no. 1299 -（1）HMAS-L）.

分布：亚洲 [中国：新疆（H. Magn. 1940，p. 121; Moreau ct Moreau，1951，p. 189），土耳其，苏联：乌兹别克，伊朗，伊拉克，阿富汗，巴基斯坦，喜马拉雅北部]，欧洲，非洲，美洲。

黑腹变种　　改级新组合

Var. ***regalis***（H, Mang.）Wei, **stat. nov.**

Basion. *Lecanora regalis* H. Magn. 《Lichens from Central Asia》, **1**:122（1940）.

Typus：新疆博格多山，alt. 2100m, 1928 VIII 10, D. Hummel（S）.

本变种不同于原变种者在于衣体下表面黑色，上表面边缘及裂缝处可见黑色镶边，其它部分鲜黄绿色，不呈枯草黄色，不带黄褐色色度，表面可见细微皱褶。子囊盘较稀少，果托局部往往黑色。

Usnic acid & zeorin.（MCT）

Specimen Examined:（见上引模式标本）。

由于未进行薄层分析测定，只做了显微化学检测，故难以肯定本变种与原变种化学成分之异同，但是，它们都具有盾脐鳞所特有的泽渥萜（zeorin)，形态特征也基本一致，因此，将 H.Magnusson 的新种降级为盾脐鳞一变种。

3. 红　脐　鳞

Rhizoplaca chrysoleuca（Sm.）Zopf, *Ann. Chem.* **340**: 276-309（1905）. Leuckeret et al. *Nova Hedwigia* **28**（1）：73（1977）. H. Hertel & Zhao, *Lichenologist* 14（2）： 150（1982）.

Basion.: *Lichen chrysoleucus* Sm. *Trans. Linn. Soc.* London. **1**:82（1791）.

=*Lecanora chrysoleuea* Ach.《Lich. Univ. 》1810: 411.

=*Lichen rubinus* Villars,《Hist. Plant. Dauphin.》3977（1789)（non Lichen rubinus Lam. Flora Franc. 1:77（1788）.

=*Lecanora rubina*（Vill.）Ach.《Lich. Univ.》**1810**: 412.

=*Squamaria rubina*（Vill.）Elenk. In Vain.《Lich. e Caucaso et ePenins. Taur.》**1899**: 284.

本种因所含化学成分及其组合差异而分为六个化学小种；本文又增加了三个新的化学小种。

Chem. st. I: Usnic acid only:

Specimens Examined:

河北：小五台山，1964，魏江春 no. 2067.**西藏：**珠峰，1966，魏江春，陈健斌， nos, 1308, 1338，1374，1771. **新疆：**天山，1978，王先业 nos. 839-（1），843。

分布：亚洲（中国，蒙古），欧洲（苏联欧洲部分），美洲（美国）。

Chem. st. II: Usnic & placodiolic acids:

Specimen Examined:

西藏：定日，加布拉山，姜恕，赵从福 no. Q 75。

分布:亚洲（中国，尼泊尔，印度，蒙古）,欧洲，美洲（美国)。

Chem. st. III: Usnic & pseudoplacodiolic acids:

Specimen Examined:

河北：小五台山，1964，魏江春 no. 2038.

分布：亚洲（中国，蒙古，巴基斯坦，尼泊尔喜马拉雅），北欧及美洲（美国）。

Chem. st. IV: Usnic & psoromic acids:

Specimen Examined:

新疆:天山，alt. 2500 m, 1978 VIII 10，no. 1299。

分布：亚洲（中国，巴基斯坦）。

Chem. st. VII: Usnic, placodiolic & lecanoric acids（Chem. st. nov.）:

Specimen Examined:

西藏：珠峰，绒布德寺，alt. 5000m, 1966，魏江春，陈健斌 no. 1338-（1）.

Chem. st. VIII: Usnic, pseudoplacodiolic & lecanoric acids（Chem. st. nov.）.

Specimens Examined:

河北：小五台山，1964，魏江春 no. 2087-（1）. **陕西：**太白山，1963，马启明，宗毓臣 no, 317.**吉林：**长白山，(采集人不明）1963 XI 2。

Chem. st. IX: Usnic & lecanoric acids（Chem. st, nov.）:

Specimen Examined:

吉林：长白山北坡，瀑布与天池之间石上，魏江春 no. 2943.

4. 垫　脐　鳞

Rhizoplaca melanophthalma（Ram.）Leuckert et Poelt, *Nova Hedwigia* **28**（1）：72（1977）.

Basion.: *Lichen melanophthalmus* Ram. in Lam. & DC. 《F1.France》, ed. 5, 2:377（1805）.

= *Lecanora chrysoleuca opaca* Ach. 《Lich. Univ.》**1810**:41.

= *Rhizoplaca opaca*（Ach.）Zopf, *Mitt.-Ann. Chem*. **340**:291（1905）.

衣体垫状，下表面以脐状物固着于基物，脐周褐色，叶周缘蓝黑色。子囊盘盘面淡色至黑色，但从不呈桔红色。

Chem. st. III: Usnic & psoromic acids:

Specimen Examined:

西藏:那曲，罗布习卡，1976 VI 10，赵魁义 no. 36-（2）.

分布：亚洲（中国，乌兹别克，土耳其，伊拉克，伊朗，阿富汗，巴基斯坦，喜马拉雅两北部），欧洲，美洲。

Chem. st. IV: Usnic & lecanoric acids:

Specimen Examined:

西藏：定日，加布拉山，1967，姜恕，赵从福 no. Q75-（1）.

分布：亚洲（中国，阿富汗）。

参 考 文 献

[1] Asahina,Y. and Sato, M.in Asahina,Y.1939.《Nippon Inkwasyokubutu Dukan》 1939: 709.

[2] Hertel, H. and Zhao C. -F. 1982. Lichens from Changbai Shan-Some additions to the lichen flora of North-East China. *Lichenologist* 14（2）：139.

[3] Jatta, A. 1902. Licheni cinesi raccoti allo Shen-si negli anni 1894-1898 dal. rev. Padre Missionario G. Giraldi. *Nuov. Giorn. Bot. Italiano* 2（9）：474.

[4] Leuckert, C., Poelt, J. und Hähnel, G. 1977. Zur Chemotaxonomie der Eurasischen Atten der Flechtengattung Rhizoplaca. *Nova Hedwigia* 28（1）：71.

[5] Magnusson, H, 1940. Lichens from Central Asia **1**: 122.

[6] Moreau et Moreau. 1951. Lichens de Chine. *Rev. Bryol. et Lichenol.* **20**: 289.

[7] Paulson, R. 1928. Lichens from Yunnan, *Journ. Bot.* **66**: 37.

[8] Sato, M. 1940. East Asiatic Lichens. *Journ. Jap. Bot.* **16**（8）：498.

[9] ——.1981. Notes on the cryptogamic flora of Prov. Shansi, North China（III）Lichenes. *Miscellanea Bryologica et Lichenologica* 9（3）：64.

[10] 魏江春、陈健斌。1974。珠穆朗玛峰地区地衣区系资料。《珠穆朗玛峰地区科学考察报告》（生物学与高山生理）1974，174。

[11] Zahlbruckner, A. 1930. Lichenes in Handel-Mazzetti, Symbolae Sinicae, III: 172.

A preliminary study of the lichen genus rhizoplaca from China

WEI Jiang-chun

（*Institute of Microbiology, Academia Sinica, Beijing*）

ABSTRACT Three species of umbilicate lichen genus *Rhizoploca* have been known in the Holarctic region so far, such as *Rh. melanophthalma*（Ram.）Leuckert et Poelt, *Rh. chrysoleuca*（Sm.）Zopf, and *Rh. peltata*（Ram.）Leuckert et Poelt.（Leuckert et al. 1977, p. 72-73）. Five taxa of the genus in China are given in the present paper. Among them one taxon has been changed in status from specific to the name of variety [*Rh. peltata* var ***regalis***（H. Magn.）Wei], one species is new to China（*Rh. melanophthalma*）, and one is new to science（***Rh. huashanensis*** Wei）. Key to the species is given:

1. Disc greenish, green-yellowish, brownish, dark brown to black, but never orange.
   2. Disc brownish, but never black; thallus below coarse, containing zeorin ································ *Rh. peltata*
   2. Disc greenish, green-yellowish, brown to black; thallus below not coarse, lacking zeorin
      3. Thallus polyphyllous, cushion ································ *Rh. melanophthalma*
      3. Thallus monophyllous, petate ································ *Rh. huashanensis*
1. Disc orange································ *Rh. chrysoleuca*

KEY WORDS *Rhizoplaca*; *Rh. melanophthalma*; *Rh. chrysoleuca*; *Rh. peltata*; *Rh. peltata* var. *regalis*; *Rh. huashanensis*

魏江春：中国脐鳞属地衣的初步研究 图版 I

Wei Jiang-Chun: A preliminary Study of the Lichen genus *Rhizoplaca* from China Plate I

华脐鳞 新种

Rhizoplaca huashanensis Wei sp. nov.

华脐鳞地衣体及子囊盘（左边两片衣体为下表面）

Thalli with apothecia of *Rhizoplaca huashanensis* Wei (two thalli on the left)

Wei Jiang-Chun: A new isidiate species of *Hypogymnia* in China Plate I

魏江春：袋衣属一裂芽新种 图版 I

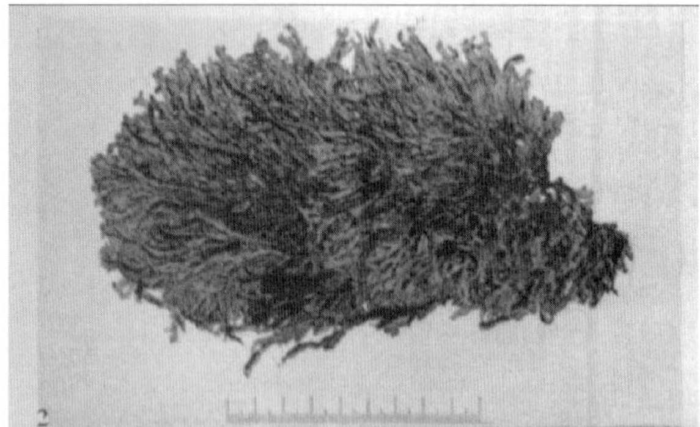

横断山袋衣

Hypogymnia hengduanensis Wei

图 2 康定亚种

Fig. 2 ssp. *kangdingensis* Wei

图 1 原亚种

Fig. 1 ssp. *hengduanensis*

（比例尺每格为 1mm scale in mm）

I am especially indebted to Dr. O. Eriksson for the arrangement made during my visit to Sweden which was a part of the exchange programme between the Academia Sinica and the Royal Swedish Academy of Sciences, and to Prof. N. Lundqvist（S）for the help offered during my stay in the Herbarium of Swedish Museum of Natural History in Stockholm.

魏江春：袋衣属一裂芽新种 **图版 II**

Wei Jiang-chun: A New isidiate species of *Hypogymnia* in China **Plate II**

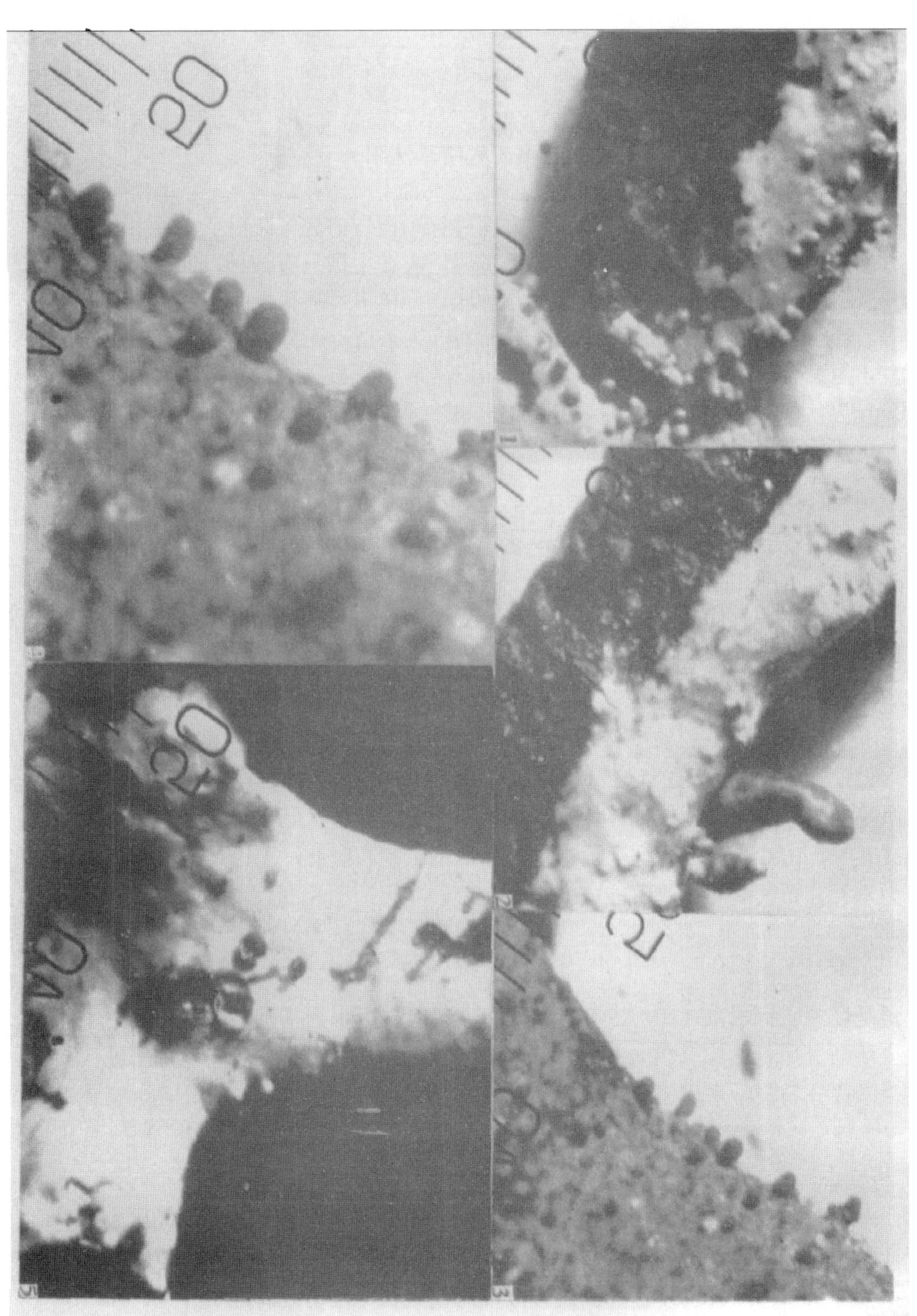

***Hypogymnia* *hengduanensis* Wei**

Fig. 1—4 ssp. *hengduanensis*

1. Isidia (Scale in 312·5 μm)

2. Adventitious lobules (Scale in 312.5 μm)

3. Isidia (Scale in 250 μm)

4. Isidia (Scale in 125 μm)

Fig. 5 ssp. *Kangdingensis* Wei (With a few isidia, scale in 250 μm)

中国石蕊科地衣研究之一
——筛蕊属的订正研究

魏江春　陈健斌　姜玉梅
（中国科学院微生物研究所，北京）
陈　锡　龄
（中国科学院林业土壤研究所，沈阳）

摘要　本文对产于中国的筛蕊属地衣在形态与化学方面进行了比较研究，订正了文献中一些错误鉴定及分布方面的错误记载。研究结果表明，筛蕊属地衣在中国只有一种，即聚筛蕊：*Cladia aggregata*（Sw.）Nyl.。其主要特征为拟果柄不肿胀，分枝多，筛孔少而排列不整齐以及只含有巴巴酸等。至于云南的变种 *straminea* 实际上为一生态型，无分类学意义。浙江的 *Cetraria aculeata* Fr.为聚筛蕊的错误鉴定。根据现有知识，本种分布区的北界在朝鲜为北纬38°，在日本为北纬 37°，在中国为北纬 31°，而非北纬 44°（黑龙江虎林）。

关键词　筛蕊属；聚筛蕊；巴巴酸

筛蕊属地衣在中国仅有一种，即聚筛蕊。关于本种在中国的最早记载由法国的 Hue（1898）所提供。这个记载是以 R. P. David 于 1869 年采自四川和 Abbe Delavay 于 1887-1892 年采自云南的标本为基础的。二十世纪二十至三十年代，本种在中国的分布主要由法国的 Harmand（1928），奥地利的 A. Zahlbruckner（1930，1933），德国的 Sandstede（1932）和日本的朝比奈泰彦及佐藤正已等外国人所报道。他们涉及的产地包括云南、四川、贵州、福建、台湾及江西等六省。自本世纪七十年代以来，中国地衣学工作者，如王贞蓉与赖明洲（1973，1976），魏江春与陈健斌（1974），芬兰的 T. Ahti 与赖明洲（1979），刘瑞棠与赖明洲等（1980），魏江春（1981），魏江春与姜玉梅（1981）以及魏江春等（1982）关于本种在台湾、西藏、云南、贵州、江西等省的记录陆续做过报道。至于本种在黑龙江虎林（陈锡龄等，1981）及吉林长白山（魏江春等，1982）的记载是基于错误的标本记录签。因此，无论是吉林长白山或黑龙江虎林作为本种分布区的北界尚不能成立。

材料与方法

本文所用材料是以本所标本室（HMAS-L）收藏的中国地衣标本为基础，同时还利用了中国科学院林业土壤研究所标本室（FPI）的部分标本。对上述标本从形态与化学方面进行了比较研究。在化学成分测定中采用了显色反应法及薄层色谱法。

问题与讨论

形态与化学：筛蕊属地衣以其拟果柄所特有的筛孔型表面而与石蕊科其他各属相区别。聚筛蕊的主要特征是拟果柄不膨胀，表面筛孔较少而排列不整齐，分枝繁多而往往相互缠绕；无子囊盘者拟果柄较细而枝梢渐尖锐；有子囊盘者拟果柄较粗壮，枝梢钝而近于齐头。体内含有巴巴酸（barbatic acid）。

本文于 1984 年 2 月 2 日收到
在考证聚筛蕊在黑龙江虎林地区的产地时，承王战教授提供重要情况；秦仁昌教授，郑儒永同志以及陈心启同志对本文提供宝贵意见；唐雯同志清绘分布图，在此一并深致谢忱。
本文原载于《真菌学报》, 4（1）: 55-59, 1985.

根据形态与化学成分的比较研究表明，所用的全部中国标本均为聚筛蕊一种。至于具有淡白色拟果柄的变种（var. *straminea*）以及具有其他细微差异的个体或群体，如拟果柄粗壮，表面粗糙，色泽暗褐；或拟果柄纤细，表面光滑，色泽淡绿或淡褐等，均含有本种特有的巴巴酸。至于有子囊盘者由于拟果柄较粗壮而枝梢钝圆至齐头，初见略似稀筛蕊 [*Cladia inflata*（F. Wills.） D. Gall.]，但不同之处在于后者拟果柄表面筛孔极少甚至缺乏，拟果柄明显肿胀，含有黑茶渍素 （Atranorin）及富马原岛衣酸（Fumarprotocetraric acid.）。因此，本种上述形态方面的细微差异可能与生境与生长发育阶段有关，无分类学意义。

错误鉴定：朱彦承（1935）以 *Cetraria aculeata* Fr.为名称报道刘慎谔（no. 6739 in HMAS-L）采自浙江杭州莫干山（朱彦承误为天目山）的标本，实为聚筛蕊之错误鉴定。此外，保存在本所标本室的由老一辈植物学家所采集的本种标本均被前人误定为 *Cetraria aculeata* Fr.

分布问题：本属地衣系热带及亚热带种类。它们主要分布于南半球，个别种类可分布至北半球。仅聚筛蕊一种的分布区可延伸至北回归线以北的中国，日本，朝鲜，尼泊尔，锡金以及印度北部等地（布万里子，1962）。本种分布区的北界在朝鲜为北纬 38°，在日本为北纬 37°，在中国为北纬 31°（图 1）。

刘慎谔所采的聚筛蕊标本（no. 13400, in FPI）附有“黑龙江虎林水克至穆稜河”的标签（陈锡龄等，1981）。上述产地位于北纬 44°左右，与迄今已知的本种分布区北界，北纬 38°左右的朝鲜金刚山更加偏北 6 度之多。但是，在工作中，作者发现刘慎谔于 1945 年在云南昆明所采的本种标本编号为 13398 （HMAS-L），与他在黑龙江虎林地区所采本种标本号 13400 （FPI）仅 2 号之差，而且这两号标本号牌无论在纸质、形状、大小、字体、字号、以及系在号牌上的线绳等都完全相同。同时，这两号同种标本的群体外形也基本一 致，一见便知它们可能来自同一生境。此外，据查实，刘慎谔从 1950 年 4 月才开始在东北地区采集植物标本。他所采的第一号东北植物标本是来自哈尔滨农学院林场，而最后一 号标本，亦即 10034 号标本，则来自黑龙江小兴安岭丰林地区。刘慎谔既未到黑龙江虎林地区采集过植物标本，在他的东北植物标本中又无大于 10034 的编号。因此，可以断定，刘慎谔的 13400 号标本并非采自虎林，而是同他的 13398 号标本一起采自云南昆明。至于刘慎谔的 13400 号标本附有黑龙江虎林的标签，很可能是后人在标本整理工作中的失误。

筛 蕊 属

Cladia **Nyl.**

Recogn.Ram.1870, p. 69.

=*Cladonia subgen. Cenomyce* Ser. *Ochrophaeae*, a. *Clathrinae* （Müll. Arg.） Vain. Monogr. Clad. Univ. XI: 223 （1887）.

聚筛蕊

Cladia aggregata （Sw.） Nyl.

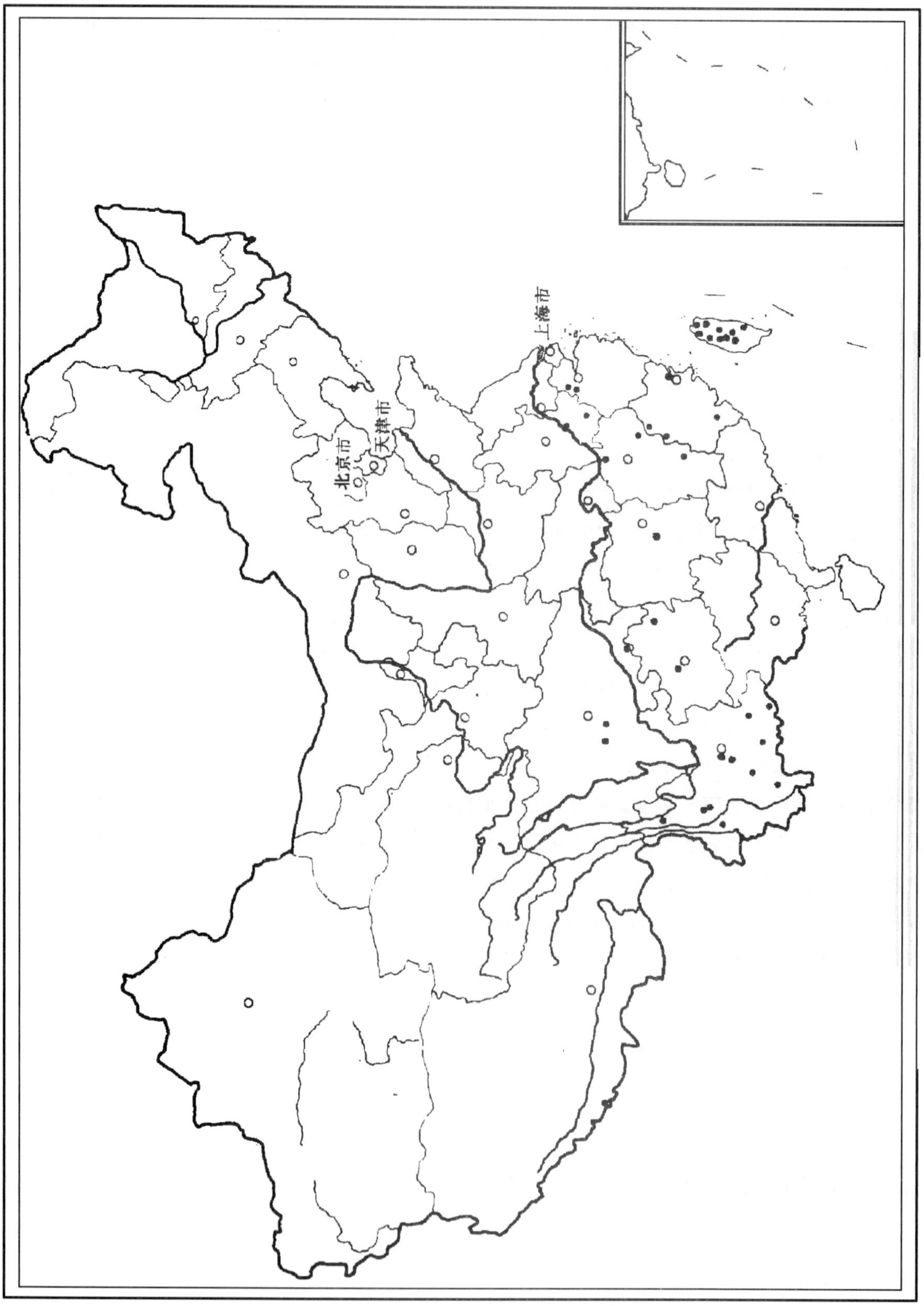

图 1 聚筛蕊（•）在中国的分布图

Fig. 1 A map of distribution of *Cladia aggregata* （Sw.） Nyl. （·） in China

Recogn. Ram, 1870, p. 69 （167）.

=*Lichen aggregatus* Sw. Nov. Gen. 1788, p.147.

=*Cladonia aggregata* （Sw.） Ach, Vet. Acad. Nya Handl.T.XVI:68 （1795）. Vain. Vain. Monogr. Clad. Univ. 1:224 （1887）.

=*Cladonia aggregata* f. *straminea* Müll.Arg. Fl.:162 （1879）; Vain, monogr. Cl.1:230 （1887）.

Cetraria aculeata sensu Tchou T. Y.: Contributions from the Inst. of Bot. National Acad, of Peiping, 3 （6）: 311 （1935）. non （Schreb.） Fr. Syst. Orb. Veget, 1825, p. 239.

M. P-.

TLC: Barbatic acid only.

Specimens Examined:

浙江：莫干山，1929.8，刘慎谔 no. 6739 （HMAS-L,误为 *Cetraria aculeata*，见 Tchou Yen-tch'eng，1935，p. 311），天目山：刘慎谔 no. 6787 （HMAS-L，误为 *Cet. aculeata*），赵继鼎、徐连旺 no. 6261-（1），6293. **安徽**：黄山，刘慎谔、钟补求 nos. 2468, 2756（HMAS-L，误为 *Cet. aculeata*），赵继鼎、徐连旺 nos. 5166， 5167，5894-（1），魏江春 no. 3764-（6）；铜陵，赵继鼎等 nos. 61，63，71，72，77，78，317。**江西**：庐山，赵继鼎 等 nos. 410， 497，魏江春 nos. 3100，3123，3139；铅山县，王庆之、袁 nos. 6，19，3140. **福建**：武夷山，王庆之、袁 no. 125；王庆之 nos, 750，790；光泽县，王庆之、袁×× no. 375，王庆之 no. 372，福建省研究院植物标本 no. 125；南靖，1200m，林尤兴 1974，10. **湖南**：衡山，19651.2.魏江春 no. 3046 **广西**：赵继鼎、徐连旺 nos. 8892，8988，8989，8991，9051， 9205， 9437。**四川**：蛾眉山，赵继鼎、徐连旺 nos. 8154，9286；南川，熊洛华，周子村 no. 90275；二郎山，蒋兴麐 no. 34894，**贵州**：樊净山，魏江春 nos. 38，173，804-（1），825 等;贵阳，魏江春，无号，1963，7. 25.王庆之 no. 139-b。**云南**:昆明，魏江春 nos. 2888-（1），2899，2992，2973，2889，2936，2972-（1），刘慎谔 nos. 13398，（HMAS-L） 13400 （FPI），16885，17045，赵继鼎、陈玉本 nos. 4707，1980，2123，2125，1545，1546， 1549，4835，姜玉梅 nos. 1973-（1）， 1547， 1549；大理，魏江春 nos. 2765（1），2766，2782 （2），2802；维西，王启无 nos. 21463， 21464。下关，姜玉梅 nos. 469（1），497，516，486-（1）；马关，王庆之 nos. 385，386；温泉，韩树金等 no. 359；保山，王庆之 no. 1296；丘北，王 nos. 815，842，845，850；个旧，赵继鼎、陈玉本 nos. 2299，2319，2300；思茅，赵继鼎、 陈玉本 nos. 2992，2543。**西藏**：樟木口岸，魏江春、陈健斌 no. 561。

参　考　文　献

[1] Ahti, T.and Lai Ming-Jou.1979. The lichen genera *Cladonia*, *Cladina* and *Cladia* in Taiwan. *Ann.Bot. Fennici* 16: 234.

[2] Asahina, Y. 1950. Lichens of Japan I. Genus *Cladonia* pp. 255.

[3] Asahina, Y. and Sato, M. in Asahina, Y. 1939.朝比奈泰彦:日本隐花植物图鉴（东京）。

[4] 陈锡龄（Chen X. L.）、赵从福、罗光裕。1981。东北地衣名录。东北林学院学报 3: 129。

[5] Culberson,C. F. 1969. Chemical and botanical guide to lichen products. University of North Carolina press, Chapel Hill pp. 628.

[6] __________. 1972, Improved conditions and new data for the identifilation of lichen products by a standardised thin-layer chromatographic method. *J. Chromatogr.* 72: 113-125.

[7] __________. Kristinsson, H. D. 1970. A standardized method for the identification of lichen products, J. Chromatogr. 46: 85-93.

[8] Galloway, D. J. 1977. Additinal notes on the lichen genus *Cladia* Nyl., in New Zealand. *Nova Hedwigia* 28（2-3）: 475.

[9] Harmand,A.1928. Lichen d'Indo-Chine recueillis per M.V.Demange. *Annal. Cryptog. Exot.* **1**（4）:322.

[10] Hue, A. M. 1898. Lichenes Extra-Europaei. *Nouv. Arch, Mus. Hist Nat.* 3（10）: 258.

[11] Liu Tang-shui, Lai Ming-jou and Lin Han-shi. 1980. The Cladoniform Lichens in Taiwan. Quart. J. Taiwan Mus. **33**（1+2）: 14.

[12] Nuno, M. 1962. Chemism of *Cladonia* Subgenus *Clathrina*（Müll. Arg.）Vain, *Journ. Jap. Bot.* 37（3）:77-80.

[13] Sandstede, H. 1932. Cladoniaceae A. Z. I. Pflanzenar. **3**: 71.

[14] Tchou Yen-tch'eng. 1935. Note preliminaire sur les lichens de Chine. *Contr. Inst, Bot. Nat. Acad. Peiping* 3: 311.

[15] Wang Yang Jen-Rong and Lai Ming-Jou, 1973. A checklist of the Lichens of Taiwan. *Taiwania* 18（1）: 89.

[16] __________, and__________1976. Additions and corrections to the lichen flora of Taiwan. *Taiwania* 21（2）; 227.

[17] 魏江春（Wei J. C.）。 1981。中国地衣标本集。植物研究 1（3）：87。

[18] 魏江春（Wei J. C.）、陈健斌、1974。珠穆朗玛峰地区地衣区系资料——珠穆朗玛峰地区科学考察报告（生物与高山生理）。176。科学出版社。北京。

[19] Wei Jiang-Chun, Jiang Yu-Mei. 1981. A biogeographical analysis of the lichen flora of Mt. Qomolangma Region in Xizang. in Proceedings of symposium on Qinghai-Xizang（Tibet）plateau（Beijing, China）1147.

[20] 魏江春（Wei J. C.）等。1982。《中国药用地衣》。科学出版社。北京。

[21] Zahlbruckner, A. 1930. Lichenes in Handel-Mazzetti. *Symbolae Sinicae* **3**: 130.

[22] Zahlbruckner, A. 1933. Flechten der Insel Formosa. Fedde Repertorium sp. nov. **33**: 46.

Studies on lichen family cladoniaceae in China I. a revision of *Cladia* NYL.

WEI JIANG-CHUN CHEN JIAN-BIN JIANG YU-MEI

(*Institute of Microbiology, Academia Sinica, Beijing*)

CHEN XI-LING

(*Institute of Forestry and Soil Sciences, Academia Sinica, Shenyang*)

ABSTRACT A comparative study in morphology and chemistry of *Cladia* Nyl. from China, as well as some revisions of misidentification are made in the present paper. All the collections examined belong to the same species, *Cladia aggregata*（Sw,）Nyl., which is characterized mainly by having not inflated pseudopodetia, numerous slender branches, and fewer and irregularly distributed perforations on the surface of pseudopodetia, and containing barbatic acid in it. The specimens with light colour were treated by some authors as *variety straminea*. This is, in fact, dependent upon its habitat and developmental stage. Moreover, the northeastern limits of this species are at 31°N in China, 37°N in Japan, and 38°N in Korea respectively. No specimen has been collected from Hulin of Heilongjiang prov. and Mt. Changbai Shan of Jilin prov. That was based on wrong lebals. The record of *Cetraria aculeata* Fr. in Zhejiang prov.（Tchou, 1935, p. 311）is a misidentification of *Cladia aggregata*（Sw.）Nyl.

KEY WORDS *Cladia*; *C. aggregata*; Barbatic acid

Studies on lichen family cladoniaceae in China II. the lichen genus *Cladina* Nyl.

WEI JIANG-CHUN　CHEN JIAN-BIN　JIANG YU-MEI

(*Institute of Microbiology, Academia Sinica, Beijing*)

ABSTRACT Nine species of *Cladina* from China are reported in the present paper. Among them *Cladina imshaugii* is a new to China together with *Cladina aberrans*. In addition, a taxonomical revision of some taxa of the genus from China is made. All of the specimens examined are tested by TLC in addition to morphological studies.

KEY WORDS Reindeer lichens; *Cladina*; *Cladina aberrans*; *Cladina imshaugii*

The lichen genus *Cladina* was established by Nylander in 1866. According to Nylander's original descriptions of *Cladina* it contained not only reindeer lichens, but also the species of the *Unciales* group of *Cladonia* and those of *Cladia*. Now, the genus *Cladina* is used only for true reindeer lichens. A good world wide monograph of this group treated as a subgenus of *Cladonia* is given by T. Ahti in 1961.

Eight species of *Cladina* from China have been reported in the scattered floristic and taxonomic literature as the members of *Cladonia*. They are *Cladina arbuscula, C. grisea, C. mitis, C. portentosa, C. pseudoevansii, C. rangiferina, C. stellaris, C. tenuis*, and some infraspecific taxa of them.

During the study of the lichen flora of China the senior author (Wei J. C.) of the present paper has made the lichenological expedition to many different areas of the country since 1963. He found a good stands of reindeer lichens in Dahinganling near Mangui in Neimongol of China as wintering ranges of about 1,000 local reindeers (*Rangifer tarandus*). Here is the only locality of reindeers in China. The good developed carpet of reindeer lichens can be found everywhere in the larch forest near Mangui.The reindeer lichens, however, are distributed in the alpine areas of the whole China as well.

The reindree lichen, i.e. *Cladina* spp. are the best known lichen group as a feeding plants for reindeer particularly in wintertime and also as one of the lichen anbotiiite resources.

In the present paper 9 species of the genus are reported. Seven of them have been known for China so far. Two species, such as *Cladina aberrans* and *Cladina imshaugii* are new to China.

As for Cladina portentosa and its infraspecific taxa previously reported under the name *Cladonia impexa* from Jilin (Moreau ct Moreau,1951), Shaanxi (Jatta, 1902; A. Z. 1930), Yunnan (Harm. ? as f. *laxiuscula*), and Sichuan (A. Z. 1934, as f. *spumosa*) we have not seen any specimens from China so far. It has to remain to be revised in future.

We are indebted to Prof. T.Ahti for giving a great help in the study of this group during the senior author's stay in Helsinki, and to Drs. N. Lundqvist (S), Y. Mäkinen (TUR), R. Moberg (UPS), and O.Vitikainen (H) for giving him access to all the facilities during his stay in these herbaria in 1982.

This paper was originally published in *Acta Mycologica Sinica*, 5 (4): 240-250, 1986.

CLADINA Nyl.（LURUI SHU）

Flora **12**:179（1866），&Not. Sällsk. F. Fl. Fenn. Förhandl. nova ser. **5**:110（1866）. =*Cladoina* G. H. Web. subg. *Cladina*（Nyl.）Leight. em. Vain. Leight. The lichen-flora of Great Britain, Ireland and the Channel Islands 1871:66; Vain. Medd. Soc. F. Fl. Fenn. 14:31（1866）; T. Ahti, Ann. Bot. Soc. 'Vanamo' 32（1）: 3（1961）.

Type species: *Cladina rangiferina*（L.）Nyl.（*Lichen rangiferinus* L.）

Key to species

1. Branching isotomic
  2. Branching predominantly dichotomic, main stems absent, P- ························ 1. *C. psendoevansii*
  2. Branching predominantly tetrachotomic or tetrachomictrichotomic, P+ or-
    3. P-, psoromic acid absent······························· 2. *C. stellaris*
    3. P+, psoromic acid present······························3.*C.aberrans*
1. Branching anisotomic
  4. Branching predominantly dichotomic
    5. Yellowish-grey, usnic acid present
      6. Main stem rather robust, resembles *C. arbuscula* ························5.*C.imshaugii*
      6. Main stem thin ······························ 4.*C.tenuis*
    5. Ash-grey, usnic acid absent ······························ 8.*C. grisea*
  4. Branching predominantly tetrachotomic or tetrachotomic trichotomic
    7. Yellowish-grey, usnic acid present
      8. P+ ······························ 7.*C.arbuscula*
      8. P- ······························6.*C. mitis*
    7. Ash-grey, usnic acid absent ······························9.*C.rangiferina*

1. *Cladina pseudoevansii*（Asahina）Hale et W. Culb.（Niquelurui）

Bryologist **73**（3）:510（1970).

Basionym: *Cladonia 'pseudoevansi'* Asahina, Journ. Jap. Bot. **16**:187（1940).

Lectotype from Japan, central Hondo, collected by Hirose in X/1923. **Isolectotype** in H（!).

This specics is very similar to both *Cladina stellaris* and *Cladina evansii*. But it is easy to distinguish from the former species by the noticeable dichotomy, and from the latter one by the distribution different and lack of atranorin in it. It has been found only in Mt. Changbai Shan from China so far.

Lichen substances: usnic and perlatolic acids.（TLC）

Reaction: P-, K-, KC+.

Specimens examined:

On the ground among grasses in forests or in mountain tundra.

JILIN: Fusong county, south slope of Mt. Changbai Shan: on roadside between Wei dong station and the post, of the frontier of China at about 1400m, 19/VII/1950, collected by Chang Y. L. et al. no. 270, & Zhou Y. L. et al. no, 1432; near Wei dong station at about 1600-1700m, 5-9/ VTII/1983, collected by Wei J. C. and Chen J. B. nos. 6621, 6623. 6696, 6780-（1). Changbai county, near Hengshan station at about 2000 m, 1/

VIII/1983, collteced by Wei J. C. & Chen J. B. no. 6399- (1)

Ahti T. (1961) and Chen et al. (1981) mention this species under genus name *Cladonia* only from Changbai mountain of Jilin prov. Actually it is found just from the south slope of the mountain. So called "Changpai-san""Chang-pei-shan", and "Pehtausan" in the literature(T. Ahti, 1961,p.27)actually reffer to the same place, i.e. Mt. Changbai shan (the chinese, name) which is situated on the border between China and Korea. "Peh-tau-san"is the korcan name of "Changbai shan".

2. *Cladina stellaris* (Opiz) Brodo (Quelurui)

Bryologist **79** (3) :363 (1976).

Basionym: *Cenomyce stellaris* Opiz

Synonym: *Cladonia alpestris* (L.) Rabh.

This lichen is very easy to recognize by its isotomic tetrachotomy predominant with trichotomy.

Lichen substances: usnic and perlatolic acids with a trace of pseudonorrangiformic acid. (TLC)

Reaction; P-, K-, KC+.

Specimens examined:

On the ground among grasses or between rocks or on mosscovered rocks in forests or mountain tundra.

NEIMONGOL: Dahinganling: Mangui: near the rail way station, 11 / IX / 1977, Wei J,C.no.3364- (1), near Molgan, alt. 800m, 12/ IX /1977, Wei J.C. nos. 3378, (2), 3387, 3388.

JILIN prov. south slope of Changbai mountain, Changbai county: in Mt. Hongtou shan, alt. 1900 m, 28/VII/1983, Wei J.C. & Chen J.B. nos. 6256, 6266, 6274, 6281; near Hengshan station, alt, 2000-2100 m, 1-3/VIII/1983, Wei J.C. & Chen. J.B. nos. 6367, 6399, 6553, 6554, 6575, 6579, & alt. 1300 m, Wei JC & Chen JB nos. 6445, 6499. Fusong county: near the post of the frontier of China, alt. 1600 m, 3/VIII/1983, Wei J.C.& Chen J.B. no. 6724. north slope of Changbai mountain: Antu county: in tundra, VIII /1977, Wei J.C. no. 2868, alt. 2250 m, 13/ VIII /1983, Wei J.C. & Chen J.B. no. 6871, near Tianchi, alt. 2100 m, VIII /1977, Wei J.C. 2936- (2), near Wenquan (hot spring), 12/ VIII /1983, Wei J.C. & Chen J.B. no. 6794.

HEILONGJIANG: Dahinganling, Gulian forestry centre, 3/IX/1977, Wei J.C. no. 3222- (1), in Qianshao tree farm, 3/IX/1977, Wei J.C. no. 3270.

SHAANXI: near Baoji town, 29/VI/1959, collector from the research institute of Chinese medicine, no. 1850; Taibai mountain near Baxiantai & Erliguan, alt. 3400 m,1963, Ma QM & Zong YC (no number), near Pingansi, alt. 2640 m, 2/VI/1963, Wei JC et al. no. 2536-(2), Baxiantai, Sanyiehai & Huangcisi 4-5/VI/1963, Wei J.C, et al. nos. 2543- (2), 2702- (4).

T. Ahti (1961, p. 51) states about this species: "Sandstede's (1932) record from Shensi (=Shaanxi) is very uncertain". Our work verifies the certainty of Sandstcde's record.

This species as *Cladonia alpesiris* was previously reported in Neimongol (Sato, 1952; Ahti, 1961; Chen et al. 1981), Jilin (A. Z. 1934; Ahti, 1961; Chen et al. 1981), Heilongjiang (Asahina, 1952), Shaanxi (Jatta, 1902; A. Z. 1930; Sandstd. 1932; Wei et al. 1982) & China (Prov. not indicated by A. Z. 1934). So, the information about distribution of this species in China given by us in this paper has not gone beyond the previous records reported by the above mentioned authors.

3. *Cladina aberrans* (des Abb.) Hale et W. Culb. (Huangquelurui)

Bryologist **73** (3) :510 (1970). (lapsu "*abberans*").

Basionym; *Cladonia alpestris* f. *aberrans* des Abb. Bull. Soc. Sci. Bretagne **16** (2) :93 (1939).

This species is similar to *Cladina stellaris*（Opiz）Brodo, but different from which in containing psoromic acid and in distributions: *Cladina stellaris* is distributed in Europe, Asia, North America and Groenland, but this species is distributed along the coasts of Pacific Ocean: North-East Asia, North America and Groenland. It is totally absent in Europe.（T. Ahti, 1961, pp. 13, 51, 52, and fig. 7）. The species is new to China.

Lichen substances: Usnic and psoromic acids（3270-（1））, and sometimes perlatolic acid in addition to above mentioned lichen substances（3378, 3387,（1））.（TLC）

Reaction: P+ yellow, K-, KC+ yellow.

Specimens examined:

On the ground or fallen wood in the larch forest.

NEIMONGOL: Dahinganling, Gulian, 4/IX/1977, Wei JC no. 3270-（1）（usnic and psoromic acids）.

T. Ahti（1961, pp. 13, 51-53, fig.7）gives us a good distribution map of this species and states that this species has its main distribution and the centre of abundance on the northern Pacific coasts, such as North Korea, Japan, Asiatic Russia, U. S. A. Alaska and its Islands, the west coast of Canada. However, it also occurs continuously along the Arctic coast of Canada to Atlantic, such as the Northwest Territories of Canada Baffin island, the Southwestern coast of Groenland. The more southern East American localities represent isolated pocket of the main area, such as Michigan and Pensylvania etc.

The senior author of the paper collected the specimens of this species from Dahinganling area of China in 1977. We have not seen it anywhere in China except Dahinganling so far. According to the knowledge of its distribution area it could be found in Changbai mountain as well.

4. *Cladina tenuis*（Flk.）Hale et W. Culb.（Xilurui）

Bryologist **73**（3）: 510（1970）.

Basionym: *Cladonia rangiferina*（L.）wigg. g. *tenuis* Flk, De Cladoniis difficillimo lichenum genere commentatio nova, **1828**:164.

Synonym: *Cladonia tenuiformis* Ahti, Ann. Bot. Soc. Vanamo **32**（1）: 63 1961）. **Type** from North Korea, Futenpo collected by Asahina in 28/VII/1934. no. 734. **isotype** in H（!）

Lichen substances: Usnic and fumarprotocetraric acids,（TLC）

Reaction: P+ red, K- or weakly yellow, KC+ yellow.

Specimens examined:

On the ground.

JILIN: Changbai mountain, south slope: Fusong county near the post of the frontier of China in Weidong station, alt. 1600 m, 5/VIII/1983, Wei J. C. and Chen J. B. nos. 6595, 6618,6626.

YUNNAN: Kunming, Liou T. N. no.13400, as Cladonia tenuiformis in FPI

XIZANG: Zayul county, alt. 4150 m, 12/VIII/1973, Zhang J.W.（no number）; Milin county , alt. 3000m , 23/VIII/1974, Zong Y.C.no. 57.

ANHUI: Mt. Huang shan; alt. 1610 m, 22/VIII/1962, Zhao J. D. and Xu L. W. no, 5712-（1）.

JIANGXI: Mt. Lu shan, alt, 950 m, 13/11/1965, Wei J. C. no. 3212.

TAIWAN: collected by T.Suzuki in 1935; from Taiping mountain collected by S. Asahina in 1936, & collected by Koponen, nos. 17089, 17283; collected by Lai, nos. 5952, 6851,7828, 7846, 8542, 10345, 9838. All of these specimens are preserved in H（!）.

This species is reporte from Taiwan by A. Zahlbruckaer（1933）, Asahina（1950）, Ahti（1961）, Lamb（1963）, Wang & Lai（1973, 1976）, Ahti & Lai（1979）, Liu et al.（1980）, and from Liaoning and Heilongjiang prov. by Chen et al（1981）, from Yunnan as v. *tenuis*（Ahti, 1961）& as f.*flavicans*（A. Z. 1934）as well. So the report from Xizang（Tibet）, Anhui, & Jiangxi is the first time for this species.

The record of this species as Cladonia tenuiformis in Hulin county of Heilongjiang prov.（Chen et al. 1981, p. 134）is based on the specimen collected by Liou T • N • from Yunnan prov. but not from Heilongjiang prov.（see Wei et al. 1985. p. 56, 58, 59.）

5. *Cladina imshaugii* （Ahti）in Beiheft 79 Zur nova Hedwigia, Festschrift J. Poelt, 1984, p. 50.
（Yalinlurui）

Basionym: *Cladonia imshaugii* Ahti, Ann. Bot. Soc. 'Vanamo'**32**（1）:113（1961）.

Holotype from the west Indies collected by H. A. Imshaug no. 23493 in H（!）.

This species is similar to *Cladina arbuscula*（Wallr.）Hale et W. Culb. in appearance at first sight, but different from it in having dichotomic branches predominant. The species is new to China.

Lichen substances: Usnic and fumarprotocetraric acids.（TLC）

Reaction： P+ red, K-, KC+ yellow.

Specimens examined:

On the soil covered rocks, and on the ground.

JILIN: Changbai mountain, north slope, tundra, VTII/1977, Wei J. C. no. 2880-（5）.

GUIZHOU: Fanjing mountain near Sanwangdian, alt. 1920-2200 m，21-23/IX/ 1963,
Wei J. C. nos. 804,822.

6. *Cladina mitis* （Sandst.）Hale et W. Culb.（Ruanlurui）

Bryologist **73**（3）: 510（1970）.

Basionym: *Cladonia mitis* Sandst. 1918 Clad, exsicc. 55.

This species is close to *C. arbuscula* in having branches of anisotomic trichotomy predominant, and those of dichotomy and tetrachotomy are also common, but different from it in having usually slender to some robust main stem and rather sparsely branched, erect or multilaterally deflexed, less browned crown and in lacking fumarprotocetraric acid, and containing usnic acid constant.

Two（Wei & Chen）of us found some good stands of a population of this species in the open lawn near forest by the frontier post of China in Weidong station with more robust main stem（more than 1 mm in diam.）and rather sparse and erect branching. The inner medulla of it seems to be usually as thick as the outer one or somewhat thicker.（Wei J. C. & Chen J.B. no. 6619）

Lichen substance: Usnic acid only（TLC）.

Reaction: P-, K-, KC+ yellow.

Specimens examined:

On the ground among grasses in forest or in the open lawn near forest.

JILIN: Changbai mountain, north slope, Antu county: Baishan station, alt. 2000 m, 26/VII/1960, Yang Y. C. et al. no. 26; South slope, Fusong county: near the frontier post of China in Weidong station, alt. 1550-1600 m, 5-7/VIII/1983, Wei J. C. & Chen J. B, nos. 6593, 6599, 6619, 6624-（1），6688; near Motianyu, alt. 2500 m, 22/ VII/1950, Zhou Y. L. et al. no. 1552 b.

Lichen substances: Usnic acid & unknown atranorin-like substance.（TLC）

Reaction： P-, K, KC+ yellow.

Specimens examined:

JILIN: Changbai mountain, north slope, Antu county: in mountain tundra, 1977, Wei JC no, 2866-（8）B; south slope, Changbai county: Hanjiagou by the Hengshan stationr alt. 1300 m, 2/VIII/1983, Wei J. C. & Chen J. B. no. 6488.

Lichen substances; Usnic acid & unknown substance in class 3（TLC: solvent syst. C），

Reaction: P-, K-,KC+ yellow.

Specimens examined:

JILIN: Chipufen mountain, 11/IX/1931, Kung JW, no. 2262 (det. by Tchou YT as Cladonia sylvatica Nyl. see Tchou, 1935, p. 305).

This species was reported before by some authors from Jilin (Ahti, 1961; Chen et al. 1981), Shaanxi (Wei et al. 1982, p. 21), Xizang (Wei & Chen, 1974), & Taiwan (Ahti, 1961; Wang & Lai; Ahtit & Lai, 1979; Lin et al. 1980).

7. *Cladina arbuscula* (Wallr.) Hale et W. Culb. (Linlurui)

Bryologist **73** (3) :510 (1970).

Basionym: *Platellaria foliacea* v.m. *arbuscula* Wallr. Naturgeschichte der Säulchen-Flechten; oder monographischer Abschluss Uber die Flecbten-Gattung Cenomyce Acharii. 1829:169.

This species resembles *C. mitis* in ramification, but differs from it in having more robust main stem and densely branched and often unilaterally deflexed, conspicuously browned crown, and in usually producing fumarprotocetraric acid in it.

Lichen substances: Usnic & fumarprotocetraric acids. (TLC)

Reaction: P+ orange, K-, KC+ yellow.

Specimens examined:

On the ground among mosses in forest or on mountain tundra.

JILIN: Changbai mountain, north slope, Antu county: on tundra, 1977, Wei J. C. no. 2868- (2) ; near Wenquan (hot spring), alt. 1850 m, 12/VIII/1983, Wei J. C. & Chen J. B. no. 6794- (1). south slope, Changbai county: in Hanjiagou by the Hengshan station, 1300 m, 2/VIII/1983, Wei J. C. & Chen J. B. no. 6478.

YUNNAN: Kunming: Hill Xi shan, Huatingsi, Zhao J. D. & Chen Y. B. no. 1559; Jindian, alt. 1900 m, Wei J. C. no. 2972. Dali: Mt. Caeng shan, Beijing University, no. 177 a.

XIZANG: Milin county, Zong Y, C. no. 85. Tsona county, alt. 4500 m, 11/VIII/ 1974, Zong Y. C. no. 53 (pr. p.). Zham county, alt. 4200 m, 13/VIII/1973, Zhang J. W. (no. number).

The following specimens contain an unknown atranorin-like substance in addition to usnic and fumarprotocetraric acids:

Lichen substances: usnic and fumarprotocetraric acids, and unknown atranorin-like sub.stance. (TLC)

Reaction: P+ orange, K-, KC+ yellow.

Specimens examined:

NEIMONGOL: Dahinganling, Mangui county: near the railway station of Mangui, 11-12 /IX/1977, Wei J. C. nos, 3334- (1), 3373- (1) ; in Morgan near Alonggeya river, Wei J, C. no. 3378- (1).

JILIN: Changbai mountain, north slope: on mountain tundra,VTII/1977.Wei J.C.no.2866- (8) -A.

HEILONGJIANG: Gulian forestry centre, by the bridge over Laocao (Laochao) river, 3/IX/ 1977, Wei J. C. nos. 3224, 3238; in Qianshao tree farm, 4/IX/1977, Wei J. C. no. 3270- (2).

SHAANXI: Taibai mountain, 1963, Wei J, C. et al. (.no number).

XIZANG: Tsona county: Pohang mountain, alt. 4500 m, 11/VIII/1974, Zong YC no. 53 (pr.p.).

Cladina arbuscula as *Cladonia sylvatica* or *Cladonia arbuscula* is reported by different authors for Neimongol (Sato, 1952), Liaoning (A. Z, 1934; Chen et al. 1981), Jilin (Tchou, 1935; Moreau et Moreau, 1951; Chen et al. 1981), Sichuan (A. Z. 1934), Yunnan (Hue, 1889, 1898; A. Z. 1930; Tchou, 1935), and Taiwan (A. Z. 1933; Wang & Lai, 1973). Ahti (1961) mentions subsp. *beringiana* under *Cladonia arbuscula* from.

Yunnan, and Wang & Lai from Taiwan (1976) as well. *Cladonia sylvatica* f. *sphagnoides* was reported by Asahina from Neimongol in 1952. So, this species previously not reported from Xizang (Tibet).

8. *Cladina grisea* (Ahti) Trass (Dongfanglurui)

Basionym: *Cladonia rangiferina* ssp. *grisea* Ahti, Ann, Bot, Soc, 'Vanamo'32 (1) : 96 (1961).

Type from Japan, Tokai district, Lake Yamanake, on the foot of Mt. Fuji collected by Asahina in 16/VIII/1952, **isotype** ia H (!)

This species is easily recognized by its very slender branchsystem, by predominant dichotomy with common trichotomy, and by the much more pale colour. *Cladina grisea* is probably an endemic species restricted to the East Asia.

Lichen substances: Fumarprotocetraric acid & atranorin. (TLC)

Reaction: P+ red, K+ yellow, KC-.

Specimens examined:

On the ground among mosses in forest, and on fallen wood.

NEIMONGOL: Dahinganling, Mangui, near railway station, 11/IX/1977, Wei JC no. 3338.

HEILONGJIANG: Daxinganling, Gulian forestry centre, by the big bridge over Laochao river about 9 km from Gulian, 3/IX/1977, Wei J. C. nos. 3224- (1), 3228- (1), 3229-A, 3233.

JILIN: Changbai mountain, north slope, Antu county: near Baishan station, 7/VIII/ 1977, Wei J. C. no. 2753; on the ground of mountain tundra, VIII/1977, Wei J. C. no. 2866- (8) -B. South slope, Changbai county: near Hengshan station, alt. 1350m, 30/ VII/1983, Wei J. C. & Chen J.B. nos 6321- (1), 6351, 6352, 6353. Fusong county: by the frontier post of China, alt. 1500m, 7/VIII/1983, Wei J. C. & Chen J. 6. no. 6687 near Weidong station, 19/VII/1950, Liu TN et al. no. 1431a (pr.p.)

SHAANXI: Taibai mountain, Baxiantai, near Erliguan, 1963, Ma Q. M. & Zong Y. C. no. 63; Yuewangchi, alt. 2900m, 3/VIII/1963, Ma Q, M. & Zong Y. C. no. 352.

YUNNAN: VII/1941, Wang H. C. no. 1083a.

GUIZHOU: Fanjing mountain, near Sanwangdian, alt. 1920m, 21/IX/1963, Wei J. C. no. 804- (4), 805.

XIZANG: 1973, Zhang J. W. (no number).

JIANGXI: Mt. Lu shan, near Wulaofeng, alt. 950m, 13/II/1965, Wei J. C. no. 3212.

This species was previously reported from China only in Hebei (Ahti, 1961), and Taiwan (Ahti, 1961; Ahti & Lai, 1979; & Liu et al. in 1980). Now, we have shown a larger area of its distribution in China.

9. *Cladina rangiferina* (L.) Nyl. (Lurui)

Flora 12:179 (1866).

Basionym: *Lichen rangifernus* L. Species plantarum, 1753:1153.

Synonym: Cladonia rangiferina (L.) G. H. Web.

This species is the most widespread lichen among Cladina, very easy to recognize by its bluish-grey to pale grey colour, and anisotomic tetrachotomic predominant branching pattern with rtichotomic and dichotomic ones.

Lichen substances: fumarprotocctraric acid & atranorine. (TLC)

Reaction: P+ orange, K+ yellow, KC-.

Specimens exaimined:

On the ground and rotten stump.

NEIMONGOL: Dahiaganling, Mangui, near the railway station, 11/IX/1977, Wei J. C. nos. 3321, 3344; in

Morgan near Alonggeya river, alt. 800m, 12/IX/1977, Wei J. C. nos. 3373-（1）-（1）3374-（·1）. 3384, 3385-（5），3387.

HEILONGJIANG: Dahinganling, Gulian, in Qianshao tree farm, 4/IX/1977, Wei JC no. 3272-(2); by the big bridge over Laochao river about 9 km from Gulian, 3/IX/ 1977, Wei J. C. nos. 3229, 3233.

JILIN: Changbai mountain, north slope, Antu county: from ice stadium to the lake'Tianchi', Wei J. C. nos. 2868.-（3），2935 south slope, Fusong county: near Weidong station（Sipingjie），19/VII/1950, Liu TN et al. no. 1431 a(pr. p.); by the frontier post of China near Weidong station, alt. 1600m, 5/VIII/1983, Wei J. C. & Chen J. B, nos. 6592, 6598, 6622, 6670, 6687. Changbai county: near Hengshan station, alt. 1300-1350 m, 27/VII/-2/VIII/1983,Wei J. C. & Chen J. B. nos. 6156, 6162, 6351; alt. 2050 m,31 /VIII/1983,Wei J. C. & Chen J. B. nos. 6362, 6363, 6374; Hongtou mountain, alt. 1900m, 28/VII/I983, Wei J. C. & Chen J. B. no. 6284.

SICHUAN: Emei mountain, 9/VII/1960, Ma Q. M. et al. no. 280.

YUNNAN: Mt. Caeng shan, 1885, 1889, Delavay, nos. 1559（=H-Vam. no. 12681, it is mixed with Thamnolia subuliformis, Cladonia amaurocraee, and Cl. furcata），4180（= H-Vain, no. 12682, it is mixed with Cl. amaurocraea and Cl. furcata），4180（ =H-Vain, no, 12683），& H-Vain. nos. 12690, 12692（with one podetium of Cl. amaurocraca），12689（Yentzihay, 1888, Delavay），12691（Lopin mountain, 1883, Delayay）. All of above cited specimens collected by Delavay from the same localities, as above mentioned, are preserved in TUR（!），and one specimen collected by Wei J. C. no. 2774-（1）from Dali, Mt. Caeng shan in 29/XII/1 1964 is preserved in HMAS-L.

XIZANG: Zham, alt, 3780-3950m, 20/V/1966, Wei J. C. & Chea J. B., nos. 981、984，1014-（1）; Zayul, alt. 4260m, 29/VI/1973, Zheng D no. 73-09.

In China this species was as *Cladonia rangiferina* previously reported from Neimongol（Sato, 1952; Chen et al. 1981），Liaoning（Chea et al, 1981），Jilin（A. Z, 1934, Tchou, 1935; Chen et al. 1981），Heilongjiang（Chen et al. 1981），Shaanxi（Jatta 1902; A. Z, 1930; Wei. et al. 1982），Hubei（A. Z. 1930），Yunnan（Hue, 1887 & 1898; A. Z. 1930），Xizang（Wei & Chen, 1974; Wei & Jiang, 1981），and Taiwan（A. Z. 1933; Wang & Lai, 1973）. Asahina and Sato（1952）mentioned it for Neimongol as ssp. and var. *rangiferina*, and Ahti（1961）for Sichuan, Yunnan and Xizang as well.

Literature cited

[1] Ahti, T. 1961. Taxonomic studise on Reindeer lichens（*Cladonia*, subgenus *Cladina*）. *Ann, Bot, Soc. "Vanamo"* **32**（1）: 1-160.

[2] ——. & Lai, M. J, 1979, The Lichen Genus *Cladonia*, *Cladina* and *Cladia* in Taiwan. *Ann.Bot. Fennici* **16**: 228-236.

[3] Asahina, Y. 1939. Nippon Inkwasyokubutu Dulan pp. 713.

[4] ___________.1950. Lichnns of Japan I. Genus *Cladonia* pp.1-255, Tokyo.

[5] ___________.1952. An addition to the Sato's Lichenes Khinganenses（Bot.Mag. Tokyo, **65**: 172）. *Journ. Jap. Bot.* **27**（12）: 373.

[6] Chen, X. L., Zhao, C. F. & Luo G. Y. 1981. A list of lichens in NE China. *Journ, North-Eastern Forestry Inst.* **3**: 127-135.

[7] Hue, A, M. 1889. *Lichenes Yunnanenses* a cl. Delavay praesertim annis 1886-1887, collectos exponit A. M. Hue（1）. series secunda. *Bull. Soc. Bot. France*, **36**: 158-176.

[8] ___________. 1898. Lichenes Extra-Europaei. *Nouv. Arch. Mus. Hist. Nat* III（10）: 213-280.

[9] Jatta, A. 1902. Licheni cinesi raccolti allo Shen-si negli anni 1894-1898 dal rev. Padre Missionario G. Giraldu *Nuov. Giorn. Bot. Italiano*, ser. 2, IX, 460-481.

[10] Lamb, I.M.1963. Index Nominum Lichenum inter annos 1932 et 1960 divulgatorum. New York.

[11] Liu, T. S, Lai, M. J. & Lin, H. S. 1980. The Cladoniform Lichens in Taiwan. *Quarterly Journ, of the Taiwan* Museum **33**（1,2）: 1-35.

[12] Moreau, F et Moreau, Mme, F.1951. Lichens de Chine. *Rev. Bryol.* et *Lichenol*, **20**: 183-199.

[13] Sandstede, H. 1932. Cladoniaceae A. Z. I. Die Pflanzenareale, Dritte Reihe, Hef 6 Dr. Ludwig Diels und Dr. G. Samuelsson beraus-gegeben von Dr. E. Hannig und Dr. H. Winkler, pp, 63-71, maps 51-60.

[14] Sato, M. M. J952. Lichenes Khinganenses: or a list of lichens collected by prof. T. Kira in the Great Khingan Range, Manchuria, *Bot. Mag*. Tokyo, **65**（769-770）: 172-175, f. 1.

[15] Tchou, Y.T.1935.Note preliminaire sur les lichens de Chine.*Contr. Inst. Bot. Nat. Acad.Peiping*,**3**: 299-322.

[16] Wang-Yang J.R.Lai, M.J.1973. A Checklist of the Lichens of Taiwan. *Taiwania* **18**（1）:83-104.

[17] Wang-Yang J. R.& Lai ,J. M,1976. Additions and corrections to the Lichen flora of Taiwan. *Taiwania* **21**（2）: 226.

[18] Wei, J. C. & Chen, J. B. 1974. Materials for the lichen flora of the Mount Qomolangma region in Southern Xizang,China Report on the Scientific investigations（1966-1968）in Mt.Qomolangma,district, Science Press, Beijing pp. 173-182（in Chinese）.

[19] Wei, J. C; Chen, J. B., Jiang, Y. M.& Chen, X. L.1985. Studies on Lichen Family Cladoniaceae in China I. A Revision of *Cladia* Nyl, *Acta Mycologica Sinica* **4**（1）: 55-59.

[20] Wei, J.C. & Jiang,Y.M. 1981. A Biogeographical Analysis of the Lichen Flora of Mt.Qomolangma Region in Xizang. Procecdings of Symposium on Qinghai-Xizang（Tibet）Plateau（Beijing, China）, pp. 1145-1151.

[21] Wei, J. C., Wang, X. Y., Wu, J. L., Wu, J. N., Chen, X. L. & Hou, J. L. 1982. Lichenes Officinales Sinenses, pp. 1-65, Science Press.

[22] Zahlbruckner, A. 1930. Lichenes in Handel-Mazzctti, Symbolae Sinicae **3**:1-254.

[23] Zahlbruckner, A. 1933. Flechten der Insel Formosa, in *Fedde's Repertorium* sp. *nov*. **33**: 22-68.

[24] ___________.1934. Nachträge zur Flechtenflora Chinas. *Hedwigia* **74**: 195-213.

中国石蕊科地衣研究之二——鹿蕊属

魏江春　陈健斌　姜玉梅

（中国科学院微生物研究所，北京）

摘要　鹿蕊属地衣（又称驯鹿地衣）广布于北极以及低纬度高寒地带；在我国多见于北方林带以及高山苔原带。在大兴安岭林区本属地衣往往是一定林型的伴生成员，又为大兴安岭驯鹿群的理想天然牧场，还是地衣抗菌素的重要资源之一。本属地衣在国内外文献记载中我国约有 8 种以及若干种下单位，其中（*Cladina portentosa*）一种我们尚未见到。本文记载了 9 种，其中黄雀鹿蕊（*Cladina aberrans*）与亚林鹿蕊（*Cladina imshaugii*）两种我国新记录。

此外，对于前人已发表的错误鉴定进行了订正。

关键词　驯鹿地衣；鹿蕊属；黄雀鹿蕊；亚林鹿蕊

本文于 1985 年 8 月 26 日收到。

Notes on lichen genus *Lobaria* in China

Wei Jiang-chun* Chen Jian-bin Jiang Yu-mei
(*Institute of Microbiology, Academia Sinica, Beijing, China*)

ABSTRACT Seventeen species including 2 infraspecific taxa and 4 chemical races from China are reported in this paper. Among them two species, *Lobaria ferax* and *Lobaria tuberculata* are new to China. Several taxa have been revised in the present paper on basis of the original specimens cited by different authors in literature. Most Chinese records of *Lobaria pulmonaria* in literature are based on misidentification.

KEY WORDS *Lobaria*; *L. ferax*; *L. pulmonaria* ; *L. tuberculata*; Chemical race

The lichens of *Lobaria* from China have been reported by some 33 different authors in about 21 scattered literature with 48 taxa belonging to 31 species. According to Wei's unpublished AN ENUMERATOIN OF THE LICHENS IN CHINA they can be placed in 25 species, 3 varieties and 8 forms. Among them 20 species are reported from Taiwan, 13 from Yunnan, 10 from Xizang, 9 from Sichuan, 6 from Shaanxi, 3 from Gansu, 2 from Hebei, Jilin, Heilongjiang, Hubei, Hunan, Jiangxi, and 1 from Zhejiang and Fujian respectively.

The results of this study show that the following Chinese records are based on erroneous identification; *Lobaria meridionalis* (H. Magn. 1940, p. 45) , *Lobaria pulmonaria* (Tchou Y. T., 1935, p. 319, cited as *Sticta pulmonacea* Ach.) , *Lobaria pulmonaria* var. *hypomelo*, (Del.) Cromb. and var. *papillaris* (Del.) Hue (H. Mag. 1940, p. 46) , and *Lobaria pulmonaria* var. *hypomelaena* A. Z. (nom. illeg. : A. Z. 1930, p. 84, and Kryptogam. Exs. Ed. a Mus. Hist. Nat. Vind. no. 2752 in S & HMAS-L) , etc. However, the specimens of the latter (Kryptogam. Exs. no. 2752) seem to be a mixture of more than two different species. One of them preserved in W has been designated as a holotype of *Lobaria yunnanensis* (Yoshim. 1971, p. 282) , but the specimens of the same number preserved in S and HMAS-L respectively contain gyrophoric acid. The latter one (in HMAS-L) bring isidia and lobules mainly along margins of thallus, is treated in the present paper as *Lobaria spathulata*, but the former one (in S) as *L . gyrophorica* owing to the absence of isidia or lobules. Above mentioned misidentified records have been revised in the present paper on basis of original specimens cited by different authors in literature.

In this paper 17 species including 2 infraspecific taxa and 4 chemical races from China are reported. Two of them, *Lobaria ferax* and *L. tuberculata* are new to China, and one chemical race under *Lobaria kazawaensis* is a new to science.

The senior author is indebted to directors of the herbaria in Sweden Prof. Dr. N. Lundqvist (S) and Dr. R. Moberg (UPS) for the kind help offered during his stay in their herbaria in 1982, which was part of the exchange programme between the Acaemia Sinica and the Royal Swedish Academy of Sciences.
This paper was originally published in *Acta Mycologica Sinica*, supplement I : 363-378, 1986.

LOBARIA (Schreb.) Hoffm. 1796, p. 138.

=*Lichen* (sect,) *Lobaria* Schreb. L. 1791, P. 768.

Lectotype, *Lichen pulmonarius L.* = *Lobaria pulmonaria* (L.) Hoffm.

Key to Chinese species of *Lobaria*

1. Soredia present.

Thallus reticulate-ridged, margin and ridges more or less sorediate; algae green; Medulla KC-, stictic, constictic, and norstictic acids present ·········· 11. *Lobaria pulmonaria*

1. Isidia and/or lobules present.

2. Thallus reticulate-ridged.

3. Medulla K+, P + .

4. Thallus with green algae .

5. Medulla KC+, gyrophoric and norstictic acids present ·········· 5. *Lobaria isidiophora*

5. Medulla KC-, stictic, constictic, and norstictic acids present ·········· 8. *Lobaria meridionalis*

4. Thallus with blue-green algae; Medulla KC-,triterpenoids present ·········· 14. *Lobaria isidiosa*

3. Medulla K-, P-.

6. Thallus with green algae; Medulla KC+, gyrophoric acid present.

7. Congyrophoric acid present·········· 7. *Lobaria tuberoulata*

7. Congyrophoric acid absent ·········· 6. *Lobaria spathulata*

6. Thallus with green or blue green algae; Medulla KC-, gyrophoric acid absent.

8. Thallus with green algae ·········· 12. *Lobaria kazawaensis*

8. Thallus with blue-green algae ·········· 16. *Lobaria retigera*

2. Thallus not reticulate-ridged, lobules present; gyrophoric and congyrophoric acids present ·········· ·········· 17. *Lobaria arassior*

1. Soredia, isidia, and lobules absent.

9. Thallus with green algae.

10. Thallus reticulate-ridged.

11. Medulla K+, P + .

12. Medulla KC+ , gyrophoric acid present ·········· 3. *Lobaria orientalis*

12. Medulla KC-, gyrophoric acid absent·········· 9. *Lobaria ahinensis*

11. Medulla K-,P-,

13. Medulla KC+, gyrophoric acid present·········· 4. *Lobaria gyrophorica*

13. Medulla KC-, gyrophoric acid absent ·········· 10. *Lobaria yunnanensis*

10. Thallus not reticulate-ridged.

14. Thallus scrobiculate ·········· 2. *Lobaria ferax* f. *stenophyllodes*

14. Thallus not scrobiculate································ 1. *Lobaria disoolor* var. *inaotiva* f. *subadsoripta*

9. Thallus with blue-green algae.

15. Medulla K+ , P+ ·· 13. *Lobaria pseudopulmonaria*

15. Medulla K-, P- ·· 15. *Lobaria kurokawae*

Sect. **Lobaria**

Lectotype: *Lobaria pulmonaria*（L.）Hoffm.

1. *Lobaria discolor*（Bory in Del.）Hue

var. *inactiva*（Asah.）Yoshim;

f. ***suhadscripta*** Yoshim. 1971, p. 266.

Type locality: Japan.

Reaction: Th. K+ ; M. KC+.

TLC: gyrophoric acid.

Specimens examined:

YUNNAN: on trees in the forest near Puvin, 1/VI /1959, collected by V. N, Sukachev, no. 1046. ZHEJIANG: Mt. Tianmu shan, Laodian, at about 1000 m, on moss covered rocks, 2/IX/1982, Zhao & Xu, no. 6387.

Distribution: TAIWAN（A, Z.1933, p.30; Yoshim. 1971, p. 266; Wang & Lai,1973, p. 92）.

2. *Lobaria ferax* Vain.

f. *stenophyllodes*（Vain.）Yoshim. 1971, p. 271.

=*Lobaria ferax var. stenophyllodes* Vain. 1913, p.133

Type locality: Philippines.

Reaction: Th. K-; M. KC+.

TLC: gyrophoric acid.

This form is different from f. ferax in the negative thalline reaction with K.

Specimens examined:

YUNNAN: 24/VI/1957, Polianskii, no.1041; Maguan county, Mt. Laojuin shan, on trees, at about 1700m, 6/VI/1959,Wang Q.Z.,no.426.

New to China.

3. ***Lobaria orientalis***（Asah.）Yoshim. 1969, p. 75.

=*Lobaria pulmonaria* var. *orientalis* Asah. 1949, p. 67.

Lectotype locality: Japan.

Lobaria meridionalis sensu H. Magn. 1940, p. 45, non Vain.

Sticta pulmonacea sensu Tchou, 1935, p. 319, non Ach.

Lobaria chinensis sensu Wei & Jiang, 1986, p. 23, non Yoshim.

This species is close to *Lobaria yunnanensis* Yoshim. , but differs from the latter by the presence of gyrophoric and norstictic acids.

Reaction: Th.K-, M. K+, KC+, P+.

TLC: gyrophoric and norstictic acids as constant substances, in addition, stictic and constictic acids present in chemical race I.

Chemical race I:

Specimens examined:

GANSU: Zagaling, on bark of tree, at about 1800m, 29/X/1930, collected by D. Hummel, in S(!), det. by H. Magn. as Lobaria meridionalis Vain(TLC not done). SICHUAN: Quantian county, Mt. Erlang shan, on rocks near

Yaiban river, at about 1950m, 18/VII/1950, Jiang X. L. , no, 34890; Mt. Emi shan, Luidongping, at about 2500m, on Abies sp. 19/Ⅷ/1963, Zhao & Xu, no. 8263; Baoxing county, Qu G. L. , no. 3531. YUNNAN: Lijiang, Mt. Yulong shan, on trees, 3/XI/1958, Han S. J. & Chen L.Y., no. 5042 -（1）. XIZANG: Lulanggou, on tree trunk in the forest of *Abies* sp. at about 3400m, 22/Ⅷ/1974, Li W. H., no. 2234. HUBEI: Shennongjia region, Jangjunzhai, alt. 2150m, 1984, Chen J. B., nos. 1193（2）, 11933-（1）; Qianjiaping, alt. 1900-1950 m, 1984, Chen J. B. , nos. 11658, 11662; Dajiuhu, alt. 2450 m, 1984, Chen J. B. , no. 10521. ZHEJIANG: Mt. Tianmu shan, 27/VII/ 1930, Liou T. N. , nos. 6792, 6813, determened by Tchou Y. T. and Liou T. N. respectively as *Sticta pulmonacea* Ach.

We have not seen the specimens nos. 6801, 6815 cited by Tchou Y. T. in his paper（1935, P. 319）, but we have studied the specimen no. 6813 determined by Tchou Y. T. as *Sticta pulmonacea* Ach, The specimen no. 6815（Tchou Y. T. , 1935, P. 319）seems to be a lapse of the no. 6813, or both nos. 6813 & 8815 are the conspecific lichen: *Lobaria orientalis*, because the above two specimens（nos. 6813, 6815）were collected by the same collector at the same time from the same locality: Zhejiang prov. Mt. Tianmu shan, 27/VII/ 1930, and determined by the same person: Tchou Y. T. as *Stica pulmonacea* Ach.

Distribution: GANSU（H. Magn. 1940, p. 45）, SICHUAN, YUNNAN（Yoshim. 1971, p. 273）, XIZANG（Wei & Jiang, 1981, p. 1146 & 1986, p.22), TAIWAN(Yoshim. 1971, p. 273; Wang & Lai, 1973, p. 92 & 1976, p. 227）.

4. ***Lobaria gyrophorica*** Yoshim. 1971, p. 257.

Type locality: Taiwan, Ilan, S. Kurokawa, No. 794（holotype）in TNS &（isotype）in NICH（not seen）.

Lobaria pulmonaria f. *hypomelaena*(Del.)Hue. Kryptogam. Exs. Ed. a Mus. Hist. Nat. Vind. no. 2752(S!), nom. illeg.

This species is similar to both *Lobaria orientalis* and *L . yunnanensis* by the morphological characters but differs from the former by the absence of stictic, constictic and norstictic acids, and from the latter by the presence of gyrophoric acid.

Reaction: Th. K-; M. K-, KC+, P-.

TLC: gyrophoric acid.

Specimens examined:

SICHUAN: Baoxing county, Qu G. L., no. 2669; Mt. Daba shan, Kangjiawan, on twigs, 13/IX/1958, Yu Y. N. & Xing Y. S., no. 1623. YUNNAN. Lijiang, Mt. Yulong shan, at about 3600 m, on bark of trees, collected by Handel-Mazzetti, distributed as Kryptog. Exs. Ed. a Mus. Hist. Nat. Vind. no. 2752（S!）; Dali, Mt. Cang shan, at about 3000 m, 5IX/1959, Wang Q.Z. no. 1197, determined by Zhao as *Lobaria pulmonaria*（L.）Hoffm. HUBEI: Shennongjia region: Jiangjunzhai, alt. 2150 m, 1984, Chen J. B. , nos. 11931, 11932, 11933; Houzishi, alt. 2650 m, 1984, Chen J. B. , no. 10446.

Distribution: TAIWAN（Yoshim. 1971, p. 275; Wang & Lai, 1973, p. 92）.

5. ***Lobaria isidiophora*** Yoshim. 1971, p. 276.

=*Lobaria meridionalis* Vain, *minor* Ras. 1949, p. 82, non *Lobaria minor*（Nyl.）Vain. 1926, p. 16.

Type locality: Philippines.

Reaction: Th. K-; M. K+, KC+, P+.

TLC: norstictic and gyrophoric acids, in addition, stictic and constictic acids in chemical race I, and lackig in chemical race II.

This species is characterized by the positive medullary reaction with KC, but sometimes, by the negative medullary reaction with it owing to low concentration of gyrophoric acid.

Chemical race I:

Selected specimens examined:

JILIN: Mt. Changbai shan, south slope, from Huapihezi to Liujiatangzi, on bark of tree, alt. 1100 m, 18/Ⅶ/1950, Zhou Y. L. , Liou T. N.,Wu Z. Y. , no. 1396. SHAANXI. Mt. Taibai shan: Yuewangchi, on *Picea* sp. at about 3080m, 6/Ⅶ/1963, Ma & Zong, no. 369; Nantianmen, at about 3090 m, on a dead tree 2/Ⅶ/1963, Ma & Zong, no. 338; On the ground near Dadian at about 2340 m, 18/Ⅶ/1963, Ma & Zong, no. 74; Tiziya, on rocks, 14/Ⅶ/1963, Ma & Zong, no. 3098. SICHUAN: Kangding, west of the town, on rather dry hills with shrubs, 11/Ⅺ/1922, H. Smith, nos. 5135（pr.p.）, 5137（pr.p.）in UPS（!）; northern Sichuan, Dongrego, Hang-lungssu, in humid Picea-Abies forest, at about 3500m, 22/Ⅻ/1922, H. Smith, no. 5120 in UPS（!）（TLC not done）; Mt. Daba shan, Beiping district, on Quercus sp. 2/Ⅸ/1958, Yu Y. N & Xing Y. s., no. 1283; Miyaluo, Muozigou, at about 2950 m, on tree & rotten wood, 12/Ⅸ/1960, Wang C. M, et al. no. 908. YUNNAN: Lijiang: Heibaishui forest centre, at about 3100m, on coniferous tress, 10-12/Ⅺ/1980, Jiang Y. M. , nos.78-（1）, 128, 129, 314, 333-（1）; Mt. Yulong shan, at. about 3000m, on trees, 7/Ⅻ/1960, Zhao & Chen, no. 3962. GUIZHOU: Kali, Xijiang, Legongping, at about 1350 m, 19/V/1959, Expedition to southern Guizhou, no. 2014 ; Mt. Fanjing shan, Huguosi, at about 1100 m, on tree trunk, 2/ⅫI/1963, Wei J. C., no. 392; Jiandaoxia near Huguosi, on *Cyclobanopsis* sp. at about 1360 m, 16/Ⅸ/1963, Wei J. C. , no. 777 & on tree trunk of Fagus sp. alt. 1800 m,14/Ⅸ/1963, Wei J. C. , no. 722; Daling, on tree trunk of Fraxinus sp. in forest, 22/Ⅷ1963, Wei J. C. , no. 458 & on bark of Betula sp. in forest, alt. 1460 m, 24/Ⅷ/1963, Wei J. C. , no. 515, XIZANG: Zham, Quxiang, at about 3480 m, 20/ Ⅴ/1966, Wei & Chen. no. 1045; Linzhi, Lulang tree farm, at about 3200 m, in the forest of Abies sp. 1974, collected by Li W. H. ; Ni Z. C. , no. 5-（2）from Xizang in 18/Ⅵ/1973. HUBEI: Shennongjia region: Mt. Dashen-nongjia, Shen-nongding（top of the Mountain）, alt. 2900m, 1984, Chen J. B. , no. 10350; Mt. Jiemeifeng, alt. 2940 m, 1984, Chen J. B. , no. 11292; at the foot of the mountain, alt. 2700 m, 1984, Wei & Chen, no. 10670; Xiaolongtan, Changaiwu,on bark of the coniferous trees, alt. 2300m, Wei, no. 10965; Qian jiaping. alt. 1900m, Chen.J.B, no. 11568; Mt. Xiaoshen-nongjia, alt. 2600 m, 1984.. Chen J. B.,no. 11060; Dajiuhu, alt. 2750 m, 1984, Chen J. B. , no. 10531; Xiaojiuhu, alt. 2000-2550 m, 1984, Chen J. B. , no. 10538. ANHUI: Ex Herb, of Fan Memoral Institute of Biology, collected by Wang.

Chemical race II:

Specimen examined:

JILIN: Mt. Changbai shan, south slope, Mt. Hongtou shan, on Abies sp. at about 1500-1700 m, 27/VII/1983, Wei & Chen, no. 6206.

Distribution: JILIN（Chen et al. 1981, p. 150）, SHAANXI（Wu & Liu, 1976. p. 67, as *Lobaria pulmonaria* var. *meridionalis*）, SICHUAN, YUNNAN（Yoshim. 1971, p. 276）., XIZANG（Wei & Jiang, 1981, p. 1146 & 1986, p. 23）, TAIWAN（Yoshim. 1971, p. 276; Wang & Lai, 1973, p. 92）.

6. ***Lobaria spathulata***（Inum.）Yoshim. 1971, p. 278.

=*Lobaria meridionalis* var. *spathulata* Inum. 1943, p. 221.

Neotype locality: Japan.

=*Lobaria pul monaria* f. *hypomelaena*（Del.）Hue, in Kryptog. exs edit,a Mus.Hist.Nat.Vind. no, 2752 in HMAS-L.（!）nom. illeg.

Thallus with a few cylindrical isidia and lobules mainly along margins.

Reaction. Th. K-; M. K-, KC-（owing to low concentration of gyrophoric acid.）

TLC: gyrophoric acid only.

Specimens examined:

SICHUAN: Markang, Wangjiagou, at about 3700 m, 18/VI/1958, collector not known. YUNNAN: Kryptogamae exsiccatae editae a Mus. Hist. Nat. Vindobon, no. 2752〔HMAS-L, as *Lobaria pulmonaria* f. *hypomelaena*(Del.)Hue〕, collected by Handel-Mazzetti from Lijiang, Mt. Yulong shan, at about 3600 m, on bark of trees. HUBEI: Shen-nongjia region: Mt. Xiaoshen-nongjia, alt. 2600-2670 m, 1984, Chen J. B. , no. 11066; Xiaolongtan, alt.2300-2350m, 1984, Chen J. B. , no. 10071; Jiangjunzhai, alt. 1800 m, Chen J. B. , no. 11888; In the vicinity of Houzishi, alt. 2620 m, 1984, Chen J.B., no. 10431.

Distribution. TAIWAN（Inum. 1941, p. 216; Asahina, 1949, p. 67; Yoshim. 1971, p. 278; Wang & Lai, 1973, p. 92）.

7. ***Lobaria tuberculata*** Yoshim. 1971, p. 280.

Type locality: Japan.

This species is close to *Lobaria isidiophora*, but differs from the latter by the negative medullary reaction with K and P respectively, and by the absence of stictic, constictic and norstictic acids and presence of congyrophoric acid.

Reaction: Th. K-; M. K-, KC+, P-.

TLC: gyrophoric, congyrophoric acids.

Specimens examined:

JILIN: Mt. Changbai shan, south slope, at about 1350 m near Hengshan station, on Abies sp. 30/VII/1983, Wei & Chen, nos. 6341, 6342, and at about 1400 m, on twigs of Picea sp.3/VIII/1983, Wei & Chen, no, 6568-（1）. HUBEI: Shen-nongjia region, Mt. Xiaoshen-nongjia, alt. 2508 m, 1984, Chen J. B., no. 11047.

New to China.

8. ***Lobaria meridionalis* Vain.** 1913, p. 128.

Lectotype locality: Philippines.

Reaction: Th. K-; M. K+, KC-, P+.

TLC: stictic, constictic, and norstictic acids.

Specimens examined:

GANSU: Zagaling, at about 1800m, 1930, collected by D. Hummel, determined by H. Magn. as Lobaria pulmonaria f. papillaris（Del.）Hue in S（!）（TLC not done）. SICHUAN: Mt. Emi shan, Jinding, at about 3160 m, on Abelia sp.17/VIII/1963, Zhao & Xu, no. 7738. YUNNAN: Lijiang, Mt. Yulong shan, at about 3600 m, 8/XII/

1960, Zhao & Chen, no. 4199. HUBEI: Shen-nongjia region, Qianjiaping, at about 1800 m, Chen J. B. , no. 11646; Shanshuping , alt. 1700 m, 1084, Chen J. B..no, 11968.

Distribution: GANSU（H. Magn. 1940, p. 46）, SICHUAN（A. Z. 1930, p. 84）, YUNNAN（A. Z. 1930, p. 84 & 1934，p. 198）, XIZANG（Wei & Jiang, 1981, p. 1146）, TAIWAN（Yoshim. 1971, p. 284;Wang & Lai, 1973, p. 92 & 1976, p. 227）.

9. ***Lobaria chinensis*** Yoshim. 1971, p. 281.

Type locality: China, Taiwan prov., Taidong. Shussuiha, H. Masuda s. n. in TNS（holotype）and in NICH（isotype）, not seen.

Reaction: Th. K-; M. K+, KC, P+.

TLC: stictic, constictic, and norstictic acids.

Specimens examined:

XIZANG: Bomi county, on trees, at about 3200 m, 10/Ⅶ/1976, Zong & Liao, no. 296. HUBEI: Shen-nongjia region, Mt. Jiemeifeng, alt. 2940 m, 1984, Chen J. B. , no. 11261; Dakengwuo, alt. 2700 m, 1984, Chen J. B. , no. 10360.

Distribution: SICHUAN, YUNNAN（Yoshim. 1971, p. 281）, TAIWAN（Yoshim. 1971, p. 281; Wang & Lai, 1973, p. 92; Ikoma, 1983, p. 70）.

10. ***Lobaria yunnanensis*** Yoshim. 1971, p. 282.

Type locality: China, Yunnan prov. Mt. Yulong shan near Lijiang, 1924, Handel-Mazzetti no. 2315（in W）, distributed as Kryptogam. exs. ed. Mus. Hist. Nat. Vindobon no. 2752（not seen）.

Reaction. Th. K-; M. K-, KC-,P-.

TLC: no lichen substance.

Specimens examined:

SICHUAN: Baoxing county, Qu G. L. , no. 3265. YUNNAN: Lijiang: Mt. Yulong shan, at about 3000 m, on trees, 9/Ⅻ/1960, Zhao & Chen, no. 4465; Baishuihe, on the ground, 12/XI/1980, Jiang Y. M. , no. 358 -（1）. HUBEI: Shen-nongjia region, Jiang-junzhai, alt. 2150 m, 1984; Chen J. B ., no. 11931-（1）.

Distribution: SICHUAN, YUNNAN（Yoshim. 1971, p. 282）.

11. ***Lobaria pulmonaria***（L.）Hoffm. 1796, p. 146.

=*Lichen pulmonarius* L. 1753, p. 1145.

Lectotype: no. 1273-103b, L.（lacking locality and other data, see Yoshim, 1971, p. 286）.

=*Sticta pulmonaria*（L.）Schaer.

=*Sticta pulmonacea* Ach.

This species is only a known sorediate one of *Lobaria in* China.

Reaction: Th. K-; M. K+, KC-, P+.

TLC: norstictic acid, in addition, stictic and constictic acids in Chemical race I.

Specimens examined:

JILIN: Mt. Changbai shan: north slope, Antu county, at about 1600 m, on rotten wood near Baishan tree farm,

10/Ⅷ/1960, Yang Y,C., Yuan J.Y. ,Yuan F, S., no. 47; south slope, Changbai county, Mt. Hongtou shan, on tree trunk of Abies sp. 27/Ⅶ/1983, Wei & Chen, nos. 6211, 6516, 6558.

This species has been found by us only in Mt. Changbai shan of Jilin province in China so far. However, some authors, such as Baroni（1894）, Elenkin（1901）, Hue（1901）, Jatta（1902）, Sasaoka（1919）, Paulson（1928）, A. Zahlbruckner（1930）, Tchou Y. T.（1935）, Asahina & Sato in Asahina（1939）, H. Magnusson（1940）, and Chen et al.（1981）reported *Lobaria pulmonaria* or its infraspecific taxa respectively from Hebei, Heilongjiang, Shaanxi, Gansu, Sichuan, Hunan, Jiangxi, Yunnan, Zhejiang and Taiwan provinces in China. Some original specimens cited by above mentioned author s from China have been reexamined in the present study. The results of it are shown on the Table 1.

As for Asiatic *Lobaria pulmonaria* or its infraspecific taxa Yoshimura（1971, p. 287）states. "However, sorediate specimens were collected only from Sakhalin and northern Korea, as mentioned by Sato（1943）and Asahina（1949）, and have not been collected from most of Asia including Japan, China（see Yoshim. 1969）, Himalayas, and southeastern Asia; most Asiatic records（e. g. Chopra, 1934, Zahlbruckner, 1930）seem to be based on erroneous identification. " According to Sato, Asahina, Yoshimura, and our study sorediate species *Lobaria pulmonaria*（L.）Hoffm. in China has been known only in Jilin province so far, and may probably be also in Heilongjiang province as reported by Chen et al.（1981, p. 151）though, the original specimen cited by Chen et al. was not available for the present study and even in the vicinity of it. Most Chinese records by above mentioned different authors from Hebei, Shaanxi, Gansu, Sichuan, Hunan, Jiangxi, Yunnan, Zhejiang and Taiwan provinces seem to be based on erroneous identification.

12. ***Lobaria kazawaensis***（Asah.）Yoshim. 1971, p. 291,

=*Lobaria sachalinensis* Asah. var. *kazawaensis* Asah.

Type locality: Japan.

This species is similar to *Lobaria isidiophora*, but from the latter differs by the negative.

Table 1 The records of *Lobaria pulmonaria* & its infraspecific taxa in China & some revised results of them

Province	Names of lichens reported by different authors	Revised names of lichens
Hebei	*Sticta pulmonacea* Ach.（Tchou. 1935, p.319）	= *Lobaria retigera*
Gansu	*Lobaria pulmonaria Hoffm.*	
	var. *hypomela*（Del .）Cromb.（H.Magn .1940, p.46）.	=*Lobaria retigera*
	var. *papillaris*（Del.）Hue（H.Magn. 1940, p.46）,	=*Lobaria meridionalis*
Yunnan	Lobaria pulmonaria	
	var. *hypomelaena* Del.）Cromb.（Kryptog.exs.	
	no.2752 in HMAS-L）.	=*Lobaria spathulata*
	（no.2752 in S.）	=*Lobaria gyrophorica*
Zhejiang	*Sticta pulmonacea* Ach.（Tchou, 1935, p.319）.	=*Lobaria orientalis*
Heilongjiang	*Lobaria pulmonaria*（L.）Hoffm.（Chen et al.	The specimen was not avai-
	1981, p.151）	label for the present study
Shaanxi	*Sticta pulmonaria* Biroia（Baroni, 1894, p.48）	not available

续表

	Sticta pulmonacea Ach.（Jatta, 1902, p.468）	not available
	Lobaria pulmonacea Hoffm.（A.Z. 1930, p.84）	not available
Sichuan	*Lobaria pulmonaria*	
	var. *hypomelaena*（Del.）Cromb.（A.Z. 1930, p.84）.	not available
Hunan	*Lobaria pulmonaria*	
Jiangxi	var.*hypomelaena*（Del.）Cromb.（A.Z. 1930 .	
Shaanxi	p.84）	not available
Yunnan	*Lobaria pulmonaria* Hoffm.（A.Z.1930, p.84）	not available
Taiwan	*Lobaria pulmonaria*（Sasaoka, 1919, p. 180）.	not available

medullary reaction with K, KC, and P. In addition, the upper surface of the thallus of the present species is dull and often covered with a thin pruina.

Reaction: Th. K-; M. K-, KC-,P-.

TLC: Triterpenoids, in addition, no lichen substance in chemical race Ⅱ.

Chemical race Ⅰ:

Specimen examined:

SHAANXI: Mt . Taibai shan, Yuewangchi , at about 3080 m, 6 / Ⅷ / 1963, Ma & Zong , no. 371.

Chemical race Ⅱ:

Speqimens examined:

SHAANXI: Mt. Taibai shan, near ping-ansi, at about 2800 m, 28/IV/1937,Wang T. P., no, 6613. HUBEI: Shen-nongjia region, Yuanling, alt,1700-1850 m ,1984, Chen J.B., nos. 11873, 1180.

Distribution: XIZANG（Wei & Jiang, 1986, p.24）

13. ***Lobaria pseudopulmonaria*** Gyeln.

Acta Fauna FL. Univ; ser, 2, 1（5-6）: 6（1933）〔fide Yoshim, 1971,p.291〕.

Type locality: Java.

This species is similar to *Lobaria kurokawae* in morphology but differs from the latter by the positive medullary reaction with K and P respectively.

Reaction: Th. K-; M. K+, P+.

TLC: norstictic acid & triterpenoids, in addition, stictic and constictic acids present in chemical race I.

Chemical race I:

Specimens examined:

SICHUAN: Between Taining（Ngata）and Maoniu（Ndrome）, on trunk of Juniperus sp. at about 3900 m, 30/Ⅸ/1934，H. Smith, no. 14090（UPS !）.（TLC not done）. YUNNAN: Dali, Mt. Cang shan, l/Ⅻ/1940, Liou T. N., no. 17516. HUBEI: Shen-nongjia region, Mt. Jiemeifeng, alt. 2950 m, 1984, Chen J. B. , no. 11423-（1）. ANHUIs Mt. Huamg shan, on the rocks near Yungusi, 18/Ⅷ/1962, Zhao & Xu, no. 5276.

Distribution. YUNNAN（Yoshim. 1971, p. 291）, XIZANG（Wei & Jiang, 1986, p. 22）, TAIWAN（Inum. 1941, p. 214; Lamb, 1963, p. 405; Yoshim. 1971, p. 291; Wang & Lai. 1973, p. 92）.

14. ***Lobaria isidiosa***（Muell. Arg.）Vain. 1913, p. 129.

= *Stictina retigera* f. *isidiosa* Muell. Arg. 1882, p. 30c.

Lectotype locality: Java.

This species is similar to *Lobaria retigera* in morphological characters but differs from the latter by the positive medullary reaction with K & P respectively.

Reaction: Th. K-; M. K+, KC-, P+.

TLC: norstictic acid and triterpenoids, in addition, stictic and constictic acids present in chemical race I.

Chemical race I:

Specimens examined:

JILIN: Mt. Changbai shan, south slope, Fusong county, on the ground in forest near Weidong frontier station at about 1600 m, 7/Ⅷ/1983, Wei & Chen, no. 6663. YUNNAN: Pingbian,Adakou, Mt. Laojian shan, at about 2380 m, 23/VI/1956, China-USSR Expedition to Yunnan, no. 3933. XIZANG: Zhamkouan, at about 2480 m, on rocks in forest, 6/Ⅴ/1966, Wei & Chen, no. 101, & on moss covered rocks, 11/Ⅴ/1966, Wei & Chen, no. 476. HUBEI: Shen-nongjia region, Mt. Dashen-nongjia, Shen-nongding, alt 2950 m, 1984, Chen J.B., no. 10249; Mt. Jiemeifeng, alt. 2950 m, 1984, Chen J.B. no. 11432.

Distribution: HEBEI（Moreau et Moreau, 1951, p, 186）, HEILONGJIANG（Chen et al. 1981, p. 151）, SICHUAN（A. Z. 1930, p. 84）, YUNNAN（Hue, 1901, p. 39; A. Z. 1930, p. 84; Yoshim. 1971, p. 295）, XIZANG（Wei & Jiang, 1986, p. 24）, HUBEI（Muell. Arg. 1893, p. 236）, TAIWAN（Sasaoka, 1919, p. 180; A. Z. 1933, p. 31; Yoshim. 1971, p. 295; Wang & Lai, 1973, p. 92）.

15. ***Lobaria kurokawae*** Yoshim. 1971, p. 292.

Type locality: Japan.

This species is characterized by the reticulately ridged thallus lacking isidia and soredia with blue green algae in it.

Reaction: Th. K-; M. K-, KC-, P-.

TLC: triterpenoids.

Selected specimens examined:

SHAANXI: Meixian county, Mt. Taibai shan, on soil covered rocks: near Wengongmiao, Ⅸ/1938, Hsia W. Y. , no. 4956; Ping-ansi, alt. 2800m, 30/Ⅸ/ 1955, Wang F. Z. no.94 & alt. 2760m, 22/Ⅶ/1963, Ma & Zone, no. 186; Huxian county, Laoyu Xihe Muziping, 24/Ⅷ/1951, Guo B. Z. , no. 665. SICHUAN: north-western Sichuan: 9/Ⅸ/1923, H. Smith nos. 5119, 5243（UPS!）; at about 3500m, 18-24/IX/1922, H. Smith nos. 5103, 5225（UPS!）; alt. 4000-4300m, 21/Ⅶ/1922, H. Smith no. 5053（UPS !）; Kangding, west of the town, on rather dry hills with shrubs, alt. 3000m, 11/Ⅺ/1922, H. Smith nos. 5137（pr. p.）& 5138（UPS!）（TLC not done）; Songpan, Conggangling, alt. 3700m, in Picea forest, Wang T. P., no. 7857. XIZANG: Zham, Quxiang, west of Deqintang, alt. 3550m, Wei & Chen, no. 960. HUBEI: Shen-nongjia region: Xiaolongtan, alt. 2200m, 1984, Chen J. B. , no. 10037; Qianjiaping, alt. 1850m, 1984, Chen J. B. , no. 11580-（2）; In the vicinity of Houzishi, at about 2660m, 1984, Chen J. B. , no. 10440, TAIWAN: Ilan county, Mt. Nanhudashan, over mosses, at about 2600-3000m, 21/I /1964, S. Kurokawa, no. 1145（as Lobaria retigera）in S（!）.

Distribution: SHAAXI（Wu & Liu, 1976, p. 63, as *Lobaria retigera*; Wu J. L. , 1981, p. 162; Wei et al. 1982, p. 19）, SICHUAN（Elenk. 1901, p. 39; Yoshim. 1971, p. 297）, YUNNAN（Yoshim. 1971, p. 297）, XIZANG（Wei & Jiang, 1981, p. 1146 & 1986, p. 22）, TAIWAN（Yoshim. 1971, p. 297; Wang & Lai, 1973, p. 92）.

16. ***Lobaria retigera***（Bory）Trey.

Lichenotheca Veneta, 1839. p. 75〔fide A. Z. Cat. Lich. Univ. 3: 321（1925)〕. Yoshim. 1971, p. 298.

=*Lichen retiger* Bory Voyag. Princip. lies Mers de Afrique 1 : 392(1804), nom. nud,; ibid . 3 :101(1804), cum descr.（fide A. Z. ibid. 3: 321（1925）& Yoshim. 1971, P. 298.）

Type locality; Bourbon（Reunion I., Mascarenes）.

Sticta pulmonacea sensu Tchou Y. T., Contr. lnst. of. Bot. National Academy of Peiping, 3（6）: 319（1935）, non Ach

Selected specimens examined:

HEBEI: Mt. Wuling shan, at about 1390m, 1/Ⅵ/1931, Liou K. M. , no. 91, determined by Tchou Y. T. as Sticta pulmonacea Ach, JILIN: Mt. Changbai shan, south slope, in Betula forest near Hengshan station, 3/Ⅷ/1983, Wei & Chen, no. 6518. HEILONGJIANG: Muling, Sanxinshan tree farm, on tree trunk of Betula sp. in Picea-Abies forest, 24/Ⅶ/1977, Wei J. C., no. 2659. GANSU: Zagaling, at about 1800m, 29/Ⅹ/1930, collected by D. Hummel, determined by H. Magn. as *Lobaria pulmonaria* var. *hypomela*(DD1.)Cromb. in S!(TLC not done). SICHUAN: Kangding, west of the town, on rather dry hills with shrubs, alt. 3000m, 11/ XI/1922, H. Smith no. 5137(pr. p., together with *Lobaria isidiophora* & *L. kurokawae*) in UPS !（TLC not done）. Mt. Emi shan, top of the mountain（Jinding）, on Abies sp. alt. 3160m, 17/Ⅷ/1993 , Zhao & Xu, no. 7630. YUNNAN: Dali, Mt. Cangshan, on the way from Zhonghefeng to Xiangshiya, 5/X/1946, Liou T. N., no. 21211; Baoshan county, wang Q. Z., no. 1290; Lijiang, Heibaishui forest centre, Baishui, on soil covered rocks, 11/XI/1980, Jiang Y. M., no. 255; Gaoligong shan, Baihualing tree- farm, alt. 1800m, on *Rhododendron* sp. 11/Ⅻ/1980, Jiang Y. M. ,no. 8.66-（4）. XIZANG: Zham, on rocks, 15/ V/ 1966, Wei & Chen, no. 749; Cuona, Maraaxiang, alt.3000m, on bark of trees, 10/Ⅷ/1974, Zong Y.C. , no.46.HUBEI: Shen-nongjia region: Mt. Dashen-nongjia, Shen-nongding, alt. 2800-3000m, 1984, Chen J. B., nos. 10305, 10314-（1）; Mt. Jiemeifeng, alt. 2550-2940m, 1984, Chen J. B., nos. 11365, 11398; Xiaolongtan, at about 2300m, 1984, Wei, no. 10940; Mt. Xiaoshennongjia,alt. 2600m, 1984, Chen J. B. , no. 11065; Xujiazhuang, alt. 1400m, 1984, Chen J. B. , no. 12017. GUANGXI: Huaping forest region, Cujiang, on rocks, alt. 1200m, 18/Ⅷ/1964, Zhao & Xu, no. 9256 & Hongtan, on trees, alt. 1000m, 17/Ⅷ/1964, Zhao & Xu, no. 9080. Zhejiang; Mt. Tianmu shan, 27/Ⅷ/1930, Liou T. N., no. 6811. FUJIAN: Chong-an, Wang Q. Z., no. 224.

Distribution: SHAANXI(Jatta, 1902, p, 468; A. Z. 1930, p. 84; Wu & Liu, 1976, p.67, as *Lobaria isidiosa*), GANSU（A.Z. 1934, p.198; H.Magn. 1940,p. 46）, YUNNAN（Hue, 1887, p22 & 1889, p. 167 & 1901, p. 38; Paulson, 1928, p. 317; A. Z. 1930, p. 84）, XIZANG(Wei & Jiang, 1986, p. 23）, JIANGXI(Yoshim. 1971, p. 298）, FUJIAN（A. Z. 1930, p. 84）, TAIWAN(Yoshim. 1971, p. 298; Wang & Lai, 1973, p. 92）.

Sect. **Ricasolia**（De Not.）Vain. 1890, p. 194.

=*Ricasolia* DeNot. Giorn. Bot. Ital. 2（pt. 1）. 178（1846）[fide Yoshim. 1971, p. 301].

Lectotype: *Ricasolia amplissima*（Scop.）DeNot.

=*Lobaria amplissima*（Scop.）Eorss.

17. ***Lobaria crassior* Vain**. 1921, p. 64.

Type locality: Japan.

=*Lobaria buiensis* Ras. 1940, p. 144.

Type locality: TAIWAN, Mt. Bui, on bark of trees, A. Yasuda 679.

Reaction: Th. K-; M. K-, KC+, P-.

TLC: gyrophoric & congyrophoric acids.

Specimens examined:

ZHEJIANG: Mt. Tianmu shan, at about 1200-1500 m, on the way from Laodian to Xiaiirending, on bark of trees, 31/Ⅷ/1962, Zhao & Xu, no.6214-a.ANHUI:Mt.Huang shan, Liou T. N., no. 2515.

Distribution. TAIWAN（Ras. 1940, p. 144; Yoshim. 1971, p. 312; Wang & Lai, 1973, p. 92）.

Literature Cited

（1）Asahina, Y. 1947. Lichenologische Notizen. 65. Revision of Japanese species of *Lobaria* sect, *Lobarina*, *Journ. Jap. Bot.* **21**: 83-85.

（2）Asahina, Y. 1949. Lichenologische Notizen. 72. Varieties, forms and related species of *Lobaria pulmonaria* from eastern Asia. *Journ. Jap, Bot.* **23**: 65-68.

（3）Baroni, E. 1894. Sopra alcuni licheni della china racolti nella provincia dello Schan-si settentro nale *Bull. Soc. Bot, Ital.* pp. 47-49.

（4）Chen, X. L. Zhao, C.F. & Luo, G. Y. 1981. A list of lichens in NE China. *Journ. NE Forestry Inst.* **4** : 150.

（5）Elenkin, A. 1901. Lichenes Florae Rossiae et regionum confinium orientalium Fasc. 1. *Acta Horti Petropoliiani* **19**: 1-52.

（6）Hoffmann,G.F.1796. Deutschland Flora oder botanisches Taschenbuch fur das Iahr 1795, **2**:1-200.

（7）Hun, A. M. 1887. Lichenes Yunnanenses a claro Delavay anno 1885 collectos, et quorum nova species a celeb. W. Nylander descriptae fuerunt, exponit A. M. Hue. *Bull. Soc. Sot, France* **34**: 16-24.

（8）______________, 1889. Lichenes Yunnanenses a cl. Delavay praesertim anris 1886-1887, collectos exponit A. M. Hue（1）. series secunda. *Bull. Soc.Bot, France* 36 . 158-176.

（9）__________________, 1901. Lichenes extra-Europaei a pluribus collectoribus ad Museum Parisiense missie. *Nquv, Arch. Mus. Hist. Nat.*（Paris）**4**（**3**）: 21-146.

（10）Inumaru, S. 1941. *Lobariae* novae japonia. *Acta Phytotax. Geobot.* **10**（**3**）: 214-216.

（11）__________________, 1943. *Lobaria* nonnullae novae iaponicae. *Acta Phytotax. Geobot.* **10**（**3**）: 214-216.

（12）Jattaf A. 1902. Licheni cinesi raccolti alio Shen-si negli aaai 1894-1898 dal. rev. Padre Missio-nario G. *Giraldi. Nuov, Giorn,* Bot. Italiauo ser. 2, 9 : 460-481.

（13） Lamb, I. M. 1963. Index Nominum Lichenum pp. 403-406.

〔14〕Linnaeus, C. von, 1753, Species Plantarum, vol. 1-2.

〔15〕Magnusson, H. 1940. Lichens from Central Asia, 1, p. 45-46.

〔16〕Moreau, F. et Moreau, Mme. F. 1951. Lichens de Chine.*Bev.Bryol.et Lichenol*, 20. 183-199.

〔17〕Muell. Arg. J. 1882. Lichenologische Beitrage, 15. *Flora* **65**: 219-333.

〔18〕______________________.1893. Lichenes chinenses Henryani a cl. Dr. Aug. Henry, anno 1889 in China media lecti quos in herbario Kewensi determinavit. *Bull, Herb, Boiss.* **1**: 235-236.

〔19〕Paulson, R. 1925. Lichens of Mount Everest. *Journ. Bot.* **63**:189-193.

〔20〕____________________.1928. Lichens from Yunnan. *Journ. Bot. Brit. & For.* **66**: 313-319.

〔21〕Rasanen, V. 1940. Lichenes ab Yasuda et aliis in Japonia collecti（Ⅱ）. *Journ, Jap. Bot.* **16**:139-153.

〔22〕Sasaoka, H. 1919. Lichens of Taiwan. *Trans. Nat. Hist. Soc. Formosa* **8**: 179-181.（in Japanese）

〔23〕Sato, M. 1943. Notes on the Japanese lung lichens *Acta Phytotax. Geobot.* **13**: 238 -241.

〔24〕Tchou, Y. T. 1935. Note preliminaire sur Les Lichens ds Chine. *Contr. lust. Bot, Natu, Acad, Peiping* **3（6）**: 319.

〔25〕Vainio, E. A. 1890. Etude sur la classification naturelle et la morphologie des lichens du Bresil, I *Acta Soc. Fauna FL. Fenn*, **7（1）**: i-xxix, pp. 1-247.

〔26〕____________________.1913. *Lichenes insularum* Philippiaarum, II. *Philippine Jour. Sci.Sect. c*, **8（2）**: 99-137.

〔27〕______________________. 1921. Lichenes ab A. Yasuda in Japonia collecti continuatio, I . *Bot, Mag , Tokyo* **35**: 45-79.

〔28〕Wang-Yang J. R. & Lai, M. J. 1973. A checklist of the lichens of Taiwan. *Taiwania*, **18（1）**: 83-104.

〔29〕______________________&_____________.1976. Additions and corrections to the lichen flora of Taiwan, ibid.**21（2）**: 226.

〔30〕Wei, J. C, An enumeration of the lichens in China（to be published）.

〔31〕Wei, J. C. & Jiang, Y. M. 1986. Lichens of Xizang, Science Press, Beijing, p. 21.

〔32〕Wu, C. N. & Liu, A. T. 1976. What is the Chinese medicinal herb. Lao Long Pi. t *Acta Phytotax. Sin.* **14（2）**: 66-68.（in Chinese）

〔33〕Wu, J. L. 1981. Medicinal Lichens in Qin Ling Mountain. *Acta Pharmaceutica Sinica* **16（3）**: 161.

〔34〕Yoshimura, I. 1969. Lichenological notes, 2-6. *Journ. Hattori Bot.* **32**: 67-78.

〔35〕______________. 1971. The Genus *Lobaria* of Eastern Asia. *Journ. Hattori Bot .Lab.* **34** : 231-364.

〔36〕Zahlbruckner, A. 1925. Cat. Lich. Univ. **3** : 321.

〔37〕________________ . 1930. Lichenes in Handel-Mazzetti. *Symb, Sin.* **3** : 83-84.

〔38〕_________________.1933. Flechten der Insel Formosa in Fedde, *Repertorium* sp. *nov*, **33**: 22-68.

〔39〕_________________.1934. Nachtrage zur Flechtenflora Chinas. *Hedwigia* **74**:195-2:3.

中国肺衣属简志

魏江春　　陈健斌　　姜玉梅

（中国科学院微生物研究所，北京）

摘要　中国肺衣属地衣迄今已有 33 位作者在 21 篇文章中记载了 48 个分类单位。这些分类单位分隶于 31 种。根据魏江春尚未发表的"中国地衣综览"，这些地衣可归纳为 25 种，3 个变种及 8 个变型；其中分布于台湾的 20 种，云南 13 种，西藏 10 种，四川 9 种，陕西 6 种，甘肃 3 种，河北、吉林、黑龙江、湖北、湖南、江西各为 2 种，浙江及福建各为一种。

本文报道了中国肺衣属地衣 17 种，其中包括两种我国新记录（亚平叶肺衣，瘤芽肺衣）及一个新的化学小种。同时，我们对于前人报道中的部分错误鉴定进行了复查和订正。研究结果表明，下列记载均以错误鉴定为基础：南肺衣（*Lobaria meridionalis*: H. Magn. 1940, p. 45）肺衣（*L. pulmonaria*: 朱，1935, *p*, 319, 以 *Stica pulmonacea* 为名），肺衣下的几个变种：var. *hypomela*（Del.）Cromb.与 var. *papillaris*（Del.）Hue（H. Magn. 1940, p. 46）以及变种 var. *hypomelaena*:（nom. illeg. A. Z. 1930, p. 84 与孢子植物标本集 no. 2752 等）.看来，孢子植物标本集 no. 2752 实为一不同种类之混杂物。因为保存在奥地利维也纳自然历史博物馆植物标本室的同号标本已被指定为云南肺衣（*L. yunnanensis*）的主模式（Yoshim. 1971），而分别保存在瑞典斯德哥尔摩自然历史博物馆孢子植物标本室及中国科学院微生物研究所真菌地衣标本室的同号标本均含有三苔色酸；而且后者地衣体边缘有裂芽及小裂片。因此，这两份同号标本在本文中被分别定名为三台色肺衣（*L.gyrophorica*）及匙芽肺衣（*L. spathulata*）两种。至于真正的肺衣（*L. pulmonaria*）在中国迄今只见于吉林长白山，可能还分布于黑龙江及其邻近地区。

关键词　肺衣属；亚平叶肺衣；肺衣；瘤芽肺衣：化学小种

本文于 1986 年 6 月 20 日收到

Notes on isidiate species of *Hypogymnia* in Asia

Wei Jiang-chun*

(*Institute of Microbiology, Academia Sinica, Beijing*)

ABSTRACT Seven isidiate species of *Hypogymnia* are reported in this paper. Among them *Hypogymnia duplicatoides* (Oxn.) Rassad., *H. mundata* (Nyl.) Rassad., and *H. subcrustacea* (Flot.) Kurok. are new to China, and a new subsection *Enteromorpha* is established.

KEY WORDS *Hypogymnia*; Subsection *Enteromorpha*; *H. duplicatoides*; *H. mundata* f. *sorediosa*; *H. subcrustacea*; Isidiate species

Seven isidiate species of *Hypogymnia* have been known by lichenologists, namely *H. austerodes* (Nyl.) Räs., *H. duplicatoides* (Oxn.) Rassad., *H. hokkaidensis* Kurok ., *H. subcrustacea* (Flot.) Kurok., *H. zeylanica* (R. Sant.) Awasthi et K. P. Singh, and *H. hengduanensis* Wei belonging to subgenus *Hypogymnia*, and *H. mundata* (Nyl.) Rassad. belonging to subgenus *Solidae* (Bitt.) Krog. The first of above mentioned species is considered as the only isidiate member of the section Subobscura Rassad. *H. duplicatoides* and *H. subcrustacea* have been put into the section *Hypogymnia* (Rassad. 1967, p. 291, as section *Metaphysodes*), and the remaining two species *H. hokkaidensis* and *H. zeylanica*, in my opinion, can be placed in the same section too. *H. hengduanensis* should be placed in the new subsection *Enteromorpha* under the section *Delavayi* Rassad. The last species *H. mundata* (f. *mundata* having isidia, which do not break up into soredia, and f. *sorediosa* having isidia, which break up into soredia) belongs to section *Mundata* Rassad. under the sugenus *Solidae*.

All the isidiate species of the genus can be found in Asia, and most of them are endemic to Asia, such as *H. duplicatoides* (to the North East Asia), *H. hengduanensis* (to Sichuan prov. and its adjacent regions in China), *H. hokkaidensis* (to Japan), and *H. zeylanica* (to the subcontinent of South Asia).

HYPOGYMNIA (Nyl.) Nyl.

Lich. Envir. Paris, 1896, p. 39.

Basionym: *Parmelia* subgenus *Hypogymnia* Nyl. Flora, 64 : 537 (1881).

---

* I am greatly indebted to Dr. I. M. Brodo & Mr. Wang (CANL) for sending the specimens on loan, and to Prof. Dr. Lundqvist, N (S), Dr. Moberg, R (UOS). Dr. Makinen, Y. (TUR) & Dr. Vitikainen, O. (H) for giving me access to all the facilities during my stay in their herbaria and sending the specimens on loan.

This paper was originally published in *Acta Mycologica Sinica*, supplement I: 379-385, 1986.

Type species: *Lichen physodes* L.
=*Parmelia physodes* (L.) Ach.
=*Hypogymnia physodes* (L.) Nyl.

Key to isidiate species of *Hypogymnia*

1. Medulla of thallus with hollow inside.
  2. Thallus almost subfruticose, rather soft, with strongly inflated, longer, subisotomic, dichotomic branched lobes, containing barbatic acid ······ 5. *H. hengduanensis*
  2. Thallus more or less connivent, rosetted foliose,with very weakly inflated or depressed and shorter lobes, lacking barbatic acid.
    3. Thallus brown grey or brown or dark brown on the upper surface. Isidia sometimes break up into soredia ······ 6. *H. austerodes*
    3. Thallus whitish mineral grey or greenish grey on the upper surface.
      4. Thallus rosetted, H. physodes' type, non perforate in the lower surface.
        5. Medulla P+ red, containing physodalic acid ······ 3. *H. suborustaoea*
        5.Medulla P–, lacking physodalic acid ······ 2. *H. hokkaidensis*
      4. Thallus with more or less longer lobes, perforate in the lower surface.
        6. Medulla P+red, containing physodalic acid. Lobes more wider without small lateral lobes ······ 1. *H. duplioatoides*
        6. Medulla P–, lacking physodalic acid. Lobes rather narrow with numerous small lateral lobules ······ 4. *H. zeylanioa*

Medulla of thallus solid inside; thallus evernioid ······ 7. *H. mundata*

I. subgenus ***Hypogymnia***
=*Parmelia* subgenus *Hypogymnia*, section *Tubulosae* Bitt. *Hedwigia*, 40:199 (1901).
=*Hypogymnia* subgenus *Tubulosae* (Bitt.) Krog, Nyt. Mag, f. *Naturvidensk*. 88 : 75 (1951).

A. Section *Hypogymnia*
=*Hypogymnia* subgenus *Tubulosae*, section *Metaphysodes* Rassad, Nov. System.Plant,non vascul, 1967, p. 290.

1. ***Hypogymnia duplicatoides*** (Oxn.) Rassad. (Pl. Ⅰ-1; Pi. Ⅱ-1)

Notul. Syst. e Sect. Cryptog. Inst. Bot. nomine V. L. Komarovii Acad, Sci. URSS, 11 : 5 (1956).

Basionym: *Parmelia duplicatoides* Oxn. Rhurn. Iast. Botan. ANURSR (Journ. Inst. Botan. Acad. Sci. URSR), **18-19**: 222 (1938).

Type locality: USSR, Khabarovsk, Muchenj, 18/Ⅶ/1927. (not seen)

The Chinese collection from Heilongjiang prov. matches well with the original description of Oxner (1938) and that of Rassadina (1971) except for the positive reaction to K in the medulla. Thus the medulla gives negative reaction to K according to Rassadina (1971), but to my own test with the Chinese specimen the medulla gives positive reaction; at first it turns tawny and then to orange-tawny.

Th. K+ yellow, P–; M. P + lemon yellow and then turns orange red, K+tawny and then turns orange-tawny.

TLC: Atranorin, physodic and physodalic acids.

Specimen examined:

CHINA: Heilongjiang prov. Muling Sanxinshan tree farm, on trunk of larch tree in the larch forest, 22/Ⅶ/1977, Wei JC no. 2577 (HMAS-L).

2. ***Hypogymnia hokkaidensis*** Kurok.（Pl. Ⅰ-2; Pl. Ⅱ-2）

in Kurok, et Nakanishi, Mem. Natl. Scl. Mus. Tokyo, 4 : 62（1971）.

Type: JAPAN, Hokkaido, Namuro prov., Ochiishi., on trunk of *Picea glehrii*. elevation about 50 m, 2/Ⅸ/1965, Coll. S. Kurokawa no. 65826（no seen）. S. Kurok. Lich. Rar. et Critici Exs. no. 167（isotype）in UPS（!），H（!），& TUR（!）.

This Species is very similar to *H. subcrustacea* but differs from that in lacking physodalic acid in it. The lobes slightly convex, mineral grey and glossy on the upper surface. The black lower surface exceeds the upper one.

Th. K + yellow: M. K−, KC + red, P −.

TLC: Atranorin, physodic and 3-Hydroxy-Physodic acids.

Specimens examined:

See above mentioned isotypes preserved in UPS, H, & TUR.

3. ***Hypogymnia subcrustacea***（Flot.）Kurok.（Pl. Ⅰ-3, Pl. Ⅱ-3）

Misc. bryol. lichen., Nichinan 5（9）: 130（1971）.

Basionym: *Imbricaria physodes* f. *subcrustacea* Flot. in Koerb. Lichenogr. Germ. Specimen, 1846, p. 11.

=*Parmelia physodes* var. *subcrustacea*（Flot.）A. Z. Cat. Lich. Univ. 10: 530（1940）.

=*Hypogymnia physodes* f. *subcrustacea*（Flot.）Rassad. Nov. System. Plant, non vascul. 1967, P. 291.

This species is close to *H. hokkaidensis*, but differs from that in containing dhysodalic acid in it. The lobes usually plane or slightly concave and weakly glossy on the upper surface. The upper surface exceeds the black lower surface.

Th. K + yellow and then turns red: M. P + red.

TLC: Atranorin, physodalic, physodic, and 3-Hydroxy-physodic acids.

Specimens examined:

CHINA: Heilongjiang prov. Muling, Sanxinshan tree farm, on *Larix* sp. 22/Ⅶ/1977, Wei J. C. no. 2590(pr. p.).

JAPAN: Mt. Myoho, Mitumine, Chichibu, Musasi prov. , 6/V/1956, collected by S. Kurokawa, no. 56133（UPS!）: Izu prov. Mt. Amagi, on bark of Cryptomeria japonica, collected by A. Yasuda, 4/Ⅸ/1922, ex Herb. Räs.no.649（H!）， misidentifiedb y Räs. in 1936 as *Parmelia physodes*（L.）Ach, f. *subisidioides* Merrill.

4. ***Hypogymnia zeylanica***（R. Sant.）Awasthi et K.P.Singh（Pl. Ⅰ-4, P1. Ⅱ-4）Geophytology 1（2）: 100（1971）.

Basionym: *Parmelia zeylanica* R, Sant. Bot. Notiser 325（1942）.

Type. SRI LANKA（Ceylon），in Mt. pedrotallagalla, 1897, E. Almquist（S!）.

This species has well developed oval, cylindrical and sometimes branched isidia, which are similar to those in *H. hengduanensis*, but differs in the following respects: lobes short（about 15 mm long）and flat, usually broad in axilla, and barbaic acid is absent.

Th. K + yellow; M. K + yellow, C−, KC + red or yellow, P− .

TLC: Atranorin, physodic and 3-Hydroxy physodic acids.

Specimens examined:

SRI LANKA（Ceylon）: Nuwara Eliya Distr., Horton Plains, between Farr Inn and "World's End". Montane forest, on tree-trunk, 14/Ⅱ/1974, collected by Nils Lundqvist no. 9066（UPS!）.

INDIA: Vezda, Exs. no. 1390（collected by Hale from India, no. 43871（UPS!）.

B. Section *Delavayi* Rassad.

Subsection *Enteromorpha*, subsect. nov.

Type species: *Parmelia enteromorpha* Ach. =*Hypogymnia enteromorpha*（Ach.）Nyl.

（type specimen in H!）

Thallus mollis, laxe adnatus, lobis linearibus, libris, longioribus, valde inflatis vel interdum plus minusve planis in superfice superlore.

Except for the type species of the subection, *H. duplicata* (Ach.) Rassad., *H. imshaugii* Krog, and *H. heterophylla* Pike, etc. can be treated as a member of the subsection. The species *H. hengduanensis* can be placed here as well.

5. ***Hypogymnia hengduanensis*** Wei

Acta Mycologica Sinica 3 (4) : 214 (1984).

This species is very similar to *H. enteromorpha* at first sight, but differs from it by having globose, oval, and cylindrical isidia and lobules on the upper surface of the lobes, and containing barbatic acid together with atranorin and sometimes an unidentified substance, and lacking protocetraric, physodic, physodalic, and diffractaic acids in it.

Th. K + yellow; M. K + yellow, C−, KC−, P−.

TLC: Barbatic acid, atranorin, unidentified substance.

5a. ssp. ***hengduanensis***

Ibid. 3 (4) : 215 (1984). Pl. I , fig. 1; Pl. II figs. 1—4.

Type: CHINA, Sichuan prov., Kangding, on trunk of *Betula* sp. alt. 3700m, 19/X/1934, Harry Smith no. 14078 (holotype in UPS! & isotype in HMAS-L).

This subspecies is characterized by the fact that the lower surface of the lobes exceeds the upper one with a conspicuous black rim, and has numerous isidia and lobules and almost dark brown even black medulla, and contains atranorin and barbatic acid in it.

Specimens examined:

CHINA: Sichuan, Mt. Emi shan, Jinding, ou *Abies* sp. alt. 3160m, 17/Ⅷ 1963, Zhao, J. D. & Xu L. W. no. 7885. Yunnan, Lijiang, Heibaishui forest center, about 50 km from Sandawan, alt. 3140m, on *Rhododenron* sp., 10/XI/1980, Jiang Y. M. no. 131- (1) (paratype) in HMAS-L.

5b. ssp. ***kangdingensis*** Wei

Ibid. 3 (4) : 215 (1984). Pl. I , fig. 2; Pl. II , fig. 5.

Type: CHINA, Sichuan, Kangding, on *Rhododendron* sp. alt. 3700m 22/Ⅶ/1934, Harry Smith 14006 (holotype in UPS! & isotype in HMAS-L).

This subspecies differs from the subspecies *hengduanensis* in that the upper surface of the lobes is exceeding the lower one without a conspicuous black rim, and has few and sparse isidia, and a white exterior medulla and a dark brown even black interior one, contains an unidentified substauce in addition to atranorin and barbatic acid in it.

C. Section *Subobscura Rassad.*

Nov. System. Plant, non yascul. 1967, p. 296.

Type species: *Parmelia subobscura* Vain. =*Hypogymnia subobscura* (Vaiin,) Poelt

6. ***Hypogymnia austerodes*** (Nyl.) Räs.

Ann. Bot. Soc. Zool. Bot. Fenn. 'Vanamo' 18 (1) : 13 (1943).

Basionym: *Parmelia austerodes* Nyl. Flora 65 : 537 (1881).

=*Parmelia physodes* f. *subisidioides* Merrill, Bryologist, 11. 86 (1908), syn. nov.

Lectotype: on old logs at Laggan, Alta, 26/VI/1904, no. 3635 in CANL (!), selected by Wei J. C.

This species is similar to *H. bitteri* (Lynge) Ahti in many ways: thallus connivent, rosetted foliose, brown grey to brown or dark brown, and more or less rugose on the upper surface, and sometimes the black lower surface of the lobes exceeds the upper surface and resulted in a conspicuous black rim. But it differs from the latter in lacking globose soralia and having numerous verruciform or cylindrical isidia on the upper surface.

Th, K + yellow; M. K−, P−.

TLC: Atranorin and physodic acid.

Specimens examined:

CHINA: Xizang, Nyalam, Mt. Himalaya, alt. 3850—3940 m, on moss covered rock in a bitch forest, 22/Ⅵ/1966, Wei, J. C. & Chen JB nos. 1820—（2）， 1865—（2）， 1886. Xinjiang, Mt. Nan shan near ürümqi city, alt. 2760m, on moss covered rock, 24/ Ⅴ/1977, Wang, X. Y, no. 137; Mt. Tian shan, north glacier, alt. 3200m, on mossy rock, 23/Ⅶ/1978, Wang, X. Y. no. 959 . Mt. Tian shan, Xiata hot spring, alt. 2700m, on rotten wood, 27Ⅶ & 3/Ⅷ/ 1978, Wang, X. Y. nos. 1024 & 1103.

Ⅱ. Subgenus ***Solidae***（Bitt.）Krog

Nyt. Mag. f. Naturvidenks. 88 : 76（1951）.

Basionym: *Parmelia* subgenus *Hypogymnia* section *Solidae* Bitt. Hedwigia 40: 72（1901）.

Section ***Mundata*** Rassad.

Nov. System. Plant, non vascul. 1967, p. 299.

This section contains the only species *H. mundata*（Nyl.）Rassad. The f. *mundata* has isidia which do not break up into soredia, but the f. *sorediosa*（Bitt.）Rassad. has isidia which are always broken up into soredia.

7. ***Hypogymnia mundata***（Nyl.）Rassad.

f. *sorediosa*（Bitt.）Rassad.

Notul. Syst. e Sect. Crptog. Inst. Bot. nomine V. L. Komarovii Acad. Sci. URSS, 11: 11（1956）.

Basionym: *Parmelia mundata* f. *sorediosa* Bitt., *Hedwigia*, 49 : 255（1901）.

This taxon is new to China.

Th. K + yellow; M. K−, KC−, P + yellow.

TLC: Atranorin, physodic and physodalic acids.

Specimens examined:

CHINA: Jilin prov. Mt. Changbai shan, north slope, near Ice Stadium, on bark of *Larix* sp. alt. 1800m, 5 & 16/Ⅷ/1977, Wei, J. C. nos. 2720, 3022, 3050—（1）.

Literature cited

〔1〕Kopaczevskaya, E. G., Makarevicz, M. F. Oxner, A. N. and Rassadina, K. A. 1971. Pertusaria- ceae，Lecanocaceae and Parmeliaceae in Handbook of the Lichens of the USSR 1. p. 285.

〔2〕Ohlsson, K. E. 1973, New and Interesting Macrolichens of British Columbia. *The Bryologist* **76（3）**: 371.

〔3〕Pike, L.H. and Hale, M. E. 1982.Three new species of *Hypogymnia* from western north America（Lichenes: Hypogymniacea）*Mycotaxon* **XVI（1）**: 157.

〔4〕Wei, J. C. An enumeration of the lichens in China（to be *published*）.

〔5〕——.1984. A new isidiate species of *Hypogymnia* in China. *Acta Mycol, Sin.* **3（4）**: 214-216.

〔6〕——.and Jiang, Y. M. 1986. Lichens of Xizang, Science Press, Beijing, p. 21.

〔7〕Zhao, J. D. 1964. A preliminary study on Chinese Parmelia. *Acta Phytotax. Sin.* **9** : 166.

〔8〕——.Xu, L. W. and Sun, Z. M. 1982. Prodromus Lichenum Sinicorum. Science Press, Beijing, p.8.

亚洲袋衣属裂芽种类札记

魏江春

（中国科学院微生物研究所，北京）

摘要 迄今已知本属裂芽种类为 7 种，而且在亚洲均有分布，其中多数为亚洲所特有，如针芽袋衣、横断山袋衣、瘤芽袋衣及南亚袋衣。

本文报道了亚洲袋衣属裂芽种类 7 种，其中针芽袋衣、实心袋衣、亚瘤芽袋衣 3 种为中国新记录。此外，描述了一新亚组：长指形亚组。

关键词 袋衣属；长指形亚组；针芽袋衣；实心袋衣；亚瘤芽袋衣；裂芽种类

本文于 1986 年 6 月 20 日收到。

WEI Jiang-chun: Notes on isidiate species of *Hypogymnia* in Asia plate I

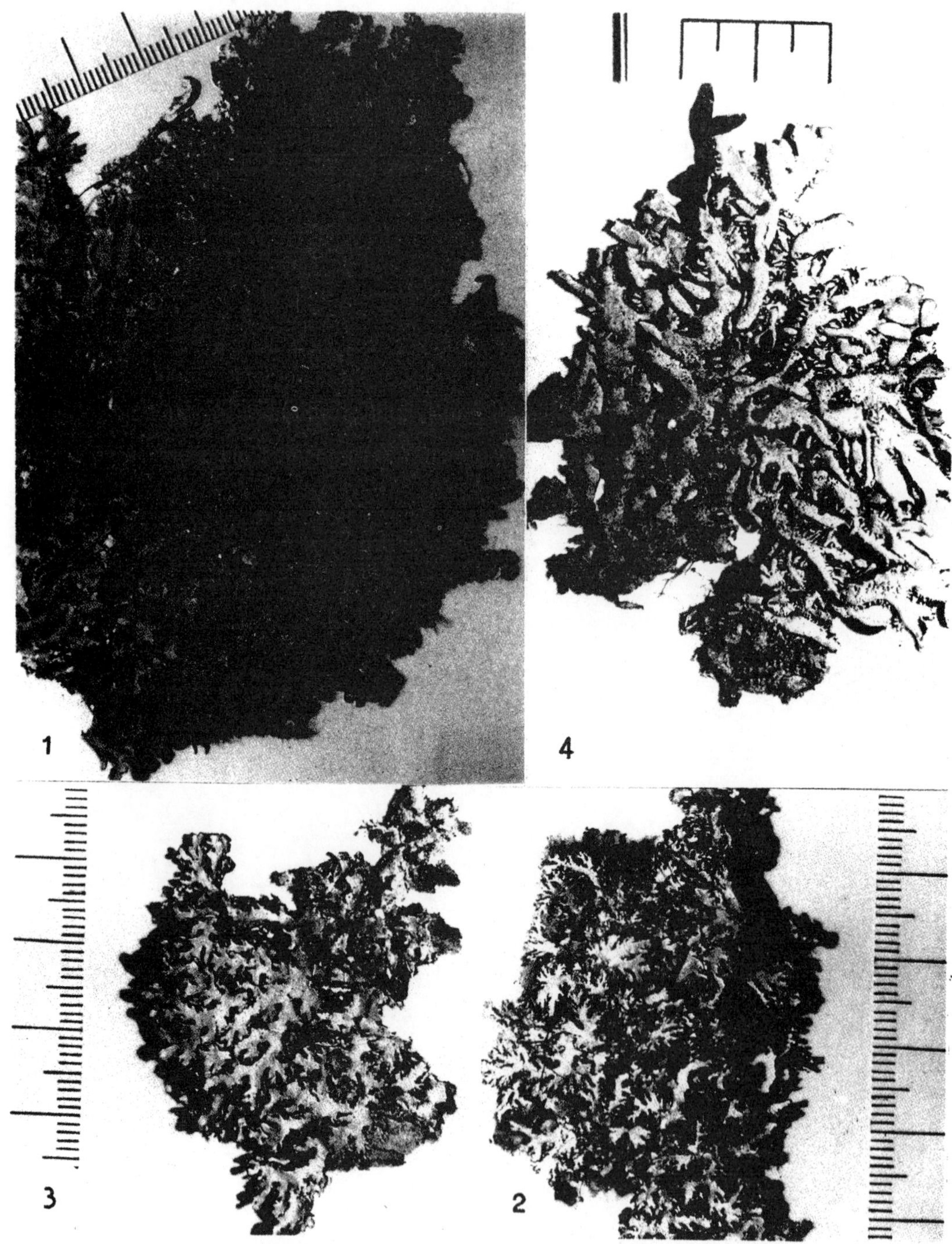

Figs. 1. *Hypogymnia duplicatoides* 2. *H. hokkaidensis*
3. *H. subcrustacea* 4. *H. zeylanica*
（scale in mm）

WEI Jiang-chun: Notes on isidiate species of *Hypogymnia* in Asia plate II

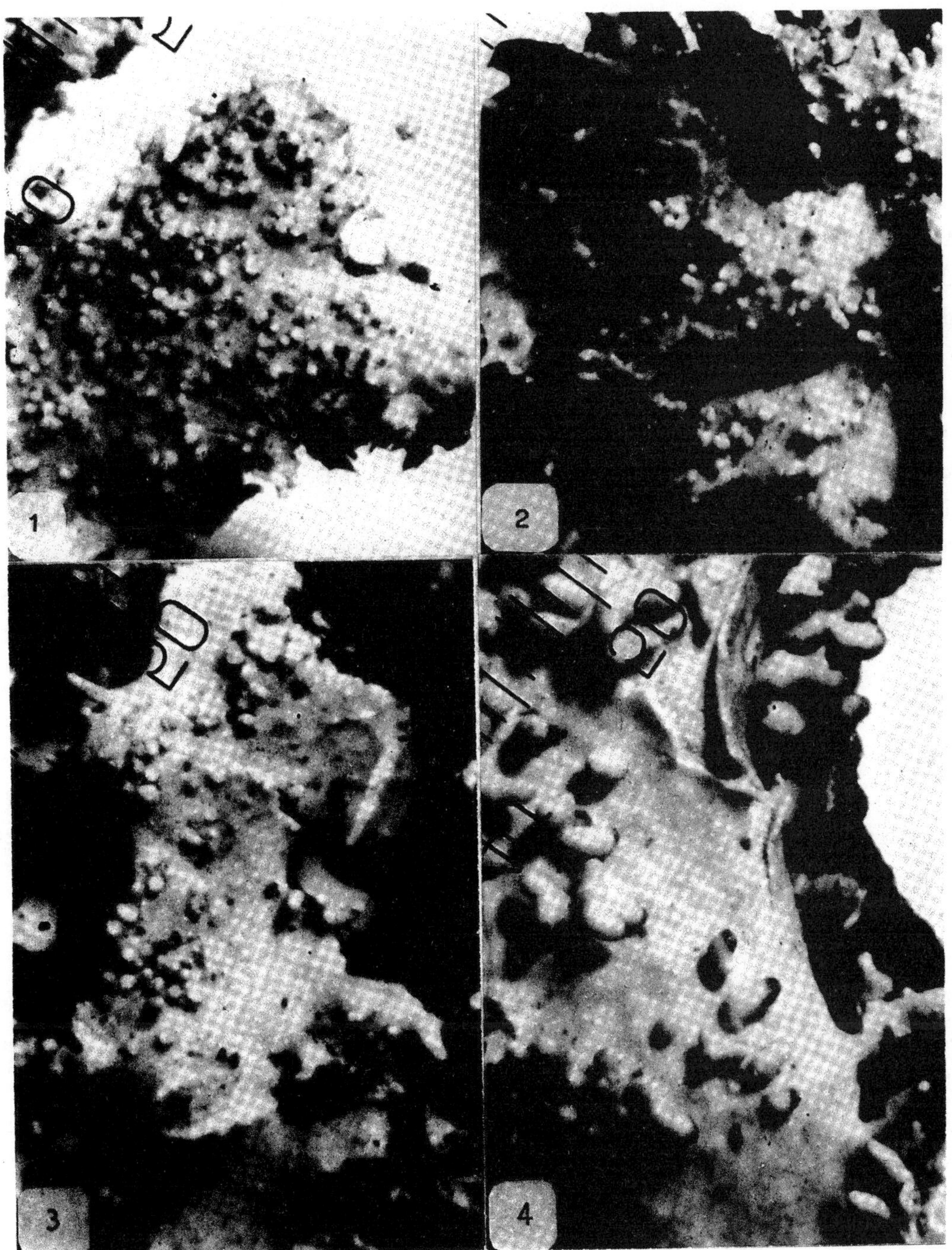

Figs. 1. Isidia of *H. duplicatoides* 2. Isidia of *H. hokkaidensis*
3. Isidia of *H. subcrustacea* 4. Isidia of *H. zeylanica*
(scale in 156 μm)

A conspectus of the lichenized Ascomycetes Umbilicariaceae in China

WEI Jiang-chun[①] **JIANG Yu-mei**

Systematic Mycology & Lichenology Laboratory
Institute of Microbiology, Academia Sinica, Beijing

ABSTRACT Henssen（1974）distinguished five apothecial types: biatorine, lecideine, lecanorine, zeorine, and superlecideoid. Both the biatorine and lecideine types belong to the lecideine type in a broad sense, and the superlecideoid, the lecanorine and zeorine to the lecanorine type in a broad sense.

The apothecium structure of some species of Umbilicariaceae has been investigated by light microscopy （LM）. According to the apothecium structure and the origin of the amphithecium, it is clear that the apothecium of *Umbilicaria virginis* and *U. rigida* with a thalline rim around the disc lacking algae in it belong to the lecanorine but not to the lecideine type in a broad sense. In conseque nce the term "superlecideoid" or "superlecideine" might better be replaced by "sublecanorine" for the apothecium type of the Umbilicariaceae.

Fourty six taxa belonging to 37 species occur in China, including 1 species new to science（*Umbilicaria pseudocinerascens*）, and 11 taxa, including 3 varieties, belonging to 11 species new to China are reported. A taxonomic revision of some taxa has also been made, and keys to species are provide.

KEYWORDS Apothecium; sublecanorine; Umbilicariaceae; *Umbilicaria pseudocinerascens*

INTRODUCTION

The species of *Umbilicaria* are generally called " Shi-er"（石耳）in Chinese, which means the ear-like plants on rocks. However, all the ear-like plants on rocks including the species of *Dermatocarpon* and some of the Collemataceae on rocks in addition to those of the Umbilicariaceae are called "Shi-er" in ancient Chinese literature and by the people（Wei *et al.*, 1982）.

Lichens of the Umbilicariaceae are important as food for man in China and also have antitum properties（Wei *et al.*, 1982）. Therefore, a systematic study of this group is of great importance not only of theoretical importance, but also of practical importance.

The structure of the apothecia in this group has been observed and described by different authors（Scholander, 1934; Frey, 1936; Henssen, 1970）, and used by some as a criterion in the arrangement of genera, 4 by Scholander （1934）and 5 by Llano（1950）.

---

① Wei Jiang-chun = Vej Tzian-czunj (used in 1960—1980)

This paper was originally published in *Mycosystema*, 1: 73-106, 1988.

The Umbilicariaceae was studied extensively by Frey (1931, 1933, 1936, 1949, etc.) who treated it as monotypic. Scholander (1934) disposed all the species of the Umbilicariaceae in four groups at the rank of genus on the basis of discus structure(tetratypic family). The sections *Lasalliae* and *Anthracinae* of *Umbilicaria* Hoffm. have been raised by Llano (1950) to generic rank according to the characteristics of the spores and thalli in addition to disc structure, and some nomenclature problems have been treated. So the five genera of the family: *Umbilicaria*, *Agyrophora*, *Omphalodiscus*, *Actinogyra*, and *Lasallia* were proposed(quinqutypic family). Two of the genera in the family, *Umbilicaria* and *Lasallia*, are accepted by Wei (1966) (bitypic family) and this has been followed by most recent authors (Cannon, Hawksworth, Pike, 1985; Moberg, 1982; Poelt, 1969; Poelt and Vězda, 1977, 1981; Santesson, 1984, etc.); and *Agyrophora* (with smooth disc calling leiodisc), *Omphalodiscus* (with disc having central sterile column or fissures calling omphalodisc), and *Actinogyra* (with disc having actinoid fissures calling actinodisc), and *Umbilicaria* (with disc having centric or eccentric gyri calling gyrodisc) are adopted as subgenera of *Umbilicaria*; in addition, *Lasallia* with smooth disc and *Pleogyra* with gyrate disc as subgenera of *Lasallia*. The monotypic family's supporter Frey, in the later period of his research changed his viewpoint and began to support the bigtypic family (Frey and Poelt, 1977; Poelt, 1977, on the basis of Frey's manuscript). Recently, three genera, *Lasallia* Mérat, *Llanoa* Dodge, and *Umbilicaria* Hoffm., in the family have been accepted by Eriksson and Hawksworth (1986 a,b,1987 a,b).

Thirty five species of Chinese Umbilicariaceae have been reported so far. Among them 24 species and 3 forms of *Umbilicaria* recorded by 27 different authors from 14 provinces in 31 scattered publications, and 11 species of *Lasallia* by 10 authors from 14 provinces in 13 scattered papers (Wei, An enumeration of the lichens in China).

During a study on the Umbilicariaceae in East Asia, a conspectus of the family from China has been accomplished.

MATERIALS AND METHODS

Apothecium preparations for LM investigation were made from herbarium materials. The following specimens, including some from abroad, were examined; *Umbilicaria virginis* var. *lecanocarpoides* (Nyl.) Wei & Jiang, SICHUAN, Daofu near Kangding, 1934, H. Smith 14006 (UPS); *U. lyngei* Schol., SHAANXI, Mt.Taibai shan, Baxiantai, 1963, collected by Wei JC; *U. leiocarpa* DC., JAPAN, Kurokawa, Lich. exs. no.291 (TUR-20229), no.292 (TUR-20230). ,*Lasallia pustulata* (L.) Mérat, SWEDEN, near Stockholm, 1982, J.C. Wei, Thol, *et al.*; *U. muehlenbergii* (Ach.) Tuck., HEILONGJIANG, Gulian, 1977, J.C. Wei; *Lasallia mayebarae* var. *sinensis* (Wei) Wei, YUNNAN, Lijiang, Mt. Yulong shan, 1964, J.C. Wei 2590; *L. rossica* Dombr.: USSR, the Buryat ASSR, (isotype, LE), And CHINA, JILIN, Mt. Changbai shan, 1983, J.C. Wei & J.B. Chen 6251. Some others have also been observed such as *Umbilicaria rigida* (Du Rietz) Frey, USSR, Savicz, Lich. Ross, no.141; *U. subglabra* Nyl., USSR, Savicz, Lich. Ross. no.162(S); *U. africana* (Jatta) Wei, Zaire(central Africa), Kivu prov. Vezda, Lich. Select. Exs. no.1371 (S); *U. zahbruckneri* Frey, Ecuador, L 223 (S); *U. cylindrica* (L.) Del., CHINA, Mt. Changbai shan, J.C.Wei 2973— (2); and *Lasallia brigantium* ssp. *hispanica* Frey, Portugal, C. Tavares 1325 (S).

Except those borrowed from abroad specimens examined were mainly collected by the senior author, and part of them by other collectors and being preserved in HMAS-L.

Chemical data were obtained by thin-layer chromatography (TLC) with the methods described by Culberson and Kristinsson(1970), and Culberson(1972), and modified by Menlove(1974), using E. Merck TLC aluminium sheets silica gel 60 F254. Microcrystal tests were performed using the methods described by Asahina(1936-1940) when necessary.

RESULTS

Excellent specimens of *Umbilicaria virginis* Schaerer with well developed apothecia of the lecanorine type are found among the extensive collections of Harry Smith from Sichuan province, China made in 1934 and preserved in UPS (Pl.I: a,b). As seen from the sections, a thalline rim lacking algal cells was found to develop around the disc of the apothecium (Pl.I: c, d; II: a, b; V). This led to the examination of all available umbilicariaceous material with the result that most of the species in the family, such as *Umbilicaria virginis* var. *lecanocarpoides* (Pl.I; II: a, b; V), *U. lyngei* (Pl.II: c, d), *U. leiocarpa* (Pl.II: e, f), *Lasallia pustulata* (Pl.III: a, b), *L. brigantium* ssp. *hispanica*, *L. rossica*, *Umbilicaria rigida*, *U. subglabra*, *U. africana*, and *U. zahlbruckneri*, etc., with leiodisc or omphalodisc apothecia bear apothecia of the lecanorine type. The others with actinogyric apothecia bear apothecia of the lecideine type. Intermediates between the above two types are observed in the subgenus *Umbilicaria*. Sometimes it is difficult to recognize, as in the case here the thalline margin of the apothecium is carbonized, but sections often show them to be apothecia with a thalline rim lacking algal cells in the amphithecium as those of *U. cylindrica* (Pl.III: d).

This conspectus of the family in China is a part of a study on the Umbilicariaceae from East Asia. Fourty six taxa belonging to 37 species including 1 species new to science (*Umbilicaria pseudocinerascens*), and 11 taxa belonging to 11 species new to China are reported here. A taxonomic revision of some taxa described from China has also been made, and keys to species are provided.

DISCUSSION

According to Hue (1907), apothecia with a thalline rim lacking algal cells around the disc belong to the lecanorine type. He considered that the presence on the sides of the apothecium of cortical tissue derived from the thallus was sufficient grounds for regarding an apothecium as lecanorine (Hue, 1907; Lamb, 1948). Regarding lecanorine apothecia, Yoshimura (1971) indicated: "in most species of *Lobaria*, the algal cells exist just below the cortex of the thalloid exciple, from which the algal cells are almost absent in the Lobaria retigera complex." Sections of apothecia of *Lobaria kurokawae* from this group show them to be of the same apothecium type as that in *U. virginis* var. *lecanocarpoides* (P l . III : c). However, the term "superlecideoid" was proposed by Frey (1936) for the similar apothecium type of *U. rigida*. Letrouit-Galinou (1973) grouped the basic apothecial types in discolichens as graphidian (p.79. Figs. 27-29), prelecanorine (p.79, Figs. 30,31), and lecanorine (p.80, Figs. 32-34). Henssen (1974) distinguished five apothecial types: biatorine, lecideine, lecanorine, zeorine, and superlecideine. Both the biatorine and lecidcine types belong to the lecidcine type in a broad sense, and the superlecideine, the lecanorine and zeorine to the lecanorine type in a broad sense. According to the apothecium structure and the origin of the amphithecium, it is clear that the apothecium type of *Umbilicaria virginis* and *U. rigida*, belong to the lecanorine and not to the lecideine type in a broad sense. In consequence the term " superlecideoid" might better be replaced by "sublecanoroid" or "sublecanorine" for the apothecium type of the Umbilicariaceae (most of the species), and for those of the species of *Lobaria retigera*-complex, etc. (Fig. 1), and of *Pannaria pycnophora* (Nyl.) Müll. Arg. (A. Eahlbr. 1926, P. 180; R. Sant, 1944, p. 16), etc.

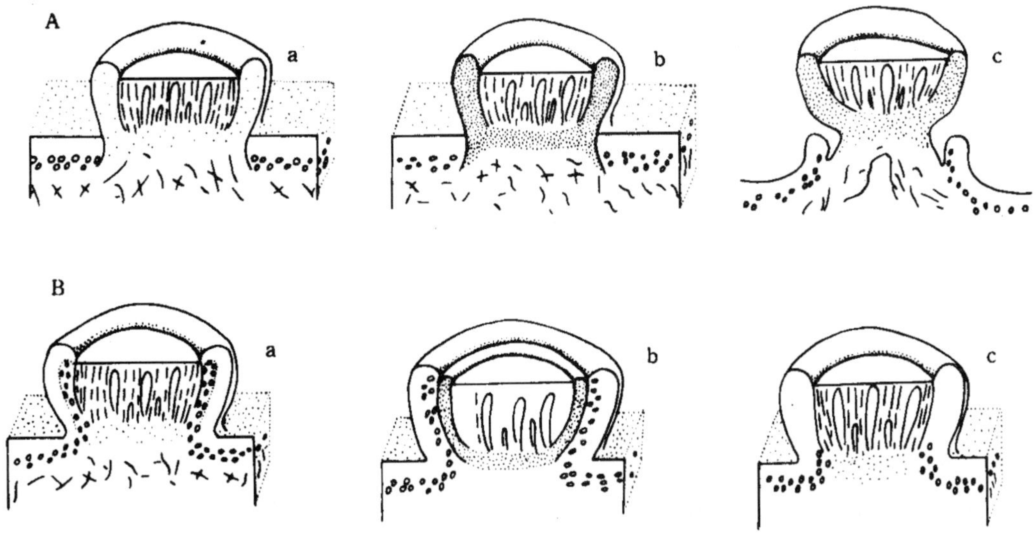

Fig. 1. Apothecial types

A.lecideine type in a broad sense:

a. biatorine. b. lecideine. c. apothecium of *U. haplocarpa* (taken from Frey, 1949, with a few changes)

B.lecanorine type in a broad sense:

a. lecanorine, b. zeorine. c. sublecanorine (= superlecideoid or superlecideine)

(A-a.b.and B-a.b: taken from Henssen, 1974, with a few changes; B-c: drawn by the present authors)

(Sketch figurs)

Llano (1950) was not clear why Du Rietz (1925) suggested that *U.lecanocarpoides* (i.e. *U. virginis* var. *lecanocarpoides*) could be compared with *U. cylindrica*. However, apothecium sections of both *U. virginis* var. *lecanocarpoides* and *U. cylindrica* show them to be similar in the apothecium type with respect to amphithecium. Indeed, it seems logical for Du Rietz to have suggested that *U. virginis* var. *lecanocarpoides* could be compared with *U. cylindrica*.

Apothecia of *Umbilicaria haplocarpa* Nyl. lack a thalline rim around the disc, but have edges of the perforated thallus surrounding it as a collar. This seems to be a degenerated or primitive form of zeorine type. As regards apothecia of the lecanorinc type in *Llanoa cerebriformis* (Dodge and Baker) Dodge, no conclusion can be drawn before checking the type specimen.

ENUMERATION OF TAXA
UMBILICARIACEAE Fée

Fée, Essai Cryptog. Ecorc. Officin. LXX, 1824 [1825] (as *Umbilicariaea*). Lindsay, Popular Hist. Brit. Lich.173, 1856 (as Umbilicariaceae Lindsay). Frey, Hedwigia, 69:219 (1929). Scholander, Nyt Mag. Naturvid. 75: 1 (1934). Wei, Acta Phytotax. Sin. 11 (1) : 4 (1966).

Typus: *Umbilicaria* Hoffm.

UMBILICARIA Hoffm.

Hoffm. Descr. Adumbr. PI. Lich. 1（1）: 8（1789）.

Wei, Acta Phytotax. Sin. 11（1）:4（1966）.

Type: *Umbilicaria exasperata* Hoffm.

[i.e. **Umbilicaria proboscidea**（L.）Schrad.]

Key to species from China

1. Thallus with rhizines.
 2. Apothecia common.
  3. Apothecia with gyrate disc.
   4. Apothecia adnate to immersed, never stipitate.
    5. Apothecia rounded, on rocks.
     6. Rhizines cylindrical.
      7. Lower surface of the thallus black, plane, with dense rhizines of short, very thin and simple form ······ ············ *U. badia*
      7. Lower surface of the thallus light, bullate to lacunose, with sparse rhizines ············ *U. herrei*
     6. Rhizines ligulate to irregularly cylindrical ············ *U. tylorhiza*
    5. Apothecia irregularly triangle to quadrilateral; lower surface of the thallus black, with rich, very short, richly branched and black rhizines, on trees ············ *U. yunnana*
   4. Apothecia adnate to stipitate.
    8. Apothecia stipitate, lower surface of the thallus light coloured to pink, with rhizines of the same colour ············ *U. cylindrica*
    8. Apothecia adnate, not stipitate.
     9. Disc gyrate with numerous perpendicular and slight cracks; lower surface of the thallus brown, weakly granulate like that in *U. rigida*, without or occasionally with individual rhizines ············ *U. pseudocinerascens*
     9. Disc gyrate, without the perpendicular cracks.
      10. Lower surface especially near the margin. with rich and black rhizines ············ *U. indica*
      10. Lower surface without rhizines, or occasionally with sparse or individual rhizines.
       11. Lower surface of the thallus rcd-yellow to purple-red ············ *U. hypococcinea*
       11.Lower surface of the thallus brown grey to black brown ············ *U. proboscidea*
  3. Apothecia of the omphalodisc type.
    12. Disc usually plane and smooth, sometimes with a fissure in it ············ *U. virginis*
    12. Disc with columns or fissures.
     13. Disc with fissures; upper surface of the thallus elevated, with folds and weak ridges in centre, margins Jaciniate; lower surface dark about umbo, without rhizines, sometimes near margin with individual grossus bristle like rhizines ············ *U. formosana*
     13. Disc with domed columns; upper surface of the thallus without conspicuous ridges and folds in the centre; lower surface black, partly densely rhizinous ············ *U. spodochroa*
 2. Apothecia rare.
   14. Lower surface of the thallus black.
     15. Rhizines capitate ············ *U. trabeculata*

15. Rhizines non capitate.
16. Rhizines richly branched.
17. Rhizines numerous, very thin, short, richly branched; thallus weakly leathery; upper surface red brown and dull ······ *U. esculenta*
17. Rhizines sparse, richly branched at the tip; thallus thin, membranous to fragile; upper surface brownish, dull to almost shiny ······ *U. caroliniana*
16. Rhizines sparsely branched.
18. Rhizines long and thick; upper surface of the thallus without squamules ······ *U. vellea*
18. Rhizines short and very thin; upper surface of the thallus with squamules ······ *U. thamnodes*
14. Lower surface of the thallus brown, with some rose tint ······ *U. hirsuta*

1.Thallus without rhizines.
19. Lower surface of the thallus with lamellae or trabeculae.
20. Apothecia actinodisc ······ *U. muehlenbergii*
20. Apothecia gyrodisc ······ *U. torrefacta*
19. Lower surface of the thallus without lamellae or trabeculae.
21. Upper surface of the thallus with folds.
22. Upper surface of the thallus with elevated ridges and folds in the ccntre.
23. Apothecia of the omphaldisc type.
24. Margins of the thallus irregularlly laciniate, sometimes with rhizinoid cilia ······ *U. formosana*
24. Margins of the thallus not laciniate, without-rhizinoid cilia ······ *U. kraschenninnikovii*
23. Apothecia gyrate; upper surface of the thallus with a umbo in the centre, covered by white pruinae ······ *U. proboscidea*
22. Upper surface of the thallus with fine reticulate folds.
25. Apothecia of the omphalodisc type; upper surface of the thallus dark grey, with some olivaceous brown colour; the reticulate folds fading near the margins and the ridges of them blunt ······ *U. decussata*
25. Apothecia of the leiodisc type; upper surface of the thallus almost dark, without conspicuously olivaceous brown colour; the reticulate folds are extending up to margins and the ridges of them acute ······ *U. lyngei*

21. Upper surface of the thallus without folds.
26. Thallus isidiate ······ *U. deusta*
26.Thallus not isidiate.
27. Apothecia always present and rich.
28. Thallus over 5 mm in diameter.
29. Apothecia concentrically gyrate with some perpendicular and slight cracks; lower surface of the thallus brown to dark brown ······ *U. pseudocinerascens*
29. Apothecia not concentrically gyrate without perpendicular cracks, lower surface of the thallus yellow redish to purple-redish ······ *U. hypococcinea*
28. Thallus about 5 mm in diameter.
30. Lower surface of the thallus black，near the margins with thick and black rhizines and rhizinoid cilia ······ *U. nanella*

30. Lower surface of the thallus white to pale rose, with rhizines of the same colour; margins of the thallus uptrend ···························· *U. cylindrica* var. *tornata*

27. Apothecia absent or very poor and single if occasionally present.

31. Upper surface of the thallus reticulate with slight cracks.

32. Lower surface of the thallus brown, weakly granular about the umbo ···························· *U. rigida*

32. Lower surface of the thallus not granular ···························· *U.leiocarpa*

31. Upper surface of the thallus not reticulate and lacking slight cracks.

33. Thallus about 5 mm in diameter, elevated near the margins but the margins recurved; upper surface red-brown, dull; lower surface black, not granular ···························· *U. kisovana*

33. Thallus above 15—20 mm in diameter; upper surface black brown; lower surface black, weakly granular or weakly reticulate with slight cracks about the umbo ···························· *U. cinerascens*

Subgenus Umbilicaria

Type: **Umbilicaria proboscidea**（L.）Schrad.

1. **Umbilicaria badia Frey**

Ber. Schweiz. Bot. Ges. 59:453（1949）.

Chemistry: gyrophoric and umbilicaric acids.

Specimens examined:

YUNNAN: Lijiang, Mt.Yulong shan, Ganhaizi, on rocks, 20/III/1943, Wang HC, no. 3506. XIZANG: Zham, alt. 2400—2700 m, 9—11/V/1966, Wei JC & Chen JB, nos, 413, 414, 418, 534, 567; Quxam valley, alt. 3200—3550 m, 18—19/V/1966, Wei JC & Chen JB, nos. 837, 932（HMAS-L）.

Literature record: Xizang（Wei JC & Jiang YM, 1986, p. 100）.

2. **Umbilicaria caroliniana Tuck.**

Proc. Amer. Acad. Arts & Sci. 12:167（1877）.

Chemistry: gyrophoric acid.

Specimen examined:

NEIMONGOL: Da Hinggan Ling, Ergun Zuoqi（Genhe）, Mt. Oklidui shan, on rocks, alt. 1530 m, 26/VII/1983, Zhao CF, no. 2963（HMAS-L, IFP）.

Literature record: Liaoning（Chen *et al.* 1981, p. 157）.

3.**Umbilicaria cinerascens**（Arnold）Frey

Hedwigia 69:250（1929）.

=*Gyrophora cinerascens* Arnold, Verhandl. Zool. -Bot. Ges. Wien 25: 438（1875）.

=*Umbilicaria cinerascens* f. *laciniata* Frey, in Rabh. Krypt. - FI. IX, Abt. 4.（1）: 380（1933）, cum Icon.

Chemistry: gyrophoric acid and'an unidentified triterpenoid.

Specimens examined:

SHAANXI: Mt.Taibai shan, on rocks, Inst. of Chinese Tranditional Medicine, Shaanxi Branch of Chinese Academy of Medicine, no. 2319; Baxiantai, Paomaliang, on rocks, 28/VII/1938, Liou TN & Tsoong PC, no. 4258, & alt. 3600 m, on rocks, 30/VII/1963, Ma & Zong, no. 303; Baxiantai, Eryehai, on rocks, alt. 3500m, 4/VI/1963, Wei JC *et al.* no. 2718-（3）, and a no numbered collection（=f. *laciniata* Frey）from the same locality.

New to China.

4.**Umbilicaria cylindrica** (L.) Del.

Bot. Gall. 2: 595 (1830).

=*Lichen cylindricus* L., Sp. Pl. 1144 (1753).

=*Umbilicaria cylindrica* var. *denudata* (Turn.) Frey in Rabh. Krypt-Fl., 9,4 (1) :330 (1933).

=*Gyrophora proboscidea* var. *denudata* Turn., Specim. Lichenogr. Brit. 219 (1839).

4.1. var. **cylindrica**

Chemistry: no lichen substances.

Specimens examined:

JILIN: Mt. Changbai shan, north slope, near Tianchi, on rocks, alt.2100m, 13/VIII/1977, Wei JC, no. 2973-(2) and alt. 2350 m, 14/VIII/1977, Wei JC, nos. 2984, 2992; southwest slope, Fusong county, alt.2050m, on rocks, 8/VIII/1983, Wei JC & Chen JB, no.6761. SHAANXI: Mt. Taibai shan, Baxiantai, 4/VI/1963, collected byWei JC.

Literature record: Jilin (Chen *et al.* 1981, p. 157).

4.2. var. **tornata** (Ach.) Nyl. (Fig. 2)

Lich. Scandin. 1861:117.

=*Gyrophora tornata* Ach., Kgl. Vetensk-Akad. Nya Handl. 1808:274.

Chemistry: gyrophoric acid.

Specimens examined:

JILIN: Mt. Changbai shan, north slope, alt. 2400 m, on rocks, 14/VIII/1977, Wei JC, no. 2984. SHAANXI: Mt. Taibai shan, Baxiantai, alt. 3660m, on rocks, 3/VI/1963, collected by Wei *et al.* & 31/VII / 1963, Ma & Zong, no.395- (1).

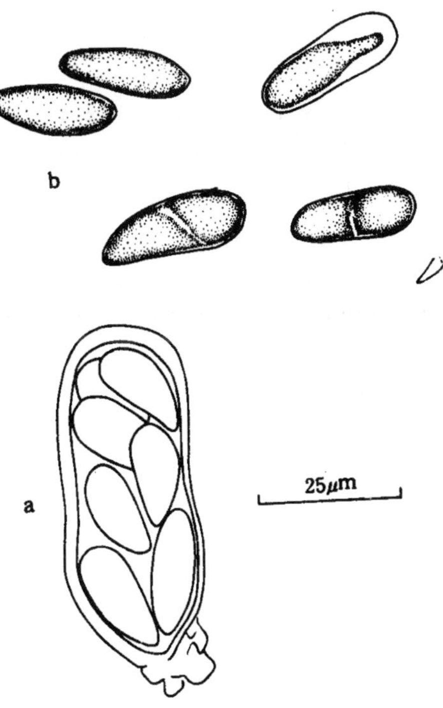

Fig. 2. *Umbilicaria cylindrica* var. *tornata*

a. ascus with ascospores. b. ascospores.

New to China.

5. **Umbilicaria deusta**（L.）Baumg.

Fl. Lips. 571（1790）.

=*Lichen deustus* L., Sp. Pl. 1150（1753）.

Chemistry: gyrophoric and umbilicaric acids.

Specimens examined:

JILIN: Mt. Changbai shan, southwest slope, Fusong county, on the way to Tianchi, alt. 2050—2250m, on rocks, 8/VIII/1983， Wei JC & Chen JB, nos. 6755,6758,6760,6762,6766, 6767, 6772; Hongtou shan, alt. 1950m, on rocks, 8/VIII/1980, Hu YC, no.2240; north slope, near Lumingfeng, alt. 2400m, on rocks, 25/VIII/ 1979, Hu YC, no.1339; near Heifengkou, alt. 2100—2400m, Hu YC, nos. 1498, 1940; Houcegou near Guanqiai, alt. 2200m, on rocks, 3/IX/1979, Hu YC, no. 2115; near Wenquan, alt. 1860m, on rocks, 3/IX/1981, Hu YC, no.2737; on west slope of the river near Wenquan, alt. 1850m, on rocks, 12/VIII/1983, Wei JC & Chen JB, no. 6845, & alt. 1880m, Hu YC, no. 4667.

New to China.

6.**Umbilicaria esculenta**（Miyoshi）Minks

Mem. Herb. Boiss. 21:46（1900）.

=*Gyrophora esculenta* Miyoshi, Bot. Centralbl. 56:161（1893）.

Chemistry: gyrophoric acid.

Specimens examined:

LIAONING: Kuandian county, on top of Baishilazi, alt. 1230m, on rocks, 26/IX/1963, Nan Yingao, no. 6314（p. p.）， and alt. 1220m, 1/IX/1964, Chen XL, no. 2766. JILIN: Mt. Changbai shan, south slope, alt. 1450m, on rocks, 17/VIII/1982, Hu YC, nos. 3514, 3528; Huichun county, Mt. Dama-an shan, alt. 580m, on rocks, 30/VII/1984, Hu YC, no. 5880; Wangqing county, Mt. Tulaopuoding shan, alt. 350—1000m, on rocks, Hu YC, nos. 6069, 6124, 6171, 6188; Jaohe county, Qianjinxiang, Baishilazi shan, alt. 780m, on rocks, 12/VIII/1984, Hu YC, no. 6264. HEILONGJIANG: Raohe county, Hualazi, on rocks, 10/IX/1952, Wang Guang-Zheng, no. 642. YUNNAN: collected by Delavay in 1885, H-Nyl. 31523（H! pr. p.）. JIANGXI: Mt. Lu shan, on rocks，31/III/1960, Zhao JD *et al.* nos. 455, 457, 458, 459, 460, 467, 468, 471; alt. 1 150-1170m, 11-14/II/1965, Wei JC, nos. 3116, 3118, 3150, 3245. ANHUI: Mt. Huang shan, near Banshansi, alt. 800m, 24/VIII/1962, Zhao JD & Xu LW, nos. 5913, 5918. HUNAN: Zhangjiaje, on rocks, 1984, collected by Zhou. ZHEJIANG: Mt. Tianmu shan, near Laodian, alt. 1000—1500m, on rocks, 31/VIII 1/IX/1962, Zhao JD & Xu LW, nos. 6189, 6300. GUANGXI: Jinxiu county, on rocks, X/1981, collector unknown.

Literature record: Jilin, Heilongjiang（Chen *et al.* 1981, p. 158）, Jiangxi（Wei JC, 1981, p. 89; Wei et al. 1982, p. 28）， Anhui, Zhejiang（Lu, 1959, p. 174）.

7. **Umbilicaria herrei** Frey

Ber. Schweiz. Bot. Ges. XLV: 219（1936）.

Chemistry: gyrophoric and umbilicaric acids.

Specimens examined:

JILIN: Mt. Changbai shan, north slope, alt. 2100—2200m, on rocks, 10, 13/VIII/1977, Wei JC, nos. 2886, 2942, & near Bingchang（ice studium）, alt. 2200m, 10/VIII/1977, Chen XL, no. 4687（as Umbilicaria hyperborea by Chen et al. 1981, p. 158）， near Tianwenfeng, alt. 2200m, on rocks, 10/VIII/1977, Hu YC, no. 121; near waterfall, alt. 2100m, on rocks, 13/VIII/1977, Hu YC, nos. 218-B, 221; slope on the east of the river near Wenquan, alt. 1850m, on rocks, 12/VIII/1983, Wei JC and Chen JB, no. 6806; slope on the west of the river near

Wenquan, alt. 1850m, on rocks, 12/VIII/1983, Wei JC and Chen JB, nos. 6815, 6846, on the way to Long menfeng near a small waterfall, on rocks, 13/VIII/1983, collected by Wei JC. and near Wenquan, alt. 1800m, on rocks, 26/VIII/1983, Hu YC, nos. 4773, 4775; south slope on the way from Hengshan station to Tianchi, on rocks, alt. 2050m, 31/VII/1983, Wei JC and Chen JB, no. 6387; southwest slope, Fusong county, Weidong station, on the way to Tianchi, alt. 2060m, on rocks, 8/III/1983, Wei JC and Chen JB, no. 6763.

This is the first report for China.

Chen et al. （1981, p. 158） reported this lichen as *U. hyperborea* from Jilin province.

8. **Umbilicaria hirsuta** （Sw.） Ach.

Svensk. Vetensk.-Akad. Nya Handl. 15:97 （1794）.

=*Lichen hirsutus* Sw. in Westr. Vetensk-Akad. Nya Handl. 1793,47.

=*Umbilicaria hirsuta* var.*vuoxaensis* （Wei） Wei,in Wei and Jiang. Acta phytotax. Sin. 20: 499 （1982）

=*Cyrophora hirsuta* var. *vuoxaensis* Wei, Notul. Syst. E Sect. Crypt. Inst. Bot. Nom. Komarovii Acad. Sci. URSS I5:8 （1962）.

Chemistry: gyrophoric acid.

Specimen examined:

XIZANG: Nyalam, alt. 3720m, on rocks, 15. VI/1966, Wei JC & Chen JB, no. 1648 （1）, and a no numbered collection from the same locality.

Literature record: Xizang （Wei JC & Chen JB, 1974, p. 177; Wei JC & Jiang YM, 1981 p. l 146, & 1986, p.99-100）.

9.**Umbilicaria hyperborea** （Ach.） Hoffm.

Descr. Adumbr. Pl. Lich. 3 （4） : 9-10. Pl. 71, fig. 1-5 （1800）.

=*Lichen hyperboreus* Ach., Vetensk.-Akad. Nya Handl. 15:89 （1794）.

Chemistry: gyrophoric and umbilicaric acids.

Specimen examined:

NEIMONGOL: Da Hinggan Ling, Ergun Zuoqi （Genhe）, Mt. Oklidui shan, alt. 1530m, on rocks, 26/VII/1983, Zhao CF, no. 2964.

Literature record: Neimongol, Da Hinggan Ling （Sato, 1952, p. 174）.

Report of this species from Mt. Changbai shan （Chen et al. 1981, p. 158, collectors no. 4687） was based on erroneous identification. see *U. herrei*）.

10. **Umbilicaria hypococcinea** （Jatta） Llano （Fig. 3）

Monogr. of the *Umbilicariaceae*, 191 （1950）（lapsu “*hypococcina*”）.

=*Gyrophora hypococcinea* Jatta. Nouv. Giorn. Bot. Ital. N.S. 9:473 （1902）.

Type: SHAANXI, Mt. Guangtou shan, collected by Giraldi.

One collection from the original preserved in S has been seen by the senior author of this paper. So that it may probably be an isotype of the species.

= *Gyrophora versicolor* Räs., Arch. Soc. Zool. Bot. Fenn. ‘Vanamo’, 3:78 （1949）.

Type: SHAANXI, Mary Strong Clements, no. 4021 （= TUR-36392, holotype） in TUR （!） and isotype in H （!）.

=*Umbilicaria nepalensis* Poelt, Khumbu Himal, 6/3:426 （1977）.

Type: L 262, not seen.

Thecium 74—86μm thick; ascus clavate, 69 × 17μm, octosporous; spores elliptical, hyaline, 11—15 × 7—

10μm; paraphyses simple, unseptate; hypothecium 37—49μm thick.

Description of the spores for this species are given here for the first time (from the specimen of Wei et al. no.2718- (4) in HMAS-L).

Chemistry: gyrophoric and umbilicaric acids.

Specimens examined:

SHAANXI: Mt. Taibai shan, on rocks, herb, from Inst, of Chinese Tranditional Medicine, no. 2318; near Mingxingsi, on rocks, alt. 2640m, 2/VI/ 1963, Wei JC et al. no. 2518; Baxiantai, on rocks, 1963, Wei JC et al. no. 2718- (4); Xizushitai, 3500m, VI/1956, Yang JX, no. 5; Domugong, alt. 2840m, 22/VII/1963, Ma & Zong, no. 140; Pingansi, alt. 2700m, on rocks, 23/VII/1963, Ma and Zong, no. 166; Fangyangsi, alt. 2950—3000m, on rocks, 26—28/VII/1963, Ma and Zong, nos. 226, 281; Yuewangchi, alt. 3080m, on rocks, 6/VII/1963, Ma and Zong, no. 380; Huxian county. Mt. Guangtou shan, Xihekou, on rocks, alt. 2000m, 2/VII/ 1964, Wei JC, no. (topotype); Shaanxi, Mary Strong Clemens no. 4021(=TUR-36392), holotype for *G. versicolor* Räs. preserved in TUR, and the isotype in H.

Literature record: Shanxi(Lu, 1959, p. 175), Shaanxi(Jatta, 1902, p.473; A.Z. 1927, p.718 and 1930, p. 138; Räs. 1949, p.78; Wu JL, 1981, p.161; Wei JC, 1981, p.89; Wei JC et al. 1982, p.28), Xizang(Wei JC and Chen JB, 1974, p. 177; Wei JC and Jiang YM, 1981, p. 1146, and 1986, p.98).

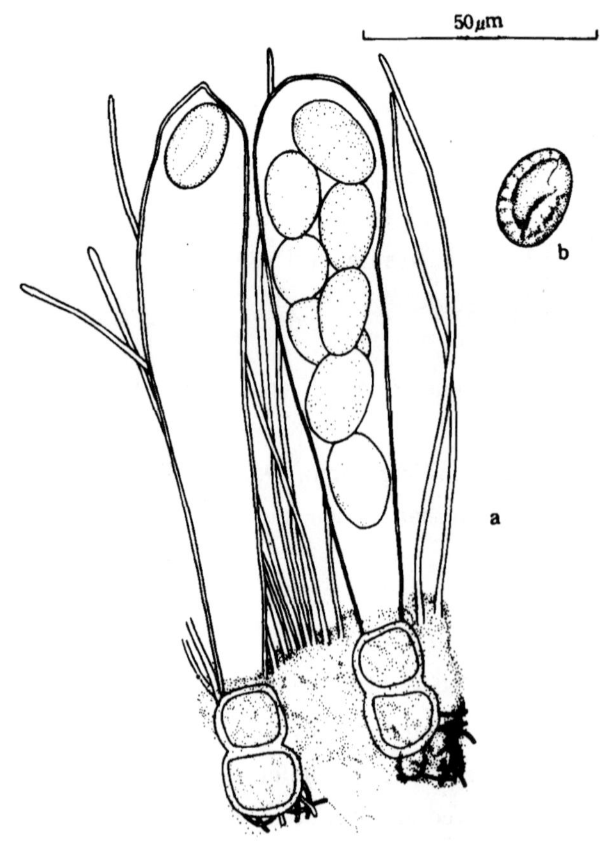

Fig. 3. *Umbilicaria hypoccocinea*

a. asci wich ascospores; b. ascospore.

11. **Umbilicaria indica** Frey

Ber. Schweiz. Bot. Ges. LIX: 456 (1949) (as nom. nov.).

=*Umhilicaria papillosa* Nyl., Synops. Lich., 2:11 (1863), non DC. 1805.

Literature record: Yunnan (Poelt, 1977, p.420), Xizang (Wei JC and Chen JB, 1974, p. 177, cited as Umbilicaria polyrrhiza; Wei JC and Jiang YM, 1986, p. 100).

Taxonomic revision of this species remains to be done in the near future.

12. **Umbilicaria kisovana** (Zahlbr. ex Asahina) Zahlbr.

Cat. Lich. Univ. 10: 405 (1940).

=*Gyrophora kisovana* A.Z. ahlbr. apud Asahina. Journ. Jap. Bot. 7:326 (1931).

Chemistry: gyrophoric acid.

Specimen examined:

JIANGXI: Mt. Lu shan, Xianrendong, alt. 850m, on rocks, 12/2/ 1965, Wei JC, no. 3179.

Literature record: Shanxi (Sato M, 1981, p.64).

13. **Umbilicaria nanella** Frey and Poelt (Pl. IV-d; Fig.4)

in Poelt, Khumbu Himal 6/3:425 (1977).

Chemistry: gyrophoric acid.

Specimens examined:

SHAANXI: Mt. Taibai shan, Baxiantai, on rocks, 3/VI/1963, collected by Wei JC, and in the same localities, alt. 3620m, Wei et al. nos. 2784, 2785; Paomaliang, on rocks, alt. 3600—3660 m, 30—31 /VI/ 1963, Ma and Zong, nos. 239, 395. YUNNAN: Dali, Mt. Cangshan, Zhonghefeng, on rocks, alt. 3970m, 3/1/ 1965, Wei JC, no. 2824.

Literature record: Sichuan(Poelt, 1977, p.425), Yunnan(Paulson, 1928, p.317; Zahlbr. 1930, p. 138), Xizang (Paulson, 1925, p. 193; Wei JC and Jiang YM, 1981, p.l 146, and 1986, p.5).

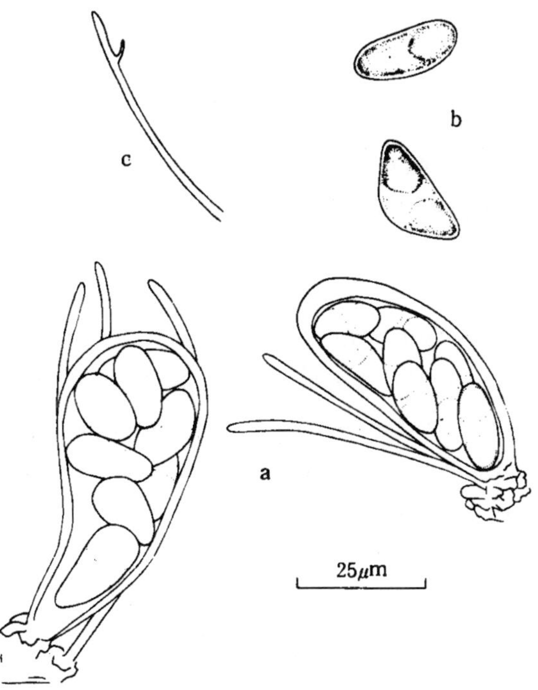

Fig. 4. *Umbilicaria nanella*

a. asci with ascosporcs and paraphyses; b. ascospores. c. simply branched paraphysis.

14. **Umbilicaria proboscidea**（L.）Schrad.

Spicil Fl. Germ. 1: 103（1794）.

=*Lichen proboscideus* L., Sp. Pl. 1150（1753）.

Chemistry: gyrophoric and umbilicaric acids.

Specimens examined:

JILIN: Mt. Changbai shan,north slope, alt. 2120m, on rocks, 13/VIII/ 1983, Wei JC and Chen JB, no. 6876-（1）. YUNNAN: Dali, Mt. Cang shan, Zhonghefeng, on rocks, alt. 4000m, 3 /I / 1965,Wei JC, nos. 2814, 2820, 2829. XIZANG: Nyalam, alt. 3850m, 12/VI/ 1966, collected by Wei JC and Chen JB.

Literature record: Yunnan（Hue, 1900, p.118; Zahlbr. 1930, p. 138）, Xizang（Wei JC and Chen JB, 1974, p.177; Wei JC and Jiang YM, 1981, p. 1147: pr. p., and 1986, p.99）.

15. **Umbilicaria pseudocinerascens** Wei and Jiang sp. nov.（Pl.IV-a, b, c; Fig. 5）

Umbilicariae cinerascentis affinis, a qua imprimis differt paginis inferioribus thallorum haud atris, versus peripheriam lixiviis prope umbem rufis verruculosis, sine rhizina vel versus marginem interdum cum rhizinis singularibus, apotheciis numerosis cum concentricis gyris in quibus fissuris perpendicularibus.

Thallus acidum gyrophoricum et umbilicaricum continens.

Typus: YUNNAN: Dali, Mt. Cang shan, Zhonghefeng, in rupibus, 3/1/ 1965, alt. 3850m, Wei JC, no. 2827（holotypus in HMAS-L et isotypi in UPS, H）, 2829-（3）.

Thallus monophyllous polyphyllous, of medium size, about 1.2—3.2cm in diametre, coriaceous, more or less waved, on the upper surface rough, not shining, brown, blackish to almost black, sooty, but usually dark grey owing to its often being covered with a pruina around the umbilicus more or less ridged, or reticulated, in periphery more or less elevated, lacking soredia or isidia; on the lower surface smooth, without rhizine or occasionaly with single rhizine, in some parts micro-papillate, something like those in *Umbilicaria rigida*, near the umbilicus often light flesh-coloured to dark brown, near the periphery dark brown to almost black and sooty, shiny, covered with a very thin pruina, with more thick umbilicus of 5—9mm, rarely 2—3mm in diametre.

Thallus (205) 333-（317）358μm thickt necral zone about 7μm thick; upper cortex brown, scleroplechtenchymatous, 14—27μm thick; algae layer 29—52μm thick; alga cell 9—11μm in diameter; medulla 103—187μm, more or less jellied parallel mycelia colourless, 27—98μm, lower cortex brown, scleroplechtenchymatous, 22-31μm thick.

Apothecia common, distributed near the periphery of thallus, without gyrus in juvenile, and concentric in large discs with perpendicular to disc gyrus fiscures; epithecium dark brown, 24—32μm thick; hymenium brownish, 29—37μm thick; hypothecium brownish, 12—15μm asci clavate, immature, 34—37×10—17μm.

Chemistry: gyrophoric and umbilicaric acids.

Specimens examined:

YUNNAN: Dali, Mt. Cang shan, Zhonghefeng, alt. 3850—3970m, on rocks, 3/I/ 1965, Wei, nos. 2815, 2821, 2829-（3）. All specimens examined are preserved in HMAS-L.

16. **Umbilicaria thamnodes** Hue

Nouv. Arch. Mus. Hist. Nat. 4（2）: 121（1900）.

Literature record: Yunnan（Hue, 1887, p.24, 1889, p. 170, and 1900, p. 121; Zahlbr. 1930, p. 138）, Xizang（Wei JC and Chen JB, 1974, p. 177, cited as *U. esculenta*; Wei JC and Jiang YM, 1981, p.l 146, p. 100）.

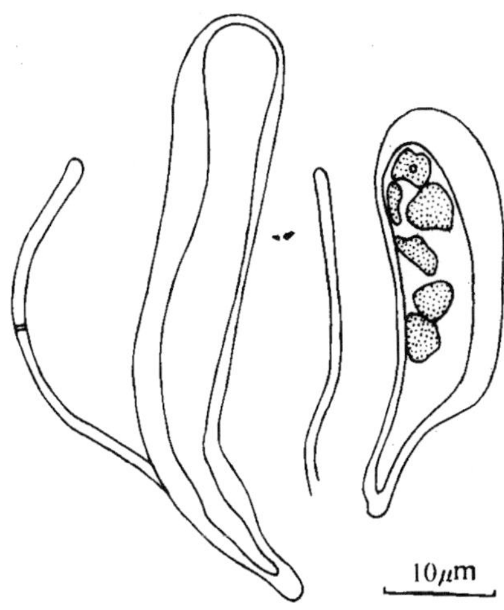

Fig. 5. *Umbilicaria pseudocincrascens* asci with paraphyses and immature ascospores.

16.1. f. **minor** Hue

Nouv. Arch. Mus. Hist. Nat: 4（2）: 122（1900）, tab.V, Fig. 3. Zahlbr. in Handel-Mazzetti, Symb. Sin. 3: 130（1930）.

Taxonomic revision of this species and its infraspecific taxa are still required.

17. **Umbilicaria torrefacta**（Lighft.）Schrad.

Spicil. Fl. German 1: 104（1794）.

=*Lichen torrefactus* Lightf., Fl. Scotica 2:862（1777）.

Chemistry: gyrophoric and umbilicaric acids.

Specimen examined:

JILIN: Mt. Changbai shan, south slope, Fusong county, on rocks in tundra zone near Laohubei, collected by Wei JC and Chen JB in 1983.

New to China.

18. **Umbilicara trabeculata** Frey and Poelt

in Poelt, Khumbu Himal 6/ 3: 429（1977）.

Chemistry: gyrophoric acid.

Specimens examined:

SHAANXI: Mt. Taibai shan, Huangcisi, alt. 2240m, on rocks, 5/VI/ 1963, Wei JC et al. no. 2548; Yuewangchi, alt. 3080m, on rocks, 6/ VIII/ 1963, Ma and Zong, no. 379. XIZANG: Zham, Quxam, alt. 3500m, on rocks, 19 / V / 1966, Wei JC and Chen JB, no. 887; alt. 3800—3910m, 12,13/VI/ 1966, Wei JC and Chen JB, nos. 1487, 1522; Nyalam, collected by Wei JC and Chen JB at about 3700m, on rocks, VI/ 1966（with apothecia）, and at about 3900m, 15/ VI / 1966, Wei JC and Chen JB, nos. 1610, 1618.Literature record; Xizang（Wei JC and Chen JB, 1974, p. 177（cited as *U. murina*）; Wei JC and Jiang YM, 1981, p.l 146, and 1986, p.101）.

19. **Umbilicaria tylorrhiza** Nyl.

Notiser ur Sallsk. Fauna et Flora Fennica Forhandl. 8: 122（1866）.

=*Gyrophora tylorhiza*（Nyl.）Nyl. apud Hue, Revue de Botan.5, 1886-87,14 et 163.

Chemistry: gyrophoric acid and a yellow pigment.

Specimen examined:

YUNNAN: collected by Delavay in 1885, H-Nyl. 31523（1）.

During the examination of Nylander's specimen we found that a part with apothecia belongs to the present species, but another to *U. esculenta*.

Literature record: Yunnan（Hue, 1887, p.23 and 1900, p.115; Zahlbr. 1930, p.138; Llano, 1950, p.264, fig.7）.

20. **Umbilicaria vellea**（L.）Ach.

Vetensk. -Akad. Nya Handl. 15:101（1794）.

=*Lichen velleus* L., Sp.Pl.l 150（ 1753）.

20.1.f. **vellea**

Chemistry: gyrophoric acid.

Specimens examined:

HEBEI: Mt. Xiaowutai shan, on rocks of the mountain top, alt. 2800m, Wei JC, no. 2096. JILIN: Mt. Changbai shan, north slope near Xiao Tianchi, alt. 1700m, on rocks, 12/ VIII/ 1977, Wei JC, no. 2908; south slope, on the way from Hengshan station to Tianchi, alt. 2050m, on rocks, 31/VII/ 1983, Wei JC and Chen JB, nos. 6370-（1），6371. SICHUAN: Kangding, on rocks in alpine region between Kangding and Zheduo, collected by Potanin in 1893, in Elenkin- Lich. Fl. Ross. IV, 1904, no. 151（pr. p. = 15 -b-1）. XIZANG: Nyalam, alt. 3920m, on rocks, 13/VI/1966, Wei JC and Chen JB, no. 1541.

Literature record: Jilin（Chen et al. 1981, p. 158）.

20.2.f. **leprosa**（Schaerer）Zahlbr.

Cat. Lich. Univ. 4:740（1927）.

=*Umbilicaria vellea* var.*cinereorufescens* f.*leprosa* Schaerer, Enum.Critic.Lich. Europ1850, p.25.

Chemistry: gyrophoric acid.

Specimens examined:

HEBEI: Mt. Xiaowutai shan, on rocks of the mountain top, alt. 2800 m, Wei JC, 2096-1. JILIN: Mt. Changbai shan, nortb.slope, on the way to Longmenfeng near a small waterfall, on rocks, 13/VIII / 1983, collected by Wei JC and Chen JB.

New to China.

21. **Umbilicaria yunnana**（Nyl.）Hue

Nouv. Arch. Mus. Hist. Nat. 4（12）: 117（1900）.

= *Gyrophora yunnana* Nyl. Bull. Soc. Bot. France 34:23（1887）.

Type: China, Yunnan super truncus arborum in silvis Hoang-lipin supra Tapintzc 2000 m.s.m.

2.V / 1885, R P Delavay 1600（H-Nyl.31740, holotype !）.

Chemistry: gyrophoric acid.

Specimens examined:

YUNNAN: Lijiang, Mt. Yulong shan, alt. 3000m, Zhao JD & Chen YB, nos. 4009, 4042, 4048, 4449; Mt. Xiang shan by the Lijiang town, on Quercus sp. alt. 2400-2600m, 24/XII/ 1964, Wei JC, nos. 2717,2718, 2719, 2720, 2721, 2722, 2723, 2724, 2725, 2726, 2727, 2728, 2729, 2749; Weixi county, 19/VI/ 1958, Han & Chen, no. 5226; Mt. Gaoligong shan, Baihualing tree farm, alt. 1750-2500m, 8-9/XII/1980, Jiang YM, nos. 673-（2），710, 719, 737,748,764.

Literature record: Yunnan（Hue, 1887, p.23 & 1889, p.170 & 1891, p.30 & 1900, p.117; Nyl. 1887, p. 135; Zahlbr. 1927, p.742 & 1930, p.138 & 1932, p. 496; Ras. 1952, p.81; Lamb, 1963, p.266; Wei JC, 1981, p.89）.

Subgenus **Actinogyra**（Schol.）Schade ex Mot.
Flora Polska, Porosty 4（2）: 77（1964）.
=*Actinogyra* Schol., Nyt Mag. Naturvid. 75:28（1934）.
Type: **Umbilicaria muehlenbergii**（Ach.）Tuck.

22. **Umbilicaria muehlenbergii**（Ach.）Tuck.

Enum. N. Am. Lich. p.55,1845.

=*Gyrophora muehlenbergii* Ach., Lich. Univ. 227（1810）.

=*Actinogyra muehlenbergii*（Ach.）Schol., Nyt Mag. Naturvid. 75:28（1934）.

Chemistry: gyrophoric acid.

Specimens examined:

NEIMONGOL: Da Hinggan Ling, Mangui, 25 km., in the south east from Mangui, on rocks, 13 / IX/ 1977, Wei JC, no. 3400; HEILONGJIANG: Gulian, near railway station Xilinji, on rocks, 5/IX/ 1977, Wei JC, nos. 3278, 3296.

Literature record: Neimongol（Chen et al.1981, p. 158）, Heilongjiang（Wei JC, 1981, p.90; Chen et al. 1981, p. 158）.

Subgenus **Omphalodiscus**（Schol.）Schade ex Wei & Jiang comb. nov.
Schade, Nova Acta Leopoldina, n.f.17: 202（1955）.（comb. inval.）
=*Omphalodiscus* Schol. Nyt Mag. f. Naturvid. 75:23（1936）.
Type: **Umbilicaria decussata**（Vill.）Zahlbr.

23. **Umbilicaria decussata**（Vill.）Zahlbr.

Cat. Lich. Univ. 8:490（1942）.

=*Lichen decussatus* Vill., Hist. Flant. Dauphine 3:964, pl. 55（1789）.

Type: As illustrated by Villars（1789）, from France, Dauphine（Hautes Alpes）.

=*Omphalodiscus decussatus*（Vill.）Schol., Nyt Mag. Naturvid. 75.23（1934）.

Chemistry: gyrophoric acid.

Specimens examined:

JILIN: Mt. Changbai shan, north slope near Tianchi, alt. 2400m, on rocks, 14/VIII/ 1977, Chen XL, no. 4825（as *U. ieiocarpa* cited by Chen et al. 1981）; near Tianwenfeng, alt. 2400m, 12/VII/ 1980, Hu YC, no. 1605.

New to China.

24. **Umbilicaria formosana** Frey

Hedwigia 71:115（1931）.

Type: China, Taiwan, Mt. Niitaka region, S.Sasaki（TNS-holotype!）

=*Omphalodiscus formosanus*（Frey）Schol., Nyt Mag. Naturvid. 75: 24（1934）.

=*Omphalodiscus krascheninnikovii* var. *formosanus*（Frey）Llano, Monogr.Umbilicariaceae, 88（1950）.

Chemistry: gyrophoric acid.

Specimens examined:

JILIN: Mt.Changbai shan, north slope, near Tianwenfeng, alt. 2600m, on rocks, 14/VIII/ 1977, Wei JC, no. 2987, & alt. 2400 m, 12/VII/1980, Hu YC, no. 1604. SHAANXI: Mt. Taibai shan, Baxiantai, on rocks, 3-4/VI/ 1963, Wei JC et al. no. 2718-（2）. YUNNAN: Dali, Mt. Changbai shan, Zhonghefeng, alt. 3900-3970m, on rocks, 3/I/1965, Wei JC, nos. 2816, 2828, 2829-（2）, XIZANG: Nyalam, alt. 3830m, on rocks, 12/VI/1966, Wei JC & Chen JB, no.1437.

Literature record: Shaanxi（Wei JC, 1981, p.90）, Taiwan（Frey E, 1931, p.115; Zahlbr. 1932, p.491 & 1933, p.49; Sato,1950, p. 170; Lamb, 1963, p.727; Wang & Lai, 1973, p.97）.

25. **Umbilicaria krascheninnikovii**（Savicz）Zahlbr.

Cat. Lich. Univ. 10:405（1939）.

=*Gyrophora krascheninnikovii* Savicz Bull. Jard. Imp. Bot. Pierre Grand 14:117（1914）.

Type: Siberia, "Kamczatka, supra massam sulphuream ad rupes vulcanicas montis（sopkae）Krascheninnikovii（in ripis lacus Kronontzkoje）abundanter lectae, 1909, V.P. Savicz 6412（13）"（LE, not seen）.

=*Omphalodiscus krascheninnikovii*（Savicz）Schol' Nyt Mag. Naturvid. 75: 24（1934）.

Chemistry: gyrophoric add.

Specimens examined:

JILIN: Mt. changbaishan, South slope, Changbai county, Mt. Hongtou shan, alt. 1880m, on rocks, 28/7/1983, Wei & Chen no. 6325; on the way from Heng shan station to Tianchi, on rocks, Wei & Chen, no. 6370; north slope, near Tianwenfeng, on rocks, 14/8/1977, Wei J C, no. 2982; alt. 2100 m. on rocks, 13/8/1977, Wei no. 2942-（1）; alt. 2120 fm, 13/8/1983, Wei & Chen, nos. 6876. XIZANG: Nyalam, Jiangning, on top of the mountain, alt. 4450 m, on rocks, 16/6/1966, Wei & Chen, no. 1669. SHAANXI: Mt. Taibai shan, Baxiantai, on rocks, 3/6/1963, Wei JC, et al. G.B.

Literature record: Jilin（Wei, 1981, p.90; Chen et al. 1981, p. 158）.

During our reexamination of the type specimen of *Gyrophora hultenii* Du Rietz and some specimens under the names of *Oprophora polaris* Schol., *Umbilicaria krascheninnikovii*（Savicz）Zahlbr. and *U. formosana* Frey respectively we found that either they are the same polymorphous species or two, three or more different taxa, and therefore they require revision in the future on the basis of reexamination of all the type specimens of above mentioned species.

26. **Umbilicaria spodochroa** Ehrh. ex Hoffm.

Deutschl. Fl. 3:113（1796）.

=*Lichen spodochrous* Ehrh. Crypt. Exs. no.317（1795）. Nomen nudum.

=*Omphalodiscus spodochrous*（Ehrh. ex Hoffm.）Schol., Nyt Mag. Naturvid. 75: 26（1934）.

Type: Ehrhart, Crypt. Exs. no.317（not seen）.

Chemistry: gyrophoric acid.

Specimen examined:

SICHUAN: On rocks in alpine region between Kangding and Zheduo, collected by Potanin（1893）, in Elinkin7s exs. "Lich. Fl Ross." IV, 1904, no. 151（pr. p. = 151-a, 151-b-2, 151 -c）, as *Gyrophora spodochroa*（Ehrh.）Ach.（UPS!）.

Literature record: Sichuan (Elenk. 1904, p.85), Yunnan (Hue, 1887, p.23), Hubei (Muell. Arg. 1893, p.235; Zahlbr. 1930, p. 138).

27. **Umbilicaria virginis** Schaerer

Biblioth. Univ. Geneve. 36:153 (1841).

=*Omphalodiscus virginis* (Schaerer) Schol., Nyt. Mag. Naturvid.75:25 (1934).

Literature record: According to Poelt (1977, p.431) a specimen of this species collected by Strachey from Xizang (Tibet) is preserved in BM, the infraspecific taxon to which it belonged was not indicated.

27.1. var. **virginis**

Chemistry: gyrophoric and norstictic acids.

Specimen examined:

XINJIANG: Mt. Tian shan, Wang, no 946.

Literature record: Xinjiang (Wang XY 1985, p.344).

27.2. var. **lecanocarpoides** (Nyl.) Wei & Jiang, comb. nov. (P1.I,V)

=*Umbilicaria lecanocarpoides* Nyl., Flora 43:418 (1860).

Chemistry:gyrophoric acid.

Specimen examined:

SICHUAN: Kangding, Daofu, southern mountain, Mt.Yara, north glacier valley, on granites, alt.4700m, 29/ VIII/ 1934, H.Smith, 14006 (UPS!).

New to China.

Subgenus **Agyrophora** Nyl.

Flora 247 (1878).

Type: **Umbilicaria leiocarpa** DC.

[*Umbilicaria leiocarpa* DC., in Lam. & DC., Flore Franc. ed.3.3:410 (1805).

The records under this name given by Chen et al. (1981, p. 158, specimens: Chen nos. 4824, 4825) were based on erroneous identifications (see *U. krascheninnikovii*, and *U. decussata* respectively).]

28. **Umbilicaria lyngei** Schol. (Pl.II:c, d)

Nyt Mag. Naturv. 75: 19 (1934).

= *Agyrophora lyngei* (Schol.) Llano, Monogr. Umbilicariaceae, 57 (1950).

Chemistry: gyrophoric and norstitic acids.

Specimen examined:

JILIN: Mt. Changbai shan, north slope, alt. 2400m, on rocks, 14/VIII/1977, Wei JC, no. 2983. SHAANXI: Mt. Taibai shan, Baxiantai, alt. 3900m, on rocks, 3/VI/1963, collected by Wei JC; Wen-gongmiao, on rocks, alt. 3380m, 29/VII/1963, Ma & Zong, no. 296.

New to China.

29. **Umbilicaria rigida**（Du Rietz）Frey

Hedwigia 71:117（1931）.

=*Gyrophora rigida* Du Rietz, Arkiv for Bot 19（12）: 3-4（1926）.

=*Agyrophora rigida*（Du Rrietz）Llano. Monogr. of the *Umbilicariaceae*, 58（1950）.

Chemistry: no lichen substances.

Specimen examined:

SHAANXI: Mt. Taibai shan, Baxiantai, on rocks, 3/VI/1963, collected by Wei JC.

New to China.

Lasallia Mérat

Mérat, Nouv. Fl. Paris ed. 2 (1):202（1821）.

Type: **Lasallia pustulata**（L.）Merat（= *Lichen pustulatus* L.）

Key to species from China

1. Apothecia absent or very poor and single if occasionally present.
    2. Lower surface of the thallus black or purple-brown, only black about the umbo.
        3. Lower surface black ………… *L. asiae-orentalis*
        3.Lower surface purple-brown. about umbo black ………… *L. xizangensis*
    2. Lower surface of the thallus pale brown.
        4. Upper surface sorediate;pustules fenestrate ………… *L. pertusa*
        4. Upper surface isidiate; pustules whole………… *L. sinorientalis*
1. Apothecia always present and rich.
    5. Lower surface of the thallus dark brown or purple-brown, black about the umbo.
        6. Upper surface of the thallus brown, neither granulated nor reticulated; lower surface dark brown ………… *L. daliensis*
        6. Upper surface grey brown to purple brown, covered with thick pruina, weakly granulated or reticulated; lower surface brownish and dark brown colours ………… *L. daliensis* var. *caengshanensis*
    5. Lower surface black brown to black
        7. Apothecia gyrate; on barck ………… *L. mayebarae*
        7. Apothecia of the leiodiscus type; on rock.
            8. Apothecia adnate; upper surface of the thallus not covered with pruina but with recurved Squamules; lower sur face black, grossly granulated ………… *L. pensylvanica*
            8. Apothecia elevated, more or less stipitate in appearance; upper surface of the thallus covered with pruina and applanate squamules; lower surface dark brown, never black, delicately granulated ………… *L. rossica*

Subgenus **Lasallia**

Type: **Lasallia pustulata** （L.） Mérat

1. **Lasailia asiae-orientalis** Asahina

Journ. Jap. Bot. 35:99 （1960）.

1.1.var. **asiae-orientalis**

Type: TAIWAN: Mt. Ali shan, 25/XII/1925,Y. Asahina, no. F-377（TNS!）. *Umbilicaria pustulata* non Hoffm. sensu, Zahlbr. in Handel-Mazzetti, Symb. Sin. III:138 （1930）, in Feddes Repertorium 33:49 （1933）; Asahina, Journ. Jap. Bot. 7:103 （1931）.

Chemistry: gyrophoric acid and an unidentified substance in Rf-class 2 of solvent system C.

Specimens examined:

YUNNAN: Lijiang, on rocks, 11/XI/1958, Han SJ & Chen LY, no. 5091; near Yungbei & Yungning, Mt. Guamao shan, at about 3075m, 29/VI/1914, Hande-Mazzetti, no.627 （Handel "Mazzetti Iter Sinense 1914-18, Sumptibus Academiae Scientiarum Vindobonensis susceptum. No.3318）(UPS!）; Mt. Yulong shan: Muzhugou, on rocks, alt. 2800m, 6/XII/1960, Zhao JD & Chen YB, no. 3885, & Ganhaizi, alt. 3020m, 18/XII/1964, Wei JC, no. 2617; Yuhucun, alt. 2820m, 20, 22/XII/1964, Wei JC, nos. 2663 （1） & 2714; Dali: Mt Cang shan, alt. 2050m, 27/XII/1964, Wei JC, no. 2750. XIZANG: Zham, alt. 3500m, 21/V/1966, Wei JC & Chen JB, no. 1081; alt. 3650m, VII/1966, Wei JC, & Chen JB, no. 510; TAIWAN: Mt. Ali shan （see type）.

Literature record: Sichuan （Zahlbr. 1930, p. 138）, Yunnan （Zahlbr. 1930, p.138: Handel-Mazzetti, no.3318; Wei JC, 1982, p.20）, Xizang （Wei JC, 1982, p.20; Wei JC & Jiang YM, 1981, p. 1146, & 1986, p. 103）, Taiwan （Zahlbr. 1933, p.49; Sato, 1937, p.298; Lamb, 1963, p.283; Wang & Lai, 1973, p.91 & 1976, p.228; Wei JC, 1982, p.20）.

1.2. var. **fanijingensis** Wei

Acta Mycol. Sin. 1:20 （1982）.

Type: GUIZHOU: Mt.Fanjing shan, Jiandaoxia, 23/IX/1963, Wei JC, no. 830 （HMAS-L）.

Chemistry: gyrophoric acid and an unidentified substance in Rf-class 2 of solvent system C.

Specimen examined: see type.

1.3. var. **major** Wei & Jiang

Acta Phytotax. Sin. 20:500 （1982）.

Type: XIZANG: Zham, Quxam, on rocks, alt.3500m,19/V/1966,Wei JC & Chen JB, no.884 （HMAS-L）.

Chemistry: gyrophoric acid.

Specimen examined: see type.

2. **Lasallia daliensis** Wei
Acta Mycol. Sin. 1:21（1982）.

2.1.var. **daliensis**
Type:YUNNAN: Dali, Mt. Cang shan, alt. 3300m, 29/XII/1964, Wei JC， no, 2778（holotype, HMAS-L）.
Chemistry: gyrophoric acid and an unidentified substance in Rf-class 2 of solvent system C.
Specimen examined:
Lijiang, Mt. Yulong shan, Ganhaizi, alt. 3000 m, 13/XII/1964, Wei. JC, no. 2421.

2.2. var. **caeongshanensis**（Wei）Wei, comb. nov.
=*Lasallia caeonshanensis* Wei JC, Acta Mycol. Sin. 1（1）:22（1982）.
Type: YUNNAN:Dali, Mt.Cang shan, alt. 3300m, on rocks, 29/XII/1964, Wei JC, no. 2758（HMAS-L）.
Chemistry: gyrophoric acid and an unidentified substance in Rf-class 2 of solvent system C.
Specimen examined: see type.

3. **Lasallia pensylvanica**（Hoffm.）Llano
Monogr. of the Umbilicariaceae 42（1950）.
= *Umbilicaria pensylvanica* Hoffm.,Descr. Adumb. Pl. Lich. 3（4）: 5-6（1801）.
Chemistry: gyrophoric acid.
Specimens examined:
JILIN: Mt. changbai shan, south slope, Mt. Hongtou schan: alt. 1950m, 5/VIII/1980,Hu YC, no. 2149; alt. 1970m, 16/VIII/1982, Hu YC, nos. 3386, 3432; alt. 1880m, 28/VII/1983, Wei JC & Chen. JB no.6250; alt. 1980—2000m, Hu YC, nos. 4218，4230,4232.
Literature record: Heilongiang（Chen et al. 1981, p. 158），Taiwan（Wang & Lai, 1976, p.228）.

4. **Lasallia pertusa**（Rassad.）Llano
Monogr. of the **Umbilicariaceae**, 48（1950）.
=**Umbilicaria pertusa** Rassad., Comp. Rand. Acad. Sci. URSS, 14（A）: 348（1929）， fig. 1-2.
=**Lasallia pensylvanica** var. **pertusa**（Rassad.）LIano, Hvalrad Skr.48: 122（1965）.

4.1. f. **pertusa**
Chemistry: gyrophoric acid.
Specimens examined:
XINJIANG: By the river near Buljin tree farm, alt. 1300 m, 13/IX/1982, Zhao CF no. 2458. XIZANG: Nyalam, on rocks, alt. 3700—3720m, 29/IV & 11, 15/V/1966, Wei JC & Chen JB, nos. 92, 1429, 1648.
Literature record: Xizang（Wei JC & Chen JB, 1974, p. 177; Wei JC, 1981, p.90; Wei JC & Jiang YM, 1981, p.l146 & 1986, p. 102）.

4.2. f. **squamulifera** Wei & Jiang
Acta Phytotax. Sin. 20:499（1982）.
Type: XIZANG: Lasa, on rocks, alt. 3750m, 3/IV/1966, Wei JC & Chen JB, no, 15（holotype, HMAS-L）; Gyrong, alt. 2880m, 14/VI/1975, Zong YC, no. 112(paratype, HMAS-L).
Literature record: Xizang（Wei JC & Jiang YM, 1986, p. 102）.

5. **Lasallia rossica** Dombr.
Nov. Syst. Plant, non vascul. 15; 180（1978）.

Type: USSR, the Buryat ASSR, VII/1965, collected by V. M. Burkova (isotype, LE!).

Chemistry: gyrophoric acid.

Specimens examined:

JILIN: Mt. Changbai shan, south slope, on top of Mt. Hongtou shan in Changbai county, alt. 1950m, 5/VIII/1980, Hu YC, nos. 2185, 2239; alt. 1960—1970m, 16/VIII/1982, Hu YC, nos. 3433, 3438, 3473; alt. 1980m, 28/VII/1983, Hu YC, no. 4216; alt. 1880m, 28/VII/1983, Wei JC & Chen JB, no. 6251.

New to China.

6. **Lasallia sinorientalis** Wei

Acta Mycol. Sin. 1:23 (1982).

Type: JIANGXI: Mt. Lu shan, Guling, alt. 1170 m, 14/II/1965, Wei JC, 3246 (HMAS-L).

=*Umbilicaria fokiensis* Merril ex Llano. Monogr. of the Umbillicariaceae, 31(1950)(Nom. inval., Art. 34).

Chemistry: gyrophoric acid and an unidentified substance in Rf - class 2 of solvent system C.

Specimens examined:

SHAANXI: Huxian county, Mt. Guangtou shan, alt. 2700m, 2/VII/1964, Wei JC, no. 77. JIANGXI: Mt. Lu shan, 1/IV/1960, Zhao JD et al. no. 472; alt. 1150—1250m, 11—12, 14/II/1965, Wei JC nos. 3115, 3117, 3121, 3246. ANHUI: Mt. Huang shan, near Banshansi, alt. 800m, 24/VIII/1962, Zhao JD et al. no. 5913(p.p.); near Yupinglou, on rocks, 2/XI/1980, Wei JC, no. 3766. FUJIAN: IV-VI/1905, Dunn, no. 3938 (FH!).

Literature record: Shaanxi, Jiangxi (Wei JC, 1982, p.24), Anhui (Lu, 1959, p. 176), Fujian (Llano, 1950, p. 31, cited as *Umbilicaria fokiensis* Merril; Lamb, 1963, p.727; Wei JC, 1982, p.23).

7. **Lasallia xizangensis** Wei & Jiang

Acta Phytotax. Sin. 20:500 (1982).

7.1 var. **xizangensis**

Type: XIZANG: Zham, Quxam, on rocks, alt. 3250m,16/V/1966, Wei JC & Chen JB, no.881 (HMAS-L).

Chemistry: gyrophoric acid.

Literature record: Xizang (Wei JC & Jiang YM,1986, p. 104).

7.2 var. **acuta** Wei et Jiang

Ibid.20 (4) :500 (1982).

Type: XIZANG: Zham, Quxam, alt. 3500m, 19/V/1966, Wei JC & Chen JB, no.888 (HMAS-L).

Chemistry:

Literature record: Xizang (Wei JC & Jiang YM, 1986, p. 104).

Subgenus **Pleiogyra** Wei

Acta Phytotax. Sin. 11:5 (1966).

Type: **Lasallia mayebarae** (Sato) Asahina

8. **Lasallia mayebarae** (Sato) Asahina

Joum. Jap. Bot. 35:101 (1960).

=*Umbilicaria mayebarae* Sato. Journ. Jap. Bot. 25: 168 (1950).

8.1 var. **mayebarae**

Umbilicaria pustulata non Hoffm., sensu Asahina, Journ. Jap. Bot. 7:102（1931）.

Type:TAIWAN: Mt. Ali shan, 9/III/1930, K.Mayebara（TNSI）.

Literature record: Taiwan（Asahina, 1931, p. 102 & 1960, p.101; Sato,1950, p. 168; Wang & Lai, 1973, p.91 & 1976, p.228; Ikoma, 1983, p. 109）.

8.2. var. **sinensis**（Wei）Wei（Fig. 6）

Bull. Bot. Res.1（3）:90（1981）.

=*Lasallia sinensis* Wei, Acta Phytotax. Sin. 11:5（1966）.

Type: YUNNAN: Lijiang, Mt. Yulong shan, Ganhaizi, on Pinus densata Masters, XII/ 1964, Wei JC, no. 2590（HMAS-L）.

Chemistry: gyrophoric acid and an unidentified substance in Rf-class 2 of solvent system C.

Specimen examined: see type.

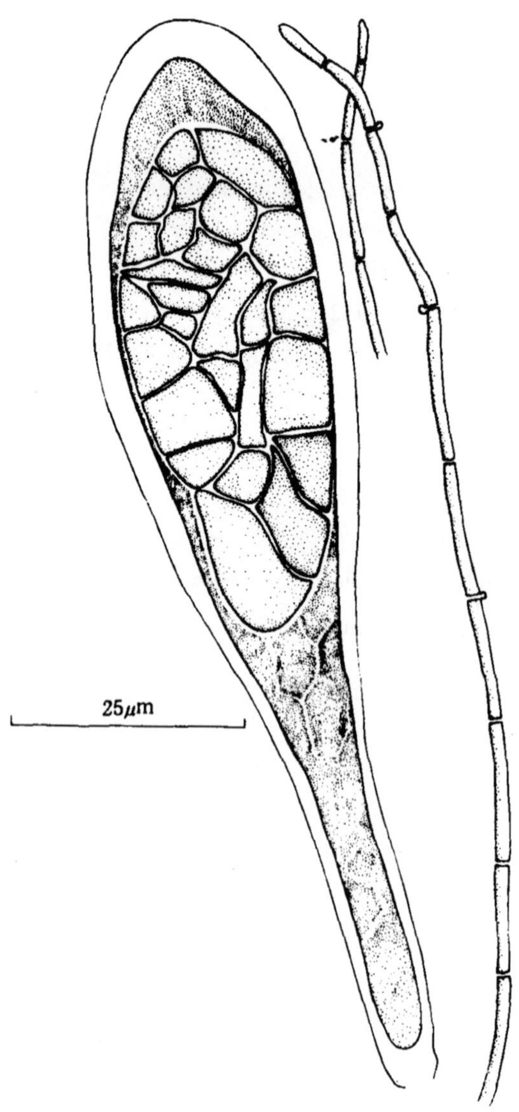

Fig. 6. *Lasallia mayebarae* var. *sinensis* asci with paraphyses and a large ascospore.

ACKNOWLEDGEMENTS

We arc very grateful to Professor T. Ahti, Professor D.L. Hawksworth, and Professor R. Santesson for reading the manuscript and giving valuable advice; and to Dr. N.S. Golubkova (LE) , to Dr. H. Kashiwadani (TNS), Professor N. Lundqvist(S), and Dr. Y. Maekinen(TUR), Dr. R. Moberg(UPS), and Dr. O.Vitikainen (H)for sending type specimens on loan. Our thanks are also due to Ms.Yuan Lancui for printing the photographs; and to Ms. Han Zhe-fang and Miss Zhu Xiangfei for inking the line drawing.

LITERATURE CITED

Asahina Y (1931) The Raiken's Soliolquy on Botanical Science, 35 (illust.). *Journ. Jap. Bot.* 7 (4) : 102

Asahina Y (1936-40) Mikrochemischer Nachweis der Flechtenstoffe. *Journ. Jap. Bot.* vols. **12,13,14,15,16**

Asahina Y (1960) Lichenologische Notizcn (§ 160-163). *Journ. Jap. Bot.* 35 (4) :97

Cannon PF, Hawksworth DL & Shcrwood-Pike MA (1985) The British Ascomycotina An Annotated Checklist. Kew, CMI,pp.1- 302

Chen XL, Zhao CF,Luo GY (1981) A list of lichens in north-eastern China. *Journ. North-eastern Forestry Inst.* 3:127-135:4:150-160

Culberson CF (1969) Chemical and botanical guide to lchen products, pp. 1-628, Chapel Hill

Culberson CF (1972) Improved conditions and new data for the identification of lichen products by a standardized thin-layer chromatographic method. *J. Chromatogr.* 72:113-125

Elenkin A (1904) Lichenes Florae Rossiae et regiooum confinium orientaliuim Fasc. 1. *Acta Horti Petropolitanl* 19:1-52

Eriksson O, Hawksworth DL (1986,a) An alphabetical list of the generic names of ascomyctes. *Systema Ascomycetum* 5:3-111

Eriksson O, Hawksworth DL (1986,b) Outline of ascomycetes. *Systema Ascomycetum* 5:185-324

Eriksson O, Hawksworth DL (1987,a) An alphabetical list of the generic names of the ascomycetes. *Systema Ascomycetum* 6:1-109

Eriksson O, Hawksworth DL (1987,b) Outline of ascomycetes. *Systema Ascomycetum* 6:259-337

Frey E (1933) Umbilicariaceae. In Rabenhorst, Krypt-Fl. edn 2,9,4 (1) :203-424

Frey E (1936) Vorarbeiten zu einer Monographic der Umbilicariaceen. *Bericht. Schweiz Bot. Ges.* 45:198-230

Frey E (1949) Neue Beitrage zu einer Monographie des Genus *Umbilicaria* Hoffm. *Bericht. Schweiz. Bot. Ges.* 59:427-470

Frey E, Poelt J (1977) Die Gattung *Lasallia* (Flechten des Himalaya 13). Mit 5 Abbildungen ... Khumbu Himal 6/3:387-395

Henssen A (1970) Die Apothecienentwicklung bei *Umbilicaria* Hoffm. emend. Frey. Dtsch. Bot. Ges., Neue Folge, Nr.4: 103-126

Henssen A, Jahns HM (1973) [1974] Lichenes. Eine Einfuhrung in die Flechtenkunde. Stuttgart: Thieme

Hue AM(1887)Lichenes Yunnanenses a claro Delavay anno 1885 collectos, et quorum novae species a celeb.W. Nylander descriptae fuerunt, exponit Hue AM. *Bull. Soc. Bot. France* 34:16-24

Hue AM(1889)Lichenes Yunnanenses a cl. Delavay praesertim annis 1886-1887, collectos exponit Hue AM(1). series secunda. *Bull. Soc. Bot France* 36:158-176

Hue AM (1900) Lichenes Extra-Europaei. *Nouv. Archiv.du Mus. Paris*, ser. 4, 2:112 Hue AM (1907) Trois *Lichens nouveaux. Bull. Soc. Bot. France* 54:418

Ikoma Y (1983) Macrolichens of Japan and adjacent regions. pp.120, Japan, Tottori City

Jatta A(1902)Licheni Cinesi raccolti allo Shen-Si negli anni 1894-1898 dal rev. Padre Missionario Giraldi G. *Nuov. Giorn. Bot. Ital.*, ser. 2,IX: 460-481

Krog H (1973) On *Umbilicaria pertusa* Rassad. and some related lichen species. *The Bryologist* 76:550-554

Lamb IM (1948) New, rare or interesting lichens from the southern hemisphere I. *Lilloa* 14:225-230

Lamb IM (1963) Index Nominum Lichenum inter annos 1932 et 1960 divulgatorum New York

Llano G（1950）A Monograph of the Lichen Family Umbilicariaceae in the western hemisphere Navexos pp 1 -831, Washington

Llano G（1965）Review of Umbilicaria Hoffm. and the *Lasallias*. *Hvalradets Skr*. 48:112-124

（Lu DA）陆定安（1959）Notes on Chinese Lichens, 2.Umbilicariaceae. Acta Phytotax. Sin. 8:173-180

Moberg R 1982 Lavar. Stockholm pp. 1-237

Paulson（1925）Lichens of Mount Everest. J. Bot. 63:189-193

Paulson（1928）Lichens from Yunnan. Ibid. 66:313-319

Poelt J（1969）Bestimmungsschlussel Europaischer Flechten. 3301 Lchre, Vcrlag von J .Cramer Poelt J（1977）Die Gattung *Umbilicaria*（Umbilicariaceae).（Flechten des Himalaya 14). *Khumbu Himal* 6 / 3:397-435

Poelt J und Vezda A（1977）Bestimmungsschlussel Europaischer Flechten Erganzungsheft I. J.Cramer, FL-9490 Vaduz

Poelt J und Vezda A（1981）Bestimmungsschlussel Europaischer Flechten Erganzungsheft II. J.Cramer, FL-9490 Vaduz

Rassadina KA（1929）Une nouvellc espece d'*Umbilicaria* de Siberie. *Umbilicaria pertusa* sp. nov. *Compt. r. Acad. Sci. VRSS*14（A）:348-350

Räsänen V（1949）Lichenes novi IV. *Arch. Soc. Zool. Bot. Fernncae "Vanamon"* 3:78

Santesson R.（1944）Contributions to the Lichen Flora of South America. Arkiv For Botanik. Band 31A（7）: 15-18

Santesson R.（1984）The Lichens of Sweden and Norway. Stockholm and Uppsala pp. 1-333

Sato MM（1937）Enumeration lichenum Insulae Formosae（III). Journ. *Jap. Bot.* 13:595-599

Sato MM（1950）Notes on some remarkable Umbilicariae collected in Far Eastern Asia. *Journ. Jap. Bot.* 25:165

Sato MM（1952）Lichenes Khinganenses: or a list of lichens collected by Prof. T. Kira in the Khingan Range, Manchurica. *Bot. Mag.Tokyo* 65（769-770）: 172-175

Sato MM（1981）Notes on the cryptogamic flora of prov. Shanxi, North China（III).

Lichenes *Miscellanea Bryologica et Lichenologica* 2（3）:64-65

Scholander PF（1934）On the apothecia in the lichen family Umbilicariaceae. *Nyt Mag. Naturvid.* 75:1-3

Wang-Yang JR, Lai MJ（1973）A checklist of the lichens of Taiwan. *Taiwania* 8:83-104

Wang-Yang JR, Lai MJ（1976）Additions and corrections to the lichen flora of Taiwan. *Taiwania* 21:226

（Wang XY）王先业（1985）天山托木尔峰地区的地衣.天山托木尔峰地区的生物新疆人民出版社 pp.328-353

（Wei JC）魏江春（1966）A new subgenus of *Lasallia* Mer. em. Wei. *Acta Phytotax. Sin.* 11:1-8

（Wei JC）魏江春（1981）Lichenes Sinenses Exsiccati（Fasc. 1:1-50). *Bull. Bot. Research* 1:81-91

（Wei JC）魏江春（1982）Some new species and materials of *Lasallia* Mer.em. Wei. *Acta Mycol. Sin.* 1:19-26

Wei JC An enumeration of the lichens in China.（to be published）

（Wei JC,Chen JB）魏江春,陈健斌（1974）珠穆朗玛峰地区地衣区系资料.珠峰地区科考报告（生物与高山理）科学出版社 1974:174

Wei JC, Jiang YM（1981）A biogeographical analysis of the lichen flora of Mt. Qomolangma region in Xizang.

Proceedings of Symposium on QinghaiXizang（Tibet）plateau, pp 1145-1151, Beijing, China

（Wei JC, Jiang YM）魏江春，姜玉梅（1982）New materials for lichen flora from Xizang. Acta *Phytotax. Sin.* 20:496-501

（Wei JC, Jiang YM）魏江春,姜玉梅（1986）Lichens of Xizang. Science Press, Beijing, China

（Wei JC, Wang XY，Chen XL, et al.）魏江春，王先业，陈锡岭等（1982）Lichenes Officinalese Sinenses. Science Press, Beijing, China

（Wu JL）吴金陵（1981）Medicinal lichens in Qin Ling Mountain. *Acta Pharmaceutica Sinica* 16（3）:161-167

Yoshimura I（1971）The genus *Lobaria* of Eastern *Asia. Journ. Hattori Bot. Lab.* 34:231-364

Zahlbruckner A（1930）Lichenes. In: Handel-Mazzetti, Symb. Sin. 3:1-254

Zahlbruckner A（1932）Cat. Lich. Univ. 8:496

Zahlbruckner A（1933）Flechten der Insel Formosa. *Feddes Repertorium,* 33:49

中国地衣型子囊菌石耳科纲要

魏江春　姜玉梅
中国科学院微生物研究所
真菌地衣系统学开放研究实验室
北京

摘要：该纲要为东亚地衣型于囊菌石耳科研究的一部分。对于该科一些种的子囊盘进行了研究。

作者从保存在瑞典乌普萨拉大学标本室的中国四川标本中发现了一种具有发育良好的“茶渍型”子囊盘的石耳。但是在显微镜下发现该茶渍型子囊盘的果托虽然是由地衣体皮层所形成，但是其中缺乏藻细胞.在对该科其它种类检查后表明，绝大多数具有平盘型和柱盘型子囊盘的种类均为这一类型。

Henssen（1974）根据不同作者的观点归纳了五种子囊盘的类型，即蜡盘型，网衣型，茶渍型，双缘型，以及超网衣型.从广义上看，蜡盘型及网衣型只具固有盘壁，而茶渍型及双缘型则具体质盘壁。至于超网衣型实际应属于具有体质盘壁的茶渍型，而不属于具固有盘壁的网衣型。因此，对于这一类型子囊盘来说“超网衣型”改称为“亚茶渍型”则更为确切。

该文报道了四十六个分类单位，隶属于三十七个种，其中包括一个新种：拟灰石耳；隶属于十一种的十一个分类单位为中国新记录。对于某些分类群进行了分类学订正，并附有分种检索表。

关键词：子囊盘；石耳科；亚茶渍型；拟灰石耳

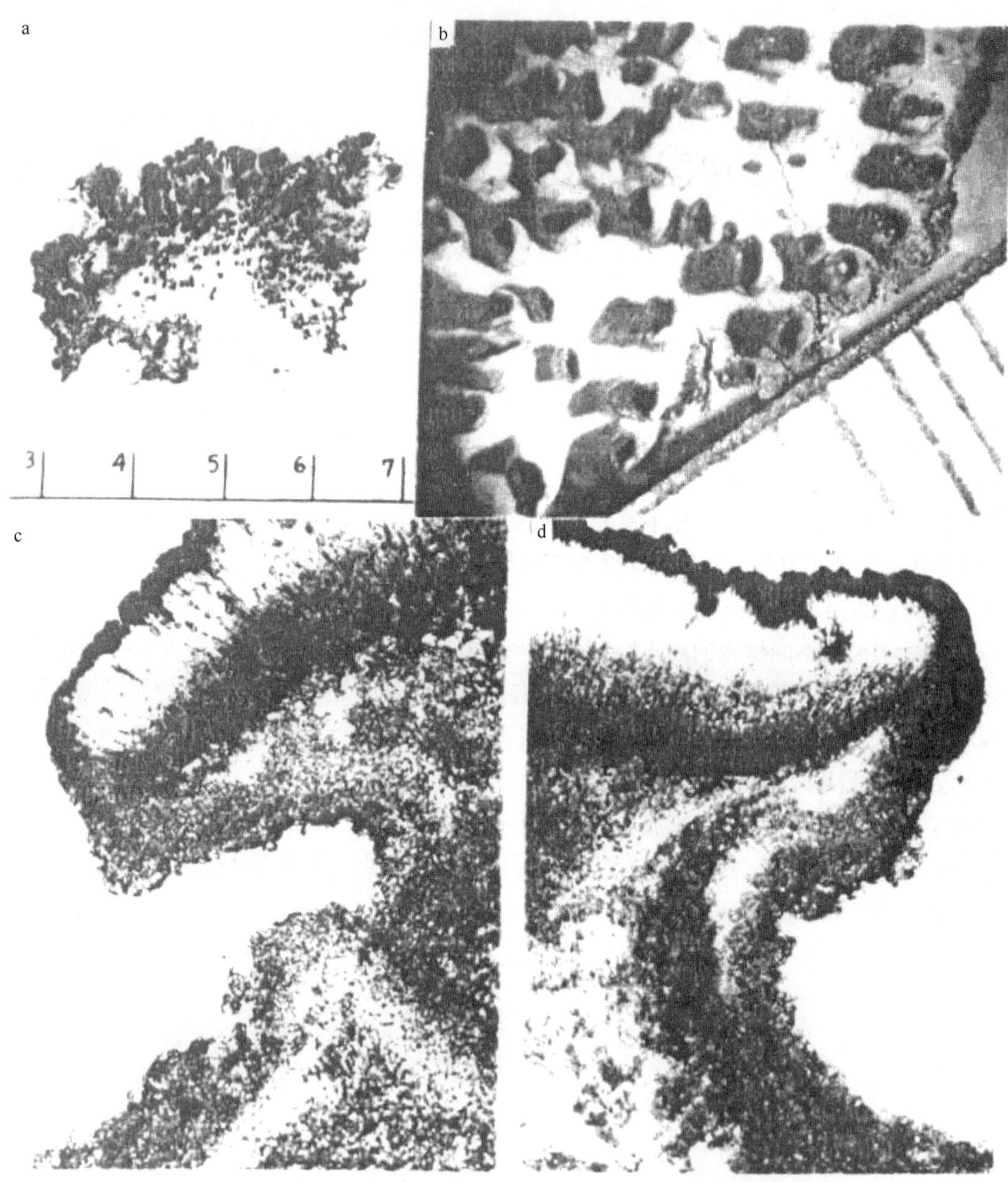

Pl. I. *Umbilicaria virginis* var. *lecanocarpoides.*（Smith H no. 14006, collected from Sichuan）

a. Thallus with apothecia, scale in cm; b. Thallus with apothecia, scale in mm; c-d. Section of apothecia（200 ×）.

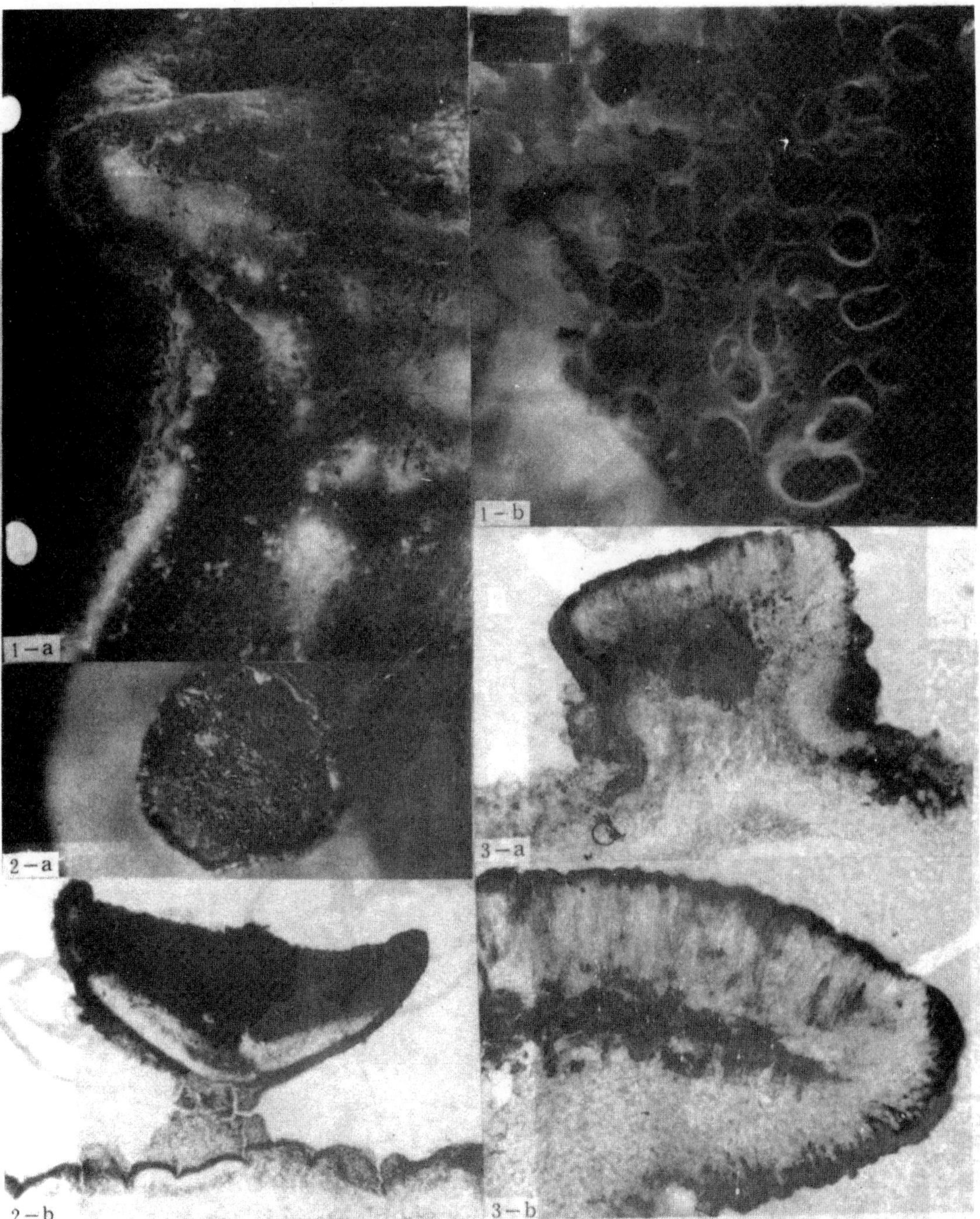

Pl. II. 1. *Umbilicaria virginis* var. *lecanocarpoides*（Smith H no.14006）

a. Apothecium（SEM. scale in 0.05μm）； b. The structure of the thalline exciple of apothecium, scale *in* o.5μm.

2. *Umbilicaria lyngei*（Wei JC collected from Mt. Taibai shan, Shaanxi）

a. Apothecium（30 ×）； b. Section of an apothecium（55 ×）.

3. *Umbilicaria leiocarpa*（Kurok. Exs. no.292 in TUR-20, from Japan）

a-b. Section of an apothecium（63×）.

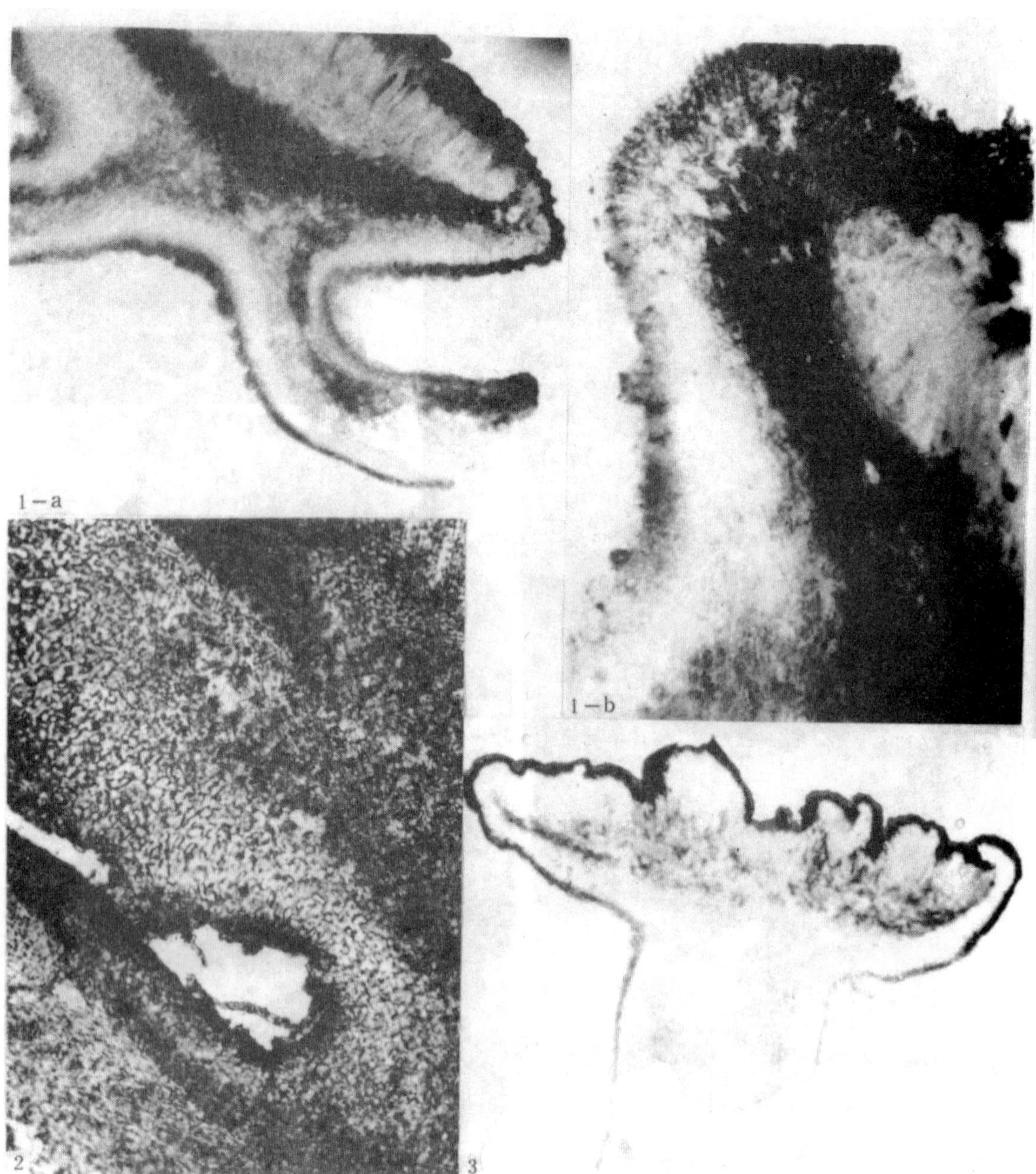

Pl. III. 1. *Lasallia pustulata*（Wei JC et al. collected from Sweden）Sections of apothecia:a. 63 ×; b. 160 ×
2. *Lobaria kurokawae*（collected from Shaanxi）Section of the thalline exciple of an apothecium（200 ×）.
3. *Umbilicaria cylindrica*（Hu YC no.2514, collected from Mt. Changbai shan, Jilin）Section of an apothecium with a long stalk（63 ×）.

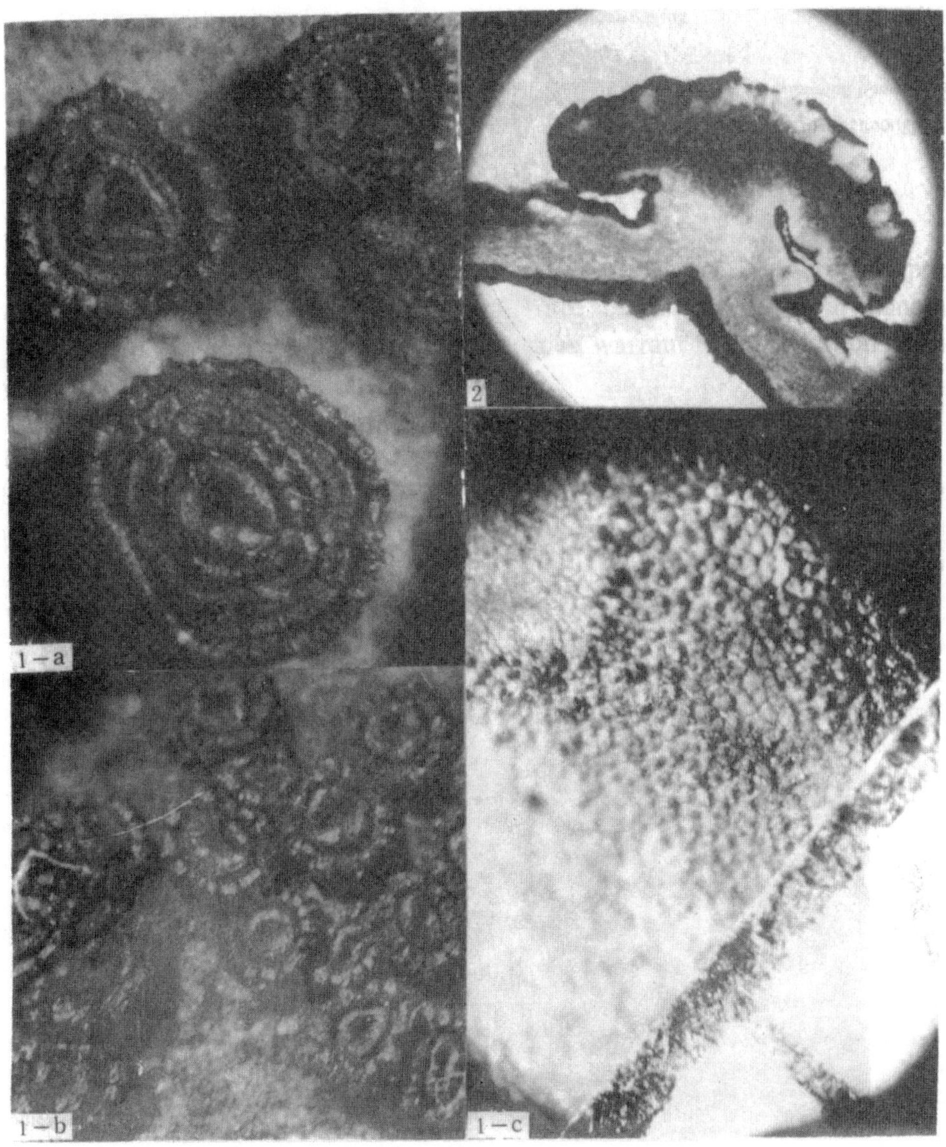

Pl. IV. *Umbilicaria pseudocinerascens*（Wei JC no.2827, collected from Yunnan）

a-b. Apothecia（30 ×）; c. Under side of the thallus with papillae.

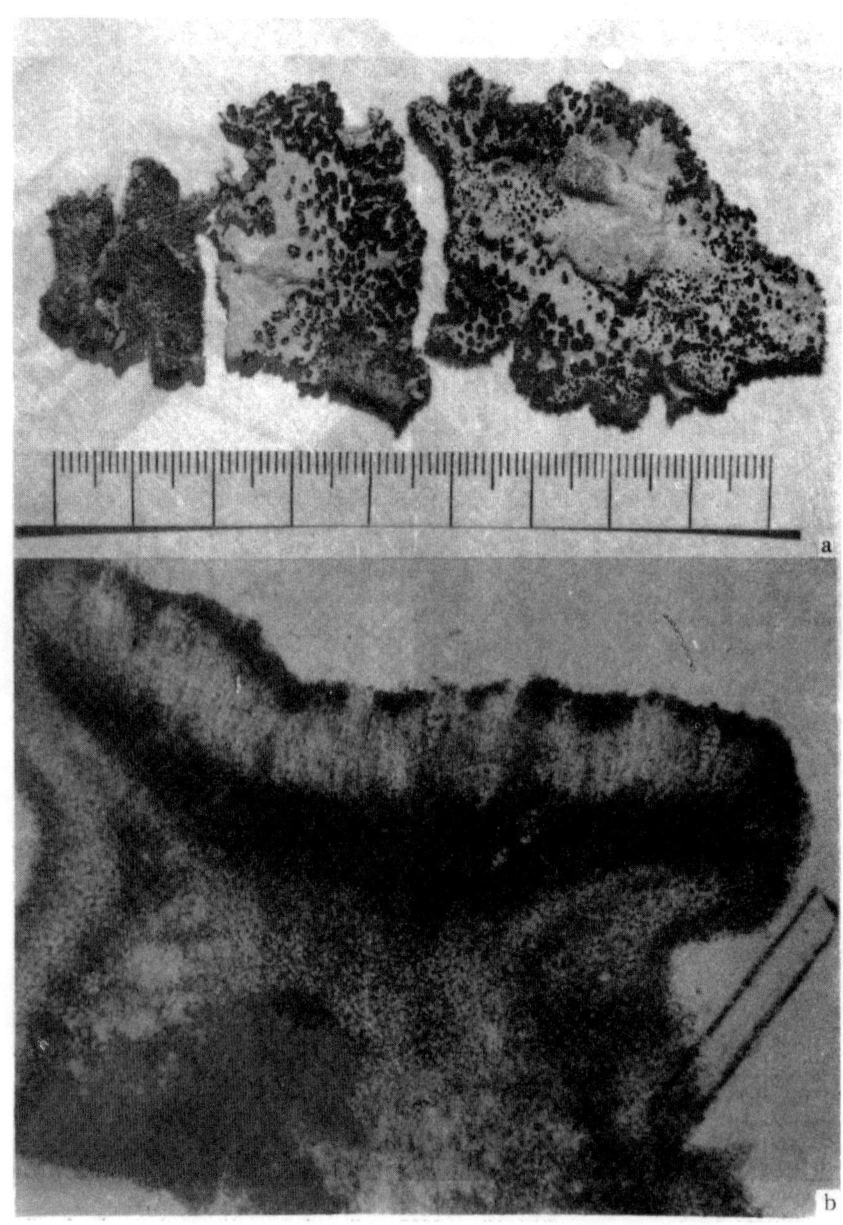

Pl. V. *Umbilicaria virginis* var. *lecanocarpoides*（Smith H no. 14006）

a. Thalli with apothecia（in colour）; b. Section of an apothecium with a thalline exciple（in colour）.

The status of the genus Llanoa Dodge and the delimitation of the genera in the Umbilicariaceae (Ascomycotina)

WEI Jiang-chun **JIANG Yu-mei**
Systematic Mycology & Lichenology Laboratory
Institute of Microbiology, Academia Sinica, Beijing

ABSTRACT: A brief history and the current situation of generic classification in the lichenized ascomycete family Umbilicariaceae is considered. A revised family with two genera has been recognized by most recent authors, but in recent years a monotypic genus *Llanoa* described by Dodge in 1968 has been accepted by some authors in addition to *Lasallia* and *Umbilicaria*. The type species of *Llanoa*, *Umbilicaria cerebriformis* Dodge & Baker [= *Llanoa cerebriformis* (Dodge & Baker) Dodge], is very close morphologically to *Umbilicaria decussata* (Vill.) Zahlbr, and the genus *Llanoa* Dodge therefore needed to be reexamined and its status clarified on the basis of the type material. A reexamination showed the type species of *Llanoa* to be a synonym of the well-known *Umbilicaria decussata*, and so the monotypic genus *Llanoa* is a synonym of the genus *Umbilicaria* Hoffm.In addition,a cluster analysis based on 35 species and 52 morphological characters favoured the revised concept of the family.

KEY WORDS: Ascomycotina; Umbilicariaceae; *Umbilicaria*;*Lasallia*;*Umbilicaria decussaia*; cluster analysis

INTRODUCTION

Eighty lichen species under the generic name "*Lichen*" were accepted by Linnaeus' in the "Species Plantarum" (1753). Seven of these species belong to the Umbilicariaceae and were transferred by Hoffmann (1789—1801) to a new generic name, *Umbilicaria*. The type species of the genus is *Umbilicaria exasperata* Hoffm, now recognized as *Umbilicaria proboscidea* (L.) Schrader. Acharius (1803, pp.100-110) proposed *Gyrophora* instead of the older name Umbilicaria for 15 species having gyrate apothecia, and placed the two species with a plane apothecial disc (*Umbilicaria pustulata* and *U. pensylvanica*) in *Lecidea* under the infrageneric taxon *Omphalaria* (loc. cit., pp.85-86). The genus *Umbilicaria*, as the single member of the family Umbilicariaceae, had been accepted by most lichenologists for a long time. It was extensively studied by Frey(1931,1933,1936, 1949,etc.)until 1967, who treated it as a monotypic family; he also described many new taxa within it. He treated the species of *Umbilicaria*, based on the characteristics of ascospores, as three subgenera: *Lasallia* (Mèrat) Endlicher (ascospores large and muriform, 1-2 per ascus), *Gyrophoropsis* (Elenkin & Savicz) Frey (ascospores small, muriform, 8 per ascus), and *Gyrophora* (Ach.) Frey (ascospores small, simple, 8 per ascus).

This paper was originally published in *Mycosystema*, 2: 135-150, 1989.

Frey's three subgeneric units within the genus *Umbilicaria* correspond respectively to: *Umbilicaria* Hoffm. sensu Körber [= subgenus *Lasallia*（Mérat）Endlicher], *Gyrophora* Ach. sensu Korber [= subgenus *Gyrophora*（Ach.）Frey], and *Gyrophoropsis* Elenkin & Savicz,（1911）[= *subgenus Gyrophoropsis*（Elenkin & Savicz）Frey].

According to the superficial appearance of the disc, Scholander（1934）divided the family into four genera: *Umbilicaria* Hoffm.（with plane discs containing both *Lasallia* and *Agyrophora* groups）, *Omphalodiscus* Scholander（1934）（with almost plane discs generally having a central sterile column or fissures）, *Gyrophora* Ach.（1803）（with gyrate discs）, *Actinogyra* Scholander（1934）（with actinogyrate discs）.

On the basis of the above concepts Llano（1950）in a monographic treatment segregated *Agyrophora*（Nyl.）Nyl.（1896）（= section *Anthracinae* with 8-spored asci and non-pustulate thalli）from *Umbilicaria* sensu Scholander（1934）, and reinstated the earlier validly published name *Lasallia* Mérat（1821）instead of misapplying the name *Umbilicaria* Hoffm. sensu Körber. He had also eliminated, in the same monograph, *Gyrophora* Ach.（1803）as a synonym for the earlier legitimate name *Umbilicaria* Hoffm.（1790）. The family then included the genera *Lasallia* Mérat, *Agyrophora*（Nyl.）Nyl., *Omphalodiscus* Scholander, *Umbilicaria* Hoffm., and *Actinogyra* Scholander.

The family name was proposed by Flotow（1850）and Körber（1855）, consisting of the genera *Gyrophora* Ach. sensu Körber（=*Umbilicaria* Hoffm.）and *Umbilicaria* Hoffm. sensu Körber（= *Lasallia* Mérat）. Poelt（1962）used the simple two-genus system of *Umbilicaria* and *Lasallia* in his first keys to European lichens. A detailed revision of the family including *Umbilicaria* Hoffm. emend. Wei [= *Agyrophora*（Nyl.）Nyl. + *Omphalodiscus* Scholander + *Umbilicaria* Hoffm. + *Actinogyra*（Nyl.）Nyl. in Llano's classification（1950）] and *Lasallia* Mérat emend. Wei consisting of both subgenera *Lasallia* and *Pleiogyra* Wei was made by Wei（1966）. The *Umbilicaria-Lasallia* system has been followed by most recent authors（Frey, 1967; Frey and Poelt, 1977; Poelt, 1969; Poelt and Vezda, 1977, 1981; Moberg, 1982; Hawksworth et al., 1983; Santesson, 1984; Cannon et al., 1985; Krog and Swinscow, 1986; Filson, 1987, etc.）

Twenty years ago a new genus *Llanoa* was described by Dodge（1968）for *Umbilicaria cerebriformis* Dodge & Baker（1938）[= *Llanoa cerebriformis*（Dodge & Baker）Dodge], and this has been accepted by Eriksson & Hawksworth（1986 a,b; 1987 a,b and 1988）in addition to *Umbilicaria* Hoffm. And *Lasallia* Mérat as it has not been critically studied, although James（in Hawksworth et al., 1983, p.223）suggested that it be a synonym of *Umbilicaria*

However, the type species of the genus *Llanoa*, *Umbilicaria cerebriformis*, is very close morphologically to *Umbilicaria decussata*（= *Omphalodiscus decussatus*）, of which it was treated as a variety by Llano（1950, p.83）prior to the publication of the genus *Llanoa*. Recently, *U. cerebriformis* has been placed by Filson（1987, p.340）as a synonym under *U. decussata* It is unfortunate that the type material of *U. cerebriformis* was not available to James or Filson Therefore a reexamination and clarification of the genus *Llanoa* Dodge is presented in the present paper based on the original material.

REEXAMINATION AND CLARIFICATION OF THE STATUS OF THE GENUS LLANOA DODGE

1. Original descriptions

1a. Genus *Llanoa* Dodge

"**Llanoa** Dodge, gen. nov.

Charcotia Dodge, B.A.N.Z.A.R.E. Reot. B 7:150. 1948, non Hue, 1915.

Type: *Umbilicaria cerebriformis* Dodge & Baker.

Thallus foliosus, monophyllus, gompho centrali, sine rhizinis; cortex fastigiatus pseudoparenchymaticus; algae Trebouxia; medulla hyphis pachydermeis; cortex inferior fastigiate pseudoparenchymaticus. Apothecia minuta, submarginalia, emergentia, lecanorina, disco laevi; amphithecium bene evolutum; parathecium carbonaceum, integrum; hypothecium hyalinum; asci pachydermei; ascosporae octonae, hyalinae, ellipsoideae, biloculares. Spermogonia immersa, perifulcrum nigro-brunneum, pseudoparenchymaticum; spermatiophorae septatae, moniliformes, rariter ramosae; spermatia lateralia, cylindrica.

Thallus foliose, monophyllous, attached by a central gomphus, underside smooth to subverrucose without rhizinae; upper cortex of fastigiate pseudoparenchyma; algae Trebouxia; medulla of thick-walled hyphae; lower cortex of one or two layers of dark brown to black spherical cells. Apothecia minute, submarginal, emergent, lecanorine, disc smooth; amphithecium well developed; parathecium entire, carbonaceous; hypothecium hyaline; asci thick-walled; 8-spored; ascospores hyaline, ellipsoid, bilocular. Spermogonia immersed; perifulcrum of dark brown pseudoparenchyma; spermatiophores septate, moniliform, rarely branched; spermatia lateral, cylindric." Dodge, Nova Hedwigia 15:310-311（1968）.

1b. Type species of the genus

"**Llanoa cerebriformis**（Dodge & Baker）Dodge, comb. nov.

Umbilicaria cerebriformis Dodge & Baker, Ann. Mo. Bot. Gard. 25:562. 1938.

Charcotia cerebriformis Dodge, B.A.N.Z.A.R.E. Rept. B 7:150. 1948." Dodge, Nova Hedwigia15:311. 1968.

"**Umbilicaria cerebriformis** Dodge & Backer, sp. nov. Pl. 45, Figs. 107—110; pl. 63, Fig. 412. Type: Marie Byrd Land, Edsel Ford Range, Skua Gull Peak, P. Siple & S. Corey 72W-15.

Thallus monophyllus, ad. 2.5cm.diametro, rugosus cerebriformisque, minute areolatus, marginibus crispatis,elevatis, subpruinosis, flavus neapolitanus aut olivaceo-alutaceus, nigricans, subtus laevis, dilute ochraceo-alutaceus, nigricans, sine rhizinis; cortex superior ad 20μm crassitudine, subfastigiatus, decompositus pseudoparenchymaticusve, non obscurascens, cellulis 3—4μm diametro, laxe dispositis, strato gelifacto ad 60μm crassitudine, fracto tectus; stratum gonidiale 40 — 50μm crassitudine, 25 — 60μm sub cortice, cellulis protococcoideis, ad 9μ diametro, in coloniis parvis; medulla frequenter 140—175μm crassitudine aut ad 400μ, hyphis 2 — 4μm diametro, ramosis anastomosantibusque, verticaliter super algas dispositis, et densius periclinalibus in strato 30—40μm crassitudine ad corticem inferiorem; cortex inferior 30—40μm crassitudine, pseudoparenchymaticus, obscurus.

Apothecia juvenilia, nigra, sessilia subimmersave; cortex 35μm, fastigiatus intus cortici thallino similis, hyphis cellulis isodiametricis nigris, 15μm diametro terminatis; parathecium tenue, hyalinum, hyphis tenuibus dense compactis; hypothecium circa 60μm crassitudine; thecium circa 60μ altitudine; paraphyses tenues, 1μm diametro; asci juveniles clavati, vaginati.

Spermogonia ampulliformia juventute, dein applanata irregulariaque, ostiola parva; murus carbonaceus, niger, cellulis parvis isodiametricis; spermatiophorae flexuosae, tenues, septatae; spermatia bacilliformia, recta, brevia.

King Edward VII Land: Rockefeller Mts., Mt.Helen Washington, P. Siple, F. A. Wade, S. Corey & O. D. Stancliff HW-10, HW-11, HW -12a, HW -13.

Marie Byrd Land: Edsel Ford Range, Skua Gull Peak, P. Siple & S. Corey 72 W-15, type; Mt. Grace Mckinley, P. Siple, F. A. Wade, S. Corey & O. D. Stancliff Mck-6, Mck-8." [Annals of the Missouri Botanical Garden, 25:562-564（1938）]

2. Type study

2a. Type specimens（holotype and isotypes）examined: Marie Byrd Land: Edsel Ford Range, Skua Gull Peak, P. Siple & S. Corey 72 W-15（FH!）.

The type specimens of this species were preserved in the C.W. Dodge Herbarium, recently transferred to Farlow Herbarium（FH）of Harvard University.

The original description, however, fits the type specimen except for the apothecia which were not found on it. The type specimen is undoubtedly closely related to *Umbilicaria decussata*（Vill.） Zahlbr. according to the morphological characteristics of thallus. Indeed, it seems logical for Llano to have suggested that *Umbilicaria cerebriformis* could be treated as a variety of *Omphalodiscus decussate*,i.e. *Umbilicaria decussata*（Llano, 1950, p.83）.

Some pycnidia have been found and examined in the type specimen. The conidiophores are septate, 3.5—7.0 × 2.0—2.5 μm, and produce conidia intercalarily, which are rod-shaped, 2.5—5.0 × 1.0—1.5μm（Fig. 1a, b）.

2b. Paratype specimens examined: King Edward VII Land: HW-11, HW-12a, HW-13 S-2（mistaken for HW-15 in herbarium label）, HW-13 S-3（FH!）; Marie Byrd Land: Mck-6, Mck-8（FH!）.

The original description of the thallus is basically correct for all the above paratype specimens. However, no lecanorine apothecia containing bilocular ascospores were found in any of the specimens examined. On the contrary, we have found some omphalodisc apothecia agreeing with those of *Umbilicaria decussata* in two of the paratype specimens examined（HW-13 S-2 & S-3; Pl. 1. nos. 1,2）. A description of them is given as follows:

Apothecia small, 0.3—0.5 mm in diameter, leiodisc to omphalodisc; epithecium 5.5—14.5 μm tall; thecium 27.0—50.5 μm tall; hypothecium 23.5— 30.5 μm tall; paraphyses simple, only rarely and simply branched, septate, about 2μm thick; asci 21.5—43.0×12.5—15.5μm containing 8 spores per ascus; ascospores unicellular, ellipsoid, hyaline, 9.0—12.5 × 4.0—5.0μm（Fig. 2 a, b）.

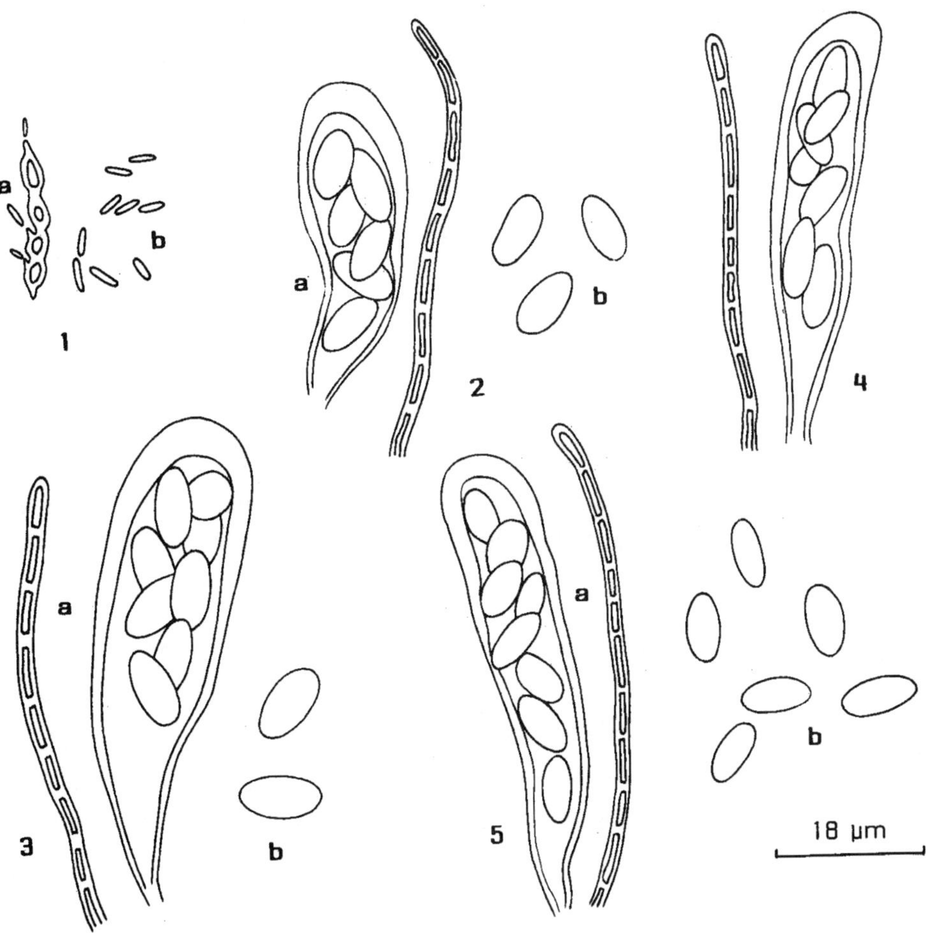

Figs. 1-5 1. Gonidiophore and conidia of *Umbilicaria cerebriformis*（72W-15, FH）: a. conidiophores, b. **conidia.**2. Ascus and ascospores of *Umbilicaria cerebriformis*（HW-13 S-2, FH）: a. ascus containing ascopores with a paraphysis, b. ascospore. 3. Ascus and ascospores of *Charcoiia cerebriformis*（Harrisson no. 58, FH）: a. ascus containing ascospores with a paraphysis, b. ascospores. 4. Ascus containing ascospores with a paraphysis of *Umbilicaria decussata* collected by Hu Shun Shi from eastern Antarctica（HMAS-L）. 5. Ascus and ascospores of *Umbilicaria decussata* from Turkey, Vězda, Lich. Sel. Exs. no. 1460（S）: a. ascus containing ascospores with a paraphysis, b. ascospores.

3. Study on specimens bearing the name *Charcotia cerebriformis* Dodge

Specimens examined: B.A.N.Z.A.R.E.Flora of King George V Land: Cape Denison, Commonwealth Bay, 67°00'S—142°36'E on rocks, January 5,6,1931,No. 536-15（FH!）. Australasian Antarctic Expedition, Flora of George V（Adelie）Land: on ridges of gneissic rocks: Commonwealth Bay 66°59'S,142°36'E, 60 ft. above sea level, 3 March 1912, Hunter no. 67（FH!）; 67°36'S, 142°36'E, 60 ft. above sea level, Cape Denison, 30 March 1912,Hunter no. 3（FH!）; Cape Denison, 67°S, 21 December 1912, no. 48（bearing both the names *Umbilicaria cerebriformis* and *Charcotia cerebriformis* FH!）. Australasian Antarctic Expedition, Flora of Queen Mary Land: L / VIII. Alligator Nunatak, 66°36'S, 97°30' E Jan. 2, 1913, Harrisson No. 58（FH!）; L/V Possession Nunatak, 66°45'S, 98°30'E, Harrisson, Dec. 25, 1912, F. Wild no. 62-1（FH!）; L/VI. Hippo Nunatak, 66°25' S, 98°E, Harrisson, Dec. 29,1912, F. Wild no. 64（FH!）; Cape Denison, no. 1049-1（FH!）.

The thalli of all the above-mentioned specimens basically correspond to those of *Umbilicaria decussata*. We would not find any lecanorine apothecia with bilocular ascospores in them. However, some omphalodisc apothecia containing 8- spored asci were found in four of above examined specimens（no. 536-15; Hunter nos. 3,

67; Harrisson no. 58 FH!; Pl. 1,nos. 3,4）. The following is a description of them:

Apothecia leiodisc to omphalodisc, 0.5—1.0 mm in diameter; paraphyses simple, rarely and simply branched, septate, about 2 μm in diameter; asci 16.0—18.0 × 43.0—54.0μm, 8-spored; ascospores unicellular, ellipsoid, hyaline, 3.5—5.5 × 8.0—10.0μm（Fig. 3 a, b）.

4. Study on some specimens of *Omphalodiscus decussatus* var. *cerebriformis*（Dodge & Baker）Llano collected from Antarctica

Specimens examined: Prince Olav coast: Lang Holvde, Soya coast on rocks, elevation about 5 m, 15 m, and 80 m. January 31, and February 1, 1968, H. Kashiwadani nos. 3660, 3695, 3697,3698,3932,4293, 299（TNS!）.

We could not find any apothecia in the above specimens.

5. Study on some specimens collected by Hu from Antarctica near the Australian station Casey（HMAS-L）

Apothecia leiodisc, subomphalodisc to pleiogyrodisc（= disc ecentrical multigyrate）containing octosporous asci, 0.5—1.5 mm in diameter; paraphyses simple, rarely and simply branched, septate, about 2μm in diameter; asci 39.5—52.0× 11.0—14.5μm; ascospores unicellular, ellipsoid, hyaline, 9.0—11.0 × 3.5—4.5μm（Fig. 4）.

6. Study on a specimen collected from Turkey

Specimen examined: A. Vězda: Lichenes Selecti Exsiccati no.1460. *Umbilicaria decussata*（Vill）. Zahlbr. Turcia asiatica. Prov. Bursa: Ulidag, 20 km in austro-orient. a Bursa, alt. 2000m s.m. Ad parietes rupium graniticarum. -31.VIII. 1976. Leg. K. Kalb et G Plobst（S!）

Apothecia omphalodisc containing octosporous asci, 0.5—1.5 mm in diameter; paraphyses simple, rarely and simply branched, septate, about 2μm in diameter; asci 52.0—65.0×11.0—12.5μm; ascospores unicellular, ellipsoid, hyaline. 7.0—9.0×3.5—4.0μm（Fig.5 a b）.

CLUSTER ANALYSIS

It seems that the methods of numerical taxonomy, as pointed out by Heywood, "will be used in future only for special classificatory problems or to illustrate some particular principle, rather than as a routine technique."（Heywood, 1976, p.49）. For the generic classification of this family, cluster analysis is used in the present paper. Thirty-five taxa of the family and 52 morphological characters were used for the analysis. The similarity of the taxa in the family was calculated by "Great Wall 0520C-H" personal computer according to the principle and methods of numerical taxonomy using the Microbiological Information Numerical Taxonomy System（MINTS）.

A. Morphological characters

1. apothecia emergent; 2. apothecia stipitate; 3. apothecia sessile; 4. leiodisc: apothecium with smooth disc; 5. gyrodisc: apothecium with disc having centric gyri; 6. actinodisc: apothecium with disc having actinoid fissures; 7. omphalodisc: apothecium with disc having central sterile column or fissures; 8.sublecanorine: apothecium with a thalline rim lacking algal cells around the disc; 9.lecideine; apothecium without a thalline rim around the disc; 10. octosporous asci: an ascus contains 8 spores; 11. monosporous asci: an ascus contains 1 or rarely 2 spores; 12. spore muriform; 13. spore unicellular; 14. ascus > 65 × 30μm;15. ascus < 65× 30μm; 16. spores > 40×20μm; 17. spores < 30× 15μm 18. thallus pustulate; 19.thallus non-pustulate; 20. thallus > 5 mm in diameter; 21. thallus < 5 mm in diameter; 22. upper surface of thallus strongly wrinkled; 23. upper surface of thallus smooth; 24. upper surface of thallus areolate; 25. upper surface of thallus with applanate squamules; 26. sorediate; 27. isidiate; 28. thallus without rhizines; 29. lower surface of thallus weakly papillose-areolate; 30. rhizines cylindrical; 31. rhizines long and strongly branched（> 1 mm long）; 32. rhizines carpet like（with kinky hair-fine branched rhizines）; 33. rhizines short and strongly branched（< 1 mm long）; 34. upper surface of thallus green grey; 35. upper surface of thallus dark brown; 36. upper surface of thallus red brown; 37. upper surface of thallus

olive-brown; 38. lower surface of thallus light coloured; 39. lower surface of thallus dark brown to black; 40. lower surface of thallus smooth; 41. lower surface of thallus coarsely verrucose; 42. lower surface of thallus finely verrucose; 43. lower surface of thallus with lamellae or trabeculae; 44. lower surface of thallus red; 45. lower surface of thallus grey to brown; 46. upper surface of thallus with vermiform rugi; 47. thallus without rhizines or sometimes with one or a few rhizines; 48. lower surface coarsely aerolate; 49. thallus with cylindrical, apically forked, ball-tipped rhizines; 50. pleiogyrodisc: apothecium with disc having eccentric gyri; 51. near the margins with thick and black rhizines and rhizinoid cilia; 52. upper surface of the thallus with recurved squamules.

B. Species examined

1. *Umbilicaria esculenta*, CHINA: Jiangxi, Wei, no. 3150（HMAS-L）; Anhui, Zhao & Xu, no. 5913（HMAS-L）. KOREA: Mt. Sok-li, collected by S. Kim in Sept. 24, 1959（H）. JAPAN: Hiroshima, M. Oshio, no. 2389（CANL）; Musashi, Kurokawa, Lich. Rar. et Crit. Exs. no. 16（S）.
2. *Umbilicaria mammulata*, USA:Virginia, M.E. Hale, no. 17858（S）.
3. *Umbilicaria rigida*, USSR: the Peninsular Kola, Savicz, Lichenotheca Ross. no. 141（S）. NORWAY: Opland, collected by S. Ahlner on July 23, 1949（S）.
4. *Umbilicaria leiocarpa*,NORWAY: Hordaland, collected by F.E. Hasselrot on Aug. 27,1939（S）.
5. *Umbilicaria haplocarpa*, SOUTH AFRICA: Basutoland, Div. Qachas Nek:Black Mts about 11000 ft., on rocks of flat summit. 2. 1963. leg. L. Kofler（HMAS-L）.
6. *Umbilicaria virginis* var. *lecanocarpoides*, CHINA: Sichuan,H. Smith, no. 14006（UPS）.
7. *Umbilicaria virginis* var. *virginis*, CHINA: Xinjiang, Wang, no. 946（HMAS-L）.
8. *Umbilicaria hypococcinea*, CHINA: Shaanxi, Wei et al. no. 2518（HMAS-L）.
9. *Umbilicaria cylindrical*, CHINA: Jilin, Wei, no. 2992（HMAS-L）JAPAN: Kai, collec- ted by A. Yasuda on July 27, 1921（H）.
10. *Umbilicaria hirsuta*, CHINA: Xizang, Wei & Chen, no. 1648-（1）（HMAS-L）.
11. *Umbilicaria muehlenbergii*, CHINA: Heilongjiang, Wei, no. 3296（HMAS-L）; Nei Mongol, Wei, no. 3400（HMAS-L）
12. *Umbilicaria yunnana*, CHINA: Yunnan, Wei, no. 2749（HMAS-L）.
13. *Umbilicaria crustulosa*, ITALY: Tyrolen, collected by I. Nordin on April 13, 1962（S）.
14. *Umbilicaria deusta*, CHINA: Jilin, Wei & Chen, no. 6755（HMAS-L）.
15. *Umbilicaria pseudocinerascens*, CHINA: Yunnan, Wei, no. 2827（HMAS-L）.
16. *Umbilicaria lyngei*, CHINA: Jilin, Wei, no. 2983（HMAS-L）; Shaanxi, collected by Wei on Jun. 3, 1963（HMAS-L）.
17. *Umbilicaria decussata*, CHINA: Jilin, Hu, no. 1605（HMAS-L）. TURKEY: Bursa, Vezda, Lich. Select. Exs. no. 1460（S）.
18. *Umbilicaria vellea*, CHINA: Jilin, Wei, no. 2908（HMAS-L）.
19. *Umbilicaria hyperborea*, CHINA: Nei Mongol, Zhao CF no. 2964（HMAS-L）.
20. *Umbilicaria herreri*, CHINA: Jilin, Wei, no. 2886（HMAS-L）.
21. *Umbilicaria proboscidea*, CHINA: Yunnan, Wei, no. 2829（HMAS-L）.
22. *Umbilicaria polyrrhiza*, USA: Idaho, Gray, J, & Nancy E. Schroeder, no. LI692（H）.
23. *Umbilicaria nanella*, CHINA: Shaanxi, Wei, no. 2784（HMAS-L）.
24. *Umbilicaria torrefacta*, CHINA: Jilin, collected by Wei & Chen from Changbai Shan in 1983（HMAS-L）.
25. *Umbilicaria spodochroa*, CHINA: Sichuan.Elenkin, Lich. Fl. Ross. IV, 1904,no. 151（UPS）.
26. *Umbilicaria caroliniana*, CHINA: Nei Mongol, Zhao CF, no. 2964（HMAS-L, IFP）.
27. *Lasallia pustulata*, SWEDEN: Stockholm, collected by Wei, Thol et al. in 1982（HMAS-L）.

28. *Umbilicaria koidzumii*, JAPAN: Kai, collected by H. Koidzumi, ex herb. V. Räsänen, no. 395（H）.
29. *Lasallia mayebarae*, CHINA: Yunnan, Wei, no. 2590（HMAS-L）; Taiwan, collected by K. Mayebara on March 9, 1930（TNS）.
30. *Lasallia daliensis*, CHINA: Yunnan, Wei, no. 2421（HMAS-L）.
31. *Lasallia pensylvanica*, CHINA: Jilin, Wei & Chen, no. 6250（HMAS-L）.
32. *Lasallia rossica*, CHINA: Jilin, Wei & Chen, no. 6251（HMAS）. USSR: Buryat ASSR, collected by V.M. Burkova in July, 1965（isotype, LE）.
33. *Lasallia papulosa*, USA: Massachusetts, Vezda, Lich. Select. Exs. No. 1488（S）.
34. *Umbiliraria badia*, CHINA: Xizang, Wei & Chen, no. 567（HMAS-L）.
35. *Umbilicaria arctica*, SWEDEN: Jamtland, collected by R. Santesson（S）. E. GREENLAND: Lich. Groenl. Exs. no. 47（S）.

C.Results of the analysis

The dendrogram of the cluster analysis shows that two distinct groups have been divided, namely one group including taxon numbers 27,30,31, 32, 33, 29 is *Lasallia*, and another including the remainder of the numbers corresponding to *Umbilicaria*（Fig. 6）.

DISCUSSIONS

The original description of the apothecia of the genus. *Llanoa* was given by Dodge（1968, pp.310-311）as follows: “Apothecia minuta, submarginalia, emergentia, lecanorina, disco laevi; amphithecium bene evolutum; parathecium carbonaceum, integrum; hypothecium hyalinum; asci pachydermei; ascosporae octonae, hyalinae, ellipsoideae, biloculares.” The key characters of the genus, in fact, were lecanorine apothecia containing definitely bilocular ascospores.

However, after careful reexamination we did not find any lecanorine apothecia containing bilocular ascospores as described by Dodge（1968, pp.310-311）, in all the type specimens（holotype and isotypes: no. 72W-15; paratypes: nos. HW-11, HW- 12a, Mck6, Mck8 FH!）, and even in the specimens bearing the name *Charcotia cerebriformis*（Dodge & Baker）Dodge（Harrisson nos. 62-1, 64, Mawson no. 1049-1 FH!）, and in the specimen from Cape Denison, 67 ° S bearing both the names *Umbilicaria cerebriformis* and *Charcotia cerebriformis*（no, 48 FH!）, and in the specimens collected from Antarctica and determined as .*Omphalodiscus decussatus* var. *cerebriformis*（Dodge & Baker）Llano（Kashiwadani nos. 3660, 3695, 5697, 3698, 3932, 4293, 4299 TNS!）. On the contrary, the omphalodisc apothecia containing unicellular 8-spored asci, which are peculiar to *Umbilicaria decussata*, were found in two of the paratype specimens（nos. HW-13 S-2 & S-3 FH!）, in four of the specimens bearing the name *Charcotia cerebriformis*（no. 536-15; Hunter nos. 3, 67; Harrisson no. 58 FH!）, and in one specimen collected by Hu from Antarctica and determined by the present authors as *Umbilicaria decussata* preserved in HMAS-L. In addition, the pycnidia containing donidia characteristic of *Umbilicaria* were found in holotype and isotype specimens（no. 72W-15 FH!）.

The thalli and apothecia of all the specimens（including types and non types）examined morphologically correspond to *Umbilicaria decussata*（Vill.）Zahlbr. From the above-mentioned information we conclude that *Lianoa cerebriformis*（Dodge & Baker）Dodge（=*Umbilicaria cerebriformis* Dodge & Baker）is a synonym of *Umbilicaria decussata*, and therefore the generic name *Lianoa* Dodge is a synonym of *Umbilicaria* Hoffm.

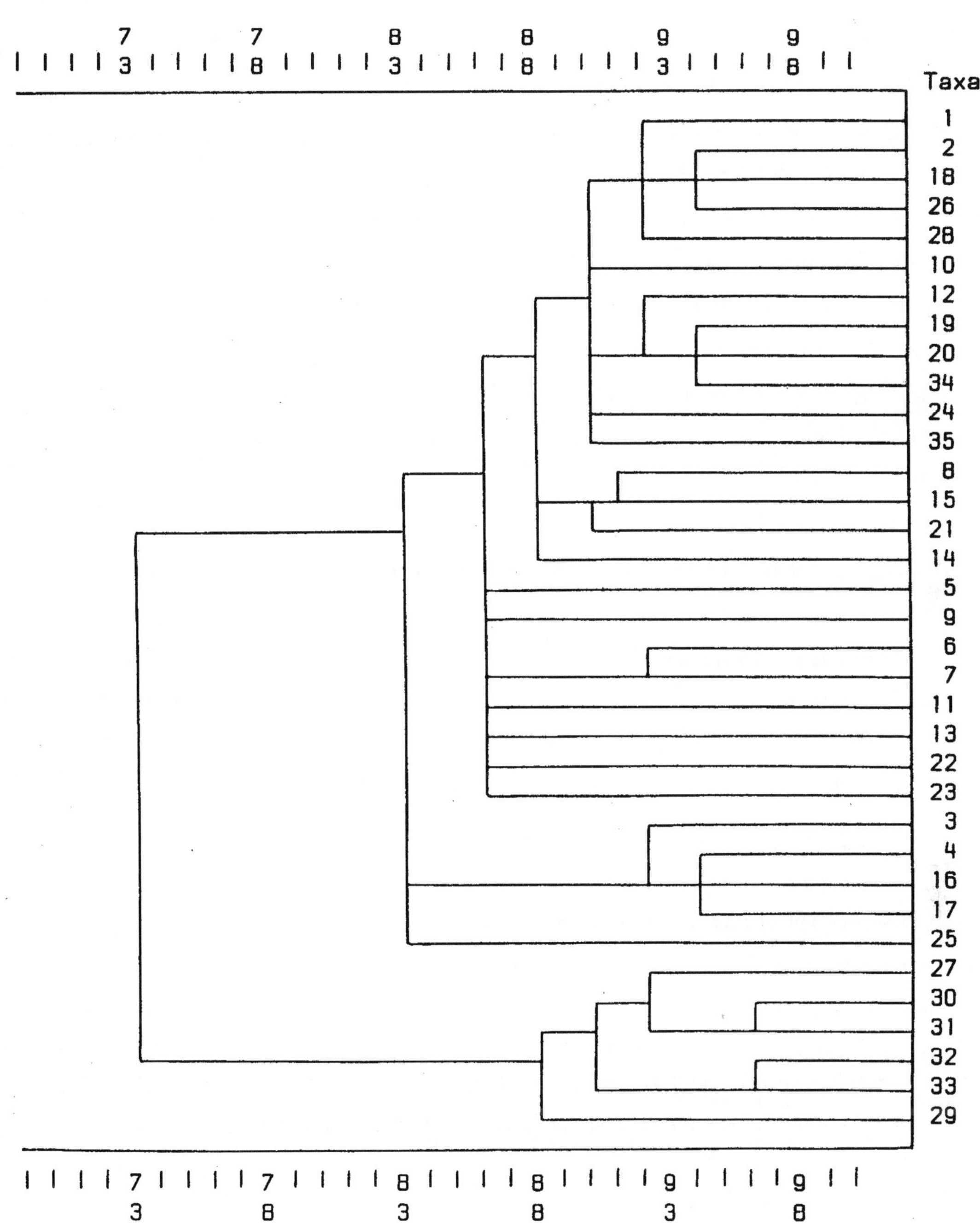

Fig.6 The dendrogram of the results of a cluster analysis of the taxa in the Umbilicariaceae based on the percentage of similarity.

Regarding the generic classification of the Umbilicariaceae, Wei（1966, p.3）pointed out that if one did not confine one’s outlook to a single feature but widen one’s horizons, one could expect to find characters in associated suites or groups. For example, every species of *Lasallia* has large and monosporous（rarely bisporous）asci with large, muriform and brown ascospores. At the same time species of *Lasallia* also have pustulate thalli without rhizines, and their lower surface is rough and verruciform. Others, belonging to *Umbilicaria*, contain smaller and 8-spored asci bearing smaller ascospores, and most of them are unicellular and hyaline, only a few later become muriform and brown. They have non-pustulate thalli, and most of them have rhizines（a few lack rhizines when the lower surface of thalli is smooth）. The concept of the Umbilicariaceae as containing two genera is based on a group or association of characters consisting of one principal and a series of related characters, not a single one.

Finally, the result of the cluster analysis also gives objective positive support to this concept of the family.

The generic classification of the family is recognized in the present paper as follows:

UMBILICARIACEAE Chev.

Chev., Fl. gen. env. Paris: 640, 1826（as Umbilicarieae）.

Fée, Essai Cryptog. Ecorc. Officin. LXX, 1824 [1825]（as Umbilicariees）, nom. inval.（Art. 18.4）.

Type genus: *Untbilicaria* Hoffmann.

UMBILICARIA Hoffmann

Hoffmann, Descr. Adumbr. PI. Lich. 1（1）: 8（1789）.

Type species: *Umbilicaria proboscidea*（L.）Schrad., Spicil Fl. Germ. 1:103（1794）.

=*Lichen proboscideus* L., Sp. Pl. 1150, 1753.

=*Umbilicaria exasperata*

Hoffmann, Descr Adumbr. Pl. Lich. 1（1）:8（1789）.

Subgenus: **Umbilicaria**

Type species: see that of the genus.

Subgenus: **Agyrophora** Nyl., Flora 247, 1878.

Type species: *Umbilicaria leiocarpa* DC.

=*Agyrophora*（Nyl.）Nyl., Lich. Environs Paris 43, 1896.

Type species: *Agyrophora atropruinosa*（Schaerer）Nyl., Lich. Environs Paris 43, 1896.

=*Umbilicaria leiocarpa* DC., in Lam. & DC. Flore Franc., ed. 3, 2:410（1805）.

=*Agyrophora leiocarpa*（DC.）Gyelnik, Ann. Mycol. 30:444（1932）.

Subgenus: **Omphalodiscus**（Scholander）Schade ex Wei & Jiang, Mycosystema 1:89（1988）.

Schade, Nova Acta Leopoldina, n.f. 17:202（1955）（comb, inval, Art. 33.2）.

Type species: *Umbilicaria decussata*（Vill.）Zahlbr.

=*Omphalodiscus* Scholander, Nyt Mag. Naturvid. 75:23（1934）.

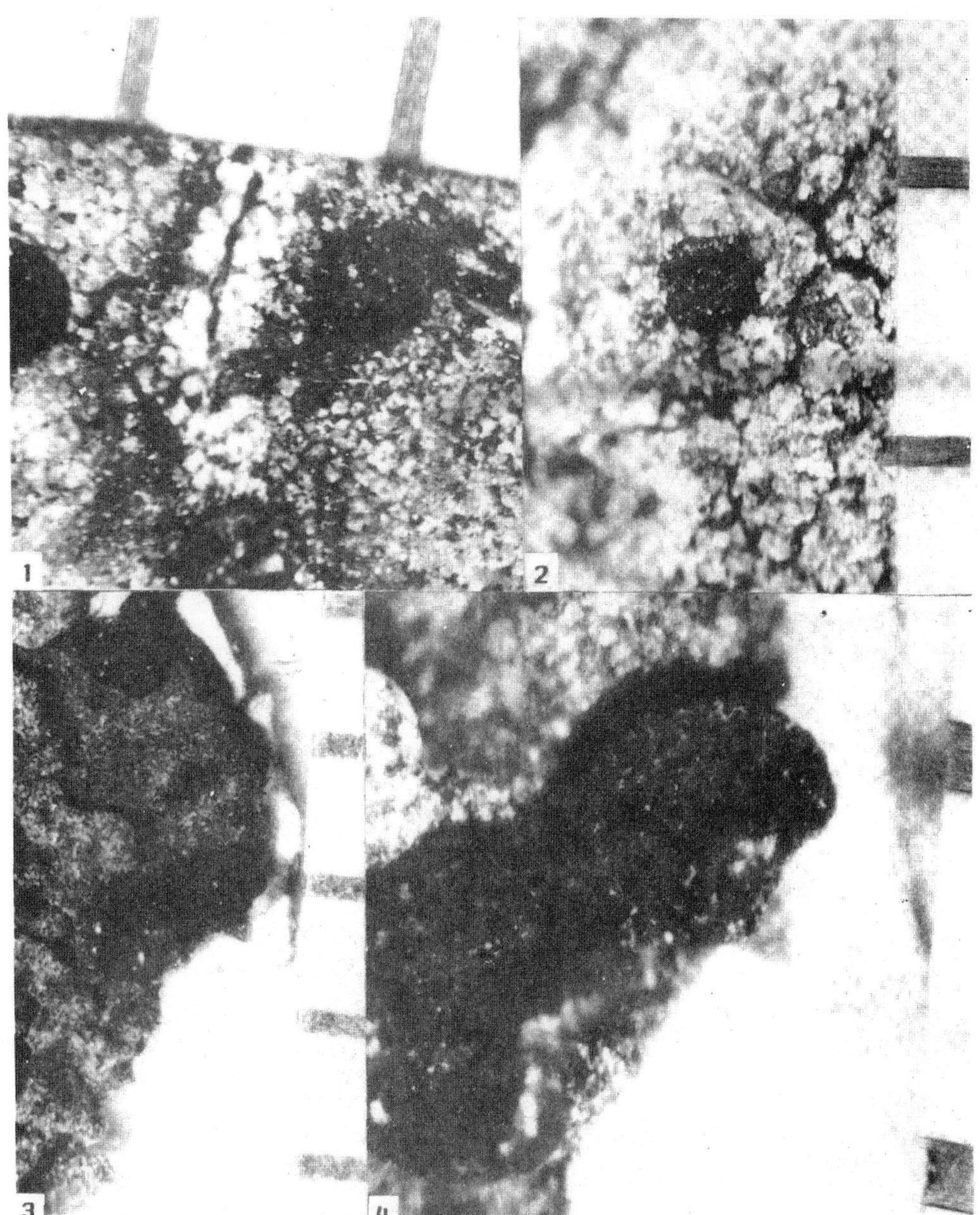

Plate 1 (Scales in 1 mm). 1. Apothecia of *Umbilicaria cerebriformis* (paratype no. HW-13 S-2, FH). 2. Apothecium of *U. cerebriformis* (paratype no. HW-13 S-3, FH). 3. Apothecia of *Charcotia cerebriformis* (Harrisson no. 58, FH). 4. Apothecia of *C. cerebriformis* (Harrisson no. 58, FH).

Type species: *Lichen decussatus* Vill., Hist. Plant. Dauphine 3:964, pl.55 (1789).

Type: as illustrated by Villars in Hist. Plant, Dauphine 3:964, pi. 55 (1789).

=*Umbilicaria decussata* (Vill.) Zahlbr., Cat. Lich. Univ. 8:490 (1942).

= *Omphalodiscus decussatus* (Vill.) Scholander, Nyt Mag. Naturvid. 75:23 (1934).

Llanoa Dodge, Nova Hedwigia 15:310 (1968).

Type species: *Umbilicaria cerebriformis* Dodge & Baker, Ann. Mo. Bot. Gard. 25:562 (1938).

Type: Marie Byrd Land, Edsel Ford Range, Skua Gull Peak, P. Siple & S. Corey 72W-15 (FH!).

= *Charcotia cerebriformis* (Dodge & Baker) Dodge, B.A.N.Z.A.R.

E.Rept. B7: 150（1948）.

= *Omphalodiscus decussata* v. *cerebriformis*（Dodge & Baker）Llano,

A Monograph of Umbilicariaceae 83,1950.

= *Llanoa cerebriformis*（Dodge & Baker ）Dodge,Nova Hedwigia 15:311（1968）.

=*Umbilicaria decussata*（Vill.）Zahlbr.

Subgenus: **Actinogyra**（Scholander）Schade ex Mot., Flora Polska, Porosty 4（2）: 77（1964）.

Type species: *Umbilicaria muhlenbergii*（Ach.）Tuck.

= *Actinogyra* Scholander, Nyt Mag. Naturvid. 75:28（1934）.

Type species: *Actinogyra muehlenbergii*（Ach.）Scholander. Nyt Mag. Naturvid. 75:28（1934）.

= *Gyrophora muehlenbergii* Ach., Lich. Univ. 227, 1810.

= *Umbilicaria muehlenbergii*（Ach.）Tuck., Enum. N. Am. Lich. 55, 1845.

LASALLIA Mérat

Mérat, Nouv. Fl. Paris ed. 2（1）: 202（1821）.

Type species: **Lichen pustulatus** L., Sp. Pl. 1150, 1753.

=**Lasallia pustulata**（L.）Mèrat, Nouv. Fl. Paris, ed. 2,1: 202（1821）.

Subgenus:**Lasallia**

Type species: see that of the genus.

Subgenus:**Pleiogyra** Wei, Acta Phytotax. Sin. 11:5（1966）.

Type species: *Lasallia mayebarae* (Sato）Asahina, Journ. Jap. Bot. 35:101（1960）.

= *Umbilicaria mayebarae* Sato, Journ.Jap.Bot. 25:168（1950）.

= *Lasallia sinensis* Wei, Acta Phytotax. sin 11:5（1966）.

=*Lasallia mayebarae* var. *sinensis*（Wei）Wei, Bull. Bot. Res. 1（3）: 90（1981）.

ACKNOWLEDGEMENTS

We are greatly indebted to Prof. Pfister（FH）, Dr. Kashiwadani（TNS）, Prof. Lundqvist（S）and Dr. Brodo and Mr. Wang（CANL）for sending on loan the type specimens and the specimens collected from Antarctica and other places related to our study, and to Prof. Hu for kindly providing her specimens collected from Antarctica. We also wish to thank Prof. Tang for valuable discussing during the study, and to express our sincere gratitude to Dr. Llano for reading the manuscript and for encouragement, especially to Prof. Hawksworth and Prof. Pfister for critical reading the manuscript and giving valuable comments and suggestions and for kindly revising the English language of the manuscript, and to Prof. Santesson and Prof. Poelt for their kind help in obtaining pertinent literature. Deep thanks are also expressed to Dr. Hale for proving information as to the place where the Dodge Herbarium was deposited. We are also obliged to Mr. Zhao and Mr. Ma for calculating the similarity of the taxa by the computer, to Ms. Yuan for printing the photographs, and to Ms. Han for inking the line drawings.

REFERENCES

Acharius E（1803）Methodus qua omnes detectos lichenes secundum organa carpomorpha ad Genera, Species et Varietates redigere atque observationibus illustrarc（Methodus Lichenum）. IV-LV, 1-393 Dodge CW（1968）Lichenological Notes on the Flora of the Antarctic Continent and the Subantarctic Islands. VII and VIII. *Nova Hedwigia* **15**: 285-332

Dodge CW（1973）Lichen Flora of the Antarctic Continent and Adjacent Islands Phoenix Publishing, Canaan, New Hampshire 03741

Dodge CW, Baker GE（1938）The second Byrd Antarctic expedtion - Botany II. Lichens and lichen parasites. Ann. Mo. Bot. Gard. **25**:515-718.

Elenkin A, Savicz VP(1911)Enumeratio lichenum in Sibiria orientali a cl. Sczcgolev anno 1903 lectorum. *Trav. Mus. Bot. Acad. Sci. St. Petersb.* **8**:26-9.

Eriksson O, Hawksworth DL（1986 a）An alphabetical list of the generic names of ascomycestes. *Systema Ascomycetum* **5**:3-111.

Eriksson O, Hawksworth DL（1987 a）An alphabetical list of the generic names of ascomycestes. *Systema Ascomycetum* **6**: 1-109.

Eriksson O, Hawksworth DL（1986 b）Outline of the ascomycetes 1986. *Systema Ascomycetum* **5**: 185-324.

Eriksson O, Hawksworth DL（1987 b）Outline of the ascomycetes 1987. *Systema Ascomycetum* **6**:259-337.

Eriksson O, Hawksworth DL（1988）Outline of the ascomycetes 1988. *Systema Ascomycetum* **7**:119-315.

Filson RB（1987）Studies in antarctic lichens 6; Further notes on Umbilicaria. *Muelleria* **6**: 335-347.

Frey B（1931）Weitere Beitrage zur Kenntnis der Umbilicariaceae. *Hedwigia* **71**: 94-119.

Frey E（1933）Umbilicariaceae. Rabenh. Krypt.-FJ. 9,**4**（1）: 203-424.

Frey E（1936）Vorarbeiten zu einer Monographie der Umbilicariaceen. *Ber. Schweiz Bot. Ges.* **45**:198-230.

Frey E（1949）Neue Beitrage zu einer Monographie des Genus *Umbilicaria* Hoffm. *Ber. Schweiz Bot. Ges.* **59**: 427-470.

Frey E(1967)Die Lichenologischen Ergebnisse der Forschungsreisen des Dr. Hans *Ulrich Stauffer. Bot. Jb.* **86**:209-255(not seen).

Frey E, Poelt J（1977）Die Gattung *Lasallia*（Flechten des Himalaya 13）. Mit 5 Abbildungen ... Khumbu Himal 6/3: 387-395.

Hawksworth DL Sutton BC and Ainsworth GC（1983）Ainsworth & Bisby ' s Dictionary the Fungi. 7th edition. Kew: Commonwealth Mycological Institute, pp. 1-445 Heywood VH（1976）Plant Taxonomy. Edward Arnold, pp. 1-63.

Hoffmann DGF（1790-1801）Descriptio et Adumbratio Plantarum e classe cryptogamica Linnaei Lichenes dicuntur. I-III.

Krog H, Swinscow TDV（1986）The lichen genera *Lasallia* and *Umbilicaria* in East Africa. *Nord. J. Bot.* **6**:75-85.

Linne（1791）Genera Plantarum ed. 8:768 no. 1688.

Llano G（1950）A Monograph of the Lichen Family Umbilicariaceae in the Western Hemisphere. Navexos pp. 1-831. Washington.

Llano G（1965）Review of *Umbilicaria* Hoffm. and the *Lasallias. Hvalradets Skr.* **48**: 112-124.

Matyka J（1964）Porosty（Lichenes）. **4**（2）: 7.

Moberg R（1982）La var. Stockholm, pp. 1-237.

Poelt J(1962)Bestimmungsschlussel der Hoheren Flechten von Europa. Mitt. *bot. Staatssammlung Munchen.* **4**:301-571(not seen).

Poelt J（1969）Bestimmungsschlussel Europaischer Flechten. Lehre: J. Cramer.

Poelt J（1977）Die Gattung *Umbilicaria*（Umbilicariaceae）（Flechten des Himalaya 14）. Khumbu Himal 6/3:397-435.

Poelt J, Vězda A（1977）Bestimmungsschlussel Europaischer Flechten Erganzungsheft I. J. Cramer, Vaduz.

Poelt J, Vězda A（1981）Bestimmungsschlussel Europaischer Flechten Erganzungsheft II. J. Cramer, Vaduz.

Santesson R（1984）The lichens of Sweden and Norway. Stockholm and Uppsala, pp. 1-333.

Scholander PF（1934）On the apothecia in the lichen family Umbilicariaceae. *Nyt Mag. Naturvid.* 75:1 -3.

Siple PA（1938）The second Byrd Antarctic expedition-Botany I. Ecology and Geographical Distribution. *Ann. Mo. Bot. Gard.* **25**:467-514.

Wei JC（as Vej TC）（1966）A new subgenus of *Lasallia* Mérat emend. Vej. *Acta Phytotax. Sin.* **11**:l-8.

Wei JC（1982）Some new species and materials of *Lasallia* Mérat emend. Wei. *Acta Mycol. Sin.* **1**:19-26.

Wei JC, Jiang YM（1988）A conspectus of the lichenized ascomycetes Umbilicariaceae in China. *Mycosystema* **1**:73-106.

子囊菌亚门石耳科属级分类及拉诺属地位的研究

魏江春 姜玉梅
中国科学院微生物研究所
真菌地衣系统学开放研究实验室
北京

摘要：关于子囊菌亚门石耳科的属级分类问题地衣学家持有不同的观点。这些观点使属级分类出现了单属系统，二属系统，三属系统以及四属系统与五属系统。现在新二属系统，即石耳属-疱脐衣属系统已被越来越多的地衣学家和子囊菌学家所接受。但是，因为由 Dodge（1968）描述的单种属拉诺属（*Llanoa*）的模式种与早已为人熟知的网脊石耳很相似，而近年来，这一单种属又被一些作者所接受。因此，我们不得不对该属模式种所依据的原始材料进行必要的复查。复查结果表明，拉诺属的模式种实为网脊石耳的异名。因此，拉诺属也就成为石耳属的异名。

此外，基于石耳科中三十五种地衣和五十二项特征的聚类分析结果也支持新的二属系统。

关键词：子囊菌亚门；石耳科；石耳属；疱脐衣属；网脊石耳；聚类分析

A preliminary study on *Everniastrum* from China

Yu-mei Jiang and Jiang-chun Wei
Institute of Microbiology, Academia Sinica, Beijing, China

ABSTRACT

Seven species of Everniastrum from China are reported in the present paper. One of them is new to science: Everniastrum subsorocheilum. This study is based on more than 200 specimens collected mostly from nine provinces of southern China, and preserved in HMAS-L. Chemical data were obtained mainly by TLC（thin-layer chromatography）and sometimes by MCT（microcrystal tests）when necessary.

中國條衣屬的初步研究

姜玉梅　魏江春
中國科學院微生物研究所

摘要：本文報導了中國條衣屬地衣七個種，其中新種一個，即亞粉芽條衣。該新種是發現在中國的第一個不含水楊嗪酸（Salazinic acid）的條衣屬種類。文中附有迄今已知的中國條衣屬分種檢索表。

條衣屬（*Everniastrum* Hale ex Sipm.）地衣為作者在進行西藏地衣研究中著重訂正的地衣類群之一。鑒于本屬地衣在醫藥及日用化學工業方面具有開發價值，以及現代地衣分類學的迅速發展，有必要對中國本屬地衣從形態學、化學（顯色反應、微量結晶法與薄板層析法）以及分佈學等方面進行訂正，以利於地衣資源的開發利用。

迄今為止，本屬地衣在全世界已描述的約 27 種，其中分布在中美洲的墨西哥高原地帶者為 12 種以上，分布在亞洲為 6 種，即黑樹發條衣（*E. alectorialicum*）、條衣（*E. cirrhatum*）、短根條衣（*E. nepalense*）、灌根條衣（*E. rhizodendroideum*）、粉芽條衣（*E. sorocheilum*）及裂芽條衣（*E. vexans*）。這六種地衣在中國均有記載，而且主要分布在南方各省區，其中多以梅衣屬（*Parmelia* Ach.）Den.的種類報導（Hue，1899；Jatta，1902；A.Zahlbr.，1930，1933，1934，1940；Sato，1933；H. Magnusson，1940；Asahina，1947，1951，1952；Moreau et Moreau，1951；Zhao，1964；Zhao et al.，1982；Wang-Yang et Lai，1973；Wei，1981，1990；Wei et Chen，1974；Wei et Jiang，1981，1982，1986；Culb. & C. Culb.，1981）。

本文原载于 *Acta Bryolichenologica Asiatica*, 1（1-2）: 43-52, 1989.

本文所用標本計 200 餘份，主要采自全國南方 9 省區，保存在中國科學院微生物研究所真菌地衣標本室（HMAS-L）。賴明洲博士則提供了其采自台灣省區之標本。

作者衷心感謝 R. Harris 博士閱讀文稿並提供寶貴意見；感謝苑蘭翠女士為本文拍攝照片。

條衣屬

Everniastrum Hale ex Sipman

Mycotaxon 26; 237（1986）.

Everniastrum Hale, Mycotaxon 3: 345（1976）.（*nom. nud.*）

本屬與其鄰近屬的主要區別首先在於它具有外皮孔和擬柵欄狀組織的內皮層（在掃描電鏡下觀察）。其次本屬所有種類在皮層裡都含有黑茶漬素（atranorin），並且幾乎半數以上種類在髓層裡含水楊嗪酸（salazinic acid）。分布在中國的種類多數都含有水楊嗪酸（salazinic acid）。

中國條衣屬分種檢索表

1.裂片上表面具粉芽或裂芽

2.僅具粉芽

3.髓層 K-，不含水楊嗪酸（salazinic acid）………………………………………… 6.亞粉芽條衣 *E. subsorocheilum*

3.髓層 K+，含水楊嗪酸（salazinic acid）…………5.粉芽條衣 *E. sorocheilum*

2.僅具裂芽 ……………………………………………7.裂芽條衣 *E.vexans*

1.裂片上表面既無粉芽，又無裂芽

4.假根單一，偶有簡單分枝

5.裂片下表面具假根

6.不含黑樹發酸（alectorialic acid）……………………3.短根條衣 *E.nepalense*

6.含黑樹發酸（alectorialic acid）…………… 1.黑樹發條衣 *E.alectorialicum*

5.裂片邊緣具緣毛型假根 ……………………………………2.條衣 *E.cirrhatum*

4.假根灌木狀分枝 ………………………… 4.灌根條衣 *E.rhizodendroideum*

1．黑樹發條衣

Everniastrum alectorialicum（Culb. & C. Culb.）Sipm., Mycotaxon 26:239（1986）.

Basionym:*Cetrariastrum alectorialicum* Culb. & C. Culb., Bryologist 84（3）: 297（1981）.

地衣體下表面僅具有稀疏假根，含黑樹發酸（alectorialic acid）.

Th: K+ yellow. M: K+ yellow to red, P+ orange red.

TLC: salazinic, alectorialic, protolichesterinic（+）& galbinic（+）acids in addition to atranorin.

基物：灌木枝。

產地：XIZANG: Zham, alt. 2450 m, 14/V/1966, Wei & J. B. Chen, no. 675 -1.

分布：YUNNAN（Culb. & C.Culb. 1981, p. 297）.

2. 條衣

Everniastrum cirrhatum（Fr.）Hale ***ex*** Sipm., Mycotaxon, 26: 235（1986）.

Basionym: *Parmelia cirrhata* Fr., Syst. Orb. Veg.1: 283（1825）.

=*Everniastrum cirrhatum*（Fr.）Hale, Mycotaxon, 3（3）: 347（1976）.（*nom.inval.*）

=*Cetrariastrum cirrhatum*（FR.）Culb. & C. Culb., Bryologist, 84（3）: 283（1981）.

地衣體由狹葉狀裂片組成；裂片稍內捲，寬約 1-3 mm；上表面灰白、灰綠至灰褐色，無粉芽或裂芽；下表面裸露，暗褐至黑色，頂端淡褐色，偶有零星假根，裂片邊緣具早一或簡單分枝的緣毛型假根，長 2-4mm。

Th: K+ yellow. M: K+ yellow to red, P+ orange red.

TLC: salazinic, protolichesterinic, galbinic（+）, protocetraric（+）acids in addition to atranorin.

基物：樹上。

產地：SICHUAN: 1982-1983, Su J.J.,無號.YUNNAN: Lijiang, alt. 3000-3600 m, 6-10/ XII/ 1960, J.D. Zhao & Y.B. Chen, nos. 3973-1, 3980 ,4036, 4123, 4170, 4182, 4191, 4192, 4193a, 4212b, 4331,4385, 4405, 4685; 4/IX/1958, S. J. Han, no. 5049a; alt. 2900-3250 m, 10-13/XI/ 1980, Y.M. Jiang, nos. 80-la, 90-5, 100, 125-2, 182-4, 340-1, 419-1,438-2, 440-1; alt. 3000 m, 22/IX/1987, Wei, nos. 9160-1, 9174-1, 9201- 1, 9271, 9277, 9279; and X.Q. Gao, nos. 2377, 2437, 2486-1, 2513; Bao- shan: Gaoligong shan, alt. 2330-2570 m, 9-11/XII/1980, Y.M. Jiang, nos. 757-2, 773-1, 807-4, 875-7, 926-la; alt. 2100 m, 23/IX/1959, Q.Z. Wang, no. 1292; Pingbian, Daweishan, alt. 1900 m, 25/VI/1956, 中蘇隊, no.4469; Dengchuan, Shiniuping, 18/IV/1949, H.C. Wang, no.1925a; Ma- guan, alt. 1700 m, 6/VI/1959, Q.Z. Wang, no. 425. GUIZHOU: Kaili, alt. 1300 m, 20/V/1959, 黔南隊,

no. 1839; Fanjing shan, alt. 1940 m, 25/IX/ 1963, Wei, no. 851. XIZANG: Nyalam, alt. 2450-3860 m, 6-31/V/1966, Wei & J. B. Chen, nos. 119, 133, 141, 181, 264, 306, 483, 492, 517-1, 543, 570, 625-lb, 746-1, 891-2, 994-3, 1039-2; alt. 2500 m, 14/V/1966, Wei & J. B. Chen,無號; 7 號樣地, 6/V/1966, Wei & J. B. Chen,無號;V/ 1966, S. Jiang & C. F. Zhao, no. Q11; Quxam, alt. 3550-3650 m, 18-19/ V/1966, Wei & J. B. Chen, nos. 867-2, 935-1, 936-2. FUJIAN: Wuyishan, alt. 1800-1850 m, 23-24/IV/1988, X. Q. Gao, nos. 2790, 2812. TAIWAN: Morrison, alt. 2700 m, Lai 10350.

分布：SHAANXI（A. Zahlbr. 1930, p. 195）, SICHUAN, YUNNAN（A. Za-hlbr. 1930, p.195 & 1934, p.210; Zhao, 1964, p. 145; Zhao *et al*. 1982, p. 15）, XIZANG（Wei & Jiang, 1986, p. 54）, ANHUI, ZHEJIANG（Zhao, 1964,p. 145; Zhao *et al.* 1982, p. 15）, TAIWAN（Asahina, 1952, p. 49）.

3.短根條衣

Everniastrum nepalense（Tayl.）Hale ex Sipm., Mycotaxon, 26: 235（ 1986）.

Basionym: *Parmelia nepalensis* Tayl., Lond. Journ. Bot., 6: 172（1847）.

=*Everniastrum nepalense*（Tayl.）Hale, Mycotaxon, 3（3）: 348（1976）.（*nom. inval.*）

=*Cetrariastrum nepalense*（Tayl.）Culb. & C. Culb., Bryologist, 84（3）: 301（1981）.

Parmelia kamtschadalis non（Ach.）Eschw. *apud* Martius: Zhao, Acta Phytotax. Sin. 9（2）: 144（1964）; Zhao, Xu, & Sun, Prodr. Lich. Sin. 1982, p. 14.

Parmelia cirrhata non Fr.: Zhao, Acta Phytotax. Sin. 9（2）: 144（1964）; Zhao, Xu, & Sun, Prod. Lich. Sin. 1982, p.15.

地衣體裂片狹葉狀，稍內捲或平坦；上表面無粉芽或裂芽；下表面具短小的黑色假根，長度一般不超過 1.5 mm.

Th: K+ yellow. M: K+ yellow to red, P+ orange red.

TLC: salazinic, protolichesterinic, protocetraric（+）& galbinic（+）acids, in addition to atranorin.

本種作為香精油原料已具有實際開發意義（Sun *et al*., 1983）。在該種中又發現了一個新的縮酚酮類化合物一條衣素 Cetrariastrumin（Sun et *al*., 1984）。

基物：樹上或灌木枝上。

產地：YUNNAN: Kunming, 2/IX-21-25/X, 2-4/XI/1960, J. D. Zhao & Y.B.Chen, nos. 162, 989, 994a, 996, 1981, 1986a, 2046, 2048, 2051, 2055a, 2078-1, 2239, 2240a, 2247; Lijiang, alt. 2800-3100 m, 6-8/XII/ 1960, J. D. Zhao & Y. B. Chen, nos. 3854a, 3881, 4014, 4086, 4123a, 4305; alt. 2400-3140 m, 9-13/XI/1980, Y.M. Jiang, nos. 37,80-1, 119- 6a, 191-3, 207, 231, 414-1, 423, 440; alt. 3050 m, 5/XII/1964, Wei, no .2496; alt. 3000 m, 22-23/IX/1987, Wei, nos. 9175-1,9237-1, 9272, 9273, 9274, 9275, 9277, 9286; Dali, alt. 2900 m, l/IX/1959, Q. Z. Wang ,no. 1169-b; Yunfeng, alt. 1650 m, ll/X/1959, Q. Z. Wang, nos. 2, 2a,3594; 1935-1936, Q. W. Wang, no. 21158a; Baoshan, alt. 2100 m, 23/IX/ 1959, Q. Z. Wang, no. 1294b; alt. 1250-2570 m, 8-10/XIl/1980, Y. M. Jiang, nos. 670-1, 681, 690-3, 807-4-1; Puer, Q. W. Wang,無號.XIZANG: Zham, alt. 2700-3360 m, 4-9/V/1966, Wei & J. B. Chen, nos. 306-2, 439-1; Zham, 7/VI/1966, Wei & J. B. Chen,無號;Quxam, alt.3650m,20/V/1966, Wei & J. B. Chen,無號;alt. 3570-3800 m, 17-19/V/ 1966, Wei& J. B.Chen, nos. 758-2, 789, 905; 1975, S. K. Chen, no. 53; Yadong, alt. 2780m, 4/IV/1975, Y. C. Zong, nos. 7, 8-4; Gyrong, alt. 3100 m, XI/1967, S. Jiang & C. F. Zhao, no. Q135-la; Zayu, alt. 2200 m & 4500 m, 15, 18/VII/1973, G. W. Zhang,無號二份。TAIWAN: Chiayi, Alishan, alt. 2275 m, Lai 10108.

GANSU（Magnusson, 1940, p. 128）, SICHUAN（A. Zahlbr. 1930, p. 195）,YUNNAN（Hue, 1899, p. 136; A. Zahlbr. 1930, p. 195; Zhao, 1964, p. 144; Zhao *et al.* 1982, p. 14）, XIZANG（Wei & Jiang, 1986, p. 53）, TAIWAN（A. Zahlbr. 1933, p.58; Asahina, 1952, p. 48）.

4.灌根條衣

Everniastrum rhizodendroideum（Wei et Jiang）Sipm., Mycotaxon, 26: 235（1986）.

Basionym: *Cetrariastrum rhizodendroideum* Wei et Jiang, Acta Phyto- Tax. Sin. 20（4）: 416（1982）.

地衣體類似條衣，裂片邊緣密生灌木狀分枝的緣毛型假根；表面裸露或具稀疏的同型假根。

Th: K+ yellow. M: K+ yellow to red.

TLC: atranorin, salazinic, protolichesterinic, galbinic（+）acids.

基物：樹上。

產地：XIZANG: Gyrong, alt. 3380 m, l/XI/1967, S. Jiang & C. F. Zhao, no. Q133; alt. 3300 m, 15/VI/1975, W. H. Li, no. 75-194; Quxam, alt. 3450 m, 21/V/1966, Wei & J. B. Chen, no. 1100; Nyalam, alt. 3650- 3740 m, 17/V, 14/VI/1966, Wei & J. B. Chen, nos. 762, 1555-1.

分布：XIZANG（Wei & Jiang, 1986,p. 54）.

5.粉芽條衣

Everniastrum sorocheilum（Vain.）Hale *ex* Sipm., Mycotaxon, 26: 235（1986）.

Basionym: *Parmelia sorocheila* Vain., Hedwigia, 38: 123（1899）.

=*Everniastrum sorocheilum*（Vain.）Hale, Mycotaxon, 3（3）: 349（1976）.（nom. Irival.）

=*Cetrariastrum sorocheilum*（Vain.）Culb. & C. Culb., Bryologist, 84（3）: 292（1981）.

地衣體較小；裂片狹葉狀；上表面灰白、灰綠至灰褐色，末端或近末端具成片的粉芽堆，具粉芽的裂片略狭且兩側邊緣往往稍內捲，末端略尖細且下捲，無粉芽的裂片寬短而平坦，兩側邊緣不下捲，末端平截，具少量邊緣假根；下表暗褐至黑色。

Th: K+ yellow. M: K+ yellow to red.

TLC: atranorin, salazinic, protolichesterinic, galbinic（+）acids.

基物：樹上。

產地：YUNNAN: Lijiang, alt. 3150 m, 16/XII/1964, Wei, no. 2546-3; alt. 3000 m, 22-23/IX/1987, Wei, nos. 9280, 9281, 9282, 9283, 9284, 9285; alt, 3000-3400 m, 10-12/XI/1980, Y. M. Jiang, nos. 91-5, 108-3, 307, 308-3, 341-1, 346-1, 436-1. XIZANG: Nyalam, alt. 3550 m, 19/V/ 1966, Wei & J. B. Chen, nos. 926, 929-1, 947, 1108-1 et alt, 3645 m, 14/VI/1966, Wei & J. B. Chen, 無號; Zham, alt. 3350 m, ll/V/1966, Wei & J. B. Chen, no. 455; Gyrong, alt. 3100 m, XI/1967, S. Jiang & C. F. Chao, no. Q135-1.

分布：XIZANG（Wei & Jiang, 1986, p. 54）, TAIWAN（Asahina, 1952,p. 50）.

6. 亞粉芽條衣 新種（圖版 1）

Everniastrum subsorocheilum Jiang & Wei, *sp. nov.*（Plate 1）

Everniastrum sorocheilum tangit, ob lobis latioribus et breviori- bus et ob absentia acidi salazinici in thall is.

Thailus ca. 3 cm longus, lobis ca. 2-3 mm longis et ca. 1 mm latis ,in axilla latioribus, planiusculis sed ad apicem saepe involutis; supra glaucis, apicem versus soralis granulosis cinereis et griseis conferruminatis tectis est; rhizinis marginatibus rarioribus, ca. 0.5- 1.0 mm longis, simplicibus; subtus nigris, arrhizis, interdum rhizinis singulis.

Apothecium et pycnidium non visum.

Thalli acidum protolichesterinicum et atranorinum continenti.

YUNNAN, Lijiang, in corticibus arborum: Ganheba, alt, 3000 m, 22/ IX/1987, J. C. Wei, no. 9170 et Yuhucun, alt. 2700 m, 25/IX/1987, J.C.Wei, no. 9247（typus）in HMAS-L conservantur.

本種類似於粉芽條衣（*Everniastrum sorochei lum*），但不同之處在於裂片較寬短，不含水楊嗪酸（salazinic acid）。

地衣體長約 3 cm，裂片長約 2-3 mm，寬約 1 mm，分腋處較寬，多次二叉分裂，平坦，末端往往內捲；上表面灰綠色，末端或近末端覆以灰綠至污灰色粉芽層，粉芽層由大量顆粒狀小粉芽堆密集匯成一片，有時粉芽層越遇末稍達 1 cm；邊緣假根稀少，長約 0.5-1.0 mm，早一或簡單分枝；下表面黑色，頂端往往褐色，平滑，裸露，老熟部分具明顯網狀棱脊，偶見個別假根；子囊盤及分生孢子器未見。

Th:K+ yellow. M:K-, C-, KC-, P -.

TLC: atranorin, protolichesterinic acid.

基物：樹上。

產地：YUNNAN: Lijiang, Ganheba, alt. 3000 m, J.C. Wei, no. 9170 et Yuhucun, alt. 2700 m, 25/IX/1987, J. C. Wei, no. 9247（模式），保存在中國科學院微生物所真菌地衣標本室（HMAS-L）。

7.裂芽條衣

Everniastrum vexans（A. Zahlbr.）Hale ex Sipm., Mycotaxon, 26: 235（1986）.

Basionym: *Parmelia vexans* A.Zahlbr. in Feddes, Repertorium, 33:35（1933）.

=*Everniastrum vexans*（A. Zahlbr.）Hale, Mycotaxon, 3（3）: 350（1976）.（*nom. inval.*）

=*Cetrariastrum vexans* A. Zahlbr. ex Culb. & C. Culb., Bryologist, 84（3）: 294（1981）.

地衣體達 12 cm 長; 裂片狹葉狀，二叉或近承叉分枝，寬 0.3-0.5 mm，稍內捲或平坦，頂端不同程度淺裂，邊緣往往具密生小裂片，小裂片頂端有時呈圓柱形；上表面灰白、灰綠或灰褐色，具單一或多分枝的裂芽，裂芽長度 0.2-0.5mm，頂端淡褐色；邊緣假根豐富，黑色，2-4 mm 長，單一或簡單分枝，下表面黑色，頂端淡褐色，裸露。

TH: K+yellow. M: K+ yellow to red, P+ orange red.

TLC: atranorin, salazinic, protolichesterinic, galbinic（+）acids.

基物：樹上、石頭。

產地：SICHUAN: Emei，alt. 1500-2500 m,10-19/VIII/1963, J.D. Zhao & Xu, nos. 6845, 6906, 6915, 6937, 7102, 7117, 7158, 7172, 7190, 7224, 8266. GUIZHOU: Fanjingshan, alt. 1570 m, 6/IX/1963, Wei, no. 699.ANHUI: Huangshan, alt. 1800 m, 19-23/VIII/1962, J. D. Zhao & Xu, nos. 5265, 5408, 5849; Huangshan, l/XI/1980, Wei, no. 3764-3. ZHEJIANG: Tianmushan, alt. 1000 m, l/IX/1962, J. D. Zhao & Xu, no. 6280. TAIWAN: Miaoli,

Chungshehshan, alt. 2750 m, Lai 9922; Taichung, Anmashan, alt. 2275 m, Lai 10024; Chiayi, Alishan, alt. 2275m, Lai 10189.

分布：TAIWAN（A. Zahlbr. 1940; Asahina, 1952, p. 49）.

REFERENCES

Asahina, Y., 1951. Lichenes Japoniae novae vel minus cognitae（2）. Journ. Jap, Bot. 26（4）: 97.

Asahina, Y., 1952. Lichens of Japan II: 48. Tokyo.

Hue, A.M., 1887. Lichenes Yunnanenses a clar. Delavay anno 1885 collectos, et quorum novae species a celeb. W. Nylander descriptae fuerunt, exponit A.M. Hue. Bull. Soc. Bot. France 34: 16.

Hue, A.M., 1889. Lichenes Yunnanenses a cl. Delavay praesertim annis 1886-1887, collectos exponit A.M. Hue（1）. ibid. 36: 158.

Hue, A.M., 1899. Lichenes extra-europaei a pluribus collectoribus ad Museum Parisieuse missi. Nouv. Archiv. du Museum IV, ser. 4（1）:27.

Ikoma, Y., 1983. Macrolichens of Japan and Adjacent Regions, p. 120. Japan, Tottori City.

Jatta, A., 1902. Licheni cinesi raccolti allo Shen-si negli anni 1894-1898 dal. rev. Padre Missionario G. Giraldi. Nuovo Giorn. Bot. Italian., ser. 2, IX: 460.

Magnusson, H., 1940. Lichens from Central Asia 1: 128. Stockholm.

Paulson, R., 1928. Lichens from Yunnan. Journ. Bot. London 66: 313.

（Sun, H.D., Lin, Z.W., Ding, Q.K., & Lou, J.F.）孫漢董、林中文、丁清凱、婁加風，1983.兩種新的地衣香料-中國橡苔Ⅰ號和中國橡苔Ⅱ號。雲南植物研究 5: 310。

（Sun, H.D. , Lin, Z.W., & Lou, J.F.）孫漢董、林中文、婁加風，1984. 星冰島衣素的結構-尼泊爾星冰島衣中的一個新縮酚酮類化合物。雲南植物研究 6: 329。

Wang-Yang, J.R., & Lai, M.J., 1973. A checklist of the lichens of Taiwan. Taiwania 18（1）: 83. 出版社，北京。

（Wei.J.C., & Jiang, Y.M.）魏江春、姜玉梅，1986. 西藏地衣。科學出版社，北京。

Wei, J .C., & Jiang, Y.M., 1981. A Biogeographical Analysis of the Lichen Flora of Mt. Qomolangma Region in Xizang. Proceedings of Symposium on Qinghai-Xizang（Tibet）Plateau, p. 1145. Beijing,China.

Zahlbruckner, A., 1930. Lichenes in Handel-Mazzetti. Symbolae Sinicae 3: 1.

Zahlbruckner, A., 1933. Flechhten der Insel Formosa. Feddes, Reper-tor iurn 31: 194 .

Zahlbruckner, A., 1934. Nachtraege zur Flechten Flora Chinas. Hedwigia 74: 95.

Zahlbruckner, A., 1940. Catalogus Lichenum Universalis. 10: 540.

（Zhao, J.D.）趙繼鼎，1964.中國梅花衣屬的初步研究。植物分類學報 9（2）:139.

（Zhao, J.D.，Xu，L.W.，& Sun，Z.M.）趙繼鼎、徐連旺、孫增美，1982. 中國地衣初編。科學出版社，北京。

图版 1.　亞粉芽條衣

Plate 1.　*Everniastrum subsorocheilum* Jiang et Wei（scale in mm）

Variations in ITS2 sequences of nuclear rDNA from two *Lasallia* species and their systematic significance

NIU Yong-chun WEI Jiang-chun
Systematic Mycology & Lichenology Laboratory
Institute of Microbiology, Academia Sinica, Beijing

ABSTRACT: The internal transcribed spacer 2（ITS2）of nuclear rDNA was amplified by PCR from total DNA extracted from sublecanorine apothecia of *Lasallia papulosa*（Ach.）
Llano and *L. rossica* Dombr. lacking algal cell, and their sequences were determined. A systematic analysis based on a combination of the molecular and morphological characters with chemical data and distribution patterns is given.
KEY WORDS: PCR; rDNA; ITS2 sequence; *Lasallia papulosa; L. rossica*

Reports on DNA extracted from lichens were given by Blum and Kashevarov（1986）for the first time; and then by Ahmadjian et al.（1987）. But the detailed methods for rapid extraction of DNA from lichens were described by Armaleo and Clerc（1991）.

A lichen is usually a symbiosis between one fungus（mycobiont）and one lower photosynthetic organism（a green algal or cyanobacterial photobiont）. The sequence analysis of the DNA of the fungal component is complicated with the mixed DNA of the green algal component. Three ways for obtaining the DNA of the fungal component were used by different authors: 1. PCR amplification of the DNA from the fungal hyphae damaged by chitinoclastic enzymes（Armaleo and Clerc, 1991）. 2. PCR amplification of the nuclear 18S rDNA of the fungal component using the nonalgal primer pairs NS17UCB - NS24UCB（Gargas and Taylor, 1992）. 3. The DNA of the lichenized fungi can be extracted from cultured mycobionts（Armaleo & Clerc, 1991）.

The DNA amplified by PCR from the sublecanorine apothecia of *Lasallia papulosa* and *L. rossica* which lack algal cells, was employed in the present work.

Lasallia rossica " appears to be an intermediate taxon between *L. papulosa* and *L. pensylvanica*（Hoffm.）Llano because of its dark brown lower surface of thalli. The rest of its characteristics basically correspond to *L. papulosa.* In addition, many collections of *L. papulosa* examined from North America also have a dark brown lower surface of thalli as well. Thus, whether this is in reality a single species remains an open question."（Wei & Jiang, 1993）In order to carry out a systematic analysis based on the combination of their molecular and morphological characters with chemical data and distribution patterns, the internal transcribed spacer 2（ITS2）sequences of the nuclear rDNA from *L. papulosa* and *L. rossica* were obtained using primer pairs 5.8SR and ITS4, and determined by ABI automatic DNA sequencing system. The ITS2 sequences of the nuclear rDNA evolve

This paper was originally published in *Mycosystema*, 6: 25-29, 1993.

faster than others, they may also vary among species within a genus or even among populations (White et al., 1990).

MATERIALS AND METHODS

Specimens. *Lasallia papulosa* was collected by Wei from GSMNP, Bullhead Trail, Tennessee, Sevier Co., alt. c. 200 m, U.S.A., March 4, 1990, and *L. rossica* was collected by Chen and Jiang from Arxan, Mt. Da Hinggan Ling, China in 1991.

DNA extraction. Dried apothecia lacking algal cell from herbarium specimens for DNA extraction were selected under a dissecting microscope and rinsed with sterilized distilled water. The apothecia (15-20 mg) were ground to powder in liquid nitrogen using a pre-chilled mortar and pestle. The powder was dispensed in small amounts into 1.5 ml microfuge tubes. Then 400 μl of 1% CTAB extraction buffer was added to each sample, and samples incubated at 65℃ for 30-60 minutes in a water bath; samples were extracted once with an equal volume of phenol / chloroform. RNase A was added to the aqueous contents of each tube and these were incubated at 65℃ for 30 minutes, before two extractions with equal volumes of chloroform / isoamyl alcohol. DNA was then precipitated from the aqueous phase by addition of exactly 2 volumes of ethanol. The precipitate was washed twice with ice cold 70% ethanol, dried in a vacuum, and resuspended in 20-40μl of TE (Bruns et al., 1990).

Fragment amplification by PCR. The 5.8SR (5'-TCGATGAAGAACGCAGCG-3' and ITS4 (5'-TCCTCCG CTTATTGATATGC-3' oligonucleotide primer pairs were used for the PCR to amplify the fragment including ITS2 region between 5.8S and 25S rDNA (Vilgalys & Hester, 1990; White et al., 1990). Template DNAs were diluted by 300 or 1200-fold. The reaction mixtures were overlaid with mineral oil and were subjected to 35 cycles on the COY Temp Cycler (Model 60, COY Laboratory Products Inc.) under the following temperature profile: 1 minute (2 minutes for the first cycle) at 94℃, 40 sec at 54℃, and 1.30 minutes (5 minutes for the last cycle) at 72°C. The PCR products were purified by Magic PCR Preps DNA Purification System (Promega Corp.) and then inserted into EcoR V sites of pBluescript SK plasmids. Ligated plasmids were transformed into the bacteria strain XL-1 blue, and the white colonies selected on IPTG/X-Gal/ampicillin plates were further identified for recombination by restriction analysis and PCR of plasmid DNA.

Sequencing. The recombinant plasmids were extracted by the alkaline lysis method, and sequencing was carried out using ABI automatic DNA sequencing system.

RESULTS

A total of 308bp of the sequence including ITS2 between 5.8S and 25S nuclear rDNA from *L. papulosa* and *L. rossica* were determined. Partial sequences of them (198bp) are shown in Fig.1. Fourty-eight (c.24%) variable positions exist between *L. papulosa* and *L. rossica*.

```
L. papulosa CGGCATTCCG GGGGGCATGC CTGTCCGAGCGT CATTGCAC    40
L. rossica  T . . T.................... T ............... A ................

L.papulosa CCCTCAAGCT CCGCTTGGT GTTGGGCCCCC GTCCCCCGGG    80
L. rossica  T ....... C . T.......A   ..... TGT. .......... ..............

L. papulosa ACGCGCCCGAAAGCGATTGGCGGCGCGGTCCGACTTCGAG    120
L. rossica  G. . . . T ........T T.G T . . A.C . T .......... A...............

L.papulosa CGTAGTAGTGAC TCCAAAC CCGCTCCGGA AGCCGGCAGG    160
L.rossica ........ A . ...... T ...... AA...... .... T ................TC.......

L.papulosa TCCGCCCCGG TCAGACAACC CGGTTGCACA CTTCG-AC    198
L.rossica  C T G . G . .-A . C . . . T ............A . . C . T - A T T .T
```

Fig. 1. Partial sequences of ITS2 of nuclear rDNA from *L. papulosa* and *L. rossica.*

DISCUSSION

A total of 187 base pairs were aligned from ITS2 of some *Lentinus* species by Hibbett and Vilgalys (1993). From the data given by Hibbett and Vilgalys we know that between the two strains of *Lentinus lepideus* there are 5 (2.67%) variable positions in ITS2 sequence, and among the three strains of *L. ponderosus* 2 (c.1%), whereas between *L. strigosus* and *L. trigrinus* 42 (22.46%) variable positions. These show that within-species sequence variability in ITS2 is lower (<3%), but interspecific sequence variability much higher (c.22%). A similar result exists in comparison with above mentioned two *Lentinus* species and two *Lasallia* species. The molecular evidence supports that *Lasallia* rossica should be a distinct species from *L. papulosa.*

The two species are morphologically very similar to each other except for characters 4-6 in the following table.

A combined analysis of the molecular and the morphological characters with their chemical data and distribution patterns strongly supported the idea that *L. rossica* is a separate species.

L. papulosa	*L. rossica*	*L. pensylvanica*
1. Apothecia more or less stipitate; often covered with redish pigment.	Apothecia more or less stipitate; usually without redish pigment.	Apothecia adnate, without redish pigment.
2. Upper surface of th. amber to brownish, often covered with whitish pruina, and showing grey-brown.	Upper surface of th. amber, brown to dark brown, usually covered with a thick whitish pruina, and showing grey-brown, peripherally dark brown.	Upper surface of th. dark brown, usually without pruina, only occasionally covered with a thin whitish pruina.

续表

3. Th. without squamules or fissure edges and th. margins with a few dark brown applanate squamules.	Th. without squamules or fissure edges and th. margins with a few dark brown applanate squamules.	Th. without squamules or fissure edges and th. margins with a few dark brown curved squamules.
4. Lower surface of th. greenish brown to whitish brown, smooth or delicately verrucose.	Lower surface of th. dark brown, delicately verrucose.	Lower surface of th. sooty black, coarsely verrucose.
5. Th. containing lecanoric gyrophoric, & umbilicaric acids.	Th. containing lecanoric & gyrophoric acids.	Th. containing lecanoric & gyrophoric acids.
6. Noto-boreal species.	North Asian species.	Arcto-holarctic species.

ACKNOWLEDGEMENTS

The authors are indebted to Dr. R. Vilgalys（Duke）and Prof. G. M. Tang（IM）for criticalreading the manuscript of this paper and giving valuable comments.

REFERENCES

Ahmadjian V, Chadegani M, Koriem AM, and Paracer S（1987）. DNA and protoplast isolation from lichens and lichen symbionts. *Lichen Physiol. Biochem.* 2:1-11.

Armaleo D and Clerc P(1991). Lichen chimeras: DNA analysis suggests that one fungus forms two morphotypes. *Experimental Mycology* 15:1-10.

Blum OB, and Kashevarov GP（1986）. DNA homologies as proof of the legitimacy of the establishment of the lichen genus *Lasallia* Merat(Umbilicariaceae). *Dok. Akad. Nauk Ukr.S.S.R.* Ser. B(12): 61-64(in Russian).

Bruns TD, Fogel R, and Taylor JW（1990）. Amplification and sequencing of DNA from fungal herbarium specimens. *Mycologia* 82（2）: 175-184.

Gargas A and Taylor JW(1992). Polymerase chain reaction(PCR)primers for amplifying and sequencing nuclear 18S rDNA from lichenized fungi. *Mycologia,* 84（4）:589-592.

Hibbett DS and Vilgalys R（1993）. Phylogenetic relationships of *Lentinus*（Basidiomycotina）inferred from molecular and morphological characters. *Syst. Bot.* 18（3）:409-433.

Vilgalys R and Hester M（1990）. Rapid genetic identification and mapping of enzymatically amplified ribosomal DNA from several Cryptococcus species. Journal of Bacteriology 172（8）: 4238-4246.

Lee SB and Taylor W（1990）. Isolation of DNA from fungal mycelia and single spores. Chap. 34 in PCR Protocols — A Guide to Methods and Applicatio ds., Innis MA, Gelfand DH, Sninsky JJ, and White TJ. Academic Press, New York.

Wei JC and Jiang YM（1993）. The Asian Umbilicariaceae. International Academic Publishers, Beijing.

White TJ, Bruns TD, Lee S, and Taylor JW（1990）. Amplification and direct sequencing of fungal ribosomal RNA genes for phylogenetics, in PCR Protocols - A Guide to Methods and Applications, eds., Innis MA, Gelfand DH, Sninsky JJ, and White TJ. Academic Press, New York.

两种疱脐衣的核 rDNA ITS2 序列变异及其系统学意义

牛永春 魏江春
中国科学院微生物研究所
真菌地衣系统学开放研究实验室，北京 100080

摘要：本文从分子系统学角度为石耳科两个疑难种的分类处理提供了佐证。真菌 DNA 是从淡腹疱脐衣及露西疱脐衣不含藻细胞的亚茶渍型子囊盘中提取后，用聚合酶链式反应（PCR）技术对于核中核糖体脱氧核酸（rDNA）的转录间区 2（ITS2）片段进行了扩增，并进行了核苷酸序列测定。在分子水平与形态特征以及化学与分布学相结合中进行了比较系统学分析。

关键词：聚合酶链式反应；核糖体脱氧核酸；转录间区 2；核苷酸序列测定；淡腹疱脐衣；露西疱脐衣

石耳科 rDNA 多型性分析及其系统学意义

魏江春 牛永春

中科院真菌地衣系统学开放研究实验室北京

（中国科学院生物分类区系特别支持费资助项目）

摘要：通过 PCR 技术扩增核 rDNA 的特定片段。对该特定片段 DNA 进行多型性分析表明，石耳科、地卷科、鸡皮衣科、黄枝衣科之间在遗传上有明显间断。为石耳科新二属系统及疱脐衣属两个形态疑难种从分子水平上提供了佐证。

关键词：PCR 技术；核 rDNA；多型性分析

茶渍目（Lecanorales）是子囊菌全系统中最大的地衣型子囊菌目。广义的茶渍目包括全部地衣型盘菌，含 23 科。根据表型性状分析表明，该目实属多源性类群。近年来，根据子囊顶部结构的微观与超微观结构比较研究及子囊开裂方式，一些科，如地卷科（Peltigeraceae）、鸡皮衣科（Pertusariaceae）、黄枝衣科（Teloschistaceae）等从茶渍目中划分出去，成为与茶渍目并列的地卷目 （Peltigerales）、鸡皮衣目（Pertusariales）、黄枝衣目（Teloschistales）（图 1）。

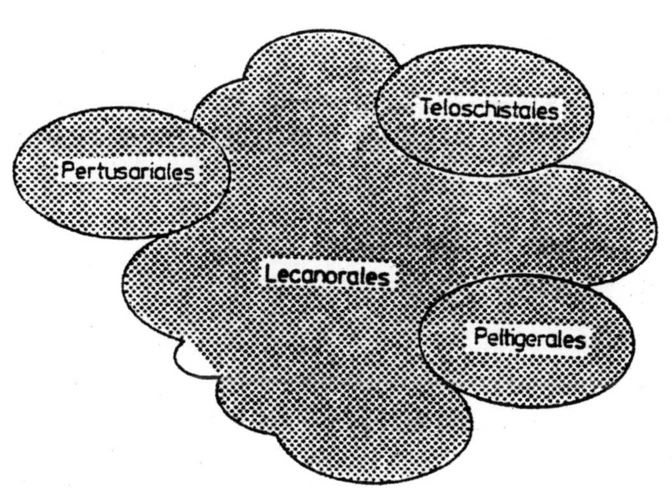

图 1 茶渍目及其分类趋势

根据我们对石耳科（Umbilicariaceae）表型性状的研究，其中包括对子囊顶部微观与超微观结构的观察表明，该科与茶渍科（Lecanoraceae）以及被从茶渍目中划分出去的地卷目、鸡皮衣目及黄枝衣目处于类似地位（Wei & Jiang, 1993）。为了从分子水平上进一步验证，我们对上述 5 个科代表种的 rDNA 多型性进行了比 较分析。属于核菌纲的皮果衣（*Dermatocar pon miniatum*）作为类群外比较也进行了验证。

本文原载于《中国科学院真菌地衣系统学开放实验室年报》: 8-12, 1994.

一、材料与方法：

1. 材料：以下各种干标本作为研究材料：

（1）	*Umdilicaria mammulata*	（石耳科）	魏江春采自美国田纳西州烟山（1990）
（2）	*U.esculenta*	（石耳科）	胡玉琛采自中国长白山地区
（3）	*U. formosana*	（石耳科）	陈健斌、贺青采自中国陕西太白山（1992）
（4）	*Lasallia rossica*	（石耳科）	陈健斌、姜玉梅采自中国内蒙阿尔山（1991）
（5）	*L. mayebarae*	（石耳科）	魏江春采自中国云南雪山地区（1964）
（6）	*Lecanora* sp.	（茶渍科）	陈健斌、贺青采自中国陕西太白山（1992）
（7）	*Nepnroma tropicum*	（地卷科）	陈健斌、贺青采自中国陕西太白山（1992）
（8）	*Xanthoria parientina*	（黄枝衣科）	陈健斌、贺青采自中国陕西太白山（1992）
（9）	*Dermatocarpon miniatum*	（瓶口衣科）	陈健斌、贺青采自中国陕西太白山（1992）

2．方法：

（1）从上述各种标本不含藻细胞的亚茶渍型子囊盘中提出微量总 DNA。通过 PCR 技术，利用 5. 8SR（TCGATGAAGAACGCAGCG ）及 LR3 （CCGTGTTTCAAGACGGG）作为引物（R. Vilgalys 实验室提供），对于 rDNA 中的 ITS2RNA 及 25SRNA 的一部分基因（即 rDNA）片段进行定位扩增（图 2），其长度约 0.88kb。然后对于扩增产物进行限制性内切酶（Hinfl, NC11 及 Aval） （White et al.,1990） 位点多型性分折。

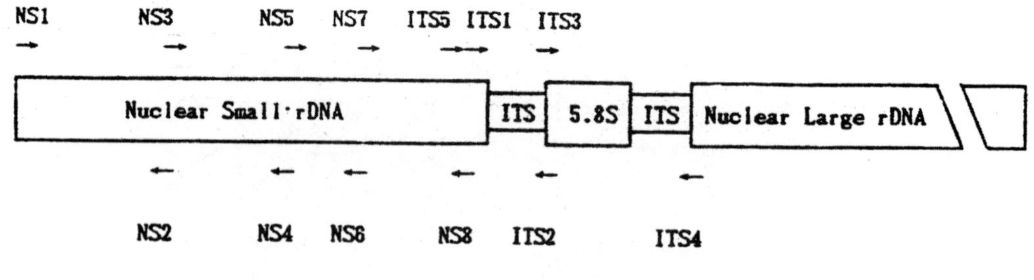

图 2 核 rDNA 序列示意图

二、结果与讨论：

1. 对石耳科的两个属中的四个种（即材料 1-4 号）进行酶切，其结果为：

（1）在 Hinfl 酶切下，石耳科的两属四种（图 3）的酶切位点完全一致。这说明石耳科作为一个独立的自然类群在基因型上的表现。

（2）在 Ncil 酶切下，石耳属的两个种，即（*U. mammulata, U. esculenta,* 各有一个长度相等的限制性片段；疱脐衣属的两个神（*L. rossica, L . mayebarae*）各有一个长度相等的限制性酶切片（图 3）。而上述两个属的限制性酶切片段之间的长度是不相等的。因为它们的酶切位点彼此是不同的，因而，其碱基序列也是相异的。这说明石耳属与疱脐衣属分别为两个独立的自然类群。这个结果与主作者在 1996 从表型性状论述过的石耳科新二属系统的观点是一致的（Wei, J. C., 1966）。

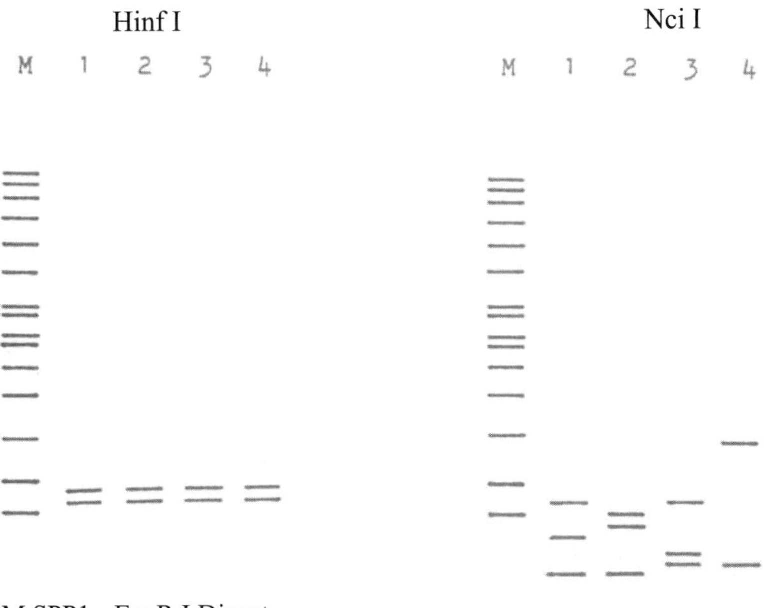

M SPP1 - EcoR I Digest

1 *Umbilicaria mammulata* 2 *U. esculenta*

3 *Lasallia rossica* 4 *L. mayebarae*

图 3 石耳科四种地衣 5.8SR - LR3 核 rDNA 序列限制性酶切片段电泳图谱

2. 对石耳科的 5 个种，茶渍科的茶渍衣（*Lecanora* sp.），地卷科热带肾盘衣（*Nephroma tropicum*），黄枝衣科的石黄衣（*Xanthoria parientina*）及作为类群外的核果类地衣-皮果衣（*Dermatocarpon miniatum*）所进行的 Aval 酶切结果如下：

石耳科的 5 个种（材料序号 1-5）都具有两个以上的共同酶切位点，而茶渍科、地卷科、黄枝衣科等彼此间以及与石耳科之间均无共同酶切位点。作为外类群的皮果衣同样也与上述各种无共同酶切位点（图 4）。这一结果说明石耳科以及上述各科都分别为独立的自然类群，因为他们彼此在遗传距离上都有明显的间断。

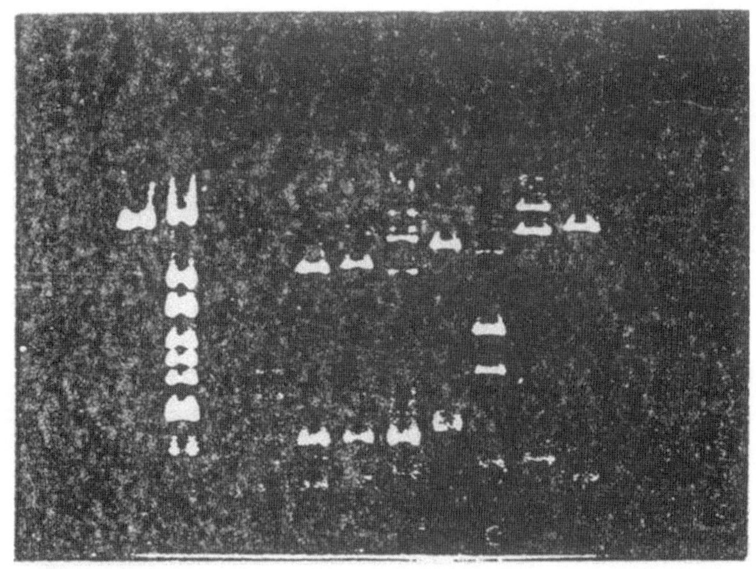

图 4 5. 8SR - LRS rDNA 序列限制性酶切电泳图谱（AvaL）

从左至右：1. 对照，2. Marker, 3 - 7:石耳科：*Umbilicaria mammulata*， *U. esculenta*，*U. formosana*， *Lasallia rossica*，*L. mayebarae*，

8. 茶渍科，9. 地卷科，10. 黄枝衣科，11. 瓶口衣科

三、初步结论：

从基因角度通过核糖体 DNA 限制性酶切位点多态性分析（RFLP）初步得出下列结论：

1. 茶渍目中的石耳科、茶渍科、地卷科、黄枝衣科在遗传距离上彼此均有明显间断，表明它们在分类上各自独立成群。这一结果与表型分类结果一致。
2. 石耳科内的两个属之间的遗传距离也有明显间断，表明它们各自成群，而不是一个类群。这一结果支持了我们关于石耳科新二属系统的观点。这一观点我们在 1966 年从表型角度进行过详细论述。在 1989 年（Mycosystema）关于否定 *Llanoa* 属一文又进行了数量分析。其结果也支持新二属系统。
3. 至于上述各科在分类上应为科级或目级。尚待进一步探索。

文 献

Hafellner, J.（1988）Principles of classification and main taxonomic groups《CRG Handbook of Lichenology》Vol. III, pp. 41 - 52.

Wei, J. C.（= Vej. J. C.）（魏江春）（1966）. A new subgenus of *Lasallia* Merat emend. Vej. Acta Phytotax. Sin. 11（1）: 1-8.

Wei, J. C& Jiang, Y. M.（1993）. The Asian Umbilicariaceae（Ascomycota）International Academic Publishers. Beijing.

White, T. J., Bruns, T. D., Lee, **S.**, and Taylor, J. W.（1990）.

Amplification and direct sequecing of fungal ribosomal RNA genes for phylogenetics, in Innis MA, Geifand, D. H., Sninsky, J. J, and White, T. J . PCR Protocols A Guide to methods and Appiications. ets., Academic Press, New York.

РОССИЙСКАЯ АКАДЕМИЯ НАУК

БОТАНИЧЕСКИЙ ИНСТИТУТ ИМ. В. Л. КОМАРОВА

ВЭЙ ЦЗЯН-ЧУНЬ

Анализ систематики и географии лишайников Сем. *Umbilicariaceae* Восточной Азии

СПЕЦИАЛЬНОСТЬ: 03.00.24-МИКОЛОГИЯ

ДИССЕРТАЦИЯ
НА СОИСКАНИЕ УЧЕНОЙ СТЕПЕНИ
ДОКТОРА БИОЛОГИЧЕСКИХ НАУК

С. Петербург

1995 г.

Посвящяется
Ста десятилетию рождения
Всеволода Павловича Савича
（1885—1972）

Работа выполнена в Открытой Лаборатории Систематической Микологии и Лихенологии Института Микробиологии Китайской Академии Наук.

Официальные оппоненты – Профессор, Доктор Биологических Наук М. М. Левитин; Доктор Биологических Наук Н. В. Седельникова; Профессор, Доктор Биологических Наук Н. С. Годубкова.

Ведущее учреждение – Санкт Петербургский Государственный Университет.

Защита состоится 24 мая 1995 г. в 15.00 час. на Заседании Специадизированного Совета Д CO_2. 46.01 по защите диссертаций на соискание ученой степени Доктора Биологических Наук при Ботаническом Институте им. В. Л. Комарова РАН по адресу: 197376 С. Петербург, ул. Проф. Попова 2, зал Ученого Совета.

С диссертацией можно ознакомиться в библиотеке Ботанического Института РАН.

Отзывы на автореферат просьба присылать в двух экземплярах по адресу Специализированного Совета.

Автореферат разослан апреля 1995 г.

Ученый секретарь

Специализированного Совета О. Я. Чаплыгина

THE RUSSIAN ACADEMY OF SCIENCES
THE BOTANICAL INSTITUTE NAMELY V. L. KOMAROV

WEI JIANG-CHUN

AN ANALYSIS OF THE SYSTEMATICS AND GEOGRAPHY OF LICHEN FAMILY UMBILICARIACEAE FROM EASTERN ASIA

Dedicated to the 110th birthday of
Vcevolod Pavlovicz Savicz
（1885—1972）

Special field: 03.00.24-Mycology

A dissertation of the requirement for the Doctor of biological Sciemces
from the Ruassian Academy of Sciences

The work was finished by the author
in the Open Laboratory of Systematic Mycology & Lichenology
Institute of Microbiology
Chinese Academy of Sciences, Beijing, China

S. Petersburg
1995.

ОГЛАВЛЕНИЕ

ВВЕДЕНИЕ

Лишайники семейства Umlilicariaceae, как известно, эпилитные лихенизированные грибы, распространены в основном в областях высокой широты и высокогорых районах низкой широты. Евразиатский материк - самый большой материк мира. Азия занимает подавляющую часть этого материка. Величаемое " Крышей мира" Цинхай-тибетское плато и 14 горных вершин высотою более 8 тыс. м над ур. м тоже расположены в Азии. Многогорная азия с разнообразными климатами является замечательным местом для исследования лишайников Сем. Umlilicariaceae.

В экосистеме на скалах роль лишайников Сем. Umlilicariaceae как укрытие для других организмов и субстрат для облигатных грибов-паразитов, а также как монитор тяжелых металлов значительна（J. C. Wei & Y. M. Jiang, 1993）.

Из лишайников Сем. Umlilicariaceae съедобный *Umbilicaria esculenta*（Miyoshi）Minks, один из эндемичных видов Восточной Азии, называется " шиэр", т. е. уховидным лишайником на скалах в Китае, "шикки" в Корее, и "Ивантаке" в Японии. Данный лишайник не только как деликатес в народе, но и содержит полисахарид GE-3, который удерживает рост саркомы 180 у белой мыши（Shibata et al., 1968）; к тому же сульфатный препарат противодействует активности вируса СПИДа（Hirabayashi et al., 1989）. Кроме того, вопрос эволюции лишайников Сем. Umlilicariaceae, характерных единственным свойством в экологическом, химическом и географическом отношении, постоянно привлекает внимание лихенологов и систематистов. Эволюционная биология является не только важным вопросом научной теории, но и имеет важное
реальное значение для накопления и получения информации о биоресурсах.

В исследовании эволюционной биологии палеонтологические данные являются самыми убедительными прямыми доказательствами. Однако, палеонтологические данные о находках лишайников крайне скудны. Поэтому в исследовании эволюции лишайников приходится ограничиваться лишь некоторыми косвенными доказательствами. На современном этапе развития наук косвенными доказательствами исследования эволюции лишайников могу служить результаты полученные из анализа макроскопических свойств лишайников, а также из исследования их микроскопических структур. В отношении макроскопических свойств их можно судить о результатах анализа из современного географического распространения.

Что касается исследования микроскопических структур, то это имеет в виду сравнительный анализ скорости геновариции лишайников при помощи техники молекулярной биологии.

Настоящая работа посвящается анализу систематики и географии лишайников Сем. Umlilicariaceae Восточной Азии.

Физико-географическое положение

Говоря о Восточной Азии, мы имеем в виду чисто географическое понятие. Сюда входят: Китай, Гималаи, Монголия, восточная Россия, Корея, Япония и зоны островов юго-восточной Азии, что не имеет ничего общего с фолористическими понятием " Восточная Азия".

Китай является главной составной частью Восточной Азии. Здесь очень сложный ландшафт; горы, холмы и нагорья занимают третью часть всей территории страны. Западная часть выше восточной. На юго-западе известная " Крыша мира" - Цинхай-Тибет плато является высочайшей частью. От нагорья Памир находящегося к северо-западу от Цинхай-Тибет плато тянутся высокогорные хребты, например: Алтай, Тяньшань, Куньлунь и Гималаи. Первый пик мира－Джомолунгма, 8848 м над ур. м., находится на южной стороне Тибета, на границе с Непалом. На севере от Цинхай-Тибет плато и к востоку от него рельеф спускается к нагорьям и впадинам, высотой до 2–1 тыс. м над ур. м., например,

Юньнань-Гуйчжоуское нагорье, нагорье Внутренней Монголии, и впадины Сычуань, Тарим, Джунгария. От хребта Куньлунь на восток тянется хребет Циньлин, который служит водоразделом двух крупных речных систем Чанцзян и Хуанхэ, имеющие огромное значение для географии растений и лишайников. Чанбайшань является самым большим горным хребтом восточной части Китая. Район Хендуаньшань – переходная полоса от Цинхай-Тибет плато к Юньнань-Гуйчжоускому нагорью и Сычуаньской впадине. Тут ландшафт наклоняется от северо-запада к юго-востоку, здесь много высоких гор и ущелий, между главными горами протекают реки с севера на юг, самая высокая вершина по сравнению с ущельем высится на 6 тыс. м. Хинган Большой, Тайханшань, Ушань и Юньнань-Гуйчжоуское нагорье с восточной стороны переходят к холмам высотой не более тыс. м над ур. м. и равнине менее 200 м над ур. м., такие как Северо-восточная равнина, Северо-китайская равнина, равнина среднего и нижнего течения Чанцзян и юго-восточные холмы. Это нижайшая часть. Тайвань и Хайнань являются крупнейшими островами среди островов Китая В Восточном Море и Южно-Китайском Море в указанном порядке. Лодкообразный остров Тайвань тянется с севера к югу на более 400 км. Широта его примерно 150 км. Ландшафт его пересечен высокими горами с северо до юга вышиной в 3–4 тыс. м над ур. м. Восточные и западные склоны высоких гор, а также их северные и южные концы опускаются к низине.

В западной, средней и северной частях Монгольского нагорья горные хребты Алтай, Ханъай, Танну, Восточный Саян, Гобиалтай, Кэнтэ образут горный район; рельеф тут постепенно снижается с запада на восток. Большинство вершин хребтов достигают 2800 м над ур. м. Восточная и юго-восточная часть Монгольского нагорья образует сравнительно ровное нагорье высотой в тысячу метров над ур. м.

Самым крупным полуостровом восточной части Азии является Корея, это многогорная полоса; рельеф тут на севере выше, на юге ниже, восток выше, запад ниже. Горные места и нагорье занимают почти три четверти всей территории страны. Большинство холмов тянутся с северо-востока на юго-запад, это ветвь хребта Чанбай.

Побережье Тихого Океана восточной части Азии вытягивается дугою в море и тут соответствует с дугообразными архипелагами,

Например, о. Рюкю, Японский архипелаг, Курильские острова, которые обращены к материку. Это доказывает, что в свое время они были соединены с азиатским материком.

Япония многогорная страна, берега ее извилисты, горные места занимают почти три четверти всей территории. В северной части Хоккайдо и Хонсю большинство холмов имеют направление с юга на север. В южной части Сикоку и Хонсю холмы идут с востока на запад. Эти холмы встречаются в середине о. Хонсю, образуя так называемый “ Центральный горный узел”, это самый высокий рельеф всей Японии. Среди них прославленная гора Фундзияма образует высочайший пик высотою в 3776 м над ур. м.

Остальные страны Юго-восточной Азии состоят и полуостровов и островов, меньшую часть занимают высокие горы, а большинство занимают равнины, холмы и низкогорья.

Северо-восточная часть Азии состоит из так называемого Ангарского древнего материка. В конце эозойской эры горообразующее движение и складчатая система севера и юга расширили прибрежные районы о. Байкала, районы забайкалья, горные районы Саянского хребта, Ханъай и Кэнтэ. Последнее образование северной и восточной частей Азии завершилось в каменноугольный период. С тех пор края этого материка были потоплены морем. В третичный период море потопило лишь район Анадыр и о. Сахалин, но к западу оно потопило всю западную часть Сибири, а степи Киргизстана превратились в острова, и отделены Тургайским заливом и Уралом. В четвертичный период малая часть района Уссури и о. Сахалин была затоплена. В восточной части Азии последнее горообразовательное движение произошло в конце третичного и в начале четвертичного периода, хребты Китая повысились. Особенностью климата восточной части Азии является то, что сезонные ветры зимой дуют с материка в море, а летом – с моря на материк. Поэтому летом многой дождей. Зимой из-за зимних ветров температура резко падает, вследствие чего среднемесячная температура материка восточной части Азии ниже чем в других частях Азии той же широты.

Краткая история исследования

В первом издании Species Plantarum Линнея（1753）под названием "Lichen" как род отмечено 80 видов. Спустя 37 лет после издания Species Plantarum Гофман（Hoffman, 1790-1801）под названием *Umbilicaria* описал новый род с новым видом *U.exasperata* Hoffm（1790）. Затем Mérat（1821）описал Род *Lasallia* основан на *Lichen pustulatus* L.（1753）. Fée（1824）и Chevalier（1826）на основе упомянутого выше рода создали сем. Umbilicariaceae. Koerber（1855）на основе Acharius（1803）излишнего родового названия "*Gyrophora*" для *Umbilicaria* разделил сем. Umbilicariaceae на два рода *Umbilicaria* и *Gyrophora*. Эту номенклатурную ошибку Llanno（1950）в своей монографии лишайников сем. Umbilicariaceae западного полушария уже исправил. Таким образом восстановил законное родовое название *Lasallia*, созданное Mérat, и выбрал вид *Umbilicaria exasperata* Hoffm（1790）как лектотил для рода *Umbilicaria* Hoffm（1790）. Но, к сожалению, *U. exasperata* было приведено как синоним *U. proboscidea* auct. На самом деле на основе образцов и записей Гофмана можно заметить *U. hyperborea*（1794）является синонимом *U. exasperata*（Hoffman, 1825; Wei, 1993; J. C. Wei & Y. M. Jiang, 1993）, и никакого отношения к *U. proboscidea* auct не имеет; тем более последнее видовое название тоже спутано последующими учеными（ Wei, 1993; J. C. Wei & Y. M. Jiang, 1993）.

В течение продолжительного времении семейство Umbilicariaceae принято лихенологами как монотипное, например Frey（1929-1949）и др. Лишь на основании особенностей строения дисков апотециев данное семейство было разбито Шоландером（Scholander, 1934）на 4 рода, т.е. *Umbilicaria* Hoffm（1790）sensu Koerb（1855）, *Omphalodiscus* Schol.（1934）, *Gyrophora* Ach.（1803）и *Actinogyra* Schol.（1934）. На основе этой четырехродовой системы Llanno（1950）обновил ранее законное родовое название *Lasallia* Mérat（1821）, исправил неправильное употребление родового название *Umbilicaria* Hoffm（1790）, и разбил данное семейство на 5 родов: *Lasallia* Mérat, *Agyrophora* Nyl., *Omphalodiscus* Schol., *Umbilicaria* Hoffm. и *Actinogyra* Schol.（Llano, 1950）.

Poelt（1962）и Wei（1966）самостоятельно рассматривали Umbilicariaceae как *Umilicaria* Hoffm – *Lasallia* Mérat система. Первый автор использовал ее лишь в ключах для определения видов Европейских лишайников（Poelt 1962）, другой же автор заложил эту систему с тщательным анализом их морфологии и анатомии Wei（1966）. Лишайники сем. Umbilicariaceae начная с Гофмана（Hoffmann, 1790）изучаются уже более двух веков. Однако, нельзя не удивляться : названия четырех самых распространенных, часто встречающихся видов *Umilicaria* постоянно путают（ Wei, 1993; J. C. Wei & Y. M. Jiang, 1993）. На самом же деле, *Lichen cylindricus* L, в Species Plantarum Линнея（ 1753）не имел в виду лишайник сем. Umbilicariaceae, а лишайник сем. Parmeliaceae: *Parmotrema perforatum*（Jocq.）Massal. *Lichen deustus* L. является базионимом лишайника *U. proboscidea* auct., а *Lichen proboscideus* L. Базионимом лишайника *U. cylindrica* auct. , однако *U. flocculosa* Hoffm является законным названием лишайника *U. deusta* auct., а *U. exasperata* Hoffm – законным названием лишайника *U. hyperborea*（ Wei, 1993; J. C. Wei & Y. M. Jiang, 1993）.

Новый род *Llanoa* описан и опубликован Доджем（Dodge, 1968）на основе лишайника *U. cerebriformis* Dodge & Baker（1938）. Этот род был призван микологами Эриксом и Гоксуорсом（Eriksson and Hawksworth 1985–1991）. Наши исследования об оригинальних материалах типа *U. cerebriformis* показали, что леканоровые апотеции и двухклеточные споры, которые были основой нового рода, не существуеют и вид *U. cerebriformis* был обработан как синоним вида *U. decussata*（Vill.）Zahlbruckner, и поэтому этот род *Llanoa* Dodge был обработан как синоним рода *Umilicaria*（J. C. Wei & Y. M. Jiang, 1989）.

При исследовании лишайников сем. Umbilicariaceae Восточной Азии, первый новый вид под названием "*U. papillosa*" был описан Ниландером(Nylander, 1869)в 19 веке. Так как название *U. papillosa* является более поздним омонимом, поэтому Фрей (Frey, 1949) заменил его новым названием "*U. indica*". Восемнадцать лет спустя Ниландер (Nylander, 1887) еще описал первый для семейства Umbilicariaceae во всем мире эпифитный вид *U. yunnana* на основе материалов из Китая. Известный съедобный лишайник из сем. Umbilicariaceae Восточной Азии *U. esculenta* описан и опубликован Мийошем (Miyoshi, 1893) на основе Японских материалов.

Начиная с 20 века, Хуе (Hue, 1990) прежде всего обнаружил и описал новый вид *U. thanmodes* из Юньнаня Китая. Вслед за этим Ятта (Jatta, 1902) обнаружил новый вид из образцов, собранных Гиральдием (Giraldi) с горы Циньлин Китая и описал его под названием "*Gyrophora hypococcinea*". После того как Савич (1914) описал два новых вида под названием "*G. krascheninnikovii*" и "*G. pulvinaria*" из Камчатского полуострова, и монография о лишайниках сем. Umbilicariaceae данного полуострова еще была опубликована (Савич, 1922). В своей монографии он записал 9 видов. В том числе один вид *G. spodochroa* (= *U. spodochroa*) [оригинальный образец №.199 (11) хранится в LE, и №.5085 (11) хранится в LE и US] и на самом деле является ошибочным определением другого вида *U. vellea* (J. C. Wei & Y. M. Jiang, 1993). Такой же лишайник под названием *U. spodochroa* из провинции Хунаня Китая(оригинальный образец №.6184 хранится в BM и FH), опубликованный Мюллером(Müll. Arg., 1893)фактически является ошибочным определением известного съедобного лишайника *U. esculenta* (Wei, 1991; J. C. Wei & Y. M. Jiang, 1993).

Лишайник *L. pertusa* из запада оз Байкал как новый вид под названием *U. pertusa* описан Рассадиной (Рассадина, 1929). Кроме того, другой эндемичный лишайник Восточной Азии как новый вид под названием "*G. kisovana*" был назван Цальбрукнером и опубликован Асахиной(Asahina, 1931). В настоящее время точно установлено, что описанный и опубликованный Фрем (Frey, 1931) лишайник *U. formosana* распространен не только в Китае, но и в Африке (Wei, рукопись не опубликована) и Америке (J. C. Wei & Y. M. Jiang, 1993). Однако эндемичный вид Японии *U. koidzumii*, названный Ясудуем и опубликованный Сатом (Sato, 1935), за пределами Японии до сих пор не обружен. Изучая лишайники восточной Индии Фрей (Frey, 1949) обнаружил и описал новый вид *U. badia*.

Во второй половине этого века произрастающий в Китае на Тайване другой эпифитный вид *L. mayebarae* под названием "*U. mayebarae*" опубликован Сатом (Sato, 1950). Затем Вэй (Wei, 1966) в Юньнане, Фрей и Поельт (Frey & Poelt, 1977) на Гималаях тоже обнаружили этот вид. Благодаря тому, что этот вид имеет гирозный диск с концентрически расположенными бороздками и скаладками и на основе его как новый подрод *Pleiogyra* в роде *Lasallia* (Wei, 1966). Другой новый вид на Тайване Китая *L. asiae-orientalis* описан Асахиной (Asahina, 1960).

На Гималаях Фрей и Поелт (Frey & Poelt, 1977) в статье Рода *Lasallia* отметили два вида, при этом приложили ключ для определения видов данного рода на основе литературных материалов. В статье гималайские лишайники *Umbilicaria* Поельт (Poelt, 1977) отмечает 15 видов. Из них Фрей и Поельт совместно описали новый вид *U. nanella* из Гималаев и Китая. Другой вид в данной же статье *U. nepalensis* описан Поельтом. Что касается разновидности *U. decussata* var. *rhizinata*, то она отнесна к разновидности *U. aprina* var. *halei* (J. C. Wei & Y. M. Jiang, 1993). Кроме того, автор описал морфологические характеры таллоконидии из 11 таксонов *Unbilicaria* и обсуждал значение для классификации *Umbilicaria*.

Близкий к *L. pensylvanica* и *L. papulosa* вид *L. rossica* был обнаружен и описан Домбровской (1978) из России. Голубкова (1981) отметила в Монголии 11 видом лишайников под родовым названием "*Umbilicaria*", два из них являются членами рода *Lasallia*, т. е. *L. pensylvanica* и *L. pertusa*. Вслед за этим она же в своей книге "Анализ Флоры лпшайников Монголии" провела географический анализ 13 видов лишайников сем. Umbilicariaceae (Голубкова, 1983). В своей статье относительно лишайников сем. Umbilicariaceae Монголии Бязров (1986) томечает 15 видов, 11 из них были новые для Монголии.

Необходимо подчеркнуть, что работа посвященная географическому анализу лишайников сем. Umbilicariaceae на территории прошлого Советского Союза（Голубкова и Шапиро, 1979）имеет большое значение для нашего исследования.

При исследовании рода *Lasallia* из Китая（ Wei, 1982）3 новый вида и одну новую разновидность были описаны. В статье конспекта лишайников сем. Umbilicariaceae Китая（J. C. Wei & Y. M. Jiang, 1988）37 видов и один из них новый для науки были отмечены. Среди Монгольских материалов Вэй и Бязров（Wei & Biazrov, 1991）обнаружили и впервые опубликовали для Азии *U. angulata* и *U. phaea*, которые были приняты как эндемичные виды Америки, причем провели первоначальный анализ их Американско-азиатской Дизъюнкции.

Вслед за этим Вэй и Цзян（J. C. Wei & Y. M. Jiang, 1992）еще опубликоввали 8 видов сем. Umbilicariaceae, 4 из них новые для науки, 3 для Китая, и один для Китая и Монголии. Те же авторы в монографии Азиатских лишайников сем. Umbilicariaceae（J. C. Wei & Y. M. Jiang, 1993）охватили 60 видов, 9 из них были описаны и опубликованы авторами как новые для науки: *L. daliensis*, *L. sinorientalis*, *L. xizangensis*, *U. altaiensis*, *U. minuta*, *U. subumbilicarioides*, *U. pseudocinerascens*, *U. squamosa*, *и U. taibaiensis*; 3 для Азии: *U. africana*, *U. angulata*, *и U*, *phaea*; 10 для Китая: *L. papulosa*, *U. badia*, *U. flocculosa*, *U. herrei*, *U. hirsuta*, *U. lyngei*, *U. nylanderiana*, *U. subglabra*, *U. torrefacta*, *U. virginis v. lecanocarpoides*, и один для Китая（*U. aprina* v. *aprina* и v. *halei*）и Монголии（*U. aprina* var. *halei*）. Эти 60 видов относятся к 6 подродам и 5секциям. Кроме того, авторы еще впервые записали таллоконидия из видов *U. indica*, *U. koidzumii*, *U. virginis и U. yunnana*. В монографии при исследовании лишайников применен комплексный метод анализа с помощью морфологии, анатомии, химии, географии, а также Макро-, Микро- и ультрамикроскопические подходы. В монографии представлены ключи для определения каждого вида и рода а также и почти каждый вид снабжен анатомическими рисунками и снимками их внешнего вида, и в том числе некоторыми электронно-микроскопическими снимками. Кроме того, в ближайшее время автор данной статьи в образцах, собранных в Цзилине и Шаньси Китая обнаружил и описал интересный новый вид *U. loboperipherica*, который, повидимому, является одним из двух викарных видов. Викарный вид *U. Loboperipherica* распространен в Китае, а другой, *U. freyi*, в Средиземноморском побережье（Wei et al. 1995）.

К настоящему времени в сем. Umbilicariaceae описано 26 новых видов, на материалах, собранных из Азии, из них 19 видов относится к роду *Umbilicaria*, 7 видов к роду *Lasallia*. Это *Umbilicaria altaiensis*, *U. badia*, *U. esculenta*, *U. indica*, *U. kisovana*, *U. koidzumii*, *U. loboperipherica*, *U. minuta*, *U. nanella*, *U. nepalensis*, *U. pseudocinerascens*, *U. pulvinaria*, *U. squamosa*, *U. subumbilicarioides*, *U. taibaiensis*, *U. thamnodes*, *U. yunnana*, *Lasallia asiae-orientalis*, *L. daliensis*, *L. mayebarae*, *L. pertusa*, *L. rossica*, *L. sinorientalis*, и *L. xizangensis*. Среди них 7 видов из рода *Umbilicaria* и 3 из рода *Lasallia* описано автором этой статьи. Что касается вида *U. zollingeri* Groenhart, основенного на материалах Явы, то он рассмотрен как синоним *U. africana*（J. C. Wei & Y. M. Jiang, 1993）. На материалах Гималаев описана разновидность *U. pensylvanica* var. *truncicola* Frey и *U. jingralensis*, из-за того что их оригинальные материалы не были исследованы автором и поэтому не вошли в статью.

Кроме упомянутых выше работ, относительно лишайников сем. Umbilicariaceae Восточной Азии опубликованы другие работы разных авторов: Андреев, 1978; Asahina, 1936-37, 1955, 1960; Awasthi, 1960, 1965; Баранов и Смирнов, 1931; Будаева, 1989; Бязров и др., 1983; Chen at al., 1981; Chopra, 1934; DuRietz, 1929; Еленкин, 1901, 1902, 1903, 1904, 1912; Еленкин, и Савич, 1911; Голубкова и Савич, 1978; Городков, 1935; Hue 1889, 1907; Ikoma, 1956, 1983; Kamiya, Kim, 1981; Kato, Kumagai, 1960; Kim, 1979, 1981; Lobaeashi, 1957; Krog, 1973; Kurokawa, 1960; Llano, 1950; Локинская, 1970. Lu, 1959; Макарова, 1973; Макрый, 1990; Микулин, 1990; Nylander, 1888; Окснер, 1926, 1993; Paulson, 1925, 1928; Räsanen, 1949, 1950, 1952; Рассадина, 1934, 1936; Sato, 1937, 1950, 1952, 1958, 1960, 1961, 1981; Савич, 1926, 1950,

Schubert et al., 1969; Schubert, Klement, 1971; Седельникова, 1978, 1985, 1990; Седельникова и Седельников, 1979; Семенов, 1922; Singh, 1964; Suza, 1925; Takamizu, 1957; Vezda, 1965; Wang-Yang, Lai, 1973,1976; Wang, 1985; Wei, 1981, 1991; Wei & Chen 1974, J. C. Wei & Y. M. Jiang, 1981, 1982, 1993; Wei et al., 1982, Wu J. L., 1981, Yamanaka & Yoshimura, 1961; Yoshimura, 1974; Yoshimura & Yamanaka, 1961; Zahlbruckner, 1911, 1930, 1932, 1933; Цогт, 1981. О них не будем говорить подробно.

При исследовании по форме очень схожих двух видов рода *Lasallia*, *L. papulosa и L. rossica* был применен комплексный анализ морфологии и молекулярной систематики, что позволило выявить границы между ними и установить, что это объективно существующие самостоятельные виды（Niu & Wei, 1993）.

МАТЕРИАЛЫ И МЕТОДЫ

Данная работа основана на личных исследованных материалах. Все эти материалы, в том числе лично собранные в основном хранятся в Гербариях лишайников как Китая: HMAS-L., так зарубежных: BM, CANL, COLO, DUKE, E, FH, H, HINU*, HMAS-L, IFP, LE, MIN, MOSCOW**, MW, NY, S, TENN, TNS, TUR, UPS и US.

В исследовании морфологии и анатомии были использованы анатомический микроскоп и световой микроскоп германской фирмы “OPTON”, а также электронный микроскоп и сканирующий элекстонный микроскоп японской фирмы “Hitachi”.

При химическом изучении главным оброзом использована тонкослойная хроматография（Culberson, C. F. & Kristinsson, 1970; Culberson, C. F., 1972）.

При географическом анализе лишайников сем. Umbilicariaceae Восточной Азии карты ареалов в основном основаны на исследованных автором материалах.

РЕЗУЛЬТАТЫ И ИХ ОБСУЖДЕНИЯ

1. Морфолого-географическое понятие о семействе, роде и виде лишайников сем. Umbilicariaceae.

На базе многолетнего исследования лишайников сем. Umbilicariaceae автор с точки зрения морфологии, анатомии, химии и географии представляет комплексное понятие семейства, рода и вида лишайников данного семейства.

1-1. Понятие о семействе лишайников сем. Umbilicariaceae

В отношении морфологии и анатомии лишайники сем. Umbilicariaceae характеризуются умбилико-листоватыми слоевищами（Рис. 1）, сублеканоровыми апотециями（Рис. 2）и сумками с амилоидными купольно-крышевидными апикальными аппаратами（Рис. 3）; в физиолого-биохимическом отношении к находящимися в постоянном симбиозе с водорослями видов *Trebouxia*, и производящими вещества Тридепсидов（Tridepsides）.

и депсидонов（Depsidones）ацетат-полималонатного происхождения（Acetate-polymalonate pathway）（Culbersons & Johnson, 1977; Hale, 1983; Рис. 4）; в географии свойственными распространенимми в областях высокой широты и в высокогорных районах низкой широты（Рис. 5）.

* Гербарий Уеонбуг национального университета и южной Кореи.

** Гербарий Института эволюционной Морфологии и Экологии животных Российской Академии Наук в Москве.

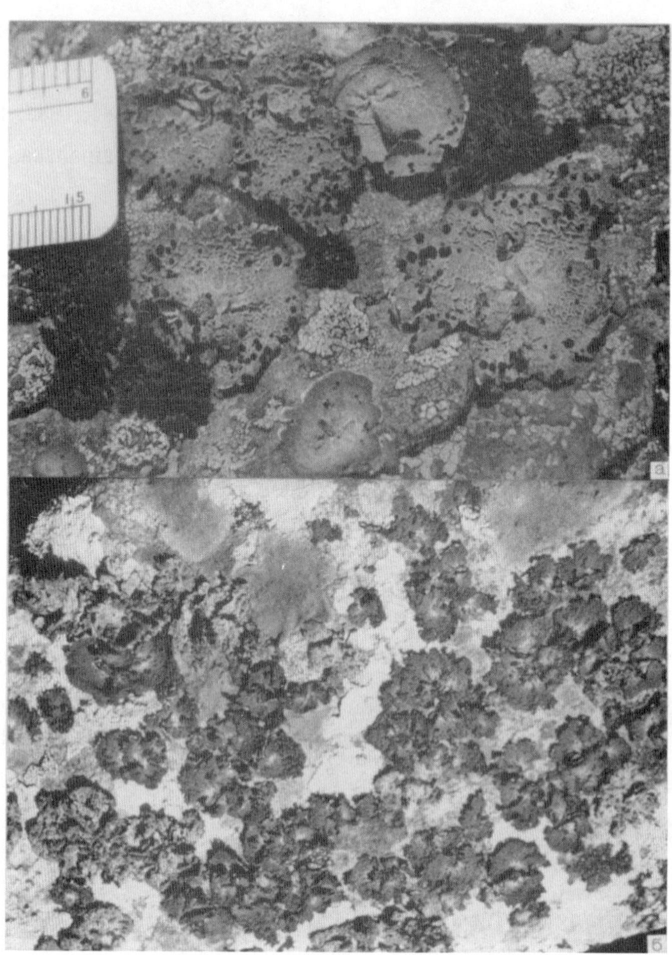

Рис. 1 Умбиликолистоватые слоевища у лишайников сем. Umbilicariaceae: а. *Umbilicaria formosana* Frey（в масштабе 1 шкала = 1mm）б. *Llasallia pensylvanica*（Hoffm.）Llano（коричневая）и *L. rossica* Dombr,（Серая）.

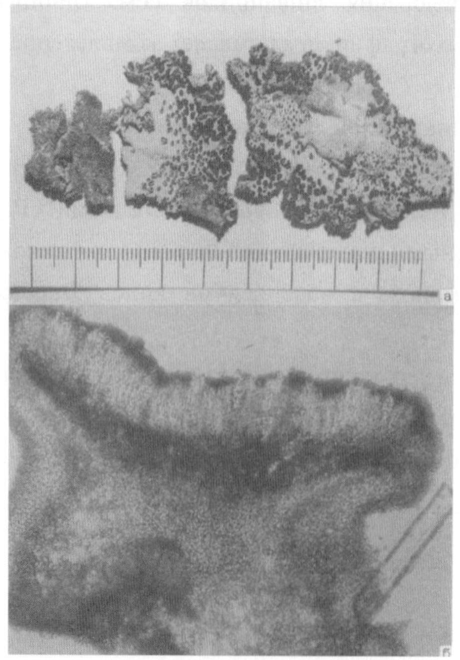

Рис. 2 Сублеканоровый апотеций у вида *Umbilicaria virginis* var. *lecanocarpoides*（Nyl.）J. C. Wei & Y. M. Jiang.

а. Слоевище с многочисленными апотециями сублеканорового типа（в масштабе 1 шкала = 1mm）б. Вертикальный разрез через сублеканорового апотеция данной разновидности（CM）. б. Схема строения апотеция сублекано-рового типа у лишайников сем. Umbilicariaceae.

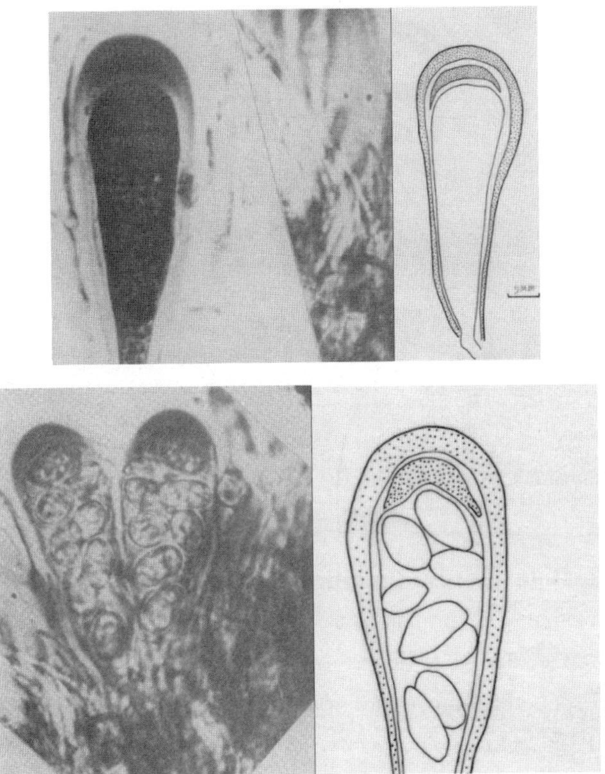

Рис. 3 Апикальный аппарат сумок у лишайников сем. Umbilicariaceae.

Сверху: *Lasallia pustulata*（L.）Merat（Налево: Фото под СМ, Направо: Рисован под СМ）.

Снизу: *Umbilicaria virginis* Schaer.（Налево: Фото под СМ, Направо: Рисован под СМ）.

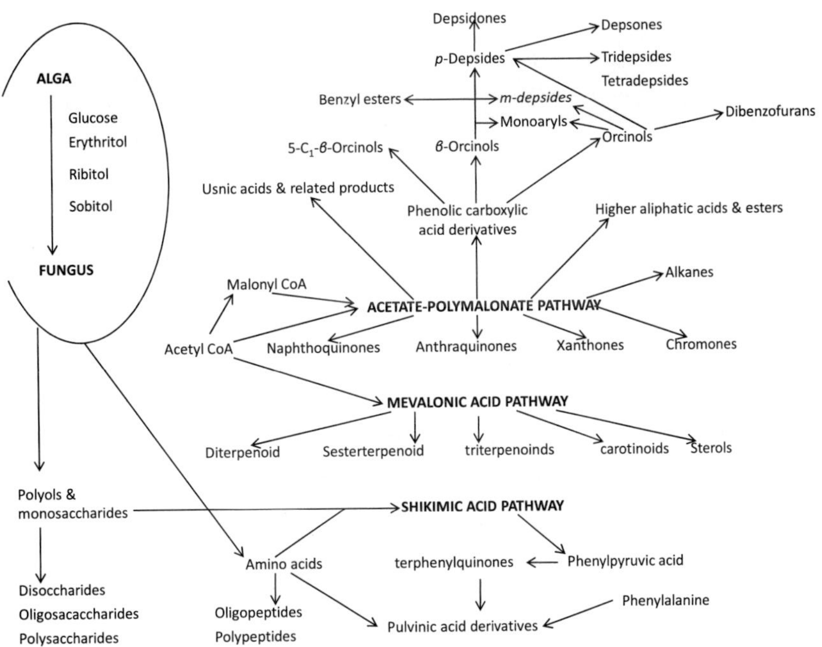

Рис. 4 Схема ацетат-полималонатного пути лишайниковых веществ（по: Culbersons & Johnson, 1977; Hale, 1983）.

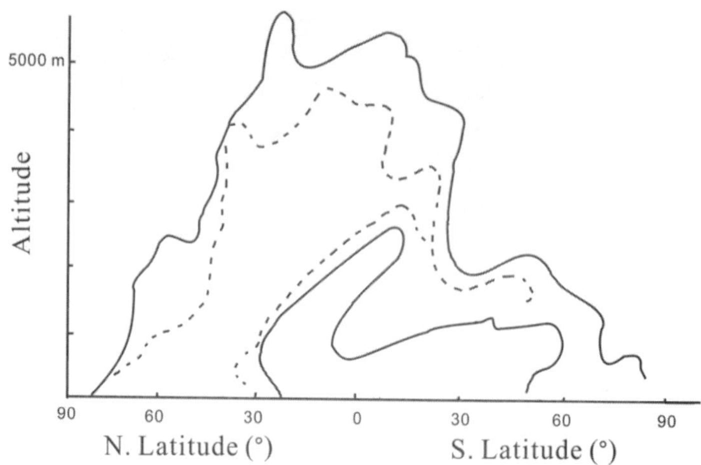

Рис. 5 распространение лишайников сем. Umbilicariaceae на разных широтах и высотах над ур. м. земного шара.

——*Umbilicaria* . ‒‒‒‒*Lasallia*

1-2. Понятие о роде лишайников сем. Umbilicariaceae.

Род *Lasallia* характеризован слоевищями с пузыревидными вздутиями（Рис. 6）без ризин и таллоконидиев, и сумками, содержащими по одной коричневой и муральной споре（Рис. 7）вместе с нотобороральными распространеними, в основном и сопровождает распространение горных областей голарктических зон（Рис. 5 и 12）. Однако род *Umbilicaria* характеризуется слоевищями без пузыревидных вздутий（Рис. 1-а）и иногда с ризинами и таллоконидиями（Рис. 8）, а также сумками, содержащими по 8 спор（Рис. 9）и вместе с биполярными и высокогорными, нотобореальными, и высокогорными распространениями（Рис. 5 и 13）.

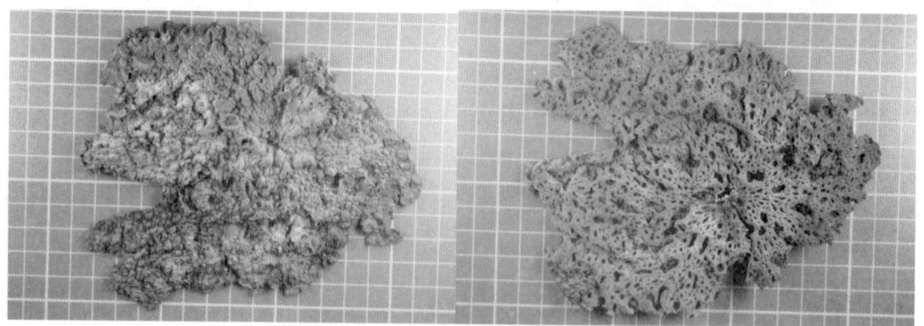

Рис. 6 Верхняя（налево）и нижняя（направо）поверхность пузыревидного слевища *Lasallia pustulata*（L.）Merat（ в масштабе 1 шкала = 1 cm）

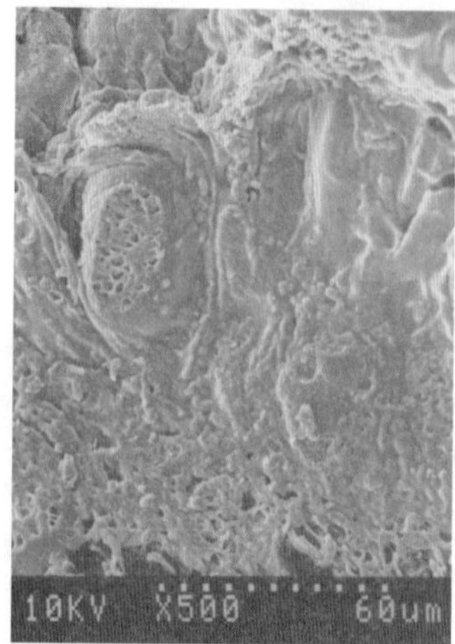

Рис. 7 Большая широкоэллипсоидная сумка по одной широкоэллипсоидной споре（*Lasallia rossica* Dombr., X500, СЭМ）.

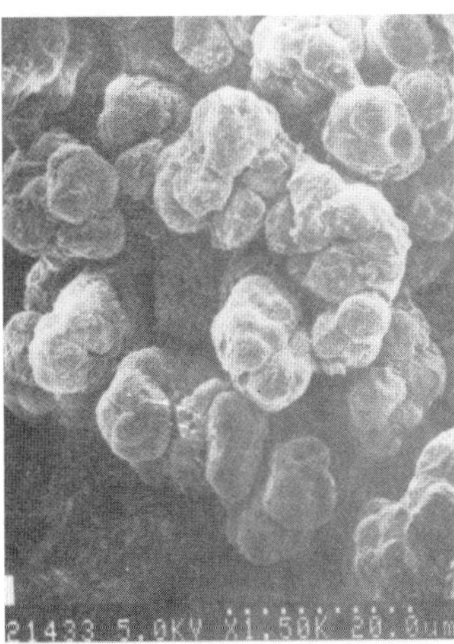

Рис. 8 Таллоконидии на нижней поверхности слоевища *Umbilicaria koidzumii* Yasuda ex Sato（СЭМ）.

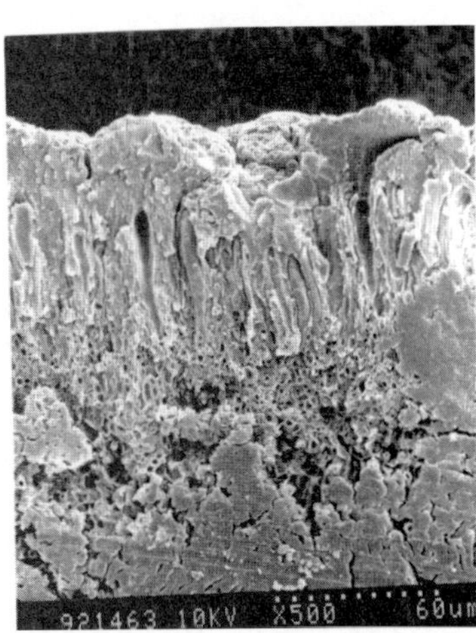

Рис. 9 Удлиненно-булавовидные сумки содержащие по 8 спор（*Umbilicaria subglabra*, X500, СЭМ）.

Рис. 10 Ровные диски у вида *Lasallia rossica* Dombr.
（Род *Lasallia*, подрод *Lasallia*（в масштабе 1 шкала = 1 cm）

Рис. 11 Гирозный диск у вида *Lasallia mayebarae*（Sato）Asahina.
（в масштабе 1 шкала = 1 cm）

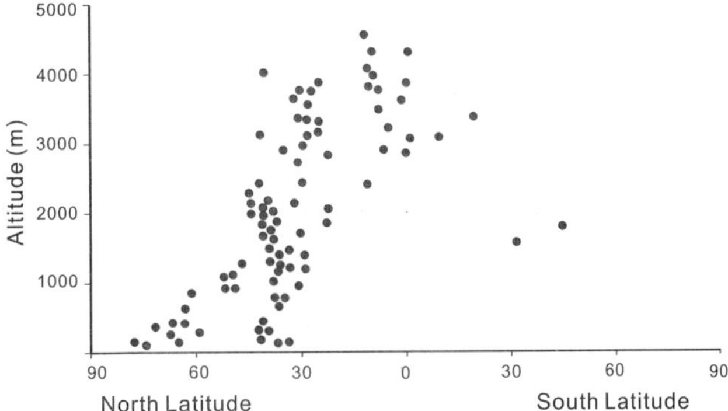

Рис. 12 Распространение лишайников рода *Lasallia* на разных широтах и высотах над ур. м. земного шара.

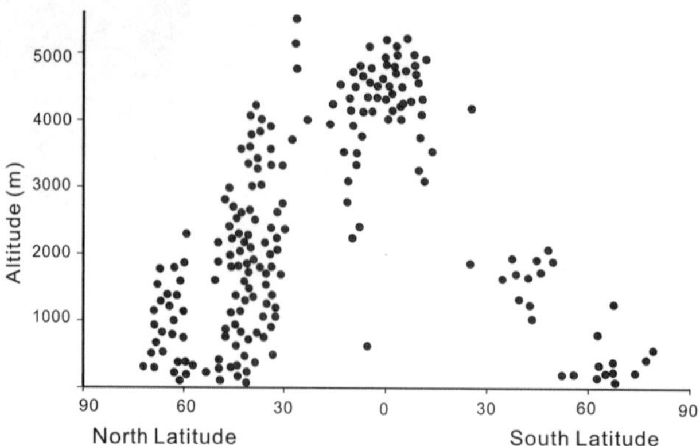

Рис. 13 Распространение лишайников рода *Umbilicaria* на разных широтах и высотах над ур. м. земного шара.

1-3. Понятие о подроде лишайников сем. Umbilicariaceae.

Каждый род лишайников сем. Umbilicariaceae может быть разделен при помощи соединенных характеров на два или четыре подрода. Так что при роде *Lasallia* у представителей подрода *Lasallia* диск ровный, гладкий, без бороздок и папилл（Рис. 10）и сопровождает нотобореальное распространение（Рис. 14）; у вида подрода *Pleiogyra* гирозный диск с не концентрически расположенными бороздками и складками（Рис. 11）вместе с Китайско-гималайским распространением（Рис. 15）. При роде *Umbilicaria* у представителей подрода *Umbilicaria* гирозный диск с концентрически расположенными бороздками и складками（Рис. 16）, и сопровождает арктическое и высокогорное и горное распространение（Рис. 20）; у видов подрода *Agyriphora* лиск ровный, гладкий без бороздок и папилл（Рис. 17）, вместе с арктическим и нотобореальным распространением（Рис. 21）; у трех видов подрода *Actinogyra* бороздки и складки на диске размещаются звездчато и разветвляются на периферии диска коротко-дихотомически（Рис. 18）и сопровождают нотобореальное распространение(Рис. 22); а у представителей подрода *Omphalodiscus* диск с папиллами, иногда с щелью（Рис. 19）и вместе с арктическими, антарктическим и высокогорным распространением низшей широты（Рис. 23）.

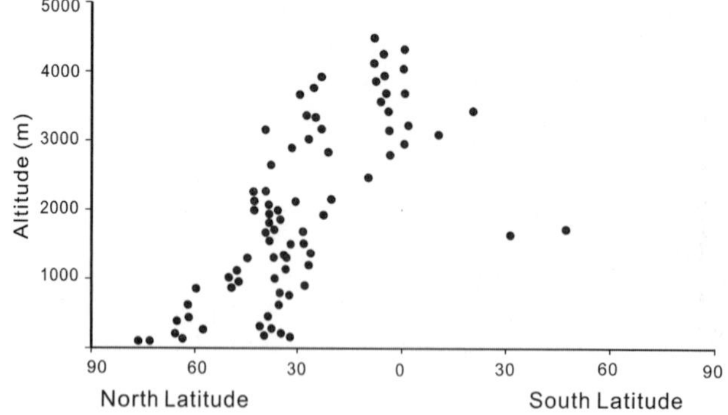

Рис. 14 Распространение лишайников подрода *Lasallia* на разных широтах и высотах над ур. м. земного шара.

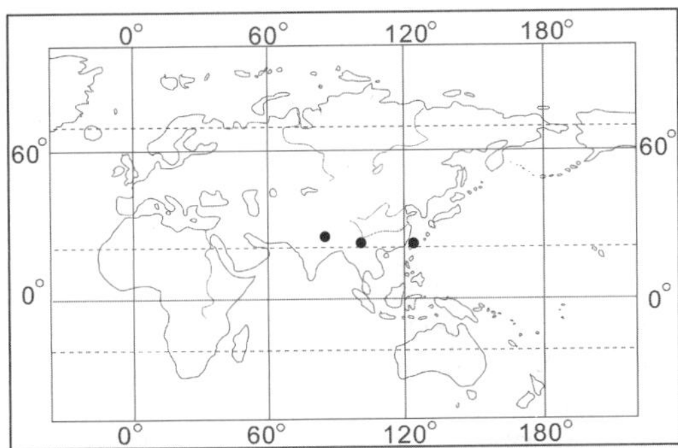

Рис. 15 Распространение на земном шаре *Lasallia mayebarae*（Sato）Asahina.

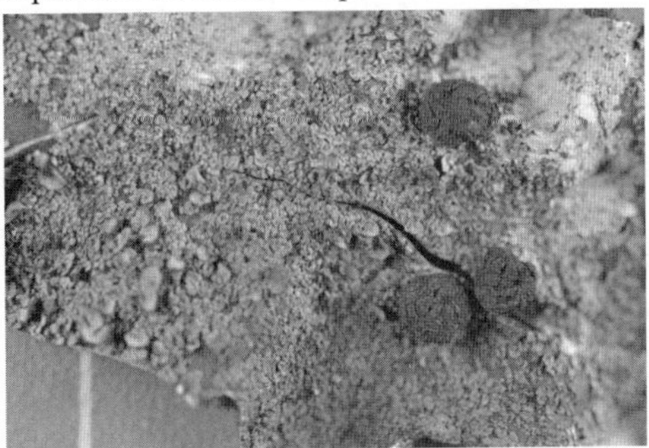

Рис. 16-а Гирозные диски у вида *Umbilicaria sguamosa* J. C. Wei & Y. M. Jiang（род *Umbilicaria* подрод *Umbilicaria*）.: *Umbilicaria squamosa*（holotype）Wei 2779（13362）.（в масштабе 1 шкала = 1 cm）

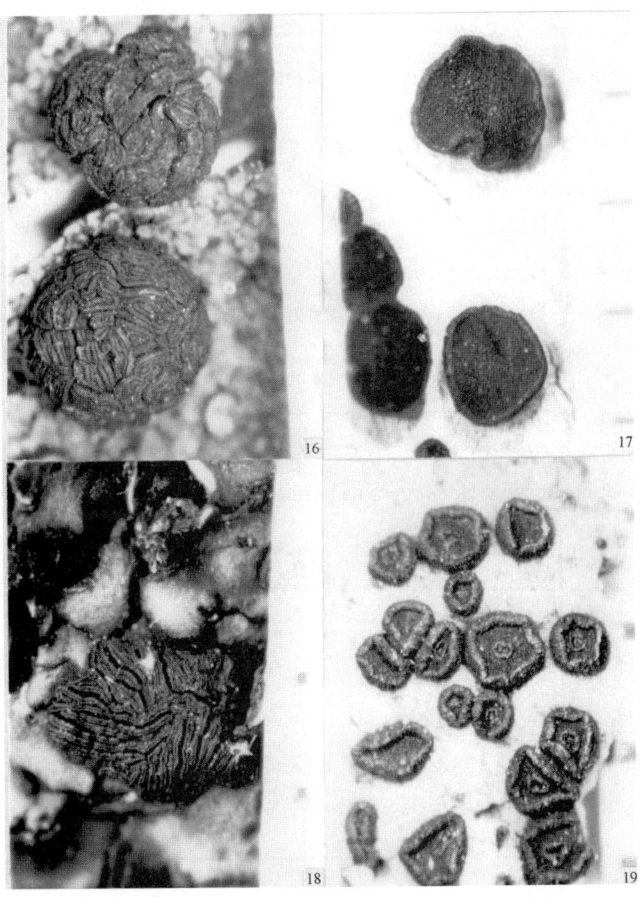

Рис. 16-б Сверху налево: Гирозные диски у вида *Umbilicaria sguamosa* J. C. Wei & Y. M. Jiang（род *Umbilicaria* подрод *Umbilicaria*）.:

Рис. 17 Сверху направо: Ровные диски у вида *Umbilicaria subglabra*（Nyl.）Harm.（в масштабе 1 шкала = 1 mm）

Рис. 18 Внизу налево: Актиногирозный диск у вида *Umbilicaria pulvinaria* Savicz.（подрод Actinogyra）.（в масштабе 1 шкала = 1 mm）

Рис. 19 Внизу направо: Омфалодиск у вида *Umbilicaria krascheninnikovii*（Savicz）Zahlbr.（подрод *Omphalodiscus*）.

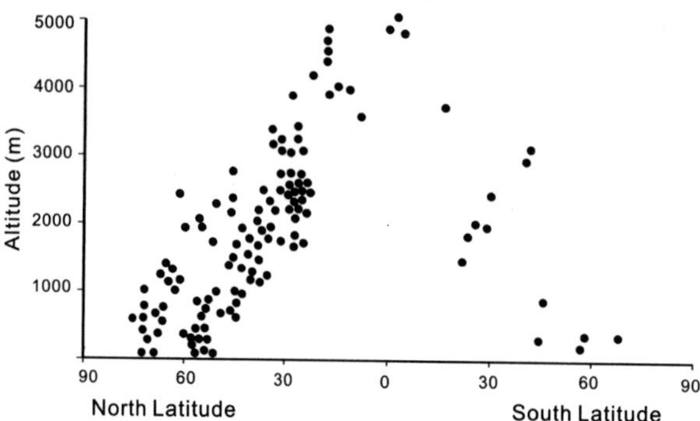

Рис. 20 Распространение подрода *Umbilicaria* на разных широтах и высотах над ур. м. Земного шара.

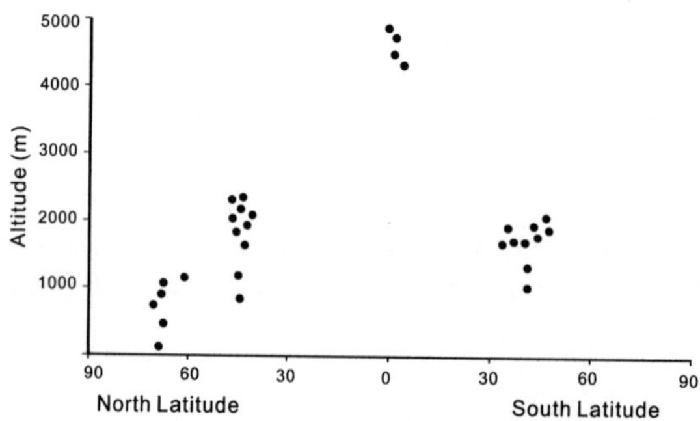

Рис. 21 Распространение подрода *Agyrophora* на разных широтах и высотах над ур. м. Земного шара.

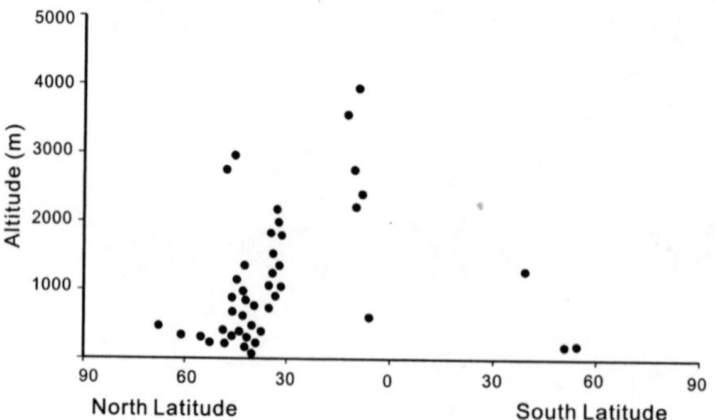

Рис. 22 Распространение подрода *Actinogyra* на разных широтах и высотах над ур. м. Земного шара.

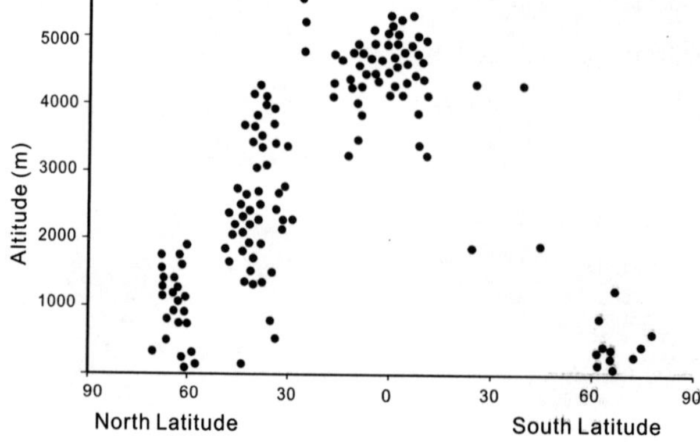

Рис. 23 Распространение подрода *Omphalodiscus* на разных широтах и высотах над ур. м. Земного шара.

1-4. Понятие о виде лишайников сем. Umbilicariaceae

Теоретически каждый биовид имеет свой особенный генетический пул. Поэтому фенотипическое выражение в морфологических или иных внешне раздичных признаках каждого биовида является результатом экспрессии генов, его отражением или можно представить как в какой-то степени расширение генетического пула. Благодаря тому что экспресия генов непременно подвергается влиянию разных факторов, а также разницы доминантности и рецессивности, поэтому все это так или иначе отражается на внещних признаках. Вот почему в отношении внешних признаков лишайников необходимо проводить комплексный анализ морфологии, анатомии, химии и географии с тем, чтобы уменьшить недоразумение о виде из-за разницы в экспрессии генов. С этой точки зрения, классификаия с использованием математических методов возможно имеет много преимуществ как в практическом, так и в теоретическом отношениях.

Поэтому при исследовании понятия о виде лишайников сем. Umbilicariaceae придерживаемся комплексного метода морфологии, анатомии, химии и географии. При условии, если четыре области не дают единого результата, то если и по крайней мере две из них приводят к единому виду, то можно признать его объективное существование. Фактически, данный Комплексный принцип применими к классификациям всех таксонов различных рангов. Благодаря развитию молекулярной биотехники, комплексный анализ соединения фенотипа и генотипа возможно как раз станет золотым ключиком и откроет двери в сущность вида.

2. Видовой состав лишайников сем. Umbilicariaceae Восточной Азии и их распростанение

Данная статья заключает 62 видов лишайников сем. Umbilicariaceae Восточной Азии, которые принадлежат двум родам, 6 подродам и 5 секциям（Табл.1）.

Табл 1. Видовой состав лишайников Сем. *Umbilicariaceae* Восточной Азии

Род	подрод	секция	Число видов	% от общего число видовидов	
Umbilicaria Hoffm.	*Umbilicaria*	*Umbilicaria*	32	51.6	83.9
		Gyrophoropsis（Elenk. & Savicz）A. Zahlbr.	4	6.5	
	Agyrophora Nyl.	*Agyrophora*	4	6.5	
	Omphaloodiscus（Schol.）Schade ex J. C. Wei & Y. M. Jiang	*Omphalodiscu*	7	11.3	
		Spodochroa Schol ex Llano	2	3.2	
Lasallia Merat	*Lasallia*		9	14.5	16.1
	Pleiogyra Wei		1	1.6	
Всего			62		

Лишайники сем. Umbilicariaceae распространение в разных странах или районах Восточной Азии（Табл. 2）. В таблице они располагаются по провинциям в Китае.

Табл 2. Лишайники сем. Umbilicariaceae в разных странах и районах Восточной Азии и разных провинциях Китая.

Вид \ район	T	S	Y	G	X	XJ	J	N	JX	SX	SH	A	F	GX	Z	H	HN	HL	L	HB	HM	IN	KI	TU	AF	TA	NP	SK	M	K	FE	AM	JV	JP	KR	B
Lasallia asiae-orientalis	+	+	+	+	+																						+									
L. deliensis			+		+																							+								
L. papulosa						+																							+	+						
L. pensylvanica						+	+	+																					+	+	+					
L. pertusa					+	+															+								+		+					
L. pustulata																													+							
L. rossica							+																							+	+		+	+		
L.sinorientalis									+	+		+	+																					+		
L. xizangensis				+																																
L. mayebarae	+	+																									+									
Umbilicaria leiocarpa																													+					+		
U. lyngei					+	+				+																			+		+					
U. rigida										+																			+					+		
U. subglabra					+																									+		+				
U. africana																																	+			
U. allaiensis					+																															
U. aprina v. *aprina*					+																		+	+	+											
U. aprina v. *halei*						+																							+							
U. decussata						+																+	+						+	+		+				
U. formosana	+	+		+		+			+																			+		+		+				
U. kraschenninikovi																														+				+		
U. virginis v.					+																	+							+			+				
U. virginis v. *lecanocarpoides*		+																			+		+				+			+						
U. crustulosa																							+					+								
U. spodochroa		+																																		
U. arctica																												+		+						

Вид \ район	T	S	Y	G	X	X J	J	N	J X	S X	S H	A	F	G X	Z	H	H N	H L	L	H B	H M	I N	K I	T U	A F	T A	N P	S K	M	K	F E	A M	J V	J P	K R	B
U. badia		+			+															+	+															
U. canescens																													+							
U. cinerascens									+																				+							
U. cylindrica					+		+		+																				+	+	+	+		+		
U. esculata		+					+	+				+		+	+	+	+	+	+														+	+		
U. exasperata							+																						+		+		+			
U. flocculosa					+		+																						+	+	+		+	+		
U. grisea																														+						
U. havaasii																														+						
U. herrei				+			+																													
U. hirsuta				+	+																			+					+	+	+					
U. hypococcinea				+					+	+																										
U. indica		+																		+	+					+		+								
U. kisovana							+	+		#																							+	+		
U. loboperipherica						+													+																	
U. minuta				+																																
U. nanella	+	+							+																											
U. nepalensis				+																						#										
U. nylanderiana					+																								+	+	+					
U. phaea																													+							
U. polyphylla																													+							
U. probescidea auct																							+						+	+			+			
U. pseudocinerascens		+																																		
U. squamosa		+		+																+		+						+								
U. subumbilicarioides				+																		+						+								
U. taibaiensis										+																	+	+								
U. thamnodes			+		+					+																										
U. torrefacta							+																						+		+			+		+
U. tylorhiza			+		+					+																		+						+		
U. vellea		+	+		+	+	+																		+				+	+	+	+		+		

Вид \ район	T	S	Y	G	X	XJ	J	N	JX	SX	SH	A	F	GX	Z	H	HN	HL	L	HB	HM	IN	KI	TU	AF	TA	NP	SK	M	K	FE	AM	JV	JP	KR	B
U. yunnana		+	+		+																															
U. angulata								+																					+							
U. caroliniana																													+		+			+		
U. cinereorufescens																													+		+			+		
U. koidzumii																																		+		
U. muchlenbergii								+										+													+			+		
U. polyrhiza																													+		+					
U. pulvinaria																															+					

Примечание к таблице 2:

A=Аньхой, AF=Афганистан, B=Берингов пролив, F=Фуцзянь, FE=Дальневосточная Россия, G=Гуйчжоу, GX=Гуанси, H=Хубэй, HB=Хэбэй, HL=Хэйлунцзян, HM=Гималаи, HN=Хунань, IN=Индия, J=Цзилинь, JP=Япония, JV=Ява, JX=Цзянси, K=Казахстан, KI=Киргизстан, KP=Корея, L=Ляонин, M=Монголия, N=Внутренняя Монголия, NP=Непал, S=Сычуань, SH=Шаньси, SK=Сикким, SX=Шэньси, T=Тайвань, TA=Таджикистан, TU=Турция, X=Тибет, XJ=Синьцзян, Y=Юньнань, Z=Чжецзян, +=на основе изученных автором образцах, £=на основе литературных материалов.

3. Пропагулы и их распространение лишайников сем. Umbilicariaceae Восточной Азии.

Пропагулы являются единицы размножения лишайников включая половые（такие как аскоспоры）и вегетационные（такие как таллоконидии, изидии, лобули, паралобули, таллилы（thallyles）, соредии, парасоредии, и фрагментации）.

Размеры ареала вида лишайников зависит от экологических условий и взаимодействия живых существ, но главное это способность размножения и миграции лишайников и приспособления к среде. Для лишайников необходимо иметь сравнительно устойчивый субстрат. Пропагулы лишайников сем. Umbilicariaceae Восточной Азии помимо соединения аскоспор с соответстующими клетками водорослей（фотобионтов）*Trebouxia* sp., имеют целый ряд вегетативных пропагулов. Благодаря тому что эти вегетативные пропагулы чаще всего сосуществуют с соответствующими грибами（микобионтами）и водорослей и вместе распраcтраняются, поэтому возможно пропагулы для размножения и распространения лишайников являются главными. Теперь рассмотрим пропагулы лишайников сем. Umbilicariaceae Восточной Азии и их пути распространения в отдельности следующие：

3-1. Аскоспоры

Зрелые акоспоры, окропленные дождем и росой, при определенной температуре под давлением набухания сами выпрыскиваются из сумки. Встретив подходящие клетки водорослей, при соответствующей среде, аскоспора начинает прорастать и образует общее тело с подходящими микобионтом и фотобионтом, превращаясь в новый молодой лишайник.

Благодаря тому, что значение пикноконидий для размножения у лишайников до настоящего времени польностью не выяснено, поэтому не остановимся на этом вопросе.

Среди видов лишайников сем. Umbilicariaceae Восточной Азии, имеющих аскоспоры следующие: *U. leiocarpa*, *U. arctica*, *U. proboscidea auct*, *U. rigida*, *U. cylindrica auct.*, *U. taibaiensis*, *U. altaiensis*, *U. exasperata*, *U. torrefacta*, *U. formosana*, *U. hypococcinea*, *U. angulata*, *U. crustulosa*, *U. nanella*, *U. caroliniana*, *U. spodochroa*, *U. phaea*, *U. muehlenbergii*, *и U. pulvinaria*.

3-2. Таллоконидии

Таллоконидии（Hestmark, 1990; J. C. Wei & Y. M. Jiang, 1993）– это толстостенные клетки, образованные от клеток гиф нижнего корового слоя лишайников рода *Umbilicaria* и на них ризинах（Рис. 8, 24）. Обычно они содержат коричневые пигменты. Эти таллоконидии могут сами произрастать（Hestmark, 1990）. Поэтому, встретив ссответстующие фотобионты, при благоприятной среде образуют маленькое слоевище и вырастает в новый, молодой лишайник. Среди видов лишайников сем. Umbilicariaceae Восточной Азии, следующие виды имеют таллоконидие:

3-2-1. Виды, имеющие только таллоконидии:

U. Lyngei, *U. nylanderiana*, *U. polyphylla*, *U. aprina*, var. *aprina*, *U. cinereorufescens*, *U. cinerascens U. polyrhiza*, *и U. havaasii*.

3-2-2. Виды, имеющие апотеции и акоспоры, а также таллоконидии:

U. subglabra, *U. nepalensis*, *U. africana*, *U. aprina* var. *halei*, *U. pseudocinerascens*, *U. decussata*, *U. subumbilicarioides*, *U. krascheninnikovii*, *U. thamnodes*, *U. virginis* var. *virginis*, *U. tylorhiza*, *U. virginis* var. *lecanocarpoides*, *U. vellea*, *U. badia*, *U. yunnana*, *U. koidzumii*, *U. indica*, *U. minuta*.

3-3. Изидии

Изидии – это маленькие выросты верхней поверхности слоевища（Рис. 25）. Они покрываются коровым слоем и содержат те же водоросли, что и все остальное слоевище. Оторванные от слоевища, в условиях благоприятных для роста, изидии развываются в новые слоевища.

Изидии встречаются у лишайников сем. Umbilicariaceae Восточной Азии из следующих видов:

3-3-1. Виды, имеющие только изидии:

L. asiae-orientalis, *L. pustulata*, *L. sinorientalis*, *L. xizangensis*.

3-3-2. Вид имеющие изидии, а также апотеции и акоспоры.

L. daliensis

3-4. Лобули

Лобули（Рис. 26）имеют вид маленьких чешуек и располагаются как на верхней поверхности, так и по краям слоевища. Они как изидии играют роль пропагул. У лишайников сем. Umbilicariaceae Восточной Азии лобули встречаются из следующих видов.

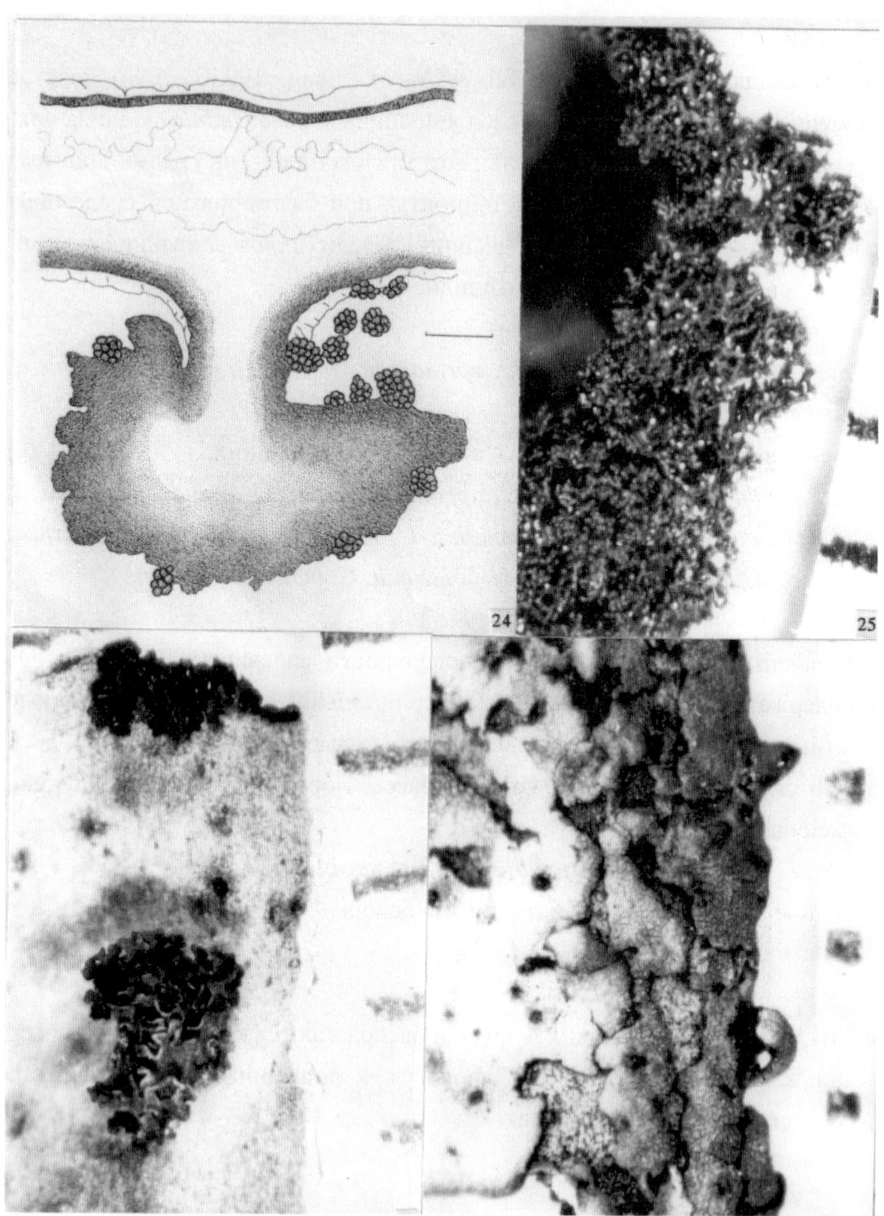

Рис. 24 Сверху налево: Таллоконидии у вида *Umbilicaria tylorhiza* Nyl.

Рис. 25 Сверху направо: Изидии у вида *Lasallia asiae-orientalis* Asahina（в масштабе 1 шкала = 1 mm）

Рис. 26-а Внизу налево: Лобулиу вида *Lasallia rossica* Dombr.（в масштабе 1 шкала = 1 mm）

Рис. 27-а Внизу направо:Паралобули у вида *Umbilicaria loboperipherica* J.C.Wei et al.（в масштабе 1 шкала = 1 mm）

3-4-1. Следующий вид имеет только лобули.

U. flocculosa.

3-4-2. Виды, имеющие не только лобули, но и апотеции и акоспоры:

L. papulosa, *L. pensylvanica*, *L. rossica*, *L. mayebarae*

3-5. Паралобули

Коровой слой верхней поверхности слоевища по периферии у вида *Umbilicaria loboperipherica* очень хрупкий и легко лопается, и части осколочков часто падают на верхнюю поверхность личного или другого слоевища лишайников, а также и на другие субстраты, в условиях благоприятных для роста и образуют новые слоевище. Оставшаясь часть корового слоя приподнята и постепенно образуется нижняя поверхность, подобная лобулу. Эти паралобули（Wei et al., 1995; Рис. 27-а, -б）часто сближаются и образуют нижнюю поверхность, и часто, как и изидии и лобули, выполняют роль пропагул. Виды лишайников сем. Umbilicariaceae Восточной Азии, имеющие паралобули, следующие:

3-5-1. Вид, имеющий только паралобули:

U. Loboperipherica

3-5-2. Вид, имеющий паралобули, а также таллоконидии:

U. esculenta

Рис. 26-б Лобулиу вида *Lasallia rossica* Dombr.（в масштабе 1 шкала = 1 cm）

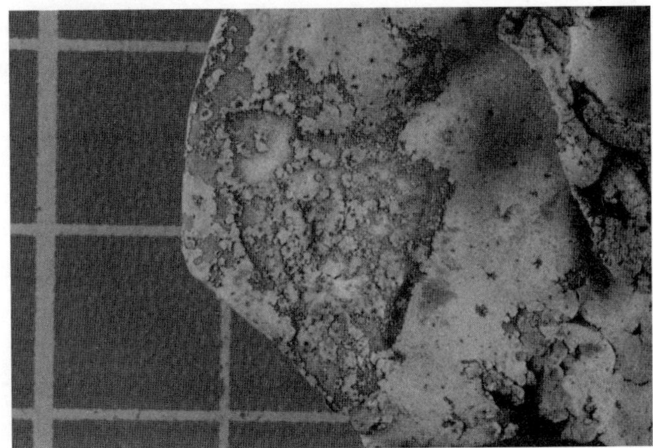

Рис. 27-б Паралобули у вида *Umbilicaria loboperipherica* J.C. Wei et al.（в масштабе 1 шкала = 1 cm）

3-6. Таллилы（thallyles）

Таллилы（J. C. Wei & Y. M. Jiang, 1993; Рис. 28）растут на нижней поверхности слоевища на кончике продолговатого ризоида в виде минислоевища, диаметр ее 1 mm. В морфологии и анатомии эти минислоевища подобны молодым слоевищам *Umbilicaria* и являются одной из форм пропагул. Такие таллилы среди лишайников сем. Umbilicariaceae Восточной Азии обнаружены на слоевищах *U. squamosa*. Кроме таллилов этот лишайник еще имеет другие.

Рис. 28 Таллилы（thallyles）у вида *Umbilicaria squamosa* J.C. Wei & Y.M. Jiang.（в масштабе 1 шкала = 1 cm）: *Umbilicaria squamosa*（holotype）Wei 2779（13362）.

3-7. Соредии

Середи представляют собой порошковые массы, которые образованы сплетением грибных нитей, заключающих внутри зеленые водоросли *Trebouxia* sp. Эти соредии при благоприятных условиях могут развиваться дальше и образовать новые слоевища. Это один из главных вегетативных пропагул.

Вид *Lasallia pertusa* является единственным лишайником из сем. Umbilicariaceae Восточной Азии, который имеет соредии（Рис. 29）.

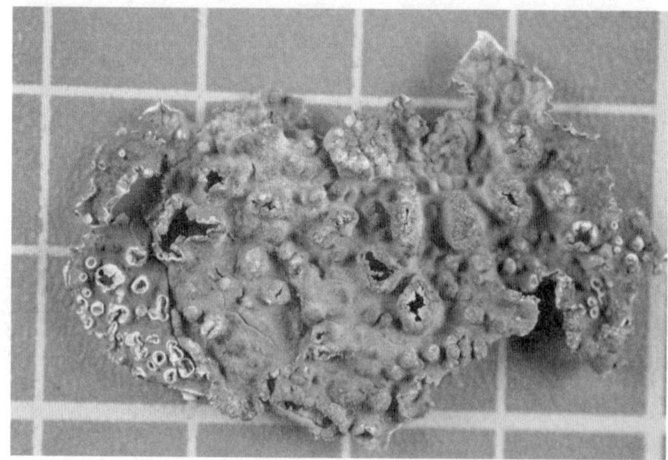

Рис. 29 Соредии в виде соралей у вида *Lasallia pertusa*（Rassad.）Llano.（в масштабе 1 шкала = 1 cm）

3-8. Парасоредии

Парасоредии（Codogno et al., 1989）представляют собой порошковые массы, которые образованы разъеденными верхними коровыми слоями. Среди лишайников сем. Umbilicariaceae Восточной Азии такие вегетативные пропагулы обнаружены только у *U. hirsuta*（Рис. 30）.

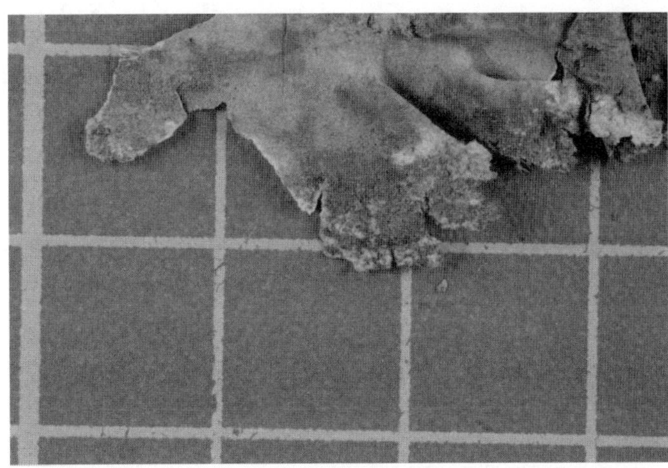

Рис. 30 Парасоредии у вида *Umbilicaria hirsuta*（Sw.）Ach.（в масштабе 1 шкала = 1 cm）

пропагулы, такие как паралобули, апотеции и аскоспоры.

3-9. Фрагментации.

Что касается тех видов, у которых отсутствуют апотеции и аскоспоры, не имеют других вегетативных пропагул, то таким лишайникам приходится размножаться и распространяться с помощью фрагментации. Фактически, у видов, имеющих все вышеуказанные пропагулы, для фрагментации играют роль вегетативных пропагул, и может служить распространителем и размножителем.

Очевилно, размножение лишайников из сем. Umbilicariaceae Восточной Азии главным образом осуществляестся путем вегетативных пропагул. В сухих условиях эти вегетативные пропагулы часто бывают хрупкими и легко измельчаются. Поврежденные когтями животных или механическими воздействиями, слоевища и их вегетативные пропагулы измельчаются или отпадают и благодаря ветрам, дождям или когтям животных распространяются и размножаются.

4. Географические элементы, типы ареалов и их происхождение лишайников из сем. Umbilicariaceae Восточной Азии.

Каждый вид лишайников размножается по-своему и распространяется своим путем, и в зависимости от способности акклиматизироваться в определенное время занимает определенное пространство, таким образом образуются разные ареалы для разных видов.

В отношении размеров ареалов вообще говоря ареалы у лишайников больше чем у высших растений. Например, ареалы видов лишайников по размерам подобны размерам ареалов родов или даже семейств высших растений. Это возможно связано с широкой экологической амплитудой лишайников и их способом размножения и распространения, имеющих вегетативные пропагулы. Так как у вегетативных пропагул отсутствует обмен генами и рекомбинации генов, поэтому лишайники генетически более консервативные, чем высшие растения. Таким образом в определенное время виды лишайников обычно занимают большее пространство чем виды высших растений. За исключением тех случаев, если среда резко изменяется или под воздействием ультра фиолетовых лучей и других факторов происходит генмутация и вслед за этим морфологическая дифференциация и географическая изоляция, происходят викарные виды и викарные ареалы. Кроме того, лишайники, это грибы в симбиозе с водорослями, как жизнь-поддерживающая система, имеют сильную способность присопособления к среде, в особенности к сухости. Это играет значительную роль в расширении ареала. Так что размеры ареалов видов лишайников обычно подобны размерам ареалов родов или даже семейств высших растений. Поэтому в географическом анализе лишайников главным образом берутся ареалы видов.

Далее проанализируем и обсудим типы ареалов видов лишайников и их происхождения из сем. Umbilicariaceae Восточной Азии.

4-1. Географические элементы и их типы ареалов

В настоящее время в составе лишайников сем. Umbilicariaceae Восточной Азии насчитывается 62 вида.

Ввиду многообразия ареалов видов лишайников сем. Umbilicariaceae Восточной Азии автор разбивает их на 7 географических элементов, в том числе группа эндемичных видов, свойственных только для Азии. В каждом элементе или группе выделены несколько типов ареалов（Табл. 3）.

Табл. 3 Географические элементы и их типы ареалов видов лишайников сем. Umbilicariaceae Восточной Азии

Географический элемент	Тип ареала	число видов	% от бощего числа видов
арктический	Еврамериканско-гренландский	1	3.2
	Кольский	1	
Аркто-Альпиский	Восточноазиатско-североамериканский	2	22.6
	Голарктический	7	
	Мультирегиональный	5	
Биполярный	Аркто-альпино-антарктический	2	3.2
Гипоарктомонтанный	Старомировой	1	18
	Восточноазиатско-северно-американский Восточноазиатско-западная часть СеверноцАмерики	1	
	Восточноазиатско-западная часть СеверноцАмерики	2	
	Средиземноморско-сычуанский	1	
	Средиземноморско-восточноазиатский	1	
	Голарктический	2	
	Мультирегиональный	3	
Ното-бореальный	Евразиатско-австралазиатский	2	8
	Евразиатско-южноамериканский	2	
	Средиземноморско-восточноазиатско-австралийский	1	
Альпийский	Азиатско-Африканский	1	5
	Восточноазиатско-восточно-африканскко-мексиканский	1	
	Восточноафриканско-яванско-колумбиатский	1	
Эндемичные виды Восточной Азии	Эндемики Китая	4	40
	Эндемики Китайско-гималаев	13	
	Эндемики Гималаев	1	
	Эндемики Китая, Кореи и Японии	2	
	Эндемики Китая и Японии	1	
	Эндемики Японии	1	
	Эндемики северо-восточной Азии	1	
	Эндемики Камчатки и Японии	1	
	Эндемики Камчатки и Якутии	1	
	Всего	62	

Ⅰ. Арктический элемент

Этот элемент охватывает виды, ареалы которых распространены в основном лишь в Арктике. К этому элементу относятся 2 вида из сем. Umbilicariaceae Восточной Азии, что составляют 3.2% от общего числа видов.

Ⅰ-1. Еврамериканско-гренландский тип ареала

К этому ареалу относится *U. havaasii*. Его ареал главным образом находится на севере Арктического круга Скандинавского полуострова и Гренландии. Этот ареал простирается даже на юг за Арктический круг, более того, типичные образцы этого вида найдены в Монголии, в районе Южной Гоби-Алтая и южной Аляски на тех же широтах, что и в Монголии. Таким образом это фактически Аркто-южноаляско-монгольская дизъюнкция（Рис. 31）.

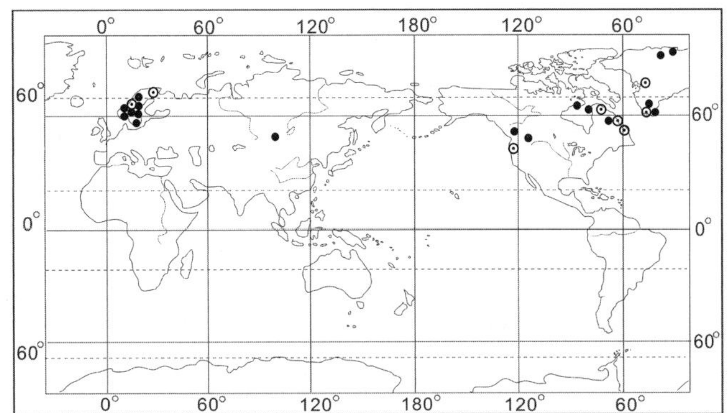

Рис. 31　Распростронение на земном шаре *Umbilicaria havaasii* Llano（● работа основана на личных исследованных материалах. ⊙ на основании Литератур.）

Ⅰ-2. Кольский тип ареала

Этот тип характерен для *U. canescens*. Благодаря распространению данного вида в Монголии（Бязров, 1986）его считают также Кольско-монгольской дизъюнкцией（Рис. 32）.

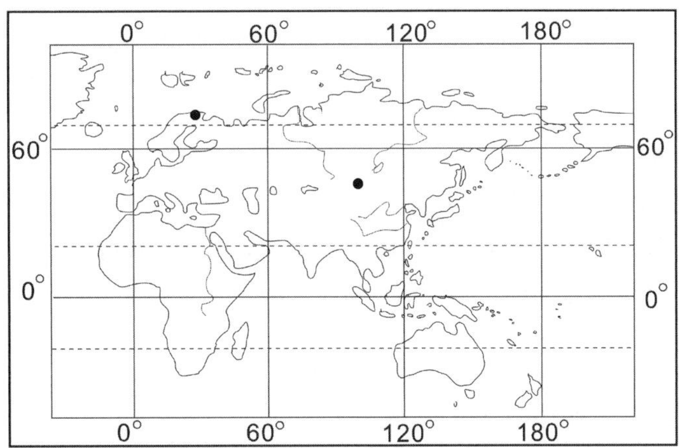

Рис. 32　Распространение на земном шаре *Umbilicaria canescens* Dombr.

Ⅱ. Аркто-альпийсккий элемент

Этот элемент в основном охватывает виды, ареал которых находится в Арктике, а также в высокогорных районах Голарктики Европы, Азии и Северной Америки.

К этому элементу относится 14 вида из сем. Umbilicariaceae Восточной Азии. Они занимают 22.6% от общего числа видов.

Ⅱ-1. Восточноазиатско-североамериканский тип ареала

К этому ареалу относится *U. caroliniana* и *L. pensylvanica*. Его восточноазиатская часть ареала находится в районах Якутии и Амура России, а также на Северо-востоке Китая и Японии, другая－в Северной Америке, а простирается на север в Арктический круг Аляски, на юг прерывисто распространяется на Северную Каролину.

Ⅱ-2. Голарктический тип ареала

Сюда входят виды, ареал которых расположен на Арктике и на альпах Голарктики. Принадлежащие этому ареалу следующие виды:

U. arctica, *U. cinereorufescens*（Рис. 33）. *U. crustulosa*, *U. leiocarpa*, *U. lyngei*, *U. rigida*, и *U. torrefacta*.

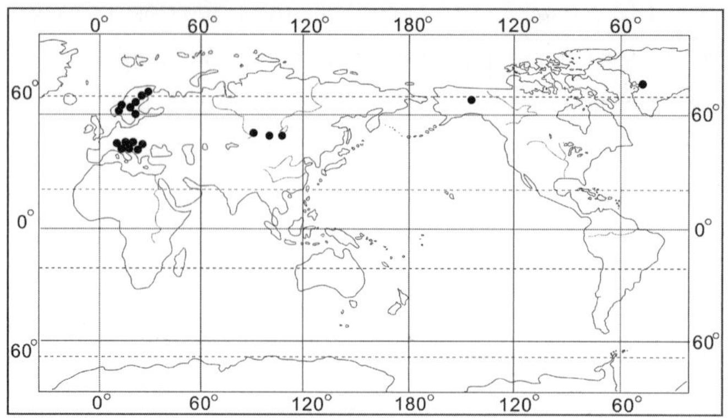

Рис. 33 Распространение на земном шаре *Umbilicaria cinereorufescens*（Schaer.）Frey.

Ⅱ-3. Мультирегиональный тип ареала

Виды лишайников очень широко распространены, в их ареалы входят даже несколько географических областей растений и лишайников. Эти виды являются т. н. космополитными или мультирегиональным（polychores или pluri-regional species, Eig, 1931）.

Данный ареал расположен в Арктике и распространяется даже на южное полушаре. Таких видов лишайников сем. Umbilicariaceae Восточной Азии существует пять: *U. cylindrica* auct, *U. exasperata*, *U. proboscidea auct*., *U. vellea*, *U. virginis*.

Ⅲ. Биполярный элемент

Этот элемент объединяет виды, одна часть ареала которых в Антарктике, другая в Арктике, а третья – в высокогорных низшей широты. К этому элементу принадлежит только 2 вида из сем. Umbilicariaceae Восточной Азии и они составляются 3.2% от общего числа видов.

Ⅲ-1. Аркто-альпино-антарктический тип ареала

Этот тип свойствен 2 виду: *U. aprina и U. decussata*.

U. decussata и *U. aprina*（Рис. 34）являются типичными представителями этого типа ареала. Однако у разновидностей *U. aprina* var. *aprina* и var. *halei*, и произошел географический викарный ареал. Ареал первой разновидности распространен на Антарктике, в высокогорных районах Африки, Средней Азии и Европы. Ареал последней разновидности ограничен в районах восточной Азии（Китай, Гималаи и Монголия）и в Арктике Северной Америки（Ривер-Клайд）.

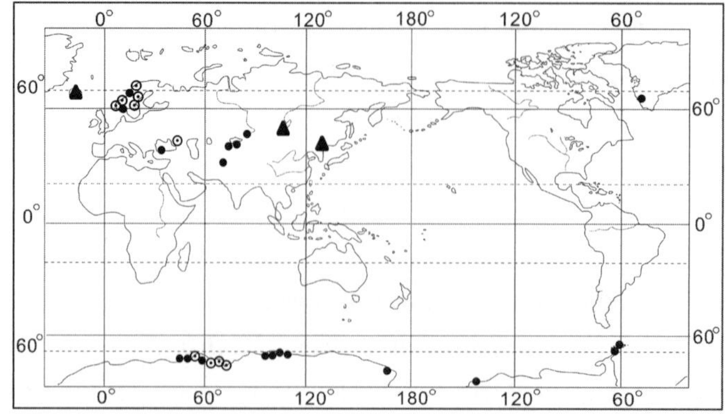

Рис. 34 Распространение на земном шаре *Umbilicaria aprina* Nyl.

[var. *aprina*（●, ⊙）, var. *halei* Llano（▲）].

（●, ▲ работа основана на личных исследованных материалах. ⊙, Δ на основании Литератур.）

Ⅳ. Гипоарктомонтанный элемент

Данный элемент в основном охватывает виды, которые распространяются в районах между Арктической зоной и тайгой. К югу может тянуться до района инзких гор. К этому элементу относятся 11 видов из лишайников сем. Umbilicariaceae Восточной Азии. Они занимаются 18% от общего чилса видов.

Ⅳ-1. Старомировой тип ареала

Принадлежащий этому ареалу *L. pustulata* имеет два центра современного ареала: один в Европе, другой в Африке（Рис. 35）.

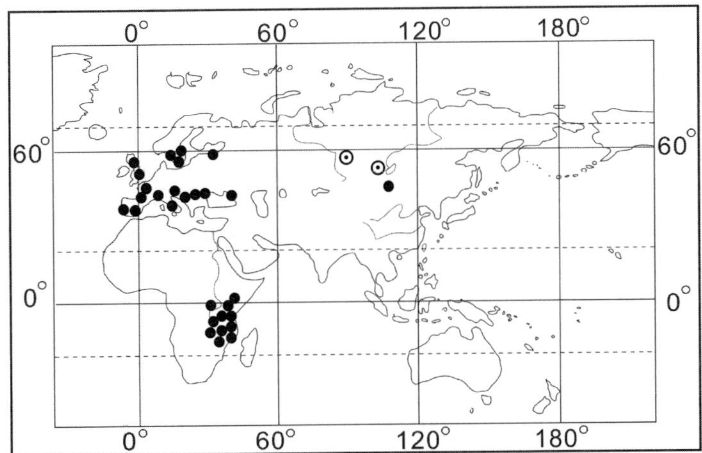

Рис. 35 Распространение на земном шаре *Lasallia pustulata*（L.）Merat.

（● работа основана на личных исследованных материалах. ⊙ на основании Литератур.）

Этот вид, опубликованный в Китае（Zahlbruckner 1930, 1922）и Японии（Asahina 1931）, фактически является ошибочным определением *L. aisae-orientalis* или *L. sinorientali*s（J. C. Wei & Y. M. Jiang, 1993）. Поэтому распространение этого вида в Монголии（Бязров 1986）, возможно является южно-восточной окраиной на Евразиатском материке. Что касается распространения этого вида в Северной Америке（Llano, 1950）, из-за того, что оригинальные материалы цитированные Llano（1950）не были иссоедованы для этой работы.

Ⅳ-2. Восточноазиатско-североамериканский тип ареала

Этот тип объединяет 3 вида: *U. angulata*, *U. phaea*, *и U. muehlenbergii.*

Для этого ареала имеется два подтипа: 1. Подтип между Восточной Азии и западной частью Северной Америки. К этому относятся виды *U. angulata* и *U. phaea*; последний тянется даже в южную Америку. 2. Подтип между Восточной Азией и восточной частью Северной Америки. К нему принадлежит вид *U. muehlenbergii.*

Ⅳ-3. Средиземноморско-сычуанский тип ареала

Этот тип охватывает *U. spodochroa*. Он главным образом распространен на северном побережье Средиземного Моря и южном побережье Скандинавского полуострова, а также в Китае в районе Сычуаньского Кандин（Рис. 36）.

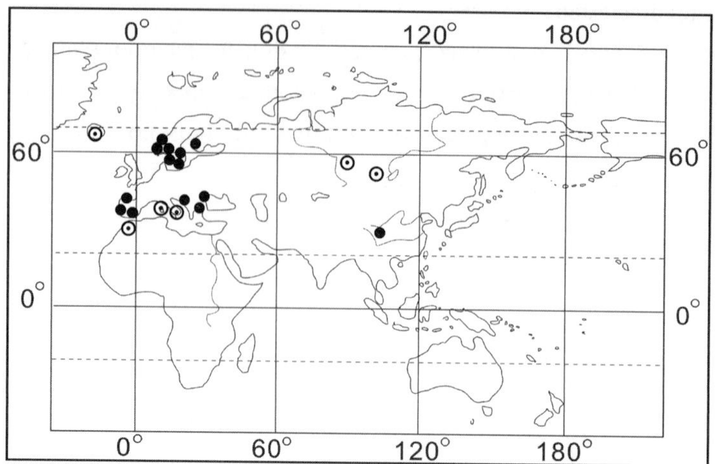

Рис. 36 Распространение на земном шаре *Umbilicaria spodochroa* Ehrh. Ex Hoffm.
（● работа основана на личных исследованных материалах. ⊙ на основании Литератур.）

Ⅳ-4. Средиземноморско-восточноазиатский тип ареала

Этот тип характерен для *U. tylorhiza*, ареал которого в основном размещен в Скандинавии Европы, и в Гималаях и Циньлине Китая, кроме того встречается и в Японии（Рис.37）.

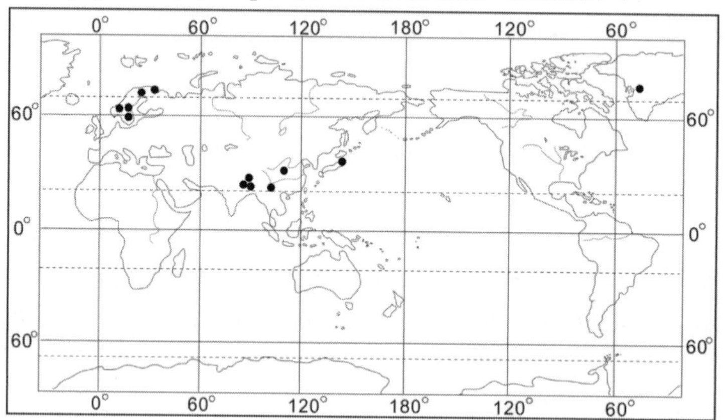

Рис. 37 Распространение на земном шаре *Umbilicaria tylorhiza* Nyl.

Ⅳ-5. Голарктический тип ареала

Этот тип свойствен 2 виду: *U. flocculosa* и *U. hirsuta*.

Ⅳ-6. Мультирегиональный тип ареала

Этот тип ареала объединяет 3 вида: *U. herrei*, *U. polyphylla*, и *L. papulosa*.

Ⅴ. Нотобореальный элемент

Этот элемент входят виды северного умеренного пояса и южного умеренного пояса. Этот элемент объединяет 5 видов, что составляется 8% от общего числа видов.

Ⅴ-1. Евразиатско-австралазиатский тип ареала

Этот ареал значительно сильнее дизъюнктирован, например, у *U. nylanderiana* и *U. subglabra*, одна часть ареала которых в основном находится в центральной Азии и Монголии, другая – в Европе, третья – в Австралии и Новозеландии южного полушария（Рис. 38）.

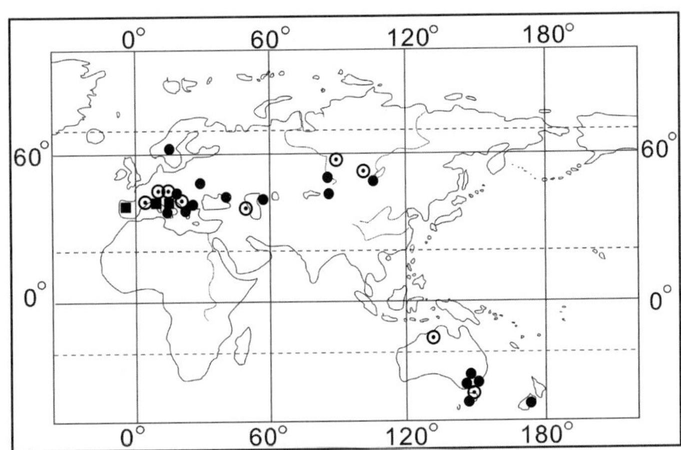

Рис. 38 Распространение на земном шаре *Umbilicaria subglabra* Nyl.（var. *subglabra*: ●работа основана на личных исследованных матерриалах, ⊙ на основании Литератур; var. *pallens*（Nyl.）Frey ■работа основана на личных исследованных матерриалах.）

Ⅴ-2. Евразиатско-южноамериканский тип ареала

Этот тип свойствен 2 вида: *U. grisea U. polyrhiza*.

Лишайник *U. grisea* главным образом распространен на побережном поясе Скандинавского полуострова, а также и в Монголии. Однако в последние годы типичные образцы данного вида обнаружены нами в коллекциях лишайников, собранных из Чили Южной Америки, а также и в юго-восточной части Австралии. Таким образом данный вид распространен не только в северном полушарии, но и в южном. Так что ареал его разъединен на три части: часть Европы, Азии и южной Америки（Рис. 39）.

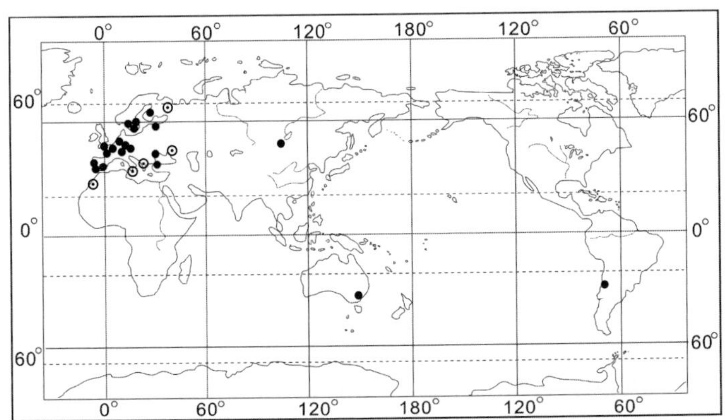

Рис. 39 Распространение на земном шаре *Umbilicaria grisea* Hoffm

（ ● работа основана на личных исследованных матерриалах, ⊙ на основании Литератур.）

Ⅴ-3. Средиземноморско-восточноазиатско-австралийский тип ареала

У вида U. cinerascens, ареал которого находится в основном на северном побережье Средиземного моря, и Монголии, в Циньлине Китая и в Австраоии южного полушария（Рис. 40）.

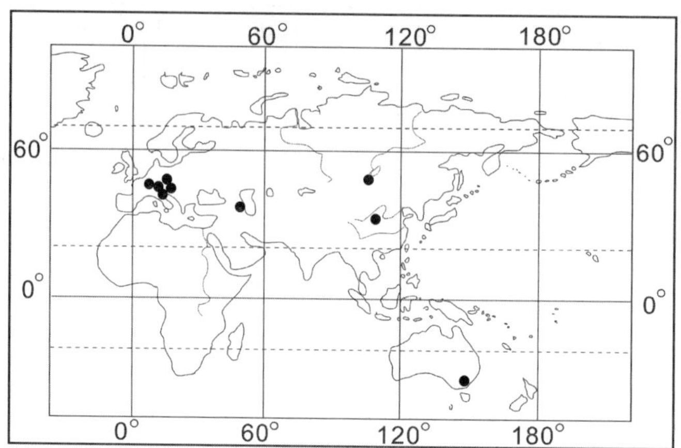

Рис. 40 Распространение на земном шаре *Umbilicaria cinerascens*（Arnold）Frey.

Ⅵ. Альпийский элемент

Этот элемент наблюдается у видов, ареал которых находится в высокогорьях умеренной области земного шара. К этому элементу относятся 3 вида из сем. Umbilicariaceae Восточной Азии. Они занимают около 5% от общего чилса видов.

Ⅵ-1. Азиатско-африканский тип ареала

Современный ареал *L. pertusa* находится на побережье оз. Байкала Сибири и в Алтае и Гималаях（Wei & Jiang, 1993）, а также в Эфиопии Восточной Африки.（Krog, 1973; Krog & Swinscow, 1986; Рис. 41）.

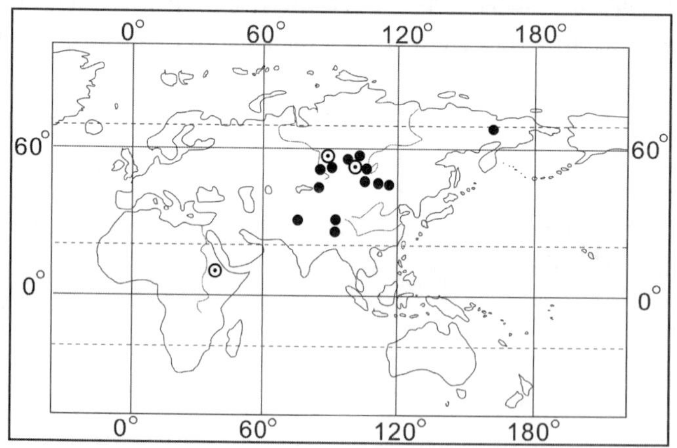

Рис. 41 Распространение на земном шаре *Lasallia pertusa*（Rassad.）Llano

（ ● работа основана на личных исследованных матерриалах, ⊙ на основании Литератур.）

Ⅵ-2. Восточноазиатско-восточноафриканско-мексиканский тип ареала

Работая в Гербариях лишайников при Британском Музее Натуральной Истории（BM）автор обнаружил много образцов вида *U. formosana* среди африканских материалов, определенных Крогом как *U. umbilicarioides* и *U. decussata*. Оригинальный материал, синтип вида *U. bigleri* описанного фреем из Мексики найден нами в Фарлов（Farlow）Гербарии лишайников при Горвардском Университет（США）. Исследование показывает что, он фактически идентичен *U. formosana*（J. C. Wei & Y. M. Jiang, 1993; рис. 1.а）. Поэтому ареал вида *U. formosana* не ограничен районами Китая（Тайвань, Циньлинь）и Гималаев, распространяется и в южной Азие, Эфиопии Восточной Африки, и в Мексике центральной Америки. Таким образом появляется дизъюкция Восточноазиатско- восточноафриканско-центральноамериканская（Рис.42）.

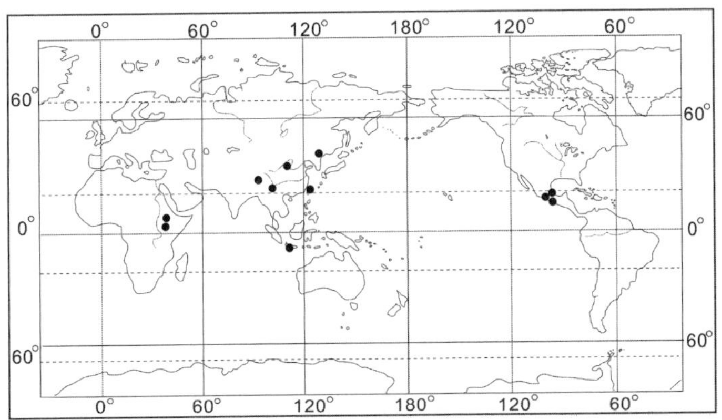

Рис. 42 Распространение на земном шаре *Umbilicaria formosana* Frey.

Ⅵ-3. Восточноафриканско-яванско-колумбиатскаий тип ареала

Этот тип характерен для вида *U.africana*.

Вида *U. zollingeri* обнаружен и описан Гронхартом（Groenhart）из материалов, найденных в высокогорьях Явы. Исследовав почти все оригинальные образцы, было установлено, что по морфологии и анатомии он ничем не отличается от вида *U .africana*. Кроме того, на основании опубликованных материалов（Sipman & Topham, 1992）, вид *U. africana* найден и в Колумбии Северной части Южной Америки. Таким образом, вид *U. africana* распространен не только в Африке(эфиопии), но и в Азии(Яве), и Южной Америке（Колумбии）. Это типично тропическая дизъюкция в высокогорных районах Азии （Рис.43）.

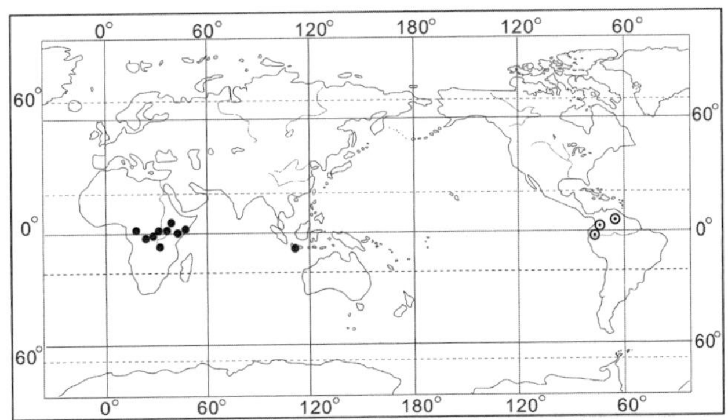

Рис. 43 Распространение на земном шаре *Umbilicaria africana*（Jatta）Krog & Swinscow

（ ● работа основана на личных исследованных матерриалах, ⊙ на основании Литератур.）

Ⅶ. Эндемичные виды Восточной Азии

В настоящее время в составе эндемичных видов Восточной Азии насчитывается 25 видов, что составляет более 40% от общего числа видов.

Ⅶ-1. Эндемики Китая

К ним относятся следующие виды: *U. altaiensis* в Аоиае Синьцзяна, *U. Loboperipherica* в Хэбэе и Цзилине, *U. pesudocineraccens* в Юньнане, *U. taibaiensis* в Шэньси на Тайбайшане.

Ⅶ-2. Эндемики Китайско-гималаев

Под Китайско-гималаи имеется в виду распространение видов от Гималаев до Юньнаня, Сычуаня и даже горных районов Циньлин в Шэньси.

К ним относятся следуюшие виды: *U. badia*(Рис. 44), *U. hypococcinea*(Рис. 45), *U. minuta*, *U. nanella*, *U. nepalensis*, *U. squamosa*（Рис. 46）, *U. subumbilicarioides*（Рис. 47）, *U. thamnodes*（Рис. 48）, *U. yunnana*（Рис. 49）, *L. asiae-orientalis*（Рис. 50）, *L. daliensis*（Рис. 51）, *L. mayebarae*（Рис. 51）, и *L. xizangensis*. Среди них *U. nanella* и *U. hypococcinea*, их ареал тянется на север восток до Утайшаня Шаньси. Что касается *L. mayebarae* и *U. yunnana*, то они являются чисто эпифитными лишайниками из сем. Umbilicariaceae. Вид *L. mayebarae* на соснах впервые обнаружен и описан на материалах, собранных в Тайване Китая. Затем Вэй（Wei, 1966）Собирал его в Юньнане Китая. Позже Поельт（Poelt, 1977）нашел его в Гималаях（Рис. 52）.

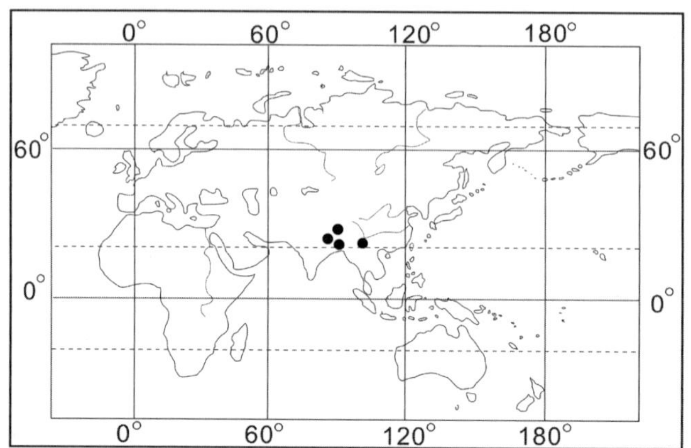

Рис. 44 Распространение на земном шаре *Umbilicaria badia* Frey

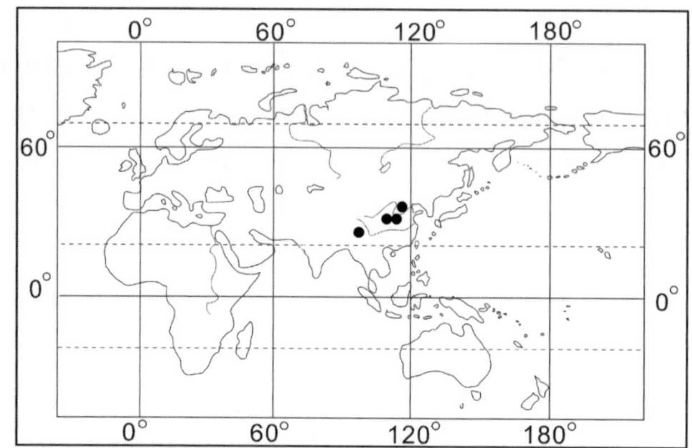

Рис. 45 Распространение на земном шаре *Umbilicaria hypococcinea*（Jatta）Llano

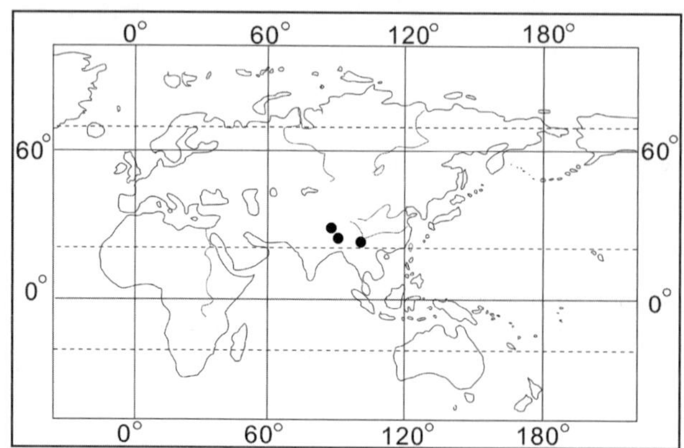

Рис. 46 Распространение на земном шаре *Umbilicaria squamosa* J. C. Wei & Y. M. Jiang

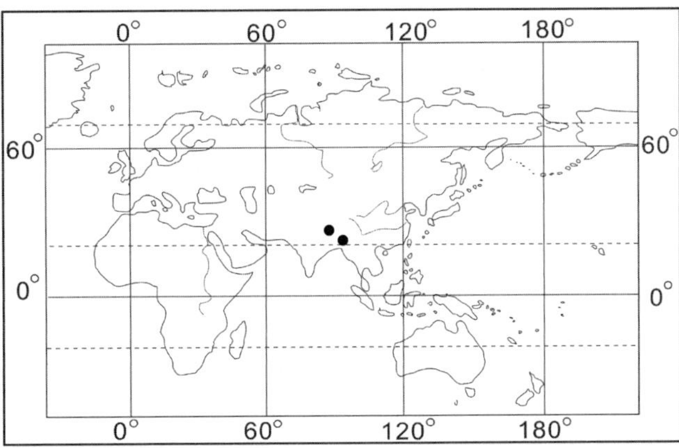

Рис. 47 Распространение на земном шаре *Umbilicaria subumbilicarioides* J. C. Wei & Y. M. Jiang

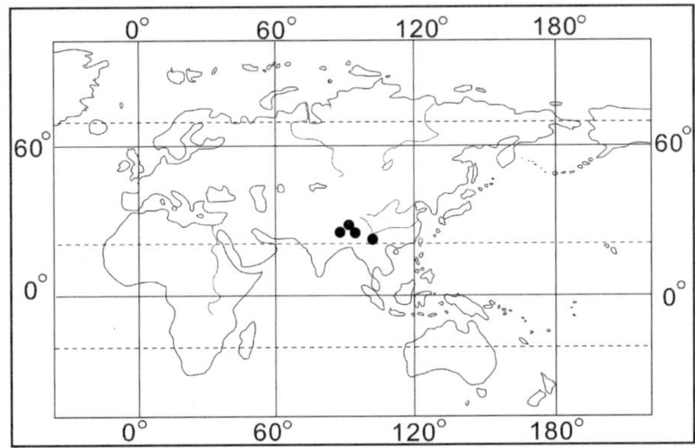

Рис. 48 Распространение на земном шаре *Umbilicaria thamnodes* Hue

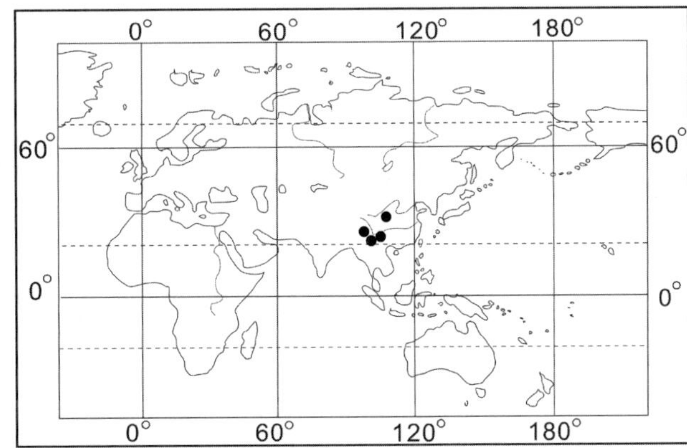

Рис. 49 Распространение на земном шаре *Umbilicaria yunnana*（Nyl.）Hue

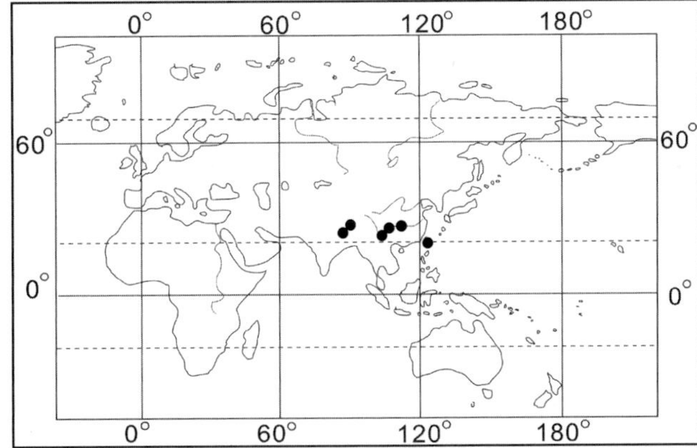

Рис. 50 Распространение на земном шаре *Lasallia asiae-orientalis* Asahina

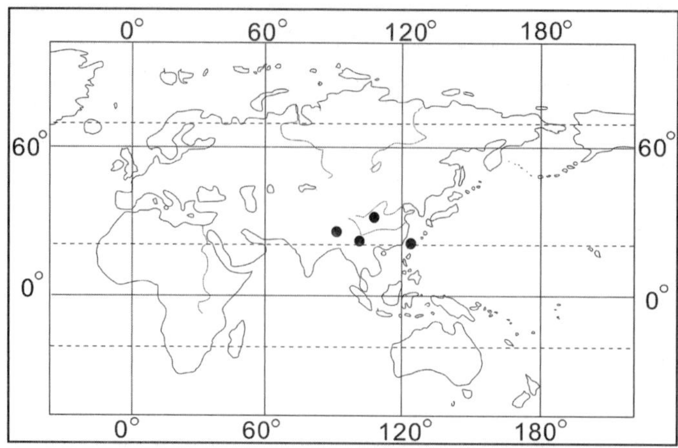

Рис. 51 Распространение на земном шаре *Lasallia daliensis* Wei

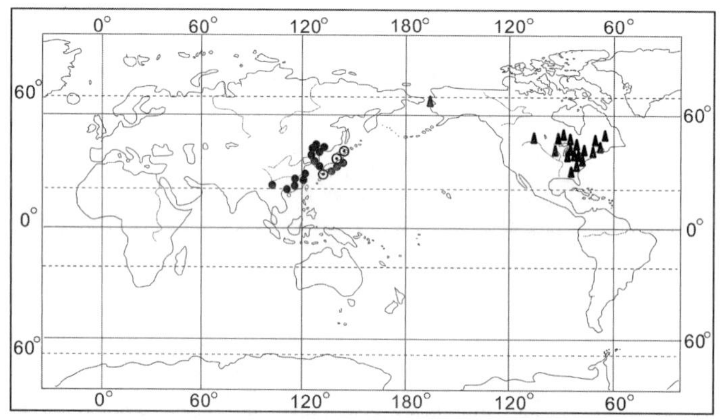

Рис. 52 Распространение на земном шаре *Umbilicaria esculenta*（Miyoshi）Minks.

（●, ⊙ и *U.mammulata*（Ach.）Llano（▲）.）

（●, ▲ работа основана на личных исследованных матерриалах, ⊙ на основании Литератур.）

Вследствие того, что лишайник *Lasallia mayebarae* является единственным видом имеющим гирозный диск с концентрически расположенными бороздками и складками, поэтому создан подрод Pleiogyra（Wei, 1966; Рис. 53）.

Другой интересный и эпифитный лишайник *U. yunnana* на дубах（Рис. 54）распространен только в пределах Юньнаньского нагорья, Сычуаня Китая и Гималаев（Рис. 49）.

Ⅶ-3. Эндем Гималаев

Типичный вид *U. indica* ограничен узким районом Гималаев, но до сих пор его не обнаружили в Юньнане и Сычуане Китая.

В составе эндемичных видов для Гималаев, Китайскогималаев, и Китая насчитывается 18 видов, что составляет более 29% от общего числа видов.

Ⅶ-4. Эндемики Китая, Кореи и Японии

Этот тип объединяет 2 вида: *U. esculenta*（Рис. 52）и *U. kisovana*（Рис. 55）

Известный миру съедобный лишайник *U. esculenta*, его ареал занимает горные районы восточного Китая, так Хуаншань Аньхоя, Тяньмушань Чжэцзяня, Лушань Цзяньси, а также Гуанси, Хунань, еще северо-восточный горный район, так Ляонин, Цзилинь, Хэйлунцзянь（J. C. Wei & Y. M. Jiang, 1993）, а также Корея（Sato, 1961; Kamiya & Kim, 1981; J. C. Wei & Y. M. Jiang, 1993）и горные места Японии（Ikoma, 1956; Kato & Kumagai, 1960; Kobaeashi, 1957; Kurokawa, 1960; Sato, 1958, 1960, 1961; Takamizu, 1957; Yamanaka & Yoshimura, 1961; Yoshimura & Ymanaka, 1961; J. C. Wei & Y. M. Jiang, 1993）, на севере редко бывает на дальновосточной России, на западе встречается в Юньнане Китая（J. C. Wei & Y. M. Jiang, 1988, 1993）. Вид *U. kisovana* распространен в некоторых горных местах восточного и северного Китая, а также в

Кореи и Японии.

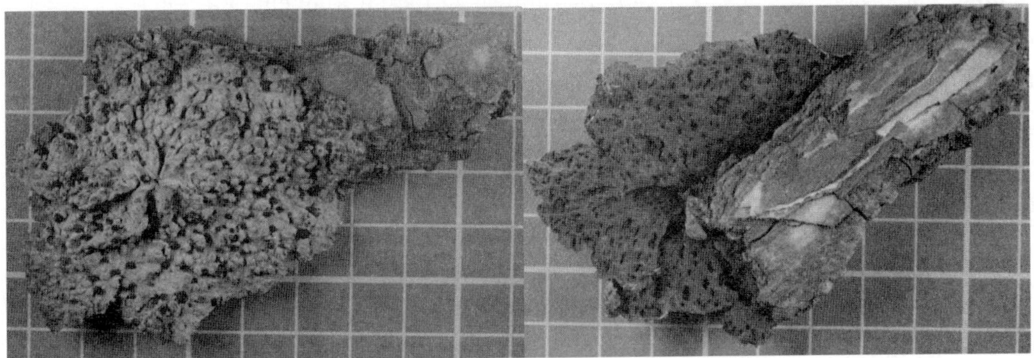

Рис. 53 чисто эпифитный лишайник их сем. Umbilicariaceae *Lasallia mayebarae*（Sato）Asahina на стволе сосны в Юньнане Китая: Налево: верхняя поверхность слоевища, Направо: нижняя поверхность слоевища.（в масштабе 1 шкала = 1 cm）

Рис. 54 Эпифитный лишайник из сем. Umbilicariaceae *Umbilicaria yunnana*（Nyl.）Hue на дубе в Юньнане Китая.（в масштабе 1 шкала = 1 mm）

а. Сверху: Дубовый лес, где растет *Umbilicaria yunnana*.

б. Снизу: *Umbilicaria yunnana* на стволе дуба.

Ⅶ-5. Эндем Китая и Японии

Лишайник *L. sinorientalis*, ареал которого ограничен Китаем и Японей（Рис. 56）.

Ⅶ-6. Эндем Японии

Этот тип свойствен виду *U. koidzumii*.

Ⅶ-7. Эндем Северновосточной Азии

Вид *L. rossica* распространен в основном в северновосточной Азии от восточной России до Монголии и северного Китая.

Ⅶ-8. Эндем Камчатки и Японии

Исследование вида *U. krascheninnikovii* на основе огромного количества материалов в том числе голотипа показало, что данный вид представляет собой эндем Камчатки и Японии. Что касается публикации этого вида в других местонахождениях, то он по-видимому является ошибочным определением вида *U. formosana*.Ⅶ-9. Эндем Камчатки и Якутии

Этот тип характерен для *U. pulvinaria*.

4-2. Происхождение ареалов

Вопрос о происхождении ареалов лишайников является чрезвычайно сложной и трудной темой. Потому что все это касается возраста видов, их экологических и биологических типов, их условия существования и способа размножения. Из-за отсутствия ископаемых лишайников для доказательства вопрос о происхождении и образовании ареалов лишайников сем. Umbilicariaceae Восточной Азии становится еще более трудным.

Анализ разъединенных или дизъюнктивных и викарных ареалов лишайников сем. Umbilicariaceae Восточной Азии имеет большое значение для понимания их происхождения. Поэтому остановимся здесь лишь на некоторых викарных и разъединенных ареалах в том числе и некоторых реликтах.

Ⅰ. Дизъюнкция

Разъединенное или дизъюнктивное распространение лишайников не только связано с происхождением вида и его возрастом, но и тесно связано с геологическими событиями, поэтому биогеографы и лихенологи проявляют чрезвычайный интерес к анализу явлений прерывистого распространения видов лишайников.

А. Межконтинентальная дизъюнкция

Ⅰ-1. Восточноазиатско-северноамериканская дизъюнкция

К этому ареалу относятся виды лишайников *U. caroliniana*, *U. muehlenbergii и L. pensylvanica* обсуждены Ляном(Llano, 1950)и Голубковый(1983). Распространение *U. angulata* и *U. phaea* в Монголии（Wei & Biazrov, 1991）впервые обнаружены нами за пределами Америки. Таким образом дизъюнкция видов лишайников сем. Umbilicariaceae Восточной Азии и Северной Америки возросло от 3 до 5 видов. Очевидно, дизъюнкция данных 5 видов была непрерывна тогда, когда материк Северной Америки и материк Евразии еще не разъединились. Современный центр ареала *U. angulata и U. phaea* оказался в Северной Америке, а распространение этих двух видов в Восточной Азии（Монголии）всего лишь край ареала данных видов, подобно *U. phaea*, обнаруженному в Чили Южной Америки. Разница лишь в том, что существование *U. phaea* в Южной Америке произошло наверно после сближения материков северной и южной Америки; а существование *U. angulata* и *U. phaea* в Восточной Азии объясняется наверно географическим реликтом.

Ⅰ-2. Азиатско-африканская дизъюнкция

Где наконец место происхождения ареала лишайника *L. pertusa*, Африка или Азии? Это очень трудный вопрос, который был анализирован Голубковой（1983）. Однако можно полагать, что место происхождения ареала этого вида, по крайней мере, не Африка. Потому что Эфиопские горы Восточной Африки существовали задолго до образования Гималаев, а разъединение Индии с Африкой произошло тоже раньше окончательного формирования Гималаев. Если утверждать, что Африка является местом происхождения, тогда становится невозможным то, что его ареал остался только в пределах Эфиопских гор Восточной Африки, тем более что в южной части Африки и на горах Индии также возможно сушествование ареала этого вида. Однако, он существует в Азии и широко разбросан на северо-востоке Китая и на обширном районе Алтая в Китае, Монголии и восточной Сибири. Что касается распространения его на Гималаях, то естественно, что это произошло только после образования Гималаев. Необходимо упомянуть также о том, что в Монголии данный вид не только прикрепляется к каменистым субстратам как эпилитный лишайник, но и переходит к обитанию на стволах деревьев как эпифитный. Таким образом широкое распространение этого вида в Монголии и такое разнообразие в экологии встречаюшиеся в Монголии показывают, что очень может быть ареал этого вида произошел в Монголии среди популяций вида *L. pustulata* после того, как появилась мутация, дефференциация, и географический викарный вид. А как тянулся ареал данного вида с Гималаев через Аравийскую пустыню в Восточную Африку? Возможно это можно объяснить наступлением и отступлением ледников. Однако, был ли на Аравийской пустыне ледниковый период? Это пока спорный вопрос.

Ⅰ-3. Азиатско-африкано-американская дизъюнкция

Ареал вида *L. papulosa* является известным（Рис. 57）. Что касается вида *U. africana*, то его обнаружили в ближайшие годы в Америке（Sipman & Topham, 1992）и Азии（J. C. Wei & Y. M. Jiang, 1993）. А вид *U. formosana* распространен не только в Китае, но и в Америке（J. C. Wei & Y. M. Jiang, 1993）и в Африке（Wei,1995; Рис. 43）.

Прерывистый ареал вида *U. africana* очень возможно образовался в то время когда континент Южной Америки и Африка еше не отделились. Потом после отделения континента Южной Америки и Африки и после появления островов юго-восточной Азии привели к тому, что уже образовавшийся сплошной ареал данного вида разорвался на три части, образуя нынешнюю Восточно-африкаско-яванско-колумбийскую дизъюнкцию. Дизъюнкция этого вида одной и той широты возможно послужит косвенным доказательством отделения древнего материка Годвана. Прерывистые три части ареала лишайника *U. formosana* в Азии на Китайских хребтах горы Чанбайшань в Циньлине и в Гималаях, в Мексике Центральной Америки и в Восточной Африке возможно сформировались после образования Гималаев（около 45 млн. лет тому назад）и до отделения древних материков Северной Америки и Евразии（около 43 млн. лет тому назад）. На западном полушарии ареал данного вида потянулся на юг в Мексику; на восточном полушарии с хребта гор Гималаев мигрировал в Восточную Африку. Эта миграция возможно произошла после того, как древние материки Африки и Евразии снова соединились（около 17 млн. Лет тому назад）, а также и после сближения материков северной и южной Америки.

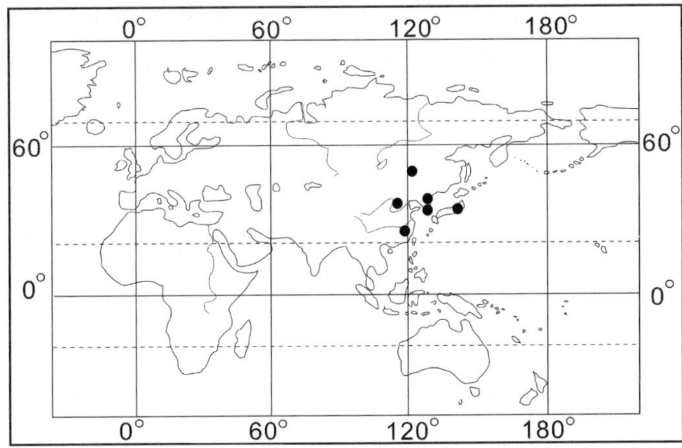

Рис. 55　Распространение на земном шаре *Umbilicaria kisovana*（Zahlbr. Ex Asahina）Zahlbr.

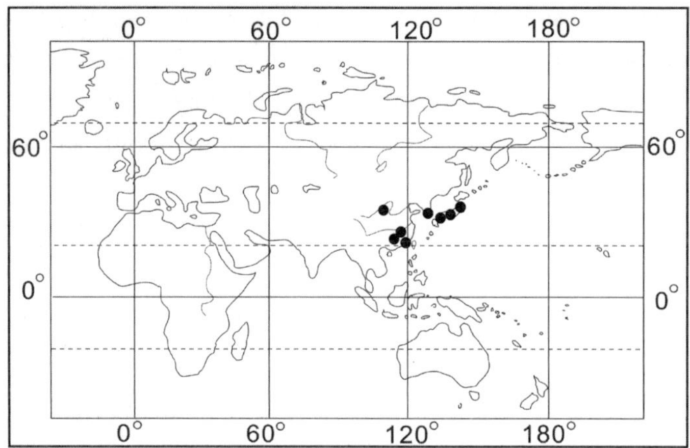

Рис. 56 Распространение на земном шаре *Lasallia sinorientalis* Wei

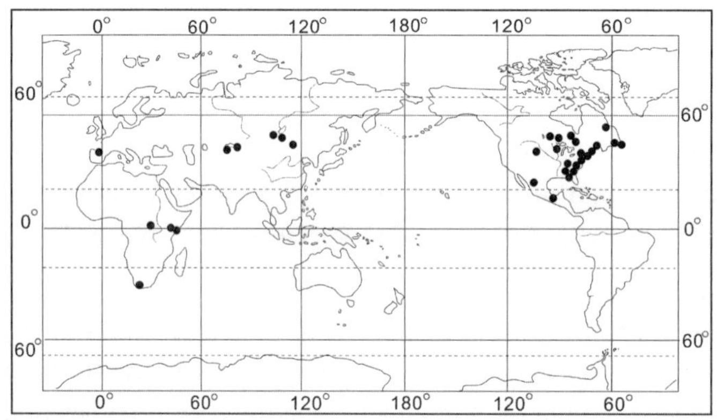

Рис. 57 Распространение на земном шаре *Lasallia papulosa*（Ach.）Llano

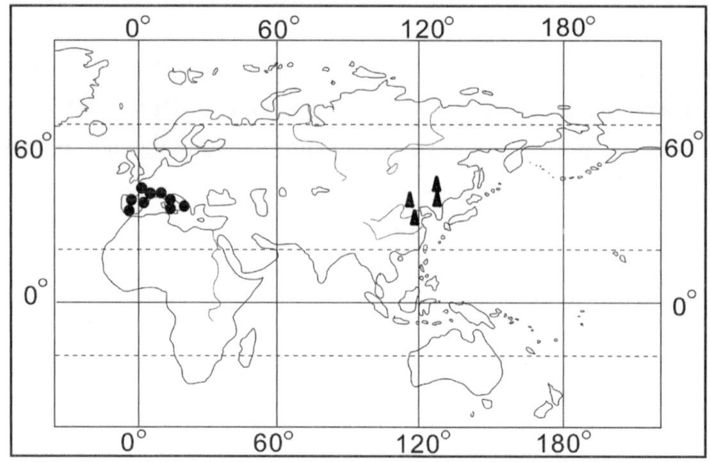

Рис. 58 Распространение на земном шаре *Umbilicaria freyi* Codogno et al.（●）*U. loboperipherica* J.C. Wei et al.（▲）

Широкое распространение прерывистых ареалов *L. papulosa* в Америке, Африке и северной части Азии возможно произошло до периода распада Пангеи（Pangaea）.

Ⅰ-4. Биполярная дизъюнкция

К этой дизъюнкции относятся виды *U. cinerascens*, *U. grisea*, *U. nylanderiana* и *U. subglabra*. Образование этой дизъюнкции видимо произошло в тот период, когда существовала Пангея. Потом после разъединения Пангеи все ареалы данных видов постепенно сократились при неблагоприятных условиях и поэтому становились реликтовыми.

Ⅰ-5. Аркто-восточноазиатская дизъюнкция

Эта дизъюнкция свойствена ареалу видов разъединенных голарктической областью так, что одна часть ареала размещается в Арктике, а другая – в Монголии, Китае или Японии. Такова дизъюнкция ареалов у вида *U. artica* – Арктика и Монголия, *U. lyngei* – Арктика, Китай и Монголия, *U. rigida* – Арктика, Китай, Монголия и Япония.

Ⅰ-6. Кольско-монгольская дизъюнкция

Эта дизъюнкция наблюдается у вида *U. canescens*. Таким образом, часть его ареала размещена в Кольском полуострове, а другая – в Монголии.

Ⅰ-7. Аркто-южноаляско-монгольская дизъюнкция

Такая дизъюнкция встречается у вида *U. havaasii*, часть ареала которого размещена в Арктических районах, а часть – в Монголии, а третья в южной Аляске.

В настоящее время указанные высше арктические виды, т.е. *U. arctica*, *U. canescens*, *U. havaasii*. *U. lyngei*, и *U. rigida*, найдены в нескольких местах Восточной Азии, так что в данном случае они несомненно являются остаточными реликтами из-за наступления и отступления ледников в четвертичном периоде. Такие реликты называют ледниковыми реликтами или как Шретер（Schroeter）их называл " Ледниковыми мигрантами"（Вульф, 1932）.

Ⅰ-8. Средиземноморско-сычуаньская дизъюнкция

Изолированное распространение вида *Umbilicaria spodochroa* в Сычуане Китая наверно является реликтом Древнесредиземноморской флоры, который остался после поднятия Гималаев и отступления Тетис（древнесредиземного моря）.

Ⅰ-9. Скандинавско-гималайская дизъюнкция

У вида *U. tylorhiza* имеется два центра современного ареала. Один находится на полуострове Скандинавии, другой – в Гималаях. У первого сосед на Гренландии, у второго – на островах Японии（Рис.37）. Происхождение такого типа прерывистых ареалов несомненно идет от Лавразии. Однако, учитывая центры нынешнего ареала этих двух типов, очень трудно установить, где же центр ареала происхождения. Однако, если рассматриваем эти центры нынешнего ареалов как реликтовые ареалы, тогда возможно это тесно связано с поднятием Гималаев и отступлением Средиземного моря. Можно себе представить, этот вид в свое время был распространен на побережье Древнесредиземного моря, но из-за горообразовательного движения Гималаев и отступления Древнесредиземного моря, часть лишайников этого вида гибла, а часть его в разных местах выжила как реликт Древнесредиземноморской флоры, и при благоприятных условиях снова разросся и образовал нынешний прерывистый ареал.

Б. Внутриазиатская дизъюнкция

Ⅰ-10. Китайско-гималайская дизъюнкция

Прерывистое распространие эпифитного лишайника на древесных стволах сосны（*Pinus* sp.）– *L. mayebarae* в Гималаях, Юньнане（*Pinus densata*）и Тайване Китая показывает, что непрерывный ареал этого вида образовался после горообразовательного движения Гималаев（45 млн лет тому назад）и до отхода Тайваня от материка. Этот вид является одним из представителей двух эпифитных видов в сем. Umbilicariaceae. Другой – *U. yunnana* на древесных стволах дуба（*Quercus* sp.）.

Лишайник *L. asiae-orientalis* имеет прерывистый ареал в Китае на Тайване, в Сычуане, Юньнане, и Тибете, а также и в Непале Гималаев. Он образовался возможно также, как и *L. mayebarae*.

Благодаря тому что Тайвань и Гималаи совместно обладают двумя видами *Lasallia*, *L. mayebarae* и *L. asiae-orientalis* показывает, что Тайвань и Гималаи были объединеным материком.

Ⅰ-11. Китайско-японская дизъюнкция

Известный миру вид *U. esculenta* широко распространен в горных районах Китая, Корея и на островах Японии. *U. kisovana*, растущий на востоке и севере Китая, в Кореи и Японии. *L. sinorientalis* распространен в восточной части Китая и Японии. У этих трех видов имеется подобный прерывистый ареал. Такая дизъюнкция оформилась наверно до 13 миллионов лет тому назад.

Благодаря тому, что данные три вида распространены в восточной части Китая и Японии, это говорит о том, что Японские острова и северная половина древнего материка Китая тоже были соединенным континентом.

Ⅱ. викарные ареалы

“ под викарными или замещающими ареалми понимают ареалы, большею частью взаимно друг друга исключающие, принадлежащие близким в родственном отношении и отличающимися лишь немногими, но характерными признаками, видами связанными происхождением от одной основной формы. В основе возникновения викарных ареалов лежит процесс образования географических рас.”（Вульф, 1932）.

Изучение викарных ареалов тесно связано с дефференциациями и образованиями видов.

Ⅱ-1. Восточноазиатский и Северноамериканский викарный ареал

Ареал викарного вида *U. esculenta* ограничен лишь в пределе Восточной Азии. Ареал другого викарного вида *U. mammulata* распространен только в пределе северной Америки（Culberson, 1973; Wei & Biazrov, 1991; J. C. Wei & Y. M. Jiang, 1993）. Эти два викарных вида по-видимому возникли благодаря изоляции и разобщению двух частей некогда обшего ареала. Эти викарные ареалы образованы наверно до 43 миллионов лет, т.е. до того, как отделился древний материк Северной Америки и Евразии.

Ⅱ-2. Евразиатско-африканско-восточноазиатский викарный ареал

Вид *L. pustulata* широко распространен в Европе и Африке, его Евразиатский ареал на юго-востоке простирается за пределы Монголии（Бязров, 1986; Wei & Biazrov, 1991; J. C. Wei & Y. M. Jiang, 1993）. Однако, он не попал в горные районы в пределах Китая, Кореи и Японии. Лишайник *L. asiae-orientalis* как викарный вид распространен на Гималаях и в Тайване Китая; другой – *L. sinorientalis* ограничен в восточной части Китая и в Японии. Все эти викарные виды отличаются от общих особенностей *L. pustulata* тем, что они имеют слабые пузыревидные вздутия. Вид *L. sinorientalis* отличается от вида *L. asiae-orientalis* тем, что первый характеризуется светлой нижней поверхностью слоевища и зернистыми и чешуйчатыми изидиами на верхней поверхности слоевища, а второй – темной нижней поверхностью слоевища и игольчатыми изидиами на верхней поверхности слоевища.

Ⅱ-3. Африко-американский и Азиатский викарный ареал

Эпифитный лишайник *L. mayebarae* на Гималаях и на Тайване Китая очень может быть является тоже викарным видом, образованным из какой-нибудь популяции вида *L. papulosa*.

Ареал у вида *L. papulosa* находится главным образом в северной Америке и в восточной Африке, а также тянется до северной Азии（Рис.57）. Ареал вида *L. rossica* ограничен лишь в северновосточной Азии, как раз совпадает с окраиной ареала вида *L. papulosa* в Азии. Поэтому можно полагать, что *L. rossica* может быть является другим викарным видом, образованным из другой популяции вида *L. papulosa*.

Ⅱ-4. Средиземноморский и Восточноазиатский викарный ареал

Что касается Средиземноморского лишайника *U. freyi* и морфологически близкого к нему восточно-Китайского вида *U. loboperipherica*, то они может быть являются тоже викарными видами（Рис. 58）. Оба вида характеризуются паралобулями на периферических местах верхней поверхности слоевища. Однако первый отличается от второго светлой поверхностью внизу слоевища без ризины, но от второй – черной поверхностью внизу слоевища с многочисленными ветвистыми ризинами（Рис. 59）. Образование этих викарных ареалов повидимому тесно связано с горообразовательными движениями Гаималаев и отступлением Древнесредиземного моря.

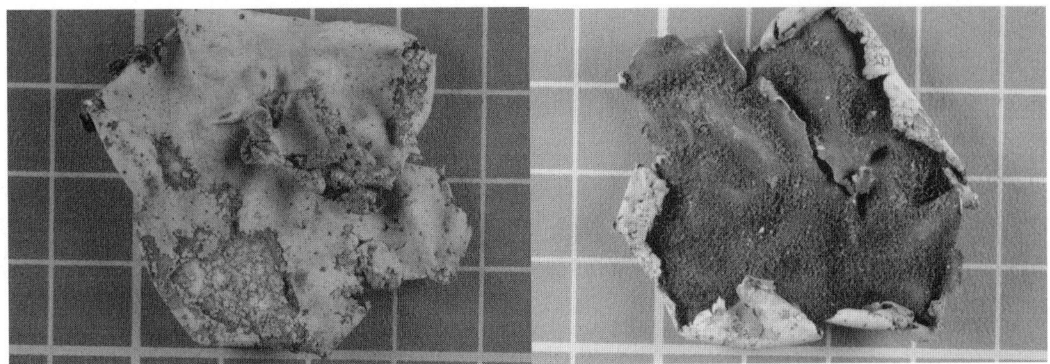

Рис. 59 *Umbilicaria loboperipherica* Wei et al.（в масштабе 1 шкала = 1 cm）

Верхняя поверхность（налево）и нижняя поверхность со многочисленными ризинами（направо）слоевища

ВЫВОДЫ

На основании приведенных результатов и их обсуждения можно прийти к следующим выводам:

С географической точки зрения лишайник сем. Umbilicariaceae Восточной Азии распространяются в областях высокой широты и в альписких и субальписких поясах высокогорных районов средних и низких широт. Из них род *Lasallia* характерен узкой амплитудой температуры, распространяется в альписких и субальпийских поясах высокогорных районов средних и низких широт, лишь один вид в Арктике. Что касается рода *Umbilicaria*, то он располагается на широкой амплитуде температуры и в областях высшей широты, а также и в альписких и субальпийских поясах высокогорных районов низких широт.

В настоящее время в составе лишайников сем. Umbilicariaceae Восточной Азии насчитывается 62 вида, относящийся к 2 родам, 6 подродам, и 5 секциям. Из них 52 вида принадлежит к роду *Umbilicaria*, 10 к роду *Lasallia*. При анализе их современного распространения все виды в данной работе распределяются между 7 географическими элементами. Это Арктический элемент（2 вида, что составляет 3.2% от общего числа видов）, Аркто-альпийский（14 видов, 22.6%）, Биполярный（2 вида, 3.2%）, Гипоарктомонтанный（11 видов, около 18%）, ното-бореальный（5 видов, 8%）, Альпийский（3 вида, около 5%）, и эндемичные для восточной Азии виды.

В составе 62 видов сем. Umbilicariaceae Восточной Азии насчитывается 25 видов свойственных Восточной Азии и занимают 40% от общего числа видов. В составе 25 видов, свойственных Восточной Азии, насчитывается 18 видов, свойственных Китайско-гималаям, что составляет 72% от общего числа видов.

Относящиеся к свойственному району резкие изменения моря и материка в свое время предоставили благоприятные условия для мутации, дифференцияции видов（эндемов или мигрантов）, для образования викарных видов и их ареалов, в конечном счете для происхождения богатых эндемов; иными словами, все упомянутые высше специфические явления являются глубоким отпечатком, оставленным перемещением моря и континента Гималаев.

Эндемичные виды Гималаев, Юньнаня и Тайваня *L. mayebarae* и *L. asise-orientalis*, и эндемичные виды восточной части Китая, Кореи и Японии *U. esculenta* и *U. kisovana* показывают, что более тесные связи существовали и между островом Тайваня и юго-западной частью Китая и Гималаями, а также и между островами Японии и северной частью Китайского континента. Единственный эндем Японии *U. koidzumii* обозначает, что отделение островов Японии от северной части Китайского континента произошло раньше, чем отделение Тайваня от юго-западной части Китая и Гималаев.

Восточноазиатско-североамериканская дизъюнкция охватывает виды *U. angulata*, *U. carolineana*, *U. muehlenbergii*, *U. phaea*, *и L. pensylvanica*. Можно предполагать, что каждый из этих видов имел свой непрерывный ареал около 43 миллионов лет тому назад, когда континенты северной Америки и Евразии были объединены. После отделения континента северной Америки от континента Евразии первоначально сплошной ареал каждого вида упомянутых выше был разорван на две части в виде нынешней дизъюнкии.

Распространение видов *U. havaasii*, *U. arctica*, *U. rigida* и *U. lyngei* в высокогорных районах Китая, Монголии и Японии наверно является ледниковыми реликтами или мигрантами, которые видимо были образованы наступлением и отступлением ледников в плейстоцене четветичного периода. Прерывистый ареал *U. subglabra* расположен на побережьях Средиземного Моря, Азии и Австралии, видимо, образовались намного раньше реликтового вида *U. spodochroa* в Средиземноморской сычуанской дизъюнкции.

СИСТЕМАТИЧЕСКИЙ СПИСОК ЛИШАЙНИКОВ СЕМ. UMBILICARIACEAE ВОСТОЧНОЙ АЗИИ

Семейство

UMBILICARIACEAE F.F.Chevalier

Fl.gen.env. Paris:640（1826）, [as ' Umbilicarieae']. Hawksworth David, Family names, in Index of fungi supplement. 73（1989）. Fee, Essai Cryptog. Ecorc. Officin. LXX, 1824（as Umbilicarieae）.[nom inval., Art 32.1（b）; see Art 18.4.]

Тип сем. *Umbilicaria* Hoffmann

Найбольшие виды на каменистых субстратах, только два вида из китая на коре деревьев.

Ключ для опрделения рода

1a. Слоевище с пузыревидными вздутиями（Рис. 6）без ризин и таллоконидиев; сумка содержащая по одной коричневой и муральной споре（Рис. 7）вместе с нотобороеальными распространеними, в основном и сопровождает распространение горных областей голарктических зон（Рис. 5 и 12）·········· род ***Lasallia***

1b. Слоевище без пузыревидных вздутий（Рис. 1-а）; , иногда с ризинами и таллоконидиями（Рис. 8）; сумка содержащая по 8 спор（Рис. 9）и вместе с биполярными и высокогорными, нотобореальными, и высокогорными распространениями（Рис. 5 и 13）··· род ***Umbilicaria***

Род

LASALLIA Mérat

Mérat, Nouv. Fl. Paris ed. 2（1）: 202（1821）.

Тип: *Lasallia pustulata*（L.）Mérat（1821）

≡ *Lichen pustulatus* L.（1753）.

Ключ для опрделения подрода

1a. Апотеции с ровными дисками·· подрод ***Lasallia***

1b. Апотеции с гирозными дисками ··· подрод ***Pleiogyra*** Wei

Ключ для опрделения видов

1a. Апотеции с гирозными дисками ······ 10. ***L. mayebarae***

1b. Апотеции с ровными дисками

2a. Слоевище с соредиями в виде соралей и без изидиев ······ 5. ***L. pertusa***

2b.Слоевище с изидиями и без соредиев

3a. Слоевище с игольчатыми изидиями

4a. Слоевище со слабо развитыми пустулами, нижняя поверхность слоевища тёмная ······ 1. ***L. asiae-orientalis***

4b. Слоевище с сильно развитыми пустулами, нижняя поверхность слоевища светлобуроватая ······ 6. ***L. pustulata***

3b. Слоевище с зернистыми и чешуйчатыми изидиям

5a. Нижняя поверхность слоевища темнобурая и чёрная ······ 4. ***L. pensylvanica***

5b. Нижняя поверхность слоевища вокруг гомфа чёрная

6a. Нижняя поверхность слоевища феолетовая ······ 9. ***L. xizangensis***

6b. Нижняя поверхность слоевища темнобурая ······ 2. ***L.daliensis***

5c. Нижняя поверхность слоевища светлобуроватая

7a.Слоевище со слабо развитыми пустулами ······ 8. ***L. sinorientalis***

7b. Слоевище с сильно развитыми пустулами

8a. На поверхности и апотеции со красными пигментами ······ 3. ***L. papulosa***

8b. На поверхности и апотеции без красных пигментов ······ 7. ***L. rossica***

Подрод ***Lasallia***

1. **Lasallia asiae-orientalis** Asahina

Journ. Jap. Bot. 35:99（1960）, cum icon.

Тип: Китай, Тайвань, гора Али шань, 25/Ⅻ/1925, Y, Asahina, No F-377（TNS!）.

=*Lasallia asiae-orientalis* var. *fanjingensis* Wei, Acta Mycol. Sin. **1**（**1**）: 20（1982）.

Тип: Китай, Гуэчжоу, гора Фаньцзин шань, Цзианьдаосия, 23/Ⅸ/1963, Вэй, No.830（HMAS-L!）.

Ошибочное определение:

Umbilicaria pustulata auct. non（L）Hoffm.: A. Zahlbr.in Handel-Mazzetti, Symb. Sin. III: 138（1930）, and in Fedde, Reportorium 33:49（1933）. Asahina, Journ. Jap. Bot. 7:103（1931）.

На скалах.

Материалы исследованные:

Китай: Сычуань, Мули, 2950м, 8/MIII/1915, Dr.Heinr, Frh. V. Handel-Mazzetti（Diar. Nr. 1364）(E!). Юньнань, Лицзян, 11/XI/1958, Хан（Han）и Чен（Chen）, No.5091（HMAS-L!）; около юунбей и юунин, гора Гуамао шань, 3075м, 29/Ⅵ/1914, Handel-Mazzetti, No.627, Acad. Sci. Vindob. No. 3318（UPS!）; гора Юйлон шань, мучжугоу, 2800м, 6/Ⅻ/1960, Чжао и Чен, No.3885, и Гайхайцзи, 3020м, 18/Ⅻ/1964, Вэй, No.-,№., Юйхуцунь, 2820м, 20,22/Ⅻ/1964 Вэй, Nos. 2663-1 и 2714; Дали, Гора Цанщань, 2050м, 27/Ⅻ/1964, Вэй, No.2750（HMAS-L!）. Тибет, Чжам, 3500м, 21/Ⅴ/1966, Вэй и Чен, No.1081; 3650м, VII/1966, Вэй и Чен, No.510（HMAS-L!）. Тайвань, гора Али шань, см. Тип; Тайчжун, гора Анма шань, 23/Ⅻ/1972, Лай, No.6859（US!）; Тайчжун, гора Силва, 11 /Ⅸ/1976, Лай, No.8072, часть из них（US!）.

Непал: Центральные Гималаи, Лантан, 3800м, 11/Ⅸ/1986, ex Plantae Graecenses, Lich. 392, Lasallia sinensis, L-84319（Crypt. Herb. COLO-411989!）.

2. **Lasallia daliensis** J.C.Wei

Acta Mycol. Sin. 1（1）: 21（1982）.

Тип: Китай, Юньнань, Дали, гора Цан шань, 3300м, 29/Ⅻ/1964, Вэй, No.2778（HMAS-L, Голотип!）

2.1. var. **daliensis**

На скалах.

Материалы исследованные:

Китай: Юньнань, Лицзян, гора Юйлон шань, Ганхайцзи, 3000м, 13/Ⅻ/1964, Вэй, No.2421（HMAS-L!）. Сычуань, Кандин, восточная гора от города Кандин, 2600м, 27X?/1934, Harry Smith, No.14003（UPS!）. Тайвань, Наньтоу, Ба-тон-гуан, 23/Ⅲ/9163, S.Nakanishi, No. 13583（TNS-22121!）.

Сикким: 8000 pel 2099, Herb. Ind. Or Hook P. V. Thomson, как U. Papulosa（BM, FH!）.

2.2. Var. **caeongshanensis**（J.C.Wei）J.C.Wei

J. C. Wei & Y. M. Jiang, The Asian Umbilicariaceae, p. 28, 1993.

= *Lasallia caeongshanensis* Wei, Acta Myool. Sin.（C1）: 22（1082）.

Тип: Китай, Юньнань, Дали, гора Цан шань, 3300м, 29/1964, Вэй, No.2758（HMAS-L!）.

=*Lasallia asiae-orientalis* Asahina £ 7 major J. C. Wei & Y. M. Jiang, Acta Phytotax. Sin. 20（4）:500（1982）.

Тип: Китай, Тибет, Чжам, Кюсям, 3500м, 19/Ⅴ/1966, Вэй и Чен, No.884（HMAS-L!）.

На скалах.

Материалы исследованные:

Китай, Тибет, Чжам, Кюсям, 3500м, 19/Ⅴ/1966, Вэй и Чен, No.884（HMAS-L!）.

3. **Lasallia papulosa**（Ach,）Llano

Monograph of the Umbilicariaceae 32（1950）.

≡ *Gyrophora papulosa* Ach. Lich. Univ. 226（1810）.

Тип: Северная Америка, Мюхленберг, H-Ach. No 570（H!）.

На скалах и изредка на деревьях.

Материалы исследованные:

Китай: Синцзян, гора Алтай, около оз. Халас, 2500м, 5/Ⅷ/1986, Гао, Nos, 1958-1-2090（HMAS-L!）.

Монголия: Дзабханский аймак, на ветках дерева, 1/Ⅶ/1076, Бязров, Nos. 6575, 7468; Ара-Хангайский аймак, на берегу реки Чулата, на коре дерева Larix sibirica, 22/Ⅷ/1970, Бязров, No. 6570; на берегу реки Цайцайрлайг, на скалах, 21/Ⅶ/1979, Бязров, No. 5736 и 7/Ⅸ/1979, Бязров, No.6190; Буланский аймак, гора Цайцайрлайт-ула, на Butula sp., 29/Ⅵ/1972, Бязров, Nos. 5108, 5109, 6689（Moscow!）.

Казахская республика: восточная часть страны, 1600м, 13/Ⅶ/1982, Домбровская, No.11, как Lasallia rossica Dombr.（HMAS-L!）, 800м, 19/Ⅵ/1963, Heinrich Aasamaa, No.7433（H!）.

4. **Lasallia pensylvanica**（Hoffm.）Llano

Monograph of the Umbilicariaceae 42（1950）.

≡ *Umbilicaria pensylvanica* Hoffm. Descr. Adumb. Pl. Lich. 3（4）-5-5, pl. 69, Fig. 1-2, 1801.

Неотип: Северная Америка, Мюленбере, H-Ach. No. 571（H!）.

На скалах.

Материалы исследованные:

Киатй: Цилин, Гора Чан Бай шань, южный склон, гора Хонтоу шань, 1950м, 5/Ⅷ/1980, Ху, No.2149;1970м, 16/Ⅷ/1982, Ху, Nos.3386, 3432; 1880м, 28/Ⅶ/1983, Вэй и Чен, No.6250; 1980-2000м, Ху, Nos.4218,4230,4232（HMAS-L!）. Внутреняя Монголия, Хинганлин, гора Цзмгуан шань, 1600м, 2/Ⅷ/1991, Чен и Цзян, Nos. A-812, A-817 A-820（HMAS-L!）. Синцзян, гора Алтай, около оз. Халас, 2000м, 4/Ⅷ/1986,

Гао, No1987（HMAS-L!）.

Монголия: Ара-Хангайский аймак, 15/Vi/1971, Бязров, No.5768（Moscow!）; 2100м, 28/Ⅷ/1971, Бязров, No.4228（Moscow!）; 7/Ⅸ/1979, Бязров, No. 6186（Moscow!）. Следующие образцы были определены как Lasallia rossica в гербарий: Ара-Хангайский аймак, 2130м, 16/Ⅷ/1972, Бязров, No. 6690; 30/Ⅵ/1976,Бязров, No.3176（с ризинами）, 2000м, 1/Ⅶ/1980, Бязров, No. 4350, 4356; Булганский аймак, 29/Ⅵ/1972, Бязров, Nos. 5110, 5103; 24/Ⅶ/1977, Бязров, No.6580; Дзабханский аймак, 2000м, 11/Ⅶ/1976,Бязров, No. 6579（Moscow!）.

Казахская республика: Туркестан, Кокчетав, 13/Ⅶ/1896, A. Tordlagei（LE!）.

Япония: Хонсю, 2400м, 27/Ⅶ/1976, Kashiwadani（No. 13413）
и K. Yoshida in Kurok: lich, Rar.et cr. Exs. 293（COLO,DUKE,FH,LE,S TENN, US!）; Musashi: недалеко от Токио, гора Кинпу, 2400м, 25/Ⅷ/1960, Asahina, Nagashi, Kurok., и Nuno [Kurok. & Kashiw. Lich. Rar.et Cr. Exs. Nos. 316, 317]（COLO,DUKE,FH,LE,S TUR, UPS, US!）; гора Охиама, H. Shibuichi（No. 4719）[Kurok. & Kashiw. Lich. Rar.et Cr. Exs. 696]（COLO,DUKE,FH,LE,S TENN, US!）, 1800м, гора Нагайамадаке, Хоккайдо, 22/Ⅷ/1971, I. Yoshimura（No. 12395）[Lich. Jap. Exs. No.39]（COLO,DUKE,H,LE,TUR, US!）. Хоккайдо, гора Томурауши, 1935, M. Sato（FH!）. Недалеко от фукушима, 3000м, 1898, Faurie No.863（LE!）, изидесан, 3000м. Faurie No. 859, pr.p.（LE, pr, p.min.MIN!）.

Россия, Чукотка, Северный Анюиск, 14/Ⅶ/1978, М. П. Андреев（HMAS-L!）. Северный Урал, 4/Ⅷ/1962, М. Тороева（H,TUR!）. Сибирь, около Реки Ангара, 1/Ⅹ/1907, Мартианов, No.171 в гербарий Vain 25 как U.caucasica Lojka（TUR!）. Севердловск район, 24/Ⅵ/19, А. Касаринова, S 25920, C.M. COLO146277（COLO!）. 16/Ⅸ/1926, А. Окснер（COLO, UPS!）. Горноалтайская авто-область, 900-1000м; 20/Ⅷ/1978, T.S.Eoias, W.Weber, C.S.Tomb I.M.Крашноборов（COLO!）> Урал, 12/Ⅸ/1823, Гербарий Chr. Steven（Hi!）. Урал, 22/Ⅵ/1847, Branth из Гербарии Акад, Петроп（UPS!）. Урал, 1900, Vasili Sokolor（H!）>

5. **Lasallia pertusa**（Rassad.）Llano
Monograph of the Umbilicariaceae 48（1950）.

≡ *Umbilicaria pertusa* Rassad. Comp. Rand. Acad. Sci. URSS,14（A）;348（1929）, Fig. 1-2. - *Lasallia pensylvanica* var. *pertusa*（Rassad）Llano, Hvalrad Skr. 48:122（1965）.

Тип: Россия, Сибирь, оз. Байкал, западное побережье, между 53 и 54 сверной широты, долина Онгурен, на валунах в лиственничному лесу, 10/Ⅶ/1928, К.А. Рассадина, No.289（LE!）.

=*Lasallia rossica* var. *pertusoides* Dombr. Nov.syst.plant.non Vascul. 15:186（1978）.

Тип: Россия, Чукотский нац. округ, Басс. Р. Белой, левогоо притока р. Андырь, юрские отложения по р. Инмуам, между усть-хвуком и устьем р. Мухоморной, 23/Ⅷ/1929, В. Б. Сочава（LE!）

=*Lasallia pertusa*（Rassad.）Llno, f. *squamulifera* J. C. Wei & Y. M. Jiang, Acta Phytotax. Sin. 20（4）:499（1982）.

Тип: Китай, Тибет, Лхаса, 3759м, 3/Ⅳ/1966, Вэй и Чен, No.15（HMAS-L, голотип）. Цзилон, 2880м, 14/Ⅳ/1965, Цзон, No.112（HMAS-L паратип!）.

На скалах и иногда на коре дерева.

Материалы исследованные:

Китай, Синцзян, на берегу реки недалеко от леспромхоза Булцзин, 1300м 13/Ⅸ/1982, Чжао, No. 2458（HMAS-L!）. Тибет, Ниаоам, на скалах, 3700-3720м, 29/Ⅳ/1966 и 11/Ⅴ/1966, Вэй и Чен, Nos.92,1429, 1648（HMAS-L!）.

Гиималаи: Панги, Д. Сколисчжа（D. scoliczha）（US!）.

Монголия: Ара-Хангайский аймак, 17/Ⅶ/1970, Бязров, No.6685（Moscow!）, 13/Ⅶ/1977, Бязров, No.3300（Moscow!）; 21/Ⅷ/1979, Бязров, Nos.6586 и на коре Larix sibirica, nos. 4387, 6587, 6589, 6599（Moscow!）, 1/Ⅶ/1980, на коре Salix sp., Бязров, No.4325（Moscow!）; 21/Ⅶ/1980, Бязров, No.4383（Moscow!）. Булганский аймак, на коре Butula sp., 29/Ⅵ/1972, Бязров, No.5088（Moscow!）; на скалах, 29/

Ⅵ/1972, Бязров, No.3199 и на коре Betula sp. No.6686（Moscow!）; на скалах, 26/Ⅶ/1977, Бязров, No.3246（Moscow!）. Центральный аймак, гора Чутаг-Чауранула, на скалах в лесу Larix sibirica, 15/Ⅶ/1974, Голубкова и Цогт, No.1058（LE!）; на скалах в лесу Betula sp., 18/Ⅷ/1974, Голубкова и Цогт, No. 2056（LE!）> Чобгол, 3/Ⅷ/1972, Голубкова и Цогт, No. 681（LE!）; западная часть страны, 2/Ⅷ/1972, Цогт, No. 606（LE!）; Дзабханский аймак, на скалах, 29/Ⅵ/19, Бязров, No.,6585（Moscow!）; на скалах, 4/Ⅶ/1976, Бязров, No.6583（Moscow!）.

Россия: Сибирь, оз. Байкал, Западное побережье, между 53 и 55 северной широты. О. Ольхон, на скалах, 28/Ⅵ/1928, Рассадина, No.199（LE, Синтип!）; местонахождение по гербарию Nos. 25,26, под названием Umbilicaria pustulata, в "lapidosis Daburiae, 1823, Гербарий Куцнецова（LE ）, Синтип!）. Красноярск, 56 с. Ш. 11/Ⅵ/1876, M. Brenner, No. 302f как U.pensylvanica（S!）. Минск, на скалах, 1901, Martianoff（H.TUR!）, 24/Ⅶ/1926, W. Reverdatta No.1（H!）. Иркутская область, оз. Байкал, на скалах, 1928, Рассадина, No.51 как U.pensylvanica f.pertus Rassa.（MIN!）. "Gouv. Tenisejeu, SLus Tenosej Auf Gestein", 30/Ⅵ/1926, W. Renerdoffi, No. 33（FH!）. Ю.з. Сибирь, Алтай горы, Горно-Алтайская автономская область, недалеко от Хабарвска, 900-1000м, 20/Ⅷ/1978, T.S.Elias, W. Weber, C.S. Tomb и I.M. Красноборов, как U.pustulata（COLO!）.

Непал: Центральные гималаи, Лантаи, 3600м, 8/Ⅸ/1986, Poelt, L-84318, C.H. COLO-411988（COLO!）.

6. **Lasallia pustulata**（L.）Merat

Nouv. Fl. Paris.ed.z, 1;202（1821）.

≡ *Lichen pustulatus* L., Sp. Pl. 1150（1753）.

Лектотип: Ш веция, 1272-201:52, Fl. Suec. 969（LINN!）.

На скалах.

Материалы исследованные:

Монголия: Ара-Хангайский аймак, 22/Ⅷ/1979, Бязров, No. 5923. Под этим номерам два конверта: в одном из них имеются многие образцы с многочисленными цилиндрическими палочковидными изидиями, в другом конверте образцы с многочисленными шаровидными изидиями（Moscow!）.

Литературные материалы под названием U. Pustulata для Китая, Японии и других стран Азии обычно имеют в виду другие виды, напр. Lasallia asiae-orientalis, L.sinorientalis, и даже L. mayebarae и т.д. Материал L. pustulata из Кореи（Kim, 1972:158）был не исследован для этой работы.

7. **Lasallia rossica** Dombr.

Nov. Syst. Plant. Non vascul. 15:180（1978）.

Изотип: Россия, восточная Сибирь, район Буриат, Ⅷ/1965, В. М. Буркова（LE!）.

≡ *Umbilicaria rossica*（Dombr.）Golubk. in Nov. Syst. Plant. non Vasc. , 1978, 15:180.

На скалах и валунах.

Материалы исследованные:

Китай: Цзилин, гора Чанбай шань, южный склон, на вершине горы Хонтоу шань, 1950м, 5/Ⅷ/1980, Ху, Nos. 2185, 2239; 1960-1970, 16/viii/1982, Ху, Nos. 3433,3438,3473; 1980м, 28/Ⅶ/1983, Ху, Хо. 4216; 1880м, 28/Ⅶ/1983, Вэй и Чен, No. 6251（HMAS-L!）.

Монголия: Ара-Хангайский аймак, 22/Ⅶ/1970, Бязров, 6684（Moscow!）, 2350-2900м, 4/Ⅶ/1977, Бязров, No. 3270（Moscow!）, 2480м, 28/Ⅷ/1971, Бязров, No. 7467（Moscow!）. Дзабханский аймак, 2000м, и 2450м, 1/Ⅶ/1976, Бязров, Nos.6577, 6578（Moscow!）. На юге от Улан-Батор, Huneck, 27/Ⅶ/1978（HMAS-L!）.

Япония: Хоккайдо, 1/Ⅷ/1935, Sato（H!）.

Россия: Якутская республика, 25/Ⅷ/1956, В. Иванова, 153/2, *Lasallia rossica* var. *rossica* f. *septentrionalis* Dombr.（LE, Голотип!）. Дальневосточный, Хабаросвская область, 27/Ⅵ/1981, T. Randlare,

No81（DUKE!）. Сибирь, Алещкино, Пальбин（LE!）; Красноярск, 56 с.ш. 11/Ⅵ/1876, M. Brennder, No.2966（S!）. Красноярск, 11/Ⅵ/1876, 296（S!）. Южный Питоран, 790м, 31/Ⅶ/1969, В. Куваев, No. 126-3（LE!）. Амурская область, 19/Ⅶ/1910, Н. Прохоров, О. Куцнева No. 581（LE!）. Урал, Ⅷ/1929, К. Инрщина（LE!）. Сибирск, недалеко от Черемохавка, 1905, Палибин（FP!）. Ю.З. Сибирь, Алтай горы, горно-алтайская Автономная Область, 1750-1950м, 19/Ⅷ/1978, T.S.Elias, W. Weber, C.S.Tomb и I. Краспоборов, L-68071, C.H. COLO-319353（COLO!）. Сибирь, Гербарий Swartzii（S!）.

Армения: Кавказ, теберидинский Заповедник, 1500м, 14/Ⅶ/1978, J. Liska [Vezda: Lich. Sel. Exs. 1648]（S!）.

8. **Lasallia sinorientalis** J. C. Wei

Acta Myco. Sin, 1（1）:23（1982）.

Тип: Китай, Цзянси, гора Лу шань, Гулин, 1170м, 14//1965, Вэй, No. 3246（HMAS-L!）.

На скалах.

Материалы исследованные:

Китай: Шэньси, Хусиан, гора Гуантоу, 2700м, 2/Ⅶ/1964, Вэй, No. 77（HMAS-L!）. Цзянси, гора Лу шань, 1/Ⅳ/1960, Чжао и др., no. 472; 1150-1250м, 11/12, 14/11/1965, Вэй, Nos. 3115, 3117, 3117, 3121, 3246（HMAS-L!）. Анхой, гора Хуан шань, 800м, 24/Ⅷ/1962, Чжао и др., no. 5931; около Юйпинлоу, 2/XI/1980, Вэй, No. 3766（HMAS-L!）. Фуцзянь, IV-VI/1905, Dunn, No. 3938（FH, MIN!）.

Япония: Хонсю, K. Yoshida（4969）[Kurok. & Kashiw. Lich. Rar.et Cr. Exs. 516], как L. Asiae-orientalis（COLO, FH, LE, TENN, US!）. Недалеко от Токио, 1300м, 6/Ⅴ/1956, Kurowaka, No.56127（TNS!）; так же, 1260м, 21/XI/1977, Kashiwadani, No. 14605（TNS!）и 1300м, 20/Ⅶ/1969, H. Shibuichi, No. 3689（TNS!）. Синан, 21/Ⅶ/1922, Koidzumi, No. 71620（TNS!）. Синан, 2122/Ⅴ/1922, Koidzumi, No. 71686（TNS!）. Ямат, иосино-гун, 1680м, 23/Ⅶ/1969, Kashiwadani, Nos. 5980, 6071; Гора Санио, 1700м, 19/Ⅷ/1953, Tanaka; гора Саниогадаке, 10/Ⅵ/1952, Togashi, TNS-22114（TNS!）. Сикоку, 1600м, 13/XI/1972, Kashiwadani, No. 10354; 1700м, Kashiwadani,13/XI/1972, 1780м, 3/Ⅷ/1968, Kashiwadani, Nos.5292, 5371; 26/Ⅷ/1933, Fujkawa: TNS-22117; Toca, 19/Ⅶ/1929, Miyazaki; 19/Ⅶ/1929, Toshinaga: TNS-22118; 23/Ⅷ/1931, Fujikawa: TNS-22120; Ава, 13/Ⅷ/1956, Toshinaga: TNS-22115; 3/Ⅷ/1954, Inobe: Herb. Asahina, No. 350, TNS-22116, и 1900м, 13/Ⅷ/1969, Kashiwadani, Nos. 6921, 6841（TNS!）. Nishijima shrine, 1650м, 23/Ⅷ/1960. I. Y. 3113（US!）.

9. **Lasallia xizangensis** J. C. Wei & Y. M. Jiang

Acta Phytotax. Sin. 20（4）:500（1982）.

Тип: Китай, Тибет, Чжам, Кюсям, на скалах, 3240м, 16/Ⅴ/1966, Вэй и Чен, No.881（HMAS-L!）.

9.1. Var. **xizangensis**

9.2. Var. **acuta** J. C. Wei et Y. M. Jiang

Ibid. 20（4）:500（1982）.

Тип: Китай, Тибет, Чжам, Кюсям, 3500м, 19/Ⅴ/1966, Вэй и Чен, No.888（HMAS-L!）.

Подрод **Pleiogyra** J. C. Wei

Acta Phytotax. Sin. 11（1）:5（1966）.

Тип: Lasallia mayebarae（Sato）Asahina（1960）.

= Umbilicaria mayebarae M. Sato（1950）.

10. **Lasallia mayebarae**（Sato）Asahina

Journ. Jap. Bot. 35（4）: 101（1960）.

≡ *Umbilicaria mayebarae* SatoJourn. Jap. Bot. 25:168（1950）.

Тип: Китай, Тайвань, гора Али шань, 9/III/1930, K. Mayebara（TNS!）.

Ошибочное определение:

Umbilicaria pustulata sensu Asahina. Journ. Jap. Bot. 7:102（1931）, non Hoffm.

10.1. Var. **mayebarae**

На коре сосны.

Материалы исследованные:

Китай: Тайвань, гора Али шань, 9/III/1930, K. Mayebara（TNS!）; Илань, гора Тайпин шань, 12/VIII/1936, S. Asahina（TNS-21971!）.

10.2. Var. **sinensis**（J. C. Wei）J. C. Wei

Bull. Bot. Res. 1（3）:90（1981）.

≡ *Lasallia sinensis* J. C. Wei, Acta Phytotaxa Sin. 11（1）:1（1966）.

Тип: Китай, Юньнань, Лицзян, гора Юйорн шань, на стволах сосны（Pinus densata）, Вэй, No.2590（HMAS-L!）.

Материалы исследованные:

Китай, Юньнань, Лицзян, гора Юйорн шань, Ганхайцзи, на стволах сосны（Pinus densata）, XII/1964, Вэй, Nos. 2539, 2540, 2589, 2590（тип）, 2591, 2596（HMAS-L!）.

Род

Umbilicaria Hoffmann

Hoffm. Descr. Adumbr. Pl. Lich. 1（1）:8（1789）.

Тип:*Umbilicaria exasperata* Hoffm.（1790）.

Ключ для опрделения подродов, секции и видов

1a. Апотеции с ровными дисками ······ 1. Подрод ***Agyrophora***

2a. Верхняи поверхность слоевища гладкая, без ребров

3a. Верхняя поверхность слоевища в центральной части беловато-серая ······ 4. ***U. subglabra***

3b.Верхняя поверхность слоевища в центральной части не беловато-серая

4a. Нижняя поверхность слоевища шагревевидно-бородавчатая кричневая, у гомфа более светлая ······ 3. ***U. rigida***

4b.Нижняя поверхность слоевища гладкая, черная, не шагревевидно-бородавчатая, покраю более светлая; верхняя поверхность слоевищамелко трещиневатозернистая, ареолировано потрескавшаяся ······ 1. ***U. leiocarpa***

2b. Верхняя поверхность слоевища со слабо возвышающимиися ребрами складок сетчатого рисунка ······ 2. ***U. lyngei***

1b. Апотеции с не ровными дисками

5a. Апотеции с сосочками или трещиной в центральной части дисков ······ 2. Подрод ***Omphalodiscus***

6a. Аскоспоры одноклеточные и бесцветные ······ Секция ***Omphalodiscus***

7a. Апотеции сублеканоровые с ровными дисками насимыми трещиной в центральной части дисков, Нижняя поверхность слоевища светло-розовая, с млогочисленными светло-розовыми ризинами ······ 12. ***U. virginis***

7b. Апотеции лецидеевые

8a. Слоевище сверху волнистое или складчатое

9a. Слоевище сверху волнистое

10a. Слоевище по краям волнистое, нижняя поверхность коричневая ………………… 11. ***U. krascheninikovii***

10b. Верхняя поверхность слоевища слабо волнистое, на нижной поверхности с беловатыми или кoричеватыми ризинами ……………………………………………………… 7. ***U. aprina***（8. *U. canescens*）

9b. Слоевище сверху складчатое

11a.Слоевище с расходящися от центра лучинистыми складками, нижняя поверхность светлая с чёрными пятнами без ризины, по краям щетинистая, диски трещиковатые …………………………… 10. ***U. formosana***

11b. Слоевище с расходящися складками на сетчатом рисунке, верхняя поверхность слоевища черно-серая и темно-коричневая ……………………………………………………………………… 9. ***U. decussata***

8b. Слоевище сверху не волнистое и не складчатое

12a. Слоевище по краям не волнистое, нижняя поверхность светло-розовая со светло-розовыми ризинами6. ***U. altaiensis***

12b. Верхняя поверхность слоевища не волнистое, на нижной поверхности и по краям с длинными ризинами ……………………………………………………………………………………5. ***U. africana***

6b. Аскоспорые муральные, бесцветные и коричневые ………………… Секция ***Spodochriae*** Schol.ex Llano

13a. Верхняя поверхность слоевища без складок от центра, нижняя поверхность слоевища чёрная с частичными и густыми ризинами …………………………………………………………… 14. ***U. spodochroa***

13b. Верхняя поверхность слоевища со складками от центра, нижняя поверхность слоевища светло-розовая со светло-розовыми ризинами……………………………………………… 13. ***U. crustulosa***

5b. Апотеции с гирозными или лучеобразными дисками

14a. Апотеции с гирозными дисками ……………………………………………………3. Подрод ***Umbilicaria***

15a. Аскоспорые одноклеточные и бесцветные ……………………………………………Секция ***Umbilicaria***

16a. Слоевище маленькое размером, не свыше15 mm в диаметре

17a. Нижняя поверхность слоевища чёрная и гладкая

18a. Апотеции обычно присутствующие；слоевище кожистое, толстое и жёсткое, не свыше 5 мм и до 10 мм в диаметре，внизу с периферическими чёрными и толстыми ризинами и по краям с черными и толстыми ресничками ……………………………………………………………………………… 31. ***U. nanella***

18b. Апотеции обычно отсуствующие; слоевище тонкое, не свыше15 mm в диаметре, сверху по периферням выпуклое, и по краям спускающее вниз………………………………………28. ***U. kisovana***

17b. Нижняя поверхность слоевища светлая

19a. Апотеции более или менее погуженные в слоевище, аскоспоры субшаровидные; слоевище около 5 мм в диаметре，сверху плоское, внизу коричневатое, без ризины ……………………………… 30. ***U. minuta***

19b. Апотеции сидячие не погуженные в слоевище, аскоспоры длинно яйцвидные; слоевище 5-10 мм в диаметре，сверху по краям поднимающеесь к верху; внизу коричневатое, с одноцветными ризинами ……………………………………………………………………………………………………40. ***U. taibaiensis***

16b. Слоевище среднее размером, свыше 15 mm в диаметре

20a. Нижняя поверхность слоевища беловатая и коричневатая, с беловатолакунозно-складками и одноцветными ризинами; апотеции треугольные или многоугольные ……………………42. ***U. torrefacta***

20b. Нижняя поверхность слоевища не лакунозно-складчатая

21a. Апотеции обычно присутствующие

22a. Апотеции более или менее погуженные в слоевище

23a. Нижняя поверхность слоевища чёрная, с густыми，тонкими，нежнымии，и короткими ризинами ……………………………………………………………………………………………………… 16. ***U. badia***

23b. Нижняя поверхность слоевища коричневая или чёрно коричневая, сильно бородавчатая, без ризины ……………………………………………………………………………………………………… 34. ***U. phaea***

22b. Апотеции сидячие，не погуженные в слоевище

24a. Апотеции нерегулярнно ланцетовидные, и треугольные;

25a. Слоевище на скалах, кожистое, толстое и жёсткое, сверху пузыревидно-бородавчатое, внизу светлое и вокруг гомфа черное, без ризины ·· 15. ***U. arctica***

25b. Слоевище на коре дерева, сверху плоское, внизу черная с крчавыми, корокими, густыми и много ветвистыми ризинами ··45. ***U. yunnana***

24b. Апотеции округлые

26a. Нижняя поверхность слоевища с ризинами,

27a. Нижняя поверхность слоевища только с язычковыми и нерегулярнно головчатыми ризинами ·· 43. ***U. tylorhiza***

27b. Нижняя поверхность слоевища с цилиндическими ризинами

28a. Апотеции сидячие и на ножке, нижняя поверхность слоевища светло-розовая со светло-розовыми ризинами ··18. ***U. cylindrica*** auct.

28b. Апотеции не на ножке

30a. Диск с много перпендикулярный трещин; нижняя поверхность слоевища шагревевидно-бородавчатая кричневая, как у *U. rigida*, без ризины или с единичной ризиной······················ 37. ***U. pseudocinerascens***

30b. Диск без перпендикулярных трещин

31a. Нижняя поверхность слоевища бородавчатая

32a. Нижняя поверхность слоевища сильно бородавчатая, с редкой ризиной, по краям густой ·· 39. ***U. subumbilicarioides***

32b.Нижняя поверхность слоевища слабо бородавчатая,

33a. Нижняя поверхность слоевища с густыми и тонкими и краткими ризинами, и без таллилов（thallyles）, сверху без чешуевидных паралобулиев ·· 41. ***U. thamnodes***

33b. Нижняя поверхность слоевища с толстыми и длинненькими ризинами и таллилами（thallyles）, сверху с чешуевидными паралобулиами ·· 38. ***U. squamosa***

31b. Нижняя поверхность слоевища ровная, не бородавчатая, чёрно-коричневая, с густыми и черно-коричневыми ризинами·· 27. ***U. indica***

26b. Нижняя поверхность слоевища обычно без ризины или случайно с единичной ризиной, светлая или розово-красная

34a. Нижняя поверхность слоевища обычно без ризины или случайно с единичной ризиной

35a. Нижняя поверхность слоевища жёлто-красная и розово-красная ························26. ***U. hypococcinea***

35b. Нижняя поверхность слоевища светлая, не жёлто-красная и розово-красная

36a. Верхняя поверхность слоевища складчатая

37a. Верхняя поверхность слоевища без беловатых налётов, нижняя поверхность слоевища коричневая и темно коричневая, без ризины ··20. ***U. exasperata***（hyperborea）

37b. Верхняя поверхность слоевища на серндинн покрытая беловатыми налётами; нижняя поверхность слоевища серо-коричневая и чёрно-коричневая, обычно без ризины и случайно с единичной ризиной ··36. ***U. proboscidea*** auct.

36b. Верхняя поверхность слоевища не складчатая

38a. Нижняя поверхность слоевища светло коричневатая, волдыреобразная, или снабжённая рядом лакун с редкими ризинами ·· 24. ***U. herrei***

38b. Нижняя поверхность слоевища беловато телесная, к периферии покрывая беловатыми налётами, инагда вокруг гомфа чёрная, с единичной ризиной ··32. ***U. nepalensis***

21b. Апотеции обычно отсуствующие

39a. Верхняя поверхность слоевища складчатая, кожистое, серо-коричневая и чёрно-коричневая, покрытая тонко беловатыми налётами, внизу чёрная ······ 33. ***U. nylanderiana***

39b. Верхняя поверхность слоевища не складчатая

40a. Верхняя поверхность слоевища серо зелёная

41a. Верхняя поверхность слоевища к периферии соредиозная; внизу светло коричневатая, с одноцветными ризинами, и случайно без ризиной ······ 25. ***U. hirsuta***

34b. Нижняя поверхность слоевища обычно без ризины

41b. Верхняя поверхность слоевища к периферии не соредиозная; внизу чёрная, бородавчатая, без ризины ······ 22. ***U. grisea***

40b. Верхняя поверхность слоевища коричневатая, коричневая и чёрно-коричневая

42a. Нижняя поверхность слоевища покрытая пушинками или ризинами

43a. Нижняя поверхность слоевища вокруг гомфа беловато телесная, к периферии темно-коричневая и почти чёрная, с чёрными пушинками, где покрытая таллоконидиами ······ 23. ***U. havaasii***

43b. Нижняя поверхность слоевища чёрная, покрытая густыми чёрными ризинами

44a. Слоевище тонкое, сверху коричневая; ризины чёрные, краткие, тонкие и ветвистые ······ 19. ***U. esculenta***

44b. Слоевище толстое, кожистое, сверху серо-зелёное и серо-коричневое, внизу чёрное

45a. Слоевище сверху к периферии покрытое паралобулиами ······ 29. ***U. loboperipherica***

45b. Слоевище сверху к периферии без паралобулиев ······ 44. ***U. vellea***

42b. Нижняя поверхность слоевища без ризины

46a. Верхняя поверхность слоевища с лобулиами ······ 21. ***U. flocculosa***

46b. Верхняя поверхность слоевища без лобулиев

47a. Верхняя поверхность слоевища серо-коричневая ······ 17.***U. cinerascens***

47b. Верхняя поверхность слоевища желто коричневая и краснокоричневая ······ 35. ***U. polyphylla***

15b. Аскоспорые муральные бесцветные и коричневые ······ ***Muriformes*** Llano

48a. Апотеции нерегулярнно многоугольные ······ 46. ***U. angulata***

48b.Апотеции округлые

49a. Слоевище монофильное, тонкое, сверху коричневая

50a. Слоевище внизу чёрные, с чёрными, длиными, толстыми и не ветвистыми ризинами ······ 47. ***U. caroliniana***

50b. Слоевище внизу темно-коричневая, с густыми, тонкими и не ветвистыми ризинами ······ 49. ***U. koidzumii***

49b. Слоевище монофильное и полифильное, толстое, кожистое, сверху серо-зелёное и серо-коричневое, внизу чёрное без ризины, с крупными наростами ······ 48. ***U. cinereorufecens***

14b. Апотеции с лучеобразными дисками ······ 4.Подрод ***Actinogyra***

51a. Нижняя поверхность слоевища лакунозно-складчатая

52a.Слоевище монофильное ······ 50. ***U. mülenbergii***

52b.Слоевище полифильное, подостланное ······ 52. ***U. pulvinaria***

51b. Нижняя поверхность слоевища не лакунозно-складчатая, светвистыми и головчатыми ризинами, где покрытая таллоконидиами ······ 51. ***U. polyrhiza***

Подрод **Agyrophora** Nyl.

Flora 247（1878）.

Тип:*Umbilicaria leiocarpa* DC,（1805）

1. **Umbilicaria leiocarpa** DC. in Lam.et DC.

Flora Franc.ed. 3.3:410（1805）.

Тип: France, Pyrenees, Port Madamette, coll.C. Ramond sb Lichen infundibuliformis

≡ *Agyrophora leiocarpa*（DC.）Gyelnik, Ann. Mycol. 30:444（1932）.

На скалах.

Материалы исследованные:

Япония: Хонсю, Етчу, гора Етчусава-даке, 2400м, 27/Ⅶ/1976. Синано, гора Кинпу, Kashiwadani（No. 13381）, и K. Yoshida [S. Kurok.:lichi. Ro.et C. Esx. No. 291]（BM, FH, H, S, UPS!）; Синано, гора Кинпу, минами-Саку-гун, 25/Ⅶ/1972, Shibuichi（No.4667）и K. Yoshida [S. Kurok.:lichi. Ro.et C. Esx. No. 292]（BM, FH, H, TUR, US!）; гора Yakusi, 27/Ⅶ/1929, H. Нага, No. 342（UPS!）. Синано, гора Сироума, A. Yasuda, No.392（H!）.

Монголия: Ара-Хангайский аймак, 2507 м, 8/Ⅷ/1971, Бязров, No. 5755（Moscow!）.

2. **Umbilicaria lyngei** Schol.

Nyt. Mag. Naturv. 75:19（1934）.

Тип: Северо-восточная Гренландия: Кар Humboldt, Aug. 3,1930, coll. Schol.（non vis.）.

≡ *Agyrophora lyngei*（Schol.）Llano, Monograph of the Umbilicariaceae, 57（1950）.

На скалах.

Материалы исследованные:

Китай: Цзилин, гора Чанбай шань, северный склон, 2400м, 14/Ⅷ/1977, Вэй, No. 2983（HMAS-L!）, и недалеко от Тянь Вен фун, 2400м, Ху, No. 1605（HMAS-L!）. Шэньси, гора Тайбай шань, Басиантай, 3900м, 3/Ⅵ/1963, Вэй（HMAS-L!）3380м, около Вен-гунмиао, 3380м, 9 /Ⅶ/1963, Ма и Цзун, No. 296（HMAS-L!）. Синцзян, гора Алтай, около оз. Халас, 2500м, 5/Ⅷ/1986, Гао, No.2075-2（HMAS-L!）.

Монголия: Убсу-Нурский айнак, 9/Ⅶ/1976, Бязров, No.5930（Moscow!）.

Россия: Восточная сибирь, гора Буторан, 25/Ⅶ/1985, М.Р.Исурбенко（LE!）. Сибирская Арктика, Таймыр полуостров, 4/Ⅶ/1947, Б. Тихомиров（как U.rigida）(H!）.

3. **Umbilicaria rigida**（DU Rietz）Frey

Hedwigia 71:117（1931）.

Тип: Швеция, 1100м, 3/Ⅷ/1911, G.E.Du Rietz（ UPS! Голотип）.

≡ *Gyrophora rigida* Du Rietz, Arkiv Bot. 19（12）:3（1925）.– *Agyrophora rigida*（Du Rietz）Llano, Monograph of the Umbilicariaceae, 58（1950）

= *Umbilicaria coriacea* Imsh.

На скалах.

Материалы исследованные:

Китай: Шэньси, гора Тайбай шань, Басиантай, 3/Ⅵ/1963. Вэй,（HMAS-L:9271!）

Япония: Хонсю, Синано, A. Yasuda, ex hreb. Ras. 392（H!）.

Монголия: Ара-Хангайский аймак: 25/Ⅷ/1970, Бязров, No.5925; 2700м, 12/Ⅶ/1976, Бязров, No.5926; 22/Ⅸ/1979, Бязров, No. 5924（Moscow!）.

4. **Umbilicaria subglabra**（Nyl.）Harm.

Lich. France 4:707（1909）.

≡ *Gyrophora subglabra* Nyl., Lich. Env. Paris, 135（1896）.

Тип: France, Pyrenees, Cebennis, coll. Mende（H-Nyl. 31666! Голотип）.

На скалах.

Материалы исследованные:

Китай: Синцзян, Чжао, No.277（HMAS-L!）; гора Алтай, около оз. Халас, 2000м, 4/Ⅷ/1986, Гао, No. 1939（HMAS-L!）и 2500м, 5/Ⅷ/1986, Гао, No. 2075（HMAS-L!）.

Казахская Республика: Восточная часть страны, 1600м, 13/Ⅶ/1982, Домбровская, No. 14, как *U. kraschennikovii*（HMAS-L!）.

Армянская Республика: Степанованский район, 1956, V.G.Nilogosjan [Савич: Lichenoheca Rossica No.162]（BM, H, S, US!）. Кавказ, Красносельск район, 2500м, 24/Ⅴ/1976, Vezda [Vezda: Lich. Sel. Exs. 1415]（BM, COLO, H!）.

Подрод

Omphalodiscus（Schol）Schade ex J. C. Wei & Y. M. Jiang

Mycosystema 1:89（1988）.

≡ *Omphalodiscus* Schol. Nyt Mag. f.Naturvid. 75:23（1936）.

Тип: *Umbilicaria decussata*（Viil.）A. Zalhbr.（1942）.

≡ *Lichen decussatus* Vill.（1789）.

Секция **Omphalodiscus**

5. **Umbilicaria afracana**（Jatta）Krog & Swinscow

Nord. J. Bot. 6（1）:79（1986）.

≡ *Gyrophora haplocarpa* var. *africana* Jatta, Ann. Bot. 6:408（1909）.

Тип: Africa, Ruwenzori, veyante ovest, Duboin 4-4500m, ex Herb.

C. sbarbaro, FH, синтип（MIN!）. - *Omphalodiscus africanus*（Jatta）Llano, Monogr of the Umbilicariaceae 1950, p.99.

= *Umbilicaria zollingeri* Groenh. Reinwardtia 1（1）:33（1950）.

Тип: Ява, Гора Ардзуно, Гроенхалт 422（ L!голотип）.

На скалах.

Материалы исследованные:

Ява: Гора Ардзуно, 3325m, 20/Ⅵ/1932, Groenhart, No.161; 27/Ⅲ/1937, Groenhart, Nos. 7659, 7660, 7661, 7662, 7663, под названием *U. zollingeri*（L!）.

6. **Umbilicaria altaiensis** J. C. Wei & Y. M. Jiang

Mecosystema 5:73（1992）.

Тип: Киатй, Синцзян, гора Алтай шань, около оз. Халас, 2500м, 5/Ⅷ/1986, Гао, No. 2069（HMAS-L, голотип!）.

На скалах.

Материалы исследованные:

Китай: Синцзян, гора Алтай шань, около оз. Халас, 2500м, 5/Ⅷ/1986, Гао, No. 20695（HMAS-L!）.

7. **Umbilicaria aprina** Nyl.

Synopsis Lich. 2:12（1869）.

Тип: Африка, эфиопия, гора Дедсин（Dedschen）, 4600м, W. Schimper（H-Nyl. 31742, голотип!）.

= *Umbilicaria syagrodema* Nyl., Flora 13:418（1860）, nom.nud.

= *Umbilicaria cristata* Dodge & Baker, Ann. Missouri Bot. Grad. 25:565（1938）.

Тип: Антарктика, гора Хелен Вашингтон, 78 05' ю.ш.- 155 20' з.д., 1934, P. Siple, F.A. Wade, S. Corey и D. Stadmcliff, No. HW-la（FH!）.

= *Umbilicaria spongiosa* Dodge & Baker, Ann. Missouri Bot. Grad. 25（2）: 566-567（1938）. - *Gyrophora spongiosa*（Dodge & Baker）Golubk. & Savicz, Novitates Systematicae Plantarum non vascularium 1966:259（1966）.

Тип: Second Byrd Antarctic Expedition, Flora of Marie Byrd Land, on gray slate, arkosic sandstone, sericiteorthoclase schist, Lichen Peak, 76°55' S-145°20' W, 1934, coll. P. Siple, F. A. Wade, S. Corey and O.D. Stancliff, No. 73-9（FH!）.

=*Gyrophora korotkeviczii* Golubk. & Savicz, Novitates Systematicae Plantarum non vascularium 1966:261-263（1966）.

Тип: Antarctida orientalis, Litus Pravda, Bunger hills ad saxa March 3,1957, coll. E-S. Korotkevicz（LE!）.

7.1. Var. **aprina**

На скалах.

Материалы исследованные:

Китай: Синцзян, гора Алтай шань, около оз. Халас, 2500м, 5/Ⅷ/1986, Гао, No. 2089（HMAS-L!）.

Таджикстан Республика: Памир-Алтай, Аини Район, 3500-3600м, 24/Ⅶ/1985, V. Vasak, как *U. krascheninuikovii*, L-83138（COLO!）; Аини район, 3700-4100м, 25/Ⅶ/1985, V. Vasak, как *U. krascheninuikovii*, L-88149（COLO!）; там же 3500-3600м, 23/Ⅶ/1985, V. Vasak, как *U.krascheninuikovii*, L-83137（COLO!）.

Афнанистан: Кабул, 500м, 68 56" В. и 34 37" с. 2,540v, 1977, D. Podlech Pod57/7, [Lichenotheca Afghanica, No.25]（BM!, LE!）; No.16, c.2,550m, Steile Silikatfelsen, ±N-NW-Exposition, May 17, 1977（BM!, LE!）; No. 58（LE!）.

Турция: Арарат, ± 4,000м, 1972, coll. Esther Buchi, Herb. Rosmarie Honegger 729d（H!）.

7.2. Var.**halei** Llano

Journ. Washington Acad. Science 46（6）:183（1956）.

Тип: Канада, Баффинова земли, 1950, Hale no. 450 in H-Llano.

= *Umbilicaria decussata* var. r*hizinata* Frey & Poelt, Khumbu Himal 6:419（1977）.

Тип: Moranen des Lobuche-Gletschers bes Lobuche, 4950-5000m. L247a（M, голотип）.

На скалах.

Материалы исследованные:

Китай, Цзилин, гора Чанбай шань, 2300-2600м, 14/Ⅷ/1977, Вэй, No. 1981; около Тянь Вен Фун, 19/Ⅷ/1987, Вэй, No.9011; недалеко от маленького водопада между Гуанцзия и Лунменфун, 13/Ⅷ/1983, Вэй и Чен, No. 6875-2（HMAS-L!）.

Монголия: Хубсугульский аймак, на побережье оз. Хусугул, 23/Ⅶ/1902, А. А. еденкин（LE!）.

8. **Umbilicaria canescens**（Dombr.）Golubk.

In Handbook of the lichens of the U. S. S. R 5:115（1978）.

≡ *Gyrophora canescens* Dombr. В кн. Хибин. Л. 1970:131.（in Lichens of Khibini, 1970, p. 131.）

Голотип: Кольский полуостров, Хибинский горный массив, слкон южной акспозиции горы Вудъяврчорр, на высоте около 380м над ур.м., на голых камнях крупнокаменистой осыпи, № 6, 5 VI 1961,

А. В. Домбровская, Ботанический институт им. В. Л. Комарова, отдел низших растений.

На скалах.

Материалы исследованные:

Монголия: Ара-Хангайский аймак, 2200м, 17/Ⅶ/1970, Бязров, No. 5937（Moscow!）.

Примечание: Этот вид может быть другой синоним вида *Umbilicaria aprina* Nyl. на основании комплексных характеров слоевища. Заклчительный вывод об этом будет оставаться за выводом молекулярной систематики.

9. **Umbilicaria decussata**（Vill.）Zahlbr.

Cat. Lich. Univ. 8:490（1942）.

≡ *Lichen decussatus* Vill. In Hist. Plant. Dauphine 3:964, pl.55（1789）. - *Omphalodiscus decussatus*（Vill.）Schol. Nyt Mag. Naturid. 75:23（1934）. Llano, A Monograph of the Umbilicariaceae, 78（(1950）.

Тип:（Iconotype）: As illustrated by Villars（1789）, from France, Dauphine（Hautes Alpes）.

= *Umbilicaria ptychophora* Nyl., Flora 52:388（1869）

Тип: Caucasus orientalis, 1861, leg. Ruprecht（H-Nyl.31631 in H!）.

= *Umbilicaria reticulata* Nyl., Flora 52:389（1869）. - *Gyrophora reticulate*（Nyl.）Th. Fr. Lichenogr. Scand. 1871:166.[166（1871）]

= *Umbilicaria cerebriformis* Dadge & Baker, Ann. Missouri. Bot Card. 25:562（1938）.

Тип: Antarctica, Marie Byrd Land, Edsel Fdscl Ford Range, Skua Gull Peak, coll. P. Siple & S. Corey 72 W-15（FH!）.

= *Charcotia cerebriformis*（Dodge & Baker）argoms Dodge, b. A. N. Z. A. R. E. Rept. B7:150（1984）.

≡ *Omphalodiscus decussatus* var. *cerebriformis*（Dodge & Baker）Llano, A Monograph of the Umbilicariaceae 83（1950）. - *Llano cerebriformis*（Dodge & Baker）Dodge, Nova Hedwigia 15:311（1986）.

= *Umbilicaria spongiosa* Dodge & Baker, Ann, Missouri Bot. Gard. 25（2）;566（1938）.

Тип: Antarctica, Marie Byrd Land, Edsel Ford Range, Lichen Peak, coll. P. Siple & S. Corcy No. 73-9,2nd Byrd Antarc. Exp. 1933-35（FH!）.

≡ *Omphalodiscus spongiosus*（Dodge & Baker Llano, A Monograph of the Umbilicariaceae 1950, p.91.

= *Umbilicaria antarctica* Frey & Lamb, var. *subvirginis* Frey & Lamb, Trans. Br. Mycol. Soc. 22（3-4）:272（1938）.

Тип: IN BERN（not seen）.

≡ *Umbilicaria spongiosa* Dodge & Baker, var. *subvirginis*（Frey & Lamb）Dodge, B. A. N. Z. Antarc. Exp. Rep. See.b. 7:148（1938）.

Ошибочное определение

Umbilicaria lyngei sensu Biazrov: Bull, Mosc. O. Va Ispeifatelei Prirodei. Otd. Biol. 91（6）:100（1986）, non Schol.

На скалах.

Материалы исследованные:

Китай, Цзилин, гора Чанбай шань, северный склон, недалеко от Тяньчи, 2400м, 14/Ⅷ/1977, Чен, No. 4825（IFP,HMAS-L!）; недалеко от Тяньвенфун, 2400м, 12/Ⅶ/1980, Ху, No.1605（HMAS-L!）.

Монголия: Хубсугульский аймак, на побережье оз. Хубсугул, 23/Ⅶ/1902, А. А. еленскин（LE!）. Убсунур, 2600м, 28/Ⅵ/1973, Бязров, No. 2969, и 2680м, 9/Ⅶ/1976, Бязров, No. 6565（Moscow!）; гора Муиву-Сардеика, 23/Ⅶ/1902, А. А. еленскин（LE!）. Дзабзанский аймак, 2650м, 12/Ⅶ/1976, Бязров, No. 5927, и 2563м, 25/Ⅵ/1978, Бязров, No. 6567; 25/Ⅵ/1978, Бязров, No. 6569, и 2680м, 9/Ⅶ/1976, Бязров, No. 6576（Moscow!）.

Россия: Западная часть Чукотского полуострова, Анюйск, 600м, 22/Ⅶ/1977, М. П. Андреев（LE!）, о.

Новая Земля, коллектор из неизвестен（LE!）.

Турция: Бурса, Улидаг, 20км к ю.-востоку от Бурсы, 2000м, 31/Ⅷ/1976, K. Kalb и G. Plost

Киргизская республика: Туркестан Терски Алатау, 7/Ⅷ/1896, V. F. Brotherus No. 34（H!）.

Армянская республика: кавказ, 2150м, 21/Ⅴ/1976, Jelinkova, около оз. Севан, 12/Ⅷ/1980, A. A. Абрамиан и A. Vezda d [

Тянь-Шань: Точное местонахождение не известно, 8/Ⅶ/1902, B. Саноцников（LE!）.

10. **Umbilicaria formosana** Frey

Hedwigia 71:115（1931）.

Тип: China, Taiwan（Formosa）, Mt. Morison（Niitaka）region, Oct. 11, 1927, coll. S. Sasaki.（MIN! синтип）.

≡ *Omphalodiscus formosana*（Frey）Schol. Nyt Mag. Naturvid. 75:24（1934）. - *Omphalodiscus krascheninnikovii* var. *formosanus*（Frey）Llano, A Monograph of the Umbilicaria, 88（1950）. - *Gyrophora formosana*（Frey）Sato, Journ. Jap. Bot. 25（8）: 170（1950）.

= *Umbilicaria bigleri* Frey, Ber. Schweiz. Bot. Ges. 59:458（1949）. Syn.nov.

Тип:（Лектотип）: Мексика, гора Попокатепетл 4000м, 7/Ⅵ/1945, Dr. Walter Kiener No. 18588（FH!）.

Ошибочное определение:

Umbilicaria krascheninnikovii sensu Wei: Bull. Bot. Res. 1（3）: 90（1981）, non（Savicz）Llno.

На скалах.

Китай: Цзилин, гора Чанбай шань, северный склон, недалеко от Тяньвенфун, 2100м, 13/Ⅷ/1977, Вэй, No. 2942-1（HMAS-L!）; 2600м, 14/Ⅷ/1977, Вэй, No. 2987, и 2400м, 12/Ⅶ/1980, Ху, No. 1604, и 2200м, 10/Ⅷ/1977, Ху, No. 120, и 2600м, 14/Ⅷ/1978, Ху, No. 4690 и 20/Ⅷ/1984, Ху, No. 6383, недалеко от водопада, 2100м, 13/Ⅷ/1977, Ху, No.218; с.з.склон, Луминфун, 2550-2580м, 20 /Ⅷ/1984, Ху, Nos. 6409, 6411; ю. Склон, Ганбахецзи, 2050м, 31/Ⅶ/1983, Ху, No. 4272; гора Хунтоу шань, 1880м, 28/Ⅶ/1983, Вэй и Чен, No. 6325（HMAS-L!）; 2010м, 28/Ⅶ/1983, Ху, No. 4234（HMAS-L!）; на дороге от Хен шань станции до Тяньчи, Вэй и Чен, No. 6370（HMAS-L!）; с.з.склон, Вэйдун станция, Лаохубей, 2200м, 8/Ⅷ/1983, Ху, Nos. 4594, 4599, 4601, 4602（HMAS-L!）.. Шэньси, гора Тайбай шань, Басиантай, 3-4/Ⅵ/1963, Вэй и др., No. 2719-2, и один образец с маркой "G.B."（HMAS-L!）; Хусиан, гора Гуантоу щань, 2960м, 2/Ⅵ/1964, Вэй , No. 79（HMAS-L!）. Юньнань, Дали, гора Цан шань, Чжунхефун, 3900-3970м, 3/Ⅰ/1965, Вэй, Nos. 2816, 2828, 2829-2（HMAS-L!）. Тибет, Ниелам, 3830м, 12/Ⅵ/1966, Вэй и Чен, No. 1437（HMAS-L!）; Ниелам, Цзянин, 4450м, 16/Ⅵ/1966, Вэй и Чен, No. 1669（HMAS-L!）. Тайвань, гора Тапачиан, 19/Ⅶ/1932, S. Sasaki, No. R-412a ex Herb.y. Asahina（TNS-21970!）.

Сикким: 4000м, 31/Ⅴ/1960,

Россия: Камчатка, Арат Шинокая Сопка, 700м, 21/Ⅶ/1920, Eric Hulten; и Иелировск район, 6/Ⅷ/1985, A. Мукулин, No. 1-653-d（LE!）. Центральная сибирь, гора Путоран, 9/Ⅶ/1984, M. П. Исурбенко（LE!）.

Армянская респубоика: Степанаван, 13/Ⅷ/1963, B. Г. Nilogocuah（LE!）.

11. **Umbilicaria krascheninnikovii**（Savicz）A. Zahlbr.

Cat. Lich. Univ. 10:405（1939）.

≡*Gyrophora krascheninnikovii* Savicz Bull. Jard. Imp. Bot. Pierre le Grand 14:117（1914）. - *Onphalodiscus krascheninnikovii*（Savicz）Schol. Nyt Mag. Naturvid. 75:24（1934）.

Тип: Камчатка, на скалах, гора Вулкан Красильниковы（сопка）, Савич, No. 6412（LE!）.

= *Gyrophora hultenii* D. Rietz, Arkiv for Bot. 22（13）:14（1929）.

Тип: Syd-Kamchatka: Aroatocha-vulkanon, 700m, July 31, 1920, leg. Eric Hulten, Svenska Kamtchatka-expedtionen 1920-1922. Lichenes n:678（UPS!）. - *Umbilicaria hultenii*（Du Rietz）A. Zahlbr., Cat. Lich. Univ. 8:492（1932）.

На скалах.

Материалы исследованные:

Россия: Камчатка（ см.тип）.

Япония: Хоккайдо, 10/Ⅷ/1932, F. Fujikawa ex Herb. Y. Asahina（TNS-21968!）.

12. **Umbilicaria virginis** Schaer.

Biblioth. Univ. Geneve. 36:153（1841）.

≡ *Gyrophora virginis*（Schaer.）Frey, Hedwigia 69:222（1929）. -*Omphalodiscus virginis*（Schaer.）Schol. Nyt. Mag. Naturvid. 75:25（1934）.

=*Umbilicaria rugifera* Nyl. Lich. Scand. 1861, p. 117.

Тип: Sibiria orientalis（Sievers)）H-Nyl. 31633!）.

12.1. Var. **virginis**

На скалах.

Материалы исследованные:

Киатй: Синцзян, Тань шань, Ван No. 479（HMAS-L!）.

Монголия: Ара-Хангайский аймак, 2500м, 28/Ⅷ/1971, Бязров, No. 4233; 3500м, 13/Ⅶ/1976, Бязров, No. 7532, 16/Ⅷ/1972, Бязров, No. 6691; 2/Ⅷ/1973, Бязров, No. 5795（Moscow!）. Убсунурский аймак, гора Дульг, 2680м, 9/Ⅶ/1976, Бязров, No. 6576（Moscow!）. Дзабаханский аймак, 3500м, 13/Ⅶ/1876, Бязров, No. 6573; 2650м, 12/Ⅶ/1876, Бязров, No. 2447（Moscow!）. Хубсугульский аймак, на берегу оз. Хубсугу, гора Муйву-Сардейскам, 23/Ⅶ/1902, А. А. Еленкин（LE!）.

Киргизская респубоика: Туркестан, Алатау, 22-24/Ⅵ/1896,V.F. Brotherus, NO. 60（H!）; Алтау, 9/Ⅶ/1907, В. Семенова（LE!）; гора Саур, 1904, В. Семенова（LE!）. Терск, Алатау, 8/Ⅷ/1896,V.F. Brotherus, NO. 39（H!）. Тянь шань, 8/Ⅶ/1902, В. Саноцников（LE!）. Тянь шань, 3800м, 11/I/1982, Е. Д. Циракова（MW!）.

Армянская республика: около Севена, 3000м, 13/Ⅷ/1980, А. А. Абрамиан（LE!）. С. З. Гималаи: Кангра район, 15/Ⅶ/1952, O. A. Hoeg, ex Herb. D. D. Awasthi, No. !%$）(DUKE,H!）.

Примечание:

Согласно Поельту（Poelt, 1977:431）, образец этого таксона, собранный из Тибета, хранится в BM. Однако он не был найден в период моей остановки в BM, и поэтому не был исследованы для этой работы.

12.2. Var. **lecanocarpoides**（Nyl.）J. C. Wei & Y. M. Jiang

Mycosystema, 1:19（1988）.

≡ *Umbilicaria lecanocarpoides* Nyl. Flora 43:418（1860）& Synops. Lich. 2:7（1863）.

Тип（Lectotype）: Гималаи, ex Herb. C. Sbarbaro, Hook fil., et Thomson, Nr, 2090（H!）.

На скалах.

Материалы исследованные:

Киатй: Сычуань, Кандин, Даву, южная гора Яара, 4700м, 29/Ⅷ/1934, H. Smith, No. 14006（UPS!）.

Сикким: alt. 15,000ft., Herb. Ind.or Hook.fil. & Thomson No.2090（BM, FH, H, UPS!）, alt. 13,000ft, Herb. ind.orHook.fil. Thomson No.2096 and alt. 19,000ft, No.2100（BM!）.

Гималаи: Hook, fil 1857, *U. lecanocarpoides* Nyl.（MIN!）.

Россия: Дальний Восток, О. Врагел, 72 с.ш. и 180 в.д. 20/Ⅷ/1938, Б. Н. Городков, No. 28（LE!）.

Киргизская республика: Алатау, 1896, coll. V. F. Brotherus No.60 pr.p.（H!）.

Секция **Spodochriae** Schol.ex Llano

A Monograph of the Umbilicariaceae 1950, p.97.

Lectotype: *Umbilicaria crustulusa*（Ach）Frey（1931）.

13. **Umbilicaria crustulosa**（Ach）Frey

Hedwigia 71:110（1931）.

≡ *Gyrophora crustulusa* Ach., Lichenogr. Univ. 673（1810）.

Тип: Hispania, H-Ach. 577（H!）.

На скалах.

Материалы исследованные:

Монголия: Ара-Хангайский аймак, 19/Ⅵ/1971, Бязров, No. 5752（Moscow!）.

Турция: Бурса, Улубаг, 2000m, 1976, coll. K. Kalb G. Plobst. 111（DUKE!）

14. **Umbilicaria spodochroa** Ehrh. Ex Hoffm.

Deutschl. Fl. 3:113（1798）.

≡ *Lichen spodochrous* Ehrh. Crypt. Exs.no. 317(1795).nom.nud. Ach. Lichenogr. Suec. Prodr. 149(1798). - *Omphalodiscus spodochrous*（Ehrh.ex Hoffm.）Schol. Nyt Mag. Naturvid. 75:26（1934）. Llano, Monograph of the Umbilicariaceae, 101（1950）.

Тип: Ehrhart, Chr. G., Crypt. Exs. 317, in Hoffm. Deutschl. Fl. 113（1796）, not seen.

На скалах.

Материалы исследованные:

Китай: Сычуань, в высокогорных районах между Кандином и Чжедуом, Потанин（1893）- в эксиккате Еленкина: Elenkin, Lich, Fl. Ross. IV, 1904, No 151（часть из них: 151-a, 151-b-2 151-c）, как *Gyrophora spodochroa*（Ehrh.）Ach.（UPS!）and No. 151（FH!）.

В своей монографии о лишайниках сем. Umbilicariaceae Камчатского полуострова Савич записал 9 видов. В том числе один вид под названием *G. spodochroa*（= *U. spodochroa*）[Оригинальный образец No. 199（11）хранится в LE, и No. 5985（11）хранится в LE, и US] на самом деле является ошибочным определением другого вида *U.vellea*（J. C. Wei & Y. M. Jiang, 1993）.

Подрод **Umbilicaria**

Секция **Umbilicaria**

15. **Umbilicaria arctica**（Ach.）Nyl.

Herb. Mus. Fen. 84（1859）.

≡ *Gyrophora arctica* Ach. Method. Lich. 106（1803）. "Icon. Tab.nostra 2.f.6. Habitat in alpibus Norland. Norveg.e Lyngentind. WAHLENBERG." - *Gyrophora proboscidea* var. *arctica*（Ach.）Ach. Synops. Lich 65（1814）. - *Gyromium proboscideum* var. *arcticum*（Ach.）Wahlbg. Flora Lappon, 483（1812）.

Тип: Ланпония, собран Вахленбергом（Wahlenberg）из Лапландии（H-Ach. 562, лектотип!）.

На скалах.

Материалы исследованные:

Монголия: Ара-Хангайский аймак, 2507m, 15/Ⅵ/1971, Бязров, No. 5739（Moscow!）.

Россия: Дальний Восток, Западная Чукотка, Анюйск, 16/Ⅶ/1977, М. Андреев, No. 27（LE!）. Центральный Урал, Ⅷ/1929, К. Игошена（LE!）.

16. **Umbilicaria badia** Frey

Ber. Schweiz. Bot. Ges. 59:453（1949）.cum icon.

Тип: India Orientalis, 9012a（BM! Holotype）.

= *Gyrophoropsis dwaliensis* Räs. Arch. Soc. Zool.- Bot. Fenn. Vanamo, 6（2）:80（1952）.

Тип: N. W. Himalaya, near Dwali（on route to Pindari Glacier）, Almora distr., on stone, alt. 8500 ft. May 25, 1950, coll. D. D. & A. M. Awasthi No. 717（H! Holotype and isotype, CANL! Isotype）

На скалах.

Материалы исследованные:

Китай, Юньнань, Лицзян, гора Юйлун шань, Ганхайцзи, 20/III/1943, Wang, No.3506（HMAS-L!）. Тибет, Чжам, 2400-2700м, 9-11/Ⅴ/1966, Вэй и Чен, Nos. 413, 414, 418, 534, 567; Кюсям, 3200-3550м, 18-19/Ⅴ/1966, Вэй и Чен, Nos. 837, 932（HMAS-L!）.

Индия: Восточная Индия, коллектор их неизвестно, 9012a（BM!）.

С. З. Гималаи: около Двали, Альмора район, 2800м, 25/Ⅴ/1950, D. D. & A. M. Awasthi（ CANL, H!）.

17. **Umbilicaria cinerascens**（Arnold）Frey

Hedwigia 69:250（1929）.

≡ *Gyrophora cinerascens* Arnold, Verhandl. Zool. Bot. Ges. Wien 25:438（1875）.

Тип: An Glimmerwanden: a）unwei der Jamthaler Hutte bei Galtur im Paznaunthale, Aug. 10, 1893; [Exs. Arnold 1579a]（UPS! Lectotype, and S! Isolectotype

На скалах.

Материалы исследованные:

Китай: Шэньси, гора Тайбай шань, институт Китайской медицины, No. 2319（HMAS-L!）; гора Тайбай шань, Басиантай, Паомалиан, 28/VI/1938, Liou TN & Tsoong PC, No. 4258, 3600m, 30/Ⅶ/1963, Ма и Цзун, No. 303, Басиантай, 3500м, 4/VI/1963, Вэй и др., No. 2718-3（HMAS-L!）.

Монголия: Ара-Хагайский аймак, 1850м, 13/Ⅷ/1970, Бязров, No. 5749（Moscow!）.

18. **Umbilicaria cylindrica** auct.

Название ошибочное употребления:

Umbilicaria cylindrica sensu: Delise in Duby, Bot. Gall. 2:595（1830）. Freyin Rabh., Kryppt. Flora, 9（4/1）:320（1932）. Llano, A monogr. Of the Umbilicariaceae in the western hemisphere 119（1950）. Thomson, American Arctic Lichens 1. The Macrolichens 447（1984）. J. C. Wei & Y. M. Jiang, Mycosystema, 1:80（1988）.

Umbilicaria neocylindrica J. C. Wei & Y. M. Jiang,

Supplement to Mycosystema 5:4-8（1993）.

= *Lichen proboscideus* L., Sp. Pl. 1150（1753）.

Тип: Linn 1273-204: 53（LINN!）, Fig. 61-A.

На скалах.

Материалы исследованные:

Китай: Цзилин, гора Чанбай шань, северный склон, около Тяньчи, 2100м, 13/Ⅷ/1977, Вэй, No. 2973-2, и 2350м, 14/Ⅷ/1977, Вэй, Nos. 2983, 2984, 2992（HMAS-L!）; на дороге от Тяньчи до водопада, 2100м, 1977, Чен, No. 4781（HMAS-L!）; недалеко от Тяньвенфун, 18-19/Ⅷ/1987, Вэй, Nos. 9010, 9113（HMAS-L!）; ю.з. Скло, фусун, 2500м, 8/Ⅷ/1983, Вэй и Чен Северный склон, 2400м, 14/Ⅷ/1977, Вэй, No. 2984, как разновидность tornata（HMAS-L!）. Шэньси, гора Тайбай шань, Басиантай 4/VI/1963, Вэй, образец без номера（HMAS-L!）. Синцзян, гора Алтай, около оз. Халас, 2500м, 5/Ⅷ/1986, Гао, Nos. 2050, 2054, 2067, 2081, 2084, 2184（HMAS-L!）.

Монголия: Хубсугульский аймак, около оз. Хубсугул, 23/Ⅶ/1902, А. А. Еленкин（LE!）.

Япония: Хоккайдо, гора Томурауси, 1/Ⅷ/1935, Sato, No. 18（MIN!）, Хоккайдо, 14/Ⅶ/1936, Sato（MIN!）, гора Онтак, VII/1889, M. Miyoshi（MIN!）.

Россия: Дальний Восток, Камчатка, Ⅷ/q908, Савич, Nos, 2232, 5639, 5839（LE!）, и 6334（LE, UPS!）. О. Врангел, 2/Ⅸ/1938, Б. Н. Городков（LE!）. Западная Чукотка, 27/VI/1977, и 1/Ⅸ/1978, М. Андреев（LE!）; и 3/Ⅸ/1978, М. Андреев（HMAS-L-3923!）; Магаранский район, 21-22/Ⅶ/1971, Макарова, Nos. 216, 308, и 8/Ⅷ/1979, Макарова, No. 151（LE!）. Полярный Урал, 7/Ⅶ/1909, В. Сукачев, и 9/Ⅷ/192, Н. Городков, No. 63（LE!）. Сибирь, Томск, 20/Ⅶ/1913, И. Шулбга（LE!）. Ю.з. Сибирь, гора Алтай, Горно-Алтайская Авто.

Область, 1750-1950 м, 19/Ⅷ/1978 T. S. Elias, W. Weber, C. S. Tomb, и И. М. Красноборов

Казахская республика: Восточная часть страны, 14/Ⅶ/1982, Домбровская（HMAS-L!）.

Армянская республика: около оз. Севан, 10/Ⅵ/1980 А. А. Абрамиан（LE!）; С тепанаван, VII/1964, В. Никогосиан（LE!）; степановански, 1965, В. Никогосиан [Savicz, Lichenotheca Ross. 161]（H!）.

19. **Umbilicaria esculenta**（Miyoshi）Minks

Mem. Herb. Boiss. 21:46（1900）.

≡ *Gyrophora esculenta* Miyoshi, preserved in the Bot. Inst. Of Tokyo Univ.（not seen）.

Ошибочное определение:

Umbilicaria spodochroa sensu Müll. Arg.in Bull. Herb. Boissier, 1:235（1893）, non Ehrh.ex Hoffm.

Gyrophora cirrhosa sensu A. Zahlbr.in Handel Mazzetti, Symbolae Sinicae 3:138（1930）, non（Hoffm.）Vain.

На скалах.

Материалы исследованные:

Китай: Лиаонин, Куандиан, Байшилацзи, 1230м, 26/Ⅸ/1963, Нан, No. 6314; 1220m, 1/Ⅸ/1964, Че, No. 2766, и 1200 м, Ⅷ/1964, Чен, No. 2766（IFP, HMAS-L!）. Цзилин, гора Чанбай шань, южный склон, 1450м, 17/Ⅷ/1982, Ху, Nos. 3514, 3528, и Хуйчунь, гора Дама-ан шань, Ванцин, гора Тулаоподин, 350-1000м, Ху, Nos.6069, 6124, 6171, 6188, и Цзиаохе, Байшилацзи шань, 780м, 12/Ⅷ/1984, Ху, Nos. 6264（HMAS-L!）. Юньнань, Delavay in 1885, H-Nyl. 31523（H! Pr.p.）. Цзянси, гора Лу шань, 1960, Чжао и др., Nos. 455, 457, 458, 459, 460, 467, 468, 471（ HMAS-L!）; гора Лу шань, 1150-1170m, 1965, Вэй, Nos. 3116, 3118, 3150, 3245（ HMAS-L!）; гора Лу шань, 1990, Ма Цзин-Я（ HMAS-L!）. Аньхой, гора Хуан шань, около Баншаньси, 800м, Чжао и Ху, Nos. 5913, 5918（ HMAS-L!）; между Байбньтъин и юйпинфен, 1740-1680м. Nov. 2, 1988, coll. O. Eriksson（UM!）. Хунань, Чжанцзияцзие, 1984, Чжоу（ HMAS-L!）, Пинцзян, 1200м, коллектор не известный, No. 8502191（HMAS-L!）. Хубэй, 1885-1888, Aug. Henry No.6184（BM, FH!）. Чжэцзян, гора Тяньму шань, окол Лаодъиан, 1000-1500м, Чжао и Ху, Nos. 6189, 6300（HMAS-L!）. Гуанси, Цзиньск коллетор неизвестный（HMAS-L!）.

Корея: Гора Сок-ли, Чун-букдо, 1959: Сенихай Ким（H!）. Канвон, 1700м, 1985, Yun Sil Park Nos.701, 765, 1453, 1533, 1555, 1616, 1673, 1681, 1690, 1748（DUKE!）. Канвон, гора Охдае национальный парк, 37°42' с.ш. и 128°32' в.д., 1986, Yun Sil Park Nos.989（DUKE!）. Киунсан, гора Кая национальный парк, 35°20' с.ш. и 128°10' в.д. 1200м, 1986, Yun Sil Park No. 1151（DUKE!）; гора Чуисе, 35°26' с.ш. и 129°6' в.д., 900м, 1986, Yun Sil Park Nos. 1101, 1104（DUKE!）. Киунсан и Чунлия, гора Чири национальный парк, 35°20' с.ш. и 127°40' в.д., 1600м, 1986, Yun Sil Park Nos. 2117, 2154, 2402, 2499, 2600, 2633, 2754, 2762（DUKE!）. Киунсан и Чунчон, гора Собак, 36°57' с.ш. и 128°30' в.д., 1200м, 1986, Yun Sil Park Nos. 2922, 2959, 2961, 3109, 3231, 3183(DUKE!); гора Чири, 1976, coll. Park, E. T.в гербарий "Jeohbug" национального университета, Nos. L-684, L-685, L-686, L-687, L-688, L-689, L-690, L-258(HJNU!); Деогиу, 1982, D. Y. Lee Nos. L-1, L-1486, L-693, L-2246, L-2247, L-2248, L-2249, L-2274, L-2275, L-2361, L-2362, L-2374, L-2417, L-2250, L-2469, L-2493, L-2520, L-2560, L-2728（HJNU!）.

Япония: Сикоку, Тоса, 1941, T. Yoshinaga No.1（US!）. Тоса, гора Куйси около Кахи, 1960, .I.Y. 3733（US!）. Kuroutsu-Kyo, 550m, 1960, Masayoshi Oshio, No. 2389（CANL!）. Гора Роиогами, 1960, A. Shiraishi [S. Kurok.: Lich. Rar.et Crit. Exs. 16 as *Gyrophora esculenta*]（FH, LE, S, TUR, US!）. Нонсю, 1959, . Shiraishi [S. Kurok. & H. Kashiw. .: Lich. Rar.et Crit. Exs. 422,(FH, LE, S, TUR, US!). Нонсю, Синано, 1200m, 1959, M. Nuno[Kurok. & Kashiw: Lich. Rar.et Crit. Exs. 315（FH, LE, S, TUR, US!）. Ацусаяма, Синано, 1958, Kurok. No. 58530（US!）. Хонсю, 300m, 1972, M. Togashi [Kurok.: Lich. Rar.et Crit. Exs. No.（US!）. Флора Японии, Эчиго Echigo, Sept. 26, 1907(MIN662624!). Япония, Miyoshi 1894, [Arnold. Lich. Exs. 1613](BM, FH, MIN, S, UPS!）. Синано, Y, Asahina（FH!）. Хиросима, 500m, 1964, M. Nakanishi & M. Oshio（US!）; Япония,

Miyoshi H-Nyl. 31525（pr.p.maj.）（H!）. Аки, гора Мисен, О. Мииацзима, 1964, M. Hale, No. 29552（US!）.

Примечание:

Лишь два экземпояра слоецишa U.esculeta, растущие на коре дерева Fagus crenata, были найдены Накаимой（Hakajima）в горе Цукуба（Tsukuba）в Японии（Sato, 1961:46）.

20. **Umbilicaria exasperata** Hoffmann

Descr.et Adumbr. Plant. Lich 1:7-8（1790）.

Lectotype（11lustration）: Tab. 2, Fig. 1（1790）.

Authentic specimen or neotype: H-Hoffmann, 8596- Umbilicaria exasperata（MW!）.

= *Gyrophora proboscidea*（L.）Ach. [var.] в G.exasperata（Hoffm.）

Ach., Method. Lich. 105（1803）.

= *Lichen hyperboreus* Ach., Kgl. Vetensk. Akad. Nya Handl. 15: 89（1794）.

Type: H-Ach.（H!）.

≡ *Gyrophora hyperboreus*（Ach.）Ach. Method. Lich. 104（1083）.

As a synonym of *U. exasperata* treated by Hoffmann in Herbarium Vivum, sive collectio plantarum siccarum, Caesareae Universitatis Mosquensis, Mosquae, 1825:454, No. 8596（MW!）. - *Umbilicaria hyperborea*（Ach..）Hoffm., Descr.et Adumbr. Plant. Lich. 3:9（1801）.

На скалах.

Материалы исследованные:

Китай: Внутреняя Монголия, Хинган Большой, Геньхе, гора Оклидуэ, 5130м, 26/Ⅶ/1983, Чжао, No. 2964（HMAS-L!）.

Монголия: Ара-Хангайский аймак, 2300м, 22/Ⅶ/1970, Бязров, No. 4229, и 1/Ⅷ/1973, Бязров, No. 5773; Байан Хонгор, 3030м, 18/Ⅷ/1972, Бязров, No. 5774（Moscow!）.

Россия: Урал, Туменская обл., Куваев（LE!）. Восточная Сибирь, гора Путоран, 30/Ⅶ/1984, М. П. Исурбенко（LE!）. Якутия, Шулга（LE!）. Якутия, Томпр, 3/Ⅷ/1956, В. И. Иванова, No. 116/16（LE!）. Центральный Урал, 3/Ⅶ/1962, М. Сторотева（TUR!）

Камчатка, экспед, No. 6335, Опред No.3 под названием Grophora hyperborea（Ach.）Ach. F. Cerebelloides Sabicz, 1909, В. П. Савич, и Экспед. Nos. 6276, 6444, Опред No. 15, f. *sublaevigata*, Савич（LE!）.

Япония: Хонсю, 2400м, 1976, Kashiw（no. 13434）K. Yoshida [Kurok. Kashiw. Lich. Rar.et Cr. Exs. 599]（FH, LE!）, и Хоккайдо, около теплого минерального источника, 1450м, 1980, N. Shibuichi（No.6210）и K. Yoshida [Kurok. Kashiw. Lich. Rar.et Cr. Exs. 499]（COLO, FH, LE, TUR!）.

21. **Umbilicaria flocculosa**（Wulf.）Hoffm.

Duetschl. Flora 110（1796）.

≡ *Lichen flocculosus* Wulf.in Jachuin, Collect. Bot. 3:99, Tab. 1.fig. 2（1789）. - *Gyrophora flocculosa*（Wulf.）Koerb. Syst. Lich. Germ. 95（1855）

Iconotype: Jacq. Collect. Bot. 3:99, Tab.1.fig. 2（1789）.

Название ошибочного употребления:

Gyrophora deusta auct.

Umbilicaria deusta auct.: Frey in Rbh. Krypt.-Flora, 9:361（1933）. Llano, A monograph of the Umbilicariaceae in the western hemisphere 124（1950）. Thomson, American Arctic Lichens 1. The macrolichens 447（1984）. J. C. Wei & Y. M. Jiang, Mycosystema, 1:81（1988）.

На скалах и валунах:

Материалы исследованные:

Китай: Цзилин, гора Чанбай шань, ю.з. Склон, фусун, Вэйдун станция, по дороге к Тяньчи,

2050-2250м, 8/Ⅷ/1983, Вэй и Чен, Nos. 6755, 6758, 6760, 6762, 6766, 6767, 6772（HMAS-L!）; С. Склон, около Луминфен, 2400м, 25/Ⅷ/1979, Ху, No. 1339, Хэйфункоу, 2100-2400м, Ху, Nos. 1498, 1940; Хоуцегоу недалеко от гуанция, 2200м, 3/Ⅸ/1979, Ху, Nos. 2115; около Теплого минерального источника（Венчуань）, 1860м, Ⅸ/1981, Ху, No. 2737; на Западном берегу реки около теплого минерального источника, 1850м, 12/Ⅷ/1983, Вэй и Чен, No. 6845, и 18880м, Ху, No. 4667（HMAS-L!）; Синьцзян, гора Алтай, около оз. Халас, 1400-2500м, 4/Ⅷ/1986, Гао, Nos. 1988, 2000-1, 2007, 2066, 2080, 2091, 2147（HMAS-L!）.

Казахская республика: восточная часть страны, в окрестности города Лениногорск, 1000-1250 м, 9-10/Ⅶ/1982, Домбровская, Алтай

Монголия: Ара-Хайгайский аймак, 2300м, 22/Ⅶ/1980, Бязров, No. 2341（Moscow!）.

Япония: Хоккайдо, Теплый Минеральный источник

Камикава районн, ю.в.склон, 1950м, 5/Ⅶ/1970, Тимо Копонен, No. 14984（HMAS-L!）.

Россия: Чукотка, 14/Ⅶ/1971, Макарова（HMAS-L!）, и Западная часть, 16 /Ⅶ/1977, М. Андреев, No. 2041; 1909, В. Савич, Nos. 6234, 6786, и 700м, 15 /Ⅶ/1981 А. Минеция, No. 1-186-d（LE!）.

22. **Umbilicaria grisea** Hoffm.

Deutschl. Flora 111（1976）.

Lllustration: In Kgl. Vetensk. Arkad. Nya Handl. 15: Tab. II, fig. 3a-c（1794）.

Неотип: H-Ach. No.543 "*Gyrophora murina*, Suecia".（H!）.

= *Lichen murinus* Ach., Lichenogr. Succ. Prodrom 143（1798）. - *Gyrophora murina*（Ach.）Ach. Method. Lich. 110（1803）. - *Umbilicaria murina*（Ach.）DC.in Lam & DC., Flore Franc., ed. 3, 2:412（1805）.

На скалах.

Материалы исследованные:

Монголия: Ара-Хангайский аймак, 15/Ⅷ/9172, Бязров, No. 5931（Moscow!）.

23. **Umbilicaria havaasii** Llano

A Monography of the Umbilicariaceae 1950, p. 136.

≡ *Gyrophora fuliginosa* Havaas in Lynge, Stud. Lich. Fl. Norway, p. 96. - *Umbilicaria fuliginosa*（Havaas）Eahlbr. Cat. Lich. Univ. 8:491（1932）, non Pers. 1810.

Тип: Norway, coll. Havaas, preserved in the Bot. Mus., Univ. Of Olso（not seen）.

=*Gyrophora wenckii* Muell. Arg. Flora 50:433（1867）, pr.p. min.

На скалах.

Материалы исследованные:

Монголия: Гоби-алтайский аймак, 9/Ⅸ/1970, Бязров, No. 5928（Moscow!）.

24. **Umbilicaria herrei** Frey

Ber. Schweiz. Bot. Ges. 45:219（1936）.

Тип: U.S.A. Washington, Mt. Baker, alt. 2000m, coll. Herre（not seen）.

= *Umbilicaria hyperborea*（Ach.）Hoffm.var. *radicicula*（Zetterst.in Lynge）Hasselr. Acta Phytogeography suecica 15:43（1941）.

На скалах.

Материалы исследованные:

Китай, Цзилин, гора Чанбай шань, с. Склон, 2100-2200м, 10/Ⅷ/1977, Вэй, Nos. 2886, 2942 и около Бинчан（Каток）, 2200м, 10/Ⅷ/1977 Чен, No. 4678, и около Тяньвенфен, 2200м, 1977, Hu YC, No.121; недалеко от водопада, 2100м, 1977, Hu YC, Nos. 21-B, 221; по дороге от Тяньчи к водопаду, 2100м, 1977, Чен, No. 4780（Moscow!）; на восточной побережье реки около Венчуань, 1850м, 1983, Вэй и Чен, No.6806; на западной побережье реки около Венчуань, 1850м, 1983, Вэй и Чен, No.6815, 6846, по дороге к Лунменфен около маленького водопада, 1983, Вэй , и около Вечуань 1800м, 1983, Ху, Nos. 4773, 4775, ю.

Склон по дороге от Хеншань станции к Тяньчи 2050м, 1983, Вэй и Чен, No.6378（HMAS-L!）; ю.з. Склон, фусон, Вэйдун станция, по дороге к Тяньчи, 2060м, 1983, Вэй и Чен, No.6763, Тибет, Нианам, 3750-3870м, 1966, Вэй и Чен, Nos. 1440, 1535, 1816（HMAS-L!）.

25. **Umbilicaria hirsuta**（Sw.）Ach.

Svensk. Vetensk. Akad. Nya Handl. 15:97（1794）.

≡ *Lichen hirsutus* Sw.in Westr. Vetensk. Akad. Nya Handl. 1793:47.

Неотип: Sweden: H-Ach. No. 582, *Gyrophora hirsuta*, Suecia.（H!）.

= *Umbilicaria hirsuta* var. *vuoxaensis*（Wei）Wei in J. C. Wei & Y. M. Jiang, Acta phytotax. Sin. 20（4）:499（1982）. - *Gyrophora hirsuta* var.*vuoxaensis* Wei, Notul. Syst. E Sect. Crypt. Inst. Bot. Nom. Komarovii Acad. Sci. URSS 15:8（1962）.

Тип: россия, Ленинградская обл. Вэй（Голотип: LE, изотип,（HMAS-L!）.

На скалах.

Материалы исследованные:

Китай: Тибет, Ниалам, 3720м, 1966, Вэй и Чен, No.1648-（1）, Синцзян, гора Алтай, около оз. Халас, 2000-2500м, 1986, Гао, Nos.1938, 1956, 1958, 2079, 2112（HMAS-L!）.

Алтай: 1909, V. Vereshagin（LE!）.

Казахская республика: Восточная часть страны, недалеко от города Лениногорск, гора Белькин, 880м, 8.VII/1982, Домбровская, No.5（HMAS-L!）.

Монголия: 47°с.ш., 104-5°в.д., 1925, и. Красильнико, Б. Цаматкинно и В. Н. Красильников（LE!）. Ара-Хагайский аймак, 10/Ⅸ/1975, Бязров, No. 1518, и 1/Ⅷ/1979 Бязров, No. 6592（HMAS-L!）.

Афганистан: Кабул, 34°37', с.ш. и 68°56' в.д., 2550м, 21/Ⅵ/1970,

Россия: Южный Урал, Башкирия, 15/Ⅵ/1955 К. Н. Игошина（LE!）.

Тип: Китай, Шэнси, Гуантоу шань, 1894. G. Giraldi Specim.orig.ex herb. Sbarbars 195. IX.)（Isotype or Isosyntype in S!）. Herb. C. Sbarbaro 9042 9042 C. L. U G.hypococcinea Jatta, In Shen2si（Shaanxi）, coll. Giraldi（FH!）.

26. **Umbilicaria hypococcinea**（Jatta）Llano.

Monography of the Umbilicariaceae, 191（1950）（lapsu "hypococcina"）.

≡ *Gyrophora hypococcinea* Jatta, Nouv. Giorn. Bot. Ital. N. S. 9:473（1902）.

= *Gyrophora versicolor* Räs. Arch. Soc. Zool. Bot. Fenn. ' Vanamo', 3:78（1949）.

Тип: Китай, шаньси, гора Вутай шань, M. S. Clements, No. 4021（=TUR-36392! Holotype）and isotype（H, FH, MIN!）.

На скалах.

Материалы исследованные:

Китай: Шаньси, гора Вутай шань, 1000м, 17-14/Ⅸ/1913, G. K. Merril（FH!）. Шаньси, гора Вутай шань, институт Китайской медицины, коллектор неизвестный, No.2381（HMAS-L!）:; около минсинси, 2640м, 2/Ⅵ/1963, Вэй и др., No.2518（HMAS-L!）; Басиантай, 1963, Вэй и др., No.2718（HMAS-L!）; Сицушитай, 3500м, /Ⅵ/1956, Yang, No. 5（HMAS-L!）; Домугун, 2840м, 22/Ⅶ/1963, Ма и Цзун, No. 140（HMAS-L!）; Пин-Ан-Си, 2700м, 23/Ⅶ/1963, Ма и Цзун, No. 166; Фан-янси, 2950-3000м, 26-28/Ⅶ/1963, Ма и Цзун, Nos. 226, 281（HMAS-L!）; Юйеванчи, 3080м, 6/Ⅶ/1963, Ма и Цзун, No. 380（HMAS-L!）; Хусиан, гора Гуантоу шань, Сихекон, 2000м, 2/Ⅶ/1964, Вэй, No.75[Wei i lich. Sin. EXS.40]（HMAS-L!）; Шэньси, Mary Strong Clemos No. 4021（TUR-36392, голотип; H, изотип; FH!）. Тибет, гора Джомолунгма, 5650м, 28/Ⅴ/1966, Вэй и Чен, No.1165（HMAS-L!）; 5000м, VII/1966, Вэй и Чен（HMAS-L!）; 5000м, 2/Ⅵ/1966, Вэй и Чен No. 1361-1（HMAS-L!）; Нианам, 3900м, 5/Ⅷ/1966, Вэй и Чен No. 1617（HMAS-L!）.

27. **Umbilicaria indica** Frey

Ber. Schweiz. Bot. Gez. 59:456（1949）（ut nom.nov）.

= *Umbilicaria papillosa* Nyl. Synops. Lich. 2:11（1869）, non DC. 1805.

Изотипы:

Гималаи: Hoo. No. 2043（H-Nyl. 31529, 31530, in H!）.

Сикким: Herb. Ind. Hook. & Thomson 2089（BM!）.

Сикким: Herb. Ind. Hook. & Thomson 2092（BM!）.

= *Gyrophora himalayensis* Räsanen, Arch. Soc. Vanamo, 5（1）:25（1950）.

Тип: India orientalis: Himalaya orient., Sikkim, 2500m s.m. Ad rupem humidam, leg. June 1947, D. D. Awasthi.ex herb. V. Rasanen（H!）.

На скалах и валунах.

Материалы исследованные:

Китай: Юньнань, Вэйси, гора Билуо Сиуе шань, Восточный склон, 3450м, 13/Ⅶ/1981, Ван, Сиао, Су, No.459（HMAS-L!）.

Индия: Дажилин район, Восточные гималаи, 3600м, VI/1948, D. D. Awasthi. No.4（H!）. India, coll. Hooker Thomson no. 2155（NY!）.

Сикким: Восточные гималаи, 2400м, 1947, D. D. Awasthi. No.2（H!）; Herb. Ind.or Hook. Fil & Thomson No.2094（BM!）, and Herb. Ind.or Hook. Fil & Thomson 2093（FH, LE!）. Herb. Ind.or Hook. F. Thomson 2089, изолектотип（FH, NY!）. С. З. Гималаи: Дхакури ex Herb. D. D. Awasthi. No.631（US!）. Непал: Восточный непал, 1953, D. D. Awasthi（No. 2312）, L-36400, Crypt. H. COLO-175769（COLO!）.

28. **Umbilicaria kisovana**（A. Zahlbr.ex Asahina）A. Zahlbr.

Cat. Lich. Univ. 10:405（1940）.

≡ *Gyrophora kisovana* A. Zahlbr.ex Asahina, Journ. Jap. Bot. 7:326（1931）.

Iconotype: Figs.: A photograph of the species habit, a drawing picture of the species habit, and a drawing picture of ascus and species in Journ. Jap. Bot. 7:326（1931）.

На скалах.

Материалы исследованные:

Китай: Цзянси, гора Лу шань, Сианженьдун. 850м, 11-12/Ⅱ/1965, Вэй, Nos. 3141, 3179（HMAS-L!）. Внутренняя Монголия, Геньхе, 1500м, 16/Ⅷ/1985, Гао, No. 1626（HMAS-L!）.

Корея: Кионги, гора Мунги, 37 24' с.ш., 127 25' в.д., в смешанных лесах, 1200м, Yun Sil Park, No. 843（DUKE!）; гора Ионгмум, 37 30' с.ш., 127 30' в.д., 1000м, 30/V-1985, Yun Sil Park, Nos. 148, 4235（DUKE!）. Киунгсан и Чунлла, гора Чири Национальный парк, 35 20' с.ш., 127 40' E: 1700м, 10/Ⅷ/1986, Yun Sil Park, Nos. 2390, 2752（DUKE!）; Гора Каиа Национальный парк, 35 20' с.ш., 128 10' в.д., 1300м, 28/Ⅵ/1986, Yun Sil Park, No. 1194（DUKE!）. Канвон, гора Сорак национальный парк, 38 10' с.ш., 128 30' в.д., 1000м, 10/Ⅶ/1986, Yun Sil Park, No. 1707（DUKE!）. Гора Чири, 10/Ⅷ/1986, Deuk-Young Lee, Nos. L-2273 2360, 2373（DUKE!）.

Япония: Оосуги-дани（Таке-гун）, 1955, M. Togasi [Lich. Jap. Exs. 124]（UPS!）. Кисо, Синано, Asahina（US!）. Титибу, Мусаси, 1300м 1956, Kurok. No.56129（US!）. Хонсю, гора Танигава, 1934, Asahina No.4（US!）. Аки, гора Мисен, о. Мийажима, 1964, M. Hale No.29522（US!）. Гора Маруцусаса, Токусима, 1961, M. Sato Ex Herb, L-37（H!）; Оосуги-дани（Таке-гун）, 1955, M. Togasi [Lich. Jap. Exs. 124]（H!）.

29. **Umbilicaria loboperipherica** J. C. Wei , Y. M. Jiang & S. Y. Guo

Mycosystema 8-9:66（1995-1996）.

Species nova *U. hirsutae* affinis, a qua imprimis differt lobulis squamosis oriundis ex corticibus ascendentibus in peripheria thallorum et sine parasorediis; ab U. Freyi distincta subtus atra et rhizinomorphis ramosissimis numerousis.

Typus: Provincia Jilinensis, Comitatus Wangqingensis, Mt. Tulaopodingensis. Ad rupes, Aug. 9, 1994, leg. Jiang, Y. M. Et Guo, S. Y. , No. 94572（HMAS-L!）.

Тип: Китай, Цзилинь, Ванцин, гора Тулаоподинцзи, на скалах, 9/Ⅷ/1994, Цзян и Гуо, No. 94572（HMAS-L!）.

На скалах.

Материалы исследованные:

Китай: Хэбэй, гора Сиаовутай шань, 2800м, Вэй, No. 2096-1（HMAS-L!）. Цзилинь, Анту, гора Чанбай шань, северный склон, по дороге к Лунменфен недалеко от маленького водонада, 13/Ⅷ/1983, Вэй и Чен（HMAS-L!）; Ванцин, Тяньчаолин, гора Тулаоподинцзи, Цзян и Гуо, Nos. 94545, 94569, 94570, 94571, 94572（тип）, 94573（HMAS-L!）.

30. **Umbilicaria minuta** J. C. Wei & Y. M. Jiang

In Wei Jiang-Chun & Jiang Yu-Mei, The Asian Umbilicariaceae（Ascomycota）, Inernational Academic Publishers Beijing, 1993:142.

Тип: Китай, Тибет, гора Джомолунгма, Ронбуси（28 1'с. ш. 86 8' в.д.）, 4850m, 1966, Вэй и Чен, No. 1370（HMAS-L:6342, голотип, 6343, изотип!）.

31. **Umbilicaria nanella** Frey & Poelt

In Poelt, Khumbu Himal 415（1977）.

Type: Khumbu Himal. Hohe westlich uber Gorak Shep, 5540m, L264（M, not seen）.

Название ошибочного употребления:

Gyrophora cylindrica var. *tornata* sensu Paulson, Journ. Bot. 63:193（1925）& 66:317（1928）, non（Ach.）Arnold（1874）.

Gyrophora tornata sensu Zahlbr.in Handel-Mazzetti, Symb. Sin. 3:318（1930）, non Ach.（1808）.

Gyrophora cylindrica var. *fimbriata* Sensu Zahlbr.in Handel-Mazzetti, Symb. Sin. 3:318（1930）, non Ach.（1810）.

На скалах.

Материалы исследованные:

Китай Шэньси, гора Тайбай шань, Басиантай, 3620м, 3/Ⅵ/1963, Вэй и др., Nos. 239, 395（HMAS-L!）. Юньнань, Дали, гора Цан шань, Чжонхефен, 3970м, 3/Ⅵ/1965, Вэй, No. 2824（HMAS-L!）. Сычуань, Мули,

32. **Umbilicaria nepalensis** Poelt

Khumbu Himal 6/3:426（1977）.

Тип: Khumbu Himal: Moranen des Lobuche- Cletschers bei Lobuche, 5000m, Gneisblocke, L262（M, not seen）.

На скалах.

Материалы исследованные:

Китай: Тибет, гора Джомолунгма, северный склон, 5530м, 28/Ⅴ/1966, Вэй и Чен, No. 1224（HMAS-L!）.

33. **Umbilicaria nylanderiana**（A. Zahlbr.）H. Magn.

Lich. Sel. Scand. Exs. 252（1937）.

≡ *Gyrophora nylanderiana* A. Zahlbr., Cat. Lich. Univ., 4:720（1927）.

= *Gyrophora heteroidea* & *cirrugata* Ach., Lich. Univ. 219（1810）.

= Umbilicaria corrugata Nyl., Lich. Scand., 119（1861）, non Hoffm. 1794.

Тип: *Gyrophora heteroidea* Ach. [var.] & corrugata, H-Ach. ①177a-183（BM!, lectotype）.

Ошибочное определение:

Umbilicaria leiocarpa sensu: Biazrov, in（Bul. Mosc. O-Va lspeitatelei prirodei.otd. Biol.）Bull. Moscow Soc. Experim. Nature, ser. Biol. 91（6）:100（1986）, non DC.

На скалах.

Материалы исследованные:

Китай: Синьцзян, гора Алтай, около оз. Халас, 2500м, 5/Ⅷ/1980, Гао, Nos. 2075-1, 2085, 2092; около оз. Халас, 2500м, 5/Ⅷ/1986, Гао, Nos. 2075（HMAS-L!）.

Монголия: Убсунурский аймак, 2680м, 9/Ⅶ/1976, Бязров,

Россия: Сибирь, около реки Табат, н. Мартианов（H-Vain. 182=TUR382!）. Южный Урал, Баширия, 15/Ⅵ/1955, К. Н. Игошина.

Армяния: Около оз. Севан, 2600м, V/1980, коллектор неизвестный（LE!）.

34. **Umbilicaria phaea** Tuck.

Lich. Calif. 15（1866）.

Lectotype: U.S.A.: California, 1864, coll. H. N. Bolander No.11, stuck on the right part of the herbarium paper, Tuck.-Herb（FH!）, and the isolectotype is in [Exs. Reliq. Tuck. No.94]（FH!）.

На скалах.

Материалы исследованные:

Монголия: Ара-Хангайский аймак, 22/Ⅶ/1970, Бязров, No.6590（MOSCOW!）.

35. **Umbilicaria polyphylla**（L.）Baumg.

Fl. Lips. 571（1790）.

≡ *Lichen polyphyllus* L., Sp.pl.1150（1753）.

Iconotype: Dill. Hist. Musc. Tab. 30, Gig. 129

Lectotype: 1273:208（LINN!）.

=*Gyrophora stygia* Hook.fil. Taylor. London Jour. Bot. 3:638*1844）.

Тип: Falkland Islands, Antarctica, 1839-1843, J. Dillwyn Llewelyn. Recd. 1889（BM!）.

На скалах.

Материалы исследованные:

Монголия: Ара-Хангайский аймак, 28/Ⅷ/1971, Бязров, №. 4230, и 13/Ⅶ/1976 Бязров, №. 5777, и 2/Ⅷ/1976, Бязров, №. 5776（Moscow!）.

36. **Umbilicaria proboscidea** auct.

Название ошибочного употребления:

Lichen proboscideus auct.

Umbilicaria proboscidea auct. non（L.）Schrad.:Frey in Rabh., Krypt.-Flora, 9（4/1）: 336（1933）. Llano, A monograph ot the Umbilicariaceae in the western hemisphere 157（1950）. Thomson, American Arctic Lichens1. The Macrolichens 456（1984）. J. C. Wei & Y. M. Jiang, Mycosystema, 1:86（1988）.

Umbilicaria neoproboscidea J. C. Wei & Y. M. Jiang.

Supplement to Mycosystema 5:12（1993）.

=*Lichen deustus* L., Sp. Pl. 1150（1753）.

Лектотип: Linn 1273-206:54, Fl. Suec.970（LINN）.（Fig. 86A-a

На скалах и валунах.

Материалы исследованные:

Монголия: Ара-Хангайский аймак, 2400м, 1971, Бязров, Nos. 4234, 5757, 5779（MOSCOW!）. Хубсугульский аймак, недалеко от оз. Хубсугул, 1902, Еленкин（LE!）.

Япония: Хонсю, Синано, гора Кимпу, Минами-саку-гун, 2550м, 1972, H. Shibuichi（4668）и K. Yoshida [Kurok. Lich. Rat. Et Cr. Exs. 196]（COLO, FH, TENN, TUR, UPS!）. Хонсю, Мусаси, гора Охима, K. Yashida（1619）[Kurok. Lich. R. Et C. Exs. 1542]（COLO, FH!）. Хонсю, гора Ширане, Никко, 2570м, 1981, H. Shibuchi（6513）[Kurok. H. Kashiw. Lich. R. C. Exs. 697]（COLO, FH!）. Хоккайдо, [Kurok. Lich. Rat. Et Cr. Exs. 250]（COLO, FH, H, TUR!）. Хоккайдо, Камикава, гора Дайсе натуральный парк, 1750-1800м, 6/Ⅷ/1970, Копонен, №. 15157（HMAS-L!）; гора Дайсе натуральный парк, 1900м, 6/Ⅷ/1970, Копонен, №. 15041（HMAS-L!）. Хоккайдо, [Lich. Jap. Exs. 50]（COLO, TENN, TUR!）.

Россия: Чукотка, западная часть, 3/Ⅸ/1978, M. Андреев（LE, HMAS-L!）и 28/Ⅵ/1977, M. Андреев как U.arctica（HMAS-L, 3871!）. Челиябински район, гора Урма, 8/Ⅶ/1959, цецелева, No.2（COLO!）. Полярный Урал, 9/Ⅷ/1925, Городков, №. 63（LE!）. Урал, туменск, Куваев（LE!）, и 20/Ⅶ/1915б Городков（LE!）. Восточная Сибирь, гора Путоран, 3Ⅷ/1984, Цурбенко（LE!）. Иркуцк, 9/Ⅶ/1956, Иванова（LE!）. Камчатка, VIII/1908, Савич, №5852（№.78）（LE!）. Северный Краспоярск, Путоран, 815м, 10/Ⅷ/1969, Куваев, №. 167-1（LE!）.

Киргизская республика: Алатау, 1/Ⅸ/1956, Викторов, №. 29（COLO!）.

37. **Umbilicaria pseudocinerascens** J. C. Wei & Y. M. Jiang

Mycosystema, 1:86（1988）& Pl. IV-a, b, c; fig.5.

Тип: Китай, Юньнань, Дали, гора Цан, Чжунхефен, 3/I/1965, 3850m, Вэй, №. 2827（HMAS-L, голотип; H, UPS, изотип）, 2829-3（HMAS-L!）.

38. **Umbilicaria squamosa** J. C. Wei & Y. M. Jiang

Mycosystema, 5:80（1992）

Тип: Китай, Юньнань, Дали, гора Цан, 3300m, 29/Ⅻ/1964, Вэй, №. 2779 9HMAS-L! Голотип）.

Ошибочное определение:

Umbilicaria esculenta sensu Wei & Chen, in Report on the scientific Investigations（1966-68）in Mt. Qomolangma Region, ser. Biology & Physiology, Science Press, Beijing, 1974, p.177, non（Migyoshi）Minks.

Umbilicaria thamnodes sensu J. C. Wei & Y. M. Jiang, in Proceedings of Symposium on Qinghai-Xizang（Tibet）plateau, p.1146, Lichens of Xizang, 1986,p.100 & Mycosystema, 1:86（1988）, non Hue.

На скалах.

Материалы исследованные:

Китай: Юньнань, Дали, гора Цан шань, 3300м, 29/Ⅻ/1964, Вэй, Nos. 2778-1, 2779（тип）. Тибет, Ниалам, Чжам, 3700м, 4,11/Ⅴ/1966, Вэй и Чен, Nos.289 и 451（HMAS-L! ）; Ниалам, Кюсям, 3200-3580м, 18, 19, 21/Ⅵ/1966, Вэй и Чен.

Гималаи: 12000 ft, 2091 Herb. Ind.or Hook. Fil & Thomson, U.calvescens v. Hypomelaena Nyl.might Hook fil 1857[FH! &LE（1868）!].

Сикким: Herb. Ind.or Hook. Fil & Thomson 2092, coll. Hooker & Thomson as U.papillosa Nyl.（NY!）and as *U. calvescens* v. *hypomelaena* Nyl.（FH!）.

Индия: Hooker & Thomson, no.2093, pr.p.（NY!）.

39. **Umbilicaria subumbilicarioids** J. C. Wei & Y. M. Jiang

Mycosystema 5:84（1992）.

Тип: Китай, Тибет, Чжам, 3600м, 11/Ⅴ/1966, Вэй и Чен, №. 470-1（HMAS-L! ）.

На скалах.

Материалы исследованные:

Сикким: Миготан, 3900м, 1960, H. Hara, H. Kanai, G. Murata, M. Togashi & T. Tuyama（TNS!）.

Индия: Hooker & Thomson, no.2093, pr.p.（NY!）.

40. **Umbilicaria taibaiensis** J. C. Wei & Y. M. Jiang

Mycosystema 5:85（1992）.

Тип: Китай, Шэньси, гора Тайбай шань, Басиантай, 3660м, 3/Ⅵ/1963, Вэй（HMAS-L, голотип; FH, H, TUR, UPS, US, изотип!）.

Ошибочное определение:

Umbilicaria cylindrica var. *tornata* sensu J. C. Wei & Y. M. Jiang, Mycosystema, 1:80（1988）, Fig. 2 in p. 81.non（Ach.）Nyl.

На скалах.

Материалы исследованные:

Китай: Шэньси, гора Тайбай шань, 3660м, 31/Ⅶ/1963, Ма и Цзон, №. 391-1（HMAS-L! ）.

41. **Umbilicaria thamnodes** Hue

Nouv. Arch. Mus. Hist. Nat. 4（2）:121（1900）.

Тип: Китай, Юньнань, гора Луопин шань и Яньцзи Хай, 1885, coll. Delavay No.1571（H-Nyl.31456!）. Delavay No.1573（H-Nyl.31451!）. Delavay No.1574（H-Nyl.31460!）.

= *Umbilicaria thamnodes* f. *minor* Hue, Nouv. Arch. Mus. Hist. Nat. 4（2）:122（1900）, tab. V, fig.3. A. Zahlbr.in Handel-Mazzetti, Symb. Sin. 3:130（1930）.

На скалах.

Материалы исследованные:

Китай: Шэньси, гора Тайбай шань, Юйе Ван чи, 6/Ⅷ/1963, Ма и Цзон, №. L-379-1（HMAS-L!）. Юньнань, гора Юйлун шань, Лицзян, 3600м, 20/Ⅵ/1914, Handel-Mazzetti, №.672（US, MIN!）; Тибет, Чжам, 3560м, 1966, Вэй и Чен, №.645（US, MIN!）; Цзаюй, 3400м, 1982, Су, №.4324（HMAS-L!）.

Сикким: Альпиский район, 12,000 ft, G.D. H. 2092 Herb. Ind. Or Hook.fil & Thomson（BM!）. Гамотан, 3900м, 1960, H. Hara, H. Kanai, G. Murata, M. Togas T. Tytana（TNS!）.

Непал: 3400м, 1967, D. H. Nicolson, No.3323（US!）. Около Функи, 3250, 1981, S. Remus/ M. Menzel, No.102（NY!）.

42. **Umbilicaria torrefacta**（Lightf.）Schrad.

Spicil. Fl. German 1:104（1794）.

≡ *Lichen torrefactus* Lightf. Fl. Scotica 2:862（1778）.

На скалах.

Материалы исследованные:

Китай: Цзилинь, гора Чанбай шань, южный склон, Фусон, около Лаохубай, 2200м, 8/Ⅷ/1983, Вэй и Чен, №. 6759（HMAS-L!）.

Монголия: Ара-Хангайский аймак, 22/Ⅶ/1970, Бязров, №.4232, 28/Ⅶ/1970, Бязров, №.5759, 5780,

5781（MOSCOW!）.

Россия: Сибирь, Камчатка, VI/1908, Савич, Nos.110, 2205（LE!）. 5024, 5649（H!）. Сибирь, Almquist, Herb. Vain. 95（TUR!）. Челиябински ряйон, гора Урма, 8/Ⅶ/1959, Цецелева（COLO!）Чукотка, 22/Ⅶ/1967, Цимофия（HMAS-L!）, и 22/Ⅶ/1977, М. Андреев（LE!）, и 19/Ⅷ/1977, М. Андреев, Норкина, и Петровский, №. 18178（HMAS-L!）.

Япония: Хоккайдо, [Kurok. & Kashiw., Lich. Rar.et. Cr. Exs. 447, 448, 544]（COLO, FH, TUR!）.

43. **Umbilicaria tylorhiza** Nyl.

Notiser ur Sallsk. Fauna et Flora Fennica Forhandl. 8:122（1866）.

Тип: Lapponia orientalis, Kantalahti, 1861, coll. Felba（H-Nyl.31524!）.

≡ *Gyrophora tylorhiza*（Nyl.）Nyl in Hue, Revue de Botan, 5:14 et 163（1886087）, & in Bull. Soc. Bot. France 34:23（1887）. A. Zahlbr. In Handel-Mazzetti, Symbolae Sinicae 3:138（1930）. Llano, A monograph of the Umbilicariaceae, 1950,p.264,fig.7.

= *Umbilicaria trabeculata* Frey & Poelt, in Poelt, Khumbu Himal 6（3）:429（1977）. Syn.nov.

Тип: Khumbu Himal: Hohe westlich Lobuche, 5050-5100m, SteilundUberhangflache eines gro Ben, windverfegten Gneisblocks, L248（M, not seen）.

*Ошибочное опреде*ление:

Umbilicaria murina sensu J. C. Wei & J. B. Chen: in Report on the Scientific Investigations（1966-1968）in Mt. Qomolangma district, Science Press, Beijing, 1974, p.177.

На скалах.

Материалы исследованные:

Китай: Шэньси, гора Тайбай шань, 2240м, 5/Ⅵ/1963, Вэй и др., No.2548（HMAS-L!）; Юйеванчи, 3080м, 6/Ⅷ/1963, Ма и Цзон, No.379（HMAS-L!）; Тибет Чжам, Кюсям, 3500м, 19/Ⅴ/1966, Вэй и Чен, No.887（HMAS-L!）; 3800-3910м, 12-12/Ⅵ/1966, Вэй и Чен, Nos. 1084-1487, 1522, Ниалам, Вэй и Чен, и 3700-3950м, 12-15/Ⅵ/1966, эй и Чен, Nos. 90б1485-2, 1522, 1610, 1618, 1648-2（HMAS-L!）; Юньнань, 1885, Delavay, [H-Nyl. 31523（pr.p.min.）!].

Сикким: Миготан-ннаиатан, 3300м, 1/Ⅵ/1960, Hara et al/（TNS!）.

Гамотан, 3900м, 1960, M. Togashi（TNS!）.

Япония: Хонсю, Кай, гора Кита-Дак около Котар-Гоиа, 3000м, 4/Ⅷ/1967, M. Togashi No.22122（TNS!）; Рикучу, гора Иват-сан, 1400м, 1971, H. kashiwadani, NO.8017a（TNS!）; Етчу, гора Етчусава-дак, 2400м, 1976, H. Kashiwadani Nos.13411, 13442（TNS!）; Синано, гора Тенгу-дак, Ятсуг-дак, 1958, S. Kurokawa, No.58233（TNS!）; Синано, гора Кимпу, 2560м, 1965, S. Kurokawa, No.65256（TNS!）; Ямато, гора Санжога-дак, Охмине, 1700м, Y. Tanaka, No.205（TNS!）.

44. **Umbilicaria vellea**（L.）Ach.

Vetensk.- Akad. Nya Hadnl. 15:101（1794）pr.p.

≡ *Lichen velleus* L. Sp.pl.1150（1753）. - *Gyrophora vellea*（L.）Ach. Meth. Lich. 1803.p. 109.

Тип: Linn 1273-199:51, Fl. Suec. 986（LINN! Lectotype）.

Ошибочное определение: *Gyrophora spodochroa* sensu Savicz, 1922, p.109 for Kamchatka.

На скалах.

Материалы исследованные:

Китай: Хэбэй, гора Сиаовутай шань, 2800м, 1964 Вэй, No. 2096（HMAS-L!）. Цзилин, гора Чанбай шань, северный склон, Сиао-тянь-чи, 1700м, 12/Ⅷ/1977 Вэй, No.2908 и Чен, No. 4730-2（HMAS-L!）; Около Катока, 1800м, 26/Ⅷ/1977, Ху, No. 521（HMAS-L!）; Южный склон, гора Хонтоу шань, 1900м, 27/Ⅶ/1983, Вэй и др., No. 6249（HMAS-L!）; Чанбай шань, по дороге от Хен-шань станции к тянь-чи, 2050м, 31/Ⅶ/1983,

Вэй и Чен, Nos. 6370-1, 6371（HMAS-L!）. Синьцзян, гора Алтай, около оз. Халас, 2500м, 5/Ⅷ/1986, Гао, No. 2073（HMAS-L!）; Сычуань, Кандин, на высокогорынх районах между Кандиным и Джедуо, Потанин в 1893. [Elenkin: Lich.fl. Ross. IV, 1904, No. 151]（часть из них: 151-6-1;（HMAS-L!）. Юаньнань, в окрестности Кунмина, 26 с.ш., 2400м, 14/Ⅲ/1914, Handel-Mazzetti,（Diar. Nr. 146）in Sump. Acad. Sci. Vindob. Nr590（W!）. Синцзян; гора Алтай, около оз. Халас, 2000м, 4/Ⅷ/1986, Гао, No. 1957（HMAS-L!）. Тибет, Ниалам, 3920м, 13/Ⅵ/1966, Вэй и Чен, No. 1541（HMAS-L!）.

Монголия: Ара-Хангайский аймак, 25/Ⅷ/1970, Бязров, No.5751; 2500м, 1/Ⅷ/1973, Бязров, No.5784, Баянхонгор, 9/Ⅷ/1973, Бязров, No.5782 и Убсунурский аймак, гора Дулг, 2680м, 99/Ⅶ/1976, Бязров, No. 6572（Moscow!）.

Алтай, 16/Ⅵ/1903, П. Крейшовас, как G. Spodochroa（LE!）.

Россия: Западная Сибирь, Свердновск, 1962, М. Стороева（H!）. Центральная Сибирь, гора Путоран（LE!）. Восточная Сибирь, Минусинк, как G. Spodochroa（LE!）. Камчатка, 1909 как G. Spodochroa Савич, Nos. 199, 5085（LE, US!）, Савич, No. 5137, как G. Spodochroa（MW!）.

Япония, Хонсю, Синано, гора Охияма, 2200м, 1979, H. Shibuichi（No.5854）K. Yoshida [Kurok. Kashiw. Lichi. Rar. Et Cr. Exs. 600]（DUKE, FH, LE!）.

Афганистан: Кабул, 2550м, M. Steiner [Lichenotheca Afghanica Nos. 15,29]（WM!）.

Казахстан: Алтай, Лениногоскист, 900м, 1963, Heinrich Aasamaa, No.7388（Loc.5435）（H!）. Карагандаст, Калкаралински, 899м, Heinrich Aasamaa, No.7418（Loc.5438）（H!）. Алма-Атасу, Ала-Тау, 1963, Heinrich Aasamaa, No.7447（Loc.5441）（H!）.

Армяния: Красносельск, Кавказ, 2500м, 1976, Vezda [Lich. Sel. Exs. 1413]（COLO,H!）.

45. **Umbilicaria yunnana**（Nyl.）Hue

Nouv. Arch. Mus. Hist. Nat. 4（12）:117（1900）.

≡ *Gyrophora yunnana* Nyl. In Hue, Bull. Soc. Bot. France 34:23（1887）, and Flora 70:134-135（1887）. - *Gyrophoropsis yunnana*（Nyl.）Ras. Arch. Soc. Zool. Bot. Fenn'. Vanamo'6（2）:81（1952）.

Тип: Китай, Юньнань, in Nylander's Herb., Univ. Of Helsingfors. Yunnan supe trunsos arborum in silvis Hoang-li-pin supra Tapin-tze 2000 m.s.m.2/5/1885, coll. R. P. Delavay No. 1600（H-Nyl.31740 in H! Holotype!, and H-Vain. Nos. 111, 112 in TUR!）.

Эпифитный вид на дубах, иногда на коре других деревьев, лишь изредка в виде исключения переходит к обитанию на мшистых камнях.

Материалы исследованные:

Китай: Юньнань, 3000м, 1885, J. M. Delavay（H-Nyl. 31739）; Кунмин, Handel-Mazzetti,（Diar. Nr. 146）in Sump. Acad. Sci. Vindob. Nr590（E!）; Сан-янпань, Handel-Mazzetti 2526. A. Zahlbr. Ex BDI（MIN, US!）; Лицзян, гора Юйлон шань, 3000м, Чжао и Чен, Nos. 4009, 4042, 4048, 4449; Лицзян, Юйфенси, го коре сосны Вэй, Nos. 9203, 9236, 9238, и на коре орех, Вэй, Nos. 9210, 9212, 9232（HMAS-L!）. Гора Сиан шань, на коре дуба 2400-2600м, Вэй, Nos. 2717, 2718, 2719, 2720, 2721, 2722, 2723, 2724, 2725, 2726, 2727, 2728, 2729, 2749（HMAS-L!）; гора Гаолигон шань, восточный склон, 1900м, на мшистых Камнях, 17/Ⅶ/1982, Су, No. 2033, и 2100м, на стволе орезе Су, No. 2179（HMAS-L!）; Сычуань, Кандин, 2600м, 1934, Harry Smith No.14002（UPS!）3000m, Harry Smith No.4005（UPS!）. Тибет, Цаюй, 2400м, 1982, Су, No. 4192（HMAS-L!）.

Секция **Muriformes** Llano, A monograph of the lichen family Umbilicariaceae in the Western Hemisphere, 1950:176.

=*Gyrophoropsis*（Elenk. & Savicz）A. Zahlbr.

46. **Umbilicaria angulata** Tuck.

Synops. Lich. New Engl. 74（1848）.

Тип: США, California, Monterey, 1875, coll. Menzies in 1842, H-Tuck. Sheet No. 994（FH! Holotype）.

≡ *Gyrophora angulata*（Tuck.）Herre, Contr. U. S. Nat. Herb. 13:318（1911）.

= *Umbilicaria semitensis* Tuck., Gen. Lich, 31（1872）.

Тип: U. S. A., California, Yosemite Valley, 1876, coll. H. N. Bolander（265）, on granite rocks, alt. 7-8000 ft., H-Tuck.（FH!）.

На скалах.

Материалы исследованные:

Монголия: Ара-Хангайский аймак, 2000м, 17/Ⅶ/1970, Бязров, No. 6591（Moscow!）.

47. **Umbilicaria caroliniana** Tuck.

Pro. Amer. Acad. Arts & Sci. 12:167（1877）.

Тип: 1845, coll. M. A. Curtis No. 88, in H-Tuck（FH!）.

≡ *Gyrophoropsis caroliniana*（Tuck.）Elenkin & Savicz, Trav. Mus. Bot. Acad. Petersbourg 8:38（1910）. - *Gyrophora caroliniana*（Tuck.）Schol. Nyt. Mag. Naturvid. 75:28（1934）.

Umbilicaria mammulata Tuck. Proc. Amer. Acad. Arts. & Sci. 1:261（1848）, non Ach.

Тип: U. S. A., North Carolina, coll. S. B. Buckley（5）, 1858, in H-Tuck.（FH!）.

На скалах.

Материалы исследованные:

Китай: Внутреняя Монголия, Хинган большой, Генхе, гора Оклидуэ, 1530м, 26/Ⅶ/1983, Чжао ц.ф. No. 2963（HMAS-L, IFP!）.

Монголия: Ара-Хангайский аймак, 2, 25/Ⅷ/1970, Бязров, Nos. 4236, 5748（MOSCOW!）. Япония: Хонсю, Синано, гора Кимпу, 1965, S. Kurokawa No.65172 [Kurok. Lich. Rar. Et Cri. Exs. No.99]）BM, COLO, E, FH, LE, S, TUR!）; Хонсю, Ишикари, гора Томуроуси, 1700м, 10/Ⅶ/1969, M. TOGASHI [Kurok. & Kashiw. Lich. Rar. Et Cri. Exs. No. 446]（S, TUR, FH, COLO!）; гора Комагатак Кисо, 1926, Y. Asahina [Lich. Japan Exs. No.49]（BM!）. Жилидесан, 3000м, 1898, Faurie No. 859, pr.p.（LE, MIN!）. Хоккайдо, гора Томурауси, Sato No.19（FH, MIN!）.

Россия: Якутская республика, Якутия, Ⅶ/1903, Шоголев, No. 1（LE!）; 25/Ⅷ/1956, Иванова 153, 3/Ⅷ/1956, Иванова No. 116/9; Якутия, Томпонск район, 1955, Добрекова как G. Pulvinaria; Якутия, Алдан, 63°с.ш. 132°в.д., 18/Ⅸ/1956, Куваев, No. 242-7（LE!）. Якутия, Жуго, 24/Ⅵ/1939, Шелудиакова（TUR!）. Амурская обл. 1500м, 19/Ⅶ/1910, Прохоров и Куценева, Nos. 580, 584, и 23/Ⅶ/1915, Прохоров и Куценева（LE!）, Амурская обл., 1915, Куценева, No. 2229（us!）. Сибирь, 19/Ⅶ/1932, Савич, No. 110（LE!）; южно-восточная Сибирь, Байкальский район, 1915, Александров, на восточном побережье оз. Байкал（LE!）; на западном побережье оз. Байкал, 25/Ⅵ/1928, Рассадина, No. 194（LE!）; 10/Ⅶ/1928, Рассадина（LE!）.

48. **Umbilicaria cinereorufescens**（Schaer.）Frey

Hedwigia 71:109（1931）.

≡ *Umbilicaria vellea* γ. *spadochroa* ε. *cinereorufescens* Schaer. Enum. Crit. Lich. Eur. 1850,p. 25.

Тип: Switzerland, in Schaerer's Herb. Inst. Of syst. Bot., univ. Pof Geneva

На скалах.

Материалы исследованные:

Монголия: Ара-Хангайский аймак, 2400м, 19/Ⅶ/1971, Бязров, No. 5750（Moscow!）.

Япония: Miyoshi [H-Nyl. 31525（pr.p.）in H!].

Россия: Восточная Сибирь, Красноярск, 56°с.ш., 11/Ⅵ/1876, Бреннер（Brenner）, No. 348b（H!）. Сибирь,

Горно-Алтайск（52°с.ш., 86°в.д.）, 360m, 1978, T. S. Elias, W. Weber, S. S. Tomb.

И. М. Красноборов, L-68248, Crypt-H. Colo. 324399（COLO!）.

49. **Umbilicaria koidzumii** Yasuda ex Sato

Jour. Jap. Bot. 11（5）:314（1935）.

Тип: Japan, prov. Kai, Mt. Komagadake, on non calcareous rocks, coll. H. Koidzumi, July 27, 1921 in Herb. Univ. Imp. Tokyo, No. 234（not seen）.

= *Umbilicaria japonica* Räs., Jap. Journ. Bot. 16:82（1940）.

Синтип: Japan, Shinano prov., Mt. Nishi-Komagatake（688）, Kai prov.（394pr.p.）, rupicola.

На скалах.

Материалы исследованные:

Япония: Кай, гора Хиташи-Камагадак, 27/Ⅶ/1921, H. Koidzumi. 395 and 3（ex herb. V. Ras. In H!）. Хонсю, Синано, гора Ятсуга-дак, 1928, M. Ogata, ex Herb. Y. Asahina（TNS!）.

Подрод **Actinogyra**（Schol.）Schade ex Mot.

Flora Polska, Porosty（Warswa）4（2）:77（1964）.

= *Actinogyra* Schol. Nyt Mag. Naturvid. 75:28（1934）.

Тип: *Umbilicaria muehlenbergii*（Ach.）Tuck.

50. **Umbilicaria muehlenbergii**（Ach.）Tuck.

An Enumer. North Americ. Lich. 1845, p.55. A. Zahlbr. Cat. Lich. Univ. 4:719（1927）.

≡ *Gyrophora muehlenbergii* Ach. Lich. Univ. 227(1810). - *Actinogyra muehlenbergii*(Ach.)Schol. Nyl Mag. Naturvid. 75:28（1934）.

Тип: America Septentr. Muehlenberg. H-ACH. 572（H! Holotype）.

На скалах.

Материалы исследованные:

Китай: Внутреняя Монголия, Хинган Большой, Мангуэ, 13/Ⅸ/1977, Вэй, No. 3400m, Хэйлунцзян, Гулиань, около станции железной дороги Силинцзи, 5/Ⅸ/1977, Вэй, Nos. 3278, 3296（HMAS-L!）.

Япония: Хонсю, Мутсу [Kurok. & Kashiw, Lich. Rar. Et Cr. Exs. 251, 252]（COLO, FH, S, TUR!）.

Россия: Южная Сибирь, Алтай, 6/Ⅷ/1931, Штыберг, No. 359（LE!）. Восточная Сибирь около Байкала, Савич, и между 53°и 55°с.ш., 10/Ⅶ/1928, Рассадина, и 13/Ⅶ/1909, No. 1200（LE!）. Читинская область, 8/Ⅶ/1943, Сергиевская, S18325（COLO!）. Красноярский край, 28/Ⅶ/1988, Weber, Vurray, Красноборов,（Lich. Exs. No.698]（COLO!）. Южная Сибирь, гора Алтай, 360м, 17, 19/Ⅷ/1978, t.s. Elias, w. Weber, c.s. Tomb, Красноборов, L-68054, L-68227（COLO!）. Амурская обл., 15/Ⅷ/1911, Соколов, No. 1050（LE!）. Якут, 12/Ⅵ/1951, Лебедева（LE!）, и Томпо, 24/Ⅸ/1954, Киваев, из Гербариав Якутской Акад. Наук, Nos. 69, 70（LE!）.

51. **Umbilicaria polyrhiza**（L.）Ach.

Vetensk.-Akad. Nya. Handl. 15:92（1794）.

≡ *Lichen polyrhizos* L. Sp. Pl. 1753, p.1151.-*Actinogyra polyrhiza*（L.）Schol. Nyt Mag. Naturvid. 75:28（1934）.

Iconotype: Cill. Hist. Musc. Tab. 30, fig. 130.

На скалах.

Материалы исследованные:

Монголия: Дзабханский аймак, сомон Баян-Ула, хребет тарбагатай, перевал Солонгойт, сопка западнее дороги, на скале ниже вершины, 18/Ⅵ/1973, Бязров, No. 5932. Убсунурский аймак, сомон Цаган-Хайрхан, плоская вершина горы Дулга в системе хребта Хан-Хухэй, 2928м, высокогорный кобрезиевый луг с большими площадями крупнообломочных каменных полей, 9/Ⅶ/1976, Бязров, No. 5933（Moscow!）.

Россия: Амурская обл., 1899, Иванов（LE!）.

52. **Umbilicaria pulvinaria**（Savicz）Frey

Hedwigia 71:114（1931）.

≡ *Gyrophora pulvinaria* Savicz, Bull. Jard Imp. Bot. Pierre Grand 15:117（1914）.

Лектотип: Камчатка, In regionibus alpinis ad rupes monits ”Krasnyi Jarczik” et montium "Poperecznyje" prope pagum :Naczika”,（1908）, coll. Savicz No.5579（LE! Lectotupe, selected here; and isolectotupes in LE, MW, Hand UPS!; syntype from the same locality, No.5699 in LE and US!）.

Таксоны не были исследованы для этой работы

1. *Umbilicaria pensylvanica* var. *truncicola* Frey

Bericht. Schweiz. Bot. Ges. 59:445（1949）.

Type: India orientalis, Lachen, 11000 pedes, on trunks（BM）.

Гималаи

Оригинальный образец этой разновидности не был найден мной в BM во время моей остановки там. Поэтому он не мог быть исследован для этой работы.

2. *Umbilicaria jingralensis* M. B. Nagarkar & P. G. Patwardhan

Curr. Sci. 50（24）: 1075（1981）.

Гималаи

Материал этого вида не был исследован для этой работы.

БЛАГОДАРНОСТЬ

Глубоко благодарен хранителям следующих гербариев: BM, E, LINN в Британии; CANL в Канаде; HJNU в Кореи; LE, Moscow, MW в России; COLO, DUKE, FH, MIN, NY, TENN, US в США; H, TUR в Финлянции; S, UME, UPS, в Швеции; и TNS ы Японии за то, что они любезно дали взаймы гербарные материалы для этой работы.

За финансовую поддержку, которая давала мне возможность исследовать образцы лишайников сем. Umbilicariaceae в их гербариях в 1989-1990 г. Благодарю доктора Кальберсон（CulbersonLDUKE）, Голубкову（LE）, Хольмгрен（Holmgren: NY）, Петерсен（Petersen: TENN）, Пфистер（Pfister: FH）, Скоr（Skog: US）, Вебер（Weber: COLO）и Ветмор（Wetmore: MIN）.

Очень признателен хранителям следующих гербариев докторам Ахти и Витикаинен（Ahti, Vitikainen: H）. Коппинс（Coppins: E）, Эрикссон（Eriksson: UME）, Галловай（Galloway: BM）, Ярвис（Jarvis: LINN）, Лундквист（Lungqvist: S）, Макинен（Makinen: TUR）, и Моберг（Moberg: UPS）, где я работал в 1982 г. и 1989 г. Согласно плану академического обмена между Академией Наук Китая и сдедующими Академаиями Наук : Шведской Королавской Академией Наук, Академией Наук финляндии, и Британским Королевским Обществом Каждого в отдельности.

В заключение выражаю благодарность дама Цзян и д. Лю за большую помощь в приготовлении рисунков, фотоснимков и рукописи в компьютере, а также и моей жене Хуан за постояную поддержку и помощь.

ЛИТЕРАТУРА

Андреев М. П. 1978. Лишайники стационара <Абориген>（тенькинский район магданской области）. – Бот. Журн. Т. 63, No. 11.

Баранов В. И., Смирнов М. П. 1931. Пихтовая тайга на предгорьях Алтай – Учен.зап. Перм.гос. Ун-та. Отделение 4. Естествознание, Вып. 1.

Будаева С. Э. 1989. Лишайники лесов Забайкалья. Новосибирск <Наука>. Сибирское отделение.

Бязров Л. Г. 1986. Дополнениея к флоре лишайников Хангая（МРН）. 1. Семейство Umbilicariaceae Fee – Бюл. Моск. Ова Испыт. Прир. Отд. Биол.т. 91. Вып. 6.

Вульф Е. В. 1932. Введение в историческую географию Растений. Ленинград

Голубкова Н. С. 1981. Конспект флоры лишайников Монгольской Народной Республики. Л.,（Сер.<Биологические ресуры и природные условия Монгольской Народной Республики>. Т. XVI）.

Голубкова Н. С. 1983. Анализ флоры лишайников Монголия.

Ленинград <Наука>. Ленинградское отделение.

Голубкова Н. С., Савич В. П.1978. Сем. Umbilicariaceae – в кн.: Определитель лишайников СССР. Л., вып. 5.

Голубкова Н. С., Шапиро И. А. 1979. Виды рода *Umbilicaria* Hoffm.em. Frey в СССР и их хемотаксономическое изучение. – Нов. Сист. Низш. Раст. Т.16.

Гордоков Б. Н. 1935. Геоботанический и почвенный очерк

Пенжинского района Дальневосточно кран. –тр.

Дальневосточ.лиф. АНСССР, сер,бот.

Домбровская А. В. 1978. Lasallia rossica Dombr. – Новый вид лишайнико из Советского Союза и его таксоны. – Нов. Сист. Низш. Раст. Т. 15.

Еленкин А. А. 1901. Список лишайников, собранных Д. В. Ивановым на <<гольцах>> в Восточной Сибири（Амурская и Приморская области）в 1899 г. – Тр. Спб. Бот. Сада, т. 19, вып.г.

Еленкин А. А. 1902. Краткий предварительный отчет о споровых, собранных в Саянских горах летом 1902 г. – изв.имп. Спб. Бот. Сада, т. 2, вып. 6.

Еленкин А. А. 1903. О замещающих видах. Ⅰ, Ⅱ. – изв.имп. Спб. Бот. Сада, т. 3, вып. 1-2.

Еленкин А. А. 1912. Список лишайников, собранных Б. А. Федченко в 1909г. на Дальнем Востоке. – тр. Спб. Бот. Сада, т.31.

Еленкин А. А., Савич В. П. 1911. Список лишайников, собранных Ир. М. Щегловым в Якутской и приморской областях по хребту Джигджуру（Становому）и его отрогам между Нелькном и Аяном в 1903 г. – тр. Бот. Музей АН, вып. 8.

Куваев В. Б. 1956. Растильность Восточного Верхоянья – в кн.: Растительность Крайнего Севера и ее освоение. М., Л., вып. 2.

Локинская м. А. 1970. Наиболее распространенные виды лишайников на Северо-Востоке СССР. – в кн.: Водоросли и грибы Сибири и Дальнего Востока, Новосибирск.

Макапрва И. И. 1973. Лашайники Иультинского района Чукотского национального округа – Нов. Сист.низш. Раст. Т.10.

Макрый Т. В. 1990. Лишайники Байкальского хребта. <<Наука>>, Сибирское отделение.

Микулин А. Г. 1990. Определитель лишайников полуострова Камчатка. Владивосток.

Окснер А. Н. 1926. Лишайники, собранные Ю. Ю. Каневским в 1916 г.в Забайкалье. – Укра. Бот. Журн., т.з.

Окснер А. Н. 1939. Лишайники бассейна рек Индигирки, Яны, Лены и Южного Прибайкалья. – Журн. Ин-та Бот. АНУсср, No23（31）.

Рассалина К. А. 1929. Новый вид *Umbilicaria* из Сибири – *Umbilicaria pertusa* sp. n. – доклады Академии Наук СССР. 14（A）: 348б рис. 1-2.

Рассалина К. А. 1934. Лишайники, собранные В. Б. Сочава в 1929 г. в Анадырском Крае. – Тр. БИН, сер. Ⅱ, вып. 2.

Рассалина К. А. 1936. Лихенологический очерк Байкальских берегов. – Тр. БИН, сер. Ⅱ, вып. 3.

Савич В. П. 1926. Тобольские лишайники, собранные Б. Н. Городковым в 1915 г. – Тр. Бот. Музея АНСССР, вып. 19.

（Савич）Savucz V. P. 1930. Lichenotheca Rossica, Dec. Ⅲ – Bull. Jard. Bot. Princ. URSS, T. 29, Liv. 1-2.

Савич в. П. 1950. Конспект лишайников к флоре Umbilicariaceae в СССР. – Бот. Матер. Отд. Споровых Раст. БИН АН СССР.

Седельникова Н. В. 1978. Географические связи в лихенофлоре Кузнецкого нагорья и Арктики. – В кн.: Систематика и география растений Сибири. Новосибирск.

Седельникова Н. В. 1985. Лихенофлора нагорья Сангилен.

Новосибирск, Издательство <<Наука>>, Сибирское отделение.

Седельникова Н. В. 1990. Лишайники Алтая и Кузнецкого нагорья Новосибирск, <<Наука>>, Сибирское отделение.

Седельникова Н. В., Седельникова В. П. 1979. Роль лишайниковых синузий в высокогорных фитоценоза северной части Алтае-Саянской горной области. – Бот. Журн. Т. 64: No.5.

Семенов Б. С. По томи и ее притокам. – Сибирская природа, Омск, No.1.

Томин М. П. 1926. Список Лишайников южно-Уссурийского Края. – Изв. Южно-Уссур. Отд. Гос. Русск. Геогр. О-ва, No.12.

Цогт У. 1981. Напочивенные лишайники лишайники Монголии.（Систематика, география, экология, жизненные формы, географические элементы и хозяйственное значенис）. – Вкп.: Исследования флоры и растительности МНР. Улан-Батор, т.3.

Acharius E. 1803. Methodus Lichenum. - Stockholm.

Asahina Y. 1931. The Raiken's Soliloquy on Botanical Science, XXXV. - Journ. Jap. Bot. 7（4）:102.

Asahina Y. 1936-40. Mikrochemischer Nachweis der Flechtenstoffe. Journ. Jap. Bot. Vols. 12, 13, 14, 15, 16

Asahina Y. 1955. Lichenes in H. Kihara: Fauna and Flora of Nepal Hmalay 1: 43-63.

Asahina Y. 1960. Lichenologische Notizen（§ 160-612）. Journ. Jap. Bot. 35（4）: 97-102.

Awasthi D. D. 1961. Some foliose and fruticose lichens from Assam and North East frontier Agency og India. Prov. Ind. Ac. Sc. 54:24-44.

Awasthi D. D. 1965. Catologue of the lichens from India. Beih. 3. Nova Hedwigia 17

Awasthi D. D. 1988. A key to the macrolichens of India and Nepal. - Journ Hattori Bot. Lab. 65: 207-302.

Chen X. L., Zhao C. F. & Luo G. Y. 1981. A list of lichens in north-eastern China（in Chinese, summary in English）. Journ. North-eastern Forestry Inst. 3: 127-135; 4: 150-160.

Chevalier F. F. 1826. Fl. Gen. Env. Paris, [as "Umbilicarieae"] cited from Hawksworth & David, Family names, Index of fungi supplement（1989）.

Chopra G. 1934. Lichens of the Himalayas. Part I. Lichens of Darjeeling and the Sikkim Himalayas. Lahore

Codogno M., Poelt J., und Puntillo D. 1989 *Umbilicaria freyi* spec. Nova und der Formenkreis von *Umbilicaria hirsuta* in Europa（Lichenes, Umbilicariaceae）. Pl. Syst. Evol. 165:55-69.

Culberson W. & C. 1970. A phylogenetic view of chemical evolution in the lichens. Bryologist 73（1）:1-31.

Culberson C. F. Culberson W. L. & Johnason A. 1977. Second supplement to Chemical and Botanical Guide to Lichen Products. St. Louis, Mo: American Bryological and Lichenological Society, Missouri Bot. Garden.

Culberson C. F. & Kristinsson H. -D. 1970. A standardized method for the identification of lichen products. J. Chromatog. 46（1）:85-93.

Culberson C. F. 1921. Improved conditions and new data for the identification of Lichen products by a standardized thin-layer chromatographic method. J. Chromatogr. 72:113-125.

Culberson W. L. 1972. Disjunctive distributions in the lichen forming fungi, Annales of the Missouri Botanical Garden 59（2）:165

Dodge C. W. 1968. Lichenological Notes on the Flora of the Antarctic Continent and the Subantarctic Islands. Ⅶ and Ⅷ. Nova Hedwigia 15:285-332

Dodge C. W & Bker G. E. 1938. The second Byrd Antarctic ex[edition - Botany Ⅱ. Lichens and lichen parasites. Ann. Mo. Bot. Gard. 25:515-718.

Du Rietz, G. E. 1929. The Lichens of the Swedish Kamtchatka - Expeditions. Arkiv or Bor. 22（B）:14.

Eig, A., 1931. Les elements et les groupes phytogeographiques auxiliaires dans la flore palestinienne, 1. Texte; Ⅱ. Tableaux analitiques（Feclde's Report, Spec Nov., Beiheftc, vol. 63）.

Erikkson O. And Hawksworth D. L. 1985. Outline of the Ascomycetes. Systema ascomycetum. Vol. 4,pp. 1-79.

Erikkson O. And Hawksworth D. L. 1986. An alphabetical list of the generic names of ascomycetes. Systema Ascomycetum vol.5, part 1, pp.3-111.

Erikkson O. And Hawksworth D. L. 1986. Outline of the ascomycetes. Systema Ascomycetum. Vol. 5, Part 2, pp. 1-185.

Erikkson O. And Hawksworth D. L. 1987. A alphabetical list of the generic names of the ascomycetes in Systema Ascomycetum vol.6, part 1, pp.1-109.

Erikkson O. And Hawksworth D. L. 1987. Outline of the ascomycetes. Systema Ascomycetum.,vol. 6, Part 2, pp. 259-337.

Erikkson O. And Hawksworth D. L. 1988. Outline of the ascomycetes. Systema Ascomycetum. Vol. 7, Part 2, pp. 119-315.

Erikkson O. And Hawksworth D. L. 1990. Outline of the ascomycetes. Systema Ascomycetum. Vol. 8, Parts 2, pp. 119-318.

Erikkson O. And Hawksworth D. L. 1991. Outline of the ascomycetes. Systema Ascomycetum. Vol. 9, Parts1- 2, pp. 39-271.

Fee 1824. Essai Cryptog Ecorc. Officin. LXX.

Frey E. 1929. Beitrage zur Biologie, Morphologie und Systematik der Umbilicariaceen. Hedwigia 69:219-252.

Frey E. 1931. Umbilicariaceae in Rabh. Krypt. -Fl. 2. Aufl. 9 IV. Abt. I. Halfte: 203-424.

Frey E. 1936. Vorarbeiten zu einer Monographie der Umbilicariaceen Bericht. Schweiz. Bot. Ges. 45:198-230.

Frey E. 1936a. Die geographische Verbreitung der Umbilicariaceen und einiger alpiner Flechten. Bericht. Schweiz. Bot. Ges. 46:412-444.

Frey E. 1949. Neue Beitrage zu einer Monographie des Genus *Umbilicaria* Hoffm. Ibid. 59:427-470.

Frey E. Poelt J. 1977. Die Gattung *Lasallia*（Flechten des Himalay 13）. Mit 5 Abbildungen...Khumubu Himal 387-395.

Hale M. 1983. The Biology of Lichens. Edward Arnold.

Haller A. von.（1768）Historia Stirpium indigenarum Helvetiae, Ⅲ. - Berne.

Hawksworth D. L. & David J. C. 1989. Family names. Index of Fungi Supplement

Hestmark G. 1990. Thalloconidia in the genus *Umibilicaria*. Nord. J. Bot. 9: 547-574.

Hirabayashi K., Iwata S., Ito, M., Shigeta, S., Narui, T., and Shibata, S. 1989. In Chem. Pharm. Bull. 37:2410.

Hoffmann D. G. F. 1790, 1974, 1801. Descriptio et Adumbratio Plantarium e classe crytogamica Linnaei Lichenes dicuntur. Ⅰ-III.

Hue A. M. 1887. Lichenes Yunnanenses a claro Delavay anno 1885 collectis, et quorum novae species a celeb. W. Nylander descriptae fuerunt, exponit Hue AM. Buell. Soc. Bot. France 34:16-24.

Hue A. M. 1889. Lichenes Yunnanenses a cl. Delavay praesertim annis 1886-1887, collectos exponit Hue AM（1）. Series secunda. Bull. Soc. Bot. Bot. France, 36:158-176.

Hue A. M. 1900. Lichenes Extra- Europaei in Nouv. Archiv. Du Mus. 4 ser. Tom. 2,p. 112.

Huneck S., Ahti T., Cogt U. Et al., 1992. Zur Berbreitung und Chemie von Flechten der Mongolei. III. Ergebnisse der Mongolisch -Deutschen Biologischen Expedition sait 1962 Nr. 217. Nova Hedwigia 54（3-4）:277-308.

Ikoma Y..1956. Phytogeographical and ecological notes on *Gyrophora esculenta* in Chugoku district. - Misc. Bryol. Lichenol. 1-3:2-3.

Ikoma 1983. Macrolichens of Japan and adjacent regions, pp. 120, Japan, Tottori City.

Jatta A. 1902. Licheni Cinesi raccolti allo Shen-Si negli anni 1894-1898 dal rev. Padre Missionario Giraldi G. Nuov. Giorn. Bot. Italiano ser. 2, IX, 460-481.

Kamiya T. & Kim, J. S. 1981. Ecological Study of *Gyrophora esculenta* Miyoshi on Mt. Jiri in Korea Journ. Phytogeogr and Taxonomy. 29（2）:112-114.

Kato K. And Kumagai T. 1960. *Gyrophora esculenta* found in Akita prefecture. -Misc. Bryol. Lichenol. 2:18-19.

Kim S. H. 1972. Lichens of Korea vol. 9（1）:145-185（1972）.

Kim S. H. 1979. Studies on the Lichens in Korea cc Bul. Kong Ju Teachers' college 15:259-268.

Kim S. H. 19781. Floral Studies on the Lichens in Korea 279-304.

Kobayashi M. 1957. Geographical range of *Gyrophora esculenta* in Prefecture Fukushima. -Misc. Bryol. Lichenol. 1-9:2

Koerber G. W. 1855. Systema Lichenum Germaniae. Breslau.

Krog H. 1973. On *Umbilicaria* pertusa Rassad. And some related lichen species. The Bryologist 76:550-554.

Krog H. & Swinscow T. 1986. The lichen genera *Lasallia* and *Umbilicaria* in East Africa. Nord. J. Bot. 6（1）:75-85.

Kurokawa S. 1960. Notes on *Gyrophora esculenta* Miyoshi. -Misc. Bryol. Lichenol. 2:16-17.

Linnaeus, 1753. Species Plantarum. 1140-1156.

Llano G. 1950. A Monogr. Of the Lichen Family Umbilicariaceae in the western hemisphere Navexos Pp.1-831 Washington.

（Lu D. A.）陆安定 1959. Notes on Chinese Lichens, 2. Umbilicariaceae. Acta Phytotax. Sin. 8（2）:173-180.

Merat 1821. Nouv. Fl. Paris ed. 2（1）:202.

Miyoshi 1893, in Bot. Centrabl. 56:161.

Nagarkar M. B. & Patwadhan P. G. 1981. A new species of *Umbilicaria* form ladak, India. Curr. Sei. 50（24）: 1075-1706.

Niu Y. C. & Wei J. C. 1993. Variations in ITS2 sequences of nuclear rDNA from two Lasallia species and their systematic significance. - Mycosystema 6:25-29

Nylander W. 1860. - Flora, 43:418.

Nylander W. 1869. Synopsis Meth. Lich. 2:11.

Nylander W. 1887. Addenda nova ad lichenographiam europaeam. Flora. 70（9）:129-136

Nylander W. 1888. Enumeratio Freti Benringii. - Bull. Soc. Linn. Norm., ser. 4, N. 1.

Paulson 1925. Lichens of Mocut Everest J. Bot. 63:189-193.

Paulson 1928. Lichens from Yunnan 1bid. 66:313-319.

Poelt J. 1962. Bestimmungsschlussel der Hoheren Flechten vo Europa. Mitt. Bot. Staatssammlung Munchen. 4:301-571.

Poelt J. 1977. Die Gattung *Umbilicaria*（Umbilicariaceae），（Flechten des Himalay 14）. Khumbu Himal 6（3）:397-435.

Räsanen Ⅴ. 1949. Lichenes novi Ⅳ. Arch. Soc. Zoo. Bot. Fennicae “Vanamo” 3:78.

Räsanen V. 1950. Lichenes novi Ⅵ. Ibid. 5:25.

Räsanen V. 1952. Lichenes novi Ⅶ. Ibid. 6:80.

Sato M. M. 1937. Enumeration lichenum Ins. Formosae（Ⅲ）. Journ. Jap. Bot. 13（8）:595-599.

Sato M. M. 1950. Notes on some remarkable *Umbilicaria* collected in Far Eastern Asia. Journ. Jap. Bot. 25（8）:165.

Sato M. M. 1952. Lichenes Khinganenses: or a list of lichens collected by Prof. T. Kira In the Khingan Range, Manchurica. Bot. Mag. Tokyo 65（769-770）:172-175.

Sato M. M. 1958. Distribution and ecology of *Gyrophora esculenta* Miyoshi. - Journ. Jap. Bot. 33:110-115.

Sato M. M. 1960. Distribution of *Gyrophora esculenta* Miyoshi in the Coastal Region of Nourth-Eastern Honshu, Japan. -Misc. Bryol. Lichenol. 2:46-49.

Sato M. M. 1961. Range of the Japanese Lichens（Ⅶ）. - Bull. Fac. Arts and Sci., Ibaraki Univ., Nat. Sci., No. 12:46-48.

Sato M. M. 1981. Notes on the cryptogamic flora of prov. Shanxi, North China（Ⅲ）Lichens Miscellanea Bryologica et Lchenologica 2（3）:64-65.

Scholander P. F. 1934. On the apothecia in the lichen family Umbilicariaceae. Nyt Mag. Naturvid. 75:1-3.

Schubert R. & Klement O. 1971. Beitrag zur Flechtenflora der Mongolishen Volksrepublik. - Fedd. Rep., Bd 82, H. 3-4.

Schubert R., Klement O. & Schamsran Z. 1969. Beitrag zur Flechtenflora der Umgebung on Ulan-bator（Mongolische Volksrepublik）. -Fedd. Bd 79, H. 6.

Shibata S. 1989.in Chem. Pharm. Bull. 37:2410.

Shibata S., Nishikawa Y., Takeda T., and Tnaka M. 1968.in Chem. Pharm. Bull. 16:2362.

Singh A. 1964. Lichens of India-Bulletin of the National Botanic Gardens No. 93. Lucknow, India.

Sipman H. J. M. & Topham. P. 1992. The genus *Umbilicaria*（lichenized ascomycetes）in Colombia. Nova Hedwigia 54（1-2）:63:75.

Suza J. A. 1925. A sketch of the distribution of lichens in Moravia with regard to the Conditions in Europe. - Publ. Fac. Sci. Univ. ,asar. Vol. 55.

Takamizu N. 1957. Geographical range of *Gyrophora esculenta* in Tokyo Pref. - Misc. Bryol. Lichenol. 1-12:3-4.

Topham. P. B., Seaward M. R. D. & Bylinska E. A. 1992. *Umbilicaria propagulifera* new to northern hemispliere. - Lichenologist 14:47-52.

Verzda A. 1965. Flechten aus NW- Mongolei. - Acta Mus. Siles., ser. A. T. 14. 18（1）:83-104.

Wang-Yang J. R. & Lai M. J. 1976. Additions and corrections to the lichen flora of Taiwan. Taiwania 21（2）:226.

（Wang X. Y.）王先业 1985. 天山托木尔峰地区的地衣。天山托木尔峰地区的生物新疆人民出版社. pp. 328-353.

（Wei J. C. ）魏江春 1966. A new subgenus of *Lasallia* Mer.em. Wei. Acta Phytotax. Sin. 11（1）:1-8.

（Wei J. C.）魏江春 1981. Lichenes Sinenses Exciccati（Fasc. 1:1-50）. Bull. Bot. Research1（3）:81-91.

（Wei J. C.）魏江春 1982. Some new species and materials of *Lasallia* Mer.em. Wei. Acta Mycol. Sin. 1（1）:19-26.

Wei. J. C. 1991. An enumeration of Lichens in China. International Academic Publishers, Beijign, pp. 1-278.

Wei. J. C. 1993. The Lectotypification of Some Specles in the Umbilicariaceae described by Linnaeus or Hoffmann. Supplement to Mycosystema 5:1-17.

Wei. J. C. 1993. Current progress of systematics of lichenized fungi. Proceedings of the first Korea- China Joint Ceminar ofr Mycology Dec. 2-5, 1993, pp. 5-12, Seoul, Korea.

Wei. J. C. & Biazrov L. G. 1991. Some disjunctions and vicariisms in the Umbilicariaceae（Acomycotina）. Mycosystema 4:65-72.

（Wei. J. C. & Chen J. B.）魏江春、陈建斌 1974. 珠穆朗玛峰地区地衣区系资料. 珠峰地区可靠报告（生物与高山理）科学技术出版社 1974:174

Wei. J. C. & Jiang Y. M. 1981. A biogeographical analysis of the lichen flora of Mt. Qomolangma region in Xizang. Proceedings of Symposium on Qinghai-Xizang（Tibet）plateau pp. 1145-1151, Beijing, China.

（Wei. J. C. & Jiang Y. M. ）魏江春、姜玉梅 1982. New materials for lichen flora from Xizang. Acta Phytotax. Sin. 20（4）:496-501.

（Wei. J. C. & Jiang Y. M. ）魏江春、姜玉梅 1986. Lichens of Xizang. Science Presss. Beijing, pp. 1-30.

Wei. J. C. & Jiang Y. M. 1988. A conspectus of the ascomycete Umbilicariaceae in China. Mycosystema 1（1）:73.

Wei. J. C. & Jiang Y. M. 1989. The status of the genus *Llanoa* Dodge and the delimitation of the genera in the Umblicariaceae（Ascomycotina）. Mycosystema 2:135-150.

Wei. J. C. & Jiang Y. M. 1992. Some species new to science and distribution of Umbilicariaceae（Ascomycotina）. Mycosystema 5:73-88.

Wei. J. C. & Jiang Y. M. 1993. The Asian Umbilicariaceae. International Academic Publishers Beijing, pp. 1-217.

Wei. J. C. & Jiang Y. M. & Guo S. Y. 1995. A new species of *Umbilicaria* and its vicarism（Ascomycota）, Mycotaxon（in press）.

（Wei J. C.et al）魏江春等 1982. Lichenes Officinalese Sinenses. Science Press, Beijing.

（Wu J. L.）吴金陵 1981. Medicinal lichens in Qin Ling Mountai. Acta pharmaceutica Sinica 16（3）:611-167.

Yamanaka T. And Yoshimura I. 1961. Sime observations on the *Gyrophora esculenta* commynity. - Journ. Jap. Bpt. 36:193-200.

Yoshimura I. 1974. Lichen flora of Japan in colour Hoikusha Publishing Co., Ltd. Pp. 1-349.

Yoshimura I. And Yamanaka T. 1961. Notes on plants attached to *Gyrophora esculenta*, - Misc. Bryol. Lichenol. 2:98-99.

Zahlbrukner A. 1930. Lichenes in Handel-Mazzetti, Symb. Sin. 3:1-254.

Zahlbrukner A. 1932. Cat. Lich. Univ. 8:496

Zahlbrukner A. 1933. Flechten der Insel Formosa in Fedde, Repertorium, Fasc.

Диссертация на соискание ученой степени доктора биологических наук, Российская Академия Наук, Ботанический Институт им. В. Л. Комарова, С. Петербург, 1995.

Taxonomic revision of six taxa in the lichen genus collema from China

Hua-jie LIU Jiang-chun WEI*

Key Laboratory of Systematic Mycology & Lichenology
Institute of Microbiology, Chinese Academy of Sciences, Beijing, 100080

ABSTRACT: Six taxa of the lichen genus *Collema* belonging to 5 species are reported in the present paper with the figures of type specimens of 5 revised taxa. The reexamination of the type and paratype specimens of 5 taxa shows *Collema brevisporum*, *C. multipartitum* var. *granulosum*, *C. corniculatum*, *C. tianmuense* and *C. pulcellum* var. *multipartitum* to be synonyms of *C. leptaleum* var. *leptaleum*, *C. multipartitum*, *C. pulcellum* var. *subnigrescens*, *C. subconveniens* and *C. subnigrescens* respectively.

KEY WORDS: Collemataceae, Lecanorales, lichenized ascomycetes

Introduction

Twenty-eight taxa of *Collema* belonging to 22 species were reported from China by about 20 authors up to the early 1990's (Wei, 1991; Wu, 1987). Afterwards, Jiang (1992) described 12 taxa belonging to 12 species new to science and reported 13 taxa belonging to 9 species new to China (Jiang, 1993). About 5 years later, Abbas & Wu (1998) added one more taxon to the *Collema* from China. To date 54 taxa of *Collema* belonging to 40 species are reported from China. During recent study on the genus from China we found that 5 taxa, *Collema brevisporum*, *C. multipartitum* var. *granulosum*, *C. corniculatum*, *C. tianmuense* and *C. pulcellum* var. *multipartitum* described by Jiang (1992) are synonyms of those taxa which have been reported from China by the following authors: Zahlbruckner (1933a, 1933b, 1940), Asahina (1952), Wang Yang & Lai (1973), Degelius (1974), Wu (1987), Thrower (1988), Chen et al. (1989), Wei (1991), Jiang (1993) and Abbas & Wu (1998).

Materials and Methods

The specimens examined in the present paper are deposited mainly in HMAS-L. The measurements were carried out using a compound microscope with hand sections mounted in water. Some terms of morphology and anatomy are used

*Corresponding author

This paper was originally published in *Mycotaxon*, 86: 349-358, 2003.

after Degelius (1954, 1974). The literature records for China are mainly based on Wei (1991).

Results and Discussions

1. Collema leptaleum Tuck. var. leptaleum

Proceed. Amer. Acad. Arts and Sc. **6**: 263 (1864).

= *C. brevisporum* Z. G. Jiang, Journal of Hebei Normal University **16**(3): 83 (1992). syn. nov. Type: China, Jilin, Changbaishan, 1370 m, 1960.8.19, Y. C. Yang et al., 52 (HMAS-L). Paratype: China, Zhejiang, Tianmushan, 1000 m, 1962.9.2, J. D. Zhao, 6399 (HMAS-L). Fig. 1 & 2.

This variety is characterized by: 1) thallus subfoliose to crustose, upper side irregularly rugose; 2) isidia and pruina absent; 3) end lobes plane, not swollen, < 5 mm wide; 4) thallus and thalline exciple without pseudocortex; 5) proper exciple euparaplectenchymatous; 6) spores bacillar and 4-celled; 7) corticolous.

Selected specimens examined: JILIN, Mt. Changbaishan, 1977.8.8, J. C. Wei, 2791-1, 2793-3. HEILONGJIANG, Dailing, Liangshui, 420 m, 1975.10.12, J. C. Wei, 2280. YUNNAN, Xichou, Fadou, 1450 m, 1991.11.16, J. B. Chen, 5015-1.

Literature records for China: TAIWAN (Zahlbruckner, 1933b, p.26 & 1940, p.247; Wang Yang & Lai, 1973, p. 90; Degelius, 1974, pp.102 & 107), HONG KONG (Thrower, 1988, p.84).

The reexamination of the type specimens of *C. brevisporum* showed that the following characteristics were present: Thallus subfoliose to subcrustose, 3-4 cm diam., rounded to somewhat irregularly shaped, usually adnate. Upper side dark olive-green to blackish, glabrous, smooth or sparsely to densely and irregularly rugose, partly glossy, epruinose, not fenestrate, without isidia, partly with isidia-like outgrowths. Lobes sparse, plane, margin entire, somewhat crisped, not swollen, 1-3 mm wide. Apothecia sparse to dense, 1 mm diam., sessile, constricted at the base.

Thallus 75-112 μm thick, without pseudocortex on both sides.

Thalline exciple without pseudocortex. Proper exciple euparaplectenchymatous, 37-61 μm thick in center. Spores bacillar, straight to curved, usually 4-celled, 25-51×2-5 μm.

The wrinkles on upper side of *C. leptaleum* var. *leptaleum* were apparently confused with the ridges of the *Nigrescens* group by Jiang (1992), who put his new species *C. brevisporum* (= *C. leptaleum* var. *leptaleum*) in the *Nigrescens* group rather than in the *Leptaleum* group.

Fig 1-2. Thallus of the type specimen of *C. brevisporum*. 1. Upper side. 2. Lower side. Scale = 1 mm.
Fig. 3. Fragments of the thallus of type specimen of *C. multipartitum* var. *granulosum*. Scale = 1 mm.
Fig 4-5. Thallus of the paratype specimen of *C. pulcellum* var. *multipartitum*. 4. Upper side. 5. Lower side. Scale = 1 mm.
Fig. 6. Thallus of the type specimen of *C. corniculatum*. Scale = 1 mm.

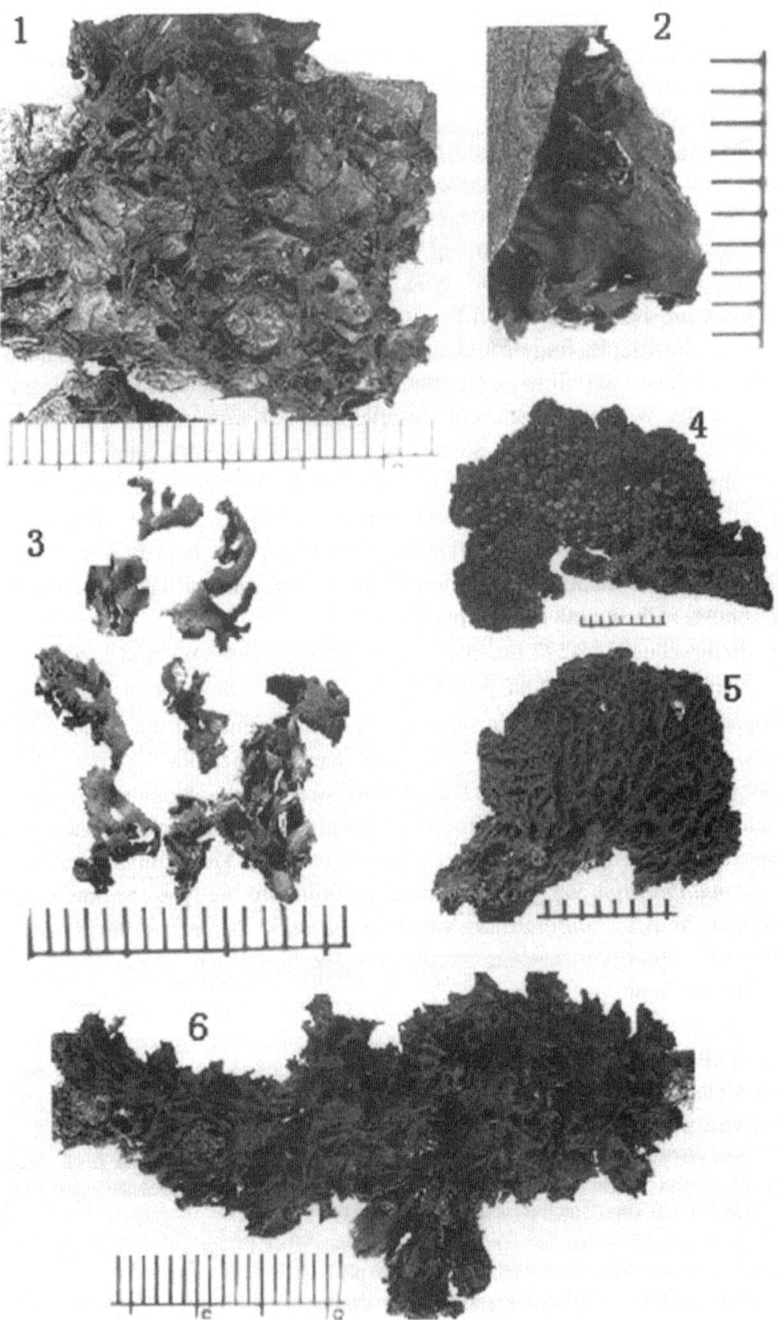
1
2
3
4
5
6

2. Collema multipartitum Smith

in Smith & Sowerby, Engl. Bot. : 36 (1814).

= *C. multipartitum* Smith var. *granulosum* Z. G. Jiang, Journal of Hebei Normal University **16**(3): 86 (1992). syn. nov. Type: China, Hebei, Xiaowutaishan, Xitai, 2000 m, 1986.6, Z. G. Jiang, 3583-1 (HMAS-L). Fig. 3.

This species is characterized by: 1) thallus foliose; 2) lobes repeatedly furcate, < 5 mm wide and convex near the margin; 3) thallus and thalline exciple without typical pseudocortex; 4) proper exciple euparaplectenchymatous; 5) spores linear-oblong, 4-celled; 6) saxicolous.

The reexamination of the type specimen of *C. multipartitum* var. *granulosum* showed that the following characteristics were present: Thallus foliose, very fragile, fragmented, deeply and richly lobate, without isidia and lobules. Upper side olive-green, glabrous, smooth, epruinose, partly glossy, not plicate, without ridges and pustules. Lower side paler, with sparse rhizines. Lobes numerous, somewhat repeatedly furcate and extended, margin not swollen or somewhat so, entire, end lobes 0.5-1 mm wide, with upper surface flattened to somewhat convex and lower surface flattened to somewhat concave. Apothecia sparse, superficial, sessile with constricted base, 1-1.5 mm diam; disc dark red, plane to somewhat irregularly convex, epruinose, with smooth and entire margin.

Thallus about 255-272 μm thick, without pseudocortex on both sides.

Thalline exciple without pseudocortex, 204-221 μm thick in the center. Proper exciple euparaplectenchymatous, 48-58 μm thick in the center, with cells globular to subglobular, > 8.5 μm diam. Hymenium 133 μm thick. Subhymenium 17-24 μm thick. Spores 8 per ascus, linear-oblong to somewhat bacillar, 4-celled, straight to somewhat curved, with usually obtuse ends, not constricted at the septa, colourless, (27) 34-37 (41) × (3.5) 5-7 μm. Saxicolous.

C. multipartitum var. *granulosum* is, according to the type specimen, not different from *C. multipartitum* var. *multipartitum*. In fact, no infraspecific taxon of *C. multipartitum* was accepted by Degelius (1954, 1974).

New to China.

3. Collema pulcellum Ach.

Syn. Meth. Lich.: 321 (1814).

3.1. var. pulcellum

C. pulcellum Ach. var. *multipartitum* auct. origin. p. p. non Z. G. Jiang (type), Z. G. Jiang, Journal of Hebei Normal University **16**(3): 86 (1992). Paratype: China, Hubei, Shennongjia, 2200 m, 1984.7.3, J. B. Chen, 10027 (HMAS-L). Fig. 4 & 5.

Collema pulcellum Ach. var. *leucopeplum* auct. non. (Tuck.) Degel., Z. G. Jiang, Journal of Hebei Normal University **17**(3): 72 (1993) (p. p.).

This variety is characterized by: 1) thallus of *nigrescens* type, mainly pustulate; 2) isidia and pruina absent; 3) thallus and thalline exciple without pseudocortex; 4) proper exciple euparaplectenchymatous; 5) spores acicular to somewhat linear fusiform, usually 6-celled, not constricted at the septa, often 51-65 × 3.5-5 μm.

Selected specimens examined: NEI MONGOL, Arxan, Bailang, 1300 m, 1991.8.18-19, J. B. Chen & Y. M. Jiang, 711-6 & A-664. HEILONGJIANG, Dailing, Liangshui, 500 m, 1975.10.2, J. C. Wei, 2065. JILIN, Mt. Changbaishan, 1100 m, 1977.8.8, J. C. Wei, 2807, 2810-1(Jiang (1993) as *C. pulcellum* var. *leucopeplum*). SHAANXI, Mt. Taibaishan, 2260 m, 1963.6.10, J. C. Wei et al., 2262. HEBEI, Mt. Xiaowutaishan, Zhongtai, 1934.7.17, Q. W. Wang, 982. ANHUI, Yuexi, Yaoluoping, 1500 m, 2001.9.5, H. J. Liu, 453; Jinzhai, Tiantangzhai, 1400 m, 2001.9.12, H. J. Liu, 668. HUNAN, Mt. Hengshan, 1000 m, 1964.8.31, J. D. Zhao & L. W. Xu, 9923. YUNNAN, Lijiang, Mt. Yulongshan, 3250 m, 1981.8.6, X. Y. Wang et al., 6542. FUJIAN, Pucheng, Jiumu, 600 m, 1960.8.8, Q. Z. Wang et al., 566. XINJIANG, Altay Shan, A. Abbas, 9800139.

Literature records for China: Northeast China (Degelius, 1974, p. 176), YUNNAN (Degelius, 1974, p. 176; Jiang, 1993, p.72), FUJIAN (Jiang, 1993, p.72).

The reexamination of the paratype specimen of *C. pulcellum* var. *multipartitum* showed that the following characteristics were present: Thallus of *nigrescens* type, foliose, 4-5 cm diam. Upper surface olive-green to blackish, glabrous, smooth, mainly pustulate, without isidia and pruina. Lobes sparse, entire, slightly downturned, somewhat crisped. Apothecia on pustules and ridges, numerous, dense, 1-1.5 mm diam., without pruina.

Thallus 90-120 μm thick, without pseudocortex on both sides.

Thalline exciple without pseudocortex. Proper exciple euparaplectenchymatous. Spores 8 per ascus, acicular to somewhat linear fusiform, 6 (-8)-celled, not constricted at the septa, colourless, (51) 58-61 (68) × 3.5-5 μm.

Jiang (1992) included two disparate elements in his original circumscription of *C. pulcellum* var. *multipartitum,* and indicated as difference between his new taxon and *C. pulcellum* var. *pulcellum* the presence of 12-16-celled spores. However, the authors were unable to find such spores when reexamining the paratype specimen of *C. pulcellum* var. *multipartitum.*

3.2. var. subnigrescens (Muell. Arg.) Degel.

Symb. Bot. Upsal. **20**(2): 173 (1974).

= *C. corniculatum* Z. G. Jiang, Journal of Hebei Normal University **16**(3): 85 (1992). syn. nov. Type: China, Yunnan, Menghai, Manjie Reserve, 1350-1400 m, 1981.1.3, Jiang Y. M., 1114-5 (HMAS-L). Fig. 6.

C. complanatum auct. non Hue, Z. G. Jiang, Journal of Hebei Normal University 17(3): 72 (1993).

C. pulcellum Ach. var. *leucopeplum* auct. non. (Tuck.) Degel., Z. G. Jiang, Journal of Hebei Normal University 17(3): 72 (1993) (p. p.).

C. pulcellum Ach. var. *pulcellum* auct. non Ach., Z. G. Jiang, Journal of Hebei Normal University 17(3): 72 (1993) (p. p.).

This variety, a member of the *Nigrescens* group, is often confused with other taxa of the same group, especially the main type and *C. complanatum.*

It differs from the main type in: 1) thallus mainly ridged; 2) spores acicular to somewhat fusiform, usually 6-8-celled, often 34-65 × 5-8.5 μm. It differs from *C. complanatum* mainly in the spores, which are shorter (< 65 μm), wider and acicular to somewhat fusiform in the former, but longer (> 65 μm), narrower and acicular in the latter.

Selected specimens examined: HEILONGJIANG, Huzhong, 500 m, 1984.9.1, X. Q. Gao, 310 (Jiang, 1993 as *C. pulcellum* var. *leucopeplum*). SHAANXI, Mt. Taibaişhan, 2700 m, 1988.7.10, C. H. Ma, 107. QINGHAI, Ledu, 2800 m, 1959.9.9, J. C. Xing & Q. M. Ma, 1528 (Jiang, 1993 as *C. pulcellum* var. *leucopeplum*). HUBEI, Shennongjia, 1800 m, 1976.7.8, P. C. Wu, 371-3. SICHUAN, Mt. Gongga Shan, 2650 m, 1982.7.2, X. Y. Wang et al., 8729; Mt. Emeishan, 2200 m, 1963.8.14, J. D. Zhao & L. W. Xu, 7325. GUIZHOU, Mt. Fanjingshan, 1963.8.21, J. C. Wei, 358-2. YUNNAN, Simao, 1960.11.19, J. D. Zhao & Y. B. Chen, 3144; Xishuangbanna, 650 m, 1980.12.21, Y. M. Jiang, 976-1 (Jiang, 1993 as *C. pulcellum* var. *pulcellum*), 570 m, 1982.8.28, L. S. Wang, 82-2790 (KUN 3453); Lijiang, 2900-3000 m, 1993.9.14, L. S. Wang, 93-13645 (KUN 13896); Judian, 1850 m, 1989.2.10, L. S. Wang, 89-440 (KUN 11110); Zhongdian, 3500 m, 1994.9.21, L. S. Wang, 94-15001 (KUN 15124). XIZANG, Zayü, Ridong, 3400 m, 1982.9.13, J. J. Su, 4643. HONG KONG, Cape d'Agailer, 1992.10.22, S. Thrower, 3262 (Jiang, 1993 as *C. complanatum*).

Literature records for China: HEILONGJIANG (Asahina, 1952, p.375; Degelius, 1974, p.176), YUNNAN (Degelius, 1974, p.176), HUBEI (Chen et al., 1989, p.420), TAIWAN (Wang Yang & Lai,1973, p.90). HONG KONG (Thrower, 1988, p.86).

The reexamination of the type specimen of *C. corniculatum* showed that the following characteristics were present: Thallus of *nigrescens* type, foliose, rounded, 3 cm diam. Upper side olive-green, glabrous, smooth, somewhat glossy, with regular ridges, without isidia and pruina. Lobes sparse, entire and somewhat crisped in margin. Apothecia not seen.

Thallus 82-99 μm thick, without pseudocortex or partly with poorly developed primitive pseudocortex.

Jiang (1992) separated *C. corniculatum* from all other taxa of *Collema* by the presence of the marginal corniculate lobules. But no such lobules were found by the authors on the type specimen.

4. Collema subconveniens Nyl.

Lich. Novae Zeland.: 8 (1888).

= *C. tianmuense* Z. G Jiang, Journal of Hebei Normal University **16**(3): 84 (1992). syn. nov. Type: China, Zhejiang, Tianmushan, Laodian, on mossy rock, 1000 m, 1962.9.1, J. D. Zhao, 6265 (HMAS-L). Fig 7 & 8.

This species is characterized by: 1) thallus of *japonicum* type; 2) ridges and pustules usually absent; 3) often without isidia; 4) lobes > 5 mm wide,

Fig. 7-8. Thallus of the type specimen of *C. tianmuense*. 7. Upper side. 8. Lower side. Scale = 1 mm.

Fig. 9-10. Thallus of the type specimen of *C. pulcellum* var. *multipartitum*. 9. Lower side. 10. Upper side. Scale = 1 mm.

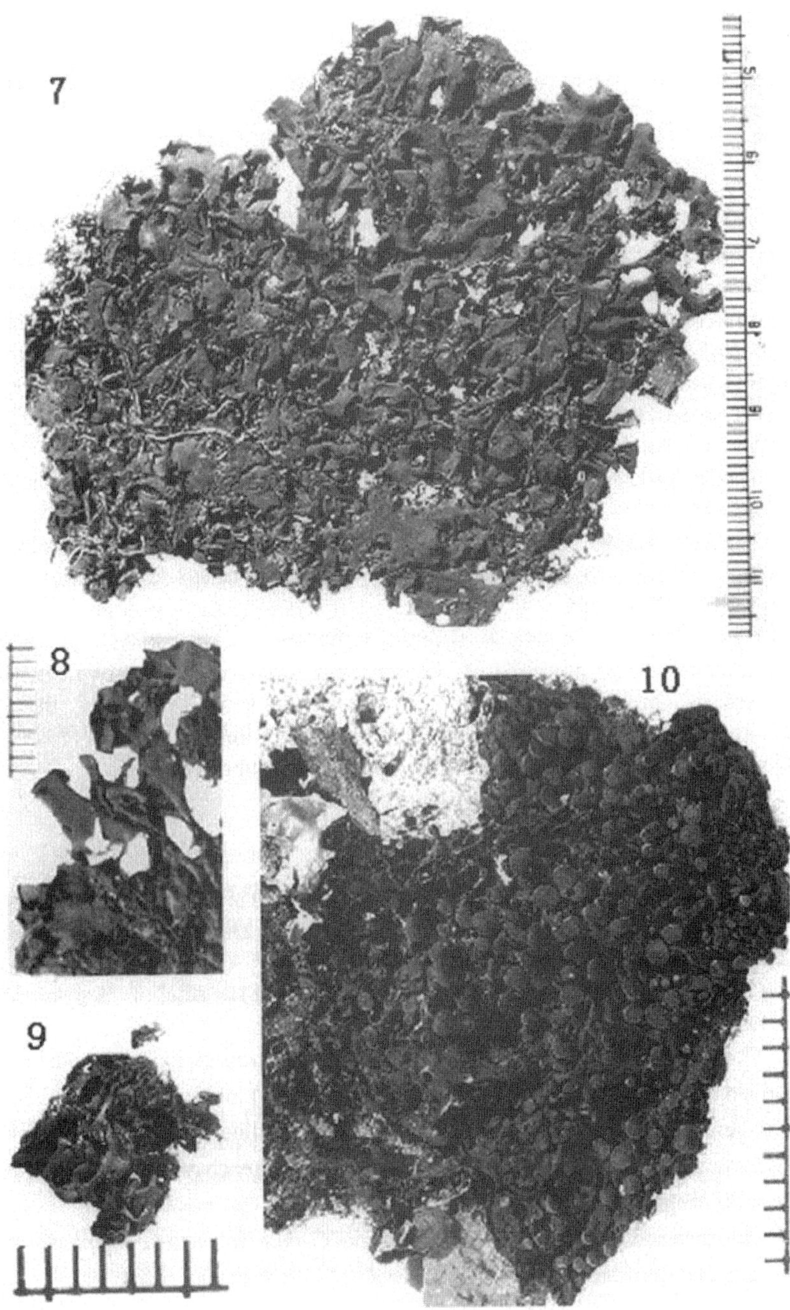
7
8
9
10

not swollen; 5) thallus and thalline exciple with typical pseudocortex; 6) proper exciple euthyplectenchymatous; 7) spores fusiform to oblong, submuriform, 17-44 × 7-12 μm.

Selected specimens examined: SHAANXI, Mt. Taibaishan, 3100 m, 1992.7.28, J. B. Chen & Q. He, 6132. HUBEI, Shennongjia, Bancang, 1200 m, 1984.8.17, J. B. Chen, 11805-2. YUNNAN, Lijiang, Mt. Yulongshan, 3050 m, 1964.10.5, J. C. Wei, 2532-1.

Literature records for China: HUBEI (Chen et al., 1989, p.421, misidentification of *Leptogium* sp.), XINJIANG (Abbas & Wu, 1998, p. 62).

The reexamination of the type specimen of *C. tianmuense* showed that the following characteristics were present: Thallus of *japonicum* type, foliose, rounded, 6-7 cm diam. Upper side olive-green to dark olive-green, glabrous, smooth, without ridges and pustules, epruinose, glossy, here and there with sparse squamiform isidia. Lower side olive-green, with sparse and white rhizines. Lobes rounded, 1-1.5 cm wide, margin entire, not crisped, not swollen, plane to downturned. Apothecia numerous, dense, 1-2 cm diam., without pruina, constricted at the base.

Thallus 80-120 μm thick, partly with primitive pseudocortex, without typical pseudocortex.

Thalline exciple without pseudocortex. Proper exciple euthyplectenchymatous, thin, 8.5-10 μm thick. Spores fusiform, 6-8-celled to submuriform, 31-34(41) × 8.5-10 μm.

Jiang (1992) indicated as differences between *C. tianmuense* and *C. subconveniens* the presence of submuriform spores and an euthyplectenchymatous proper exciple, and the rocky substrate. However, these characters are no deviation from the species definition by Degelius (1974). The authors were unable to find any other significant difference.

5. Collema subnigrescens Degel.

Symb. Bot. Upsal. **13**(2): 413 (1954).

= *C. pulcellum* Ach. var. *multipartitum* Z. G. Jiang, Journal of Hebei Normal University **16**(3): 86 (1992) (p.p.). Type: China, Hubei, Shennongjia, 2200 m, J. B. Chen, 11298 (HMAS-L). Fig. 9 & 10.

C. pulcellum Ach. var. *subnigrescens* auct. non. (Muell. Arg.) Degel., Z. G. Jiang, Journal of Hebei Normal University **17**(3): 72-73 (1993) (p. p.).

This species is characterized by: 1) thallus of *nigrescens* type, mainly ridged on upper side; 2) without isidia and pruina; 3) lobes > 5 mm wide, not swollen, not crisped, plane; 4) thallus and thalline exciple often with pseudocortex; 5) proper exciple usually euthyplectenchymatous; 6) spores usually irregularly clavate, usually 5-6-celled.

This species can be easily distinguished from *C. pulcellum* and *C. complanatum* by its pseudocorticate thalline exciple and euthyplectenchymatous proper exciple.

Selected specimens examined: HUNAN, Mt. Hengshan, Cangjingdian, 960 m, 1964.9.2, J. D. Zhao & L. W. Xu, 10315. SICHUAN, Huanglong, 3250-3300 m, 2001.9.23, Z. T. Zhao, et al., S172 & S178. YUNNAN, Lijiang, Mt. Yulongshan, 3050 m, 1964.12.15, J. C. Wei, 2497 (Jiang, 1993 as *C. pulcellum* var. *subnigrescens*). XIZANG, Zham, 2580 m, 1966.5.8, J. C. Wei & J. B. Chen, 391-2.

Literature records for China: SHAANXI (Wu, 1987, p.55).

The reexamination of the type specimen of *C. pulcellum* var. *multipartitum* showed that the following characteristics were present: Thallus of *nigrescens* type, foliose, 5 cm diam. Upper surface olive-green to dark olive-green, glabrous, smooth, partly glossy, mainly ridged, pustulate marginally, without isidia and pruina. Lower side olive-green, smooth, not glossy, rhizines sparse. Lobes sparse, rounded, ca. 1 cm broad, not swollen, entire, not crisped to slightly so. Apothecia on ridges and pustules, numerous, dense, ca. 1 mm diam., without pruina.

Thallus 85-102 μm thick, without pseudocortex on both sides.

Thalline exciple with typical pseudocortex. Proper exciple euthyplectenchymatous to somewhat subparaplectenchymatous. Spores 8 per ascus, clavate to acicular, usually 6-10-celled, colourless, not constricted at the septa, (51) 54-65 (85) × (3.5) 5-7 μm.

According to the reexamination of the type specimen, *C. pulcellum* var. *multipartitum* (type: J. B. Chen, 11298) is identical with *C. subnigrescens*, though the spores are 6-10-celled, unusual for *C. subnigrescens*.

Acknowledgment

The authors are greatly indebted to both the Drs. Harrie Sipman and Richard P. Korf for reading and improving the manuscript. The authors' thanks are also given to Mr. Wang Li-song for sending some specimens on loan. This study was financially supported by National Natural Science Foundation of China (39899400).

References

Abbas, A. & J. N. Wu. 1998. Lichens of Xinjiang, Sci-Tec & Hygiene Publishing House of Xinjiang (K), Urumqi (in Chinese).

Asahina, Y. 1952. An addition to the Sato's Lichenes Khinganenses (Bot. Mag. Tokyo, 65:172). *Journ. Jap. Bot.* **27**(12): 373-375.

Chen, J. B., J. N. Wu & J. C. Wei. 1989. Lichens of Shennongjia, in Fungi and Lichens of Shennongjia. World Publishing Corp, Beijing (in Chinese).

Degelius, G. 1954. The lichen genus *Collema* in Europe. Morphology, Taxonomy,

Ecology. *Symb. Bot. Upsal.* **13**(2): 1-499.

Degelius, G. 1974. The lichen genus *Collema* with special reference to the extra-European Species. *Symb. Bot. Upsal.* **20**(2): 1-215.

Jiang, Z. G. 1992. New taxa of lichen genus *Collema* Weber from P. R. China. *Journal of Hebei Normal University (Natural Science)* **16**(3): 83-87 (in Chinese).

Jiang, Z. G. 1993. New recorded species of the genus *Collema* Weber from P. R. China and Hong Kong region. *Journal of Hebei Normal University (Natural Science)* **17**(3): 69-73 (in Chinese).

Thrower, S. L. 1988. Hong Kong Lichens. An Urban Council Publication. Hong Kong.

Wang Yang, J. R. & M. J. Lai. 1973. A checklist of the lichens of Taiwan. *Taiwania* **18**(1): 83-104.

Wei, J. C. 1991. An Enumeration of lichens in China. International Academic publishers. Beijing.

Wu, J. L. 1987. Iconography of Chinese Lichen. Zhong Guo Zhan Wang Chu Ban She. Beijing (in Chinese).

Zahlbruckner, A. 1933a. Flechten der Insel Formosa. *Feddes, Repert.* **31**: 194-224.

Zahlbruckner, A. 1933b. Flechten der Insel Formosa. *Feddes, Repert.* **33**: 22-68.

Zahlbruckner, A. 1940. Cat. Lich. Univ. 10. Leipzig. reprinted by Johnson Reprint Corporation, New York, N. Y., 1951.

The systematic position of *Gymnoderma* and *Cetradonia* based on ssu rDNA sequences

QI-MING ZHOU[1], JIANG-CHUN WEI[1], TEUVO AHTI[2], SOILI STENROOS[2] AND FILIP HÖGNABBA[2]

ABSTRACT. Phylogenetic relationships of *Gymnoderma coccocarpum* and *Cetradonia linearis* were investigated using nucleotide sequence data of the SSU rDNA. Our results show that *Gymnoderma* and *Cetradonia* form a clade, which is a sister group of *Pycnothelia* and *Carassea*. As a result, both *Gynmoderma* and *Cetradonia* are clearly members of the family Cladoniaceae and the family name *Cetradonia*ceae has to be abandoned. The SSU rDNA of *Gymnoderma* contains seven group I and one spliceosomal intron, the positions and lengths of which are discussed.

KEY WORDS: introns, lichens, *Cetradonia*, *Gymnoderma*, SSU rDNA

INTRODUCTION

The lichen genus *Gymnoderma* Nyl. currently includes two definite species, *G. coccocarpum* Nyl. (type species) and *G. insulare* Yoshim. & Sharp. A taxonomic study of *Gymnoderma* based on multiple morphological characteristics was made by Wei and Ahti (2002) before any molecular data were available of the type species, and the genus was then placed in the Cladoniaceae. Furthermore, they also segregated *G. lineare* (A. Evans) Yoshim. & Sharp to a distinct genus *Cetradonia*. Stenroos & DePriest (1998a) had by then published a molecular phylogeny, which rather surprisingly placed *Cetradonia* (as "*Gymnoderma*") as basal to the genus *Stereocaulon* and selected members of the Parmeliaceae. Following these results a new family Cetradoniaceae was erected by Wei & Ahti (2002) for the genus. Phylogenetic analyses of the Cladoniaceae have not included *Gymnoderma* s. str., but only its segregates, *Neophyllis* F. Wilson and *Cetradonia* J. C. Wei & Ahti (Stenroos & DePriest, 1998a; Wedin, Döring & Ekman, 2000; Stenroos et al., 2002a, 2002b). Therefore, the phylogenetic position of *Gymnoderma* itself remained unclear until very recently. The SSU sequence of *G. coccocarpum*, produced for the present study, was released prior to its publication and it was used in the analyses presented by Peršoh, Beck & Rambold (2004). In their cladogram *Gymnoderma* correctly appeared in the Cladoniaceae, close to *Pycnothelia*. It was also reported that its ascus tip structure is deviating from the other studied genera in the family.

The goal of the present study is to clarify and confirm the relationships of both *Gymnoderma* and *Cetradonia*. The fact that the earlier used SSU rDNA sequence of *Cetradonia linearis* (Stenroos & DePriest, 1998a; as *Gymnoderma lineare*, GBANAF

[1] Systematic Mycology & Lichenology Laboratory, Institute of Microbiology, Academia Sinica, Bei-yi-tiao #13, Zhong-guan-cun, Beijing 100080, P. R. China.

[2] Botanical Museum, P.O. Box 7, FI-00014 Helsinki University, Finland.

* The corresponding author's e-mail: soili.stenroos@helsinki.fi

The result, based on SSU rDNA sequences of lichen-froming fungus *Cetradonia linearis* (A. Evans) J. C. Wei & T. Ahti, made by J. C.Wei's Lab, is contrary to the data, made by S. Stenroos' team, and showed that the fungus is fallen into the family Cldoniaceae. Because S. Stenroos' data from the same fungus lacks 185bp (1998a), and She also mistook the SSU rDNA sequences of phycobiont *Trebauxia erici* from *C. linearis* for the sequences of its fungus (in communication with J. C. Wei by E-mail, 2005). The manuscript of this paper wrote by J. C. Wei as corresponding author. S. Stenroos wanted to be a co-author through T.Ahti with J.C.Wei's permission. It was surprising that S. Stenroos behind J. C. Wei's back replaced J. C. Wei as corresponding author! The moral standards for scientists are honest, but some is dishonest!

This paper was originally published in *J. Hattori Bot. Lab.*, No.100: 871-880, 2006.

Table 1. GenBank accession numbers and the current family classification of all species used in the phylogenetic analysis. Species with sequences produced for the present study are marked with an asterisk.

Species	Accession Number	Current family
Bunodophoron scrobiculatum	U70958	Sphaerophoraceae
Caloplaca demissa	AF515609	Teloschistaceae
*Cetradonia linearis**	AY648095	not defined
Cladia aggregata	U72713	Cladoniaceae
Cladonia rangiferina	AF184753	Cladoniaceae
Cladonia sulphurina	AF241544	Cladoniaceae
*Gymnoderma coccocarpum**	AF523362	Cladoniaceae
Heterodea muelleri	AF184754	Cladoniaceae
Lecanora intumescens	AF091586	Lecanoraceae
Lepraria umbricola	AF517836	Stereocaulaceae
Metus conglomeratus	AF184755	Cladoniaceae
Neophyllis melacarpa	AF117981	Sphaerophoraceae
Parmelia saxatilis	AF117985	Parmeliaceae
Pilophorus strumaticus	AF184758	Cladoniaceae
Pycnothelia papillaria	AF085474	Cladoniaceae
Sphaerophorus globosus	AF117983	Sphaerophoraceae
Stereocaulon paschale	AF140236	Stereocaulaceae
Thamnolia vermicularis	AF085472	Icmadophilaceae

085470) turned out to be partially erroneous, further encouraged us to revisit the present topic.

Materials and Methods

Specimens

The voucher data for the sequenced taxa is as follows: *Gymnoderma coccocarpum*; China, Hunan Province, Zhang-yi, Mt. Mangshan, alt. 1170 m, July 9, 2001 *J. B. Chen et al. 20912*; HMAS-L). *Cetradonia linearis*; U.S.A., North Carolina, Jackson Co., July 15, 2002 *J. C. Lendemer no. 405a* and the collection was distributed in Lendemer, Lich. East. N. Amer. Exsicc. 5 (H, HMAS-L). The latter specimen was analyzed in Beijing by Q. M. Zhou in Wei's team and its duplicate in Helsinki by F. Högnabba in Stenroos's team.

In addition to the above mentioned samples, 17 other species, mostly belonging to the Lecanorales, were selected for the analyses. The sequence data for these samples were obtained from the GenBank (see Table 1 for species and GenBank accession numbers).

DNA extraction

A rapid one-tube genomic DNA extraction was used for *Gymnoderma coccocarpum* (Lee & Taylor 1990; Steiner et al. 1995) modified by us as follows: Two dried apothecia were cleaned under a dissecting microscope. Algal-free periclinal sections containing the hymenia were cut out by hand and transferred directly into a 1.5 ml eppendorf tube. The material was grinded with a pestle in liquid nitrogen until a fine powder was obtained. 3 μl deionized sterile water was added and the tube was immediately stored at −20°C. The total DNA of *Cetradonia linearis* was extracted from both the lichen thallus and an axenic culture of its mycobiont with a modified CTAB method (Rogers & Bendich, 1988).

PCR amplification and sequencing

The fungal SSU rDNA was amplified by using the following primers (Gargas & Taylor, 1992): NS17UCB (5′-CATGTCTAAGTTTAAGCAA-3′) and NS20UCB (5′-CG-TCCCTATTAATCATTACG-3′), NS21UCB (5′-GAATAATAGAATAGGACG-3′) and NS24UCB (5′-AAACCTTGTTACGACTTTTA-3′). After initial denaturation at 95°C for 3 min, amplification was run for 30 cycles (30 sec denaturation at 94°C, 40 sec annealing at 52°C, 2 min 30 sec extension at 72°C) with Taq Plus Polymerase (SABC), followed by 1 cycle of 8 min extension at 72°C. The obtained products were verified by electrophoresis on 1% agarose gels. Purified PCR products were sequenced with primers NS17UCB, NS20UCB, NS21UCB, and NS24UCB. All the sequences were obtained with an ABI 3700 automatic sequencer.

In order to avoid the possible epiphytic contaminants and ensure the specificity, DNA extraction from two copies of the specimens and one culture of the mycobiont of *Cetradonia linearis* were tested. The SSU rDNA sequences from the lichen thallus and the culture were identical. In addition, the duplicate specimen (H) of *Cetradonia linearis* was analyzed as well, and the identity of the sequence was thereby confirmed.

Phylogenetic analysis

The fungal SSU rDNA sequence of *Gymnoderma coccocarpum*, *Cetradonia linearis* as well as the 17 sequences downloaded from the GenBank were used for the analyses. *Thamnolia vermicularis* of the Icmadophilaceae as well as *Sphaerophorus*, *Bunodophoron* and *Neophyllis* of the Sphaerophoraceae were used as outgroup taxa. Major insertions, when present in the sequences, were excluded prior to the analyses.

Phylogenetic hypotheses essentially depend on homology assumptions made during the aligning process and therefore ambiguous alignments are not an option. The SSU rDNA is variable in length, and it is therefore impossible to obtain an unambiguous alignment of this gene with conventional methods. We therefore chose to use direct optimization (Wheeler 1996) as implemented in the program POY, version 3.0.6 (Gladstein & Wheeler 1997). In direct optimization the aligning process and tree search are done simultaneously. This can be done for sequences of unequal length because, similarly to base changes, also insertions and deletions are treated equally as transformations (Schulmeister et al. 2002). In the present analyses all transformations were equally weighted, following the argumentation by Frost et al. (2001), Grant (2003) and Grant & Kluge (2003).

The POY analyses were performed at CSC (Scientific Computing Ltd., Espoo, Finland), using the IBMSC parallel supercomputer IBM eServer Cluster 1600 system. Sixteen processors were used for the analyses. The command line was as follows: poi -parallel -solospawn 17 datafile -molecularmatrix 111 -maxtrees 3 -holdmaxtrees 30 -random 100 -multibuild 16 -nodiscrepancies -ratchettbr 5 -ratchettrees 2 -treefuse -fuselimit 30 -check-

Table 2. Positions and lengths of introns in the SSU rDNA gene of *Cetradonia linearis* and *Gymnoderma coccocarpum*. Positions corresponds the 5′ flanking nucleotide in both *Escherichia coli* and *Saccharomyces cerevisiae* (the latter in parentheses).

Intron positions	*Cetradonia*	*Gymnoderma*
114 (98)	192	207
471 (532)	188	204
516 (563)	280	399
789 (1001)	—	252
989 (1210)	760	—
1046 (1262)	—	226
1199 (1428)	593	422
1210 (1439)	—	224
1231 (1462)	—	112

slop 10 -fitchtrees -noleading -norandomizeoutgroup -indices -impliedalignment. Parsimony jackknife values (Farris et al. 1996) were calculated using POY with the following settings: -replicates 1000 -jackboot -notbr -jackfrequencies all -jacktree -time>jack.out.

Results and Discussion

The SSU rDNA sequences obtained from *Cetradonia linearis* and *Gymnoderma coccocarpum* in this study consisted of 3705 bp and 3765 bp, respectively. However, about 1750 bp-length product would generally be obtained when primers NS17USB-NS24USB were used to sequence fungal SSU rDNA. The SSU of both *C. linearis* and *G. coccocarpum* were therefore notably longer than normally, which is due to multiple sequence insertions (introns) in the gene. There are eight introns in the SSU rDNA of *G. coccocarpum* and five in that of *C. linearis*. The positions and lengths of these introns are shown in Table 2. All introns in *Cetradonia* and seven of those in *Gymnoderma* are of group I type. Most of the positions now reported are already known from other lichen-forming fungi (DePriest & Been 1992, Gargas et al. 1995, Stenroos & DePriest 1998b). However, the positions 471 and 989 (corresponding the 5′ flanking nucleotide in *Escherichia coli*) are reported here as new. We noticed, however, that Lutzoni & Reeb (unpubl.) have found the intron 471 in their sample of *Hymenelia epulotica* (AF279393, intron length 238 bp).

In addition to the group I introns *Gymnoderma* also possesses one spliceosomal intron (for characteristics of spliceosomal introns, see Battacharya et al. 2000), which is 112 nucleotides in length and is located at the position 1231 (corresponding the position in *Escherichia coli*, and the position 1462 in *Saccharomyces cerevisiae*). This position is a new one and adds to the following list of spliceosomal intron positions previously reported: 265, 296, 297, 300, 330, 331, 332, 390, 393, 513, 673, 882, 883, 939, 943, 1129, 1416, 1510 (as summarized by Battacharya et al. 2000). However, the position 1129 reported by Cubero et al. (2000) from *Physconia detersa*, *P. distorta*, *P. elegantula*, *P. perisidiosa*, and *P. venusta* appears to be erroneously determined. In fact, this position turned out to be

Table 3. Species with a spliceosomal intron at the position 1231 (corresponding *Escherichia coli*)

Species	Length	GenBank No.
Dimelaena oreina	75	AJ421683
Gymnoderma coccocarpum	112	AF523362
Physconia detersa	59	AJ240509
Physconia distorta	60	AJ240508
Physconia elegantula	63	AJ240507
Physconia kurokawae	59	AJ240510
Physconia perisidiosa	63	AJ240512
Physconia venusta	64	AJ240511

identical with the one we are now reporting from *Gymnoderma*, i.e. the position 1231. In Table 3 we list the lengths of this intron in each of the species we known it from. The GenBank accession numbers accidentally mixed up in Cubero et al. (2000) are also corrected here, and *Physconia kurokawae* as well as *Dimelaena oreina* are included in the list.

Spliceosomal introns are infrequent in ascomycete rDNA. Their positions are known to almost exclusively contain only spliceosomal introns but not group I introns that are numerous in other positions along the SSU of many ascomycete fungi. The position 943 is an exception as it may house both spliceosomal and group I introns (Gargas et al. 1995, Battacharya et al. 2000, Cubero et al. 2000; see also Grube et al. (1999) for the adjacent position 940). However, according to a preliminary screening we think that introns reported from position 943 should be re-evaluated and some of them possibly repositioned.

Introns are common and numerous along the SSU rDNA of many members of the lecanoralean fungi. Although introns up to five per sample are common (e.g. DePriest 1993), larger numbers are rarely found. The sample of *Gymnoderma coccocarpum* turned out to be rather exceptional in that it houses introns in as many as eight positions in its SSU rDNA. Gargas et al. (1995) and Battacharya et al. (2002) reported eight intron positions in *Lecanora dispersa* and *Physconia perisidiosa*, respectively. Stenroos (unpubl.) found the same number in *Cladonia portentosa*. This amount of introns approximately doubles the total length of the whole SSU gene in the case of *Gymnoderma* and almost so in *Cetradonia* and *Lecanora dispersa* (Gargas et al. 1995 reported a total length of 3350 bp for the latter).

The SSU rDNA sequence of *Cetradonia linearis* includes two exceptionally long insertions, the lengths of which are 760 nucleotides (position 990/1210, corresponding to *Escherichia coli* and *Saccharomyces cerevisiae*, respectively) and 593 nucleotides (position 1199/1428). Long introns have been reported in lecanoralean fungi before but rarely: DePriest (1993) and Beard & DePriest (1996) reported group I introns of ca. 400 nt in *Cladonia*. Introns longer than those now found in *Cetradonia* are known from other organisms, such as slime molds, red algae and amoebae (see summary in Gargas et al. 1995). It is necessary to study the structure of the two *Cetradonia* introns in order to determine the reason for the exceptional length. Major size variation in some group I introns has been explained by the insertion of internal open reading frames (Dujon 1989) or by the recombination of introns (Eddy & Gold 1991), for instance, but these reasons have not explained the excessive intron length in studies by DePriest (1993), at least.

The distribution of introns is sporadical, which is a result of rare gains followed by extensive losses in descendants during the evolutionary history of lichen-forming fungi (Battacharya et al. 2002). However, it has been shown that introns at a particular position are homologous and can therefore usually be used in evaluating phylogenies (Stenroos & DePriest 1998b, Myllys et al. 1999, Thell 1999, Thell & Miao 1999, Grube et al. 1999, Myllys et al. 2001, Battacharya et al. 2002). Some taxonomic groups are more rich in introns than others. Cladoniaceae, for instance, is rich in introns, but not uniformly. The genera *Cetradonia, Cladonia, Cladia, Gymnoderma, Heterodea,* and *Notocladonia* contain them, whereas introns appear to be absent from *Carassea, Metus* and *Pilophorus*. Spliceosomal introns appear in *Cladonia, Heterodea, Gymnoderma* and *Notocladonia* (Stenroos et al., unpubl.).

Poy analyses including 19 taxa resulted in two equally parsimonious optimizations (and hence trees) of 339 steps. In the strict consensus tree shown in Fig. 1 the Cladoniaceae is monophyletic and includes the genera *Heterodea, Cladia, Pilophorus, Metus, Cladonia, Gymnoderma, Cetradonia, Carassea* and *Pycnothelia*. Cladoniaceae is sister to the Stereocaulaceae, which includes *Stereocaulon* and *Lepraria*. These results are in accordance with

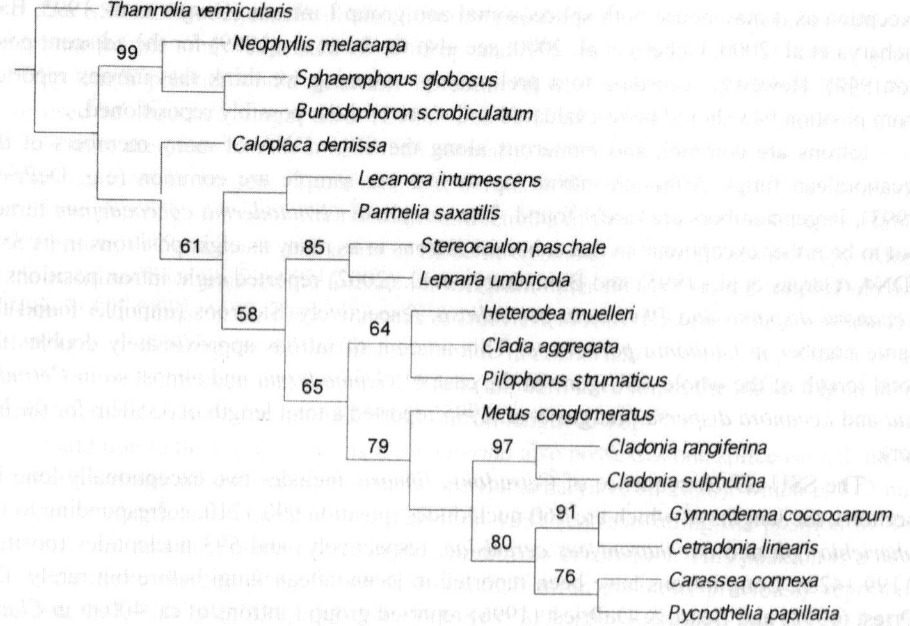

Fig. 1. Strict consensus of two equally parsimonious optimizations based on SSU rDNA sequence data. Tree length 339 steps; CI 32; RI 85. Jacknife support values are shown at nodes.

analyses based on larger data sets (Stenroos et al. 2002a, Myllys et al. 2005, for instance). *Cetradonia linearis* groups with *Gymnoderma coccocarpum* first and this topology is supported by a Jacknife value 91.

The position of *Cetradonia linearis* well outside of the Cladoniaceae suggested by Stenroos & DePriest (1998a; AF08540) is based on a sequence deviating rather notably from our sample of the same taxon. This difference in sequence is probably due to a contamination when amplifying the rear segment of the SSU of the sample AF085470. In fact, when using the BLAST search available in GenBank (http://www.ncbi.nlm.nih.gov/blast/) it turned out that the rear segment of AF085470 showed the highest similarity with *Capnobotryella* spp. of mitosporic Ascomycota. When this segment was removed and the remaining portion was tested with BLAST, the search yielded mostly Cladoniaceae, and particularly our present sample of *Cetradonia* (AY648095). The remaining SSU, from which the rear segment was removed, was also tested in a preliminary phylogenetic analysis, which confirmed a correct placement of this portion as sister to *Gymnoderma coccocarpum* within the Cladoniaceae (results not shown).

Our results suggest that both *Gymnoderma*, represented by *G. coccocarpum*, and *Cetradonia*, represented by *C. linearis*, should be placed in the family Cladoniaceae. Therefore the family name Cetradoniaceae needs to be reduced to a synonym of the Cladoniaceae. However, we would keep the two genera *Cetradonia* and *Gymnoderma* separate for now. This view cannot be tested with sequence analyses due to the lack of more material, but it is supported by the following characteristics. The asci in *Cetradonia* are very narrow, with uniseriate spores, while in *Gymnoderma* the asci are broader, with thicker wall and biseriate spores. Peršoh et al. (2004) found that *G. coccocarpum* has amyloid tholi lacking an inamyloid axial body, although all the other species of the Cladoniineae clade in their study have a tube-like amyloid reaction pattern in the ascus apex (*Porpidia* type s. lat.). *Cetradonia* seems to belong to the latter group.

The total list of genera included in the Cladoniaceae is still incomplete. As demonstrated in Table 4 the Cladoniaceae provisionally includes 14 genera and the status of three of them still need to be confirmed with proper methods. It is to be noted that the Australian species *Ramalea cochleata* was recently segregated into *Notocladonia* S. Hammer (Ham-

Table 4. Genera currently assigned to the Cladoniaceae. Those marked with an asterisk (*) are still awaiting for DNA analyses and their inclusion in the Cladoniaceae is provisional.

Genus	Number of species	Genus	Number of species
*Calathaspis**	1	*Heterodea*	2
*Calycidium**	1	*Metus*	3
Carassea	1	*Notocladonia*	1
Cetradonia	1	*Pilophorus*	11
Cladonia	ca. 450	*Pycnothelia*	2
Cladia	14	*Squamella**	1
Gymnoderma	2 (4?)	*Thysanothecium*	2

mer 2003), and on morphological grounds the rest of *Ramalea* is not expected to belong to the Cladoniaceae. *Calycidium* Stirt. from New Zealand is provisionally listed here following the suggestion by Wedin (2002). Based on material seen by the author Ahti in New Zealand, as well, *Calycidium* is not very different from *Gymnoderma*.

A poorly known African species known as *Heterodea madagascarea* Nyl. presents an additional problem (Wei & Ahti 2002). It resembles *Gymnoderma coccocarpum* but is not identical. H. Krog and E. Timdal recollected it in 1991 in Madagascar but the material was too old to yield usable DNA. The final generic placement of *Heterodea madagascarea* therefore remains unresolved. Yet another problem is the status of *Cladonia botryocephala* Hepp (Wei & Ahti 2002: 28), which may be one more species of *Gymnoderma*, but additional fresh material is required to clarify its position.

ACKNOWLEDGEMENTS

We are particularly indebted to Professor J. B. Chen and Dr. J. C. Lendemer for providing the fresh specimens of *Gymnoderma coccocarpum* and *Cetradonia linearis*, respectively. Special thanks must be given also to Prof. Y. M. Jiang for supplying the culture collection of the mycobiont isolated from *Cetradonia linearis*. The present work was supported by The National Natural Science Foundation of China (30499340) and a grant from the Academy of Finland (211172).

REFERENCES

Battacharya, D., F. Lutzoni, V. Reeb, D. Simon, J. Nason and F. Fernandez. 2000. Widespread occurrence of spliceosomal introns in the rDNA genes of Ascomycetes. Mol. Biol. Evol. 17: 1971–1984.

Battacharya, D., T. Friedl and G. Helms. 2002. Vertical evolution and intragenic spread of lichen-fungal group I introns. J. Mol. Evol. 55: 74–84.

Beard, K. and P. T. DePriest. 1996. Genetic variation within and among mats of the reindeer lichen, *Cladina subtenuis*. Lichenologist 28: 171–182.

Cubero, O. E., P. D. Bridge and A. Crespo 2000. Terminal sequence conservation identifies spliceosomal introns in ascomycete 18S RNA genes. Mol. Biol. Evol. 17: 751–756.

DePriest, P. T. 1993. Small subunit rDNA variation in a population of lichen fungi due to optional group-I introns. Gene 134: 67–74.

DePriest, P. T. and M. D. Been. 1992. Numerous group I introns with variable distributions in the ribosomal DNA of a lichen fungus. J. Mol. Biol. 228: 315–321.

Dujon, B. 1989. Group I introns as mobile genetic elements: facts and mechanistic speculations—a review. Gene 82: 91–114.

Eddy, S. R. and L. Gold. 1991. The phage T4 *nrdB* intron: a deletion mutant of a version found in the wild. Genes Dev. 5: 1032–1041.

Farris, J. S., Albert, V. A., Kallersjö, M., Lipscomb, D. & Kluge, A. G. 1996. Parsimony jackknifing outperforms neighbor-joining. Cladistics 12: 99–124.

Frost, D. R., M. T. Rodrigues, T. Grat and T. A. Titus. 2001. Phylogenetics of the lizard genus *Tropidurus* (Squamata: Tropiduridae: Tropidurinae): direct optimization, descriptive efficiency, and sensitivity analysis of congruence between molecular data and morphology. Mol. Phyl. Evol. 21: 352–371.

Gargas, A. and J. W. Taylor. 1992. Polymerase chain reaction (PCR) primers for amplifying and sequencing nuclear 18S rDNA from lichenized fungi. Mycologia 84: 589–592.

Gargas, A., P. T. DePriest and J. W. Taylor. 1995. Positions of multiple insertions in SSU rDNA of lichen-forming fungi. Mol. Biol. Evol. 12: 208–218.

Gladstein, D. S. and W. C. Wheeler. 1997. POY: the optimization of alignment characters. Program and documentation. American Museum of Natural History, New York. (ftp.amnh.org/pub/molecular).

Grant, T. 2003. Against sensitivity analysis in phylogenetic systematics. *In*: Muona, J. (ed.), Abstracts of the annual meeting of the Willi Hennig Society. Cladistics 19: 148–163.

Grant, T. and A. Kluge. 2003. Data exploration in phylogenetic inference: scientific, heuristic, or neither. Cladistics 19: 397–418.

Grube, M., B. Gutmann, U. Arup., A. de los Rios, J.-E. Mattsson and M. Wedin. 1999. An exceptional group-I intron-like insertion in the SSU rDNA of lichen mycobionts. Current Genetics 35: 536–541.

Hammer, S. 2003: *Notocladonia*, a new genus in the Cladoniaceae. Bryologist 106: 162–167.

Lee, S. B. and J. W. Taylor. 1990. Isolation of DNA from fungal mycelia and single spores. *In*: M. A. Innis, D. H. Gelfand, J. J. Sninsky and T. J. White (eds), PCR protocols—a guide to methods and applications: 282–287. Academic Press, San Diego.

Myllys, L., F. Högnabba, K. Lohtander, A. Thell, S. Stenroos and J. Hyvönen. 2005. Phylogenetic relationships of Stereocaulaceae based on simultaneous analysis of beta-tubulin, GAPDH and SSU rDNA sequences. Taxon 54: 605–618.

Myllys, L., K. Lohtander, M. Kallersjö and A. Tehler. 1999. Sequence insertions and ITS provide congruent information on *Roccella canariensis* and *R. tuberculata* (Arthoniales, Euascomycetes) phylogeny. Mol. Phyl. Evol. 12: 295–309.

Myllys, L., K. Lohtander and A. Tehler. 2001. ß-tubulin, ITS and group I intron sequences challenge the species pair concept in *Physcia aipolia* and *P. caesia*. Mycologia 93: 335–343.

Peršoh, D., A. Beck and G. Rambold. 2004. The distribution of ascus types and photobiontal selection in Lecanoromycetes (Ascomycota) against the background of a revised SSU nrDNA phylogeny. Mycol. Progr. 3: 103–121.

Rogers, S. O. and A. J. Bendich. 1988. Extraction of DNA from plant tissues. Plant Molecular Biology Manual A6:1–10. Kluwer Academic Publishers, Dordrecht.

Schulmeister, S., W. C. Wheeler and J. M. Carpenter. 2002. Simultaneous analysis of the basal lineages of Hymenoptera (Insecta) using sensitivity analysis. Cladistics 18: 455–484.

Steiner, J. J., C. J. Poklemba, R. G. Fjellstrom and L. F. Elliott. 1995. A rapid one-tube genomic DNA extraction process for PCR and RAPD analyses. Nucl. Acids Res. 23: 2569–2570.

Stenroos, S. and P. T. DePriest. 1998a. SSU rDNA phylogeny of cladoniiform lichens. Amer. J. Bot. 85: 1548–1559.

Stenroos, S. & P. T. DePriest (1998b) Small insertions at a shared position in the SSU rDNA of Lecanorales (lichen-forming Ascomycetes). Current Genetics 33: 124–130.

Stenroos, S., J. Hyvönen, L. Myllys, A. Thell and T. Ahti. 2002a. Phylogeny of the genus *Cladonia* s.lat. (Cladoniaceae, Ascomycetes) inferred from molecular, morphological, and chemical data. Cladistics 18: 237–278.

Stenroos, S., L. Myllys, A. Thell and J. Hyvönen. 2002b. Phylogenetic hypotheses: Cladoniaceae, Stereocaulaceae, Baeomycetaceae, and Icmadophilaceae revisited. Mycol. Progr. 1: 267–282.

Thell, A. 1999. Group I intron versus ITS sequences in phylogeny of cetrarioid lichens. Lichenologist 31: 441–449.

Thell, A. and V. Miao. 1999. Phylogenetic analysis of ITS and group I intron sequences from European and North American samples of cetrarioid lichen. Ann. Bot. Fenn. 35: 275–286.

Wedin, M. 2002. The genus *Calycidium* Stirt. Lichenologist 34: 63–69.

Wedin, M., H. Döring and S. Ekman. 2000. Molecular phylogeny of the lichen families Cladoniaceae, Sphaerophoraceae, and Stereocaulaceae (Lecanorales, Ascomycotina). Lichenologist 32: 171–187.

Wei, J. C. and T. Ahti. 2002. *Cetradonia*, a new genus in the new family Cetradoniaceae (Lecanorales, Ascomycota). Lichenologist 34: 19–31.

Wheeler, W. C. 1996. Optimization alignment: the end of multiple sequence alignment in phylogenetics? Cladistics 12: 1–9.

A study of the pruinose species of *Hypogymnia* (*Parmeliaceae*, Ascomycota) from China

Xin-Li WEI and Jiang-Chun WEI

Abstract: Six pruinose species of *Hypogymnia* are reported in this paper, including one new species *Hypogymnia pruinoidea*. The type of *Hypogymnia pseudopruinosa* was found to be a mixture with *H. laccata*. *Hypogymnia pseudopruinosa* is therefore typified with a lectotype, and the description of *H. pseudopruinosa* is revised. Distributions of the six pruinose species are given and discussed. Comments on differences and similarities between pruinose species of *Hypogymnia* are made. Diagnostic characters of each species, and a key to the pruinose species of *Hypogymnia* in China, are also provided.

Key words: *H. pruinoidea*, *H. pseudopruinosa*, lichen substances, pruina

Accepted for publication 6 *June* 2012

Introduction

Although over 100 species of *Hypogymnia* (Nyl.) Nyl. have been reported worldwide, according to the latest statistics from Index Fungorum, only five species have pruina on the upper surface of the lobes (Wei & Jiang 1980; Chen 1994; Wei & Wei 2005). So far, all of these species are endemic to China. Thus, during our studies on the taxonomy of Chinese species of *Hypogymnia*, the pruinose species group attracted our attention. A new pruinose species collected from Mt. Taibaishan, Shaanxi Province, is described in this paper. The type of *Hypogymnia pseudopruinosa* X. L. Wei & J. C. Wei (Wei & Wei 2005) was found to be a mixture with *H. laccata* J. C. Wei & Y. M. Jiang, necessitating the lectotypification of *H. pseudopruinosa*.

X-L. Wei and J-C. Wei (corresponding author): State Key Laboratory of Mycology, Institute of Microbiology, Chinese Academy of Sciences, Beijing 100101, China. Email: weijc2004@126.com

Materials and Methods

Specimens treated here are preserved in the Lichen Section of Herbarium Mycologicum Academiae Sinicae (HMAS-L). A dissecting microscope (ZEISS Stemi SV 11), compound microscope (OPTON III), and scanning electron microscope (SEM, FEI, Quanta 200) were used to study morphology. Colour test reagents (10% aqueous KOH and concentrated alcoholic *p*-phenylenediamine) and thin-layer chromatography (TLC) were used for the detection of lichen substances (Culberson & Kristinsson 1970; Culberson 1972).

Taxonomy

Lichen thalli often have a whitish, flour-like surface covering called pruina. It is sometimes treated as a diagnostic character to delimit lichen species (Giralt *et al.* 2001), although in some genera it is considered to be highly variable and unreliable. To date, six pruinose species of *Hypogymnia*, *H. pruinoidea* X. L. Wei & J. C. Wei, sp. nov, *H. lijiangensis* J. B. Chen, *H. pruinosa* J. C. Wei & Y. M. Jiang, *H. pseudopruinosa* X. L. Wei & J. C. Wei, *H. subfarinacea* X. L. Wei & J. C. Wei, and *H. subpruinosa* J. B. Chen, all endemic to China, have been found and reported.

This paper was originally published in *The Lichenologist*, 44(6): 783-793, 2012.

Key to the pruinose species of *Hypogymnia* in China

1 Soredia present . . . . **H. subfarinacea**
Soredia absent 2

2(1) Medulla PD+ orange-red, containing physodalic acid in medulla 3
Medulla PD− 4

3(2) Lobes divergent, with black borders; holes mainly on axils, rarely on lower surface . . . . **H. lijiangensis**
Lobes crowded, without black borders; holes on lobe tips. . . . . . . **H. subpruinosa**

4(2) Pruina layer dense, with a sharp demarcation between pruinose and epruinose regions; containing alectoronic acid in thallus . . . . **H. pruinosa**
Pruina layer thin; lacking alectoronic acid in thallus 5

5(4) Thallus texture cartilaginous to slightly papery; lobes subdichotomously branched; holes mainly present on lobe tips . . . . **H. pseudopruinosa**
Thallus texture cartilaginous; lobes isodichotomously branched; holes present on lobe tips, axils, and lower surfaces . . . . **H. pruinoidea**

The Species

Hypogymnia lijiangensis J. B. Chen

Acta Mycologica Sinica **13** (2)**:** 109 (1994); type: China, Yunnan, Lijiang county, 1981, *X. Y. Wang, X. Xiao, J. J. Su* 6991 (HMAS—holotype).

(Figs 1A, 2A, 3A)

This species is characterized by discrete lobes with black borders, pruina limited to lobe tips, and presence of physodalic acid in the thallus. It is similar to *H. subfarinacea* X. L. Wei & J. C. Wei and *H. subpruinosa* J. B. Chen in the location of pruina and chemistry, but clearly differs in morphology. *Hypogymnia lijiangensis* has separated lobes, holes mainly in the axils, lacks soredia on the upper surface, and has a cartilaginous texture; *H. subfarinacea* has granular soredia on the upper surface, being easily separated from the other pruinose species of *Hypogymnia*, cartilaginous texture but slightly papery; and *H. subpruinosa* has crowded lobes with holes only on the lobe tips.

Chemistry. Thallus cortex K+ yellow; medulla K+ pale reddish brown, PD+ orange-red. TLC: atranorin, physodic acid, physodalic acid, 3-hydroxyphysodic acid, protocetraric acid.

Known distribution. Sichuan, Tibet and Yunnan provinces of China (Fig. 4).

Selected specimens examined. **China:** *Sichuan*: Xiangcheng, Mt. Wumingshan, 1983, *J. J. Su, H. A. Wen, B. Li* 6036 (HMAS-L). *Tibet*: Riwoqê county, 1976, *Y. C. Zong, Y. Z. Liao* 256-3 (HMAS-L). *Yunnan*: Lijiang county, Yulong Snow Mountain, alt. 3060 m, 1981, *X. Y. Wang, X. Xiao, J. J. Su* 5124 (HMAS-L); Shangri-La county, Mt. Tianbaoshan, alt. 3700 m, 1981, *X. Y. Wang, X. Xiao, J. J. Su* 5189 (HMAS-L).

Hypogymnia pruinoidea X. L. Wei & J. C. Wei sp. nov.

MycoBank No.: MB 564892

Species characterized by isodichotomously branched lobes, pruinose lobe tips and upper surface, and abundant perforations present on lobe tips, axils, and lower surface.

Type: China, Shaanxi, Mt. Taibaishan, alt. 2800 m, on *Abies* trunk, 3 August 2005, *X. L. Wei* 1727 (HMAS—holotype).

(Figs 1B, 2B, 3B)

Thallus foliose, up to 6 cm wide, texture cartilaginous, loosely adnate, with flat, crowded, mostly isodichotomously branched, hollow lobes 0·5–1·0 mm wide and 0·5–2·0 mm long, apices blunt; upper surface greyish green, rugose, dull, thinly pruinose or pruina limited to the lobe tips, with a sharp demarcation between the pruinose and epruinose

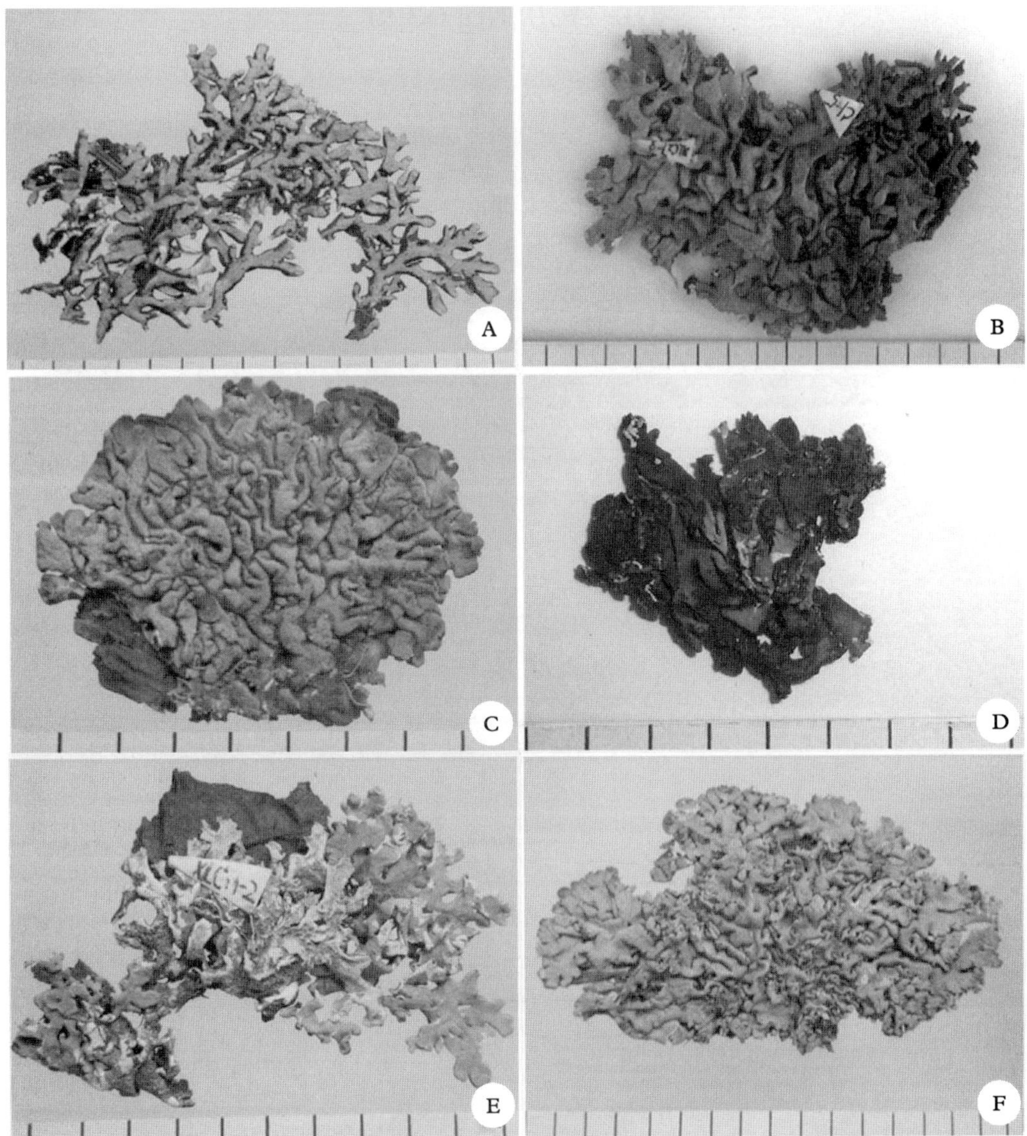

FIG. 1. *Hypogymnia* species. A, *H. lijiangensis* (holotype); B, *H. pruinoidea* (holotype); C, of *H. pruinosa* (holotype); D, *H. pseudopruinosa* (lectotype); E, *H. subfarinacea* (holotype); F, *H. subpruinosa* (holotype). Scale = 5 mm increments.

region; *soredia* and *isidia* lacking; lobules present; lower surface black, pale brown at lobe tips, rugose, dull, perforations present on many lobe tips, axils, and lower surfaces, the holes not rimmed; *medulla* hollow, both the ceiling and floor of the cavity white to pale dirty brown.

Apothecia not seen.

Pycnidia mostly at lobe tips, immersed to slightly protruding, black, mostly punctiform, aggregated; *conidia* fusiform, simple, 6·0–7·5 × 1·0 μm.

Chemistry. Reagent tests: cortex K+ yellow, P−; medulla K+ pale reddish brown, KC+

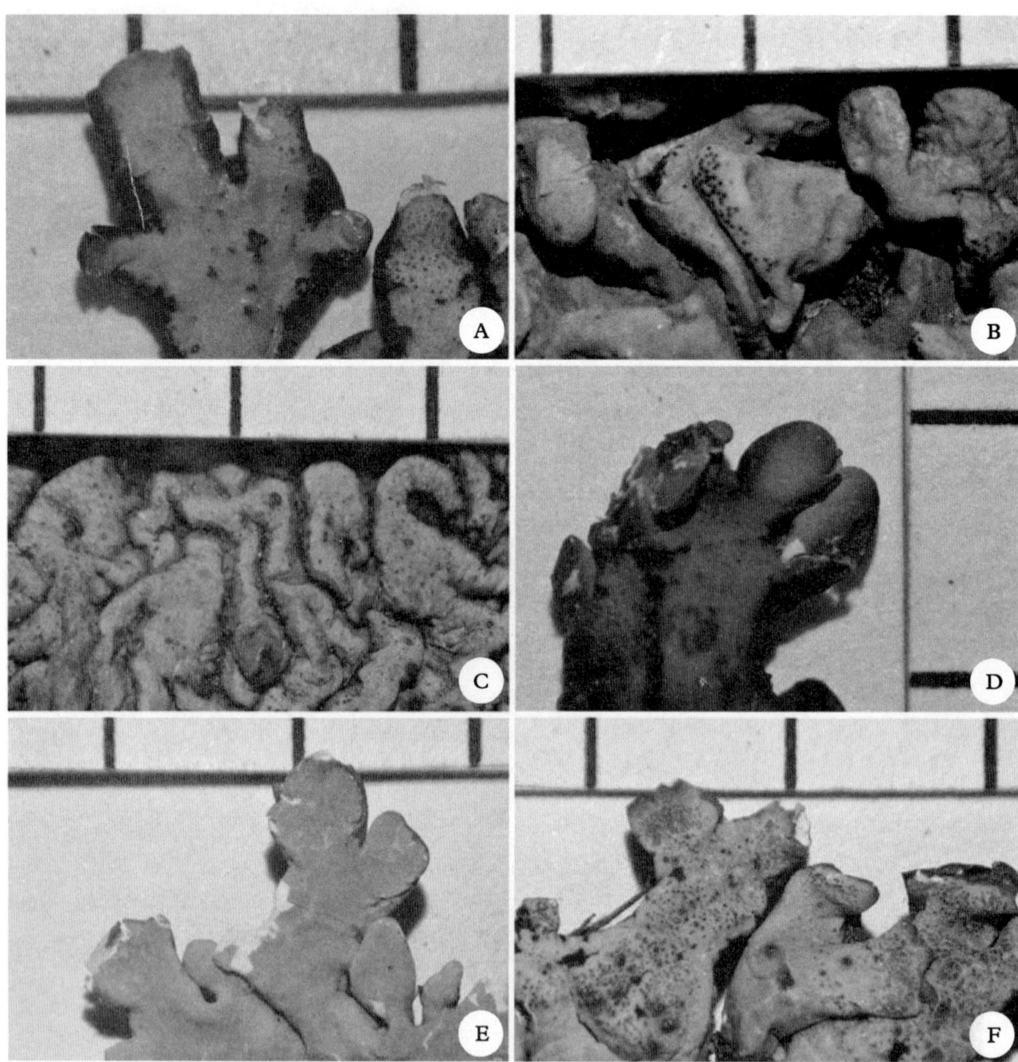

FIG. 2. Pruina of *Hypogymnia* species. A, *H. lijiangensis* (holotype); B, *H. pruinoidea* (holotype); C, *H. pruinosa* (holotype); D, *H. pseudopruinosa* (lectotype); E, *H. subfarinacea* (holotype); F, *H. subpruinosa* (holotype). Scale = 5 mm increments.

pale pink, P−. All specimens contain atranorin, physodic and 3-hydroxyphysodic acids, most specimens also contain vittatolic acid.

Etymology. The epithet 'pruinoidea' refers to the pruinose lobe tips and upper surface in this species.

Habitat, distribution and substratum. *Hypogymnia pruinoidea* occurs on bark and wood, usually of *Abies* in Mt. Taibaishan, Shaanxi province (Fig. 4). The high-elevation mountain habitats are cool and moist, supporting lichen-rich forests, woodlands and subalpine areas. These areas are rich in *Hypogymnia* species such as *H. arcuata* Tchaban. & McCune, *H. austerodes* (Nyl.) Räsänen, *H. flavida* McCune & Obermayer, *H. hypotrypa* (Nyl.) Rass., and others.

Remarks. Morphologically the new species resembles *H. pseudopruinosa*, but differs in

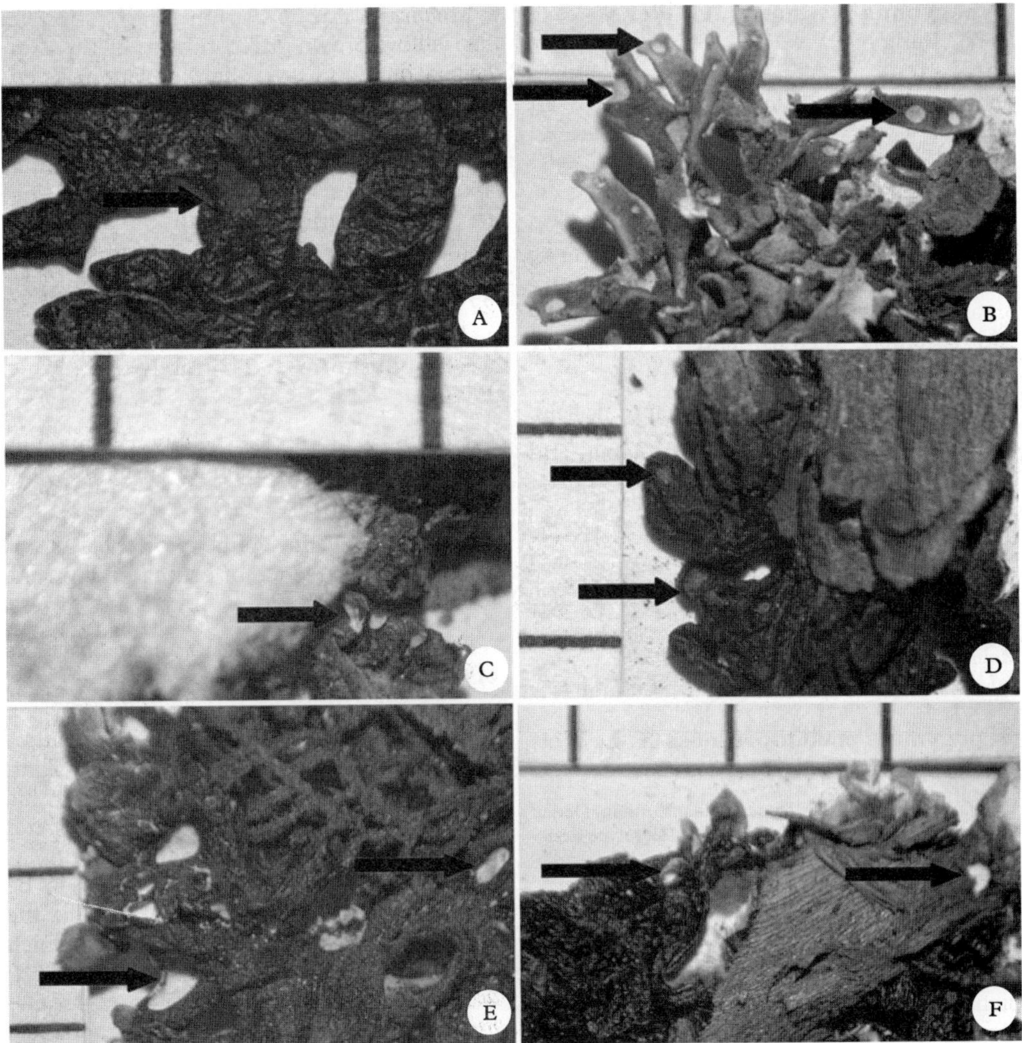

FIG. 3. *Hypogymnia* species, location of holes indicated by arrows. A, *H. lijiangensis* (holotype); B, *H. pruinoidea* (holotype); C, *H. pruinosa* (holotype); D, *H. pseudopruinosa* (lectotype); E, *H. subfarinacea* (holotype); F, *H. subpruinosa* (holotype). Scale = 5 mm.

having a thallus with a more cartilaginous texture, isodichotomously branched lobes, and abundant perforations present on lobe tips, axils, and lower surface, whereas *H. pseudopruinosa* is characterized by a thallus with a cartilaginous to slightly papery texture, subdichotomously branched lobes, and perforations mainly present on lobe tips. The new species also resembles *H. pruinosa*, but differs by perforations sometimes grouped on the lower surface and containing physodic and 3-hydroxyphysodic acids. *Hypogymnia pruinosa* is characterized by perforations mainly at lobe tips, and contains alectoronic acid.

Specimens examined. **China:** *Shaanxi Province*: Mt. Taibaishan, 2800 m, on trunk of *Abies* tree, 2005, *X. L. Wei* 1728, 1729; 3136 m, on trunk of *Abies* tree, 2005, *X. L. Wei* 1662, 1663, 1664 (HMAS-L).

Hypogymnia pruinosa J. C. Wei & Y. M. Jiang

Acta Phytotax. Sin. **18** (3): 386 (1980); type: China, Tibet, Changdu, 1976, *Y. C. Zong, Y. Z. Liao* 215 (HMAS—holotype; HMAS—isotype).

(Figs 1C, 2C, 3C)

This species is characterized by almost circular thalli, wide and short lobes, a layer of dense pruina on the upper surface, presence of a sharp demarcation between pruinose and epruinose regions, and in containing alectoronic acid in the thallus.

Chemistry. TLC: atranorin, alectoronic acid.

Distribution. Shaanxi, Sichuan, Tibet, and Yunnan provinces of China (Fig. 4).

Selected specimens examined. **China:** *Shaanxi*: Mei county, Mt. Taibaishan, 2005, *X. L. Wei* 1624 (HMAS-L). *Sichuan*: Kangding county, Minya Konka, alt. 3600 m, 1982, *X. Y. Wang, X. Xiao, B. Li* 9438 (HMAS-L). *Tibet*: Linzhi county, alt. 3011m, 2004, *X. L. Wei* 1041 (HMAS-L). *Yunnan*: Lijiang county, Yulong Snow Mountain, alt. 3100 m, 1981, *X. Y. Wang, X. Xiao, J. J. Su* 4829 (HMAS-L).

Hypogymnia pseudopruinosa X. L. Wei & J. C. Wei

Mycotaxon **94:** 155 (2005); type: China, Yunnan, Dêqên county, 1981, *X. L. Wei* 7606 (HMAS—lectotype designated here).

(Figs 1D, 2D, 3D)

Thallus foliose, tightly appressed, texture cartilaginous to slightly papery; lobes subdichotomously branched, 1–2 mm wide, 5 mm long; dense layer of pruina limited to the lobe tips; upper surface greyish green to dark brownish yellow, partly black, rugose; *isidia*, *soredia* and lobules lacking; lower surface black, brown near the apices, rugose, and perforations mainly present on the lobe tips; *medulla* hollow, the ceiling of the cavity white to pale brown, and floor of the cavity white to dark brown.

Upper cortex prosoplectenchymatous, pale yellow, 12·0–14·5 μm thick; algal layer 10·0–14·5 μm thick, consisting of green, subspherical cells of 7·5–9·5 μm diam.; hyphae in medulla colourless, *c.* 2 μm diam.; lower cortex prosoplectenchymatous, pale yellow, 10 μm thick.

Apothecia rare, 1–3 mm diam., stipitate; *disc* yellow-brown or red-brown, glossy, concave at first and then slightly plane with entire and thin margin; *epithecium* brown, 7–9 μm thick; *hymenium* colourless, 32–36 μm thick; *asci* clavate, 23·5–25·0 × 8–14 μm, 8-spored; spores simple, colourless, ellipsoid to nearly spherical, 5·5–7·0 × 3·5–4·0 μm; *paraphyses* linear, septate, 2 μm wide, slightly swollen at the tips; *hypothecium* colourless, 27–36 μm thick.

Pycnidia common, brown to black; *conidia* weakly bifusiform, *c.* 2·5 (rarely to 5 μm) × 1·0 μm.

Chemistry. Cortex K+ yellow, P–; medulla K+ yellow, P–; containing atranorin and physodic acid.

Etymology. The epithet '*pseudopruinosa*' refers to the presence of pruina only on the lobe tips.

Distribution. Shaanxi and Yunnan provinces of China (Fig. 4).

Remarks. This species resembles *H. macrospora* (Zhao 1964; Wei 1991) at first sight, but differs by having a dense layer of pruina limited to the tips of the lobes, holes not rimmed, and smaller ascospores. The holotype of *H. pseudopruinosa* was found to be mixed with *H. laccata*. Some lobes of the two species overlapped each other, which led to some phenotypic characters of *H. laccata* being wrongly assigned to *H. pseudopruinosa*, whereas some diagnostic characters were actually based on both species. The diagnostic characters mistakenly attributed to *H. pseudopruinosa* include the glossy upper surface and PD+ medulla (physodalic acid), which were actually based on *H. laccata*. A lectotype is chosen here (ICBN Art. 9.9) from the separated true *H. pseudopruinosa* portion of the original holotype and isotype. The *H. laccata* in the mixture has been separated out as specimen no. 7606-1. When this species was published as new, it was based on a single specimen from Yunnan Province but, during our recent research, at least four additional specimens were collected from Shaanxi province.

FIG. 4. Distribution of the six pruinose species of *Hypogymnia*. *H. lijiangensis* (○), *H. pruinoidea* (■), *H. pruinosa* (●), *H. pseudopruinosa* (Δ), *H. subfarinacea* (▲), *H. subpruinosa* (□).

Selected specimens examined. **China:** *Shaanxi*: Mei county, Mt. Taibaishan, 1988, *C. H. Ma* 074 (HMAS-L); 2011, *X. L. Wei* w11139, w11143, w11150 (HMAS-L).

Hypogymnia subfarinacea X. L. Wei & J. C. Wei

Mycotaxon **94:** 156 (2005); type: China, Sichuan, Nanping county, Jiuzhai Gou, 1983, *X. Y. Wang & X. Xiao* 10582 (HMAS—holotype).

(Figs 1E, 2E, 3E)

This species superficially resembles *H. farinacea* Zopf, but differs in having more separated lobes, the pruina limited to the lobe tips, and containing physodalic acid; *H. farinacea* contains atranorin, chloroatranorin, physodic acid, 3-hydroxyphysodic acid, and 2′-*O*-methylphysodic acid (McCune 2002).

Chemistry. Cortex K+ yellow, P−; medulla K+ yellow, P+ orange-yellow → orange-red. TLC: atranorin, physodic acid, physodalic acid, 3-hydroxyphysodic acid, and protocetraric acids.

Distribution. Sichuan and Yunnan provinces of China (Fig. 4).

Specimens examined. **China:** *Yunnan*: Lijiang county, Yulong Snow Mountain, alt. 2900 m, 1981, *X. Y. Wang, X. Xiao, J. J. Su* 4892 (HMAS-L); *ibid.*, alt. 3100 m, 1981, *X. Y. Wang, X. Xiao, J. J. Su* 6591 (HMAS-L).

Hypogymnia subpruinosa J. B. Chen

Acta Mycologica Sinica **13** (2): 107 (1994); type: China, Yunnan, Zhongdian county, 1981, *X. Y. Wang, X. Xiao, J. J. Su* 7094 (HMAS—holotype).

(Figs 1F, 2F, 3F)

This species is characterized by wide and short lobes with round tips, strongly rugose upper surface, pruina limited near to the lobe tips, and presence of physodalic acid in the thallus. It is similar to *H. pruinosa* in external appearance, but the latter has pruina all over the upper surface and contains alectoronic acid.

TABLE 1. *Comparison of phenotypic characters of the six pruinose species of* Hypogymnia *from China*

	H. lijiangensis	*H. pruinoidea*	*H. pruinosa*	*H. pseudopruinosa*	*H. subfarinacea*	*H. subpruinosa*
Texture						
Cartilaginous	+	+	+	+	+	+
Papery	−	−	−	+	+	−
Lobes*						
Crowded	−	−	+	−	−	+
Isodichotomous	−	+	−	−	−	−
Subdichotomous	+	−	+	+	+	+
Soredia						
Present	−	−	−	−	+	−
Holes (on)						
Lobe tip	−	+	+	+	−	+
Axil	+	+	−	−	+	−
Lower surface	−	+	−	−	+	−
Medulla						
Ceiling colour†	+	+	+	+	+	+
Floor colour 1	−	+	−	−	−	−
Floor colour 2	+	−	+	+	+	+
PD+	+	−	−	−	+	+

* Isodichotomous: isodichotomously branched; subdichotomous: subdichotomously branched

† Ceiling colour: white to pale brown in all species; floor colour 1: white to pale brown; floor colour 2: pale to dark brown

Chemistry. Cortex K+ yellow; medulla K+ pale reddish brown, PD+ orange-red. TLC: atranorin, physodic acid, physodalic acid, 3-hydroxyphysodic acid, and protocetraric acids.

Distribution. Heilongjiang, Shaanxi, Sichuan, Tibet, and Yunnan provinces of China (Fig. 4).

Selected specimens examined. **China:** *Heilongjiang*: Dailing, s. n. (HMAS-L). *Shaanxi*: Mt. Taibaishan, alt. 3280 m, 2005, *X. L. Wei* 1986 (HMAS-L). *Sichuan*: Maerkang county, Mt. Mengbishan, alt. 3700 m, *X. Y. Wang & X. Xiao* 11503 (HMAS-L). *Tibet*: Linzhi county, alt. 3010 m, *X. L. Wei* 1036 (HMAS-L). *Yunnan*: Shangri-La county, Mt. Tianbaoshan, alt. 3800 m, 1981, *J. J. Su, X. Xiao, X. Y. Wang* 6808 (HMAS-L).

Discussion

Of the six pruinose species of *Hypogymnia*, only *H. subfarinacea* has granular soredia on the upper surface, easily separating it from the other five. Presence or absence of physodalic acid, with the result of the medulla being PD+ orange-red or PD−, respectively, is sometimes a significant diagnostic character to delimit species of *Hypogymnia*. Two pruinose species *H. lijiangensis* and *H. subpruinosa*, have a PD+ medulla. *Hypogymnia lijiangensis* has separate rather than contiguous lobes with black borders, and holes mainly in the axils, rarely on the lower surface; *H. subpruinosa* has crowded lobes, without black borders, and holes only on the lobe tips. The three species with a PD− medulla are chemically distinct: alectoronic acid is present in *H. pruinosa*, but absent in the other two species, *H. pseudopruinosa* and *H.pruinoidea*. *Hypogymnia pseudopruinosa* superficially resembles the epruinose species, *H. macrospora*, but differs by having a dense layer of pruina limited to the tips of the lobes and smaller ascospores. Compared with *H. pruinoidea*, the thallus of *H. pseudopruinosa* has a cartilaginous to papery texture, subdichotomously branched lobes, and perforations mainly present on lobe tips, whereas *H. prui-*

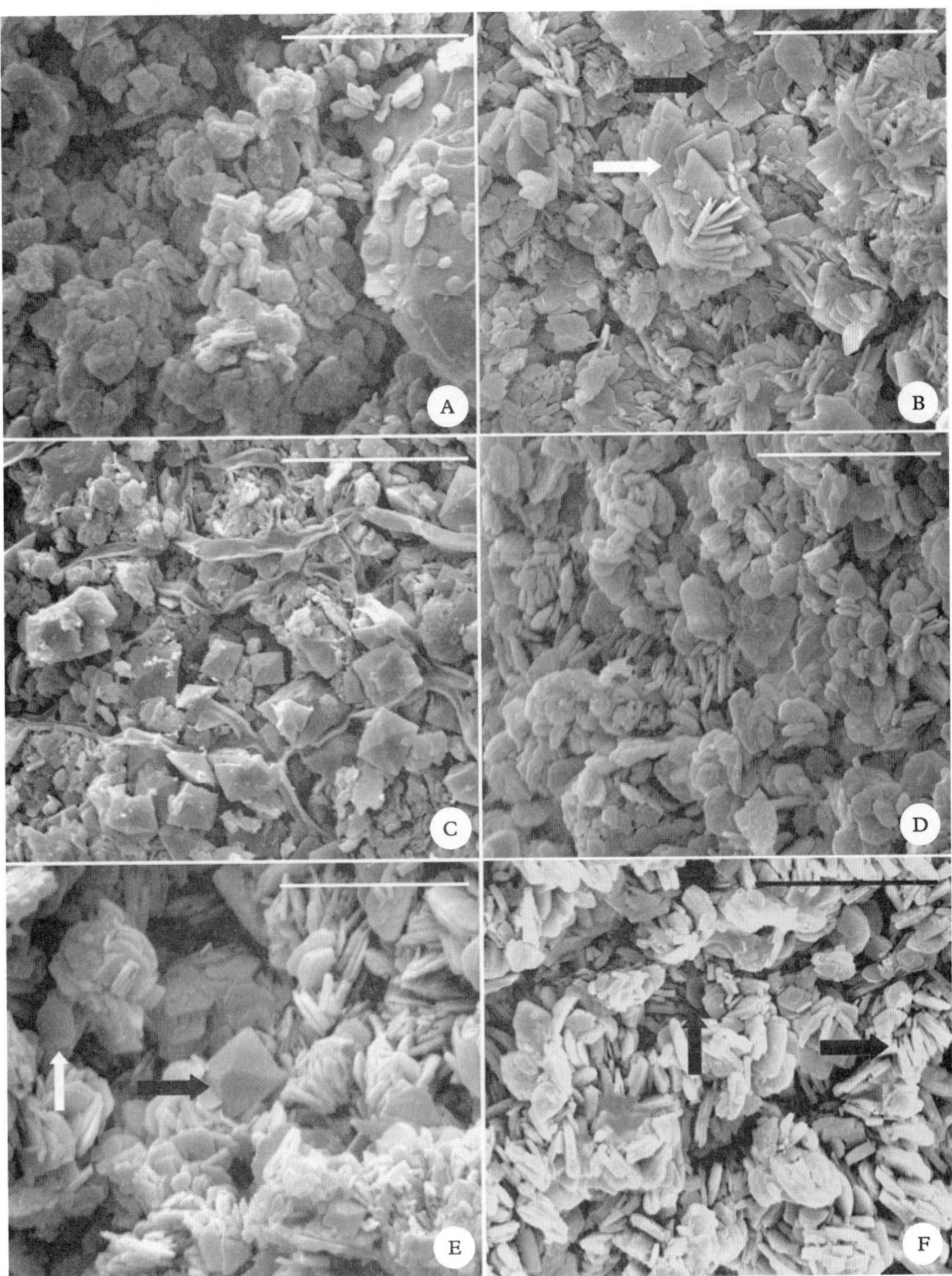

FIG. 5. SEM photographs of pruina on *Hypogymnia* species. A, whewellite in *H. lijiangensis* (holotype); B, weddellite of crystal type 3 (white arrow) and whewellite (black arrow) in *H. pruinoidea* (holotype); C, weddellite of crystal type 2 in *H. pruinosa* (holotype); D, whewellite in *H. pseudopruinosa* (lectotype); E, weddellite of crystal type 2 (black arrow) and whewellite (white arrow) in *H. subfarinacea* (holotype); F, weddellite of crystal type 3 (arrow on right) and whewellite (upper arrow) in *H. subpruinosa* (holotype). Scales: A–D & F = 10 μm; E = 5 μm.

noidea has a thallus with a more cartilaginous texture, isodichotomously branched lobes, and abundant perforations present on lobe tips, axils and lower surface. Table 1 compares six pruinose species of *Hypogymnia* more clearly, including the main different phenotypic characters, such as thallus texture, lobes, holes, soredia, and medulla PD reactions.

Among the six pruinose species of *Hypogymnia*, all except *H. pruinoidea* occur in south-western China, and two species, *H. pruinosa* and *H. subpruinosa*, extend to north-western and (or) north-eastern China. *Hypogymnia pruinoidea* is found only in Shaanxi province, north-western China. Located on a low latitude plateau, south-western China arises from south-east Tibet, crosses over west Sichuan, and expands to central and north Yunnan. A combination of complex landforms with advantageous moisture conditions in south-western China results in an extremely abundant biodiversity, with many endemic species of animals and plants. *Hypogymnia* species are also abundant in south-western China. Over 80% of known *Hypogymnia* species in China can be found in this area, and about 18% of the species are endemic. South-western China can therefore be regarded as a main centre of speciation for *Hypogymnia*. Mount Taibaishan is the highest peak of Qinling, Shaanxi province, which is the watershed of Yangtze River and Yellow River. As the highest mountain in the eastern half of China, it has a very complex geography and climate. The marked difference in elevation leads to a vertical distribution of climate, animals, and plants, and many new species of plants endemic to Taibaishan have been reported. Over 30% of known *Hypogymnia* species in China can be found in this area, including the new endemic species, *H. pruinoidea*.

The common character of the six species of *Hypogymnia* treated in this paper is having pruina. Pruina is a whitish, flour-like surface covering, consisting primarily of superficial deposits, most commonly calcium oxalate (Büdel & Scheidegger 1996). There are two kinds of calcium oxalate found in lichen pruina, weddellite and whewellite, and four types of crystal structure have been reported based on SEM study. Types 1–3 belong to the weddellite, and type 4 belongs to the whewellite (Wadsten & Moberg 1985). In our SEM study, three types of crystal structure have been confirmed on the pruina of the six pruinose *Hypogymnia* species: weddellite type 2, type 3, and whewellite. *Hypogymnia pruinoidea* has two crystal types, weddellite type 3 and whewellite (Fig. 5A); *H. lijiangensis* has only one crystal type, whewellite (Fig. 5B); *H. pruinosa* has only one crystal type, weddellite type 2 (Fig. 5C); *H. pseudopruinosa* has only one crystal type, whewellite (Fig. 5D); *H. subfarinacea* has two crystal types, weddellite type 2 and whewellite (Fig. 5E); and *H. subpruinosa* has two crystal types, weddellite type 3 and whewellite (Fig. 5F). It is postulated that the formation of calcium oxalate on the surface of lichens is caused by a need for the lichen to dispose of an excess of calcium, and there seems to be some relationship between different types of crystal structure and habitat, especially with regard to different conditions of acidity and water content (Wadsten & Moberg 1985). However, the taxonomic significance of different crystal structures of calcium oxalate is not yet clear.

In conclusion, based on the above mentioned phenotypic characters, presence or absence of soredia and black borders, location of pruina and thallus perforations, and chemistry, all the six pruinose species can be easily separated. All the phenotypic characters used to define the six pruinose species of *Hypogymnia* are commonly used in the taxonomy of *Hypogymnia*, while their objectivity and the phylogenetic placement of each species based on them will require molecular methods.

This project was supported by the Main Direction Program of Knowledge Innovation of the Chinese Academy of Sciences (KSCX2-EW-Z-9). We thank Dr Irwin M. Brodo for his careful modification of English grammar in the manuscript. Professor Bruce McCune of Oregon State University is thanked for the significant discussion about *Hypogymnia* with the first author, and he and one anonymous reviewer are also thanked for their valuable suggestions on the manuscript. Professor Jianbin Chen gave constructive comments on an earlier version of our paper. Mr Chunli Li helped take SEM photograghs of pruina. The authors are also grateful to Ms. Hong Deng for giving considerable assistance during the studies in HMAS-L.

References

Büdel, B. & Scheidegger, C. (1996) Thallus morphology and anatomy. In *Lichen Biology* (T. H. Nash III, ed.): 37–64. Cambridge: Cambridge University Press.

Chen, J. B. (1994) Two new species of *Hypogymnia* (Nyl.) Nyl. (Hypogymniaceae, Acomycotina). *Acta Mycologica Sinica* **13** (2): 107–110.

Culberson, C. F. (1972) Improved conditions and new data for the identification of lichen products by a standardized thin-layer chromatographic method. *Journal of Chromatography* **72:** 113–125.

Culberson, C. F. & Kristinsson, H. (1970) A standardized method for the identification of lichen products. *Journal of Chromatography* **46:** 85–93.

Giralt, M., Mayrhofer, H., van den Boom, P. P. G. & Elix, J. A. (2001) *Rinodina turfaceoides*, a new corticolous, blastidiate species from the Iberian Peninsula. *Lichenologist* **33:** 97–102.

McCune, B. (2002) *Hypogymnia*. In *Lichen Flora of the Greater Sonoran Desert Region* Vol. 1 (T. H. Nash III, B. D. Ryan, C. Gries & F. Bungartz, eds): 228–238. Tempe, Arizona: Lichens Unlimited, Arizona State University.

Wadsten, T. & Moberg, R. (1985) Calcium oxalate hydrates on the surface of lichens. *Lichenologist* **17:** 239–245.

Wei, J. C. (1991) *An Enumeration of Lichens in China*. Beijing: International Academic Publishers.

Wei, J. C. & Jiang, Y. M. (1980) Species novae lichenum e Parmeliaceis in regione xizangensi. *Acta Phytotaxonomica Sinica* **18:** 386–388.

Wei, X. L. & Wei, J. C. (2005) Two new species of *Hypogymnia* (Lecanorales, Ascomycota) with pruinose lobe tips from China. *Mycotaxon* **94:** 155–158.

Zhao, J. D. (1964) A preliminary study on Chinese *Parmelia*. *Acta Phytotaxonomica Sinica* **9:** 139–166.

第三章　新分类群
Taxa New to Science

疱脐衣属的一新亚属*

魏 江 春

(中国科学院微生物研究所)

A NEW SUBGENUS OF LASALLIA MERAT EMEND. VEJ

VEJ TZJAN-CZUNJ

(Institute of Microbiology, Academia Sinica)

1964 年，我们在云南丽江玉龙山的西康油松（*Pinus densata* Masters）树皮上发现了一个在脐衣科（Umbilicariaceae）的分类上有意义的新种。这一发现，促使我们在确定这一新种的分类等级时，不得不对该科內各分类单位及等级进行检查和讨论。

Linnaeus 在他的 "Species Plantarum"（1753）一书中以 *Lichen* 为属名，描述了 80 种地衣。其中属于脐衣科的有 7 种。后来，D. G. F. Hoffmann(1790—1801)改以 *Umbilicaria* 为属名[属的模式种为 *Umbilicaria exasperata* Hoffm.＝*Umbilicaria proposcidea*（L.）Schrad.——子囊盘表面环纹状，子囊內含八孢，原植体无疱状凸起]，描述了脐衣科的 11 个种。1803 年，Acharius 在他的 "Methodus Lichenes" 一书中将该科植物以 *Gyrophora* 为属名[属的模式种为 *Gyrophora glabra* Ach.＝*Umbilicaria polyphylla*（L.）Baumg.——子囊盘表面环纹状，子囊內含八孢，原植体无疱状凸起]又进行了描述。这样一来，*Gyrophora* Ach.（1803）一属名就成了 *Umbilicaria* Hoffm.（1790—1801）的异名。后来，Flotow（1850）及 Koerber（1855）依子囊孢子特性及原植体有无疱状凸起为根据，把脐衣科划为 2 个属（即 *Umbilicaria*——子囊盘表面平坦，子囊內含单孢，原植体具疱状凸起；*Gyrophora*——子囊盘表面环纹状，子囊內含八孢，原植体无疱状凸起），从而在认识植物方面，由外部形态进入內部结构，在该科植物的分类研究中作出了他们的贡献。但是，不幸的是，他们将 *Umbilicaria* Hoffm. 及其异名 *Gyrophora* Ach. 作为该 2 属的两个属名。因而，在该科的属名方面造成了一定的混乱。为了消除混乱，根据国际植物命名法规，*Gyrophora* Ach. 一属名应予废弃。

E. Frey 对该科植物进行了比较详细而深刻的研究和分析之后，在 1931 年提出了一个分类系统，和 Hoffmann 一样，将该科的全部种类都归入一属，即脐衣属（*Umbilicaria* Hoffm.）。但是，E. Frey 在该属內又根据子囊孢子特性承认并组合了 3 个亚属，即疱脐衣亚属（subg. *Lasallia* Endlich.——孢子砖墙式，褐色，原植体具疱状凸起），拟环纹亚属[subg. *Gyrophoropsis*（Elenk. et Savicz）Frey——孢子随后呈砖墙式结构，原植体无疱状凸起]及环纹亚属［subg. *Gyrophora*（Ach.）Frey——孢子单胞，无色，原植体无疱状凸

* 本文承戴芳澜、邓叔羣、王云章三位教授审阅，并提供宝贵意见；部分照片系在孙荣钦等同志协助下摄制而成，特此深致谢忱。

魏江春 Wei Jiang-Chun，曾用名 Vej Tzjan-czunj。

本文原载于《植物分类学报》，11（1）：1-10，1966.

起]。在后一亚属内，E. Frey 根据原植体在形态上的微小差异，又建立与组合了5个组，即 sect. *Velleae* Frey, sect. *Rugifrae* Frey, sect. *Polymorphae* Frey, sect. *Glabrae* Frey 及 sect. *Anthracinae* Frey。

对该科植物的子囊盘表面形态进行了仔细的研究之后，P. F. Scholander (1936) 大胆地提出了一个新的分类方案，即仅仅根据子囊盘的表面形态，把该科植物划为4个属，即脐衣属 [*Umbilicaria* (Hoffm.) Schol.——子囊盘表面平坦]，柱盘脐衣属 (*Omphalodiscus* Schol.——子囊盘表面具不孕的柱体或裂隙)，环纹脐衣属 (*Gyrophora* Ach. emend. Schol.——子囊盘表面呈同心圆式的环纹)及辐射盘脐衣属 (*Actinogyra* Schol.——子囊盘表面呈放射状条纹，无共同的盘缘)。包括在 *Umbilicaria* (Hoffm.) Schol. 一属中的全部种类都具有平坦的盘面。但是，就其他特性来看，其中既包括有砖墙式褐色孢子类型的种，也包括有单胞无色孢子类型的种；既有单孢子囊型者，又有八孢子囊型者；既有疱状原植体型的种类，又有非疱状原植体型的种类。这说明，子囊盘表面的形态特征，并沒有眞正反映出该科植物的內在特性，亦即与其內在特性沒有任何相关性。但是，Scholander 的热烈支持者 G. A. Llano (1950) 把 *Umbilicaria* (Hoffm.) Schol. 一属中的种类再分为2个属，即疱脐衣属 (*Lasallia* Merat——单孢子囊型，孢子砖墙式，原植体具疱状凸起)及平盘脐衣属 (*Agyrophora* Nyl. 八孢子囊型，孢子单胞无色或随后呈砖墙式褐色，原植体无疱状凸起)。因此，根据 Llano 的观点，脐衣科內就包括有5个属，即疱脐衣属 (*Lasallia* Merat)，平盘脐衣属 (*Agyrophora* Nyl.)，柱盘脐衣属 (*Omphalodiscus* Schol.)，脐衣属 (*Umbilicaria* Hoffm.) 及辐射盘脐衣属 (*Actinogyra* Schol.)。很明显，Llano 的观点是把稳定性強的(即主要的)特征与稳定性差的(即次要的)特征同等地看作为属的分类依据。此外，在平盘脐衣属，柱盘脐衣属及脐衣属等3属中，把是否为砖墙式孢子作为根据，分别建立了2个组。这就是经过 Llano 略加修改后的 Scholander 的系统。尽管 E. Frey 根据大量事实来反对 Scholander 的系统所依据的基础，但是，不能因此而完全否定 Scholander 在这方面的重要发现。

在前人研究的基础上，V. P. Savicz (1950) 根据 Flotow (1850), Koerber (1855) 及 Elenkin (1906) 的观点，把该科植物分为2属，即脐衣属 [*Umbilicaria* (Hoffm.) Koerb.] 及环纹脐衣属 [*Gyrophora* (Ach.) Emend. Savicz]。又根据子囊盘表面特征及孢子有无砖墙式结构，把环纹脐衣属中的种类分归5个亚属，即 subg. *Gyrophoropsis* (Elenk. et Savicz) Savicz (盘面环纹状，孢子呈砖墙式结构)，subg. *Agyrophora* (Nyl.) Savicz, subg. *Omphalodiscus* (Schol.) Savicz, subg. *Eugyrophora* (A. Zahlbr.) Savicz 及 subg. *Actinogyra* (Schol.) Savicz。

在对国內外大量材料进行对比研究的基础上，我们认为，仅仅把很不稳定的子囊盘表面的形态特征作为该科分属的唯一标准表明，Scholander 试图用最简单的方法来认识和阐明复杂的地衣有机体。

尽管由 Llano 略加修改后的 Scholander 的系统，克服了一些表面性的缺点，但是，Llano 只是单纯地做到了对该科植物进行较为彻底的区分。在这里，他把每个单一的特征都同等地，不分主次地作为分属的标准。

因此，如果我们完全接受 Scholander 的系统，那末，我们的新种(子囊盘表面呈非同心圆式的多环状褶纹，从不呈现平坦的盘面，单孢子囊型，孢子砖墙式，褐色，原植体具疱状凸起）将不得不被嵌入 *Gyrophora* Ach. emend. Schol.（子囊盘表面呈环纹状，八孢子囊型，孢子单胞，无色，或随后呈砖墙式，褐色，原植体无疱状凸起）一属内。如果完全接受经过 Llano 修改后的这个系统，那末，我们的新种势必成为一个新属。

关于由 Scholander 所发现的子囊盘表面特征，根据我们的对比观察，证实了 E. Frey 所据以反对 Scholander 系统的事实，即各类型盘面之间的相似性比较明显，但是，这种相似性多表现在子囊盘个体发育的幼小阶段，而少见于成熟阶段中。就在这种情况下，E. Frey 根本拒绝考虑子囊盘表面特征的分类价值。在他自己的系统中，不仅在属的等级上，就是在亚属及组的等级上，也根本不把盘面特征作为分类依据。因此，如果说，Scholander 在该科的分属方面，沿着盘面特征这条狭窄的小胡同走到了一个极端，那末，E. Frey 却沿着相反的一条小胡同走到了另一个极端。这样，他们在该科的分属界限上，都表现了相当的主观性和片面性。各据一个单一的特征，建立各自的系统。缺乏在对该科植物进行全面而综合的对比研究和分析中，抓住主要特征及其相关特征，而非各取其一，从而得出较为全面的，更接近于自然系统的分类方案。

如果我们不是单纯地从子囊盘的外部形态来探讨各类群之间的系统关系和分属界限，而是透过外部特征去观察和研究其内部特性，那末，我们就会发现，一部分种类仅具单孢子囊（稀为双孢子囊），孢子特大，砖墙式多胞，褐色；同时，这一群种类的原植体也正好都具有疱状凸起。另一部分种类，则具八孢子囊（稀为六孢子囊），孢子相应地较小；同时，这一群种类的原植体也正好不具疱状凸起。由此看来，原植体有无疱状凸起这个外部特征，在一定程度上反映了该科植物各类群的内在特性。因此，我们基本上赞成 Flotow（1850），Koerber（1855），Elenkin（1906）及 Savicz（1950）的观点，根据上述界限，把脐衣科的植物承认为 2 个属，即疱脐衣属（*Lasallia* Merat emend. Vej）与脐衣属（*Umbilicaria* Hoffm. emend. Vej）。

我们将单孢子囊型与八孢子囊型看作是 2 个属的自然界限的标准之一，是因为这个特性较子囊盘表面特征更为稳定，尤其是这个特性，系该科植物的性过程中的内在特性。而且，这个特性的相关特征，表现在原植体有无疱状凸起上。因而，不仅在营养器官的外部形态上，而且也在性器官的内在特性上，都明显地反映出这 2 个属之间的自然界限。此外，除了原植体有无疱状凸起这个特征是性器官的内在特性的相关特征之外，而且，原植体特征在地衣植物的分类中，通常具有相当重要的意义。因此，在该科的系统分类中，同样不能忽视关于原植体有无疱状凸起这个比较稳定的形态特征。这样看来，把子囊内含孢子数目及原植体有无疱状凸起当作属所显示出来的重要特征可能比较符合客观实际。

其次，关于盘面特征在分类上虽有其一定的意义，但是，因为这个特征在稳定性方面不如前两个特征那样强，以及它与原植体有无疱状凸起和子囊内含孢子数目等特征性状的不一致性，亦即，它没有反映出该科植物的内在特性，因而，使我们无法接受 Scholander 的系统。但是，我们赞成 Savicz 的观点，把盘面特征当作亚属的分类标准。不过，我们并不赞成他关于建立与组合 *Gyrophoropsis*（Elenk. et Savicz）Savicz 一亚属的观点。因为，具砖墙式孢子的种类中，既有环纹状盘面的，也有平盘和柱盘类型的种类。要是接受这个观点，那末其他 4 个亚属的建立基础将不复存在。此外，在八孢子囊型的种类中，砖多胞这一特征在稳定性方面还远不如盘面特征为強。比方，人们在 *Umbilicaria yunnana* (Nyl.) Hue 的子囊中偶见有砖墙式多胞的孢子 (Frey, 1931, p. 103)。但是，我们在对该种的大量标本所进行的研究中，到目前为止，倘未遇到砖墙式的子囊孢子。虽然不能因此就否定前人的发现，但是，至少可以说明，这种特性在稳定性方面比盘面特征还要小。因此，在这一点上，我们赞成 Llano 的意见，把这种特性隶属于盘面特征之下，作为分组的界限。我们从附表中可以轮廓地比较脐衣科各分类系统之间的异同。下面是我们赞成的分类系统提要：

脐衣科 Umbilicariaceae

I. 疱脐衣属 **Lasallia** Merat emend. Vej——原植体具疱状凸起；子囊盘表面平坦至多中心的环状褶纹；子囊內含单孢(偶见双孢)，孢子特大，砖墙式多胞，褐色。

1. 疱脐衣亚属 subg. **Lasallia**——盘面平坦。

2. 环纹疱脐衣亚属 subg. **Pleiogyra** Vej——盘面呈非同心圆式的多环状褶纹。

II. 脐衣属 **Umbilicaria** Hoffm. emend. Vej [syn. *Gyrophora* (Ach.) emend. Savicz]——原植体无疱状凸起；子囊盘表面平坦，具不孕的中柱或裂隙，呈环纹或辐射状条纹；子囊內含八孢（偶见六孢），孢子相应地较小，单胞无色或后期有时呈砖墙式多胞。

1. 平盘脐衣亚属 subg. **Agyrophora** (Nyl.) Savicz——盘面平坦。

(1) 平盘脐衣组 sect. **Agyrophora** (syn. sect. *Anthracinae* emend. Llano)——孢子单胞无色。

(2) 褐孢脐衣组 sect. **Dichroae** Llano——孢子后期有时呈砖墙式结构，褐色。

2. 柱盘脐衣亚属 subg. **Omphalodiscus** (Schol.) Savicz——盘面呈不孕的中柱或裂隙。

(1) 柱盘脐衣组 sect. **Omphalodiscus** (syn. sect. *Decussatae* Schol.)——孢子单胞无色。

(2) 多胞脐衣组 sect. **Spodochroae** Schol.——孢子后期呈砖墙式结构，褐色。

3. 脐衣亚属 subg. **Umbilicaria**——盘面呈同心圆式的环纹；孢子单胞无色或后期呈砖墙式多胞，褐色。

(1) 脐衣组 sect. **Umbilicaria** (syn. sect. *Simplices* Llano)——孢子单胞无色。

(2) 砖孢脐衣组 sect. **Gyrophoropsis** (Elenk. et Savicz)A. Zahlbr. (syn. sect. *Muriformes* Llano)——孢子后期呈砖墙式多胞，褐色。

4. 辐射盘脐衣亚属 subg. **Actinogyra** (Schol.) Savicz——子囊盘无共同边缘，盘面呈辐射状条纹。

疱 脐 衣 属

Lasallia Merat emend. Vej

Merat, Nouv. Fl. Envir. Paris, edit. 2, vol. I, 1821, p. 202; Llano, A Monogr. of the Lich. Fam. *Umbilicariaceae* in the West. Hemisph. 1950, p. 27.——*Umbilicaria* sect. *Lasallia* Endlich., Gen. Pl. 1836, p. 13; A. Zahlbr., Cat. Lich. IV, 1927, p. 745.——*Umbilicaria* subg. *Lasallia* Frey in Hedwigia, 1931, 71, p. 106 et *Umbilicariaceae* in Rabenh. Krypt.-Fl. B. IX, Ab. IV, Haf. I, 1933, p. 208.

Thallus foliaceus, papulosus, ad substratum umbilico adfixus, subtus erhizoideus.

Apothecia haud gyrosa vel semper gyroso-plicata, sessilia vel stipitata, marginata vel non marginata; sporae solitariae (rarissime binae), muriformi-polyblastae, demum cinnamomeae.

原植体叶状，具疱状凸起；借助于中央脐固着于基物上，下表面假根缺如。

子囊盘表面无环纹或经常具环状褶纹，无柄或有柄，共同盘缘有或无；子囊单孢型（偶见双孢型），孢子砖墙式多胞，随后呈褐色。

1. **疱脐衣亚属**

Subg. **Lasallia**

Apothecia haud gyrosa (rarissime gyroso-plicata), lecideina, simplicia, marginata (vel rarissime non marginata).

子囊盘表面无环纹（偶见微弱的环纹），平坦，具共同的盘缘（或偶见无共同盘缘者）。

2. **环纹疱脐衣亚属** 新亚属

Subg. **Pleiogyra** Vej, subg. nov.

Apothecia semper gyroso-plicata, econcentrica, marginata vel interdum non marginata.

Typus subgeneris *Lasallia sinensis* Vej est.

子囊盘表面经常呈多环状褶纹，具共同的盘缘或偶见无共同盘缘者。

亚属的模式种：*Lasallia sinensis* Vej。

中华疱脐衣 新种

Lasallia sinensis Vej, sp. nov.

Diagnos. Thallus foliaceus, monophyllus, papulosus, vulgo eisidiifer vel solitario-isidiifer (squamiformis), ad substratum umbilico adfixus, subtus ater, erhizoideus. Apothecia frequentia; discus non-planiusculus, semper gyroso-plicata, econcentrica; sporae solitariae, muriformi-polyblastae, cinnamomeae.

Descriptio. Thallus foliaceus, monophyllus, papulosus, ca. 4—9 cm diam. in margine plus minusve laciniatus, superne dense pruinosus, griseolus, plumbeus, peripheriam versus cinnamomeus, badius, pullus, superne interdum plus minusve reticulatus [ut in *Umbilicaria decussata* (Vill.) Frey] vel subrugosus [ut in *Umbilicaria corrugata* (Ach.) Arn.], vulgo eisidiifer vel solitario-isidiifer (squamiformi, pullus atrans), subtus erhizoideus, ater, pullo-ater, epruinosus, peripheriam versus interdum subreticulatus; thallus ca. 86.38—493.60 μ crassus; extra zona necralis ca. 7.41—37.15 μ crassa; cortex superne paraplectenchymatus, brunneus, ca. 17.29—37.05 μ crassus; stratum gonidiale continuum, ca. 37.05—74.10 μ crassum; medulla laxa, ca. 29.64—74.10 μ crassa; cortex inferior sclero-plectenchymatus ca. 81.51—243 μ crassus; inferior zona necralis ca. 2.47—29.64 μ crassa.

Apothecia frequentia, ca. 1—2.5 mm in diametro, sessilia vel stipitata, marginata vel interdum non marginata; discus non planiusculus, gyroso-plicatus, econcentricus, juvenilius interdum fissurus centralis instructus [ut in subg. *Omphalodiscus* (Schol.) Savicz] demum convexus, ater; parathecium atrum, ca. 17.29—61.70 μ crassum; hypothecium ca. 111.15 μ crassum; thecium ca. 86.45 μ crassum; paraphyses filiformes, hyalinae, septatae; asci clavati, ca. 61.75—86.45 (98.80) × 9.88—32.11 (61.75) μ (membrana ca. 2.47—7.4 μ crassa); sporae solitariae, obovatae, muriformi-polyblastae, cinnamomeae, ca. 44.46—

74.10(98.80) × 19.76—41.99(61.75) μ.

Pycnidia subglobosa, pullata; pycniosporae simplices, hyalinae, ellipticae, ca. 4.94 × 1.25—1.50 μ.

Yunnan: Yulun-shan (Li-tzjan), Gan-hai-tze, ad corticem Pini. Vej Tzjan-czunj: *2539*, *2540* (isotypus), *2589*, *2590* (holotypus), *2591*,*2596*. In Herbario Mycologico Instituti Microbiologici Academiae Sinicae conservatur.

鉴别特征: 原植体单叶状，具疱状凸起，通常不具珊瑚芽或偶见个别的鳞片状珊瑚芽，借助于中央脐固着于基物上，下表面黑色，假根缺如。子囊盘常见，盘面经常呈非同心圆型的多环状褶纹；子囊内含单孢，孢子砖墙式多胞，褐色。

描述: 原植体单叶状，具疱状凸起，叶片直径约4—9厘米，周缘多少呈浅裂状，上表面覆盖有白色霜层，呈灰色，铅灰色，并带有褐色色度，近周缘处则呈褐色，栗褐色至暗棕色，有时表面多少呈现网格状脉纹[类似于 *Umbilicaria decussata* (Vill.) Frey] 或微弱的褶皱[类似于 *Umbilicaria corrugata* (Ach.) Arn.]，通常不具珊瑚芽或偶见个别的黑棕色鳞片状珊瑚芽，下表面假根缺如，黑棕色至黑色，无霜层，近周缘处有时呈现微弱的网格状脉纹；原植体厚约86.38—493.60微米；上死层厚约7.41—37.15微米；上表皮为假薄壁组织，褐色，厚约17.29—37.05微米；藻层连续不断，厚约37.05—74.10微米；髓层疏松，厚约29.64—74.10微米；下表皮为密丝组织，厚约81.51—243微米；下死层厚约2.47—29.64微米。

子囊盘常见，直径约1—2.5毫米，无柄或具短柄，具共同的盘缘或有时缺如，盘面呈非同心圆型的多环状褶纹，幼期有时具中央裂隙[类似于 Subg. *Omphalodiscus* (Schol.) Savicz]；盘面后期凸起，黑色；盘壁黑色，厚约17.29—61.70微米；子实下层厚约111.15微米；子实层厚约86.45微米；侧丝线状，具横膈膜；子囊棍棒状，大小约61.75—86.45(98.80) × 9.88—32.11(61.75)微米(膜厚约2.47—7.40微米)；子囊内含单孢，孢子椭圆形；砖墙

表1 脐衣科 (Umbilicariaceae) 各分类系統之比較

Hoffmann (1790—1801)	Acharius (1803)	Koerber (1855)	Frey (1931—1933)
(A) ……	……	*Umbilicaria* (genus) ……	…… *Lasallia* (subg.)
(C) ……	……	……	……*Gyrophoropsis* (subg.)
(A)—(F) ……	……	……	*Umbilicaria* (genus)
(B) *Umbilicaria* (genus)	*Gyrophora* (genus)……	*Gyrophora* (genus) ……	…… *Gyrophora* (subg.)
Scholander (1936)	Llano (1950)	Savicz (1950)	Vej (1965)
(A) *Umbilicaria* (genus) (A + D)	*Lasallia* (genus)……	*Umbilicaria* (genus) ……	**Lasallia** (genus) *Lasallia* (subg.)
(G) ……	……	……	…… *Pleiogyra* (subg.)
(B) *Gyrophora* (genus)	*Umbilicaria* (genus) ……	*Gyrophora* (genus) ……	**Umbilicaria** (genus)
		Eugyrophora (subg.)……	……*Umbilicaria* (subg.)
(C) ……	……	*Gyrophoropsis* (subg.) ……	…… [*Gyrophoropsis* (sect.)]
(D) ……	*Agyrophora* (genus)……	*Agyrophora* (subg.)……	…… *Agyrophora* (subg.)
(E) *Omphalodis* cus (genus)	*Omphalodiscus* (genus)……	*Omphalodiscus* (subg.)……	…… *Omphalodiscus* (subg.)
(F) *Actinogyra*(genus)	*Actinogyra* (genus) ……	*Actinogyra* (subg.) ……	…… *Actinogyra* (subg.)

注: 表中 (A),(B),(C)……等系各等级单位模式种之代号。

式多胞，褐色，大小约 44.46—74.10 微米 (98.80) × 19.76—41.99(61.75) 微米。

分生孢子器近球形，暗色；分生孢子单胞，无色，椭圆形，大小约 4.94 × 1.25—1.50 微米。

云南：丽江，玉龙山，乾海子，西康油松 (*Pinus densata* Masters) 树皮上。魏江春：*2539*, *2540* (副模式), *2589*, *2590* (模式), *2591*, *2596*。均保存于北京中国科学院微生物研究所真菌标本室。

Summary

In the present paper a new species of *Lasallia* Merat is described and placed under a new subgenus *Pleiogyra,* characterized by having econcentrical multi-gyrate apothecia. In consequence of this finding, the conception of plane disc in the genus *Lasallia* of *Umbilicariaceae* becomes untenable.

As a result of careful examination of rich materials, the writer considers the difference in disc found by Scholander among all species of the family is of certain significance for classification. Unfortunately, Scholander's classification is based solely on this rather variable morphological character. If his view be adopted the new species will have to be inserted into the genus *Gyrophora* Ach. emend. Schol. If accepting Llano's division of *Umbilicariaceae* into five genera, namely *Lasallia, Agyrophora, Omphalodiscus, Umbilicaria* and *Actinogyra,* the new species will have to be considered a new genus of the family.

The present writer considers both Scholander's and Llano's systems as being artificial, because the former author emphasizes too much the unstable morphological appearance of disc and the later simply divides the family into five genera simultaneously on the basis of the stable and unstable characters of the various groups. In contrast with Scholander's and Llano's systems Frey's classification neglects wholly the difference in the characters of disc. Such treatment is likewise unsound. The present writer supports the view held by Flotow, Koerber, Elenkin and Savicz (1950). Nevertheless, he does not agree with Savicz in erecting the subgenus *Gyrophoropsis* (Elenk. et Savicz) Savicz, because muriform spores are present not only in the group having gyrate discs, but also in the groups having plane and omphalic discs of the Genus *Umbilicaria* Hoffm. emend. Vej. The recognition of this subgenus must therefore deny the basis on which the other four subgenera are established.

The system of **Umbilicariaceae** held by the writer is as follows:

I. Genus **Lasallia** Merat emend. Vej …… Thallus pustulate; disc plane to econcentrical multi-gyrate; spores 1(2) per ascus, quite large, muriform.

1. Subgenus **Lasallia**...Disc plane.
2. Subgenus **Pleiogyra** Vej……Disc econcentrical multigyrate.

II. Genus **Umbilicaria** Hoffm. emend. Vej (syn. *Gyrophora* Ach. emend. Savicz)* …Thallus non-pustulate; disc plane and smooth or plane with central sterile column (or fissure) or with gyri; spores 8 per ascus, simple, hyaline or becoming brown muriform, small.

* The name of genus Gyrophora Ach. (1803) being a synonym of Umbilicaria Hoffm. (1790—1801) is therefore illegitimate.

1. Subgenns **Agyrophora** (Nyl.) Savicz……Disc plane.
   (1) Section **Agyrophora** (syn. sect. *Anthracinae* emend. Llano). ……Spores simple and hyaline.
   (2) Section **Dichroae** Llano……Spores becoming brown muriform.
2. Subgenus **Omphalodiscus** (Schol.) Savicz……Disc with central sterile column or fissure.
   (1) Section **Omphalodiscus** (syn. sect. *Decussatae* Schol.)……Spores simple and hyaline.
   (2) Section **Spodochroae** Schol.……Spores becoming muriform.
3. Subgenus Umbilicaria...Disc with concentrical gyri; spores simple and hyaline or becoming muriform.
   (1) Section **Umbilicaria** (syn. sect. *Simplices* Llano)……Spores simple and hyaline.
   (2) Section **Gyrophoropsis** (Elenk. et Savicz) A. Zahlbr. (syn. sect. Muriformes Llano)……Spores becoming muriform.
4. Subgenus **Actinogyra** (Schol.) Savicz……Disc with radial gyri, proper margin absent.

参 考 文 献

[1] Elenkin, A. A. Lichenes florae Rossiae Mediae, Pars 1,1906, pp. 52—64.
[2] Frey, E. Weitere Beiträge zur Kenntnis der *Umbilicariaceae*, in Hedwigia, 71, 1931, pp. 94—119.
[3] ——————, *Umbilicariaceae* in Rabenhorst's Kryptogamen-Flora, IX, 1933, pp. 203—426.
[4] ——————, Vorarbeiten zu einer Monographie der *Umbilicariaceae*, in Berichte der Schweizerischen Botanischen Gesellschaft, Band. 45, 1936, pp. 198—230.
[5] ——————, Neue Beiträge zu einer Monographie des Genus *Umbilicaria* Hoffm., Nyl. in Berichte der Schweizerischen Botanischen Gesellschaft. Band. 59, 1949, pp. 427—470.
[6] Hoffmann, D. G. F. Descriptio et Adumbratio Plantarum e classe cryptogamica Linnaei Lichenes dicuntur. I—III, 1790—1801.
[7] Linnaeus. Species Plantarum, 1753, pp. 1140—1156.
[8] Llano, G. A. A Monograph of the Lichen Family *Umbilicariaceae* in the Western Hemisphere, 1950.
[9] Savicz, V. P. Conspectus Lichenum Ad Floram Umbilicariacearum in URSS, in Notulae Systematicae e Sectione Cryptogamica Instituti Botanici nomine V. L. Komarovii Academiae scietiarum URSS, 1950, pp. 97—108.
[10] Scholander, P. F. On the Apothecia in the Lichen Family *Umbilicariaceae* in Nytt Magasin for Naturvidenskapene Bind, 75, 1936, pp. 1—31.
[11] Zahlbruckner, A. Catalogus Lichenum Universalis, IV, 1927, pp. 675—754.

植物分类学报 第十一卷 图版一
Acta Phytotaxonomica Vol. XI, Pl. 1

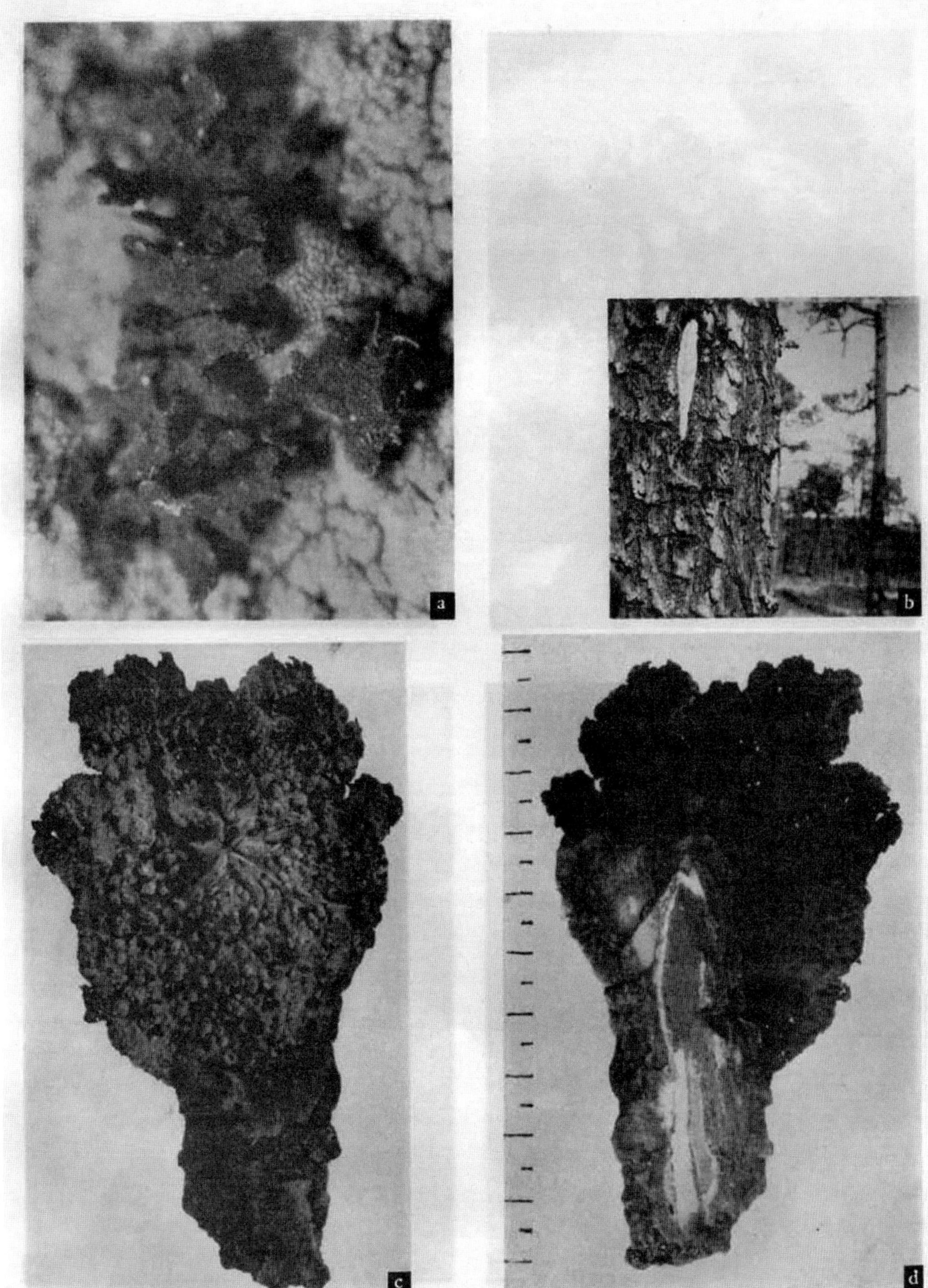

a. 原植体上表面的鳞片状珊瑚芽(×55.5)。
b. 生长在 *Pinus densata* Masters 树皮上的 *Lasallia sinensis* Vej 的野外景观(×1/12)。
c. *Lasallia sinensis* Vej 的原植体上表面(带有一小块树皮;自然大小)。
d. *Lasallia sinensis* Vej 的原植体下表面(同上)。

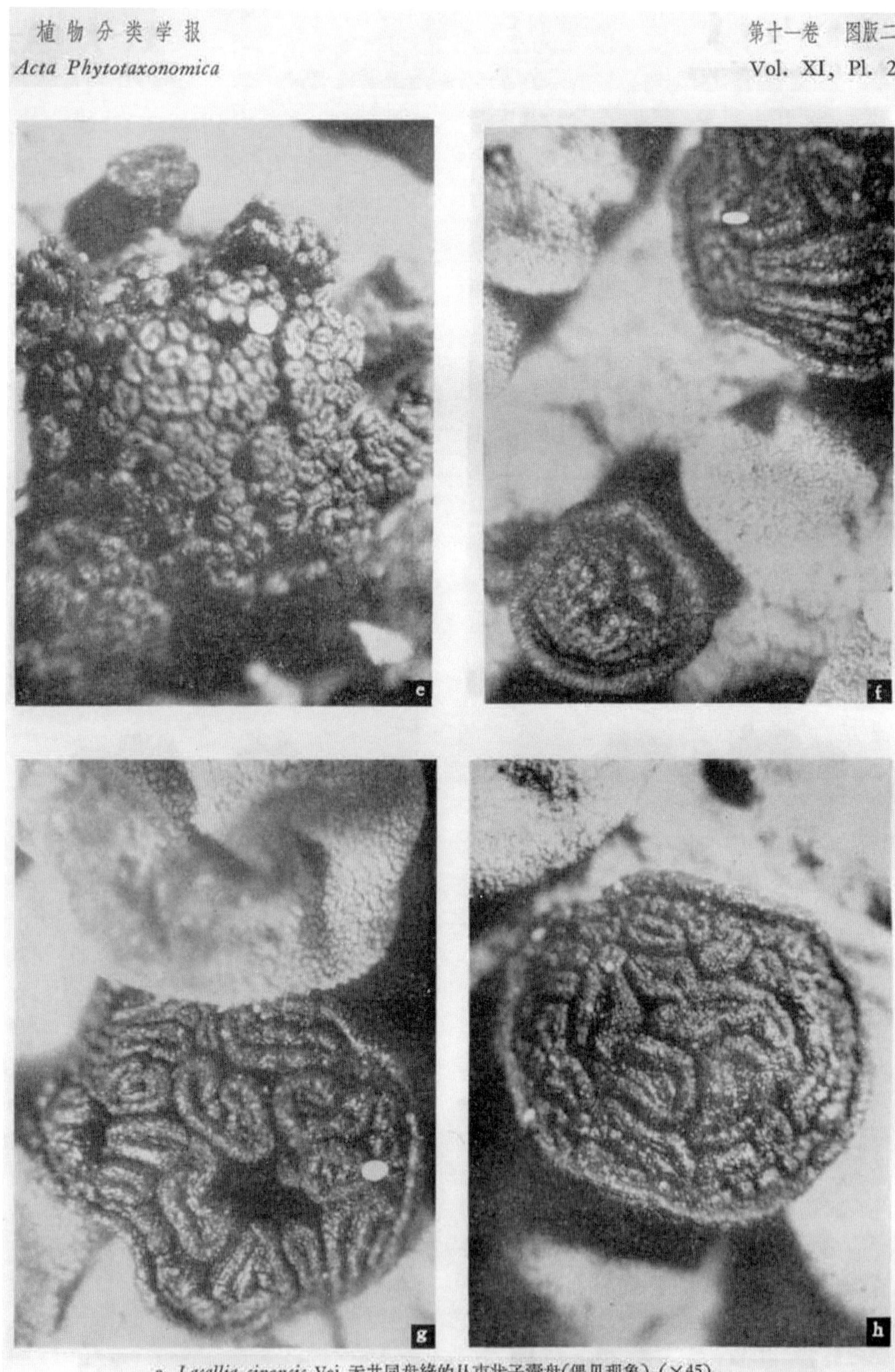

e. *Lasallia sinensis* Vej 无共同盘缘的丛束状子囊盘(偶见现象),(×45)。
f. *Lasallia sinensis* Vej 左下角为未成熟的子囊盘;右上角为较成熟的子囊盘(×45)。
g—h. *Lasallia sinensis* Vej 典型的环纹状盘面的子囊盘及明显的疱状凸起(×45)。

西藏梅衣科地衣新种*

魏江春　姜玉梅
（中国科学院微生物研究所）

1.**中华袋衣**　新种　图版 8:1

Hypogymnia sinica Vej et Jiang, sp. nov. Pl. 8—1.

Species *H. hypotrypellae* (Asah.) Rassad. similis est, sed differt pagina supera thalli grisea vel viridulo-grisea, lobis planis, ostiolis infra apicem loborum sitis, med. K +, P—.

Thallus foliaceus, dichotomus, lobis longis et planis, non inflatis vel subinflatis, ca. 1.5—4 mm latis, apicibus rotundis, supra griseis vel viridulo-griseis, apicem versus brunneis, sed non viridi-flavis vel sulphureis, opacis, interdum pro parte corticibus fragilibus vel subsorediosis, subtus aquilis, subnitidis, apicem versus rubiginosis, ostiolis infra-apicalibus subrotundis, ca. 0.5—2 mm diametro.

Cortex superus prosenchymaticus, brunneus, ca. 20.80—31.20 μ crassus; cortex inferus prosenchymaticus, brunneus, ca. 29.12—49.92 μ crassus; stratum algaceum ca. 24.96—41.60 μ crassum; cellulae algae subglobosae, viridiflavae, ca. 6.24—12.48 μ diametro; hypha medullae ca. (1.66) 2.08—4.16 (6.24) μ diametro. Apothecia ignota.

Cortex superus K+ luteus; medulla K+ fulvescens, C-, KC-, P-; atranorinum, chloroatranorinum et acidum non identificatum continens.

Xizang: Nyalam, Qüxiang, DeQingtang, ad corticem Betulae sp., alt. 3660 m, 1966 V 21, leg. Vej T. C. et Chen J. B. 1110 (holotypus) et ad corticem Rhododendroni sp. leg. Vej T. C. et Chen J. B. 1117, 1118-(1), in HMAS** (Beijing) conservatur.

本种类似黄粉袋衣，但衣体上表面灰色至灰绿色，而非黄绿色或硫磺色，下表面孔洞位于裂片近末端处而不位于近裂腋处，髓层 K+，P—。

西藏：聂拉木，曲香，德青塘附近，桦树皮上，海拔 3,600米，1966、V、21，魏、陈 1110（模式标本）；杜鹃灌木上，魏、陈 1117、1118-①，保存于中国科学院微生物研究所真菌标本室（缩写：HMAS，下同），北京。

2.**霜袋衣**　新种　图 1:1 图版 8:2-3

Hypogymnia pruinosa Vej et Jiang, sp. nov. Fig. 1—1; Pl. 8—2,3.

Thallus ad marginem lacinulatus; lobus latus, brevis et non acclivis; pagina infera substrato valde adhaerens; supra pruina alba crassa obtectus; ostiolo apice lobi inter paginam superam et paginam inferam sito.

Thallus foliaceus, subrotundus, rigidus, ad marginem lacinulatus, lobis latis, brevibus et non acclivibus, axillis latis, substrato valde adhaerens pagina infera, supra valde rugosus, vermicularis, fulvus, ad elevationem pruina, margo quae perspicuus, alba crassa obtectus (Pl. 8—3), isidiis sorediisque destitutus, punctis atris aspersis, instructus, subtus ater, subrugosus, ad marginem loborum brunneus, ostiolo apice lobi inter paginam superam et paginam inferam sito.

Cortex superus prosenchymaticus, ca. 21.50—25.80 μ crassus; cortex inferus prosenchymaticus ca. 10.40—14.56 μ crassus; stratum algaceum ca. 20.80—35.35 μ crassum; cellulae algae ca. 7.28—11.44 μ diametro, virides; stratum medullare ca. 3.12—4.16 μ crassum; hypha medullae ca. 2.49—3.12 μ diametro.

Apothecia numerosa, ad centrum thalli aggregata; disci rubiginosi vel aquilis; ascis octosporis, sporis subglobosis., ca. 3.12—4.16μ diametro.

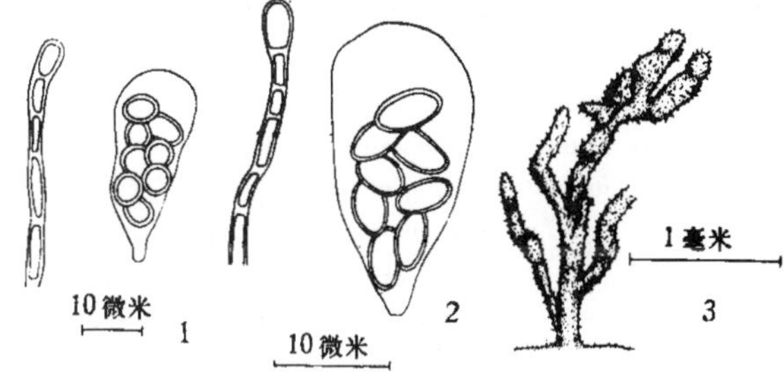

图1　1.霜袋衣 **Hypogymnia pruinosa** 的子囊及子囊孢子；2.蜡光袋衣 **Hypogymnia laccata** 的子囊及子囊孢子；3.珊茸梅衣 **Parmelia subverruculifera** 的珊瑚状裂芽及其表面的茸毛。

* 简荔同志描绘插图，苑兰翠同志拍摄部分照片，在拉丁文修改中王文采同志提供了宝贵意见，在此一并致谢。

**Herbarium Mycologicum Instituti Microbiologici Academiae Sinicae.

魏江春 Wei Jiang-Chun，曾用名 Vej Tzjan-czunj。
本文原载于《植物分类学报》，18（3）：386-388，1980.

Cortex superus K+ luteus; medulla K—, C—, KC+ aurantiaca, P—; atranorinum, chloroatranorinum et acidum identificatum continens.

Xizang: Changdu, corticola, 1975 VI 1, leg. Zong Y. C. et Liao Y. Z. 215 (holotypus), 217-(1), in HMAS (Beijing) conservatur.

本种衣体近圆形叶状，质地略坚硬，边缘浅裂，裂片宽短而紧贴于基物，上表面腊肠状隆起处覆以浓厚的白色粉霜层，霜层边缘界限清晰（图版8:3），孔洞位于裂片顶端而不位于下表面，与其他种易于区别。

西藏：昌都，杉树皮上，1976. VI、1，宗、廖215（模式标本）、217-①，保存于 HMAS，北京。

3. **蜡光袋衣** 新种 图1:2，图版8:4。

Hypogymnia laccata Vej et Jiang, sp. nov. Fig. 1—2; Pl. 8—4.

Species *H. bitteri* (Lynge) Ahti et *H. subobscurae* (Vain.) Poelt similis est, sed differt thallo isidiis sorediisque destituto, supra griseo-viridi, pro parte griseo-viridi griseo-albo et brunneo, valde laccato, ad medium valde rugoso, ad marginem plano, med. P +, chloroatranorinum continens.

Thallus foliaceus, ca. 8 cm. diametro, substrato adhaerens pagina infera, supra griseo-viridis, pro parte griseo-viridis, griseo-albus et brunneus, discolor, subreticulato-foveatus, valde laccatus, pro parte striis et maculis nigris instructus, lobis ca. 1—2 mm latis, apice latioribus, planioribus et subinflatioribus, saepe brunneis et reticulato-foveolatis confusis, ad marginem saepe atris, ad medium valde rugosis, subtus ater, apice brunneus, ostiolo subrotundo magno instructus.

Cortex superus prosenchymaticus,, ca. 8.32—16.64 μ crassus, fulvescens, supra se strato hyalino ca. 2.08—4.16 μ crasso instructus; stratum algaceum ca. 20.80—43.68 μ crassum; cellulae algae subglobosae, ca. 10.40—16.64 μ diametro; hypha medullae ca. 2.08—5.20 μ diametro; cortex inferus prosenchymaticus, aquilus, ca. 12.48—16.64 μ crassus.

Pycnidia subglobosa, ca. 29.12—31.20 μ diametro, pycnidiosporis bacillaribus, ca. 5.20—6.24 × 1.04 μ, hyalinis.

Apothecia ca. 1—8 mm diametro, discis rubiginosis vel aquilis, epithecio brunneo, ca. 6.88—8.32 μ crasso, hymenio 34.40—43.00 μ crasso, hypothecio hyalino ca. 30.10—34.40 μ crasso, ascis pyriformibus, ca. 24.96—29.12 × 8.32—11.44 μ octosporis, sporis ellipsoideis hyalinis, ca. 4.16—6.03 × 2.08—3.32 μ.

Cortex superus K+ lutescens; medulla K+ fulvescens, C—, KC+ rubescens, P+ lutea, pro parte interdum aurantiaca; chloroatranorinum et acidum physodicum continens.

Xizang: Zuogong, corticola, alt. 4400 m, 1976 VIII 31 leg. Zong Y. C. et Liao Y. Z. 506 (typus) in HMAS (Beijing) conservatur.

本种近似粉球袋衣及亚暗袋衣，但既无粉芽，又无裂芽，上表面具有强烈蜡样光泽，质地较硬，灰绿色，灰白色、淡灰绿色及褐色彼此相杂成花斑，衣体中央处褶皱较明显，边缘较平坦，髓层P+，含有 Chloroatranorin。

西藏：左贡，杉树上，海拔 4,400 米，1976 VIII 31，宗、廖 506（模式标本），保存于 HMAS，北京。

4. **珊茸梅衣** 新种 图1:3，图版8:5—6

Parmelia (Melaenoparmelia) **subverruculifera** Vej et Jiang, sp. nov. Fig. 1—3; Pl. 8—5, 6,

Species *P. verruculiferae* Nyl. similis est, sed isidiis densis granuliformibus, bacillaribus et coralliformibus, ca. 103—258 μ diametro et ad 2 mm longis differt.

Thallus foliaceus, supra brunneus, valde elevatus et concavus, valde rugosus, isidiis densis granuliformibus, bacillaribus et coralliformibus ca. 103—258 μ diametro et ad 2 mm longis (Fig. 1—3; Pl. 8: 6), tota superficie pilis densis hyalinis setaceis ca. 24—25.80 × 4.30—6.88 μ tectus, subtus ater, ad marginem atro-brunneus, rhizinis concoloribus.

Thallus ca. (107) 129—223.60 (258) μ crassus; cortex superus pseudoparenchymaticus, ca. 6.20—12.90 μ crassus; stratum algaceum ca. 25.80—51.60 μ crassum; cellulae algae virides, ca. 8.32—10.40 μ diametro; medulla ca. 68.80—180.60 μ crassa; hypha medullae 2.08—5.20 μ diametro; cortex inferus pseudoparenchymaticus ca. 6.20—12.90 μ crassus. Apothecia ignota.

Medulla K—, C+ rubescens, KC+ subrubescens, P—.

Acidum lecanoricum et non identificatum continens.

Xizang: Nyalam, ad corticem Betulae sp. alt. 3840 m, 1966 VI 22, leg. Vej T. C. et Chen J. B. 1899-(1) (typus) in HMAS (Beijing) conservatur.

本种衣体上表面褐色，在实体显微镜下观察时可见表面密布无色透明的刚毛状茸毛，类似于茸梅衣及瘤茸梅衣，但不同之处在于本种上表面除了颗粒状裂芽外，还密布大量棒状至多分枝的珊瑚状裂芽，其长度可达 2 毫米；裂芽表面同衣体表面一样密布无色透明的茸毛（插图1:3；图版8—6）。

西藏：聂拉木，桦树皮上，海拔3,840米，1966、VI、22，魏、陈1899-①(模式标本)，保存于HMAS，北京。

5.**藏岛衣** 新种 图版8—7

Cetraria xizangensis Vej et Jiang, sp. nov. Pl. 8—7.

Species *C. pinastri* (Scop.) Röhl. primo adspectu maxime similis est, sed ab ea differt essentialiter sorediis sordido-albis, medulla alba; cortice et medulla K +, medulla P +.

Thallus foliaceus, suborbiculatus vel irregularis, ca. 1—3 cm. diametro, substrato adhaerens rhizinis, non nisi ad marginem lacinulatus et subacclivis, lobis rotundis interdum pro parte crenulatis, supra stramineis et subunduloso-inaequabilibus, apicem versus sorediosis, sed juventute eis carentibus pruinis albis praeditis sorediis minute granularibus sordido-albis, zonis sorediorum ab apicibus loborum ad media ca. 0.5—2 mm latis, subtus flavidis vel brunneolis, interdum ad marginem subaquilibus, nitidis ubi flavidis et convexis, rhizinis aspersis brunneis vel aquilis.

Thallus ca. (94.64) 139.36—176.80 (195.60) μ crassus; cortex superus pseudoprenchymaticus, brunneolus ca. 20.80—27.04 μ crassus; stratum algaceum ca. 20.80—49.92 μ crassum; cellulae algae ca. 11.44—16.64 μ diametro, virides; stratum medullare ca. 20.80—97.76 μ crassum; hypha medullae ca. 3.12—4.16 μ diametro; cortex inferus pseudoparenchymaticus ca. 16.64—20.80 μ crassus. Apothecia ignota.

Cortex superus K + flavens; medulla alba, K+ flavens, C—, KC—, P+ flavens →aurantiaca; acidum usneicum continens.

Xizang: Nyalam, silvicola, ad corticem Betulae sp., alt. 3,900 m, 1966 VI 22, leg. Vej T. C. et Chen J. B. 1899 (Typus) in HMAS (Beijing) conservatur.

本种初看时近似于黄花岛衣，但不同之处在于粉芽为污白色，髓层白色，皮层K+，髓层K+，P+，不含吴尔品酸 (Vulpinic acid)。

西藏：聂拉木，林内桦树皮上，海拔3,900米，1966、VI、22，魏、陈1899(模式标本)，保存于HMAS，北京。

SPECIES NOVAE LICHENUM E PARMELIACEIS IN REGIONE XIZANGENSI

Vej Tzjan-czunj et Jiang Yu-mei

(*Institutum Microbiologicum Academiae Sinicae, Beijing*)

Summary

Five new species of the lichen family Parmeliaceae, namely, *Hypogymnia sinica* Vej et Jiang, *H. pruinosa* Vej et Jiang, *H. laccata* Vej et Jiang, *Parmelia subverruculifera* Vej et Jiang and *Cetraria Xizangensis* Vej et Jiang, all collected from Xizang (Tibet) Autonomous Region are reported. The lichen substances existed in these new species are determined by microchemical tests.

中国孢脐衣属的新种与新资料

魏 江 春

（中国科学院微生物研究所，北京）

摘要 本文报道了中国孢脐衣属四个新分类单位，其中包括三个新种：华东孢脐衣（*Lasallia sinorientalis* Wei），大理孢脐衣（*L. daliensis* Wei），苍山孢脐衣（*L. caeonshanensis* Wei）及东亚孢脐衣一新变种：樊净山变种（*L. asiae-orientalis* var. *fanjingensis* Wei）。此外，还从东亚孢脐衣的模式标本中首次发现了该种的子囊盘，并作了补充描述。

作者认为，根据有关中国地衣区系的现有知识，迄今尚未发现分布于中国的孢脐衣（*Lasallia pustulata* (L.) Merat）。而华东孢脐衣可能是孢脐衣在亚洲的地理替代种。

关键词 孢脐衣；大理孢脐衣；苍山孢脐衣；华东孢脐衣；东亚孢脐衣樊净山变种

孢脐衣属（*Lasallia* Merat emend. Wei）是石耳科（*Umbilicariaceae*）地衣中界限清楚的一个自然属。本属地衣已经描述的计有十四种；中国有三种；本文又描述了三个新种。因此，迄今，本属计有十七种；中国有六种。

关于石耳科地衣的分类系统我们曾进行过概括性的讨论（魏江春，1966）。近年来，不论是 Hoffmann-Frey 的单属系统（HOFFMANN, 1790—1801; FREY, 1931），还是 Scholander-Llano 的多属系统（Scholander, 1936; Llano, 1950），或是经作者修订了的新二属系统（魏江春，1966），都受到不同作者的赞成。不过，在此值得指出的是，多年来一直坚持旧二属系统的 Savicz（1950）在他学术研究的后期却改变了观点，放弃了旧二属系统，转而支持 Hoffmann-Frey 的单属系统（GOLUBKOVA et al., 1978）。与此相反的是单属系统的支持者与修订者 Frey 近年来却放弃了他的单属系统，转而赞成经我们修订了的新二属系统（Frey & POELT, 1977）。

至于中国的孢脐衣属地衣，尚缺乏专门系统的研究，只有零散的记载，其中最早的报道要算是法国的 Hue（1900）。他记载了采自云南的孢脐衣（*Umbilicaria pustulata* Hoffm. = *Lasallia pustulata* (L.) Mer.）一种。后来，Zahlbruckner 又以上述名称报道了云南、四川（1930）以及台湾（1933）的孢脐衣。佐藤月二（SATO, T. 1937）也报道了台湾产的同一种地衣。后来，朝比奈泰彦（ASAHINA., 1960）将 Zahlbruckner 报道的台湾孢脐衣描述为新种，即东亚孢脐衣（*Lasallia asiae-orientalis* Asahina）。佐藤月二所报道的台湾孢脐衣，从文中所提供的标本照片来看，也近似于东亚孢脐衣。而 Hue 及 Zahlbruckner 所指的云南、四川孢脐衣究为何种，尚待今后订正。不过，我们从采自云南及四川的大量孢脐衣属标本中迄今尚未发现孢脐衣（*L. pustulata*），而东亚孢脐衣却屡见不鲜。朝比奈泰彦（1931）在其"蕾轩独语"中也曾以"*Umbilicaria pustulata*"的名称报道了台湾产的孢脐衣。后来，佐藤正己（SATO, M. 1950）将它描述为新种，即 *Umbilicaria mayebarae* Sato = *Lasallia mayebarae* (Sato) Asahina. (1960)。本文作者于 1966 年将采自云南丽江西康油松树皮上的孢脐衣以"*Lasallia sinensis*"为名称描述为新种（魏江春 1966），后又改级为 *L. mayebarae* 的变种（魏江春 1981）。此外，根据其环纹状子囊盘特征，在孢脐衣属内建立了一新亚属：环纹孢脐衣亚属（Subg. *Pleiogyra* Wei），并在对石耳科的分类系统进行全面检查和

本文于 1982 年 4 月 2 日收到。

本文承秦仁昌教授及王云章教授审阅指正，郑儒永同志提供宝贵意见，苑兰翠同志拍摄部分照片，在此一并致谢。

本文原载于《真菌学报》, 1(1): 19-26, 1982.

修订的基础上，提出了新的二属系统(魏江春，1966)。

如果说东亚疱脐衣及中华疱脐衣是在澄清中国究竟有无 *L. pustulata* 种的过程中描述的，那么，本文所报道的新种之一，华东疱脐衣，即为这一过程中所描述的第三个新种。Llano (1950) 在他的西半球石耳科专著中将 Merrill 在标本签上手写的未经合格发表的名称 *Umbilicaria fokiensis* 的中国标本 (Dunn, no. 3938, IV—VI 1905) 当作 *Lasallia pustulata* 处理，而陆定安(1959)根据 Llano 的观点，将采自安徽黄山的两号标本也作为 *L. pustulata* 予以报道。我们对于采自安徽黄山、江西庐山、浙江天目山以及陕西秦岭光头山的大量疱脐衣标本与陆安定文中所引用的两号原始标本以及 Merrill 的标本 (Dunn, no. 3938)进行了比较研究之后证实，它们都是同一种类，而且并非真正的疱脐衣 (*L. pustulata*)，因而作为新种予以描述。

东亚疱脐衣 图版 I-1

Lasallia asiae-orientalis Asahina, *Journ. Jap. Bot.* **35**(4): 99—10 1(1960).

Holotypus: **Taiwan**, Mt. Alishan, 25XII 1925, Y. Asahina, no. F-377(TNS!)

朝比奈泰彦在描述本种时未发现子囊盘。我们于 1978 年在本种模式标本中发现了少量子囊盘。子囊盘网衣型，盘面平坦，有时具不规则形突起物或裂隙，类似于柱盘型盘面，直径 0.6—1.5mm，盘缘圆形或稍弯曲。囊层被厚约 10.40—16.64μm，子囊层厚约 76.96—83.20μm，囊层基厚约 129—146μm，子囊 20.80—104.00 × 8.32—31.00μm，内含一个孢子。孢子广矩圆形，初期无色，随后棕色，砖壁式多胞，52.00—90.00 × 25.80—45.00 μm

原 变 种

var. **asiae-orientalis**

地衣体直径 1—4.5cm。

基物：岩石。

产地：**台湾**阿里山，1925 XII 25，Y. Asahina, no. F-377(TNS!)。**云南**丽江，玉龙山下千海子，3,020m 1964 XII 18，魏，no. 2617(HMAS*)；玉胡村，2,820m 1964 XII 20，魏，no. 2663-① 及 2714 (1964 XII 22) (HMAS)；大理，苍山，2,050m 1964 XII 27，魏，no. 2750(HMAS)。**西藏**樟木口岸，3,650m 1966 V11，魏、陈，no. 510 及 1081 (3,500m 1966 V 21)，(HMAS)。

分布：日本、朝鲜、中国。

梵净山变种 新变种

var. **fanjingensis** Wei, var. nov.

Varietas nostra a var. *asiae-orientali* thallo pertenui superne coffeato subtus pernigro differt.

Guei-zhou: Mt. Fanjing-shan, Jian-dao-xia, 23 IX 1963, Wei, no. 830(**Holotypus**) in HMAS conservatur, Biejing (Peking)。

---

* Herparium Mycologicum Instituti Microbiologici Academiae Sinicae, Beijing

我们的变种不同于原变种之处在于衣体很薄，上表面咖啡褐色，下表面深黑色。

基物：岩石。

产地：贵州樊净山，尖刀峡顶部，1963 IX 23，魏，no. 830(HMAS)。

大理疱脐衣 新种 图版 I-2

Lasallia daliensis Wei, sp. nov.

Species similis *Lasalliae sinorientali* sed subtus perobscuris dissimilis et similis *L. asiae-orientali* sed pagina superiore sine microstructura areolata recedit.

Thallus monophyllus, ad 2.5—4.5cm diametro, coriaceus, rigidus, pagina superiore nicotianus, bistraceus vel atrocastaneus, saepe pruinosus, glabrus, sine microstructura areolata, opacus, interdum centro undulatus; pustulis rarioribus conspicuis, ca. 2—4mm latitudine et 0.5—1mm altitudine, saepe ad apicem penetrantibus inactinodispositis, centro absentibus ve lrarissimis; isidiis atro-brunneis, coralloideis ramosis in fasciculum globulosum vel linearem aggregati, sup erficialibus margineque dispositis; subtus granulosus, vel papillaris, atrofuscus, centro ater, ad peripheriam saepe pruinosus, opacus, laculosus, sine rhizinis.

Thallus ca. 292.40—555.30μm crassus; cortex superus paraplectenchymaticus, ca. 34.40—43.00μm crassus, prope superficiem brunneus; strato gonidiali continuo ca. (17.70) 24.70—(34.40)49.40μm crasso, cellulis algaceis ca. 6.24—8.32μm diametro; medulla ca. 189.20—258.00μm crassa, hyphis ca. 2.08—4.16μm diametro; cortice infero scleroplectenchymatico, ca. 20.80(43.00)—31.20(51.60)μm crasso, prope paginam inferam brunneo, papillis ca. 31.20—94.60μm altis, 79.04—131.04μm latis.

Apothecia vulgata, lecideina, ad 2.5mm diametro, margine rotundata vel flexuosa, prope peripheriam dispersa centro destituta; epitheciis atro-fuscis, ca. (6.24) 22.88—24.68 (61.70) μm crassis; theciis brunneolis ca. (72.80) 123.40—160.42 (292.40) μm crassis; hypotheciis atro-fuscis, ca. 46.80—308.50μm crassis; ascis pyriformis ca. 33.28—114.40 × 20.80—49.40μm(membrana ca. 2.08—3.12μm crassa); sporis solitariis, late oblongis, primum decoloribus dein fuscescentibus, muriformi-multilocularibus, ca. (22.88) 45.76—112.32 × (12.48) 24.96—52.00μm.

Pycnidia subglobosa, atro-fusca, nigra; pycnidiosporis simplicibus hyalinis, bacillaribus, ca. 16.64 × 4.16μm; pycnidiophoris ca. 3.70—4.16μm diametro.

Acidum gyrophoricum continens. (MCT)

Yunnan: Lijiang, Mt. Yulungshan, Ganhaizi, alt. 3,000m. 13 XII 1964, Wei J. C. 2421; Dali, Mt. Caeonshan, alt. 3,300m 29 XII 1964, Wei J. C. 2778 (**Holotypus**) in HMAS consernvantur, Beijing (Peking). Saxicola.

本种类似于华东疱脐衣，但是不同之处在于下表面暗色；类似于东亚疱脐衣，但是不同之处在于上表面无微粒状网块结构。

衣体单叶型，直径为 2.5—4.5cm，革质，坚硬；上表面烟草棕色，茶褐色或暗栗褐色，往往覆有粉霜层，无微粒状或网状小区块型结构，光滑而无光泽，有时中央波状隆起；疱状突起明显而稀少，直径 2—4mm 左右，较大者顶部凹凸不平或有穿孔，不呈放射状排列，中央部分缺乏或极少有疱状突起；裂芽黑褐色，珊瑚状多分枝，成球状簇群或线条状排列，分枝末端呈头状膨大，着生于上表面及边缘；下表面具乳头状突起或颗粒状结构，暗棕色，中央暗黑色，近周缘地带往往覆有粉霜层，无光泽，具明显的凹穴，假根缺如。

衣体厚约 292.40—555.30μm；上皮层为假薄壁组织，厚约 34.40—43.00μm，靠近上表面部分褐色；藻层连续型，厚约 (17.20)24.70—(34.40)49.40μm，藻细胞直径约 6.24—8.32μm；髓层厚约 189.20—258.00μm，髓层菌丝直径约 2.08—4.16μm；下皮层硬壁组织，厚约 20.80(43.00)—31.20(51.60)μm，靠下表面部分褐色，乳头状突起高约 31.20—94.60μm，宽约 79.04—131.04μm。

子囊盘亚常见，网衣型，分散于近周缘地带，直径可达 2.5mm；盘缘圆形或弯曲形；囊层被暗棕色，厚约 (6.24)22.88—24.68(61.70)μm；子囊层厚约 (72.80)123.40—160.40(292.40)μm；囊层基黑棕色，厚约 46.80—308.50μm；子囊梨形，33.28—114.40 × 20.80—49.40μm（膜厚约 2.08—3.12μm）；子囊内含单孢，宽矩圆形，未成熟时无色，成熟后变为褐色，砖壁式多胞，(22.88)45.76—112.32 × (12.48)24.96—52.00μm。

分生孢子器亚球形，暗棕色，黑色；器孢子单胞，无色透明，杆状，16.64 × 4.16μm；器孢子梗粗棒状，直径 3.70—4.16μm。

衣体内含三苔色酸。

基物：岩石。

产地：**云南**丽江，玉龙山，干海子，3,000m 1964 XII 13，魏，no. 2421；大理，苍山，3,300m 1964 XII 29，魏，no. 2778（主模式）保存于 HMAS。

苍山疱脐衣 新种 图版 I-3

Lasallia caeonshanensis Wei, sp. nov.

Species f. *majori Lasalliae asiae-orientalis* similis sed subtus pallidis, brunneolis, circa umbonem ater differt.

Thallus monophyllus, ca. 4—6cm diametro, coriaceus, rigidus, superne atro-cinereus, atro-cinerofuscus et atro-violaceus vel sordide violaceus, saepe pruinosus, crystallinus, micrositructura areolata, opacus; saepe centro actino-undulato-elevatus et interdum subrugosus; pustulis nconspicuis, minoribus, rarioribus et inactinodispositis, centro absentibus vel rarissimis isidiis atro-fuscis, coralloideis ramosis apice globulosis, superficialibus margineque dispositis, centro absentibus vel rarissimis; subtus granulosus vel papillaris, brunneolus, interdum atro brunneo-confusus, circa umbonem ater, ad peripheriam saepe subpruinosus, lacunis conspicuibus.

Thallus ca. 258—459μm crassus; cortex superus paraplectenchymaticus, ca. 20.80—34.40(60.20)μm crassus, ad superficiem brunneus; stratum gonidiale continuum, ca. 25.80—52.00μm crassum; medulla superior contans ex hyphis valde laxe intricatis, ca. 43.00—86.00μm crassa, medulla inferior constans ex hyphis parallelis ad superfiem valde compactis, ca. 215—258μm crassa; cortex inferus paraplectenchymaticus, brunneus, ca. 25.80—43.00μm crassus.

Apothecia vulgata, lecideina, pseudomphalodisca, ad 2mm diametro, dispersa, interdum glomerata vel margine valde flexuosa, ad peripheriam disposita, centro absentia, epithecia atro-fusca, ca. 10.40—24.96μm crassa; thecia ca. 87.36—137.28um crassa; hypothecia brunneola, ca. 10.80—33.28μm crassa; paraphyses 72.80—124.80 × 3.12μm; asci pyriformes, 76.96—124.80 × 33.28—62.40μm; sporae solitariae, late oblongae, primum decolores dein fuscescentes, muriformi-multiloculares, 49.92—108.16 × 27.04—52.00μm.

Acidum gyrophoricum continens. (MCT)

Yunnan: Dali, Mt. Caeonshan, alt. 3,300m 29 XII 1964 Wei J. C. no. 2758(**Holotypus**) in HMAS conservatur, Beijing (Peking).

Saxicola.

本种近似于东亚疱脐衣大叶变种,但是不同之处在于下表面淡色,浅棕色,中央脐周围黑色。

衣体单叶型,直径 4—6cm,革质,坚硬;上表面暗灰色至暗灰褐色以及暗紫色或污紫色,往往覆有粉霜层,解剖镜下观察为晶体状,表面具微粒状或微网状结构,无光泽,中央脐处往往呈放射状大波浪型隆起,有时具微弱的皱纹;疱状突起不明显,小而稀少,不呈放射状排列,中央脐附近无疱状突起或极少见;裂芽丛头状,暗棕色,珊瑚状分枝,分枝末端呈头状膨大,散布于上表面及边缘,中央缺乏或极少;下表面颗粒状结构或密布乳头状小突起,浅褐色,暗褐色混杂在一起,中央黑色,周缘往往稍覆有粉霜层,凹穴比上表面相应的疱状突起更为明显可见。

衣体厚约 258.00—459.00μm;上皮层为假薄壁组织,厚约 20.80—34.40(60.20)μm,靠近上表面部分褐色;藻层连续不断型,厚约 25.80—52.00μm;上髓层由极疏松的交错菌丝所组成,厚约 43.00—86.00μm,下髓层由与表面相平行的极为紧密结合的菌丝所组成,厚约 215—258μm;下皮层假薄壁组织,褐色,厚约 25.80—43.00μm。

子囊盘常见,网衣型,假柱盘型,直径达 2mm,散生型,有时若干个聚集而生,或盘缘极弯曲,散布于近周缘地带,中央缺如。囊层被暗棕色,厚约 10.40—24.96μm;子囊层厚约 87.36—137.28μm;囊层基浅褐色,厚约 10.80—33.28μm;侧丝直径约 3.12μm,子囊梨形,76.96—124.80 × 33.28—62.40μm;子囊内含单孢;孢子宽矩圆形,初期无色,随后变为褐色,砖壁式多胞,49.92—108.16 × 27.04—52.00μm。

衣体内含三苔色酸。

基物:岩石。

产地:**云南**大理,苍山,向阳垂面,3,300m,1964 XII 29,魏,no. 2758(主模式)(HMAS!)。

华东疱脐衣 新种 图版 I-4

Lasallia sinorientalis Wei, sp. nov.

Species nostra *Lasalliae pustulatae* similis sed thallis minoribus, pustulis sparsis debilibus inconspicuisque, isidiis granulatis plano-compressis squamiformibus ramosissimisque, in extremitatibus auctis atque globosis differt.

Thallus monophyllus, ca. 2—3.5(6)cm diametro; superne atro-brunneus aterque, opacus, interdum pruinosus, centro elevatus; pustulis sparsis debilibus inconspicuisque; ad peripheriam et marginem isidiis granulatis plano-compressis squamiformibus ramosissimisque, in extremitatibus auctisatque globosis; subtus lacunosus, brunneolus, interdum in umbilico atro-brunneus, rhizinis destitutus, opacus, granulis planis.

Thallus ca. 160.43(in pustulis)—370.20μm crassus. Cortex superne paraplectenchymatus, ca. 43.00—51.60μm crassus, in dimidio superiore brunneolus ca. 14.56—27.04μm crassus, in dimidio inferiore hyalinus ca. 17.20—31.20μm crassus. Stratum gonidiale ca. 18.72—43.00 μm crassum. Medulla ca. 68.80—154.80μm in dimidio superiore laxa, ca. 41.60—77.40μm crassa, in dimidio inferiore compactus, ca. 43.00—77.40μm crassa. Cortex inferior scleroplectenchymatus, ca. 24.96—43.00μm crassus, in dimidio superiore hyalinus, ca. 12.48—25.80 μm crassus, in dimidio inferiore brunneolus, ca. 10.40—17.20μm crassus.

Apothecia ignota.

Acidum gyrophoricum continens. (MCT)

Shaanxi: Mt. Guongtou shan, alt. 2,700m 2 VII 1964, Wei J. C. no. 77.

Jiangxi: Mt. Lushan, alt. 1,250m 11 II 1965, Wei J. C. no. 3115, alt. 1, 150 m Wei J. C. no. 3117, alt. 1,170m Wei J. C. no. 3121; Wulaofeng, alt. 1, 200m 12 II 1965, Wei J. C. no. 3211; Guling, alt. 1, 170m 14 II 1965, Wei J. C. no. 3246 (**Holotypus**) in HMAS conservantur, Beijing (Peking).

Fujian: Dunn, no. 3938, IV—VI 1905(FH!).

Saxicola.

我们的种类似于疱脐衣，但不同之处在于衣体较小，疱状突起稀疏而微弱，裂芽呈小颗粒状，扁平小鳞状分枝，末端呈球状膨大。

衣体单叶型，直径约 2—3.5(6)cm；上表面暗褐色及浅黑色，无光泽，有时覆有粉状霜层，中央脐处隆起；裂芽颗粒状，压扁状的鳞片形并具分枝，末端呈球状膨大；下表面具凹穴，浅褐色，有时中央脐处暗褐色，无假根，无光泽，颗粒状结构平坦。

衣体厚 160.43μm (疱状突起处)—370.20μm，上皮层假薄壁组织，厚约 43.00—51.60 μm，其中上半层浅褐色，厚约 14.56—27.04μm，下半层无色透明，厚约 17.20—31.20μm，藻层厚约 18.72—43.00μm，髓层厚约 68.80—154.80μm，其中上半层疏松，厚约 41.60—77.40 μm，下半层紧密，厚约 43.00—77.40μm，下皮层为硬壁组织，厚约 24.96—43.00μm，其中上半层无色透明，厚约 12.48—25.80μm，下半层浅褐色，厚约 10.40—17.20μm。

子囊盘未见。

衣体内含三苔色酸。

基物：岩石。

产地：**陕西**秦岭，光头山，2,700m，1964 VII 2，魏，no. 77。**江西**庐山，1,250m，1965 II 11，魏，no. 3115；1150m，魏，no. 3117、3119；1，170m，魏，no. 3121；五老峰，1,200m，1965 II 12，魏，no. 3211；牯岭，1,170m，1965 II 14，魏，no. 3246 (主模式) (HMAS!). **福建**，Dunn，no. 3938，IV—VI 1905(FH!).

参 考 文 献

[1] Asahina, Y. (朝比奈泰彦)。1931. 蕾轩独语(其三十五)。植物研究杂志 **7**(4): 102。
[2] Asahina, Y. 1960. Lichenologische Notizen. *J. Jap. Bot.* **35**(4): 97.
[3] Frey, E. 1931. Weitere Betrage zur Kenntnis der *Umbilicariaceae Hedwigia* **71**: 94.
[4] Frey, E. & Poelt, J. 1977. Die Gattung *Lasallia* (Umbilicariaceae) (Flechten des Himalaya 13.) *Khumbu Himal.* **6**(3): 387.
[5] Golubkova, N. S., Savicz, V. P., Trass. H. H. 1978. Handbook of the Lichens of the U. S. S. R. 89.
[6] Hoffmann, D. G. F. 1790—1801. Descriptio et Adumbratio Plantarum e classe cryptogamica Linnaei Lichenes dicuntur I—III.
[7] Hue, A. 1900. Lichenes Extra-Europaei. *Nouv. Arch.* du *Mus. Hist. Nat.* **4**(2): 112.
[8] Llano, G. A. 1950. A Monograph of the Lichen Family Umbilicariaceae in the Western Hemisphere. Navexos P—831, Washington. D. C.
[9] Llano, G. A. 1965. Review of *Umbilicaria* Hoffm. and the *Lasallias Hvalradets* skr. **48**: 112.
[10] 陆定安。1959。中国地衣杂录 2——脐衣科 (Umbilicariaceae)。植物分类学报 **8**(2): 176。
[11] Sato, M. (佐藤正已). 1950. Notes on some remarkable *Umbilicariae* collected in Far Eastern Asia. *J. Jap. Bot.* **25**(8). 165.
[12] Sato, T. (佐藤月二). 1937. On the *Umbilicaria pustulata* Hoffm. *J. Jap. Bot.* **13**(4): 298.
[13] Savicz, V. P. 1950. Conspectus Lichenum Ad Floram Umbilicariacearum in URSS. *Not. Syst.* e *Sect. Crypt.* Inst. Bot. nom. V. L. Komarovii Acad. Sci. URSS. 97.
[14] Scholander, P. F. 1936. On the Apothecia in the Lichen Family Umbilicariaceae. *Nytt. Mag. Natur.* **75**: 1.
[15] Wang-Yang, J.-R. & Lai. M.-J. (杨王贞容、赖明洲). 1976. Additions and corrections to the Lichen Flora of Taiwan. *Taiwania* **21**(2): 228.
[16] 魏江春。1966。疱脐衣属的一新亚属。植物分类学报 **11**(1): 1。
[17] 魏江春。1981。中国地衣标本集(第一辑: 1—50)。植物研究 **1**(3): 81。
[18] Zahlbruckner, A. 1930. Lichenes in Handel-Mazzetti's. *Symbol. Sin.* **3**: 138.
[19] Zahlbruckner, A. 1923. Flechten der Insel Formosa in Fedde. *Repert. Spec. nov.* **33**: 49.

SOME NEW SPECIES AND MATERIALS OF LICHEN GENUS *LASALLIA* FROM CHINA

WEI Jiang-chun*

(Institute of Microbiology, Academia Sinica, Beijing)

ABSTRACT In this paper are reported 4 new taxa of *Lasallia* from China, of which 3 are new species: ***Lasallia daliensis*** Wei, ***L. sinorientalis*** Wei, and ***L. Caeonshanensis*** Wei and 1 as new variety: *L. asiae-orientalis* var. ***fanjingensis*** Wei.

G. K. Merrill used the herbarium name "*Umbilicaria fokiensis*" for a specimen (no. 3938, FH) collected by Dunn from Fujian prov. of China in 1905. Many years later, when G. A. Liano (1950) published his monograph, he cited "*U. fokiensis*" (nom. inval) as a part of the species *Lasallia pustulata* (L) Merat. Lu Ding-an (1959) also identified the specimen collected from Mt. Huang-shan, Anhui prov., China as *L. pustulata*. However, after a comparison of the above mentioned specimens with specimens from Mts. Huang-shan (Anhui prov.), Tian-mu-shan (Zhejiang prov.), Lu-shan (Jiangxi prov.), and Guangtou-shan (Shanxi prov.), the present author has come to the conclusion that, although they all belong to the same species, but are not conspecific with *L. pustulata* by having sparse and weak pustules and smaller thalli, etc. Hence, the Chinese materials are described by the author as a new species: *Lasallia sinorientalis* sp. nov., while *L. pustulata* is believed to be an entirely different species hitherto not found in China. It seems to the author that the new species may be a vicarious species of *L. pustulata*.

The apothecia of *L. asiae-orientalis* Asahina, which had been overlooked by Asahina are now found and described for the first time by the author from the type specimen.

EKY WORDS *Lasallia; L. daliensis; L. caeonshanensis; L. sinorientalis; L. asiae-orientalis* var. *fanjingensis*

---

* 1. =Vej Tzjan-czunj

2. The author takes this opportunity to express his grateful thanks to Dr. Syo Kurokawa, Dr. Hiroyuki Kashiwadani (TNS), Dr. Donald H. pfister (FH), Dr. Mason E. Hale (US), Dr. I. I. Abramov and Dr. N. S. Golubkova (LE) for the loan of type specimens.

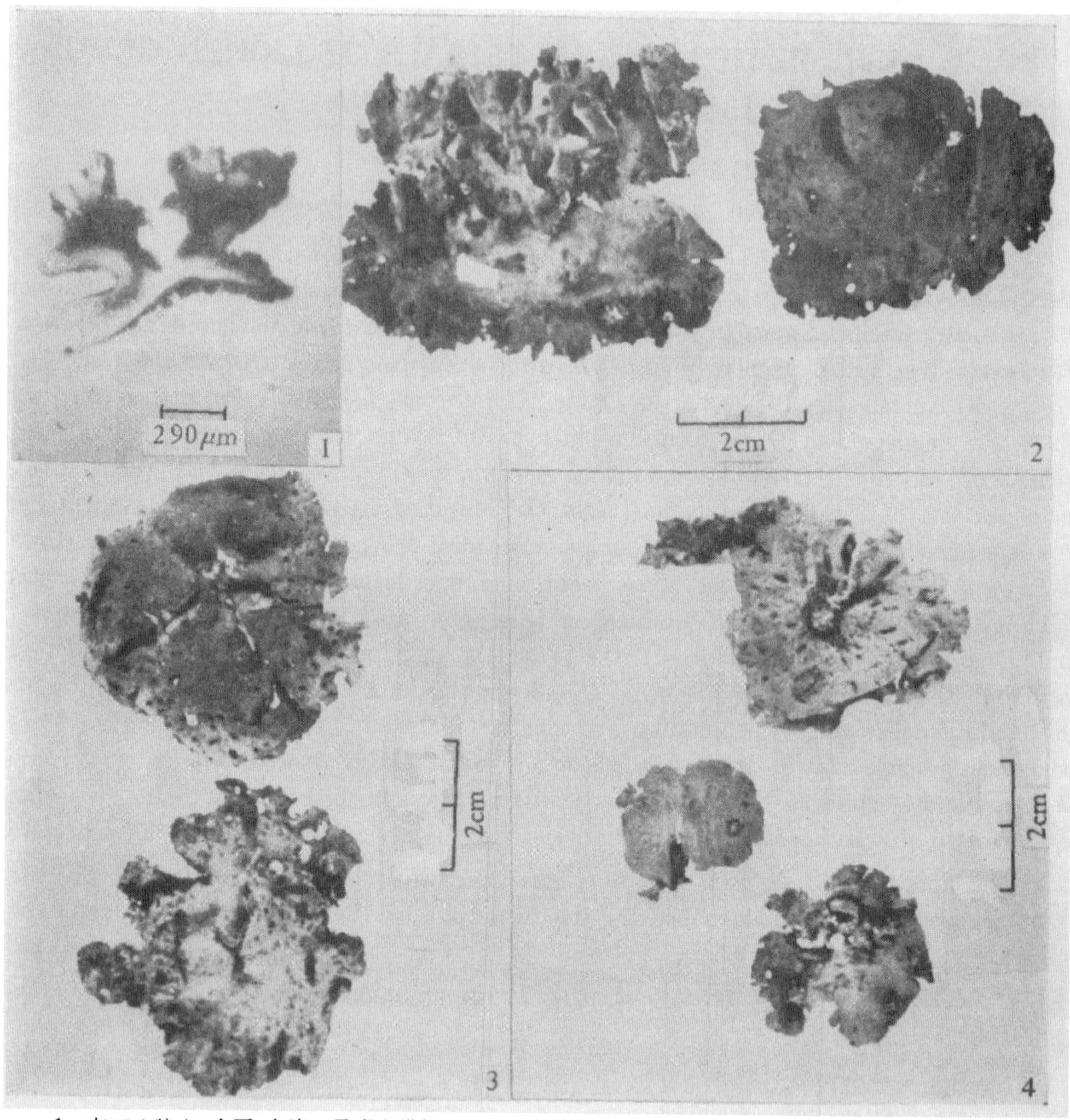

1. 东亚痂脐衣，中国，台湾。子囊盘纵切面。
Lasallia asiae-orientalis Asahina. Taiwan, China. Y. Asahina no. F-377, (TNS) holotypus. Longitudinal section of an apothecium.

2. 大理痂脐衣，中国，云南。上表面具子囊盘(左)；下表面具脐(右)。
Lasallia daliensis Wei, Yunnan, China. Wei no. 2778, (HMAS) holotypus. Left specimen showing dorsal surface with apothecia; right specimen showing ventral surface with umbilicus.

3. 苍山痂脐衣，中国，云南。下表面具脐(上)；上表面具子囊盘(下)。
Lasallia caeonshanensis Wei, Yunnan, China. Wei no. 2758, (HMAS) holotypus. Upper specimen showing ventral surface with umbilicus; lower specimen showing dorsal surface with apothecia.

4. 华东痂脐衣，中国，江西。下表面具脐(上)；上表面具子囊盘(下)。
Lasallia sinorientalis Wei, Jiangxi, China. Wei no. 3246,(HMAS) holotypus. Upper specimen showing ventral surface with umbilicus; lower specimens showing dorsal surface with apothecia.

西藏地衣的新类群与新资料

魏江春　　姜玉梅
（中国科学院微生物研究所）

本文报道了十种西藏地衣，其中三个新种，三个新的种下单位，一个种及一个变种的新组合。作者还首次描述了金丝绣球的子囊盘。此外，本文还涉及到一些有关的命名法问题。我们发现，*Gyrophora hypococcina* Jatta (A. Zahlbruckner, 1927,1934) 及*G. hypocrocina* Jatta (A. Zahlbruckner, 1930) 的种加词实际上乃是原缀法 *G. hypococcinea* Jatta (1902)的缀法错误或印刷错误。因此，按照国际植物命名法规第73条的规定，原缀法必须予以保留。这样，红腹石耳的正确学名应该是 *Umbilicaria hypococcinea* (Jatta) Llano, 而不是 *U. hypococcina* (Jatta) Llano 1950 (陆定安 1959)。

本文所依据的标本均保存于中国科学院微生物研究所真菌标本室(缩写为 HMAS)。

梅衣科 Parmeliaceae

1. **灌根条衣**　新种　图版 2:1

Cetrariastrum rhizodendroideum Wei et Jiang, sp. nov.

Species nostra *Cetrariastro cirrhato* similis est, sed a *C. cirrhato* rhizinis ramosissimis differt. Apothecia ignota.

Xizang: Zhangmukouan, Quxiang, alt. 3740 m, corticola, 17. V. 1966, Wei J. C. et Chen J. B. No. 762 (typus); Gyirong Xian, Gyirong, corticola, alt. 3300 m, 1975, Li W. H. No. 194, in HMAS conservatur.

本种类似于条衣 *Cetrariastrum cirrhatum*，但是，不同之处在于具有灌丛状分枝的假根。

西藏：樟木口岸，曲香，海拔 3740 米，小灌木上，1966 年 5 月 17 日，魏江春、陈健斌 762 (模式标本)；吉隆，海拔 3300 米，高山栎上，1975 年，李文华 194。

松萝科 Usneaceae

2. **金丝刷**

Letharielia cladonioides (Nyl.) **Krog**, Norw. J. Bot. **23**: 93. 1976 p.p. —— *Chlorea cladonioides* Nyl. Synops. Lich. **1**: 276. 1860. —— *Usnea cladonioides* DR. Svensk Botanisk Tidskr. **20**: 91. 1926; Mot. Lich. Gen. Usnea Stud. Monogr. Pars Syst. 44. 1936—38. —— *Usnea reticulata* DR. Svensk Botan. Tidskr. **20**: 91. 1926. —— *Usnea sernanderi* Mot. Lich. Gen. Usnea Stud. Monogr. Pars Syst. 46. 1936—38.

本种与曲金丝 *L. flexuosa* 不同之处在于地衣体灌丛状分枝繁茂；主枝与分枝均较笔直，不呈明显的弧形弯曲，有时只在尖细分枝末梢处微弯曲，尤其当末梢处密布粉芽时则弯曲更明显。曲金丝主枝及分枝均呈明显的弧形弯曲，分枝稀疏，形态上易与本种相区

本文原载于《植物分类学报》, 20 (4): 496-501, 1982

别。但是，Krog 只根据化学成分相同并强调了形态学上的相同特征，却忽视了形态学上的相异特征而并为一种。

基物：树干及灌丛枝。

产地：西藏，察隅，海拔 4260 米，1966 年 6 月 29 日，郑度 73-08；米林，海拔 4200 米，1972 年 7 月 4 日，中国科学院植物研究所（无号）；林芝，海拔 4400 米，1974 年，李文华 39。

分布：四川、西藏（Mot. 1936—38，p. 44）.

3. **曲金丝** 新组合

Lethariella flexuosa (Nyl.) Wei, comb. nov. —— *Chlorea flexuosa* Nyl. Synops. Lich. **1**: 267. 1860. (Isotypus: J. D. Hooker, no. 1735 (H. herb. Nyl. 36698)). —— *Usnea flexuosa* DR. Svensk Botan. Tidskr. **20**: 91. 1926. —— *Usnea hookeri* Mot. Lich. Gen. Usnea Stud. Monogr. Pars Syst. 45. 1936—38. —— *Lethariella cladonioides* Krog, Norw. J. Bot. **23**: 93. 1976 p.p.

地衣体分枝稀疏，主枝与分枝呈明显弧形弯曲，枝体近圆柱形，表面网状棱脊钝圆，主枝直径约 1 毫米，幼小分枝针芽状，无粉芽，往往覆以微薄的白色粉霜，无粉霜处则呈蜡样光泽。

基物：灌木。

产地：西藏，那曲，海拔 5800 米，1976 年 6 月 10 日，赵魁义 35；巴青，海拔 5000 米，1976 年 8 月 30 日，赵魁义 50。

分布：西藏（Mot. 1936—38，p. 45—46）.

4. **金丝绣球** 图版 2:2

Lethariella cashmeriana Krog, Norw. J. Bot. **23**: 91. 1976.

本种衣体分枝繁茂、密集而坚硬，呈绣球状，分枝末梢不尖锐而较钝圆，体内含有三苔色酸（gyrophoric acid）。我们在西藏标本中首次发现大量子囊盘及分生孢子器。

子囊盘侧生，直径 1—13 毫米，盘面褐色至暗褐色，基本无光泽或稍具光泽，幼小盘面深陷于果托中，呈茶碗形；老熟的盘面不深陷于果托中，呈浅盘形；果托外壁与衣体同色，局部桔红色，灰白色至污白色，具明显的纵向刀刃型棱脊，纵向棱脊之间具微弱的横向棱脊，构成网格状棱脊。

囊层被红棕色，厚度 4.16—6.24 微米；子囊层淡色至淡褐色，厚度 43.00—60.20 微米；囊层基淡褐色，半透明，厚度 77.40—94.60 微米；侧丝有分隔，基本无分枝，有时末梢具一次二叉式分枝，顶端微膨大，直径约 2.08 微米（顶端膨大处直径 3.12—4.16 微米），长约 33.28—37.44 微米；子囊棍棒状，35.36—41.60 × 14.56 微米，内含 8 个孢子；孢子无色透明，椭圆形，3.12—4.16 × 4.16—6.64 微米。

基物：柏树。

产地：西藏，昌都，大若卡林场，火烧迹地，柏树林内，海拔 4300 米，1976 年 7 月 2 日，李文华 76-95；类乌齐，海拔 4000 米，1976 年 7 月 2 日，宗毓臣、廖寅章 255；然乌，海拔 4300 米，1976 年 8 月 15 日，宗毓臣、廖寅章 477；左贡，海拔 4400 米，1976 年 8 月 31 日，宗毓臣、廖寅章 493。

5. **中华金丝** 新种 图版 2:4

Lethariella sinensis Wei et Jiang, sp. nov.

Species nostra *L. zahlbruckneri* (DR.) Krog similis est, sed a qua superficiebus thallorum valde rugosis reticulatis et substantiis inaequalibus differt.

Apothecia et Pycnidia ignota.

Ad ramulos Thujae sp.

Xizang: Qamdo, alt. 4300 m, 2. VII. 1976, Li Wen-hua 76-95-(1) (Typus); Riwoqe, alt. 4200 m, 2. VII. 1976. Zong Yu-chen et Liao Yin-zhang 260. in HMAS conservatur.

本种近似于金丝带(图版 2:3),但是,不同之处在于丝状体表面呈明显的网状稜脊,而后者则基本无此特征(图版 2:3);此外,二者所含化学成分相异(图版 2:5)。子囊盘及分生孢子器未见。

基物:柏树枝。

产地:西藏,昌都,大若卡林场火烧迹地,海拔 4300 米,1976 年 7 月 2 日,李文华 76-95-(1)(模式标本);类乌齐,海拔 4200 米,1976 年 7 月 2 日,宗毓臣、廖寅章 260。

石耳科 Umbilicariaceae

6. **红腹石耳**

Umbilicaria hypococcinea (Jatta) Llano, Monogr. Umbilicariaceae, 191. 1950. —— *Gyrophora hypococcinea* Jatta, Nouv. Giorn. Bot. Ital. N. S. **9**: 473. 1902. —— *Gyrophora versicolor* Räs. Arch. Soc. Zool. Bot. Fenn. "Vanamo", **3**: 78. 1949 (Typus in H). Lamb, Lndex Nom. Lich. 265. 1963. —— *Umbilicaria nepalensis* Poelt, 17 Khumbu Himal **6/3**: 426. 1977 (Typus in L, 262, non visus).

本种分布于我国陕西秦岭(模式标本产地为秦岭光头山),山西五台山及西藏珠穆朗玛峰地区。根据我们对 Räsänen, V. (1949) 曾描述的新种 *Gyrophora visicolor* Räs. 的模式标本所进行的对比研究证实,Räsänen 的新种实际上就是红腹石耳的异名。而根据 Poelt (1977) 对产于喜马拉雅的新种 *Umbilicaria nepalensis* 所做的形态描述及其分布高度(3600—5100 米)来看,与红腹石耳几乎完全一致。红腹石耳在珠穆朗玛峰地区的分布高度为 3700 米(聂拉木)至 5650 米(珠峰中绒布)。因此,我们将 Poelt 的新种也列为红腹石耳的异名。

关于红腹石耳的种加词的拼法,A. Zahlbruckner (1930) 在 Handel-Mazzetti 的"Symbolae Sinicae"第 3 卷中拼写为"*hypocrocina*",而在其"世界地衣名录"(Catalogus Lichenum Universalis) 第 4 卷 (p. 718) 与第 9 卷索引中以及 Llano (1950) 在其西半球的石耳科专著中均拼写为"*hypococcina*"。后来,陆定安 (1959, p. 175—176) 仅凭间接推理法,便从 A. Zahlbruckner 的上述两种拼法中选用了后者。经我们查对,Jatta (1902) 在发表红腹石耳新种时,其种加词的原拼缀为"*hypococcinea*"。这个原拼缀法,无论在印刷上或拼缀法上均无错误,而这样的错误却出现在 A. Zahlbruckner 的上述两种拼法中。因此,根据列宁格勒 (1978) 国际植物命名法规条款 73·1 的规定,红腹石耳种加词的原拼缀"*hypococcinea*"必须予以保留。

7. **粗根石耳** 稀根变种 新组合

Umbilicaria hirsuta (Sw.) Ach. **var. vuoxaënsis** (Wei*) Wei, comb. nov. —— *Gyrophora hirsuta* (Sw.) Ach. var. *vuoxaensis* Wei*, Not. Syst. e sect. Crypt. Inst. Bot. nom. V. L. Komarovii Acad. Sci. URSS, **15**: 8. 1962 (Holotypus in LE, Isotypus in HMAS).

基物：波曲河岸水边岩石上。

产地：西藏，聂拉木，海拔 3700—3720 米，1966 年 6 月 15 日，魏江春、陈健斌 1632 及无号标本。

本变种西藏标本的小生境(水边岩石上)与模式标本产地者(苏联列宁格勒州)基本一致。由此看来，本变种的存在，可能是一定生态特征在地衣形态上的具体反映。

8. **孔疱脐衣**

Lasallia pertusa (Rassad.) Llano, Monogr. Umbilicariaceae, 48. 1950. —— *Umbilicaria pertusa* Rassad. Comp. Rend. Acad. Sci. URSS. **14**(A): 348. 1929.

原变型 f. ***pertusa***

基物：岩石。

产地：西藏，聂拉木，海拔 3700—3720 米，1966 年 4 月 29 日，6 月 11 日、15 日，魏江春、陈健斌 92、1429、1648。

小鳞变型

f. **squamulifera** Wei et Jiang, f. nov.

Forma nostra a f. *pertusa* squamis orbicularibus differt.

Saxicola.

Xizang: Lhasa, alt. 3750 m, 3. IV. 1966, Wei J. C. et Chen J. B. 15 (Typus); Gyirong, alt. 2880 m, 14. VI. 1975, Zong Y. C. 112 in HMAS conservatur.

本变型不同于原变型之处在于具有圆饼形至小鳞片形裂芽，它们位于疱状突起顶端穿孔的边缘与疱状突起侧面以及疱状突起之间的叶面上。

基物：岩石上。

产地：西藏拉萨，海拔 3750 米，1966 年 4 月 3 日，魏江春、陈健斌 15 (模式标本)；吉隆，海拔 2880 米，1975 年 6 月 14 日，宗毓臣 112。

9. **东亚疱脐衣**

Lasallia asiae-orientalis Asahina, Journ. Jap. Bot. **35**(4): 99. 1960. —— *Umbilicaria pustulata* (non Hoffm.) A. Z. Fedde, Repert. 33: 49. 1933; Asahina, Journ. Jap. Bot. **7**: 103. 1931.

原变型 f. ***asiae-orientalis***

地衣体直径为 1—4.5 厘米。

基物：岩石。

产地：西藏，樟木口岸，海拔 3650 米，1966 年 5 月 11 日，魏江春、陈健斌 510；曲香，海拔 3500—3580 米，1966 年 5 月 19 日、21 日，魏江春、陈健斌，880-(1)、1070-(1)，

---

* = Vej.

1081；米林，海拔3000米，1974年8月23日，宗毓臣61。

分布：台湾省（A. Z. 1933, p. 49; Asahina, 1960, p. 99; 1973, p. 91）。

大叶变型

f. **maior** Wei et Jiang, f. nov.

Forma nostra a f. *asiae-orientalis* thallo majore, ca. 4—6 cm diametro, duriore et foveolis majoribus in pagina thalli inferiore ca. 4 mm diam. differt.

Saxicola.

Xizang: Zhangmukouan, Quxiang, alt. 3500 m, 19. V. 1966, Wei J. C. et Chen J. B. 884 (Typus) in HMAS conservatur.

本变型与原变型不同之处在于衣体较大，直径为4—6厘米，质地较坚硬；上表面紫褐色至污紫色，除上表面珊瑚状裂芽外，边缘具流苏型珊瑚状裂芽；下表面凹穴较大，直径可达4毫米。

基物：岩石。

产地：西藏，樟木口岸，曲香，海拔3500米，1966年5月19日，魏江春、陈健斌884（模式标本）。

10. **藏疱脐衣** 新种 图版2:6

Lasallia xizangensis Wei et Jiang, sp. nov.

Species nostra *Lasalliae asiae-orientali* similis est, sed a qua structuris granulosis nullis in superficie supera et foveolis profundioribus in superficie inferiore thalli.

本种接近于东亚疱脐衣，但不同之处在于上表面在实体显微镜下未见微粒状结构；下表面小凹穴较深。

原变种 var. **xizangensis.**

Thallus magnus, ca. 12 cm diametro, ca. 266.60—516.00 μ crassus, subtus granulis obtusis ca. 43—77 μ altis praeditus. Conidiophorum ca. 4.16 μ diametro; conidia 3.12—4.16 (5.20) × 1.04 μ.

Ad rupes.

Xizang: Zhangmukouan, Quxiang, alt. 3250 m, 16. V. 1966, Wei J. C. et Chen J. B. 881 (Typus) in HMAS conservatur.

地衣体大而厚，直径约12厘米，厚约266.60—516.00微米；下表面颗粒状突起钝而平，高约43—77微米；分生孢子梗直径约4.16微米；分生孢子3.12—4.16（5.20）× 1.04微米。

基物：岩石。

产地：樟木口岸，曲香，海拔3250米，1966年5月16日，魏江春、陈健斌881（模式标本）。

尖粒变种 新变种

var. **acuta** Wei et Jiang, var. nov.

Varietas haec ab varietate typica differt thallis minoribus ca. 2.5—5 cm diametro et tenuibus ca. 163.20—240.80 μ crassis; sutus granulis acutis, ca. 103.20—129.00 altis. Conidiophorum ca. 3.12 μ diametro; conidia ca. 2.08—4.16 × 0.64 μ.

Ad rupes.

Xizang: Zhangmukouan, Quxiang, alt. 3500 m, 19. V. 1966, Wei J. C. et Chen J. B. 888 (Typus) in HMAS conservatur.

本变种不同于原变种之处在于衣体小而薄，直径约 2.5—5 厘米，厚约 163.20—240.80 微米；下表面颗粒状突起尖锐，高约 103.20—129.00 微米。分生孢子梗直径 3.12 微米；分生孢子约 2.08—4.16 × 0.64 微米。

基物：岩石。

产地：西藏，樟木口岸，曲香，海拔 3500 米，1966 年 5 月 19 日，魏江春、陈健斌 888 (模式标本)。

参 考 文 献

[1] 陆定安，1959：植物分类学报，第 8 卷，第 2 期，第 173 页。
[2] Jatta, A., 1902: Nouv. Giorn. Bot. Ital. N. S. 9: 473.
[3] Llano, G. A., 1950: A Monograph of the Family Umbilicariaceae in the Western Hemisphere Washington, D. C.
[4] Zahlbruckner, A., 1927: Catalogus Lichenum Universalis, IV.
[5] ——————, 1934: Catalogus Lichenum Universalis, IX.
[6] ——————, 1930: In Handell-Mazzetti, Symbolae Sincae, III.

NEW MATERIALS FOR LICHEN FLORA FROM XIZANG

WEI JIANG-CHUN* JIANG YU-MEI

(*Institute of Microbiology, Academia Sinica*)

Abstract

In the present paper, ten lichen species from Xizang are reported. Three of them are new species and three new infraspecific taxa. One species and one variety are made as new combinations. The apothecium of *Lethariella cashmeriana* Krog is described for the first time.

This paper also deals with nomenclatural problems of certain species. The authors found that the specific epithets of *Gyrophora hypococcina* Jatta (A. Zahlbruckner, 1927, 1934) and *G. hypocrocina* Jatta (A. Zahlbruckner, 1930) were either orthographic or typographic errors for *G. hypococcinea* Jatta (1902) (original spelling). According to the Article 73:1 of the International Code of Botanical Nomenclature [Leningrad Code (1978)], the epithet (*hypococcinea*) must be retained and the name *Umbilicaria hypococcinea* (Jatta) Llano must, therefore be used substituting for *U. hypococcina* (Jatta) Llano (1950; Lu, 1959).

The authors wish to express their sincere appreciation to Prof. T. Ahti and Dr. O. Vitikainen (H) and Dr. Roland Moberg (UPS) for the loan of type specimens.

* =Vej Tzjan-czunj.

A new isidiate species of *Hypogymnia* in China

WEI Jiang-chun

(*Institute of Microbiology, Academia Sinica, Beijing*)

ABSTRACT *Hypogymnia hengduanensis* Wei, a new species containing barbatic acid and having isidia, is described based on Haary Smith's collections from Sichuan Prov. of China in 1934.

KEY WORDS *Hypogymnia hengduanensis*; barbatic acid; isidia

The new species is very interesting in both chemistry and morphology. It is the first one, that contains barbatic acid as a constant substance in the genus, and the only one, that bears real isidia in *Hypogymnia enteromopha* group. The description of the new species is based on Harry Smith's collections, who made two journeys in China, particularly in northeastern and western Sichuan (1921—23, 1934—35). I have found in **UPS** some interesting lichens in his extensive collections, including the new species described here.

Hypogymnia hengduanensis Wei, sp. nov. (Pl. I, Fig. 1—5)

Species haec *Hypogymniae enteromorphae* (Ach.) Nyl. primo habitu similis est, sed ab ea differt isidiis globulosis ovatis cylindricis et a speciebus omnibus *Hypogymniae* praeter specie *H. imshaugii* Krog thallo acidum barbaticum continenti.

Thallus laxe adnatus, ca. 19 cm longus, lobis linearibus libris, cavis, inflatis, ca. 1—2 mm diametro et ca. 3 cm longis, subisotomis, dichotomis, in axillis vulgo contractis; supra convexus, interdum depressus, griseolo-viridis, nitidus cum aliquot maculae denigratae vel interdum ex parte medullis fuscis nudis sed in partibus juvenilibus albis, isidiis globosis ovalibus, cylindricis, margine vel interdum supra cum lobulis adventitiis cavis inflatis, dorsiventralibus; subtus rugosus, niger, apicem versus brunneus vel brunneolus, nitidus, sparse perforatus (ca. 1.5 mm diametro).

Stratum corticatum pseudoparenchymaticum, cerinum, ca. 8.00—15.00μm crassum. Stratum algaceum ca. 31.00—54.00μm crassum. Stratum medullosum arachnoideum, ca. 48.00—159.00 μm crassum, juxta stratum algaceum incoloratum sed parte cetera brunneum, fuscum et nigrum.

Apothecia ignota.

Cortex K + luteus.

I am grateful to Dr. R. Moberg, director of the Herbarium of Uppsala University, for so kindly providing facilities for working at the Herbarium, being of great help during my stay there, and sending on loan the specimens, to Prof. R. Santesson for his kind encouragement and great help, and to Dr. L. Tibell for giving good advice about the chemistry of the new species. My visit to the Herbarium of Uppsala University was a part of exchange programmes between Academia Sinica and the Royal Swedish Academy of Sciences in 1982. I am also indebted to Prof. Ching R. C. & Dr. Chen X. Q. for reading the manuscript and revising the language of it.

This paper was originally published in *Acta Mycologica Sinica*, 3（3）: 214-216, 1984.

Medulla K + lutescens, C-, KC-, P-.

TLC: Thallus acidum barbaticum, atranorinum, et interdum sub stantiam ignotam continens.

At first sight the new species is very similar to *H. enteromorpha*, from which differs by having globose, oval, and cylindrical isidia on the upper surface of the lobes and containing barbatic acid together with atranorin, and sometimes an unidentified substance, and lacking protocetraric, physodic, physodalic, and diffractaic acids.

The new species is the first one that produces barbatic acid as a constant substance. Although the same substance is also found in *H. imshaugi*, it is an accessory. In addition, the latter has no isidia and contains physodic and physodalic acids together with atranorin.

Subsp. ***hengduanensis*** (Pl. I-1; Fig. 1—4)

Thallus lobis nigromarginatis, pagina infera paginam superam superant, medulla fusca et nigra, isidiis cylindricis numerosis, acidum barbaticum et atranorinum continens.

Holotypus: SICHUAN, Kangding regio, Yülinkong, Gomba La, ad truncum *Betulae* sp., alt. 3700m, 19 Oct. 1934, Harry Smith no. 14078 in **UPS** et isotypus in **HMAS-L** conservantur.

YUNNAN: Lijiang, Heibaishui, alt. 3140m, ad *Rhododendron* sp. 10 Nov. 1980, Jiang Yu—Mei no. 131—(1) in **HMAS-L** conservatur.

This subspecies is characterized by that the lower surface of the lobes is exceeding the upper surface with a conspicuous black rim, having numerous isidia and almost dark brown even black medulla, and containing atranorin and barbatic acid in it.

Subsp. ***kangdingensis*** Wei, ssp. nov. (Pl. I-2; Fig. 5)

Subspecies haec a ssp. *hengduanensi* pagina supera lobi paginam inferam superanti, isidiis sparsis, strato medulloso conspicuo juxta stratum algaceum, acidum barbaticum, atranorinum, et substantiam ignotam continenti differt.

Holotypus: SICHUAN, Kangding, Yülingkong, Gomba La, ad truncum *Rhododendron* sp., alt. 3700m, 22 Jul. 1934, Harry Smith no. 14061 in **UPS** et isotypus in **HMAS-L** conservantur.

This subspecies differs from ssp. *hengduanensis* by that the upper surface of the lobes is exceeding the lower surface without a conspicuous black rim, having fewer and sparse isidia and a white exterior medulla and a dark brown even black interior one, and containainning an unidentified substance together with atranorin and barbatic acid in it.

袋衣属一新种

姜玉梅　魏江春
（中国科学院微生物研究所，北京）

摘要　本文描述了袋衣属一新种：云南袋衣。其主要特征为裂片长，末端及边缘常具大量小裂片。裂片上表面无光泽，具丘疹；下表面呈明显的网状褶皱，中央呈暗玫瑰色至暗褐色，近末端呈淡玫瑰色至淡褐色，具稀疏而较大的孔洞。

关键词　袋衣属；云南袋衣；网状褶皱；具丘疹

作者在云南丽江玉龙山地区发现了袋衣属中一种迄今未被描述的地衣。经过形态学及化学特性的研究之后，现在描述如下：

云南袋衣　新种　图1—2

Hypogymnia yunnanensis Y.M. Jiang & Wei, sp. nov. Figs. 1—2

Typus: YUNNAN, Lijiang, Heibaishui, in arborum truncis Pini sp., alt 3000m

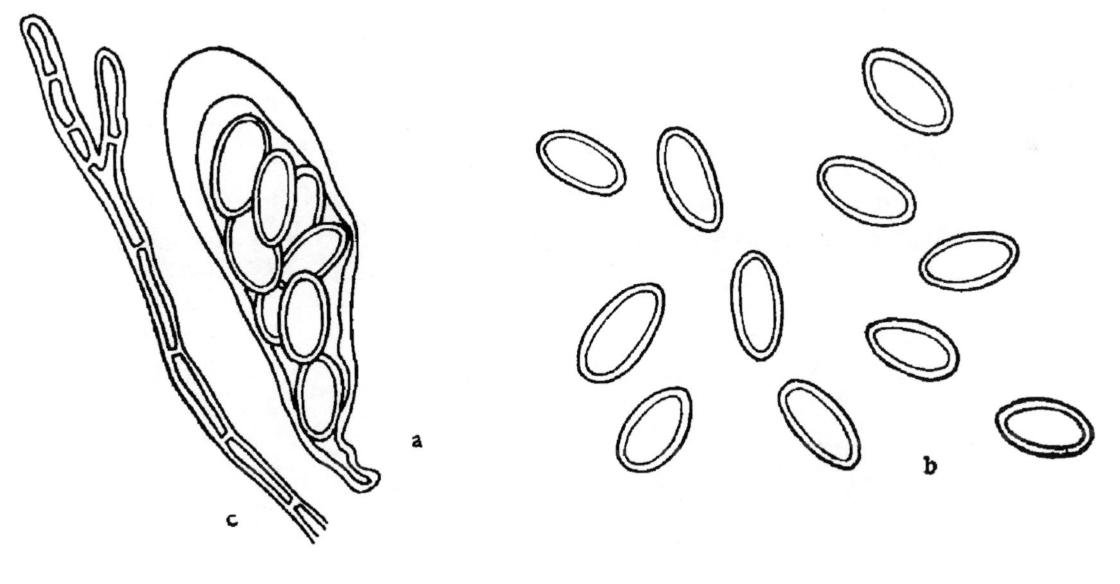

图1　云南袋衣的子囊，子囊孢子及侧丝

a. 子囊及子囊孢子　b. 子囊孢子　c. 侧丝

Fig. 1　An ascus with ascospores, ascospores and a paraphysis of *Hpogymnia yunnanensis* Y.M. Jiang & Wei

a. An ascus with ascospores b. ascospores c. a paraphysis

1989-07-14 收稿

作者衷心感谢苑兰翠同志为本文拍摄照片，感谢韩者芳同志为本文描绘插图

本文原载于《真菌学报》，9（4）：293-295, 1990.

11/XI/1980, Jiang YM, no. 286 (HMAS-L).

Species lobis supra papillosis et subtus plicatis reticulatis a congeneribus diversa; medula P—, KC+, atranorinum et acidum physodalicum continentia.

Specimina examinata: YUNNAN. Lijiang, Heibaishui, in fruticum ramulis, alt. 3140m,10/XI/1980, Jiang YM nos. 67,134-2 (HMAS-L).

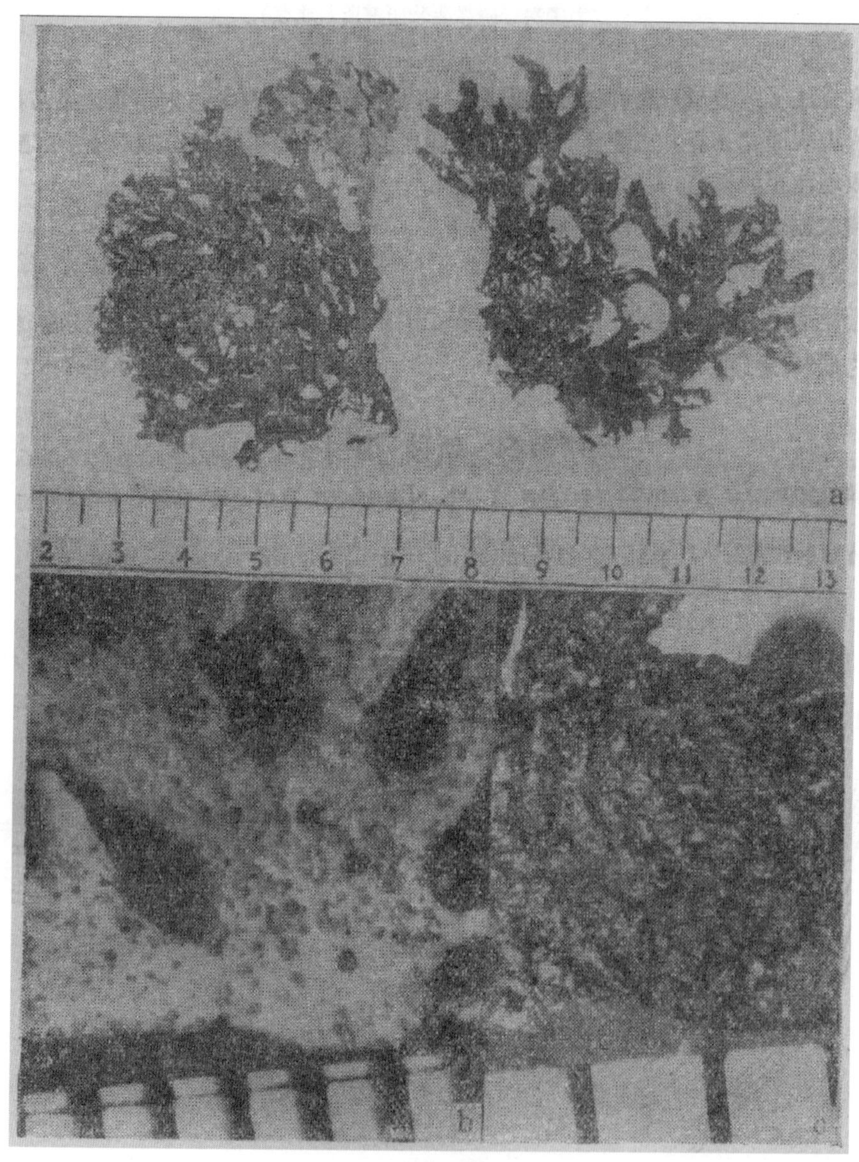

图 2 云南袋衣

a. 地衣体(比例尺每格为 5mm) b. 地衣体上表面(比例尺每格为 1mm)
c. 地衣体下表面(比例尺每格为 1mm)

Fig. 2 *Hypogimnia yunnanensis* Y. M. Jiang & Wei

a. Thallus (scale in 5mm) b. Upper surface of the thallus (scale in 1mm) c. lower surface (scale in 1mm)

本种以其丘疹状的上表面和明显网状褶皱的下表面与本属其它种类区别。地衣体由大量裂片组成。裂片长为 2—13mm，宽为 (0.5)2—3.5(5)mm，中空，近二叉式分枝，裂腋处稍缢缩，末端宽圆而略鼓起，边缘常具不定小裂片；上表面灰绿色并稍带淡褐色色度，不平坦，具丘疹状突起，无明显光泽，无粉芽及裂芽；下表面呈明显的网状褶皱，并具光泽及明显而稀疏的孔洞；裂片近基部暗玫瑰色至暗褐色，近末端处则呈淡玫瑰色至淡褐色。

子囊盘直径 (1)4—7(13)mm，具肿胀而带纵沟的短柄，长 1—3.5mm，盘面淡褐至褐色，具光泽。

上皮层拟厚壁组织，淡黄褐色，厚度为 18—23μm；藻层厚为 16—36μm，绿藻，亚球形，直径 7—12μm；髓层菌丝直径为 1.5—4.5μm；下皮层为拟厚壁组织，厚为 9—14μm。

子实上层淡褐色，厚为 3.5—7μm；子实层厚为 34—37.5μm，子实下层淡黄色，厚为 9—12.5μm；子囊倒卵形，27—32×7—10.5(12.5)μm. 内含 8 孢；孢子椭圆形，无色透明，5—7 × 3.5—4.5μm。

分生孢子器亚球形，未成熟。

皮层：K + 黄色；髓层：K + 黄色→红色，C−，KC + 红色，P−。

体内含黑茶渍素 (atranorin) 和袋衣甾酸 (physodalic acid)。

产地：云南丽江黑白水，距三大湾 50km. 云南松树干上，海拔：3000m, 11/XI/1980, 姜玉梅，no.286(Typus, HMAS-L)；灌丛枝上，海拔：3140m, 10/XI/1980, nos. 67,134-2 (HMAS-L).

A NEW SPECIES OF *HYPOGYMNIA*

JIANG YU-MEI WEI JIANG-CHUN

(Institute of Microbiology, Academia Sinica, Beijing)

ABSTRACT A new species of *Hypogymnia*, *H. yunnanensis* Y. M. Jiang et Wei, is described in the present paper. The new species is different from all the members of the genus by the papuliferous upper surface of the lobes and reticulated folds in the lower side of lobes. A detailed description of morphology, anatomy, and chemistry is given.

KEY WORDS *Hypogymnia*; *H. yunnanensis*; Reticulated folds; Papuliferous

Some species new to science and distribution in Umbilicariaceae (Ascomycota)

WEI Jiang-chun JIANG Yu-mei
Systematic Mycology & Lichenology Laboratory
Institute of Microbiology, Academia Sinica, Beijing

ABSTRACT: Eight species of Umbilicariaceae are reported in the present paper. Amongst them are four species new to science, *Umbilicaria altaiensis* sp. nov., *U. squamosa* sp. nov., *U. subumbilicarioides* sp. nov., and *U. taibaiensis* sp. nov.; three species, *U. nylanderiana, U. subglabra,* and *Lasallia papulosa* new to China, and one species *U. aprina* new to both China and Mongolia.

KEY WORDS: Umbilicariaceae; *Umbilicaria*; *Lasallia*; lichens; China; Mongolia

No monograph of the Umbilicariaceae from the Asia were published. The earlier studies of the family from the Asia were scattered in a series of publications made by different authors. A special paper regarding the family from Kamchatka was published by Savicz(1922). The papers "Die Gattung Lasallia" (Frey & Poelt, 1977) and "Die Gattung Umbilicaria"(Poelt, 1977) from Himalaya were published. Eleven species new to Mongolia were reported by Biazrov (1986). Two species new to Asia were reported by Wei & Biazrov(1991). Thirty seven species of the Umbilicariaceae from China were given in " A Conspectus of the lichenized ascomycetes Umbilicariaceae in China"(Wei & Jiang, 1988).

In the study on the Asian Umbilicariaceae four undescribed species were descovered. These four species have been named *U. altaiensis, U. squamosa, U. subumbilicarioides,* and *U. taibaiensis* respectively. All these species are provided with the Latin diagnoses, the English descriptions and figures, etc. The other three species, reported in the present paper, are new to China, and the last one, *U. aprina* is new to both China and Mongolia.

All the above mentioned eight species are described not only from the angle of morphology, anatomy, distribution and chemistry, but also from the angle of the holomorph, i. e. the teleomorph and the anamorph, if they exist and are found. For the genus *Umbilicaria* the thalloconidium state also exists in addition to the pycnidium state in the anamorph (Poelt, 1977; Hestmark, 1990).

Umbilicaria altaiensis Wei & Jiang, sp. nov. Figs. 1, 2.

TYPE: CHINA, Xinjiang, Mt. Altai shan, by the Halas lake, on rocks, alt. 2500m, Aug. 5, 1986, Coll. Gao X.Q. No. 2069(HMAS–L!).

This paper was originally published in *Mycosystema*, 5: 73-88, 1992.

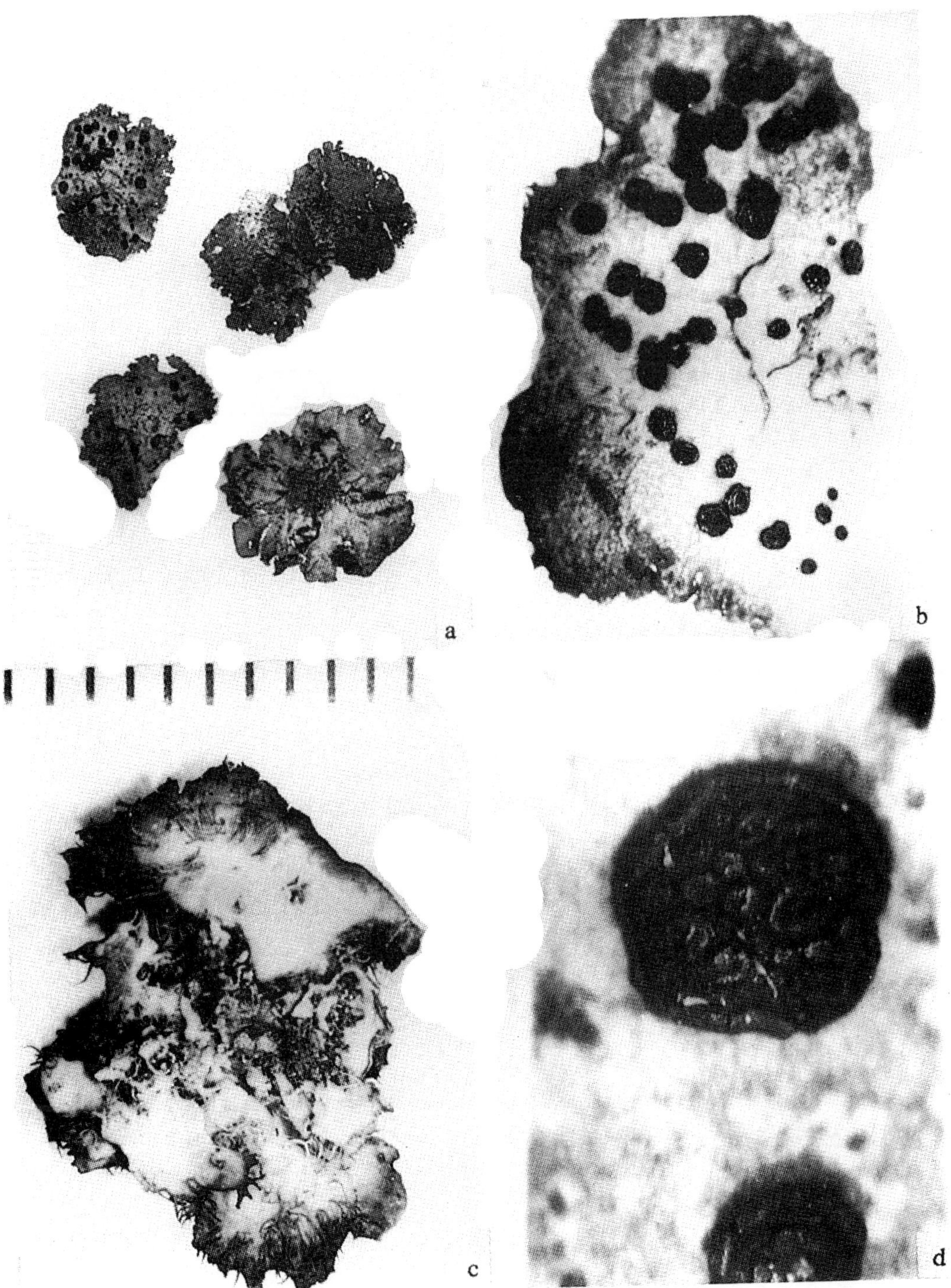

Fig. 1 *Umbilicaria altaiensis* Wei & Jiang, the holotype specimen. Gao XQ, No. 2069(HMAS–L), scale in mm. a. upper surface (above 4 thalli) and lower surface (below on the right 1 thallus) b. a part of a thallus (upper surface) c. lower surface of a thallus. d. Apothecia.

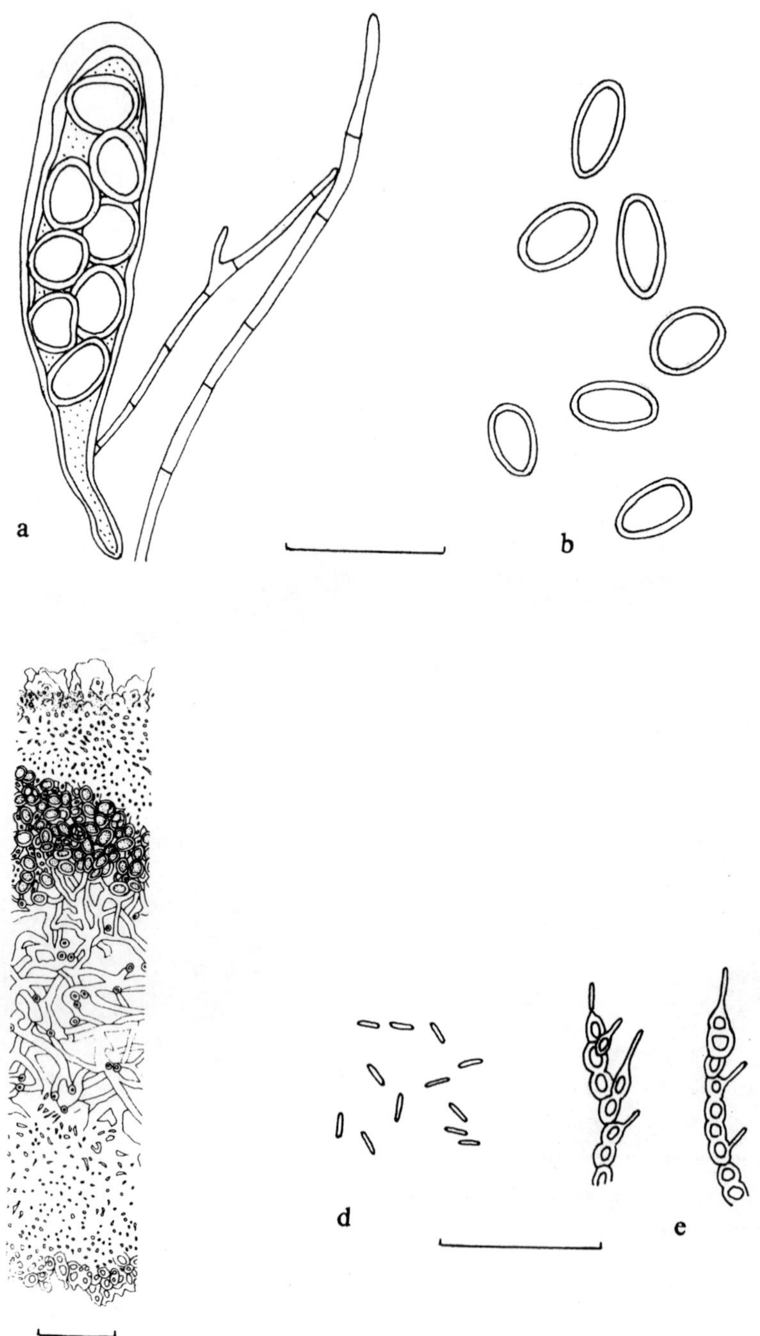

Fig. 2 *Umbilicaria altaiensis* **Wei & Jiang (2069).** a. Ascus containing spores with paraphyses; b. Ascospores; c. Thallus section; d. Pycnoconidia; e. Conidiophores with pycnoconidia. a,b,d,e: the scale in 18μm; c: the scale in 46.4μm.

DIAGNOSIS:

Umbilicariae cylindricae auct. primo adspectu maxime similis, sed apotheciis omphalodiscis; *U. crustulosae* similis, sed paginiis superis thallorum valde lacunosis et ascosporis unicellularibus hyalinis parvis (c. 9 × 7 μm).

DESCRIPTION:

Thallus monophyllous, rigid, orbicular, 1.5–2.5 cm diam.; margins whole, sometimes irregularly lobed, weakly revolute; upper surface irregularly and strongly lacunose, rugose and weakly elevated over umbo, covered with a thick layer of white pruina, grey brown in colour, in parts grey-green to dark-brown with tints of olive green visible by dissecting microscope; lower surface pale rose, greyish, avellaneous, marginally greyish brownish with concolor simple to branched rhizines.

Thallus 200–425 μm thick; necral zone 4.5–23 μm thick; upper cortex consisting of a thin brown outer layer of 1.5–7 μm thick and a thick colourless and much mucilaginous inner layer of 28–74.5 μm thick; algal layer 42–69.5 μm thick; medulla extremely loose, 55–230 μm thick; lower cortex consisting of a thick colourless and much mucilaginous inner layer of 23–83.5 μm thick and a greyish brownish outer layer of 14–37 μm thick, which consists of scleroplechtenchymatous tissues.

Apothecia numerous, up to 1.5 mm diam., adnate to weakly stipitate; typically omphalodisc pattern (disc with numerous columns or buttons); epihymenium 7–16 μm thick; hymenium 63–90μm thick; subhymenium 14.5–36 μm thick; paraphyses septate, simple to poorly branched, 1–2 μm diam.; asci 53–68.5 × 9–18 μm; spores ellipsoid, 7–11.5 × 5.5–7 μm. Pycnidia 260–400 × 220–255 μm; pycnoconidia bacilliform, 2.5–3.5 × 1 μm.

Algae: Trebouxia sp.

CHEMISTRY: no lichen substances.

SPECIMEN EXAMINED:

CHINA: Xinjiang, Mt. Altai shan, by the Hanas lake, alt. 2500m, on rocks, Aug. 5, 1986, Gao XQ, No. 2065 (pr. p.).

COMMENTS: The diagnostic characteristics for this species are: Thallus at first sight very similar to that of *Umbilicaria cylindrica* auct. but differing in that the thallus margin is weakly revolute downwards, and the apothecia typically omphalodiscs, the latter is similar to those of *U. crustulosa*, but differs in the strongly lacunose upper surface and the unicellular, hyaline and smaller ascospores. In addition, the new species has both the extremely thick mucilaginous colourless inner layer of the upper cortex and that of the lower cortex. The medulla is extremely loose.

Umbilicaria aprina Nyl.

Synops. Lich. 2:12(1869).

TYPE: AFRICA, Ethiopia (=Abyssinia), Mt. Dedschen, alt. 14,200 ft, coll. W.Schimper, H-Nyl. 31742! (holotype).

TYPE OBSERVATIONS: There are four fragments of thalli in Nylander's herberium of this species. The first is about 2 cm diam., rigid and fragile, with the upper surface sandy colour, faintly corrugated or vermiform visible by the unaided eye, and weakly granular under the dissecting microscope; lower surface black; rhizines cylindrical or slightly flattened, horn–shaped, simple or rarely dich- or occasionally trichotomously branched, black or often white, or white only at the tip.

The second (about 1 cm diam.) is similar morphologically to the first one, with a stalkes of about 3 mm long and usually about 2 mm thick and thicker (about 4 mm) at the base; upper surface reticulately ridged on the corresponding place to the umbilicus of the lower surface.

The third is also similar to the first, but is different in having no clear vermiform upper surface and by brownish color on periphery of the lower surface of the thallus.

The fourth is the largest one, 3.5 cm long and 2.5 cm wide; upper surface dark brown to black-brown and granulated, lightly vermiform on the margin only; lower surface black, the margin paler with almost white rhizines.

= *Umbilicaria syagroderma* Nyl. Flora 13:418(1860), nom. nud.

Type: See the type of *Umbilicaria aprina*Nyl. (H-Nyl. 31742 H!).

= *Umbilicaria cristata* Dodge & Baker, Ann. Missouri Bot. Gard. 25:565 (1938).

Type: Second Byrd Antarctic Expedition, Flora of King Edward VII Land, on pink granite, Mt. Helen Washington, Rockefeller Mt., 78 ° 05 ′ S – 155 ° 20 ′ W. 1934, coll. P.Siple, F.A.Wade, S.Corey and O.D.Stancliff No.HW-1a (type, FH!).

= *Umbilicaria spongiosa* Dodge & Baker, Ann. Missouri Bot. Gard. 25(2):566–567 (1938).

Type: Second Byrd Antarctic Expedition, Flora of Marie Byrd Land, on gray slate, arkosic sandstone, sericiteorthoclase schist, Lichen Peak, 76 ° 55 ′ S – 145 ° 20 ′ W, 1934, coll. P. Siple, F.A.Wade, S.Corey and O.D.Stancliff, No.73–9 (Type, FH!).

≡ *Gyrophora spongiosa* (Dodge & Baker) Golubk. & Savicz, Novitates Systematicae Plantarum non vascularium 1966: 259.

= *Umbilicaria antarctica* v. subvirginis Frey & Lamb, Trans. Br. Mycol. Soc. 22: 272–273 (1939).

Type: preserved in BERN (not seen).

= *Gyrophora korotkeviczii* Golubk. & Savicz, Novitates Systematicae Plantarum non vascularium. 261–263(1966).

Type: Antarctida orientalis, Litus Pravda, Bunger hills ad saxa March 3, 1957, leg.E-S. Korotkevicz (LE!)

var. **aprina**

DESCRIPTION: Thallus 1.5–2.5 cm diam.; upper surface pale sandy to dark brown, vermiform, and in parts reticulately ridged; lower surface pallid or pinkish to partly brown or black with pale rhizines and a long stalked umbilicus; umbilicus c. 2 mm long and no more than 1 mm thick.

Thallus 167–371 μm thick; upper cortex consisting of a brown outer layer, 7–9 μm thick, and a colourless inner layer, 7–27 μm thick; algal layer 18.5–69.5 μm thick; medulla 97–185.5 μm thick, including the layer of loose hyphae under the algal layer and the layer of concentrated and mucilaginous hyphae; lower cortex 23.5–52 μm thick, consisting of a colourless inner layer and a brown outer layer.

Apothecia absent.

Pycnidia 111–329 × 88–246 μm, pycnidial wall 4.5–9 μm thick,; conidia bacilliform, colourless, 2.5–5.5 × 1 μm.

Thalloconidia also present, hyaline and colourless to brownish, but not brown, distributed on both the lower surface of thallus and on the surface of rhizines; conidia subrounded 5–10.5 μm diam., conidial wall 1.5–2.50 μm thick.

CHEMISTRY: lecanoric and gyrophoric acids.

SPECIMENS EXAMINED:

CHINA: Xinjiang, Mt. Altai, on rocks by the Halas lake, alt. 2,500 m, Aug. 5, 1986, Gao X.Q. No.2089 (HMAS-L!).

AFGHANISTAN: [Lichenotheca Afghanica, No.25], Prov. Kabul: Paghman- Gebirge oberhalb des Ortes Paghman, c.500 m unterhalb der Talgabelung Chap- Darrah und Rast- Darrah, 68° 56′ E / 34° 37′ N, c.2,540 m, - Auf Silikatfelsen, April 10, 1977, coll. D.Podlech Pod 57 / 7 (BM!, LE!), and No.16, c.2,550 m, Steile Silikatfelsen, ± N-NW-Exposition, May 17, 1977 (BM!, LE!) and No.58 (LE!).

TURKEY: Ararat, ± 4,000 m, 1972, coll. by E sther Buchi, Herb. Rosmarie Honeger 720d (H!)

OTHER SPECIMENS EXAMINED:

Antarctica: Lichenes Antarctici Exsiccati,[issued by Rex B. Filson, National Herbarium of Victoria (MEL), No.19,] Antarctica, Mac. Robertson Land: west side of Nawson Rock, growing on pebbles and flat rock faces in sheltered positions. Feb. 1, 1974, Coll. Rex B. Filson No.14820 (UPS!, BM!). Eastern Antarctica, Oazis Bunger, Coll. V.S.Korotkevicz, Jan. 29, 1956 & Jan. 9, 1957 (LE!); Haswell insula, 1956, Coll. O. Vjalov, No.60 (LE!) Bunger Hills Oasis-2-station 66° 18′ S / 100° 45′ E, south of station, on stone and rocks, Feb. 11, 1989, Coll. M. Andreev, No.890907 (LE!) and on rock near a nest of Pagodroma nivea Feb. 22, 1989, Coll. V. Bulavintsev No.893101 (LE!).

LITERATURE RECORD FOR ASIA: Afghanistan(Poelt, 1977, p.415)

COMMENTS: It seems that the colour of the lower surface of the thallus is dependent upon the developmental stage of the thalloconidia distributed on it. It is light when immature, and dark when mature.

var. **halei** Llano,

Journ. Washington Acad. Science 46(6):183(1956)

TYPE: CANADA, N.W.T.,Baffin Island, head of Clyde Fiord, on exposed gneiss boulders, August 26, 1950, Coll. M. E. Hale, Jr., no. 450 herb. Llano, holotype; not seen.

= *Umbilicaria decussata* var. *rhizinata* Frey & Poelt, Khumbu Himal 6:419(1977).

Type: Moranen des Lobuche-Gletschers bes Lobuche, 4950–5000 m,(L247a). (not seen).

DESCRIPTION:

Thallus orbicular, 1.5–3cm diam.; upper surface blackish grey, but not sandy, grey and rugose over umbo and undulately elevated; lower surface on inner part around umbo covered with a circular, sooty, usually mottled patch terminating in black flecks, and contrasting sharply with the brown or grey brown outer lower rim.

Thallus 273.5–324.5 μm thick, including necral zone, colourless, 9–19 μm thick; upper cortex 30.5–57.5 μm thick, consisting of colourless inner layer, 18–34 μm thick, and brown outer layer, 9–12.5 μm thick; algal layer, 36–68.5 μm thick; medulla 97.5–139 μm thick; and lower cortex scleroplechtenchymatous, 30.5–68.5 μm thick consisting of colourless inner layer, 18–59 μm thick and brown outer layer, 9–20 μm thick.

Apothecia omphalodiscs, c.1 mm diam., epihymenium 5.5–14.5 μm thick; hymenium 54–72 μm thick; subhymenium brownish, 36–54 μm thick; hypothecium dark brown, 27–72 μm thick; paraphyses branched and septate, c. 2 μm wide; asci when immature, 41–45 × 9–14 μm.

Thalloconidia distributed on lower surface of thallus and on the surface of rhizines, conidia subglobose, brown, 5.5–1 μm diam., cell wall 1–3.5 μm thick.

CHEMISTRY: gyrophoric acid.

SPECIMENS EXAMINED:

CHINA: Jilin, Mt. Changbai shan, north slope, on rocks among lichens *Umbilicaria formosana* Frey in alpine tundra zone, alt. 2300–2600 m, Aug. 14, 1977, Coll. Wei, J.C. No. 2981 (HMAS-L !), and on rocks near the peak Tian Wen Feng , Aug. 19, 1987, Coll. Wei, J.C. Nos. 9011, 9012-(1) (HMAS-L!).

MONGOLIA: Hubsugul Lake (Kosogol Lake), on top of Mt.Muivu-Sardeika, July 23, 1902, coll. A. A. Elenkin (LE!).

OTHER SPECIMENS EXAMINED:

Norway: Oppland, Oystre slide, Valdresflya mellom riksvegen og Rasletjern, MP 87–89 06 (1617 IV), ± level plateau tundra with streams boulders and out crops, alt.1390–1400 m(BM!); Oppland, Vang, Ostsida av Bitihorn, MN 89 ° 95 ′ (1617 IV), On rock outcrops and boulders, alt.1170 m (BM!). S.Iceland: Alftavatn, Skaftar tunga, on lava bolock, Coll. unknown (BM!). Central Highlands: Storisandur Area, 1 km NNW of Blafell Hill, 700 m, on basalt stones in Cairn, Aug. 20, 1968, coll. Hordux Kristisson No.24253 (DUKE!).

LITERATURE RECORD FOR ASIA: Himalaya (Poelt, 1977, p.419).

COMMENTS: The specimens examined completely correspond to the descriptions of both var. *halei* Llano of *U. aprina* Nyl., and the var. *rhizinata* Frey & Poelt of *U. decussata*(Vill.) A. Zahlbr. although the types of them were not available for the study. The key differences between the two varieties are: var. aprina has a sandy upper surface to the thallus, lacks apothecia and distributed in Antarctica and Central and western Asia in addition to the type locality in Africa, while the var. *halei* Llano is of blackish grey upper surface, sometimes bearing apothecia and distributing in North East Asia and North Europe and North America.

New to both China and Mongolia.

Umbilicaria nylanderiana (A. Zahlbr.) H. Magn.

Lich. Sel. Scand. Exs.no.252(1937).

≡*Gyrophora nylanderiana* A. Zahlbr., Cat. Lich. Univ., 4: 720(1927).

≡*Gyrophora heteroidea* [var.] *δ corrugata* Ach., Lich. Univ. 219(1810).

≡*Umbilicaria corrugata* Nyl., Lich. Scand., 119(1861), non Hoffm. 1794.

TYPE: *Gyrophora heteroidea* [var.] *δ corrugata*, Ach. 177a–183 (BM! lectotype).

MISIDENTIFICATION: *Umbilicaria leiocarpa* sensu: Biazrov, in Bull. Mosc. O-Va lspeitatelei prirodei. otd. Biol. 91(6): 100(1986), non DC.

DESCRIPTION:

Thallus monophyllous, coriaceous, rigid, up to 5.5 cm diam., finely rugose, sometimes irregularly reticulate in parts, deeply incised-torn and irregularly lobed, perforate, grey-brown to dark grey brown, covered with a thin layer of white grey pruina; lower surface plane, smooth, sooty-black covered with a thick layer of sooty-black thalloconidia, the small and young lobes with a marginal narrow zone brownish to grey-brown, lacking the thalloconidia, the centre with an umbilicus about 6 mm diam.

Thallus 271–636 μm thick, including upper cortex of 17–29 μm thick (the upper brown layer 6–11 μm thick; the next colourless layer 9–14 μm thick); algal layer 21.5–59.5 μm thick; medulla consisting of a layer of loose hyphae directly below the algal layer, 12–412 μm thick, and a layer of compacted and mucilaginous hyphae above the lower cortex, 47–177 μm thick; lower cortex including a colourless layer, 7–18 μm thick and a brown layer, 9–27 μm thick.

Apothecia absent.

Pycnidia 106.5–125 × 79–116 μm.; conidia immature.

Thalloconidia arising on lower surface; conidia subround, brown, 7–11 μm diam., cell wall 2–3.5 μm thick.

CHEMISTRY: lecanoric, gyrophoric and umbilicaric acids.

SPECIMENS EXAMLNED:

CHINA: Xinjiang, Mt. Altai, Hanas lake, alt. 2500 m, on rocks, Aug. 5, 1986, coll. Gao, Nos. 2075–1, 2085, 2092. Xinjiang, Mt. Altai, near Hanas lake, alt. 2,500 m, on rocks, Aug. 5, 1986, Gao X. Q. Nos. 2075 (pr. p.) & 2092(HMAS–L).

MONGOLIA: Ubsunursk, Mt. Dulg, alt. 2680 m, on rocks, July 9, 1976, Coll. Biazrov, No. 6568 as *U. leiocarpa* (Moscow!). ①

RUSSIA: Siberia, by Tabat river, coll. N. Martianoff (H-Vain. 00182 in TUR 382!).

LITERATURE RECORD FOR ASIA: Mongolia(Biazrov, 1986, p. 100, as U. leiocarpa).

COMMENTS: This species is similar to *U. exasperata* (syn. *U. hyperborea*) but differs in having sooty black thalloconidia covering the lower surface of the thallus.

New to China.

Umbilicaria squamosa Wei & Jiang, sp. nov. Fig. 3A, 4.

TYPUS: CHINA, Yunnan, Dali, cangshan, alt. 3300 m, on rocks, Dec. 29, 1964, Wei J. C. No. 2779 (holotypus HMAS-L!)

MISIDENTIFICATION:

Umbilicaria esculenta sensu Wei & Chen, in Report on the scientific Investigations(1966–68) in Mt. Qomolangma Region, ser. Biology & Physiology, Science Press, Beijing, 1974, p.177.

Umbilicaria thamnodes sensu Wei & Jiang, in Proceedings of Symposium on Qinghai- Xizang (Tibet) plateau, p.1146, Lichens of Xizang, 1986, p.100 & Mycosystema, 1:86(1988), non Hue.

DIAGNOSIS:

Species nova *Umbilicariae esculentae* primo aspectu maxime simila, sed thallis coriaceis rigidis crassis et supra hic

① Moscow = Herbarium of Institute of Evolutionary Morphology and Ecology of Animals, Academy of Sciences of Russia (the former U.S.S.R.)

inde corticibus refringentibus squamiformibus et rhizinis non intricatis.

DESCRIPTION:

Thallus monophyllous, coriaceous, rigid and thick, 3–11 cm diam., the upper side black-violet or dirty violet, covered with a very thin chalk-white pruina, locally with a broken up and elevated cortex occuring squamules, locally with the white medulla visible where the cortex has peeled off, and afterwards becoming black-violet; below black, finely granulose with scattered black warts, and locally bare or with scattered, cylindrical, branched and short (1–1.5 m long), but not entangled rhizines.

Apothecia rare, gyrodiscs, 4 mm diam; asci immature.

On rocks.

CHEMISTRY: lecanoric and gyrophoric acids.

SPECIMENS EXAMINED:

CHINA: Yunnan, Dali, Cangshan, alt. 3300 m, on rock, Dec. 29, 1964, Coll. Wei J. C. Nos. 2778–1, 2779: (HMAS-L! type). Xizang, Nyanam, Zham, alt. 3,700 m, May. 4 & 11, 1966, Coll. Wei & Chen,nos. 289 & 451(HMAS-L!); Nyalam, Quxiam, alt. 3,200–3,580 m, June 18, 19 &21, 1966, Coll. Wei & Chen nos. 830, 880 & 1070(HMAS-L!).

HIMALAYA: alt. 12000ft, 2092 Herb. Ind. or Hook. fil & Thomson, U. calvescens v. hypomelaena Nyl. Hook fil 1857 [FH! & LE (1868)!].

COMMENTS: *U. esculenta* is most similar at first sight but this species has coriaceous, rigid and thick thalli, with the upper side locally with broken up and elevated patches of cortex occuring as squamules, and below with cylindrical and not entangled rhizines.

A Himalayan species

Umbilicaria subglabra (Nyl.) Harm.

Lich. France 4:707(1909).

≡*Gyrophora subglabra* Nyl., Lich. Env. Paris, 135(1896).

TYPE: FRANCE, Pyrenees, Cebennis, coll. by Mende (H-Nyl. 31666, holotype, H!).

TYPE OBSERVATION:

There is only one thallus in Herb. Nylander 31666. Thallus 3 × 4 cm diam., peripherally incised to lacerate forming truncate torn lobes; upper surface black brown, covered with a thick layer of white pruina becoming white with minute radial fissures in centre over umbo, grey at margin; lower surface smooth, pallid around umbo and grey brown in the other parts covered with a thick layer of black brown thalloconidia giving a black lower surface, peripherally grey-brown in a narrow zone lacking thalloconidia.

Apothecia absent.

DESCRIPTION:

Thallus 3 × 4 cm, upper surface covered with a thick layer of white pruina, showing sordid white with radial lines in centre over umbo, grey-brown to black brown in other parts; lower surface smooth, covered with black brown to black at the thalloconidia, appearing black with a narrow grey -brown zone at the peripherary, umbo 6 mm diam.

CHEMISTRY: lecanoric (trace) and gyrophoric acids.

U. umbilicarioidis primo adspectu maxime similis, sed a qua differt corticibus superis palisadoplectenchymatis,

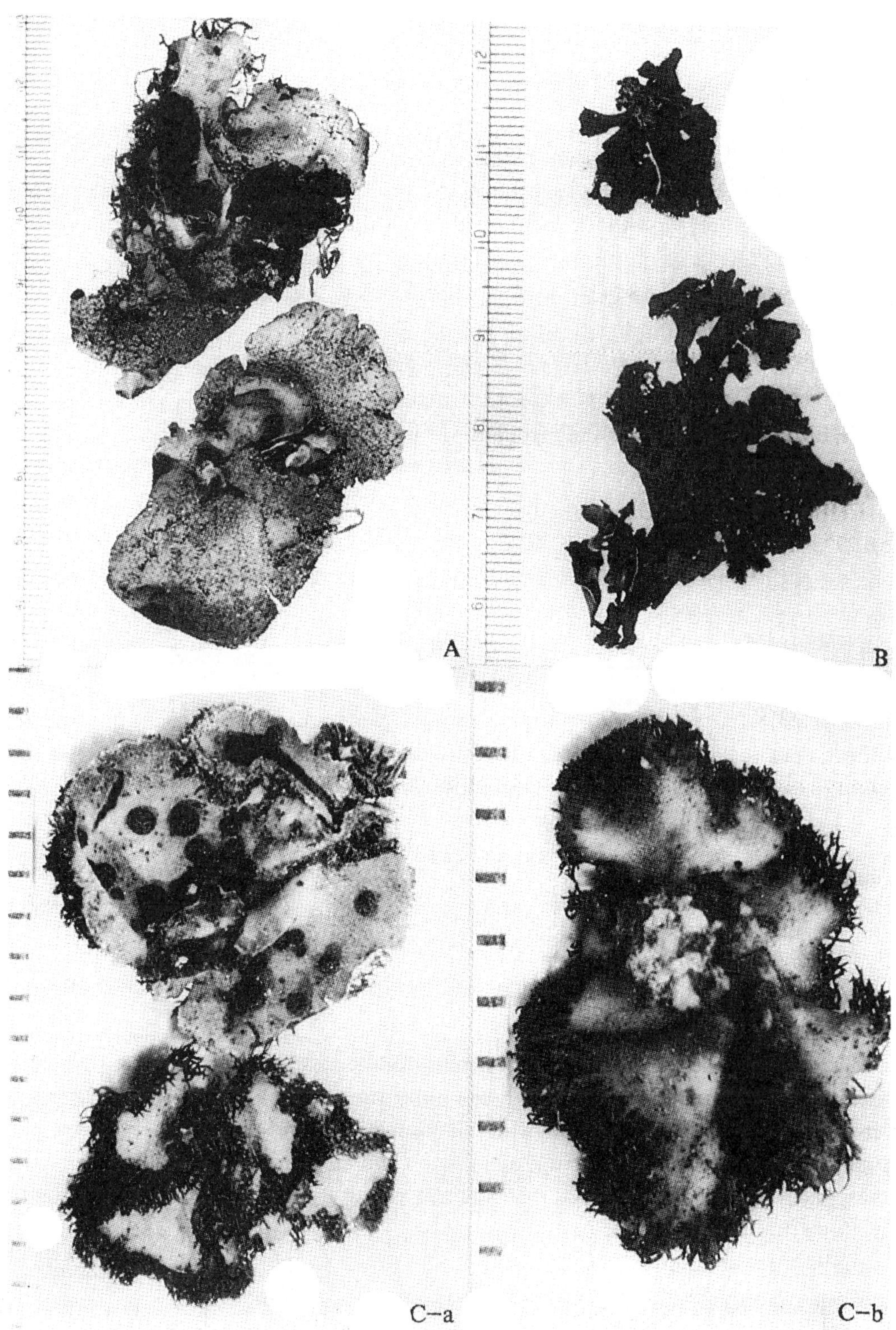

Fig. 3 A. *Umbilicaria squamosa* Wei & Jiang, the holotype specimen (Wei, No. 2779 in HMAS–L), scale in mm. B. *Umbilicaria subumbilicarioides* Wei & Jiang, the holotype specimen (Wei & Chen, No. 470 in HMAS–L), scale in mm. C. *Umbilicaria taibaiensis* Wei & Jiang, the holotype specimen collected by Wei from Mt. Taibai shan, Baxiantai, June 3, 1963(HMAS–L)., the scale in mm; a. the upper surface; b. the lower surface.

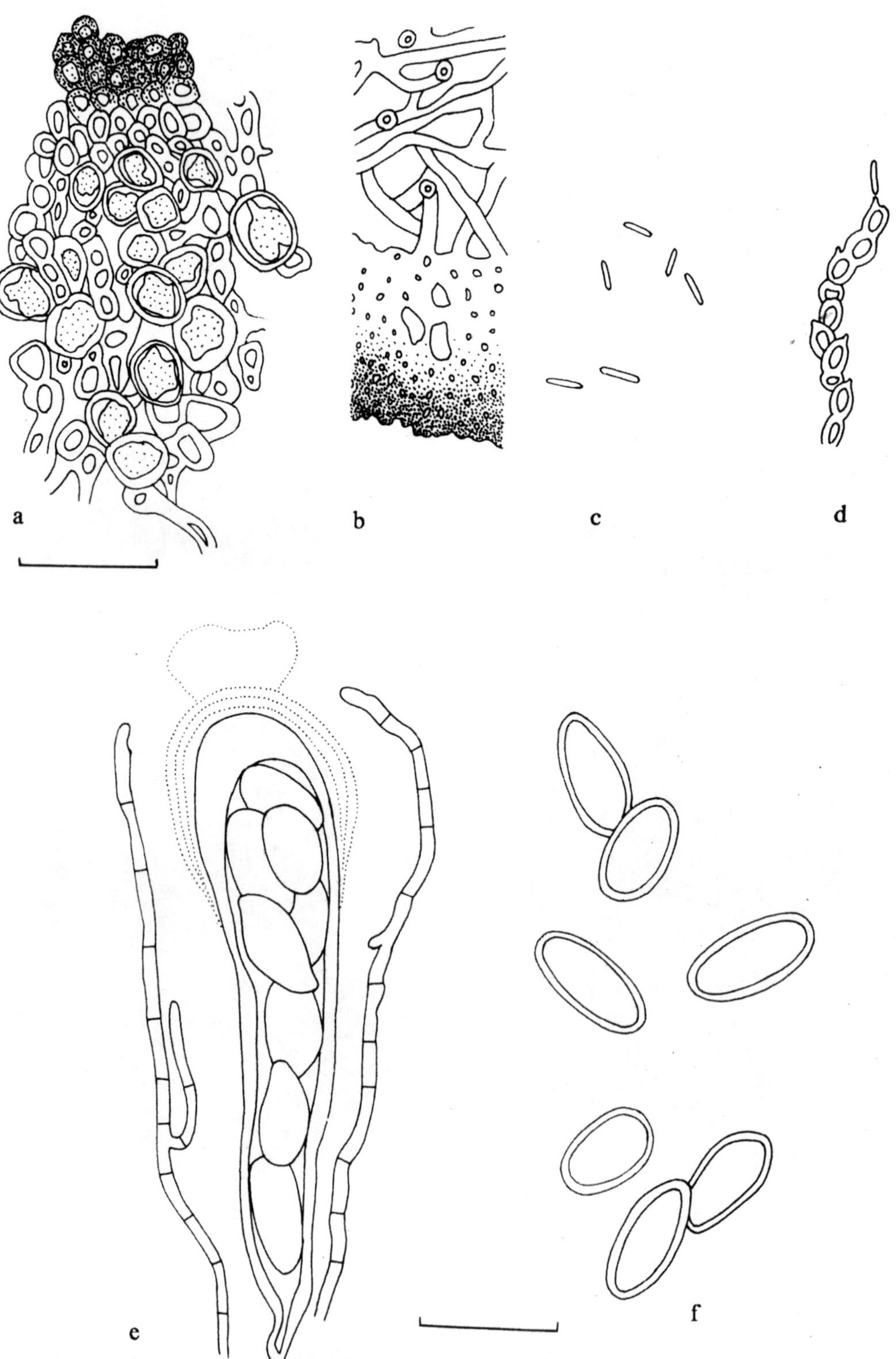

Fig. 4 *Umbilicaria squamosa* Wei & Jiang (2779). a. Thallus section (upper cortex and algal layer); b. Thallus section (medulla and lower cortex); c. Pycnoconidia; d. Conidiophore with a pycnoconidium; e. Ascus containing spores with paraphyses; f. Ascospores. The scale in 18μm.

SPECIMENS EXAMINED:

CHINA: Xinjiang, Zhao C. F. No. 2771(HMAS-L); Mt. Altai shan, Hanas lake, alt. 2500 m, on rocks, Aug. 5, 1986, Coll. Gao X.Q. No. 2075 (HMAS-L!).

OTHER SPECIMENS EXAMINED:

Switzerland: Graubunden, Fimbertal 2420 m. Leg. M. Steiner 8. 1960 [Poelt, Lich. Alp. no. 232] (HMAS–L–5026). Graubunden. Bernina-Hospiz, reg. alp. 30–31. VII. 1923 Ernst Hayren ex Mus. Bot. Univ. Helsinki (H) (HMAS–L. 3933!). Tanzania: Kilimanjaro, Saddle between Kibo and Mawenzi, in the southern part of the saddle, on boulders at the path from peters hut. Alt. 4470 m. 18 June 1948. O. Hedberg no. 1266d. Dupl. ex Herb. UPS(L–04292) (HMAS–L–5458!). Russia: Transcaucasus, Armenia, districtus Stepanovansky, cacumen trajecti Puschkinsky, ad rupes. 1956, leg. V. G. Nikogosjan [Savicz, Lichenotheca Rossica No. 162] (S!).

LITERATURE RECORD FOR ASIA: Altai (Golubkova et al. 1978, p. 96).

COMMENTS: The diagnostic character for this species is the smooth and neither reticulately rugose nor areolate upper surface of the thalli with radial lines from the point over umbo to margins.

New to China.

Umbilicaria subumbilicarioides Wei & Jiang. sp. nov. Fig. 3B, 5

TYPUS: CHINA, Xizang(Tibet), Zham, alt. 3660 m, on rocks, May 11, 1966, Coll. Wei & Chen No. 470–1(HMAS–L holotypus).

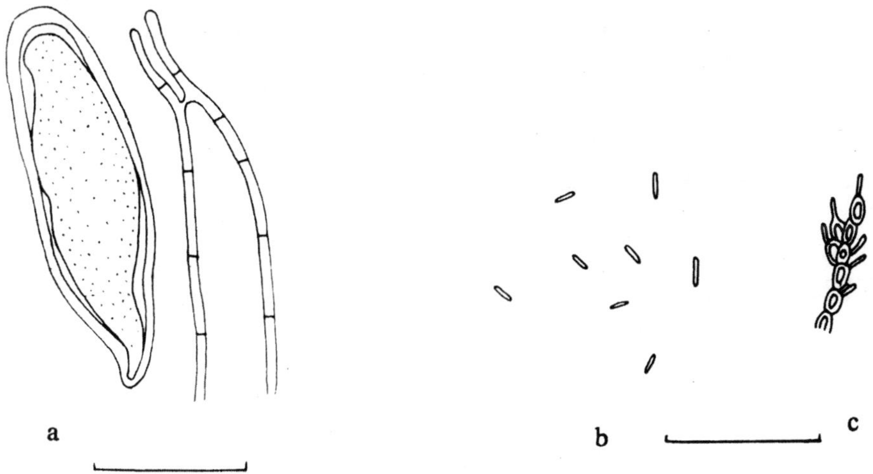

Fig. 5 *Umbilicaria subumbilicarioides* Wei & Jiang(470), the scale in 18μm. a. The immature ascus with paraphyses; b. Pycnoconidia; c. conidiophore with pycnoconidia.

DIAGNOSIS:

U. umbilicarioidi primo adspectu maxime similis, sed a qua differt corticibus superis palisadoplectenchymatis, stratis algarum continuis, corticibus inferinis scleroplectenchymatis cum cavitis cellularum minutis, rhizinis brevibus c.0.5–1.5 mm diametro longis, thalloconidiis ingularibus evolutis debiliter et acidis lecanoricis gyrophoricis cum substantia incoginta in Rf class 3 in B.

DESCRIPTION:

Thallus 3.5 × 4.5 cm, deeply lobed peripherally; upper surface smooth, weakly lacunose, in parts visibly rugose with the dissecting microscope, dark grey-brown to dark red brown, covered with a thin layer of white pruina, with some black spots and small perforations, black-brown rhizines growing thickly along the perforated rim and the spot margin; lower surface black, strongly verrucose, with black and branched short rhizines (0.5–1.5 mm long) mainly in the peripheral zone, and extended marginally, bare around the umbo.

Thallus 99.5–278.5 μm thick; upper cortex palisade plectenchymatous 14–25.5 μm thick, the outer layer brown, 4.5–9 μm thick, the inner layer colourless, 9–18.5 μm thick; algal layer continuous, 18.5–51 μm thick; medulla mucilaginous, 37–148.5 μm thick; lower cortex 18.5–46.5 μm thick, the inner layer colourless, 9–32 μm thick, the outer layer brown, 9–23 μm thick.

Apothecia with gyrose discs, c. 2.5 mm diam.; epihymenium 14–18.5 μm thick; hymenium 36–54 μm thick; subhymenium c. 18 μm thick; paraphyses septate, c.1.5 μm thick, the tip slightly inflated; asci when immature, 34–43 × 12.5–14.5 μm.

Pycnidia 125–255 × 107–213 μm; conidia bacilliform, 1.5–3.5 × 1 μm.

Thalloconidia arising on the lower suface; conidia dark brown, a few, single, only occasionally aggregated in several cells, subround, 5.5–11.5 × 3.5–7 μm, cell wall 1–3.5 μm thick.

Algae: Trebouxia.

CHEMISTRY: lecanoric, gyrophoric and umbilicaric acids.

COMMENTS: This new species is at first sight most similar to *U. umbilicarioides,* but differs from that in the palisade plectenchymatous upper cortex, continuous algal layer, scleroplectenchymatous lower cortex with very small cell cavities, short rhizines, c. 0.5–1.5 mm long, poorly developed single cell thalloconidia, and containing lecanoric and gyrophoric acids with an unidentified substance in Rf class 3 in B.

Umbilicaria taibaiensis Wei & Jiang, sp. nov.

TYPUS: CHINA, Shaanxi, Mt. Taibai shan, Baixiantai, alt. 3660 m, on rocks; June 3, 1963, Coll. Wei J.C. (HMAS-L, holotype; UPS, H, TUR, FH, US, Isotype).

MISIDENTIFICATION: *Umbilicaria cylindrica* auct. var. *tornata* sensu Wei & Jiang, Mycosystema, 1:80 (1988), Fig. 2 in p. 81, non (Ach.) Nyl.

DIAGNOSIS:

Species nova *Umbilicariae cylindricae* var. *tornatae* (Ach.) Nyl. primo adspectu maxime similis, sed thalli crassitudine 2 / 3 var. *tornatae* partes aequantes, sporis tertia parte major, et acido gyrophorico.

DESCRIPTION:

Thallus consisting of numerous subround small thalli of 0.5–1 cm diam. forming a rosette

3–5 cm diam. Each small thallus consists of 3–5 deeply divided lobes, weakly elevated marginally, and each lobe looking like a miniature teaspoon, with very finely black linear margins and simple or dichotomously branched black-brown cilia; upper surface dark brown with a slight tint of violet, covered with a thin layer of white pruina and a weakly greyish tint, dull, rarely and weakly undulating; lower surface brownish to pale brownish, black brown to blackish around umbo, smooth, not granulate, occasionally weakly granulate at the dark margins; umbilicus c. 5 mm diam., equalling 1 / 2 part of the small thallus, with brown to dark brown rhizines on the dark brown peripheral zone of the lobes.

Thallus 148–222 μm thick; upper cortex dark brown, 12.5–17.5 μm thick; algal layer 49.50–98.50 μm thick, and absent under the apothecia; medulla 24.5–74 μm thick; lower cortex brownish, 12.5–19.5 μm thick.

Apothecia scattered on the upper surface of thallus, 0.5–1 mm diam., with gyrose discs; epihymenium dark brown to black-brown, 12.5–24.5 μm thick; hymenium brownish, 49.5–86.5 μm thick; subhymenium brown-black, 49.5–61.5 μm thick; hypothecium brownish, 37–49 μm thick; asci obovoid, octosporous, 37–54 × 15–20 μm; ascospores long ovoid, unicellular in the juvenile stage and some bicellular later, hyaline, colourless, 14.5–22 × 7.5–12.5 μm.

CHEMISTRY: gyrophoric acid.

SPECIMEN EXAMINED:

CHINA: Shaanxi, Mt. Taibai-shan, alt. 3660 m, on rocks, July 31, 1963, Coll. Ma & Zong. No. 395–1 (HMAS–L).

COMMENTS: At first sight this species is most similar to *U. cylindrica* auct. var. *tornata* (Ach.) Nyl., but differs from the latter in the larger ascospores and containing gyrophoric acid.

Lasallia papulosa (Ach.) Llano

Monograph of the Umbilicariaceae 32(1950).

≡*Gyrophora papulosa* Ach., Lich. Univ. 226(1810).

TYPE: AMERICA septentrionalis, Leg. Muehlenberg, H-Ach. No. 570(H! holotype).

TYPE OBSERVATION:

Thallus similar to that of *Lasallia pensylvanica*, but differing in the lower surface of the thallus being a light brownish to brown, black around umbilicus, between umbilicus and peripheral zone almost whitish; and usually covered with a thin layer of a red pigment on the apothecia.

DESCRIPTION:

Thallus monophyllous, strongly pustulate, c. 2 cm diam.; upper surface greyish brown to brownish; lower surface greyish brown to whitish brown.

Thallus 471–813 μm thick; upper cortex consisting of a brown layer, 9–21.5 μm thick, and a colourless layer, 20–48.5 μm thick; algal layer 36–77.5 μm thick; medulla 232–559 μm thick, the part, close to algal layer, grey-brown, that close to the lower cortex colourless; lower cortex 51–111 μm thick, inner layer of cortex colourless and the outer one grey brown.

Apothecia sublecanorine, leiodiscs, c.1 mm diam.; epihymenium 9–18 μm thick; hymenium reddish, 86.5–126 μm thick; subhymenium yellowish, 23–51 μm thick; hypothecium brown, 55.5–171.5 μm thick; a zone between subhymenium and hypothecium yellowish brown, 23–60 μm thick; paraphyses septate, branched, 1–2 μm thick, at tip inflated, 2.5–3.5 μm thick; Asci monosporous, occasionally containing two spores; 75.5–108 × 29–39.5 μm. Ascospores large, brown, multiseptately muriform, 36–65 × 18–32.5 μm.

Pycnidia 65–45 × 30.5–74 μm; conidia bacilliform, colourless, 4.5–8 × 1 μm.

CHEMISTRY: lecanoric, gyrophoric and umbilicaric acids.

SPECIMENS EXAMINED:

CHINA: Xinjiang, Mt. Altai, by the Halas lake, alt. 2500 m, on rocks, Aug. 5, 1986, Coll. Gao Nos. 1958–1, 2090.(HMAS–L!)

MONGOLIA: Dzabhan, on wood twigs, July 1, 1976, Coll. Biazrov, nos. 6575, 7468; Ara-Khangai on the left bank of Chulata river, on tree bark of *Larix sibirica*, Aug. 22, 1970, Coll. Biazrov. no. 6570, 17km from southern Somon, on the left bank of Caicairlaig River, on rock, July 21, 1979, Coll. Biazrov, no. 5736 & Sept. 7, 1979, Coll. Biazrov, no. 6190; Bulgan, Dashinchilen residential area, Mt. Caicairlat-Ula, southern residential area, on Betula sp., Jun. 29, 1972, Coll. Biazrov, nos. 5109, 6689(Moscow!).

KAZAKSTAN: Eastern Kazakstan, alt. 1600 m, July 13, 1982, Coll. A.V. Dombrovskaja, No. 11 [as L. rossica Dombr]. (HMAS–L–5651!).

LITERATURE RECORD FOR ASIA: Russia (Golubkova et al. 1978, p. 136 for Yakut).

COMMENTS: This species rare-northern Asia, and mainly charaterized by its light coloured to brownish lower surface of thallus. The species is mainly distributed in North America and Africa. It is mainly attached to rocks in North America, but in Africa and northern Asia it is often on tree bark as well. In addition, it is iteresting that the specimen from China is redish in hymenium.

ACKNOWLEDGEMENTS

We are very grateful to the directors and curators of the herbaria where the senior author worked and or the specimens have been sent on loan for this study. Many thanks to Prof. Hawksworth for critical reading and giving lots of valuable suggestions and revising the English language. Thanks also to Ms Han ZF for inking the line drawing, and to Ms Yuan LC for preparing the photographs and to Ms Deng H for preparing the manuscript.

REFERENCES

Biazrov LG (1986) Additions to the lichen flora in Khangai (Mongolian People's Republic). 1. The family Umbilicariaceae Fee [Bul. Mosc. O–va I specitatelei Prirodei. Otd. Biol.] **91(6)**:99 (in Russian)

Frey E, Poelt J (1977) Die Gattung Lasallia in Khumbu Himal, Band 6 / **3**: 387–395, Uunchen

Hestmark G (1990) Thalloconidia in the genus Umbilicaria in Nord. *J. Bot.* **9**:547–574

Lawrey JD (1984) Biology of Lichenized Fungi Praeger

Poelt J (1977) Die Gattung Umbilicaria in Khumbu Himal, Band 6 / **3**:397–435

Savicz VP (1922) De Umbilicariaceae e Kamczatka notula. *Not. Syst. Inst. Crypt. Hort. Bot. Petrop.* **1**:102–109

Wei & Jiang (1988) A conspectus of the lichenized ascomycetes Umbilicariaceae in China. *Mycosystema* **1**:73–106

Wei & Biazrov (1991) Some disjunctions and vicariisma in the Umbilicariaceae (Ascomycotina). *Mycosystema* **4**:65–72

石耳科（子囊菌门）的新种和新分布

魏江春 姜玉梅

中国科学院微生物研究所真菌地衣系统学开放研究实验室

北 京

摘要：本文报导了石耳科的八个种，其中新种四个，即阿尔泰石耳，鳞芽石耳、亚石耳及太白石耳；三个种是中国的新分布，即皱石耳、亚光石耳及淡腹疱脐衣；一个种即白根石耳为中国及蒙古的新分布。

关键词：石耳科；石耳属；疱脐衣属；地衣；中国；蒙古

A new species of *Everniastrum* containing diffractaic acid

YU-MEI JIANG and JIANG-CHUN WEI*

Abstract: A new species, *Everniastrum diffractaicum*, the only member of the genus containing diffractaic acid, is described. It has much larger ascospores than those of any other member of the genus.

Introduction

The genus *Everniastrum* Hale ex Sipman (Lecanorales: Parmeliaceae) comprises some 29 species (Culberson & Culberson 1981; Sipman 1986; Jiang & Wei 1989; Chen *et al.* 1989) in the world, and from it 13 lichen acids, one pigment, and one group of fatty acids are recorded (Culberson 1969, 1979; Culberson *et al.* 1977). Eight species of the genus are reported from China (Jiang & Wei 1989; Chen *et al.* 1989), and six lichen substances are known from them.

Lichen substances are of considerable importance in the systematics of *Everniastrum*. Therefore, the Chinese specimens were investigated chemically during studies on this genus. All the species in China except *E. subsorocheilum* contain salazinic acid. Diffractaic acid was found in collections of *Everniastrum* from Yunnan, which is the first record of this compound in the genus. Because Culberson & Culberson (1981) showed that all members of the *E. cirrhatum* group contain group 1 fatty acids, including either protolichesterinic acid, alloprotolichesterinic acid or some compounds similar to them, it was necessary to re-examine whether *E. sinense* contains any such fatty acids.

Materials and Methods

The species described below is based on material collected by Wang Han-chen from Yunnan province, Dali district in 1941.

Both microcrystal tests (MCT) (Asahina 1936; Yoshimura 1974) and thin-layer chromatography (TLC) (Culberson & Kristinssen 1970; Culberson 1972) of acetone extracts of thalli were carried out.

Key to the species of *Everniastrum* from China

1	Soredia, isidia or lobulets present, rhizines marginal	2
	Soredia, isidia or lobulets absent, rhizines laminal or marginal	5
2(1)	Soredia absent	3
	Soredia present	4
3(2)	Isidia present	**E. vexans**
	Lobulets present	**E. sinense**

*Systematic Mycology & Lichenology Laboratory, Institute of Microbiology, Academia Sinica, Beijing 100080, People's Republic of China.

0024–2829/93/010057+04 $03.00/0

This paper was originally published in *Lichenologist*, 25（1）: 57-60, 1993.

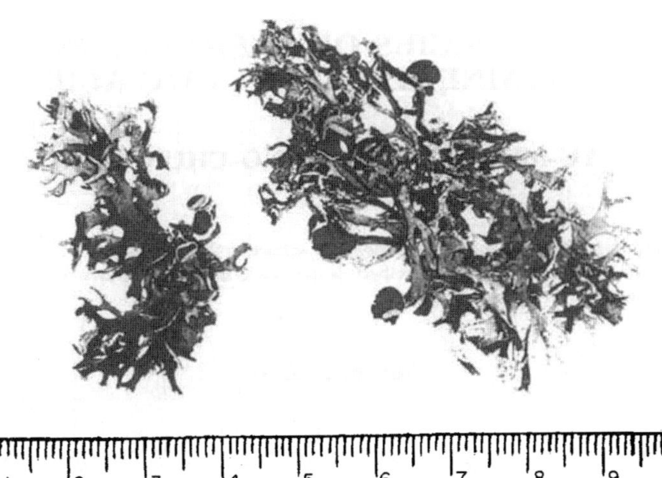

FIG. 1. *Everniastrum diffractaicum* (holotype). Scale in cm.

4(2) Medulla K−, salazinic acid absent . . . . **E. subsorocheilum**
Medulla K+, salazinic acid present . . . . . . **E. sorocheilum**
5(1) Rhizines unbranched or occasionally poorly branched 6
Rhizines richly branched, short . . . . **E. rhizodendroideum**
6(5) Rhizines mostly laminal 7
Rhizines only marginal . . . . **E. cirrhatum**
7(6) Rhizines short, *c.* 1 mm long, diffractaic acid absent 8
Rhizines long, *c.* 2 mm long, diffractaic acid present **E. diffractaicum**
8(7) Alectorialic acid absent . . . . **E. nepalense**
Alectorialic acid present . . . . **E. alectorialicum**

The Species

Everniastrum diffractaicum Y. M. Jiang & Wei sp. nov.

Species sporis majoribus [(19·5–)23·5–28·5(–32·5) × 7·0–10·5(–12·5) μm] et substantia (acido diffractaico) a congeneribus diversa.

Typus: Yunnan, Dali, Mt. Longquanfeng, ad ramulos *Ribis* sp. (Grossulariaceae), Julius 1941, *Wang, H. C.* 1065a (HMAS-L—holotypus) et 1084b (HMAS-L—paratypus).

(Figs 1 & 2)

Thallus linear-lobate and spreading; *lobes* 2–6 mm long and 0·5–1·5 mm wide, repeatedly dichotomously branched, flattened to subinvolute. *Upper surface* grey to whitish or pale brown, without soredia and isidia. *Lower surface* black, sometimes strongly wrinkled, but brownish to brown and plane at apices of young lobes, sparsely rhizinate or sometimes bare on lower surface, richly rhizinate only at margins; rhizines 1–2·5 mm long, simple to branched, and black.

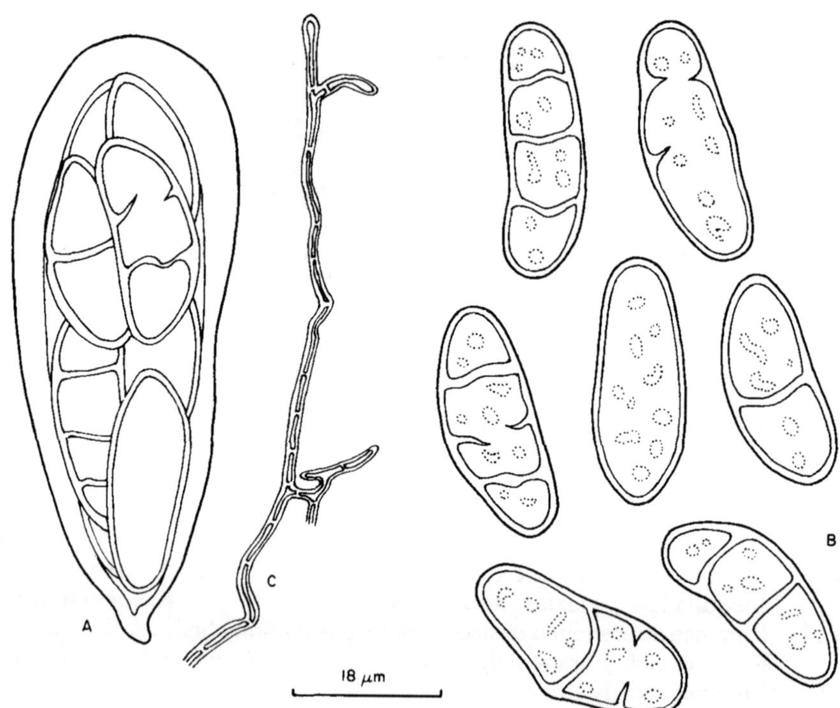

FIG. 2. *Everniastrum diffractaicum*. A, Ascus with ascospores; B, ascospores; C, a paraphysis.

Apothecia abundant, discs brown, rounded, (1–)2–6(–7) mm in diam. *Epithecium* 10·5–16·0 µm thick, brown; *hymenium* 84·5–104·5 µm thick, colourless to slightly yellowish or brownish; *hypothecium* 23·5–39·5 µm thick, yellowish to brownish, K–. *Asci* broadly elongate, 61·0–73·5 × 27·0–30·5 µm, 8-spored. *Ascospores* usually simple, colourless to yellowish or brownish, oblong to reniform, sometimes with 1–3 irregular transverse septa, (19·5–)23·5–28·5(–32·5) × 7·0–10·5(–12·5) µm.

Pycnidia black, numerous. *Conidia* bacilliform, 5·5–7·0 × 1 µm.

Chemistry: Cortex K+ yellow. Medulla K+ yellow to orange, C–, KC+ blood red, PD+ orange. Atranorin, protolichesterinic acid, diffractaic acid, salazinic acid, and galbinic acid (trace).

Note: These collections were identified and published before by Zhao (1964: 144) and Zhao *et al.* (1982): 15) as *Parmelia cirrhata*, non Fr.

Everniastrum diffractaicum is distinguished from other species in the genus by the presence of diffractaic acid in the medulla and by its larger ascospores. The range of ascospore sizes in the genus are recorded by Sipman (1986) to be (7–)10–16(–28) × (4–)6–7(–11) µm.

This species is, so far, known only from the type locality.

The chemicals found in Chinese *Everniastrum* species are shown in Table 1.

TABLE 1. *Chemical constituents of* Everniastrum *species from China*

Species	Atranorin	Diffractaic acid	Protolichesterinic acid	Alectorialic acid	Salazinic acid	Galbinic acid	Protocetraric acid
E. alectorialicum	+*	–	(+)	+	+	(+)	–
E. cirrhatum	+	–	+	–	+	(+)	(+)
E. diffractaicum	+	+	+	–	+	(+)	–
E. nepalense	+	–	+	–	+	(+)	(+)
E. rhizodendroideum	+	–	+	–	+	(+)	–
E. sinense	+	–	–	–	+	–	–
E. sorocheilum	+	–	+	–	+	(+)	–
E. subsorocheilum	+	–	+	–	–	–	–
E. vexans	+	–	+	–	+	(+)	–

*+ = Present, – = absent, (+) = present in only trace amounts.

We wish to thank Dr C. F. Culberson and A. Johnson for offering kind help and confirming the identification of the diffractaic acid. We also thank Ms L. C. Yuan for printing the photographs, and Ms Z. F. Han for making the line drawings. The project was supported by the National Natural Science Foundation of China.

REFERENCES

Asahina, Y. (1936) Mikrochemischer Nachweis der Flechtenstoffe (II Mitteil.) *Journal of Japanese Botany* **12:** 859–872.

Chen, J. B., Wu, J. N. & Wei, J. C. (1989) Lichens of Shennongjia. In *Fungi and Lichens of Shennongjia*: 386–493. Beijing: World Publishing Co.

Culberson, C. F. (1969) *Chemical and Botanical Guide to Lichen Products.* Chapel Hill: University of North Carolina Press.

Culberson, C. F. (1970) Supplement to 'Chemical and Botanical Guide to Lichen Products' *Bryologist* **73:** 177–377.

Culberson, C. F. (1972) Improved conditions and new data for the identification of lichen products by a standardized thin-layer chromatographic method. *Journal of Chromatography* **72:** 113–125.

Culberson, C. F., Culberson, W. L. & Johnson, A. (1977) Second Supplement to 'Chemical and Botanical Guide to Lichen Products'. St Louis: Missouri Botanical Garden. American Bryological and Lichenological Society.

Culberson, C. F. & Kristinsson, H. (1970) A standardized method for the identification of lichen products. *Journal of Chromatography* **46:** 85–93.

Culberson, W. L. & Culberson, C. F. (1981) The genera *Cetrariastrum* and *Concamerella* (Parmeliaceae): a chemosystematic synopsis. *Bryologist* **84:** 273–314.

Jiang, Y. M. & Wei, J. C. (1989) A preliminary study on *Everniastrum* from China. *Acta Bryolichenologica Asiatica* **1:** 43–52.

Sipman, H. (1986) Notes on the lichen genus *Everniastrum* (Parmeliaceae). *Mycotaxon* **26:** 235–251.

Yoshimura, I. (1974) *Lichen Flora of Japan in Colour*. Hokusha Publishing Co.

Zhao, J. D. (1964) A preliminary study of Chinese *Parmelia. Acta Phytotaxonomica Sinica* **9:** 144.

Zhao, J. D., Xu, L. W. & Son, Z. M. (1982) *Prodromus Lichenum Sinicorum*. Beijing: Science Press.

Accepted for publication 8 *March* 1992

A new species of *Umbilicaria* (Ascomycota)

WEI Jiang-chun JIANG Yu-mei GUO Shou-yu
Systematic Mycology & Lichenology Laboratory
Institute of Microbiology, Academia Sinica, Beijing 100080

ABSTRACT: A new species of the lichen- forming genus *Umbilicaria* from China, *U. loboperipherica* sp. nov. is described. The new species is characterized by numerous squamose lobules derived from the ascending cortex on the periphery of thallus, and a black lower surface with numerous and richly branched rhizines; it is confined to northern China. The vicarism of the new species is also discussed briefly and a key including closely related species is given in addition to a Latin diagnosis, English description with illustrations.

KEY WORDS: *Umbilicaria*; *U. loboperipherica*; new species

In previous studies on the Asian Umbilicariaceae (Wei & Jiang, 1993) some specimens collected from Mt. Xiaowutai in Hebei province, China, and also from Mt.Changbai in Jilin province, were treated as under *U. vellea* f. *leprosa* (Schaer.) Zahlbr. (Wei & Jiang, 1988). Recently, many additional specimens of this lichen were collected by Jiang & Guo from Mt. Tulaopoding, Tianqiaoling, in Wangqing county, Jilin province. After a careful examination of the morphology, anatomy, and geographical distribution we came to the conclusion that this lichen is undescribed and should be described as a remarkable new species, quite distinct from those already known to us.

The new species is somewhat similar to *U. freyi* (Codogno et al., 1989) in having numerous squamose lobules derived from the ascending cortex on the periphery of the thallus, but differs in other characteristics. As a matter of fact, *U. vellea*, f. *leprosa*(Schaer.) Zahlbr. (Llano, 1950) is of a somewhat sorediose upper thallus.

MATERIAL AND METHODS

The new species described below is based on specimens collected from China and preserved in the Lichen Section of "Herbarium Mycologicum Instituti Microbiologici Academiae Sinicae" (HMAS-L).

A dissecting microscope OPTON, and research microscope OPTON III were used for morphological and anatomical studies. Both microcrystal test (MCT) (Asahina, 1936, 1937a, b),

This paper was originally published in *Mycosystema*, 8-9: 65-70, 1996

and thin-layer chromatography (TLC) (Culberson & Kristinsscn, 1970; Culberson, 1972; White and James, 1985) of acetone extracts of thalli were carried out.

Taxonomy

Umbilicaria loboperipherica Wei, Jiang & Guo, sp. nov. (Pl.1) Species nova *U. hirsutae* et *U. velleae* f. *leprosae* affinis, a qua imprimis differt lobulis squamosis oriundis ex corticibus ascendentibus in peripheria thallorum (paralobulis) et sine parasorediis * ; ab *U. freyi* distincta subtus atra et rhizinis ramosissimis numerosis.

Typus: Sina, Provincia Jilinensis, Comitatus Wangqingensis, Mt. Tulaopodingensis, ad rupes, 9 Aug. 1994, leg. Jiang, Y. M. et Guo, S. Y., 94572(HMAS–L–holotypus).

Thallus thick and rigid, usually 3–4 (rarely 6–8) cm diam, irregularly torn-lacerate and revolute downwards on the margins; upper surface smooth, dark green when wet and grey or dark grey to brown-grey when dry, without parasoredia① but with numerous small and squamose lobules derived from the ascending cortices on the periphery of the thallus, can be called "paralobules;" lower surface black, with numerous and richly branched rhizines and a thin umbo 2–3 mm diam.

Thallus 120–218μm thick; upper cortex consisting of colourless to greyish palisade plectenchymatous tissues, the upper surface of the cortex sometimes brownish and 5–10μm thick at the margin, greyish and 18–28μm thick near the umbo; algal layer 30–55(–65)μm thick; medulla consisting of an upper loose layer 20–50μm thick, and a compact mucilaginous lower layer 10–95μm thick; lower cortex scleroplectenchymatous, consisting of colourless inner layer of 9–18μm thick and brown outer layer of 10–15μm thick.

Apothecia not seen.

Pycnidia, 115–140 × 175–190μm; conidia bacilliform, hyaline, 3–4 × < 1μm.

Chemistry: lecanoric and gyrophoric acids.

The new species is similar to *U. hirsuta* and *U. vellea* f. *leprosa* (Schaer.) Zahlbr., from which it differs in the absence of parasoredia and the pressence of small squamose lobules derived from the ascending cortex (paralobules)on the periphery of thallus. It differs from *U. freyi* in having a black lower surface with numerous and richly branched rhizines.

The upper surface of the naked medullary layer is darkened and forms a brown structured plectenchyma, recalling a secondary cortex without an algal layer under it.

It appears that both *U. freyi* and *U. loboperipherica* are vicarious species; the former distributed in the Mediterranean area and the latter in China(Fig. 1). The origin of these two vicarious species could have been caused by the upheaval of the Himalayas and the retreat of the ancient Mediterranean sea.

---

①The term "parasoredia" was proposed by Codogno et al. (1989) for the so- called soredia in *U. hirsuta*. They differ from true soredia by originating from the structured plectenchyma in the upper thallus layer, while true soredia are from unstructured medullary plectenchyma.

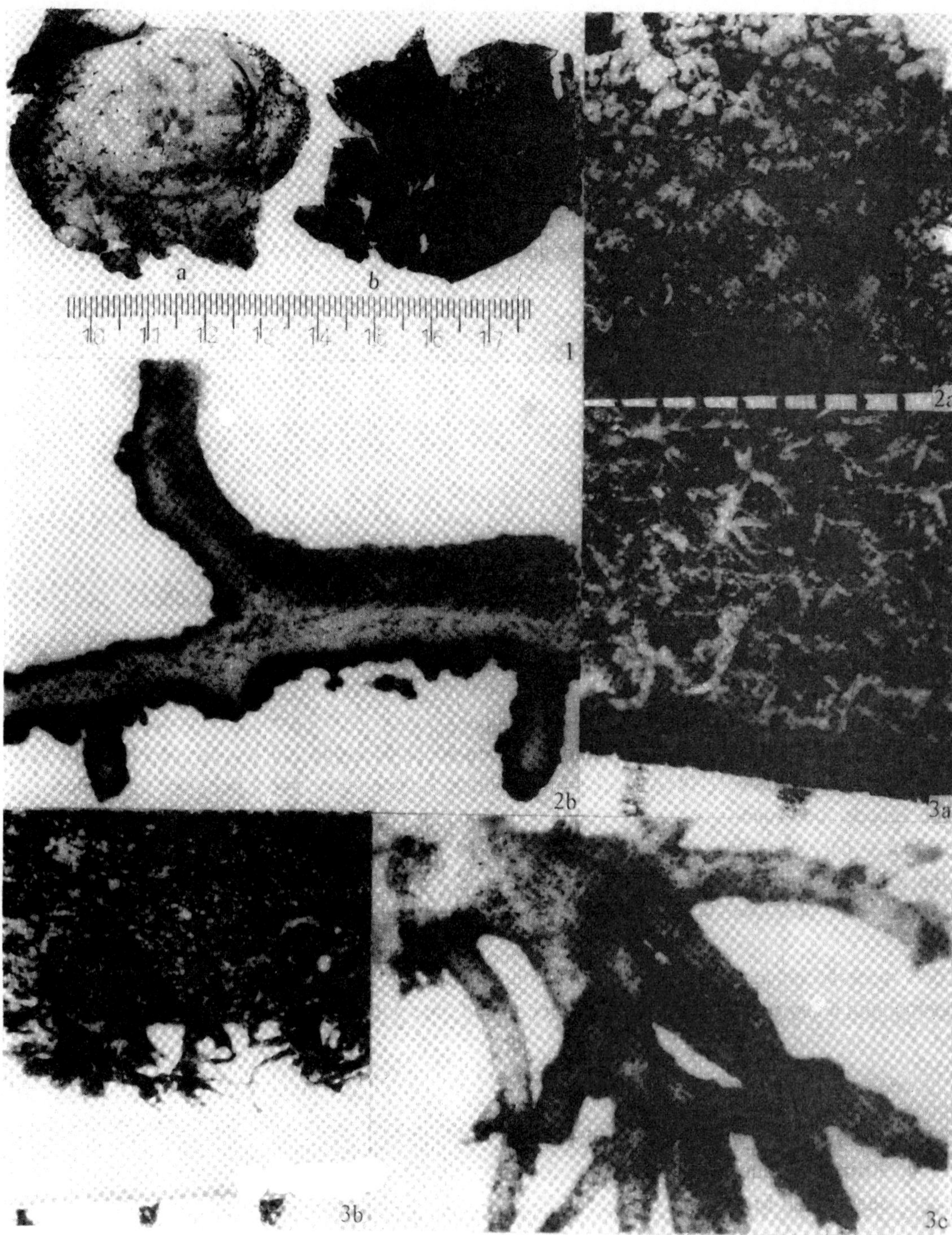

Plate: I. 1. Habit of *U. loboperipherica* (scale in mm): (a). upper surface of a thallus; (b). lower surface of a thallus. 2, Numerous squamose lobules derived from the ascending cortex on periphery of a thallus: (a). Habit of the squamose lobules under the dissecting microscope (scale in mm); (b). section of a squamose lobule under the research microscope (× 6.3). 3. Rhizines: (a). surface view under the dissecting microscope (scale in mm); (b). profile under dissecting microscope (scale in mm); (c). profile under the research microscope (× 6.3).

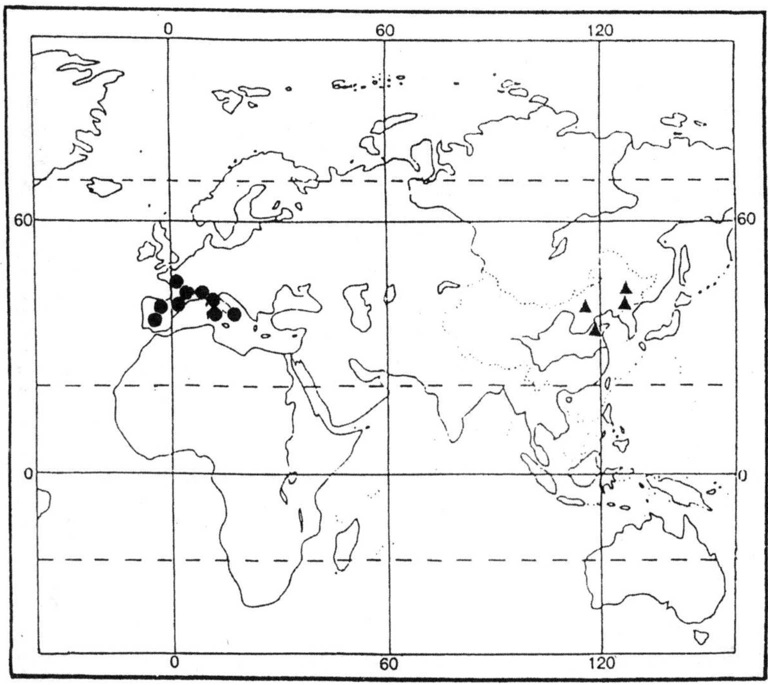

Fig. 1 The distributional areas of the vicarious species *Umbilicaria loboperipherica* ▲ and *U. freyi.* ●.

Specimens examined

China: Hebei province, Mt. Xiaowutai, on rocks of mountain top, alt. 2800 m, 16 Aug. 1964, coll. Wei, 2096–1(HMAS–L). Jilin province, Antu county, Mt. Changbai, north slope, on the way to Longmenfeng near a small waterfall, on rocks, 13 Aug. 1983, coll. Wei & Chen (HMAS–L); Wangqing county, Tianqiaoling, Mt. Tulaopoding, on rocks, 9 Aug. 1994, coll. Jiang Y. M. and Guo S. Y., Nos. 94514, 94515, 94545, 94569, 94570, 94571, 94572(Type HMAS–L)., 94573,94574

This remarkable new species is quite distinct from the most closely related members of the genus, and a key to the new species and its related ones will facilitate recognition.

Key to New and Related Species

1. Rhizines covered with thalloconidia
    2. Rhizines numerous, very thin and short, richly branched; thallus weakly leathery, upper surface red-brown and dull . . . . *U. esculenta*
    2. Rhizines long and thick, simple, poorly branched
        3. Rhizines cylindrical, but swollen on the base caused by covering with black thalloconidia and / or mixed with stubby or oval ones covered with thalloconidia; upper surface of thallus pale green-grey in all tints of mouse-grey to grey-brown or brown; widely distributed . . . . *U. vellea*
        3. Rhizines fine cylindrical and club-shaped, neither swollen on the base nor mixed with

stubby or oval ones; upper surface of thallus cinamon buff, light brownish olive, or isabella colour, when light coloured the upper surface is stained light brown, endemic to N, America . . . . *U. mammulata*

1. Rhizines lacking thalloconidia
   4. Thallus on periphery parasorediate; widely distributed . . . . *U. hirsuta*
   4. Thallus on periphery lacking parasoredia
      5. Thallus on periphery paralobulose; upper surface mouse-grey or brownish
         6. Lower surface of thallus brownish lacking rhizines or occasionally with few or individul and short rhizines in parts; NW coastal mountains of the Mediterranean . . . . *U. freyi*
         6. Lower surface of thallus black, with numerous and richly branched rhizines; Hebei and Jilin provinces of China . . . . *U. loboperipherica*
      5. Thallus occasionally paralobulose locally on the upper surface but not on periphery; upper surface black-violet or dirty violet, covered with a very thin chalk-white pruina; Himalayas . . . . *U. squamosa*

ACKNOWLEDGEMENTS

We are very indebted to Dr. Clifford M. Wetmore for reviewing the manuscript, to Ms Deng for preparing the manuscript on computer, and to Ms Yuan for developing and printing the photographs.

REFERENCES

Asahina Y. 1936. Mikrochemischer Nachweis der Flechtenstoffe (I). J. of Jap. Bot. 12(7): 516–525.

Asahina 1937a, Mikrochemischer Nachweis der Flechtenstoffe (III. Mitteil.). J. of Jap. Bot. 13(7): 529–536.

Asahina Y. 1937b. Mikrochemischer Nachweis der Flechtenstoffe (IV. Mitteil.). J. of Jap. Bot. 13(12): 855–861.

Codogno M., Poelt, J., und Puntillo, D. 1989. Umbilicaria freyi spec. nova und der Formenkreis von Umbilicaria hirsuta in Europa (Lichenes, Umbilicariaceae). Pl. Syst. Evol. 165: 55–69.

Culberson C. F. 1972. Improved conditions and new data for the identification of lichen products by a standardized thin-layer chromatographic method. J. Chromatography 72: 113–125.

Culberson C. F. & Kristinsson, H. 1970. A standardized method for the identification of lichen products. J. Chromatography 46: 85–93.

Llano G. 1950. A monograph of the lichen family Umbilicariaceae in the western hemisphere, Navexos, Washington.

Wei J. C. and Jiang, Y. M., 1988. A conspectus of the lichenized ascomycetes Umbilicariaceae in China. Mycosystema 1:73–106.

Wei J. C. and Jiang, Y. M., 1993. The Asian Umbilicariaceae (Ascomycota). Mycosystema Monographicum, International Academic Publishers, Beijing, pp. 1–217.

White F. J. and James P. W. 1985. A new guide to microchemical techniques for the identification of lichen substances. Bull. British Lichen Soc. 57(suppl.), pp.1–47.

石耳属一新种(子囊菌门)

魏江春 姜玉梅 郭守玉
中国科学院微生物研究所真菌地衣系统学开放实验室, 北京 100080

摘要: 本文描述了地衣型石耳属一新种，周鳞石耳。新种脐叶体上表面以周边皮层碎片上翘而形成的大量准鳞芽，类似于淡腹鳞石耳，但是，不同之处在于新种脐叶体下表面黑色，覆以大量同色而多分枝的假根以及具有不同的地理分布，从而呈现为地理替代现象。文中为新种提供了拉丁文特征提要、英文描述、图片与地理分布图以及新种及其相关种的检索表。

关键词: 石耳属; 周鳞石耳; 新种

Two new taxa of the lichen genus *Collema* from China*

LIU Hua-Jie　　WEI Jiang-Chun**

(Key Laboratory of Systematic Mycology & Lichenology, Institute of Microbiology, Chinese Academy of Sciences, Beijing 100080)

ABSTRACT: Two taxa new to science, *Collema sichuanense* and *C. substipitatum* var. *gonggashanense*, are described and illustrated.

KEY WORDS: *Collema sichuanense*, *Collema substipitatum* var. *gonggashanense*, Collemataceae, Lecanorales, Ascomycota

In our studies on some collections of *Collema* deposited in HMAS-L, two new taxa, *Collema sichuanense* and *C. substipitatum* var. *gonggashanense* were encountered and are described and illustrated.

Some terms of morphology and anatomy are used after Degelius (1954, 1974).

Collema sichuanense H. J. Liu & J. C. Wei, sp. nov. (Fig. 1, 3, 4, 5); *C. nigrescenti* primo adspectu maxime simile, a quo sporis minoribus differt.

Typus: SICHUAN, Zoige County, Tiebu, 2800 m, 1983.6.21, X. Y. Wang & X. Xiao, 10993 (HMAS-L). Paratypus: SICHUAN, Aba, 3100 m, 1983.6.28, X. Y. Wang, 11343 (HMAS-L).

Thallus foliose, membranaceous, rounded or somewhat irregularly rounded, ca. 3-5 cm diam., not fenestrated; upper side olive-green to dark olive-green, glabrous, mainly regularly and densely pustulate, with short and often pustules-like ridges, matt to somewhat glossy, without isidia and lobules, epruinose; lower side with depressions corresponding to pustules on upper side; lobes few and rounded to somewhat extended, 5-10 mm wide, plane, margin entired and not swollen, not crisped or slightly so.

Thallus 60-102 ì m thick; both sides with pseudocortex.

Apothecia numerous and dense, mainly on pustules and ridges, sessile with constricted base or somewhat stipitate, 1-1.5 mm diam.; disc plane to somewhat convex when old, concave when young, red to dark red, glabrous and smooth, not glossy to somewhat so, epruinose; margin of disc not prominent, concolourous with thallus, smooth and entired.

Apothecia zeorine type; thalline exciple with pseudocortex consisted of two to several layers of somewhat globular cells; proper exciple euthyplectenchymatous, (10) 13.5-17 ì m thick in center; hymenium 81.5-112 ì m thick; subhymenium 27-37.5 ì m thick; spores 8 per ascus, linear-fusiform with acute ends, commonly 6-celled, not constricted at septa, colourless, straight, (27) 30.5-37.5 (44) × 3.5-5 ì m.

Corticolous.

Notes: This species is characterized by: 1) thallus is similar to that of *C. nigrescens*; 2) thallus and apothecia epruinose; 3) thallus and thalline exciple with pseudocortex; 4) proper exciple euthyplectenchymatous; 5) spores 6-celled, linear-fusiform; 6) corticolous.

Owing to the type of thalline and proper exciples, *C. sichuanense* is closely related to *C.subnigrescens*,

* Supported by the National Natural Science Foundation of China (39899400).

** Corresponding author

Received:2003-04-16, accepted: 2003-05-03

This paper was originally published in *Mycosystema*, 22（4）: 531-533, 2003.

C. nigrescens, *C. nepalense* and *C. ryssoleum*, from which it differs chiefly in the shorter and narrower, linear-fusiform spores. It looks more like *C. nepalense*, also with mainly pustulate thallus and fusiform spores, but the latter lichen has pruinose and little apothecia as well as broader spores.

Collema substipitatum Zahlbruck. var. ***gonggashanense*** H. J. Liu & J. C. Wei, sp. nov. (Fig. 2, 6); differt a var. substipitato praecipue excipulo proprio euparaplectenchymatico.

Typus: SICHUAN, Mt. Gonggashan, Dongpo, Yanzigou, 2650 m, in *Salici sp.*, 1982.7.2, X. Y. Wang, X. Xiao & B. Li, 8729-1 (HMAS-L).

Thallus subcrustose to subfoliose, membranaceous, irregular in shape, up to 3.5 cm diam., adnate to substrate, fenestrated; upper side dark olive-green to blackish, glabrous, smooth, matt, without isidia and lobules, epruinose, strongly rugose and sparsely pustulate; lower side dark olive-green, with sparse and white rhizines; lobes few and rounded, > 5 mm wide, margin entired, not swollen, not crisped.

Thallus 68-122.5 ì m thick, without pseudocortex; hyphae running in all directions, branched, I-.

Apothecia few and sparse, superficial, sessile with constricted bases to stipitate, up to 3 mm diam.; disc dark red, plane to somewhat convex, smooth, glossy, epruinose; margin of disc prominent, glossy, smooth, entired.

Apothecia zeorine type; thalline exciple without pseudocortex; proper exciple euparaplectenchymatous (cells thin-walled, isodiam. ones commonly < 8 ì m diam.), colourless, in its central part 20.5-30.5 ì m thick, in marginal part thinner; hymenium 75-110 ì m thick, I+ blue; subhymenium 17-35 ì m thick, I+ blue; Spores 8 per ascus, acicular, straight to curved, commonly 6-celled, not constricted at septa, colorless, (61) 68-75 (81.5) × 3.5-5 ì m.

Notes: The new variety differs from var. *substipitatum* in the euparaplectenchymatous proper exciple, and is similar to *C. leptaleum* var. *leptaleum* as to the thalline and proper exciples as well as habitus of the thallus, but differing from the latter in the shape and size of the spores. From *C. shiroumanum*, also has the subcrustose to subfolious thallus with the rugose upper side, it diifers in the different proper exciple and spores.

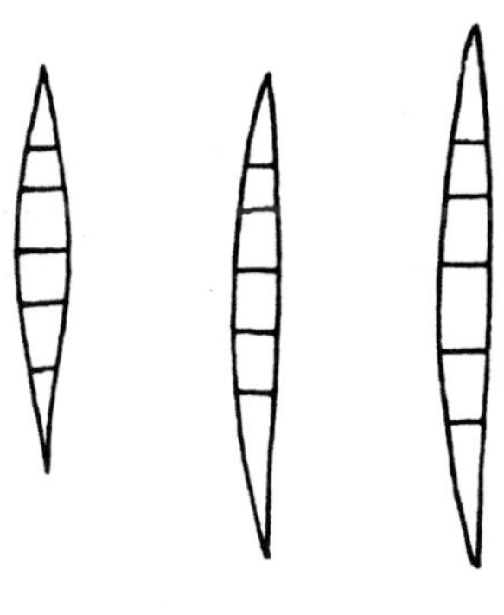

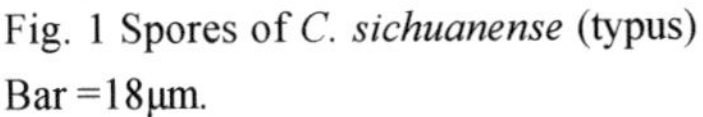

Fig. 1 Spores of *C. sichuanense* (typus) Bar =18μm.

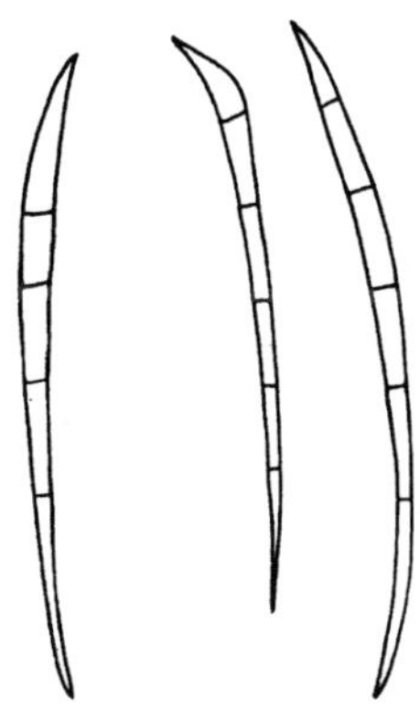

Fig. 2 Spores of *C. substipitatum* var. *gonggashanense* (typus) Bar =18μm.

Fig. 3. Thalli with apothecia of *C. sichuanense* (typus): the upper side; Fig. 4 Portion of preceding to show the apothecia more distinctly; Fig. 5 Thallus of *C. sichuanense* (typus): the lower side with depressions; Fig. 6 Thallus of *C. substipitatum* var. *gonggashanense* (typus). Scales = 1 mm.

[REFERENCES]

Degelius G, 1954. The lichen genus *Collema* in Europe. Morphology, taxonomy, ecology. *Symb. Bot. Upsal.* **13** (2): 1-499

Degelius G, 1974. The lichen genus Collema with special reference to the extra-European Species. *Ibid.* **20** (2): 1-215

胶衣属地衣的新分类群

刘华杰　魏江春

（中国科学院微生物研究所真菌地衣系统学重点实验室，北京，100080）

摘 要：本文描述了产于我国四川的胶衣属 (*Collema*) 两个新分类群：四川胶衣 (*Collema sichuanense*) 与短柄胶衣贡嘎山变种 (*Collema substipitatum* var. *gonggashanense*)。文中除了拉丁文特征提要以及英文描述以外，还附有新分类群的外形照片与孢子形态线条图。

关键词：子囊菌门，茶渍目，胶衣科，胶衣属，四川胶衣，亚柄胶衣贡嘎山变种

中图分类号：Q939.5　**文献标识码**：A　**文章编号**：1007-3515（2003）04-0531-0533

Cetradonia, a new genus in the new family *Cetradoniaceae* (Lecanorales, Ascomycota)

Jiang-chun WEI and Teuvo AHTI

Abstract: A taxonomic revision of the lichen genus *Gymnoderma* Nyl. sensu Yoshimura & Sharp (*Cladoniaceae*) based on an integrative analysis of polyphasic characters supports the segregation of a new genus *Cetradonia* J.-C.Wei & Ahti in a new family *Cetradoniaceae* J.-C.Wei & Ahti. The monotypic genus and family are based on *Cladonia linearis* Evans. In addition, two species are recognized in genus *Neophyllis* F. Wilson and two species are retained in the genus *Gymnoderma* Nyl. of the *Cladoniaceae*.

Introduction

During investigations of East Asian–North American disjunct taxa of lichen-forming fungi, the genus *Gymnoderma* Nyl. emended by Yoshimura and Sharp was studied in detail. The genus *Gymnoderma*, based on *G. coccocarpum* Nyl., was originally established by Nylander (1860, 1869). Yoshimura and Sharp (1968) retained it as a distinct genus, emphasizing the short and solid podetia without symbiotic algae as the principal distinguishing morphological character. On the basis of this single character, they transferred the North American endemic species *Cladonia linearis* Evans to *Gymnoderma* as *G. lineare* (Evans) Yoshim. & Sharp and described the second species from Japan, *G. insulare* Yoshim. & Sharp. Subsequently, and again on the basis of the same single character, Yoshimura (1973) transferred the Australian *Neophyllis melacarpa* (F. Wilson) F. Wilson to *Gymnoderma*. Thus, a total of four species, *G. coccocarpum*, *G. insulare*, *G. lineare*, and *G. melacarpum* (F. Wilson) Yoshim. were accepted in the genus *Gymnoderma* by Yoshimura (1973). However, the genus as circumscribed by Yoshimura and Sharp has a markedley disjunct distribution in East Asia, Australasia and North America and it is doubtful whether it is homogeneous, but it has been difficult to find many well defined criteria to distinguish or delimit more than one genus. Therefore, the delimitation of the genus was re-examined after a careful study of morphological, anatomical, geographical and chemical characteristics.

J.-C. Wei: Systematic Mycology & Lichenology Laboratory, Academia Sinica, Zhong-guan-cun, Bei-yi-tiao 13, Beijing 100080, China.
T. Ahti: Department of Ecology and Systematics, P.O. Box 47, Fin-00014, University of Helsinki, Finland.

Materials and Methods

This study is based on herbarium material from Yunnan in China, Japan, Malaysia, India, Indonesia, Australia, and the Appalachian Mountains of North America located in the following herbaria: H, HMAS-L, MIN, MSC, NY, TENN and US; and fresh material collected from the Appalachian Mountains by one of us (J-CW).

Morphology and anatomy were studied by light microscopy and by SEM (Eriksson 1981). Spores for measurements were mounted in water.

The standardized thin-layer chromatography (TLC) (Culberson & Kristinsson 1970; Culberson 1972; White & James 1985) was used for identification of the lichen substances in combination with microcrystal tests (MCT) (Asahina 1936, 1937; Evans 1943).

This paper was originally published in *Lichenologist*, 34（1）: 19-31, 2002.

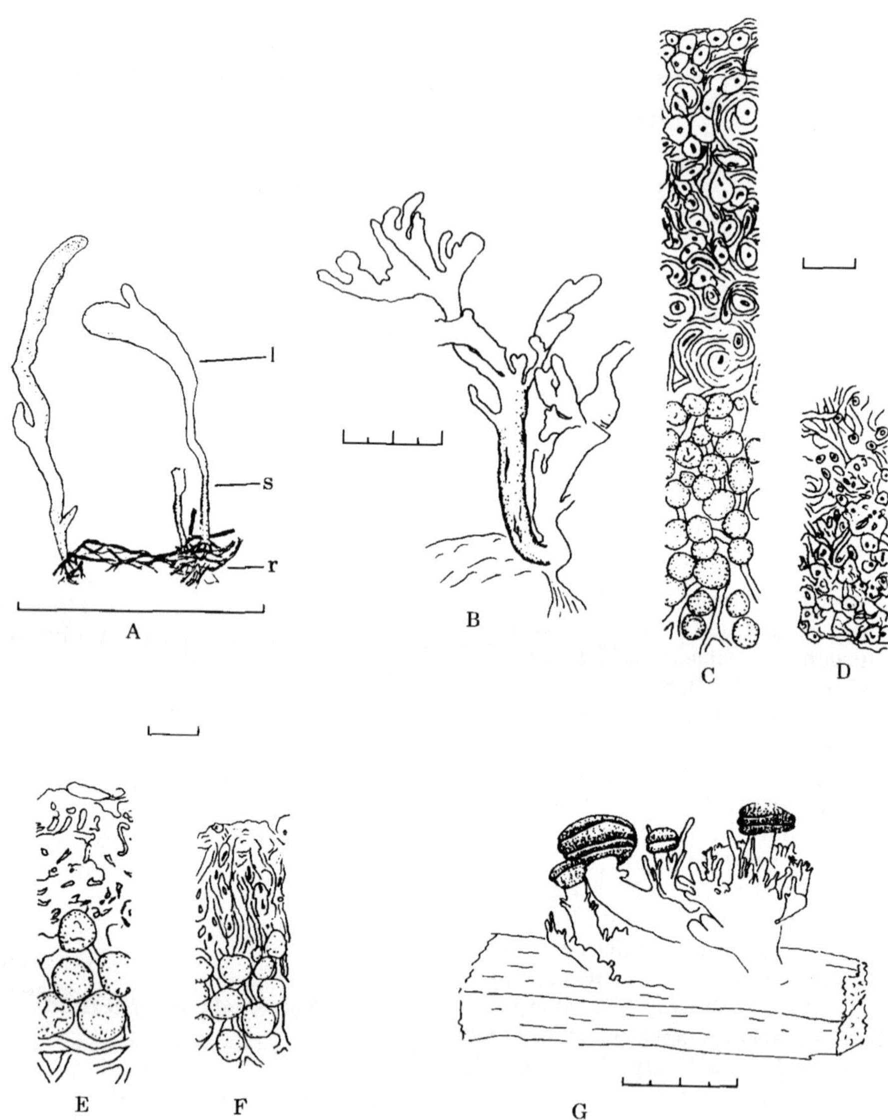

FIG. 1. A–D *Cetradonia linearis*. A & B, general habit (r=rhizines, s=phyllopodia, l=linear lobes); C, cross section of upper cortex with algal layer of linear lobe; D, cross section of lower cortex of linear lobe. E, *Cladonia yunnana*, cross section of upper cortex with algal layer of primary thallus. F, *Gymnoderma coccocarpum*, cross section of upper cortex with algal layer of primary thallus. G, *Neophyllis melacarpa*, general habit. Scales: A=7 mm; B=2 mm; C–F=20 μm; G=2 mm.

Results and Discussion

A careful examination of fresh and herbarium collections of *Gymnoderma lineare* shows that the thallus is cylindrical (fruticose) at the base becoming flat and strap-shaped towards the apex (Figs 1A & B, 3A). It is completely corticate at this stage (Fig. 1C &D). In contrast, *G. coccocarpum* (Figs 1F & 4), the type species, as well as *Cladonia yunnana* (Vain.) Abbayes ex J.-C.Wei & Y. M. Jiang (Figs 1E, 5) and most other

TABLE 1. *Comparison of* Cetradonia *with* Gymnoderma *s. str. and allied genera*

Character	*Cetradonia*	*Gymnoderma*	*Cladonia*	*Neophyllis*
Shape of primary squamules	Strap-shaped, branched, cylindrical at base	Fan-shaped, crenulate	Various, rarely strap, or fan-shaped	Small, multifid, coralloid and flat lobules
Ventral side of thallus	Corticate, not veined	Ecorticate, veined	Ecorticate, rarely slightly corticated (Figs 1E & 5), rarely veined	Fruticose—ecorticate, coralloid and flat lobules—corticated
Podetia	Solid; often lacking algae	Present or absent, solid, corticate lacking algae	Hollow, usually containing algae, corticate or ecorticate	Solid, corticate, lacking algae, with coralloid and flat lobules
Height of podetia	(0) 0·3–2 mm	0–0·5 mm	(0) 0·5–150 mm	0·5–3 mm
Origin of podetia and conidomata	Ventral side of squamules	Margin or subapical ventral side of squamules	Dorsal side of squamules, rarely marginal	Dorsal side or margin of squamules
Hymenial discs	Dark brown to black	Light to dark brown	Brown to dark brown, red or waxy	Black, rarely brownish
Phenols	Atranorin	Unknown compounds	Various including atranorin, not melacarpic acid	Melacarpic acid, etc
Aliphatic fatty acids	Lichesterinic acids (3 compounds)	Absent	Present but lichesterinic acids very rare	Absent
Ecology	On moist rocks	On tree bark	On various substrata	On rotting logs or tree bark, rarely on soil
Distribution	N. America	S.E. Asia, Australasia, E. Africa (?)	Cosmopolitan	Australasia

Cladonia species produce small foliose to squamulose thalli that are ecorticate below. *Gymnoderma lineare* also differs from both *G. coccocarpum* and *Cladonia* spp. in other characters such a its chemistry, distribution and habitat. (see Table 1). *Gymnoderma coccocarpum* appears to be more closely similar to *Cladonia* spp. than to *G. lineare*. Strong support for this view is provided by molecular data. Using sequence data from the nuclear SSU rDNA, Stenroos & DePriest (1998) found *G. lineare* fell well outside the monophyletic group formed by *Cladonia* and *Pycnothelia*. The evidence suggests that *G. lineare* is sufficiently distinct to merit placement in a separate genus.

The genus *Neophyllis* F. Wilson should also be recognized as distinct from *Gymnoderma* in ascus structure and ascoma ontogeny (Table 2), and from *Cladonia* in the type of conidiophores (Table 3). It seems more closely related to *Austropeltum* Henssen *et al.* (Verdon & Elix 1986; Henssen *et al.* 1992; Döring 1992; Döring *et al.* 1999; Wedin & Döring 1999). Phylogenetic analysis, based on the nuclear SSU rDNA sequences, strongly suggests that *Neophyllis* and *Austropeltum* are more closely related to the *Sphaerophoraceae* than to genera in the *Cladoniaceae* or *Stereocaulaceae* (Wedin and Döring 1999; Doring & Wedin 2000).

It is evident that the genus *Gymnoderma sensu* Yoshimura and Sharp (1968) and Yoshimura (1973) is polyphyletic and heterogeneous. The name *Gymnoderma* should be retained for the type species, *G. coccocarpum* and its presumed relatives. Therefore, *Cladonia lineare* Evans is here transferred to a new distinct genus *Cetradonia* J.-C. Wei & Ahti and the genus *Neophyllis* is retained for

TABLE 2. *The principal differences in the ascoma development and ascus type between* Neophyllis *and* Gymnoderma *s. str.*

Character	*Neophyllis*	*Gymnoderma*
Ascoma development	Ascogonia twisted and coiled. Primordia form a shallow band below the upper cortex. Boundary tissue produced (Döring *et al.* 1999).	Ascogonia linear to curved. Primordia develop in a vegetative outgrowth of the medulla. Boundary tissue not produced (Jahns 1970: Döring 1990).
Ascus type	Strongly amyloid tube structure (Döring *et al.* 1999).	Amyloid ascus tholus (Verdon & Elix 1986; Döring *et al.* 1999).

TABLE 3. *The principal differences in the type of conidiophores between* Neophyllis *and* Cladonia

	Neophyllis	*Cladonia*
Type of conidiophores	The conidiophores produce small, bacilliform conidia terminally and intercalary. Correspond to Vobis type VI (Vobis 1980).	Conidiophores with a main axis, ± dichotomously branched with terminal conidiophore cells. Correspond to Vobis type IV (Gluck 1899: Vobis 1980).

N. melacarpa. Thus, only two species of *Gymnoderma* (*G. coccocarpum* and *G. insulare*), both endemic to E. Asia, are now accepted in the amended genus. The monotypic genus *Cetradonia* is endemic to the southern Appalachian Mts., in eastern North America, and the genus *Neophyllis* is endemic to Australia, Tasmania and New Zealand. As a result, the three genera *Gymnoderma*, *Neophyllis* and *Cetradonia* are endemic to different continents and islands (Fig. 2).

Our phenotypic analysis of several characters, and the phylogenetic analysis based on SSU rDNA data (Stenroos & DePriest 1998) show that *Cladonia linearis* is probably a phylogenetically separate group within the Lecanorales. According to that analysis *Cladonia linearis* is distant from not only the family *Stereocaulaceae* (*Stereocaulon* and *Pilophorus*) but also the cladoniiform lichens (Ahti 1982), such as the families *Cladoniaceae*, *Cladiaceae*, and especially distant from the families *Siphulaceae*, *Baeomycetaceae*, and other families within the order Lecanorales. Therefore, we propose not only a new genus *Cetradonia*, but also a new family *Cetradoniaceae*, based on *Cladonia linearis*.

Key to the Genera

1. Primary thallus strap-shaped, branched, corticate below; podetia present, occasionally absent, solid, (0)0·3–2 mm long, often lacking algae, originating from ventral side of strap-shaped primary thallus. . . . . **Cetradonia**
   Primary thallus squamulose, ecorticate below 2

2(1). Podetia usually present (very rarely absent), hollow, (0)0·5–150 mm long, normally containing algae, originating dorsally, rarely marginally, from primary squamules . . . . **Cladonia**
   Podetia absent or present, 0–3 mm long, solid, normally lacking algae, originating marginally or dorsally. 3

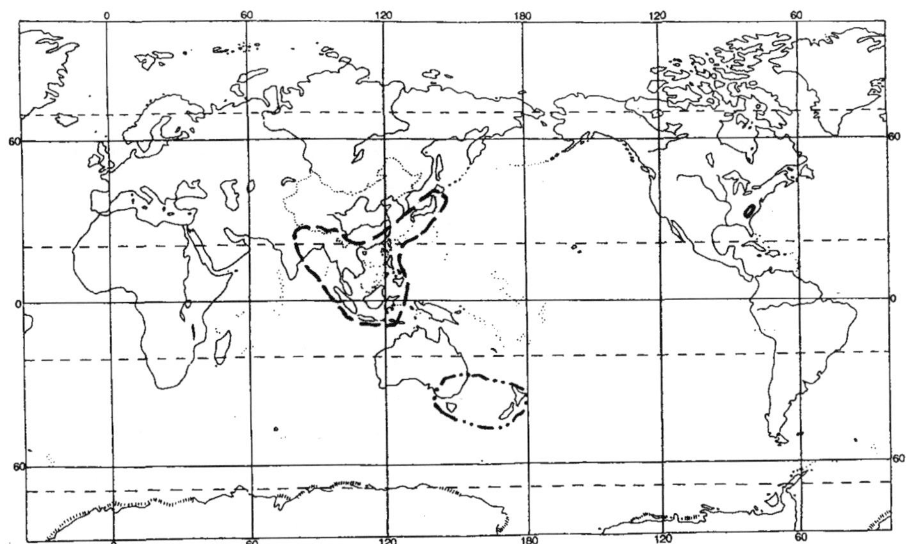

FIG. 2. Distribution map of the genera *Cetradonia* (——), *Gymnoderma* (– – – –) and *Neophyllis* (– · · – · · –).

3(2). Primary thallus squamulose, fan-shaped; podetia absent or present, very short (0–0·5 mm), originating from margins of squamules . . . . . . **Gymnoderma**

Primary thallus minutely fruticose, consisting of terete lobes, multifid; podetia present and conspicuous, 0·5–3 mm long, originating from dorsal side or margin of squamules, growing coralloid or flat primary thallus lobules . . . **Neophyllis**

Cetradoniaceae J.-C. Wei & Ahti, fam. nov.

Thallis primariis utrinque corticatis constantibus et podetiis fere cylindricis denigratis et squamulis linearibus complanatis a membris familiarum omnium Lecanoralium diversa.

Type genus: *Cetradonia* J.-C.Wei & Ahti

Cetradonia J.-C. Wei & Ahti, gen. nov.

Thallis primariis utrinque corticatis constantibus et podetiis fere cylindricis denigratis et squamulis linearibus complanatis a membris generis omnium Cladoniacearum diversa. Apothecia biatorina nigra. Acidum lichesterinicum, acidum protolichesterinicum et (probabiliter) alloprotolichesterinicum et atranorinam continens.

Type species: *Cetradonia linearis* (Evans) J.-C. Wei & Ahti

(Figs 1A–D, 2 & 3)

Thallus persistent, well-developed, forming tufts, with often nearly cylindrical basal stalks and linear, flattened, strap-shaped, suberect, unbranched to sparingly branched, cartilaginous lobes, 5–25 mm long, 0·5–1 mm broad, basally united to other thalli by horizontal rhizomorphs; upper surface thickly corticate, greyish green to pale yellowish brown or olive-green, with cartilaginous stereome inside; lower surface thinly corticate, not veined, but occasionally transversely rugulose in lower parts, and furrowed and pitted near tips, apically white, cream-coloured, becoming black towards base. *Podetia* originating from upper quarter of lower surface, usually subapical, dirty white, (0)0·3–2 mm tall, corticate, solid, without or when well-developed with algae.

Ascomata with clustered hymenial discs, clusters to 2 mm broad, single discs 0·1–1 mm broad, flat to spherical, dark brown to black, occasionally forming several discs successively on top of each other (as in *Cladia aggregata*); *Asci* very narrow and spores

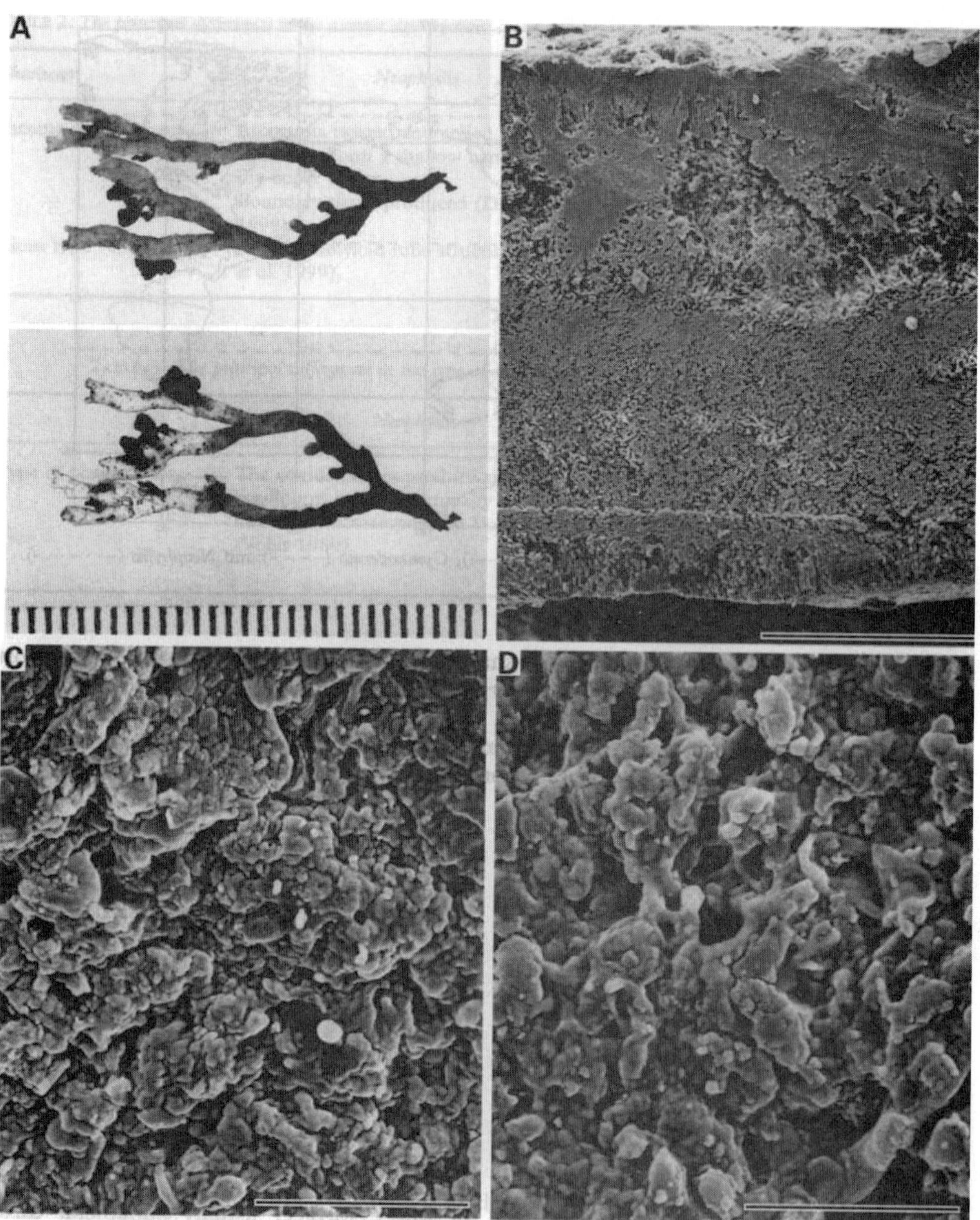

FIG. 3. *Cetradonia linearis*. A, general habit, upper (top) and lower (bottom) surface of linear lobes; B–D SEM micrographs of linear lobe; B, section; C, upper surface; D, lower surface. Scales: A in mm; B=150 μm; C & D=10 μm.

uniseriate, 8–10 × 2–2·5 μm, oblong, apical structure as in *Cladonia*. *Conidiomata* originating on basal thalli, either on lower side or on margins, dark brown, spherical to irregular, 100–150 μm diam, stalked, often 2–12 forming on the same branch, successively (see drawing in Jahns 1970: fig. 29), stalks up to 2 mm long; *conidia* not observed.

Chemistry. Contains atranorin and lichesterinic acids (3 compounds). Culberson (in Vězda 1980) reported atranorin and lichesterinic acid.

Notes. Cetradonia differs from *Cladonia* by its solid podetia, which originate on the ventral side of the primary thallus (for ontogeny, see Jahns 1970). The podetia are reported to lack algae, but in well-developed examples (e.g., in *Schofield* 10514) algae are present. The new genus differs from *Gymnoderma* by its stiff, linear and strap-shaped thalli and narrow asci. The primary thalli, which are corticate on the lower side, contain a central, cartilaginous stereome, thereby differing from both *Cladonia* and *Gymnoderma*.

Distribution. Cetradonia is endemic to the southern Appalachian Mts, eastern North America (Fig. 2).

Cetradonia linearis (Evans) J.-C.Wei & Ahti, comb. nov.

Cladonia linearis Evans, *Bryologist* **50**: 46 (1947).—*Gymnoderma lineare* (Evans) Yoshim. & Sharp, *Amer. J. Bot.* **55**: 639 (1968); type: United States, Tennessee, Sevier Co., Roaring Fork, Mt. LeConte, 1933, *K. Baur & M. Fulford* 39 (US, not seen, possibly lost).

(Figs 1A–D, 2 & 3)

Notes. Cetradonia linearis differs from *Gymnoderma coccocarpum* (the type of *Gymnoderma*) and *G. insulare* in the well-developed cortex on the lower surface of the strap-shaped primary thallus and the well-developed podetia arising from its lower surface, black apothecia, three lichesterinic acids, and a distribution restricted to the southern Appalachians in eastern North America (Fig. 2).

Cetradonia linearis is listed as an endangered species in the United States (many new additional localities, not listed here, have been recorded in recent inventories by environmental authorities).

Distribution. Confined to the Southern Appalachians (Kentucky, Ohio, North Carolina, Tennessee, Georgia) in eastern North America (Thomson 1968).

Selected specimens examined: **U.S.A.:** *North Carolina*: Yancey County, Camp Alice in the Black Mts., 5850 ft, 1972, *J. P. Dey* 1410 (H). *Tennessee*: Sevier Co., Clingman's Dome, 1959, *W. B. Schofield* 10514, [Vězda Lich. Sel. Exsicc. no. 1694](H); Great Smoky Mts. National Park, rock, stream bed just below Ramsey Cascades, *J. Drumke* (TENN). *Georgia*: Rabun Co.; Chattahoochee Natl. Forest, Rabun Bald, 1115–1420 m, 1997, *W. R. Buck* 32688 (NY).

Neophyllis F. Wilson

J. Linn. Soc., Bot. **28**: 372 (1891); type species: *Neophyllis melacarpa* (F. Wilson) F. Wilson.

Phyllis F. Wilson, *Victorian Naturalist* **6**: 68 (1889), *nom. illeg.*

(Figs 1G & 2)

Thallus squamulose. *Podetia* cylindrical, short, coralloid, and solid, with white medulla. *Photobiont Trebouxia* sp.

Ascomata brownish black, capitate, 0·5–2 mm wide, on top of podetia. *Ascus* 8-spored. *Ascospores* simple and hyaline.

Notes. This lichen is similar to some species of *Sphaerophorus* but without the spore mass (mazaedium) or to some *Stereocaulon* spp. but without cephalodia. It seems to be more similar morphologically to *Cetradonia* than to other lichenized fungi.

An analysis, based on SSU rDNA sequences, suggested that *Neophyllis* is more closely related to *Austropeltum* or *Sphaerophorus* than to genera of either *Cladoniaceae* or *Stereocaulaceae* (Wedin & Döring 1999).

More work on the systematic position of this genus is needed.

Neophyllis melacarpa (F. Wilson) F. Wilson

J. Linn. Soc., Bot. **28**: 372 (1891).—*Phyllis melacarpa* F. Wilson, *Victorian Naturalist* **6**: 68 (1889).—*Gymnoderma melacarpum* (F. Wilson) Yoshim. *J. Jap. Bot.* **48**: 287 (1973); type: Australia, Victoria, Mt Black Spur, 1888, *F. Wilson* (NSW—lectotype, designated by Filson (1986); BM, G—isolectotypes).

Psora dactylophylla Müll. Arg., *Bull. Herb. Boissier* **1**: 135 (1893).—*Lecidea dactylophylla* (Müll. Arg.)

Zahlbr., *Cat. Lich. Univ.* 4: 867 (1925); type: Australia: Victoria, Mt. Macedon, F. Wilson (G—holotype).

Phyllopsora melanocarpa Müll. Arg., *Hedwigia* 34: 28 (1895).—type: Australia: Victoria, *F. Wilson* 150 (G—holotype; W—isotype).

(Fig. 1G & 2)

Thallus squamulose, sublinear, irregularly to pinnately lobed, rounded at apex, sinuous at margins, small, 1·5 × 0·5 mm, the upper surface pale greenish brown to olive-green, the lower surface cream or yellowish brown, often with lobules on the upper surface or margins. *Podetia* cylindrical, simple, dense, sometimes flattened, to 3 mm tall.

Ascomata on top of podetia, with convex to hemispheric, shining, blackish brown to black discs about 1 mm wide. *Ascospores* ellipsoidal, simple, 5–13 × 4–7 μm.

Chemistry. Fumarprotocetraric, melacarpic, grayanic acids (major), congrayanic and 4-*0*-dimethylgrayanic acids (minor) (Filson 1992).

Notes. *Neophyllis* is usually (e.g. Filson 1992) considered to include two species, *N. melacarpa* and *N. pachyphylla*. The species *N. melacarpa* differs from *N. pachyphylla* by growing on rock or soil, by having larger apothecia (to 2 mm wide), and melacarpic acid only. However, Döring (1992) showed that these species differ only slightly in their chemistry (both, however, contain the rare dibenzofuran melacarpic acid). The older name *N. pachyphylla* assumes priority over the commonly used *N. melacarpa*, in case they are united.

Because there are very few specimens of *N. pachyphylla*, Döring *et al.* (1999) suggested that two species are maintained until the status of *N. pachyphylla* has been clarified.

Habitat and distribution. On rotting logs and tree bases in wet sclerophyll forests up to 1300 m (Filson 1992). Confined to SE Australia, Tasmania and New Zealand (Yoshimura 1973; Yoshimura & Kurokawa 1976; Filson 1992).

Selected specimens examined: **Australia**: *New South Wales*: SE of Rossi, 19 Dec. 1967, *W. A. Weber & D. McVean*, [Lich. Exsicc. COLO 246] (H); 6 km E of Captains Flat, 1985, *D. Verdon* 5618, [Elix: Lich. Australas. Exsicc. 111] (H). *Tasmania*: Near Boyd River, 1984, *G. Kantvilas* 493/84, [Vězda: Lich. Sel. Exsicc. 2020] (H).—**New Zealand:** *Wellington:* 18 km S of Otaki, Kapahapanui Hut, 1990, *L. Tibell* 19035, [Lich. Sel. Exsicc. Upsal. 128] (H).

Neophyllis pachyphylla (Müll. Arg.) Gotth. Schneid.

Bibl. Lichenol. 13: 168 (1979).—*Psora pachyphylla* Müll. Arg., *Flora* 70: 319 (1887); type: Australia, Victoria, Mt. William, *D. Sullivan* 86 (G—holotype).

Thallus squamulose. Squamules thick, densely pressed together and imbricate. *Podetia* similar to *N. melacarpa*.

Ascomata to 2 mm wide, terminal or subterminal on short podetia, convex, shining, brownish black. *Ascospores* oblong-ovoid, 11–12 × 5–7 μm.

Chemistry. Melacarpic acid.

Habitat and distribution. On rock or a clay bank. Known only from the type locality and Pigeon House Range, N.S.W. (Filson 1992).

Gymnoderma Nyl.

Flora 43: 546 (1860) *nom. cons.*; type species: *Gymnoderma coccocarpum* Nyl.

(Figs 1F, 3 & 4)

Thallus squamulose; squamules broadly fan-shaped, only the upper side corticate, upper surface of lobes of squamules convex or not convex.

Ascomata almost sessile on the margin of the primary thallus without podetia or sometimes with very short, solid podetia without symbiotic algae. *Asci* broader than in *Cetradonia*; spores biseriate; ascus tip structure similar to *Cladonia*.

Notes. Galloway (1985) claimed that *Gymnoderma* was invalidly published by Nylander (1860). However, the name is conserved on the basis of this publication, as the descriptive words 'genus *Psoromeorum* (analogus *Eriodermati*)' can be interpreted by

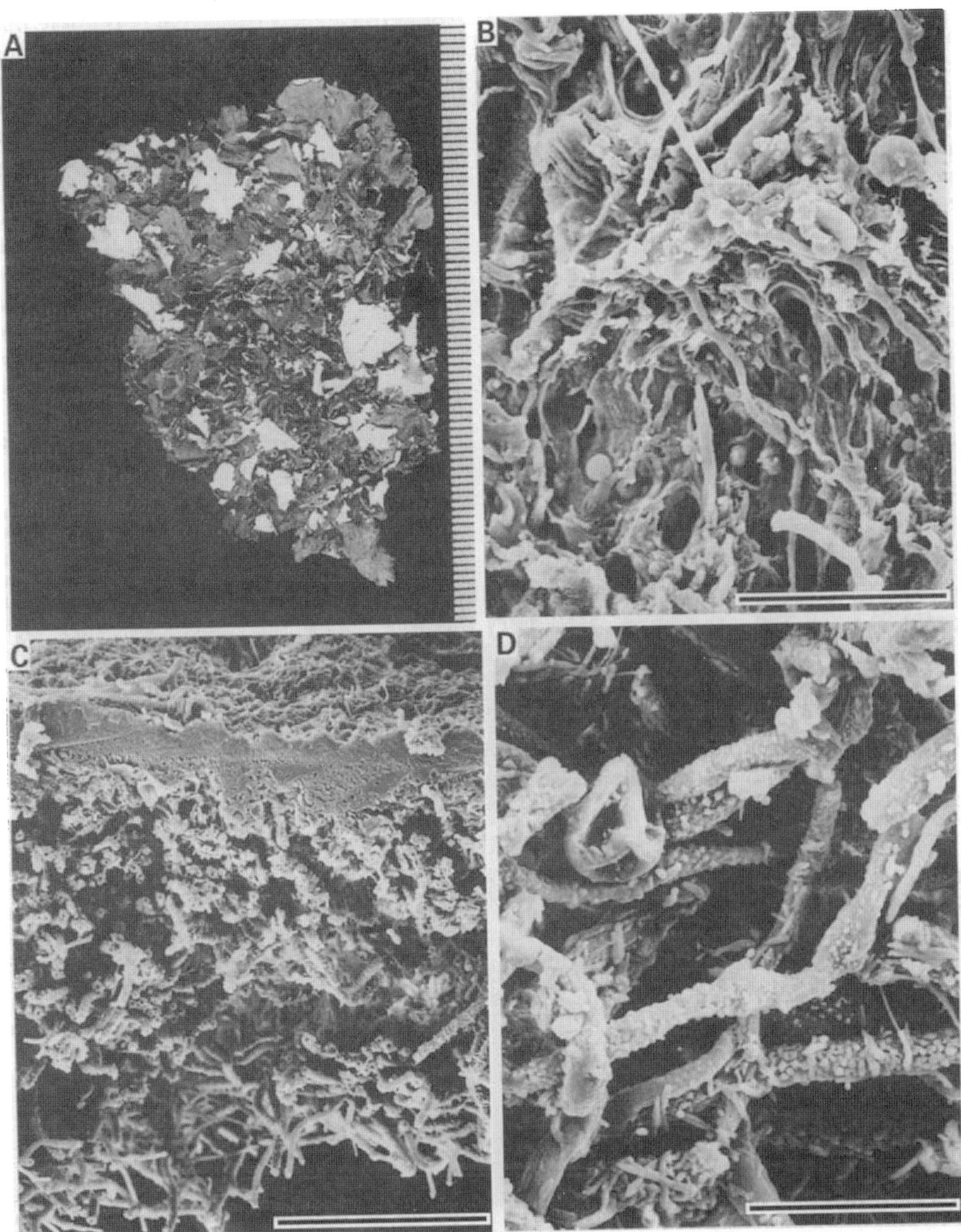

FIG. 4. *Gymnoderma coccocarpum*. A, general habit of primary thallus; B–D, SEM micrographs of the thallus; B, upper surface; C, section; D, lower surface. Scales: A, in mm; B & D=10 μm; C=50 μm.

specialists as not having apothecia which are biatorine, unlike in the other *Psoromei*, but exactly like *Erioderma* which has biatorine apothecia even though this genus is placed in the non-biatorine *Pannariacei*: see discussions in Nylander (1860: 545; 1869). Therefore, we regard the publication of *Gymnoderma* Nyl. 1860 as perfectly valid.

Gymnoderma is a distinct member of the *Cladoniaceae*.

Distribution. The two species of *Gymnoderma* accepted here occur in Asia; an additional species is found in Africa.

Gymnoderma coccocarpum Nyl.

Flora **43**: 546 (1860).—*Cladonia coccocarpa* (Nyl.) Evans, *Bryologist* **50**: 50 (1947); type, Sikkim, Tonglo ('Tongloo'), *J. D. Hooker & T. Thomson* 2124 (PC—lectotype, designated by Ahti (1993):102; BM, H-NYL 30796 & 30953, LE, NY, US—isolectotypes).

?=*Cladonia botryocephala* Hepp in Zollinger, *Syst. Verzeichn.* 5: 9 (1854); *Natuurk. Tijdsschr.* **6**: 143 (1854); type: Indonesia: Java, Tjikoja, *Zollinger* 447b (G, PC—Delise-syntype).

(Figs 1F & 4)

Primary *squamules* large, up to 4 cm long, 0·8 cm wide, upper surface of lobes not convex, but here and there depressed or grooved.

Chemistry. Unidentified medullary substance (UV+, KC+ rose) (Yoshimura & Sharp 1968).

Habitat and distribution. On bark of trees. Confined to E. Asia.

Notes. The type material (very poor) of an older name *Cladonia botryocephala* Hepp may be conspecific with *G. coccocarpum*, requiring that the epithet *botryocephalum* should be adopted. However, the syntypes available have much dissected primary thalli, indicating that they may be a distinct species of *Gymnoderma* [a description also provided by Vainio (1887)]. If *C. botryocephala* eventually proves to be synonymous with *G. coccocarpum*, then a conservation proposal should be made to conserve the name *coccocarpum*. *Gymnoderma coccocarpum* was adopted in the Names in Current Use list (Ahti 1993), and because it is possible that this list will be sanctioned later, conservation of *G. coccocarpum* would be an appropriate action. However, the distinct possibility that *G. botryocephala* is a separate species based on one depauperate specimen cannot be excluded.

Jahns and van der Knapp (1973) synonymized *Heterodea madagascarea* Nyl. (Nylander 1888) with *Gymnoderma coccocarpum*. However, according to Krog and Timdal (*in litt.*), it is a distinct species of *Gymnoderma*, which will be treated elsewhere. Therefore, *Gymnoderma* does occur in Africa.

Gymnoderma coccocarpum is listed as a threatened species in Japan and China.

Specimens examined. **China:** *Yunnan*: Maguan county, Mt. Laojunshan, on tree bark, 1700 m, 1959, *O. Z. Wang* 392 (HMAS-L!); Mt Gong shan, Dulong jiang, alt. 1400 m, on rotten wood among mosses, 1982, *J. J. Su* et al. 2880 (HMAS-L). *Hunan*: Zhang-yi, Mt Mang shan, alt. 1170 m, on rotten wood, 2001, *J. B. Chen* et al. (HMAS-L).—**Taiwan:** Hwalien Co., Hopin Forest Station, 1971, *M. J. Lai* 2020 (US).—**Sikkim**: Mainomcha, *J. D. Hooker & T. Thomson* 2101 (NY).—**Malaysia**: *Sabah*: along Mesilau Trail on the plateau from Mesilau River to Kundason, elev. *c.* 1600 m, Kinabalu National Park, Aug. 1964, *M. E. Hale* 28174 (US).

Literature reports: Himalaya (Yoshimura & Sharp 1968; Yoshimura 1982); China [Guangxi ('Kinangri', may be mistaken for Kwangsi=Guangxi), Yoshimura & Sharp 1968], Yunnan (Yoshimura 1982), Taiwan [Formosa], (Sato 1941; Zahlbruckner 1956; Yoshimura 1982)], Thailand, Borneo (Yoshimura & Sharp 1968; Yoshimura 1982), Java (Groenhart 1954; Yoshimura & Sharp 1968; Sato 1941; Zahlbruckner 1956; Yoshimura 1982), Philippines (Yoshimura 1982), Japan (Kashiwadani & Gradstein 1982; Yoshimura 1982), Malaysia (Kashiwadani & Gradstein 1982).

Gymnoderma insulare Yoshim. & Sharp

Amer. J. Bot. **55**: 638 (1968). type: Japan, Honshu, Prov. Kii, Mt. Koya, on trunk of *Cryptomeria japonica*. 800 m, 7 Nov 1957, *S. Kurokawa* 57277 [in Kurokawa, Lich. Rar. Crit. Exsicc. no. 8] (TNS—holotype; DUKE, M, NICH, H, TENN, UPS, US—isotypes).

Primary squamules small, less than 1 cm long, and 0·4 cm wide, upper surface of lobes convex, looking like a miniature upside down spoon.

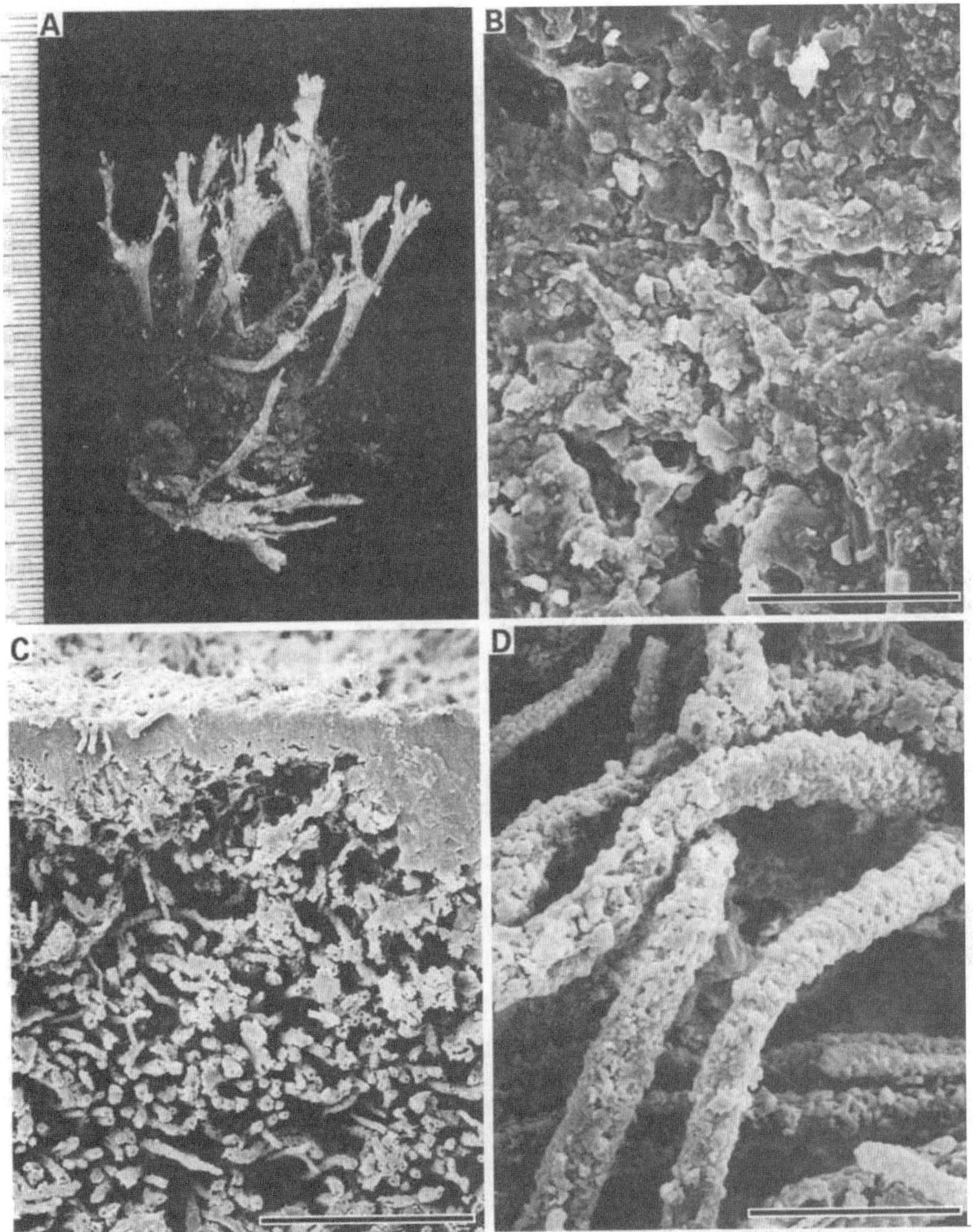

FIG. 5. *Cladonia yunnana*. A, general habit of primary thallus; B–D, SEM micrographs of the thallus; B, upper surface; C, section; D, lower surface. Scales: A, in mm; B & D=10 μm; C=50 μm.

Chemistry. Didymic acid and atranorin.

Notes. This species is distinguished from *G. coccocarpum* by having smaller primary squamules and having the upper surface of squamule lobes convex, which look like miniature upside down spoons, and by containing atranorin and didymic acid.

Habitat and distribution. On bark of trees. Known only from Japan.

Specimens examined. **Japan:** *Honshu*: Prov: Kii, Mt. Koya, Okunoin, on bark of *Cryptomeria japonica*, 19 Oct 1952, *Y. Asahina*, Lich. Jap. Exsicc. o. 107 (H); *I. Yoshimura* 72435, [Lich. Jap. Exsicc. no. 38] (H, MIN, TENN). *Kyushu*: Prov. Ohsumi: Yakusugi-land, Yakushima Island on bark of *Cryptomeria japonica*; 1000 m, 1984, *H. Shibuichi* (no. 7683) & *K. Yoshida* [Kurokawa and Kashiwadani, *Lich. Rar. Crit. Exsicc.* no. 621] (H, TENN).

J.-C. Wei thanks Dr G. Mueller and Dr Qiuxin Wu for facilitating field work at Highlands, North Carolina, herbarium work in The Field Museum, Chicago, and for the loan of specimens from various herbaria. He also thanks Dr C. M. Wetmore for support in the University of Minnesota's herbarium and for borrowing specimens from different herbaria, the curator of the Michigan State University Herbarium for the opportunity to examine its lichen collections, and also the curators of the following herbaria for the loan of collections: TENN and US. Dr G. Mueller and Prof. D. Pfister kindly improved the English in the manuscript, Prof. C. W. Smith gave valuable comments and suggestions, and Dr S. Y. Guo prepared the distribution map. The study by J.-C. Wei was supported in part by the National Natural Science Foundation of China (39670002), the Special fund of the Committee for Systematic Biology of CAS, the Bureau of International Cooperation of CAS, and U.S. National Science Foundation (INT95-13874). It was undertaken in cooperation with the Field Museum of Natural History, Chicago, Illinois, U.S.A. The work by T. Ahti was supported by grants from the Academy of Finland.

References

Ahti, T. (1982) Evolutionary trends in cladoniiform lichens. *Journal of the Hattori Botanical Laboratory* **52:** 331–341.

Ahti, T. (1993) Names in current use in the Cladoniaceae (lichen-forming ascomycetes) in the ranks of genus to variety. *Regnum Vegetabile* **128:** 58–106.

Asahina, Y. (1936) Mikrochemischer Nachweis der Flechtenstoffe (1). *Journal of Japanese Botany* **12:** 516–525.

Asahina, Y. (1937) Mikrochemischer Nachweis der Flechtenstoffe (IV). *Journal of Japanese Botany* **13:** 855–861.

Chester, D. O. & Elix, J. A. (1980) A new dibenzofuran and diphenyl ether from the lichen *Gymnoderma melacarpum*. *Australian Journal of Chemistry* **33:** 1153–1156.

Culberson, C. F. (1972) Improved conditions and new data for the identification of lichen products by a standardized thin-layer chromatographic method. *Journal of Chromatography* **72:** 113–125.

Culberson, C. F. & Kristinsson, H. (1970) A standardized method for the identification of lichen products. *Journal of Chromatography* **46:** 85–93.

Döring, H. (1990) Morphologisch-ontogenetischen Studien an australasiatischen Flechten. Unpublished "Diplomarbeit", Universitat Marburg.

Döring, H. (1992) Developmental morphology of *Neophyllis pachyphylla* (Müller Arg.) G. Schneider, (synonym: *Neophyllis melacarpa* (F.Wils.) F.Wils.). In *International Association for Lichenology Symposium 2 Abstract Volume* (I. Kärnefelt, ed.): 12. Lund: University of Lund.

Döring, H., Henssen, A. & Wedin, M. (1999) Ascoma development in *Neophyllis melacarpa* (Lecanorales, Ascomycota), with notes on the systematic position of the genus. *Australian Journal of Botany* **47:** 783–794.

Döring, H. & Wedin, M. (2000) Homology assessment of the boundary tissue in fruiting bodies of the lichen family Sphaerophoraceae (Lecanorales, Ascomycota). *Plant Biology* **2:** 361–367.

Eriksson, O. (1981) The family of bitunicate ascomycetes. *Opera Botanica* **60:** 1–220.

Evans, A. W. (1943) Microchemical studies on the genus *Cladonia* subgenus *Cladina*. *Rhodora* **45:** 417–438.

Filson, R. B. (1986) Index to type specimes of Australian lichens: 1800–1984. *Australian Flora and Fauna Series* **4:** 1–317.

Filson, R. B. (1992) *Neophyllis*. *Flora Australia* **54:** 145–146.

Galloway, D. J. (1985) *Flora of New Zealand Lichens*. Wellington: Hasselberg.

Glück, H. (1899) Entwurf zu einer vergleichenden Morphologie der Flechten-Spermegonien. *Verhandlungen des Naturhistorisch-Medizinischen Vereins zu Heidelberg, N.F.* **6**(2): 81–216.

Groenhart, P. (1954) Malaysian lichens IV. *Reinwardtia* **2:** 385–402.

Henssen, A., Döring, H. & Kantvilas, G. (1992) *Austropeltum glareosum* gen. et sp. nov., a new lichen from mountain plateaux in Tasmania and New Zealand. *Botanica Acta* **105:** 457–467.

Jahns, H. M. (1970) Untersuchungen zur Entwicklungsgeschichte der Cladoniaceen unter besonderer Berücksichtigung des Podetien-Problems. *Nova Hedwigia* **20:** 1–177.

Jahns, H. M. & van der Knapp, P. (1973) Die Flechtengattung *Heterodea* Nyl. Systematik und Ontogenie der Fruchtkörper. *Herzogia* **2:** 437–451.

Kashiwadani, H. & Gradstein, S. R. (1982) Notes on *Baeomyces sanguineus* Asah. and *Gymnoderma coccocarpum* Nyl. *Miscellanea Bryologica et Lichenologica* **9:** 79–81.

Nylander, W. (1860) De lichenis nonnullis europaeis. *Flora* **43:** 545–547.

Nylander, W. (1869) *Synopsis Lichenum* **2**: 1–64. Paris: L. Martinet.

Nylander, W. (1888) *Lichenes Novae Zelandiae*. Paris: Schmidt.

Sato, M. (1941) Cladoniales (1). In *Nova Flora Japonica*, 7. (T. Nakai & M. Honda, eds): Tokyo: Sanseido.

Stenroos, S. & DePriest, P. T. (1998) SSU rDNA phylogeny of cladoniiform lichens. *American Journal of Botany* **85:** 1548–1559.

Thomson, J. W. (1968) ['1967'] *The Lichen Genus Cladonia in North America*. Toronto: University of Toronto Press.

Vainio, E. A. (1887) Monographia Cladoniarum universalis. 2. *Acta Societatis pro Fauna et Flora Fennica* **10:** 1–499.

Verdon, D. & Elix, J. A. (1986) *Myelorrhiza*, a new Australian lichen genus from North Queensland. *Brunonia* **9:** 193–214.

Vězda, A. (1980) *Lichenes Selecti Exsiccati* Fasc. **68:** 1676–1700. Pruhonice.

Vobis, G. (1980) Bau und Entwicklung der Flechten-Pycnidien und ihrer Conidien. *Bibliotheca Lichenologica* **14:** 1–141.

Wedin, M. & Döring, H. (1999) The phylogenetic relationship of the Sphaerophoraceae, Austropeltum and Neophyllis (lichenized Ascomycota) inferred by SSU rDNA sequences. *Mycological Research* **103:** 1131–1137.

White, F. J. & James, P. W. (1985) A new guide to microchemical techniques for the identification of lichen substances. *British Lichen Society Bulletin* 57(Suppl.): 1–47.

Yoshimura, I. (1973) Notes on *Gymnoderma melacarpum*, comb. nov. *Journal of Japanese Botany* **48:** 283–288.

Yoshimura, I. (1982) Distribution of *Gymnoderma coccocarpum* Nyl. and *G. insulare* Yoshim. & Sharp. *Bulletin of Kochi Gakuen Junior College* **13:** 83–86.

Yoshimura, I. & Sharp, A. J. (1968) A revision of the genus *Gymnoderma*. *American Journal of Botany* **55:** 635–640.

Yoshimura, I. & Kurokawa, T. (1976) Chemical substances of *Gymnoderma melacarpum*. *Bulletin of Kochi Gakuen Junior College* **7:** 51–53.

Zahlbruckner, A. (1956) Flechtenflora von Java, 2. *Willdenowia* **1:** 433–528.

Three new taxa of *Stereocaulon* from China

MAN-RONG HUANG & JIANG-CHUN WEI*

*Corresponding author: weijc@95777.com

The Systematic Mycology & Lichenology Laboratory, Institute of Microbiology, Chinese Academy of Sciences, Beijing 100080, China

Abstract—Three taxa of the lichen genus *Stereocaulon* from China, namely *Stereocaulon intermedium* var. *gracile*, *St. kangdingense* and *St. sorediiphyllum* are described in this paper as new to science. The diagnoses in Latin and descriptions and remarks in English are given. In addition, each taxon is provided with a photograph.

Key words—*Stereocaulon intermedium* var. *gracile*, *St. kangdingense*, *St. sorediiphyllum*

During studies of the lichen genus *Stereocaulon* in China, some interesting specimens drew our attention. After an examination of them in detail three taxa are considered as new to science and described in the present paper.

1. ***Stereocaulon intermedium* (Savicz) H. Magn. var. *gracile* M. R. Huang & J. C. Wei,** *var. nov.* **FIGURE 1**

Varietas nostra a var. intermedio *pseudopodetiis decumbentibus et apotheciis nec apicalibus et lateralibus differt.*

TYPE—CHINA. SICHUAN: E. SLOPE OF MT. GONGGA, alt. 2400 m, on sandy soil, 23.VI.1982, *Wang X. Y., Xiao X. & Li B., no. 8841* (HMAS-L, **HOLOTYPE!**).

DESCRIPTION—Primary thallus evanescent in early stage. Secondary thallus with decumbent and caespitose stalks, i.e., pseudopodetia compactly caespitose, tuft-like, decumbent, slightly dorsiventral, cylindrical, slightly twisted, not tapered, gracile, 2.5-5 cm long, 0.5-1 mm thick, moderately branched, irregular, main stems conspicuous, completely decorticated, heavily or sometimes rather thinly tomentose, felted, whitish, not ligneous, without soredia, firmly or sometimes rather loosely attached to substrate; phyllocladia numerous, grain-like to mainly terete-coralloid, 0.1-0.2 mm thick, ash-gray, with paler tips, usually not over 1 mm long, but abundantly branched in an antler-like manner. Cephalodia abundant, sessile, globose to subglobose, 0.1-0.5 (-0.8) mm in diameter, with white pruina in surface or half-immersed in tomentosa, poorly corticate, containing *Nostoc* sp.

Apothecia abundant, terminal and lateral, with pseudolecanorine margin in earlier stage, disappearing later; disc brown, flattened or slightly convex; hymenium hyaline, 40-50 μm thick, both hypothecium and central cone colorless; asci 31-47 μm × 7-10 μm; spores 8 per ascus, colorless, clavate, 3-septate, 27-37 μm × 2-3 μm.

This paper was originally published in *Mycotaxon*, 90（2）: 469-472, 2004.

Pycnidia not observed.

CHEMISTRY—K+ yellow, P+ pale yellow; TLC: atranorin, lobaric acid.

OTHER SELECTED SPECIMENS EXAMINED — **CHINA.** SICHUAN: MT. EMEI, alt. 2800 m, on rock, 18.VIII.1963, *Zhao J. D. & Xu L. W., no. 8144*; **Wolong**, alt. 2200 m, on rock, 24.VIII.1982, *Wang X. Y., Xiao X & Li B, no. 9635*. YUNNAN: **Gongshan**, alt. 3300 m, on humus, 26.VII.1982, *Su J. J., no. 2620*; **Weixi**, E. SLOPE OF MT. BILUO, alt. 3300 m, on rock, 13.VII.1981, *Wang X. Y., Xiao X. & Su J. J., no. 4642*. XIZANG: **Nyalam**, alt. 3670 m, on soil, 14.VI.1966, *Wei J. C., Chen J. B. & Zong Y. C., no. 1571*. **INDIA.** ANCHAL PRADESH: **Chamoli Distr.**, BETWEEN WAAN AND BHUNA, alt. 11500 feet, on soil over rock with mosses and foliose hepatics, *A. Singh, no. 91599-1* (FH!), which was segregated from the holotype specimen of *St. paradoxum* I. M. Lamb.

REMARKS —The variety is distinguished from *St. intermedium* var. *intermedium* (FH-syntype!) mainly by lateral and terminal apothecia and decumbent, gracile and thickly tomentose pseudopodetia. It is widespread in southwestern China.

2. ***Stereocaulon kangdingense*** **M. R. Huang & J. C. Wei,** *sp. nov.* **FIGURE 2**

Species St. esterhuyseniae *similis a qua pseudopodetiis non lignosis tomentulosis et cephalodiis presentibus differt.*

TYPE — **CHINA.** SICHUAN: **Kangding,** SHADE, alt. 3300 m, on soil, 18.X.1999, *Chen L. H., no. 990108* (HMAS-L, **HOLOTYPE!**).

DESCRIPTION—Primary thallus evanescent in early stage. Pseudopodetia compactly caespitose, decumbent, obviously dorsiventral, cylindrical, 1-1.5 cm long, 0.5-1 mm thick, expanded into flattened and flabellate ends with digitately divided margins, sparsely dichotomously branched; main stems inconspicuous, completely decorticated on ventral side, covered with thin and smooth tomentosum, well corticate on dorsal side, especially the parts near the tips, not ligneous, without soredia, firmly attached to substrate; phyllocladia scarce, replaced by flabellate pseudopodetial apices or poorly differentiated cortex and adhered to pseudopodetia, crenate-lobulate. Cephalodia abundant, sessile, globose to verrucose, not over 0.5 mm in diameter, bluish, but pruinose in most cases, poorly corticate, containing *Nostoc* sp.

Apothecia and pycnidia not observed.

CHEMISTRY—K+ yellow, P+ red; TLC: atranorin, stictic acid, norstictic acid.

ADDITIONAL SPECIMEN EXAMINED — SICHUAN: NW SLOPE OF MT. GONGGA, alt. 2950 m, on rock, 11.VIII.1982, *Li B., Wang X.Y. & Xiao X., no. 0374*.

REMARKS—It seems to be very close to *St. esterhuyseniae* I. M. Lamb (CANL-holotype!). However, it can be recognized by presence of cephalodia, thin tomentum and non-ligneous pseudopodetia. What is more, the latter species is found only in South Africa so far (Lamb 1953, 1977). The resemblance of these two species suggests that they are a pair of vicarious species.

3. ***Stereocaulon sorediiphyllum*** **M. R. Huang & J. C. Wei,** *sp. nov.* **FIGURE 3**

Species St. coniophylli *similis a qua phyllocladiis granulosis et pseudopodetiorum apicibus non subfoliosis praecipue differt.*

TYPE — **CHINA.** JILIN: MT. CHANGBAI, alt. 1750 m, on rock, 9.IX.1997, *Guo S. Y., no. 1105* (HMAS-L, **HOLOTYPE!**).

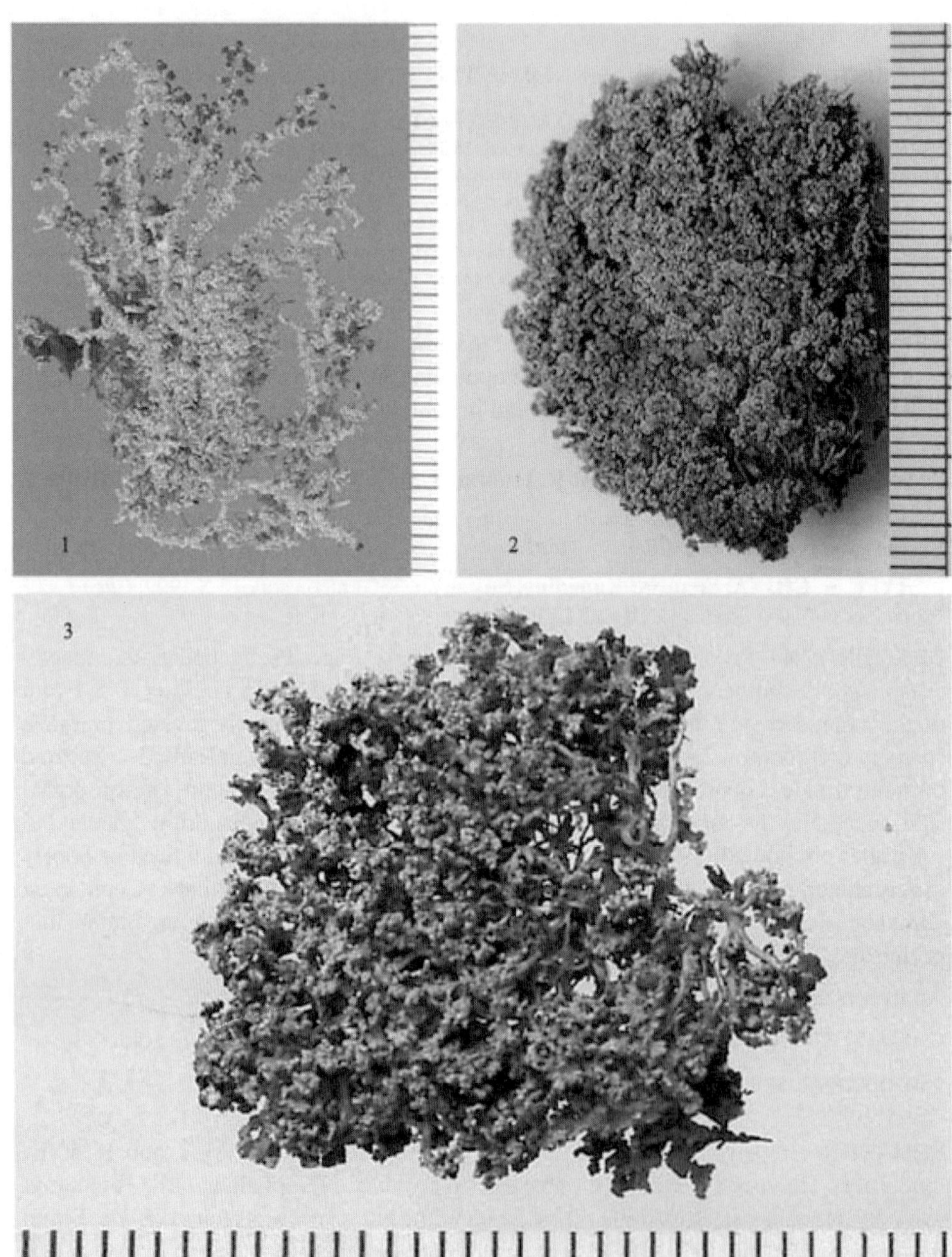

Fig. 1. *Stereocaulon intermedium* var. *gracile* (holotype). Scale=1 mm. Fig. 2. *Stereocaulon kangdingense* (holotype). Scale=1 mm. Fig. 3. *Stereocaulon sorediiphyllum* (holotype). Scale=1 mm.

DESCRIPTION — Primary thallus evanescent in early stage. Pseudopodetia compactly caespitose, erect, not dorsiventral, cylindrical, dwarfish, only 0.5-1.5 cm high, ca. 1 mm in diameter, abundantly branched, mostly corymbose; main stems inconspicuous, completely decorticated, glabrous, not tomentose, yellowish, obviously ligneous, some pseudopodetial ends flattened and subfoliose and brown, corticate on upper surface, and ecorticate, completely sorediate on lower surface, like but not as much as in *St. coniophyllum* I. M. Lamb, firmly attached to substrate; phyllocladia numerous, grain-like, ca. 0.2 mm in diameter, aggregated at tips of pseudopodetia and their branchlets, but rare or absent in other parts, whitish, thinly pruinose; some phyllocladia are sorediate. Cephalodia rare, sessile, subglobose, 0.5-1.5 mm in diameter, brown, protosacculate, containing *Nostoc* sp.

Apothecia and pycnidia not observed.

CHEMISTRY—K+ yellow, P+ pale yellow. TLC: atranorin, lobaric acid.

ADDITIONAL SPECIMEN EXAMINED — SHAANXI: MT. TAIBAI, on rock, 5.VI.1963, *Wei* J. C. et al., *s.n.*

REMARKS—The new species is closely related to *St. coniophyllum,* from which it is distinguished by bearing numerous grain-like phyllocladia and by the basically nonsubfoliose apices of pseudopodetia.

Acknowledgements

The project was financially supported by the National Natural Science Foundation of China no. 30270005. The authors are very grateful to the keepers of the herbaria FH and CANL for loan of type specimens, and to Ms. H. Deng for checking some place names of China.

Literature Cited

Lamb, IM. 1953. New, rare or interesting lichens from the southern hemisphere II. Lilloa 26: 401-438.

Lamb, IM. 1977. A conspectus of the lichen genus *Stereocaulon* (Schreb.) Hoffm. J. Hattori Bot. Lab. 43: 191-355.

Two new species of *Hypogymnia* (*Lecanorales*, Ascomycota) with pruinose lobe tips from China*

XINLI WEI [1,2] JIANGCHUN WEI *,1

*weijc2004@126.com

[1]Systematic Mycology & Lichenology Laboratory
Institute of Microbiology, Academia Sinica
Zhong-guan-cun, Bei-yi-tiao#13, Beijing 100080, China

[2]Graduate School of Chinese Academy of Sciences, Beijing 100039, China

Abstract—Two new species of lichens are described from China, viz. *Hypogymnia pseudopruinosa* (in Yunnan) and *H. subfarinacea* (in Sichuan and Yunnan). They are characterized by means of morphology and secondary chemistry. Latin diagnoses, English descriptions, and habitus photographs are provided.

Keywords—*H. macrospora*, *H. farinacea*, lichen substances

Until recently, forty species of *Hypogymnia* (Nyl.) Nyl. have been reported from China (Wei 1991; Chen 1994; McCune & Obermayer 2001; McCune & Tchabanenko 2001; McCune et al. 2003). During our studies on the lichen flora of China, two new species of the genus were collected from Yunnan and Sichuan Provinces, which are described in this paper. The gross morphology and anatomy were examined using the dissecting microscope (ZEISS Stemi SV 11) and compound microscope (OPTON Ⅲ). The lichen substances were detected by colour reagents and thin-layer chromatography (Culberson & Kristinsson 1970; Culberson 1972; White & James 1985).

Hypogymnia pseudopruinosa X.L. Wei & J.C. Wei, sp. nov. **Plate I: A-B**

Hypogymniae macrosporae similis, sed sporis minoribus, lobi pruina obductis et atranorina deest.

Type: China. Yunnan, Dèqên county, alt. 4100 m, on dead branches of *Sabina* sp., X. Y. Wang, X. Xiao & J. J. Su 7606, 29 August 1981 (holotype, HMAS-L).

Thallus foliose, tightly appressed; with subdichotomously branched lobes of 1-2 mm wide and 5 mm long; upper surface dark brownish-yellow, partly black, with black margin, rugose, glossy, lacking isidia, soredia and lobules but with dense layer of pruina limited to the lobe tips; lower surface black, brown near the apices, rugose, glossy, and with round perforations at the lobe tips and on the lower surface.

* The project was supported by the National Natural Science Foundation of China (30499340)

This paper was originally published in *Mycotaxon*, 94（1）: 155-158，2005.

Upper cortex prosoplectenchymatous, pale yellow, 12-14.5 μm thick; algal layer 10-14.5 μm thick, consisting of green, subspherical cells of 7.5-9.5 μm in diameter; hyphae in medulla colorless, about 2 μm in diameter; lower cortex prosoplectenchymatous, pale yellow, 10 μm thick.

Apothecia rare, 1-3 mm in diameter, stipitate; disc yellow brown or red brown, glossy, concave at first and then slightly plane with entire and thin margin; epithecium brown, 7-9 μm thick; hymenium colourless, 32-36 μm thick; asci clavate, 8-14 x 23.5-25 μm, 8-spored; spores simple, colourless, ellipsoid to nearly spherical, 3.5-4 × 5.5-7 μm; paraphyses linear, septate, 2 μm wide, slightly swollen at the tips; hypothecium colourless, 27-36 μm thick. Pycnidia not seen.

Chemistry. Cortex K-, C-, P-; medulla K+ yellow, C-, P+ orange yellow → orange red; containing physodalic and physodic acids, and a pale spot in R_f class 6 (solvent system C).

Comments: The new species resembles *H. macrospora* (J.D. Zhao) J.C. Wei (Zhao 1964; Wei 1991) at first sight, but differs by having a dense layer of pruina limited to the tips of the lobes and smaller ascospores, and by lacking atranorin.

Hypogymnia subfarinacea X.L. Wei & J.C. Wei, **sp. nov.** **Plate I:** C-E

Habitu cum Hypogymnia farinacea optime congruens, sed differt lobis pruinosis et acidum physodalicum continens.

Type: China. Sichuan Province, Nanping County, Jiuzhai Gou, alt. 2151 m, on trunk of *Tsuga* sp., 10 June 1983, X. Y. Wang & X. Xiao 10582 (holotype, HMAS-L).

Thallus foliose, loosely appressed, with subdichotomously branched and separated lobes of 2 mm wide; upper surface gray, dull, slightly rugose to smooth with some pieces of upper cortex in the old lobes disintegrating, lacking isidia and lobules but bearing granular soredia coalescent in more or less sacciform structures, with thin layer of pruina limited to the lobe tips; lower surface black, pale brown near the apices, wrinkled, glossy, with large, round perforations of 2 mm in diameter.

Upper cortex prosoplectenchymatous, pale yellow, 14.5 μm thick; algal layer 20.5-22.5 μm thick, consisting of green and subspherical cells of 3-4 μm in diameter; hyphae in medulla colorless, septate, 1-2 μm in diameter; lower cortex prosoplectenchymatous, pale yellow, 12-14.5 μm thick. Apothecia and pycnidia unknown.

Chemistry. Cortex K-, C-, P-; medulla K+ yellow, C-, P+ orange yellow → orange red, containing physodalic, physodic, 3-hydroxyphysodic (conphysodic), protocetraric acids, and atranorin.

Other Material Examined: **China, Yunnan:** Lijiang County, Mt. Yulong shan, alt. 2900 m, on the bark of *Quercus* sp., 8 August 1981, X. Y. Wang, X. Xiao & J. J. Su 4892 (HMAS-L); alt. 3100 m, on the ground, 8 August 1981, X. Y. Wang, X. Xiao & J. J. Su 6591 (HMAS-L).

Comments: The new species resembles *H. farinacea* Zopf at the first sight, but differs by more separated lobes, presence of pruina limited to the lobe tips, and in containing physodalic acid.

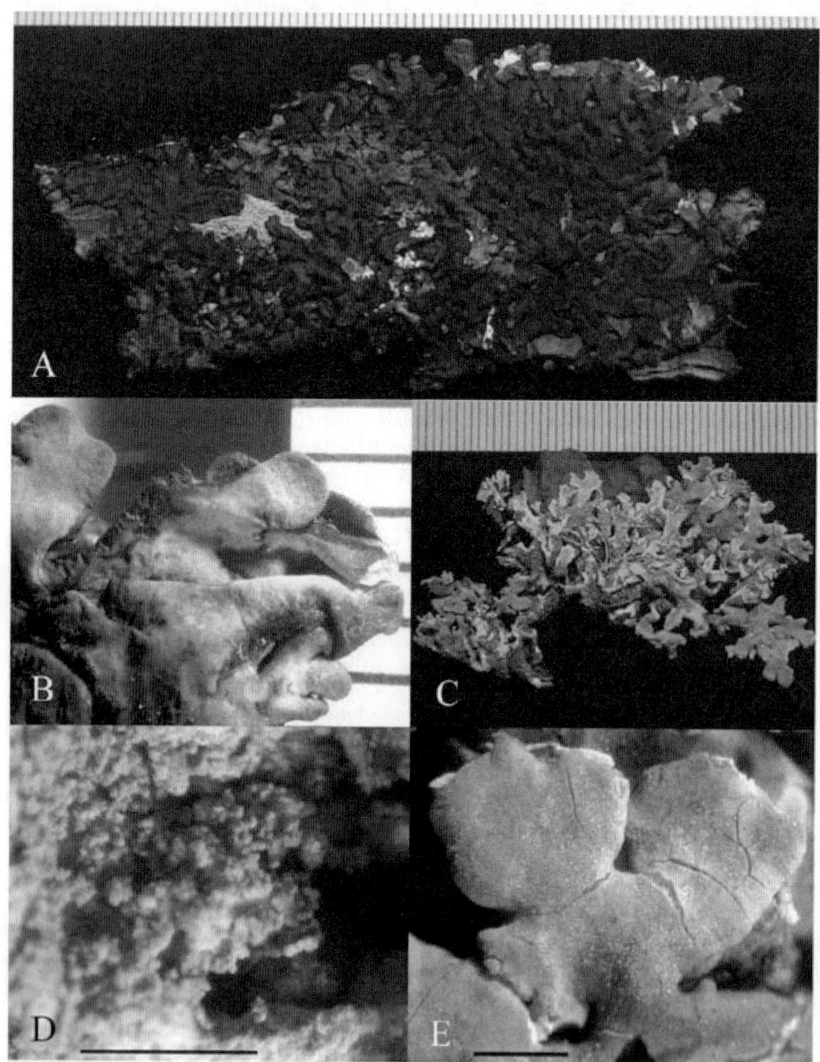

Plate I. A. *Hypogymnia pseudopruinosa*, Wang et al. *7606* (holotype in HMAS-L), showing general appearance of thallus. B. *H. pseudopruinosa*, Wang et al. *7606*, showing the pruina limited to the margin of lobes. C. *H. farinacea*, Wang et al. 10582 (holotype in HMAS-L), showing general appearance of thallus. D. *H. farinacea*, Wang et al. 10582, showing the pruina limited to the margin of lobes. E. *H. farinacea*, Wang et al. 10582, showing the soredia on the upper surface.

A-C: Scale in mm; D-E: Scale bar = 1 mm.

Acknowledgements

We are indebted to Prof. Teuvo Ahti and Dr. Harrie Sipman for giving valuable comments on the manuscript. Special thanks due to Ms. H. Deng for giving considerable assistance during the studies in HMAS-L. The authors are also grateful to Mr. Q. M. Zhou, and Drs. M. R. Huang and S. Y. Guo for kind help in taking the pictures.

Literature Cited

Chen JB. 1994. Two new species of *Hypogymnia* (Nyl.) Nyl. (*Hypogymniaceae*, Acomycotina). Acta Mycologica Sinica 13 (2): 107-110.

Culberson CF. 1972. Improved conditions and new data for the identification of lichen products by a standardized thin-layer chromatographic method. Journal of Chromatography 72: 113-125.

Culberson CF, Kristinsson H. 1970. A standardized method for the identification of lichen products. Journal of Chromatography 46: 85-93.

McCune B, Martin EP, Wang LS. 2003. Five new species of *Hypogymnia* with rimmed holes from the Chinese Himalayas. The Bryologist 106 (2): 226-234.

McCune B, Obermayer W. 2001. Typification of *Hypogymnia hypotrypa* and *H. sinica*. Mycotaxon 79: 23-27.

McCune B, Tchabanenko S. 2001. *Hypogymnia arcuata* and *H. sachalinensis*, two new lichens from east Asia. The Bryologist 104 (1): 146-150.

Wei JC. 1991. An Enumeration of Lichens in China. International Academic Publishers.

White FJ, James PW. 1985. A new guide to microchemical techniques for the identification of lichen substances. British Lichen Society Bulletin 57 (Suppl.): 1-41.

Zhao JD. 1964. A preliminary study on Chinese *Parmelia*. Acta Phytotaxonomica Sinica 9: 139-166.

A new genus and species *Rhizoplacopsis weichingii* in a new family Rhizoplacopsidaceae (Ascomycota)

ZHOU Qi-Ming　　WEI Jiang-Chun*

(Key Laboratory of Systematic Mycology and Lichenology, Institute of Microbiology, Chinese Academy of Sciences, Beijing 100080)

ABSTRACT: A new genus and species *Rhizoplacopsis weichingii* in a new family Rhizoplacopsidaceae within Umbilicariales, Lecanoromycetes, Ascomycota are described in this paper. The new species is similar morphologically to species of *Rhizoplaca*, from which it is different by lecideine apothecia located mainly in the center of the thallus among several peltate scales. Also, its ascus apex structure belongs to the Umbilicaria-type. The phylogenetic analyses of *Rhizoplacopsis weichingii* and other umbilicate lichens based on the molecular data support our conclusion to describe the new family Rhizoplacopsidaceae.

KEYWORDS: Internal transcribed spacer (ITS), large subunit (LSU) ribosomal DNA, morphological characters, small subunit (SSU) ribosomal DNA, umbilicate lichens

新科 Rhizoplacopsidaceae 中的新属及新种 *Rhizoplacopsis weichingii*

周启明　　魏江春*

（中国科学院微生物研究所真菌地衣系统学重点实验室，北京 100080）

摘 要：本文描述了位于新科盾叶科 Rhizoplacopsidaceae（Umbilicariales，Lecanoromycetes，Ascomycota）中的新属盾叶属 *Rhizoplacopsis* 和新种蔚青盾叶 *Rhizoplacopsis weichingii*。蔚青盾叶取名于著名中国真菌学家王云章教授之字"蔚青"，作为庆祝教授百岁生日之纪念。该新种在外形上与 *Rhizoplaca* 属地衣极为相似，但位于地衣体上的子囊盘却为网衣型。此外，它的子囊顶器结构非常接近于 *Umbilicaria*-type。基于分子数据，对 *Rhizoplacopsis weichingii* 及其它相关地衣进行的系统发育分析支持成立新属盾叶属 *Rhizoplacopsis* 和新科盾叶科 Rhizoplacopsidaceae。

关键词：ITS，LSU rDNA，SSU rDNA，形态特征，具脐地衣

中图分类号：Q939.5　　**文献标识码：**A　　**文章编号：**1672-6472（2006）03-0376-0385

1 INTRODUCTION

An interesting new lichen species was recently discovered in China. At first sight the thalli of this new species are similar to those of *Rhizoplaca* and can thus be called "rhizoplacoid". However, the ascus apex structure of the lichen belongs to the *Umbilicaria*-type which was described by Janex-Favre (1974) and Wei & Jiang (1993). The interesting new species is different from any other known lichens and described here as a new genus within a new family.

In order to clarify the systematic position of the new taxon, and to examine the relationships among

The project was supported by the National Natural Science Foundation of China (30070002, 30570008).

*通讯作者 Corresponding author's E-mail: weijc2004@126.com

Received:2006-05-22, accepted: 2006-06-20

This paper was originally published in *Mycosystema*, 25（3）: 376-385, 2006.

other umbilicate lichens, especially *Rhizoplaca* and *Umbilicaria*, phylogenetic analyses based on the small-subunit ribosomal DNA (SSU rDNA), the internal transcribed spacer region (ITS1, 5.8S and ITS2), and the large subunit ribosomal DNA (LSU rDNA) are presented here.

2 MATERIALS AND METHODS

2.1 Morphological, anatomical and chemical studies

The specimens examined for the new lichen were collected from Tulaopodingzi hill in Wangqing county, Jilin province, China. A Leica DM LD microscope was used for the light microscopy. Squash mounts, hand sections, and sections made on a freezing microtome equipped with a dissecting microscope were routinely examined. The ascus apex structure was examined in the light microscope after staining with Lugol's solution. Standardized thin-layer chromatogaphy (TLC: Culberson & Kristinsson, 1970; Culberson, 1972; White & James, 1985) in combination with microcrystal tests (MCT: Asahina, 1936, 1937; Evans, 1943) have been used to identify secondary metabolites in the specimens examined.

2.2 DNA extraction

Herbarium specimens of 6 lichen species used for DNA extraction are listed in Table 1 and deposited in the Lichen Section of the Herbarium Mycologicum Academiae Sinicae (HMAS-L). The DNA was obtained from dried thalli. Before extraction, both upper and lower cortices of the thalli were carefully cleaned to avoid contamination. Total DNA was extracted from individual thalli using a modified CTAB method (Rogers & Bendich, 1988).

2.3 PCR-amplification and sequencing

The primers used for amplification are listed in Table 2. The sequences of the primers are based on Gargas & Taylor (1992), White et al. (1990), or were obtained from: http://www.biology.duke.edu/fungi/mycolab/primers.htm. After initial denaturation at 95 °C for 3 min, polymerase chain reaction (PCR) amplification was run for 30 cycles (30 sec. denaturation at 94 °C, 45 sec. annealing at 52 °C or 48 °C, 2 min extension at 72 °C) with Taq Plus (Taq + Pfu) Polymerase (Sangon), followed by 1 cycle of 8 min extension at 72 °C. All PCR products were verified by electrophoresis on 1% agarose gels using the Gel Extraction Mini Kit (SABC). The PCR products less than 900bp were sequenced directly, and those more than 900bp were ligated into pGEM-T vector (Promega) after purification. According to the protocol enclosed in Promega's kit, competent cells of *E. coli* strain DH5α were prepared with calcium chloride and transformed with the ligased plasmid. The positive clones were selected for sequencing. Sequencing was carried out by Shanghai Genecore Biotechnologies. Walking reactions were performed with the primers designed on the sequences that had been sequenced if needed. All sequences were obtained with an ABI 3700 automatic sequencer.

2.4 Phylogenetic analysis

The SSU rDNA, LSU rDNA, and ITS (including ITS1, 5.8S rDNA and ITS2) sequences determined in this study and those from GenBank (Table 3) were aligned with DNAMAN4.0 (Lynnon Biosoft) and manually optimized. The phylogenetic analyses were executed with the software MEGA2 (Kumar et al., 2001). The Kimura 2-parameter model was chosen and the gaps were retained initially but excluded in the pairwise distance estimation. The phylogenetic trees were evaluated according to the minimum evolution (ME) method. Close-neighbor-interchange (CNI) was used to examine the neighborhood of the neighbor-joining (NJ) trees to find the potential ME trees. Reliability of the inferred trees was tested by 1000 bootstrap replications.

3 RESULTS AND DISCUSSION

The thallus of the new lichen is very similar to that of *Rhizoplaca* species, but the ascus apex structures are significantly distinct from each other. Many studies have previously demonstrated that the thallus growth forms of lichens can vary considerably even within groups of closely related taxa, e.g., *Lecanora* and *Rhizoplaca* (Arup & Grube, 1998, 2000), and cladoniiform lichens (Stenroos & DePriest, 1998). These studies clearly show that the growth form of lichens is of minor importance in comparison with ascomatal characters of the lichenized fungi. However, even a careful analysis of morphological, anatomical and chemical characters would suggest that the new lichen cannot be included in *Rhizoplaca* as part of the Lecanoraceae, but instead appears very close to the Umbilicariaceae according to the type of the ascus apex structures. In order to clarify this systematic hypothesis, molecular data were successfully used in our study.

Molecular data are capable of identifying variability at many different levels in the taxonomic hierarchy. In general, SSU and LSU rDNA are relatively conserved and therefore available to determine relationships at higher taxonomic ranks, extending from phyla and orders to families and genera (e.g., Gargas et al., 1995; Platt & Spatafora, 2000; Wiklund & Wedin, 2003). In contrast, data from the ITS region has been extensively used for lower level phylogenies, such as species delimitation and population studies (e.g., Goffinet & Miądlikowska, 1999; Crespo et al., 2002). These general statements are, however, not necessarily applicable for all different lichen groups. For example, Lumbsch et al. (2001) also used ITS data to study the phylogeny in the Agyriales.

We obtained sequences of partial SSU rDNA (NS21-NS24) varying in length between 883 bp in *Rhizoplaca huashanensis* and 1782 bp in the new lichen, partial LSU rDNA between 2033 bp in the new lichen and 2696 bp in *Rhizoplaca chrysoleuca*, and 549 bp in the new lichen and 849 bp in *Rhizoplaca huashanensis* for the ITS region (including partial SSU and LSU rDNA sequences). All sequences determined in this study were accessioned in GenBank (Table 1). Also, group I introns which cause length variety of the SSU and LSU rDNA for all samples were sequenced, however, these insertions were excluded from the alignments. The final data matrix for SSU rDNA contained 920 characters. Though the total lengths of LSU rDNA fragments determined in this study were about 1800bp (after excluding group I introns), the sequence alignment for all taxa produced a matrix of only 1237 characters for LSU rDNA because the LSU rDNA sequences data of other lichenized fungi in GenBank were mostly less than 1300bp. And the alignment of the ITS sequences resulted in a data set of 537 characters.

In the analysis of the SSU rDNA data, the Umbilicariaceae appear as monophyletic with 64% bootstrap support. The new lichen groups seem sister taxon of the Umbilicariaceae, and this topology has 55% bootstrap support. The species of *Rhizoplaca* form a distinctly separate group with 94% bootstrap support, located far from the new lichen (Fig. 1). In the LSU rDNA analysis, all Umbilicariaceae form a monophyletic group with a bootstrap support of 68%; the new lichen still appears as a sister taxon to the family. In comparison with taxa that do not belong to the Umbilicariaceae there is strong bootstrap support (95%) for this sister group relationship of the new lichen with the Umbilicariaceae clade. *Rhizoplaca huashanensis* and *R. chrysoleuca* group together with 100% bootstrap support and form a sister group to the remaining taxa belonging to the Lecanorales (Fig. 2). The monophyletic group Umbilicariaceae has 78% bootstrap support and the new lichen is still the sister taxon to it in the analysis using ITS data. This topology is supported with 96% bootstrap value. However, the species of *Rhizoplaca* do not form a monophyletic group (Fig. 3).

The results of the phylogenetic analyses based on SSU rDNA, LSU rDNA and ITS data show an obvious congruence supporting our hypothesis that the new lichen appears to be quite distantly related to the Lecanorales, but clustered close to the Umbilicariaceae as a sister clade in all the trees and this relationship was supported by bootstrap support (Figs. 1-3). The bootstrap values are not very high in the analysis with partial SSU rDNA data, which suggests that SSU rDNA is much conserved and there are not enough variant sites when only half-length sequences were used. The new lichen appears much more closely related to the Umbilicariaceae than to other taxa, and it can not be included in the Umbilicariaceae or other families. So it is reasonable to treat the new lichen as a new genus named *Rhizoplacopsis* and a new family named Rhizoplacopsidaceae. Our results also indicate that the character of rizoplacoid thallus is a not homologous character in lichens and the growth forms have lower taxonomic value than they were given.

The clades of the Umbilicariaceae in three trees have moderate bootstrap support, which seems to be caused by lacking variant sites because the bootstrap support value of the Umbilicariaceae was much higher when the full-length SSU rDNA sequences data were used in the phylogenetic analysis (data not shown here). Our analyses provide strong bootstrap support for that the Umbilicariaceae together with the Rhizoplacopsidaceae can be raised to an order as Umbilicariales. The taxonomic and nomenclatural treatment of them about the Umbilicariales would be given in a separate publication.

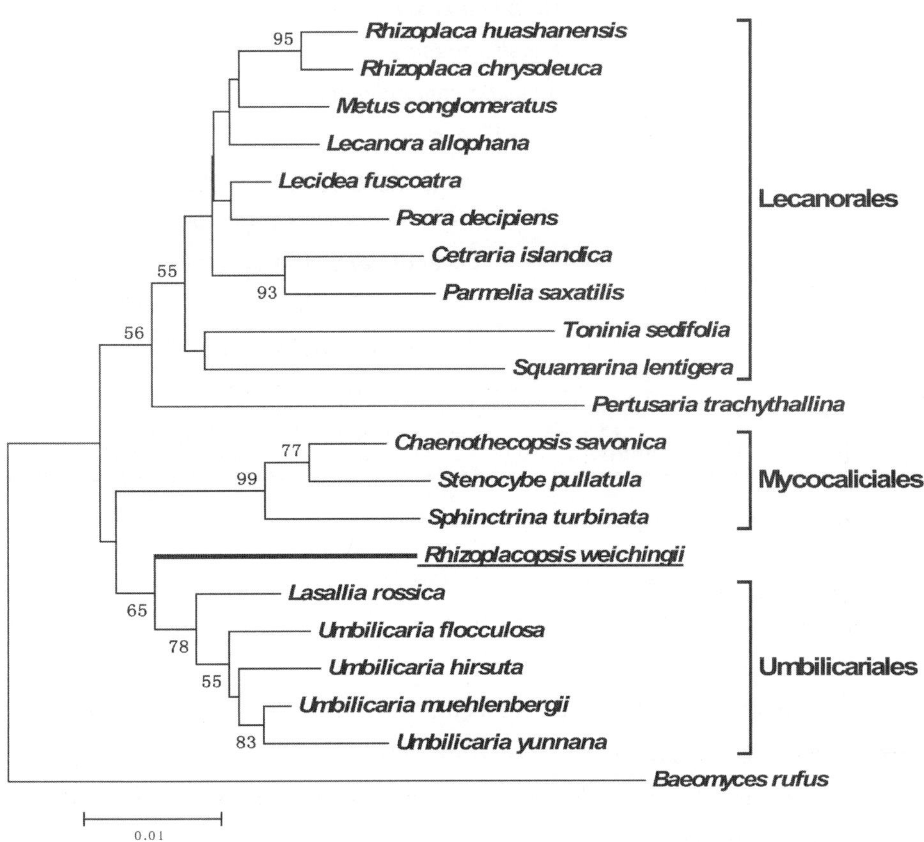

Fig. 1 The minimum evolution (ME) tree inferred from partial SSU rDNA data

The reliability of the inferred tree was tested by 1000 bootstrap replications; the numbers in each node represent bootstrap support values. Only bootstrap values greater than 50% are shown

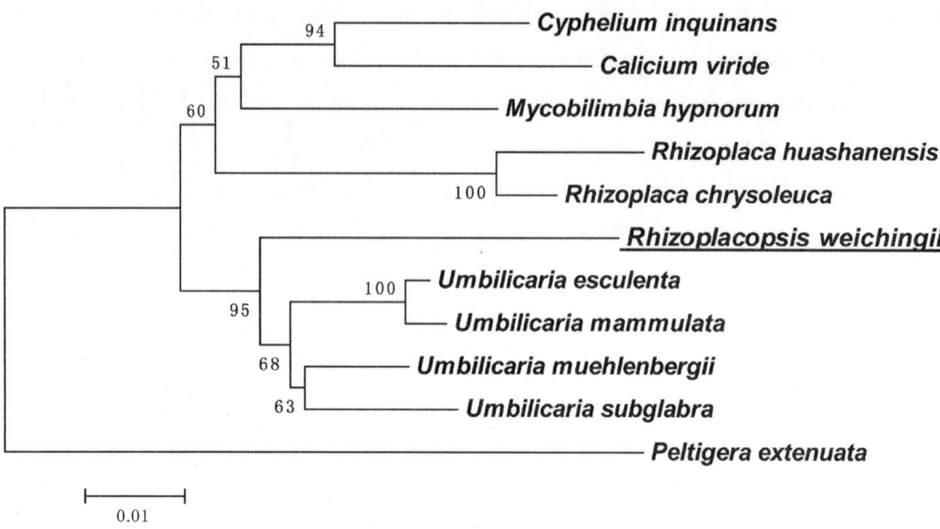

Fig. 2 The minimum evolution (ME) tree inferred from partial LSU rDNA data

The reliability of the inferred tree was tested by 1000 bootstrap replications. Only bootstrap values greater than 50% are shown

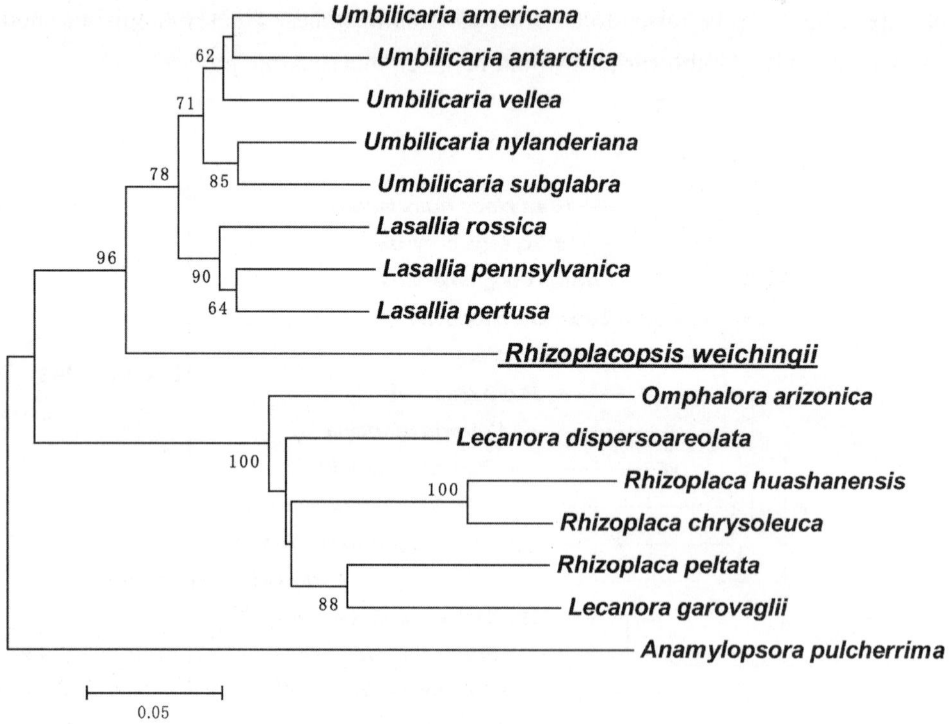

Fig. 3 The minimum evolution (ME) tree based on ITS and 5.8S rDNA data

The number of bootstrap replicates was 1000. Only bootstrap values greater than 50% are shown

The new lichen is treated and described here as follows:

Rhizoplacopsidaceae J. C. Wei & Q. M. Zhou, fam. nov.

Thallus rhizoplacoideus, apotheciis lecideinis in centro thalli inter squamas peltatas; structura ad apicem ascorum ut in *Umbilicariaceis*.

TYPUS: *Rhizoplacopsis* J. C. Wei & Q. M. Zhou

Thallus rhizoplacoid; consisting of several peltate scales with lecideine apothecia located in the thallus center; asci with an apical structure that belongs to the *Umbilicaria*-type.

Rhizoplacopsis J. C. Wei & Q. M. Zhou, gen. nov.

Genus novum *Rhizoplaco* (Zopf) Leuckert et al. similare, a qua apotheciis lecideinis, in centro thalli inter squamas peltatas; structura ad apicem ascorum ut in *Umbilicariaceis* differt.

TYPUS: *Rhizoplacopsis weichingii* J. C. Wei & Q. M. Zhou

The new genus has a thallus very similar to the genus *Rhizoplaca*, but has apothecia that are lecideine. These apothecia are located in the thallus center among two or more peltate scales. The ascus apex is similar to that of the Umbilicariaceae.

Rhizoplacopsis weichingii J. C. Wei & Q. M. Zhou, sp. nov. (Figs. 4, 5)

Etymology: The specific epithet "weichingii" of the new species is given in compliment to Professor Yun-Chang Wang, a famous Chinese mycologist, using his alternative name "weiching" to celebrate his 100th birthday.

TYPUS: Habitat in rupibus in colli Tulaopodingensi in provincia Jilinensi in Sina, 8-VI-1996, Wei et Jiang no. 45 (holotypus in HMAS-L hic designatus, isotypi in FH, H, UPS, US).

Diagnosis: Species nova *Rhizoplaco melanophthalmae* similaris, a qua apotheciis lecideinis, nigris, structura ad apicem ascorum ut in *Umbilicariaceis* differt.

Description: Thallus umbilicate, forming rosettes, consisting of several peltate scales, much thicker in the center, rigid, 2-3 or 4 cm in diam., with a broadly lobate margin; upper surface with well developed cortex, gray-brown to greenish gray, smooth, glossy, areolated; lower surface dark to dark brown, rough, fissured, the white medulla becoming exposed where the cortex is peeling off, with a narrow marginal zone where the cortex remains intact, similar in color to the upper surface; attached with broad zone of white rhizinate structures extending 1.0-1.5 (-2.5) cm in diam. Thallus 660-1732 μm in cross section: upper cortex 42-70 μm, algal layer 56-97 μm, and medullary layer 530-1472 μm thick.

Apothecia lecideine, black, 2-3 (-4) mm in diam., more or less immersed between thallus scales in the center of the rosettes.; a pseudothalline margin lacking symbiotic algae can occasionally be observed in immature apothecia; epithecium dark green to dark blue, 3.5-5.5 μm tall; hymenium hyaline, 36-63 μm tall; hypothecium hyaline, 21.5-27.0 μm tall; without algal cells below; paraphyses sparsely septate, sometimes simple to sparsely branched, dark green to dark blue and slightly expanded at the tips, 1-1.5 μm in diam.; asci 8-spored, 34-38 × 9-11 μm; ascospores ellipsoidal, hyalin 6-7 × 3.5-4.0 μm.

Chemistry: Only Lecanoric acid detected here.

Specimens examined: On rocks of Tulaopodingzi hill in Wang-qing county, Jilin province, 9-VIII-1994,Y. M. Jiang & S. Y. Guo, no.94496; 8-VI-1996, J. C. Wei, Y. M. Jiang & Y. Z. Wang, nos.42, 45 (type), 48, 102, 140, 151, 165, 173, 185, 238 (HMAS-L).

Fig. 4 *Rhizoplacopsis weichingii*.– General growth morphology, upper (top) and lower (bottom) surface of the peltate thalli; the lower surface of the peltate thalli with white rihizinate structures that attach the peltate thalli to the substrate. (Scale bar = 1 cm)

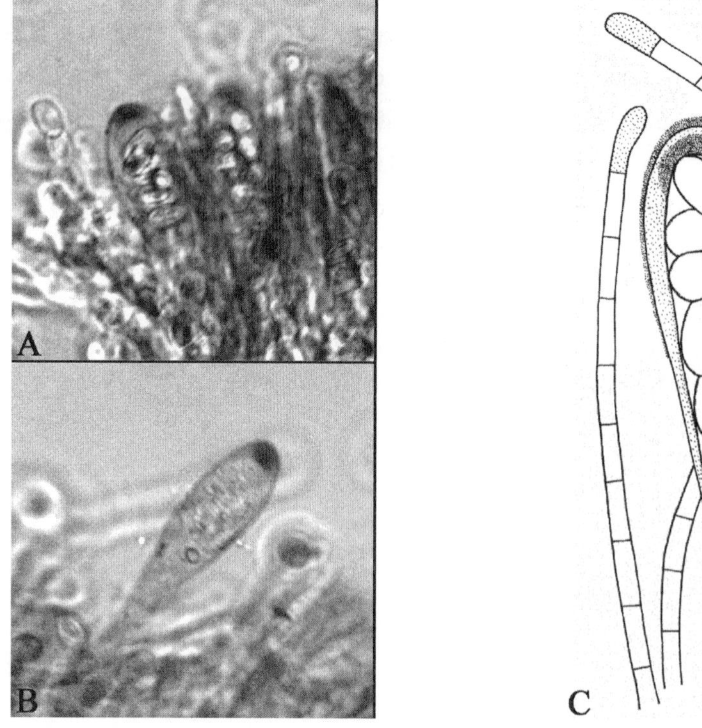

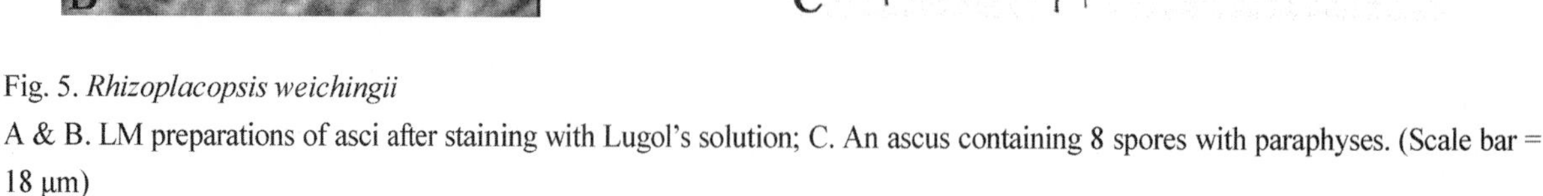

Fig. 5. *Rhizoplacopsis weichingii*

A & B. LM preparations of asci after staining with Lugol's solution; C. An ascus containing 8 spores with paraphyses. (Scale bar = 18 μm)

ACKNOWLEDGMENTS We are indebted to Drs. H. Sipman and O. Eriksson for their comments and suggestions on the manuscript.

Table 1 List of the lichen species and DNA sequence accession numbers determined in this study

		Accession No.		
		SSU rDNA	ITS	LSU rDNA
Rhizoplaca chrysoleuca	Guo & Huang, 144, 2001	AY530888	AY304157	AY648104
Rhizoplaca huashanensis	Wei et al., 18357, 1998	AY530885		AY648105
Rhizoplacopsis weichingii	Zhou & Wei, 03036, 2003	AY530886		AY648106
Lasallia sineorientalis	Wei et al., 119, 2000	AY648109		
Umbilicaria loboperipherica	Wang et al., Aer236b, 2002	AY648113		
Umbilicaria thamnodes	Wang et al., WY178, 2003	AY648116		

Table 2 The primers used for amplifications

Species	Primers for SSU rDNA	Primers for ITS	Primers for LSU rDNA
Rhizoplaca chrysoleuca	NS21-NS24	ITS5-ITS4	SR0R-LR8
Rhizoplaca huashanensis	NS21-NS24	SR6R-LR1	SR0R-LR8
Rhizoplacopsis weichingii	NS21-NS24	ITS5-ITS4	SR0R-LR8
Lasallia sineorientalis	NS21-NS24		
Umbilicaria loboperipherica	NS21-NS24		
Umbilicaria thamnodes	NS21-NS24		

Table 3 Accession numbers and current family and ordinal classification of related lichens, of which SSU rDNA, ITS or LSU rDNA sequences were used for phylogenetic comparison.

Species	Accession Number	Systematic position
Species used in SSU rDNA phylogenetic analysis		
Anamylopsora pulcherrima	AF119501	Agyriales; Anamylopsoraceae
Baeomyces rufus	AF113718	Lecanoromycetes; Baeomycetaceae
Cetraria islandica	AF117986	Lecanorales; Parmeliaceae
Chaenothecopsis savonica	U86691	Mycocaliciales; Mycocaliciaceae
Lasallia rossica	AF088238	Umbilicariales; Umbilicariaceae
Lecanora allophana	AF515608	Lecanorales; Lecanoraceae
Lecidea fuscoatra	AF088239	Lecanorales; Lecideaceae
Metus conglomeratus	AF184755	Lecanorales; Cladoniaceae
Parmelia saxatilis	AF117985	Lecanorales; Parmeliaceae
Peltula radicata	AF336903	Lichinales; Peltulaceae
Pertusaria trachythallina	AF088242	Pertusariales; Pertusariaceae
Psora decipiens	AF184759	Lecanorales; Lecideaceae
Sphinctrina turbinata	U86693	Mycocaliciales; Sphinctrinaceae
Squamarina lentigera	AF088250	Lecanorales; Ramalinaceae
Stenocybe pullatula	U86692	Mycocaliciales; Mycocaliciaceae
Toninia sedifolia	AF091591	Lecanorales; Ramalinaceae
Umbilicaria subglabra	AF088253	Umbilicariales; Umbilicariaceae

续表 3

Species	Accession Number	Systematic position
	Species used in LSU rDNA phylogenetic analysis	
Calicium viride	AY453638	Lecanorales; Caliciaceae
Cyphelium inquinans	AY453639	Lecanorales; Caliciaceae
Mycobilimbia hypnorum	AY533005	Lecanorales; Porpidiaceae
Peltigera extenuata	AY266034	Peltigerales; Peltigeraceae
Umbilicaria esculenta	AY645334	Umbilicariales; Umbilicariaceae
Umbilicaria mammulata	AY645335	Umbilicariales; Umbilicariaceae
Umbilicaria muehlenbergii	AY645336	Umbilicariales; Umbilicariaceae
Umbilicaria subglabra	AY645339	Umbilicariales; Umbilicariaceae
	Species used in ITS phylogenetic analysis	
Anamylopsora pulcherrima	AF274089	Agyriales; Anamylopsoraceae
Lasallia pennsylvanica	AF096202	Umbilicariales; Umbilicariaceae
Lasallia pertusa	AF096203	Umbilicariales; Umbilicariaceae
Lasallia rossica	AF096201	Umbilicariales; Umbilicariaceae
Lecanora dispersoareolata	AF070016	Lecanorales; Lecanoraceae
Lecanora garovaglii	AF189718	Lecanorales; Lecanoraceae
Omphalora arizonica	AY340867	Lecanorales; Parmeliaceae
Rhizoplaca peltata	AF159936	Lecanorales; Lecanoraceae
Umbilicaria americana	AF096218	Umbilicariales; Umbilicariaceae
Umbilicaria antarctica	AF096213	Umbilicariales; Umbilicariaceae
Umbilicaria nylanderiana	AF096205	Umbilicariales; Umbilicariaceae
Umbilicaria subglabra	AF096200	Umbilicariales; Umbilicariaceae
Umbilicaria vellea	AF096208	Umbilicariales; Umbilicariaceae

[REFERENCES]

Arup U, Grube M, 1998. Molecular systematics of *Lecanora* subgeneus *Placodium*. *Lichenologist*, **30**: 415-425

Arup U, Grube M, 2000. Is *Rhizoplaca* (Lecanorales, lichenized Ascomycota) a monophyletic genus? *Can J Bot*, **78**: 318-327

Asahina Y, 1936. Mikrochemischer Nachweis der Flechtenstoffe (I). *J Jap Bot*, **12**: 516-525

Asahina Y, 1937. Mikrochemischer Nachweis der Flechtenstoffe (III.Mitteil). *J Jap Bot*, **13**: 529-536

Crespo A, Molina MC, Blanco O, Schroeter B, Sancho LG, Hawksworth DL, 2002. rDNA ITS and β-tubulin gene sequence analyses reveal two monophyletic groups within the cosmopolitan lichen *Parmelia saxatilis*. *Mycol Res*, **106**: 788-795

Culberson CF, 1972. Improved conditions and new data for the identification of lichen products by a standardized thin-layer chromatographic method. *J Chromatogr*, **72**: 113-125

Culberson CF, Kristinsson HD, 1970. A standardized method for the identification of lichen products. *J Chromatogr*, **46**: 85-93

Evans AW, 1943. Microchemical studies on the genus *Cladonia* subgenus *Cladina*. *Rhodora*, **45**: 417-438

Gargas A, Taylor JW, 1992. Polymerase chain reaction (PCR) primers for amplifying and sequencing nuclear 18S rDNA from lichenized fungi. *Mycologia*, **84**: 589-592

Gargas A, DePriest PT, Grube M, Tehler A, 1995. Multiple origins of lichen symbioses in fungi suggested by SSU rDNA phylogeny. *Science*, **268**: 1492-1495

Goffinet B, Miądlikowska J, 1999. *Peltigera phyllidiosa* (Peltigeraceae, Ascomycotina), a new species from the southern Appalachians corroborated by its sequences. *Lichenologist*, **31**: 247-256

Janex-Favre MC, 1974. Lontogenie et la structure des apothecies de l'*Umbilicaria _ylindrica. Rev Bryol Lichenol*, **40**: 9-88

Kumar S, Tamura K, Jakobsen IB, Nei M, 2001. MEGA2: Molecular evolutionary genetics analysis software., Arizona State University, Tempe, Arizona, USA

Lumbsch HT, Schmitt I, Doring H, Wedin M, 2001. ITS sequence data suggest variability of ascus types and support ontogenetic characters as phylogenetic discriminators in the Agyriales (Ascomycota). *Mycol Res*, **105**: 265-274

Platt JL, Spatafora JW, 2000. Evolutionary relationships of nonsexual lichenized fungi: molecular phylogenetic hypotheses for the genera *Siphula* and *Thamnolia* from SSU and LSU rDNA. *Mycologia*, **93**: 475-487

Rogers SO, Bendich AJ, 1988. Extraction of DNA from plant tissues. *Plant Molecular Biology Manual.* **A6**: 1-10

Stenroos S, DePriest PT, 1998. SSU rDNA phylogeny of cladoniiform lichens. *Am J Bot*, **85**: 1548-1559

Wei JC, Jiang YM, 1993. The Asian Umbilicariaceae (Ascomycota). Mycosystema Monographicum Series No.1. International Academic Publishers Beijing, China.

White FJ, James PW, 1985. A new guide to microchemical techniques for the identification of lichen substances. *Bulletin of the British Lichen Society*, **57** (Supplement): 1-41

White YJ, Bruns T, Lee S, Taylor J, 1990. Amplication and direct sequencing of fungal ribosomal RNA genes for phylogenetics. *In* Innis MA, Gelfand DH, Sninsky JJ, White TJ (eds.) PCR protocols: a guide to methods and applications. Academic Press, San Diego, California. 315-322

Wiklund E, Wedin M, 2003. The phylogenetic relationships of the cyanobacterial lichens in the Lecanorales suborder Peltigerineae. *Cladistics*, **19**: 419-431

A new order Umbilicariales J. C. Wei & Q. M. Zhou（Ascomycota）

ZHOU Qi-Ming　　WEI Jiang-Chun*

(Key Laboratory of Systematic Mycology and Lichenology, Institute of Microbiology, Chinese Academy of Sciences, Beijing 100080)

Abstract: The phylogenetic position of the lichen family Umbilicariaceae is investigated using nucleotide sequences of the nuclear SSU rDNA region. Sequences of 6 species were obtained in this study and aligned to those of other lichenized and non-lichenized fungal species from GenBank. The result indicates that the Umbilicariaceae cannot be included in the order Lecanorales and the order Umbilicariales is supported by both the molecular and morphological data. The new order Umbilicariales within Lecanoromycetes, Ascomycota is described in this paper accordingly.

Keywords: Molecular systematics, small subunit (SSU) ribosomal DNA, Umbilicariaceae

子囊菌的一个新目 Umbilicariales J.C. Wei & Q.M. Zhou

周启明　　魏江春*

（中国科学院微生物研究所真菌地衣系统学重点实验室, 北京 100080）

摘 要：本文基于核糖体 SSU rDNA 序列对石耳科 Umbilicariaceae 的系统地位进行了研究。将所获得的石耳科地衣中 6 个种的 SSU rDNA 序列与 GenBank 中其它地衣型及非地衣型真菌的相关序列进行比对用于系统发育研究。结果表明长期以来系统地位不够明确而暂时被置于茶渍目 Lecanorales 的石耳科不能被包括在茶渍目中，分子数据支持成立石耳目 Umbilicariales。基于分子数据并结合形态学和解剖学特征描述了新目 Umbilicariales J.C. Wei & Q.M. Zhou（Lecanoromycetes， Ascomycota）。

关键词：SSU rDNA，分子系统学，石耳科

中图分类号：Q939.5　　**文献标识码**：A　　**文章编号**：1672-6472（2007）01-0040-0045

INTRODUCTION

The lichen family Umbilicariaceae had been provisionally placed in the order Lecanorales. Then, Poelt (1973) treated this family as a suborder Umbilicariineae under the Lecanorales, while he pointed out the fact that this family seemed to be an ancient group and was apart from others by sharp discontinuities. It did not fit in closely with the Lecideaceae, and it seemed more convenient to give this family a separate place in the classification.

Some researches based on the SSU rDNA sequences of *Lasallia rossica* and *Umbilicaria subglabra* indicated that the Umbilicariaceae should be excluded from the Lecanorales (DePriest &

*通讯作者 Corresponding author's E-mail: weijc2004@126.com

Received:2006-05-22, accepted: 2006-07-03

This paper was originally published in *Mycosystema*, 26（1）: 40-45, 2007.

Gargas, 1996; DePriest et al., 1997; Stenroos & DePriest, 1998), and other studies based on LSU rDNA or mt SSU rDNA sequences also confirmed that the family Umbilicariaceae differ from Lecanorales (Kauff & Lutzoni, 2002; Lumbsch et al., 2004). However, these studies didn't reveal the phylogeny or systematic position of the Umbilicariaceae, and only one or two species of the Umbilicariaceae were used in the analysis. In order to clarify the systematic position of the Umbilicariaceae in the Ascomycota, the SSU rDNA sequences data obtained from other 6 species of the Umbilicariaceae are presented in this paper.

MATERIALS AND METHODS

The lichen samples selected for this study are listed in Table 1.

Table 1. Information for samples included in this study

Species	Specimen voucher	Accession No.
Lasallia pensylvanica	Abbas, 106, 1997	AY648108
Umbilicaria flocculosa	Abbas, 9600303, 1996	AY648111
Umbilicaria hirsuta	Abbas, 9600460-1, 1996	AY648112
Umbilicaria muehlenbergii	Wang et al., Aer236a, 2002	AY648115
Umbilicaria vellea	Guo, 1102, 1997	AY648117
Umbilicaria yunnana	Zhou et al., 99084, 1999	AY648118

Before the DNA extraction, the lichen thalli were cleaned and the upper cortex with algal layer was scraped using a razor blade under the dissecting microscope. The lower cortex was cleaned carefully as well in order to avoid DNA pollution. Total DNA was extracted from individual thallus using the modified CTAB method (Rogers & Bendich, 1988). PCR reactions were performed with primers pairs NS17/NS20 and NS21/NS24 (Gargas & Taylor, 1992) in a DNA Thermal Cycler (Biometra) as follows: initial denaturation at 95 ?C for 3 min, followed by 30 cycles of 40 sec denaturation at 94 ?C, 40 sec annealing at 48 ?C, 2 min extension at 72 ?C. The amplification was ended with a final 8 min extension at 72 ?C. All the PCR products were verified by electrophoresing on 1% agarose gels and ligated into pGEM-T vector (Promega) after being purified using Gel Extraction Mini Kit (SABC). According to the protocol enclosed in Promega's kit, prepared competent *E.coli* DH5a cells with calcium chloride and transformed them with the ligation mixes. After incubated overnight at 37 ?C in LB plates with ampicillin/IPTG/X-Gal, positive clones were selected to sequence by Shanghai Genecore Biotechnologies with T7 promoter sequencing primer (5'-TAATACGACTCACTATAGGG-3') and SP6 promoter sequencing primer (5'-GATTTAGGTGA CACTATAG-3').

The SSU rDNA sequences acquired in this study and those from GenBank (Table 2) were aligned with DNAMAN4.0 (Lynnon Biosoft). The phylogenetic analysis was executed with software MEGA2 (Kumar et al., 2001). We took Kimura 2-parameter model, and retained gaps initially while excluding them in the pairwise distance estimation. The neighbor joining (NJ) method was used in constructing the phylogenetic tree and the reliability of the inferred tree was tested by 1000 bootstrap replications.

Table 2 Information for other samples obtained from GenBank

Species	GenBank Accession No.	Species	GenBank Accession No.
Amphisphaeria umbrina	AF225207	*Pertusaria trachythallina*	AF088242
Bunodophoron australe	AF184749	*Pilophorus strumaticus*	AF184758
Caloplaca demissa	AF515609	*Placopsis gelida*	AF119502
Capronia mansonii	X79318	*Pseudophacidium ledi*	AF315623
Cetraria islandica	AF117986	*Pseudoplectania nigrella*	Z27408
Chaenothecopsis savonica	U86691	*Psora decipiens*	AF184759
Chytridium confervae	M59758	*Rhizocarpon geographicum*	AF088246
Cladia aggregata	U72713	*Rhytidhysteron rufulum*	AF201452
Cladonia rangiferina	AF184753	*Solorina crocea*	X89220
Coprinopsis cinerea	M92991	*Sphaerotheca cucurbitae*	AB033482
Cyanodermella viridula	U86583	*Sphinctrina turbinata*	U86693
Dibaeis baeomyces	AF113713	*Squamarina lentigera*	AF088250
Diploschistes thunbergianus	AF274112	*Stenocybe pullatula*	U86692
Fusarium culmorum	AF548073	*Stereocaulon ramulosum*	AF088251
Gyalecta ulmi	AF088237	*Stylodothis puccinioides*	AY016353
Gymnoascus reesii	AJ315169	*Taphrina alni*	AJ495834
Lasallia rossica	AF088238	*Thamnolia vermicularis*	AF085472
Lecanora allophana	AF515608	*Tilletia caries*	U00972
Lecidea fuscoatra	AF088239	*Toninia sedifolia*	AF091591
Leotia viscosa	AF113715	*Trapelia placodioides*	AF119500
Medeolaria farlowii	AF234840	*Trichoderma viride*	AF548104
Neophyllis melacarpa	AF117981	*Umbilicaria subglabra*	AF088253
Neurospora crassa	AY046271	*Urnula hiemalis*	Z49754
Ochrolechia parella	AF274109	*Usnea florida*	AF117988
Parmelia saxatilis	AF117985	*Xanthoria elegans*	AF088254
Peltigera neopolydactyla	X89218	*Xylaria carpophila*	Z49785
Peltula radicata	AF336903	*Zoophthora anglica*	AF368524
Penicillium italicum	AF548091		

RESULTS AND DISCUSSION

Almost the full-length sequences of SSU rDNA from 6 species were obtained in this study. There were two group I introns in *Umbilicaria yunnana* and one in *Umbilicaria flocculosa*, these introns were excluded from the alignments with DNAMAN4.0.

The SSU rDNA data indicate that all the species examined in the Umbilicariaceae form a monophyletic group with 81% bootstrap support. This group lies systematically in a separate position which is far especially from the Lecanorales (Fig. 1).

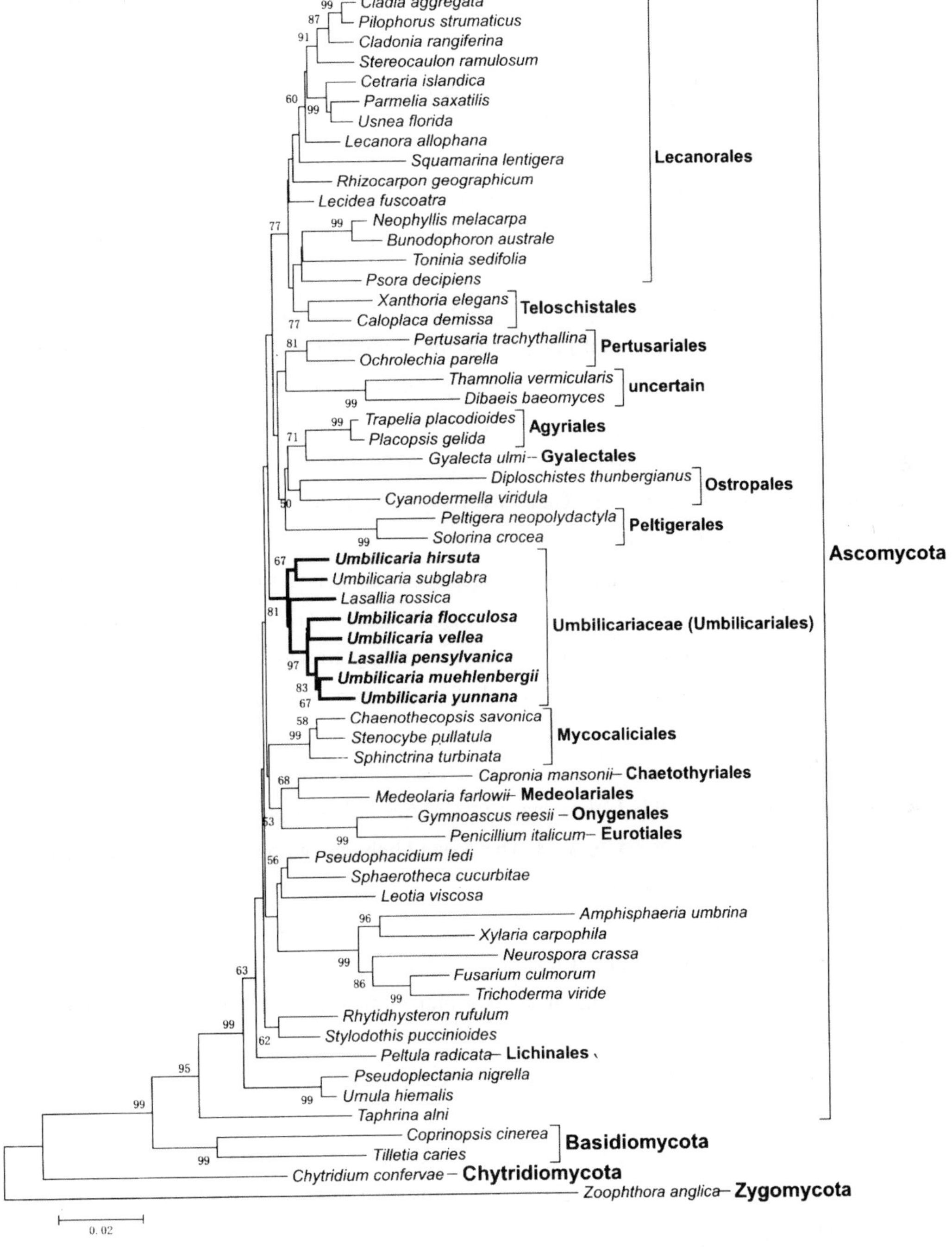

Fig. 1 The NJ tree inferred from SSU rDNA data. The reliability of the inferred tree was tested by 1000 bootstrap replications; the numbers in each node represent bootstrap support values. Only bootstrap values greater than 50% are shown

The morphological, anatomical and molecular data show that the Umbilicariaceae is an isolated taxon and cannot be included in any known orders in the Ascomycota. So the Umbilicariaceae have to be described as a new order Umbilicariales. Another family Rhizoplacopsidaceae (Zhou & Wei, 2006) can be also placed into this order. As to the family Elixiaceae (Lumsch, 1997), it remains to be carefully examined in the future.

Umbilicariales J.C. Wei & Q.M. Zhou Ordo novus

Typus: Umbilicariaceae Chevall.(1826)

Thallus umbilicatus. Apothecia lecideina vel sublecanorina, sessilia vel leviter stipitata, nigra ad castanea; paraphyses parce ramosae interdum tumidae ad apicem; asci cum structura apicis asci ut in typo Umbilicariae, 1-8-spori.

The members of the new order have umbilicate thallus with apothecia of lecideine or sublecanorine type (Wei & Jiang, 1988) containing the asci with ascus apex structure of *Umbilicaria* type (Janex-Favre, 1974; Wei & Jiang, 1993). The phylogenetic trees based on nuclear SSU rDNA, LSU rDNA, and mt SSU rDNA sequences showed that the Umbilicariales located systematically in a separate position which is differs from that of the Lecanorales and other known orders.

Umbilicariaceae F.F. Chevalier, Fl. Gen. Env. Paris:640 (1826).

Typus: *Umbilicaria* Hoffmann

Thallus foliose, more or less leathery, umbilicate, corticated on both sides, pustulate or non-pustulate. Apothecia lecideine to sublecanorine, with smooth, gyrate discs bearing a common margin to radially gyrate discs lacking a common margin. Asci monosporous or disporous to octosporous. Ascospores simple and hyaline to brown and muriform. On rocks, only few species on tree bark.

Umbilicaria Hoffmann, Descr. Adumbr. Pl. Lich. 1(1):8 (1789).

Typus: *Umbilicaria exasperata* Hoffmann, Descr. Et Adumbr. Plant. Lich. 1: 7-8 (1790), Illustration: Tabula 2, Fig. 1-2.[=*Lichen hyperboreus* Ach.,1794 =*Umbilicaria hyperborean* (Ach.) Hoffmann (1801)]; Wei, J.C. & Jiang, Y.M. The Asian Umbilicariaceae (Ascomycota). International Academic Publishers, Beijing, 119-123 (1993).

Thallus non-pustulate, with or without rhizines. Apothecia lecideine to sublecanorine, with smooth or gyrate discs with a common margin to radially gyrate discs without a common margin. Asci octosporous. Ascospore simple and hyaline or muriform and brown. On rocks, only one species from China on tree bark.

Lasallia Merat, Nouv. Fl. Paris ed. 2 (1): 202 (1821).

Typus: *Lasallia pustulata* (Linn.) Merat (1821).

Thallus pustulate, lacking rhizines. Apothecia with smooth to gyrate discs having a common margin. Asci monosporous to disporous. Ascospores muriform and brown. On rocks, only one species sometimes on tree bark.

Rhizoplacopsidaceae J.C. Wei & Q.M. Zhou In Q.M. Zhou, J.C. Wei, *Mycosystema*, **25**:376-385 (2006)

Typus: *Rhizoplacopsis* J.C. Wei & Q.M. Zhou (2006)

Thallus rhizoplacoid, consists of several peltate scales. Apothecia with lecideine to sublecanorine discs with a common margin which are growing in the center of the thallus among the peltate scales. Asci octosporous. Ascospores simple, hyaline.

Rhizoplacopsis J.C. Wei & Q.M. Zhou In Q.M. Zhou, J.C. Wei, *Mycosystema*, 25: 376-385 (2006)

Typus: *Rhizoplacopsis weichingii* J.C. Wei & Q.M. Zhou (2006)

Characters as given for the family.

The following group is excluded in this paper:

Elixiaceae Lumbsch J. Hattori Bot. Lab. 83:1 (1997)

Typus: *Elixia* Lumbsch (1997)

Elixia Lumbsch (1997)

Typus: *Elixia flexella* Lumbsch J. Hattori Bot. Lab. 83:1 (1997)

This monotypic family based on a crustose genus and species *Elixia flexella* was treated by Lumbsch (1997) as a member of the suborder Agyriineae in the Lecanorales. However, the phylogenetic analysis based on LSU rDNA data showed that the Umbilicariaceae together with the Elixiaceae form a well-supported group and are different from the Lecanorales (Lumbsch et al., 2004). Whether the Elixiaceae can be treated as a member of the Umbilicariales, it remains to be carefully studied in the future.

[REFERENCES]

DePriest PT, Gargas A, 1996. Origins of the lichen association in the fungi: phylogenetic analyses of nuclear small subunit ribosomal DNA sequences. Third meeting of the International Association for Lichenology. Progress and Problems in Lichenology in the Nineties Abstracts: 100. Salzburg: University of Salzburg.

DePriest PT, Stenroos S, Ivanova NV, Gargas A, 1997. Origins of the lichen association in the fungi: phylogenetic analyses of nuclear small subunit ribosomal DNA sequences. Invited presentation, Bridging the gap between phylogeny and classification of lichenforming Ascomycetes: Annual AIBS Meeting, Montreal, Canada. American Journal of Botany **84** (suppl.): 9

Gargas A, Taylor JW, 1992. Polymerase chain reaction (PCR) primers for amplifying and sequencing nuclear 18S rDNA from lichenized fungi. *Mycologia*, **84**:589-592

Janex-Favre MC, 1974. Lontogenie et la structure des apothecies de l'Umbilicaria cylindrica. Extr.: Rev. Bryol. Et Lichenol., T. XL, Fasc.

Kauff F, Lutzoni F, 2002. Phylogeny of Gyalectales and Ostropales (Ascomycota, Fungi): among and within order relationships based on nuclear ribosomal RNA small and large subunits. *Mol Phylogenet Evol*, **25**:138-156

Kumar S, Tamura K, Jakobsen IB, Nei M, 2001, MEGA2: Molecular evolutionary genetics analysis software, Arizona State University, Tempe, Arizona, USA

Lumbsch HT, 1997. Systematic studies in the suborder Agyriineae (Lecanorales). *The J Hattori Bot Lab*, **83**: 1-73

Lumbsch HT, Schmitt I, Palice Z, Wiklund E, Ekman S, Wedin M, 2004. Supraordinal phylogenetic relationships of Lecanoromycetes based on a Bayesian analysis of combined nuclear and mitochondrial sequences. *Mol Phylogenet Evol*, **31**: 822-832

Poelt J, 1973. Classification. *In*: Ahmagjian V, Hale ME (eds.) The Lichens. Academic Press, New York. pp.599-632

Rogers SO, Bendich AJ, 1988. Extraction of DNA from plant tissues. *Plant molecular biology manual.* **A6**: 1-10

Stenroos S, DePriest PT, 1998. SSU rDNA phylogeny of cladoniiform lichens. *Am J Bot*, **85**: 1548-1559

Wei JC, Jiang YM, 1988. A conspectus of the lichenized ascomycetes Umbilicariaceae in China. *Mycosystema*, **1**: 73-106

Wei JC, Jiang YM, 1993. The Asian Umbilicariaceae (Ascomycota). Beijing: International Academic Publishers.1~

Zhou QM, Wei JC, 2006. A new genus and species *Rhizoplacopsis weichingii* in a new family Rhizoplacopsidaceae (Ascomycota). *Mycosystema*, **25**: 376-385

Graphis fujianensis, a new species of *Graphidaceae* from China

Ze-Feng Jia[1] & Jiang-Chun Wei[2*]

*zfjia2008@163.com * weijc2004@126.com*

[1] *College of Life Sciences, Shandong Agricultural University Taian, 271018, China*

[2]*Key Laboratory of Systematic Mycology & Lichenology, Institute of Microbiology Academia Sinica, Beijing,100080, China*

Abstract—*Graphis fujianensis*, from Wuyi Mountain in Fujian, China, is described as new to science.

Key words— curved ascospores, stictic acid

Introduction

During a study of the lichen family *Graphidaceae* (*Ostropales*, *Ascomycota*) in China, an interesting saxicolous species new to science was found. It is placed in the genus *Graphis* as delimited by Staiger (2002).

Material and methods

The type specimen was collected in the forest Wanmulin in Wuyishan, Fujian province, Southern China.

A dissecting microscope (TECH XTS-20) and a research microscope (OLYMPUS CHB-213) were used for the morphological and anatomical studies. Manual cross-sections of lirellae were examined in tap water. The lichen compound was detected by thin-layer chromatography (TLC) (Culberson & Kristensson, 1970; Culberson, 1972).

Taxonomic description

Graphis fujianensis Z.F Jia & J.C. Wei, **sp. nov.** FIGURE 1

MycoBank MB 511505

Differt a G. saxicola ascosporis minoribus curvatibus, sine marginibus thallinis apotheciorum et acidum sticticum cotinenti.

* *Corresponding author*

This paper was originally published in *Mycotaxon*, 104: 107-109, 2008.

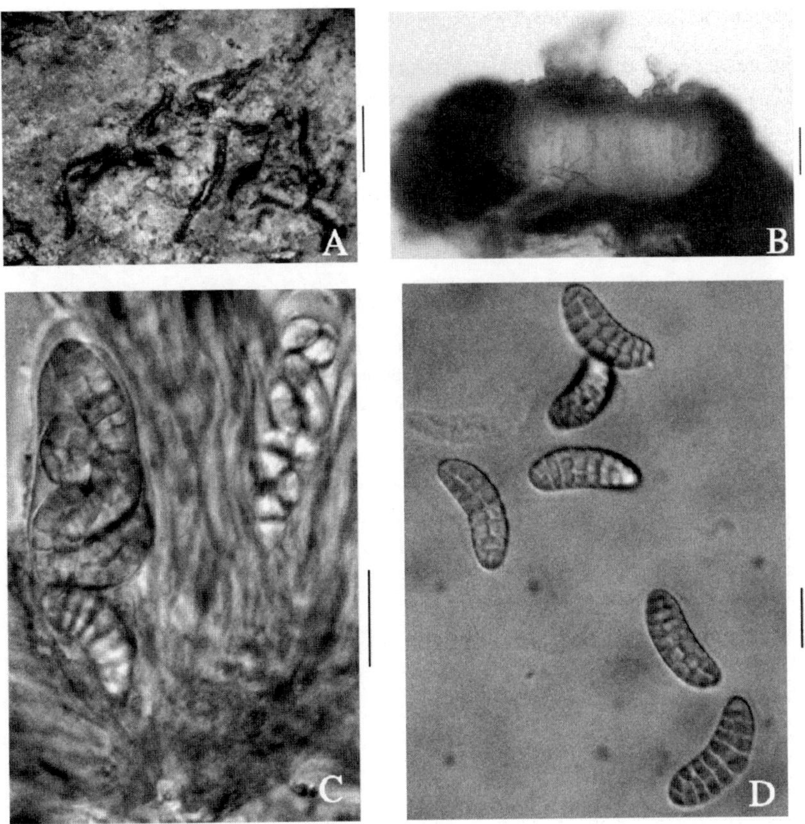

Fig. 1 *Graphis fujianensis*. A. Habit (bar = 1 mm); B. Apothecium cross section (bar = 50 μm); C. Two asci containing ascospores (bar = 20 μm); D. Curved, muriform ascospores (bar = 20 μm)

Typus: CHINA. Fujian Provincia, Jian-ou, Wanmulin, alt. 580m, ad saxa. 2/VI/2007, *Li Jing FJ1182*, (holotypus in LHS; isotypus in HMAS-L.).

ETYMOLOGY: The specific epithet "fujianensis" is derived from the type locality in Fujian province.

THALLUS crustose, pale grayish-white, thin, saxicolous, with dull surface. APOTHECIA lirelliform, black, conspicuous, initially fissurine, becoming sub-immersed to sessile, with slightly raised margins, curved and sinuous, lips sometimes opening slightly, 0.5-4 mm long, 0.1-0.25 mm wide, not grooved, without thalline margins; PROPER EXCIPLE completely carbonized; HYMENIUM hyaline, not inspersed, 100-120 μm tall, clear, I-ve, EPITHECIUM 20-28.5 μm tall, brownish; PARAPHYSES up to 1.5 μm wide, septate, slightly widened at apices;

ASCI cylindric, 70-80 × 13-20 μm, 8-spored; ASCOSPORES oblong-ellipsoid to slightly curved, hyaline, 17.5-33 μm long, 9-13 μm wide, muriform, 8-10 × 1-3-locular, I+ blue.

CHEMISTRY: C-, K+ yellow, P+red; stictic acid (TLC).

The new species differs from *Graphis saxicola* (Müll. Arg.) A.W. Archer (1999, 2005) by the smaller and curved ascospores, the lack of thalline margins and the presence of stictic acid.

Acknowledgments

The authors thank Alan W. Archer and Harrie J. Sipman for presubmission review.

Literature cited

Archer AW. 1999. The lichen genera *Graphis* and *Graphina* (*Graphidaceae*) in Australia. 1. Species based on Australian Type specimens. Telopea 8(2): 273-295.

Archer AW. 2005. New combinations and synonymies in the Australian *Graphidaceae*. Telopea 11(1): 59-78.

Culberson CF. 1972. Improved conditions and new data for the identification of lichen products by a standardized thin-layer chromatographic method. Journal of Chromatography 72: 113-125.

Culberson CF, Kristensson H. 1970. A standardized method for the identification of lichen products. Journal of Chromatography 46: 85-93.

Staiger B. 2002. Die Flechtenfamilie *Graphidaceae* Studien in Richtung einer natürlicheren Gliederung. Bibliotheca Lichenologica 85: 1-526.

A new species of *Melanelixia* (*Parmeliaceae*) from China

Hai-Ying Wang [1,2], Jian-Bin Chen [1] & Jiang-Chun Wei[1*]

*lichenwhy@Yahoo.com.cn *weijc2004@126.com*

[1]*Key Laboratory of Systematic Mycology & Lichenology*
Institute of Microbiology, Chinese Academy of Sciences
Beijing 100101, P.R. China

[2]*Graduate University of Chinese Academy of Sciences*
Beijing 100049, P.R. China

Abstract — A new *Melanelixia* species characterized by the presence of cortical hairs and wart-like isidia, *M. subvillosella*, is described from China. *Melanelia subverruculifera* is reported as a new synonym of *Melanelixia villosella*. A key to the nine known *Melanelixia* species is also provided.

Keywords — Asia, lichens, taxonomy

Introduction

The lichen genus *Melanelixia* O. Blanco et al. in the *Parmeliaceae* was segregated from *Melanelia* Essl. based on molecular as well as chemical and morphological data (Blanco et al. 2004). The genus is morphologically characterized by pored or fenestrate epicortex and lecanoric acid as the primary medullary constituent (Blanco et al. 2004, Esslinger 1977). The genus presently includes eight species known in the world: *M. albertana* (Ahti) O. Blanco et al., *M. fuliginosa* (Fr. ex Duby) O. Blanco et al., *M. glabra* (Schaer.) O. Blanco et al., *M. glabroides* (Essl.) O. Blanco et al., *M. huei* (Asahina) O. Blanco et al., *M. subargentifera* (Nyl.) O. Blanco et al., *M. subaurifera* (Nyl.) O. Blanco et al., and *M. villosella* (Blanco et al. 2004). All these *Melanelixia* species except *M. albertana* and *M. subaurifera* have been reported in China (Wei 1991, Abbas et al. 1998, Omar et al. 2004, Chen & Esslinger 2005).

During our study of the lichen flora of China, an interesting species of *Melanelixia* new to science was found. Here we present a brief Latin diagnosis, an extended description, and key to the presently known species of the genus. Five photographs for the new species are provided as well. In addition, as we studied the new species and related taxa, we discovered that *Melanelia subverruculifera* is a synonym of *Melanelixia villosella*.

This paper was originally published in *Mycotaxon*, 104: 185-188, 2008.

Materials and methods

The material examined was collected from Northeast China and is preserved in HMAS-L (Herbarium Mycologicum Academiae Sinicae-Lichenes). For the morphological study of the material, a dissecting microscope (Zeiss Stemi SV 11) has been used and the anatomical study was carried using a compound microscope (Zeiss Axioscop 2 plus). For the ultrastructural observations, a scanning electron microscope (Fei Quanta 200) has been used. Lichen substances were identified using the standardized thin layer chromatography techniques of Culberson (1972).

Key to *Melanelixia* species

1a. Cortical hairs absent 2
1b. Cortical hairs present 5

2a. Thallus with isidia, soredia or both 3
2b. Thallus without isidia or soredia 4

3a. Isidia fine, 0.1-0.25 mm long and 0.02-0.06 mm in diameter, seldom branched; soredia usually present .... ***M. subaurifera***
3b. Isidia larger, mostly 0.2-1mm long and 0.05-0.1 mm in diameter, often branched; soredia never present .... ***M. fuliginosa***

4a. Corticolous; upper surface of lobes with obscure to distinctive pseudocyphellae; often with the orange pigment rhodophyscin in the lower medulla .... ***M. huei***
4b. Saxicolous; upper surface of lobes without pseudocyphellae; without rhodophyscin .... ***M. glabroides***

5a. Thallus with soredia 6
5b. Thallus without soredia 7

6a. Soralia marginal and strongly labriform .... ***M. albertana***
6b. Soralia laminal and marginal, the laminal ones arising from small, ± hemispherical pustules .... ***M. subargentifera***

7a. Thallus without isidia .... ***M. glabra***
7b. Thallus with isidia 8

8a. Isidia cylindrical, often branched and with obvious hyaline hairs at their tips .... ***M. villosella***
8b. Isidia wart-like, non-branched and without hairs on their tips .... ***M. subvillosella***

Taxonomic description

Melanelixia subvillosella H.Y. Wang & J.C. Wei, sp. nov. FIG. 1

MycoBank MB 511544

Species haec a M. glabra verrucis et a M. villosella sine trichomis in pagina isidiorum differt.

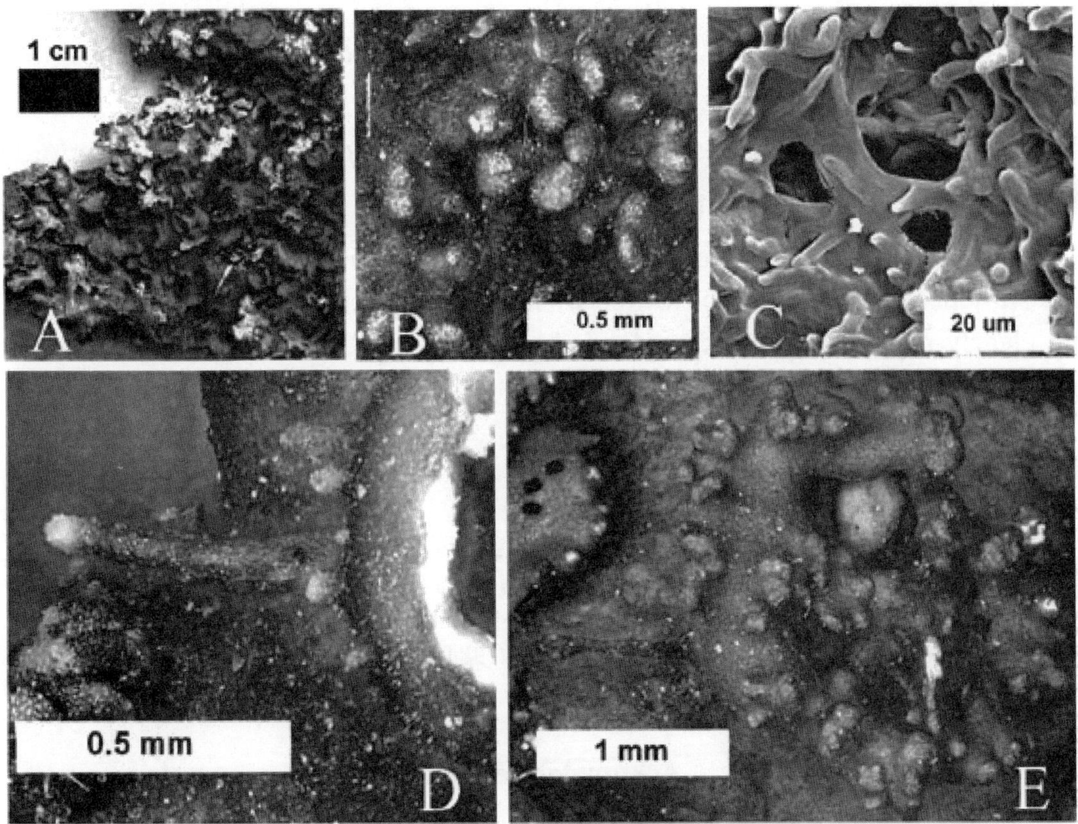

Fig.1 Photographs of *Melanelixia subvillosella*. A. Thallus. B. Isidia. C. SEM photograph, showing the pored and coarse epicortex. D. Rhizinae and hairs on an apothecial margin. E. Rhizina, papilla and lobules on an apothecial margin.

Holotypus a China, Jilin provincia, Mt. Changbaishan, alt. 1100m, ad corticem arborum, X.D. Lu 0211, 24 Jun. 1985 in HMAS-L (077798) conservatur.

Expanded description — Thallus foliose, 10 cm in diameter (Fig. 1A). Lobes 1-3 mm broad, more or less flat, elongated, sometimes crenulate on the margin, contiguous to imbricate. Upper surface olive-brown to dark-brown; rather shiny at lobe ends, inward becoming dull; with obvious cortical hairs, especially on the lobe ends and the apothecial margins (Fig. 1D); without pseudocyphellae or soredia; isidiate (Fig. 1B), the isidia arising as hemispherical papillae, and then slightly elongating into short claviform isidia (0.2-0.1×0.5-1 mm); lobulate, the lobules short and rounded, marginal to laminal. Lower surface black, often paler at the margin; smooth to wrinkled, dull to slightly shiny; abundantly rhizinate, the rhizines simple to branched, concolorous with the lower surface, to 2 mm long. Lobes 130-160 μm thick, upper cortex 8-12 μm thick; lower cortex 8-11 μm thick. Epicortex pored (pore, 6-16 μm in diameter) and coarse (Fig. 1C). Apothecia common, sessile to short stipitate, concave, to 4 mm in diameter; margin entire when small, soon becoming crenate-papillate and lobulate (Fig. 1E), often with rhizinae (Fig. 1D,E); hymenium 54-70 μm thick, subhymenium 50-70 μm thick; spores 8, globose to ellipsoid, 8-10×5-7 μm, spore wall 1 μm thick. Pycnidia unknown.

CHEMISTRY — Cortex K–, HNO_3–; medulla C+ red, K–, PD–; containing lecanoric acid.

DISTRIBUTION AND SUBSTRATE — At present, *M. subvillosella* is known only from the type locality; on bark.

COMMENTS — The apothecia with rhizinae and the coarse epicortex distinguish *M. subvillosella* from all the other *Melanelixia* species. The new species can be clearly separated from *M. glabra* by having wart-like isidia and from *M. villosella* by lacking the cortical hairs on the surface of isidia.

Melanelixia villosella (Essl.) O. Blanco, A. Crespo, Divakar, Essl., D. Hawksw. & Lumbsch, Mycol. Res. 108(8): 882 (2004)

≡ *Parmelia villosella* Essl., J. Hattori. Bot. Lab. 42: 95 (1977)

= *Parmelia subverruculifera* J.C. Wei & Y.M. Jiang, Acta Phytotax. Sin. 18: 386 (1980) = *Melanelia subverruculifera* (J.C. Wei & Y.M. Jiang) J.C. Wei, Enum. Lich. China: 153 (1991). (syn. nov.) Holotype: CHINA. Tibet, Nielamu, alt. 3840m, 22 Jun 1966, J.C. Wei & J.B. Chen 1899-1 (HMAS-L).

Acknowledgements

The project was financially supported by the National Natural Science Foundation of China (30499340). The authors would like to thank the keeper of the HMAS-L Ms Deng Hong for giving assistance during this study. The authors thank Teuvo Ahi, Theodore L. Esslinger, and Syo Kurokawa for expert presubmission review.

Literature cited

Abbas A, Wu JN. 1998. Lichens of Xinjiang. 178 pp. Sci-Tech & Hygiene Publishing House of Xinjiang, Urumqi.

Blanco O, Crespo A, Divakar PK, Esslinger TL, Hawksworth DL, Lumbsch HT. 2004. *Melanelixia* and *Melanohalea*, two new genera segregated from *Melanelia* (*Parmeliaceae*) based on molecular and morphological data. Mycological Research 108(8): 873-884.

Chen JB, Esslinger TL. 2005. *Parmeliaceae* (*Ascomycota*) lichens in China's mainland IV. *Melanelia* species new to China. Mycotaxon 93: 71-74.

Culberson CF. 1972. Improved conditions and new data for the identification of lichen products by a standardized thin-layer chromatographic method. Jour Chromatogr. 72: 113-125.

Esslinger TL. 1977. A chemosystematic revision of the brown *Parmeliae*. Jour. Hattori Bot. Lab. 42: 1-211.

Omar Z, Keyimu A, Abbas A. 2004. New Chinese records of the lichen genus *Melanelia*. Acta Bot. Yun. 26: 385-386.

Wei JC. 1991. Enumeration of lichens in China. 278 pp. International Academic Publishers. BeiJing.

The new lichen species *Endocarpon crystallinum* from semiarid deserts in China

Jun Yang & Jiang-Chun Wei*

weijc2004@126.com
Key Laboratory of Systematic Mycology & Lichenology
Institute of Microbiology, Chinese Academy of Sciences
No. A3 Datun Road, Chao-yang District, Beijing 100101, P.R.China

Abstract — The new lichen species *Endocarpon crystallinum* was collected from semiarid deserts in China and is described in the present paper. Research on the morphology, anatomy, and chemistry of the species has been carried out. The species is unique in the genus by the thick upper cortex that is broken into columnar polygonal elements. A similar cortex structure is however known from a few, unrelated, other desert lichens and appears to be an adaptation to that climate.

Keywords — *Ascomycota*, cortex structure, *Verrucariaceae*

Introduction

During research on lichen species and their genetic diversity in the arid and semiarid deserts of northwestern China, numerous species of *Endocarpon* have been found. The genus is mainly characterized by a squamulose thallus, immersed perithecioid ascomata with hymenial algae, and asci with (1–)2(–8) muriform ascospores. Among the specimens collected was one that could not be identified with any of the known species, and it is consequently described below as new to science. The genus *Endocarpon* belongs to the *Verrucariaceae* (Kirk et al. 2001), *Verrucariales*, *Chaetothyriomycetidae*, *Eurotiomycetes*, *Ascomycota* (Geiser et al. 2006).

Materials and methods

The specimens examined were collected by either J. Yang and E.R. Zhang or T. Zhang from the deserts in the Shanxi province in northwestern China. Hand sections and squash preparations were made for anatomical study. The morphological and anatomical examinations were performed using the

* Corresponding author.

This paper was originally published in *Mycotaxon*, 106: 445-448, 2008.

dissecting microscope Leica MZ8 and the biological microscope Leica DMR respectively. For chemical analysis, the standardized thin-layer chromatograph (TLC) method for identification of lichen products was used, applying solvent C (Culberson 1972, Culberson & Kristensson 1970, White &James 1985).

Results and discussion

Endocarpon crystallinum J.C.Wei & Jun Yang, **sp. nov.** FIG. 1

MYCOBANK MB512364

Species nova cortice supero crasso quo fracto in elementa similia crystallis, sed constantia ex textura paraplectenchymata hyalina a congeneribus diversa.

Typus: China, Shanxi provincia: Yang-gao comitatus, Xiejiatun pagus, ad arenam, September 23, 2004, J.Yang & E.R. Zhang SX-28 (holotypus) in HMAS-L conservatur.

ETYMOLOGY: The specific epithet "crystallinum" refers to the upper cortex of the thallus, which is covered with a broken hyaline layer reminiscent of crystals.

EXPANDED DESCRIPTION — Thallus terricolous, squamulose, consisting of small squamae, covered with a thick layer of hyaline tissue that is broken into columnar polygonal elements resembling crystals (FIG. 1 A); squamae solitary, with dark brown to black peripheral margins (FIG. 1 A), or occasionally contiguous, never imbricate, pale to yellowish brownish, rounded, elongate or irregular, shallowly to deeply lobate, 2–4(–7) mm broad.

Upper cortex consisting of two layers: the upper thick paraplectenchymatous layer of hyaline tissue that is broken into columnar polygonal elements resembling crystals in the shape of peaks is (30–)50–175 μm tall, the elements (50–)62.5–275(–475) μm wide at the base, 17.5–175(–225) μm wide at the top; the lower brownish paraplectenchymatous layer is 34–80(–110) μm tall. (FIG. 1 B)

Algal layer 38–100(–112)μm tall, consisting of subglobose and bright green cells, 5–7.5 × 5–10 μm, without visible medulla (FIG. 1 B).

Lower cortex consisting of obscure plectenchyma, dark brown to black, about 50 μm tall, with brown to black and branched rhizines of 1.5–2.5 mm long (FIG. 1 C).

Ascomata perithecioid, subglobose, uniloculate, immersed in thallus, solitary, 1–6 per squama, with brown to dark brown apex of 0.25 mm in diam. bearing a depressed ostiole lighter in color than the surrounding tissue; centrum obpyriform, 150–490 μm wide; excipulum brown to dark brown, 105–400 μm thick at the base and sides, pale brown at the apex near the ostiole; gelatinizing periphysoids 75–100 μm long, simple to sparingly branched; hymenial gel Lugol's I + blue; hymenial algal cells globose to oval, green, 2.5–5 μm in diam. or 2.5–3.5 × 3.5–5 μm (FIGS. 1D, E).

Asci fissitunicate, bisporous (FIG. 1E), clavate to cylindro-clavate, 40–60 × 15.5–21.5 μm; ascospores muriform, yellowish to brownish, elongate-ellipsoid

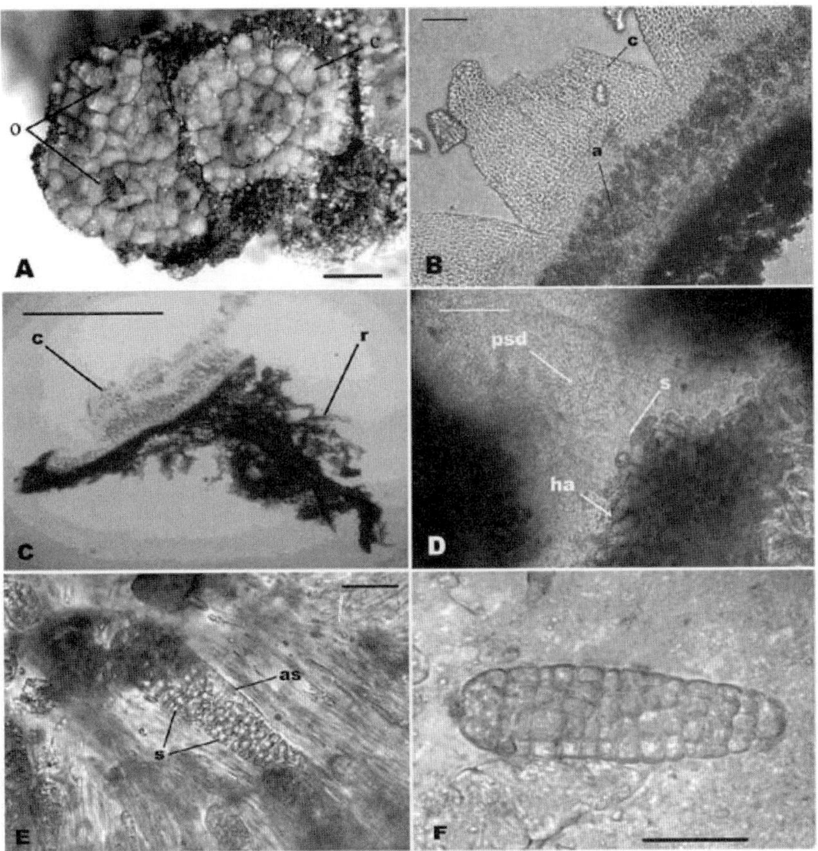

FIG. 1. A. *Endocarpon crystallinum*: Upper surface of squamae covered with minute crystal-like broken cortex elements and a few perithecioid ascomata of which only the ostioles are visible (bar = 1mm); B. Cross section of a squama (bar = 50μm); C. Cross section of a squama with branched rhizines (bar=1mm); D. The hymenial algal cells and ascospores within a perithecioid ascoma (bar = 100μm); E. An ascus containing 2 spores within a perithecioid ascoma; note that the upper spore is much broader and shorter than the lower spore (bar = 20 μm); F. A muriform ascospore (bar =20μm).

(a = algal layer; as = ascus; c = crystals; ha = hymenial algal cells; o = ostiole; psd = periphysoid; r= rhizines; s=spore; t=thallus)

to subcylindrical, with 2–4 transverse divisions and 9–15 longitudinal divisions, the upper spore much broader and shorter than the lower spore; upper spore: 50–65(–87.5) × 12.5–15μm; lower spore: (32.5–)40–50 ×15–17.5μm (FIGS. 1 E,F).

Conidiomata not seen.

CHEMISTRY: no lichen substances detected with TLC.

ADDITIONAL SPECIMENS EXAMINED: China, Shanxi prov.: Yang-gao county: Xiejiatun, on the wall of soil, September 23, 2004, J.Yang & E.R. Zhang SX-32 (HMAS-L); west of the Yang-gao county town 2 km, on slope of soil, April 20, 2007, J. Yang & T. Zhang, SX-093, SX-113 (HMAS-L); Ningxia Autonomous region, Xiang mountain of Shapotou region, soil among rocks; April 17, 2007, J. Yang & T. Zhang SPT-202 (HMAS-L).

COMMENTS: This new species resembles in its upper cortex structure several other unrelated lichens from desert habitats. Examples are *Psora crystallifera* and several *Toninia* species (Timdal 1991), e.g. *T. albilabra* and *T. sculpturata*. The congruence is so striking that Fig. 2E in Timdal (1991), taken from *Toninia albilabra*, is indistinguishable from FIG. 1B below, taken from the new species, which belongs to a different order (and even class). This structure is undoubtedly an adaptation to desert climate.

The new species differs from other members currently accepted in the genus (Breuss 2002, Harada 1993, McCarthy 1991) by the thick upper cortex that is broken into columnar polygonal elements resembling crystals, but made of hyaline paraplectenchymatous tissue.

Acknowledgments

This work was financially supported by the National Natural Science Foundation of China (NSFC, No.30270005) and the Ministry of Science and Technology of the People's Republic of China (MOST, No.2007AA021405). The authors are most grateful to Dr. A. Aptroot and Dr. H. Sipman for reading the manuscript and giving the valuable comments and suggestions, and also to Mr. E.R. Zhang and Mr. T. Zhang for accompanying with the first author to do the field work.

Literature cited

Breuss O. 2002. *Endocarpon*. pp. 181-187 in: Nash TH, Ryan BD, Gries C & Bungartz F. Lichen flora of the greater Sonoran desert region. Volume 1. Arizona State University, Tempe.

Culberson CF. 1972. Improved conditions and new data for the identification of lichen products by a standardized thin-layer chromatographic method. J. Chromatogr. 72:113-125.

Culberson CF, Kristensson H. 1970. A standardized method for the identification of lichen products. J. Chromatogr. 46:85–93.

Geiser DM, Gueidan C, Miadlikowska J, et al. 2006. *Eurotiomycetes: Eurotiomycetidae* and *Chaetothyriomycetidae*. Mycologia 98(6):1053–1064.

Harada H. 1993. A taxonomic study of the lichen genus *Endocarpon* (*Verrucariaceae*) in Japan. Nova Hedwigia 45:335–353.

Kirk PM, Cannon PF, David JC, Stalpers JA. 2001. Ainsworth & Bisby's Dictionary of the Fungi (9th Edition) CABI Publishing, Egham.

McCarthy PM. 1991. The lichen genus *Endocarpon* Hedwig in Australia – Lichenologist 23(1): 27–52.

Timdal E. 1991. A monograph of the genus *Toninia* (*Lecideaceae, Ascomycetes*). Opera Botanica 110: 1–137.

White FJ, James PW. 1985. A new guide to microchemical techniques for the identification of lichen substances. Brit. Lich. Soc. Bull. 57 (Suppl.): 1–41

A new isidiate species of *Graphis* (lichenised *Ascomycotina*) from China♣

Ze-Feng Jia[1] & Jiang-Chun Wei[2]*

zfjia2008@163.com & weijc2004@126.com

[1]*College of Life Sciences, Shandong Agricultural University Taian 271018, China*

[2]*Key Laboratory of Systematic Mycology & Lichenology, Institute of Microbiology Chinese Academy of Sciences, Beijing 100101, China*

Abstract — This paper describes a third isidiate species with trans-septate spores in the genus *Graphis* sensu Staiger. The new species is from Guangdong Province of China. It is characterized by an isidiate thallus, a completely carbonized exciple with entire lips, trans-septate ascospores, and the absence of lichen substances.

Key words — *Graphis guangdongensis, Graphidaceae, Ostropales, Ascomycota*

Introduction

The members of the *Graphidaceae* (*Ostropales*, lichen forming fungi) are widely distributed in pantropical areas of the world. In the genus *Graphis*, sensu Staiger (Staiger 2002), there are two known isidiate species with trans-septate spores, viz. *G. patwardhanii* C.R. Kulk. 1978 and *G. isidiza* Adaw. & Makhija (Adawadkar & Makhija 2004).

The new isidiate species of the genus, which differs from the two species mentioned above, was collected in China and is described as new to science in the present paper.

Material and methods

The material examined was collected from Guangdong in southern China. A dissecting microscope (TECH XTS-20 and AIGO Digital Viewer GE-5) and a

♣ The project was supported by the National Natural Science Foundation of China (No. 39899400).

* Corresponding author

This paper was originally published in *Mycotaxon*, 110: 27-30, 2009.

compound microscope (OLYMPUS CHB-213) were used for the morphological and anatomical studies. Measurements and illustrations were taken from manual cross-sections of lirellae in tap water. The chemistry was determined by thin-layer chromatography (TLC) (Culberson & Kristensson 1970, Culberson 1972, White & James 1985).

Fig. 1 *Graphis guangdongensis*. A. Habit (bar =1mm); B. A cross section of an apothecium (bar =100 mm); C. An ascus containing ascospores (bar = 50μm); D. Ascospores (bar = 50 μm).

Taxonomy

Graphis guangdongensis Z.F. Jia & J.C. Wei, **sp. nov.** FIGURE 1

MYCOBANK MB 514121

Species nova G. patwardhanii *similis sed excipulo omnino carbonaceo et labello integro differt.*

HOLOTYPE: CHINA. provincia Guangdong, comitatus Fengkai, Heishiding, 23°27′N, 113°30′E, alt. 250 m, in cortice arboris latifoliae. X/28/1998. coll. Shou-Yu Guo 2185 (**holotypus** in HMAS-L *X024581*, isotypus in LHS).

ETYMOLOGY: The specific epithet "guangdongensis" refers to the place name of the province Guangdong, China, the type locality of the new species.

THALLUS corticolous, pale white to milky white, unevenly thickened, tightly attached to the substratum, distinctly isidiate. Isidia small, globose, or more or less digitiform, 100–150 µm tall. APOTHECIA elongate, 0.5–2.5 mm long, 0.25–0.3 mm wide, simple or rarely slightly branched, erumpent, black, curved and straight, rounded to pointed at the ends, not striate, scattered over the thallus. PROPER EXCIPLE conspicuous, completely carbonized. EPITHECIUM 8–10.5 µm thick, grayish. HYPOTHECIUM brownish, 20–41.5 µm tall HYMENIUM colorless, clear, 160–190 µm tall, I–. PARAPHYSES unbranched, filiform, septate, up to 1.5 µm wide ASCI cylindrical, 110–132 × 22–33 µm, 2-4-spored. ASCOSPORES colourless, long fusiform, 18–20 transverse septate, (65–)75–110 × (8–)11–15.5 µm, I+ blue.

CHEMISTRY: C–, K–, P–; no lichen compounds detected.

REMARKS: The new species is similar to both *Graphis patwardhanii* and *G. isidiza* in having an isidiate thallus and producing trans-septate spores. *Graphis patwardhanii* differs in having a laterally carbonized, 3–6-striate exciple while *G. isidiza* has smaller (21–34 × 4–8 µm) ascospores and contains the lichen compounds constictic acid and stictic acid (Adawadkar & Makhija 2004). *Graphis guangdongensis* is so far known only from the type material.

Acknowledgments

The authors are grateful to Dr. Alan W. Archer and Dr. Urmila Makhija for reading and giving valuable comments and suggestions.

Literature cited

Adawadkar B, Makhija U. 2004. A new isidiate species of *Graphis* from India. The Lichenologist 36(6): 361–363.

Culberson CF, Kristensson H. 1970. A standardized method for the identification of lichen products. Journal of Chromatography 46: 85–93.

Culberson CF. 1972. Improved conditions and new data for the identification of lichen products by a standardized thin-layer chromatographic method. Journal of Chromatography 72: 113–125.

Staiger B. 2002. Die Flechtenfamilie *Graphidaceae*. Studien in Richtung einer natürlicheren Gliederung. Bibliotheca Lichenologica 85: 1–526.

White FJ, James PW. 1985. A new guide to microchemical techniques for the identification of lichen substances. British Lichen Society Bulletin 57: 1–41.

A new species, *Thalloloma microsporum* (*Graphidaceae, Ostropales, Ascomycota*)

ZE-FENG JIA[1] & JIANG-CHUN WEI[2*]

*zfjia2008@163.com * & weijc2004@126.com*

[1]*College of Life Sciences, Shandong Agricultural University
Taian 271018, China*

[2]*Key Laboratory of Systematic Mycology & Lichenology, Institute of Microbiology
Chinese Academy of Sciences, Beijing 100101, China*

Abstract — A new corticolous species of *Thalloloma* from the Qinling Mountains in Shaanxi Province of China is described. The fungus is characterized by the small ascospores and cinnabarine lips.

Key words — lichen, morphology, taxonomy

Introduction

During a study of the lichen family *Graphidaceae* (*Ostropales, Ascomycota*) from Shaanxi Province, China, a corticolous species of *Thalloloma* was found in dry deciduous forests of the Qinling Mountains. It is new to science and described as *Thalloloma microsporum*. The genus as delimited by Staiger (2002) has not been reported from China before.

Material and methods

A dissecting microscope (TECH XTS-20) and a light microscope (OLYMPUS CHB-213) were used for the morphological and anatomical studies. Measurements and illustrations were taken from the manual cross-sections of fruitbodies in tap water. The lichen substance was detected and identified by thin-layer chromatography (TLC) (Culberson & Kristensson 1970, Culberson 1972).

Taxonomy

Thalloloma microsporum Z.F. Jia & J.C. Wei, **sp. nov.** FIGURE 1

MYCOBANK MB 512502

Species nova similis T. hypolepto, *a quo labellis fere cinnabarinis et ascosporis minoribus.*

HOLOTYPE: CHINA. Shaanxi Province, Qinling mountains, Banqiaogou, 33°88′N, 108°01′E, alt. 1520 m, on cortices of cortice *Zelkova serrata* (Thunb.) Makino. 29-

* Corresponding author

This paper was originally published in *Mycotaxon*, 107: 197-199, 2009.

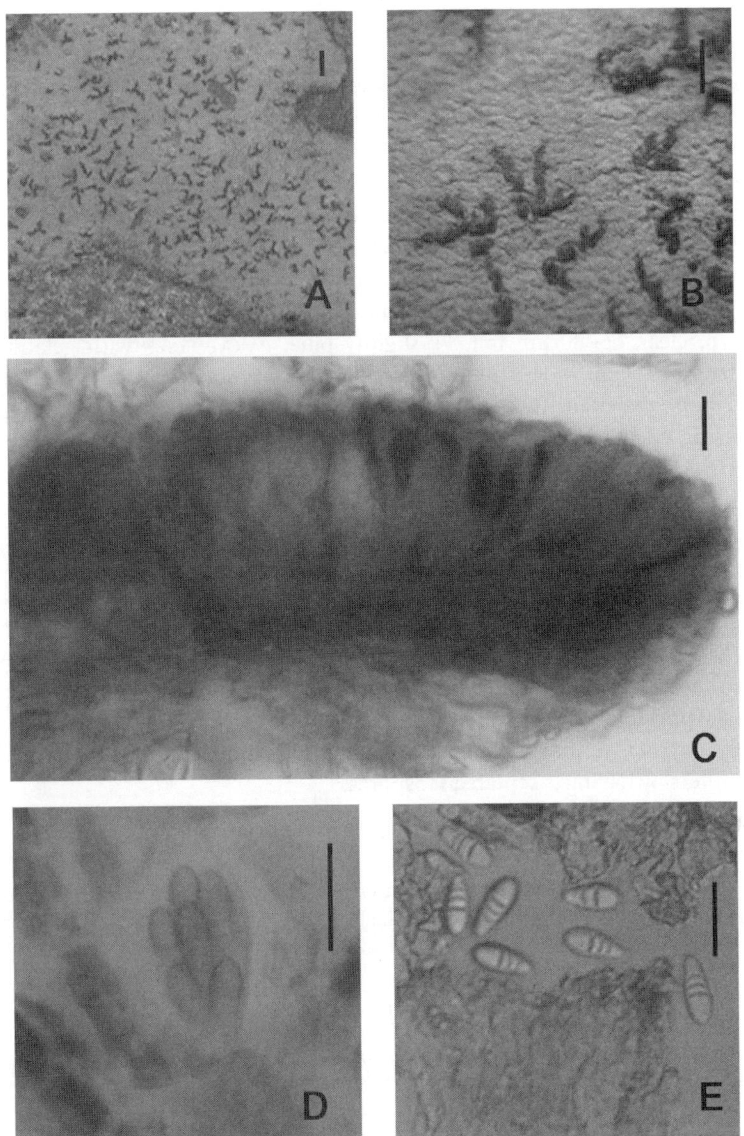

FIG. 1 *Thalloloma microsporum*. A, B. Thallus with apothecia; C. Apothecium cross section; D. An ascus containing ascospores; E. Ascospores.
Bars: A = 2 mm; B = 1 mm; C, D & E = 20 μm.

VII-2005, Ze-feng Jia SQ380 (holotype in LHS; isotype in HMAS-L.); paratypes: ibid., on cortices of *Pinus armandii* Franch. 29-VII-2005, Jia Ze-feng SQ374, SQ375 (LHS, HMAS-L).

ETYMOLOGY: The specific epithet "*microsporum*" refers to the small ascospores; *micro* is from Greek, small; *spora* in Greek, a seed.

Description: THALLUS corticolous, pale white to grayish-white, thin, 0.1–0.2 mm thick, with a dull surface. APOTHECIA elongate, rarely branched, apparently brown, conspicuous, sub-immersed to sessile, curved and sinuous, often with opening cinnabarine lips because of the reddish brown pigment isohypocrellin (Fig. 1A,B), 0.5–1.5 mm long, 0.1–0.25 mm wide, not grooved, without a distinguishable margin, surface of discs slightly granulous, grayish; PROPER EXCIPLE inconspicuous, not carbonized; EPITHECIUM 13–22 µm thick, brownish; HYPOTHECIUM red-brown, 15–30 µm tall; HYMENIUM slightly brown, clear, 66–80 µm tall, I+ slightly blue; PARAPHYSES with gelatinized wall, up to 1.5 µm wide, septate, enlarged at apices; ASCI club-shaped, 33–42 × 13–20 µm, 8-spored; ASCOSPORES ellipsoid with one end narrower and slightly pointed, 3-septate, hyaline, 15.5–20 × 5.5–8.0 µm, I+ blue, the largest upper cells 5.5–6.5 × 5.0–6.0 µm, the lower end cells 3.5–4.0 × 2.5–3.5 µm, the two middle rectangular cells 4.5–5.5 × 1.5–3 µm (Fig. 1E).

CHEMISTRY: C–, K–, P–; contains isohypocrellin (reddish brown pigment).

The new species is characterized by the opening with reddish lirellae, due to the presence of isohypocrellin, and small, 3-septate, hyaline ascospores. It is similar to *Thalloloma hypoleptum* (Nyl.) Staiger, but differs in the red brown to nearly cinnabarine lips and smaller ascospores (15.5–20 × 5.5–8.0 µm vs. 20–30 × 6.0–8.0 µm) (Staiger 2002). It also resembles *T. cinnabarinum* (Fée) Staiger and *T. rhodastrum* (Redinger) Staiger in containing isohypocrellin (Staiger 2002) but differs by having 3-septate ascospores.

Acknowledgments

The authors are deeply grateful to Dr. Alan W. Archer and Prof. W.Y. Zhuang for serving as pre-submission reviewers and their valuable comments. This project was supported by the National Natural Science Foundation of China (No. 39899400).

Literature cited

Culberson CF. 1972. Improved conditions and new data for the identification of lichen products by a standardized thin-layer chromatographic method. Journal of Chromatography 72: 113–125.

Culberson CF, Kristensson H. 1970. A standardized method for the identification of lichen products. Journal of Chromatography 46: 85–93.

Staiger B. 2002. Die Flechtenfamilie *Graphidaceae* Studien in Richtung einer natürlicheren Gliederung. Bibliotheca Lichenologica 85: 1–526.

A new subspecies of *Gyalidea asteriscus* from China[1]

Jun Yang & Jiang-Chun Wei[2]

rain_man_yj@hotmail.com & weijc2004@126.com
Key Laboratory of Systematic Mycology & Lichenology
Institute of Microbiology, Chinese Academy of Sciences
No. 1 Beichen West Road, Chao-Yang District, Beijing 100101, P. R. China

Abstract — The genus *Gyalidea* is reported for the first time from China. *Gyalidea asteriscus* subsp. *gracilispora* from deserts of northern China is described as new to science. Latin diagnosis, English description, and illustrations are given for the new taxon. A new combination, *G. asteriscus* subsp. *nigrescens*, is also made.

Key words — lichens, *Asterothyriaceae*, new Chinese record, taxonomy

Introduction

During the lichen study of arid and semiarid deserts from northern China, numerous specimens of apothecia with star-shaped margins (Figs. 2a–b) containing polysporic asci (Fig. 2f) were collected from soil and microbiotic crust in the arid land of Hebei, Shanxi, Ningxia, Gansu, and Qinghai. The lichen examined is close to both *Solorinella asteriscus* and *S. nigrescens* (Thor 1984) in habit and chemistry, but differs in its wider paraphyses, smaller asci, and narrower ascospores. *S. nigrescens* has been reduced to subspecific rank as *S. asteriscus* subsp. *nigrescens* (Vězda, Lumbsch & Øvstedal, 1990).

The genus *Solorinella* has recently been transferred to the genus *Gyalidea* based on phenotypically phylogenetic analysis, and the species *S. asteriscus* has been recombined as *G. asteriscus* (Aptroot & Lücking 2003). Our collections from the desert in China represent a new taxon, *G. asteriscus* subsp. *gracilispora*. In keeping with these recent nomenclatural changes, we transfer *Solorinella asteriscus* subsp. *nigrescens* to *Gyalidea* under the new combination, *G. asteriscus* subsp. *nigrescens*.

The lichen genus *Gyalidea* (*Asterothyriaceae, Ostropales, Ostropomycetidae*) is reported for the first time from China.

---

[1]This work was financially supported by the National Natural Science Foundation of China (NSFC, No.30270005) and the Ministry of Science and Technology of the People's Republic of China (MOST, No.2007AA021405).

[2]Corresponding author

This paper was originally published in *Mycotaxon*, 109: 373-377, 2009.

Material and methods

The specimens studied were collected from the microbiotic crusts in the arid land of Hebei, Shanxi, Ningxia, Gansu, and Qinghai. The dissecting microscope (Leica MZ8) and compound microscope (Zeiss Axioskop 2 plus) were available for morphological and anatomical studies, and color test and standardized TLC was used for detecting of lichen substances (Culberson & Kristinssen 1970, Culberson 1972, Culberson & Johnson 1982, White & James 1985).

Taxonomy

Gyalidea asteriscus (Anzi) Aptroot & Lücking, Biblioth. Lichenol. 86: 67 (2003).

Basionym. *Solorinella asteriscus* Anzi, Catal. Lich. Sondr. p. 37 (1860).

Gyalidea asteriscus subsp. ***gracilispora*** Jun Yang & J.C. Wei, **subsp. nov.**

MycoBank MB 513530 Figs. 1, 2.

Subspecies nova habitu et substantia cum Gyalidea asteriscus *subsp.* asteriscus *et* G. asteriscus *subsp.* nigrescente *optime congruens, sed differt paraphysibus crassioribus, ascis brevioribus et parvulioribus, ascosporis gracilioribus.*

Type collection: Hebei, Mt. Xiaowutai, south of Jinghekou management area, N39°56′, E 114°56′, alt. 1190 m, on soil, 15 April 2005, Hai-Ying Wang & Xin-Li Wei, 3058 (**holotypus** - HMAS-L).

Etymology: Latin: "*gracilispora*" = narrow ascospore.

Apothecia with deep concave discs of dark brown to black color and of 1.1–1.8 mm in diam. bearing white star-shaped margins consisting of 4–8 triangles or cones of white lobelet. Margins star-shaped, white, paraplectenchymatous, lobelets 10–150 μm wide at top and 300–350 μm wide at bottom, 200–300 μm high; epithecium brownish, 25–30 μm thick; hymenium K−, I−, hyaline to pale brownish, 105–145μm thick; hypothecium hyaline to pale brownish, 25–35 μm high; paraphyses hyaline, unbranched, septate, 2–3 μm thick; asci cylindrical with apex structure of *Tremolecia*-type, polysporic, 52.5–72.5(−82.5)

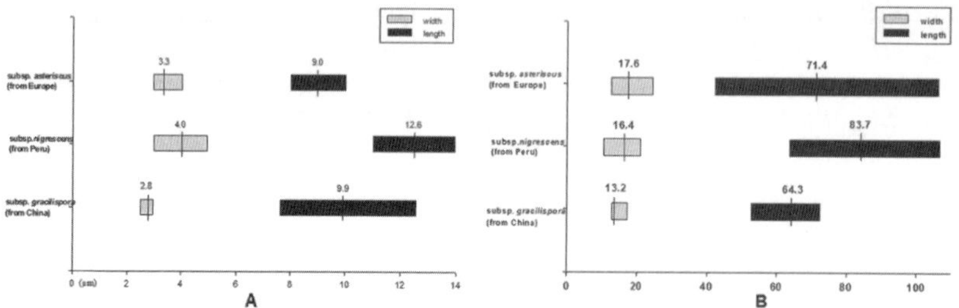

Fig. 1. Comparisons of ascospore (A) and ascus (B) size among three subspecies of *G. asteriscus*. Data for subsp. *asteriscus* and subsp. *nigrescens* are from Thor (1985, "1984").

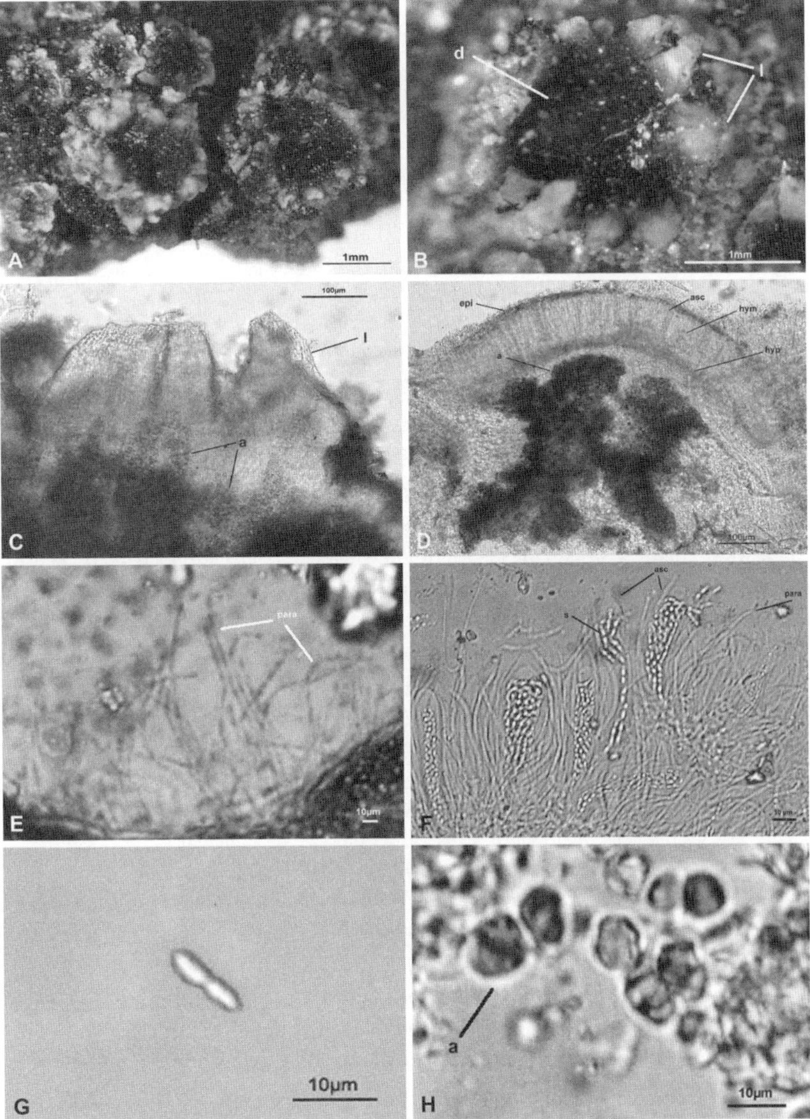

Fig. 2. A. The habit of *G. asteriscus* subsp. *gracilispora* (bar = 1 mm); B. One apothecium with star = shaped lobes at margin (bar = 1 mm); C. Cross section of a star = shaped lobe (bar = 100 µm); D. Cross section of an apothecium (bar = 10 µm); E. Septate paraphyses in cotton blue (bar = 10 µm); F. Asci containing spores (bar = 10 µm); G. An ascospore; H. Algal cells.

(a = algal cells; asc = ascus; d = disk of apothecium; epi = epithecium; hym = hymenium; hyp = hypothecium; l = lobe; para = paraphyses; s = spores)

× (10–)12.5–17.5 μm; ascospores hyaline, oblong ellipsoidal, 2-celled, 7.5–12.5(–17.5) × 2.5–3(–3.5) μm.

CONIDIOMATA not seen.

CHEMISTRY: no lichen substances detected by TLC; all parts C–, K–, KC–, PD–.

PHYCOBIONT belonging to *Chlorococcaceae*, 7.5–10 μm diameter.

HABIT: on soil of the microbiotic crusts in arid land.

ADDITIONAL SPECIMENS EXAMINED: **HEBEI:** Mt. Xiaowutai, south of Jinghekou management area, N39°56′, E 114°56′, alt. 1190 m, on soil, 15 April 2005, Hai-Ying Wang & Xin-Li Wei, 3041, 3043, 3052, 3053, 3056 (HMAS-L); Fengning County, Xiaobazi Village, on soil, 24 April 2004, XBZ038 (HMAS-L). **SHANXI:** Yanggao County, Xiejiatun Village, on soil, Jun Yang & Tao Zhang, 23 September 2004, SX039 (HMAS-L). **NINGXIA:** Zhongwei, Shapotou, on microbiotic crust, 6 August 2003, Jiang-Chun Wei & Jun Yang, SPT372 (HMAS-L). **GANSU:** Weiyuan County, Mt. Junshan, 28 October 2004, on soil, En-Ran Zhang, GS055, GS067 (HMAS-L). **QINGHAI:** Gonghe County, Qiabuqia, alt. 2910 m, on soil, 10 September 2005, Man-Rong Huang & Jun Yang QH030 (HMAS-L).

REMARKS: The new subspecies is identical to *G. asteriscus* subsp. *asteriscus* and subsp. *nigrescens* in habit and chemistry but differs in its wider paraphyses, thinner ascospores, and smaller asci.

We recognize three *G. asteriscus* subspecies: subsp. *asteriscus* from Europe, subsp. *gracilispora* from China, and subsp. *nigrescens* from Peru:

Gyalidea asteriscus subsp. ***nigrescens*** (G. Thor) Jun. Yang & J.C. Wei, **comb. nov.**

MYCOBANK MB 514070

BASIONYM. *Solorinella nigrescens* G.Thor, Nord. J.Bot. 4(6): 823 (1985, "1984").

≡ *Solorinella asteriscus* subsp. *nigrescens* (G.Thor) Vězda, Lumbsch & Øvstedal, Nova Hedwigia 50(3–4): 528 (1990).

Acknowledgments

The authors express their thanks to Dr. Xin-Li Wei, Dr. Man-Rong Huang, Mr. Hai-Ying Wang, Mr. Tao Zhang, and Mr. En-Ran Zhang for collecting the materials examined in this study from different arid regions of northern China. They also thank Drs. Wen-Ying Zhuang, André Aptroot and Shaun Pennycook for reading the manuscript and valuable suggestions.

Literature cited

Aptroot A, Lücking R. 2002. Proposal to conserve *Gyalidea* (lichenized fungi: *Asterothyriaceae, Ostropales*) against an additional name, *Solorinella*. Taxon 51 (3): 565.

Aptroot A, Lücking R. 2003. Phenotype-based phylogenetic analysis does not support generic separation of *Gyalidea* and *Solorinella* (*Ostropales*: *Asterothyriaceae*). Bibliotheca Lichenologica 86: 53–78.

Culberson CF. 1972. Improved conditions and a new data for the identification of lichen products by a standardized thin-layer chromatographic method. J. Chromatography 72: 113–125.

Culberson CF, Johnson A. 1982. Substitution of methyl tert.-butyl ether for diethyl either in the standardized thin-layer chromatographic method for lichen products. J. Chromatography 238: 483–487.

Culberson CF, Kritinssen H. 1970. A standardized method for the identification of lichen products. J. Chromatography 46: 85–91.

Thor G. 1985 ("1984"). A new species of *Solorinella* (*Asterothyriaceae*) from Peru. Nord. J. Bot. 4(6): 823–826.

White FJ, James PW. 1985. A new guide to microchemical techniques for the identification of lichen substances. Bul. British Lichen Soc. 57(suppl.): 1–47.

Caloplaca tianshanensis (lichen-forming *Ascomycota*), a new species of subgenus *Pyrenodesmia* from China

HURNISA XAHIDIN[1, 2], ABDULLA ABBAS[1] & JIANG-CHUN WEI[3*]

Hurnisa_xju@sina.com & weijc2004@126.com or weijc@im.ac.cn

[1]College of Life Science and Technology
[2]College of Resource and Environment Sciences
Xinjiang University, Urumqi 830046, P. R. China

[3]Key Laboratory of Systematic Mycology & Lichenology,
Institute of Microbiology, Chinese Academy of Sciences
1-3 West Beichen Road, Chaoyang District, Beijing 100101, P. R. China

Abstract — *Caloplaca tianshanensis* is described as a species new to science. It has a crustose and areolate thallus of yellowish-brown color with conspicuous cracks, bearing dark brown to black apothecia. An analysis of ITS sequences supports the affinity of the new species to subgenus *Pyrenodesmia*.

Key words — *Teloschistaceae*, peltate areoles, zeorine, isthmus

Introduction

As presently circumscribed, the subgenus *Pyrenodesmia* (A. Massal.) Boistel of the lichen-forming genus *Caloplaca* Th. Fr. (*Teloschistaceae*) contains lichens characterized by brown or black apothecia, an epihymenium that is usually K– or K+ violaceous, and a thallus that is not yellow, orange or red unlike most other *Caloplaca* spp., and lacks the K+ red reaction of the parietin complex (Tretiach & Muggia 2006).

Forty-two species of the genus *Caloplaca* were reported from China (Wei 1991). Among them 9 species belong to the subgenus *Pyrenodesmia*: *C. chrysophora* Zahlbr., *C. cupreorufa* Zahlbr. and *C. cervina* Zahlbr. from Sichuan (Zahlbruckner 1930, 1932), *C. giraldii* Jatta from Shaanxi (Jatta 1902) and Sichuan (Zahlbruckner 1930, 1931), *C. ochrotropa* Zahlbr. from Yunnan (Zahlbruckner 1930, 1932), *C. plumbeoolivacea* H. Magn., *C. circumalbata* (Delile) Wunder from Inner Mongolia (Magnusson 1944, as *C. aegyptiaca* (Müll.

*corresponding author

This paper was originally published in *Mycotaxon*, 114: 1-6, 2010.

TABLE 1. Lichen species and sequences used to generate the phylogenetic tree.

SPECIES	GENBANK #
Caloplaca albopruinosa (Arnold) H.Olivier	EF093577
	EF093578
C. albopustulata Khods. & S. Y. Kondr.	EU192150
C. alociza (A. Massal.) Mig.	EF090933
	EF090936
C. badioreagens Tretiach & Muggia	EF081039
	EF081040
C. cerina (Ehrh.Ex Hedw.) Th.Fr.	AF353958
C. chalybaea (Fr.) Müll.Arg.	AY313970
	AY313971
C. chlorina (Flot.) Sandst.	AF353959
C. concreticola Vondrák & Khodos.	EU192153
	EU192152
C. cretensis (Zahlbr.) Wunder	EF093579
C. erodens Tretiach et al.	EF090922
	EF090921
C. obscurella (J. Lahm) Th.Fr.	AY313976
	AY313977
C. peliophylla (Tuck.) Zahlbr.	AY313965
C. tianshanensis Xahidin, A. Abbas & J.C. Wei [a]	GU552277
C. transcaspica.	EU192156
C. variabilis (Pers.) Müll. Arg.	EF090926
	EF090925

Arg.) Stnr; Wunder 1974), *C. transcaspica* (Nyl.) Zahlbr. from Inner Mongolia (Magnusson 1944, as *C. paulsenii*), Gansu, Qinghai (Magnusson 1940, as *C. paulsenii*) and Xinjiang (Poelt & Hinteregger 1993)., and *C. alociza* (Massal.) Mig. from Jiangsu (Wu & Xiang 1981, as *C. agardhiana* (Flot.) Flag., 1981).

During a study of the lichen genus *Caloplaca* in China numerous samples were collected by the first two authors from the Xinjiang region. Some specimens belonging to *Pyrenodesmia* attracted our special attention and were examined in detail for morphology, anatomy, chemistry and molecular systematics. As a result, one of them, *C. tianshanensis*, is described here as new to science.

Material and methods

Material

The lichen material examined for morphology, anatomy, chemistry and molecular analyses was collected from Miaoergou on Mt. Nan-shan in the Tianshan mountain chain, Xinjiang region, in 2009.

Morphological observations

Observations and photographs were made with a dissecting microscope (Leica MZ 12), a Zeiss Axioplan compound microscope and an Axiocam digital camera with associated software. Squash mounts and hand sections were routinely examined using tap water as the mounting medium. Lichen substances were detected by TLC and MCT (Culberson & Kristinsson 1970, Culberson 1972, Orange et al. 2001).

DNA extraction, amplification, and sequencing

The dried apothecia first were checked under the dissecting microscope for well-developed fruit bodies to avoid contamination of other organisms.

Total DNA was extracted from dry apothecia following the rapid one-tube genomic DNA extraction (Steiner et al. 1995) with modifications: seven dried and cleaned apothecia were transferred directly into a 2 ml Eppendorf tube. The material was grinded with a pestle in liquid nitrogen until a fine powder was obtained. Then 150 μl TE solution was added into the tube and stirred for 2 min. until the powder was well-distributed, and immediately stored at −20°C.

Primers for PCR of the nuclear ribosomal ITS region ITS1F (Gardes & Bruns 1993) and ITS4 (White et al. 1990) were used.

The phylogenetic tree was constructed with a Bayesian approach based on the nuclear ribosomal ITS sequence data of the new species and sequences of species from the same subgenus retrieved from GenBank (Table 1).

Taxonomy

Caloplaca tianshanensis Xahidin, A. Abbas & J.C. Wei, **sp. nov.** (Figs 1, 2)

MycoBank MB 518332

Species nova similis C. peliophyllae *a qua thallo flavido-brunneo areolato cum rimis conspicuis et areolis peltatis, stipitatis in centro thalli, discis apotheciorum atris raro atrobrunneis, substantias lichenium ignotas continente differt.*

Type: China, Xinjiang, Mt. Nan-shan in Tianshan mountain chain, Miaoergou, on limestone, alt. 1280 m, April 10, 2009, A. Abbas & H. Xahidin 20090001 (**holotype** in XJU, **isotype** in HMAS-L).

Etymology: The specific epithet refers to the type locality.

Thallus crustose, 2–11 cm in diam., consisting of numerous peltate areoles of 0.7–3 mm wide and 0.4–0.6 mm thick, much thicker in central part of the thallus, yellowish brown, flat, separated by conspicuous cracks (Fig. 1a, b), with a whitish gray to light gray and very thin prothallus.

Upper cortex well developed, paraplectenchymatous, 50–175 μm thick; algal layer discontinuous (Fig. 1c).

Ascomata apothecia, orbicular to irregular in shape, immersed or somewhat prominent, 0.8–1 mm in diam., numerous, usually 1 per areole, sometimes 2 or occasionally more than 2, zeorine, with both a proper and a thalline margin; thalline margin raised and proper margin not visible when younger;

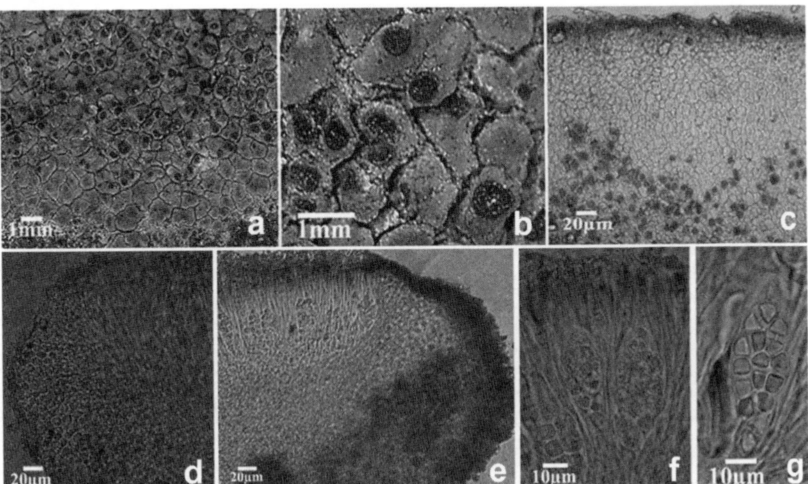

FIG.1. *Caloplaca tianshanensis*: a, b. habit; c. cross section of a peltate areole of the thallus showing the well-developed paraplectenchyma in the upper cortex; d. cross section of an apothecium showing the well-developed paraplectenchyma in the proper exciple; e. cross section of an apothecium showing the double or zeorine margin, with both thalline and proper exciple; f. cross section of the hymenium showing asci containing spores and paraphyses with beaded apices consisting of 2-5 swollen terminal cells; g. an ascus containing 8 spores with thin septa.

proper margin raised and prominent, and thalline margin lower when mature (FIG. 1e); disc dark brown to black, concave, shiny, without or with thin whitish pruina (FIG. 1a, b); hymenium 75–115 μm thick; paraphyses septate, simple, with beaded apices consisting of 2–5 swollen cells (FIG. 1f); asci 44–62 × 12–26 μm, 8-spored; spores broadly ellipsoid, polarilocular, 12–18 × 5–9 μm (FIG. 1f, g); proper exciple paraplectenchymatous (FIG. 1d); hypothecium with gray crystals, 55–90 μm thick.

CONIDIOMATA not seen.

CHEMISTRY: upper cortex K−, C−, epihymenium K−; two unknown substances were detected by TLC: one gives a spot in R_f class 5–6 by solvent systems A, B and G, and in R_f class 6 by solvent system C, grey-brown after charring; the other gives a spot in R_f class 5 by solvent systems A and G, in R_f class 2 by B, and in R_f class 2–3 by C, green after charring.

REMARKS: The new species is similar to *C. peliophylla* in its yellowish brown thallus, but different by the areolate thallus, dark brown to black apothecium discs, the presence of two unknown lichen substances, and the Asian distribution. The latter species differs in its subsquamose thallus with shiny brown apothecia, an American distribution and the absence of lichen substances (Wetmore 1994). In addition, the new species is similar to *C. transcaspica* in its crustose and

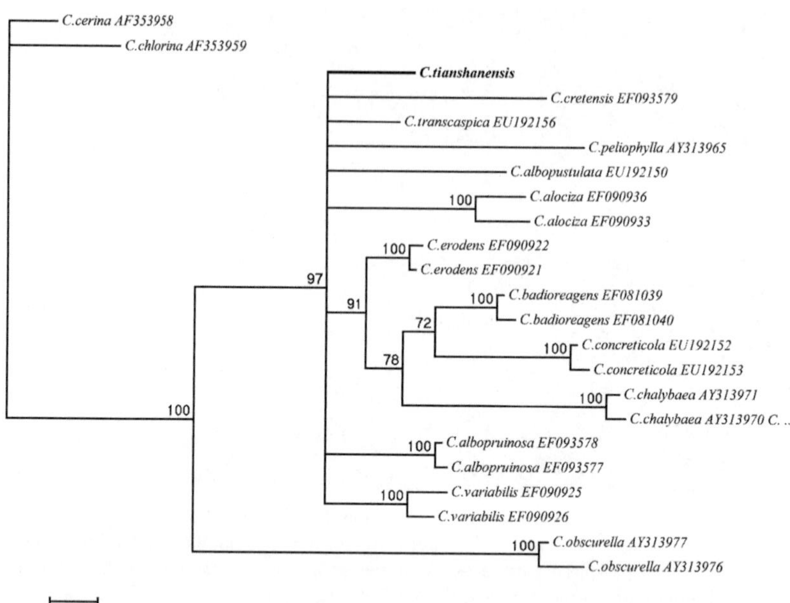

FIG. 2. Consensus tree generated by Bayesian analysis based on ITS region rDNA sequence data. *Caloplaca tianshanensis* groups with species of subgenus *Pyrenodesmia*. Bootstrap support values from 1000 replicates higher than 50% are reported at the nodes. *C. cerina* and C. *chlorina* from subgenus *Caloplaca* were selected as outgroup.

areolate thallus, but differs by its yellowish brown color, dark brown to black discs, smaller ascospores, wider isthmus in cells, and 2–5 swelling terminal cells of the paraphyses.

The ITS sequence of *C. tianshanensis* grouped with those of other 12 related species was retrieved from GenBank as a group belonging to the subgenus *Pyrenodesmia* with 100% bootstrap support. The ITS sequence of the new species *C. tianshanensis* form a distinct clade among the other 11 well recognized related species, such as *C. cretensis, C. transcaspica, C. peliophylla, C. albopustulata*, etc., with 97% bootstrap support. These results show that *C. tianshanensis* is clearly distinct from the above-mentioned well-recognized species (FIG. 2).

Acknowledgments

The first and second authors thank the National Natural Science Foundation of China for the financial support from the grants Nos: 30960003, 30750012. The authors are very grateful to Drs. H. Sipman and W.Y. Zhuang for reading the manuscript and giving valuable comments and suggestions.

Literature cited

Culberson CF. 1972. Improved conditions and new data for the identification of lichen products by a standardized thin-layer chromatographic method. Journal of Chromatography 72: 113–125. doi:10.1016/0021-9673(72)80013-X

Culberson CF, Kristinsson H. 1970. A standardized method for the identification of lichen products. Journal of Chromatography 46: 85–93. doi:10.1016/S0021-9673(00)83967-9

Gardes M, Bruns TD. 1993. ITS primers with enhanced specificity for basidiomycetes - application for the identification of mycorrhizae and rusts. Molecular Ecology 2: 113–118. doi:10.1111/j.1365-294X.1993.tb00005.x

Magnusson HA. 1940. Lichens from Central Asia. Reports from the scientific expedition to the north-western provinces of China under the leadership of Dr. Sven Hedin. The Sino-Swedish Expedition 13.

Magnusson HA. 1944. Lichens from Central Asia. Part II. Reports from the scientific expedition to the north-western provinces of China under the leadership of Dr. Sven Hedin. The Sino-Swedish Expedition 22.

Orange A, James PW, White FJ. 2001. Microchemical methods for the identification of lichens. British Lichen Society, London. doi:10.1639/0007-2745(2003)106[0345:R]2.0.CO;2

Steiner JJ, Poklemba CJ, Fjellstrom RG, Elliott LF. 1995. A rapid one-tube genomic DNA extraction process for PCR and RAPD analyses. Nucleic Acids Research 23(13): 2569–2570. doi:10.1093/nar/23.13.2569-a

Tretiach M, Muggia L. 2006. *Caloplaca badioreagens*, a new calcicolous, endolithic lichen from Italy. The Lichenologist 38(3): 223–229. doi:10.1017/S0024282906005305

Wei JC. 1991. An Enumeration of Lichens in China. International Academic Publishers, Beijing.

Wetmore CM. 1994. The lichen genus *Caloplaca* in North and Central America with brown or black apothecia. Mycologia 86(6): 813–838. doi:10.2307/3760596

White TJ, Bruns T, Lee S, Taylor J. 1990. Amplification and direct sequencing of fungal ribosomal DNA genes for phylogenies. In: PCR protocols: a guide to methods and applications. (M. A. Innis, D H. Gelfand, J. J. Sninsky & T. J. White, eds): 315–322. San Diego: Academic Press. doi:10.1002/mrd.1080280418

Wu JN, Xiang T. 1981. A preliminary study of the lichens from Yuntai mountain in Lianyungang , Jiangsu. Journal of Nanjing Normal College (Natural Science Edition) 3:1–11.

Wunder H. 1974. Schwarzfrüchtige, Saxicole Sippen der Gattung *Caloplaca* (Lichenes, *Teloschistaceae*) in Mitteleuropa, dem Mittelmeergebiet und Vorderasien. Bibliotheca Lichenologica Band 3.

Zahlbruckner A. 1930. Lichenes, in: Handel-Mazzetti, H. (ed.), Symbolae Sinicae III. Julius Springer, Wien.

Zahlbruckner A. 1931. Catalogus Lichenum Universalis 7. Borntraeger, Leipzig.

Zahlbruckner A. 1932. Catalogus Lichenum Universalis 8. Borntraeger, Leipzig.

Phyllobaeis crustacea sp. nov. from China

SHUNAN CAO[1,2], XINLI WEI[1], QIMING ZHOU[1] & JIANGCHUN WEI[1*]

[1]*State Key Laboratory of Mycology, Institute of Microbiology, Chinese Academy of Sciences, No. 1 Beichen West Road, Chaoyang District, Beijing 100101, China*

[2]*University of Chinese Academy of Sciences, No. 19A Yuquan Road, Beijing 100049, China*

*CORRESPONDENCE TO: *weijc2004@126.com*

ABSTRACT — The lichen-forming fungus genus, *Phyllobaeis,* is reported for the first time from China. A new crustose species, *P. crustacea,* is described and illustrated.

KEY WORDS — *Ascomycota, Baeomycetaceae,* chemistry, molecular systematics, morphology

Introduction

The lichen-forming fungus genus *Phyllobaeis* is known from five squamulose species (Gierl & Kalb 1993, Index Fungorum 2013) occurring in the Neotropics. During our studies on the lichen flora of China we encountered a species that showed affinities with this genus but deviated conspicuously by its crustose (rather than squamulose) thallus. We present our analysis of the morphology, anatomy, chemistry, and phylogeny of this species to clarify its taxonomy.

Materials & methods

The material used in this study was collected from Hainan, Yunnan, and Qinghai provinces and Xizang (Tibet) autonomous region of China in 2010 and 2012, and from the Antarctica in 2011. The collections examined are preserved in the lichen section of the Herbarium Mycologicum Academiae Sinicae (HMAS-L). A specimen of *Phyllobaeis imbricata* was borrowed from the former Herbarium Universitatis Amstedamensis, now in Museum Naturalis in Leiden (L). A compound microscope (ZEISS Axioskop 2 plus) and a dissecting microscope (MOTIC SMZ-168) were used for the study of morphology and anatomy. A 10% solution of potassium hydroxide (KOH), a 5% bleaching solution (sodium hypochlorite, NaOCl), concentrated alcoholic *p*-phenylenediamine (PD), Lugol's solution of Iodine, and thin-layer chromatography (TLC) (Culberson & Kristinsson 1970; Culberson 1972; White & James 1985) were used for the detection of lichen substances.

This paper was originally published in *Mycotaxon*, 126: 31-36, 2013.

TABLE 1. Twenty-two nrDNA ITS sequences used in phylogenetic analysis.

SPECIES	LOCALITY; VOUCHER SPECIMEN *	GENBANK NO. ^
Baeomyces placophyllus	Xizang, China; HMAS-L 124223	**KC414621**
	Xizang, China; HMAS-L 124222	**KC414620**
	China; —	DQ001274
Baeomyces rufus	Qinghai, China; HMAS-L 124225	**KC414623**
	Yunnan, China; HMAS-L 124226	**KC414622**
	France; —	AF448457
	France; —	AF448458
Dibaeis absoluta	Hainan, China; HMAS-L 118071	**KC414625**
	Hainan, China; HMAS-L 118073	**KC414626**
Dibaeis baeomyces	—	DQ782844
Dibaeis sorediata	Hainan, China; HMAS-L 118090	**KC414627**
	Hainan, China; HMAS-L 118097	**KC414628**
Icmadophila japonica	Japan; —	AB623070
Phyllobaeis crustacea	Hainan, China; HMAS-L 118086	**KC414614**
	Hainan, China; HMAS-L 118087	**KC414615**
	Hainan, China; HMAS-L 118089	**KC414616**
	Hainan, China; HMAS-L 118095 (holotype)	**KC414617**
	Hainan, China; HMAS-L 118096	**KC414618**
Phyllobaeis imbricata	Carchi, Ecuador; L 0790053	**KC414619**
	—	HQ650635
Placopsis contortuplicata	Antarctica; HMAS-L 124227	**KC414624**
	Antarctica; —	DQ534479

* Missing data indicated with "—". ^ New sequences are shown in bold font.

DNA extraction, PCR amplification and sequencing

The extraction procedure followed the modified CTAB method (Wang et al. 2011). PCR amplifications were performed using a Biometra T-Gradient thermal cycler. The primer pair ITS5 and ITS4 (White et al. 1990) was used to amplify the nrDNA ITS region. Reactions were carried out in 50 μl reaction volume and the components used were 1 μl total DNA, 2 μl each primer (10 μM), 1 μl Taq polymerase (rTaq DNA Polymerase, 5 U/μl), 4 μl dNTP (2.5 mM each), 5 μl amplification buffer (10×, 25 mM $MgCl_2$ contained), 35 μl ddH_2O. Cycling parameters were set to an initial denaturation at 95°C for 5 min, followed by 30 cycles of denaturation at 94°C for 40 s, annealing at 52°C for 40 s, extension at 72°C for 2 min, and a final extension at 72°C for 10 min. Negative control, without DNA template, was prepared in every series of amplification in order to minimize the possibility of contamination. Finally, PCR products were purified by gel purification kit (Biocolor BioScience & Technology Co. Ltd.). Then, PCR products were sequenced using ABI 3730 XL DNA Sequencer.

Altogether 22 nrDNA ITS sequences belonging to nine species were used for the phylogenetic analysis (TABLE 1). Fifteen samples representing seven species were sequenced by the authors, and another seven samples (belonging to six species) were downloaded from GenBank.

Phylogenetic analysis

All sequences were aligned using ClustalW 1.6 (Higgins et al. 1994). The phylogenetic analysis was executed with software Mega5.10 (Tamura et al. 2011). The Kimura-2-parameter was selected as the nucleotide substitution model, and gaps or missing data were set as pairwise deletion. The maximum likelihood (ML) method was used in constructing the phylogenetic tree and the reliability of the inferred tree was tested by 1000 bootstrap replications.

Results

The ML-tree (FIG. 1) of the ITS rDNA sequences shows that the five *Phyllobaeis crustacea* individuals cluster with 100% bootstrap support onto a separate branch, which is most closely related to *P. imbricata*. Together these two species form a common *Phyllobaeis* branch with 79% bootstrap support. *Baeomyces placophyllus* and *B. rufus* cluster together with 98% bootstrap support. *Phyllobaeis* and *Baeomyces* represent *Baeomycetaceae* with 59%

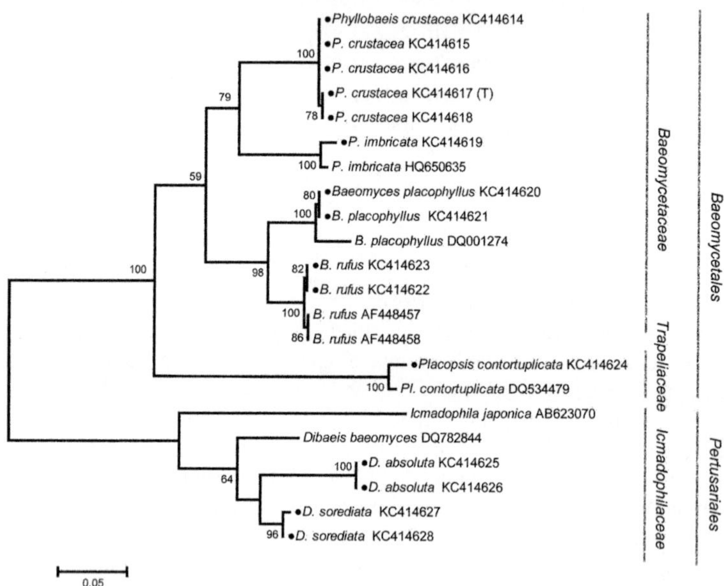

FIG. 1. ML tree based on nrDNA ITS region sequences. The species in the tree marked with "•" were sequenced by the authors. Nucleotide: K2 model, gaps or missing data were partial deletion, bootstrap = 1000. Genetic distance scale = 0.05. Numbers at nodes present the bootstrap support value (numbers <50 not shown).

bootstrap. Meanwhile, the *Baeomycetales,* containing the *Baeomycetaceae* and *Trapeliaceae* (represented by *Placopsis*), is supported by a bootstrap value of 100%, and the *Dibaeis* species (formerly included in *Baeomyces*) are clearly shown to belong to the outgroup, *Icmadophilaceae.*

Summarizing, the ITS rDNA sequences of *P. crustacea* differ significantly from the other species of *Baeomycetaceae* that are most closely related to the *Phyllobaeis* group.

Taxonomy

Phyllobaeis Kalb & Gierl, Herzogia 9:610.1993; emend. S.N. Cao & J.C. Wei

Original diagnosis (Gierl & Kalb 1993: 610): *Genus novum a genere* Baeomyces *differt thallo squamuloso, superne et infra corticato acido norstictico continente; regionibus tropicis distributum.*

Emended diagnosis: Differs from *Baeomyces* by its production of norstictic acid and its tropical distribution.

Phyllobaeis crustacea S.N. Cao & J.C. Wei, **sp. nov.** Fig. 2

Fungal Name FN570052

Differs from the other *Phyllobaeis* species by its crustose thallus.

Type: China, Hainan: Changjiang County, Mt. Bawangling, 19°16' N 109°03' E, alt. 300 m, on rock, 25 Nov. 2010, S. N. Cao CSN047 (**Holotype**, HMAS-L 118095, GenBank KC414617; **Isotype**: HMAS-L 127984).

Etymology: Latin *crustaceus,* referring to the crustose thallus.

Thallus crustose, grayish green, matt, varnish-like, tightly attached to the substrate, forming a patch of 2.5–5 cm in diameter, lacking cortical layers, irregularly delimited; algae layer continuous, algal cells green, ovoid or ellipsoid, single, 5–7.5 × 3.75–5 μm.

Apothecia pale reddish brown to brownish, round and plump, 0.3–0.5mm in diameter, short-stiped, without clear margin, scattered over the thallus; podetia whitish, 0.1–0.5 mm tall, 0.5 mm in diameter, lacking algae; hymenium 112.5–125 μm thick, I–; paraphyses simple, non-septated; asci long-clavate, 8-spored, with apex I–, 80–87.5(–92.5) × 7.5 μm; ascospores oblong or fusiform, hyaline, one-septate, 10–12.5 × 5 μm.

Chemistry: Spot tests: Thallus K+ yellow turning red, C–, KC–, P+ yellow. All specimens contain norstictic acid (TLC).

Additional material examined: CHINA, Hainan: Changjiang County, Mt. Bawangling, 19°16' N 109°03' E, alt. 300 m, on rock, 25 November 2010, S.N. Cao CSN048 (HMAS-L 118096, GenBank KC414618); CSN049 (HMAS-L 118089, GenBank KC414616); CSN050 (HMAS-L 118087, GenBank KC414615); CSN051 (HMAS-L 118086, GenBank KC414614).

FIG. 2. *Phyllobaeis crustacea* (holotype): A, habit; B, apothecia; C, ascus with eight ascospores, apex I– (arrow). Scale bars: A, B = 0.5 mm; C = 10 μm.

Acknowledgments

We are very grateful to Dr. Harrie Sipman and Dr. Robert Lücking for reviewing the manuscript. Our thanks are also given to Drs. Shaun Pennycook, Wen-Ying Zhuang, and Klaus Kalb for valuable and constructive comments on previous versions of this paper and providing an important literature, and to the curator of L who kindly sent a specimen on loan. Special thanks are due to Ms. H. Deng for giving considerable assistance during the studies in HMAS-L. This research was supported by the Ministry of Science and Technology of PRC (2006FY120100) and the Chinese Arctic and Antarctic Administration (2011GW12016).

Literature cited

Culberson CF, Kristinsson H. 1970. A standardized method for the identification of lichen products. Journal of Chromatography 46: 85–93. http://dx.doi.org/10.1016/S0021-9673(00)83967-9

Culberson CF. 1972. Improved conditions and new data for the identification of lichen products by a standardized thin-layer chromatographic method. Journal of Chromatography. 72: 113–125. http://dx.doi.org/10.1016/0021-9673(72)80013-X

Gierl C, Kalb K. 1993. Die Flechtengattung *Dibaeis*. Eine Übersicht über die rosafrüchtigen Arten von *Baeomyces* sens. lat. nebst Anmerkungen zu *Phyllobaeis* gen. nov. Herzogia 9: 593–645.

Index Fungorum. 2013. http://www.indexfungorum.org/Names/Names.asp

Tamura K, Peterson D, Peterson N, Stecher G, Nei M, Kumar S. 2011. MEGA5: Molecular evolutionary genetics analysis using maximum likelihood, evolutionary distance, and maximum parsimony methods. Molecular Biology and Evolution 28: 2731–2739. http://dx.doi.org/10.1093/molbev/msr121

Thompson JD, Higgins DG, Gibson TJ. 1994. CLUSTAL W: improving the sensitivity of progressive multiple sequence alignment through sequence weighting, position-specific gap penalties and weight matrix choice. Nucleic Acids Research. 22: 4673–4680. http://dx.doi.org/10.1093/nar/22.22.4673

Wang YY, Zhang T, Zhou QM, Wei JC. 2011. Construction and characterization of a full-length cDNA library from mycobiont of *Endocarpon pusillum* (lichen-forming *Ascomycota*). World Journal of Microbiology and Biotechnology. 27: 2873–2884. http://dx.doi.org/10.1007/s11274-011-0768-5

White FJ, James PW. 1985. A new guide to microchemical techniques for the identification of lichen substances. British Lichen Society Bulletin 57(Suppl.): 1–41.

White TJ, Bruns TD, Lee SB, Taylor JW. 1990. Amplification and direct sequencing of fungal ribosomal RNA genes for phylogenetics. 315–222, in: MA Innis et al. (eds). PCR protocols: a guide to methods and applications. Academic Press, New York.

Allocetraria capitata sp. nov. (Parmeliaceae, Ascomycota) from China[Δ]

WANG Rui-Fang[1, 2]　WANG Li-Song[3]　WEI Jiang-Chun[1, 2*]

[1] *College of Life Sciences, Shandong Agricultural University, Tai'an, Shandong 271018, China*

[2] *State Key Laboratory of Mycology, Institute of Microbiology, Chinese Academy of Sciences, Beijing 100101, China*

[3] *Key Laboratory of Biodiversity and Biogeography, Kunming Institute of Botany, Chinese Academy of Sciences, Kunming, Yunnan 650204, China*

Abstract: *Allocetraria capitata* is described as new to science from Sichuan Province, China. A diagnosis and a description are provided. Four photographs of the morphology and anatomy of the holotype are also given. The species differs from all known species in the genus by having capitate soralia on the top of lobes.

Key words: soralia, usnic acid, secalonic acid, Sichuan

同岛衣属一新种—粉头同岛衣（梅衣科，子囊菌门）[Δ]

王瑞芳[1, 2]　王立松[3]　魏江春[1, 2*]

[1] 山东农业大学生命科学学院 山东 泰安 271018

[2] 中国科学院微生物研究所真菌学国家重点实验室 北京 100101

[3] 中国科学院昆明植物研究所生物多样性与生物地理学重点实验室 云南 昆明 650204

摘　要：描述了采自中国四川的同岛衣属新种粉头同岛衣，提供了新种的特征提要与描述并附有新种原型的形态及解剖图片共 4 张。新种与该属所有已知种之区别在于其裂片末端具头状粉芽堆。

关键词：粉芽堆，松萝酸，黑麦酮酸，四川

INTRODUCTION

The genus *Allocetraria* Kurok. & M.J. Lai was described with a new species *A. isidiigera* Kurok. & M.J. Lai, and two new combinations *A. ambigua* (C. Bab.) Kurok. & M.J. Lai, *A. stracheyi* (C. Bab.) Kurok.

[Δ]This work was carried out and finished in the State Key Lab of Mycology, IM, CAS, and supported by the National Natural Science Foundation of China (No. KSCXZ-EW-Z-9).

[*]Corresponding author. E-mail: weijc2004@126.com

Received: 11-04-2013, accepted: 19-05-2013

本文原载于《菌物学报》, 33（1）: 19-22, 2014.

& M.J. Lai in 1991. The genus was originally characterized by dichotomously or subdichotomously branched and foliose to suberect or erect lobes with sparse rhizines, angular to sublinear pseudocyphellae, palisade plectenchymatous upper cortex as well as by a unique chemistry. In species of the genus, a small amount of usnic acid is produced in the upper cortex, but never atranorine (Kurokawa & Lai 1991).

Thell *et al.* (1995) described two new species, *A. flavonigrescens* A. Thell & Randlane and *A. sinensis* X.Q. Gao, and proposed three new combinations, *A. denticulata* (Hue) A. Thell & Randlane, *A. globulans* (Nyl. ex Hue) A. Thell & Randlane and *A. oakesiana* (Tuck.) Randlane & A. Thell. The genus containing above mentioned eight species was delimited by the palisade plectenchymatous cortex, asci with a very broad axial body, globose or subglobos ascospores, and filiform pycnoconidia. In the next year, other two new combinations, *A. endochrysea* (Lynge) Kärnefelt & A. Thell, and *A. madreporiformis* (With) Kärnefelt & A. Thell were added to the genus (Kärnefelt & Thell 1996).

The world distribution of nine species excluding *A. denticulata* (Randlane *et al.* 2001), and ten species (Randlane *et al.* 2004) of *Allocetraria* were analyzed.

The phylogentic analyses based on ITS1, 5.8S and ITS2 (Saag *et al.* 2002), combined molecular data (Thell *et al.* 2009), and ITS, mtSSU, nulLSU and RPB1 (Nelsen *et al.* 2011) showed that the genus *Allocetraria* containing *A. ambigua*, *A. flavonigrescens*, *A. globulans*, *A. madreporiformis*, *A. sinensis*, and *A. stracheyi* forms as a monophyletic group within the *Cetraria* clade. However, the phylogentic analysis based on the morphological data sets, including anatomical and chemical data showed that the genus *Allocetraria* was placed in a paraphyletic group (Saag *et al.* 2002).

It is known that the genotype regulates the phenotype, and the phenotype is the end product of the genotype. In other words, "the actual end product, the phenotype, is the result of a complex series of reactions between genes and internal and external environmental factors" (Heywood 1976). Therefore, the inconsistency between phenotype and genotype as a result of phylogenic studies on the genus *Allocetraria* remains to be researched in detail in the future.

During a study of *Allocetraria*, a terricolous species bearing capitate soralia was found in Mountain Que'ershan, Dege County, Northwest Sichuan. It is new to science and described as *Allocetraria capitata*.

1 MATERIALS AND METHODS

The specimen examined was collected by L.S. Wang from Northwest Sichuan in China, and is deposited in Cryptogamic Herbarium of Kunming Institute of Botany, Chinese Academy of Sciences (KUN-L) and Lichen Section of Mycological Herbarium, Chinese Academy of Sciences (HMAS-L). The dissecting microscope (ZEISS Stemi SV 11) and compound microscope (ZEISS Axioskop 2 plus) were available for study of the morphology and anatomy of this species. The lichen substances were detected by color reaction of K (10% KOH), C (5% NaOCl), and Pd (concentrated alcoholic p-phenylenediamine), and also TLC (thin-layer chromatography) (Culberson & Kristinsson 1970; Culberson 1972).

2 DESCRIPTIONS

Allocetraria capitata R.F. Wang, L.S. Wang & J.C. Wei, sp. nov. Fig. 1 A–D

Fungal Name FN570064

Diagnosis: The new species is similar to *A. isidiigera* Kurok. & M.J. Lai in habitus, but different from the latter and all other members in the genus by having capitate soralia on the top of lobes.

Fig. 1 *Allocetraria capitata* sp. nov. (holotype). A: Habit in the field (provided by L.S. Wang); B: Habit in herbarium, scale bar= 1cm (B-1 & arrow indicate the apical lobes with capitate soralia, scale bar= 2mm); C: Cross section of the thallus, scale bar= 20μm; D: Pycnoconidia, scale bar= 20μm.

Type: China. Sichuan Province, Dege County, Manigangge Village, Mt Que'ershan, alt. 4 510m, on the ground, N 31°55', E 98°56', Wang Li-Song *et al.* 07-28259, 30 August 2007 (Holotype: KUN-L 20001; isotype: HMAS-L 125690).

Etymology: The epithet 'capitata' is derived from the Latin word "capitatus", refers to capitate soralia on the top of lobes of this species.

Thallus linear foliose, suberect to erect, 2–4cm high; lobes subfruticose to fruticose, terete, subdichotomous to irregular branched, 250–450μm thick, and 0.6–3mm wide, with capitate soralia of dirty white to yellowish color on the top (Fig. 1B-1), commonly dark brown and pruinose at the apices of lobes those without capitate soralia, with black and papillate pycnidia and isidia-like lobules on the margin; upper surface convex, whitish to yellowish green in the field tending to be straw yellow in herbarium with passage of time, smooth, shiny; lower surface pale yellow to reddish brown, concave, weakly shiny and reticulately lacunose or scrobiculate, with rare marginal pseudocyphellae and simple, black rhizines of 1mm long, restricted to the margin of lobes; medulla yellowish to ochraceous.

Apothecia not seen.

Upper cortex palisade plectenchymatous, 35–45μm; algal layer 25–40μm; medulla 150–350μm (Fig. 1C); lower cortex palisade plectenchymatous to more or less paraplectenchymatous, 25–30μm (Fig. 1C); pycnoconidia colorless, filiform, one end slightly swollen, 10–17×0.5–1.5μm (Fig. 1D).

Chemistry: Cortex: K-, KC+, yellow; medulla:

K-, C-, KC-, P-; usnic acid in the cortex, protolichesterinic and lichesterinic fatty acids and secalonic acid in the medulla.

Ecology and substrate: on the ground, mixed with moss.

Acknowledgements: The authors would be grateful to Dr.Wei X.L. for giving valuable suggestions, and Ms. H. Deng, the keeper and senior Lab technician of the lichen herbarium (HMAS-L) for giving kind assistance during the studies.

[REFERENCES]

Culberson CF, 1972. Improved conditions and new data for the identification of lichen products by a standardized thin-layer chromatographic method. *Journal of Chromatography*, 72: 113-125

Culberson CF, Kristinsson H, 1970. A standardized method for the identification of lichen products. *Journal of Chromatography*, 46: 85-93

Heywood VH, 1976. Plant taxonomy. 2nd edition. Edward Arnorld, London. 1-63

Kärnefelt I, Thell A, 1996. A new classification for the *Dactylina/Dufourea* complex. *Nova Hedwigia*, 62(3-4): 487-511

Kurokawa S, Lai MJ, 1991. *Allocetraria*, a new genus in the Parmeliaceae. *Bulletin of the National Science Museum (Tokyo), Series B*, 17(2): 59-65

Nelsen MP, Chavez N, Sackett-Hermann E, Thell A, Randlane T, Divakar PK, Rico VJ, Lumbsch HT, 2011. The cetrarioid core group revisited (Lecanorales: Parmeliaceae). *Lichenologist*, 43(6): 537-551

Randlane T, Saag A, 2004. Distribution patterns of some primary and secondary cetrarioid species. *Symbolae Botanicae Upsalienses*, 34(1): 359-376

Randlane T, Saag A, Obermayer W, 2001. Cetrarioid lichens containing usnic acid from the Tibetan area. *Mycotaxon*, 80: 389-425

Saag A, Randlane T, Thell A, Obermayer W, 2002. Phylogenetic analysis of cetrarioid lichens with globose ascospores. *Proceedings of the Estonian Academy of Sciences, Biology, Ecology*, 51: 103-123

Thell A, Randlane T, Kärnefelt I, Gao XQ, Saag A, 1995. The lichen genus *Allocetraria* (Ascomycotina, Parmeliaceae). In: Daniels FJA, Schulz M, Peine J (eds.) Flechten Follmann. Contributions to lichenology in honour of Gerhard Follmann. University of Cologne, Cologne, Germany. 353-370

Thell A, Högnabba F, Elix JA, Feuerer T, Kärnefelt I, Myllys L, Randlane T, Saag A, Stenroos S, Ahti T, Seaward MRD, 2009. Phylogeny of the cetrarioid core (Parmeliaceae) based on five genetic markers. *Lichenologist*, 41(5): 489-511

第四章　中国新纪录
Taxa New to China

Studies on the lichen family Cladoniaceae in China III a new genus to China: *Thysanothecium*[①]

WEI Jiang-chun JIANG Yu-mei GUO Shou-yu
Systematic Mycology & Lichenology Laboratory
Institute of Microbiology, Academia Sinica, Beijing 100080

ABSTRACT: The occurrence of the genus *Thysanothecium* Mont. & Berk. of the lichen family Cladoniaceae in China is reported. It is new to the lichen flora of China, with the species *T. scutellatum*. The lichen species of this genus are very rare not only for China, but also for the whole world. So, a short description for the genus and species from China is given.

KEY WORDS: Cladoniaceae; lichens; *Thysanothecium*.

The recent discovery of the lichen genus *Thysanothecium* Mont. & Berk. (Cladoniaceae) in Hainan island has added a new genus to the lichen flora of China. Galloway mentioned in 1985 that this genus was known from China (Galloway, 1985). In fact, it was not known from China until the present paper. It might be a printers error to mistake Indo–China for China (Galloway, 1977; Galloway & Bartlett, 1983). Therefore, the Cladoniaceae in China contain five genera, i. e. *Cladia, Cladina, Cladonia, Gymnoderma* and *Thysanothecium.*

The so called "Swan River Lichen", which was found earliest in Australia, was described by Montagne and Berkeley as a new genus *Thysanothecium* with a new species *T. hookeri* (1846). A new species of *Cladonia, C. scutellata,* was deseribed by Elias Fries (1846) from the Australian collections in the same year. It was neglected until Galloway & Bartlett (1982) and found that *C. scutellata* is an earlier name for another species of *Thysanothecium*, which was traditionally called *T. hyalinum* (Taylor) Nyl. Also the taxa *T. hyalinum* var. *intortum* F. Wils.(1893) from Australia, *T. indicum* Harmand (1928) from Vietnam, *T. casuarinarum* Groenhart (1954) from East Java, and *T. nipponicum* Asahina (1956) from Japan proved to be taxonomic synonyms of *T. scutellatum* (Galloway & Bartlett, 1982). Thus, the genus *Thysanothecium* contains only two species, i. e. *T. hookeri* Mont. & Berk. and *T. scutellatum* (Fr.) D. Galloway. Among them, the former is distributed only in Australia, Tasmania, and New Zealand, but the latter is known from New Caledonia, Java, Vietnam and Japan, in addition to Australia, Tasmania, and New Zealand.

Thysanothecium Mont. & Berk.
Hook. Lond. J. Bot. 5:257(1846).
Type species: *Thysanothecium hookeri* Mont. & Berk.

①The project supported by the National Science Foundation.

This paper was originally published in *Mycosystema*, 7: 23-27, 1994.

Primary thallus crustose to squamuliform, consisting of small granules and squamules, which are 0.05–2 mm in diam., scattered or closely imbricate, adnate or ascending at morgins, and often finely dissected to coralloid or occasionally dissolving into a sorediate mass.

Pseudopodetia developing from primary thallus, erect, cylindrical near base, becoming flattened above, and expanded terminal umbrella-shaped to fan-shaped apothecia; cortex continuous to longitudinally fissured, smooth to warty, like toad skin.

Apothecia regularly terminal and umbrella- shaped (disc surface upwards) or often fan-shaped (disc surface downwards or facing a side); disc surface plane or convex, cerinous to brownish; asci clavate, octosporous; ascospores hyaline, ellipsoidal to bacilliform, simple or occasionally 1–septate, straight or curved.

Pycnidia black.

Phycobiont: *Trebouxia* sp.

Chemistry: usnic, barbatic and divaricatic acids.

Thysanothecium scutellatum (Fr.) D. Galloway (Pl. I, a–f; Fig. 1. a & b)

Nova Hedwigia 36: 390 (1982).

≡*Cladonia*? *scutellata* Fr., in: Lehmann, Plantae Preissianae **2**:141 (1846).

= *Thysanothecium hyalinum* (Taylor) Nyl., Mem. Soc. Sci. Nat. Cherbourg **5**:94 (1858) ('1857').

≡*Baeomyces hyalinus* Taylor, Hook. Lond. J. Bot. **6**:187(1847).

= *Thysanothecium hyalinum* f. *intortum* F. Wils., Proc. Roy. Soc. Vict. n.s. **5**:176(1893).

= *Thysanothecium hyalinum* f. *squamulosum* F. Wils., ibid. **5**:176(1893)

= *Thysanothecium indicum* Harmand, Ann. Crypt. Exot. **1**:321(1928).

= *Thysanothecium casuarinarum* Groenhart, Reinwardtia **2**:388(1954).

= *Thysanothecium casuarinarum* ssp. *nipponicum* (Asahina) Asahina, J. Jap. Bot. **32**:35(1957).

≡*Thysanothecium nipponicum* Asahina, J. Jap. Bot. **31**:65(1956).

Primary thallus crustose to squamuliform, consists of small granules and squamules, which are usually smaller than 0.3 mm or about 1 mm in diam., finely dissected at margins, scattered or often closely imbricate, adnate or ascending at margins, greenish grey with a yellowish green tint.

Pseudopodetia developing from primary thallus , erect, cylindrical, 0.5–1.5 mm in diam., 5–7 mm tall, often becoming flattened above; cortex developed well and on the surface with longitudinal nerves, fissions, and warts, like toad skin, grey-green to yellowish green, in some areas lacking algal layer and appearing cerinous and cartilaginous; towards the tip expanded to form apothecia.

Apothecia umbrella-shaped (disc surface upwards), fan-shaped (disc surface towards a side or downwards); disc surface plane to convex, cerinous to brownish; epithecium brownish, 2.5–3.5 μm tall; hymenium hyaline, 23.5–32.5 μm tall; Paraphyses simple, septate, <1 μm in diam.; hypothecium 9–14 μm tall; asci clavate, 18–27 × 7– 15 μm, octosporous; ascospores hyaline, ellipsoidal to bacilliform, 6.5–10.5 × 2–3.5 μm.

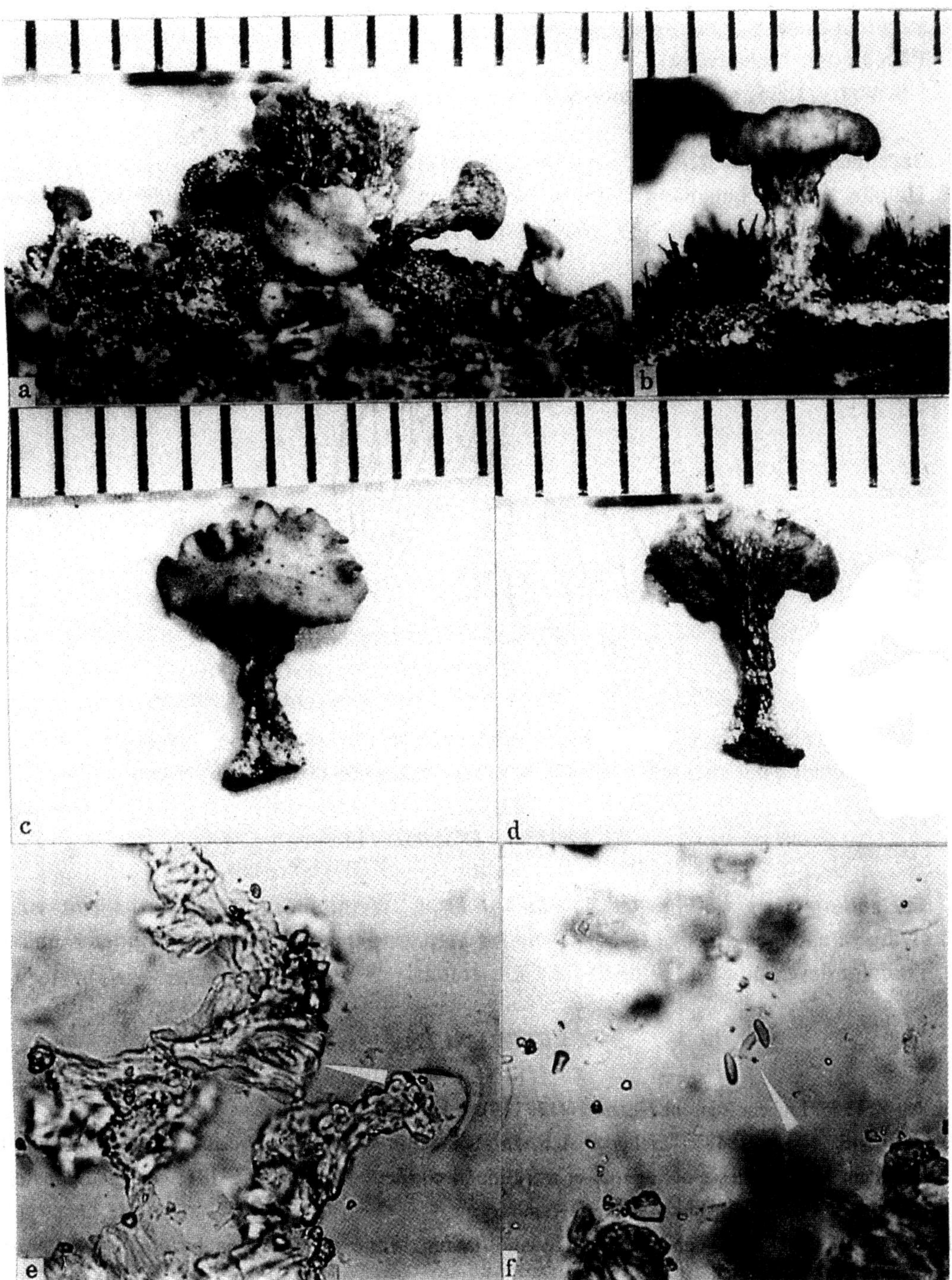

Plate I *Thysanothecium scutellatum*(Fr.) D. Galloway (the scale in mm) a. Several pseudopodetia with apothecia arising from primary thalli on rotten wood. b. A pseudopodetium with an umbrella-shaped apothecium. c. A pseudopodetium with a fan-shaped apothecium (the disc surface of the apothecium). d. A pseudopodetium with a fan-shaped apothecium (the back of the apothecium). e. An ascus containing ascospores (arrow indicates position of the ascus). ×40 f. The three ascospores discharged from asci (arrow indicates position of the ascospores). ×40

Pycnidia black.

Phycobiont: *Trebouxia* sp.

Chemistry: Usnic and divaricatic acids.

Specimens Examined:

Hainan Prov., Changjiang County, Bawangling, on rotten wood, alt.1000 m, 23 March 1993, coll. Jiang Y. M. & Guo S. Y. Nos. H−303, H−305, and H−306 **(HMAS−L)**.

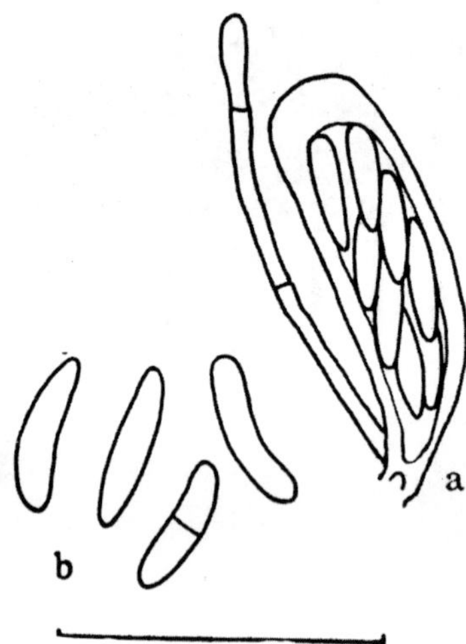

Fig. 1 An ascus and ascospores of *Thysanothecium scutellatum* (Fr.) D. Galloway (the scale in 20 μm). a. An ascus containing ascospores with a paraphysis. b. Four ascospores discharged from asci.

ACKNOWLEDGEMENTS

The authors are grateful to Dr. T. Ahti and Dr. C. Wetmore for reviewing the manuscript and giving valuable comments; to Ms. Deng for preparing the manuscript in computer; and to Ms. Yuan for developing and printing the photographs.

REFERENCES

Asahina, Y. 1956. Lichenologische Notizen (117−119). J. Jap. Bot. **31**:65−70

Fries, E. 1846. " 1846−1847" . Lichenes Ach. In: Lehmann, C. Plantae Preissianae sive enumeratio plantarum in Australasia occidentali et meridionali-occidentali annis 1834−1841 collegit Ludovicus Preiss **2**:140−145. Sumptibus Meissnerii Hamburgi.

Galloway, D. J. 1977. The lichen genus Thysanothecium Mont. & Berk., an historical note. Nova Hedwigia **28**:499−513.

Galloway, D. J. 1985. Flora of New Zealand, Lichens, pp.1−662.

Galloway, D. J., and Bartlett, J. K., 1982. The lichen genus Thysanothecium Mont. & Berk., in New Zealand. Nova Hedwigia **36**:381−398.

Groenhart, P. 1954. Malaysian lichens IV. Reinwardtia **2**:385−402.

Harmand, J. 1928. Lichens d'Indo−Chine recueillis par M. V. Demange (introduction par Dr. J. Maheu). Ann. Crypt. Exot. **1**:319−337.

Montagne, J. F. C. & Berkeley, M. J. 1846. On Thysanothecium, a new genus of-lichens. Hook. Lond. J. Bot. 5:257–258.

Nylander, W. 1858. "1857" Enumeration generale des lichens avec l'indication sommaire de leur distrbution geographique. Mem. Soc. Sci. Nat. Cherbourg 5:332–339.

Wilson, F. R. M. 1893. The lichens of Victoria. Part I. Proc. Roy. Soc. Vict. n. s. 5:141–177.

Studies on the lichen family Cladoniaceae in China Ⅳ species of *Cladonia* new to China（Ascomycota）

GUO Shou-yu WEI Jiang-chun
Systematic Mycology & Lichenology Laboratory
Institute of Microbiology, Academia Sinica, Beijing

ABSTRACT: Twelve species of the lichen genus *Cladonia* in China are reported in the present paper. Among them one is new to science: *Cl. gongganensis* S.Y.Guo & Wei, and eleven are new to China: *Cl. bacilliformis*, *Cl. beaumontii*, *Cl. clavulifera*, *Cl. cylindrica*, *Cl. farinacea*, *Cl. hokkaidensis*, *Cl. incrassata*, *Cl. macrophylla*, *Cl. mateocyatha*, *Cl. multiformis*, and *Cl. subrangiformis*. The chemistry of each species was examined by TLC, and the distribution is also given.

KEY WORDS: *Cladonia*; *Cl. gongganensis*; *New*; *Chemistry*; *Distribution*; China

Eighty-three species which contain five subspecies, twenty-seven varieties and forty-one forms have been reported from China (Wang-Yang JR & Lai MZ 1973, 1976; Ahti & Lai 1979; Liu TS et al. 1980; Chen XL et al. 1981; Wang XY, 1985; Wei JC & Jiang YM, 1986; Chen JB et al. 1989; Wei 1991). In the course of studies on the lichen family *Cladoniaceae* from Mount Hengduan region in south-western China, we found fifty-five species of *Cladonia*. Among them one species, *Cl. gongganensis* S.Y. Guo & Wei, is new to science, and eleven species are new to China. All of the specimens examined were tested using thin-layer chromatography (TLC) in addition to chemical spot tests and morphological studies.

Geographical analysis of these species shows as follows: one species (*Cl. hokkaidensis*) is endemic to East Asia; four species (*Cl. clavulifera*, *Cl. cylindrica*, *Cl. farinacea* and *Cl. mateocyatha*) are East Asia-North American disjunction (Yoshimura, 1968); one (*Cl. multiformis*) is East Asia – North America – South African species; three (*Cl. bacilliformis*, *Cl. beaumontii* and *Cl. macrophylla*) are holoarctic; one (*Cl. subrangiformis*) is Eurasian-American and one (*Cl. incrassata*) cosmopolitan as well.

1 *Cladonia bacilliformis* (Nyl.) DT. & Sarnth

Flecht-tirol, Vorarlberg und Lichstenstein. (1902).

≡ *Cladonia carneola* var. *bacilliformis* Nyl., in Nyl. et Saelan. Herb. Musei Fenn. 79(1859).

Primary squamules persistent, small. Podetia short, approximately 12 mm tall, unbranched, copless or with narrow cups; the margins of the cups with slender proliferations; abundant

This paper was originally published in *Mycosystema*, 7: 29-35, 1994.

farinose soredia; esquamulose. podetia yellow. Apothecia small, pale brown. Pycnidia brown, on the tips of podetia.

Chemistry: Cortex K－, KC＋yellow, P－. Containing usnic and barbatic acids.

Substrate: rotten wood and rocks.

Specimens examined: XIZANG: Zuogong, 4500 m, Oct.1982, Su JJ No.5347 (HMAS-L). YUNNAN: Zhongdian, 3650 m, Aug.1981, Wang XY et al No.5929 (HMAS-L).

Holoarctic species.

Comments:

This species has previously been repoted as *Cladonia cyanipes* (Sommerf.) Nyl. by Wei & Chen (1974), Wei (1981) and Wei & Jiang (1986).

The podetia of this species are usually longer, but some shorter of the Chinese collections. It is different from *Cl. cyanipes* by the closed bottom of the scypha in these Chinese specimens.

2. *Cladonia beaumontii* (Tuck.) Vain.

Acta Soc. F. Fl. Fenn. 10:455(1894).

≡ *Cladonia santensis* b. *beaumontii* Tuck., Syn. N. Am. Lich. 1:245(1882).

Primary squamules persistent. Podetia cupless; cylindrical, the surface bearing numerous squamules; esorediate. Apothecia dark brown; borne at the tips of the podetia. Pycnidia brown, also borne at the tips of podetia.

Chemistry: Cortex K－, KC－, P＋yellow; UV＋blue. Containing baeomycesic and squamatic acids.

Substrate: ground and decayed wood.

Specimens examined: YUNNAN: Deqin, 4230 m, Oct. 1982, Su JJ No. 5755 (HMAS-L); Zhongdian, 3350－3700 m, Aug.1981, Wang XY et al Nos. 5197, 5906 (HMAS-L).

Holoarctic species.

3. *Cladonia clavulifera* Vain.

in Robb., Rhodora 26: 145 (1924).

Primary squamules persistent, small. Podetia cylindrical, esorediate, with sparse squamules. Apothecia dark brown, borne at the tips of the podetia. Pycnidia on the primary squamules.

Chemistry: Cortex K－, KC－, P＋red. Containing fumarprotocetraric acid.

Substrate: rotten wood and mosses over rocks.

Specimens examined: SICHUAN: Mt. Gongga, 3200 m, July 1982, Wang XY et al No. 8583 (HMAS-L). XIZANG: Zayu, 3200 m, Oct. 1982, Su JJ No. 5062 (HMAS-L). YUNNAN: Gongshan, 2300 m, Sept. 1982, Su JJ No. 4118 (HMAS-L); Weixi, 3350 m, May 1982, Su JJ No. 0426 (HMAS-L); Zhongdian, 3700 m, Aug. 1981, Wang XY No. 5188 (HMAS-L).

East Asia－North American disjunctive species.

4. *Cladonia cylindrica* (Evans) Evans

Rhodora 52: 116 (1950).

≡ *Cladonia borbonica* f. *cylindrica* Evans, Trans. Conn. Acad. Arts. & Sci. 30:482 (1930).

Primary squamules persistent or disappearing early. Podetia cylindrical and unbrached, cupless, the tips very abruptly contracted and acute; the surface with farinose soredia. Apothecia and pycnidia brown, borne at the tips of the podetia.

Chemistry: Cortex K – , KC – , P + red. Containing fumarprotocetraric and grayanic acids.

Substrate: rotten wood.

Specimens examined: YUNNAN: Zhongdian, 3500 m, Aug. 1981, Wang XY et al No. 6092 (HMAS–L).

East asia – North American disjunctive species.

5. *Cladonia farinacea* (Vain.) Evans

Rhodora 52:95 (1950).

≡ *Cladonia furcata* var. *scabriuscula* f. *farinacea* Vain., Acta Soc. F. Fl. Fenn. 4:339 (1887).

Primary squamules persistent. Podetia slightly branched, approximately 55 mm tall, the upper part with soredia, and the basal part with squamules; podetia with interior of the cups opening into hollow podetium. Apothecia and pycnidia brown, well developed.

Chemistry: Cortex K – , KC – , P + red. Containing fumarprotocetraric acid. Substrate: soil over rocks.

Specimens examined: YUNNAN: Weixi, 3300 m, May 1982, Su JJ No. 0433 (HMAS–L); Zhongdian, 3650 m, Aug. 1981, Wang XY et al No. 5832 (HMAS–L).

East Asia – North American disjunctive species.

6. *Cladonia gonggaensis* S.Y. Guo & Wei sp. nov. (Fig. 1)

Cl. pocillo affinis sed podetio phyllocladiis orbiculatis obtecti et acidum psoromicum et thamnolicum continenti notabilis.

Typus: CHINA Sichuan province, Mount Gongga, Alt. 3750 m, July 30, 1982, Coll. Wang XY et al No. 9189 (HMAS–L – holotypus).

Primary squamules persistent, middle-sized, 10 mm long and 4 mm wide. the lower surface white, and the upper surface yellow-brown.

Podetia growing from the upper surface of the primary squamules and developing into broad cups, cortex bearing small squamules, esorediate.

Apothecia and pycnidia brown, on the margins of the cups.

Chemistry: Cortex K + yellow, KC – , P + yellow. Containing psoromic and thamnolic acids.

Substrate: the ground and rotten wood.

Specimens examined: SICHUAN: Mt. Gongga, 3600 m, Aug.1982, Wang XY et al No. 9297 (HMAS–L). YUNNAN: Deqin, 3500 m, Aug. 1981, Su JJ Nos. 5720, 5724, 5727 (HMAS–L).

Comments:

The new species in habit is similar to *Cl. pocillum* (Ach.) O.J. Rich but from which different by that the podetium is covered with orbicular phyllocladia, and contains psoromic and thamnolic acids. This combination of lichen substances is very rare in *Cladonia*.

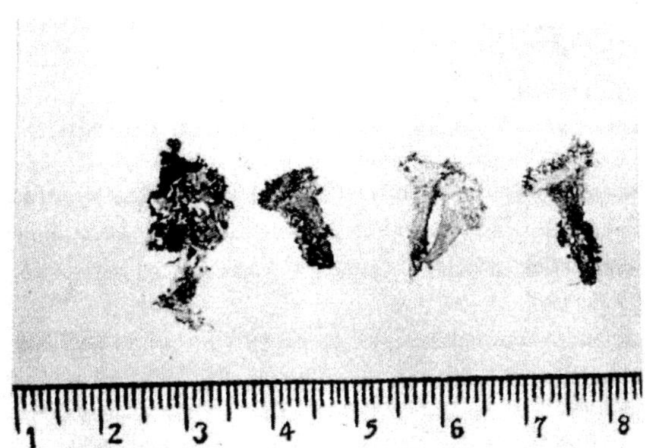

Fig. 1 Thallus of *Cladonia gonggaensis* S.Y. Guo & Wei sp. nov. (holotype)

7. *Cladonia hokkaidensis* Asah.
ATLAS of Japanese Cladoniae, 10(1971).

Primary squamules persistent, middle-sized. Podetia cylindrical, esorediate, with sparse squamules. Apothecia dark brown, borne at the tips of the podetia. Pycnidia on the primary squamules.

Chemistry: Cortex K −, KC + yellow, P −. Containing barbatic, usnic and 4-o-demethylbarbatic acids.

Substrate: ground.

Specimens examined: SICHUAN: Muli, 3800 m, June 1982, Wang XY et al No. 7986 (HMAS-L). YUNNAN: Fugong, 3850 m, May 1982, Su JJ No. 0885 (HMAS-L); Gongshan, 3600 m, July 1982, Su JJ No. 2605 (HMAS-L).

Endemic to East Asia.

8. *Cladonia incrassata* Florke
in E. Fries, Nov. Sched. Crit. 20(1826).

Primary squamules persistent, middle-sized. Podetia poorly developed, yellow. Apothecia and pycnidia red, borne at the tips of the podetia.

Chemistry: Cortex K −, KC + yellow, P −; UV + blue. Containing squamatic and usnic acids, zeorin.

Substrate: ground.

Specimen examined: SICHUAN: Mt. Gongga, 1600 m, June 1982, Wang XY et al No. 8104 (HMAS-L).

Cosmopolitan species.

9. *Cladonia macrophylla* (Schaer.) Stenham.
Ofvers Kungl. Vet-Akad. Forhandl. (4):231(1865).

Primary squamules persistent. Podetia cupless, subulate, up to 15 mm tall, approximately 1 mm diameter; simply branched, with numerous squamules. Apothecia and pycnidia brown, borne at the tips of the podetia.

Chemistry: Cortex K −, KC −, P + yellow. Containing psoromic acid. Substrate: rotten wood and mosses over rocks.

Specimens examined: YUNNAN: Gongshan, 1400 m, Aug. 1982, Su JJ Nos. 2818, 3042, 3179 (HMAS–L); Zhongdian, 3700 m, Aug. 1981, Wang XY et al No. 6064 (HMAS–L).

Holoarctic species.

10. *Cladonia mateocyatha* Robb.
Rhodora 27:50 (1925).

Primary squamules persistent or disappearing early, middle-sized. Podetia short and stout, forming very irregular cups and the cups forming proliferations marginally; podetia esorediate, sometimes with squamules, little aerolates covering on the surface. Apothecia small, brown, borne at the tips of podetia. Pycnidia brown, borne at the tips of podetia.

Chemistry: Cortex K −, KC −, P + red. Containing fumarprotocetraric acid.

Substrate: rotten wood and mosses over rocks.

Specimens examined: YUNNAN: Weixi, 3300–3500 m, May 1982, Su JJ Nos. 0451, 0816 (HMAS–L); Zhongdian, 3500 m, Aug. 1981, Wang XY et al No. 5217 (HMAS–L).

East Asia – North American disjunctive species.

11. *Cladonia multiformis* Merr.
Bryologist 12:1 (1909).

Primary squamules persistent or disappearing early. Podetia with cups, and the cups with inner sieve-like perforated membranes. Podetia branching by dichotomies, cortex covering with small squamules, esorediate. Apothecia dark brown, borne at the tips of the podetia. Pycnidia dark brown, on the tips of the branchlets or the margins of the cups.

Chemistry: Cortex K −, KC −, P + red. Containing fumarprotocetraric and ursolic acids.

Substrate: rotten wood and ground.

Specimens examined: XIZANG: Zayu, 3200–3700 m, Sept. 1982, Su JJ Nos. 4486, 4795 (HMAS–L). YUNNAN: Zhongdian, 3350–3700 m, Aug. 1981, Wang XY et al Nos. 5248, 5657, 5759, 5902, 5960 (HMAS–L).

East Asia – North America – South African species.

Comments:

This species was usually considered as endemic to America. But it has been reported from South Africa by des Abbayes (1938) and from Japan by Yoshimura (1968). Therefore, its known range is extended to Africa and Asia.

12. *Cladonia subrangiformis* Sandst.

Abh. Nat. Ver. Bremen. 25:165(1922).

Primary squamules disappering early. Podetia cylindrical, cupless, branching by dichotomies or whorls, the axils open. Podetia covering small squamules. Apothecia and pycnidia brown, borne at the tips of the branchlets.

Chemistry: Cortex K + yellow, KC − , P + red. Containing fumarprotocetraric acid and atranorin.

Substrate: rotten wood, rocks, ground, soil and mosses over rocks.

Specimens examined: XIZANG: Zayu, 3100–3400 m, Sept. 1982, Su JJ Nos. 4504, 4586, 4598 (HMAS–L); Zuogong, 2900–4300 m, Oct. 1982, Su JJ Nos. 5451, 5474, 5513 (HMAS–L). YUNNAN: Fugong, 3400–3500 m, May 1982, Su JJ Nos. 0850, 1028 (HAS–L); Weixi, 2900–3450 m, May 1982, Su JJ Nos. 0473, 0483 (HMAS–L); Aug. 1981, Wang XY et al No. 4480 (HMAS–L); Zhongdian, 3600 m, Aug. 1981, Wang XY et al No. 5758 (HMAS–L).

Eurasian-American species.

ACKNOWLEDGEMENTS

We would like to express our thanks to Prof. Clifford Wetmore for his reviewing the manuscript and some helpful suggestions on this paper.

REFERENCES

Ahti T, Lai MJ (1979) The lichen genera *Cladonia*, *Cladina* and *Cladia* in Taiwan. *Ann. Bot. Fennici* **16**:228–236

Chen JB, Wu JN, Wei JC (1989) Lichens of Shennongjia, in Fungi and Lichens of Shennongjia, World Publishing Corp., Beijing, pp.386–493 (in Chinese)

Chen XL, Zhao CF, Luo GY (1981) A list of lichens in northeastern China. *J. NE Forestry Inst.* **3**:127–135 & **4**:150–160 (in Chinese)

Culberson CF (1972) Improved conditions and new data for the identification of Lichen products by a standardized thin-layer chromatographic method. *Journ. Chromatogr.* **72**:113–121

Liu TS, Lai MJ, Lin HS (1980) The Cladoniform Lichens in Taiwan. Quart. *Journ. Taiwan Mus.* **33**:1–35

Thomson JW (1967) The Lichen Genus *Cladonia* in North America, Toronto Univ. Press, Toronto

Thomson JW (1984) American Arctic Lichens Ⅰ. The Macrolichens, Columbia Univ. Press, New York

Wang XY (1985) The Lichens of the Mt. Tuomuer Region in Tianshan. in Fauna and Flora of the Mt. Tuomuer Region in Tianshan, p.328. People Publishing House of Xinjiang

Wang-yang JR, Lai MJ (1973) A Checklist of the Lichens of Taiwan. *Tanwania* **18**(1):83–104

Wang- yong JR, Lai MJ (1976) Additions and Corrections to the Lichen Flora of Taiwan. *Taiwania* **21**(2):226

Wei JC (1981) A Biogeographical Analysis of the Lichen Flora of Mt. Qomolangma Region in Xizang. Proceedings of Symposium on Qinghai- Xizang (Tibet) Plateau, Beijing, Vol.Ⅱ: Environment and Ecology of Qinghai-Xizang plateau, p.1145

Wei JC (1991) An Eumeration of the Lichens in China, International Academic Publishers, Beijing

Wei JC, Chen JB (1974) Materials for the Lichen Flora of the Mt. Qomolangma Region in Southern Xizang, China's Report on the Scientific Investigations (1966−1968) in Mt. Qomolangma District, ser. Biology & Physiology, p.174. Science Press, Beijing (in Chinese)

Wei JC, Jiang YM (1986) Lichens of Xizang, Science Press, Beijing (in Chinese)

Yoshimura I (1968) The phytogeographical relationships between the Japanese and North American species of *Cladonia*, *J. Hattori Bot. Lab.* **31**:227−246

中国石蕊科地衣研究之四
石蕊属地衣新种及中国新记录

郭守玉 魏江春

中国科学院微生物研究所真菌地衣系统学开放研究实验室

北 京

摘要：本文报道了中国横断山地区石蕊属地衣 12 种，其中新种 1 个：贡嘎石蕊（*Cladonia gongganensis* S.Y. Guo & Wei），中国新记录 11 种：类黄粉石蕊（*Cl. bacilliformis*），比蒙氏石蕊（*Cl. beaumontii*），小棍棒石蕊（*Cl. clavulifera*），圆筒石蕊（*Cl. cylindrica*），具粉石蕊（*Cl. farinacea*），北海道石蕊（*Cl. hokkaidensis*），厚叶石蕊（*Cl. incrassata*），大叶石蕊（*Cl. macrophylla*），丛杯石蕊（*Cl. mateocyatha*），多孔石蕊（*Cl. multiformis*）和亚石蕊（*Cl. subrangiformis*）。对于每一种的化学成分及地理分布进行了测定和分析。

关键词：石蕊属；贡嘎石蕊；新种；新记录；化学；分布；中国

A lichen genus *Brodoa*, new to China in Parmeliaceae

WEI Jiang-Chun JIANG Yu-Mei

(*Systematic Mycology & Lichenology Laboratory, Academia Sinica, Beijing* 100080)

ABSTRACT: A lichen genus *Brodoa* with a species *B. oroarctica* (Krog) Goward in the Parmeliaceae (Lecanorales, Ascomycota) is reported from Altai Mountains in Xinjiang of the northwestern China. The discovery of the genus from Xinjiang is new to China. Diagnostic notes of the genus and the species collections from China are given in this paper.

KEY WORDS: *Brodoa*, *Brodoa oroarctica*, lichen, China

While preparing the lichen flora of China we discovered an unusual lichen genus *Brodoa* with a species *B. oroarctica* in the Parmeliaceae from the northwestern China, Xinjiang Autonomous Region in the Altai Mountains. The genus *Brodoa* was segregated by Goward (1986) from the *Hypogymnia*, and is equivalent to *H. intestiniformis* complex (Krog, 1974). None of which was previously reported from China (Wei, 1891).

More than 1700 lichen species belonging to 233 genera from China have been reported so far (Wei, 1991; Wei et al., 1994). The two hundred and thirty fourth genus *Brodoa* Goward in the Parmeliaceae from China is reported in this paper.

1 MATERIALS AND METHODS

This study is based on the collections from the Altai Mountains in Xinjiang Autonomous Region of northwestern China, preserved in HMAS-L. The specimens were subjected to the standardized thin-layer chromatography (TLC) in accordance with the methods described by Culberson and Kristinsson (1970), Culberson (1972), and White and James (1985). The research microscope (OPTON), and dissecting microscope (OPTON) for morphological study were available.

2 RESULTS AND DISCUSSIONS

After a careful examination in morphology, anatomy, and chemistry of the specimens, collected from Xinjiang, it shows that they were identified with *Brodoa oroarctica*, an arctic-northern boreal, circumpolar species (Krog, 1974).

Brodoa Goward, Bryologist 89:222(1986).

= *Parmelia* subg. *Hypogymnia* tax. vag. ('Grup') *Solidae* Bitter, Hedwigia 40:251(1901).

≡*Hypogymnia* sect. *Solidae* (Bitter) Dahl, Rev. Bryol. Lichenol. 21:127(1952).

Type species of the genus is *Brodoa oroarctica* (Krog) Goward.

The genus *Brodoa* was segregated from the genus *Hypogymnia* by soild and compact lobes; upper cortex paraplectenchymatous and covered in a polysaccharide-like layer; lower cortex palisade

Foundation item: supported in part by the National Natural Science Foundation of China (NNSFC 39391800).

Received: 1999-01-26

This paper was originally published in *Mycosystema*, 18（4）: 445-448, 1999.

plectenchymatous and lacking polysaccharide-like layer covering; growing on rocks only, and having an arctic-alpine distribution. It is also distinguished from the related genus *Allantoparmelia* by palisade plectenchymatous structure of lower cortex, lacking polysaccharide-like layer, and present of cortical atranorin, and from the genus *Cavernularia* by lacking cavernulae below and larger and ellipsoidal spores.

New to China.

Brodoa oroarctica (Krog) Goward (Figs.1-3)

Bryologist 89:222(1986).

≡*Hypogymnia oroarctica* Krog, Lichenologist 6:136(1974).

Type: Spitsbergen, Lomfjorden, Dvergbreen, 14 August 1931 P. F. Scholander (O, not seen).

Thallus foliose, consists of terete, soil, nodulouse, and branched lobes, 0.5-1mm diam., irregularly spreading, loosely attached to the substrate; upper surface pale grey to ash grey or brown to dark brown or in parts black, maculate, areolate, with cracks and lines; lower surface black, brown with a rose tint covered with thin layer to white pruina towards the apices.

Apothecia not found in specimens from Xinjiang.

Chemistry: Cortex of thallus K+ yellow; medulla K−, KC− redish, P−. Contains atranorin and physodic acid.

Specimens Examined: China, Xinjiang Autonomous Region, Altai Mountains on rocks, alt. 2500m, Aug. 5, 1986, coll. X. Q. Gao nos. 2087, 2146 (HMAS-L).

New to China.

Fig.1. Habit of *B. oroarctica* (Krog) Goward. lower surface of a thallus (above left), upper surface of 3 thalli (above right and below) (scale in mm).

Fig.2. *B. oroarctica*. upper surface of a thallus (scale in mm).

The species is characterized by an irregularly spreading thallus consisting of separate, terete, nodulous lobes, and loosely attached to rocks; the mdulla reacts PD−. It contains physodic acid in addition to cortical atranorin. In contrast to the species in question *B. intestiniformis* (Vill.) Goward is characterized by its numerous flattened secondary lobes, and always contains fumarprotocetraric acid in addition to cortical atranorin. Another species *B. atrofusca* (Schaer.) Goward also contains physodic acid, but morphologically has orbicular thalli adnate to rocks, with contiguous peripheral lobes. The medulla reacts PD+ orange, Both the *B. intestiniformis* and *B. atrofusca* have an arctic-alpine distribution restricted mainly within Europe. In contrast to the above mentioned two species, *B. oroarctica* has an arctic-northern boreal, circumpolar distribution. Outside of the arctic and northern boreal regions it is restricted to high mountains. It has its southern limit in the south central Scandinavian mountains in Europe (Krog. 1974). It is known from the Rocky Mountains to New Mexico in western North America (Hasselrot, 1953; Ohlsson, 1973), and from White Mountains of New Hampshire and the Adirondack Region of New York in eastern North America (Ohlsson, 1973).

Fig.3. *B. oroarctica*. Lower surface of a thallus (scale in mm).

The distribution of this species in Asia has not been recorded so far. Whether the record of *Hypogymnia intesiniformis* and *H. encausta* in Asia (Kopaczevskaja *et al.*, 1971) and that of *H. intestiniformis* in Mongolia (Golubkova, 1981) is that of *Brodoa oroarctica* remains a question. This problem can not be clarified until the above mentioned collections from Asia are checked.

[REFERENCES]

Culberson C F, 1972. Improved conditions and new data for the identification of lichen products by a standardized thin-layer chromatographic method. *J. Chromat.* 72:113~125.

Culberson C F, Kristinsson H, 1970. A standardized method for the identification of lichen products, *J. Chromat.* 46:85~93.

Golubkova N S, 1981. Conspect Flori Lishainikov Mongolskoi Narodnoi Respubliki (in Russian) (Conspect of Lichen Flora from the People's Republic of Mongolia, Leningrad 〈Science〉 press, pp. 1~200.

Goward T, 1986. *Brodoa*, a new lichen genus in the Parmeliaceae. *Bryologist* 89:219~223.

Hasselrot T E, 1953. Nordliga lavari Sydoch Mrllansverige. *Acta Phytogeogr. Suec.* 33:1~200.

Kopaczevskaja E G, *et al.*, 1971. Opredelitele Lishainikov SSSR (in Russian)(Handbook of the lichens of the U. S. S. R, Leningrad 〈Science〉 press, 1:1~410.

Krog H, 1974. Taxonomic studies in the *Hypogymnia intestiniformis* complex. *Lichenologist* 6:135~140.

Ohlsson K E, 1973. New and interesting macrolichens of British Columbia. *Bryologist* 76:366~387.

Wei J C, 1991. An Enumeration of Lichens in China. Beijing: International Academic Publishers. pp.1~278.

Wei J C, Jiang Y M, Guo SY, 1994. Studies on the lichen family *cladoniaceae* in China III: a new genus to china: Thysanothecium. *Mycosystema* 7: 25~28

White F J, James P W, 1985. A new guide to microchemical techniques for the identification of lichen substances, *Bull British Lichen Soc* 57(supp.): 1~47

The lichen genus *Pseudevernia* Zopf in China*

WEI Jiang-Chun**

(*Systematic Mycology & Lichenology Laboratory Institute of Microbiology, Academia Sinica, Beijing 100080*)

ABBAS Abdulla

(*Xinjiang University, Urumqi*)

ABSTRACT: The lichen genus *Pseudevernia* with the species *P. furfuracea* is reported for the first time from China. Descriptions are given to both the genus and the species based on Chinese collection. Thallus photographs of both the upper and lower sides are provided. The key to the known species of the genus is given as well.

KEY WORDS: *Pseudevernia furfuracea*, new record, Xinjiang, description, key to species

During the study on the lichen flora of China a specimen of the lichen genus *Pseudevernia* Zopf was found among the materials collected by Abdulla Abbas from the Natural Conservation of Kalas in the Xinjiang Uygur Autonomous Region of China. The genus is one of the members in the lichen family Parmeliaceae belonging to the Lecanorales, Lecanoromycetidae, Ascomycetes, and Ascomycota (Kirk *et al.* 2001). This genus has not been found in China before.

MATERIALS AND METHODS

The materials were collected by Abdulla Abbas (coll. no. 980100) in 1998 from Kalas, alt.1560m, Xinjiang, China. Both the lichens *Cladonia* sp. and *Pseudevernia* sp. in a pocket were found. The last one was available for this research.

The dissecting microscope (OPTON) was used for morphological studies. Both the colour reaction and thin-layer chromatography (TLC) were used for the detection of lichen substances.

DESCRIPTIONS FOR THE TAXA

The descriptions for both the genus and species are given. The article is also provided with a key to the known species of the genus in the world.

Pseudevernia Zopf

Beih. Bot. Centralbl. 14:124(1903).

≡*Parmelia* subgen.*Pseudevernia* (Zopf) Berry, Ann. Missouri Bot. Gard. 28:135 (1941).

Type species *Pseudevernia furfuracea* (L.) Zopf

Thallus foliose-shrubby, with narrow strap-shaped and dorsiventral lobes of 1-10 mm wide,

---

*The project was supported by the National Natural Science Foundation of China (39899400)

**Corresponding author e-mail: weijc@95777.com;.

Received: 2002-09-13

This paper was originally published in *Mycosystema*, 22（1）: 26-29, 2003.

dichotomously branched, with a single point of attachment. Upper surface greenish-grey, with or without isidia or soredia. Lower surface channeled, without rhizines.

Ascomata apothecia. Asci 8-spored and *Lecanora*-type. Ascospores single, colourless. Pycnidia immersed. Conidia bifusiform.

Chemistry: Thallus in cortex contains atranorin, and in medulla depside or depsidone.

The genus is similar to the genus *Evernia*, but differs by the naked, channeled, often darkens to a purplish black or mottled white lower surface, and never containing usnic acid.

Pseudevernia furfuracea (L.) Zopf (Plate 1)

Beih. Bot. Centralbl. 14:124 (1903).

≡*Lichen furfuraceus* L.,Sp. Pl. 1146 (1753).

Thallus composed of numerous strap-shaped lobes; lobes 1-10 mm wide, dichotomously branched, with the down incurved margins; upper surface gray-white, rough, with numerous isidia and lobules; lower surface channeled and mottled black, brownish and pinkish.

Apothecia not seen.

Chemistry: Cortex K+ yellow, contains atranorin (TLC); medulla Pd-, K-, KC-, C-, contains physodic acid (TLC).

Habitat:China:, Xinjiang Uygur Automonous Region, Kalas, alt.1560m, 1998, Abdula Abbas 980100-1 (HMAS-L).

The species is similar to *Evernia prunastri*, but differs by presence of isidia and gray-white upper surface, dark lower surface and lacking usnic acid.

Fig. 1 *Pseudevernia furfuracea*,

General habit of the thallus: the upper surface(left); the lower surface (right), scales=1mm

DISCUSSIONS

No record of the species from China in previous literatures was found (Wei,1991).The record of the species from China by Kotlov (1996, ed. Golubkova) seems to be very doubtful because neither the Chinese collection of the species in LE nor the literature of that can be cited for confirmation (personal communication with Kotlov, 2002).

There are six species known in the genus *Pseudevernia*. Among them *P. cladonia* (Tuck.) Hale & W.Culb., *P. consocians* (Vain.) Hale & W.Culb. and *P. intensa* (Nyl.) Hale & W.Culb. are endemic to North and Central America; *P. olivetorina* (Zopf) Zopf, *P. soralifera* (Bitt.) Zopf, and *P. furfuracea* are distributed in Europe, North Africa, and Pakistan of western Asia, and with the sporadic occurrence of the last species in tropical America (Hale, 1968), Mongolia (Golubkova, 1981) and China.

P. furfuracea is important for perfume industry. *Evernia prunastri* (L.) Ach., another most important lichen for perfume industry distributed in Europe, has been not found in China so far. In Asia it is reported only in Japan （Yoshimura，1974）.The available lichen species *Evernia mesomorpha* Nyl. for perfume industry is distributed very common in China. The famous perfume lichen *Lobaria pulmonaria* (L.) Hoffm., however, is very rare in China (Wei, 1986).

Key to the known species of *Pseudevernia* Zopf

1a.Thallus with isidia or soredia 2

1b.Thallus without isidia or soredia 4

2a.Thallus having isidia only 3

2b.Thallus having capitate soralia in addition to sparse isidia and containing physodic acid........................ *P. soralifera*

3a.Thallus in medulla contains lecanoric acid; lobes narrow; endemic to America........................ *P. consocians*

3b.Thallus in medulla contains olivetoric acid; lobes elongated; distribution mainly in North Europe .... *P. olivetorina*

3c.Thallus in medulla contains physodic acid; morphologically variable; distribution mainly in South Europe .... *P. furfuracea*

4a.Thallus very narrow, with branched lobes of about 1 mm wide; endemic to eastern North America ....*P. cladonia*

4b.Thallus robust, often rugose, leathery, and often with abundant large apothecia; in Mexico, Central Ameica and U.S. . ... *P. intensa*

ACKNOWLEDGEMENTS The authors are indebted to Dr. Y. Kotlov (St.Petersburg) for kindly checking the record of *Pseudevenia furfuracea* from China and giving a valuable information.

[REFERENCES]

Culberson C F, Kristinsson H, 1970. A standardized method for the identification oflichen products. *J. Chromatogr.* **46**: 85~93.

Culberson C F, 1972. Improved conditions and new data for the identification of lichen products by a standadized thin-layer chromatographic method. *J. Chromatogr.* **72**: 113~125.

Hale, M.E.Jr. (1968) A Synopsis of the Lichen Genus Pseudevernia – The Bryologist **71** (1): 1~11.

Kirk, P.M., P.F.Cannon, J.C.David, and J.A.Stalpers (edited) 2001. Ainsworth & Bisby's Dictionary of the fungi (Ninth Edition) prepared by CABI Bioscience. CABI International.

Kotlov, Y. (1996) *Pseudevernia* in Golubkova (ed.) Handbook of the lichens of Russia **6**: 62.

Golubkova, N.S. (1981) Conspect Flori Lishainikov Mongolskoi Narodnoi Respublic – L. Nauka, 200. (Conspect of Lichen Flora from the People's Republic of Mongolia, in Russian).

Wei J.C., Chen J.B. and Jiang Y.M. (1986). Notes on lichen genus Lobaria in China. *Acta Mycol. Sin. Suppl.*I: 370~371.

Wei, J.C. (1991) An Enumeration of Lichens in China. Beijing: International Academic Publishers. 1~278.

White F J, James P W. 1985. A new guide to microchemical techniques for the identification of lichen substancews. *Bull. Br. Lichen Soc.* **57** (Suppl.):1~41.

Yoshimura, I. (1974) Lichen Flora of Japan in colour. Hoikusha Publishing Co., Ltd., Japan. 1~349

A lichen genus *Varicellaria* Nyl. from China*

REN Qiang[1,2] ZHAO Zun-Tian[1] WEI Jiang-Chun[2*]

([1]*College of Life Sciences, Shandong Normal University,* Jinan, 250014)

([2]*Systematic Mycology & Lichenology Laboratory, Institute of Microbiology, Academia Sinica,* Beijing 100080)

ABSTRACT: A lichen genus *Varicellaria* with the species *V. rhodocarpa* is reported from China for the first time. Both the genus and species in the paper are provided with the descriptions respectively and two line drawings for the species as well. The key to all the three species of the genus is also given in the present paper.

KEY WORDS: *Varicellaria*, *V. rhodocarpa*, China

During the research of the lichen flora of China "Flora Lichenum Sinicorum" the lichen genus *Varicellaria* Nyl. with the species *V. rhodocarpa* (Körb.) Th. Fr. was found among the lichen specimens preserved in HMAS-L. The genus is one of the three genera in the lichen family Pertusariaceae belonging to the Pertusariales, Lecanoromycetidae, Ascomycetes, and Ascomycota (Kirk *et al.* 2001). Three species in the world have been known from the genus so far. The genus is reported from China for the first time and the lecanoric acid from the species *V. rhodocarpa* is detected for the first time as well.

MATERIAL AND METHODS

The material examined was collected from Shaanxi province, Mt. Taibai, near the Temple Ping' ansi, alt. 2700 m, on bark of *Betula platyphylla* Suk., 8/VI/1963, J.C. Wei 2609-1 (HMAS-L).

Both the dissecting microscope ORIENT made in China and research microscope OPTON made in Germany were available for the morphological study. Thin layer chromatography (TLC, Culberson & Kristinsson, 1970; Culberson, 1972; White & James, 1985) was used for the identification of the lichen substance.

RESULTS AND DISCUSSIOIN

Varicellaria Nyl.

Varicellaria Nyl. Mé m. Soc. Sci. Natur. Cherbourg 5: 117 & 119 (1858) (' 1857')

Thallus crustose, thin to thick; Upper surface smooth to rugose or fissured without isidia, sometimes sorediate.

Apothecia lecaorine, buried in thallus warts (fertile verrucae), sometimes breaking down at the summit and appearing coarsely sorediate; paraphyses septate, branched and reticulately anastomosed; asci cylindrical to clavate, 1 or 8-spored, with IKI+ dark blue in ascus walls; ascospores colorless, large (200- 400μm long), 2-celled and strongly constricted in the middle, usually thick-walled.

Pycnidia not seen.

Photobiont green alga (*Trebouxia*).

On a variety of substrates.

---

*Supported by the National Natural Science Foundation of China (国家自然科学基金委资助项目 no. 39899400)

*Corresponding author J.C.Wei's e-mail:weijc@95777.com

Received:2002-10-11, accepted: 2002-12-04

This paper was originally published in *Mycosystema*, 22（2）: 216-218, 2003.

New to China.

The genus is similar to *Pertusaria*, but it has huge, 2-celled ascospores (to 400μm long), strongly constricted at the septate.

The genus name was validly published by Nylander in 1858 because of short combined description (ICBN, Art. 42, 1994).

Key to the species of *Varicellaria* Known

1. Asci 1-spored; ascospores large, 200~400 × 100~140μm, with very thick ascospore wall, easily breaking into two; medulla C+red.

2. Thallus smooth first, immediately turning into soradiate completely, P+blue-gray; with small soradiate fertile verrucae of, 0.5mm in diam. ....*V. kemensis* Räsänen

2. Thallus never soradiate, P-;with non-sorediate fertile verrucae of 0.5~1.2mm in diam ....*V. rhodocarpa* (Körb.) Th. Fr.

1. Asci 8-spored; ascospores small, 10~16 × 5~8 μm, with thin ascospore wall, not easily breaking into two; medulla C- ....*V. carneonivea* (Anzi) Erichsen

Varicellaria rhodocarpa (Körb.) Th. Fr. (Figs. 1,2) Lich. Scand. 1: 322 (1871)

≡*Pertusaria rhodocarpa* Körb. System. Lich. Germ. 384 (1855)

=*Varicellaria microsticta* Nyl. Mé m. Soc. Sci. Natur. Cherbourg 5: 117&119 (1858)(' 1857')

Thallus thin, upper surface smooth to rugose or lightly fissured, shiny, epruinose, whitish to yellowish, lacking isidia, sometimes becoming sorediate or leprose, marginally indefinite, unzoned, with numerous, scattered, hemispheric, rarely joined together fertile verrucae of 0.5~1.2 mm in diam., light to strongly constricted on the base.

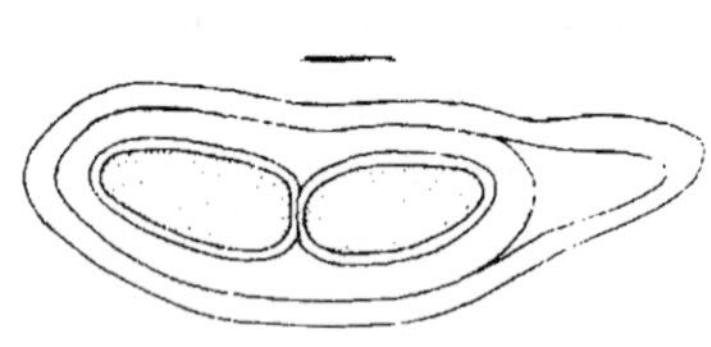

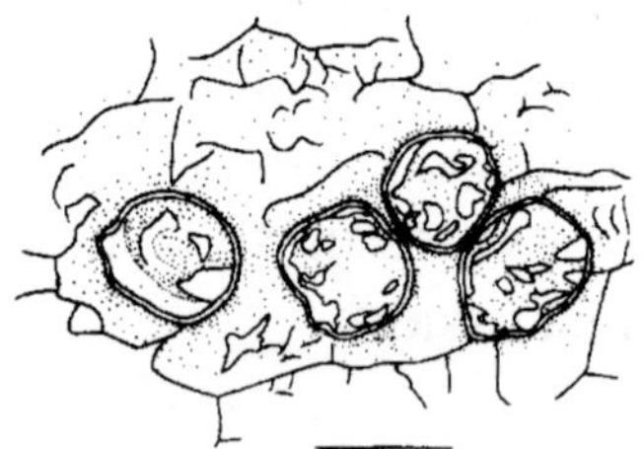

Fig. 1 *Varicellaria rhodocarpa*
Left: an ascus with a two-celled ascospore, scale bar = 46 μm
Right: upper surface of the thallus habit with apothecia, scale bar = 1 mm

Apothecia 1 or 2~3 immersed within a fertile verruca, with expanded, open, pinkish or yellowish disc, sometimes with the ascus tips showing as tiny glistening dots of 0.3~0.5 mm in diam; epithecium hyaline, 9~23μm tall; hymenium 224~271μm tall; hypothecium hyaline, 59~118μm tall; paraphysoids branched and anastomosed, 1.5~2.0 μm in diam.; asci clavate, 1-spored; ascospore hyaline, 2-celled, ellipsoid, constricted at the septum and easily breaking into two, large, 227~270 x 55~110 μm; ascospore walls 10~27 μm thick.

Pycnidia not seen.

Chemistry: Cortex K-, C-, KC-, Pd-; Medulla K-, C+red, KC+red becoming yellowish, especially fertile verrucae, containing lecanoric acid (TLC).

Habitat: on the bark of *Betula platyphylla* Suk..

Specimens examined: SHAANXI: Mt. Taibai, near the Temple Ping' ansi, alt. 2700 m, 8/VI/1963, J.C. Wei

2609-1 (HMAS-L).

New to China.

The species has been reported from Norway, Sweden, Finland, the Alps, Carpathians, Scotland, North America and Japan.

Geographical element: an arctic-alpine species.

The speices is different from *V. kemensis* by lacking soredia, and from *V. carneonivea* by 1-spored asci.

[REFERENCES]

Culberson C F, Kristinsson H, 1970. A standardized method for the identification of lichen products. *J Chromatogr* **46**: 85~93.

Culberson C F, 1972. Improved conditions and new data for the identification of lichen products by a standardized thin-layer chromatographic method. *J Chromatogr* **72**: 113~125.

Erichsen E, 1936. Pertusariaceae in Rabenhorst' s Kryptogamen-Flora von Deutschland, Œsterreich und der Schweitz. Ed. 2, 9 Band, 5 Abt., 1Teil. Leipzing, 687~699.

Kirk P M, Cannon P F, David J C, Stalpers J A, (edited) 2001. Ainsworth & Bisby' s Dictionary of the fungi (ninth edition) prepared by CABI Bioscience. CABI International.

Oshio M, 1968. Taxonomical Studies on the Family Pertusariaceae of Japan. *Journal of Science Hiroshima University*, Series B, Div. 2, **12**: 139~140.

Wei J C, 1991. An enumeration of lichens in China. Beijing: International Academic Publishers, 1~278.

White F J, James P W, 1985. A new guide to microchemical techniques for the identification of lichen substances. *Bull Br Lichen Soc* **57** (Suppl.): 1~41.

Overlooked taxa of *Stereocaulon* (Stereocaulaceae, Lecanorales) in China

by

Man-Rong Huang and Jiang-Chun Wei*

The Systematic Mycology & Lichenology Laboratory, Institute of Microbiology
Chinese Academy of Sciences, Beijing 100080

Huang, M.-R. & J.-C. Wei (2006): Overlooked taxa of *Stereocaulon* (Stereocaulaceae, Lecanorales) in China. - Nova Hedwigia 82: 435-445.

Abstract: Twelve species and two varieties of *Stereocaulon* are reported for the first time from China, viz. *Stereocaulon alpestre, St. botryosum, St. coniophyllum, St. depreaultii, St. exutum, St. foliolosum* var. *botryophorum, St. graminosum, St. himalayense, St. intermedium, St. myriocarpum* var. *altaicum, St. prostraturn, St. rivulorum, St. saxatile* and *St. sterile.* Descriptions and remarks are given.

Key words: Lichenized fungi, biodiversity.

Introduction

Stereocaulon coralloides var. *japonicum* Th. Fr. (= *St. dactylophyllum* Florke) is the first taxon in this genus reported from China by Nylander (1857). Since then several foreign and Chinese lichenologists have recorded taxa of this genus from the country (Hue 1898; Tchou 1935; Sat8 1940; Lamb 1977; Wei & Jiang 1986; Wu 1987; Thrower 1988; Chen et al. 1989; Wei 1991; Abbas & Wu 1998; Aptroot 1999; Lai 2000, 2001), viz. *St. alishanum, St. alpinum, St. apocalypticum, St. condensaturn, St. dactylophyllum, St. exutum, St. fastigiatum, St. foliolosum* var. *strictum, St. japonicum, St. macrocephalum, St. massartianum, St. myriocarpum, St. nigrum, St. octomerellum, St. octomerum, St. paschale, St. pendulum, St. pileatum, St. piluliferum, St. pomiferum, St. sasakii, St. saviczii, St. sorediiferum* var. *sorediijerum, St. sorediiferum* var. *leprosolingulatum, St. tomentosum, St. verruculigerum, St. vesuvianum* var. *nodulosum.* There are also a lot of literatures on this genus in the countries around China, especially the Himalayan area out of China. During a careful taxonomic study of this genus from China, the authors of this paper have found 14 additional species and varieties which are treated below.

'Corresponding author: weijc@95777.com

DOI: 10.1127/0029-5035/2006/0082-0435

This paper was originally published in *Nova Hedwigia*, 82（3-4）: 435-445, 2006.

Materials and methods

All Chinese specimens examined for this study are deposited in HMAS-L. Some type specimens were borrowed from FH and H. Their chemical constitution was detected using TLC in solvent system C as standardized by Culberson (1972). Color tests were carried out according to Lamb (1977), where K stands for 10% aqueous solution of potassium hydroxide and P stands for freshly prepared 0.1% alcoholic solution of para-phenylenediamine. The measurements were carried out using a compound microscope with hand sections mounted in tap water. The terminology for the morphology and anatomy follows Lamb (1977, 1978).

The taxa of *Stereocaulon* new to China

1. **Stereocaulon alpestre** (Flot.) Dombr.

Botanischeskii Zhurnal 77 (7): 98 (1992).

DESCRIPTION: Primary thallus evanescent in an early stage. Pseudopodetia compactly caespitose, tuft-like, more or less dorsiventral, cylindrical, 2-2.5 cm long, ca. 1 mm thick, with heavily tomentose surface, pale grey, rather smooth, not ligneous, without soredia, firmly attached to the substrate; phyllocladia numerous, densely distributed on the pseudopodetia, neighbouring ones joining each other into areolate, thick squamulose, 0.3-1 mm in diameter, with undulating or lobulate margin. Cephalodia rare and inconspicuous, sessile, warty to subglobose, 0.2-1 mm in diameter, cyaneous or concolorous with phyllocladia, poorly corticate, containing *Nostoc* sp.

Apothecia not observed.

CHEMISTRY: K+ yellow, P+ orange red; TLC: atranorin, stictic acid and norstictic acid.

SPECIMENS EXAMINED: Sichuan, Mt. Baimang, alt. 4450 m, on soil, 13 July 1981, coll. Wang M.Z., no. 7836.

REMARKS: The species resembles *St. alpinum* Laurer in morphology, but differs in chemistry by having stictic and norstictic acids and lacking lobaric acid.

2. **Stereocaulon botryosum** Ach.

Lamb, J. Hattori Bot. Lab. no. 43: 207 (1977)

DESCRIPTION: Primary thallus evanescent in an early stage. Pseudopodetia compactly caespitose, erect, not dorsiventral, cylindrical, 2-3 cm high, abundantly branched, polychotomous, cauliflower-like, main stems inconspicuous, completely decorticated, glabrous or slightly tomentose, not ligneous, without soredia, more or less emorient below and loosely attached to its substrate; phyllocladia verrucose, grain-like or squamulose, densely distributed in the upper parts of the pseudopodetia. Cephalodia abundant, sessile, verrucose to botryose, rufous to grey-brown, poorly corticate, containing *Stigonema* sp.

Apothecia terminal, slightly convex, up to 4 mm in diameter, dark brown, often divided into small secondary discs when aged; hymenium hyaline, 44-51 μm tall, hypothecium slightly brown in color, central cone colorless; spores clavate to fusiform, colorless, 1-3 septate, (17-)27-37 μm x 4-5 μm.

CHEMISTRY: K+ yellow, P+ pale yellow; TLC: atranorin, porphyrilic acid.

SPECIMENS EXAMINED: Jilin, Mt. Changbai. on humus, 13 August 1977, coll. Wei J.C., nos. 2950 & 2960-3-1.

REMARKS: The species is widespread in the northern hemisphere, but rarely found in China. It is characterized by its cauliflower-like habit and the presence of porphyrilic acid. Sometimes there are many black dots on its phyllocladia which are formed by a lichenicolous fungus.

3. Stereocaulon coniophyllum Lamb

J. Hattori Bot. Lab. no. 43: 268 (1977).

DESCRIPTION: Primary thallus evanescent in an early stage. Pseudopodetia not caespitose, decumbent, obviously dorsiventral, cylindrical, 1.5-3.5 cm high, 0.8-1.8 mm thick, slightly branched, typically expanded into foliose ends, completely decorticated, glabrous, rust red, ligneous, heavily sorediate on the tips of its branchlets or on the lower surface of foliose ends; phyllocladia almost completely absent or replaced by foliose ends of pseudopodetial branchlets. Cephalodia normally present, shortly stiped, subglobose to botryose, 1-5 mm in diameter, brown to cyaneous, protosacculate, containing *Nostoc* sp.

Apothecia not observed.

CHEMISTRY: K+ yellow, P+ pale yellow; TLC: atranorin, lobaric acid.

SPECIMENS EXAMINED: Jilin, Mt. Changbai, alt. 2100 m, on rock, 13 August 1977, coll. Wei J.C., no. 2951. Shaanxi, S. slope of Mt. Taibai, on rock, 5 June 1963, coll. Wei et al. Sichuan, E. slope of Mt. Gongga, on rock, 28 June 1982, coll. Wang, Xiao & Li, nos. 8429 & 8426-1.

REMARKS: This is a very interesting and rare species, distinguished by its sorediate foliose pseudopodetial apices and absence of phyllocladia.

4. Stereocaulon depreaultii Delise ex Nyl.

Lamb, J. Hattori Bot. Lab. no. 43: 215 (1977).

TYPE: Canada, Newfoundland, no exact locality given, coll. Despreaux (H-NYL, no. 39814, holotype!).

TYPE OBSERVATION: Primary thallus not observed. Pseudopodetia caespitose, fastigiate, erect, not dorsiventral, cylindrical, tapered, 2-3 cm high, diameter 1-2 mm at base, 0.5-1 mm near tips, numerously branched, irregular, di- to polychotomous, main stems inconspicuous, completely decorticated, glabrous, smooth, ligneous, yellowish in upper parts, rust red near base, without soredia; phyllocladia numerous, mostly flat grain-like to ginger tuber-like, opalescent to pale russet, diameter 0.1-0.5 mm, mainly distributed in the upper parts of the pseudopodetia. Cephalodia numerous, sessile, spherical to verrucose, botryose, 0.5-1 mm in diameter, grey brown.

Apothecia rare, mostly terminal, occasionally lateral, 1-3 mm wide, black, slightly convex, divided into secondary small discs.

DESCRIPTION: Primary thallus evanescent in an early stage. Pseudopodetia loosely caespitose, erect, not dorsiventral, cylindrical, dwarfish, only 1 cm long, 0.8 mm thick, moderately and irregularly branched, main stems conspicuous, completely decorticated, glabrous, yellowish to slightly rust red, highly ligneous, without soredia;

phyllocladia abundant, mainly present in the upper parts of the pseudopodetia, minute grain-like to wart-like, pale, usually 0.1-0.2 mm in diameter. Cephalodia rare, inconspicuous, sessile, wart-like to subbotryose, not over 0.5 mm in diameter, cinerous, poorly corticate, containing *Stigonema* sp.

Apothecia not observed.

CHEMISTRY: K+ yellow, P+ pale yellow; TLC: atranorin, lobaric acid.

SPECIMEN EXAMINED: Only one specimen was collected in China, viz. Jilin, Hunchun, alt. 580 m, on soil, 30 July 1984, coll. Lu X.D., no. 847105-2.

REMARKS: This species is mainly distinguished by its numerous minute phyllocladia. Chinese material of this species shows a certain difference from the type specimen, but they have the essentially same types of pseudopodetia, phyllocladia and cephalodia, and variations between them fall within a species.

5. **Stereocaulon exutum** Nyl.

Lamb, J. Hattori Bot. Lab. no. 43: 260 (1977).

TYPE OBSERVATION: Primary thallus not seen. Pseudopodetia erect, dorsiventral, cylindrical, tapered, 4-7 cm long, diameter 0.8-1.5 mm at the base and ca. 0.5 mm near the tip, frequently branched, mostly dichotomous, main stems inconspicuous, completely decorticated, glabrous, pale yellow or occasionally rust red at the base, ligneous, without soredia, firmly attached to its substrate; phyllocladia numerous, harmoniously and densely distributed, cylindrical-coralloid, up to 2-4 mm long, dichotomously ramulose, dorsiventral, corticate on their dorsal sides, deep olivaceous, and ecorticated on their ventral sides, whitish. Cephalodia abundant and conspicuous, usually stiped, rarely sessile on the pseudopodetia, globose to botryose, 1-2 mm in diameter, smooth.

Apothecia abundant, terminal, rufous to dark brown, 1-2 mm wide, convex and with slightly reflexed margin, occasionally several apothecia aggregated to form larger secondary discs.

DESCRIPTION: Primary thallus evanescent in an early stage. Pseudopodetia caespitose, tuft-like, erect to decumbent, sometimes dorsiventral, cylindrical, 1-3 cm long, 1-1.5 mm thick, moderately branched, harmoniously dichotomous, main stems inconspicuous, completely decorticated, glabrous, yellowish, ligneous, without soredia, firmly attached to its substrate; phyllocladia cylindrical, pale whitish, robust, up to 3 mm long, 0.2-0.3 mm thick, frequently branched, resembling branchlets of pseudopodetia. Cephalodia abundant and conspicuous, sessile or inconspicuously stiped, concolorous with the phyllocladia, mostly botryose, 0.8-1.2 mm in diameter, poorly corticate, containing *Stigonema* sp.

Apothecia not observed.

CHEMISTRY: K+ yellow, P+ pale yellow; TLC: atranorin, lobaric acid.

SPECIMENS EXAMINED: Jilin, Changbai, alt. 1270 m, on rock, 27 July 1983, coll. Wei J.C. & Chen J.B., no. 6316.

REMARKS: This species was reported by Wu (1987) from China without citing any specimens. Thus it is impossible for scientists to prove the record of it. Therefore, we list this species here as a new to China. It is mainly characterized by its long and frequently branched cylindrical phyllocladia and botryose cephalodia. The differences between the Chinese material and type specimen are normal within a species.

6. **Stereocaulon foliolosum** Nyl. var. **botryophorum** (Müll. Arg.) Lamb

J. Hattori Bot. Lab. no. 43: 267 (1977).

DESCRIPTION: Primary thallus evanescent in an early stage. Pseudopodetia fastigiately caespitose, erect, not dorsiventral, cylindrical, 5-7 mm high, 1-2 mm thick, subsimple near base, sparingly branched towards the apical region, completely decorticated, glabrous, slightly rust red, ligneous, without soredia, firmly attached to its substrate; phyllocladia digitate-squamulose, 1-3 mm long or sometimes granulose, whitish at the tips. Cephalodia sessile, subglobose to globose, 1-2 mm in diameter, protosacculate, containing *Nostoc* sp.

Apothecia terminal, brown, slightly convex, 2-3 mm wide; hymenium hyaline, 90-120 μm high; asci 90-105 μm x 12-15 μm, both hypothecium and central cone colorless; spores spirally wound to vermiform, colorless, 6-9 septate, 60-90 μm x 3-5 pm.

CHEMISTRY: K+ yellow, P+ pale yellow; TLC: atranorin, lobaric acid.

SELECTED SPECIMENS EXAMINED: Sichuan, E. slope of Mt. Gongga, alt. 3450 m, on rock, 28 June 1982, coll. Wang X.Y., Xiao X. & Li B., no. 8429-1. Xizang, Nyalam, 20 June 1966, coll. Wei J.C., Chen J.B. & Huang Z.W., no. 973-land 22 June 1966, Wei J.C. & Chen J.B. no. 1817; Zham, alt. 3500 m, on rock between mosses, 11 May 1966, coll. Wei J.C. & Chen J.B., no. 449. Yunnan, Cangshan, alt. 3900 m, on rock, 31 May 1955, coll. Wu Z.Y. et al, no. 1578.

REMARKS: This variety differs from *St. foliolosum* var. *foliolosum* by its digitate-squamulose phyllocladia. It is also morphologically similar to *St. himalayense,* but can distinguished by its larger and more septate spores.

7. **Stereocaulon graminosum** Schaer. in Moritzi

Lamb, J. Hattori Bot. Lab. no. 43: 252 (1977).

DESCRIPTION: Primary thallus evanescent in an early stage. Pseudopodetia fastigiately caespitose,decumbent,cylindrical,slightly tapered, 2-3 cm long, 0.8-2 mm thick, frequently and irregularly branched, main stems conspicuous, completely decorticated, thinly tomentose, yellowish, not ligneous, without soredia; phyllocladia grain-like to verrucose, ca. 0.1 mm in diameter, or squamulose, thick, up to 0.5 mm in diameter, occasionally with darker, depressed center and paler margin and umbilicate, those occurring near the base of the pseudopodetiaoften becoming flattened-peltate. Cephalodia subglobose, with whitish pmina on their surface, 0.2-1 mm in diameter, containing *Nostoc* sp.

Apothecia not observed.

CHEMISTRY: K+ yellow, P+ orange red; TLC: atranorin, stictic acid and norstictic acid.

SPECIMEN EXAMINED: Xizang, Zogang, alt. 4400 m, on rock, 31 August 1976, coll. Zong Y.C. & Liao Y.Z., no. Zang 494.

REMARKS: This species is characterized by its small, crowded and sometimes peltate phyllocladia, which also distinguish it from *St. vesuvianum* Pers. It was reported from Southeast Asia before, and seems to have a disjunctive distribution between the Malaysian and the Himalayan region.

8. **Stereocaulon himalayense** D.D.Awasthi & Lamb

J. Hattori Bot. Lab. no. 43: 269 (1977).

DESCRIPTION: Primary thallus evanescent in an early stage. Pseudopodetia compactly caespitose, decumbent, robust, (0.5-)1-4 cm high, 0.8-2 mm thick, abundantly branched, dichotomous, main stems usually inconspicuous, completely decorticated, glabrous, usually rust-red, ligneous, without soredia; phyllocladia numerous, granulose to coralloid, never subfoliose, whitish, 0.5-1.5 mm long, ca. 0.2 mm thick. Cephalodia abundant and conspicuous, sessile, subglobose to botryose, 1-5 mm in diameter, rufous, protosacculate, containing *Nostoc* sp.

Apothecia terminal on the pseudopodetia, with conspicuous pseudolecanorine margin in an early stage, later up to 5 mm in diameter, convex or saddle-like, brown, often divided into small secondary discs when aged; hymenium colorless, 68-119 μm high, hypothecium and central cone colorless; spores clavate, colorless, 1-4 septate, 24-36 μm X 4-6 μm.

CHEMISTRY: K+ yellow, P+ pale yellow; TLC: atranorin, lobaric acid.

SPECIMENS EXAMINED: Shaanxi, Mt. Taibai, on rock, 5 June 1963, coll. Wei et al. Sichuan, E. slope of Mt. Gongga, 27 June 1982, coll. Wang X.Y., Xiao X. & Li B., nos. 8371 & 8382; Kangding, coll. Liu Z.S., no. 1541-2. Xizang, Mt. Namjagbarwa, alt. 4450 m, on rock, 28 June 1983, coll. Mao X.L., no. 1022; Nyalam, 14 June 1966, coll. Wei J.C., Chen J.B. & Zong Y.C., no. 1599. Yunnan, Deqin, Mt. Baimang, alt. 4200 m, on rock, 28 August 1981, coll. Wang X.Y., Xiao X. & Su J.J., no. 7612-1; Zhongdian, alt. 4300 m, on rock, 8 September 1981, coll. Wang X.Y., Xiao X. & Su J.J., no. 7203.

REMARK: Sometimes this taxon is difficult to distinguish from *St. foliolosum* var. *botryophorum* by morphology, and it differs more clearly by its less septate spores. Occasionally its apothecia and cephalodia are unusually large and the hymenium very thick. It grows often together with *St. coniophyllum.*

9. **Stereocaulon intermedium** (Savicz) H.Magn.

Lamb, J. Hattori Bot. Lab. no. 43: 223 (1977).

TYPE: Russia, Kamtchatka, near Tar' inskaya Bucht, coll. V.P.Savicz, no. 2247198 (FH-syntype!).

TYPE OBSERVATION: Primary thallus not seen. Pseudopodetia compactly caespitose, erect, not dorsiventral, cylindrical, tapered, up to 5 cm high, diameter 1-2 mm at the base, and 0.5-1 mm near the tip, abundantly branched, polychotomous, main stems conspicuous, completely decorticated, thinly and finely tomentose, rather smooth, not ligneous, without soredia, firmly attached to its substrate; phyllocladia numerous, ginger tuber-like to mainly cylindrical-coralloid, ca. 1 mm long, usually one or two times branched, distributed on the middle part of the pseudopodetia. Cephalodia rare, but large, conspicuous, 1-2 mm in diameter, sessile on the pseudopodetia, verrucose, with divided surface, brown.

Apothecia not observed.

DESCRIPTION: Primary thallus evanescent in an early stage. Pseudopodetia compactly caespitose, tuft-like, erect or sometimes decumbent and dorsiventral, cylindrical, 2-

4 cm high, 1-2 mm thick, simple to slightly branched, main stems usually conspicuous, completely decorticated, subtomentose, grey-yellowish, not ligneous, without soredia, firmly or sometimes rather loosely attached to its substrate; phyllocladia grain-like, verrucose to mainly cylindrical-coralloid, dirty yellow, paler at the tips, 0.3-0.8 cm long, ca. 0.1 mm thick, usually one or two times branched, neighbouring phyllocladia sometimes fused near their base and squamulose or ginger tuber-like. Cephalodia rare, sessile, subglobose to verrucose, 0.1-1 mm in diameter, bluish grey, usually pruinose, aggregated, cortex poorly developed, containing *Nostoc* sp.

Apothecia terminal, with conspicuous, paler pseudolecanorine margin in an early stage, convex when mature, ca. 0.8 mm wide, rufous; hymenium hyaline, 51-68 μm high, both hypothecium and central cone colorless; spores clavate, colorless, 3-5 septate, 23-31 μm x 4-5 pm.

Chemistry: K+ yellow, P+ pale yellow; TLC: atranorin, lobaric acid.

Specimens examined: Hubei, Shennongjia, alt. 2950 m, on soil, 2 August 1984, coll. Chen J.B., no. I1316-1. Shaanxi, Mt. Taibai, on soil, 4 June 1963, coll. Wei *et al*, no. 2724. Sichuan, Emci Shan, 16 August 1963, coll. Zhao J.D. & Xu L.W., no. 07461; E. slope of Mt. Gongga, 23 June 1982, coll. Wang X.Y., Xiao X. & Li B., nos. 8100,8086 & 8837; Muli, alt. 2300 m, on rock, 29 May 1982, coll. Wang X.Y., Xiao X. & Li B., no. 7926; Tianquan, alt. 1950 m, on rock, 18 July 1953, coll. Jiang X.L., no. 34892; alt. 2500 m, on soil, 24 August 1963, coll. Guan K.J. & Wang W.C., no. 1893. Xizang, Lhozhag, alt. 3300 m, on soil, 1 September 1975, coll. Zong Y.C., no. Zang086; Nyalam, on soil, 18 May 1966, coll. Wei J.C. & Chen J.B., nos. 827 & 1491-5. Yunnan, 4 April 1982, Gongshan, Su J.J., no. 4032; Weixi, Mt. Biluo, 13 July 1981, coll. Wang X.Y., Xiao X. & Su J.J., nos. 4509, 4510 & 4621. Zhejiang, no exact locality and collectors were indicated.

Remark: *St. intermedium* is morphologically similar to *St. dactylophyllum* Florke, but differs by its mere blunt phyllocladia and the absence of stictic and norstictic acids.

10. **Stereocaulon myriocarpum** Th. Fr. var. **altaicum** Lamb

J. Hattori Bot. Lab. no. 43: 225 (1977).

Description: Primary thallus evanescent in an early stage. Pseudopodetia not caespitose, erect, robust, 1.5-6 cm long, 1-2.5 mm thick, abundantly and complicatedly branched, polychotomous, main stems inconspicuous, completely decorticated, glabrous or thinly tomentose, rust red near base, ligneous, without soredia; phyllocladia numerous, coralloid, ramulose or wart-like, pale to yellowish. Cephalodia abundant and conspicuous, sessile, subglobose to verrucose, warty in surface, 0.5-2 mm in diameter, concolorous with phyllocladia, poorly corticate, containing *Nostoc* sp.

Apothecia usually abundant, lateral or terminal on the pseudopodetia, lecideine, 0.4-1 mm in diameter, slightly convex; hymenium hyaline, 75-90 μm high, both hypothecium and central cone colorless; asci 50-75 μm x 10-14 μm; spores colorless, clavate, 3-5 septate, 27-35 μm x 3-5 μm.

Chemistry: K+ yellow, P+ pale yellow; TLC: atranorin, stictic acid and norstictic acid.

Specimens examined: Sichuan, Batang, alt. 4100 m, on rock, 29 July 1983, coll. Su J.J., Wen H.A. & Li B., no. 5960; W. slope of Mt. Gongga, alt. 4200 m, 29 July 1982, coll. Wang X.Y., Xiao X. & Li B., no. 9102; Muli, alt. 3900 m, on rock, 5 June 1982, coll. Wang X.Y., Xiao X. & Li B., 7967-1. Xizang,

Mêdog, alt. 4150 m, 14 August 1982, coll. Zong Y.C. Yunnan, Deqin, alt. 4500 m, on rock, 8 October 1982, coll. Su J.J., no. 5391; Weixi, 11 July 1981, coll. Wang X.Y., Xiao X. & Su J.J., no. 4557.

REMARKS: This variety is characterized mainly by its robust, glabrous and ligneous pseudopodetia.

11. **Stereocaulon prostratum** Zahlbr.

Lamb, J. Hattori Bot. Lab. no. 43: 266 (1977).

DESCRIPTION: Primary thallus evanescent in an early stage. Pseudopodetia caespitose, decumbent, dorsiventral, cylindrical, dwarfish, only 5-8 mm long, ca. 1 mm thick, moderately to abundantly and irregularly branched, main stems conspicuous, completely decorticated, glabrous, slightly rust red, more or less ligneous, without soredia, firmly attached to its substrate and half-concealed; phyllocladia numerous, cylindrical, slightly tapered toward tips, yellowish, usually 0.5-1 mm long, occasionally up to 2 mm, ca. 0.1 mm thick, simple or strongly ramulose near tips. Cephalodia numerous, conspicuous, sessile, globose to warty, grey to whitish, 0.8-2 mm in diameter; poorly corticate, containing *Stigonema* sp.

Apothecia terminal on the pseudopodetia, rufous, convex to hemispherical, ca. 1 mm in diameter; hymenium hyaline, 68-85 ym tall, hypothecium slightly brownish, central cone colorless; spores clavate, colorless, 3-septate, 24-30 μm x 4-7 ym.

CHEMISTRY: K+ yellow, P+ pale yellow; TLC: atranorin, lobaric acid.

SPECIMENS EXAMINED: Jilin, Mt. Changbai, on soil, 23 August 1984, coll. Lu X.D., no. 848313-1 and 12 August 1977, Wei J.C., no. 2932.

REMARKS: This species is rare and was only reported from Japan. It can be easily recognized by its long and densely clustered phyllocladia on the dorsal side of the pseudopodetia.

12. **Stereocaulon rivulorum** H.Magn.

Lamb, J. Hattori Bot. Lab. no. 43:228 (1977).

DESCRIPTION: Primary thallus evanescent in an early stage. Pseudopodetia caespitose, decumbent, cylindrical, tapered toward tips, frequently branched, main stems inconspicuous, completely decorticated, tomentose or glabrous in basal parts, obviously ligneous, without soredia, firmly attached to substrates; phyllocladia abundant, vermcose to thick-squamulose, subdigitate, pale whitish, many derived from the ends of phyllocladioid branchlets of pseudopodetia. Cephalodia rare, sessile, vermcose, brownish, 0.5-1 mm in diameter, poorly corticate, containing *Stigonema* sp.

Apothecia both lateral and terminal, slightly concave in an early stage, flat when mature, up to 2 mm in diameter, black; hymenium colorless, 51-70 μm tall, both hypothecium and central cone colorless; spores clavate, colorless, 4-5 septate, 32-40 μm x 3-5 ym.

CHEMISTRY: Two chemical strains were observed: Strain II: K+ yellow, P+ pale yellow; TLC: atranorin, lobaric acid. Strain III: K+ yellow, P+ orange red; TLC: atranorin, stictic acid and norstictic acid.

SPECIMENS EXAMINED: Strain II: Inner Mongolia, Altai, alt. 2500 m, on soil, 5 August 1986, coll. Gao X.Q., no. 2149. Strain III: Jilin, Changbai, 2 August 1983, coll. Wei J.C. & Chen J.B., nos. 6440,

6443 & 6501-3. Shaanxi, Mt. Taibai, 2 June 1963, coll. Wei et al, no. 2539-1. Sichuan, Emei Shan, on rock, 17 August 1963, coll. Zhao J.D. & Xu L.W., no. 7952. Yunnan, Weixi, on rock, 18 July 1981, coll. Wang X.Y., Xiao X. & Su J.J., nos. 3909 & 4498-1; alt. 3400 m, on rock, 27 May 1982, coll. Su J.J., no. 0870.

REMARKS: Characterized by its phyllocladia derived from the ends of phyllocladioid branchlets of pseudopodetia. Tønsberg (1977) distinguished 4 chemical strains in this species, but Lamb (1977) preferred to consider strain III as a distinct species. However, we follow here Tønsberg.

13. **Stereocaulon saxatile** H.Magn.

Lamb, J. Hattori Bot. Lab. no. 43: 230 (1977).

DESCRIPTION: Primary thallus evanescent in an early stage. Pseudopodetia compactly caespitose, tuft-like, decumbent, obviously dorsiventral, cylindrical, 1-3 cm long, usually 0.5 mm thick, abundantly and irregularly branched, main stems inconspicuous, completely decorticated, thinly or sometimes rather heavily tomentose, ash grey to grey, not ligneous, without soredia, usually emorient below and loosely attached to the substrate; phyllocladia ramulose-coralloid to verrucose, yellowish, not over 0.5 mm thick. Cephalodia rare or usually absent, botryose, small, not over 1 mm in diameter, cyaneous, poorly corticate, containing *Stigonema* sp.

Apothecia rare, terminal on pseudopodetia, plane, ca. 0.8 mm wide, brown; hymenium colorless, 40-55 ym tall, both hypothecium and central cone colorless; spores fusiform, colorless, 20.5-28 ym x 2-3 ym.

CHEMISTRY: K+ yellow, P+ pale yellow; TLC: atranorin, lobaric acid.

SELECTED SPECIMENS EXAMINED: Heilongjiang, Huzhong, alt. 1410 m, on humus, 3 September 1984, coll. Gao X. Q., no. 387-1; Wudalianchi, 29 July 2002, coll. Chen J.B. & Hu G.R., nos. 21888 & 21894. Inner Mongolia, Arxan, 2 August 2002, coll. Wei J.C. *et al*, no. Aer209; Ergun Zuoqi, 10 August 1985, coll. Gao X.Q., nos. 1343 & 1493. Jilin, Changbai, 28 July 1983, coll. Wei & Chen, nos. 6288 & 6854); Hunchun, alt. 560 m, on soil, 30 July 1984, coll. Lu X.D., no. 647090-1; Wangqing, 12 June 1996, coll. Wang Y.Z. et al, no. 345. Liaoning, Kuandian, Baishilazi Natural Reserve, 21 July 2000, coll. Liu H.J., no. 055.

REMARKS: This species is difficult to distinguish from *St. paschale* (L.) Hoffm. Usually, it is dorsiventral, ash-grey in color and cephalodia are rare or completely absent. It is usually found growing with *Cladonia* spp.

14. **Stereocaulon sterile** (Savicz) Lamb in Krog

Lamb, J. Hattori Bot. Lab. no. 43: 234 (1977).

DESCRIPTION: Primary thallus evanescent in an early stage. Pseudopodetia compactly caespitose, decumbent, obviously dorsiventral, cylindrical, 2-4 cm long, ca.1 mm thick, moderately branched, main stems inconspicuous, completely decorticated, glabrous, slightly rust red, or slightly tomentose, pale, whitish, ligneous, without soredia; phyllocladia numerous, grain-like to coralloid or cylindrical, occasionally squamulose with dentate lobules on the margin, pale green in color, 0.3-2 mm long, 0.1-0.2 mm thick, ramulose, obviously dorsiventral, firmly attached to the substrate. Cephalodia rare, sessile, subglobose to verrucose, not over 1 mm in diameter, concolorous with the pseudopodetia or phyllocladia, poorly corticate, containing *Nostoc* sp.

Apothecia terminal, lecidein, 1-4 mm in diameter, slightly convex, rufous; hymenium ca. 62 mm tall, both hypothecium and central cone colorless; asci 41-50 μm x 12-15 μm; spores clavate, colorless, 3-septate, 21-28 μm x 3.5-4.5 pm.

CHEMISTRY: K+ yellow, P+ pale yellow; TLC: atranorin and lobaric acid.

SPECIMENS EXAMINED: Jilin, Mt. Changbai, alt. 1850 m, on rock, 12 August 1983, coll. Wei J.C. & Chen J.B., no. 6811 and alt. 1650 m, on rock, 1 August 1985, coll. Lu X.D., no. 1014. Liaoning, Kuangdian, Baishilaizi Natural Reserve, alt. 670 m, on rock, 21 July 2000, coll. Jiang Y.M. & Huang M.R., no. 040.

REMARKS: *St. sterile* somewhat resembles *St. exutum*, but can immediately be distinguished by inconspicuous, verrucose cephalodia and grain-like phyllocladia.

Acknowledgements

This project was financially supported by the National Natural Science Foundation of China, grant no. 30270005. Our special thanks are due to the keepers of the herbaria FH and H for the kind loan of type specimens, to Dr. M.J.Lai for sending literature and to Miss H.Deng for checking Chinese locality names.

References

ABBAS, A. & J.N. WU (1998): Lichens of Xinjiang. - Sci-Tec & Hygiene Publishing House of Xinjiang (K), Urumqi. (in Chinese)

APTROOT, A. (1999): Annotated checklist of Hongkong lichens. - Tropi. Bryol. 17: 57-101.

CHEN, J.B., J.N. WU & J.C. WEI (1989): Lichens of Shennongjia. - In: Chen, J.B. & X.Q. Kong (eds.): Fungi and lichens of Shennongjia: 386-493. World Publishing Corp, Beijing. (in Chinese)

CULBERSON, C.F. (1972): Improved conditions and new data for the identification of lichen products by a standardized thin-layer chromatographic method. - J. Chromatogr. 72: 113-125.

HUE, A.M. (1898): Lichenes extra-europaei. - Nouv. Arch. Mus. Hist. Nat. (Paris) 3: 213-280.

LAI, M.J. (2000): Illustrated macrolichens of Taiwan (I). - The Council of Agriculture, the Executive Yuan, Taipei. (in Chinese)

LAI, M.J. (2001): The lichen family Stereocaulaceae of Taiwan. - Taiwan J. For. Sci. **16:** 175-180.

LAMB, I.M. (1977): A conspectus of the lichen genus *Stereocaulon*. - J. Hattori Bot. Lab. 43: 191-355.

LAMB, I.M. (1978): Keys to the lichen genus *Stereocaulon*. - J. Hattori Bot. Lab. 44: 209-250.

NYLANDER, W. (1857): Enumeration generale des lichenes, avec l' indication sommaire de leur distribution geographique. - Mem. Soc. Sci. Cherbourg **5:** 85-146.

SATÔ, M. (1940): East Asiatic lichens (II). - J. Jap. Bot. **16:** 42-47.

TCHOU, Y.T. (1935): Note preliminaire sur les lichens de Chine. - Contr. Inst. Bot. Nat. Acad. Peiping **3**:299-322.

THROWER, S.L. (1988): Hong Kong lichens. - An Urban Council Publication, Hong Kong.

TØNSBERG, T. (1977): The chemical strains in *Stereocaulon rivulorum* and their distribution. - Norw. J. Bot. 24: 213-216.

WEI, J.C. (1991): An enumeration of lichens in China. - International Academic Publishers, Beijing.

WEI, J.C. & Y.M. JIANG (1986): Lichens of Xizang. - Science Press, Beijing. (in Chinese)

WU, J.L. (1987): Iconography of Chinese lichen. - Zhong Guo Zhan Wang Chu Ban She, Beijing. (in Chinese)

Some species of Graphidaceae (Ostropales, Ascomycota) rare and new to China

MIAO Xiao-Lin[1] JIA Ze-Feng[1] MENG Qing-Feng[1] WEI Jiang-Chun[1,2*]

([1]*College of Life Science, Shandong Agricultural University, Tai'an 271018, China;* [2]*Key Laboratory of Systematic Mycology and Lichenology, Institute of Microbiology, Chinese Academy of Sciences, Beijing 100101, China*)

Abstract: Twelve species of *Dyplolabia*, *Glyphis*, *Hemithecium*, *Platygramme*, *Platythecium* and *Sarcographa* in the Graphidaceae from China are reported. Six of them are new to China. A description with two photographs to each species based on Chinese collection is given. A working key is also provided.

Key words: *Dyplolabia*, *Glyphis*, *Hemithecium*, *Platygramme*, *Platythecium*, *Sarcographa*

中国文字衣科（厚顶盘目、子囊菌门）地衣之稀有和新记录种

苗晓琳[1] 贾泽峰[1] 孟庆峰[1] 魏江春[1,2*]

（[1]山东农业大学生命科学学院 泰安 271018；[2]中国科学院微生物研究所真菌地衣系统学重点实验室 北京 100101）

摘 要： 报道了采自中国的 12 种稀见的文字衣科（厚顶盘目、子囊菌门）地衣。它们分别隶属于白唇衣属 *Dyplolabia*、刻痕衣属 *Glyphis*、半实衣属 *Hemithecium*、凸唇衣属 *Platygramme*、双实衣属 *Platythecium* 和星衣属 *Sarcographa*，其中中国新记录 6 种。文中每一种地衣均有文字描述、图片和分种检索表。

关键词： 白唇衣属，刻痕衣属，半实衣属，凸唇衣属，双实衣属，星衣属

中图分类号： Q939.5 **文献标识码：** A **文章编号：** 1672-6472（2007）04-0493-0506

One hundred and twenty one species of the family Graphidaceae from China have been accepted and reported so far (Zahlbrückner, 1930; Thrower, 1988; Wu & Qian, 1989; Wei, 1991) according to an unnatural system established by Müller (Müll. Arg., 1880, 1882). During the research on the family, 24 species of 8 genera were found by the authors based on a more natural system (Staiger, 2002). Among them 12 species belonging to 6 genera were determined to be rare and new to China and are reported as follows.

MATERIALS AND METHODS

The present study was based on the materials collected by the authors and other collectors and conserved in the Herbarium Mycologicum Academiae Sinicae-Lichenes (HMAS-L) and Lichen Herbarium of Shandong Agricultural University (LHS).

Supported by the National Natural Science Foundation of China (No.30499340)

*Corresponding author. E-mail: weijc2004@126.com

Received: 2007-04-02, Revised: 2007-06-22

This paper was originally published in *Mycosystema*, 26（4）：493-506，2007.

The study were carried out based on morphological, anatomical and chemical characteristics of the specimens. The dissecting microscope (ZEISS Stemi SV 11) and the compound microscope (OPTON III; Olympus CHB213) were used for the morphological and anatomical studies. The lichen substances were detected by the color reaction and thin-layer chromatography (TLC) (Culberson & Kristensson, 1970; Culberson, 1972; White & James, 1985). The color reaction of spores was carried out by Lugol's solution (I).

RESULTS AND DISCUSSIONS

Twelve species belonging to 6 genera, viz. *Dyplolabia* (1 species), *Glyphis* (1 species), *Hemithecium* (3 species), *Platygramme* (2 species), *Platythecium* (2 species) and *Sarcographa* (3 species) in the Graphidaceae are reported in this paper. A working key, description and short discussion to each species are provided.

KEY TO THE 12 SPECIES OF GRAPHIDACEAE FOUND IN CHINA

1a. Apothecia embedded in stromata……………………………………………………………………2

1b. Apothecia not embedded in stromata…………………………………………………………………5

2a. Stromata carbonized entirely; mature ascospores hyaline, with transverse septa only, 8-10(-12)-locular, (29-) 33-38 × 9-11μm……………………………………………………… ***Glyphis cicatricosa*** Ach.

2b. Mature ascospores brownish………………………………………………………………………3

3a. Stictic acid present, stromata oblong and raised from the thallus distinctly, hymenium with secondarily compartments…………………………………………… ****Sarcographa heteroclita***(Mont.)Zahlbr.

3b. No lichen compounds present, stromata not raised from the thallus distinctly and approximately round, hymenium without secondarily compartments………………………………………………………4

4a. The carbonization of exciple spreading to adjacent stromata; mature ascospores 4-locular, 18-20×6-7.5μm…. ……………………………………………………………… ****Sarcographa medusulina***(Nyl.)Müll.Arg.

4b. The carbonization of exciple not spreading to adjacent stromata; mature ascospores 4-locular, 15-18 × 6-8μm ……………………………………………………………………… ***Sarcographa tricosa***(Ach.)Müll.Arg.

5a. Apothecia covered by fine white pruina(C+ red)…………………………… ***Dyplolabia afzelii***(Ach.)A.Massal.

5b. Apothecia not covered by fine white pruina, or covered by pruina but lacking the color reaction mentioned above……………………………………………………………………………………6

6a. Labia weakly developed, ascospores minuteness(≤15μm)……………………………………………7

6b. Labia well developed, ascospores larger…………………………………………………………8

7a. Salazinic acid present…………………………………… ****Platythecium colliculosum***(Mont.)Staiger

7b. Norstictic acid present…………………………………… ****Platythecium dimorphodes***(Nyl.)Staiger

8a. The lateral exciple carbonized wedge-shapely……………………………………………………9

8b. Exciple uncarbonized, labia divergent and covered by well-developed thalline margin……………………10

9a. Stictic acid and its compounds present; mature ascospores with transverse septa only, 4-6-locular, 16-22(-26) × 5-8μm…………………………………………………… ***Platygramme discurrens***(Nyl.)Staiger

9b. No lichen compounds present; mature ascospores submuriform, 10-12/1-2-locular, (47-)49.5-56 ×13-16 (-18)μm.. *__Platygramme pachyspora__(Red.)Staiger

10a. Proper margin indistinctly crenate; hymenium clear; mature ascospores with distinct halo, 8/ascus, 18-22/ 6-8-locular, 84-100×20-31μm.. *__Hemithecium chapadanum__(Red.)Staiger

10b. Proper margin distinctly crenate; hymenium inspersed; mature ascospores without distinct halo...............11

11a. Mature ascospores 2-4/ascus, 18-22/6-8-locular, 84-94(-100)×20-23μm...
.... **Hemithecium chlorocarpoides** (Nyl.) Sraiger

11b. Mature ascospores 8/ascus, 10-14/3-5-locular, 45-58×12-15μm..**Hemithecium chrysenteron** (Mont.) Trevis.

Note: New records are marked with "＊".

DESCRIPTIONS AND DISCUSSIONS

Dyplolabia afzelii (Ach.) A. Massal., Neagenea Lichenum: 6 (1854).

≡*Graphis afzelii* Ach., Synopsis Methodica Lichenum: 85 (1814).

Description: Thallus corticolous, crustose, within the bark, waxy; apothecia lirelliform, prominent, with thick, white, powdery pruina, straight or curved, usually unbranched, 0.8~3.0mm long and 0.5~0.8mm wide, opening by a slit (Fig.1-A); proper exciple convergent, well-developed and carbonized (Fig.1-B); hymenium clear, 85~102μm tall; mature ascospores 8/ascus, hyaline, ovoid to elongate, 4-locular, cells lenticular, 16~18×6.5~9μm, I-.

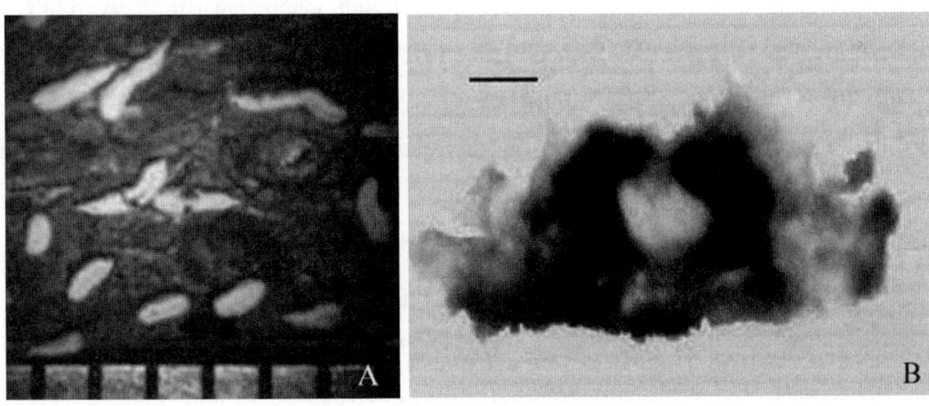

Fig.1 *Dyplolabia afzelii* (Ach.) A. Massal (HMAS-L X008029). A: Apothecia, scale bar = 1mm; B: Cross-section of an apothecium, scale bar =10μm.

Chemistry: pruina C+ red, lecanoric acid

Specimens Examined: YUNNAN: Menglun County, altitude not noted, Jiang Yu-Mei, 1127-10, 30 XII 1980 (HMAS-L X017675). GUANGXI: Shangsi County, 500m, Guo Shou-Yu, 1601, 29 XII 1997 (HMAS-L X008029).

Discussion: This species is characterized by its lirella covered by the fine white pruina with C+ reaction, the carbonized proper exciple and the typical hyaline *Graphis*-type ascospores. It was put under the genus *Graphis* by former investigators. Once Massalongo used this species as type to establish the genus *Dyplolabia* in 1854 but such a treatment was not accepted by other researchers (Kalb & Staiger, 2001). Kalb and Staiger held out Massalongo's conclusion, and put the fine white pruina with C+ reaction as the main character that can distinguish the species of the genus *Dyplolabia* from the similar genera (Kalb & Staiger, 2001). This species was reported under the Genus *Graphis* in Hong Kong (Thrower, 1988).

Glyphis cicatricosa Ach., Synopisis Methodica Lichenum: 107 (1814).

Description: Thallus corticolous, crustose, flat, waxy, pale to greenish fawn; apothecia lirelliform or spot-like, numerous, embedded in conspicuous, raised, cream-colored stromata; stromata oval to irregularly elongate, 1.0~3.5mm long, 1.0~1.5mm wide, carbonized entirely and covered with white tissue; aging lirella can be more crowded, branched and even covered the surface of the stromata region; discs red brown to dark brown, epruinose, flat to somewhat concave (Fig.2-A); proper exciple carbonized entirely, divergent, lacking a thalline cover, lateral part weakly developed while the carbonized base continuous to the carbonized stromata region entirely (Fig.2-B); hymenium inspersed, ca. 100μm tall, with brownish epithecium; mature ascospores 8/ascus, hyaline, oblong-elliptical, with transverse septa only, 8-10(-12)-locular, cells lenticular, (29-) 33~38×9~11μm, I+ violet.

Chemistry: no lichen compounds found.

Specimens Examined: YUNNAN: Xichou County, 1400m, Chen Jian-Bin, 5294, 5348, XI 1991 (HMAS-L 070755, 076884). GUANGDONG: Guangzhou City, 100m, Miao Xiao-Lin, GD35, 2 X 2006(LHS GD35).

Discussion: This is a typical *Glyphis*-species distributed mainly in tropical and temperate regions all over the world, and has been reported from China before. It is characterized by the conspicuous and carbonized stromata, crowded and open lirella, dark red brownish discs, transverse septa and hyaline ascospores as well as the absence of lichen compound. The stromata of this species is quite thick, completely carbonized and fragile. This species was reported from Shanghai, Jiangsu, Zhejiang, Taiwan and Hong Kong in China (Wei, 1991).

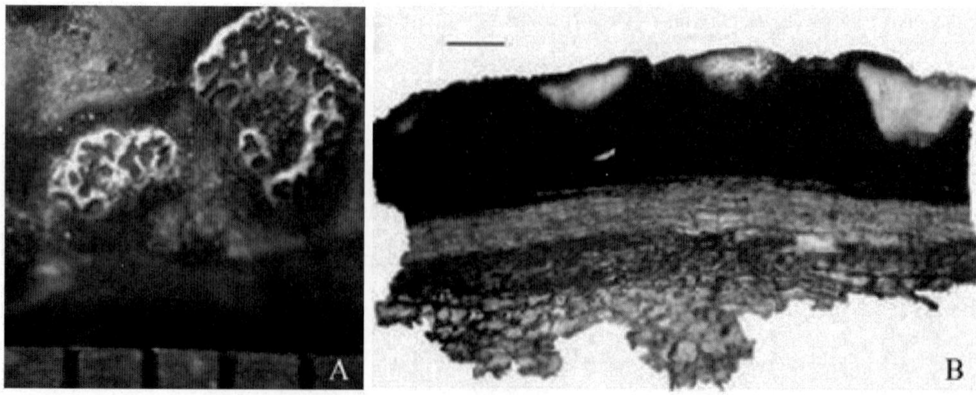

Fig.2 *Glyphis cicatricosa* Ach (LHS GD35). A: Apothecia, scale bar = 1mm; B: Cross-section of apothecia, scale bar =100μm.

Hemithecium chapadanum (Redinger) Staiger, Bib. Lichen., 85: 281 (2002).

≡*Phaeographina chapadana* Redinger, Ark. Bot. 26 A (1): 100 (1934).

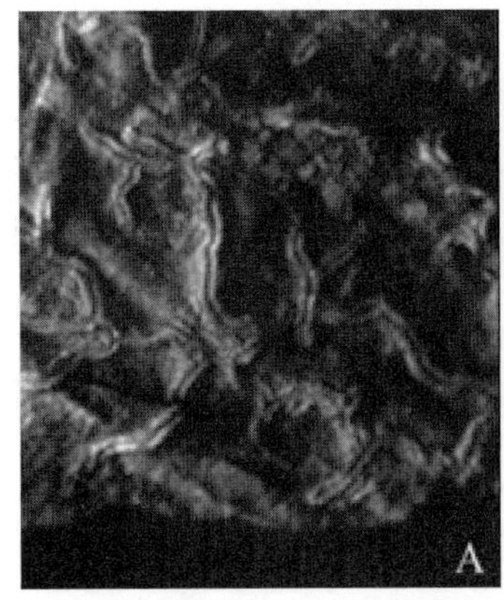

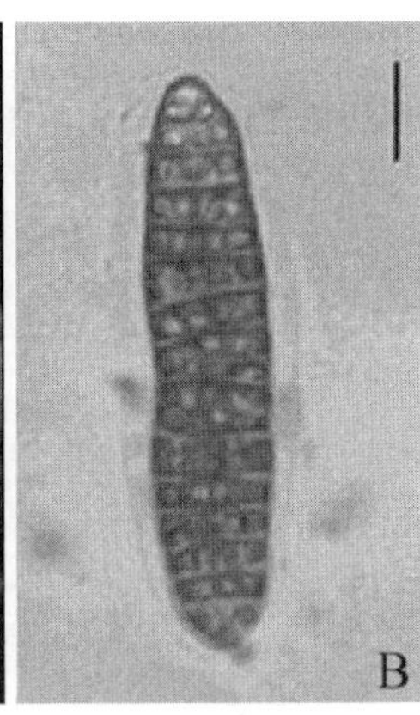

Fig.3 *Hemithecium chapadanum* (Redinger) Staiger (HMAS-L X006981-3). A: Apothecia, scale bar = 1mm; B: Ascospore with halo, scale bar =20μm.

Description: Thallus corticolous, crustose, thick, waxy, pale yellowish to brownish, with cream-colored medullary tissue that can be distinguished from cortex from transverse section; apothecia lirelliform, numerous, large-scale, 2.0~7.0mm long and 0.5mm wide, gently branched, covered entirely by the cream-colored thalline margin (crenate indistinctly); discs cannot be seen and gives a split-liked appearance (Fig.3-A); proper exciple convergent, composed of yellowish swollen and loosely twisted hypha, the basal part weakly developed, the lateral developed radially to adjacent thalline region and the terminal tissue carbonized slightly as a rule; hymenium clear, elliptical, 150~170μm tall, with thick brown epihymenium; mature ascospores 8/ascus, brownish, elongate to bullet-like, muriform, 18-22/6-8-locular, 84~100 × 20~31μm, I+ rufous, with distinct thick halo around (Fig.3-B).

Chemistry: stictic acid.

Specimens Examined: GUANGXI: Wuming County, 1250m, Chen Jian-Bin, Hu Guang-Rong & Xu Lei, 20408, Ⅵ 2001 (HMAS-L X000801); Longsheng County, 900m, Chen Jian-Bin, Hu Guang-Rong & Xu Lei, 20065, 6 Ⅵ 2001 (HMAS-L X000821). HAINAN: Mt. Wuzhi shan, 1340m, Jiang Yu-Mei & Guo Shou-Yu, H-1029, 28Ⅺ 1980 (HMAS-L X005420-1). GUANGDONG: Boluo County, 300m, Guo Shou-Yu, 1948, 14 Ⅹ 1998 (HMAS-L X006981-3). YUNNAN: Luxi City, 1340m, Jiang Yu-Mei, 559, 563, 28 Ⅺ 1980 (HMAS-L X017836-3, X 017682-1).

Discussion: As other species of this genus, this species is characterized for its large-scale and strongly convergent lirella, radical-developed proper exciple and large, muriform ascospores. The distinct halo of mature ascospores and indistinctly crenate thalline margin are characteristics distinguished this species from analogous species. This species is reported for the first time in China.

Hemithecium chlorocarpoides (Nyl.) Staiger, Bib. Lichen., 85: 283 (2002).

≡*Graphis chlorocarpoides* Nyl., Flora 49: 133 (1866);

≡*Phaeographina chlorocarpoides* (Nyl.) Zahlbr., Cat. Lich. Univ. Ⅱ: 435 (1923).

Description: Thallus corticolous, crustose, thick, with cream-colored medullary tissue that can be distinguished from cortex from transverse section, upper surface waxy, pale yellowish; apothecia lirelliform, numerous, large-scale, 1.0~4.0mm long and 0.3~0.5mm wide, gently branched, covered entirely by the cream-colored thalline margin(crenate distinctly); discs concealed and gives a split-liked appearance (Fig.4-A); proper exciple convergent, composed of yellowish swollen and loosely twisted hypha, the basal region may thick and darkened, the lateral developed radially to adjacent thalline region and the terminal tissue carbonized slightly as a rule (Fig.4-B); hymenium inspersed, elliptical, 135~150μm tall, with thick brown epihymenium; mature ascospores 2-4/ascus, brownish, elongate to bullet-like, muriform, 18-22/6-8-locular, 84~94(~100)×20~23μm, I+ rufous, without halo around.

Chemistry: stictic acid.

Specimens Examined: GUANGXI: Lingui County, 1250m, Chen Jian-Bin, Hu Guang-Rong & XU Lei, 20434, 6 Ⅵ 2001 (HMAS-L X000803); Lingui County, 1460m, Chen Jian-Bin, Hu Guang-Rong & Xu Lei, 20266, 6 Ⅵ 2001(HMAS-L X000805, X000806, X000807); Wuming County, 1100m, Guo Shou-Yu, 1283, 19 Ⅻ 1997(HMAS-L 067951). HAINAN: Mt. Wuzhi shan, Jiang Yu-Mei, H-1029, 20 Ⅳ 1993 (HMAS-L X005420-2); Mt. Diaoluo shan, 900m, Chen Jian-Bin, Hu Guang-Rong & Xu Lei, 20496, 16 Ⅵ 2001 (HMAS-L 076911); Mt. Diaoluo shan, Wei Jiang-Chun, 6 Ⅵ 2006 (HMAS-L HN3-1).

Discussion: This species is characterized by its large-scale and strongly convergent apothecia. The hymenium, proper exciple and ascospores of this species are similar to those of *Hemithecium chapadanum*, but it can be easily distinguished from *H. chapadanum* by its prominently crenate thalline margin and ascospores without any halo around (immature ascospres may have an unconspicuous halo). This species was reported from Guangdong (Zahlbückner, 1930) and Hong Kong (Thrower, 1988) in China under the genus *Phaeographina*.

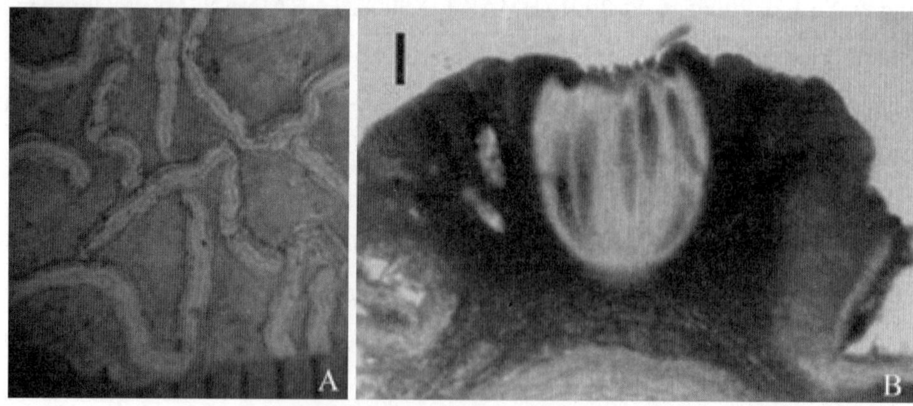

Fig.4 *Hemithecium chlorocarpoides* (Nyl.) Staiger (HMAS-L X000803). A: Apothecia, scale bar = 1mm; B: Cross section of an apothecium, scale bar = 50μm.

Hemithecium chrysenteron (Mont.) Trevis., Spighe e Paglie 1: 13 (1853).

≡*Graphis chrysenteron* Mont., Ann. Sci. Nat., Bot. 18: 270 (1842);

≡*Leucogramma chrysenteron* (Mont.) A. Massal., Atti Reale Ist. Veneto Sci. Lett. Arti ser. 3 (5): 320

(1860);

≡*Phaeographina chrysentera* (Mont.) Müll. Arg., Hedwigia 30: 52 (1891).

Description: Thallus corticolous, crustose, thick, with cream-colored medullary tissue that can be distinguished from cortex from transverse section, upper surface waxy, pale yellowish to somewhat olive green; apothecia lirelliform, large-scale, 1.0~2.5mm long and 0.3~0.5mm wide, gently branched, covered entirely by the cream-colored thalline margin (crenate distinctly); discs concealed and gives a split-liked appearance (Fig.5-A); proper exciple convergent, composed of yellowish swollen and loosely twisted hypha, the basal region may be thick and darkened, the lateral developed radially to adjacent thalline region and the terminal tissue carbonized slightly as a rule (Fig.5-B); hymenium inspersed, elliptical, 100~115μm tall, with thick brown epihymenium; mature ascospores 8/ascus, brownish, elongate to bullet-like, muriform, 10-14/3-5-locular, 45~58×12~15μm, I+ rufous, without halo around.

Chemistry: stictic acid.

Specimens Examined: YUNNAN: Luxi City, 1340m, Jiang Yu-Mei, 559, 28 Ⅺ 1980 (HMAS-L X017836-4). Yizhang County, 1200m, Jiang Yu-Mei & Yang Jun, M032, 12 Ⅸ 2002 (HMAS-L X025266, X025267) and 1150m, Jiang Yu-Mei & Yang Jun, M062, 12 Ⅸ 2002 (HMAS-L 067956). GUANGXI: Shangsi County, 300m, Guo Shou-Yu, 1560, 27 Ⅻ 1997 (HMAS-L 067954-1).

Discussion: This species is quite similar to *Hemithecium chlorocarpoides* in the appearance, proper exciple structure and chemistry. These two species can be distinguished by the height of hymenium and the size of ascospores. This species was reported under *Phaeographina* from Fujian in China (Zahlbrückner, 1930).

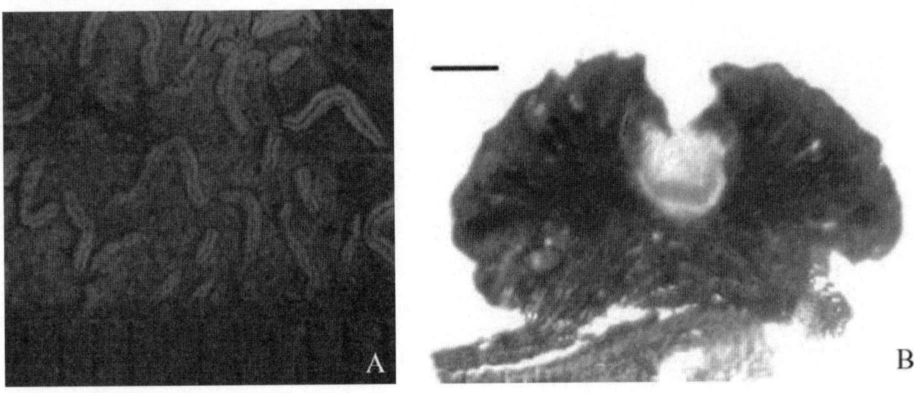

Fig.5 *Hemithecium chrysenteron* (Mont.) Trevis (HMAS-L X017836-4). A: Apothecia, scale bar = 1mm; B: Cross section of an apothecium, scale bar = 100μm.

Platygramme discurrens (Nyl.) Staiger, Bib. Lichen., 85: 361 (2002).

≡*Graphis discurrens* Nyl., Ann. Sci. Nat., Bot. Ser. 4, 19: 358 (1863);

≡*Phaeographina discurrens* (Nyl.) Müll. Arg., Flora 65: 604 (1882).

Description: Thallus corticolous, crustose, thin, upper surface flat to somewhat rough, pale yellowish; apothecia lirelliform, branched vastly, 3.0~4.0mm long and 0.1~0.2mm wide; proper margin well-developed, dark brown, slightly pruinose; discs mostly can be seen, dark brown, slightly pruinose, flat to somewhat concave (Fig.6-A); proper exciple convergent slightly, uncarbonized at basal and most of the lateral region, except the labia that carbonized wedge-shaped; crystal compounds existed in thalline margin (Fig.6-B); hymenium inspersed, 70~95μm tall, with thick brown epihymenium and obvious subhymenium (ca. 15μm); mature ascospores 8/ascus, brownish, elongate, with transverse septa only, 4-6-locular, 16~22(~26)×5~8μm, I+ rufous.

Chemistry: stictic acid (major) and constictic acid (trace).

Specimens Examined: YUNNAN: Menglun County, 560m, Jiang Yu-Mei, 1015-3, 23 XII 1981 (HMAS-L X017659) and 650m, Jiang Yu-Mei, 949, 20 XII 1981(HMAS-L X017839). FUJIAN: Mt. Wuyi shan, 250m, Jia Ze-Feng, FJ360, 27III2004 (LHS FJ360).

Discussion: This species is characterized of its wedge-shaped carbonized lateral exciple, the transversely septate mature ascospores, and the presence of stictic acid. The type specimen of this species was collected in Hong Kong by Nylander (Nylander, 1863), falling under the genus *Graphis* and transferred to the genus *Platygramme* by Staiger (Staiger, 2002).

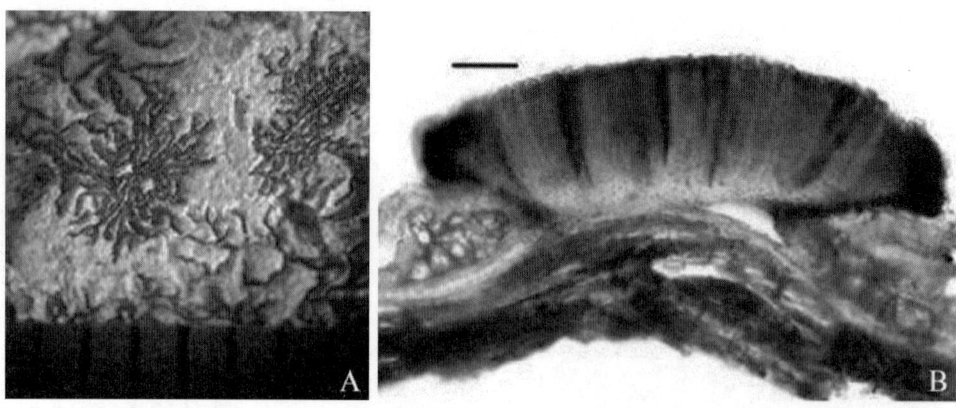

Fig.6 *Platygramme discurrens* (Nyl.) Staiger (LHS FJ360). A: Apothecia, scale bar = 1mm; B: Cross section of an apothecium, scale bar = 50μm.

Platygramme pachyspora (Redinger) Staiger, Bib. Lichen., 85: 364 (2002).

≡*Phaeographis pachyspora* Redinger, Ark. Bot. 27 A (3): 77 (1935).

Description: Thallus corticolous, crustose, thin, pale greenish to brownish; apothecia lirelliform, single, 1.5~5.0mm long and 0.2~0.4mm wide; discs can be seen when soaked, dark brown, slightly pruinose, flat to somewhat concave; proper exciple convergent when soaked, composing of yellow hypha and the lateral carbonized wedge-shaped , while the basal part weakly developed to absent (Fig.7-A); hymenium inspersed, 155~170μm tall; mature ascospores 8/ascus, brownish, elongate, submuriform, 10-12/1-2-locular (Fig. 7B), (47~) 49.5~56 × 13~16 (~18)μm, I+ rufous.

Chemistry: no lichen compounds found.

Specimens Examined: FUJIAN: Mt. Wuyi, 250m, Jia Ze-Feng, FJ 370, 29 III 2004(LHS FJ370).

Discussion: This species is characterized by the conspicuous black-brown labia, the wedge-shaped carbonized lateral exciple and the brownish, submuriform ascospores. The specimen examined here has its own characters relative to the type specimen. First, the labia of the specimen here is convergent in dry circumstance and becomes divergent after soaked; second, the apothecia of the specimen are larger than the type. To make a comprehensive view, the specimen accords with the main characteristics of this species, so it can be fallen under *Platygramme pachyspora* which is a new record from China.

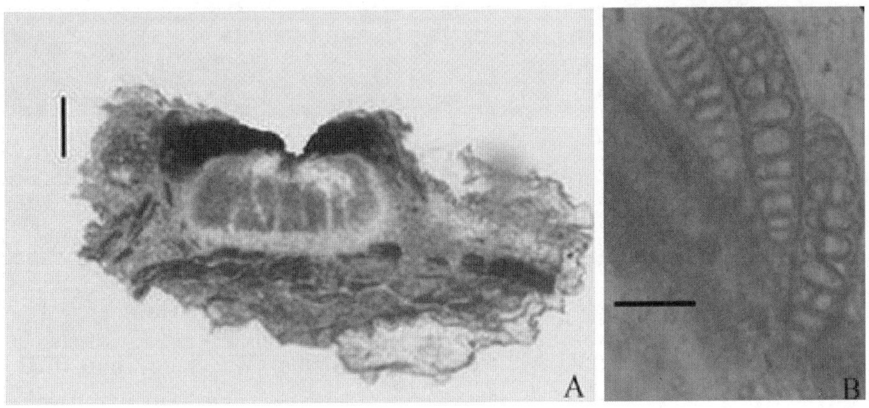

Fig.7 *Platygramme pachyspora* (Red.) Staiger (LHS FJ370). A: Cross section of an apothecium (10×10), scale bar = 100μm; B: Ascospores, scale bar = 10μm.

Platythecium colliculosum (Mont.) Staiger, Bib. Lichen., 85: 380 (2002).

≡*Sclerophyton colliculosum* Mont., Ann. Sci. Nat., Bot. ser. 3, 16: 61 (1851);

≡*Fissurina colliculosum* (Mont.) Massal., Atti. Reale Ist.Veneto Sci. Lett. Arti ser. 3 (5): 276 (1860);

≡*Opegrapha colliculosum* (Mont.) Stizen., Ber. Tatigt. (Jahrb.) St. Gallischen Naturewiss. Ges. 1861: 154 (1862);

≡*Graphis colliculosum* (Mont.) Nyl., Ann. Sci. Nat., Bot. ser. 4, 19: 367 (1863);

≡*Graphina colliculosum* (Mont.) Zahlbr., Cat. Lich. Univ. II: 404 (1923).

Description: Thallus corticolous, crustose, thick, hardness, with cream-colored medullary tissue that can be distinguished from cortex from transverse section, upper surface pale greenish to somewhat brownish, waxy and slightly warty; apothecia unconspicuous, lirelliform, elongate, branched vastly to wavy, at most 8.0mm long and 0.2~0.4mm wide; discs can be seen, jelly-like, brownish, slightly pruinose or not, flat to somewhat concave (Fig.8-A); proper margin covered by cream-colored thalline margin; proper exciple convergent, composed of yellowish to brownish swollen and loosely twisted hypha, the base region weakly developed and the lateral darkened

(Fig.8-B); hymenium clear, 75~80μm tall, subhymenium layer existed which can developed up to 40μm; mature ascospores 8/ascus, hyaline, ovoid to elongate, with transverse septa only or sometimes submuriform, 4/1(-2)-locular, 9~11.5 × 3.5~4.5μm, I+ violet.

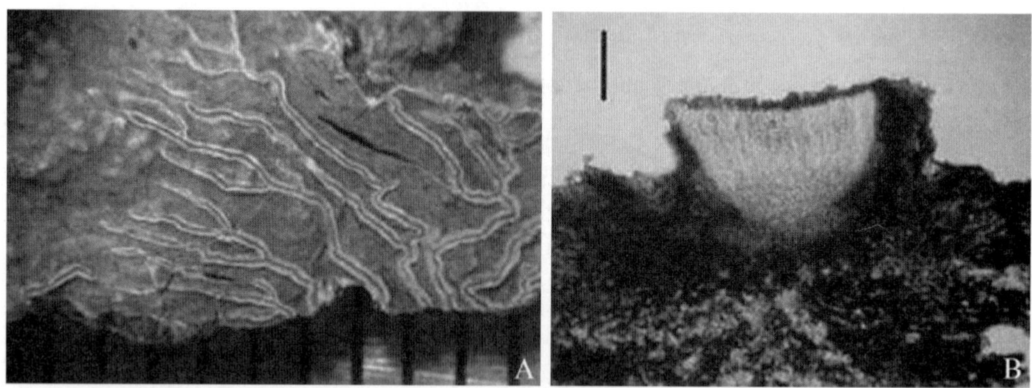

Fig.8 *Platythecium colliculosum* (Mont.) Staiger (HMAS-L 076856). A: Apothecia, scale bar =1mm; B: Cross section of an apothecium, scale bar = 50μm.

Chemistry: salazinic acid.

Specimens Examined: GUIZHOU: Mt. Fanjing shan, 1800m, Wei Jiang-Chun, 0733, 14 Ⅸ 1963 (HMAS-L X017791). HAINAN: Ledong County, 1100m, Chen Jian-Bin, Hu Guang-Rong & Xu Lei, 20773, 29 Ⅵ 2001 (HMAS-L 076856).

Discussion: This is a typical *Platythecium* species which is characterized by the thickened exciple base, tiny ascospores, the presence of norstictic acid and the isidium-like warts situated in the edge of the thallus. Thrower has reported this species under the abolished genus *Graphina* in Hong Kong, with norstictic acid as its chemical substance (Thrower, 1988). It can be confirmed by the description that it might actually be a misconception, and the present report should be the first report of *P. colliculosum* in China. Most of the ascospores of the Chinese specimens examined are aging. The systematic status of this species is presently contentious, the author approve of Staiger (2002) and accept *Platythecium colliculosum* as correct name.

Platythecium dimorphodes (Nyl.) Staiger, Bib. Lichen., 85: 383 (2002).
≡*Graphis dimorphodes* Nyl., Trans. Linn. Soc. 27: 176 (1869);
≡*Fissurina dimorphodes* (Nyl.) Nyl., Acta. Soc. Sci. Fenn. 26 (10): 23 (1900);
≡*Graphina dimorphodes* (Nyl.) Zahlbr., Cat. Lich. Univ. Ⅱ: 404 (1923).

Description: Thallus corticolous, crustose, thick, hardness, with cream-colored medullary tissue that can be distinguished from cortex from transverse section, upper surface pale gray to somewhat olive, waxy and slightly warty, with tiny cream-colored warts (isidia-like, dia. 0.1~0.3mm) situated in the edge; apothecia inconspicuous, lirelliform, elongate, single to gently branched, 0.5~2.0mm long and ca. 0.3mm wide; discs can be seen, jelly-like, brownish, slightly pruinose, flat to somewhat concave(Fig.9-A); proper exciple convergent, composed of yellowish to brownish, swollen and loosely twisted hypha, the basal region thick and darkened while the lateral developed radially into the adjacent thalline margin; hymenium clear, 85~95μm tall, subhymenium layer existed which can developed up to 90μm; mature ascospores 8/ascus, hyaline, ovoid to elongate, submuriform, 4/1-2-locular (Fig.9-B), 11~13.5× 4.5~6μm, I+ violet.

Chemistry: norstictic acid.

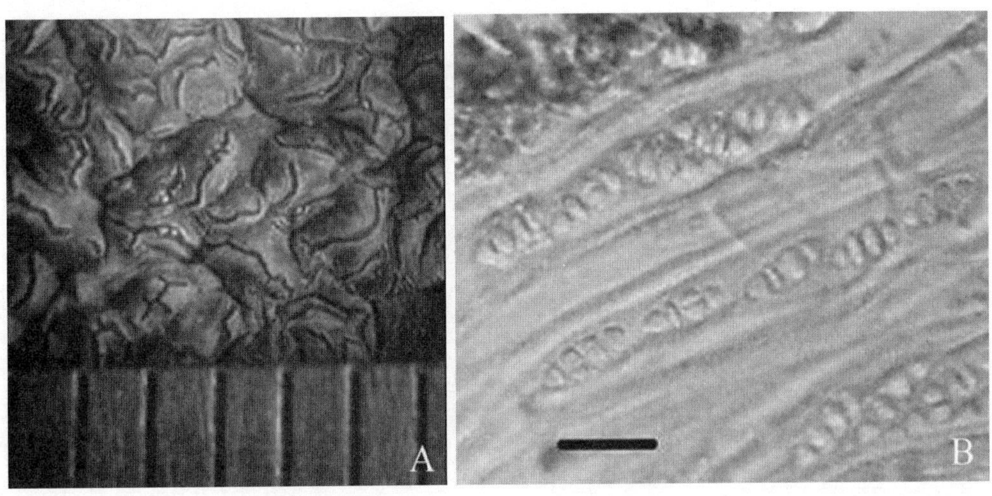

Fig.9 *Platythecium dimorphodes* (Nyl.) Staiger (HMAS-L 071762). A: Apothecia, scale bar =1mm; B: Ascospores, scale bar = 10μm.

Specimens Examined: GUIZHOU: Mt. Fanjing shan, 1300m, Wei Jiang-Chun & Zhang Tao, G420, G426, G567, G191, G444, G198, 5 Ⅷ 2004 (HMAS-L 071762, 071763,071069,071778, 071788,071790).

Discussion: This species is analogous to *Platythecium colliculosum*, the most distinctly difference between them is that *P. dimorphodes* contains norstictic acid instead of salazinic acid. Two specimens are saxicolous according to the collection record, but bark tissue is discovered distinctly as the actual substrate of the thallus during anatomic examination, so these two specimens should be corticolous and fall under such species.

Sarcographa heteroclita (Mont.) Zahlbr., Denkschr. Kaiserl. Akad. Wiss., Wien. Math.-Naturwiss. Kl., 88: 19 (1911).

≡*Glyphis heteroclita* Mont., Ann. Sci. Nat. Bot., Bot. 19: 83 (1843);

≡*Actinoglyphis heteroclita* (Mont.) Kremp., Nuovo Giorn. Bot. Ital. 7: 44 (1875);

≡*Graphis heteroclita* (Mont.) Vain., Ann. Acad. Sci. Fenn., Ser. A 15 (6): 233 (1921).

Description: Thallus corticolous, crustose, thin, flat, waxy, pale yellowish fawn; apothecia numerous, crowded, divided into polygonal pieces, leading intricate appearance, embedded abreast in conspicuous, raised, cream-colored stromata; stromata oval to elongate, 0.5~3.5mm long, 0.5~1.5mm wide, uncarbonized; discs dark brown, slightly pruinose, flat to somewhat concave (Fig.10-A); proper exciple carbonized entirely, divergent, lacking a thalline cover, lateral region weakly developed while the base possesses of increasing thickness (Fig.10-B); hymenium inspersed, 75~100μm high, having secondarily partitions, hypothecium carbonized and forming the thicker basal exciple; mature ascospores 8/ascus, pale brown, oblong-elliptical, with transverse septa only, 4(-6)-locular, cells lenticular, 15~18 × 6~8μm, I+ rufous.

Chemistry: stictic acid and constictic acid.

Specimens Examined: YUNNAN: Menglun County, 680m, Jiang Yu-Mei, 1035-10, XII 1980 (HMAS-L 025270); Xishuangbanna, 680m, Chen Jian-Bin, 6503, VII 1994 (HMAS-L 027682-1).

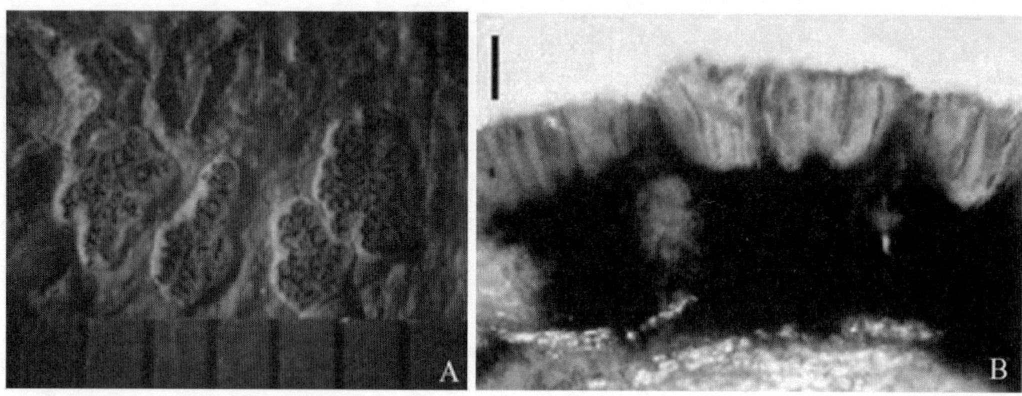

Fig.10 *Sarcographa heteroclita* (Mont.) Zahlbr (HMAS-L 027682-1). A: Apothecia, scale bar = 1mm; B: Cross section of apothecia and the region of stromata, scale bar = 50μm.

Discussion: This species is characterized by the oblong and distinctly raised stromata, the secondary compartments in hymenium that can be observed by appearance of discs, the carbonized basal exciple that cannot be interrupted by the stromata and the brownish ascospores. This species is newly recorded from China.

Sarcographa medusulina (Nyl.) Müll. Arg., Mém. Soc. Phys. Genève 29 (8): 77 (1887).

≡*Glyphis medusulina* Nyl., Acta. Soc. Sci. Fenn. 7: 485 (1863).

Description: Thallus corticolous, crustose, thin, flat, pale yellowish fawn; apothecia clustered, radial-like, strongly branched and immersed in unconspicuous, cream-colored to yellowish stromata; stromata rounded to oval, patches up to 1.6~2.0mm across; discs dark brown, slightly pruinose and sunk more or less (Fig.11-A); proper exciple carbonized entirely, divergent, lacking a thalline cover, and the carbonization of basal region radiates to the stromatic tissue in evidence (Fig.11-B); hymenium inspersed slightly, 85~95μm high, hypothecium carbonized; mature ascospores 8/ascus, pale brown, oblong-elliptical, with transverse septa only, 4-locular, cells lenticular, 18~20 × 6~7.5μm, I+ rufous.

Chemistry: no lichen compound found.

Specimens examined: GUANGDONG: Guangzhou City, Pan Xue-Ping, (HMAS-L 071765-1).

Discussion: This species can be distinguished from *Phaeographis* species by its stromata-like regions, the carbonization of hypothecium and carbonized exciple bases. It is characterized by the carbonization of exciple that can spread to adjacent regions of stromata. This report is new to China.

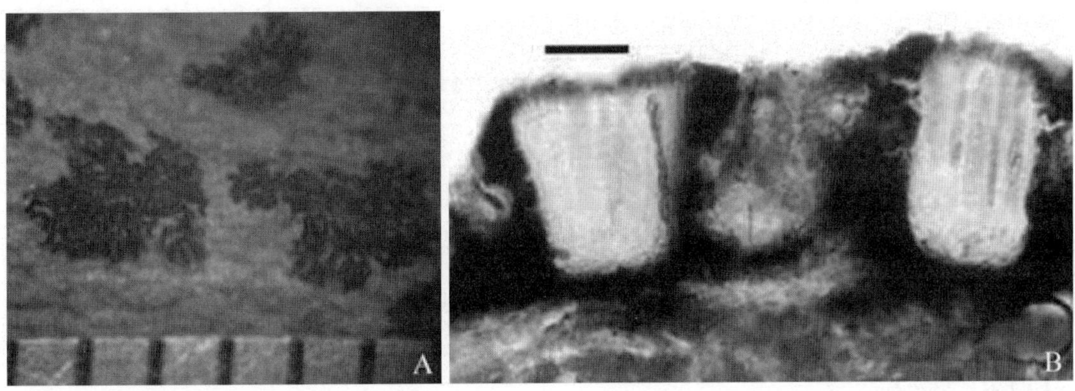

Fig.11 *Sarcographa medusulina* (Nyl.) Müll. Arg (HMAS-L 071765-1). A: Apothecia, scale bar = 1mm; B: Cross section of apothecia and carbonized stromata region, scale bar = 50μm.

Sarcographa tricosa (Ach.) Müll. Arg., Mém. Soc. Phys. Hist. Nat. Genève 29 (8): 63 (1887).

≡*Graphis tricosa* Acharius, Lichenographia Universalis: 674 (1810).

Description: Thallus flat, waxy, olive-green; apothecia clustered, crowded, radial-like, strongly branched and immersed in unconspicuous yellowish stromata; stromata rounded to oval, patches up to about 1.5mm across; discs dark brown, slightly pruinose and sunk more or less (Fig.12-A); proper exciple carbonized well at base and weak at lateral, divergent, lacking a thalline cover, the carbonization of basal region doesn't radiates to the stromatic tissue (Fig.12-B); hymenium inspersed slightly, 60~75μm high, hypothecium slightly carbonized; mature ascospores 8/ascus, pale brown, oblong-elliptical, with transverse septa only, 4-locular, cells lenticular, 15~18 × 6~8μm, I+ rufous.

Chemistry: no lichen compound found.

Specimens Examined: GUANGDONG: Guangzhou City, 80m, Miao Xiao-Lin, GD15, GD17, 2 V 2006 (LHS GD15, GD17).

Discussion: This species is characterized by its unconspicuous stromata, the carbonization of hypothecium and carbonized exciple bases. It is quite similar to *Sarcographa medusulina* by the similarity of appearance, hymenium and ascospore character, and these two species have always been misconcepted. It can only be distinguished by the transverse section of apothecia. Moreover, the carbonization of exciple can spread to adjacent regions of stromata in *S. medusulina*. This species has been reported in China (Zahlbückner, 1930; Thrower, 1988).

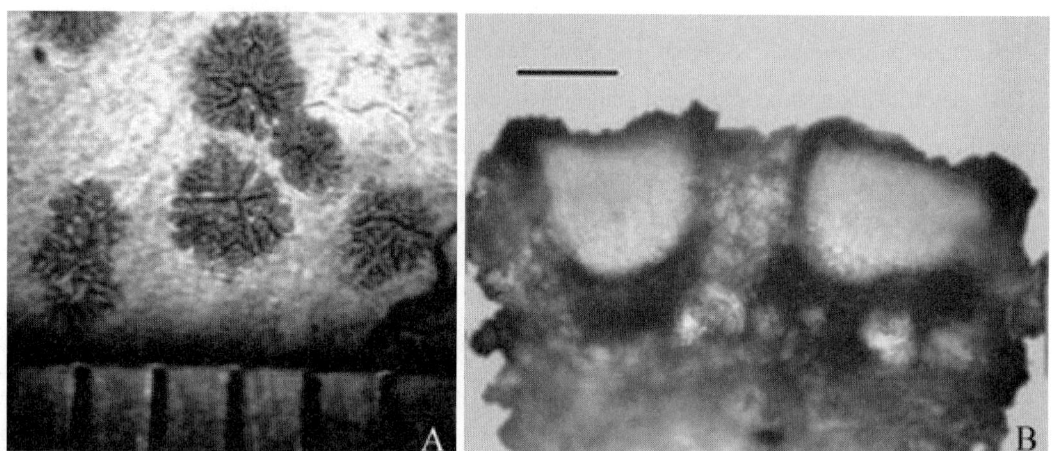

Fig.12 *Sarcographa tricosa* (Ach.) Müll. Arg (LSH GD15). A: Apothecia, scale bar = 1mm; B: Cross section of apothecia, scale bar = 50μm.

[REFERENCES]

Culberson CF, Kristensson H, 1970. A standardized method for the identification of lichen products. *J Chromatogr*, **46**: 85~93

Culberson CF, 1972. Improved conditions and new data for the identification of lichen products by a standardized thin-layer chromatographic method. *J Chromatogr*, **72**: 113~125

Kalb K, Staiger B, 2001. *Dyplolabia* Massalongo, monographie einer vergessenen flechtengattung. *J Hoppea*, **61**: 409~422

Müller A, 1880. Lichenologische beiträge 10. *J Flora*, **63**: 17~24, 40~45

Müller A, 1882. Lichenologische beiträge 15. *J Flora*, **65**: 291~306, 316~322, 326~337, 381~386, 397~402

Nylander W, 1863. Lichens. In: Triana J, Planchon JE (eds.) Prodromus florae Novo-Grannatensis ou Enumeration des plantes de la Nouvelle-Gennade avec description des especes nouvelles. *J Ann Sci Nat, Bot* ser 4, **19**: 286~382

Staiger B, 2002. Die Flechtenfamilie Graphidaceae. Bibliotheca Lichenologica 85, Berlin. 1~526

Thrower SL, 1988. Hong Kong Lichens. An Urban Council Publication, Hong Kong. 1~193

White FJ, James PW, 1985. A new guide to microchemical techniques for the identification of lichen substances. *J Bri Lich Soc Bull*, **57** (Suppl.): 1~41

Wu JN, Qian ZG, 1989.Lichens. In: Xu BS (ed.) Cryptogamic Flora of the Yangtze Delta and Adjacent Regions. Shanghai Scientific & Technical Publishers, Shanghai. 158~266 (in Chinese)

Wei JC, 1991. An Enumeration of Lichens in China. International Academic Publishers, Beijing. 1~278

Zahlbrückner A, 1930. Lichenes in Handel-Mazzetti, Symbolae Sinicae III. Julius Springer In Vienna, Austria. 1~254

[附中文参考文献]

吴继农，钱之广，1989. 地衣：徐炳升（主编）长江三角洲及邻近地区孢子植物志. 上海：上海科学技术出版社. 158~266

Graphis fujianensis, a new species of *Graphidaceae* from China

Ze-Feng Jia[1] & Jiang-Chun Wei[2*]

*zfjia2008@163.com * weijc2004@126.com*

[1] *College of Life Sciences, Shandong Agricultural University Taian, 271018, China*

[2]*Key Laboratory of Systematic Mycology & Lichenology, Institute of Microbiology Academia Sinica, Beijing, 100080, China*

Abstract—*Graphis fujianensis*, from Wuyi Mountain in Fujian, China, is described as new to science.

Key words— curved ascospores, stictic acid

Introduction

During a study of the lichen family *Graphidaceae* (*Ostropales*, *Ascomycota*) in China, an interesting saxicolous species new to science was found. It is placed in the genus *Graphis* as delimited by Staiger (2002).

Material and methods

The type specimen was collected in the forest Wanmulin in Wuyishan, Fujian province, Southern China.

A dissecting microscope (TECH XTS-20) and a research microscope (OLYMPUS CHB-213) were used for the morphological and anatomical studies. Manual cross-sections of lirellae were examined in tap water. The lichen compound was detected by thin-layer chromatography (TLC) (Culberson & Kristensson, 1970; Culberson, 1972).

Taxonomic description

Graphis fujianensis Z.F Jia & J.C. Wei, **sp. nov.** FIGURE 1

MycoBank MB 511505

Differt a G. saxicola ascosporis minoribus curvatibus, sine marginibus thallinis apotheciorum et acidum sticticum cotinenti.

* *Corresponding author*

This paper was originally published in *Mycotaxon*, 104: 107-109, 2008.

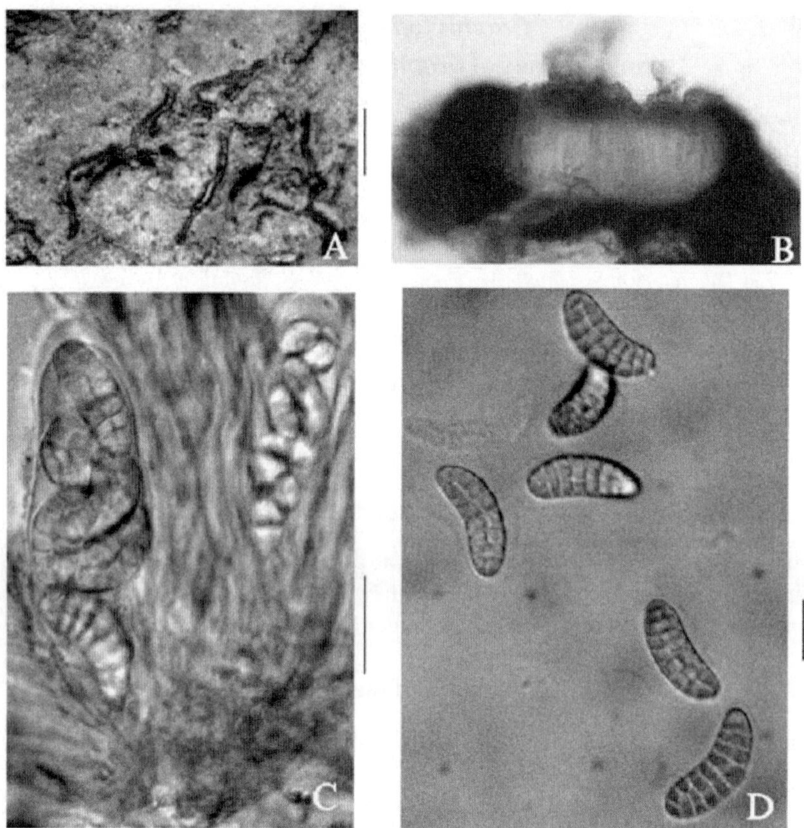

Fig. 1 *Graphis fujianensis*. A. Habit (bar = 1 mm); B. Apothecium cross section (bar = 50 μm); C. Two asci containing ascospores (bar = 20 μm); D. Curved, muriform ascospores (bar = 20 μm)

Typus: CHINA. Fujian Provincia, Jian-ou, Wanmulin, alt. 580m, ad saxa. 2/VI/2007, *Li Jing FJ1182*, (holotypus in LHS; isotypus in HMAS-L.).

ETYMOLOGY: The specific epithet "fujianensis" is derived from the type locality in Fujian province.

THALLUS crustose, pale grayish-white, thin, saxicolous, with dull surface. APOTHECIA lirelliform, black, conspicuous, initially fissurine, becoming sub-immersed to sessile, with slightly raised margins, curved and sinuous, lips sometimes opening slightly, 0.5-4 mm long, 0.1-0.25 mm wide, not grooved, without thalline margins; PROPER EXCIPLE completely carbonized; HYMENIUM hyaline, not inspersed, 100-120 μm tall, clear, I-ve, EPITHECIUM 20-28.5 μm tall, brownish; PARAPHYSES up to 1.5 μm wide, septate, slightly widened at apices;

ASCI cylindric, 70-80 × 13-20 μm, 8-spored; ASCOSPORES oblong-ellipsoid to slightly curved, hyaline, 17.5-33 μm long, 9-13 μm wide, muriform, 8-10 × 1-3-locular, I+ blue.

CHEMISTRY: C-, K+ yellow, P+red; stictic acid (TLC).

The new species differs from *Graphis saxicola* (Müll. Arg.) A.W. Archer (1999, 2005) by the smaller and curved ascospores, the lack of thalline margins and the presence of stictic acid.

Acknowledgments

The authors thank Alan W. Archer and Harrie J. Sipman for presubmission review.

Literature cited

Archer AW. 1999. The lichen genera *Graphis* and *Graphina* (*Graphidaceae*) in Australia. 1. Species based on Australian Type specimens. Telopea 8(2): 273-295.

Archer AW. 2005. New combinations and synonymies in the Australian *Graphidaceae*. Telopea 11(1): 59-78.

Culberson CF. 1972. Improved conditions and new data for the identification of lichen products by a standardized thin-layer chromatographic method. Journal of Chromatography 72: 113-125.

Culberson CF, Kristensson H. 1970. A standardized method for the identification of lichen products. Journal of Chromatography 46: 85-93.

Staiger B. 2002. Die Flechtenfamilie *Graphidaceae* Studien in Richtung einer natürlicheren Gliederung. Bibliotheca Lichenologica 85: 1-526.

A new species of *Melanelixia* (*Parmeliaceae*) from China

HAI-YING WANG [1,2], JIAN-BIN CHEN [1] & JIANG-CHUN WEI [1*]

*lichenwhy@Yahoo.com.cn *weijc2004@126.com*

[1]*Key Laboratory of Systematic Mycology & Lichenology*
Institute of Microbiology, Chinese Academy of Sciences
Beijing 100101, P.R. China

[2]*Graduate University of Chinese Academy of Sciences*
Beijing 100049, P.R. China

Abstract — A new *Melanelixia* species characterized by the presence of cortical hairs and wart-like isidia, *M. subvillosella*, is described from China. *Melanelia subverruculifera* is reported as a new synonym of *Melanelixia villosella*. A key to the nine known *Melanelixia* species is also provided.

Keywords — Asia, lichens, taxonomy

Introduction

The lichen genus *Melanelixia* O. Blanco et al. in the *Parmeliaceae* was segregated from *Melanelia* Essl. based on molecular as well as chemical and morphological data (Blanco et al. 2004). The genus is morphologically characterized by pored or fenestrate epicortex and lecanoric acid as the primary medullary constituent (Blanco et al. 2004, Esslinger 1977). The genus presently includes eight species known in the world: *M. albertana* (Ahti) O. Blanco et al., *M. fuliginosa* (Fr. ex Duby) O. Blanco et al., *M. glabra* (Schaer.) O. Blanco et al., *M. glabroides* (Essl.) O. Blanco et al., *M. huei* (Asahina) O. Blanco et al., *M. subargentifera* (Nyl.) O. Blanco et al., *M. subaurifera* (Nyl.) O. Blanco et al., and *M. villosella* (Blanco et al. 2004). All these *Melanelixia* species except *M. albertana* and *M. subaurifera* have been reported in China (Wei 1991, Abbas et al. 1998, Omar et al. 2004, Chen & Esslinger 2005).

During our study of the lichen flora of China, an interesting species of *Melanelixia* new to science was found. Here we present a brief Latin diagnosis, an extended description, and key to the presently known species of the genus. Five photographs for the new species are provided as well. In addition, as we studied the new species and related taxa, we discovered that *Melanelia subverruculifera* is a synonym of *Melanelixia villosella*.

This paper was originally published in *Mycotaxon*, 104: 185-188, 2008.

Materials and methods

The material examined was collected from Northeast China and is preserved in HMAS-L (Herbarium Mycologicum Academiae Sinicae-Lichenes). For the morphological study of the material, a dissecting microscope (Zeiss Stemi SV 11) has been used and the anatomical study was carried using a compound microscope (Zeiss Axioscop 2 plus). For the ultrastructural observations, a scanning electron microscope (Fei Quanta 200) has been used. Lichen substances were identified using the standardized thin layer chromatography techniques of Culberson (1972).

Key to *Melanelixia* species

1a. Cortical hairs absent 2
1b. Cortical hairs present 5

2a. Thallus with isidia, soredia or both 3
2b. Thallus without isidia or soredia 4

3a. Isidia fine, 0.1-0.25 mm long and 0.02-0.06 mm in diameter, seldom branched; soredia usually present .... *M. subaurifera*
3b. Isidia larger, mostly 0.2-1mm long and 0.05-0.1 mm in diameter, often branched; soredia never present .... *M. fuliginosa*

4a. Corticolous; upper surface of lobes with obscure to distinctive pseudocyphellae; often with the orange pigment rhodophyscin in the lower medulla .... *M. huei*
4b. Saxicolous; upper surface of lobes without pseudocyphellae; without rhodophyscin .... *M. glabroides*

5a. Thallus with soredia 6
5b. Thallus without soredia 7

6a. Soralia marginal and strongly labriform .... *M. albertana*
6b. Soralia laminal and marginal, the laminal ones arising from small, ± hemispherical pustules .... *M. subargentifera*

7a. Thallus without isidia .... *M. glabra*
7b. Thallus with isidia 8

8a. Isidia cylindrical, often branched and with obvious hyaline hairs at their tips .... *M. villosella*
8b. Isidia wart-like, non-branched and without hairs on their tips .... *M. subvillosella*

Taxonomic description

Melanelixia subvillosella H.Y. Wang & J.C. Wei, sp. nov. FIG. 1

MYCOBANK MB 511544

Species haec a M. glabra verrucis et a M. villosella sine trichomis in pagina isidiorum differt.

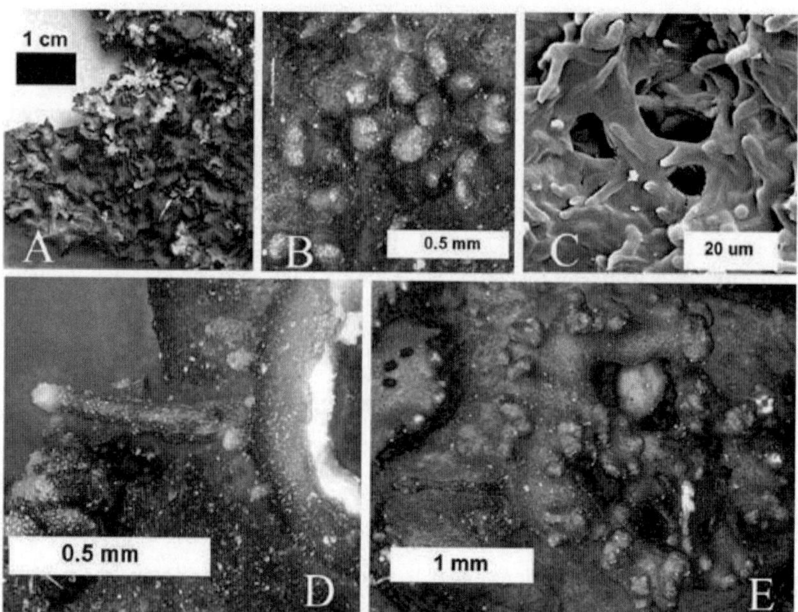

Fig.1 Photographs of *Melanelixia subvillosella*. A. Thallus. B. Isidia. C. SEM photograph, showing the pored and coarse epicortex. D. Rhizinae and hairs on an apothecial margin. E. Rhizina, papilla and lobules on an apothecial margin.

Holotypus a China, Jilin provincia, Mt. Changbaishan, alt. 1100m, ad corticem arborum, X.D. Lu 0211, 24 Jun. 1985 in HMAS-L (077798) conservatur.

Expanded description — Thallus foliose, 10 cm in diameter (Fig. 1A). Lobes 1-3 mm broad, more or less flat, elongated, sometimes crenulate on the margin, contiguous to imbricate. Upper surface olive-brown to dark-brown; rather shiny at lobe ends, inward becoming dull; with obvious cortical hairs, especially on the lobe ends and the apothecial margins (Fig. 1D); without pseudocyphellae or soredia; isidiate (Fig. 1B), the isidia arising as hemispherical papillae, and then slightly elongating into short claviform isidia (0.2-0.1×0.5-1 mm); lobulate, the lobules short and rounded, marginal to laminal. Lower surface black, often paler at the margin; smooth to wrinkled, dull to slightly shiny; abundantly rhizinate, the rhizines simple to branched, concolorous with the lower surface, to 2 mm long. Lobes 130-160 μm thick, upper cortex 8-12 μm thick; lower cortex 8-11 μm thick. Epicortex pored (pore, 6-16 μm in diameter) and coarse (Fig. 1C). Apothecia common, sessile to short stipitate, concave, to 4 mm in diameter; margin entire when small, soon becoming crenate-papillate and lobulate (Fig. 1E), often with rhizinae (Fig. 1D,E); hymenium 54-70 μm thick, subhymenium 50-70 μm thick; spores 8, globose to ellipsoid, 8-10×5-7 μm, spore wall 1 μm thick. Pycnidia unknown.

CHEMISTRY — Cortex K–, HNO_3–; medulla C+ red, K–, PD–; containing lecanoric acid.

DISTRIBUTION AND SUBSTRATE — At present, *M. subvillosella* is known only from the type locality; on bark.

COMMENTS — The apothecia with rhizinae and the coarse epicortex distinguish *M. subvillosella* from all the other *Melanelixia* species. The new species can be clearly separated from *M. glabra* by having wart-like isidia and from *M. villosella* by lacking the cortical hairs on the surface of isidia.

Melanelixia villosella (Essl.) O. Blanco, A. Crespo, Divakar, Essl., D. Hawksw. & Lumbsch, Mycol. Res. 108(8): 882 (2004)

= *Parmelia villosella* Essl., J. Hattori. Bot. Lab. 42: 95 (1977)

= *Parmelia subverruculifera* J.C. Wei & Y.M. Jiang, Acta Phytotax. Sin. 18: 386 (1980) = *Melanelia subverruculifera* (J.C. Wei & Y.M. Jiang) J.C. Wei, Enum. Lich. China: 153 (1991). (syn. nov.) Holotype: CHINA. Tibet, Nielamu, alt. 3840m, 22 Jun 1966, J.C. Wei & J.B. Chen 1899-1 (HMAS-L).

Acknowledgements

The project was financially supported by the National Natural Science Foundation of China (30499340). The authors would like to thank the keeper of the HMAS-L Ms Deng Hong for giving assistance during this study. The authors thank Teuvo Ahi, Theodore L. Esslinger, and Syo Kurokawa for expert presubmission review.

Literature cited

Abbas A, Wu JN. 1998. Lichens of Xinjiang. 178 pp. Sci-Tech & Hygiene Publishing House of Xinjiang, Urumqi.

Blanco O, Crespo A, Divakar PK, Esslinger TL, Hawksworth DL, Lumbsch HT. 2004. *Melanelixia* and *Melanohalea*, two new genera segregated from *Melanelia* (*Parmeliaceae*) based on molecular and morphological data. Mycological Research 108(8): 873-884.

Chen JB, Esslinger TL. 2005. *Parmeliaceae* (*Ascomycota*) lichens in China's mainland IV. *Melanelia* species new to China. Mycotaxon 93: 71-74.

Culberson CF. 1972. Improved conditions and new data for the identification of lichen products by a standardized thin-layer chromatographic method. Jour Chromatogr. 72: 113-125.

Esslinger TL. 1977. A chemosystematic revision of the brown *Parmeliae*. Jour. Hattori Bot. Lab. 42: 1-211.

Omar Z, Keyimu A, Abbas A. 2004. New Chinese records of the lichen genus *Melanelia*. Acta Bot. Yun. 26: 385-386.

Wei JC. 1991. Enumeration of lichens in China. 278 pp. International Academic Publishers. Beijing.

A phylogenetic analysis of *Melanelia tominii* and four new records of brown parmelioid lichens from China

Hai-Ying Wang [1,2], Jian-Bin Chen [1] & Jiang-Chun Wei [1*]

*lichenwhy@Yahoo.com.cn chenjbin@yahoo.com * weijc2004@126.com*

[1]*Key Laboratory of Systematic Mycology & Lichenology*
Institute of Microbiology, Chinese Academy of Sciences
Beijing 100101, P.R. China

[2]*Graduate University of Chinese Academy of Sciences*
Beijing 100049, P.R. China

Abstract -- The molecular analysis based on ITS nrDNA sequences indicates that *Melanelia tominii* probably belongs to *Melanelixia*. Four new records from China — *Melanelia predisjuncta, Melanohalea subelegantula, M. olivaceoides,* and *M. septentrionalis* — are reported. A key to the 21 species belonging to *Melanelixia, Melanohalea* and *Melanelia* from China is provided.

Keywords -- Asia, taxonomy, gyrophoric acid

Introduction

The lichen genus *Melanelia* (*Parmeliaceae*) was originally established by Esslinger in 1978. Two more genera, *Melanelixia* O. Blanco et al. and *Melanohalea* O. Blanco et al., were subsequently split from *Melanelia*, based on molecular as well as chemical and morphological data (Blanco et al. 2004). *Melanelixia* is characterized by having a pored or fenestrate epicortex, by lacking pseudocyphellae and by containing lecanoric acid as the primary medullary constituent (Blanco et al. 2004, Esslinger 1977). *Melanohalea* is characterized by pseudocyphellae, often on warts or isidial tips, by a non-pored epicortex, and by a medulla containing depsidones or lacking secondary compounds (Blanco et al. 2004, Esslinger 1977). The placement of the type species of *Melanelia*, *M. stygia* (L.) Essl., outside the parmelioid lichens was strongly supported in the molecular systematic studies (Blanco et al. 2004). Although *Melanelia tominii* resembles *M. stygia* morphologically, the former species contains the tridepside gyrophoric acid (usually with other tridepsides as well), and the latter contains

* *Corresponding author*

This paper was originally published in *Mycotaxon*, 107: 163-173, 2009.

the β-orcinol depsidones fumarprotocetraric and protocetraric acid (Esslinger 1977, 1992). Because molecular sequences were not available for *M. tominii*, its systematic position remained uncertain (Blanco et al. 2004).

Worldwide, *Melanelixia* includes nine known species, *Melanohalea* nineteen species and *Melanelia* still contains a heterogeneous residue of seventeen species (Esslinger 1977, 1978, 1987, 1992; Ahti et al. 1987, Egan 1987, Galloway & Jørgensen 1990, Thell 1995, Divakar et al. 2001, 2003; Blanco et al. 2004, Divakar & Upreti 2005, Wang et al. 2008). In China, *Melanelixia* includes seven species, *Melanohalea* five species and *Melanelia* five species (Wei 1991, Abbas & Wu 1998, Kurokawa & Lai 2001, Zibirnisa et al. 2004, Chen & Esslinger 2005, Wang et al. 2008).

During our study of these genera in China, four new records were discovered, namely *Melanelia predisjuncta*, *Melanohalea subelegantula*, *M. olivaceoides* and *M. septentrionalis*, and the systematic position of *M. tominii* was investigated based on its ITS sequences (including ITS1, 5.8S nrDNA and ITS2) and morphological and chemical characters. In addition, a key to 21 species belonging to *Melanelixia*, *Melanohalea* and *Melanelia* in China is provided.

Materials and methods

Morphology and Chemistry

The specimens studied are housed in HMAS-L (Lichen Section, Herbarium of the Institute of Microbiology, Academia Sinica) unless otherwise indicated. The morphology of the lichen specimens was examined using a Zeiss stereo microscope (Stemi SV 11) and Zeiss compound microscope (Axioscop 2 plus).

TABLE 1. Specimens of *Melanelia tominii* in which the morphology, chemistry or ITS sequences were studied.

HERBARIUM ACCESSION #	SPECIMEN INFORMATION	GENBANK ACCESSION #
114000	CHINA. Hebei, Mt. Wulingshan, alt. 1750m, on rock, T. Zhang & H.Y. Wang, WLS||042, May 17, 2004.	EU784154
036389	CHINA. Inner Mongolia, Mt. Arxan, alt. 1600m, on rock, J.C. Wei et al., Aer192, August 2, 2002.	EU784155
071058	CHINA. Inner Mongolia, Bairin Youqi, alt. 1800m, on rock, J.B. Chen & G.R. Hu, 21423, August 27, 2001.	EU784156
029902	CHINA. Sichuan, Mt. Gongga, alt. 3300m, on rock, X.Y. Wang et al., 8987, July 26, 1982	–
007086	CHINA. Hebei, Mt. Xiaowutaishan, alt. 2800m, on rock, J.C. Wei, 2042, August 16, 1964.	–
007087	CHINA. Tibet, Mt. Qomolangma, alt. 5000m, on rock, J.C. Wei & J.B. Chen, 1332, June 2, 1966.	–
077822	CHINA. Tibet, Chayu County, alt. 4250m, on rock, J.J. Su, 4801, September 26, 1982.	–
080967	U.S.A. Arizona, Cochise County, alt. 1830m, on rock, T.L. Esslinger, 12261, January 10, 1992.	–

Lichen substances in all specimens cited were identified using the standardized thin layer chromatography techniques (Culberson 1972). Information on the specimens of *M. tominii* studied is shown in TABLE 1.

Molecular systematics

TAXON SAMPLING — Sequence data of the ITS nrDNA of *M. tominii* were obtained from three specimens (TABLE 1). Fifteen sequences of other related taxa were downloaded from GenBank (TABLE 2). *Lecanora leptyrodes* and *L. rupicola* were used as outgroup.

TABLE 2. Species and ITS sequences downloaded from GenBank.

SPECIES	GENBANK ACC. #	SPECIES	GENBANK ACC. #
Melanelia disjuncta	AY611077	*Melanohalea elegantula*	AY611094
M. hepatizon	AF451776	*M. exasperata*	AY611081
M. stygia	AY611121	*M. olivacea*	AY611091
		M. septentrionalis	AY611093
Melanelixia fuliginosa	AY611088	*M. subelegantula*	AY611115
M. glabra	AY611114	*M. subolivacea*	AY611123
M. subargentifera	AY611098	*Lecanora rupicola*	DQ451666
M. subaurifera	AY611099	*L. leptyrodes*	AY541255

PCR AMPLIFICATION AND SEQUENCING — Total DNA was extracted by the modified CTAB method (Rogers and Bendich 1988). DNA extracts were used for PCR amplification of the ITS nrDNA with ITS1 (White et al. 1990) and 1R (TATGCTTAAGTTCAGCGGGT) as primers. PCR reactions were performed in a DNA Thermal Cycler (Biometra) as follows: initial denaturation at 95°C for 3 min, followed by 35 cycles of 30 s denaturation at 94°C, 45 s annealing at 58°C, 1 min extension at 72°C, and completed with a final 8 min extension at 72°C. Products were purified with Gel Extraction Mini Kit (SABC). Sequencing reactions were carried out by Shanghai Genecore Corp. with an ABI 3700 Sequencer. Both complementary strands of each sample were sequenced.

DATA ANALYSIS —The alignment was analyzed using the programs ClustalX 1.8.1. The aligned ITS matrix was edited manually and the flanking regions of the small subunit and large subunit rDNA were deleted through software MEGA 4 (Tamura et al. 2007). Phylogenetic analyses were conducted also in MEGA4. The phylogenetic tree was inferred using the Minimum Evolution method (Rzhetsky & Nei 1992), of which the reliability was tested by 1000 bootstrap replications (Felsenstein 1985). The evolutionary distances were computed using the Maximum Composite Likelihood method (Tamura et al. 2004). All positions containing alignment gaps and missing data were eliminated only in pairwise sequence comparisons (Pairwise deletion option).

The phylogenetic analysis of *Melanelia tominii*

Melanelia tominii (Oxner) Essl., Lichenologist 24(1): 17 (1992)
≡ *Parmelia tominii* Oxner, Zh. Bio.-Bot. Tsyklu, Kyev 1933(7–8): 171 (1933)
= *Parmelia substygia* Räsänen, Lichenes Fenniae Exs. 51 (1935)
≡ *Melanelia substygia* (Räsänen) Essl., Mycotaxon 7: 47 (1978)
= *Parmelia borisorum* Oxner, Bot. Zh., Kyyiv 1: 33 (1940)
= *Parmelia saximontana* R.A. Anderson & W.A. Weber, Bryologist 65: 236 (1963)
= *Parmelia altaica* Oxner, Ukr. bot. Zh. 27(2): 176 (1970)

ITS PHYLOGENETIC ANALYSIS — There were a total of 501 positions in the final dataset of ITS sequences. Six *Melanohalea* species, four *Melanelixia* species and four *Melanelia* species are included in the inferred tree based on ITS (FIG.1). All the *Melanohalea* species formed a monophyletic clade supported by 83% bootstrap value. All the *Melanelixia* species clustered in a clade supported by 70% bootstrap value. *Melanelia stygia* and *M. hepatizon* formed a monophyletic clade supported by 99% bootstrap value. The *Melanohalea* clade, the *Melanelixia* clade and *Melanelia disjuncta* form a large clade supported by 92% bootstrap value, while the clade comprised of *M. stygia* and *M. hepatizon* becomes the outgroup of the former three clades. Although *M. disjuncta* clustered in a clade together with *Melanohalea*, the clade has low bootstrap support (<50%) so the placement of *M. disjuncta* remains uncertain. These results are consistent with previous analyses based on polygenes (Blanco et al. 2004, Thell et al. 2002). In our study, *Melanelia tominii* 1 represents a specimen from Mt. Wuling, Hebei Province, China (herbarium accession no. 114000). *M. tominii* 2 represents two specimens from Inner Mongolia Province, China (herbarium accession nos. 036389, 071058), for which the ITS sequences are identical. *M. tominii* 1 and *M. tominii* 2 form a clade supported by 100% bootstrap value. Within the *Melanelixia* clade (70% bootstrap value), *M. tominii* and the interior clade comprised of *Melanelixia fuliginosa* and *M. subaurifera* (99% bootstrap value) form a moderately supported clade (58% bootstrap value), while another interior clade comprised of *M. glabra* and *M. subargentifera* (100% bootstrap value) becomes the outgroup of the three former clades. The fact that *M. tominii* locates within the *Melanelixia* clade in the phylogenetic tree, strongly indicates that *M. tominii* belongs to *Melanelixia* rather than *Melanelia*.

MORPHOLOGY AND CHEMISTRY — *Melanelia tominii* was included in the nominal subgenus *Melanelia* (Esslinger 1978) together with *M. stygia*, *M. disjuncta*, *M. panniformis*, *M. predisjuncta*, and *M. sorediata*. The rather small, dark, narrow lobed saxicolous thalli with the effigurate pseudocyphellae distinguish the members of this group from all the other species previously included in *Melanelia*. Further, both *M. stygia* and *M. tominii* have cylindrical to bifusiform conidia. However, *M. stygia* is chemically unique in this group by containing the β-orcinol depsidones (fumarprotocetraric and protocetraric

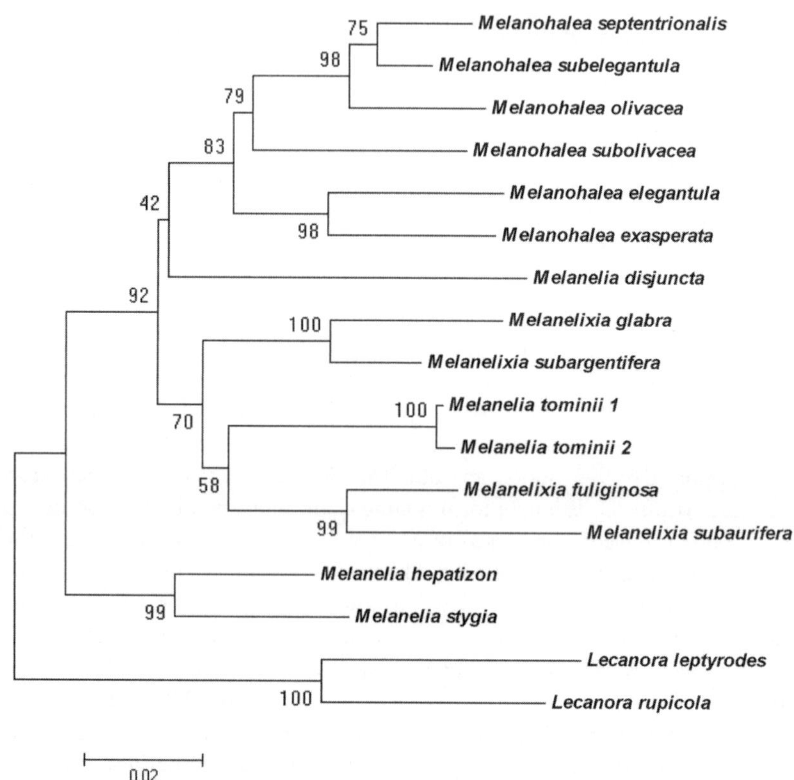

FIG. 1 The ME tree inferred from ITS data. The bootstrap values (1000 replicates) are shown next to the branches.

acids), while five of the other species contain orcinol *para*-depsides. *M. tominii* contains the tridepside gyrophoric acid, while *M. disjuncta*, *M. panniformis*, *M. predisjuncta*, and *M. sorediata* contain the depsides, perlatolic and stenosporic acids. The molecular analysis indicates that *M. disjuncta* and *M. tominii* are phylogenetically quite distant from *M. stygia*, but are closely related to *Melanelixia* and *Melanohalea* (FIG.1). *Melanohalea* species contain β-orcinol depsidones (six species contain fumarprotocetraric acid and one species norstictic acid) or lack lichen substances, while *Melanelixia* species contain the orcinol *para*-depside, lecanoric acid. Lecanoric and gyrophoric acids are closely related chemically as both derive from orsellinic acid moieties: the former is derived from two molecules of orsellinic acid, while the latter is derived from three. Thus *M. tominii* is also chemically similar to *Melanelixia*. In addition to *Melanelia tominii*, *M. microglabra*, *M. calva*, *M. fuscosorediata*,

M. glabratuloides, *M. piliferella*, *M. pseudoglabra*, and *M. subglabra* also produce gyrophoric acid as a major constituent (Esslinger 1977, Divakar et al. 2003). *M. tominii* is widely distributed in the northern hemisphere, and *M. microglabra* is known only from the type specimen in India, while the other six species are restricted to the Southern Hemisphere. As originally circumscribed (Esslinger 1978), the genus *Melanelia* comprised 46 species. The eight species with cortical hairs are now considered to belong to *Melanelixia* or the group containing gyrophoric acid, e.g. *M. fuscosorediata*, *M. piliferella*, and *M. pseudoglabra*. This evidence indicates that the species containing gyrophoric acid probably belong to *Melanelixia*, and the generic concept may be amended accordingly.

However, neither *M. tominii* nor the *Melanelia* species containing gyrophoric acid are transferred formally to *Melanelixia* in this paper, because of the paucity of our molecular data.

New records

1. ***Melanelia predisjuncta*** (Essl.) Essl., Mycotaxon 7(1): 47 (1978)
≡ *Parmelia predisjuncta* Essl., J. Hattori bot. Lab. 42: 50 (1977)

This species is characterized by the saxicolous habit, the narrow lobes (0.4–0.8 mm broad), the presence of pseudocyphellae, the lack of isidia and soredia, the acerose to slightly bifusiform conidia, the common apothecia, the black lower surface, the moderate rhizines, and the presence of perlatolic and stenosporic acids in the medulla (K–, C–, KC–, PD–). *M. predisjuncta* is superficially similar to *M. stygia*, but *M. stygia* can be readily distinguished by the more regular and distinctive pseudocyphellae, the bifusiform conidia, the thicker upper cortex (30–50 μm cf. 8–12 μm), and the absence of perlatolic and stenosporic acids. Of the three other species containing perlatolic and stenosporic acids, namely *M. disjuncta*, *M. sorediata* and *M. panniformis*, the former two have soralia, and the third species has distinctive laminal lobules. In our study only one specimen of *M. predisjuncta* was found. The specimen lacks apothecia, its upper cortex is 8–10 μm thick, lower cortex 9–12 μm thick and lobes 90–110 μm thick.

M. predisjuncta has been reported from Japan (Esslinger 1977) and Russia (Makryi 1981). New to China.

SPECIMEN EXAMINED: CHINA. **Jilin**, Hongtoushan, Mt. Changbaishan, alt. 1900 m, on rock, J.C. Wei & J.B. Chen 6251 (HMAS-L: 052098).

2. ***Melanohalea subelegantula*** (Essl.) O. Blanco, A. Crespo, Divakar, Essl., D. Hawksw. & Lumbsch, Mycol. Res. 108(8): 883 (2004)
≡ *Parmelia subelegantula* Essl., J. Hattori bot. Lab. 42: 89 (1977)
≡ *Melanelia subelegantula* (Essl.) Essl., Mycotaxon 7(1): 48 (1978)

This species is characterized by the typically corticolous habit, the moderate lobes (1–3 mm broad), the lack of pseudocyphellae, soredia and pycnidia, the rare apothecia, the black lower surface, the moderate rhizines, the absence of lichen substances (K–, C–, KC–, PD–), and the distinctive, small isidia (0.1–0.3 mm long). The isidia arise as hemispherical papillae, elongating into cylindrical isidia and then into lobules usually with rhizines and are unique in this genus. Among the five other species with isidia, *M. poeltii* contains fumarprotocetraric acid (PD + orange), *M. elegantula* has isidia with pseudocyphellae at the tips, *M. exasperatula* has hollow isidia, *M. infumata* has broader lobes [1–4(–6 mm)] and longer isidia (0.2–1 mm long), and *M. ushuaiensis* from southern South America has broader lobes (2–5 mm wide) and sparse isidia. In our study only one specimen of *M. subelegantula* was found. The specimen differs from typical *M. subelegantula* in the very dense cylindrical isidia, the very sparse lobules developing from isidia, and in the shorter rhizines (to 0.5 mm cf. to 1 mm long). The sparse lobules are short, small (60–110 × 40–100 μm), and occasionally rhizinate. The lobes of this specimen are 80–100 μm thick, upper cortex 8–12 μm thick, and lower cortex 8–12 μm thick.

Previously *M. subelegantula* was only known from western North America (Esslinger 1977). New to China.

SPECIMEN EXAMINED: CHINA. **Tibet**, Gongbogyamda county, alt. 3500m, on bark, G.R. Hu h537, 24 Jul. 2004 (HMAS-L: 077825).

3. ***Melanohalea olivaceoides*** (Krog) O. Blanco, A. Crespo, Divakar, Essl., D. Hawksw. & Lumbsch, Mycol. Res. 108(8): 883 (2004)

≡ *Parmelia olivaceoides* Krog, Norsk Polarinst. Skr. 144: 109 (1968)

≡ *Melanelia olivaceoides* (Krog) Essl., Mycotaxon 7(1): 48 (1978)

This species is characterized by the typically corticolous habit, the moderate lobes (1–4 mm broad), the lack of pseudocyphellae, isidia and pycnidia, the laminal and punctiform soralia, the granular to isidioid soredia, the rare apothecia, the black lower surface, the moderate rhizines, and the presence of fumarprotocetraric and protocetraric acids (PD + orange, K–, C–, KC–, or sometimes lacking lichen substances). *M. olivaceoides* is related to the other five species containing fumarprotocetraric, namely *M. olivacea*, *M. septentrionalis*, *M. halei*, *M. poeltii*, and *M. gomukhensis*. However, the former four species are all esorediate, and *M. gomukhensis* has distinctive pseudocyphellae. In our study only one specimen of *M. olivaceoides* was found and this specimen lacked apothecia. The lobes were 80–100 μm thick, the upper cortex 8–12 μm thick and lower cortex 8–12 μm thick.

Melanelia olivaceoides has been reported from Alaska, Canada, United States, Europe, Siberia, Japan and Russia (Esslinger 1977, Makryi 1981). New to China.

SPECIMEN EXAMINED: CHINA. **Tibet**, Gongbogyamda county, alt. 3500m, on bark, G.R. Hu h659, 24 Jul. 2004 (HMAS-L: 071066).

4. ***Melanohalea septentrionalis*** (Lynge) O. Blanco, A. Crespo, Divakar, Essl., D. Hawksw. & Lumbsch, Mycol. Res. 108(8): 883 (2004)

≡ *Parmelia olivacea* var. *septentrionalis* Lynge, Bergens Mus. Årbok 1912(10): 4 (1912).
≡ *Parmelia septentrionalis* (Lynge) Ahti, Acta Bot. Fenn. 70: 22 (1966)
≡ *Melanelia septentrionalis* (Lynge) Essl., Mycotaxon 7(1): 48 (1978)

This species is characterized by the corticolous habit, the shiny lobes (1-3 mm broad), the lack of isidia, soredia and lobules, the rare or absence of pseudocyphellae, the smooth apothecial margin, the subhymenium which is thinner than the hymenium, the acerose to slightly bifusiform conidia, the black lower surface, the moderate to dense rhizines, and the presence of fumarprotocetraric and protocetraric acid (PD+ orange, K-, C-, KC-). *M. septentrionalis* is closely related to *M. halei* and *M. olivacea*. However, *M. septentrionalis* can be separated from *M. halei* by the darker colored upper surface, the absence of papillae and lobules in the central parts of the thallus, the smaller spores (9–13 × 5.5–8.5 μm cf. 15–20 × 8–12.5 μm), the weakly bifusiform rather than cylindrical conidia, and the K- rather than K+ reaction of the medulla (yellow turning dingy orange for *M. halei*). *M. septentrionalis* differs from *M. olivacea* in its smaller thalli, the less common pseudocyphellae, the presence of numerous apothecia near the thallus periphery, and the smooth rather than crenate or tuberculate apothecial margin. In addition, *M. septentrionalis* is the only species of *Melanohalea* where the hymenium is obviously thicker than the subhymenium (2 × the thickness of the latter). In our study only one specimen of *M. septentrionalis* was found. The specimen had an obviously shiny and rugose upper surface. the lobes were 100–120 μm thick, the upper cortex 12–14 μm thick and lower cortex 10–12 μm thick. The hymenium was 70–100 μm high, and subhymenium 35–50 μm high. The spores are ellipsoid (9–11 × 6–8 μm), and spore wall 1 μm thick. Conidia are weakly bifusiform, 5–7 μm long.

M. septentrionalis has been reported from North America, Europe, Russia and India (Esslinger 1977, Öztürk 1990, Spribille & Kolb 2000, Motiejunaite 2002, Divakar & Upreti 2005). New to China.

Specimen examined: CHINA. **Heilongjiang**, Linchang, Mt. Dabaishan, alt. 1200 m, on bark, X.Q. Gao, 405, 3 Sep. 1984 (HMAS-L: 036135).

Key to *Melanelia*, *Melanelixia* and *Melanohalea* in China

1a. Thallus with soredia or isidia 2
1b. Thallus without soredia or isidia 12

2a. Thallus with isidia 3
2b. Thallus with soredia 9

3a. Medulla PD + red-orange, with fumarprotocetraric acid ...... *Melanohalea poeltii*
3b. Medulla PD –, other substances present or none 4

4a. Medulla C + rose or red 5
4b. Medulla C – 7
5a. Cortical hairs absent .... *Melanelixia fuliginosa*
5b. Cortical hairs obvious, especially on the lobe ends 6
6a. Isidia papillate, with obvious cortical hairs on the tips .... *Melanelixia villosella*
6b. Isidia cylindrical, without cortical hairs on the tips .... *Melanelixia subvillosella*
7a. Isidia mostly compressed-clavate to spathulate, hollow . *Melanohalea exasperatula*
7b. Isidia cylindrical, not hollow 8
8a. Pseudocyphellae present at the tip of isidia .... *Melanohalea elegantula*
8b. Pseudocyphellae absent; some isidia growing into lobules with rhizines
.... *Melanohalea subelegantula*
9a. Medulla C + rose or red 10
9b. Medulla C –.. 11
10a. Gyrophoric acid present, cortical hairs absent .... *Melanelia tominii*
10b. Lecanoric acid present, cortical hairs present .... *Melanelixia subargentifera*
11a. Medulla P + red-orange, with fumarprotocetraric and protocetraric acid; corticolous; soralia punctiform, laminal .... *Melanohalea olivaceoides*
11b. Medulla P–, with perlatolic and stenosporic acid; saxicolous; soralia granular to isidioid, on the ends of the ascending lateral branches ...... *Melanelia sorediata*
12a. Medulla P + orange or red-orange 13
12b. Medulla P – 16
13a. Pycnidia exogenous, medulla with stictic and norstictic acids
.... *Melanelia hepatizon*
13b. Pycnidia not exogenous, medulla with fumarprotocetraric and/or protocetraric acids 14
14a. Saxicolous; upper cortex thicker (30–50 μm); conidia bifusiform
.... *Melanelia stygia*
14b. Corticolous; upper cortex thinner (8–14 μm); conidia acerose to weakly bifusiform 15
15a. Apothecia margin smooth; hymenium twice as thick as subhymenium
.... *Melanohalea septentrionalis*
15b. Apothecia margin crenate or tuberculate; hymenium as thick as subhymenium
.... *Melanohalea olivacea*
16a. Medulla C + rose or red 17
16b. Medulla C– 20
17a. Medulla with gyrophoric acid; lobes distinctively pseudocyphellate; saxicolous .... *Melanelia tominii*
17b. Medulla with lecanoric acid 18
18a. Cortical hairs present .... *Melanelixia glabra*
18b. Cortical hairs absent 19

19a. Thallus saxicolous; pseudocyphellae absent; apothecia rare .... *Melanelixia glabroides*

19b. Thallus corticolous; pseudocyphellae present; apothecia common .... *Melanelixia huei*

20a. Corticolous usually; no lichen substances detected; papillae conical, with conspicuous pseudocyphellae at the tip .... *Melanohalea exasperata*

20b. Saxicolous; perlatolic and stenosporic acid present; without papillae 21

21a. Lobules abundant, marginal to laminal; pseudocyphellae absent or faint .... *Melanelia panniformis*

21b. Lobules absent; pseudocyphellae obvious .... *Melanelia predisjuncta*

Acknowledgements

The project was financially supported by the National Natural Science Foundation of China (30499340). The authors would like to thank the keeper of the HMAS-L Ms Deng Hong for assistance during this study. The authors thank John A. Elix and Theodore L. Esslinger for expert presubmission reviews.

Literature cited

Abbas A, Wu JN. 1998. Lichens of Xinjiang. 178 pp. Sci-Tech & Hygiene Publishing House of Xinjiang, Urumuqi.

Ahti T, Brodo IM, Noble WJ. 1987. Contributions to the lichen flora of British Columbia, Canada. Mycotaxon 28: 91–97.

Blanco O, Crespo A, Divakar PK, Esslinger TL, Hawksworth DL, Lumbsch HT. 2004. *Melanelixia* and *Melanohalea*, two new genera segregated from *Melanelia* (*Parmeliaceae*) based on molecular and morphological data. Mycological Research 108(8): 873–884.

Chen JB, Esslinger TL. 2005. *Parmeliaceae* (Ascomycota) lichens in China's mainland. IV. *Melanelia* species new to China. Mycotaxon 93: 71–74.

Culberson CF. 1972. Improved conditions and new data for the identification of lichen products by a standardized thin-layer chromatographic method. Jour Chromatogr. 72: 113–125.

Divakar PK, Upreti DK. 2005. A new species in *Melanohalea* (*Parmeliaceae, Ascomycotina*) and new lichen records from India. Lichenologist 37(6): 511–517.

Divakar PK, Upreti DK, Elix JA. 2001. New species and new records in the lichen family *Parmeliaceae* (*Ascomycotina*) from India. Mycotaxon 80: 355–362.

Divakar PK, Upreti DK, Sinha GP, Elix JA. 2003. New species and records in the lichen family *Parmeliaceae* (*Ascomycota*) from India. Mycotaxon 88: 149–154.

Egan RS. 1987. A fifth checklist of the lichen-forming, lichenicolous and allied fungi of the continental United States and Canada. The Bryologist 90(2): 77–173.

Esslinger TL. 1977. A chemosystematic revision of the brown *Parmeliae*. Jour. Hattori Bot. Lab. 42: 1–211.

Esslinger TL. 1978. A new status for the brown *Parmeliae*. Mycotaxon 7: 45–54.

Esslinger TL.1987. A new species of *Melanelia* from Nepal. Mycotaxon 28: 215–217.

Esslinger TL. 1992. The brown *Parmelia* type specimens of A. N. Oxner. Lichenologist 24(1): 13–20.

Felsenstein J. 1985. Confidence limits on phylogenies: An approach using the bootstrap. Evolution 39: 783–791.

Galloway DJ, Jørgensen PM. 1990. *Bartlettiella*, a new lichen genus from New Zealand, with notes on a new species of *Melanelia* and a new chemodeme of *Bryoria indonesica* in New Zealand. N.Z. Jour. Bot. 28(1): 10.

Kurokawa S, Lai MJ. 2001. Parmelioid lichen genera and species in Taiwan. Mycotaxon 77: 225–284.

Makryi TV. 1981. New species of lichen flora of the USSR. Botanicheskii Zhurnal 66(1): 129–132.

Motiejunaite J. 2002. Additions to the Lithuanian flora of foliose and fruticose lichens. Botanica Lithuanica 8(1): 69–76.

Öztürk S. 1990. New records of lichens for Turkey. Turkish Journal of Botany 14: 87–96.

Rogers SO, Bendich AJ. 1988. Extraction of DNA from plant tissues. Plant Molecular Biology Manual A6: 1–10. Kluwer Academic Publishers. Dordrecht.

Rzhetsky A, Nei M. 1992. A simple method for estimating and testing minimum evolution trees. Molecular Biology and Evolution 9: 945–967.

Spribille T, Kolb A. 2000. *Cetraria sepincola* new to Idaho and *Melanelia septentrionalis* new to Montana, with notes on their distribution and ecology. Evansia 17(4): 112–115.

Tamura K, Dudley J, Nei M, Kumar S. 2007. MEGA4: Molecular Evolutionary Genetics analysis (MEGA) software version 4.0. Molecular Biology and Evolution 24: 1596–1599.

Tamura K, Nei M, Kumar S. 2004. Prospects for inferring very large phylogenies by using the neighbor-joining method. PNAS 101: 11030–11035.

Thell A. 1995. A new position of the *Cetraria commixta* group in *Melanelia* (*Ascomycotina*, *Parmeliaceae*). Nova Hedwigia 60(3–4): 407–422.

Thell A, Stenroos S, Feuerer T, Kärnefelt I, Myllys L, Hyvönen J. 2002. Phylogeny of cetrarioid lichens (*Parmeliaceae*) inferred from ITS and β-tubulin sequences, morphology, anatomy and secondary chemistry. Mycological Progress 1(4): 335–354.

Wang HY, Chen JB, Wei JC. 2008. A new species of *Melanelixia* (*Parmeliaceae*) from China. Mycotaxon 104: 185–188.

Wei JC. 1991. An enumeration of liches in China. 278 pp. International Academic Publishers. Beijing.

White YJ, Bruns T, Lee S, Taylor JW. 1990. Amplification and direct sequencing of fungal ribosomal RNA genes for phylogenetics. In: PCR protocols: a guide to methods and applications. pp. 315–322. Ed. Innis MA, Gelfand DH, Sninsky JJ & White TJ. Academic Press. San Diego, California.

Zibirnisa O, Aziguli K, Abbas A. 2004. New Chinese records of the lichen genus *Melanelia*. Acta Botanica Yunnanica 26: 385–386.

A brief overview of and key to species of *Collema* from China

Huajie Liu* & Jiangchun Wei**

huajie_l@126.com weijc2004@126.com
Key Laboratory of Systematic Mycology & Lichenology
Institute of Microbiology, Chinese Academy of Sciences, Beijing 100101, China

Abstract — Forty-seven taxa of the lichen genus *Collema* belonging to 33 species are reported in the present paper. Among them, the following nine taxa are new to China: *Collema coccophorum, C. furfuraceum* var. *luzonense, C. kauaiense, C. nepalense, C. nipponicum, C. poeltii, C. polycarpon, C. subnigrescens* f. *caesium*, and *C. tenax* var. *expansum*. A key to the known species from China is given. *Collema pulchellum* var. *leucopeplum* and *C. latzelii* are excluded from the lichen flora of China.

Key words — *Collemataceae, Lecanorales,* lichenized *Ascomycetes*

Introduction

Collema is a crustose to foliose lichen genus belonging to *Collemataceae, Lecanorales, Ascomycota* (Kirk et al. 2001). It comprises about 80 species in the world (Kirk et al. 2001) and has been monographed by Degelius (1954, 1974).

The genus has been reported from China in 29 scattered publications with 59 taxa belonging to 43 species (Wei 1991; Jiang 1992, 1993; Liu & Wei 2003ab). The first species from China, *Collema limosum* was reported by Nylander & Crombie (1883). The second one, *C. coccophylloides* Nyl. (= *C. callibotrys* var. *coccophyllizum*), was recorded by Hue (1898). The two species, *Synechoblastus sublaevis* Jatta (= *C. furfureolum*) and *S. flaccidus* (Ach.) Körb. (≡ *C. flaccidum*) were reported by Jatta (1902). Zahlbruckner (1930, 1933) added six taxa, viz. *C. nigrescens, C. substipitatum, C. raishanum* Zahlbr. (= *C. leptaleum*), *C. ogatae* Zahlbr. (≡ *C. tenax* var. *ogatae*), *C. complanatum* and *C. japonicum*. Magnusson (1940) described two species from China, *C. kansuense* H. Magn. (= *C. tenax* var. *corallinum*) and *C. substellatum* H. Magn. (≡ *C. tenax* var. *substellatum*). Degelius (1974) added five taxa, viz. *C. glebulentum, C. pulchellum* var.

* Present address: College of Life Sciences, Hebei University, No. 180, Wusidong Road, Baoding, Hebei, 071002, China.

** Corresponding author

This paper was originally published in *Mycotaxon*, 108: 9-29, 2009.

subnigrescens, *C. subflaccidum*, *C. tenax* var. *vulgare* and *C. texanum*. The following 12 taxa were added to the Chinese *Collema* flora in eight papers (Wei 1991), viz. *C. pustuligerum* Hue (= *C. pulchellum* var. *pulchellum*) (Asahina 1952), *C. crispum* (Wu & Xiang 1981), *C. fasciculare* (Chen et al. 1981a), *C. peregrinum* (Ikoma 1983), *C. furfuraceum*, *C. ryssoleum*, *C. subnigrescens*, *C. tunaeforme* (Ach.) Ach. (= *C. fuscovirens*) [the above mentioned four species were recorded by Wu (1987)], *C. leptaleum* var. *biliosum*, *C. rugosum* [both in Thrower (1988)], *C. cristatum* (Wu & Qian 1989) and *C. subconveniens* (Chen et al. 1989).

Twelve taxa from China were described by Jiang (1992), viz. *C. beijingense*, *C. brevisporum*, *C. clavisporiferum*, *C. corniculatum*, *C. fanjingshanense*, *C. fusiosporum*, *C. lushanense*, *C. multipartitum* var. *granulosum*, *C. pulchellum* var. *multipartitum*, *C. sorediatum*, *C. tetrasporum* and *C. tianmuense*. The same author (Jiang 1993) reported 13 new records: *C. auriforme*, *C. callopismum*, *C. ceraniscum*, *C. cristatum* var. *marginale*, *C. latzelii*, *C. pulchellum* var. *leucopeplum*, *C. shiroumanum*, *C. tenax* var. *ceranoides*, *C. tenax* var. *crustaceum*, *C. tenax* var. *diffracto-areolatum*, *C. tenax* f. *papulosum*, *C. undulatum* var. *undulatum* and *C. undulatum* var. *granulosum*. Abbas & Wu (1998) reported *C. thamnodes* from Xinjiang. The occurrence of *Collema* in China was also reported by Magnusson (1944), Wang & Lai (1973), Chen et al. (1981b), Wu et al. (1984), Wei (1991), Abbas et al. (1996, 2001), Aptroot & Seaward (1999) and Guo (2005).

Although a large number of species were reported, there has not been any revisionary study on *Collema* from China. As part of this research, we synonymized five taxa described by Jiang (1992), recorded *C. multipartitum* as new to China (Liu & Wei 2003a), and described *C. sichuanense* and *C. substipitatum* var. *gonggashanense* (Liu & Wei 2003b). In this paper, we report 47 taxa belonging to 33 species, including nine new records, and we exclude two taxa from the Chinese lichen flora.

Materials and methods

Most of the specimens examined are deposited in the Herbarium Mycologicum Academiae Sinicae-Lichenes (HMAS-L); some are in the Herbarium of Xinjiang University (XJU) and the Herbarium, Kunming Institute of Botany, Chinese Academy of Sciences (KUN).

For the morphological and anatomical studies, a dissecting (Motic) and a compound microscope (Olympus BHA) were routinely used for all materials. Hand sections were routinely made and examined with water as the mounting medium. Terms of morphology and anatomy follow Degelius (1954, 1974). Comments on the distribution of *Collema* taxa are mainly based on Wei (1991).

Results and discussion

Forty-seven taxa belonging to 33 species are accepted in the present paper. Among them, nine taxa are new to China: *C. coccophorum*, *C. furfuraceum* var. *luzonense*, *C. kauaiense*, *C. nepalense*, *C. nipponicum*, *C. poeltii*, *C. polycarpon*, *C. subnigrescens* f. *caesium* and *C. tenax* var. *expansum*; these are marked with "***" in the taxonomic list. Both *C. pulchellum* var. *leucopeplum* and *C. latzelii* are misapplied names and are therefore excluded from the lichen flora of China.

Collema F.H. Wigg., Primit. Fl. Holsat.: 89 (1780).

Thallus crustose to foliose, gelatinized; upper side smooth, rugose to ridged or pustulate. Apothecia superficial, commonly sessile with constricted base. Thallus without cortex, homoeomerous; photobiont *Nostoc*. Apothecia zeorine; thalline exciple with or without a pseudocortex; proper exciple euthyplectenchymatous, subparaplectenchymatous or euparaplectenchymatous; ascospores 8 per ascus, 1- to multi-septate or somewhat muriform, variable in size and shape, colorless.

Corticolous, saxicolous to terricolous.

Key to the taxa of *Collema* from China

1. Thallus crustose to subfoliose, often fenestrate; lobes indistinct or absent 2
1. Thallus foliose to somewhat subfoliose, not fenestrate (but see *C. nepalense*); lobes distinct 9

2. Thallus with isidia, rugose on upper side; apothecia absent .... *C. leptaleum* var. *biliosum*
2. Thallus without isidia 3

3. Spores submuriform (with longitudinal septa) 4
3. Spores without longitudinal septa 5

4. Spores cubic .... *C. lushanense*
4. Spores ellipsoid .... *C. nipponicum*

5. Spores > 10-celled, vermiform and curved in various ways, > 65 μm long .... *C. fasciculare*
5. Spores < 10-celled, not vermiform, straight or curved 6

6. Spores bacillar and 4-celled, < 50 μm long; proper exciple euparaplectenchymatous .... *C. leptaleum* var. *leptaleum*
6. Spores not bacillar, > 4-celled 7

7. Spores dumbbell-shaped, < 50 μm long; proper exciple euthyplectenchymatous .... *C. shiroumanum*
7. Spores fusiform to acicular, > 60 μm long 8

8. Proper exciple euthyplectenchymatous to subparaplectenchymatous . . . . *C. substipitatum* var. *substipitatum*
8. Proper exciple euparaplectenchymatous . . . . . . *C. substipitatum* var. *gonggashanense*

9. Thallus rugose on both sides . . . . *C. auriforme*
9. Thallus rugose only on upper side or not rugose 10

10. Lobes swollen and sometimes plicate (at least in the tips) 11
10. Lobes not swollen and not plicate 21

11. Lobes convex, repeatedly furcate . . . . *C. texanum*
11. Lobes plane to concave 12

12. Thallus commonly < 2 cm diam.; lobes < 2 mm wide 13
12. Thallus > 2 cm diam.; lobes > 2 mm wide 16

13. Lobes erect, often forming erect clusters . . . . *C. tenax* var. *corallinum*
13. Lobes adnate to ascending, not forming erect clusters 14

14. Thallus areolate, isidiate . . . . *C. tenax* var. *diffracto-areolatum*
14. Thallus not areolate, nonisidiate 15

15. Spores 2-celled . . . . *C. coccophorum*
15. Spores 4-celled to submuriform . . . . *C. tenax* var. *crustaceum*

16. Thallus > 5 cm diam.; lobes > 5 mm wide; upper side glossy . . . . *C. tenax* var. *expansum*
16. Thallus < 5 cm diam.; lobes < 5 mm wide; upper side dull 17

17. Lobes long, up to 15 mm long . . . . *C. tenax* var. *substellatum*
17. Lobes short, commonly < 5 mm long 18

18. Upper side with erect accessory lobules . . . . *C. tenax* var. *ogatae*
18. Upper side not so 19

19. Thallus isidiate . . . . *C. tenax* f. *papulosum*
19. Thallus nonisidiate 20

20. Spores 4-celled to submuriform; proper exciple euthyplectenchymatous . . . . *C. tenax* var. *vulgare*
20. Spores 4-celled; proper exciple subparaplectenchymatous to euparaplectenchymatous . . . . *C. polycarpon*

21. Lobes > 5 mm wide and plane 22
21. Lobes < 5 mm wide, concave, plane or convex 38

22. Upper side of the thallus with regular ridges or pustules 23
22. Upper side of the thallus without regular ridges or pustule 33

23. Upper side of the thallus with isidia 24
23. Upper side of the thallus without isidia 25

24. Proper exciple euparaplectenchymatous . . . . *C. furfuraceum* var. *luzonense*
24. Proper exciple euthyplectenchymatous . . . . *C. furfuraceum* var. *furfuraceum*

25. Thalline exciple with pseudocortex; proper exciple euthyplectenchymatous 26
25. Thalline exciple without pseudocortex; proper exciple euparaplectenchymatous 31

26. Spores fusiform to ellipsoid 27
26. Spores broadly acicular to acicular 29
27. Thallus often fenestrate; apothecia pruinose; spores 4–6-celled ...... *C. nepalense*
27. Thallus not fenestrate; apothecia epruinose; spores 6-celled 28
28. Spores > 5 μm wide; saxicolous .... *C. ryssoleum*
28. Spores < 5 μm wide; corticolous .... *C. sichuanense*
29. Spores acicular, 6–10-celled, < 5 μm wide .... *C. nigrescens*
29. Spores broadly acicular, 6-celled, > 6 μm wide 30
30. Apothecia pruinose .... *C. subnigrescens* f. *caesium*
30. Apothecia epruinose .... *C. subnigrescens* f. *subnigrescens*
31. Spores > 65 μm long; upper side of the thallus mainly ridged ..... *C. complanatum*
31. Spores < 65 μm long 32
32. Upper side of the thallus mainly ridged .... *C. pulchellum* var. *subnigrescens*
32. Upper side of the thallus mainly pustulate .... *C. pulchellum* var. *pulchellum*
33. Spores with longitudinal septa .... *C. subconveniens*
33. Spores without longitudinal septa 34
34. Thallus rarely isidiate; proper exciple euparaplectenchymatous ...... *C. japonicum*
34. Thallus isidiate; proper exciple euthyplectenchymatous to subparaplectenchymatous 35
35. Thallus with squamiform isidia .... *C. flaccidum*
35. Thallus with globular to teretiform isidia 36
36. Isidia teretiform, > 1 mm long, repeatedly branched .... *C. glebulentum*
36. Isidia globular to teretiform, < 1 mm long, not branched or slightly so 37
37. Upper side with irregular ridges or pustules; isidia teretiform .... *C. rugosum*
37. Upper side without ridges or pustules; isidia often globular ...... *C. subflaccidum*
38. Thallus nonisidiate 39
38. Thallus isidiate 42
39. Lobes convex, < 1 mm wide, not crisped; proper exciple euparaplectenchymatous .... *C. multipartitum*
39. Lobes concave to plane 40
40. Lobes canaliculate at least in part 41
40. Lobes not so, strongly undulate .... *C. undulatum* var. *undulatum*
41. Lobes irregularly furcate, incised with lobules in margin .... *C. cristatum* var. *cristatum*
41. Lobes regularly furcate, often entire in margin .... *C. cristatum* var. *marginale*
42. Isidia squamiform when old; lobes crisped 43
42. Isidia globular to cylindrical when old; lobes not crisped or slightly so 44
43. Spores < 20 μm long .... *C. furfureolum*
43. Spores > 20 μm long .... *C. crispum*

44. Lobes concave, furcate .... *C. undulatum* var. *granulosum*
44. Lobes plane to somewhat concave, not branched or slightly so 45
45. Upper side of the thallus irregularly pustulate .... *C. fuscovirens*
45. Upper side of the thallus not pustulate 46
46. Proper exciple euthyplectenchymatous to subparaplectenchymatous..... *C. poeltii*
46. Proper exciple euparaplectenchymatous .... *C. kauaiense*

1. ***Collema auriforme*** (With.) Coppins & J.R. Laundon, in Laundon, Lichenologist 16(3): 228 (1984).

This species is characterized by 1) thallus with distinct wrinkles on both sides; 2) lobes 4–10 mm wide, irregularly branched; 3) proper exciple euparaplectenchymatous; 4) spores submuriform, (17–) 27–40.5 (–51) × 6.5–15.5 (–17) μm. It is muscicolous on soil-covered rocks.

Selected specimens examined: NEI MONGOL, Horqin Youyi Qianqi, 1250 m, Gao XQ, 758. HEBEI, Mt. Xiaowutaishan, 1800 m, Jiang ZG, 3627. QINGHAI, Huangcheng, 3000 m, 11.IX.1958, Ma QM, 9. XINJIANG, Yecheng, 3000 m, 8.VIII.1992, Abbas A, 92-0244.

Literature records for China: JILIN, HEBEI (Jiang 1993, p. 70), SHAANXI (Guo 2005, p. 46), QINGHAI (Jiang 1993, p. 70; Guo 2005, p. 46), XINJIANG (Guo 2005, p. 46).

2. ***Collema coccophorum*** Tuck., Proc. Am. Acad. Arts & Sci. 5: 385 (1862).***

This species is characterized by 1) thallus foliose, 3 cm diam.; 2) lobes numerous, concave, 1–2 mm wide, with swollen ends; 3) thallus and thalline exciple not pseudocorticate; 4) proper exciple euthyplectenchymatous; 5) spores fusiform to ovate, 2-celled, (17–) 20.5–25 × 6.5–9 μm. It is terricolous.

In gross morphology, *C. coccorphorum* resembles some infraspecific taxa of *C. tenax*, which differs in having 4-celled to submuriform spores.

Selected specimens examined: NEI MONGOL, Xilinhot, Inner Mongolian Grassland Ecosystem Research Station of Academia Sinica, 1150 m, 27.VII.2003, Liu HJ, 740. HUNAN, Mt. Hengshan, 1200 m, 1.IX.1964, Zhao JD & Xu LW, 10351.

3. ***Collema complanatum*** Hue, J. Bot. (Morot) 20: 85 (1906).

This species differs from other species of *Collema* in having long, 6–10-celled, acicular spores (commonly > 65 μm long). It is corticolous.

Selected specimens examined: NEI MONGOL, Ergun Zuoqi, 1200 m, 16.VIII.1985, Gao XQ, 1620. JILIN, Mt. Changbaishan, 1100 m, 7.VIII.1977, Wei JC, 2776. HEILONGJIANG, Dailing, 450 m, 2.X.1975, Wei JC, 2055. SHAANXI, Mt. Taibaishan, 2300 m, 9.VII.1988, Ma CH, 60 [The above mentioned specimens were cited by Jiang (1993) as *C. pulchellum* var. *leucopeplum*]. ANHUI, Yuexi, 1150 m, 4.IX.2001, Liu HJ, 444. JIANGXI, Mt. Lushan, Zhao JD et al., 579 [Jiang (1993) as *C. shiroumanum*]. YUNNAN, Lijiang, Mt. Yulongshan, 3050 m, 15.XII.1964, Wei JC, 2504-5 [Jiang (1993) as *C. pulchellum* var. *pulchellum*].

LITERATURE RECORDS FOR CHINA: SHAANXI (Guo 2005, p. 46), ZHEJIANG (Wu & Qian 1989, p. 194), FUJIAN (Wu et al. 1984, p. 1), TAIWAN (Zahlbruckner 1933, p. 26; Wang & Lai 1973, p. 90; Degelius 1974, p. 161), HONG KONG (Thrower 1988, p. 83; Jiang 1993, p. 72).

4. ***Collema crispum*** (Huds.) Weber ex F.H. Wigg., Primit. Fl. Holsat.: 89 (1780).

This species is characterized by the foliose thallus with crisped, < 5 mm wide lobes. It is terricolous.

SPECIMEN EXAMINED: XIZANG, Lhünzê, 3600 m, 5.VII.1975, Zang M, 1117.

LITERATURE RECORDS FOR CHINA: SHAANXI (Wu 1987, p. 53), JIANGSU (Wu & Xiang 1981, p. 2; Wu & Qian 1989, p. 194; Wu 1987, p. 53), ZHEJIANG (Wu & Qian 1989, p. 194).

5. ***Collema cristatum*** (L.) Weber ex F.H. Wigg., Primit. Fl. Holsat.: 89 (1780).

5.1. var. ***cristatum***

This variety is characterized by 1) thallus foliose, often > 5 cm diam.; 2) lobes radiating and irregularly furcate, < 2 mm wide, somewhat concave; 3) thalline exciple without pseudocortex; 4) proper exciple euparaplectenchymatous; 5) spores ellipsoid, 4-celled to submuriform, 17–28 × 6.5–12 μm. It is saxicolous to terricolous, often growing together with var. *marginale*.

SELECTED SPECIMENS EXAMINED: NEI MONGOL, Hexigten Qi, 1500 m, 29.VII.1985, Gao XQ, 1049. HEBEI, Mt. Xiaowutaishan, 2100 m, Jiang ZG, 3692. XINJIANG, Zhaosu, 2.VII.1992, Abbas A, 219, 221. Wensu, Mt. Tomur, 2600 m, 30.VI.1977, Wang XY et al., 341.

LITERATURE RECORDS FOR CHINA: BEIJING, HEBEI (Jiang 1993, p. 70), XINJIANG (Guo 2005, p. 46).

5.2. var. ***marginale*** (Huds.) Degel., Symb. Bot. Upsal. 13(2): 316 (1954).

This variety is similar in anatomy (thalline and proper exciples, spore shape and size) to var. *cristatum*, but differs in having the regularly furcate lobes. It is saxicolous to terricolous.

SELECTED SPECIMENS EXAMINED: NEI MONGOL, Hexigten Qi, 1500 m, 29.VII.1985, Gao XQ, 1035. HEBEI, Mt. Baihuashan, 1500 m, 18.VII.1964, Xu LW & Zong YC, 8461. XINJIANG, Zhaosu, 6.VII.1993, Abbas A, 930172 [Abbas & Wu (1998) and Abbas et al. (1996, 2001) as *C. fuscovirens*].

LITERATURE RECORDS FOR CHINA: BEIJING, NEI MONGOL (Jiang 1993, p. 70), XINJIANG [Abbas et al. 1996, p. 12; Abbas & Wu 1998, p. 62; Abbas et al. 2001, p. 363 (All cited as *C. fuscovirens* in the abovementioned three literatures); Guo 2005, p. 46].

6. ***Collema fasciculare*** (L.) Weber ex F.H. Wigg., Primit. Fl. Holsat.: 89 (1780).

This species differs from all other crustose to subfoliose Chinese taxa of *Collema* in having 10–16-celled, vermiform spores that are > 65 μm long. It is corticolous.

Collema fasciculare can be well distinguished from *C. complanatum* by its crustose to subfoliose thallus without ridges or pustulates on the upper side.

SELECTED SPECIMENS EXAMINED: JILIN, Mt. Changbaishan, 1100 m, 8.VIII.1977, Wei JC, 2807-1. SICHUAN, Mt. Emeishan, 2200 m, 14.VIII.1963, Zhao JD & Xu LW, 7322.

LITERATURE RECORDS FOR CHINA: HEILONGJIANG (Chen et al. 1981a, p. 134; Wu 1987, p. 53), SHAANXI (Wu 1987, p. 53).

7. ***Collema flaccidum*** (Ach.) Ach., Lichenogr. Univ.: 647 (1810).

This species is characterized by its foliose thallus with broad (> 5 mm wide) lobes, irregular pustules or ridges on the upper side, and squamiform isidia. Spores fusiform, 23.8–37.4 × 5–8.5 μm. It is corticolous.

SELECTED SPECIMENS EXAMINED: HEILONGJIANG, Mulin, 610 m, 21.VII.1977, Wei JC, 2545-1. ANHUI, Yuexi, 1670 m, 5.IX.2001, Huang MR, 655. SICHUAN, Mt. Emeishan, 2200 m, Zhao JD et al., 7399. XINJIANG, Mt. Altay Shan, 1700 m, Abbas A, 2002948.

LITERATURE RECORDS FOR CHINA: SHAANXI (Jatta 1902, p. 480; Zahlbruckner 1930, p. 76; Wu 1987, p. 53), JIANGSU (Wu 1987, p. 53), ANHUI (Wu & Qian 1989, p. 194), XINJIANG (Abbas et al. 1996, p. 12; Abbas & Wu 1998, p. 61; Abbas et al. 2001, p. 363; Guo 2005, p. 46).

8. ***Collema furfuraceum*** (Arnold) Du Rietz, Ark. Bot. 22A(13): 3 (1929).

8.1. var. ***furfuraceum***

This taxon is characterized by 1) thallus foliose, with regular ridges and pustules on upper side; 2) lobes > 5 mm wide; 3) cylindrical isidia on ridges and pustules; 4) thallus and thalline exciple often with typical pseudocortex; 5) proper exciple euthyplectenchymatous to somewhat subparaplectenchymatous; 6) spores fusiform, commonly 6-celled, 36–67 × 3.5–6.5 μm. It is corticolous.

In its isidia and lobe width, this taxon resembles *C. subflaccidum*, from which it can be separated by the regular ridges and pustules on the upper side.

SELECTED SPECIMENS EXAMINED: JILIIN, Mt. Changbaishan, 20.VIII.1977, Wei JC, 3118. SHAANXI, Mei Xian, 800 m, 12.IV.1963, Ma QM & Zong YC, 2008. ANHUI, Jinzhai, 680 m, 10.IX.2001, Huang MR, 674. HUNAN, Mt. Hengshan, 600–1000 m, 31.VIII.1964, Zhao JD & Xu LW, 9872, 10011. SICHUAN, Mt. Emeishan, 1400–3160 m, 20.VIII.1963, Zhao JD & Xu LW, 7704, 8220. XINJIANG, Mt. Tianshan, 2500 m, 7.VIII.1978, Wang XY, 1205.

LITERATURE RECORDS FOR CHINA: ZHEJIANG, ANHUI, JIANGXI (Wu 1987, p. 53), SHAANXI (Wu 1987, p. 53; Guo 2005, p. 46), XINJIANG (Guo 2005, p. 46).

8.2. var. ***luzonense*** (Räsänen) Degel., Symb. Bot. Upsal. 20(2): 179 (1974).***

This taxon differs from var. *furfuraceum* in having euparaplectenchymatous proper exciple. It is corticolous.

SPECIMEN EXAMINED: HUNAN, Mt. Hengshan, 500 m, 11.II.1965, Wei JC, 3030.

9. *Collema furfureolum* Müll. Arg., Flora 72: 142 (1889).

This species is characterized by 1) thallus foliose, without ridges and pustules; 2) end lobes 2–3 mm wide, slightly crisped; 3) isidia superficial, squamiform; 4) thallus without typical pseudocortex. It is saxicolous.

Specimen examined: SHAANXI, Mt. Taibaishan, 4.VI.1963, Wei JC et al., 2718.

Literature records for China: SHAANXI (Jatta 1902, p. 481; Zahlbruckner 1930, p. 76; Degelius 1974, p. 79; Guo 2005, p. 46), ZHEJIANG (Degelius 1974, p. 79).

10. *Collema fuscovirens* (With.) J.R. Laundon, Lichenologist 16(3): 219 (1984).

This species is characterized by the branched foliose thallus with < 5 mm wide lobes and irregularly pustulate upper side, and obovate to ellipsoid spores (14–26 × 6.5–12 μm). It is saxicolous to terricolous.

Selected specimens examined: NEI MONGOL, Horqin Youyi Qianqi, 1250 m, 3.VII.1985, Gao XQ, 729-1. SHAANXI, Mt. Taibaishan, 1200 m, 7.VII.1988, Gao XQ, 2937 [Jiang (1993) as *C. undulatum* var. *undulatum*]. SICHUAN, Huanglong, 3325 m, 25.IX.2001, Jiang YM & Zhao ZT, S220. XIZANG, Nyalam, 3350 m, 21.V.1966, Wei JC & Chen JB, 1083-1. XINJIANG, Mt. Altay Shan, Abbas A, 98-006-8a.

Literature records for China: SHAANXI (Wu 1987, p. 55), JIANGSU (Wu 1987, p. 55; Wu & Qian 1989, p. 195), XINJIANG (Guo 2005, p. 46).

11. *Collema glebulentum* (Nyl. ex Cromb.) Degel., in Magnusson, Ark. Bot. ser. 2, 2(2): 88 (1952).

The species is distinct from all other species of *Collema* by the repeatedly branched and large isidia (often > 1 mm long). It is corticolous.

Specimen examined: JILIN, Mt. Changbaishan, 16.VIII.1977, Wei JC, 3041.

Literature records for China: SHAANXI (Guo 2005, p. 46). XINJIANG (Degelius 1974, p. 144; Abbas & Wu 1998, p. 63; Guo 2005, p. 46).

12. *Collema japonicum* (Müll. Arg.) Hue, Nouv. Arch. Mus. Hist. Nat. Paris Sér. 3, 10: 220 (1898).

This species is characterized by 1) thallus foliose with > 5 mm wide lobes; 2) upper side smooth or with irregular pustules and ridges; 3) thalline exciple with pseudocortex; 4) proper exciple euparaplectenchymatous; 5) spores fusiform, 6-celled, 27–54 (–67) × 5–9 μm. It is corticolous, rarely saxicolous.

Selected specimens examined: SHAANXI, Mt. Taibaishan, 1100 m, 30.VI.1963, Wei JC et al., 2779. ANHUI, Jinzhai, 650 m, 10.IX.2001, Liu HJ, 548. HUNAN, Mt. Hengshan, 960 m, 2.IX.1964, Zhao JD & Xu LW, 10288. SICHUAN, Mt. Gonggashan, 2500 m, 1.VII.1982, Wang XY et al., 8646.

Literature records for China: SHAANXI (Guo 2005, p. 46), ANHUI, ZHEJIANG (Wu & Qian 1989, p. 195), TAIWAN (Zahlbruckner 1933, p. 26; Wang & Lai 1973, p. 90).

13. ***Collema kauaiense*** H. Magn., in Magnusson & Zahlbruckner, Ark. Bot. 31A(1): 63 (1943).***

This species is characterized by 1) thallus foliose, 2 cm diam.; 2) lobes 2–3 mm wide, margin crisped, not swollen; 3) isidia on upper side, laminal, globular to somewhat squamiform; 4) apothecia pruinose; 5) proper exciple euparaplectenchymatous; 6) spores fusiform, 4–6-celled, 23.5–34.5 × 6.5–10 μm. It is corticolous.

In having pruinose apothecia, this species resembles *C. nepalense* and *C. subnigrescens* f. *caesium*, from which it can be separated by the smaller thallus (< 3 cm diam.), narrower lobes (< 5 mm wide) and upper side without ridges and pustules.

Specimen examined: SICHUAN, Mt. Gonggashan, 2500 m, 1.VII.1982, Wang XY et al., 8638.

14. ***Collema leptaleum*** Tuck., Proc. Am. Acad. Arts & Sci. 6: 263 (1866).

14.1. var. ***leptaleum***

= *C. brevisporum* Z.G. Jiang, Journal of Hebei Normal University (Natural Science) 16(3): 83 (1992).

This variety is similar in appearance to var. *biliosum*, but differs in lacking isidia and in having numerous apothecia. It is corticolous. See Liu & Wei (2003a) for details.

Literature records for China: JILIN, HEILONGJIANG, ZHEJIANG, YUNNAN (Liu & Wei 2003a, p. 350), TAIWAN (Zahlbruckner 1933, p. 26, 1940, p. 247; Wang & Lai 1973, p. 90; Degelius 1974, pp. 102 & 107), HONG KONG (Thrower 1988, p. 84).

14.2. var. ***biliosum*** (Mont.) Degel., Symb. Bot. Upsal. 20(2): 105 (1974).

This taxon is characterized by 1) thallus crustose to subfoliose, sometimes fenestrate; 2) upper side irregularly rugose, with globular isidia; 3) apothecia absent. It is corticolous.

Selected specimens examined: JILIN, Mt. Changbaishan, Wenquan, 16.VIII.1977, Wei JC, 3048. NEI MONGOL, Horqin Youyi Qianqi, 1200 m, 6.VII.1985, Gao XQ, 892-1. SHAANXI, Mt. Taibaishan, 4.VI.1963, Wei JC et al., 2718-3. JIANGXI, Mt. Lushan, 21.IV.1960, Zhao JD, 519. YUNNAN, Weixi, 2900 m, 18.VII.1981, Wang XY et al., 3848. HONG KONG, 1.I.1973, Thrower SL, 1527.

Literature records for China: SHAANXI (Guo 2005, p. 46), HONG KONG (Thrower 1988, p. 85).

15. ***Collema lushanense*** Z.G. Jiang, Journal of Hebei Normal University (Natural Science) 16: 83 (1992).

Type(!): JIANGXI, Mt. Lushan, 3.IV.1960, Zhao JD et al., 577 (HMAS-L). Corticolous.

This species is characterized by 1) thallus subcrustaceous to subfoliose, markedly rugose, fenestrate; 2) isidia and pruina absent; 3) thallus and thalline exciple without pseudocortex; 4) proper exciple euparaplectenchymatous; 5) spores 8 per ascus, cubic, submuriform, 13.5–17 × 8.5–10 μm. It is corticolous.

This species differs from other Chinese *Collema* taxa in having cubic spores. It is endemic to China and known only from the original locality.

Literature records for China: JIANGXI (Jiang 1992, p. 83).

16. *Collema multipartitum* Sm., in Smith & Sowerby, Engl. Botan. vol. 36, tab. 2582 (1814).

= *C. multipartitum* var. *granulosum* Z.G. Jiang, Journal of Hebei Normal University (Natural Science) 16(3): 83 (1992).

This species is characterized by the foliose thallus with narrow (< 5 mm wide), repeatedly furcate, convex lobes, euparaplectenchymatous proper exciple, and 4-celled linear-oblong spores (34–37.4 × 5–6.5 μm). It is saxicolous. See Liu & Wei (2003a) for details.

Literature records for China: HEBEI (Liu & Wei, 2003a, p. 352).

17. *Collema nepalense* Degel., Symb. Bot. Upsal. 20(2): 157 (1974).***

This species is characterized by 1) thallus foliose, often fenestrate; 2) upper side regularly pustulate to somewhat ridged; 3) lobes often > 5 mm wide; 4) thalline exciple with distinct pseudocortex; 5) proper exciple euthyplectenchymatous to somewhat subparaplectenchymatous; 6) spores fusiform, 6-celled, (34–) 38–44 (–51) × (3.5–) 5–7 μm. It is corticolous.

Selected specimens examined: YUNNAN, Lijiang, 3050 m, 15.XII.1964, Wei JC, 2493-1. XIZANG, Zogang, Mt. Meilixueshan, 3200 m, 8.X.1982, Su JJ, 5501.

18. *Collema nigrescens* (Huds.) DC., in Lamarck & de Candolle, Fl. Franç., ed. 3, 2: 384 (1805).

This species is mainly characterized by its foliose thallus with regular ridges and pustules, pseudocorticate thalline exciple, euthyplectenchymatous to subparaplectenchymatous proper exciple and 6–10-celled, transversely septate spores. It differs from *C. subnigrescens* by its longer [58–87 (–102) μm vs 54–69 (–85) μm] and narrower [3.5–5 μm vs (3.5–) 5–7 μm] spores with more septa [6–10-celled vs 6-celled]. It is corticolous.

Selected specimens examined: HEILONGJIANG, Dailing, 450 m, 1.VIII.2002, Chen JB & Hu GR, 22030. NEI MONGOL, Hexigten Qi, 1950 m, 30.VII.1985, Gao XQ, 1084.

Literature records for China: SHAANXI, ZHEJIANG, ANHUI, GUANGDONG (Wu 1987, p. 54), SICHUAN (Zahlbruckner 1930, p. 76), HUNAN, YUNNAN, FUJIAN (Wu 1987, p. 54; Zahlbruckner 1930, p. 76).

19. *Collema nipponicum* Degel., Symb. Bot. Upsal. 20(2): 53 (1974).***

This species is characterized by 1) thallus subfoliose, 2–3 cm diam.; 2) lobes swollen towards margin; 3) thallus and thalline exciple without pseudocortex;

4) proper exciple euparaplectenchymatous; 5) spores ellipsoid, submuriform, 27–35 × 9–13.5 μm. It grows on soil-covered rocks.

In gross morphology *C. nipponicum* resembles some infraspecific taxa of *C. tenax*, but differs in the euparaplectenchymatous proper exciple.

Specimens examined: XINJIANG, Tomort, 2800–2900 m, 3–25.III.1977, Wang XY et al., 394, 533-1.

20. *Collema poeltii* Degel., Symb. Bot. Upsal. 20(2): 96 (1974).***

This species is characterized by 1) thallus foliose, often fragmented, 2–4 cm diam.; 2) lobes 1–2 mm wide, slightly crisped; 3) thalline exciple with typical pseudocortex; 4) proper exciple euthyplectenchymatous to subparaplectenchymatous; 5) spores fusiform, (4–)6-celled, 28–45 × 7–10 μm. It is saxicolous or terricolous.

Selected specimens examined: HEBEI, Mt. Donglingshan, 18.VIII.1957, Zhao JD, 008 [Jiang (1993) as *C. latzelii*]. HENAN, Mt. Jigongshan, 20.IX.2001, Liu HJ, 705. ANHUI, Yuexi, 1450 m, 5.IX.2001, Liu HJ, 477. YUNNAN, Xichou, 1580 m, 17.XI.1991, Chen JB, 5173.

21. *Collema polycarpon* Hoffm., Deutschl. Fl. 2: 102 (1796).***

This species is characterized by 1) thallus foliose, 2–3 cm diam.; 2) lobes 1–2 mm wide, margins slightly swollen; 3) thallus and thalline exciple without pseudocortex; 4) proper exciple subparaplectenchymatous; 5) spores fusiform, 4-celled, 17–26 × 6–8.5 μm. It is terricolous.

Specimen examined: YUNNAN, Yingjiang, 1500 m, Wang X. Y. et al., 3277.

22. *Collema pulchellum* Ach., Syn. Meth. Lich.: 321 (1814).

22.1. var. ***pulchellum***

This taxon is characterized by the foliose thallus with regular pustules on the upper side, euparaplectenchymatous proper exciple and acicular spores (commonly < 5 μm wide). It is corticolous. See Liu & Wei (2003a) for details.

Selected specimens examined: HEILONGJIANG, Tahe, 500 m, 3.VIII.1984, Gao XQ, 084. JILIN, Mt. Changbaishan, 1000 m, 25.VII.1983, Wei JC & Chen JB, 6053, 6111, 6150 [The above three specimens were cited by Jiang (1993) as *C. pulchellum* var. *leucopeplum*].

Literature records for China: Northeast China (Asahina 1952, p. 375; Degelius 1974, p. 176), NEI MONGOL, HEILONGJIANG, JILIN, HEBEI, ANHUI, HUNNAN (Liu & Wei 2003a, p. 354), YUNNAN (Degelius 1974, p. 176; Jiang 1993, p. 72; Liu & Wei 2003a, p. 354), FUJIAN (Jiang 1993, p. 72; Liu & Wei 2003a, p. 354), SHAANXI, XINJIANG (Liu & Wei 2003a, p. 354; Guo 2005, p. 46).

22.2. var. ***subnigrescens*** (Müll. Arg.) Degel., Symb. Bot. Upsal. 20(2): 173 (1974).

= *C. corniculatum* Z.G. Jiang, Journal of Hebei Normal University (Natural Science) 16(3): 85 (1992).

This taxon differs from var. *pulchellum* in having regular ridges rather than pustules on the upper side and broader spores (commonly > 5 μm wide). It is corticolous.

SPECIMENS EXAMINED: see Liu & Wei (2003a) for details.

LITERATURE RECORDS FOR CHINA: HEILONGJIANG (Liu & Wei 2003a, p. 355), SHAANXI, QINGHAI (Liu & Wei 2003a, p. 355; Guo 2005, p. 46), HUBEI (Chen et al. 1989, p. 420; Liu & Wei 2003a, p. 355), SICHUAN, GUIZHOU (Liu & Wei 2003a, p. 355), YUNNAN (Degelius 1974, p. 176; Liu & Wei 2003a, p. 355), XIZANG (Liu & Wei 2003a, p. 355), HONGKONG (Thrower 1988, p. 86; Liu & Wei 2003a, p. 355).

23. ***Collema rugosum*** Kremp., in Fenzl, Reise Österr. Novara Bot. 1: 128 (1870).

The species is characterized by its foliose thallus with > 5 mm wide lobes, distinct ridges on upper side and rugose to isidiate apothecial margin. It is corticolous.

It is similar to *C. subflaccidum*, but differs in having isidiate apothecial margins and distinct ridges on the upper side of the thallus.

SELECTED SPECIMENS EXAMINED: HEILONGJIANG, Jingpohu Lake, 30.VII.1977, Qian ZG, Herbarium no.: 021472 (HMAS-L). ANHUI, Yuexi, 950 m, 4.IX.2001, Liu HJ, 458. JIANGXI, Mt. Lushan, 2.IV.1960, Zhao JD, 519-1; GUIZHOU, Daozhen, 1600 m, 11.VI.1987, Wu, 2929. XINJIANG, Kalas, 2300 m, Abbas A, 980088.

LITERATURE RECORDS FOR CHINA: HONG KONG (Thrower 1988, p. 87; Aptroot & Seaward 1999, p. 83). XINJIANG (Guo 2005, p. 46).

24. ***Collema ryssoleum*** (Tuck.) A. Schneid., Guide Study Lich.: 181 (1898).

This species resembles *C. nigrescens* and *C. subnigrescens* in general appearance, but differs in having 4–6-celled, fusiform to ellipsoid spores (23.5–40.5 × 4.5–9.5 μm), and in being saxicolous rather than corticolous.

SELECTED SPECIMENS EXAMINED: HEBEI, Lingshou, Manshan, 1400 m, 13.VI.1986, Jiang ZG, 2000. SHAANXI, Mt. Taibaishan, 2750 m, 11.VII.1988, Ma CH, 123. SICHUAN, Mt. Emeishan, 2800 m, 18.VIII.1963, Zhao JD & Xu LW, 8140. YUNNAN, Mt. Gongshan, Dulongjiang, 2000 m, 3.IX.1982, Su JJ, 3978.

LITERATURE RECORDS FOR CHINA: SHAANXI (Wu 1987, p. 54; Guo 2005, p. 46).

25. ***Collema shiroumanum*** Räsänen, Journ. Jap. Bot. 16: 147 (1940).

This species is characterized by its 6–8-celled, (31–) 37.5–47.5 (–54.5) × 3.5–5 μm, dumbbell-shaped spores. It is corticolous.

SELECTED SPECIMENS EXAMINED: HUBEI, Shennongjia, 2250 m, 3.VII.1984, Chen JB, 10030. JIANGXI, Mt. Lushan, 3.IV.1960, Zhao JD et al., 573.

LITERATURE RECORDS FOR CHINA: HUBEI, JIANGXI (Jiang 1993, p. 73).

26. ***Collema sichuanense*** H.J. Liu & J.C. Wei, Mycosystema 22: 531 (2003).

Type(!): SICHUAN, Zoige County, Tiebu, 2800 m, 21.VI.1983, XY Wang & X Xiao, 10093 (HMAS-L). Paratype(!): SICHUAN, Aba, 3100 m, 28.VI.1983, XY Wang, 11343 (HMAS-L).

This species is characterized by 1) thallus foliose, lobes > 5 mm wide; 2) upper side regularly pustulate; 3) thallus and thalline exciple with pseudocortex; 4) proper exciple euthyplectenchymatous; 5) spores 6-celled, linear-fusiform, (27–) 30.5–37.5 (–44) × 3.5–5 μm. See Liu & Wei (2003b) for details.

LITERATURE RECORDS FOR CHINA: SICHUAN (Liu & Wei 2003b, p. 531).

27. ***Collema subconveniens*** Nyl., Lich. Nov. Zel.: 8 (1888).

= *C. tianmuense* Z.G. Jiang, Journal of Hebei Normal University (Natural Science) 16(3): 84 (1992).

This species is characterized by its foliose thallus with broad lobes (> 5 mm wide) having a distinct pseudocortex on both sides, and its submuriform spores. It is corticolous or terricolous. See Liu & Wei (2003a) for details.

LITERATURE RECORDS FOR CHINA: HUBEI, YUNNAN (Liu & Wei 2003a, p. 357), SHAANXI (Liu & Wei 2003a, p. 357; Guo 2005, p. 47), XINJIANG (Abbas & Wu 1998, p. 62; Liu & Wei 2003a, p. 357; Guo 2005, p. 47).

28. ***Collema subflaccidum*** Degel., Symb. Bot. Upsal. 20(2): 140 (1974).

This species is characterized by 1) foliose thallus with > 5 mm wide lobes; 2) upper side without regular ridges or pustules; 3) globular to occasionally cylindrical isidia on the upper side; 4) spores 6-celled, fusiform, 34–57.5 × 3.5–8.5 (–13.5) μm. It is corticolous.

SELECTED SPECIMENS EXAMINED: BEIJING, Mt. Xishan, 12.X.1961, Zhao JD & Sun ZM, 5069. JILIN, Mt. Changbaishan, 20.VIII.1977, Wei JC, 3118. NEI MONGOL, Horqin Youyi Qianqi, 1200 m, 6.VII.1985, Gao XQ, 930. SHAANXI, Mt. Taibaishan, 2260 m, 10.VI.1963, Wei JC et al., 2589. ANHUI, Jinzhai, 650 m, 10.IX.2001, Liu HJ, 552. GUIZHOU, Mt. Fanjingshan, 1220 m, 19.VIII.1963, Wei JC, 300-1. JIANGXI, Mt. Tianchishan, 800 m, 12.II.1965, Wei JC, 3090. YUNNAN, Zhongdian, 3650 m, 22.VIII.1981, Wang XY et al., 7562. XINJIANG, Mt. Altay Shan, Abbas A, 200732.

LITERATURE RECORDS FOR CHINA: HEILONGJIANG (Chen et al. 1981a, p. 134; Wu 1987, p. 55), JILIN (Wu 1987, p. 55), SHAANXI (Wu 1987, p. 55; Guo 2005, p. 47), XINJIANG (Guo 2005, p. 47), JIANGSU (Wu & Xiang 1981, p. 2; Wu & Qian 1989, p. 195; Wu 1987, p. 55), ANHUI, SHANGHAI, ZHEJIANG (Wu & Qian 1989, p. 195), JIANGXI (Degelius 1974, p. 140), FUJIAN (Wu et al. 1984, p. 1).

29. ***Collema subnigrescens*** Degel., Symb. Bot. Upsal. 13(2): 413 (1954).

29.1. f. ***subnigrescens***

= *C. pulchellum* var. *multipartitum* Z.G. Jiang, Journal of Hebei Normal University (Natural Science) 16(3): 86 (1992) (p.p.).

This form differs from f. *caesium* in having apothecia without pruina. It is corticolous. See Liu & Wei (2003a) for details.

LITERATURE RECORDS FOR CHINA: SHAANXI (Wu 1987, p. 55; Guo 2005, p. 47), HUNNAN, SICHUAN, XIZANG (Liu & Wei 2003a, p. 358).

29.2. f. ***caesium*** (Clemente) Degel., Symb. Bot. Upsal. 13(2): 417 (1954).***

This form is characterized by 1) thallus foliose, lobes > 5 mm wide; 2) upper side with regular ridges and pustules; 3) apothecia pruinose; 4) thalline exciple with pseudocortex; 5) proper exciple euthyplectenchymatous; 6) spores broadly acicular, 6-celled, (34–) 45–64 × (3.5–) 5–7 μm. It is corticolous.

SELECTED SPECIMENS EXAMINED: SICHUAN, Yanyuan, 3450 m, 26.VIII.1983, Wang LS, 83-1294 (KUN 6911). GUIZHOU, Mt. Fanjingshan, 1220 m, 19.VIII.1963, Wei JC, 300.

30. ***Collema substipitatum*** Zahlbr., in Handel-Mazzetti, Symb. Sin. 3: 76 (1930).

30.1. var. ***substipitatum***

This taxon resembles *C. leptaleum* and *C. shiroumanum* in external appearance (with subfoliose to crustose and fenestrate thallus, rugose upper side and < 5 mm wide lobes), but differs in having 6–10-celled, longer (> 40 μm long), fusiform to acicular spores.

SELECTED SPECIMENS EXAMINED: GUIZHOU, Daozhen, 1600 m, 15.VI.1987, Wu, 2933. SICHUAN, Mt. Emeishan, 2800 m, 18.VIII.1963, Zhao JD & Xu LW, 8017.

LITERATURE RECORDS FOR CHINA: YUNNAN (Zahlbruckner 1930, p. 76. Degelius 1974, p. 188); TAIWAN (Degelius 1974, p. 188).

30.2. var. ***gonggashanense*** H.J. Liu & J.C. Wei, Mycosystema 22: 532 (2003).

Type(!): SICHUAN, Mt. Gonggashan, Dongpo, Yanzigou, 2650 m, 2.VII. 1982, XY Wang et al., 8729-1 (HMAS-L).

This taxon differs from var. *substipitatum* in having a well developed euparaplectenchymatous proper exciple. See Liu & Wei (2003b) for details.

LITERATURE RECORDS FOR CHINA: SICHUAN (Liu & Wei 2003b, p. 532).

31. ***Collema tenax*** (Sw.) Ach., Lichenogr. Univ.: 635 (1810).

This species is distinguished by the marginally swollen and plicate lobes, euthyplectenchymatous proper exciple, and fusiform to ellipsoid, 4-celled to submuriform spores (15–28 × 6–10.5 μm). It is terricolous.

LITERATURE RECORDS FOR CHINA: HEILONGJIANG (Chen et al. 1981b, p. 150), SHAANXI (Wu 1987, p. 55; Guo 2005, p. 47).

The articles mentioned above recorded this species without indicating infraspecific taxa. In any case, var. *tenax* is not included here because it has not been found in the materials we have examined from China so far.

31.1. var. ***corallinum*** (A. Massal.) Degel., Symb. Bot. Upsal. 13(2): 165 (1974).

This variety is characterized by 1) thallus subfoliose to foliose, 2–3 cm diam.; 2) lobes < 2 mm wide, erect, often forming erect clusters. It is terricolous.

SELECTED SPECIMENS EXAMINED: NEI MONGOL, Xilinhot, Inner Mongolian Grassland Ecosystem Research Station of Academia Sinica, 1150 m, 27.VII.2003, Liu HJ, 751.

XINJIANG, Pishan, 3500 m, 13.VIII.1992, Abbas A, 92-0373-c [Abbas & Wu (1998) as *C. tenax* var. *substellatum*].

LITERATURE RECORDS FOR CHINA: NEI MONGOL (Degelius 1974, p. 49), GANSU (Magnusson 1944, p. 20; Degelius 1954, p. 183, 1974, p. 49), XINJIANG (Abbas et al. 1996, p. 12; Abbas & Wu 1998, p. 63; Abbas et al. 2001, p. 363; Guo 2005, p. 47).

31.2. var. ***crustaceum*** (Kremp.) Degel., Symb. Bot. Upsal. 13(2): 164 (1954).

In general habit, this variety resembles *C. tenax* var. *vulgare*, but differs in having smaller thallus (< 1 cm diam.), narrower lobes (< 2 mm wide) and smaller apothecia (< 1.5 mm diam.).

SELECTED SPECIMENS EXAMINED: HUNAN, Mt. Hengshan, 700 m, 1.IX.1964, Zhao JD & Xu LW, 10047. XINJIANG, Tomort, 2900 m, 25.VII.1977, Wang XY et al., 553.

LITERATURE RECORDS FOR CHINA: HEBEI (Jiang 1993, p. 69), XINJIANG (Guo 2005, p. 47).

31.3. var. ***diffracto-areolatum*** (Schaer.) Degel., Symb. Bot. Upsal. 13(2): 164 (1954).

This variety differs from other infraspecific taxa of *C. tenax* by the isidiate and areolate thallus.

SELECTED SPECIMENS EXAMINED: XINJIANG, Yecheng, 3000 m, 8.VIII.1992, Abbas A, 92-0293.

LITERATURE RECORDS FOR CHINA: HEBEI (Jiang 1993, p. 70), XINJIANG (Guo 2005, p. 47).

31.4. var. ***expansum*** Degel., Symb. Bot. Upsal. 13(2): 162 (1954).***

This variety is characterized by 1) thallus foliose, > 5 cm diam.; 2) lobes glossy, plane, often > 5 mm wide, with somewhat swollen and plicate margins; 3) thallus and thalline exciple without typical pseudocortex; 4) proper exciple euthyplectenchymatous; 5) spores fusiform to ellipsoid, 4-celled to submuriform, (13.5–) 17–24 (–27) × (7–) 9–13.5 μm.

It differs from all the other infraspecific taxa of *C. tenax* in the large thallus and glossy upper side.

SELECTED SPECIMENS EXAMINED: NINGXIA, Mt. Helanshan, 30.V.1984, Liu SR, 9.

31.5. var. ***ogatae*** (Zahlbr.) Degel., Symb. Bot. Upsal. 20(2): 47 (1974).

This variety is characterized by its foliose thallus with distinctly branched lobes and accessory lobules on the upper side.

SELECTED SPECIMENS EXAMINED: XINJIANG, Kuche, 2600 m, 23.V.1992, Abbas A, 92-0741; Aketao, 3000 m, 30.VII.1992, Abbas A, 92-024 [Abbas & Wu (1998) as *C. fuscovirens*].

LITERATURE RECORDS FOR CHINA: TAIWAN (Zahlbruckner 1933, p. 27; Degelius 1974, p. 47 & 49; Ikoma 1983, p. 59), XINJIANG (Guo 2005, p. 47).

31.6. var. ***substellatum*** (H. Magn.) Degel., Symb. Bot. Upsal. 20(2): 47 (1974).

This taxon differs from all the other infraspecific taxa of *C. tenax* in having long (15 mm long) and narrow (< 1 mm wide) lobes.

Specimen examined: XINJIANG, Baicheng, 2600 m, 21.V.1998, Wang XY, 6678-1.

Literature records for China: XINJIANG (Abbas et al. 1996, p. 12; Abbas & Wu 1998, p. 63; Abbas et al. 2001, p. 363; Guo 2005, p. 47), GANSU (Magnusson 1940, p. 41; Degelius 1974, p. 49).

31.7. var. ***vulgare*** (Schaer.) Degel., Symb. Bot. Upsal. 13(2): 163 (1954).

This taxon is characterized by its radiating, narrow (1–4 mm wide), plane lobes.

Selected specimens examined: HEBEI, Mt. Xiaowutaishan, 2100 m, Jiang ZG, 3696-1. NEI MONGOL, Ergun Zuoqi, Awuni, 10.VIII.1985, Gao XQ, 1467. NINGXIA, Mt. Helanshan, 1300 m, 21.V.1961, Han SJ et al., 2013. XINJIANG, Aketao, 3000–3250 m, 30.VII.1992, Abbas A, A-B, 920023, 920025, 920030, 9200109 [the specimens mentioned above from Xinjiang were cited as *C. fuscovirens* by Abbas & Wu (1998) and Abbas et al. (1996)].

Literature records for China: XIZANG (Degelius 1974, p. 49), XINJIANG (Abbas et al. 1996, p. 12, as *C. fuscovirens*; Abbas & Wu 1998, p. 63; Guo 2005, p. 47).

31.7.1. f. ***papulosum*** (Schaer.) Degel., Symb. Bot. Upsal. 13(2): 163 (1954).

This form often grows together with var. *vulgare*, from which it differs mainly in having the numerous, dense and globular isidia on the upper side.

Selected specimens examined: HEBEI, Mt. Baihuashan, 900 m, 14.IX.1978, Wei JC & Jiang YM, 3501. XINJIANG, Kuche, 3200 m, 23.V.1992, Abbas A, 92-0767 [Abbas & Wu (1998) as *C. tenax* f. *vulgare*].

Literature records for China: HEBEI (Jiang 1993, p. 69), XINJIANG (Guo 2005, p. 47).

32. ***Collema texanum*** Tuck., Am. Journ. Arts & Sci., ser. 2, 28: 200 (1859).

The species is characterized by the foliose thallus with repeatedly furcate, convex, marginally swollen, narrow (< 2 mm wide) lobes. It grows on soil-covered rocks.

Specimen examined: YUNNAN, Kunming, 1780 m, 16.I.1981, Jiang YM, 1153-1.

Literature records for China: ZHEJIANG (Degelius 1974, p. 57).

33. ***Collema undulatum*** Laurer ex Flot., Linnaea 23: 161 (1850).

33.1. var. ***undulatum***

The variety is characterized by the foliose thallus with 2–4 mm wide, repeatedly furcate, somewhat concave and undulate lobes. It is terricolous.

Selected specimens examined: XINJIANG, Kuche, 3100–3500 m, 22–30.V.1992, Abbas A, 92-0426, 92-0800.

Literature records for China: SHAANXI, GUANGXI (Jiang 1993, p. 71), XINJIANG (Guo 2005, p. 47).

33.2. var. ***granulosum*** Degel., Symb. Bot. Upsal. 13(2): 369 (1954).

This variety differs from var. *undulatum* in having globular isidia on upper side. It is saxicolous or terricolous.

SELECTED SPECIMENS EXAMINED: NEI MONGOL, Horqin Youyi Qianqi, 1500 m, 9.VII.1985, Gao XQ, 937. SHAANXI, Mt. Taibaishan, 1300 m, VII.1992, Chen JB & He Q, 5614. JILIN, Mt. Changbaishan, 1850 m, 28.VIII.1984, Lu XD, 848390-1.

LITERATURE RECORDS FOR CHINA: NEI MONGOL, HEBEI (Jiang 1993, p. 71), SHAANXI (Guo 2005, p. 47).

Taxa excluded from China

1. ***Collema pulchellum*** var. ***leucopeplum*** (Tuck.) Degel.
Symb. Bot. Upsal. 20(2): 172 (1974).

This name has been used for specimens of three taxa, *C. complanatum*, *C. pulchellum* var. *pulchellum* and *C. pulchellum* var. *subnigrescens* (see specimen citation of each taxon for details). It was reported from Nei Mongol, Jilin, Heilongjiang, Shaanxi and Qinghai (Jiang 1993).

2. ***Collema latzelii*** Zahlbr., Österr. Bot. Zeitschr. 59: 493 (1909).

This species was reported from Hebei by Jiang (1993) based on a misidentified specimen of *C. poeltii* (see specimen citation for details).

Taxa not included in this study

The following taxa were previously recorded in China. Among them, three species described by Jiang (1992), *C. fanjingshanense*, *C. fusiosporum* and *C. tetrasporum*, are doubtful because the original descriptions are short, not very informative, and not consistent with the type material. Specimens of the other taxa listed below were not available for study and the reports, therefore, could not be verified.

1. ***Collema beijingense*** Z.G. Jiang,
Journal of Hebei Normal University (Natural Science) 16: 84 (1992).

LITERATURE RECORDS FOR CHINA: BEIJING, HEBEI (Jiang 1992, p. 84).

2. ***Collema callibotrys*** var. ***coccophyllizum*** (Zahlbr.) Degel.,
Symb. Bot. Upsal. 20(2): 68 (1974).

LITERATURE RECORDS FOR CHINA: YUNNAN (Hue 1898, p. 217; Zahlbruckner 1930, p. 76).

3. ***Collema callopismum*** A. Massal., Miscell. Lich.: 23 (1856).

LITERATURE RECORDS FOR CHINA: HEBEI (Jiang 1993, p. 70).

4. ***Collema ceraniscum*** Nyl., Flora 48: 353 (1865).

LITERATURE RECORDS FOR CHINA: BEIJING, HEBEI (Jiang 1993, p. 70).

5. ***Collema clavisporiferum*** Z.G. Jiang,
Journal of Hebei Normal University (Natural Science) 16: 85 (1992).
LITERATURE RECORDS FOR CHINA: YUNNAN (Jiang 1992, p. 85).

6. ***Collema fanjingshanense*** Z.G. Jiang,
Journal of Hebei Normal University (Natural Science) 16: 86 (1992).
LITERATURE RECORDS FOR CHINA: GUIZHOU (Jiang 1992, p. 86).

7. ***Collema fusiosporum*** Z.G. Jiang,
Journal of Hebei Normal University (Natural Science) 16: 85 (1992).
LITERATURE RECORDS FOR CHINA: YUNNAN (Jiang 1992, p. 85).

8. ***Collema limosum*** (Ach.) Ach., Lichenogr. Univ.: 629 (1810).
LITERATURE RECORDS FOR CHINA: SHANGHAI (Nylander & Crombie 1883, p. 62; Degelius 1974, p. 52).

9. ***Collema peregrinum*** Degel., Symb. Bot. Upsal. 20(2): 109 (1974).
LITERATURE RECORDS FOR CHINA: TAIWAN (Degelius 1974, pp. 109–111; Ikoma 1983, p. 59).

10. ***Collema sorediatum*** Z.G. Jiang,
Journal of Hebei Normal University (Natural Science) 16: 85 (1992).
LITERATURE RECORDS FOR CHINA: ANHUI, GUANGXI, SICHUAN (Jiang 1992, p. 85).

11. ***Collema tenax*** var. ***ceranoides*** (Borrer) Degel.,
Symb. Bot. Upsal. 13(2): 162 (1954).
LITERATURE RECORDS FOR CHINA: HEBEI (Jiang 1993, p. 69)

12. ***Collema tetrasporum*** Z.G. Jiang,
Journal of Hebei Normal University (Natural Science) 16: 84 (1992).
LITERATURE RECORDS FOR CHINA: BEIJING (Jiang 1992, p. 84).

13. ***Collema thamnodes*** Tuck. ex Riddle, Bull. Torrey Bot. Club 43: 155 (1916).
LITERATURE RECORDS FOR CHINA: XINJIANG (Abbas & Wu 1998, p. 64).

Acknowledgments

The authors are indebted to Prof. Abdulla ABBAS (XJU) for sending the specimens on loan and to Ms Hong DENG (HMAS-L) for giving us access to all the facilities in the herbarium. Thanks go to both Dr. Irwin M. BRODO, Canadian Museum of Nature, and Prof. Wenying ZHUANG, Institute of Microbiology, CAS, for reading and improving the manuscript, and for acting as presubmission reviewers.

Literature cited

Abbas A, Hairet M, Anwar T, Wu JN. 2001. A checklist of the lichens of Xinjiang, China. Harv. Pap. Bot. 5: 359–370.

Abbas A, Wu JN. 1998. Lichens of Xinjiang. Sci-Tec & Hygiene Publishing House of Xinjiang: Urumqi (China). 178 pp. (in Chinese with English abstract).

Abbas A, Wu JN, Jiang YC. 1996. A study of lichens of Xinjiang. Arid Zone Research 13: 1–19 (in Chinese with English abstract).

Aptroot A, Seaward MRD. 1999. Annotated checklist of Hong Kong lichens. Trop. Bryol. 17: 57–101.

Asahina Y. 1952. An addition to the Sato's Lichenes Khinganenses (Bot. Mag. Tokyo, 65: 172). Journ. Jap. Bot. 27: 373.

Chen JB, Wu JN, Wei JC. 1989. Lichens of Shennongjia. in Mycological and Lichenological Expedition to Shennongjia, Academia Sinica (eds.) Fungi and Lichens of Shennongjia. World Publishing Corp.: Beijing (China). pp. 386–493 (in Chinese with English abstract).

Chen XL, Zhao CF, Luo GY. 1981a. A list of lichens in N. E. China (I). Journ. North-eastern Forestry Inst. 3: 127–135 (In Chinese with English abstract).

Chen XL, Zhao CF, Luo GY. 1981b. A list of lichens in N. E. China (II). Journ. North-eastern Forestry Inst. 4: 150–160 (In Chinese).

Degelius G. 1954. The lichen genus *Collema* in Europe: Morphology, taxonomy, ecology. Symb. Bot. Upsal. 13(2): 1–499.

Degelius G. 1974. The lichen genus *Collema* with special reference to the extra-European species. Symb. Bot. Upsal. 20(2): 1–215.

Guo SY. 2005. Lichens. In Zhuang WY (ed.) Fungi of Northwestern China. Mycotaxon Ltd: New York (USA). pp. 31–82.

Hue AM. 1898. Lichenes extra-Europaei. Nouv. Arch. Mus. Hist. Nat. ser. 3, 10: 213–280.

Ikoma Y. 1983. Macrolichens of Japan and adjacent regions. Tottori City (Japan). 120 pp.

Jatta A. 1902. Licheni cinesi raccolti allo Shen-si negli anni 1894–1898 dal. rev. Padre Missionario G. Giraldi. Nuovo Giorn. Bot. Italiano. ser. 2, 9: 460–481.

Jiang ZG. 1992. New taxa of lichen genus *Collema* Weber from P. R. China. Journal of Hebei Normal University (Natural Science) 16: 83–87 (in Chinese with English abstract and Latin diagnosis).

Jiang ZG. 1993. New recorded species of the genus *Collema* Weber from P. R. China and Hong Kong region. Journal of Hebei Normal University (Natural Science) 17: 69–73 (in Chinese with English abstract).

Kirk PM, Cannon PF, David JC, Stalpers JA. 2001. Ainsworth & Bisby's Dictionary of the Fungi (9th edition). CAB International: Wallingford (UK). 616 pp.

Liu HJ, Wei JC. 2003a. Taxonomic revision of six taxa in *Collema* from China. Mycotaxon 86: 349–358.

Liu HJ, Wei JC. 2003b. Two new taxa of the lichen genus *Collema* from China. Mycosystema 22: 351–353.

Magnusson AH. 1940. Lichens from central Asia I. Reports from the scientific expedition to the North-western Provinces of China under the leadership of Dr. Sven Hedin. Publication 13 (XI. Botany) 1: 1–168.

Magnusson AH. 1944. Lichens from central Asia II. Reports from the scientific expedition to the North-western Provinces of China under the leadership of Dr. Sven Hedin. Publication 22 (XI. Botany) 2: 1–68.

Nylander W & Crombie JM. 1883. On a collection of exotic lichens made in Eastern Asia by the late Dr. A. C. Maingay. J. Linn. Soc. London, Bot. 20: 62–66.

Thrower SL. 1988. Hong Kong lichens. Urban Council: Hong Kong (China). 193 pp.

Wang JR, Lai MJ. 1973. A checklist of the lichens of Taiwan. Taiwania 18: 83–104.

Wei JC. 1991. An enumeration of lichens in China. International Academic Publishers: Beijing (China). 278 pp.

Wu JL. 1987. Lichen iconography of China. China Zhanwang Press: Beijing (China). 236 pp. (in Chinese).

Wu JN, Qian ZG. 1989. Lichens. in Xu BS (ed.) Cryptogamic flora of the Yangtze Delta and adjacent regions. Shanghai Scientific & Technical Publisher: Shanghai (China). pp. 158–266 (in Chinese).

Wu JN, Xiang T. 1981. A preliminary study of the lichens from Yuntai Mountain in Lianyungang, Jiangsu. Journal of Nanjing College (Natural Science Edition) 3: 1–11 (in Chinese).

Wu JN, Xiang T, Qian ZG. 1984. Notes on Wuyi Mountain lichens (II). Wuyi Science Journal 4: 1–7 (in Chinese).

Zahlbruckner A. 1930. Lichenes. in Handel-Mazzetti H (ed.) Symb. Sin. 3: 1–254. Julius Springer: Vienna (Austria).

Zahlbruckner A. 1933. Flechten der Insel Formosa. Feddes Repert. Sp. Nov. Regni Veg. 33: 22–68.

第五章　系统演化生物学
Systematic & Evolutionary Biology

地衣的物种概念与进化论

魏 江 春

（中国科学院微生物研究所）

THE SPECIES CONCEPT IN LICHENS AND THE THEORY OF EVOLUTION

Wei Jiangchun

(The Institute of Microbiology, Academia Sinica)

地衣是一类与藻共生的、具有独特构造的专化性真菌。在地衣中，共生菌依赖共生藻借光合作用所提供的有机养分而生活。共生藻在菌丝组织包围中处于有利的弱光照下正常生活；也有利于藻细胞水分状况的改善以及对可溶性无机盐的需要。在人工培养中，湿度过大则使共生藻徒长；培养基中有机养分过多则使菌、藻各自独立生长。因此，地衣实际上是一种生态平衡系统，也是菌、藻相互依存的生命支持系统。

现代海洋、湖泊与溪流中广泛存在的短暂而松散和长期而稳定的菌、藻共生现象（这类生物也叫类地衣），生长在两栖地带岩石表面的低等地衣，以及分布于地球表面各自然带的两万多种陆生地衣，这些事实作为一个整体来看，犹如展现在我们面前的一幅由类地衣到两栖地带低等地衣向陆生地衣过渡的地衣进化趋势图。在这一进化过程中，有些类地衣向以真菌为主的菌、藻共生方向演变而成为具有独特构造的地衣；另一些则向菌、藻各自独立生活的方向演变而成为今天的陆生子囊菌和陆地藻。因为菌、藻相互依存的生命支持系统与自由生活菌、藻相比，对于干旱与缺乏有机养分的生境具有更强的适应性。因而，在海陆变迁以及子囊菌由水生向陆生进化的过程中，类地衣及地衣这种生命支持系统很可能起到了重要的桥梁作用。

种间的遗传隔离说明种间 DNA 碱基序列之明显差异，而变种间无遗传隔离，说明 DNA 碱基序列差异不大。而表型特征差异的大小，正是物种 DNA 碱基序列特异性的反映与放大。因此，在地衣分类中，确定种与变种的界限理应根据若干表型特征的组合及其差异的主次、大小与多少而进行。在地衣种的确定上，主要是考虑形态差异、地理分布、生物起源较远的化学成分的替代或同种化学成分之有无等表型特征的组合。而变种主要是考虑形态上的微小差异，生境与小地区分布以及生物起源较近的化学成分的替代或同种化学成分之有无，或有无群居现象等表型特征的组合。变种与种之间是一种动态关系。变种可能是未来的种。

一种地衣从一生态区向另一生态区的散布，或由于气候巨变，海陆变迁或历史性的其他原因，都会引起物种分化，从而伴随替代种的出现。拟菊叶黄梅（*Xanthoparmelia taractica* (Krempelh) Hale）与旱黄梅（*X. camtschadalis* (Ach.)Hale）可能就是不同生态条件下

本文原载于《进化论选集》. 北京: 科学出版社: 169-170, 1983.

物种分化的结果。而华东疱脐衣（*Lasallia sinorientalis* Wei）可能是疱脐衣（*L. pustulata* (L.)Merat）在亚洲(中国)的地理替代种。拟石耳(*Umbilicaria mammulata* Ach.)则是中国、日本特有的石耳（*U. esculenta*(Miyoshi)Minks）在北美的替代种。而黄粉袋衣(*Hypogymnia hypotrypella*(Asahina)Rassad.）与黄袋衣（*H. hypotrypa*(Nyl.)Rassad.)，金丝带（*Lethariella zahlbruckneri*(DR.)Krog）与中华金丝（*L. sinensis* Wei *et* Jiang)，金丝刷（*L. cladonioides* (Nyl.)Krog）与金丝绣球（*L. cashmeriana* Krog）以及槽枝衣原变型（*Sulcaria sulcata* f. sulcata）与黄槽枝变型（*f. vulpinoides*(A. Z.)Hawksw.）等四对种及变型的分化都是在原种分布区内出现的，而且集中分布于中国横断山脉地带及其邻近地区。这可能是由于复杂的地貌特征与南北的冷、热、干、湿气流的交替所造成的独特自然地理条件在物种变异与自然选择中所起的作用。

如果说旱黄梅是由拟菊叶黄梅决定衣体形态的等位基因突变后所分化的种，那么，旱黄梅的具有抗旱特性的管状衣体可能就是突变后的等位基因的特异性的表型特征，而原等位基因却没有这特异性。因为，基因的适应性只能在生物的表型中，而不是在基因分子中去识别。因此，旱黄梅的分化正是在物种变异的基础上自然选择的结果。

无论在地衣从水生向陆生的进化中，或在物种分化与新种形成中，在基因交换与重组或基因突变的基础上，自然选择起着重要的作用。然而，Motoo Kimura（1968, 1980）以其中性基因的观点来反对达尔文的自然选择说，实际上是在反对生物的适应性，因而也是在否认生物本身的存在。因为，从分子水平上看，中性基因也许是可以接受的，而对环境条件缺乏适应性的中性生物却是不可想象的。生物的适应性只有广狭之分，绝无有无之别。达尔文的自然选择学说是建立在对自然界的客观规律进行广泛考察与研究，以及对经济动植物和微生物的人工选择所取得的巨大成就的基础上的，因此，它不会因为分子生物学的兴起而过时，恰恰相反，它将从分子生物学中获得越来越多的证据。

粗皮松萝群体分化性状的研究

魏春江[1]　胡玉琛[2]　姜玉梅[1]　陈健斌[1]

1）中国科学院微生物研究所真菌地衣系统学开放研究实验室，北京　100080）

2）吉林省长白山自然保护区科学研究所，吉林，安图　133613）

摘要　本文对于长白山南坡高山半寒漠植被中大面积匍匐生长在地表的粗皮松萝群体从形态学、生态学、化学及分类学方面进行了初步研究。在分类学上将这一地生群体命名为匍匐变种；在种群落结构方面进行了样方调查；从生物系统学角度进行了初步分析。

关键词　粗皮松萝，匍匐变种，地生群体，高山半寒漠，生物系统学

在松萝属（*Usnea*）中，除了石萝亚属（*Neuropogon*）的种类以及从松萝属中分出来的金丝属（*Lethariella*）部分种类之外，几乎全部种类都是以树木为基物而生长的，只有个别种类偶而也能生长在岩石或土表[1]。但是，1983 年我们在长白山南坡高山半寒漠[2,3]地面芽型植被中发现了大面积匍匐生长在地表的粗皮松萝（*Usnea montis-fuj* Mot.）群体*，形成了以它为特色的地面芽型植物群落(图版 I:1,2)。这一发现，无论是在松萝属的生态学中，还是在粗皮松萝的生物系统学**中都具有重要意义。

1　环境特点

长白山地处我国吉林省东南部，与朝鲜毗邻，位于北纬 41°36′，东经 126°55′—129°。我国一侧主峰白云峰海拔为 2691 米[4]。长白山山顶(2600 米以上)年平均气温－7.3℃，最暖月(7—8 月)平均气温 8.5℃，最冷月(1 月)平均－24℃，也曾出现过－44℃的现象[5]。年降水量为 1340.4 mm，最多年分曾达 1809.1mm。山顶多雾，年平均为 264 天。年平均绝对湿度为 3.7mb，最大 17.3mb。山顶风大，≥40 米/秒的大风全年各月均有出现；年平均风速为 11.5 米/秒。山顶以西风和偏西风为主；南坡则为全年西风[6]。

2　粗皮松萝研究简史

Motyka（1936—38）[1]在其“松萝属地衣研究专著”中对于采自日本、中国及俄罗斯

---

* “群体”是来自英文“Population”一词。该词在生物学上有多种译法。“群体”的译法虽较普遍，但是，近年来“居群”的译法却更受鼓励。（见 V. H. Heywood 所著“植物分类学”中文译本第 8 页脚注）不过“居群”一词似更适用于动物。因此，本文仍延用“群体”一词，即在种的分布区内占有同一分布区或生长于相似生境的一群生物个体。“群体”之间往往由于地理的、生态的、甚至体态或颜色等方面的某些差异而呈现某种间断。因此，我们认为“群体”的译法也许不仅适用于地衣型与非地衣型真菌和绿色植物，而且也适用于动物。

** “生物系统学”是来自英文“Biosystematics”一词。这是该词的直接译法。但是由于它易与生物的“系统学”（systematics）一词相混淆而有译为“物种生物学”的。易于混淆的主要原因也许是因为国内在这方面的研究工作不多而使人们忽视了它的特定含意所致。它的特定含意着重于种内群体间及群体内分化、变异和演化过程的研究。因此在按其特定含意理解和使用时，通常是不易混淆的。因此，本文仍采用“生物系统学”的译法。

1993 03-23 收稿。

本文原载于《真菌学报》，13（3）：199-207, 1994.

远东库页岛的类似于长萝松（*Usnea longissima* Ach.)的地衣以日本"富士山"（Fujisan）名称为种加词作为新种（*Usnea montis-fuji* Mot.）予以命名和描述。主模式是朝比奈泰彦于1925年采自日本富士山,保存在奥地利维也纳自然博物馆。副模式是由不同的采集者，如 Giraldi（1896）采自中国陕西秦岭山区；Faurie 采自俄罗斯远东库页岛；Du Bois-Reymond（1914）及 Sakurai（1909）分别采自日本。这些副模式标本均保存在柏林植物博物馆。而 Almquist（1880）及 Vega Exp. 考察队分别采自日本的副模式标本保存在瑞典斯德哥尔摩国家自然历史博物馆;由 Faurie（1898)采自日本的副模式标本保存在法国港市敦刻尔克一标本室。但是,原作者 Motyka 对于上引主、副模式标本的基物均未指明。朝比奈泰彦(1956)[7]在"日本之地衣"第三册中对于本种的基物也未提及。赵继鼎等(1982)[8]在"中国地衣初编"中对于不同采集者采自辽宁、陕西、四川和湖北的粗皮松萝作了记载。陈锡龄等(1981)[9]对于采自长白山的粗皮松萝也作了报导。这些记载都明确地指出其基物为树木。更重要的是迄今为止，尚未见到粗皮松萝以土壤为基物的任何报导。

3 材料与方法

我们于1983年夏季对于长白山南坡高山半寒漠大面积匍匐生长的粗皮松萝地表群体进行了实地考察。在考察中采用了1×1米样方法（Quadrat method)。

用于形态学及化学研究的标本于1983年采自考察地区海拔为2080m至2380m地段。作为对照用标本采自吉林浑江市临江镇(1950),陕西太白山(1955),四川峨眉山以及尼泊尔(1963)（TNS 21324!）等。在形态学观察中使用光学显微镜（OPTONII）及解剖镜（OPTON)。测定采用 C. Culberson & Kristinsson[10] 和 C. Culbersor[11]标准薄层色谱方法（TLC）。

4 形态特征及化学分析

粗皮松萝地生群体及树生群体在形态上均属于丝状类型（Filamentose：朝比奈泰彦,1956, p. 8, Fig. 1, C)。但是地生群体在形态上的变异在于主茎变为宽大的屋瓦型扁枝或近圆形基盘。大量短小分枝由屋瓦型扁枝或基盘周围丛生；其中通常只有一条分枝作为初次分枝而继续蔓延伸长而形成细长的丝状体匍匐于地面(图1，2)。这些丝状体随处可生不定假根。丝状体的主茎基盘及其不定假根均牢固地固着在地面(图版II：3—6)。至于化学成分,树生群体与地生群体完全一致,即均含有松萝酸及水杨嗪酸两种。

此外，我们在野外考察中以及在所采集的大系列标本中迄今尚未发现子囊盘。这至少可以说明有性繁殖是偶见的。因此,这样大面积地生群体的形成与蔓延,主要是通过有丝分裂地衣体或其片段借助风力传播而实现的。但是，最初可能是从林区传来的树生群体的有丝分裂丝状体个体或片段在高山寒漠地带经过长期适应的结果。高山半寒漠的地生群体与林区的树生群体之间显然已在很大程度上形成了生态隔离。

这样的隔离使地生群体在形态上虽然并未发生明显的变异，但是也出现了基盘型和屋瓦型主茎以及不定假根等微小差别。这些差别究竟是高山半寒漠气候及地面生境所引起的持久饰变（dauermodification)[12],还是在偶见的有性繁殖中发生了遗传突变，难以作出明确结论。不过,由于这样的生态隔离而导致的形态分化,在本种内群体间已表现出微小的间断，因而在分类学上具有一定的意义。但是，考虑到这种形态分化及间断不能排除由生境引起饰变的可能性，以及从丝状体的基本特征及其所含化学成分等整体表型性状来看，地生群体并未超出粗皮松萝的种级界限。因此，我们将它置于粗皮松萝下一变种地位，命名为匍匐变种。

5 群落种类组成

长白山高山植被带通常被称为高山冻原带，位于长白山森林界限以上，海拔 2200—2400 米处为界又分为下部的“高山带常绿矮石楠灌丛亚带”和上部的“高山半荒漠亚

表 1 粗皮松萝地生群体群落结构记名样方

Table 1 The list quadrats of the community structure for terricolous population of *Usnea montis-fuji* Mot.

种名	样方编号			
	I	II	III	IV
	海拔			
	2300m	2200m	2080m	2080m
	盖度（%）			
1.*Usnea montis-fuji* var. *sarmentosa* 粗皮松萝 匍匐变种（松萝科）	0.2	30	10	2
2.*Chrysanthemum zawadzkii* var. *alpinum* 毛山菊（菊科）	20.0	10	3	1
3.*Dryas octopetala* var. *asiatica* 宽叶仙女木（蔷薇科）	24.0	30	3	0
4.*Rhodiola sachalinensis* 长白景天（景天科）	5	4	1	1
5.*Oxytropis anertii* 长白棘豆（豆科）	5	2	2	3
6.*Bupleurum euphorbioides* 大苞柴胡（伞形科）	1	1.5	1	0
7.*Allium senescens* 山葱（百合科）	10	3	0.5	0.1
8.*Cyperaceae* （莎草科）	20	10	3	0
9.*Cetraria laevigata* 白边岛衣（岛衣科）	0	0.5	7	0
10.*Cladonia alinii* 沟石蕊（石蕊科）	0	0.5	5	0
11.*Cladonia amaurocraea* 黑穗石蕊（石蕊科）	0	0.2	5	0
12.*Thamnolia vermicularis* 地茶（有丝分裂类）	0	0.1	0.5	0

带"[3,4]。粗皮松萝地生群体在长白山高山半寒漠的分布范围主要局限在南坡海拔 2080 米至 2380 米之间,而且以海拔 2200 米上下的盖度最大(图版 1:1, 2)。它和其它草本植物及地衣组成了铺地型植被 (matted vegetation)。这里既无越橘 (*Vaccinium* spp.),又无鹿蕊(*Cladina* spp.) 及苔藓。记名样方 (list quadrat) 结果见表1。

由于这些植物的高度基本一致,又缺乏壳状地衣种类,因而植被中缺乏明显的成层现象。从群落中的种类组成来看,多为界于中生与旱生之间而又偏于旱生的过渡型种类。这是因为山顶年降水量虽然较大,但是由于南坡全年盛行的强西风劲吹,增强了蒸发,抵消了降水量较大的特点,因而使这里的地面层植被种类以界于中生与旱生而又偏于旱生的种类为特点。此外,由于匍匐生长的丝状粗皮松萝地生群体中的每一个体明显的呈现为

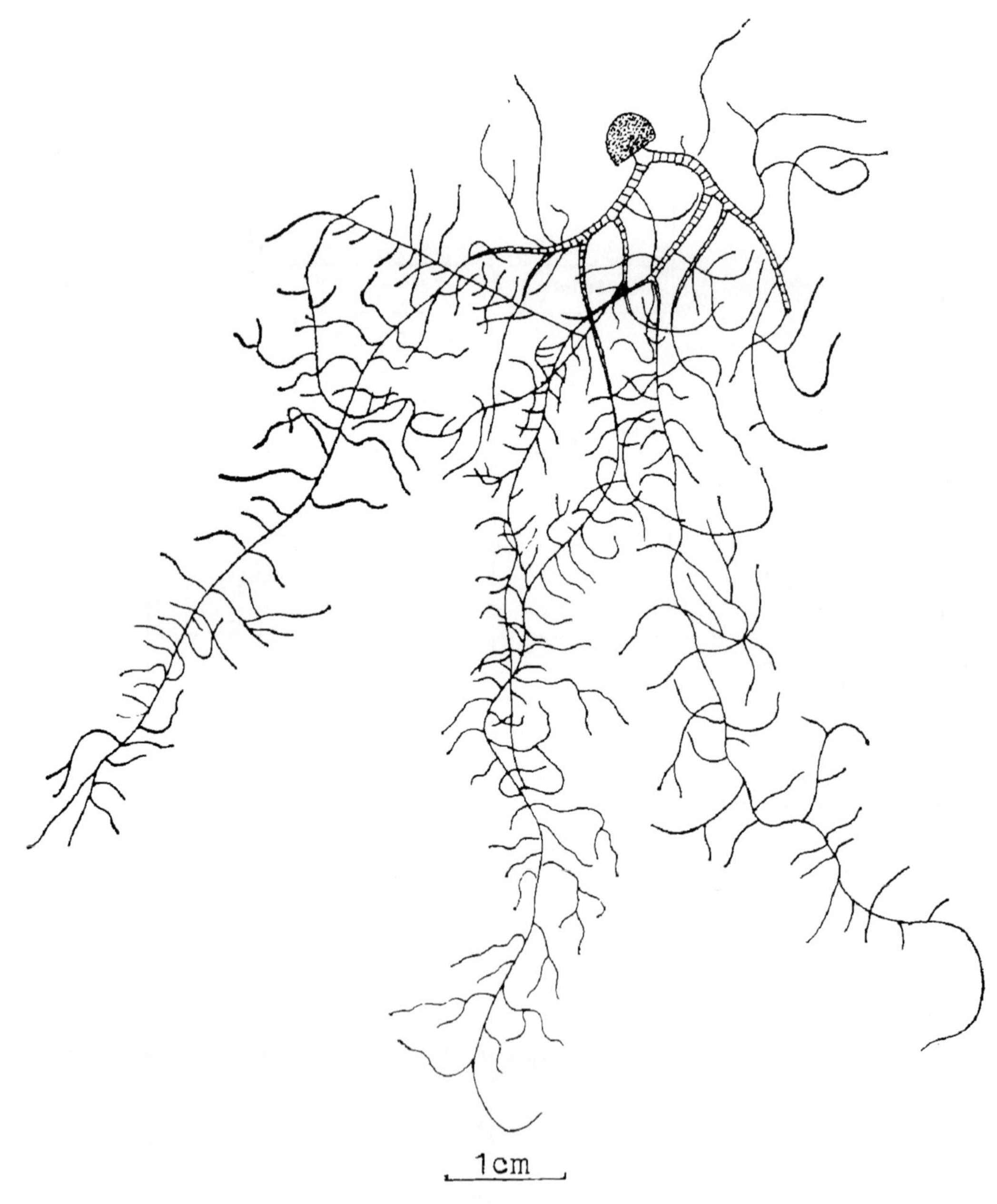

图1 粗皮松萝原变种(比例尺为一厘米)

Fig. 1 Usnea montis-fuji var. *montis-fuji* (the scale in 1 cm)

由西向东的走向，无疑是全年强劲西风所造成的必然结果。这一结果，也是当地盛行风方向的明显标志。

6 分类学处理

粗皮松萝

Usnea montis-fuji Mot.

图2 粗皮松萝匍匐变种(比例尺为一厘米)
Fig. 2 Usnea montis-fuji var. *sarmentosa* Wei et al. (the scale in 1 cm)

Lichenum Generi s Usnea Studium Monographicum，1936—1938：420.

原变种（图1；图版II:4,5）

var. ***montis-fuji*** （Fig. 1; Plate II:4,5）

地衣体枝状，主茎极短，圆柱形，直径约为1.5mm左右，直接固着于基物上；第一次分枝为等二叉式分枝，分枝比主茎稍长而细，直径约0.7mm，表面具环裂纹；第二次分枝为不等二叉式分枝，即在第二次分枝上往往不对称地又生出第三次分枝。一般情况则第三次分枝便是无环裂纹而具纤毛的细长丝状体，但有时第三次分枝仍为环裂纹型，第四次分枝才是无环裂纹而具纤毛的丝状体。

TLC：usnic＆salazinic acids.

Specimens Examined：

吉林：浑江市，临江镇，22，VI，1950，刘檏谔等 no. 94la. 陕西：太白山，放羊寺附近，海拔3200m，X，1955，汪发缵 no. 113，同一产地，海拔2800m，马启明等 no. 191。四川：峨届山，洗象池，海拔2200m，赵继鼎等 no. 7141。

匍匐变种（图2；图版I:1,2；图版II:3）

var. ***sarmentosa*** J. C. Wei et a1. var. nov. （Fig. 2；Plate I：1，2；Plate II：3）

Varietas haec a var. *montis-fnji* ramulis fasciculatis actinomorphis e basi disciformi rotunda ramosissimis prominentibus et ramis filamentosis sarmentosis ad terram in tundris montis Changbaiensis differt. Thalli acida usneica et salazinica contienti（TLC）.

Holotypus in comitato Changbaiensi in clivo australi montis Changbaiensis provinsiae Jilinensis, alt. 2200m. 3l, VII, 1983.Wei JC et Chen JB no. 64l1 in HMAS-L conservatus est.

本变种与原变种不同之处在于主茎变态为圆盘形或屋瓦状扁枝，其周缘簇生大量放射状短小分枝，其中通常只有一条分枝作为初次分枝而继续蔓延伸长而形成细长的丝状体匍匐于地面；这些丝状体随处可生不定假根。丝状体的主茎基盘及其不定假根均牢固地固着在地面；继续蔓延伸长而形成细长的这条丝状体第一次分枝为二叉式分枝，通常在这二叉式分枝上则直接生出无环裂纹而具短小纤毛的细长丝状体。

模式标本产于吉林长白山南坡长白县境内高山半寒漠，alt. 2200m，3l，VII，1983，魏江春，陈健斌，no.641l，保存在中国科学院微生物研究所真菌标本室地衣部（HMAS-L），等模式列入中国地衣标本集第二辑内。

苑兰翠洗印黑白照片；韩者芳为线条图涂墨，特此致谢。

参 考 文 献

[1] Mocyka J. Lichenum Generis Usnea Studium Monographicum. Leopoli. 1936—38。

[2] Lindsay D. C. Lichen Ecology. edited by M. R. D. seaward，Academic Press London, New York, San Fracisco. 1977,183—209。

[3] 赵大昌,森林生态系统研究，1980，**1**：65。

[4] 张风山，迟振文，裴铁番，等。森林生态系统研究，1981，**2**：179。

[5] 钱家驹,张文仲。森林生态系统研究,1980,**1**: 51。
[6] 张风山,迟振文,李晓晏。森林生态系统研究,1980,**1**: 193。
[7] Asahina, Y. 日本之地衣(第三册)。1956。
[8] 赵继鼎,徐连旺,孙曾美。中国地衣初编。北京: 科学出版社。1982。
[9] 陈锡黎,赵从福,罗光裕。东北林学院学报,1981,**4**: 150—160。
[10] Culberson C F, Kristinsson H D. *Jou Chromatogr*, 1970, **46**:85—93。
[11] Culberson C F. *Jou Chromatogr*,*a*1972, **72**:113—125。
[12] Heywood V H. Plant Taxonomy Edward Arnold. 1976。
(中文译本,北京: 科学出版社。1979)

A STUDY ON DIVERGENT CHARACTERS OF POPULATIONS IN *USNEA MONTIS-FUJI* MOT

Wei Jiangchun[1] Hu Yuchen[2] Jiang Yumei[1] Chen Jianbin[1]

(1 *Systematic Mycology & Lichenology Laboratory, the Institute of Microbiology, Academia Sinica, Beijing* 100080, *China*)

(2 *Research Institute of Sciences for Changbai Shan Natural Servation Area of Jilin province, Antu* 133613, *China*)

ABSTRACT A terricolous population of *Usnea montis-fuji* Mot. is found in the alpine cold semi-desert of Changbai mountain of Jilin province in north-eastern China. The population covers a considerably large area from alt. 2080 to 2380m in southern slope of the mountain. The strong prevailing wind trains the prostrate filamentous thalli of *Usnea montis-fuji* Mot. over the earth from west toward east. The terricolous population is characterized by its disciform or tile-shaped stem base of its filamentous thalli instead of cylindroid one in epiphytic population, and its adventitious rhizines. A preliminary analysis and comparison in biosystematics of the populations from morphology, ecology, chemistry, and taxonomy are carried out.

KEY WORDE *Usnea montis-fuji* Mot., terricolous population, *U. montis-fuji* var. *sarmentosa*, alpine could semi-desert, biosystematics

图版 I：1.长白山南坡匍匐生长在地表的粗皮松萝匍匐变种景观。**Plate I:** 1. A landscape of the terricolous *Usnea montis-fuji* var. *sarmentosa* J.C.Wei et al. trailing over the ground on south slope in the Mt. Changbai, Jilin.

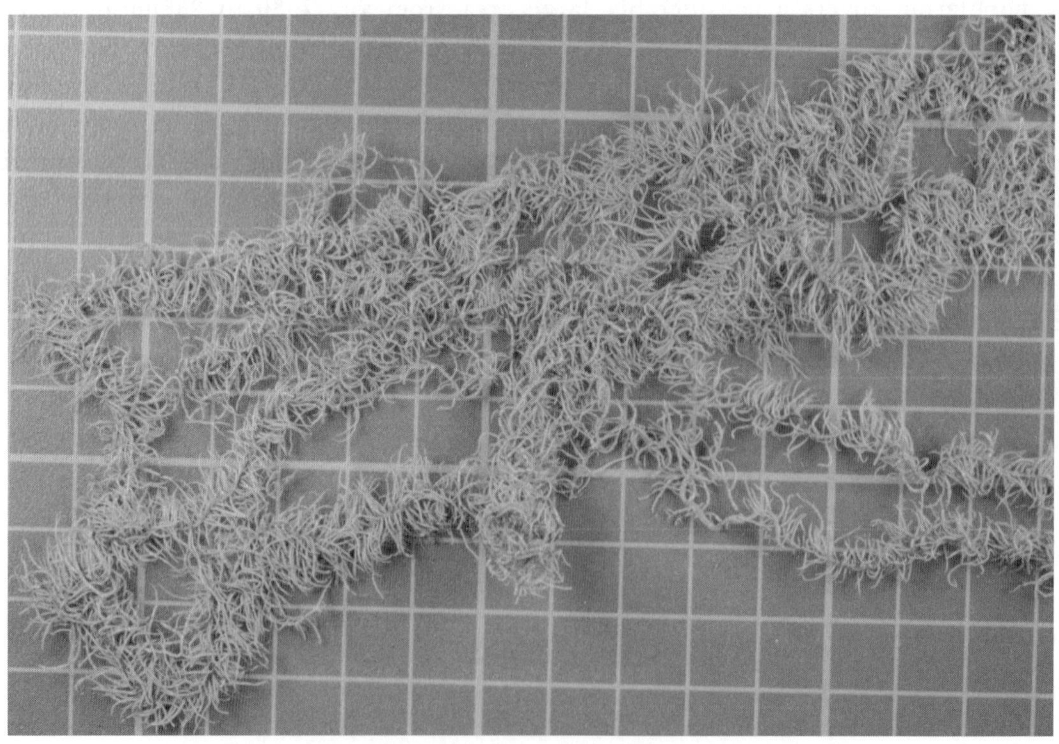

图版 I：2.匍匐生长在吉林长白山地表的.粗皮松萝匍匐变种，1983.7.31. 魏等 6411 （HMAS-L.109911-主模）标本（1 方格标尺=1 平方公分）。**Plate I:** 2. The habit of *Usnea montis-fuji* var. *sarmentosa* July 31, 1983. J. C. Wei et al. 6411（HMAS-L.109911-holotypus） trailing over the ground of Mt. Changbai, Jilin （a square scale=1cm^2）.

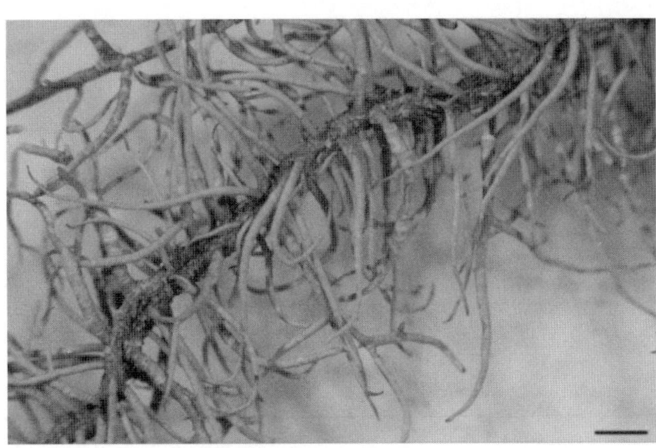

图版Ⅱ：3.匍匐生长在吉林长白山地表的粗皮松萝匍匐变种，1983.7.31. 魏等 6411（HMAS-L.109911-主模）枝状体局部特征（比例尺=1mm）。**Plate Ⅱ**：3. The habit of a part of *Usnea montis-fuji* var. *sarmentosa* July 31, 1983.J.C.Wei et al. 6411（HMAS-L.109911-holotypus） trailing over the ground of Mt. Changbai, Jilin （scale=1mm）.

图版Ⅱ：4.悬垂地固着于吉林临江三岔子 *Betula platyphylla* 枯枝上的粗皮松萝原变种枝状体局部特征（比例尺=1mm）,1950.06.22.刘慎谔等 941a（HMAS-L-13784）。**Plate Ⅱ**：4.The habit of a part of *Usnea montis-fuji* var. *montis-fuji*, June 22, 1950. Liu et al.941a（HMAS-L-13784） suspended from dead branches of *Betula platyphylla* in Sanchazi of Linjiang, Jilin （scale=1mm）.

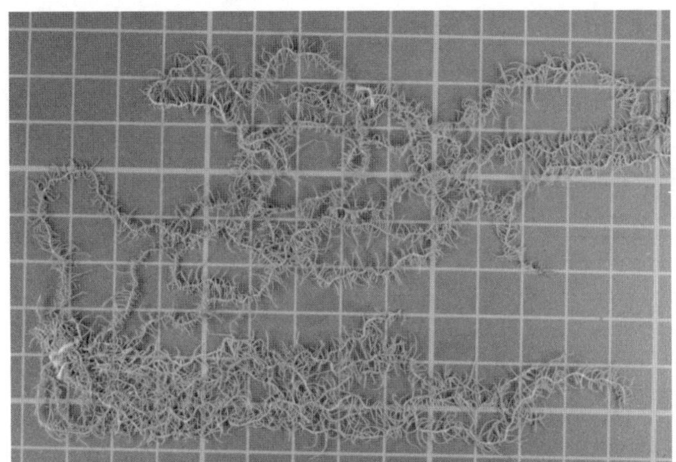

图版Ⅱ：5. 悬垂地固着于吉林临江三岔子 *Betula platyphylla* 枯枝上的粗皮松萝原变种，1950.06.22.刘慎谔等 941a（HMAS-L-13784）（1 方格标尺=1 平方公分）。**Plate Ⅱ**：5.The habit of *Usnea montis-fuji* var. *montis-fuji* June 22, 1950.Liu et al.941a（HMAS-L-13784） suspended from dead branches of *Betula platyphylla* in Sanchazi of Linjiang, Jilin （a square scale=1cm^2）.

真菌共祖起源的分子证据

付少彬 [1] 魏江春 [1,2*]

[1]河北大学生命科学学院 保定 071002

[2]中国科学院微生物研究所真菌地衣系统学重点实验室 北京 100101

摘　要：以真菌界、动物界和植物界中的 19 种各类生物作为代表，从具有进化速率不同区段结构的核糖体核酸基因大亚基全序列的 C3、C9 及 C11 三个高度保守区段中鉴定出了 4 条真菌所共有的核苷酸序列，亦即真菌界所特有的核苷酸序列。从其保守区段 C5 中鉴定出了 1 条子囊菌门所特有的保守序列；从保守区段 C7 中鉴定出了 1 条担子菌门所特有的保守序列。研究结果为各类真菌起源于一个共同祖先提供了分子证据。用于研究的美味石耳的核糖体核酸基因大亚基全序列是由本实验室所提供；其他 18 种核糖体核酸基因大亚基全序列来自 GenBank。

关键词：美味石耳，子囊菌门，担子菌门，核糖体核酸基因，特有序列

Molecular evidence for common descent of true fungi

FU Shao-Bin[1]　WEI Jiang-Chun[1,2*]

[1]*College of Life Science, Hebei University, Baoding 071002, China*

[2]*Key Laboratory of Systematic Mycology and Lichenology, Institute of Microbiology, Chinese Academy of Sciences, Beijing 100101, China*

Abstract: Four characteristic sequences of true fungi were found in the conserved regions C3, C9, and C11 of LSU rDNA. One characteristic sequence of Ascomycota and another one of Basidiomycota were identified respectively in the conserved region C5 and C7. The results showed that these are the molecular evidence for common descent of the true fungi. The whole sequence of LSU rDNA from *Umbilicaria esculenta* was made by

基金项目：国家自然科学基金（No. 30570008）

*Corresponding author. E-mail: Weijc2004@126.com

收稿日期：2008-01-14，接受日期：2008-01-31

This paper was originally published in *Mycosystema*, 27（2）: 217-223, 2008.

Wei's lab., and the other whole sequences of LSU rDNA from 18 species were obtained from GenBank.

Key words: *Umbilicaria esculenta*, Ascomycota, Basidiomycota, LSU rRNA gene, characteristic sequences

根据达尔文进化论的共同祖先学说，每一类生物都来自一个共同的祖先（Mayr 1988）。这就意味着，当今地球上如此丰富多彩的生物多样性，都是由一个共同祖先通过其基因突变和重组，引起基因组结构的变化，进而引起蛋白质中氨基酸序列的改变，最终导致表型性状的差异，从而在漫长的生物进化过程中，通过自然选择逐步分化并形成的。如此丰富多彩的后代生物往往都具有其祖先的性状，即祖征 (plesiomorphic characters)。这无疑是各类生物共同祖先起源的表型性状证据。

尽管已有大量含有以分子数据为基础的系统发育树的文章不断发表，然而，关于各类生物共同祖先起源的分子证据的文章却很少。屈良鹄在他的学位论文工作中从核 rRNA 基因大亚基 (nu-LSU rDNA) 进化速率不同的 12 个结构区[1]中发现了 C4[2], C5, C9 和 C11 为高度保守区。而且从中鉴别出了为细胞生物所共有的核苷酸序列。这些共有序列不仅表明所有细胞生物起源于共同的祖先，同时也是地球上细胞生物共同的分子特征。此外，该作者还在高度保守区 C9 中鉴定出了真核生物成员共有的核苷酸序列。该序列与原核的细菌和古菌中各自所共有的核苷酸序列存在着明显的差异（屈良鹄和陈月琴 1999）。这一结果也同样表明，所有真核生物、原核生物中的细菌和古菌都分别起源于它们各自的共同祖先。

本文的目的在于试图从核 rRNA 基因大亚基全序列中鉴定出真菌界各代表成员所共有的核苷酸序列。

1 材料与方法

1.1 DNA提取

用于 DNA 提取的美味石耳 *Umbilicaria esculenta* (Miyoshi) Minks 采自吉林通化白鸡腰国家森林公园，2006 年 8 月 5 日。采集人付少彬，采集号 F3，海拔 1200m。

采用 CTAB 法（Cubero *et al.* 1999）从美味石耳中提取总 DNA。

1.2 PCR扩增

为了获得美味石耳中 LSU rDNA 的核苷酸全序列，根据 GenBank 提供的最新数据，本实验室设计出适合扩增地衣型真菌的引物系列（见图 1 及表 1）。

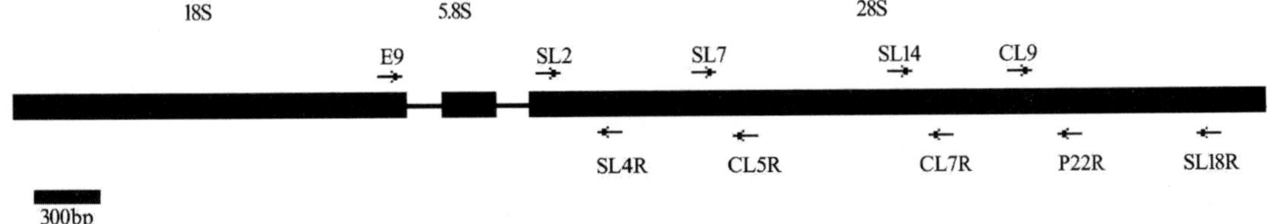

图 1 用于扩增和测序 28S 和 ITS rDNA 的引物

Fig. 1 Primers for amplification and sequencing of LSU and ITS rDNA.

[1]保守区 (Conserved region) 及高变区 (Divergent region) 镶嵌而存在于大亚基中（屈良鹄等 1999，图 1）

[2]第 4 保守区 (Conserved region 4)

表 1 由本实验室设计的引物

Table 1 Primers designed by Wei's Lab

引物名称 Primer name	引物序列 (5′→3′) Sequence (5′→3′)	引物位置以 *Saccharomyces cereviseae* 为标准 Position within *Saccharomyces cereviseae* rRNA
E9	TTGTACACACCGCCCGT	1640-1656 (18S)
SL2	CGGCGAGTGAAGCGGCA	106-122
SL4R	TTCGATCACTCTACTTGTGC	356-375
SL7	GATCGATTTGCACGTCAGAA	886-905
SL14	TAGCTCCGGACAAACCGAT	1736-1754
SL18R	CTTAGAGGCGTTCAGCCAT	3120-3139
CL5R	CTTACCAAAAATGGCCCACT	1136-1155
CL7R	GATCTATTTGCCGACTTC	1916-1934
CL9	ATGACGAGGCATTTGGCTA	2286-2304
P22R	CAATGTCAAACTAGAGTCAAGC	2426-2447

表 2 从 GenBank 数据库中下载的序列

Table 2 Sequences obtained from GenBank

种名 Species	序列号 Accession No.	系统地位 Systematic position
Aspergillus fumigatus	AAHF01000017	Ascomycota; Pezizomycotina; Eurotiomycetes; Eurotiales
Saccharomyces cerevisiae	NC_001144	Ascomycota; Saccharomycotina; Saccharomycetes; Saccharomycetales
Penicillium verruculosum	AF510496	Ascomycota; Pezizomycotina; Eurotiomycetes; Eurotiomycetidae; Eurotiales
Phaeosphaeria sp.	EF590325	Ascomycota; Pezizomycotina; Dothideomycetes; Pleosporomycetidae; Pleosporales
Umbilicaria esculenta	EU534208	Ascomycota; Pezizomycotina; Lecanoromycetes; Lecanoromycetidae; Umbilicariales
Cryptococcus neoformans	CPC5825S	Basidiomycota; Hymenomycetes; Heterobasidiomycetes; Tremellomycetidae; Tremellales
Minimedusa polyspora	DQ915476	Basidiomycota; Agaricomycotina; Agaricomycetes; Cantharellales; mitosporic Cantharellales
Amanita bisporigera	AY550243	Basidiomycota; Hymenomycetes; Homobasidiomycetes; Agaricales
Glomus intraradices	DQ273790	Glomeromycota; Glomeromycetes; Glomerales
Paraglomus occultum	DQ273827	Glomeromycota; Glomeromycetes; Paraglomerales
Spizellomycete sp.	DQ273821	Chytridiomycota; Spizellomycetales; unclassified Spizellomycetales
Orpinomyces sp.	AJ864475	Chytridiomycota; Neocallimastigales
Rhizopus oryzae	AACW02000151	Zygomycota; Zygomycetes; Mucorales
Mucor racemosus	MRARRHA	Zygomycota; Zygomycetes; Mucorales
Zea mays	AJ309824	Viridiplantae; Streptophyta; Embryophyta; Tracheophyta; Spermatophyta
Arabidopsis thaliana	X52320	Viridiplantae; Streptophyta; Embryophyta; Tracheophyta; Spermatophyta

续表 2

Mus musculus	NR_003279	Metazoa; Chordata; Craniata; Vertebrata; Euteleostomi; Mammalia; Eutheria
Rattus norvegicus	V01270	Metazoa; Chordata; Craniata; Vertebrata; Euteleostomi; Mammalia; Eutheria
Homo sapiens	NR_003287	Metazoa; Chordata; Craniata; Vertebrata; Euteleostomi; Mammalia

PCR 扩增选用 50μL 反应体系：36.5μL 无离子水；10×PCR buffer（含有 $MgCl_2$）5μL；dNTP（2.5mmol/L）4μL；上游引物（10μmol/L）和下游引物（10μmol/L）各 1μL；Taq DNA 聚合酶（5U/μL）0.5μL；模板 DNA 1μL。

PCR 反应按照如下条件进行：95℃预变性 3min；94℃变性 30s，52 复性 30s，72℃延伸 30s，共进行 37 个循环，72℃后延伸 10min。

1.3 测序及数据处理

核酸测序由北京华大基因研究中心完成。对所测得的 DNA 序列通过软件 Clustal X，DNAman4.0 及 Mega3.1 对序列进行处理，去除内含子并进行同源性分析，根据保守性将 28S 分为 12 区（图 2）。从 12 个保守区我们试图找到各个分类单元的共有序列，作为该分类单元的分子标记。保守区的位置以 *Saccharomyces cerevisiae*（NC_001144）为标准。

用于本研究的真菌界以及动物界和植物界的生物共计 19 种，其中 18 种的 28S rDNA 全序列来自 GenBank（见表 2）。

2 结果

我们从 19 种真菌和动植物的 28S rDNA 全序列中的 C3、C5、C7、C9 和 C11 四个保守区中鉴定出 6 条高度保守的核苷酸序列，其中 4 条为真菌界所共有（图 2、图 3、图 4、和图 5），1 条为子囊菌门所共有（图 6），1 条为担子菌门所共有（图 7）。

```
conservative            TCTGACGTGCAAATCGATCGTCRAAT
Saccharomyces_cerev     --------------------------
Umbilicaria_esculen     --------------------------
Phaeosphaeria_nodor     --------------------------
Aspergillus_fumigat     --------------------------
Penicillium_chrysog     --------------------------
Cryptococcus_neofor     --------------------------
Minimedusa_polyspor     --------------------------
Amanita_bisporigera     --------------------------
glomus_intraradices     --------------------------
paraglomus_occultum     --------------------------
Spizellomycete          --------------------------
orpinomyces_sp          --------------------------
Rhizopus_oryzae         --------------------------
Mucor_racemosus         --------------------------
Zea_mays                a---------------t-----tg-c
Arabidopsis_thalian     a---------------t-----tg-c
Mus_musculus            c---------------g-----cg-c
Rattus_norvegicus       c---------------g-----cg-c
Homo_sapiens            c---------------g-----cg-c
```

图 2 第 3 保守区真菌界特有序列（873-898）

Fig. 2 Peculiar sequences to fungi in 3rd conservative region (873-898).

```
conservative          CTAATTAAAACATAGCATTGYGATRKYCR
Saccharomyces_cerev   -----------------------------
Umbilicaria_esculen   -----------------------------
Phaeosphaeria_nodor   -----------------------------
Aspergillus_fumigat   -----------------------------
Penicillium_chrysog   -----------------------------
Cryptococcus_neofor   -----------------------------
Minimedusa_polyspor   -----------------------------
Amanita_bisporigera   -----------------------------
glomus_intraradices   -----------------------------
paraglomus_occultum   -----------------------------
Spizellomycete        -----------------------------
orpinomyces_sp        -----------------------------
Rhizopus_oryzae       -----------------------------
Mucor_racemosus       -----------------------------
Zea_mays              t-----------a-------c--cggt-c
Arabidopsis_thalian   t-----------a-------c---ggt-c
Mus_musculus          -ccgg-gggg-ggg-ggcccg--cact-g
Rattus_norvegicus     -ccgg-ggg..ggg-ggcccg--cacc-g
Homo_sapiens          -ccgcg-g..gggg--ccc-g-gacccgg
```

图 3 第 9 保守区真菌界特有序列（2203-2230）

Fig. 3 Peculiar sequences to fungi in 9th conservative region (2202-2230).

```
conservative          TGACTGTCTAATTAAAACAT
Saccharomyces_cerev   --------------------
Umbilicaria_esculen   --------------------
Phaeosphaeria_nodor   --------------------
Aspergillus_fumigat   --------------------
Penicillium_chrysog   --------------------
Cryptococcus_neofor   --------------------
Minimedusa_polyspor   --------------------
Amanita_bisporigera   --------------------
glomus_intraradices   --------------------
paraglomus_occultum   --------------------
Spizellomycete        --------------------
orpinomyces_sp        --------------------
Rhizopus_oryzae       --------------------
Mucor_racemosus       --------------------
Zea_mays              c------t-----------a
Arabidopsis_thalian   c------t-----------a
Mus_musculus          cag-g-c-ccgg-gggg-gg
Rattus_norvegicus     ccg-g-c-ccgg-gggg..g
Homo_sapiens          g-g-a-c-ccgcg-gg...g
```

图 4 第 9 保守区真菌特有序列（2284-2305）

Fig. 4 Peculiar sequences to fungi in 9th conservative region (2284-2305).

```
conservative          GTGAATACAAACCATGAAAGT
Saccharomyces_cerev   ---------------------
Umbilicaria_esculen   ---------------------
Phaeosphaeria_nodor   ---------------------
Aspergillus_fumigat   ---------------------
Penicillium_chrysog   ---------------------
Cryptococcus_neofor   ---------------------
Minimedusa_polyspor   ---------------------
Amanita_bisporigera   --------m------------
glomus_intraradices   ---------------------
paraglomus_occultum   ---------------------
Spizellomycete        ---------------------
orpinomyces_sp        ---------------------
Rhizopus_oryzae       ---------------------
Mucor_racemosus       ---------------------
Zea_mays              ac------g------------c
Arabidopsis_thalian   ac------g----g------c
Mus_musculus          ac-------g---g------c
Rattus_norvegicus     ac-------g---g------c
Homo_sapiens          ac-------g---g------c
```

图 5 第 11 保守区真菌特有序列（2911-2931）

Fig. 5 Peculiar sequences to fungi in 11th conservative region (2911-2931).

```
conservative            GTCTTCGGAWGGATTTGAGTAMGAGCRYAGC
Saccharomyces_cerev     -------------------------------
Umbilicaria_esculen     -------------------------------
Phaeosphaeria_nodor     -------------------------------
Aspergillus_fumigat     -------------------------------
Penicillium_chrysog     -------------------------------
Cryptococcus_neofor     c-tc-gt--t-----------a----at-ta
Minimedusa_polyspor     c-tc-gt--c-c--cg---gta----at-ta
Amanita_bisporigera     c-t--g---c----c--c-gta----aagca
glomus_intraradices     a-t--tt--ta----------a----at-t-
paraglomus_occultum     a-a-atc--ta-g--------a----ac-t-
Spizellomycete          t-t--tt--ta---c------a----at-t-
orpinomyces_sp          t-t--at--ga---c-----cta---at-tt
Rhizopus_oryzae         t-a-a-t--attgacc---acaa---actaa
Mucor_racemosus         t-taa-t--aatgacc---gtaa---actta
Zea_mays                c-tc-gt--a--g--c----tg----actc-
Arabidopsis_thalian     c-tc-ga--a--g--c----gt----atgc-
Mus_musculus            ccgcg-c-gg-agg-g---c-c----gt-cg
Rattus_norvegicus       ccgcg-c-gg-agg-g---c-c----gt-cg
Homo_sapiens            ccgcg-c-gg-agg-g---c-c----gc-cg
```

图 6 第 5 保守区子囊菌门特有序列（791-821）

Fig. 6 Peculiar sequences to Ascomycota in 5th conservative region (791-821).

```
conservative            GTGAGAACCTTGAAGACTGAAGTGGAGAAAGGTTCCATGG
Saccharomyces_cerev     ----------------------------------------
Umbilicaria_esculen     ----------------------------------------
Phaeosphaeria_nodor     ----------------------------------------
Aspergillus_fumigat     a-------t----------------g-----------c-t
Penicillium_chrysog     a-------t----g-----------g-------------t
Cryptococcus_neofor     a-------t----g-----------g----------g--t
Minimedusa_polyspor     a-------t----g-----------g-------------t
Amanita_bisporigera     a-------t----g-----------g-------------t
glomus_intraradices     a-------t------------------------------t
paraglomus_occultum     a-------t----------------------------c-c
Spizellomycete          a-------t---------------------------tg-a
orpinomyces_sp          a-------t---------------------------tg-a
Rhizopus_oryzae         a-------t------------------------------t
Mucor_racemosus         a-------t------------------------------t
Zea_mays                a-------t------g-c----a----------------t
Arabidopsis_thalian     a-------t------g-c----a--g-------------t
Mus_musculus            ac------t------g-c-----------g---------t
Rattus_norvegicus       ac------t------g-c-----------g---------t
Homo_sapiens            ac------t------g-c-----------g---------t
```

图 7 第 7 保守区担子菌门特有序列（1491-1533）

Fig. 7 Peculiar sequences to Basidiomycota in 7th conservative region (1491-1533).

3 讨论

我们从真菌界、动物界以及植物界的 19 种生物的 28S rDNA 全序列的 5 个保守区中鉴定出了 6 条保守序列。其中子囊菌门的成员共享 C5 结构区的 1 条保守序列（图 6），亦即该序列系子囊菌门的分子特征，因而它们可能起源于一个共同祖先；担子菌门的成员共享 C7 结构区的 1 条保守序列（图 7），该序列系担子菌门的分子特征，因而担子菌的成员起源于它们自己的共同祖先；真菌界的成员共享 C3，C9，C11 三个结构区中的 4 条保守序列（图 2-5），这些序列为真菌界的分子特征，从而表明各类真菌则起源于一个共同祖先。这是为细胞生物、真核生物、原核生物中的细菌和古菌提供的分子特征（屈良鹄和陈月琴 1999）之后，又为真菌界及其子囊菌门和担子菌门所提供的分子特征。这些数据作为达尔文进化论的共同祖先学说的分子证据，在生物进化上具有重要的科学意义。

至于子囊菌门和担子菌门是否应归隶为真菌界下之双核菌亚界 (Hibbett *et al.* 2007)，尚待进一步通过鉴定其共有核苷酸序列途径进行研究。

[REFERENCES]

Cubero OF, Crespo A, Fatehi J, Bridge PD, 1999. DNA extraction and PCR amplification method suitable for fresh, herbarium-stored, lichenized and other fungi. *Plant Systematics and Evolution*, **216**: 243-249

Hibbett DS, Binder M, Bischoff JF, Blackwell M, Cannon PF, Eriksson OE, Huhndorf S, James T, Kirk PM, Lücking R, Lumbsch HT, Lutzoni F, Matheny PB, McLaughlin DJ, Powell MJ, Redhead S, Schoch CL, Spatafora JW, Stalpers JA, Vilgalys R, Aime MC, Aptroot A, Bauer R, Begerow D, Benny GL, Castlebury LA, Crous PW, Dai YC, Gams W, Geiser DM, Griffith GW, Gueidan C, Hawksworth DL, Hestmark G, Hosaka K, Humber RA, Hyde KD, Ironside JE, Kõljalg U, Kurtzman CP, Larsson KH, Lichtwardt R, Longcore J, Miądlikowska J, Miller A, Jean-Marc Moncalvo, Mozley-Standridge S, Oberwinkler F, Parmasto E, Reeb V, Rogers JD, Roux C, Ryvarden L, Sampaio JP, Schüßler A , Sugiyama J, Thorn RG, Tibell L, Untereiner WA, Walker C, Wang Z, Weir A, Weiss M, White MM, Winka K, Yao YJ, Zhang N, 2007. A higher-level phylogenetic classification of the fungi. *Mycological Research*, **111**: 509-547

Mayr E, 1988. Toward a new philosophy of biology. Cambridge: Harvard University Press. 1-575

Qu LH, Chen YQ, 1999. Key to molecular taxonomy—principle and methods. *Acta Scientiarum Naturalium Universitatis Sunyatseni*, **38** (1): 1-6 (in Chinese)

[附中文参考文献]

屈良鹄，陈月琴，1999. 生物分子分类检索表—原理与方法. 中山大学学报，**38** (1): 1-6

Ascomycota has a faster evolutionary rate and higher species diversity than Basidiomycota

WANG HaiYing[1,2], GUO ShouYu[1], HUANG ManRong[3], Lumbsch H. THORSTEN[4*] & WEI JiangChun[1*]

[1]*Key Laboratory of Systematic Mycology & Lichenology, Institute of Microbiology, Chinese Academy of Sciences, Beijing 100101, China;*
[2]*Graduate University of Chinese Academy of Sciences, Beijing 100039, China;*
[3]*Beijing Museum of Natural History, Beijing 100050, China;*
[4]*Department of Botany, The Field Museum, 1400 S. Lake Shore Drive, Chicago, IL 60605, USA*

Received March 18, 2010; accepted May 20, 2010

Differences in rates of nucleotide or amino acid substitutions among major groups of organisms are repeatedly found and well documented. A growing body of evidence suggests a link between the rate of neutral molecular change within populations and the evolution of species diversity. More than 98% of terrestrial fungi belong to the phyla Ascomycota or Basidiomycota. The former is considerably richer in number of species than the latter. We obtained DNA sequences of 21 protein-coding genes from the lichenized fungus *Rhizoplaca chrysoleuca* and used them together with sequences from GenBank for subsequent analyses. Three datasets were used to test rate discrepancies between Ascomycota and Basidiomycota and that within Ascomycota: (i) 13 taxa including 105 protein-coding genes, (ii) nine taxa including 21 protein-coding genes, and (iii) nuclear LSU rDNA of 299 fungal species. Based on analyses of the 105 protein-coding genes and nuclear LSU rDNA datasets, we found that the evolutionary rate was higher in Ascomycota than in Basidiomycota. The differences in substitution rates between Ascomycota and Basidiomycota were significant. Within Ascomycota, the species-rich Sordariomycetes has the fastest evolutionary rate, while Leotiomycetes has the slowest. Our results indicate that the main contribution to the higher substitution rates in Ascomycota does not come from mutualism, ecological conditions, sterility, metabolic rate or shorter generation time, but is possibly caused by the founder effect. This is another example of the correlation between species number and evolutionary rates, which is consistent with the hypothesis that the founder effect is responsible for accelerated substitution rates in diverse clades.

evolutionary rate, amino acid substitution, nucleotide substitution, fungal evolution, species diversity, founder effect

Citation: Wang H Y, Guo S Y, Huang M R, *et al*. Ascomycota has a faster evolutionary rate and higher species diversity than Basidiomycota. Sci China Life Sci, 2010, 53: 1163–1169, doi: 10.1007/s11427-010-4063-8

It is well known to biologists that the number of species among different major groups of organisms differs dramatically from small, isolated clades, including only a single surviving species, to megadiverse groups, such as beetles, with well over 350000 described species [1]. Ages of clades, adaptive radiations, and key innovations are among the factors that have been invoked to explain these differences in species numbers among clades [2–5]. Differences in evolutionary rates among taxa are also well documented in molecular studies [6–11]; however, an exact molecular clock [12,13] seems a rare exception, if present at all. Several studies have investigated correlations of evolutionary rates with other factors, such as body size and metabolic rate [14–19], generation time [20–25], symbiotic associa-

*Corresponding author (email: tlumbsch@fieldmuseum.org; weijc2004@126.com)

This paper was originally published in *Science China Life Sciences*, Volume 53, Issue 10: 1163-1169, 2010.

tions [9,11,26] or environmental conditions [27,28]. In addition to these studies, researchers have addressed issues of correlation between evolutionary rates and rates of diversification (speciation minus extinction). A growing body of evidence supports the existence of such a correlation [29–34]. This correlation is explained by the speciation rate hypothesis [31], which is based on population biological and evolutionary theory [35–38]. If speciation takes place in small, peripheral populations, these are expected to undergo rapid genetic change as a result of genetic drift, resulting in the founder effect [35]. If the different types of speciation processes happened at different frequencies among clades, we would expect differences among these in evolutionary rates. In fact, the theory of punctuated equilibrium even predicts a general correlation between the rates of diversification and genetic change [39–41].

The Kingdom Fungi are an important and species-rich major clade of organisms that forms the sister-group of animals [42]. Terrestrial fungi represent an important part of terrestrial ecosystems, with species forming mutualistic and parasitic relationships with other groups of organisms, and representing the most important group of decomposers [43]. About 98% of all known fungal species belong to the phyla Ascomycota or Basidiomycota [44,45], sister-groups that represent the crown group of fungi [46]. The two phyla differ in the number of species; Ascomycota includes more than twice as many described species as Basidiomycota (Table 1) [45]. As the number of undescribed taxa is assumed to be much higher in Ascomycota than in the latter phylum, the difference in number of species is likely to be even higher [47,48].

Given the difference in number of species between Ascomycota and Basidiomycota, we were interested in testing whether the evolutionary rates between these phyla differ, which would be expected if the punctuated equilibrium theory was correct. We generally used sequences from whole genome sequencing projects available in GenBank. However, since data for numerous protein-coding genes are not yet available from any representative of the largest lichenized fungi class, Lecanoromycetes of Ascomycota [49–54], we generated sequences of a species of this class. We chose the lichenized fungus, *Rhizoplaca chrysoleuca*, the genetic variability of which we have previously studied [55]. We obtained DNA sequences of 21 protein-coding genes from this lichenized fungus and aligned them with orthologous sequences from GenBank for our analyses. We assembled three different datasets: two with numerous orthologous protein-coding genes (21 genes including nine fungal species, and 105 genes including 13 fungal species) and a dataset of nuclear (nu) LSU rDNA including 299 fungal species. We used these datasets to test for the presence of significant rate discrepancies between Ascomycota and Basidiomycota. We also compared the evolutionary rates among the main classes of Ascomycota (Table 1).

1 Materials and methods

1.1 Sampling and alignments

The gene fragments of orthologous proteins for *R. chrysoleuca* were obtained through sequencing at random inserts

Table 1 Number and classification of species in fungal phyla and classes (following reference [45]) included in our analyses

Class/phylum	Number of family/ species in class or phylum	Number of family/species included in the nu LSU rDNA analyses	Species included in the protein coding gene analyses	21-gene dataset	105-gene dataset
Ascomycota	327/64163	170/201	–	–	–
Dothideomycetes	90/19010	44/44	*Phaeosphaeria nodorum*	+	+
Eurotiomycetes	27/3401	17/26	*Aspergillus fumigatus*	+	+
			Coccidioides immitis	+	+
Lecanoromycetes	77/14199	40/40	*Rhizoplaca chrysoleuca*	+	–
Leotiomycetes	19/5587	15/27	*Sclerotinia sclerotiorum*	+	+
Pezizomycetes	16/1684	11/21	–	–	–
Sordariomycetes	64/10564	38/38	*Gibberella zeae*	+	+
			Magnaporthe grisea	+	+
			Neurospora crassa	+	+
Saccharomycetes (outgroup)	–	5/5	*Yarrowia lipolytica*	+	–
Basidiomycota	177/31515	91/94	*Coprinopsis cinerea*	–	+
			Cryptococcus neoformans	–	+
			Laccaria bicolor	–	+
			Malassezia globosa	–	+
			Ustilago maydis	–	+
Mucoromycotina (outgroup)	–	–	*Rhizopus oryzae*	–	+
Chytridiomycota (outgroup)	–	4/4	–	–	–

in 1000 clones of an EST library. The above work was completed by the Beijing Genomics Institute, Beijing. The cDNA fragments sequenced at random were translated into protein sequences using DNAMAN 4.0 software (Lynnon Biosoft). The protein fragments were identified through a similarity search (Blastp) in GenBank, and 21 orthologous protein fragments were selected for analysis. All 21 proteins and corresponding gene sequences have been deposited in GenBank (accession numbers HM007281–HM007304).

All nu LSU rDNA sequences of 299 fungal species and all additional protein-coding gene sequences were obtained from GenBank (Appendix Table 1 in the electronic version). The protein-coding gene sequences of *Rhizopus oryzae* were from http://www.broadinstitute.org/science/data.

The datasets were aligned separately using ClustalX 1.81 [56]. Mega 4 [57] was used to combine protein-coding gene sequences and to delete gaps present in the dataset.

1.2 Test for rate differences

The programs Baseml and Codonml (part of the PAML 3.14b package) [58] were used to test for significant rate differences. The resulting likelihoods were compared using likelihood ratio tests [59]. Tests were performed to determine whether significant departures from rate homogeneity were present between Ascomycota and Basidiomycota for the 105 protein-coding genes, the nu LSU rDNA dataset, and among the classes of Ascomycota for all three datasets. We compared two-rate models, in which two selected clades had the same rate, but a third clade had a different rate versus a three-rate model, in which all three included clades had different rates.

1.3 Calculation of evolutionary distances

Evolutionary distances in the nu LSU rDNA dataset were calculated using the maximum composite likelihood method [60] as implemented in the software Mega 4 [57]. All positions containing alignment gaps and missing data were eliminated in pairwise sequence comparisons. The final dataset included 1129 base pairs (bp). Differences of average evolutionary distances among the target groups in relation to the outgroups were examined by one-way ANOVA using SPSS 13.0 software.

The final dataset of 21 protein-coding genes of nine species included 3879 amino acids (aa), while the final dataset of 105 protein-coding genes of 13 species included 39279 aa.

1.4 Calculation of relative evolutionary rate

The relative evolutionary rates between pairwise clades were obtained based on the evolutionary distances calculated by the software Mega 4 [57]. The evolutionary distances based on the protein data or the nu LSU rDNA data were obtained respectively using the poisson correction method [61] or the maximum composite likelihood method [60].

2 Results

2.1 Significant rate differences between phyla and classes in crown fungi

We used likelihood ratio tests to determine rate differences between Ascomycota and Basidiomycota, and among the classes of Ascomycota. To identify particular clades with deviating rates, two-rate and three-rate models were compared for three selected clades in each analysis using the likelihood ratio test statistics.

For the evolutionary rate comparison between Ascomycota and Basidiomycota, we analyzed two datasets: the 105-protein coding gene dataset and the nuclear LSU rDNA dataset. The former included seven ascomycetes and five species of Basidiomycota, and one species of Mucoromycotina that was used as an outgroup (Table 1). The latter included 196 species of Ascomycota and 94 species of Basidiomycota, and four species of chytrids that were used as outgroups (Table 1). Significant rate differences were found between Ascomycota and Basidiomycota using the two datasets (Table 2). The results indicate that the evolutionary rate differs between the two phyla, with the rates in Ascomycota being consistently higher than in Basidiomycota.

The same nu LSU rDNA and 105-protein coding gene datasets, as well as the 21-protein coding gene dataset, were used for evolutionary rate comparison among the classes of Ascomycota. The 21-protein coding gene dataset included all the ascomycete samples of the 105-protein coding gene dataset, one species of Lecanoromycetes, and one species of Saccharomycetes that was used as an outgroup (Table 1). These three analyses all show a consistently higher evolutionary rate in Sordariomycetes than in the other classes of Ascomycota. In addition, the analysis of the nu LSU rDNA dataset shows that Eurotiomycetes has the second-fastest evolutionary rate, Leotiomycetes has the slowest, and three other classes (Lecanoromycetes, Dothideomycetes, and Pezizomycetes) have the third-fastest.

2.2 Evolutionary distances

The analyses of the significant rate differences by the programs Baseml and Codonml are based on a given phylogenetic tree. Because of the large number of samples and the limited number of positions in the nu LSU rDNA dataset, it is not easy to reconstruct a perfect phylogenetic tree based on this dataset. The topological structure of a phylogenetic tree will affect the result of significant rate difference analysis to a certain extent. Therefore, we compared the

Table 2 Comparison of models using PAML (*df*=1)[a)]

Groups	Data set	Ln*L* (null) 2-rate model	Ln*L* (alt.) 3-rate model	LRT	*P*	Relative rate
Ascomycota vs. Basidiomycota	105 proteins	–527398.204058	–527312.526435	85.677623	<0.0001	1∶0.824
	nu LSU rDNA	–65068.740739	–65062.100412	6.640327	<0.01	1∶0.724
Sordariomycetes vs. Eurotiomycetes	105 proteins	–527560.318578	–527402.805448	157.51313	<0.0001	1∶0.859
	21 proteins	–27015.482279	–27001.819130	13.663149	<0.001	1∶0.763
	nu LSU rDNA	–64998.459349	–64970.501967	27.957382	<0.0001	1∶0.536
Sordariomycetes vs. Lecanoromycetes	21 proteins	–24678.859250	–24662.406111	16.453139	<0.0001	1∶0.925
	nu LSU rDNA	–65040.207085	–64974.290373	65.916712	<0.0001	1∶0.435
Sordariomycetes vs. Dothideomycetes	105 proteins	–527547.627172	–527410.509802	137.11737	<0.0001	1∶0.960
	21 proteins	–24815.434902	–24805.185578	10.249324	<0.01	1∶0.963
	nu LSU rDNA	–65025.983459	–64978.705730	47.277729	<0.0001	1∶0.409
Sordariomycetes vs. Leotiomycetes	105 proteins	–527783.148246	–527434.405216	348.74303	<0.0001	1∶0.813
	21 proteins	–24305.693655	–24274.466383	31.227272	<0.0001	1∶0.792
	nu LSU rDNA	–65035.747310	–64975.958430	59.78888	<0.0001	1∶0.299
Sordariomycetes vs. Pezizomycetes	nu LSU rDNA	–65019.466588	–64978.806885	40.659703	<0.0001	1∶0.412
Eurotiomycetes vs. Lecanoromycetes	nu LSU rDNA	–65063.689716	–65056.385849	7.303867	<0.01	1∶0.782
Eurotiomycetes vs. Dothideomycetes	nu LSU rDNA	–65068.612580	–65067.109686	1.502894	>0.1	–
Eurotiomycetes vs. Leotiomycetes	nu LSU rDNA	–65062.767258	–65040.574709	22.192549	<0.0001	1∶0.603
Eurotiomycetes vs. Pezizomycetes	nu LSU rDNA	–65068.264591	–65064.268406	3.996185	<0.05	1∶0.702
Lecanoromycetes vs. Dothideomycetes	nu LSU rDNA	–65054.212621	–65051.467129	2.745492	>0.05	–
Lecanoromycetes vs. Leotiomycetes	nu LSU rDNA	–65032.323634	–65028.335008	3.988626	<0.05	1∶0.735
Lecanoromycetes vs. Pezizomycetes	nu LSU rDNA	–65053.936657	–65052.906898	1.029759	>0.1	–
Dothideomycetes vs. Leotiomycetes	nu LSU rDNA	–65050.186493	–65039.104095	11.082398	<0.001	1∶0.767
Dothideomycetes vs. Pezizomycetes	nu LSU rDNA	–65063.346948	–65063.045087	0.301861	>0.1	–
Pezizomycetes vs. Leotiomycetes	nu LSU rDNA	–65042.811015	–65035.001269	7.809746	<0.01	1∶0.887

a) A *P*-value less than 0.05 indicates a significant difference between every two groups.

evolutionary distances to the outgroups between Ascomycota and Basidiomycota, and among the classes of Ascomycota, based on the nu LSU rDNA dataset.

For the comparison between Ascomycota and Basidiomycota, we analyzed two nu LSU rDNA datasets. One was the same dataset as in the analysis of significant rate difference, and another dataset did not include the fastest-rate class of Ascomycota, Sordariomycetes. The analyses based on the two datasets all show that the evolutionary distance from ascomycetes to chytrids is significantly longer than from basidiomycetes to chytrids (Table 3). The results indicate that the significant difference in the evolutionary distances to the outgroups between Ascomycota and Basidiomycota does not result from the acceleration of individual classes of Ascomycota. A plot of the average evolutionary distance for each analyzed species (Figure 1) shows that every Ascomycota sample has a higher evolutionary rate than Basidiomycota.

For the comparison among the ascomycete classes, we used the nu LSU rDNA dataset, which included all the 196 ascomycete species used in the above analysis and five species of Saccharomycetes that were used as outgroups (Table 1). According to the results of the significant rate differences, these evolutionary distances of samples to the outgroups were divided into four groups: Sordariomycetes, Eurotiomycetes, Leotiomycetes and a group including Lecanoromycetes, Dothideomycetes and Pezizomycetes. This analysis showed that these four groups are significantly different from each other (Table 3). This result agrees with that of the significant rate differences analysis.

3 Discussion

Based on different analyses and three different datasets, this study provides clear evidence for a general acceleration in the rates of molecular evolution in Ascomycota compared with that of Basidiomycota. We also demonstrated significant differences in evolutionary rates among the classes of Ascomycota. Although this study provides the first statistical evidence for a significantly accelerated substitution rate in Ascomycota and the class Sordariomycetes, several studies have demonstrated accelerated evolutionary rates in smaller clades of basidiomycetes [11,26] and ascomycetes [9,27]. In these cases, accelerated evolutionary rates were associated with mutualism or ecological conditions. There are no consistent differences in ecological conditions between Ascomycota and Basidiomycota, or among the studied classes of Ascomycota. In addition, the three classes of Ascomycota with the richest species studied here (Dothideomycetes, Lecanoromycetes and Sordariomycetes) differ in their mutualistic associations. While the vast majority of taxa in Lecanoromycetes form lichen symbioses, the species of Dothideomycetes are rarely lichenized, and

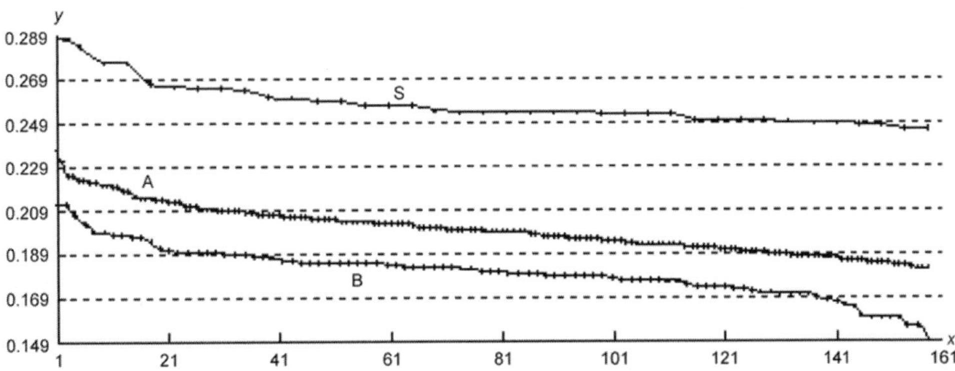

Figure 1 Each sample's average evolutionary distance to the outgroup using the nu LSU rDNA dataset. The y-axis is the evolutionary distance and the x-axis is the sample number where the samples in the corresponding group are distributed uniformly from 1 to 158 according to the descending order of evolutionary distance. S, Sordariomycetes (38 species); A, Ascomycota (158 species, Sordariomycetes absent); B, Basidiomycota (94 species). The outgroups in Chytridiomycota consist of *Boothiomyces macroporosum*, *Kappamyces laurelensis*, *Rhizophlyctis harderi* and *Rozella* sp.

Table 3 Mean value tests on each group's evolutionary distance to outgroup in the nu LSU rDNA dataset[a)]

Groups	Outgroups	n	Mean	Minimum	Maximum	Sig.	Relative rate
Ascomycota	Chytridiomycota	196	0.211	0.182	0.289	0.000	1:0.824
Basidiomycota		94	0.181	0.149	0.212		
Ascomycota (Sordariomycetes absent)	Chytridiomycota	158	0.200	0.182	0.237	0.000	1:0.804
Basidiomycota		94	0.181	0.149	0.212		
Sordariomycetes	Saccharomycotina	38	0.201	0.179	0.243	0.000	–
Eurotiomycetes		26	0.149	0.137	0.160		
Lecanoromycetes Dothideomycetes Pezizomycetes		104	0.135	0.112	0.161		
Leotiomycetes		27	0.124	0.113	0.159		

a) A significant (sig) value of less than 0.05 indicates a significant difference among the groups.

Sordariomycetes are never lichenized [62]. Hence, these reasons appear unlikely to explain the differences in evolutionary rates found in this study.

In general, increased rates of nucleotide substitution could result from different factors, such as positive selection or a general acceleration of nucleotide substitution. We found no evidence for positive selection when we compared the ratio of synonymous and non-synonymous substitutions in the 105 protein-coding gene dataset (Appendix Table 2 in the electronic version), therefore indicating a general acceleration of nucleotide substitutions in Ascomycota. This could be caused by factors such as sterility [63], metabolic rate [64], shorter generation time [65] or the founder effect [35]. The loss of sexual reproduction has been identified as a cause of accelerated evolution in endosymbiotic bacteria [63,66]; in fact, the number of asexual species is higher in Ascomycota than in Basidiomycota [67]. However, this cannot explain the differences in evolutionary rates among the classes of Ascomycota; for example, Eurotiomycetes has a high number of mitosporic fungi, but shows a lower substitution rate than Sordariomycetes. There is no evidence for general differences in metabolic rates or generation times between Ascomycota and Basidiomycota [68]. This leaves the founder effect as the most plausible cause. The fact that evolutionary rates are higher in species-rich clades at the phylum level supports this interpretation.

4 Conclusion

We presented further evidence for a correlation between the number of species and the evolutionary rate of clades. This is true for the comparison between Ascomycota and Basidiomycota. Our analyses of classes within Ascomycota indicate that the main contribution to higher substitution rates does not come from mutualism, ecological conditions, sterility, metabolic rate or shorter generation time, but possibly from the founder effect. This is consistent with the theory of punctuated equilibrium, which assumes that the main pattern of speciation is the peripatric type. This involves small isolated populations that undergo rapid genetic change as a result of genetic drift; hence, speciation processes would accelerate the evolutionary rate of a clade over time. Although the patterns of speciation could be explained by the peripatric, allopatric, parapatric or sympatric types, or the mutation of speciation genes, the fact that Ascomycota has a faster evolutionary rate and is richer in species than its sister group Basidiomycota is consistent with the theory of punctuated equilibrium.

This work was supported by the National Natural Science Foundation of China (Grant Nos. 30570008 and 30770012) and the Chinese Academy of Sciences (Grant No. KSCX2-YW-Z-041).

1 Resh V H, Cardé R T. Encyclopedia of Insects. San Diego: Academic Press, 2003. 209–229

2 Hughes C, Eastwood R. Island radiation on a continental scale: exceptional rates of plant diversification after uplift of the Andes. Proc Natl Acad Sci USA, 2006, 103: 10334–10339

3 Kristensen N P. Phylogeny of endopterygote insects, the most successful lineage of living organisms. Eur J Entomol, 1999, 96: 237–253

4 Dodd M E, Silvertown J, Chase M W. Phylogenetic analysis of trait evolution and species diversity variation among angiosperm families. Evolution, 1999, 53: 732–744

5 Bond J E, Opell B D. Testing adaptive radiation and key innovation hypotheses in spiders. Evolution, 1998, 52: 403–414

6 Arbogast B S, Edwards S V, Wakeley J, *et al.* Estimating divergence times from molecular data on phylogenetic and population genetic timescales. Annu Rev Ecol Syst, 2002, 33: 707–740

7 Britten R J. Rates of DNA-sequence evolution differ between taxonomic groups. Science, 1986, 231: 1393–1398

8 Bromham L, Penny D. The modern molecular clock. Nat Rev Genet, 2003, 4: 216–224

9 Woolfit M, Bromham L. Increased rates of sequence evolution in endosymbiotic bacteria and fungi with small effective population sizes. Mol Biol Evol, 2003, 20: 1545–1555

10 Langley C H, Fitch W M. An examination of the constancy of the rate of molecular evolution. J Mol Evol, 1974, 3: 161–167

11 Lutzoni F, Pagel M. Accelerated evolution as a consequence of transitions to mutualism. Proc Natl Acad Sci USA, 1997, 94: 11422–11427

12 Margoliash E. Primary structure and evolution of cytochrome c. Proc Natl Acad Sci USA, 1963, 50: 672–679

13 Bryson V, Vogel H. Evolving Genes and Proteins. New York: Academic Press, 1965. 97–166

14 Fontanillas E, Welch J J, Thomas J A, *et al.* The influence of body size and net diversification rate on molecular evolution during the radiation of animal phyla. BMC Evol Biol, 2007, 7: 95

15 Mooers A O, Harvey P H. Metabolic rate, generation time, and the rate of molecular evolution in birds. Mol Phylogenet Evol, 1994, 3: 344–350

16 Bromham L. Molecular clocks in reptiles: life history influences rate of molecular evolution. Mol Biol Evol, 2002, 19: 302–309

17 Bromham L, Rambaut A, Harvey P H. Determinants of rate variation in mammalian DNA sequence evolution. J Mol Evol, 1996, 43: 610–621

18 Thomas J A, Welch J J, Woolfit M, *et al.* There is no universal molecular clock for invertebrates, but rate variation does not scale with body size. Proc Natl Acad Sci USA, 2006, 103: 7366–7371

19 Martin A P, Palumbi S R. Body size, metabolic-Rate, generation time, and the molecular clock. Proc Natl Acad Sci USA, 1993, 90: 4087–4091

20 Ohta T. An examination of the generation-time effect on molecular evolution. Proc Natl Acad Sci USA, 1993, 90: 10676–10680

21 Gu X, Li W H. Higher amino acid substitution in rodents than in humans. Mol Phylogenet Evol, 1992, 1: 211–214

22 Bousquet J, Strauss S H, Doerksen A H, *et al.* Extensive variation in evolutionary rate of rbcL gene sequences among seed plants. Proc Natl Acad Sci USA, 1992, 89: 7844–7848

23 Conti E, Fischbach A, Sytsma K J. Tribal relationships in Onagraceae: implications from rbcL sequence data. Ann Mo Bot Gard, 1993, 80: 672–685

24 Laroche J, Li P, Maggia L, *et al.* Molecular evolution of angiosperm mitochondrial introns and exons. Proc Natl Acad Sci USA, 1997, 94: 5722–5727

25 Kay K M, Whittall J B, Hodges S A. A survey of nuclear ribosomal internal transcribed spacer substitution rates across angiosperms: an approximate molecular clock with life history effects. BMC Evol Biol, 2006, 6: 36

26 Zoller S, Lutzoni F. Slow algae, fast fungi: exceptionally high nucleotide substitution rate differences between lichenized fungi *Omphalina* and their symbiotic green algae *Coccomyxa*. Mol Phylogenet Evol, 2003, 29: 629–640

27 Lumbsch H T, Hipp A L, Divakar P K, *et al.* Accelerated evolutionary rates in tropical and oceanic parmelioid lichens (Ascomycota). BMC Evol Biol, 2008, 8: 257

28 Bromham L, Cardillo M. Testing the link between the latitudinal gradient in species richness and rates of molecular evolution. J Evolution Biol, 2003, 16: 200–207

29 Barraclough T G, Savolainen V. Evolutionary rates and species diversity in flowering plants. Evolution, 2001, 55: 677–683

30 Webster A J, Payne R J H, Pagel M. Molecular phylogenies link rates of evolution and speciation. Science, 2003, 301: 478–478

31 Jobson R W, Albert V A. Molecular rates parallel diversification contrasts between carnivorous plant sister lineages. Cladistics, 2002, 18: 127–136

32 Mindell D P, Sites J W, Graur D. Speciational evolution: a phylogenetic test with allozymes in *Sceloporus* (Reptilia). Cladistics, 1989, 5: 49–61

33 Mindell D P, Thacker C E. Rates of molecular evolution: phylogenetic issues and applications. Annu Rev Ecol Syst, 1996, 27: 279–303

34 Barraclough T G, Harvey P H, Nee S. Rate of rbcL gene sequence evolution and species diversification in flowering plants (angiosperms). P Roy Soc Lond B Bio, 1996, 263: 589–591

35 Mayr E. Animal Species and Evolution. Cambridge, MA: Harvard University Press, 1963. 1–797

36 Carson H L, Templeton A R. Genetic revolution in relation to speciation: the founding of new populations. Annu Rev Ecol Syst, 1984, 15: 97–131

37 Harrison R G. Molecular changes at speciation. Annu Rev Ecol Syst, 1991, 22: 281–308

38 Coyne J A. Genetics and speciation. Nature, 1992, 355: 511–515

39 Gould S J, Eldredge N. Punctuated equilibrium comes of age. Nature, 1993, 366: 223–227

40 Eldredge N, Gould S J. Punctuated equilibrium prevails. Nature, 1988, 332: 211–212

41 Webster A J, Payne R J H, Pagel M. Molecular phylogenies link rates of evolution and speciation. Science, 2003, 301: 478

42 Wainright P O, Hinkle G, Sogin M L, *et al.* Monophyletic origins of the metazoa: an evolutionary link with fungi. Science, 1993, 260: 340–342

43 Alexopoulos C J, Mims C W, Blackwell M. Introductory Mycology. 4th ed. New York: John Wiley & Sons, 1996. 1–880

44 Hibbett D S, Binder M, Bischoff J F, *et al.* A higher-level phylogenetic classification of the fungi. Mycol Res, 2007, 111: 509–547

45 Kirk P M, Cannon P F, David J C, *et al.* Ainsworth & Bisby's Dictionary of the Fungi. 10th ed. Wallingford (Oxon): CAB International, 2008. 1–771

46 Lutzoni F, Kauff F, Cox C, *et al.* Assembling the fungal tree of life: progress, classification, and evolution of subcellular traits. Am J Bot, 2004, 91: 1446–1480

47 Hawksworth D L. The fungal dimension of biodiversity: magnitude, significance, and conservation. Mycol Res, 1991, 95: 641–655

48 Hawksworth D L. The magnitude of fungal diversity: the 1.5 million species estimate revisited. Mycol Res, 2001, 105: 1422–1432

49 Spatafora J W, Sung G H, Johnson D, *et al.* A five-gene phylogeny of Pezizomycotina. Mycologia, 2006, 98: 1018–1028

50 Miadlikowska J, Kauff F, Hofstetter V, *et al.* New insights into classification and evolution of the Lecanoromycetes (Pezizomycotina, Ascomycota) from phylogenetic analyses of three ribosomal RNA- and two protein-coding genes. Mycologia, 2006, 98: 1088–1103

51 Wedin M, Wiklund E, Crewe A, *et al.* Phylogenetic relationships of Lecanoromycetes (Ascomycota) as revealed by analyses of mtSSU and nLSU rDNA sequence data. Mycol Res, 2005, 109: 159–172

52 Persoh D, Beck A, Rambold G. The distribution of ascus types and photobiontal selection in Lecanoromycetes (Ascomycota) against the background of a revised SSU nrDNA phylogeny. Mycol Prog, 2004,

3: 103–121
53 James T Y, Kauff F, Schoch C, *et al.* Reconstructing the early evolution of fungi using a six-gene phylogeny. Nature, 2006, 443: 818–822
54 Lumbsch H T, Schmitt I, Palice Z, *et al.* Supraordinal phylogenetic relationships of Lecanoromycetes based on a Bayesian analysis of combined nuclear and mitochondrial sequences. Mol Phylogenet Evol, 2004, 31: 822–832
55 Zhou Q M, Guo S Y, Huang M R, *et al.* A study of the genetic variability of *Rhizoplaca chrysoleuca* using DNA sequences and secondary metabolic substances. Mycologia, 2006, 98: 57–67
56 Thompson J D, Higgins D G, Gibson T J. Clustal W: improving the sensitivity of progressive multiple sequence alignment through sequence weighting, position-specific gap penalties and weight matrix choice. Nucleic Acids Res, 1994, 22: 4673–4680
57 Tamura K, Dudley J, Nei M, *et al.* MEGA4: molecular evolutionary genetics analysis (MEGA) software version 4.0. Mol Biol Evol, 2007, 24: 1596–1599
58 Yang Z. PAML: a program package for phylogenetic analysis by maximum likelihood. Comput Appl Biosci, 1997, 13: 555–556
59 Felsenstein J. Confidence-limits on phylogenies: an approach using the bootstrap. Evolution, 1985, 39: 783–791
60 Tamura K, Nei M, Kumar S. Prospects for inferring very large phylogenies by using the neighbor-joining method. Proc Natl Acad Sci USA, 2004, 101: 11030–11035
61 Bryson V, Vogel. H. Evolving Genes and Proteins. New York: Academic Press, 1965. 97–166
62 Lumbsch H T, Huhndorf S M. Whatever happened to the Pyrenomycetes and Loculoascomycetes? Mycol Res, 2007, 111: 1064–1074
63 Moran N A. Accelerated evolution and Muller's rachet in endosymbiotic bacteria. Proc Natl Acad Sci USA, 1996, 93: 2873–2878
64 Rand D M. Thermal habit, metabolic-rate and the evolution of mitochondrial DNA. Trends Ecol Evol, 1994, 9: 125–131
65 Kohne D E. Evolution of higher-organism DNA. Q Rev Biophys, 1970, 3: 327–375
66 Lynch M, Blanchard J L. Deleterious mutation accumulation in organelle genomes. Genetica, 1998, 103: 29–39
67 Taylor J W. Making the Deuteromycota redundant: a practical integration of mitosporic fungi. Can J Bot, 1995, 73: S754–S759
68 Kavanagh K. Fungi: Biology and Applications. New York: John Wiley & Sons, 2005. 1–239

（正文内容见1163-1169）

SCIENCE CHINA
Life Sciences

Appendix Table 1A Samples of the 105 protein-coding genes dataset

Proteins	*Aspergillus fumigatus*	*Coccidioides immitis*	*Coprinopsis cinerea*	*Cryptococcus neoformans*	*Gibberella zeae*	*Laccaria bicolor*	*Magnaporthe grisea*	*Malassezia globosa*	*Neurospora crassa*	*Phaeosphaeria nodorum*	*Sclerotinia sclerotiorum*	*Ustilago maydis*	*Rhizopus oryzae*
Vacuolar protein sorting-associated protein 26	XP_751625	XP_001247863	EAU83945	XP_776046	XP_381331	EDR10318	XP_362384	XP_001731425	EAA35892	EAT77623	XP_001592543	EAK82921	RO3G_03069
vacuolar ATP synthase catalytic subunit A	XP_748105	XP_001242551	EAU88427	XP_775506	XP_387180	EDR11023	XP_362504	XP_001732794	EAA32337	EAT81520	XP_001595698	EAK80890	RO3G_01898
pre-mRNA splicing factor (Prp8)	XP_749438	XP_001245132	EAU86045	XP_778129	XP_382712	EDR13411	XP_359533	XP_001732213	EAA33717	EAT91945	XP_001595120	EAK82591	RO3G_01072
gamma tubulin MipA	XP_752709	XP_001241990	EAU88379	XP_772622	XP_390169	EDR13746	XP_368283	XP_001729231	EAA28357	EAT82125	XP_001598808	EAK84729	RO3G_11100
Centromere/microtubule-binding protein cbf5	XP_754061	XP_001242479	EAU92842	XP_775935	XP_387382	EDR15040	XP_367607	XP_001729376	EAA31591	EAT83460	XP_001590061	EAK80867	RO3G_15034
triosephosphate isomerase	XP_753309	XP_001239740	EAU86629	XP_775227	XP_386878	EDR10156	XP_364060	XP_001732054	EAA29827	EAT76573	XP_001587441	EAK84286	RO3G_03048
translation initiation factor 2 alpha subunit	XP_754274	XP_001246338	EAU89116	XP_773976	XP_386396	EDR08279	XP_001404451	XP_001729265	EAA32669	EAT79164	XP_001585956	EAK82336	RO3G_13185
threonyl-tRNA synthetase	XP_751461	XP_001248022	EAU80442	XP_776995	XP_381424	EDR10518	XP_361008	XP_001729744	EAA30565	EAT80431	XP_001587735	EAK86112	RO3G_06614
T-complex protein 1, zeta subunit	XP_754660	XP_001248128	EAU91603	XP_773781	XP_386489	EDR09063	XP_361383	XP_001731930	EAA29712	EAT80679	XP_001591294	EAK83388	RO3G_13197
T-complex protein 1, eta subunit	XP_750420	XP_001244568	EAU92308	XP_778256	XP_386482	EDR15287	XP_360767	XP_001732682	EAA29703	EAT76803	XP_001586957	EAK81214	RO3G_10956
T-complex protein 1, delta subunit	XP_754830	XP_001248566	EAU88477	XP_774369	XP_380734	EDR14879	XP_363656	XP_001728916	EAA34893	EAT76083	XP_001597630	EAK83741	RO3G_14969
T-complex protein 1, beta subunit,	XP_749924	XP_001238899	EAU82336	XP_777942	XP_385954	EDR09515	XP_365277	XP_001730993	EAA30290	EAT88787	XP_001595013	EAK87115	RO3G_03092
T-complex protein 1, alpha subunit	XP_755937	XP_001246551	EAU88242	XP_776254	XP_385426	EDR10956	XP_366138	XP_001728987	EAA26670	EAT78684	XP_001590906	EAK82142	RO3G_05866
splicing factor 3B subunit 1	XP_755711	XP_001245212	EAU90529	XP_774063	XP_387227	EDR14089	XP_368888	XP_001731366	EAA34811	EAT88157	XP_001596870	EAK82854	RO3G_07617
small monomeric GTPase SarA	XP_750243	XP_001245464	EAU88667	XP_777426	XP_386822	EDR03200	XP_369847	XP_001731788	EDO65417	EAT85260	XP_001591874	EAK87233	RO3G_05258
signal recognition particle protein SRP54	XP_747957	XP_001242701	EAU88500	XP_775089	XP_389140	EDR14856	XP_369487	XP_001730607	EAA31407	EAT77106	XP_001595676	EAK83583	RO3G_13370
methyltransferase	XP_750530	XP_001243369	EAU88807	XP_778233	XP_385904	EDR14902	XP_363121	XP_001732469	EAA35697	EAT88653	XP_001594354	EAK82606	RO3G_15486
DUF652 domain protein	XP_749469	XP_001241339	EAU91589	XP_775101	XP_387375	EDR08878	XP_362746	XP_001731785	EAA32079	EAT89089	XP_001596852	EAK80884	RO3G_13163
60S ribosome biogenesis protein Nip7	XP_756040	XP_001239792	EAU92678	XP_777449	XP_387138	EDR15126	XP_368762	XP_001730043	EAA27993	EAT76236	XP_001589048	EAK86086	RO3G_09534
rRNA processing protein (Rrp20)	XP_750535	XP_001243372	EAU88852	XP_777870	XP_381521	EDR14785	XP_359641	XP_001730122	EAA32853	EAT88982	XP_001594483	EAK84633	RO3G_00885
RIO1 family protein kinase	XP_753477	XP_001239555	EAU93038	XP_776475	XP_389960	EDR15017	XP_367194	XP_001730876	EAA32736	EAT78642	XP_001590865	EAK86208	RO3G_04042
cytosolic large ribosomal subunit protein L7A	XP_751177	XP_001247914	EAU86554	XP_772335	XP_388654	EDQ99374	XP_362167	XP_001732654	EAA30992	EAT86044	XP_001598059	EAK80786	RO3G_14825
ribonucleotide reductase small subunit RnrA	XP_753417	XP_001239725	EAU84682	XP_772723	XP_385585	EDR13226	XP_369893	XP_001730759	EAA33584	EAT82336	XP_001595401	EAK87225	RO3G_12051
DNA replication factor C subunit Rfc4	XP_747574	XP_001248860	EAU92758	XP_774249	XP_387376	EDR15164	XP_367615	XP_001730833	EDO65226	EAT91453	XP_001598347	EAK81118	RO3G_16283
secretory pathway gdp dissociation inhibitor	XP_755448	XP_001240816	EAU92695	XP_774160	XP_386342	EDR15137	XP_367212	XP_001731580	EAA32823	EAT79915	XP_001589137	EAK81128	RO3G_03976
pyruvate kinase	XP_750636	XP_001240869	EAU93215	XP_776923	XP_387704	EDR16037	XP_362480	XP_001729750	EAA30602	EAT92314	XP_001594760	EAK81542	RO3G_08020
proteasome component Pup2	XP_755476	XP_001241019	EAU92402	XP_772516	XP_386335	EDR15938	XP_367205	XP_001731866	EAA32830	EAT91216	XP_001589129	EAK86958	RO3G_13946
phosphatidylinositol:UDP-GlcNAc transferase subunit PIG-A	XP_753063	XP_001243553	EAU82881	XP_775674	XP_381136	EDR04632	XP_361459	XP_001731545	EAA28961	EAT90228	XP_001588529	EAK82847	RO3G_07782
phenylalanyl-tRNA synthetase beta chain, cytoplasmic	XP_749491	XP_001245059	EAU84898	XP_773126	XP_381280	EDR12918	XP_361980	XP_001729085	EAA27845	EAT92087	XP_001594633	EAK82585	RO3G_11637
L-ornithine aminotransferase Car2	XP_751911	XP_001240799	EAU87983	XP_773582	XP_385722	EDR09594	XP_369877	XP_001729938	EAA27181	EAT88909	XP_001587988	EAK84399	RO3G_12123
mitochondrial processing peptidase beta subunit	XP_752790	XP_001243696	EAU85792	XP_774411	XP_381039	EDR13198	XP_361057	XP_001731874	EAA36444	EAT81329	XP_001593250	EAK86965	RO3G_04336
peptide N-myristoyl transferase (Nmt1)	XP_752019	XP_001241626	EAU88866	XP_772024	XP_389573	EDR14336	XP_367304	XP_001732829	EAA30036	EAT92391	XP_001594233	EAK83827	RO3G_06862
lysyl-tRNA synthetase	XP_750655	XP_001247536	EAU90925	XP_773851	XP_388937	EDR11645	XP_359811	XP_001731476	EAA28399	EAT82803	XP_001595938	EAK82995	RO3G_14000
lipoic acid synthetase precursor	XP_754950	XP_001246687	EAU92837	XP_773405	XP_389620	EDR15648	XP_001522320	XP_001732059	EAA35490	EAT91765	XP_001589458	EAK85510	RO3G_01981
isoleucyl-tRNA synthetase	XP_752741	XP_001243953	EAU91061	XP_778214	XP_390898	EDR11511	XP_360677	XP_001731519	EAA26778	EAT80885	XP_001585648	EAK83034	RO3G_12153
isocitrate dehydrogenase Idp1	XP_754749	XP_001240510	EAU84027	XP_777363	XP_390523	EDR10319	XP_367343	XP_001730822	EAA26614	EAT91095	XP_001593497	EAK86993	RO3G_13820
iron-sulfur protein subunit of succinate dehydrogenase Sdh2	XP_753605	XP_001241858	EAU82988	XP_774483	XP_385786	EDR13751	XP_369077	XP_001729560	EAA35916	EAT88556	XP_001594577	EAK81918	RO3G_03816
IMP dehydrogenase	XP_749494	XP_001245055	EAU87620	XP_777938	XP_381037	EDR09420	XP_361156	XP_001730522	EAA35740	EAT91916	XP_001587025	EAK83736	RO3G_14462
GTP-binding protein YchF	XP_752347	XP_001239251	EAU85656	XP_776732	XP_382940	EDR08546	XP_361880	XP_001728946	EAA35430	EAT85562	XP_001592250	EAK86999	RO3G_07275
GMP synthase	XP_748432	XP_001239961	EAU83257	XP_772762	XP_390534	EDR09233	XP_368325	XP_001732749	EAA30515	EAT84008	XP_001595541	EAK84975	RO3G_02515
glycine-tRNA ligase	XP_754040	XP_001242329	EAU86846	XP_774702	XP_387134	EDQ99825	XP_369806	XP_001729902	EAA28055	EAT83336	XP_001598549	EAK84392	RO3G_11381
succinate dehydrogenase subunit Sdh1	XP_754832	XP_001248568	EAU81873	XP_773861	XP_387537	EDR13384	XP_369076	XP_001729428	EDO65095	EAT76087	XP_001591238	EAK81956	RO3G_13321
translation initiation factor eIF-6	XP_754031	XP_001242313	EAU85399	XP_773008	XP_385604	EDR08495	XP_363745	XP_001730047	EAA29796	EAT90992	XP_001591620	EAK87150	RO3G_13353
peptide chain release factor eRF/aRF, subunit 1	XP_754064	XP_001242476	EAU88180	XP_776668	XP_387103	EDR11251	XP_369751	XP_001730088	EAA28060	EAT83625	XP_001598524	EAK85196	RO3G_00516
translation elongation factor EF-Tu	XP_752585	XP_001239381	EAU93042	XP_774180	XP_387358	EDR15015	XP_362579	XP_001729762	EAA32277	EAT82586	XP_001597967	EAK81523	RO3G_01455
translation elongation factor G1	XP_752015	XP_001241630	EAU83626	XP_776356	XP_381567	EDR02488	XP_363639	XP_001731326	EAA36106	EAT92120	XP_001589746	EAK86751	RO3G_07529
ATP dependent RNA helicase (Sub2)	XP_747814	XP_001239034	EAU92983	XP_777578	XP_388839	EDR15710	XP_001522452	XP_001729648	EAA35657	EAT76998	XP_001593842	XP_761087	RO3G_15740
DNA-directed RNA polymerase III, beta subunit,	XP_749996	XP_001242390	EAU87529	XP_773994	XP_381019	EDR09689	XP_370487	XP_001731706	EAA27959	EAT91415	XP_001595664	EAK83484	RO3G_12315
DNA replication licensing factor Mcm5	XP_748090	XP_001242563	EAU93404	XP_777467	XP_386953	EDR15422	XP_363062	XP_001730681	EAA32301	EAT77199	XP_001591048	EAK85924	RO3G_10698
diphthamide biosynthesis protein	XP_755576	XP_001245771	EAU84561	XP_777602	XP_380812	EDR14729	XP_360694	XP_001732172	EAA34196	EAT85141	XP_001588209	EAK84039	RO3G_16190
dimethyladenosine transferase	XP_749024	XP_001247217	EAU87730	XP_773062	XP_385225	EDR09459	XP_368993	XP_001729060	EAA31554	EAT76682	XP_001593673	EAK83482	RO3G_14104
dihydrolipoamide succinyltransferase	XP_755064	XP_001248359	EAU89463	XP_777242	XP_388146	EDR10293	XP_360606	XP_001729588	EAA30207	EAT78149	XP_001587717	EAK82572	RO3G_06527

(*To be continued on the next page*)

(*Continued*)

Proteins	*Aspergillus fumigatus*	*Coccidioides immitis*	*Coprinopsis cinerea*	*Cryptococcus neoformans*	*Gibberella zeae*	*Laccaria bicolor*	*Magnaporthe grisea*	*Malassezia globosa*	*Neurospora crassa*	*Phaeosphaeria nodorum*	*Sclerotinia sclerotiorum*	*Ustilago maydis*	*Rhizopus oryzae*
ATP-dependent RNA helicase	XP_750371	XP_001244087	EAU88082	XP_773820	XP_384204	EDR11265	XP_001411591	XP_001732004	EAA27679	EAT84178	XP_001592326	EAK82902	RO3G_03957
ATP-dependent RNA helicase HAS1	XP_751498	XP_001246788	EAU88940	XP_777235	XP_384526	EDR16076	XP_370007	XP_001732565	EAA29692	EAT86830	XP_001588603	EAK85282	RO3G_07575
cytochrome C1/Cyt1	XP_749957	XP_001238954	EAU90205	XP_774726	XP_380737	EDR05656	XP_360371	XP_001729525	EAA34520	EAT88545	XP_001586215	EAK85488	RO3G_08415
tRNA splicing protein (Spl1)	XP_754202	XP_001246354	EAU88735	XP_776031	XP_390225	EDR14351	XP_368477	XP_001732187	EAA29943	EAT76535	XP_001598849	EAK86015	RO3G_11728
CTP synthase	XP_748988	XP_001246057	EAU82860	XP_776312	BAA33767	EDR12113	XP_367634	XP_001729415	EAA28496	EAT79715	XP_001589075	EAK81142	RO3G_03216
cell division control protein Cdc48	XP_756045	XP_001239786	EAU90639	XP_773075	XP_385706	EDR14459	XP_359584	XP_001732423	EAA27769	EAT82668	XP_001595101	EAK81798	RO3G_07895
DEAD-box RNA helicase Dhh1/Vad1	XP_755058	XP_001248331	EAU89171	XP_776581	XP_390967	EDR08188	XP_360845	XP_001730716	EAA30775	EAT80597	XP_001588066	EAK84197	RO3G_06550
spartyl-tRNA synthetase Dps1	XP_749394	XP_001248036	EAU93029	XP_776192	XP_384730	EDR15009	XP_001522528	XP_001729396	EAA35505	EAT91850	XP_001589717	EAK85319	RO3G_04619
alanyl-tRNA synthetase	XP_747195	XP_001248510	EAU91282	XP_774788	XP_381082	EDR11794	XP_361064	XP_001731546	EAA36461	EAT88464	XP_001593079	EAK82846	RO3G_07558
adenylosuccinate synthetase AdB	XP_752913	XP_001243896	EAU88944	XP_776751	XP_385363	EDR03268	XP_367061	XP_001731021	EAA30951	EAT76814	XP_001593669	EAK84757	RO3G_08252
adenylate kinase	XP_750499	XP_001248754	EAU81921	XP_772932	XP_390913	EDR11223	XP_368186	XP_001732798	EAA27017	EAT91128	XP_001597412	EAK83143	RO3G_13367
adenosylhomocysteinase	XP_752379	XP_001239289	EAU81357	XP_775643	XP_385791	EDR10213	XP_359622	XP_001732329	EAA33210	EAT85768	XP_001596998	EAK84912	RO3G_07184
60S ribosomal protein L4	XP_748041	XP_001242607	EAU83110	XP_773160	XP_387362	EDR09812	XP_362607	XP_001728871	EAA31868	EAT80905	XP_001597207	EAK83947	RO3G_09581
60S ribosomal protein L19	XP_755133	XP_001247141	EAU93376	XP_774059	XP_390050	EDR00720	XP_001404186	XP_001729625	EAA30828	EAT89870	XP_001595641	EAK82415	RO3G_04235
60S ribosomal protein L17	XP_752811	XP_001243753	EAU86083	XP_771988	XP_382047	EDR13793	XP_364349	XP_001731924	EAA32243	EAT90522	XP_001586516	EAK86962	RO3G_01574
60S ribosomal protein L11	XP_752052	XP_001239964	EAU90374	XP_773228	XP_381257	EDR14488	XP_367849	XP_001730226	EAA36404	EAT88761	XP_001586298	EAK85026	RO3G_09340
60S ribosomal protein L1	XP_752537	XP_001243105	EAU92198	XP_775117	XP_387100	EDR08282	XP_362251	XP_001731247	EAA28529	EAT85593	XP_001598446	EAK85891	RO3G_16218
60S ribosomal protein P0	XP_750257	XP_001245443	EAU85302	XP_774147	XP_387003	EDR08463	XP_362022	XP_001728628	EAA28947	EAT77462	XP_001584828	EAK86939	RO3G_09023
40S ribosomal protein S8e	XP_750180	XP_001245539	EAU91660	XP_774458	XP_380807	EDR08803	XP_360708	XP_001731848	EAA34193	EAT77032	XP_001588292	EAK84649	RO3G_04360
40S ribosomal protein S3Ae	XP_754085	XP_001242450	EAU84719	XP_777512	XP_387656	EDR12827	XP_370422	XP_001728782	EAA26738	EAT83615	XP_001598109	EAK85901	RO3G_12122
40S ribosomal protein S0	XP_754290	XP_001244060	EAU91101	XP_773962	XP_391081	EDR11584	XP_001406293	XP_001731370	EAA27604	EAT86783	XP_001587181	EAK83011	RO3G_04736
proteasome regulatory particle subunit Rpt2	XP_753929	XP_001242204	EAU85352	XP_775283	XP_382204	EDR08470	XP_360102	XP_001730706	EAA32354	EAT83263	XP_001591603	EAK83732	RO3G_10320
proteasome regulatory particle subunit Rpt1	XP_754831	XP_001248567	EAU88428	XP_774476	XP_380735	EDR11355	XP_363655	XP_001732793	EAA34894	EAT76084	XP_001597631	EAK80891	RO3G_11342
proteasome regulatory particle subunit Rpt5	XP_750366	XP_001244098	EAU86097	XP_772095	XP_391773	EDR13340	XP_001403458	XP_001732254	EAA28255	EAT84531	XP_001596943	XP_760694	RO3G_03248
proteasome regulatory particle subunit Rpt6	XP_746652	XP_001241366	EAU82882	XP_772972	XP_381781	EDR04627	XP_363454	XP_001732083	EAA34118	EAT90722	XP_001597447	EAK81907	RO3G_12232
proteasome component Pup3	XP_752084	XP_001239915	EAU92531	XP_777884	XP_381412	EDR15819	XP_362106	XP_001731296	EAA35245	EAT89142	XP_001586346	EAK80963	RO3G_16220
proteasome component Pup1	XP_749622	XP_001247301	EAU81477	XP_776879	XP_380543	EDR09720	XP_362084	XP_001729238	EAA34801	EAT78113	XP_001584802	EAK86392	RO3G_00008
proteasome component Pre9	XP_750785	XP_001247565	EAU90883	XP_774604	XP_385541	EDR12033	XP_362705	XP_001731725	EAA29550	EAT80099	XP_001587697	EAK83098	RO3G_16334
6-phosphogluconate dehydrogenase Gnd1	XP_750696	XP_001247382	EAU92918	XP_772874	XP_381287	EDR09736	XP_369069	XP_001728891	EAA35723	EAT76717	XP_001589237	EAK83747	RO3G_01672
alpha-ketoglutarate dehydrogenase complex subunit Kgd1	XP_751665	XP_001247826	EAU85723	XP_777169	XP_384485	EDR13583	XP_001408601	XP_001731922	EAA34012	EAT78214	XP_001593012	EAK85720	RO3G_08761
translation release factor eRF3	XP_749613	XP_001247604	EAU84760	XP_774637	XP_388779	EDR12396	XP_368795	XP_001730065	EAA31180	EAT79373	XP_001596056	EAK86098	RO3G_13821
pre-mRNA splicing factor RNA helicase	XP_753488	XP_001239544	EAU81615	XP_774927	XP_390933	EDR10006	XP_361419	XP_001731936	EAA27287	EAT85898	XP_001589848	EAK82057	RO3G_08924
60S ribosomal protein L7	XP_746575	XP_001240725	EAU88749	XP_566981	XP_382718	EDR14897	XP_359540	XP_001732427	XP_962950	EAT86244	XP_001597919	XP_759725	RO3G_16080
proteasome component Pre6	XP_748127	XP_001242525	EAU88401	XP_775738	XP_387458	EDR10996	XP_369957	XP_001729803	EAA28095	EAT81628	XP_001591078	EAK86352	RO3G_10410
transketolase TktA	XP_752720	XP_001239980	EAU88185	XP_570402	XP_390174	EDR10917	XP_365769	XP_001731824	XP_961414	EAT82418	XP_001588699	XP_761114	RO3G_13067
t-complex protein 1, gamma subunit (Cct3)	XP_754318	XP_001244016	EAU92138	XP_567673	XP_386161	EDR13239	XP_370142	XP_001730709	XP_956627	EAT86746	XP_001592815	XP_762214	RO3G_03228
AAA family ATPase Rvb2/Reptin	XP_749991	XP_001242359	EAU92476	XP_566914	XP_386436	EDR15378	XP_360630	XP_001732143	XP_963328	EAT90854	XP_001597906	XP_760373	RO3G_07845
phosphoglycerate kinase PgkA	XP_752401	XP_001239296	EAU88439	XP_571947	XP_384168	EDR11030	XP_359714	XP_001728700	XP_962430	EAT85787	XP_001596913	XP_761018	RO3G_08041
translation elongation factor EF-1 alpha subunit	XP_750388	XP_001244267	EAU93024	XP_568462	XP_388987	EDR15006	XP_361098	XP_001732312	XP_964868	EAT80707	XP_001594091	XP_757071	RO3G_15351
DNA-directed RNA polymerase III largest subunit	XP_754189	XP_001246195	EAU80581	XP_571468	XP_390036	EDR04865	XP_362032	XP_001730008	XP_956571	EAT91146	XP_001588093	XP_762303	RO3G_12475
DNA polymerase delta catalytic subunit Cdc2	XP_755997	XP_001239851	EAU90154	XP_772922	XP_387351	EDR05542	XP_362488	XP_001728836	XP_961558	EAT82147	XP_001587512	XP_757605	RO3G_03762
TFIIH complex helicase Rad3	XP_752761	XP_001243936	EAU88545	XP_566880	XP_384471	EDR14295	XP_360589	XP_001732571	XP_956536	EAT82740	XP_001589884	XP_760298	RO3G_12052
deoxyhypusine synthase	XP_748168	XP_001242489	EAU84757	XP_567440	XP_380499	EDR12843	XP_364328	XP_001732398	XP_963241	EAT83590	XP_001597869	XP_756218	RO3G_10161
prolyl-tRNA synthetase	XP_755936	XP_001246550	EAU81430	XP_569307	XP_385425	EDR09277	XP_366137	XP_001729319	XP_955907	EAT78685	XP_001590907	XP_757002	RO3G_08732
arsenite translocating ATPase ArsA	XP_754484	XP_001244255	EAU93342	XP_567863	XP_390067	EDR15466	XP_364690	XP_001731054	XP_960897	EAT86334	XP_001593411	XP_759985	RO3G_11494
ribosomal protein L16a	XP_750349	XP_001244411	EAU93058	XP_568413	XP_382275	EDR15024	XP_360074	XP_001730190	XP_961587	EAT77902	XP_001591799	XP_757073	RO3G_15704
proteasome regulatory particle subunit (RpnK)	XP_749475	XP_001245086	EAU84677	XP_566594	XP_380957	EDR12356	XP_369606	XP_001732491	XP_964366	EAT92004	XP_001592900	XP_757688	RO3G_06585
GTP binding protein	XP_750305	XP_001244472	EAU84685	XP_569574	XP_387417	EDR12813	XP_360356	XP_001731642	XP_957275	EAT80854	XP_001591529	XP_759154	RO3G_15244
prohibitin	XP_755243	XP_001247251	EAU86052	XP_566440	XP_381295	EDR13797	XP_365041	XP_001730682	XP_964487	EAT82909	XP_001589275	XP_761177	RO3G_13787
AAA family ATPase Pontin	XP_751754	XP_001247704	EAU82356	XP_566529	XP_385326	EDR08315	XP_361484	XP_001730521	XP_955769	EAT76964	XP_001593913	XP_762348	RO3G_15559
DNA replication factor C subunit Rfc3	XP_748155	XP_001242497	EAU90493	XP_771880	XP_386203	EDR14578	XP_001414254	XP_001729496	XP_963397	EAT76168	XP_001597346	XP_760770	RO3G_03308
cell division control protein 2 kinase	XP_750689	XP_001247390	EAU81687	XP_569525	XP_388644	EDR06007	XP_363436	XP_001729871	XP_960117	EAT79379	XP_001596080	AAP94021	RO3G_05995
karyopherin alpha subunit	XP_755944	XP_001243434	EAU93208	XP_570493	XP_387317	EDR14943	XP_367135	XP_001730902	XP_960652	EAT82614	XP_001587480	XP_760972	RO3G_11023

Appendix Table 1B Samples of the 21 protein-coding genes dataset

Proteins	*Rhizoplaca chrysoleuca*	*Aspergillus fumigatus*	*Coccidioides immitis*	*Gibberella zeae*	*Magnaporthe grisea*	*Neurospora crassa*	*Phaeosphaeria nodorum*	*Sclerotinia sclerotiorum*	*Yarrowia lipolytica*
Malate dehydrogenase	HM007281	XP_748936	XP_001246025	XP_382637	XP_364559	XP_958408	XP_001796362	XP_001585256	XP_502909
Mitochondrial Hsp70 chaperone	HM007284	XP_755328	XP_001244995	XP_386330	XP_361717	XP_961753	XP_001803790	XP_001589111	XP_501940
Cytosolic hydroxymethyltransferase	HM007285	XP_755116	XP_001247153	XP_390049	XP_001404185	XP_960065	XP_001793492	XP_001595640	XP_503153
Translation elongation factor 1-alpha	HM007287, HM007304	XP_750388	XP_001244267	XP_388987	XP_361098	XP_964868	XP_001801902	XP_001594091	XP_501628
Peptidyl-prolyl cis-trans isomerase	HM007288	XP_749504	XP_001245045	XP_380953	XP_366228	XP_001728468	XP_001790747	XP_001593362	XP_501664
Aminopeptidase	HM007289	XP_751922	XP_001241125	XP_388516	XP_366388	XP_959172	XP_001793910	XP_001594581	XP_500893
ATP synthase F1, beta subunit	HM007290	XP_753589	XP_001242378	XP_384488	XP_360642	XP_963253	XP_001791943	XP_001591168	XP_500475
Glyceraldehyde-3-phosphate dehydrogenase GpdA	HM007291	XP_748145	XP_001242378	XP_384488	XP_360642	XP_963253	XP_001791924	XP_001591168	XP_500475
Vacuolar ATP synthase subunit d1	HM007292	XP_754283	XP_001244066	XP_391083	XP_370510	XP_956842	XP_001792129	XP_001587183	XP_505487
60S ribosomal protein L20	HM007293	XP_750201	XP_001245490	XP_381692	XP_361110	XP_963936	XP_001801850	XP_001592983	XP_505817
mRNA-nucleus export ATPase (Elf1)	HM007294	XP_747719	XP_001239067	XP_388708	XP_001522177	XP_958018	XP_001798057	XP_001593480	XP_500490
40S ribosomal protein Rps16	HM007295	XP_755383	XP_001244905	XP_384312	XP_359488	XP_964828	XP_001798536	XP_001598002	XP_500817
Transketolase TktA	HM007296, HM007301	XP_752720	XP_001239980	XP_390174	XP_365769	XP_961414	XP_001800365	XP_001588699	XP_503628
60S ribosomal protein L3	HM007298	XP_755517	XP_001245984	XP_386465	XP_360650	XP_963317	XP_001806656	XP_001598836	XP_502097
Rho GTPase Rho1	HM007299	XP_750583	XP_001244593	XP_384576	XP_367251	XP_956761	XP_001795250	XP_001595777	XP_504289
Cytochrome c	HM007300	XP_755643	XP_001245655	XP_391057	XP_370188	XP_956486	XP_001805810	XP_001598900	XP_502612
Molecular chaperone and allergen Mod-E/Hsp90/Hsp1	HM007282, HM007297	XP_747926	XP_001245288	XP_382190	XP_370262	XP_961298	XP_001791544	XP_001591945	XP_501578
Ribosomal protein S23 (S12)	HM007302	XP_752311	XP_001243230	XP_390909	XP_368178	XP_956255	XP_001795358	XP_001588322	XP_500301
Alternative oxidase AlxA	HM007303	XP_749637	XP_001247280	XP_381518	XP_001403506	XP_962086	XP_001793620	XP_001596660	XP_502637
Mitochondrial peroxiredoxin PRX1	HM007283	XP_751969	XP_001241702	XP_387712	XP_362792	XP_959621	XP_001791462	XP_001594896	XP_500241
Cobalamin-independent methionine synthase MetH/D	HM007286	XP_752090	XP_001244621	XP_391001	XP_370215	XP_957152	XP_001934603	XP_001588472	XP_503874

Appendix Table 1C Samples of the nu LSU rDNA dataset

nu LSU rDNA	Species	Family	Order	Class	Phylum
EU940084	D_Acrospermum_compressum	Acrospermaceae	Acrospermales	Dothideomycetes	Ascomycota
AY004336	D_Botryosphaeria ribis	Botryosphaeriaceae	Botryosphaeriales	Dothideomycetes	Ascomycota
AY004337	D_Microxyphium_citri	Capnodiaceae	Capnodiales	Dothideomycetes	Ascomycota
DQ678074	D_Davidiella tassiana	Davidiellaceae	Capnodiales	Dothideomycetes	Ascomycota
EU019274	D_Penidiella_columbiana	incertae sedis	Capnodiales	Dothideomycetes	Ascomycota
DQ246232	D_Mycosphaerella flexuosa	Mycosphaerellaceae	Capnodiales	Dothideomycetes	Ascomycota
AY016366	D_Piedraia_hortae	Piedraiaceae	Capnodiales	Dothideomycetes	Ascomycota
EF134949	D_Schizothyrium pomi	Schizothyriaceae	Capnodiales	Dothideomycetes	Ascomycota
GU301874	D_Teratosphaeria_jonkershoeken	Teratosphaeriaceae	Capnodiales	Dothideomycetes	Ascomycota
AY004342	D_Stylodothis_puccinioides	Dothideaceae	Dothideales	Dothideomycetes	Ascomycota
AY016359	D_Columnosphaeria fagi	Dothioraceae	Dothideales	Dothideomycetes	Ascomycota
AY541493	D_Hysteropatella_clavispora	Hysteriaceae	Hysteriales	Dothideomycetes	Ascomycota
AY541492	D_Farlowiella carmichaeliana	incertae sedis	incertae sedis	Dothideomycetes	Ascomycota
DQ678083	D_Helicomyces roseus	Tubeufiaceae	incertae sedis	Dothideomycetes	Ascomycota
DQ384104	D_Zopfia rhizophila	Zopfiaceae	incertae sedis	Dothideomycetes	Ascomycota
GU301796	D_Aliquandostipite_khaoyaiensi	Aliquandostipitaceae	Jahnulales	Dothideomycetes	Ascomycota
DQ678060	D_Elsinoe_veneta	Elsinoaceae	Myriangiales	Dothideomycetes	Ascomycota
AY016365	D_Myriangium_duriaei	Myriangiaceae	Myriangiales	Dothideomycetes	Ascomycota
DQ678081	D_Lophium mytilinum	Mytilinidiaceae	Mytilinidiales	Dothideomycetes	Ascomycota
FJ469672	D_Rhytidhysteron rufulum	Patellariaceae	Patellariales	Dothideomycetes	Ascomycota
GU301793	D_Aigialus_grandis	Aigialaceae	Pleosporales	Dothideomycetes	Ascomycota
AY538339	D_Arthopyrenia_salicis	Arthopyreniaceae	Pleosporales	Dothideomycetes	Ascomycota
DQ384102	D_Curreya pityophila	Cucurbitariaceae	Pleosporales	Dothideomycetes	Ascomycota
DQ384090	D_Delitschia_didyma	Delitschiaceae	Pleosporales	Dothideomycetes	Ascomycota
AB524621	D_Roussoella hysterioides	Didymosphaeriaceae	Pleosporales	Dothideomycetes	Ascomycota
AY016363	D_Lojkania enalia	Fenestellaceae	Pleosporales	Dothideomycetes	Ascomycota
FJ515646	D_Didymella vitalbina	incertae sedis	Pleosporales	Dothideomycetes	Ascomycota
GU301822	D_Keissleriella_cladophila	Lentitheciaceae	Pleosporales	Dothideomycetes	Ascomycota
DQ470946	D_Leptosphaeria_maculans	Leptosphaeriaceae	Pleosporales	Dothideomycetes	Ascomycota
AB521737	D_Lindgomyces_ingoldianus	Lindgomycetaceae	Pleosporales	Dothideomycetes	Ascomycota
DQ782384	D_Lophiostoma_arundinis	Lophiostomataceae	Pleosporales	Dothideomycetes	Ascomycota
DQ678065	D_Massaria platani	Massariaceae	Pleosporales	Dothideomycetes	Ascomycota

(*To be continued on the next page*)

(Continued)

nu LSU rDNA	Species	Family	Order	Class	Phylum
AY016356	D_Bimuria_novaezelandiae	Melanommataceae	Pleosporales	Dothideomycetes	Ascomycota
DQ678086	D_Montagnula opulenta	Montagnulaceae	Pleosporales	Dothideomycetes	Ascomycota
EF590318	D_Phaeosphaeria nodorum	Phaeosphaeriaceae	Pleosporales	Dothideomycetes	Ascomycota
AY004340	D_Phaeotrichum_benjaminii	Phaeotrichaceae	Pleosporales	Dothideomycetes	Ascomycota
AY004341	D_Pleomassaria_siparia	Pleomassariaceae	Pleosporales	Dothideomycetes	Ascomycota
AY544645	D_Cochliobolus heterostrophus	Pleosporaceae	Pleosporales	Dothideomycetes	Ascomycota
AY004343	D_Westerdykella_cylindrica	Sporormiaceae	Pleosporales	Dothideomycetes	Ascomycota
AY016357	D_Byssothecium_circinans	Teichosporaceae	Pleosporales	Dothideomycetes	Ascomycota
DQ678067	D_Lepidosphaeria nicotiae	Testudinaceae	Pleosporales	Dothideomycetes	Ascomycota
AB524620	D_Quadricrura_septentrionalis	Tetraplosphaeriaceae	Pleosporales	Dothideomycetes	Ascomycota
DQ678072	D_Trematosphaeria_pertusa	Trematosphaeriaceae	Pleosporales	Dothideomycetes	Ascomycota
DQ384100	D_Metacoleroa dickiei	Venturiaceae	Pleosporales	Dothideomycetes	Ascomycota
AY004339	E_Ceramothyrium_carniolicum	Chaetothyriaceae	Chaetothyriales	Eurotiomycetes	Ascomycota
AY004338	E_Capronia_mansonii	Herpotrichiellaceae	Chaetothyriales	Eurotiomycetes	Ascomycota
AJ507323	E_Phaeococcomyces_chersonesos	Herpotrichiellaceae	Chaetothyriales	Eurotiomycetes	Ascomycota
AF346420	E_Glyphium_elatum	Incertae sedis	Chaetothyriales	Eurotiomycetes	Ascomycota
DQ470987	E_Caliciopsis_orientalis	Coryneliaceae	Coryneliales	Eurotiomycetes	Ascomycota
DQ782908	E_Monascu_purpureus	Elaphomycetaceae	Eurotiales	Eurotiomycetes	Ascomycota
FJ358291	E_Xeromyces_bisporus	Elaphomycetaceae	Eurotiales	Eurotiomycetes	Ascomycota
FJ358281	E_Chromocleista_malachitea	Trichocomaceae	Eurotiales	Eurotiomycetes	Ascomycota
FJ358278	E_Aspergillus_fumigatus	Trichocomaceae	Eurotiales	Eurotiomycetes	Ascomycota
GU332517	E_Dolabra_nepheliae	Incertae sedis	Incertae sedis	Eurotiomycetes	Ascomycota
FJ358273	E_Arachnomyces_glareosus	Arachnomycetaceae	Onygenales	Eurotiomycetes	Ascomycota
FJ358275	E_Ascosphaera_apis	Ascosphaeraceae	Onygenales	Eurotiomycetes	Ascomycota
FJ358282	E_Ctenomyces_serratus	Arthrodermataceae	Onygenales	Eurotiomycetes	Ascomycota
AY062120	E_Trichophyton_interdigitale	Arthrodermataceae	Onygenales	Eurotiomycetes	Ascomycota
FJ358283	E_Eremascus_albus	Eremascaceae	Onygenales	Eurotiomycetes	Ascomycota
FJ358286	E_Leucothecium_emdenii	Gymnoascaceae	Onygenales	Eurotiomycetes	Ascomycota
FJ358284	E_Gymnoascus_reesii	Gymnoascaceae	Onygenales	Eurotiomycetes	Ascomycota
FJ358288	E_Shanorella_spirotricha	Onygenaceae	Onygenales	Eurotiomycetes	Ascomycota
FJ358287	E_Onygena_corvina	Onygenaceae	Onygenales	Eurotiomycetes	Ascomycota
DQ782909	E_Spiromastix_warcupii	Onygenaceae	Onygenales	Eurotiomycetes	Ascomycota
DQ782906	E_Anisomeridium_polypori	Monoblastiaceae	Pyrenulales	Eurotiomycetes	Ascomycota
EF411063	E_Pyrenula_aspistea	Pyrenulaceae	Pyrenulales	Eurotiomycetes	Ascomycota
DQ823103	E_Pyrgillus_javanicus	Pyrenulaceae	Pyrenulales	Eurotiomycetes	Ascomycota
EF411062	E_Granulopyrenis_seawardii	Requienellaceae	Pyrenulales	Eurotiomycetes	Ascomycota
EF643805	E_Bagliettoa_parmigera	Verrucariaceae	Verrucariales	Eurotiomycetes	Ascomycota
EF643774	E_Staurothele_drummondii	Verrucariaceae	Verrucariales	Eurotiomycetes	Ascomycota
AY640939	L_Acarospora_bullata	Acarosporaceae	Acarosporales	Lecanoromycetes	Ascomycota
DQ986774	L_Placynthiella_uliginosa	Agyriaceae	Agyriales	Lecanoromycetes	Ascomycota
AF356658	L_Baeomyces_placophyllus	Baeomycetaceae	Baeomycetales	Lecanoromycetes	Ascomycota
DQ986791	L_Candelaria_concolor	Candelariaceae	Candelariales	Lecanoromycetes	Ascomycota
AY640956	L_Maronea_constans	Fuscideaceae	incertae sedis	Lecanoromycetes	Ascomycota
AY533006	L_Hymenelia_lacustris	Hymeneliaceae	incertae sedis	Lecanoromycetes	Ascomycota
AF279376	L_Arthrorhaphis_citrinella	Arthrorhaphidaceae	Lecanorales	Lecanoromycetes	Ascomycota
AY533011	L_Sporastatia_testudinea	Catillariaceae	Lecanorales	Lecanoromycetes	Ascomycota
AY584640	L_Cladonia_caroliniana	Cladoniaceae	Lecanorales	Lecanoromycetes	Ascomycota
AY584653	L_Crocynia_pyxinoides	Crocyniaceae	Lecanorales	Lecanoromycetes	Ascomycota
AY584651	L_Lopezaria_versicolor	Incertae sedis	Lecanorales	Lecanoromycetes	Ascomycota
DQ973028	L_Ophioparma_lapponica	Ophioparmaceae	Lecanorales	Lecanoromycetes	Ascomycota
AY584641	L_Xanthoparmelia_conspersa	Parmeliaceae	Lecanorales	Lecanoromycetes	Ascomycota
DQ883798	L_Heterodermia_vulgaris	Physciaceae	Lecanorales	Lecanoromycetes	Ascomycota
DQ314902	L_Porpidia_flavocaerulescens	Porpidiaceae	Lecanorales	Lecanoromycetes	Ascomycota
AY533009	L_Psora_rubiformis	Psoraceae	Lecanorales	Lecanoromycetes	Ascomycota
AY533010	L_Rhizocarpon_geminatum	Rhizocarpaceae	Lecanorales	Lecanoromycetes	Ascomycota
AY453643	L_Leifidium_tenerum	Sphaerophoraceae	Lecanorales	Lecanoromycetes	Ascomycota
AY533002	L_Stereocaulon_vesuvianum	Stereocaulaceae	Lecanorales	Lecanoromycetes	Ascomycota
AY452504	L_Calicium_adspersum	Caliciaceae	Lecanorales	Lecanoromycetes	Ascomycota
DQ986746	L_Lecanora_contractula	Lecanoraceae	Lecanorales	Lecanoromycetes	Ascomycota
AY532990	L_Cecidonia_umbonella	Lecideaceae	Lecanorales	Lecanoromycetes	Ascomycota
DQ986762	L_Niebla_cephalota	Ramalinaceae	Lecanorales	Lecanoromycetes	Ascomycota
AF356662	L_Petractis_clausa	Gyalectaceae	Ostropales	Lecanoromycetes	Ascomycota
DQ912346	L_Coccocarpia_domingensis	Coccocarpiaceae	Peltigerales	Lecanoromycetes	Ascomycota
AY424212	L_Leptogium_gelatinosum	Collemataceae	Peltigerales	Lecanoromycetes	Ascomycota

(To be continued on the next page)

(*Continued*)

nu LSU rDNA	Species	Family	Order	Class	Phylum
AY584655	L_Lobaria_scrobiculata	Lobariaceae	Peltigerales	Lecanoromycetes	Ascomycota
AY584656	L_Nephroma_parile	Nephromataceae	Peltigerales	Lecanoromycetes	Ascomycota
AY584657	L_Peltigera_degenii	Peltigeraceae	Peltigerales	Lecanoromycetes	Ascomycota
AF356674	L_Placynthium_nigrum	Placynthiaceae	Peltigerales	Lecanoromycetes	Ascomycota
DQ917419	L_Fuscopannaria_mediterranea	Pannariaceae	Peltigerales	Lecanoromycetes	Ascomycota
AF329171	L_Ochrolechia_balcanica	Ochrolechiaceae	Pertusariales	Lecanoromycetes	Ascomycota
AF356679	L_Thamnolia_subuliformis	Icmadophilaceae	Pertusariales	Lecanoromycetes	Ascomycota
DQ782907	L_Pertusaria_dactylina	Pertusariaceae	Pertusariales	Lecanoromycetes	Ascomycota
AY584648	L_Letrouitia_domingensis	Letrouitiaceae	Teloschistales	Lecanoromycetes	Ascomycota
AY584650	L_Megalospora_tuberculosa	Megalosporaceae	Teloschistales	Lecanoromycetes	Ascomycota
AY584647	L_Teloschistes_exilis	Teloschistaceae	Teloschistales	Lecanoromycetes	Ascomycota
AF279405	L_Porina_guentheri	Porinaceae	Trichotheliales	Lecanoromycetes	Ascomycota
AY648106	L_Rhizoplacopsis_weichingii	Rhizoplacopsidaceae	Umbilicariales	Lecanoromycetes	Ascomycota
DQ782912	L_Umbilicaria_mammulata	Umbilicariaceae	Umbilicariales	Lecanoromycetes	Ascomycota
DQ470960	Le_Bulgaria_inquinans	Bulgariaceae	Helotiales	Leotiomycetes	Ascomycota
DQ470942	Le_Mollisia_cinerea	Dermateaceae	Helotiales	Leotiomycetes	Ascomycota
AY064705	Le_Neofabraea_alba	Dermateaceae	Helotiales	Leotiomycetes	Ascomycota
DQ470967	Le_Pezicula_carpinea	Dermateaceae	Helotiales	Leotiomycetes	Ascomycota
AY487098	Le_Pilidium_concavum	Dermateaceae	Helotiales	Leotiomycetes	Ascomycota
DQ470981	Le_Microglossum_rufum	Geoglossaceae	Helotiales	Leotiomycetes	Ascomycota
AY544656	Le_Chloroscypha_enterochroma	Helotiaceae	Helotiales	Leotiomycetes	Ascomycota
AY544680	Le_Crinula_caliciiformis	Helotiaceae	Helotiales	Leotiomycetes	Ascomycota
DQ470944	Le_Cudoniella_clavus	Helotiaceae	Helotiales	Leotiomycetes	Ascomycota
AF356694	Le_Fabrella_tsugae	Hemiphacidiaceae	Helotiales	Leotiomycetes	Ascomycota
AY789415	Le_Hyaloscypha_daedaleae	Hyaloscyphaceae	Helotiales	Leotiomycetes	Ascomycota
DQ227262	Le_Hyphodiscus_hymeniophilus	Hyaloscyphaceae	Helotiales	Leotiomycetes	Ascomycota
AY544674	Le_Lachnum_bicolor	Hyaloscyphaceae	Helotiales	Leotiomycetes	Ascomycota
AY544679	Le_Chaetomella_acutiseta	Incertae sedis	Helotiales	Leotiomycetes	Ascomycota
AY544644	Le_Leotia_lubrica	Leotiaceae	Helotiales	Leotiomycetes	Ascomycota
DQ470978	Le_Lambertella_subrenispora	Rutstroemiaceae	Helotiales	Leotiomycetes	Ascomycota
AY789423	Le_Mitrula_paludosa	Sclerotiniaceae	Helotiales	Leotiomycetes	Ascomycota
AY544683	Le_Monilinia_fructicola	Sclerotiniaceae	Helotiales	Leotiomycetes	Ascomycota
FJ176874	Le_Vibrissea_truncorum	Vibrisseaceae	Helotiales	Leotiomycetes	Ascomycota
AY541491	Le_Myxotrichum_deflexum	Myxotrichaceae	incertae sedis	Leotiomycetes	Ascomycota
DQ470988	Le_Pseudeurotium_zonatum	Pseudeurotiaceae	incertae sedis	Leotiomycetes	Ascomycota
AF279379	Le_Cudonia_circinans	Cudoniaceae	Rhytismatales	Leotiomycetes	Ascomycota
AY541496	Le_Spathularia_flavida	Cudoniaceae	Rhytismatales	Leotiomycetes	Ascomycota
DQ470954	Le_Meria_laricis	Rhytismataceae	Rhytismatales	Leotiomycetes	Ascomycota
DQ470949	Le_Potebniamyces_pyri	Rhytismataceae	Rhytismatales	Leotiomycetes	Ascomycota
AF356696	Le_Rhytisma_acerinum	Rhytismataceae	Rhytismatales	Leotiomycetes	Ascomycota
FJ176895	Le_Thelebolus_caninus	Thelebolaceae	Thelebolales	Leotiomycetes	Ascomycota
AY544678	P_Ascobolus_crenulatus	Ascobolaceae	Pezizales	Pezizomycetes	Ascomycota
FJ176870	P_Saccobolus_dilutellus	Ascobolaceae	Pezizales	Pezizomycetes	Ascomycota
DQ470966	P_Eleutherascus_lectardii	Ascodesmidaceae	Pezizales	Pezizomycetes	Ascomycota
DQ247799	P_Caloscypha_fulgens	Caloscyphaceae	Pezizales	Pezizomycetes	Ascomycota
AY544673	P_Gyromitra_californica	Discinaceae	Pezizales	Pezizomycetes	Ascomycota
AY544652	P_Barssia_oregonensis	Helvellaceae	Pezizales	Pezizomycetes	Ascomycota
AY544655	P_Helvella_compressa	Helvellaceae	Pezizales	Pezizomycetes	Ascomycota
AY544667	P_Disciotis_venosa	Morchellaceae	Pezizales	Pezizomycetes	Ascomycota
AY544665	P_Morchella_elata	Morchellaceae	Pezizales	Pezizomycetes	Ascomycota
AY544664	P_Morchella_esculenta	Morchellaceae	Pezizales	Pezizomycetes	Ascomycota
AY544666	P_Verpa_conica	Morchellaceae	Pezizales	Pezizomycetes	Ascomycota
DQ470948	P_Peziza_vesiculosa	Pezizaceae	Pezizales	Pezizomycetes	Ascomycota
AY544668	P_Sarcosphaera_crassa	Pezizaceae	Pezizales	Pezizomycetes	Ascomycota
AY544654	P_Aleuria_aurantia	Pyronemataceae	Pezizales	Pezizomycetes	Ascomycota
AY544661	P_Cheilymenia_stercorea	Pyronemataceae	Pezizales	Pezizomycetes	Ascomycota
DQ247806	P_Scutellinia_scutellata	Pyronemataceae	Pezizales	Pezizomycetes	Ascomycota
DQ470961	P_Rhizina_undulata	Rhizinaceae	Pezizales	Pezizomycetes	Ascomycota
EF494056	P_Phymatotrichopsis_omnivora	Rhizinaceae	Pezizales	Pezizomycetes	Ascomycota
AY544658	P_Pithya_cupressina	Sarcoscyphaceae	Pezizales	Pezizomycetes	Ascomycota
AY544647	P_Sarcoscypha_coccinea	Sarcoscyphaceae	Pezizales	Pezizomycetes	Ascomycota
FJ176877	P_Tuber_gibbosum	Tuberaceae	Pezizales	Pezizomycetes	Ascomycota
DQ438186	Y-Sporopachydermia_cereana	Dipodascaceae	Saccharomycetales	Saccharomycetes	Ascomycota
DQ518995	Y-Myxozyma_udenii	Lipomycetaceae	Saccharomycetales	Saccharomycetes	Ascomycota

(*To be continued on the next page*)

(*Continued*)

nu LSU rDNA	Species	Family	Order	Class	Phylum
EF550320	Y-Pichia_mississippiensis	Pichiaceae	Saccharomycetales	Saccharomycetes	Ascomycota
EU011594	Y-Ambrosiozyma_platypodis	Saccharomycopsidaceae	Saccharomycetales	Saccharomycetes	Ascomycota
DQ442681	Y-Trichomonascus_ciferrii	Trichomonascaceae	Saccharomycetales	Saccharomycetes	Ascomycota
AY780054	S_Camarops_amorpha	Boliniaceae	Boliniales	Sordariomycetes	Ascomycota
AY761081	S_Togninia_novaezealandiae	Calosphaeriaceae	Calosphaeriales	Sordariomycetes	Ascomycota
AY544685	S_Carpoligna_pleurothecii	Chaetosphaeriaceae	Chaetosphaeriales	Sordariomycetes	Ascomycota
AY346276	S_Coniochaetidium_savoryi	Coniochaetaceae	Coniochaetales	Sordariomycetes	Ascomycota
AF408339	S_Microthia_havanensis	Cryphonectriaceae	Diaporthales	Sordariomycetes	Ascomycota
AY818964	S_Gnomonia_gnomon	Gnomoniaceae	Diaporthales	Sordariomycetes	Ascomycota
AF408364	S_Harknessia_lythri	incertae sedis	Diaporthales	Sordariomycetes	Ascomycota
AF362567	S_Melanconis_stilbostoma	Melanconidaceae	Diaporthales	Sordariomycetes	Ascomycota
AF408386	S_Valsa_ceratosperma	Valsaceae	Diaporthales	Sordariomycetes	Ascomycota
AY490788	S_Ceriosporopsis_halima	Halosphaeriaceae	Halosphaeriales	Sordariomycetes	Ascomycota
AY686634	S_Bionectria_ochroleuca	Bionectriaceae	Hypocreales	Sordariomycetes	Ascomycota
EU334679	S_Beauveria_bassiana	Cordycipitaceae	Hypocreales	Sordariomycetes	Ascomycota
FJ176861	S_Emericellopsis_maritima	incertae sedis	Hypocreales	Sordariomycetes	Ascomycota
AB067709	S_Cordyceps_sinensis	Ophiocordycipitaceae	Hypocreales	Sordariomycetes	Ascomycota
AF218207	S_Metarhizium_anisopliae	Clavicipitaceae	Hypocreales	Sordariomycetes	Ascomycota
AF510497	S_Hypocrea_jecorina	Hypocreaceae	Hypocreales	Sordariomycetes	Ascomycota
AY489727	S_Stephanonectria_keithii	Nectriaceae	Hypocreales	Sordariomycetes	Ascomycota
AY489720	S_Niesslia_exilis	Niessliaceae	Hypocreales	Sordariomycetes	Ascomycota
DQ471018	S_Apiospora_montagnei	Apiosporaceae	incertae sedis	Sordariomycetes	Ascomycota
DQ286199	S_Glomerella_cingulata	Glomerellaceae	incertae sedis	Sordariomycetes	Ascomycota
DQ493955	S_Magnaporthe_grisea	Magnaporthaceae	incertae sedis	Sordariomycetes	Ascomycota
DQ470950	S_Papulosa_amerospora	Papulosaceae	Incertae sedis	Sordariomycetes	Ascomycota
AY780077	S_Sinosphaeria_bambusicola	Thyridiaceae	Incertae sedis	Sordariomycetes	Ascomycota
DQ522856	S_Lulworthia_grandispora	Lulworthiaceae	Lulworthiales	Sordariomycetes	Ascomycota
FJ176888	S_Gondwanamyces_capensis	Ceratocystidaceae	Microascales	Sordariomycetes	Ascomycota
DQ470979	S_Ambrosiella_xylebori	Microascaceae	Microascales	Sordariomycetes	Ascomycota
DQ470955	S_Ophiostoma_piliferum	Ophiostomataceae	Ophiostomatales	Sordariomycetes	Ascomycota
DQ286206	S_Colletotrichum_crassipes	Phyllachoraceae	Phyllachorales	Sordariomycetes	Ascomycota
AY346257	S_Annulatascus_triseptatus	Annulatascaceae	Sordariales	Sordariomycetes	Ascomycota
AF431950	S_Cephalotheca_sulfurea	Cephalothecaceae	Sordariales	Sordariomycetes	Ascomycota
AY346305	S_Zopfiella_ebriosa	Chaetomiaceae	Sordariales	Sordariomycetes	Ascomycota
AY436419	S_Lasiosphaeris_hispida	Lasiosphaeriaceae	Sordariales	Sordariomycetes	Ascomycota
AY346301	S_Sordaria_macrospora	Sordariaceae	Sordariales	Sordariomycetes	Ascomycota
FJ176892	S_Nigrospora_oryzae	mitosporic	Trichosphaeriales	Sordariomycetes	Ascomycota
AF431949	S_Cainia_graminis	Amphisphaeriaceae	Xylariales	Sordariomycetes	Ascomycota
AY780068	S_Duradens_sp	Clypeosphaeriaceae	Xylariales	Sordariomycetes	Ascomycota
DQ470964	S_Diatrype_disciformis	Diatrypaceae	Xylariales	Sordariomycetes	Ascomycota
AY544676	S_Xylaria_acuta	Xylariaceae	Xylariales	Sordariomycetes	Ascomycota
DQ911601	B_Leucoagaricus_barssii	Agaricaceae	Agaricales	Agaricomycetes	Basidiomycota
DQ071728	B_Limacella_glioderma	Amanitaceae	Agaricales	Agaricomycetes	Basidiomycota
DQ470820	B_Plicaturopsis_crispa	Atheliaceae	Agaricales	Agaricomycetes	Basidiomycota
DQ457660	B_Conocybe_lactea	Bolbitiaceae	Agaricales	Agaricomycetes	Basidiomycota
DQ071735	B_Macrotyphula_fistulosa	Clavariaceae	Agaricales	Agaricomycetes	Basidiomycota
EU600883	B_Inocybe_sp	Cortinariaceae	Agaricales	Agaricomycetes	Basidiomycota
AY820890	B_Crepidotus_versutus	Crepidotaceae	Agaricales	Agaricomycetes	Basidiomycota
AY380407	B_Rhodocybe_aureicystidiata	Entolomataceae	Agaricales	Agaricomycetes	Basidiomycota
U66435	B_Hygrocybe_citrinopallida	Hygrophoraceae	Agaricales	Agaricomycetes	Basidiomycota
DQ071709	B_Bovista_nigrescens	Lycoperdaceae	Agaricales	Agaricomycetes	Basidiomycota
DQ071741	B_Omphalotus_illudens	Omphalotaceae	Agaricales	Agaricomycetes	Basidiomycota
DQ071722	B_Pleurotus_ostreatus	Pleurotaceae	Agaricales	Agaricomycetes	Basidiomycota
AY634279	B_Pluteus_romellii	Pluteaceae	Agaricales	Agaricomycetes	Basidiomycota
DQ457661	B_Coprinopsis_atramentaria	Psathyrellaceae	Agaricales	Agaricomycetes	Basidiomycota
DQ071725	B_Schizophyllum_commune	Schizophyllaceae	Agaricales	Agaricomycetes	Basidiomycota
DQ987904	B_Tubaria_serrulata	Strophariaceae	Agaricales	Agaricomycetes	Basidiomycota
AY639434	B_Marasmiellus_palmivorus	Tricholomataceae	Agaricales	Agaricomycetes	Basidiomycota
DQ144610	B_Amyloathelia_crassiuscula	Atheliaceae	Atheliales	Agaricomycetes	Basidiomycota
AY634277	B_Auricularia_sp	Auriculariaceae	Auriculariales	Agaricomycetes	Basidiomycota
AY645056	B_Exidia_uvapsassa	Exidiaceae	Auriculariales	Agaricomycetes	Basidiomycota
DQ831025	B_Pseudohydnum_gelatinosum	Hyaloriaceae	Auriculariales	Agaricomycetes	Basidiomycota
AY684158	B_Boletellus_projectellus	Boletineae	Boletales	Agaricomycetes	Basidiomycota
EU118643	B_Leucogyrophana_mollusca	Coniophoraceae	Boletales	Agaricomycetes	Basidiomycota

(*To be continued on the next page*)

(*Continued*)

nu LSU rDNA	Species	Family	Order	Class	Phylum
DQ534669	B_Gomphidius_roseus	Gomphidiaceae	Boletales	Agaricomycetes	Basidiomycota
AY586659	B_Hygrophoropsis_aurantiaca	Hygrophoropsidaceae	Boletales	Agaricomycetes	Basidiomycota
AY645059	B_Paxillus_vernalis	Paxillaceae	Boletales	Agaricomycetes	Basidiomycota
AJ440941	B_Serpula_lacrymans	Serpulaceae	Boletales	Agaricomycetes	Basidiomycota
AY586715	B_Suillus_luteus	Suillaceae	Boletales	Agaricomycetes	Basidiomycota
DQ071746	B_Tapinella_panuoides	Tapinellaceae	Boletales	Agaricomycetes	Basidiomycota
AY885164	B_Waitea_circinata	Ceratobasidiaceae	Cantharellales	Agaricomycetes	Basidiomycota
AY885163	B_Multiclavula_mucida	Clavulinaceae	Cantharellales	Agaricomycetes	Basidiomycota
DQ863692	B_Phlebia_chiricahuaensis	Corticiaceae	Corticiales	Agaricomycetes	Basidiomycota
AJ583431	B_Gloeophyllum_abietinum	Gloeophyllaceae	Gloeophyllales	Agaricomycetes	Basidiomycota
AY647207	B_Gomphus_clavatus	Gomphaceae	Gomphales	Agaricomycetes	Basidiomycota
AY635770	B_Hydnochaete_duportii	Hymenochaetaceae	Hymenochaetales	Agaricomycetes	Basidiomycota
U66452	B_Contumyces_rosella	Incertae sedis	Hymenochaetales	Agaricomycetes	Basidiomycota
U66432	B_Loreleia_marchantiae	Incertae sedis	Hymenochaetales	Agaricomycetes	Basidiomycota
U66441	B_Sphagnomphalia_brevibasidiat	Incertae sedis	Hymenochaetales	Agaricomycetes	Basidiomycota
DQ911614	B_Auriscalpium_vulgare	Auriscalpiaceae	Polyporales	Agaricomycetes	Basidiomycota
AY635771	B_Cyphella_digitalis	Cyphellaceae	Polyporales	Agaricomycetes	Basidiomycota
DQ071727	B_Fistulina_hepatica	Fistulinaceae	Polyporales	Agaricomycetes	Basidiomycota
AJ583428	B_Donkioporia_expansa	Fomitopsidaceae	Polyporales	Agaricomycetes	Basidiomycota
DQ208413	B_Ganoderma_lucidum	Ganodermataceae	Polyporales	Agaricomycetes	Basidiomycota
AF506469	B_Scytinostroma_odoratum	Lachnocladiaceae	Polyporales	Agaricomycetes	Basidiomycota
DQ071736	B_Phyllotopsis_nidulans	Lentinaceae	Polyporales	Agaricomycetes	Basidiomycota
DQ457673	B_Hydnopolyporus_fimbriatus	Meripilaceae	Polyporales	Agaricomycetes	Basidiomycota
AY629320	B_Polyporus_squamosus	Polyporaceae	Polyporales	Agaricomycetes	Basidiomycota
AY629315	B_Pterula_echo	Pterulaceae	Polyporales	Agaricomycetes	Basidiomycota
AY629321	B_Sparassis_crispa	Sparassidaceae	Polyporales	Agaricomycetes	Basidiomycota
EU118658	B_Xenasmatella_ardosiaca	Xenasmataceae	Polyporales	Agaricomycetes	Basidiomycota
AY684166	B_Albatrellus_higanensis	Albatrellaceae	Russulales	Agaricomycetes	Basidiomycota
DQ234539	B_Bondarzewia_montana	Bondarzewiaceae	Russulales	Agaricomycetes	Basidiomycota
AF506430	B_Echinodontium_tinctorium	Echinodontiaceae	Russulales	Agaricomycetes	Basidiomycota
DQ457665	B_Epithele_typhae	Epitheliaceae	Russulales	Agaricomycetes	Basidiomycota
AF506439	B_Gloeocystidiellum_formosanum	Gloeocystidiellaceae	Russulales	Agaricomycetes	Basidiomycota
EU118625	B_Dentipellis_leptodon	Hericiaceae	Russulales	Agaricomycetes	Basidiomycota
EU118634	B_Hyphodontiella_multiseptata	Hyphodermataceae	Russulales	Agaricomycetes	Basidiomycota
EU118662	B_Phlebiopsis_flavidoalba	Meruliaceae	Russulales	Agaricomycetes	Basidiomycota
EU118621	B_Cyphellostereum_laeve	Podoscyphaceae	Russulales	Agaricomycetes	Basidiomycota
EU118670	B_Steccherinum_ochraceum	Steccherinaceae	Russulales	Agaricomycetes	Basidiomycota
DQ521406	B_Sebacina_incrustans	Sebacinaceae	Sebacinales	Agaricomycetes	Basidiomycota
AY586635	B_Bankera_fuligineoalba	Bankeraceae	Thelephorales	Agaricomycetes	Basidiomycota
AY634276	B_Polyozellus_multiplex	Thelephoraceae	Thelephorales	Agaricomycetes	Basidiomycota
AY586724	B_Typhula_phacorrhiza	Typhulaceae	Thelephorales	Agaricomycetes	Basidiomycota
AY586720	B_Trechispora_nivea	Sistotremataceae	Trechisporales	Agaricomycetes	Basidiomycota
AY745730	B_Bensingtonia_ciliata	Agaricostilbaceae	Agaricostilbales	Agaricostilbomycetes	Basidiomycota
DQ831021	B_Phleogena_faginea	Phleogenaceae	Atractiellales	Atractiellomycetes	Basidiomycota
DQ832205	B_Sakaguchia_dacryoidea	Incertae sedis	Erythrobasidiales	Cystobasidiomycetes	Basidiomycota
DQ831020	B_Naohidea_sebacea	Incertae sedis	Naohideales	Cystobasidiomycetes	Basidiomycota
AY586643	B_Cerinomyces_crustulinus	Cerinomycetaceae	Dacrymycetales	Dacrymycetes	Basidiomycota
DQ831032	B_Rhamphospora_nymphaeae	Rhamphosporaceae	Doassansiales	Exobasidiomycetes	Basidiomycota
DQ663687	B_Entyloma_calendulae	Entylomataceae	Entylomatales	Exobasidiomycetes	Basidiomycota
AY745714	B_Tilletiopsis_washingtonensis	Entylomataceae	Entylomatales	Exobasidiomycetes	Basidiomycota
DQ667151	B_Exobasidium_rhododendri	Exobasidiaceae	Exobasidiales	Exobasidiomycetes	Basidiomycota
AY745713	B_Tilletiopsis_minor	Tilletiariaceae	Georgefischeriales	Exobasidiomycetes	Basidiomycota
DQ832196	B_Rhodotorula_hinnulea	Incertae sedis	Microstromatales	Exobasidiomycetes	Basidiomycota
EF551316	B_Rhodotorula_nothofagi	Incertae sedis	Microstromatales	Exobasidiomycetes	Basidiomycota
DQ832244	B_Tilletia_controversa	Tilletiaceae	Tilletiales	Exobasidiomycetes	Basidiomycota
AY629313	B_Colacogloea_peniophorae	Incertae sedis	Incertae sedis	Microbotryomycetes	Basidiomycota
AY745728	B_Kriegeria_eriophori	Incertae sedis	Incertae sedis	Microbotryomycetes	Basidiomycota
AY646098	B_Leucosporidium_scottii	Leucosporidiaceae	Leucosporidiales	Microbotryomycetes	Basidiomycota
DQ789982	B_Microbotryum_violaceum	Microbotryaceae	Microbotryales	Microbotryomycetes	Basidiomycota
DQ831009	B_Mixia_osmundae	Mixiaceae	Mixiales	Mixiomycetes	Basidiomycota
AY646099	B_Insolibasidium_deformans	Platygloeaceae	Platygloeales	Pucciniomycetes	Basidiomycota
AY745696	B_Kuehneola_uredinis	Phragmidiaceae	Pucciniales	Pucciniomycetes	Basidiomycota
DQ419920	B_Auriculoscypha_anacardiicola	Septobasidiaceae	Septobasidiales	Pucciniomycetes	Basidiomycota
DQ831016	B_Mrakia_frigida	Cystofilobasidiaceae	Cystofilobasidiales	Tremellomycetes	Basidiomycota

(*To be continued on the next page*)

(*Continued*)

nu LSU rDNA	Species	Family	Order	Class	Phylum
EF551318	B_Guehomyces_pullulans	Incertae sedis	Cystofilobasidiales	Tremellomycetes	Basidiomycota
DQ836002	B_Trichosporon_lignicola	Incertae sedis	Cystofilobasidiales	Tremellomycetes	Basidiomycota
AY586653	B_Elmerina_holophaea	Aporpiaceae	Tremellales	Tremellomycetes	Basidiomycota
DQ645514	B_Asterotremella_humicola	Tremellaceae	Tremellales	Tremellomycetes	Basidiomycota
AY745724	B_Malassezia_pachydermatis	Incertae sedis	Malasseziales	Ustilaginomycetes	Basidiomycota
DQ846888	B_Ustanciosporium_standleyanum	Cintractiaceae	Ustilaginales	Ustilaginomycetes	Basidiomycota
DQ094784	B_Ustilago_tritici	Ustilaginaceae	Ustilaginales	Ustilaginomycetes	Basidiomycota
DQ273824	C_Kappamyces_laurelensis	Kappamycetaceae	Rhizophydiales	Chytridiomycetes	Chytridiomycota
DQ273823	C_Boothiomyces_macroporosum	Terramycetaceae	Rhizophydiales	Chytridiomycetes	Chytridiomycota
DQ273766	C_Rozella_sp.	Incertae sedis	Spizellomycetales	Chytridiomycetes	Chytridiomycota
DQ273775	C_Rhizophlyctis_harderi	Spizellomycetaceae	Spizellomycetales	Chytridiomycetes	Chytridiomycota

Appendix Table 2 Codon-based test of neutrality for analysis among sequences[a)]

	1	2	3	4	5	6	7	8	9	10	11	12	13
1. *Gibberella zeae*		−184.317	−142.743	−226.836	−117.516	−191.803	−148.869	−113.929	−137.833	−129.891	−164.784	−160.468	−180.596
2. Magnaporthe grisea	0.000		−188.521	−227.962	−117.358	−120.369	−193.850	−144.449	−125.675	−87.916	−200.894	−135.379	−186.449
3. Neurospora crassa	0.000	0.000		−177.944	−178.191	−112.182	−123.331	−113.202	−94.598	−141.345	−116.777	−100.106	−121.007
4. *Sclerotinia sclerotiorum*	0.000	0.000	0.000		−133.105	−196.832	−243.138	−177.307	−234.438	−108.635	−220.575	−133.662	−162.847
5. *Aspergillus fumigatus*	0.000	0.000	0.000	0.000		−247.560	−155.108	−111.981	−119.304	−100.577	−174.783	−94.895	−171.981
6. *Coccidioides immitis*	0.000	0.000	0.000	0.000	0.000		−156.800	−141.129	−124.953	−111.981	−135.773	−103.759	−195.702
7. *Phaeosphaeria nodorum*	0.000	0.000	0.000	0.000	0.000	0.000		−175.179	−84.938	−142.238	−124.861	−181.418	−159.492
8. *Malassezia globosa*	0.000	0.000	0.000	0.000	0.000	0.000	0.000		−191.895	−128.378	−176.426	−136.619	−131.781
9. *Ustilago maydis*	0.000	0.000	0.000	0.000	0.000	0.000	0.000	0.000		−114.146	−164.587	−143.049	−233.382
10. *Cryptococcus neoformans*	0.000	0.000	0.000	0.000	0.000	0.000	0.000	0.000	0.000		−116.655	−177.228	−157.077
11. *Laccaria bicolor*	0.000	0.000	0.000	0.000	0.000	0.000	0.000	0.000	0.000	0.000		−293.352	−260.705
12. *Coprinopsis cinerea*	0.000	0.000	0.000	0.000	0.000	0.000	0.000	0.000	0.000	0.000	0.000		−164.899
13. *Rhizopus oryzae*	0.000	0.000	0.000	0.000	0.000	0.000	0.000	0.000	0.000	0.000	0.000	0.000	

a) The probability of rejecting the null hypothesis of strict-neutrality (d_N=d_S) (below diagonal) is shown. Values of P less than 0.05 are considered significant at the 5% level and are highlighted. The test statistic (d_N–d_S) is shown above the diagonal. d_S and d_N are the numbers of synonymous and nonsynonymous substitutions per site, respectively. The variance of the difference was computed using the bootstrap method (1000 replicates). Analyses were conducted using the Nei-Gojobori method in MEGA4. All positions containing gaps and missing data were eliminated from the dataset (complete deletion option). There were a total of 39279 positions included in 105 protein-coding genes in the final dataset.

第六章　命名法

Nomenclature

The lectotypification of some species in the Umbilicariaceae described by Linnaeus or Hoffmann

LASALLIA Merat

Nouv. Fl. Paris ed. 2 (1): 202 (1821).

Type: *Lasallia pustulata* (L.) Merat (1821).

≡ *Lichen pustulatus* L. (1753).

Lasallia pensylvanica (Hoffm.) Llano

Monograph of the Umbilicariaceae 42(1950).

≡ *Umbilicaria pensylvanica* Hoffm. Descr. Adumb. Pl. Lich. 3(4):5–6, Pl. 69, fig. 1–2, 1801.

Neotype: Amer. Septent. Muehlenberg, H–Ach. No.571(H!).

TYPE OBSERVATION: Thallus smooth with squamules on the uppert surface, black below, gorssly verrucose; Apothecia leiodiscs.

DESCRIPTION: Thallus monophyllous, orbicular, undulating, rigid, to 6 cm. diam.; upper surface smooth to more or less areolately papilate, dull to shining, pustulated, warm buff to wood brown, covered with a thin layer of whitish pruina, with a few scattered black-brown curved squamules, over umbo elevated, occasionally with a few small thalli, marginally laciniated with many dark brown to black-brown toothed lacinia; lower surface sooty black, coarsely verrucose, without rhizines or occasionally with a few cylindrical rhizines up to 4 mm long, surface of the rhizines clearly papillated, with concave lacunae corresponding to the pustules of upper surface.

Thallus 354–385 μm thick; outer cortex yellowish brown, 23–13 μm thick, innercortex colourless, 4.5–18.5 μm thick; algal layer 41.5–55.5 μm thick; medulla 236.5–260 μm thick, the part next to algae brownish, below colourless; outer lower cortex brown, 9–18.5 μm thick, inner lower cortex scleroplectenchymatous, colourless. 23–46 μm thick.

Apothecia leiodisc, occasionally with a single fissure or button in the disc; epihymenium brown, 5.5–16 μm thick; hymenium colourless, 54–106 μm thick; subhymenium brownish, 18–63 μm thick; paraphysesbranched,septate, colourless, sometimes brownish, 1.5 μm thick, inflated at the tip, ellipsoid, 1.5–5.5 μm thick; asci monosporous, 61–81 × 23–30.5 μm; spores large, muriform, brown, 48.5–52 × 25–30.5 μm.

Pycnidia 148.5–204 × 143.5–172.5 μm; conidia bacilliform, colourless, 3–4 × < 1 μm.

CHEMISTRY: lecanoric and gyrophoric acids.

SPECIMENS EXAMINED:

China: Jilin, Mt. Changbai shan, south slope, Mt. Hongtou shan: alt. 1950 m, Aug. 5, 1980, coll. Hu YC, No.2149; alt.1970 m, Aug. 16, 1982, coll. Hu YC, nos. 3386, 3432; alt. 1880 m, July 28, 1983, coll. Wei JC & Chen JB no. 6250; alt. 1980–2000 m, coll. Hu YC, nos. 4218, 4230, 4232 (HMAS–L!). Neimongol, Xinganling, Bailong forest farm, Mt. Jiguan shan, alt.1600 m, on rocks, Aug. 2, 1991, coll. Chen & Jiang, nos.A–812, A–817, A–820 (HMAS–L!). Xinjiang, Mt. Altai shan, near Halas lake, alt. 2000 m, on rocks, Aug. 4, 1986, coll. Gao XQ, No.1987 (HMAS–L!).

Mongolia: Ara–Khangai, on rocks, Jun. 15, 1971, coll. Biazrov, No. 5768 (Moscow!); alt. 2100 m, Aug. 28, 1971, coll. Biazrov, No. 4228 (Moscow!); Sept. 7, 1979, coll. Biazrov, No. 6186

This paper was originally published in *Mycosystema*, supplement 5: 1-17, 1993.

(Moscow!). The folowing specimens examined were identified as Lasallia rossica Dombr. in the herbarium: Ara–Khangai; alt. 2130 m, on rocks, Aug. 16, 1972, coll. Biazrov, No. 6690 (Moscow!); Jun. 30, 1976, coll. Biazrov, No. 3178 (with rhizines, Moscow!); alt. 2000 m, July 1, 1980, coll. Biazrov, Nos. 4350, 4356 (Moscow!); Bulgan, Jun. 29, 1972, coll. Biazrov, Nos. 5110, 5103 (Moscow!), July 24, 1977, coll. Biazrov, No. 6580 (Moscow!). Zabhan, alt.2000 m, on rocks, July 11, 1976, coll. Biazrov, No. 6579 (Moscow!).

Kazakstan: Turkestan, Kokchetav, July 13, 1896, coll. A. Tordiagei(LE!). Japan: Honshu prov. Etchu: Mt. Etchusawa-dake on rocks, alt. 2400 m, July 27, 1976, coll. Kashiwadani (no. 13413) and K. Yoshida in Kurok.: Lich. Rar. et Cr. Exs. 293 (COLO, DUKE, FH, LE, S, TENN, US!); Musashi: Mt. Kinpu. on rocks, alt. 2400 m, Aug. 25, 1960, coll. Asahina, Togashi, Kurok., and Nuno, [Kurok. & Kashiw. Lich. Rar. et Cr. Exs. No. 316, 317] (COLO, DUKE, FH, LE, S, TUR, UPS, US!); Mt. Ohyama, coll. H. Shibuichi(No.4719) [Kurok & Kashiw., Lich.Rar. et Cr. Exs. 696] (COLO, DUKE, FH, LE, US!), on rocks, alt. c.1800 m, Mt. Nagayamadake,Mt. Daisetsu National Park, Prov. Ishikari, Hokkkaido, Aug. 22, 1971, Coll. I. Yoshimura (No.12395) [Lich. Jap. Exs. No.39] (COLO, DUKE, H, LE, TUR, US!). Hokkaido, Mt. Tomuraushi, July 31, 1935, coll. M.M.Sato (FH!). Herb. of the U.S. North Pacific Exploring Expedition undercommanders Ringgold and Rodgers, 1853–56 Umbilicaria pensylvanica, C. Wright (FH!). Near Fukushima alt. 3000 m, Aug. 1898, coll. Fauric No.863 (LE!), Jiidesan, alt.3000 m, Aug. 1898, Coll. Fauric, No. 859, pr. p.(LE, pr. p. min. MIN!).

Russia: Chukotka, western Chukotka, North Anuisk, July 14, 1978, coll. M. P. Andreev (HMAS–L!). North Ural, Herb. of Inst. of Biology of Ural Academy of Sciences, Aug. 4, 1962, coll. by M.Storojeva (H, Partly TUR!). Siberia, near Angara river, on rocks, Oct. 1, 1907, coll. Martianov, No. 171 in Herb. Vain 25 as U. caucasica Lojka (TUR!). Sverdlovsk region, Jun. 24, 1959, coll. A. Kasarinova, S25920, C. M. COLO–146277, Pr. P. (COLO!). Sept. 16, 1926, coll. A. Oxner (COLO, UPS!). Gorno–Altaisk Auto–oblast. SE of Ongudai on Little Ilzumen River near Khabarovka below Chikemansky Pass, alt. 900–1000 m; arid mountain steppe with low granite outcrops, also riparian woodland, Aug. 20, 1978, coll. T. S. Elias, W. Weber, C. S. Tomb & I. M. Krashnoborov (COLO!). Vom Gipfel des Urals bei Jekaterinburg, auf Granit Sept. 12, 1823, Herb. Chr. Steven (H!). Exped. Uralensis, Jun. 22, 1847, coll. Branth ex Herb. Acad. Petrop. (UPS!). Ural, 1900, coll. Vasili Sokolor (H!).

OTHER SPECIMENS EXAMINED:

[H. Lojka: Lichenotheca Universalis No.13] (S!). ExHerb. W. G. Farlow, *Umbilicaria pensylvanica* Hoffm., Jun. 13, 1883, coll. E. Faxon(S!). Lichens collected by Carolyn W. Harris in Vicinity of Chilson Lake, Essex Co, N.Y. Alt. 1200ft, *Umbilicaria pensylvanica* Hoffm (S!). N. Amer. (N. Carol.) coll. H.A. Green, July,1883, *Umbilicaria pensylvanica* Hoffm. (S!).

LITERATURE RECORDS FOR ASIA:

China (Chen et al. 1981, p.158; Wang & Lai, 1976, p.228).

Mongolia (N. Golubkova, 1981, p.137; Huneckct al., 1992, p. 296).

Japan (Sato, 1943, p.103; Yoshimura, 1974, p.119; Ikoma, 1983, p.109).

Russia (Golubkova, 1978, p.133 for Siberia an d Far East; Sedelnikova, 1985, p.89 for Siberia; Makrei, 1990, p.55 for Siberia).

Himalaya (Singh, 1964, p.178 as U. pensylvanica; Awasthi, 1965, p.126).

COMMENTS: It is easy to distinguish this species from *L. papulosa* by its sooty black lower surface of thallus and recurved squamules, while in *L. papulosa*with applanate squamules on the upper surface and light to brownish coloured lower surface of thallus.

Arctic-holarctic Multirange's.

Lasallia pustulata (L.) Merat

Nouv. Fl. Paris, ed. 2, 1:202 (1821).

≡ *Lichen pustulatus* L., Sp. Pl. 1150 (1753).

Lectotype: Sweden, 1273–201:52, Fl. Suec. 969 (LINN!).

TYPE OBSERVATION: One specimen from Sweden (Fl. Suec. 969), and two figures (Vaill. paris. 116. t. 20, f. 9 and Dill. musc. 227, t. 30 f, 131) were cited by Linnaeus in Sp. Pl. (1753). All of above cited should be considered as syntypes for this species. Three specimens (1273–201: 52, Fl. Suec. 969; 1273–202:52; and 1273–203) for this species are preserved in LINN. One of these three specimens, the specimen from Sweden (Fl. Suec. 969) cited by Linnaeus (1753, p.1150), was selected here as a lectotype.

There are two thalli glued to a herbarium paper. Thallus strongly pustulate, upper surface grey-brown, over umbo covered with a thick layer of whitish pruina, with a crystal-granulate surface, marginally smooth and plane, with black brown isidia; lower surface brownish to brown, coarsely papilated or verrucose.

DIAGNOSIS: This species is close to *Lasallia pensylvanica* (Hoffm.) Llano, but differs from the latter in having a grey-brown or brownish to brown lower surface of thallus and finely cylindrical or bacilliform and branched (Biazrov No. 59233) or granulated globoid isidia (Biazrov, No.5923, duplicate) of dark brown colour. In addition, apothecia are very rare.

SPECIMENS EXAMINED:

Mongolia: Ara–Khangai, on rocks, Aug. 22, 1979, coll. Biazrov, No.5923, the specimen with numerous, thin and bacilliform isidia, and a duplicate of this collection with granulated globoid isidia (Moscow!).

OTHER SPECIMENS EXAMINED:

Finland: Nylandia, Orimattila, Fonno. Ad rupem ad SW versus loco aprico. Oct. 23, 1948, Rainar Hakulinen [Lichenotheca Fennica No.676] (S!).

Sweden: near Stokholm, on rocks, 1982, coll. Wei J.C. et al. (HMAS–L!).

LITERATURE RECORDS FOR ASIA:

Mongolia (Schubert et. al., 1969, p.406; Golubkova, 1981, p.134; Olech, 1984, p.215; Biazrov, 1986, p.99).

Russia (Sedelnikova, 1985, p.89 for Siberia).

COMMENTS: This species differs from others of *Lasallia* in Asia byitslarger thalli and strongly convexed pustules with bacilliform isidia on the upper surface of thalli, sometimes with numerous finely globoid isidia, which is usually treated by some lichenologists as an infraspecific taxon.

The collection from Mongolia was not available for the anatomic study and chemical tests.

Bipolar multirange's.

UMBILICARIA Hoffmann

Descr. Adumbr. Pl. Lich. 1(1):8(1789).

Type: *Umbilicaria exasperata* Hoffmann

Umbilicaria cylindrica auct.

Lichen cylindricus L. in Species Plantarum (1753, p.1144) as a basionym has been incorrectly applied for the species which had been named *Lichen proboscideus* (1753, P.1150). In the Linnaean Herbarium there is no specimen bearing the name "*Lichen cylindricus*" . Delise made the combination in *Umbilicaria* based on the specimens of Schaerer's Lich. Helvet. Nos. 143–147, as cited in Duby's paper (1830). After careful examination of Schaerer's specimens, made by the author in FH, it is certain that both the description furnishedby Delise in Duby (1830) and Schaerer's specimens in Lich. Helvet. correspondcompletelywith specimens bearing "*Lichen proboscideus*" in the *Linnaean Herbarium*. Unfortunately, "Lichen cylindricus" used by Delise as basionym was recombined and misapplied as *Umbilicariacylindrica*. auct Subsequent authors followed this mistake, which should be rectified. (Princ. 2; Arts.7.2, 7.12 and 55.2).

Actually, Linnaean "*Lichen cylindricus*" was based on both Linnaeus's own previous description (Amoen. Acad. 2, 1751, p.264) and Dillenius' description and illustration (p.149, Tab. 20, fig. 42) cited in the Species Plantarum (1753, p.1144).

First, the same figure given by Dillenius was also cited in Linnaeus'(1751) previous description. From Dillenius' description and figure 42A (Fig. 60) we know that *Lichen cylindricus* exhibits a blueish grey thallus with long fimbriate imbriate cilia at margins and grows on a tree twig. Apothecia on the same figure are deep bowl on stalks, that is, they do not belong to any of the leiodiscs, omphalodiscs, gyrose or actinodiscs (Wei & Jiang, 1988, p.74), which are characteristics of the apothecia in the Umbilicariaceae.

Second, when Linnaeus (1753:1144, No.29) arranged the lichens in Species Pantarum, he placed *Lichen cylindricus* close to members belonging to Lobariaceae, Parmeliaceae, and Physciaceae, while other umbilicariaceous species (pp.1150–1151, Nos.51–56) were five pages and 21 species later. These 6 umbilicariaceous species were linked together in almost the same page (pp. 1150–51).

Finally, Linnaeus described the six umbilicariaceous species by the term "umbilicatus" and / or indicated that they were saxicolous. For *Lichen cylindricus* he neither used "umbilicatus" nor indicated that it grew on rocks, on the contrary, it grew on a tree twig recognized from the figure 42A.

According to above mentioned facts, Dillenius' figure (42A) of *Lichen cylindricus* looks like a member of *Parmotrema*, rather than *Umbilicaria*. The other figures on the same page are members of the Parmeliaceae and Physciaceae and their allies, but none from the Umbilicariaceae.

In typifying *Parmelia perforata*, Hale (1965, p.336) stated:

"It is interesting to note that *Lichen cylindricus* L. (Sp. Pl.1150①, 1753) which antedates *Lichen perforatus*, is based primarilly on 'Lichenoides foliorum laciniis crinitis' Dill. Hist. Musc. 149, p.20, fig. 42 (1742). Dillenius' specimen (OXF) is typical *P. perforata* collected in Pennsylvania. Linnaeus cited a second syntype from Lapland; this specimen (LINN) is a small *Umbilicaria*. Linnaeus himself later realized that the American specimen did not fit his concept of *Lichen cylindricus* and put this name in the umbilicate series as a synonym of *L. proboscideus* L. without further explanation. All later authors have followed Linnaeus' concept without, however, realizing that *L. cylindricus* was based in large part on *Parmelia perforata.*"

There is no doubt that *Lichen cylindricus* L. was based in large part on *Parmelia perforata* according to Linnaeus' (Amoen. acad. 2. p.264) and Dillenius' (1741–p.149) descriptions and Dillenius' fig. 42A cited by Linnaeus (1753, p.1144–45). The problem is what species the second syntype cited by Linnaeus (1753) belongs to. Hale (1965, p.336) pointed out that it belongs to "a small *Umbilicaria*" and was based on a specimen preserved in Linnaean Herbarium (LINN). As a matter of fact, in the LINN there is no specimen bearing the name "*Lichen cylindricus*". The protologue for the second syntype of *Lichen cylindricus* is Linnaeus' and Dillenius' descriptions and Dillenius' fig. 42B cited by Linnaeus (1753).

It is impossible to recognize that the second syntype belongs to a small *Umbilicaria*. Perhaps the specimen examined by Hale was typotype of fig.42B indeed, i.e. it was refferred to Dillenius' one preserved in OXF, on which the Dill. fig.42B was based. Even if that is so, *Lichen cylindricus* (pr.p. minore) can be considered as a synonym of *Lichen proboscideus* at most, as did Linnaeus later (1764, p.1617).

The problem in nomenclature and typification like this not only concerns *U. cylindrica*, but also several other species of the genus. After a careful check of the Linnaean specimens in LINN and of the original literatures it is clear that the lichenologists havebeen mistaking *Lichen flocculosus* Wulf. for *Lichen deustus* L., *Lichen* deustus L. for *Lichen proboscideus* L., and *Lichen proboscideus* L. for *Lichen cylindricus* L. These are interlocking mistakes in the nomenclature. In order to solve this problem and to revise these mistakes can choose between the following two ways:

1. Keeping the misapplied manes for name stability. However, the consequence of keeping the misapplied names certainly will make us unable to typify on the basis of the principles and rules of the ICBN unlessthe typification will be done according to the concept of the misapplied names. Evidently, this is in disagreement with the principles and rules of the ICBN.

2. These interlocking mistakes in the nomenclature should be rectified through the typification on the basis of the principles and rules of ICBN (Principle II; Art.7. 2; 7,12; 55. 2). Unfortunately, rectifying these interlocking mistakes in through the typification certainly will cause these rectified epithets to be thrown into confusion. However, the confusion consequent upon the typification can be removed by abandoning these misapplied epithets (Art. 69), and replaced by the new epithets on the basis of the same word roots of the misapplied epithets (Art.72.1). Treating by this way is both in line with ICBN, and in favour of name stability.

It seems that the second way is the better way to solve the problem and to revise these mistakes. As above mentioned the rectified name *Umbilicaria proboscidea* (L.) Schrad. [Spicil. Fl.

①Hale in his paper (1965, p. 336) mistook p. 1144 for p. 1150.

Germ. 1:103 (1794)] maybe kept (Art.55.2) on the basis of the typification. But this rectified name certainly can be thrown into confusion. Because this is one of the misapplied names in the interlocking mistakes. Therefore this misapplied name may be ruled as rejected (Art.69) and replaced by a new epithet which should be both in line with ICBN and in favour of name stability. So, a new epithet "*neocylindrica*" has been selected for *U. cylindrica* auct. instead of the rectified name *U. proboscidea*.

According to the author's careful examination of Linnaean specimens in LINN, "the specimen 1273–204:53" has been selected as lectotype of *Lichen proboscideus* L. (Principle II; Art. 7. 12).

Lichen proboscideus L., Sp. Pl. 1150 (1753).
Lectotype: Linn 1273–204:53 (LINN!).

TYPE OBSERVATION: Thallus umbilicate, orbicular to irregular, monophyllous upper surface smooth, or with a fine reticulating pattern of dark lines on a grey and pruinose background, or becoming wrinkled, dull to subshing, greyish-brown to dark grey-brown; margins crenate to torn, more or less perforated, with occasional rhizinoid spines growing upright from upper surface; lower surface smooth, only occasionally and locally weakly granulate, pale to pale pinkish-brown, with sparse to well developed rhizines, mostly peripheral, long and cylindrical or flat and branched.

Apothecia sublecanorine, stipitate; stipe and lower surface of apothecia concolorous to thallus, margin of discs often carbonized and becoming black.

Umbilicaria neocylindrica Wei, nom. nov.
≡ *Lichen proboscideus* L., Sp. Pl. 1150 (1753).
Lectotype: Linn 1273–204: 53 (LINN!).

MISAPPLIED NAME:
Lichen cylindricus auct.
Umbilicaria cylindrica sensu: Delise in Duby, Bot. Gall. 2:595 (1830). Frey in Rabh., Krypt.-Flora, 9(4 / 1):320(1932). Llano, A monograph of the Umbilicariaceae in the western hemisphere 119 (1950). Thomson, American Arctic Lichens 1. The Macrolichens 447 (1984). Wei & Jiang, Mycosystema, 1:80 (1988).

DESCRIPTION: Thallus mono- to polyphyllous, orbicular, margins lobed, sometimes several thalli develop in a rosette form, 2–3 cm diam., more or less elevated at margins; upper surface weakly lacunose, covered with a thick layer of whitish pruina, dark-grey to grey-brown; lower surface smooth, whitish to light brown, around umbo dark, with light brown to dark brown and branched rhizines mainly at margins which are often extended outwards.

Thallus 116–27.8 μm thick; neeral zone 4.5–14 μm thick; the yellow-brown outer layer of the upper cortex is usually 11.5–27.5 μm thick, the inner layer is very thin like a single hypha in thickness; algal layer 25.5–83.5 μm thick; medulla loose, 18.5–111 μm thick; the colourless inner layer of the lower cortex densely mucilaginous, 9–28 μm thick, the grey brown outer layer with uncellular structure, 9–18.5 μm thick.

Apothecia with gyrose discs, stipitate, 1–2 mm diam.; epihymenium brown, 5.5–10.5 μm thick; hymenium 36–73.5 μm thick; subhymenium 18–39.5 μm thick; paraphyses septate, c. 2 μm thick; asci when immature 19–43 × 4.5–81 μm thick.

Pycnidia 102–292 × 37–14–83.5 μm; conidia bacilliform, 2.5–3.5 × < 1 μm.

CHEMISTRY: no lichen substances.

SPECIMENS EXAMINED:

China: Jilin, Mt. Changbai shan, north slope, near Tianchi, on rocks, alt. 2100 m, Aug. 13, *1977, coll. Wei JC, No. 2973–(2) and alt. 2350 m, Aug. 14, 1977, Wei JC, Nos. 2983, 2984, 2992* (HMAS–L!); and on the way from Tianchi to waterfall, alt. 2100 m, 1977, coll. Chen X. L. No. *4781 (HMAS–L!); near the Tianwenfeng, on rocks, Aug.18, 19, 1987, coll. Wei J.C. Nos. 9010,* 9113 (HMAS–L!); southwest slope, Fusong county, alt. 2050 m, on rocks, Aug. 8, 1983, coll. Wei JC & Chen JB, No. 6761; Mt. Changbai shan, north slope, alt. 2400 m, on rocks, Aug. 14, 1977, coll. Wei JC, No. 2984. as a var. tornata (HMAS–L!). Shaanxi, Mt. Taibai shan, Baxiantai, Jun. 4, 1963, coll. Wei JC.(HMAS–L!). Xinjiang, Mt. Altai shan, by the Hanas lake, alt. 2500 m, on rocks, Aug. 5, 1986, coll. Gao X.Q. Nos. 2050, 2054, 2067, 2081, 2084, 2148 (HMAS–L!).

Mongolia: Near Hubsugul lake (Kosogol lake), on top of Mt. Muivu–Sardeika, July 23, 1902, coll. A. A. Elenkin, pr. p. (LE!).

Japan: Hokkaido, Mt. Tomuraushi, Aug. 1, 1935, Coll. M.M.Sato, No.18 (MIN!). Hokkaido, Mt. Daisetu, July 14, 1936, coll. M.M.Sato ex Herb. Univ. Imp. Tokyo (MIN!), Mt. Ontake, July 1889, ex coll. M. Miyoshi (MIN!).

Russia: Far East, Kamchatka, Aug. 1908, coll.V.Savicz, Nos.2232, 5639, 5839 (LE!) and 6334 (LE, UPS!). Vrangel Island, Sept. 2, 1938, coll. B.N. Gorodkov (LE!). Western Chukotka, Jun. 27, 1977, and Sept. 1, 1978, coll. M.P. Andreev (LE!); and Sept. 3, 1978, Coll. M. P. Andreev (HMAS–L–3923!); Magaransk area, July 21, 22, 1967, coll. E. V. Zimorskaia (HMAS–L–3884!); Chukotka, July 17, 20, 1971, coll. Makarova Nos.216, 308 and Aug. 8, 1979, coll. Makarova No. 151 (LE!). Polar Ural, July 7, 1909, coll. V. Sukachev and Aug. 9, 1925, coll. N. Gorodkov No.63pr.p. (LE!). Western Siberia, Tomsk, July 20, 1913, coll. N.J. Kuznezov, No.3252 (LE!). Eastern Siberia, Yakut prov. 1905, coll. I. Shulbga (LE!). SW Siberia, Altai mountains Gorno–Altaisk Auto. oblast. S. side of Saddle of Syeminsky Pass bwtween Shebalino and Ongudai, alt.1750–1950m, subalpine stepe- meadows scattered fixed rock outcrops. Aug. 19, 1978, coll. T.S. Elias, W. Weber, C.S. Tomb & I.M. Krasnoborov, L–68071, C. H. COLO–319353 pr.p. (COLO!).

Kazakstan: Eastern Kazakstan, July 14, 1982, coll. Dombrovskaia (HMAS–L!).

Armenia: near Sevan lake, Jun. 10, 1980, coll. A. A. Abramian (LE!); Ctepanavan, on rocks, July 1964, coll. V. Nikogosian, pr.p.(LE!). Districtus Stepanovansky, cacumen trajecti Puschkinsky, ad rupes, 1965, coll. V. Nikogosian [Savicz, Lichenotheca Ross. 161] (H!).

OTHER SPECIMENS EXAMINED:

U.K.: [Leighton: Lich. Brit. No. 95], *Umbilicaria varia* var. proboscidea, a Leight. Leight. Brit. Umbil. inedit. E. Bot. t. 522, upper fig. *Gyrophora proboscidea*, a. Turn. & Borr, Lich. Brit. 219. Schaer. Exs. 143, 146, Clova, Scotland (FH!). Moug. & Nestl. Crypt. 59. *Gyrophora cylindrica* Ach. *Lichen cylirdricus* L. *Umbilicaria proboscidea* Decand. (FH!).

Notes: The specimen of Leighton: Lich. Brit. No. 95. *Umbilicaria varia*, var. *proboscidea*, preserved in FH is corresponding to that of *Lichen proboscideus* L. (1273–204:53), preserved in LINN. *Umbilicaria proboscidea* Decand. was treated as conspecific with *Gyrophora cylindrica* auct. (≡*Lichen cylindricus* auct.) in Moug. & Nestl. Crypt. 59, preserved in FH.

LITERATURE RECORDS FOR ASIA:

China (Chen et al. 1981, p.157; Wei & Jiang, 1988, p.80; Wei, 1991, p.246).

Mongolia (Golubkova,1981, p.133).

Japan (Asahina, 1939, p.699; Sato, 1943, p.41; Yoshimura, 1974, p.120; Ikoma, 1983, p.109).

India, Darjeeling (Chopra, 1934, p.46; Paulson, 1925, p. 193; Smith, 1931, p.130; Singh, 1964, p.176; Awasthi, 1965, p.125).

Russia (Savicz, 1922, p.103 for Kamchatka; Golubkova, 1978, p.113 for Siberia; Makrei,1990, p.56 for Siberia).

COMMENTS: This species is similar to both *U. umbilicarioides* and *U. virginis* but differs from both them in lacking thalloconidia; and from the former in lacking conspicuous papillae on the under side, and from the llater in having gyose discs.

Multirange's.

Umbilicaria exasperata Hoffmann

Descr. et Adumbr. Plant. Lich. 1:7–8 (1790).

Illustration: Tabula 2, Fig. 1–2 (1790).

Neotype: H–Hoffm. 8596. *Umbilicaria exasperata* Hoffm. (*Gyrophora hyperborea* Ach.) Suec. (MW!).

TYPE OBSERVATION: Afteracareful examination of the specimenbearing *Umbilicaria exasperata* Hoffm. in Hoffmann's Herbarium in Moscow (MW!) the author came to the conclusion that Hoffmann's specimen was conspecific with *Umbilicaria hyperborea*. This specimen is selected here as a neotype for *U. exasperata* Hoffm.

≡*Gyrophora proboscidea* (L.) Ach. [var.] β G. *exasperata* (Hoffm.) Ach., Method. Lich. 105 (1803).

= *Lichen hyperboreus* Ach., Kgl. Vetensk. Akad. Nya Handl. 15:89 (1794). syn. nov.

Type: H–Ach. (H!).

≡*Gyrophora hyperborea* (Ach.) Ach. Method. Lich:104 (1803). ≡*Umbilicaria hyperborea* (Ach.) Hoffm. Deser. et Adumbr. Plant. Lich. 3:9 (1801).

DESCRIPTION: Thallus monophyllous, coriaceous, rigid and more or less pliable, over umbo weakly elevated; upper surface weakly undulating, smooth to wrinkled or strongly vermiform extending to margins, dull to slightly shiny, olive-brown to dark brown or grey-brown with olive-green tint; lower surface smooth to papillate, bullate to lacunose, brown to clove-brown or dark brown or black, without rhizines.

Thallus 134.5–324.5 μm thick; upper cortex 9–20 μm thick, the outer layer brown, 5.5–9 μm thick, the inner layer colourless, 3.5–10.5 μm thick; algal layer discontinuous, 23–49 μm thick; medulla loose in upper layer, 32.50–21μm thick, mucilaginous in lower layer, 16–60 μm thick; lower cortex 25.5–88.5 μm thick, the inner layer colourless with some brown spots,

11.5–21 μm thick, the outer layer brown, 14–28 μm thick.

Apothecia common, circular or irregular in shape, with convex discs; epihymenium 14.5–19.5 μm thick; hymenium 36–57.5 μm thick; subhymenium 12.5– 27 μm thick; paraphyses septate, not inflated at the tip, c. 1 μm thick; asci 23.5–34 × 5.5–9 μm; ascospores immature.

CHEMISTRY: lecanoric, gyrophoric and umbilicaric acids.

SPECIMENS EXAMINED:

China: Neimongol: Da Hinggan Ling, Ergun Zuoqi (Genhe), Mt. Oklidui shan, alt.1530 m, on rocks, July 26, 1983, coll. Zhao CF, No. 2964 (HMAS–L!).

Mongolia: Ara–Khangai, alt. 2300 m, on rocks, July 22, 1970, coll. Biazrov, No.4229 (MOSCOW!); & Aug. 1, 1973, coll. Biazrov, No. 5773 (MOSCOW!); Baian Khongor, alt.3030 m, Aug. 18, 1972, coll. Biazrov, No. 5774 (MOSCOW!).

Russia:Ural, Tumensk prov. coll. Kuvaev, pr.p. (LE!). Eastern Siberia, Mt. Putoran, July30, 1984, coll. M.P. Isurbenko (LE!). Yakut, coll.Shulga(LE!). Yakut, Tomop, Aug. 3, 1956, coll. V.I. Ivanova 116 / 16 (LE!). Central Ural. July 3, 1962, coll. M. Storoteva (TUR!). In Kamczatcam expeditio Rjabuschinskiana 1908–1909,V.L. Komarov et V.P. Savicz,Iter Kamczaticum I et II Exped No.6335, Determ. No.3.G.hyperborea f. cerebelloides Savicz, f. nov. Asia orientalis, kamczatca In regionibusalpinis in locis apricis (apritis?) vertosis que ad saxa 1909 coll. V. Savicz (LE!), and Exped. Nos. 6276, 6444, Determ. No.15 f. sublaevigata coll. V. Saviicz, f. nov. (LE!).

Japan: Honshu, prov. Etchu, Mt. Etchusawadake, alt. 2400 m, July 27,1976, coll. Kashiw (no.13434) & K. Yoshida [Kurok. & Kashiw. Lich. Rar. et Cr. Exs. 599] (FH, LE!),and Hokkaido, prov. Ishikari, Tokachidake, Hot Spring, Furano city, alt.1450 m, Aug. 7, 1980, coll. N. Shibuichi (No.6210) and coll. K. Yoshida [Kurok. & Kashiw. Lich. Rar. et Cr. Exs. 499] (COLO, FH, LE, TUR!).

OTHER SPECIMENS EXAMINED:

Canada: Plants of interior Ouebec–labrador Canada, Vicinity of Gerin Mountain, Lat. 55 ° 04′N, Long 67 ° 14′W alt.2800 ft. Aug. 1, 1955, Coll. Les A. Viereck, Mc Gill. Subarctic Research Lab. No.789 (FH!). Merri [Lich. Exs. No.128 (FH!).

LITERATURE RECORDS FOR ASIA:

China (Sato, 1952, p.174; Wei & Jiang, 1988, p.83; Wei, 1991, p.247).

Altai (Golubkova, 1978, p.105).

Japan (Sato, 1943, p.41; Yoshimura, 1974, p.120; Ikoma, 1983, p.109).

Russia (Savicz, 1922, p.104–105 for Kamchatka; Golubkova, 1978, p.105 for Kamchatka, Chukotka and Siberia; Makrei, 1990, p.56 for Siberia).

All above cited were reported as *U. hyperborea* or *G. hyperborea*. Record of this species from Mt. Changbai shan (Chen et al. 1981, p.158, collector's No.4687) was based on erroneous identification. (see *U. herrei*)

COMMENTS: In "A Monograph of the Lichen family Umbilicariaceae in the western hemisphere" Llano came to the conclusion that "The type species of *Umbilicaria* Hoffm. is *Umbilicaria exasperata* Hoffm. in Descr. Adumbr. Pl. Lich. 1 (1): 8. Pl. 2, fig 1–2. 1789; this is now recognized as *U. proboscidea* (L.) Schrad." (1950, p.105). It seems that above cited conclusion was drawn by Llano according to the data showing in Table 14a (1950, p.109). A part of the Taυle concerning *Umbilicaria exasperata* Hoffm. is quoted as follows:

Hoffmann 1788–1801	Acharius 1803, 1811	Scholander 1934	Llano 1950
Table Figure species	species	species	species
2 1–2 U. exasperata	G. proboscidea	G. proboscidea	U. proboscidea

In fact, Acharius treated *U. exasperata* Hoffm. as an infraspecific taxon *β. G. exasperata* under *G. proboscidea* (1803, p.105). According to Acharius treatment of *U. exasperata* Hoffm. the table 14a could be revised as follows:

Hoffmann 1788–1801	Acharius 1803	Scholander 1934	Llano 1950
Table Figure species	variety	species	species
2 1–2 U. exasperata	G. proboscidea β. G. exasperata	G. proboscidea	U. proboscidea

In that way, only from illustrations and description can one to judge whether the *U. exasperata* Hoffm. is conspecific with *U. proboscidea* remains a question. After a careful examination of the specimen of *U. exasperata*Hoffm. in Hoffmann's Herbarium in Moscow, theauthor of this study found that Hoffmann not only treated *G. hyperborea*(Ach.) Ach as a synonym of *U. exasperata* Hoffm. in 1825, but the specimen bearing *U. exasperata* Hoffm., preserved in Hoffmann's herbariumin Moscow (MW!), is also completely conspecific with *G. hyperborea* (Ach.) Ach. indeed. The specimen of this species in Hofffmann's herbarium is neotypified in the present study.

Multirange's.

Umbilicaria flocculosa (Wulf.) Hoffm.

Deutschl. Flora 110 (1796).

≡ *Lichen flocculosus* Wulf. in Jacquin, Collect. Bot 3:99, Tab. 1. fig. 2 (1789).

Iconotype: Jacq. collect. Bot. 3:99, Tab. 1, fig. 2 (1789).

≡*Gyrophora flocculosa* (Wulf.) Koerb. Syst. Lich. Germ. 95 (1855).

Misapplied name:

Gyrophora deusta auct.

Umbilicaria deusta auct.: Frey in Rabh. Krypt.–Flora,9:361 (1933). Llano, A monograph of the Umbilicariaceae in the western hemisphere 124 (1950). Thomson, American Arctic Lichens 1. The macrolichens 447 (1984). Wei & Jiang, Mycosystema, 1:81 (1988).

DESCRIPTION: Thallus umbilicate, monophyllous to polyphyllous, up to 7 cm diam., usually very thin and fragile, more or less curled and reflexed, margins lobed and crenated; upper surface undulating, isidiose to subsquamose, cinnamon buff to dark brown; lower surface bare, without rhizines, weakly lacunose with some vein-like ridges.

Thallus 116–218 μm thick; upper cortex consisting of two layers, the outer layer, brown, 7–14 μm thick, and the inner layer, colourless, 2–18.5μm thick; algal layer 23–69.5 μm thick; medulla 51–97.5 μm thick; lower cortex consisting of a colourless inner layer of 14–37 μm thick, and a brown outer layer of 11.5–18.5 μm thick.

Apothecia very rare, small, c. 1 mm diam., immature.

Pycnidia 218–255 × 153–209 μm.

CHEMISTRY: lecanoric and gyrophoric acids.

SPECIMENS EXAMENED:

China: Jilin, Mt. Changbai shan, South–West slope, Fusong county, weidong station, on rocks in tundra zone, on the way to Tianchi alt. 2050–2250 m, Aug. 8, 1983, coll. Wei & Chen, Nos. 6755, 6758, 6760, 6762, 6766, 6767, 6772 (HMAS–L!); south slope, Changbai county, Hongtou shan, alt. 1950 m, on rocks, Aug. 8, 1980, coll. Hu YC, No. 2240 (HMAS–L!); north slope, near Lumingfeng, alt. 2400 m, on rocks, Aug. 25, 1979, coll. Hu YS, No. 1339; near Heifengkou, alt. 2100–2400m, coll.Hu YC, Nos. 1498, 1940; Houcegou near Guanqiya, alt. 2200 m, on rocks, Sept. 3, 1979, coll. Hu YC, No. 2115; near Wenquan, alt. 1860 m, on rocks, Sept. 1981, coll. Hu YC, No. 2737; on west slope of the river near Wenquan, alt. 1850 m, on rocks, Aug. 12, 1983, coll. Wei JC & Chen JB, No. 6845, & alt. 1880 m, coll. Hu YC, No.4667(HMAS– L!). Xinjiang, Mt. Altai, near the Halas lake, alt. 1400–2500 m, on rocks, Aug. 4, 5, 1986, coll. Gao XL. Nos.1998, 2000–1, 2007, 2066, 2080, 2091, 2147 (HMAS–L!).

Kazakstan: Eastern Kazakstan, in the suburbs of Leninogorsk town, alt.1000–1250 m, on rocks, July 9–10, 1982, coll. Dombrovskaja, Nos. 7, 10 (HMAS–L!). Altai, Leninogoarxist 5 Km S, on-obraznoja crust alpiinrest Voondist 2000m, Jun. 25, 1963, coll. Heinrich Aasamaa No. 7415 with apothecia (H!).

Mongolia: Ara–Khangai, alt. 2300 m, on rocks, July 22, 1970, coll. Biazrov, No.4231 (MOSCOW!).

Japan, Hokkaido, prov. Ishikari, Kohgen Hot Spring, Daisetsu Mts. Aug. 23, 1968, coll. Y. Endo, [det. by S. Kurok. as *Lasallia asiae- orientalis* Asahina] (TNS!); Kamikawa Distr., kamikawa-cho, Mt. Daisetsu Nature Park, Mt. Hakuuntake. Along small creek in "alpine" meadow on SE slope near the hut,alt. 1950 m (orohemiarctic zone). Aug. 5, 1970, coll. Timo Koponen, No.14984 (HMAS–L!).

Russia, Chukotka: peninsula, July 14, 1971, coll. Makarova (HMAS–L!), and westernpart, on rocks, July 16, 1977, M.P. Andeev No.C24 (HMAS–L!). Far East, Kamchatka, 1908, coll. V. Savicz, No.2041; 1909, coll. V. Savicz, Nos. 6234, 6786 and alt.700 m, on rocks, July 15, 1981, coll. A. Minczin, No.1–186–d (LE!). Behring straits: C. Wright coll. Arakamtch etchene I. ex Herb. of the U. S. North pacific Exploring Expedition under commanders Ringgold and Rodgers, 1853–56 (FH!).

OTHER SPECIMENS EXAMINED:

Canada: Prov. Ontario, 14 mi E of Nipigon, Thunder Bay District, Aug. 1965, coll. Beth Denison No. 618 (E!).

LITERATURE RECORDS FOR ASIA:

China (Wei & Jiang, 1988, p. 81; Wei, 1991, p.246–7).

Japan (Yoshimura, 1974, p. 121).

Russia (Savicz, 1922, p.106 for Kamchatka; Golubkova, 1978, p.110 for Siberia, Chukotka and Far East; Makrei, 1990, p.56 as U. deusta for Siberia).

COMMENTS: This widespread Arctic–Holarctic species is similar to both the *Lasallia asiae-orientalis* and *L. sinorientalis* at first sight. The apothecia are very rare in *U. flocculosa* and *L. asiae-orientalis* (Wei, 1982) and have never been found in L. sinorientalis so far. This species differs from the latter two species mainly by that the lower surface smooth, never granulate, and there is not corresponding hat-shaped convex on the upper surface of the thallus though the lower surface is irregularly and weakly lacunose.

Multirange's.

Umbilicaria proboscidea auct.

Notes: The name *Umbilicaria deusta* (L.)Baumg. [Fl.Lips.571 (1790)] may be kept on the basis of the typification [Art. 55 (2)]. This name is one of the misapplied names of interlocking mistakes in nomenclature. Therefore, it is easy to cause confusion on the names. So, it may be ruled as rejected (Art.69) and replaced by a new epithet as follows (Art.72.1.; see *U. cylindrica* auct):

Lichen deustus L., Sp. Pl. 1150 (1753).

Lectotype: Linn 1273–206: 54, Fl. Suec. 970 (LINN).

TYPE OBSERVATION: Thallus umbilicate, orbicular, c.22 mm diam., monophyllous, membranaceous, rigid, thin and fragile, over umbo elevated, with a reticulate of sharp edged rugi arranged circularly around umbo and becoming veriform ridges to margins.

Apothecia regularly gyrose.

MISAPPLIED NAME:

Lichen proboscideus auct.

Umbilicaria proboscidea auct. non (L.) Schrad.: Frey in Rabh., Krypt.–Flora, 9 (4 / 1) :336 (1933). Llano, A monograph of the Umbilicariaceae inthe western hemisphere 157(1950). Thomson, AmericanArcticLichens1. The Macrolichens 456 (1984). Wei & Jiang, Mycosystema, 1:86 (1988).

Umbilicaria neoproboscidea Wei, nom. nov.

≡*Lichen deustus* L., Sp. Pl. 1150 (1753).

Lectotype: Linn 1273–206: 54, Fl. Succ. 970 (LINN).

DISCRIPTION: Thallus monophyllous, membranous, rigid and fragile, orbicular, crenated at margins, usually 4–5 cm and up to 13 cm diam.; upper surface dark red-brown to black-brown, covered with white pruina, which is more thick in place over umbo showing grey-white colours with a reticulating pattern of sharp edged rugi, which can be sometimes arranged circularly around umbo; occasionally a few rhizoidal spines project from upper surface of thallus; lower surface smooth to areolate; umbo small, c.4 mm diam., nut-brown around umbo and dark to black-brown near periphery, covered with a thick layer of white pruina showing grey to dark grey colours, usually without rhizines, occasionally with sparse simple to branched rhizines.

Thallus 34.5–348 μm thick; necral zone discontinuous, 4.5–23 μm thick; upper cortex consisting of a brown outer layer in 7–14 μm thick with larger cell cavity, and a colourless inner layer in 5.5–23 μm thick with smaller cell cavity; algal layer 14–46.5 μm thick; medulla 44–232 μm, consisting of both the loose upper layer and a compact mucilaginous lower layer; the latter not clearly differentiated from the lower cortex; lower cortex, 32.5–55.5 μm, the outer layer brown.

Apothecia typically gyrose, c.2 mm diam., with plane to strongly convex or globose discs, sparsely scattered near the peripheral zone; epihymenium brown, 7–18 μm thick; hymenium 45–64 μm thick; subhymenium yellowish-brown, 20–32 μm thick; paraphyses septate, branched, c.1 μm thick, weakly expanded at tip, c.2 μm thick; asci when immature 27–37 × 8–10 μm.

Pycnidia 176–213 × 120–199 μm; conidia bacilliform, 3.5 × < 1 μm.

CHEMISTRY: lecanoric, gyrophoric and umbilicaric acids.

SPECIMENS EXAMINED:

Mongolia: Khangai: alt.2400 m, Jun. 19, 1971, coll Biazrov No.4234, and Aug. 28, 1971, coll. Biazrov No.5757, and alt. 2928 m, July 9, 1976, coll. Biazrov No.5779 (MOSCOW!). Near Kosogol Lake, Jun. 1902, coll. A.A. Elenkin (LE!).

Japan: Honshu, prov. Shinano: Mt. Kimpu, Minami–saku–gun, on rocks; elevation 2550 m, July 25, 1972, coll. H. Shibuichi (4668), and K. Yoshida [Kurok. Lich. Rar. et Cr. Exs. 196] (COLO, FH, TENN, TUR, UPS!). Honshu, prov. Musashi, Mt. Ohyama, near Jumonji Pass, Chichibu–gun, on rocks, coll. K. Yashida (1619) [Kurol. Lich. R. et C. Exs. 542] (COLO, FH!). Honshu, prov. Shimotsuke, Mt. Shirane, Nikko. on rocks, alt. 2570 m, Aug. 4, 1981, coll. H. Shibuchi (6513) [Kurok. & H. Kashiw. Lich. R. & C. Exs. No.697] (COLO, FH!). Hokkaido, Ishikari, [Kurok, Lich. Rar. et Cr. Exs. 250] (COLO, FH, H, TUR!). Hokkaido, Kamikawa Distr.: Mt. DaisetsuNature Park,1–2.5 km S of Mt. Hakuuntake, Takanegahara. Windswept mountain ridgewith scattered Pinus pumila shrubs, alt. 1750–1800 m, (orohemiarctic zone) Aug. 6, 1970, coll. Timo Koponen, No.15157 (HMAS–L!); Mt. Daiseteu Nature Park, 1 km S of Mt. Hakuuntake, Windswept mountain ridge with polygon soil, alt. 1900 m, (orohemiarctic zone) Aug. 6, 1970, coll. Timo Koponen, No.15041 (HMAS–L!). Hokkaido, prov. Ishikari, [Lich. Jap. Exs. 50] (COLO, TENN, TUR!).

Russia: Chukotka, western Chukotka, upper reaches of the Anu Minór river, Sept. 3, 1978, coll. M.P. Andreev (LE, HMAS–L!), and Jun. 28, 1977, coll. M.P. Andreev as. U. arctica (HMAS–L, 3871!). Cheliabinsk region, Mt. Urma, July 8, 1959, coll. Tshetshelewa No.2 (COLO!). Polar Ural,Aug.9, 1925, coll.B.N. Gorodkov, No.63 pr.p. (LE!). Ural, Tumensk prov., coll. Kuvaev pr.p. (LE!), and N. Ural, Tobolsk, near N. Sosiva river, Jun. 25, 1927, coll. V.B. Sochava (LE!), and July 20, 1915, coll. B.N. Gorodkov (LE!). Eastern Siberia, Mt. Putoran, Aug. 3, 1984, coll.M.P. Zurbenko (LE!). Irkutsk, July 30, 1902, coll. A. A. Elenkin (LE!). Yakut,Tompo (N. 63 ° lat., E 135 ° long), July 9, 1956, coll. V. Ivanova(LE!). Far East,

Kamchatka, Aug. 1908, coll. V. Savicz No. 5852 (Herb. No.7 & 8)(LE!). North Krasnojarsk, Putoran, alt. 815 m, Aug. 10, 1969, coll. V. Kuvaev, No.167–1 (LE!).

Kirghizia: Jugum Talasskij Alatau, Sept. 1, 1955, coll. S. Victorov No. 29 (COLO!).

Bering Straits: C.Wright coll. Arakamtchetchene I ex Herb. of the U.S. North Pacific Exploring Expedition under Commanders Ringgold and Rodgers, 1853 = 56 (FH!).

OTHER SPECIMENS EXAMINED:

Russia: Savicz: [Lichenotheca Rossica Nos. 41, 42] (HMAS–L!). St. Peterburg (Leningrad), priozersk, by the Ladazsk lake, on rocks, Aug. 8, 1960, coll. Wei, Nos.1, 2, 6 (HMAS–L!). Lapland conservation "Zapavednik", alt.470m, on rocks, Aug. 17, 1966, coll. Dombrovskaja No.173 (HMAS–L!). Chibinsk, alt. 530 m, Jun. 17, 1961, coll. Dombrovskaja No.67 (HMAS–L!).

Sweden: Smaland, Barkeryd, Aug. 9, 1928, coll. G. Haglund (HMAS–L!). Norway: Oppland, Dovre, 1875, H. Falk (HMAS–L!).

Switzerland: [Lich. Helvet. Exs. 148.] *U. polymorpha* B. *deusta* Schaer. Spicil. p.88 (*L. deustus* Linn.), in M. Grimsel. (FH!).

LITERATURE RECORDS FOR ASIA:

China (Hue, 1900, p.118 & A. Zahlbr., 1930, p.138 for Yunnan).

Mongolia (Golubkova, 1981, p.133).

Japan (Sato, 1943, p.41; Ikoma, 1983, p. 109).

Himalaya (Hue, 1892, p.118 1900, p.118–119; Singh, 1964, p.178 as *G. proboscidea*; Awasthi, 1965, p.126).

Russia (Savicz, 1922, p.103 for Kamchatka; Golubkova,1978, p.116 for Siberia and Far East; Makrei, 1990, p.56 as *U. proboscidea* for Siberia).

The records of this species as *U. proboscidea* from Xizang, China, were based on erroneous identification (see *U. herrei* Frey).

COMMENTS: The specimen of Schaerer: Lichenes Helvetici Exsiccati No.148 (FH!) is the only one examined bearing the correct name corresponding to *Lichen deustus* Linn. (1753).

After a careful examination of the original specimens of this species and original illustration in protologue lectotypification for it has been done. Because the typification is able to cause the rectified name to be thrown into confusion,thereforethe rectifiedname has to berejected (Art.69) and replaced by a new epithet (Art. 72. 1.). Besides, we gave consideration to the habitualy use of the old misapplied name when we were selecting a word for the new epithet. So, the epithet "neoprocoscidea", which is in favour of name stability, has been selected.

Multirange's.

Umbilicaria vellea (L.) Ach.

Vetensk. –Akad. Nya Handl. 15:101 (1794) pr.p.

≡ *Lichen velleus* L. Sp.Pl.1150 (1753).

Type: Linn 1273–199:51, Fl. Succ. 968 (LINN! lectotype).

TYPE OBSERVATION: Three thalli are pasted on a sheet of herbarium paper in LINN. The first thallus above on the left is of 6 × 3 cm size, the second on the right is of 11 × 7 cm size, and the third (Ventral surface) below in the left is of 4 × 4.5 cm size.

Thallus monophyllous; upper surface pale green-grey to grey-brown or brown, covered with a thin layer of white pruina, sometimes perforated with clusters of rhizines extruding dorsally; lower surface black, at margins usually darkbrown to brown, with both long, cylindrical, brown rhizines and short, thick, club tipped black ones, usually at the basic parts of long and cylindrical brown rhizines inflated and blackened because of thalloconidia.

≡ *Gyrophora vellea* (L.) Ach. Meth. Lich. 1803. p.109.

DESCRIPTION: Thallus monophyllous, coriaceous, rigid, 5–12 cm diam; upper surface undulating, grey-green in all tints of mouse-grey; lower surface black, at margins brown, with numerous dark brown to black, long and cylindrical rhizines, at the base inflated and blackened, covered with black thalloconidia, among them scattered oval clubs, covered with black thalloconidia as well, sometimes with rich lamellar and scattered thin and cylindrical rhizines; umbo 7 × 5 mm diam.

Thallus 120.5–148.5 μm thick; necral zone 2–23 μm thick; upper cortex 14– 23 μm thick, including a brownish outer layer of 4.5–9 μm thick, and a colourless inner layer of 7–14 μm thick; algal layer 18.5–37 μm thick; medulla 23–65 μm thick; lower cortex 28–46 μm thick, including a colourless layer of 14–32.5 μm thick, and a brown outer layer of 9–18.5 μm thick.

Apothecia with gyrose discs, up to 2 mm diam.; epihymenium brown, 4.5–14μm thick; hymenium 37–65 μm thick; subhymenium colourless, 23–55.5 μm thick; asci, when immature, 19.5–30.5 × 4.5–7 μm; paraphyses c. 1 μm thick, at the tip inflated, c. 1.5 μm thick, septate, branched.

Pycnidia 120.5–148.5 × 65–102 μm; conidia bacilliform, 3–3.5 × 1–1.5 μm.

Thalloconidia distributed on the lower surface of thallus, on the surface of club shaped rhizines, and on inflated basic parts of cylindrical rhizines; conidia subrounded, brown, 5.5–9 × 4.5–7 μm..

CHEMISTRY: lecanoric (trace) and gyrophoric acids.

SPECIMENS EXAMINED:

China: Hebei, Mt. Xiaowutai shan, on rocks of the mountaintop, alt. 2800 m, 1964, coll. Wei, No.2096 (HMAS–L!). Jilin, Mt. Changbai shan, north slope near Xiao Tianchi, alt. 1700 m, on rocks, Aug. 12, 1977, coll. Wei, No.2908 and Chen XL No.4730–2 (HMAS–L!); Near the Ice Stadium, alt.1800 m, on rocks, Aug. 26, 1977, coll. Hu YC No.521 (HMAS–L!); south slope, Changbai county, on the top of the Hongtou shan, on rocks, alt. 1900 m, July 27, 1983, coll. Wei et al. No.6249 (HMAS–L!); Changbai county, on the way from Hengshan station to Tianchi, alt.2050 m, on rocks, July 31, 1983, coll. Wei & Chen JB, Nos.6370–(1), 6371 (HMAS–L!). Xinjiang, Mt. Altai shan, near Hanas lake, alt. 2500 m, on rocks, Aug. 5, 1986, coll. Gao XQ No.2073 (HMAS–L!). Sichuan, Kangding, on rocks in·alpine region between Kangding and Zheduo, coll. Potanin in 1893, [Elenkin: Lich. Fl. Ross. IV, 1904, No.151] (pr.p. = 151–b–1: HMAS–L!). Yunnan, prope vicum sanyingpanad septentr. urbis Yunnanfu, 26° lat., in regione calide temperata c.2400 m, March 14, 1914, coll. Handel–Mazzetti,(Diar.Nr.146) in Handel–Mazzetti, IterSinense 1914–1918. sumptibus Academiae Scientiarum Vindobonensis susceptum Nr.590 (E!). Xinjiang, Altai shan, by the Hanas lake, on rocks, alt. 2000 m, Aug. 4, 1986, coll. Gao XQ No.1957 (HMAS–L!). Xizang,

Nyalam, alt.3920 m, on rocks, Jun. 13, 1966, coll. Wei & Chen JB, No.1541 (HMAS L!).

Mongolia: Ara–Khangai, on rocks, Aug. 25, 1970, coll. Biazrov. No. 5751 as U. cirrosa (MOSCOW!); Ara–Khangai, alt. 2500 m, on rocks, Aug. 1, 1973, coll. Biazrov, No.5784 (MOSCOW!); Baian Khongor, Aug. 9, 1973, coll. Biazrov. No.5782 and Ubsunursk, Mt. Dulg, alt. 2680 m, July 9, 1976, coll. Biazrov, No. 6572 (MOSCOW!).

Altai: Jun. 16, 1903, coll. P. Kreishovas G. Spodochroa (LE!).

Russia: Western Siberia, Sverdlovsk, July 3, 1962, coll. M. Storojeva, ex herb. of Inst. of Biol. Ural Academy of Science (H!). Central Siberia, Mt. Putoran, Aug. 28, 1984, Aug. 9, 1985, coll. M. P. Isurbenko, as *U. hirsuta*; and Aug. 29, 1984, coll. M. P. Isurbenko, *U. vellea* together with *Dermatocarpon vellereum* Zschacke, misidentified as *U. hirsuta* (LE!). Eastern Siberia, M inusinsk flora, on rocks, as *G. spodochroa* (LE!). Far East, Kamchatka, 1909 as *G. spodochroa*, coll. *V. Savicz*, Nos.199, 5085 (LE, US!).

Japan: Honshu, prov. Shinano, Mt. Ohyama near Jymonji Pass, Minamisaku–gun, alt. 2200 m, July 12, 1979, coll. H. Shibuichi (No.5854) & K. Yoshida [Kurok. & Kashiw. Lich. Rar. et Cr. Exs. 600] (DUKE, FH, LE!).

Afghanistan: Prov. Kabul, Paghman–Gebirge, oberhalb des Ortes Paghman, an der Talgabelung Chap–Darrah und Rast–Darrah 68° 56′E 34° 37′N, c. 2550 m. Steile Silikatfelsen, ± N–exponiert. coll. M. Steiner [Lichenotheca Afghanica Nos.15, 29] (BM!).

Kazakstan: Altai, Leninogoskist 4 km S, alt. 900 m, Jun. 25, 1963, coll. Heinrich Aasamaa, No.7388 (Loc.5435) (H!). Alma–Atast 60 km NE, Ala–Tau, alt. 2000 m, Jun. 28, 1963, coll. Heinrich Aasamaa, No.7447 (Loc.5441) (H!). Karagandaast 200 km SE, Karkaralinski, alt.800 m, Jun. 19, 1963, coll. Heinrich Asamaa, No.7418 (Loc.5438) (H!).

Armenia: Distr. Krasnoselsk, Caucasus Minor, alt. 2500 m, ad saxa silicea, May 24, 1976, coll. Vezda [Lich. Sel. exs. 1413] (COLO, H!).

OTHER SPECIMENS EXAMINED:

Stirpes Crypt. Vogeso–Rhenane, Mougeot et Nestler 540. *G. spadochroa* (misidentification).

Austria: [Crypt. exs. ed. a Mus. Hist. Natur. Vindob. 4244] (H!).

LITERATURE RECORDS FOR ASIA:

China (Chen et al. 1981, p.158 for Jilin).

Mongolia (Golubkova, 1981, p.134 cited as *U.* cfr. *vellea*. This collection wasnot available for study; Biazrov, 1986, p. 99, as *U. cirrosa*; Huneck et al., 1992, p.304).

Japan (Sato, 1943, p.42).

Russia (Savicz, 1922, p.108 for Kamchatka; Golubkova, 1978, p.122 for Siberia and Far East; Makrei, 1990, p.55 for Siberia).

India (Jatta, 1905, p.179; Awasthi, 1965, p.126).

Sri Lanka (Leighton, 1869, p.165; Awasthi, 1965, p.126).

COMMENTS: This widespread bipolar species is easy to recognized by its numerous long and dark brown to black rhizines and always basally inflated and covered with sooty black thalloconidia, in addition there are oval clubs, covered with thalloconidia as well; upper surface neither sorediate nor eroded at margins.

Bipolar multirange's.

Acknowledgements

I am most grateful to Dr. Galloway, head of Lichen Section in the Cryptogamic Herbarium of the British Museum, for giving me a lot of help in different ways during my work in the Herbarium, and for arranging for my visit to the Linnean Society of London to examine the Linnaean collections. Special thanks are due to Dr. Jarvis (BM) who gave me very helpful comments and suggestions in the terms of the background for the Linnaean collections, without which this revision and lectotypification could not be made, and to Mrs. Douglas (LINN) for giving a lot of help during my stay in the Linnaean Herbarium.

I am particularly indebted to Prof. Pfister, director of the Gray and the Farlow Reference Library and Herbarium of the Harvard University, for his kind help during my study in the Farlow Library and Herbarium, for the important assistance I want to express also my thanks to Dr. Boise and Mr. Cacavio. My sincere indebt to Prof. Hawksworth (CMI) and Prof. Petersen (TENN) for the helpful discussions on the nomenclatural problems, and to Dr. Hale for the important discussions on the typification of both *Parmotrema perforatum* (Jacq.) Hale and *Lichen cylindricus* L. during my stay in US in the early April 1990.

This work was a part of the exchange programme between the Academia Sinica and the British Royal Society. My visit to the United States was supported by Hesler Foundation in University of Tennessee, by Smithsonian Institution, and in part by the Farlow Reference Library and Herbarium of the Harvard University, respectively. Special thanks are due to Dr. Topeisheva (MW) for giving a lot of help during my stay in MW to study the Hoffmann's herbarium collections and to Dr Hu (GH) for improving the English language.

References

Dillenius, J. J. 1741. Historia Muscorum.

Duby, J. E. 1830. Botanicon Gallicum. Paris

Hale, M. E. 1965. A monograph of Parmelia subgenus Amphigymnia in Contribution from the United States National Herbarium, vol. 36, part 5.

Hoffmann, G. Fr. (1825) Herbarium vivum, sive Collectio Plantarum Siccarum, Caesareae Universitatis Mosquensis, Mosquae.

Linnaeus, C. 1751. Amoenitates Academicae.

———, –. 1753 (1st ed.). Species Plantarum.

———, –. 1764 (3rd ed.). Species Plantarum.

Savage, S. (compiled and annotated). A cataloque of the Linnaean Herbarium.

第七章　地衣化学
Lichen Chemistry

Chemical revision of *Hypogymnia hengduanensis*

JIANG-CHUN WEI

Systematic Mycology & Lichenology Laboratory, Academia Sinica, Zhong-guan-cun, Bei-yi-tiao 13, Beijing 100080, China

WANG-FU BI

Experimental Center of Forest Biology, College of Biology, Beijing Forestry University, Beijing 100083, China

Abstract. *A new lichen species with isidia* Hypogymnia hengduanensis, *containing barbatic acid and atranorin, was described by Wei in 1984. This chemical revision, using the combined methods of microcrystal test (MCT), thin-layer chromatography (TLC), and high-performance liquid chromatography (HPLC) shows that diffractaic acid is the major compound in addition to barbatic acid and atranorin in* H. hengduanensis.

Lichen substances, in combination with morphology and distribution, are of great importance in the classification of lichens, especially of lichen genus *Hypogymnia* (Nyl.) Nyl. A new species with isidia, *H. hengduanensis* Wei, containing barbatic acid and atranorin was described by Wei (Wei 1984, 1986). During the study of the lichen flora of China, extensive collections of the genus *Hypogymnia* with isidia were examined. Among them, a chemical revision of *H. hengduanensis* shows that diffractaic acid is the major compound, in addition to barbatic acid and atranorin. As a consequence, the type collections of *H. hengduanensis* were reexamined by standardized thin-layer chromatography (TLC) and microcrystal test (MCT), in combination with high-performance liquid chromatography (HPLC).

MATERIALS AND METHODS

Type collection of *H. hengduanensis* Wei ssp. *hengduanensis* were examined: Sichuan Province, Kangding district, Yulinkong, Gomba La, on the tree bark of *Betula* sp., elev. 3,700 m, October 19, 1934, collected by *H. Smith 14078* (isotype, HMAS-L). Yunnan Province, Lijiang district, Heibaishui tree farm, on *Rhododendron* sp., elev. 3,140 m, November 10, 1980, collected by *Y. M. Jiang 131* (paratype, HMAS-L). Also, the type collection of *H. hengduanensis* ssp. *kangdingensis* Wei was examined: Sichuan Province, Kangding district, Gomba La, on the tree bark of *Rhododendron* sp., elev. 3,700 m, July 22, 1934, *H. Smith 14061* (isotype, HMAS-L).

These specimens were analyzed by TLC using the three-solvent-system method standardized for lichen products (Culberson 1974; White & James 1985) except that methyl *tert.*-butyl ether was used in place of diethyl ether for solvent B (Culberson & Johnson 1982) and MCT (Asahina 1936, 1937). The isotype specimen of *H. hengduanensis* ssp. *hengduanensis* was analyzed by high-performance liquid chromatography in addition to MCT and TLC.

High-performance Liquid Chromatography.—Thallus fragments were extracted first with benzene at room temperature and then four times with warm (38°C) acetone. A portion of the residue was redissolved in 3–5 drops of methanol. Small volumes (2–4 μl) were chromatographed on a precolumn (50 × 4.6 mm I.D.) and an analytical column (250 × 4.6 mm I.D.) filled with YWGCH (C_{18}, 10 μm) [Tianjin Second Factory of Chemical Reagents, China] with CH_3OH: THF: 0.5% H_3PO_4 mobile phase of 22.5:22.5:55, v:v:v, at 1 ml/min. at 1,700 psi. Representative retention times for the following different compounds were: diffractaic acid, 2.80 min. atranorin, 3.73 min. and barbatic acid, 4.06 min. Identification of atranorin, diffractaic acid, and barbatic acid was made by comparison with authentic samples for both the TLC and HPLC analyses given kindly by S. Huneck.

RESULTS AND DISCUSSION

The MCT gave abundant crystals of diffractaic acid in GE. By TLC, solvent B showed a strong spot of diffractaic acid (R_f class 6; $R_f \times 100 = 58$) and a trace of barbatic acid (R_f class 6; $R_f \times 100 = 69$), which separated just above. Diffractaic and barbatic acids ran together in solvents A (R_f class 4) and C (R_f class 6). A trace of atranorin was detected in all three solvents. It is clear that the mixture of barbatic and diffractaic acids was mistaken by the senior author for barbatic acid in the type description of *H. hengduanensis,* using solvent C only and without authentic samples for comparison.

The HPLC analysis (Fig. 1) agreed with the TLC result: Peak D is the major compound diffractaic acid, and the trace peaks A and B are atranorin and barbatic acid, respectively. By MCT, abundant crystals of the major substance, diffractaic acid, were confirmed in GE, but the low concentrations of atranorin and barbatic acid were not detected. Thus, *H. hengduanensis* contains diffractaic acid as the major compound accompanied by trace amounts of barbatic acid and atranorin.

Diffractaic acid has been reported from some species of *Hypogymnia* as an accessory compound, often in low concentration, accompanying larger amounts of the β-orcinol depsidone physodalic

This paper was originally published in *The Bryologist*, 101（4）: 556-557, 1998.

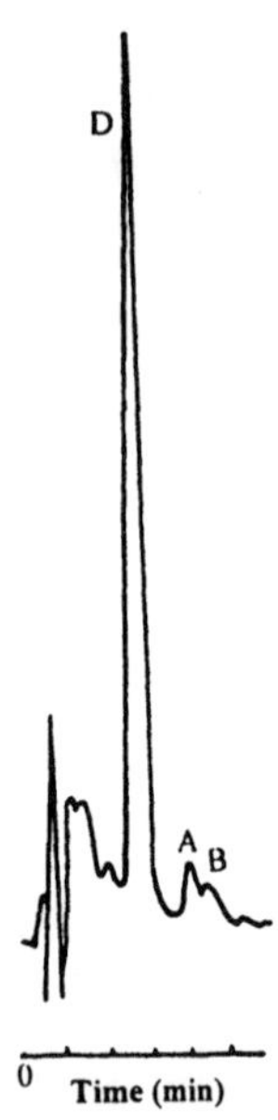

FIGURE 1. HPLC analyses of acetone extracts. The compounds are diffractaic acid (D, 2.80 min.), atranorin (A, 3.73 min.), and barbatic acid (B, 4.06 min.).

acid, which is itself usually accompanied by traces of its probable precursor protocetraric acid. Most of these species of *Hypogymnia* also contain the less closely related orcinol depsidone physodic acid. Diffractaic acid was reported as an accessory substance with atranorin and physodalic acid in *H. duplicata* (Goward et al. 1994; Pike & Hale 1982) and with atranorin, physodic, physodalic, and protocetraric acids in *H. enteromorpha* (Goward et al. 1994; Hale & Cole 1988; Pike & Hale 1982). The more variable chemistry of *H. imshaugii,* with chemotypes having or lacking physodic acid and physodalic acids (again with traces of protocetraric usually detected), also have diffractaic acid as an accessory substance (Goward et al. 1994; Hale & Cole 1988; McCune 1982). The original report of barbatic acid in this species was a misidentification based on crystal tests (Krog 1968; Pike & Hale 1982).

As in species of other genera that have high concentrations of diffractaic acid, this β-orcinol-type *p*-depside, is accompanied by a trace of barbatic acid, a possible precursor differing from it only by lacking 2-O-methylation. The fact that barbatic acid was not detected in other species of *Hypogymnia,* which contain diffractaic acid in much lower concentrations, is doubtless due to the inability of TLC to detect it in such low amounts. *Hypogymnia hengduanensis* is the first species in the genus known to contain diffractaic acid as the major medullary product.

ACKNOWLEDGMENTS

The senior author wishes to thank S. Huneck for authentic samples of lichen substances. Extremely helpful was the technical assistance from Y. M. Jiang and S. Y. Guo, for which we are grateful. Heartfelt thanks are due to Chicita Culberson for her furnished valuable editorial help. This research was supported in part by the National Natural Science Foundation of China (NNSFC39670002).

LITERATURE CITED

ASAHINA, Y. 1936. Mikrochemischer Nachweis der Flechtenstoffe I, II. Journal Japanese Botany 12: 516–525, 859–872.

———. 1937. Mikrochemischer Nachweis der Flechtenstoffe III, IV. Journal Japanese Botany 13: 529–536, 855–861.

CULBERSON, C. F. 1974. Conditions for the use of Merck silica gel 60 F254 plates in the standardized thin-layer chromatographic lichen products. Journal of Chromatography 97: 107–108.

——— & A. JOHNSON. 1982. Substitution of methyl tert-butyl ether for diethyl ether in the standardized thin-layer chromatographic method for lichen products. Journal of Chromatography 238: 483–487.

GOWARD, T., B. MCCUNE & D. MEIDINGER. 1994. The Lichens of British Columbia, Illustrated Keys. Part 1. Crown Publications, Victoria. Foliose and Squamulose Species. British Columbia Ministry of Forests.

HALE, M. E., JR. & M. COLE. 1988. Lichens of California. University of California Press, Berkeley.

KROG, H. 1968. The macrolichens of Alaska. Norsk Polarinstitutt Skrifter 144: 1–180.

MCCUNE, B. 1982. Lichens of the Swan Valley, Montana. THE BRYOLOGIST 85: 13–21.

PIKE, L. H. & M. E. HALE, JR. 1982. Three new species of *Hypogymnia* from western North America (Lichenes: Hypogymniaceae). Mycotaxon 16: 157–161.

WEI, J. C. 1984. A new isidiate species of *Hypogymnia* in China. Acta Mycologica Sinica 3: 214–216.

———. 1986. Notes on isidiate species of *Hypogymnia* in Asia. Acta Mycologica Sinica Supplement 1: 379–385.

WHITE, F. J. & P. W. JAMES. 1985. A new guide to microchemical techniques for the identification of lichen substances. British Lichen Society Bulletin (Supplement) 57: 1–47.

ms. submitted Oct. 21, 1997; accepted May 21, 1998.

第三篇　荒漠地衣生理生态学

Physiology & Ecology of Desert Lichens

沙漠生物地毯工程——干旱沙漠治理的新途径

魏江春

(中国科学院微生物研究所 真菌地衣系统学重点实验室, 北京 100080)

我国是世界上荒漠化最严重的国家之一, 荒漠化土地面积达 2.64×10^6 km^2, 占国土面积的 27.5%, 其中尤以沙漠危害最为严重, 其面积已达 8.09×10^5 km^2, 而且, 还继续扩大蔓延, 平均每年有 610 km^2 左右的沙漠出现活化, 其中有 310 km^2 土地沦为沙地。沙漠化不仅造成生态系统失衡, 而且使可耕地面积不断缩小, 对我国工农业生产和人民生活带来严重影响。我国西北干旱区沙漠和沙漠化土地, 已成为中国乃至亚太地区沙尘暴主要源地之一, 给国家社会经济造成了巨大的损失。因此, 沙漠治理是国家在生态建设和环境保护方面的迫切需求。

长期以来, 植树造林和种草是沙漠治理的主要途径, 在实践中也获得了很大效果。然而, 并非任何地区的沙漠都可以通过这一途径进行治理。在自然界, 年降水量在 600~ 800 mm 及以上者通常为森林地带, 在 400~ 600 mm 者为草原地带, 在 200~ 400 mm 者为干旱草原地带, 在 200 mm 以下者为荒漠地带。由此可见, 水分承载力是不同生态条件下植被类型存在的关键。有人研究得知, 荒漠地带由维管束植物组成的植被盖度 ≤30% (Townshend 等, 1986), 而微型生物结皮的盖度 ≥70% (Belnap 等, 1994)。因此, 在干旱荒漠地区微型生物结皮占绝对优势, 是荒漠地区天然植被中特有的生态景观。

所谓微型生物是由一些微小而耐旱的地衣、藻类和苔藓植物所组成。这些微型生物体在沙土表面形成一层厚厚的, 地毯式壳状物, 即所谓结皮(crust)。这些结皮生物通过其假根、菌丝和藻丝及其分泌物将沙土微粒缠结成团, 从而起到固定沙尘的作用, 这就是微型生物结皮(microbiotic crust)。

年降水量为 200 mm 以下的水分承载力在荒漠地区便足以支撑结皮的微型生物占优势, 并间有稀疏维管束植物的存在, 却无力支撑由维管束植物组成的优势植被。因此, 在 20 世纪 50 年代为了保护京兰铁路采取种植耐旱灌丛植物治理沙漠期间, 曾发现大片枯死的灌丛植物被大片微型生物结皮所替代的景象。从植树造林治沙角度来看, 这些微型生物结皮被认为是大片灌丛植物枯死的根源。因为微型生物结皮的结构相当致密, 在被降水饱和之后, 随后的降水便沿结皮表面泾流而损失, 起到了截流降水下渗的作用, 从而造成水分流失与蒸发, 使沙土深层水分缺乏, 导致灌丛植物根系缺水而枯死。因此, 清除截流地表降水下渗的微型生物结皮, 从而使灌丛植物免遭枯死的思路虽有一定道理。然而, 如果从水分承载力及固沙尘角度来看, 占优势地位的微型生物结皮正是干旱荒漠地区自然形成的特有生态景观。因此, 在干旱荒漠中, 微型生物结皮的存在, 不仅是自然规律的一种表现, 而且起着明显的固定沙尘作用。相反, 在年降雨量 200 mm 以下的干旱沙漠地区, 通过植树造林等传统方式进行沙漠治理是不现实的, 既违反自然规律, 也难以达到治沙目标, 除非配置引水灌溉系统, 这对于水资源奇缺的干旱荒漠地区来说是不可能的。顺应自然规律, 在干旱沙漠通过人工培植微型生物结皮以治理沙漠的设想, 是值得研究的重要课题。至于微型生物结皮可能具有的截流降水下渗的作用, 可能因参与结皮的微型生物种类不同而有别。研究发现, 凡是以 *Endocarpon pusillum* 和

本文原载于《干旱区研究》, 22 (3) : 287-288, 2005.

*Collema tenax*为优势的结皮(图1，2)，可能具有一定程度的截流降水下渗作用，而以*Psora decipens*和*Toninia* sp.为优势的结皮(图3)，则往往使沙漠表面呈现网状裂隙，有利于降水下渗。因此，在进行沙漠“生物地毯工程”的基础研究中，有可能通过相关物种的组合予以解决。

所谓沙漠“生物地毯工程”，是以干旱荒漠地区所特有的、自然形成的地毯式微型生物结皮为“模版”，通过现代生物技术途径予以“复制”，为沙漠铺上微型生物结皮式的“地毯”，即所谓利用微型生物结皮治沙的“生物地毯工程”。这一系统工程蓝图是符合自然规律的干旱沙漠治理新途径。由于微型生物结皮的形成、生长和发育及其固氮作用的结果，促进了沙漠表面有机物质的富积，为短命草本植物和微型土壤生物的繁殖创造了条件，从而导致微型生物结皮为主的沙漠生态建设及其生物多样性演替进入良性循环过程。

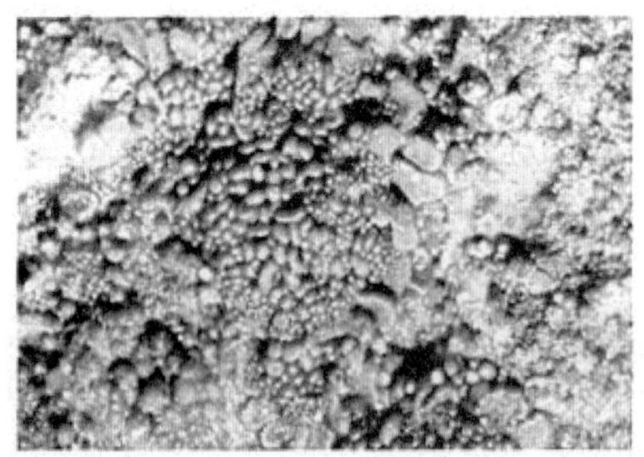

图1 宁夏中卫沙坡头治沙站以 *Endocarpon pusillum* 和 *Collema tenax* 为优势的结皮(近距照片)

图2 宁夏中卫沙坡头治沙站以 *Endocarpon pusillum* 和 *Collema tenax* 为优势的结皮(远距照片)

图3 新疆阜康北沙窝以 *Psora decipens* 和 *Toninia* sp.为优势的结皮呈现网状裂隙(近距照片)

Survival analyses of symbionts isolated from *Endocarpon pusillum* Hedwig to desiccation and starvation stress

ZHANG Tao[1,2] & WEI JiangChun[1*]

[1]*Key Laboratory of Systematic Mycology & Lichenology, Institute of Microbiology, Chinese Academy of Sciences, Beijing 100101, China;*
[2]*Graduate University of Chinese Academy of Sciences, Beijing 100049, China*

Received March 20, 2009; accepted June 28, 2009

This work deals with the survival analyses of the symbionts isolated from the lichen *E. pusillum* under desiccation and starvation stress. The mycobiont of the symbionts was under the desiccation in combination with starvation stress, and under starvation stress alone as well. The phycobiont of the symbionts was under desiccation stress alone. The experiments were detected by means of the biomass size, weight and cell density, deformity of the hyphae and cells, and metabolic activity through SEM (scanning electron microscopy), TEM (transmission electron microscopy), FM (fluorescence microscopy), spectrophotometry, and FCM (flow cytometry). The results show that the mycobiont can survive for seven months under desiccation stress in combination with starvation stress, and for eight months under starvation stress alone. The phycobiont can survive for two months under desiccation stress. It can provide a scientific basis for further research of the reproduction biology of lichens and arid desert biocarpet engineering to fix sand and carbon.

lichens, mycobiont, phycobiont, SEM, TEM, FM, FCM, reproduction, biocarpet

Citation: Zhang T, Wei J C. Survival analyses of symbionts isolated from *Endocarpon pusillum* Hedwig to desiccation and starvation stress. Sci China Life Sci, 2011, 54: 480–489, doi: 10.1007/s11427-011-4164-z

One of the central challenges of adversity-resistance biology is to understand the responses and adaptations of desert species to arid conditions. A lichen is a life-support system composed of a specific fungus (mycobiont) and a corresponding alga (phycobiont) or a cyanobacterium (cyanobiont) in an unique symbiotic relationship. "Their unique symbiotic relationship with algae has enabled these fungi to colonize and flourish in a wide range of habitats from the Antarctic continent to the rain forests of the tropics. Their ecological success in so many types of habitats depends on a number of unique structural and functional adaptations that are only now beginning to be generally appreciated by mycologists and microbiologists" [1].

Lichens are some of the most drought-resistant organisms; this characteristic has been extensively studied. In particular, levels of resistance to desiccation have been assessed using photosynthesis or respiration as indicators. Most of lichens are able to survive desiccation for several months [2–4], and a few species can survive for more than a year [4,5]. In reviewing previous studies, Lange [4] concluded that there was a general correlation between water conditions within the habitat and desiccation tolerance, i.e., the drier the habitat, the greater the desiccation tolerance of the lichen.

Although the length of the desiccation period that lichens can survive is well documented, this information is still limited for isolated symbionts. Ahmadjian and Hale [6] reviewed the previous studies of desiccation resistance of isolated symbionts, and indicated that the isolated mycobionts could survive desiccation for five weeks (three species) or six weeks (seven species), while the isolated phycobionts could survive for six weeks (13 species) or 41 weeks (two

*Corresponding author (email: weijc2004@126.com)

This paper was originally published in *Science China Life Sciences*, 54（5）: 480-489, 2011.

species).

With respect to the symbiotic relationship in lichens, the mycobionts must obtain nutrients provided by the specific phycobionts and then establish the true symbiosis. Schaper and Ott [7] found that the establishment of initial contact between mycobiont and phycobiont (e.g., formation of haustoria-like structures with hyphae branching and enveloping the phycobiont surface) appeared to take about six weeks. Mycobionts, therefore, can probably survive starvation for some time before obtaining nutrients from phycobionts. However, little is known about the limits of survival of the isolated mycobionts subjected to starvation stress.

Some non-lichenized fungi and algae are also known to be desiccation-tolerant. Microcolonial rock fungi (MCF), for example, are known to survive desiccation for 32 [8] or 56 d [9]. The alga, *Chlorella vulgaris* Beijerinck, remained viable in desiccated state for about 30 d [10], and five algae, *Microcoleus chthonoplastes* Thuret, *Phormidium bohneri* Schmidle, *Rhizoclonium crassipellitum* West & G. S. West, *Lyngbya mesotricha* Skuja and *Scytonema millei* Bornet, were observed to survive from 0.5 to 2 months [11].

Desert lichens are ecologically significant because they weaken windstorm by fixing sand and dust, and weaken global warming by fixing carbon dioxide from the atmosphere through photosynthesis. *E. pusillum* is a dominant species that is widespread in the arid desert regions of northern China. We study the stress tolerance of symbionts isolated from *E. pusillum*, which is a previous research of the arid desert biocarpet engineering to improve environments and the screening of drought-resistant genes from lichen symbionts.

1 Materials and methods

1.1 Isolation and identification of the symbionts

The lichen *E. pusillum* was collected from Shapotou Desert Experimental Research Station (37°32′N, 105°02′E; alt. 1340 m; mean annual precipitation in this area is 186 mm, and annual potential evaporation is 2800 mm [12]). Symbionts within *E. pusillum* (Figure 1A) were isolated by spore discharge [13] from the perithecia of the same lichen thalli. The ascospores of the mycobiont were synchronously discharged together with the hymenial algal cells onto the surface of 1.5% water agar in a Petri dish. When the ascospores germinated, the germinated spores and small algal colonies (Figure 1C) were transferred to tubes containing potato dextrose agar (PDA). A colony was formed from the germinated ascospores and the hymenial algal cells (Figure 1D). A small piece of the mycelia and some of the alga were then transferred to the tubes containing PDA and Bold's Basal Medium (BBM) for cultivation, respectively.

Analyses of morphology and molecular systematics were used to confirm that the isolated fungi and algae were sym-

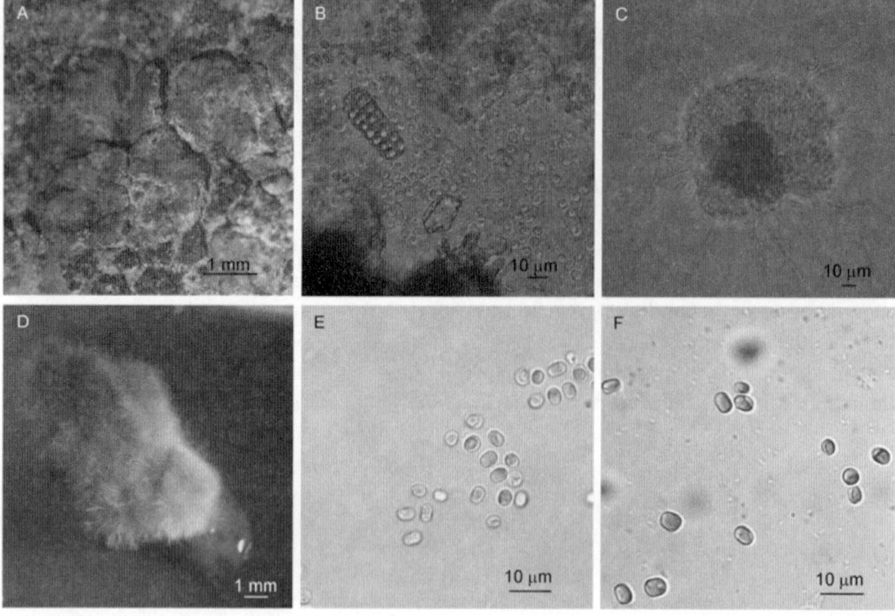

Figure 1 The lichen *E. pusillum* and its symbionts. A, The habit of *E. pusillum*. B, Ascospores with hymenial algal cells within a perithecium. C, Germinating ascospores with numerous algal cells ejected from a perithecium. D, A colony formed from the discharged ascospores and the hymenial algal cells for two months after being transferred. E, Algal cells from *E. pusillum* in the liquid BBM. F, Algal cells of *D. chodatii* (UTEX No. 1177) in liquid BBM.

bionts within lichen *E. pusillum* rather than contamination.

Total DNA was extracted using a modified CTAB method [14] from samples as follows: (i) Mycobiont: the mycobiont isolated from *E. pusillum*, *E. pusillum* and out-group species *E. crystallinum* J. C. Wei & J. Yang. Sequence data for other out-group lichen species were downloaded from GenBank (Table 1); (ii) Phycobiont: the phycobiont isolated from *E. pusillum* and *D. chodatii* (Table 2). Sequence data for out-group algal species were down-loaded from GenBank.

The ITS region (including ITS1, 5.8S rDNA and ITS2) from lichen and algal species was analyzed. Primers ITS1 and ITS4 [15] were used for PCR amplification of the mycobiont. Primers nr-SSU-1780-5′-algal and nr-LSU-0012-3′-algal [16] were used for PCR amplification of the phycobiont.

PCR products were purified and sequenced by Genewiz Inc. (Beijing, China). Phylogenetic trees were constructed using neighbor-joining method [17].

Voucher specimens of *E. pusillum* and *E. crystallinum* were deposited in Herbarium Mycologicum Academiae Sinicae-Lichenes (HMAS-L). The alga *D. chodatii* obtained from the Culture Collection of Algae at the University of Texas at Austin (UTEX) was deposited in the Laboratory of Lichen Biology at the Institute of Microbiology, Chinese Academy of Sciences.

1.2 Experimental design for survival analyses of symbionts

Colonies of the isolated mycobiont were transferred into liquid PDA. After two months, the liquid suspension was passed through a sterilized stainless steel filter with a 500-μm mesh, and mycelial pellets were collected. These pellets (diameter ≈ 1 mm) were transferred directly onto the plastic bases of Petri dishes (35 mm in diameter) without nutrients and water, and placed in desiccators over fused silica gel (relative humidity⩽10%, room temperature 20–27°C, and room light intensity). Samples were thus subjected to the stress conditions in combination with desiccation and starvation. Other mycelial pellets of the isolated mycobiont were transferred into 50-mL Erlenmeyer flasks containing sterile distilled water and placed in the laboratory (room temperature 20–27°C and room light intensity) to subject the samples to starvation stress alone.

Samples were examined and measured at regular time, using colony size and weight measurement (CSM & CWM; three replications), scanning electron microscopy (SEM, Quanta 200, FEI Co.), transmission electron microscopy (TEM, JEM-1400, JEOL Ltd.) and florescence microscopy (FM).

FM was used to determine whether the mycobiont cells were metabolically active. In metabolically active cells, fluorescein diacetate (FDA) is converted into green fluorescent fluorescein. The hyphae of the isolated mycobiont were suspended in a working solution of FDA (Merck, CB343209; stock solutions: 1 mg FDA in 1 mL acetone; working solutions: 1 μL stock FDA in 4 mL deionized water), and incubated for 30 min at room temperature. The green fluorescence of the mycobiont cells was examined using a fluorescence microscope (Axio imager A1, ZEISS Co.) with an excitation filter of 450–490 nm and a barrier filter of 520 nm.

Colonies of the isolated phycobiont were transferred into liquid BBM. After one month, 1 mL of phycobiont cells suspension ($\approx 6.5\times10^6$ cells mL^{-1}) was transferred by a sterilized transfer pipette onto the plastic bases of Petri dishes (35 mm in diameter) without medium and water, and they were placed in desiccators over fused silica gel (relative humidity⩽10%, room temperature 20–27°C, and room

Table 1 Lichen species examined and sequenced (in bold text) and downloaded from GenBank

Species	Source	GenBank accession number
Mycobiont isolated from *E. pusillum*	**Shapotou**	**HM237333**
E. pusillum	**Shapotou, HMAS-L**	**HM237334**
E. crystallinum	**Shapotou, HMAS-L**	**HM237332**
E. pallidulum (Nyl.) Nyl.	GenBank	DQ826735
Verrucaria viridula (Schrad.) Ach.	GenBank	DQ553510

Table 2 Algal species examined and sequenced (in bold text) and downloaded from GenBank

Species	Source	GenBank accession number
Phycobiont isolated from *E. pusillum*	**Shapotou**	**HM237336**
Phycobiont in *E. pusillum*	**Shapotou**	**HM237335**
D. chodatii	**University of Texas (UTEX No. 1177)**	**HQ129931**
Stichococcus bacillaris Nägeli	GenBank	AJ431678
S. mirabilis Lagerheim	GenBank	AJ431679
C. vulgaris	GenBank	FM205855

light intensity) to subject the samples to desiccation stress.

Samples were examined and measured at regular intervals by measurement of naked eyes (MNE) and cell density measurement (CDM) using a spectrophotometer (UV-2800 UV/Vis, Unico Shanghai Instruments Co.; three replications). SEM, TEM and flow cytometry (FCM) were also used to examine the samples.

FCM was used to determine whether the phycobiont cells were metabolically active when subjected to desiccation. For FDA staining of the isolated phycobiont, 1 μL of the FDA stock solutions (1 mg mL^{-1} in acetone) was mixed with 1 mL of phycobiont cells suspension and then incubated for 30 min at room temperature. The green fluorescence of phycobiont cells was examined using a flow cytometer (FCM, BD FACSAria, Becton Dickinson Co.). Excitation at 488 nm was induced by an argon laser and the emission was confined by a 515–545 nm interference filter [18]. BD FACSDiVa software (Becton Dickinson Co.) was used for instrument control and sample analysis.

1.3 Statistical analysis

Data were analyzed using one-way analysis of variance (ANOVA). Significant differences were determined by Duncan's multiple range test (P<0.05). All statistical analyses were performed using SPSS 15.0 statistic software (SPSS Inc., USA).

2 Results and discussion

2.1 Identification of the symbionts

Ascospores of the mycobiont isolated from *E. pusillum* were morphologically identical to those within a perithecium of *E. pusillum* (Figure 1B).

The phylogenetic tree (Figure 2) shows that the mycobionts isolated from *E. pusillum* (HM237333) and *E. pusillum* (HM237334) form a monophyletic group with high bootstrap support (100%). A large genetic distance was found between this clade and the outgroup, including *E. pallidulum* (DQ826735; bootstrap value=58%), *E. crystallinum* (HM237332; bootstrap value<50%) and *Verrucaria viridula* (EU553510; bootstrap value<50%). This indicates the isolated mycobiont is the true mycobiont of *E. pusillum*.

The phylogenetic tree (Figure 3) shows that the phycobiont isolated from the *E. pusillum* (HM237336), the phycobiont in the *E. pusillum* (HM237335) and *D. chodatii* (HQ129931) form a monophyletic group with high bootstrap support (100%). A large genetic distance was found between this clade and the outgroup, including *S. bacillaris* (AJ431678; bootstrap value=54%), *S. mirabilis* (AJ431679; bootstrap value<50%) and *C. vulgaris* (FM205855; bootstrap value<50%). This indicates the isolated phycobiont is the true phycobiont of *E. pusillum*, and the same as *Diplosphaera chodatii* Bialosuknia [19] (≡*Stichococcus chodatii* (Bialosuknia) Heering [20]=*Stichococcus diplosphaera* Chodat [21]), and belongs to Chaetophoraceae (Chlorophyta).

2.2 Survival analyses of the mycobiont

2.2.1 Colony size and weight measurement of the mycobiont

The growth capability of the isolated mycobiont was determined by CSM (Figures 4 and 5, Table 3) and CWM (Figure 6).

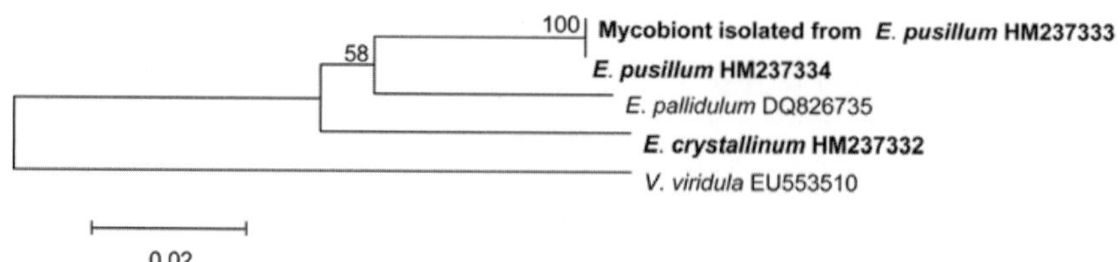

Figure 2 NJ tree based on rDNA ITS sequences of mycobionts. Bootstraps values greater than 50% are indicated. The scale bar represents the Kimura-2-parameter genetic distance. Samples in bold were examined by the authors.

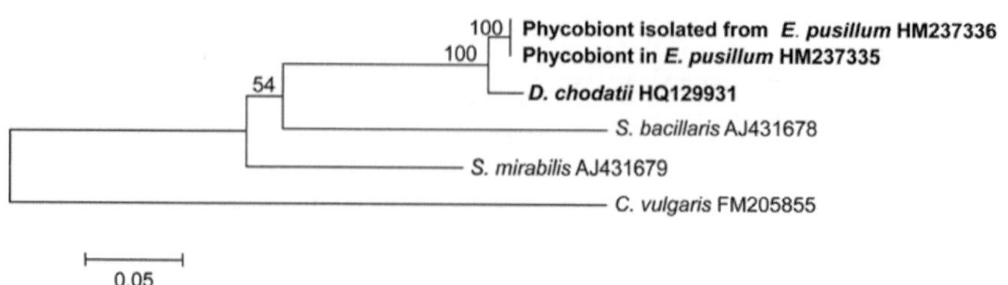

Figure 3 NJ tree based on rDNA ITS sequences of phycobionts. Bootstraps values greater than 50% are indicated. The scale bar represents the Kimura-2-parameter genetic distance. Samples in bold were examined by the authors.

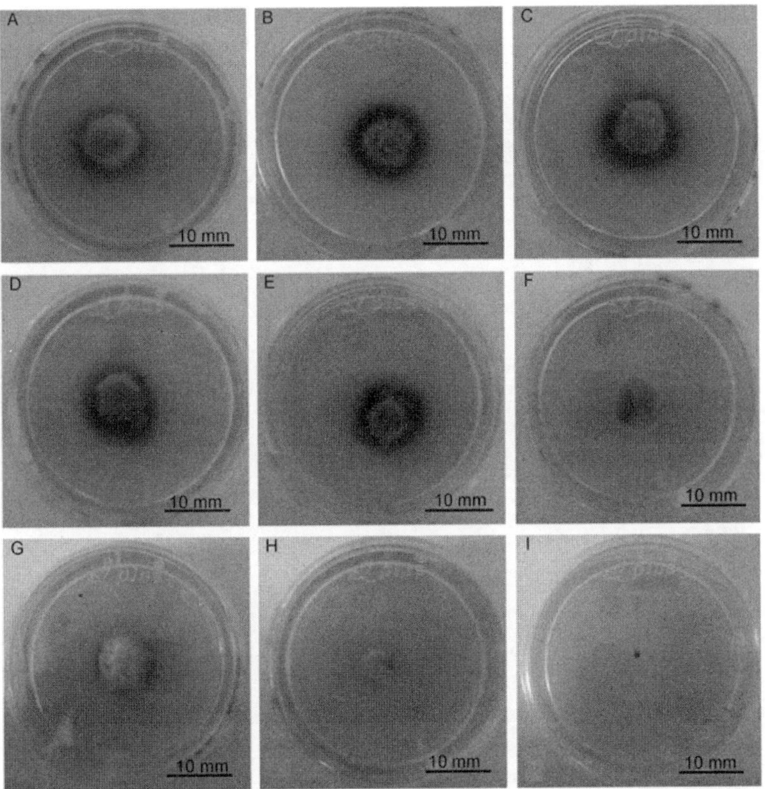

Figure 4 Colonies of the isolated mycobiont grown on PDA for two months after transfer from combined desiccation and starvation. A, Control. B, Transfer after one month of stress. C, After two months. D, After three months. E, After four months. F, After five months. G, After six months. H, After seven months. I, After eight months.

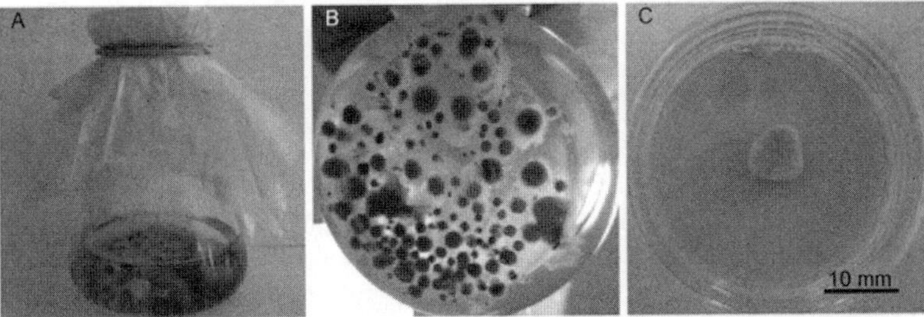

Figure 5 Growth of the isolated mycobiont under starvation stress. A and B, After eight months of growth in the 50-mL Erlenmeyer flask containing sterile distilled water. C, Colony of the isolated mycobiont on PDA in the Petri dish transferred from the Erlenmeyer flask after eight months.

The CSM of the isolated mycobiont growing on PDA for two months after transfer from stress conditions shows that the colonies are 8–9 mm in diameter after stress treatment for 1–6 months (Figure 4B–G, Table 3), and 5 mm in diameter after stress for seven months (Figure 4H, Table 3), and 0 mm in diameter after stress for eight months (Figure 4I, Table 3).

The colony grown on PDA in the Petri dish for two months after transfer from an Erlenmeyer flask was 8 mm in diameter (Figure 5C).

The result of CWM (Figure 6) was basically consistent with that of CSM (Figures 4 and 5, Table 3). It shows that the mycobiont of *E. pusillum* subjected to desiccation stress

Table 3 Colony size of isolated mycobiont grown on PDA for two months after transfer from that under the stress conditions in combination with desiccation and starvation

Periods of stress treatment	Colony size (diameter)
A: control without stress treatment	9 mm
B: one month	8 mm
C: two months	9 mm
D: three months	8 mm
E: four months	7.5 mm
F: five months	8 mm
G: six months	8 mm
H: seven months	5 mm
I: eight months	0 mm

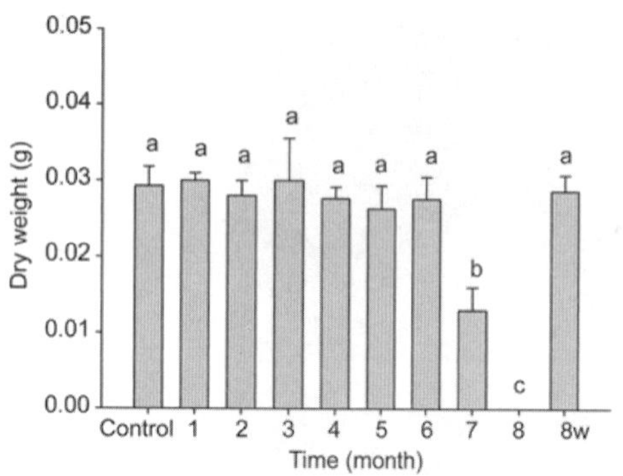

Figure 6 Dry weight of the colonies of the isolated mycobiont grown on PDA for two months. Data are presented as mean±SD. Three replicates were analyzed for each. From left to right the columns represent control, mycobiont under combined desiccation and starvation for 1–8 months and mycobiont in water alone for eight months (8w). a, well growth; b, moderate growth; c, death.

in combination with starvation stress could maintain normal viability after six months, maintain half viability after seven months and lose viability after eight months. Therefore, its survival time limit under desiccation stress in combination with starvation stress was seven months. Additionally, the isolated mycobiont subjected to starvation alone could maintain normal viability after eight months, but its survival time limit under starvation alone has yet to be explored further.

2.2.2 *Comparative morphological analysis of the mycobiont*

The morphology and anatomy of the isolated mycobiont were observed using SEM and TEM (Figure 7).

The hyphae subjected to three months of combined desiccation and starvation were deformed, but the stereoscopic hyphal contour was still visible (Figure 7D). Hyphae had

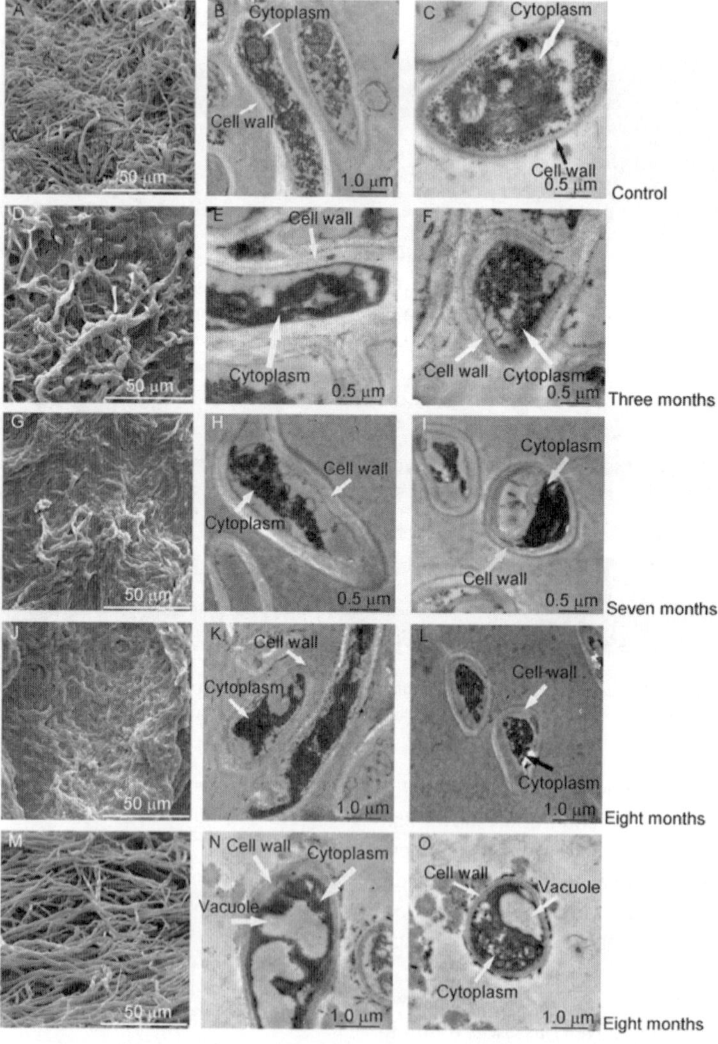

Figure 7 Comparative morphology and anatomy of isolated mycobiont under desiccation and starvation. SEM: A, control; D, after three months of combined desiccation and starvation; G, after seven months of combined desiccation and starvation; J, after eight months of combined desiccation and starvation; M, after eight months of starvation alone. TEM (longitudinal section): B, control; E, after three months of combined desiccation and starvation; H, after seven months of combined desiccation and starvation; K, after eight months of combined desiccation and starvation; N, after eight months of starvation alone. TEM (cross section): C, control; F, after three months of combined desiccation and starvation; I, after seven months of combined desiccation and starvation; L, after eight months of combined desiccation and starvation; O, after eight months of starvation alone.

undergone some plasmolysis (Figure 7E and F). After seven months, the hyphae were strongly deformed into block and the stereoscopic hyphal contour was only visible in some places (Figure 7G). Hyphae had undergone obvious plasmolysis (Figure 7H and I). After eight months, hyphae were significantly deformed into plane block, stereoscopic hyphal contour was not visible (Figure 7J), and hyphae had undergone serious plasmolysis (Figure 7K and L). After eight months of starvation stress alone, hyphae were still normal as compared with the control, but large parts of the hyphae were occupied by vacuoles (Figure 7M–O).

2.2.3 *Metabolic activity analysis of the mycobiont*

The metabolic activity of the isolated mycobiont was determined using fluorescence microscopy (Figure 8).

After being subjected to combined starvation and desiccation for 1–8 months, hyphae did not produce green fluorescence (Figure 8B–E). After being subjected to starvation for eight months, hyphae in sterile distilled water produced green fluorescence (Figure 8F). Our florescence microscopy method was not sensitive enough to detect the metabolic activity of hyphae subjected to combined starvation and desiccation for 1–3 months. Hyphae subjected to starvation for eight months remained metabolically active. The starved but hydrated mycobiont clearly had greater survival capability after eight months than the starved and dried mycobiont after 1–3 months.

2.3 Survival analyses of the phycobiont

2.3.1 *Cell density measurement of the phycobiont*

The growth capability of the isolated *D. chodatii* phycobiont was determined by MNE (Figure 9) and CDM (Figure 10).

Colonies of phycobiont were scattered across the bases of the plastic Petri dish and were clearly visible after desiccation stress for one month (Figure 9B). None of the colonies was visible on the bottom of the dish after being subjected to stress for two months, but a greenish color was visible locally (Figure 9C). After three months of stress, no colonies or greenish color were visible (Figure 9D).

The cell density decreased rapidly when the phycobiont was subjected to desiccation, and almost no cell density was detected after three months (Figure 10).

These two results were basically consistent that the isolated *D. chodatii* phycobiont could maintain viability under desiccation stress for two months, and lose viability after three months.

After one month of desiccation stress, phycobiont cells had contracted with thick cellular contour (Figure 11D) and undergone some plasmolysis (Figure 11E and F). After two months, phycobiont cells had obviously contracted without thick cellular contour (Figure 11G) and undergone obvious plasmolysis (Figure 11H and I). After three months, phycobiont cells had significantly contracted without thick cellular contour (Figure 11J), and undergone serious plasmolysis and lysis of cell walls (Figure 11K and L).

2.3.2 *Metabolic activity analysis of the phycobiont*

The metabolic activity of the isolated *D. chodatii* phycobiont was determined using FCM (Figure 12).

2.3.3 *Comparative morphological analysis of the phycobiont*

The morphology and anatomy of the isolated *D. chodatii* phycobiont were observed using SEM and TEM (Figure 11).

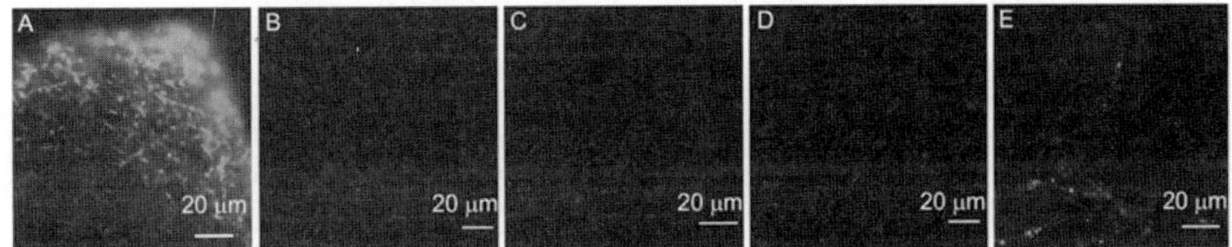

Figure 8 Metabolic activity of the isolated mycobiont under starvation and desiccation stress as detected by fluorescence microscopy. A, Control. B, After one month of combined desiccation and starvation. C, After three months of combined desiccation and starvation. D, After eight months of combined desiccation and starvation. E, After eight months of starvation alone.

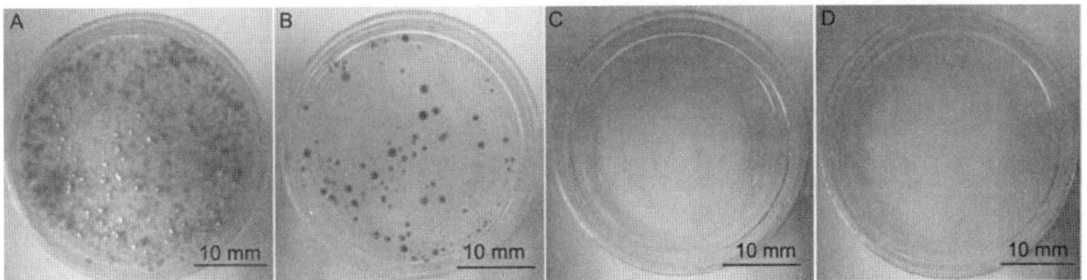

Figure 9 Colonies of the isolated *D. chodatii* phycobiont on PDA for two months after transferring from desiccation stress. A, Control. B, Transferred after one month of desiccation stress. C, After two months of desiccation stress. D, After three months of desiccation stress.

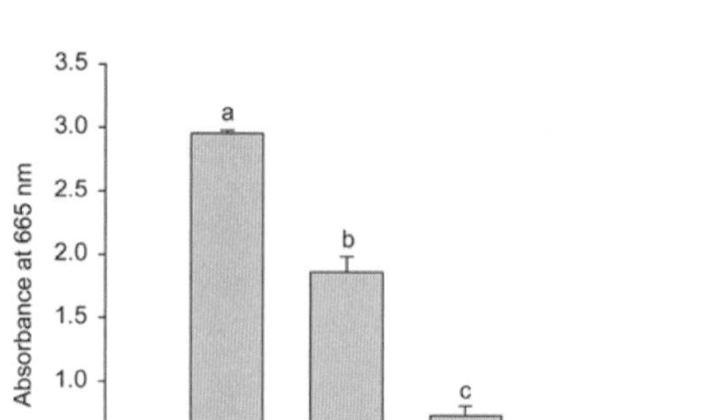

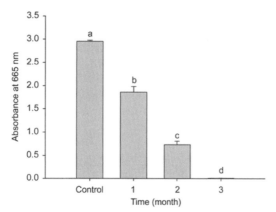

Figure 10 Cell density of 1 mL cell suspension from the isolated phycobiont *D. chodatii* grown on PDA for two months. Data are presented as mean±SD. Three replicates were analyzed for each. a, normal density; b, density reduced by nearly half; c, density reduced to 1/4; d, density reduced to zero.

Non-stressed phycobiont cells produced a low fluorescent signal if unstained with FDA (control; Figure 12A) and a high fluorescent signal if stained with FDA (control; Figure 12B). The number of green fluorescent cells decreased rapidly when the *D. chodatii* phycobiont was subjected to desiccation for 1–2 months, and almost no cells generated a fluorescent signal after three months (Figure 12C–E).

The results of metabolic activity analysis indicate that the isolated phycobiont *D. chodatii* can survive for two months under desiccation stress. *D. chodatii* demonstrates strong desiccation tolerance when compared with the other algae studied to date [10,11].

Based on the above experiments, it appears that the survival time of mycobiont isolated from *E. pusillum* subjected to combined stress of desiccation and starvation was seven months, meaning that the starved mycobiont is able to maintain viability for seven months under drought

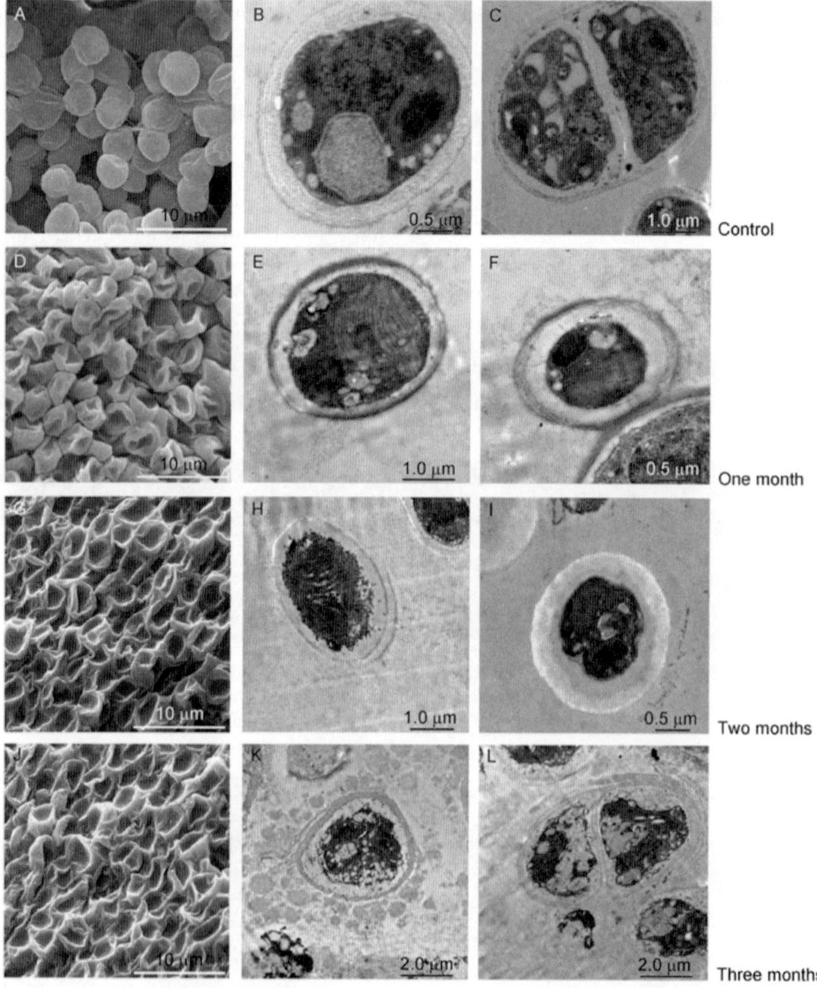

Figure 11 Comparative morphology and anatomy of isolated phycobiont under desiccation stress. SEM: A, control; D, after one month of desiccation; G, after two months of desiccation; J, after three months of desiccation. TEM: B and C, control; E and F, after one month of desiccation; H and I, after two months of desiccation; K and L, after three months of desiccation.

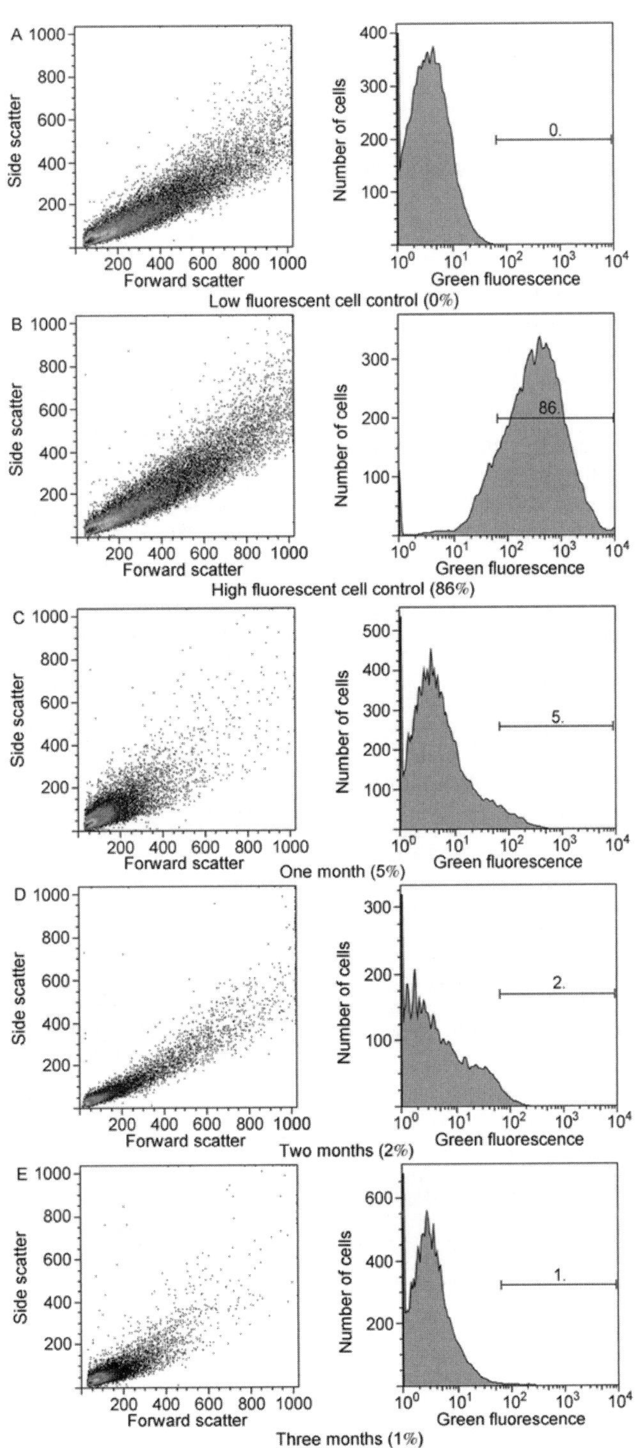

Figure 12 FCM analysis of the isolated *D. chodatii* phycobiont. Each frequency distribution histogram corresponds to 3×10^4 cells analyzed. Two-parameter light scatter plots (forward scatter, FSC, depending on the cell volume; side scatter, SSC, depending on the inner complexity of the particle) and single-histograms of green fluorescence intensity are shown. A, Low fluorescent cells control. B, High fluorescent cells control. C, After one month of desiccation stress. D, After two months of desiccation stress. E, After three months of desiccation stress.

conditions. Such strong desiccation and starvation tolerance increases the chance of symbiotic contact between the mycobiont and its phycobiont in the arid desert in seven months, thus guaranteeing the ecological dominance of *E. pusillum* in desert regions.

The survival time of its phycobiont *D. chodatii* subjected

to desiccation was only two months. However, both the synchronous release of mycobiont and phycobiont from the same lichen thallus after intermittent rainfall, and the existence of drought-tolerant mycobiont in nature tremendously increase the chances of the symbionts to form lichen *E. pusillum* and its reproduction in desert regions.

Our results provide a scientific basis for the arid desert biocarpet engineering to fix sand and carbon, which is accomplished by forming lichen crust through artificial cultivation and inoculation of symbionts isolated from *E. pusillum* in field, and also provide a referential desiccation time for drought-resistant genes that may be screened from this lichen mycobiont.

The further research will determine the survival limit of mycobionts subjected to the starvation stress for longer than eight months and that of mycobionts subjected to continuous rainfall for a period of six months or less in nature.

This work was supported by the National Natural Science Foundation of China (Grant No. 30670004) and Ministry of Science and Technology of China (Grant No.2007AA021405).

1 Lawrey J D, ed. Biology of Lichenized Fungi. New York: Praeger Publishers, 1984

2 Benkô Z, Juhász A, Pócs T, *et al.* Desiccation survival times in different desiccation-tolerant plants. Act Biol Szeg, 2002, 46: 231–233

3 Kranner I, Zorn M, Turk B, *et al.* Biochemical traits of lichens differing in relative desiccation tolerance. New Phytol, 2003, 160: 167–176

4 Lange O L. Hitze- und Trockenresistenz der Flechten in Beziehung zu ihrer Verbreitung. Flora, 1953, 140: 39–97

5 Becquerel P. Reviviscence du Xanthoria parietina desséché avec sa faune, six ans dans le vide et deux samaines á–189°, Ses conséquences biologiques. C R Acad Sci, 1948, 226: 1413–1414

6 Ahmadjian V, Hale M E, eds. The Lichens. New York: Academic Press, 1973

7 Schaper T, Ott S. Photobiont selectivity and interspecific interactions in lichen communities. I. Culture experiments with the mycobiont *Fulgensia bracteata*. Plant Biol, 2003, 5: 1–10

8 Sterflinger K, Krumbein W E. Multiple stress factors affecting growth of rock-inhabiting black fungi. Bot Acta, 1995, 108: 490–496

9 Gorbushina A A, Kotlova E R, Sherstneva O A. Cellular responses of microcolonial rock fungi to long-term desiccation and subsequent rehydration. Stud Mycol, 2008, 61: 91–97

10 Agrwal S C, Singh V. Viability of dried cells, and survivability and reproduction under water stress, low light, heat and UV exposure in *Chlorella vulgaris*. Isr J Plant Sci, 2001, 49: 27–32

11 Gupta S, Agrawal S C. Survival of blue-green and green algae under stress conditions. Folia Microbiol, 2006, 51: 121–128

12 Yoshimura I, Yamamoto Y, Nakano T, *et al.* Isolation and culture of lichen photobionts and mycobionts. In: Kranner I, Beckett R P, Varma A K, eds. Protocols in Lichenology. Culturing, Biochemistry, Ecophysiology and Use in Biomonitoring. Berlin: Springer, 2002. 3–33

13 Li X R, Kong D S, Tan H J, *et al.* Changes in soil and in vegetation following stabilisation of dune in southeastern fringe of the Tengger Desert, China. Plant Soil, 2007, 300: 221–231

14 Cubero O F, Crespo A, Fatehi J, *et al.* DNA extraction and PCR amplification method suitable for fresh, herbarium-stored, lichenized and other fungi. Plant Sys Evol, 1999, 216: 243–249

15 White T J, Bruns T, Lee S, *et al.* Amplification and direct sequencing of fungal ribosomal RNA genes for phylogenetics. In: Innis M A, Gelfand D H, Sninsky J J, *et al.* eds. PCR Protocols: A Guide to Methods and Applications. New York: Academic Press, 1990. 315–322

16 Piercey-Normore M D, DePriest P T. Algal switching among lichen symbionts. Am J Bot, 2001, 88: 1490–1498

17 Saitou N, Nei M. The neighbor-joining method: A new method for reconstructing phylogenetic trees. Mol Biol Evol, 1987, 4: 406–425

18 Hadjoudja S, Vignoles C, Deluchat V, *et al.* Short term copper toxicity on Microcystis aeruginosa and Chlorella vulgaris using flow cytometry. Aquat Toxicol, 2009, 94: 255–264

19 Bialosuknia M W. Sur un nouveau genre de pleurococcacees. Bull Soc Bot Geneve, 1909, 1: 101–104

20 Heering W. Ulotrichales, Mikrosporales, Oedogoniales. In: Pascher A, ed. Die Susswasserflora Deutschlands, Osterreichs und der Schweiz, Heft 6, Chlorophyceae 3. Jena: G. Fisher, 1914. 51–53

21 Chodat R. Monographies d'algues en culture pure. Mat Fl Crypt Suisse, 1913, 4: 1–266

Estimation of *Endocarpon pusillum* Hedwig carbon budget in the Tengger Desert based on its photosynthetic rate

DING LiPing[1,2], ZHOU QiMing[1] & WEI JiangChun[1*]

[1]*State Key Laboratory of Mycology, Institute of Microbiology, Chinese Academy of Sciences, Beijing 100101, China;*
[2]*University of Chinese Academy of Sciences, Beijing 100049, China*

Received April 18, 2013; accepted June 20, 2013; published online July 31, 2013

This study investigated the photosynthetic rate of the lichen *Endocarpon pusillum* at the Chinese Academy of Sciences Shapotou Desert Research Station and estimated its annual contribution to the carbon budget in the ecosystem. The software SigmaPlot 10.0 with "Macro-Area below curves" was used to calculate the carbon fixation capacity of the lichen. The total carbon budget (ΣC) of the lichen was obtained by subtracting the respiratory carbon loss (ΣDR) from the photosynthetic carbon gain (ΣNP). Because water from precipitation plays an important role in photosynthesis in this ecosystem, the annual carbon budget of *E. pusillum* at the station was estimated based on the three-year average precipitation data from 2009 to 2011. Our results indicate that the lichen fixes 14.6 g C m^{-2} annually. The results suggest that artificial inoculation of the crust lichen in the Tengger Desert could not only help reduce the sand and dust storms but also offer a significant carbon sink, fixing a total of 438000 t of carbon over the 30000 km^2 of the Tengger Desert. The carbon sink could potentially help mitigate the atmospheric greenhouse effect. Our study suggests that the carpet-like lichen *E. pusillum* is an excellent candidate for "Bio-carpet Engineering" of arid and semi-arid regions.

crust organisms, carpet-like lichen, carbon income, carbon loss, carbon sink

Citation: Ding L P, Zhou Q M, Wei J C. Estimation of *Endocarpon pusillum* Hedwig carbon budget in the Tengger Desert based on its photosynthetic rate. Sci China Life Sci, 2013, 848–855, doi: 10.1007/s11427-013-4526-9

The vegetation for sand control planted since 1956 in the Shapotou region of the Tengger Desert in northwest China has led to high soil water loss in the region [1–3]. The changes in soil water content were observed 10 years after the artificial vegetation was planted [4]. By contrast, ecological niches with native crust microbial communities showed little change in their soil water content [5]. In addition, these crust microbial communities have helped stabilize the sands, with relatively few incidences of sand and dust storms [6]. This suggests that carpet-like crust microbial communities can be used to maintain and revitalize the water balance of arid and semi-arid regions, via a process called "Bio-carpet Engineering" [7].

Microbiotic crusts consisting of bacteria, fungi, algae, lichens, and bryophytes colonize most terrestrial surfaces, especially in arid and semi-arid regions. These crust microbial communities are able to fix carbon and nitrogen from the atmosphere [8]. This enriches the fertility of the sand and soil, reduces sand and dust storms, and potentially contributes to mitigating the atmospheric greenhouse effect. Of the various crust organisms, lichens are usually the most common.

A lichen is a symbiotic association of a fungus (*mycobiont*) and a photosynthetic partner (*photobiont*), which may be an alga (*phycobiont*) or a cyanobacterium (*cyanobiont*). The association is a complicated arrangement in which the fungus produces a *thallus*, or body, within which the photobionts are housed [9].

*Corresponding author (email: weijc2004@126.com)

This paper was originally published in *Science China Life Sciences*, 56（9）：848-855，2013.

Therefore, the lichen-forming fungi together with photosynthetic algae and/or cyanobacteria in symbioses as lichens appear to function like a single autotrophic "organism" [10]. The relationship is a relatively stable and well-balanced symbiotic association containing both heterotrophic and autotrophic components. Indeed, the lichen can be regarded as a self-contained miniature ecosystem [11–13].

In lichens, carbohydrate acquisition is directly related to photosynthesis performed by the photobionts algae or cyanobacteria, while minerals and water are acquired through the thallus. Since mycobiont hyphae almost always make up the dominant part of the lichen biomass, it is reasonable to assume that the fungal hyphae acquire most of the water and minerals, similar to the functions of roots in plants. However, unlike plant roots, the cyanobacteria in lichens can fix nitrogen, converting N_2 to NH_3. In addition, polyols, the secondary carbon metabolites, and cell walls of lichens have been proposed as major carbon sinks [14].

The photosynthetic and respiratory CO_2 exchange rates of three soil-crust lichens, *Acarospora cf. schleicheri*, *Caloplaca volkii* and *Lecidella crystallina* have been measured in the Namib Desert in South Africa. When experimentally exposed to optimal conditions of light and hydration, the lichen soil crusts in the Namib Desert showed a photosynthetic potential almost equal to that of higher plant leaves [15]. The potential carbon sinks offered by the carpet-like lichens covering arid and semiarid deserts can make a positive contribution to mitigating the atmospheric greenhouse effect.

In this preliminary study of a "Bio-carpet Engineering" project using the carpet-like lichens to control the spread of sandy deserts, the photosynthetic rate of the lichen *Endocarpon pusillum* was measured at the Chinese Academy of Sciences (CAS) Shapotou Desert Research Station (SDRS) which is located in the Ningxia Hui Automonous Region on the southeast edge of the Tengger Desert at 37°32′N and 105°02′E with an elevation of 1339 m. The mean annual air temperature is 10.0°C with a recorded minimum temperature of −25.1°C and maximum temperature of 38.1°C. The annual sunlight exposure is 3264 h; the mean annual precipitation is 180.2 mm, and the annual evaporation is 3000 mm. These values place this region in the arid desert zone [16–18].

The lichen *E. pusillum* is one of the dominant species among the crust organisms at the CAS SDRS. However, the potential carbon sink offered by the lichen and its contribution to the global cycles of carbon have not yet been quantified. In this study, the field measurements of the photosynthetic rate of the lichen at the CAS SDRS were performed and its potential carbon fixation was calculated.

1 Materials and methods

1.1 Materials

Fresh samples of the lichen *E. pusillum* for the laboratory measurements were collected from the SDRS 2 d before measurements. Its phycobiont *Diplosphaera chodatii* Bialosuknia [19] was isolated from the hymenial algal cells of the lichen. The strain of *D. chodatii* for the study is permanently stored in Centre for General Microbiological Culture Collection (CGMCC), and the specimen of the lichen *E. pusillum* is kept in Herbarium Mycologicum Academiae Sinicae (HMAS-L). Field measurements of *E. pusillum* photosynthetic rates were conducted at the CAS SDRS.

1.2 Chlorophyll extraction for quantification

The photosynthesis of the lichen *E. pusillum* is performed by its phycobiont partner. Therefore, comparison of the photosynthetic rates between the lichen *E. pusillum* and its phycobiont *D. chodatii* is necessary. Here we will start with quantifying chlorophylls from both the lichen and the alga.

For this test, the phycobiont was maintained in an axenic liquid culture in 300 mL of Bold's Basal Medium [20] in an Erlenmyer flask and incubated in a shaking incubator (BHWY-2112, Safe, Ningbo, Zhejiang, China) with illumination. Conditions of 130 r min^{-1} at 20°C under 2000 lx were maintained for three months. The liquid culture of the phycobiont was centrifuged at 200×*g* for 5 min. The supernatant was discarded and 200 μL of the phycobiont *D. chodatii* suspension from the precipitate (8 mL) was transferred to a 5 mL centrifuge tube. A 1 cm^2 piece of the *E. pusillum* was transferred into a separate 5 mL centrifuge tube.

Chlorophyll extraction was performed with dimethyl sulfoxide (DMSO) [21]. Four milliliter of DMSO was added to centrifuge tubes containing either *D. chodatii* or the lichen *E. pusillum*. The tubes were placed in a water bath at 65°C for 40 min to extract the chlorophyll. The extracts were centrifuged at 300×*g* for 1.5 min. The supernatants were scanned with an ultraviolet spectrophotometer (DU-800) between the wavelengths of 400–700 nm with a slit of 1 nm. The absorption peaks of the chlorophyll of both the *E. pusillum* and *D. chodatii* were very similar: 2.4 at 435 nm and 1.4 at 665 nm. The result indicates that the chlorophyll content in 1 cm^2 of the lichen *E. pusillum* is equal to that in the phycobiont *D. chodatii* suspension of 200 μL. Each experiment was repeated three times.

1.3 Laboratory measurement of *E. pusillum* and *D. chodatii* photosynthetic rates

For comparison, the photosynthetic rates based on CO_2 exchange in the green algal lichen *E. pusillum* and its phycobiont *D. chodatii* were measured. CO_2 exchange was measured under controlled conditions with a Handheld Photosynthesis System CI-340 (CID Bio-Science, Inc., USA). A fresh sample of the lichen *E. pusillum* was collected from the CAS SDRS 2 d before measurement. Ten square centimeters of water-saturated lichen was placed on a water-saturated soil layer in a leaf chamber of 10 cm^2. For the phycobiont, a 2 mL suspension of cells prepared as described above was spread onto a thin sand layer in another leaf chamber. The saturated lichen was sprayed with water equivalent to >2 mm of precipitation.

The measurements of both the lichen *E. pusillum* and its phycobiont *D. chodatii* were carried out at temperatures of 5°C, 20°C and 30°C, and under red light intensities of 0, 100, 200, 400, 800, 1200, 1600, 2000 and 2400 μmol m^{-2} s^{-1} gradient, and CO_2 concentration of 600±50 ppm. Each experiment was repeated three times.

The respiratory rate of *E. pusillum* in the dark was also measured under the conditions simulated the field conditions.

1.4 Field measurements of *E. pusillum* photosynthetic rates

CO_2 exchange in the field was measured under natural conditions with a Handheld Photosynthesis System CI-340 (CID Bio-Science, Inc., USA) at the CAS SDRS. The four different vegetation zones at the CAS SDRS contain crusts of different ages. The eldest crust is in the 1956 artificial vegetation zone, and the youngest one in the 1990 zone. Measurement of *E. pusillum* photosynthetic rates was conducted in the 1956 artificial vegetation zone.

A 10 cm^2 area was transferred to the leaf chamber for measurement. The leaf chamber was not enclosed after each measurement, so the CO_2 concentration in the chamber was equal to that in the atmosphere. Each sample was measured for 2 min and repeated an average of three times per hour. The field measurements were typically conducted from 5:00 a.m. to 8:00 p.m. of June 30–July 8, October 25–November 9 in 2010, and April 5–16, December 26–28 in 2011.

1.5 Estimated carbon budget of *E. pusillum* in the field

The carbon budget (ΣC) of *E. pusillum* lichen crust was obtained by subtracting the respiratory carbon loss (ΣDR) from the photosynthetic carbon income (ΣNP) (ΣC=ΣNP–ΣDR) [22]. SigmaPlot 10.0 software with "Macro-Area below curves" was used to calculate the ΣC. Because water availability has a large influence on both photosynthesis and respiration, the ΣC of the lichen per year was calculated and estimated based on the mean annual precipitation from 2009 to 2011 (Figure 1).

2 Results and discussion

2.1 The photosynthetic rates of *E. pusillum* and *D. chodatii* in the laboratory

All experiments were performed under optimal moisture conditions. The results demonstrated that the respiratory rates of both the *D. chodatii* and *E. pusillum* were between 0 and –1 μmol m^{-2} s^{-1} at 5°C. The light saturation points (LSP) for *E. pusillum* and *D. chodatii* were 1200 and 400 μmol m^{-2} s^{-1}, respectively. The maximum net photosynthetic rate (*Pn*) for *D. chodatii* was 2.8 μmol m^{-2} s^{-1}, slightly higher than that for *E. pusillum*, which was 1.2 μmol m^{-2} s^{-1} (Figure 2A).

The respiratory rate of *E. pusillum* was –6.5 μmol m^{-2} s^{-1}, lower than that of *D. chodatii* (–1.6 μmol m^{-2} s^{-1}) at 20°C. The LSPs for *E. pusillum* and *D. chodatii* were 1600 and 1200 μmol m^{-2} s^{-1}, respectively. At 20°C, the maximum *Pn* for *E. pusillum* and *D. chodatii* were 3.5 and 6.1 μmol m^{-2} s^{-1}, higher than the maximum values at 5°C (Figure 2B).

The respiratory rate of *E. pusillum* was –11.0 μmol m^{-2} s^{-1}, much lower than that of *D. chodatii* (–4.4 μmol m^{-2} s^{-1}) at 30°C. The LSPs for *E. pusillum* and *D. chodatii* were 2400 and 2000 μmol m^{-2} s^{-1}, respectively. The maximum *Pn* for the lichen was only 1.5 μmol m^{-2} s^{-1}, close to that at 5°C, but significantly lower than that at 20°C; and the *Pn* for the phycobiont was 5.9 μmol m^{-2} s^{-1}, similar to that at 20°C (Figure 2C).

The results showed that the *Pn* of the phycobiont *D. chodatii* was always higher than that of the lichen

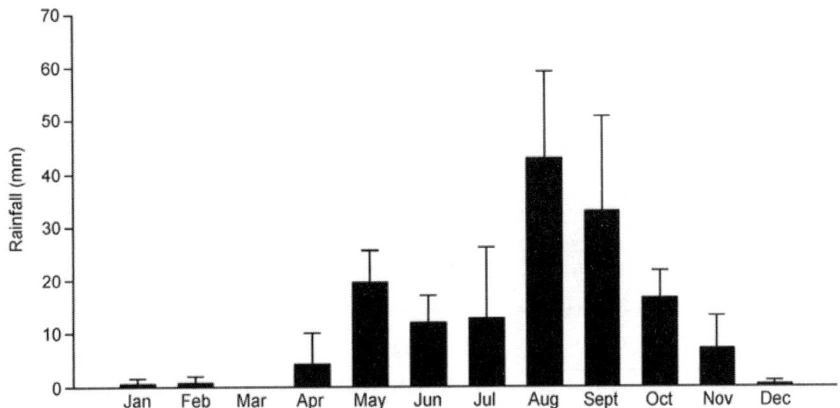

Figure 1 Average monthly rainfall at the CAS SDRS during 2009–2011.

E. pusillum, while the respiratory rate of the phycobiont always lower than that of the lichen at all three temperatures. This can be explained by the fact that the photosynthetic rate of both the lichen *E. pusillum* and the phycobiont *D. chodatii* is performed by *D. chodatii* only, but both the mycobiont *E. pusillum* and the phycobiont *D. chodatii* contribute to *E. pusillum*'s respiratory rate.

2.2 Field measurements of *E. pusillum* photosynthetic rates and carbon budget calculation

2.2.1 The results of measurements in 2010

The results of the daytime measurements of the photosynthetic rate of the lichen during the 25 days from June 30 to July 8, and from October 25 to November 9 in 2010 at the CAS SDRS indicated that both the respiratory rate (in early morning) and the net photosynthetic rate increased after rainfall (Figure 3).

Our results showed that in the early morning after a rainfall event on June 30, 2010, the respiratory rate reached –6.3 μmol m^{-2} s^{-1}, the lowest rate measured in the field. The *Pn* increased gradually after 8:00 a.m. and reached a maximum value of 4.8 μmol m^{-2} s^{-1} at 11:00 a.m. After 11:00 a.m. the *Pn* declined gradually and reached zero at 1:00 p.m. In the late afternoon, the *Pn* became negative and the respiratory rate declined to –3.0 μmol m^{-2} s^{-1} owing to

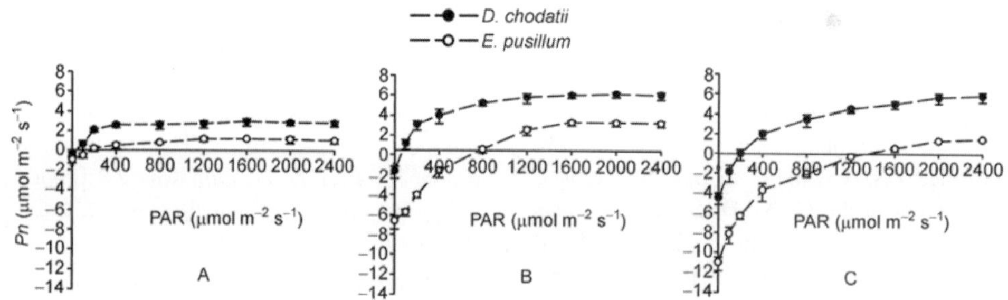

Figure 2 Response of net photosynthetic rate (*Pn*)) to photosynthetic active radiation (PAR) for both *E. pusillum* and its phycobiont *D. chodatii*, as measured at 5°C (A), 20°C (B), 30°C (C) and optimal moisture conditions in the laboratory.

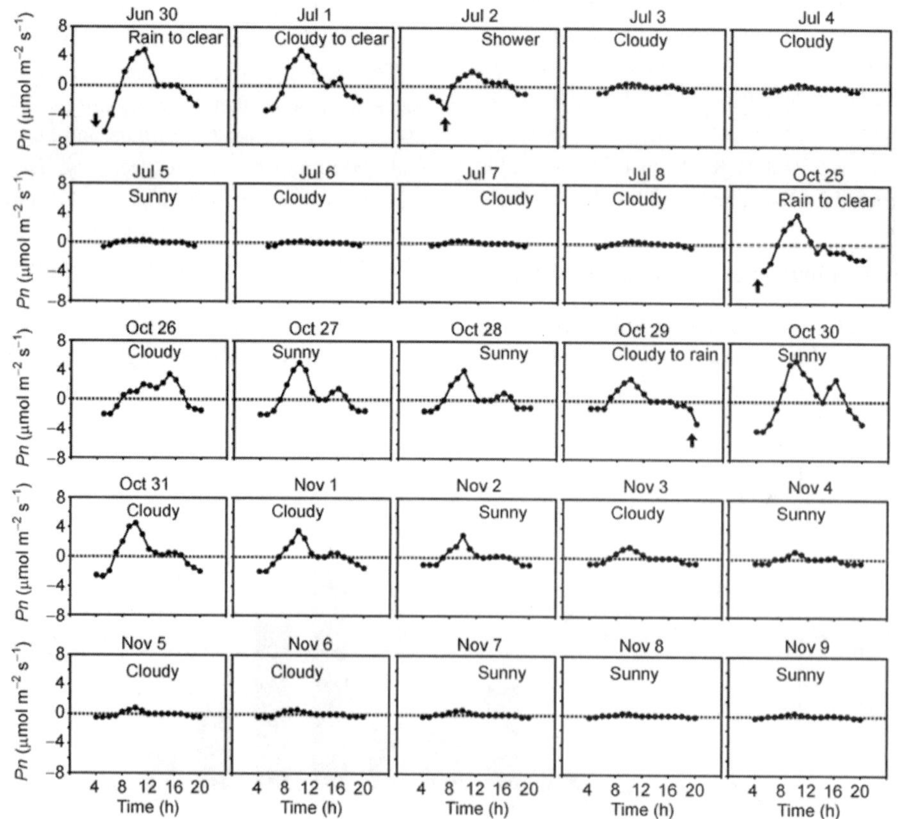

Figure 3 Field observations of the net photosynthetic rate (*Pn*) of *Endocarpum pusillum*, at the CAS SDRS over a period of 25 days from June 30 to July 8, and from October 25 to November 9 in 2010. The arrows indicate rainfall events.

the decreased light intensity. A similar result was observed during July 1–2, October 25–31, and November 1–4 (Figure 3). The *Pn* values of *E. pusillum* were close to zero in the days without any rain and the lichen thallus was almost dry.

Photosynthetic carbon fixation was calculated based on the *Pn* data from 5:00 a.m. to 8:00 p.m. (4:00 a.m.–8:00 p.m., Oct. 26–Nov. 9) on each day (Figure 3). The software SigmaPlot 10.0 was used to quadrature the difference between the upper abscissa area and the down area of the *Pn* curve. The results were then converted into mg C (Figure 4).

After rainfall of 5 mm in the summer of 2010, the carbon assimilation was 201 mg m^{-2} on June 30 and 319.7 mg m^{-2} on July 1. Although there was 0.5 mm rainfall on July 2, the carbon intake was 99.4 mg m^{-2}. The amount of carbon fixation was approximately 30.0 mg m^{-2} during the six cloudy and sunny days from July 3 to 8 (Figure 4). The total carbon assimilation of the lichen is estimated to be 650 mg m^{-2} for the period of the nine days from June 30 to July 8 of 2010 (Figure 4).

It rained on October 25 and 29 and the carbon fixation was positive in the next nine days from October 26 to November 3 except October 25 because of the greater respiratory rate. The highest peak (505.4 mg m^{-2}) of the carbon income was on October 30. Carbon sequestration became negative (−73.1 mg m^{-2}) on November 4. The total carbon assimilation of the lichen is estimated to be 2174 mg m^{-2} during the period from October 25 to November 9 of 2010, which included two rainy days (Figure 4).

2.2.2 *Field results in 2011*

The results of the daytime measurement of the photosynthetic rate of the lichen during the 15 days from April 5 to 16, and from December 26 to 28 in 2011 at the CAS SDRS showed that the *Pn* was almost zero throughout the tested periods because of the lack of precipitation, except for the rainfall event on April 5 (Figure 5).

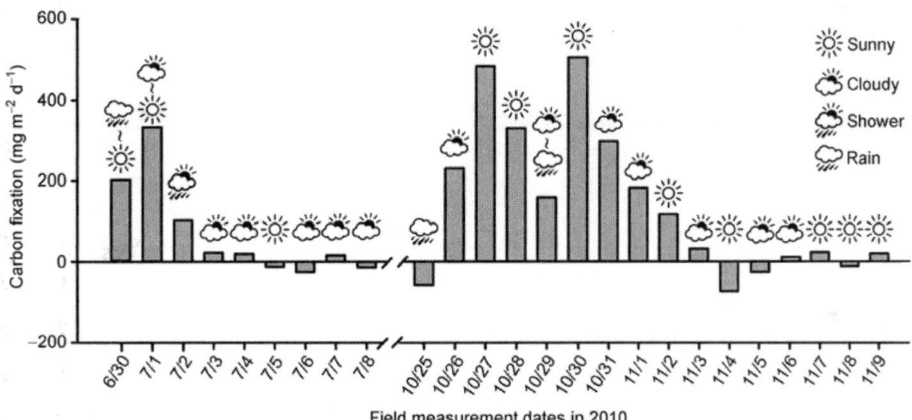

Figure 4 Field measurements of *Endocarpon pusillum*'s diel photosynthetic carbon income over the period June 30–July 8, October 25–November 9, 2010 at the CAS SDRS.

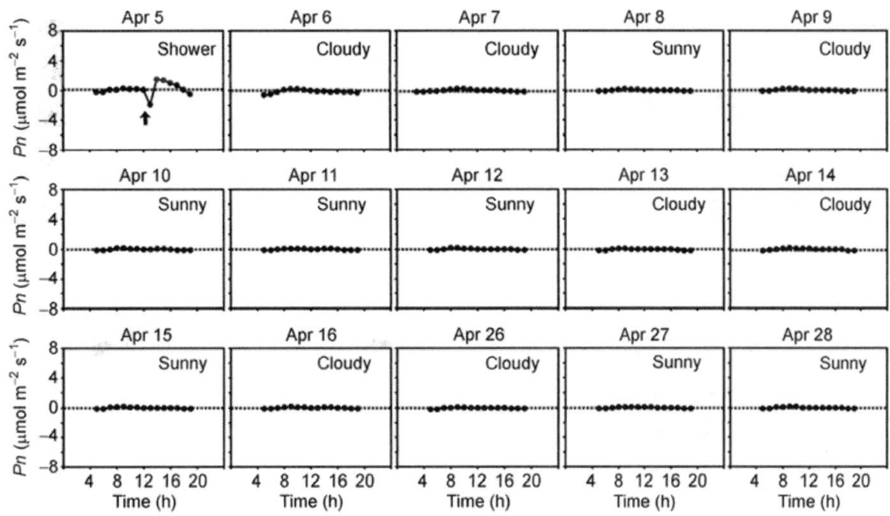

Figure 5 Field observations of the net photosynthetic rate (*Pn*) of *Endocarpon pusillum* at the CAS SDRS over the periods of April 5–16 and December 26–28 in 2011. The arrows indicate rainfall events.

There was a shower with a total precipitation of 1.0 mm on April 5, the first rain for 2011. Because of the rain, the carbon sequestration by *E. pusillum* was relatively high, at 85.2 mg m^{-2} in the two days after the spring shower. However, there was no rain for 10 days from April 7 to 16 and the total carbon sequestration was low, about 40.1 mg m^{-2} during these 10 days (Figure 6). The combined total carbon assimilation was 125.3 mg m^{-2} during the 12 days in spring.

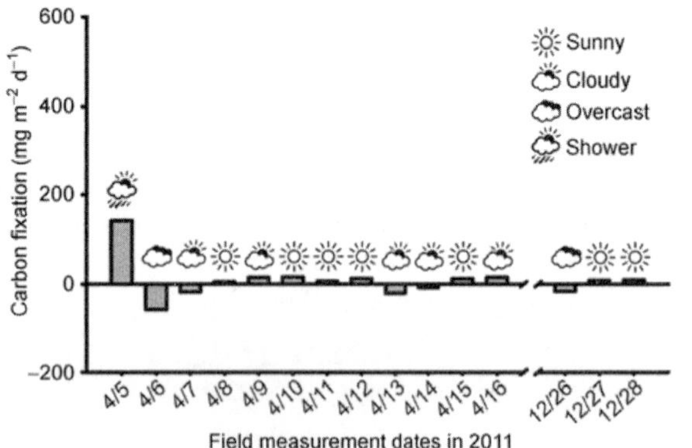

Figure 6 Field measurements of *Endocarpon pusillum*'s daytime photosynthetic carbon intake over the periods April 5–16 and December 26–28, 2011 at the CAS SDRS.

Winter temperatures at the CAS SDRS are very cold and there is almost no precipitation. The carbon sequestration rates were −16.7, 8.1, and 9.5 mg m^{-2} in the three days from December 26 to 28, and the total carbon sequestration was only 0.9 mg m^{-2}.

2.3 Respiration rate of *E. pusillum* in the dark under simulated conditions and the night respiratory carbon loss in the laboratory

Night respiration rate data were obtained in the laboratory under simulated conditions of rainfall, temperature and humidity. The simulated rainfall was given in the morning and the measurement was carried out at night.

When more than 2 mm of the simulated rainfall was given, the respiration rate of *E. pusillum* in the dark was −1 μmol m^{-2} s^{-1} at 15°C on the night of July 10 only (Figure 7). However, negative (−0.8, −0.5, and −0.2 μmol m^{-2} s^{-1} at 5°C) respiration rates were observed for three consecutive nights after the simulated rainfall on November 10 (Figure 7).

No obvious respiration was detected at night for *E. pusillum* when there was 0.5 mm of simulated rainfall on July 12 (Figure 7).

The software SigmaPlot 10.0 was also used to compute the nocturnal respiratory carbon loss. The results were then converted into mg C (Figure 8). When *E. pusillum* was saturated by spraying with water equivalent to ⩾2 mm precipitation, the respiratory carbon loss was 343.5 mg m^{-2} after the simulated rain on July 10. However, the total carbon loss was 463.5 mg m^{-2} in the three nights after simulated rain on November 10 (Figure 8).

2.4 Estimating the annual carbon budget

2.4.1 *Daytime photosynthetic carbon income (ΣNP)*

The average monthly precipitation at the CAS SDRS was relative abundant (>5 mm) in the seven months from May to November during 2009–2011 (Figure 1). In these months, the lichen was relatively moist and the carbon assimilation was 2824 mg m^{-2} in the 25 days during June 30–July 8 and October 25–November 9 in 2010. Based on this number, the total estimated carbon assimilation was 24173 mg m^{-2} for the 214 days in the seven months from May to November (Table 1).

In the five months from December 2010 to April 2011, the average of monthly rainfall was little (Figure 1), and the lichen was dormant for much of the time in the relative dry

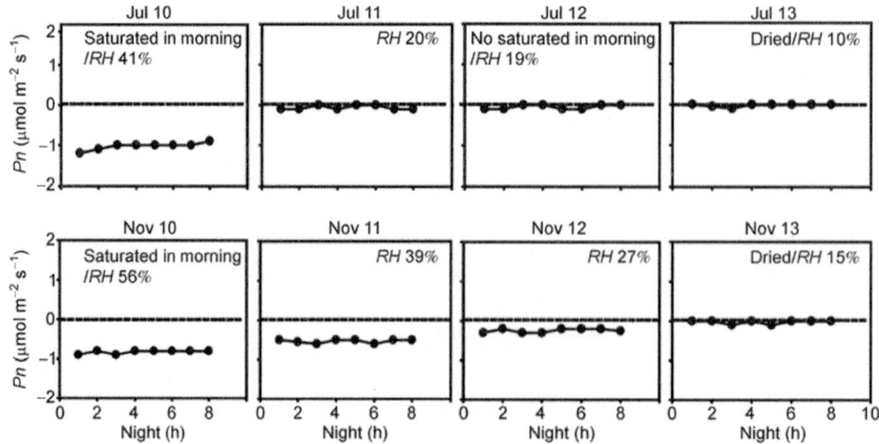

Figure 7 The nocturnal respiration rate of *Endocarpon pusillum* under laboratory simulated precipitation or dry conditions in July 10–13 and November 10–13, 2010. *RH* represents the relative humidity of the leaf chamber.

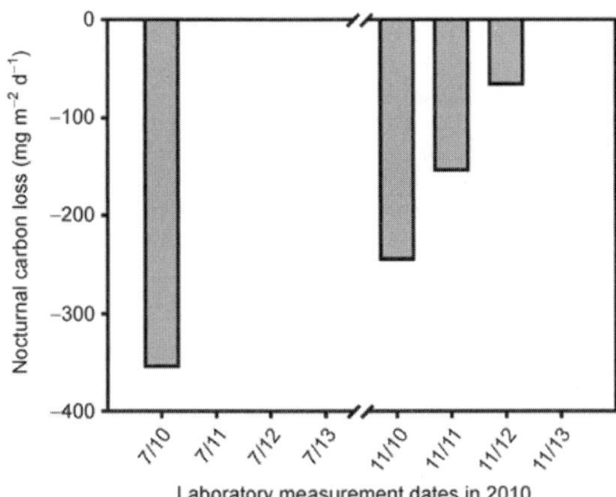

Figure 8 *Endocarpon pusillum*'s nocturnal respiratory carbon loss under laboratory simulated conditions of precipitation or dry conditions over the periods of July 10–13 and November 10–13, 2010.

periods. During this period, the carbon assimilation was similar to that of the periods of April 5–16 and December 26–28 in 2011. Because carbon assimilation was 126.2 mg m^{-2} in the 15 days, the total estimated carbon assimilation would be 1270 mg m^{-2} for the 151 days in the five months from December to April (Table 1).

Altogether, the total carbon assimilation ΣNP was about 25443 mg m^{-2} per year.

2.4.2 Nocturnal respiratory carbon loss (ΣDR)

According to the laboratory simulation experiments (Figures 7 and 8), the nocturnal carbon loss was 1271 mg m^{-2} during the period of June 30–July 8 and October 25–November 9 in 2010 at the CAS SDRS. During the 25 days, there were three rainfall events, one in the summer and two in the fall, each of which saturated the lichen thalli with water. Therefore, the carbon loss due to nocturnal respiration was 10880 mg m^{-2} in the 214 days covering the seven humid months (Table 1). At the remaining nights the respiratory rates were weak and the carbon loss can be ignored. Based on these estimates, the *E. pusillum* annual nocturnal ΣDR was 10880 mg m^{-2}.

2.4.3 Photosynthetic carbon budget (ΣC)

Annual carbon gain is calculated according to the formula ΣC=ΣNP−ΣDR [22]. Because the annual ΣNP was 25443 mg m^{-2} and the annual ΣDR was 10880 mg m^{-2}, the net annual gain of carbon by *E. pusillum* was 14563 mg m^{-2} a^{-1} (Table 1, Figure 9).

3 Conclusion

The results of the experiments show that the amount of carbon fixed by photosynthesis of the lichen *E. pusillum* at the CAS SDRS is 14.6 g m^{-2} a^{-1}. Thirty thousand square kilometers of sand dunes in the Tengger Desert could be turned into a carbon sink capable of fixing 438000 t of carbon per year, if a Desert Biocarpet Engineering project could be carried out using the lichen *E. pusillum*.

Table 1 Annual *Endocarpn pusillum* carbon budgets divided into two groups according to the average monthly rainfall[a)]

State	Field measured	Period (d)	NP (mg m^{-2})	DR(mg m^{-2})	ΣSP (d)	ΣNP (mg m^{-2})	ΣDR (mg m^{-2})	ΣC (mg m^{-2})
Relatively humid	Jun. 30–Jul. 8, 2010	9	2824	1271	214	24173	10880	13293
	Oct. 30–Nov. 9, 2010	16						
Relatively dry	Apr. 5–16, 2011	12	126.2	–	151	1270	–	1270
	Dec. 26–28, 2011	3						
Total		40	2950.2	1271	365	25443	10880	14563

a) The field measurements include net photosynthesis (NP) and simulated nocturnal carbon loss (DR) and the similar period (ΣSP) to calculate the photosynthetic carbon income (ΣNP), nocturnal respiratory carbon loss (ΣDR) and total annual carbon budget (ΣC).

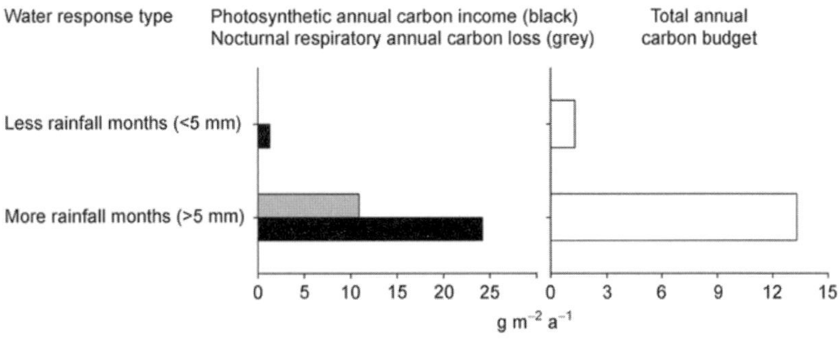

Figure 9 *Endocarpon pusillum* annual gain and loss of carbon, showing the contribution of the rainy months and dry months to the photosynthetic carbon income, nocturnal carbon loss and the carbon budget over a one-year period.

We are grateful to the directors and the staff members of the CAS SDRS for their kind help during our work in the field, and for providing local weather data. Many thanks also go to Wang ChengChong of CAS SDRS, Liu Meng and Cao ShuNan of IM CAS for helping with our field measurements, and also Chen YanChang of Zeal Quest Scientific Technology Company for providing the opportunity to use the instrument for our experiments. This work was supported by the National Natural Science Foundation of China (31070018), the Knowledge Innovation Program of the Chinese Academy of Sciences (2010-Biols-CAS-0104) and Ministry of Science and Technology of China (2011BAC07B03).

1 Xiao H L, Li X R, Duan Z H, et al. Succession of plant-soil system in the process of mobile dunes stabilization (in Chinese). J Desert Res, 2003, 23: 605–611

2 Xiao H L, Li X R, Duan Z H, et al. Impact of evolution of plant-soil system on the water environment during the mobile dunes stabilization (in Chinese). Acta Pedo Sin, 2003, 40: 809–814

3 Long L Q, Li X R. Effects of soil microbiotic crusts on seedling survival and seedling growth of two annual plants (in Chinese). J Desert Res, 2003, 23: 656–660

4 Duan Z H, Xiao H L, Li X R, et al. Evolution of soil properties on stabilized sand in the Tengger Desert, China. Geomorphology, 2004, 59: 237–246

5 Shi S, Ma F Y, Liu L C, et al. The effect on different vegetation structure to soil water contents in Shapoto region (in Chinese). J CUN, 2004, 13: 137–145

6 More W G. A Dictionary of Geography—Definitions and Explanations of Terms used in Physical Geography. London: Adam & Charles Black, 1975

7 Wei J C. The Desert Biocarpet Engineering—A new way to control the arid desert (in Chinese). Arid Zone Res, 2005, 22: 287–288

8 Elbert W, Weberb B, Büdel B, et al. Microbiotic crusts on soil, rock and plants: neglected major players in the global cycles of carbon and nitrogen? Biogeosci Discuss, 2009, 6: 6983–7015

9 Ahmadjian V. The Lichen Symbiosis. New York: John Wiley & Sons, 1993. 1–250

10 Green T G A, Nash T H, Lange O L. Physiological ecology of carbon dioxide exchange. In: Nash T H, ed. Lichen Biology. 2nd ed. Cambridge: Cambridge University Press, 2008. 152–208

11 Farrar J F. The lichen as an ecosystem: observation and experiment. In: Brown D H, Hawksworth D L, Bailey R H, eds. Lichenology: Progress and Problems. London: Academic Press, 1976. 385–406

12 Seaward M R D. Contribution of lichens to ecosystems. In: Galun M, ed. CRC Handbook of Lichenology, Vol. 2. Boca Raton: CRC Press, 1988. 107–129

13 Nash T H, ed. Lichen Biology. 2nd ed. Cambridge: Cambridge University Press, 2008. 1–8

14 Palmqvist K, Dahlman L, Jonsson A, et al. The carbon economy of lichens. In: Nash T H, ed. Lichen Biology. 2nd ed. Cambridge: Cambridge University Press, 2008. 182–215

15 Lange O L, Meyer A, Zellner H, et al. Photosynthesis and water relations of lichen soil crusts: field measurements in the coastal fog zone of the Namib Desert. Funct Ecol, 1994, 8: 253–264

16 Li X R, Shi Q H, Zhang J G, et al. Study of plant diversity changes during the artificial vegetation evolution processes in the Shapotou region (in Chinese). J Desert Res, 1998, 18: 23–29

17 Li X R, Zhang J G, Wang X P, et al. Study on soil microbiotic crust and its influences on sand-fixing vegetation in arid desert region. Acta Bot Sin, 2000, 42: 965–970

18 Li X R. Influence of variation of soil spatial heterogeneity on vegetation restoration. Sci China Ser D-Earth Sci, 2005, 48: 2020–2031

19 Zhang T, Wei J C. Survival analyses of symbionts isolated from *Endocarpon pusillum* Hedwig to desiccation and starvation stress. Sci China Life Sci, 2011, 54: 480–489

20 Deason T R, Bold H C. Phycological Studies I. Exploratory Studies of Texas Soil Algae. Austin: University of Texas Publication, 1960

21 Ronen R, Galun M. Pigment extraction from lichens with dimethyl sulfoxide (DMSO) and estimation of chlorophyll degradation. Envir Exp Bot, 1984, 24: 239–245

22 Lange O L. Photosynthetic productivity of the epilithic lichen *Lecanora* muralis: long-term field monitoring of CO_2 exchange and its physiological interpretation III. Diel, seasonal, and annual carbon budgets. Flora, 2003, 198: 277–292

荒漠地衣结皮固沙研究进展*

一、引言

沙暴（sandstorm）是由狂风将大量粗沙吹入空中所形成的一种风暴。尘暴（duststorm）是由湍流风侵入干旱多尘地域的上空而引起的。尘暴推进时，前面还有非常高大，有时可以高到3000 m的尘墙为它开路（Moore，1973）。无论是狂风、暴风或湍流风，那是气候现象，是自然规律，迄今人类尚无法控制它。但是，能否形成沙尘暴，则取决于干旱地区地面有无沙尘外露。这便涉及沙尘治理问题，是人类可以控制的，至少可以减缓。

“利用植被防治荒漠化是指增加植被对土地的覆盖，改善土壤的物理、化学特性和土壤的水分状况。生长茂密（适度）的植被可以有效地防风治沙，降低风蚀和水蚀，改善小气候，减缓大风对裸露沙地和雨滴对地面的直接打击，从而预防和治理荒漠化。”（慈龙骏等，2005）。因此，当我们面对沙漠治理任务时，通常首先就会想到种树、种灌、种草的生态工程以实施人工植被对沙地覆盖的措施。

然而，实施人工植被生态工程的前提之一便是水分平衡原则。在干旱、半干旱地区，“无论进行造林、植灌与种草首先必须考虑水分收支平衡的原则，即大气降水、径流的输入（含灌溉）与输出以及植物的蒸腾强度三方面进行估算，其收入的水分（大气降水+径流输入）应超出支出（径流输出+蒸腾）的20%，以此来决定造林植灌的密度（单位面积的株数）或种草的覆盖度”（张新时，1994）。

二、无灌溉人工植被固沙防护体系及其自然演替

为了确保包兰铁路经过腾格里沙漠东南缘沙坡头地区流沙地段畅通无阻，中国科学院与有关单位受命于20世纪50年代中期建立了“以固为主，固阻结合，机械固沙与生物固沙相结合”的人工植被固沙防护体系（赵兴梁，1988；邸醒民，2005），从而获得了巨大成功。我们以此作为案例进行具体分析。

沙坡头地区（37°32′N，105°02′E）位于腾格里沙漠东南缘，属于草原荒漠地带，是沙漠与绿洲的过渡区（李新荣等，1998）。海拔1339 m，以高大、密集的格状新月形沙丘链连绵分布而著称（李新荣等，2005）。年平均气温10℃，最低温度-25. 1℃，最高温度38. 1℃。年平均降水量180. 2 mm，年平均蒸发量为3000 mm，沙层稳定含水量仅为2%~3%。年平均风速2. 9 m/s，尘暴59天（Li，2005）。

作为人工植被固沙防护体系工程的第一步，在苏联土库曼共和国科学院院士彼得罗夫（M. P. Petrov）的建议下，将中亚荒漠固沙行之有效的措施——半隐蔽式草方格沙障用于沙坡头固沙防护体系工程，从而在稳定沙面、防止流沙移动方面起到了关键作用（杨喜林，2005），为第二步的人工栽植旱生灌木柠条（*Caragana korshenskii*）、油蒿（*Artemisia ordosica*）、花棒（*Hedysarum scoparium*）等植物创造了条件（李新荣等，2005）。

沙坡头地区无灌溉人工植被固沙防护体系建成后，“植被覆盖度最初10年为15%~25%，从第20年开始到现在，灌木盖度下降至6%~9%”。作为生物结皮组分的“大量微生物、隐花植物也进入到固沙区

---

* 本节作者为魏江春（中国科学院微生物研究所真菌学国家重点实验室）

本文原载于《中国当代生态学研究 生态系统恢复卷》（李文华主编）. 北京：科学出版社：337-346, 2013.

来”。“人工植被建立 40 年左右的固沙区，生物结皮覆盖度达到 90% 以上，厚度达 11 mm。现在，沙坡头固沙区已经形成了人工植被、一年生草本和生物结皮为主体的固沙体系”（李守中，2005）。

在对沙坡头地区人工植被蒸散耗水与水量平衡的研究中发现，人工植被的蒸散耗水量明显大于降水量（冯金朝等，1995）。人工植被前，土壤深层含水量丰富，随着人工植被的形成，人工植被下土壤深层水分越来越少；人工植被盖度越来越小，逐年衰退（Duan et al，2004；Long and Li，2003；Xiao et al，2003a；2003b）。

然而，生物结皮群落却逐年形成，在人工植被与生物结皮群落下土壤深层水分含量的比较研究表明，在生物结皮群落下，随着土壤深度越大，则含水量越大。在人工植被下，随着土壤深度越大，则含水量越小（Shi et al，2004）。

根据调查统计，1956 ~ 2007 年分 4 批建立的沙坡头地区无灌溉人工植被防护体系由人工植被为主体演变为微型生物结皮为主体的演替过程分为 4 个发育期：①17 年发育区（1990 ~ 2007 年），②26 年发育区（1981 ~ 2007 年），③43 年发育区（1964 ~ 2007 年）及④51 年发育区（1956 ~ 2007 年）。

从人工固沙植被防护体系建成至第 17 年，即①17 年发育区（1990 ~ 2007 年）的调查结果为生物结皮盖度、人工植被盖度以及裸露沙土面积之比为 30.41 ∶ 19.42 ∶ 50.25；而从体系建成至第 51 年，即④51 年发育区（1956 ~ 2007 年）的调查结果为生物结皮盖度、人工植被盖度以及裸露沙土面积之比为 61.14 ∶ 19.92 ∶ 19.07（=6.1 ∶ 2 ∶ 1.9）（图 3-7-1）（Yang and Wei，2012）。

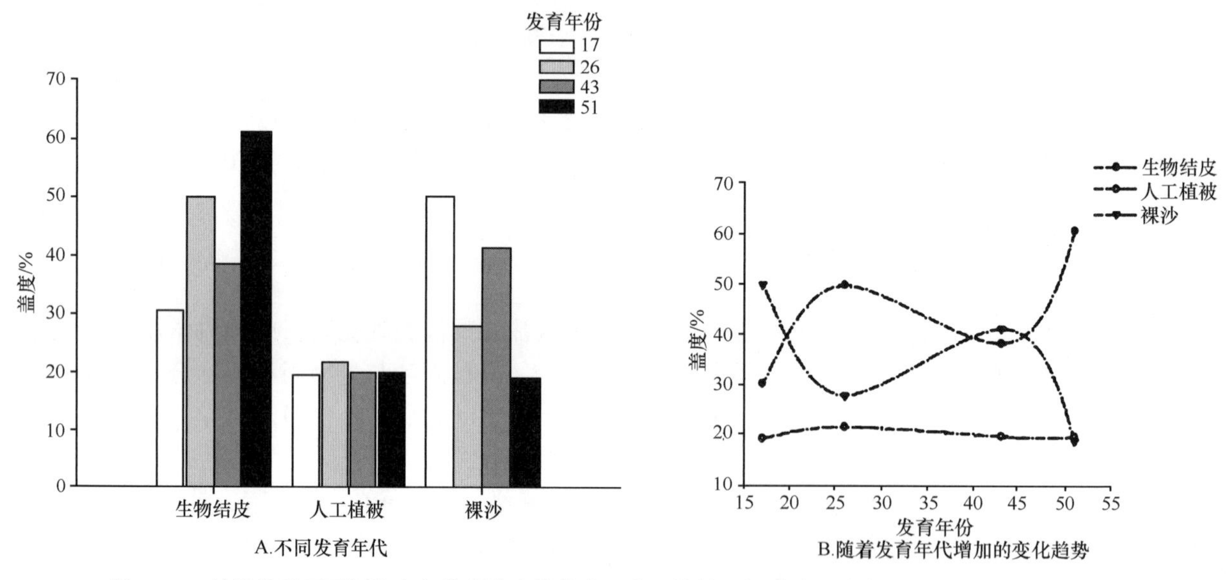

图 3-7-1 沙坡头地区不同发育年代荒漠生物结皮、人工植被和裸露沙土盖度（Yang and Wei，2012）

这一结果表明，微型生物结皮的盖度与发育年代的长短成正比，即发育年代越长，微型生物盖度越大；而裸露沙土面积与发育年代的长短则成反比，即发育年代越长，裸露沙土面积越小。至于其中的人工植被盖度则稳定在 19% 。这说明以人工植被为主体的固沙防护体系向以微型生物为主体的固沙防护体系的转变，在人工固沙植被防护体系建成后的 17 年内已经基本完成；人工植被稳定在 19% ，而微型生物结皮则继续发育至 61% ，而裸露沙土面积继续缩小至 19% 。

沙坡头地区无灌溉人工植被固沙防护体系经过 50 多年的自然演替结果表明，具有抽水机效应的人工植被的蒸腾作用加强了水分输出大于输入的矛盾，导致人工植被因干旱而逐年衰退，具有固沙功能而又耗水量小，不具有抽水机效应的一片片地毯似的，以地衣为优势的微型生物结皮却逐年发育，从而使规划中的人工植被为主体的固沙防护体系演替成以天然形成的生物结皮为主体，混有残留人工植被和裸露沙土之比为 6.1 ∶ 2 ∶ 1.92（Yang and Wei，2012），甚至可达 9 ∶ 1（李守中，2005）的固沙防护新体系。沙坡头地区无灌溉人工植被固沙防护体系的覆盖度已经逐年降至近于 6% 的水分平衡状态（李守中，2005）。

虽然沙坡头地区无灌溉人工植被固沙防护体系未能达到预期效果，然而，通过该体系的实施，在水分平衡和“物竞天择，适者生存”规律的作用下，使规划中的人工植被固沙防护体系逐年演变成符合水分收支平衡的生物结皮为主体，混有残留人工植被的固沙防护新体系。

顺应这一自然规律，通过生物技术途径，将荒漠生物结皮的漫长自然形成过程变为人工培养与接种，使之快速形成以生物结皮为主体的地毯式固沙防护体系“沙漠生物地毯工程”设想应运而生（魏江春，2005）。

三、荒漠微型生物结皮

生物结皮是指由蓝细菌（蓝藻）、地衣和藓类植物等微型生物在沙表与沙粒缠绕在一起而形成的皮壳状生物组织。研究者对生物结皮有多种称谓（图 3-7-2），然而，微型生物结皮（microbiotic crusts，MBC）已被多数研究者所采纳（Belnap et al.，2001）。

有关美国西部、澳大利亚、以色列的荒漠微型生物结皮有大量文献报道，而南美和亚洲则少有报道（Belnap et al.，2001）。实际上，荒漠微型生物结皮几乎分布于地球表面一切荒漠地带，包括高纬度的南北极地和低纬度高海拔的高山寒漠地带以及南北半球的干旱半干旱荒漠地带。但是，微型生物结皮在年平均降水量为 30 mm 左右的极干旱地区，如我国南疆干旱荒漠则难以生长。

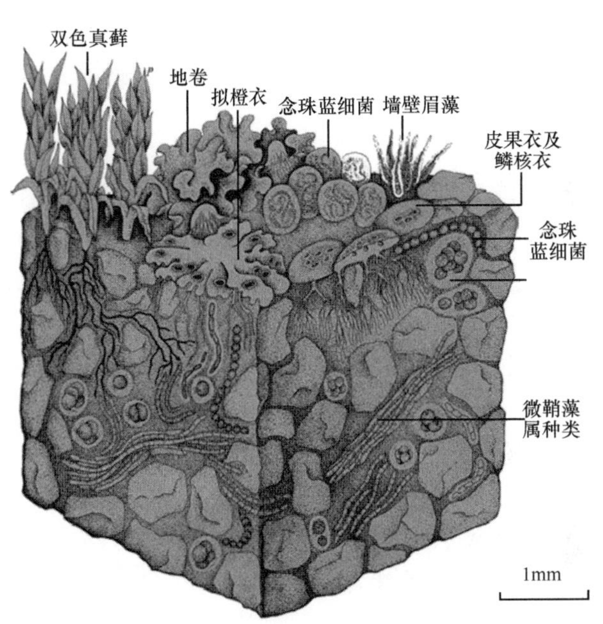

图 3-7-2 生物结皮示意图，结皮厚度为 3mm；其中结皮生物示意图未按比例（Belnap et al.，2001）

据调查研究，沙坡头地区荒漠微型生物结皮的生物组分有地衣、苔藓和蓝藻（蓝细菌）等。通过样方调查统计和分析表明，各生物组分盖度分别为蓝藻 21.1%，地衣 28.5%，苔藓 19% 和土壤 27%（图 3-7-3，Yang and Wei，2012）。

根据在四个不同恢复时间结皮区所作的 8 个样地，总计 155 个样方得到的统计数据，得出：①17 年发育区：裸沙 50%，结皮 50%，其中，地衣 12%，蓝藻 37%，藓类 1%；②26 年发育区：裸沙 21%～27%，结皮 73%～79%，其中，地衣 18%～25%，蓝藻 25%～29%，藓类 12%～20%；③43 年发育区：裸沙 39%，结皮 61%，其中，地衣 31%，蓝藻 10%，藓类 9%；④50 年发育区：裸沙 15%，结皮 85%，其中，地衣 45%～48%，蓝藻 8%，藓类 29%～30%；⑤50～51 年：裸沙 24%～25%，结皮 75%～76%，其中，地衣 23～28%，蓝藻 21%～31%，藓类 18%～33%；总平均为裸沙 27%，结皮 73%，其中，地衣 28.5%，蓝藻 21.125%，藓类 19%。

可以看到，地衣的盖度逐年稳步增加，蓝藻的盖度呈现逐年减少的趋势，而苔藓的盖度也是逐渐增多，同时，随着结皮发育成熟，裸露土壤也渐渐变少。这说明在沙坡头地区，发育初期的结皮是以蓝藻为主，同时结皮的盖度并不大，裸露土壤较多，而随着恢复年代增加，结皮逐渐成熟，此时成熟的结皮是以地衣和苔藓为主，同时结皮也能覆盖更多的土壤。结果显示微型生物结皮的形成与发育程度，虽然受自然环境复杂性的干扰，但是总的趋势和走向随着时间的延长而结皮发育越良好（Yang and Wei，2012）。

沙坡头地区荒漠微型生物结皮中当前的各类生物组分比例及其在生物演替过程中的消长趋势（图 3-7-3）表明，地衣盖度明显占优势；地衣及苔藓盖度与微型生物结皮发育年代呈正相关，即发育年代越长则盖度越大。然而，蓝藻盖度及裸露沙土面积与生物结皮发育年代呈负相关，即发育年代越长则盖度和面积越小。而地衣的盖度则与发育年代呈正相关，即发育年代越长则盖度越大。

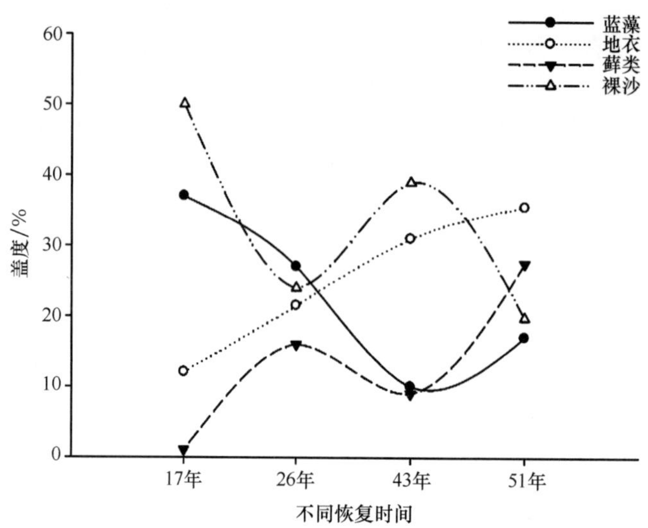

图 3-7-3 沙坡头地区荒漠微型生物结皮中各生物组分在演替中的消长（Yang and Wei，2012）

四、微型生物结皮中地衣物种及其固沙作用

为了对于“沙漠生物地毯工程”的可行性进行前期研究，有必要对于沙坡头地区荒漠微型生物结皮，尤其是其中的地衣物种组成进行了研究。

地衣是一大群专化型真菌与特定的藻类及蓝细菌处于稳定共生中的生态群落。通常所谓的地衣物种实际上只这一共生生态群落中真菌的物种。

1. 地衣种类组成

根据初步研究表明，沙坡头地区荒漠微型生物结皮中发现地衣计 36 种，隶属于 23 属，16 科和 8 目（Yang and Wei，2012；刘萌和魏江春，2012）。

沙坡头地区地衣名录：

微孢衣目 Acarosporales

微孢衣科 Acarosporaceae

糙聚盘衣

Glypholecia scabra（Pers.）Müll. Arg.

黄茶渍目 Candelariales

黄茶渍科 Candelariaceae

灰体黄茶渍

Candelariella antennaria Räsänen

莲座黄茶渍

C. rosulans（Müll. Arg.）Zahlbr.

茶渍目 Lecanorales

茶渍科 Lecanoraceae

蒙古副茶渍

Lecania mongolica H. Magn.（D. J. Galloway，1985）

加州茶渍衣

Lecanora californica Brodo

碎茶渍衣

L. argopholis（Ach.）Ach.
红鳞茶渍衣
Rhizoplaca chrysoleuca（Sm.）Zopf
梅衣科 Parmeliaceae
卷叶黄梅
Xanthoparmelia convoluta（Kremp.）Hale
鳞网衣科 Psoraceae
红鳞网衣
Psora decipiens（Hedw.）Hoffm.
树花衣科 Ramalinaceae
异色杆孢衣
Bacidia heterochroa（Müll. Arg.）Zahlbr.
珊瑚枝科 Stereocaulaceae
条斑鳞茶渍
Squamarina lentigera（G. H. Weber）Poelt
异极衣目 Lichinales
异极衣科 Lichinaceae
泥蜂窝衣
Heppia lutosa（Ach.）Nyl.
厚顶盘目 Ostropales
薄蜡衣科 Asterothyriaceae
大孢星衣亚种
Gyalidea asteriscus ssp. *gracilispora* J. Yang 和 J. C. Wei
污核衣科 Porinaceae
安奈污核衣
Porina aenea（Wallr.）Zahlbr.
疣孔衣科 Thelotremataceae
藓生双缘衣
Diploschistes muscorum（Scop.）R. Sant.
地卷目 Peltigerales
胶衣科 Collemataceae
球胶衣
Collema coccophorum Tuck.
地图衣目 Rhizocarpales
腊肠衣科 Catillariaceae
座叶泡鳞衣
Toninia sedifolia（Scop.）Timdal
黄枝衣目 Teloschistales
粉衣科 Caliciaceae
黑白黑瘤衣
Buellia alboatra（Hoffm.）Th. Fr.
美丽黑瘤衣
B. venusta（Körber）Lettau

椰子黑盘衣

Pyxine cocoës (Sw.) Nyl. Mém.

陆生饼干衣

Rinodina terrestris Tomin

黄枝衣科 Teloschistaceae

南方拟橙衣

Fulgensia australis (Arnold) Poelt

拟橙衣

F. bracteata (Hoffm.) Räsänen

荒漠拟橙衣

F. desertorum (Tomin) Poelt

凹面茸枝衣

Seirophora lacunosa (Rupr.) Frödén

柔毛茸枝衣

S. villosa (Ach.) Frödén

丽黄石衣

Xanthoria elegans (Link) Th. Fr.

瓶口衣目 Verrucariales

瓶口衣科 Verrucariaceae

旱石果衣

Endocarpon aridum P. M. McCarthy

晶石果衣

E. crystallinum Wei 和 Yang

淡白石果衣

E. pallidum Ach.

石果衣

E. pusillum Hedw.

诺氏石果衣

E. rogersii P. M. McCarthy

莲座石果衣

E. rosettum Amar Singh 和 Upreti

黑边石果衣

E. simplicatum (Nyl.) Nyl.

中华石果衣

E. sinense H. Magn.

褐点鳞衣

Placidium rufesens (Ach.) A. Massal.

2. 地衣群落组成及其固沙作用

沙坡头地区荒漠微型生物结皮中作为菌藻共生稳定共生群落的地衣之间形成的地衣群落有 3 种：球胶衣群落，球胶衣-石果衣群落，石果衣群落。根据顶极群落-格局理论，各种类型的群落，不是截然呈离散状态，而是连续变化的，因而形成连续的顶极类型，构成一个顶极群落连续变化的格局。因此，这三种地衣群落的演替关系是：球胶衣群落为先锋群落（pioneer community），其先锋种（pioneer species）和为优势种（dominant species）为球胶衣。该群落的形成是在流沙上最早形成的原生演替（primary succession）群落。其次为球胶衣-石果衣群落的出现。其建群种为球胶衣和石果衣。最后为石果衣群落，其建群种为石果衣。由于这3种群落中都没有分层现象，所以它们的建群种（constructive species）也是优势种。至于其他属的地衣在微型生物结皮中均为零星分布，很少能形成优势地位。其中球胶衣属的种类在微型生物结皮群落中的盖度高达71.9%，石果衣为27.6%，其他地衣只有0.5%（Yang and Wei，2012）。

地衣和灌木等维管束植物不同的是，它们不但不具有庞大根系将沙土深层水分通过根茎运输和叶的蒸腾作用送向大气的抽水机效应，而且地衣及其群落以其彼此相接的皮壳状地衣体所形成的地毯式生物组织（图3-7-4和图3-7-5）对沙表所起的覆盖作用，以及通过其下表假根或茸毛（图3-7-6和图3-7-7）伸向沙土中以固定沙粒，起到明显的固沙作用。

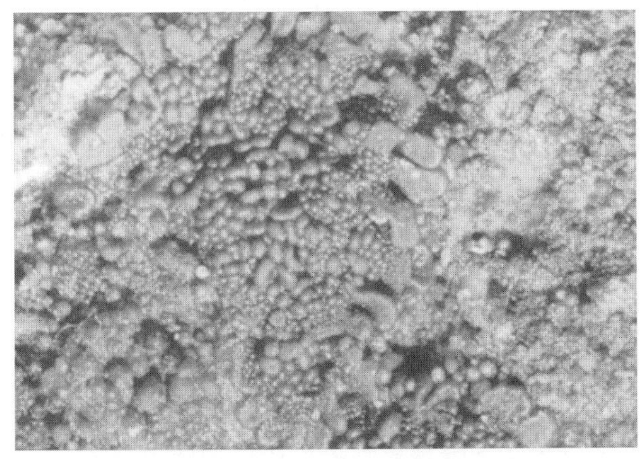

图3-7-4 球胶衣（*Collema coccophorum* Tuck.）结皮

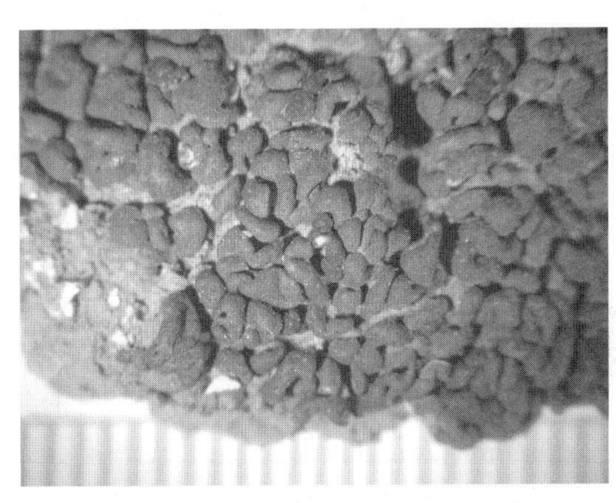

图3-7-5 座叶泡鳞衣（*Toninia sedifolia*（Scop.）Timdal）结皮

图3-7-6 球胶衣茸毛缠绕的沙粒

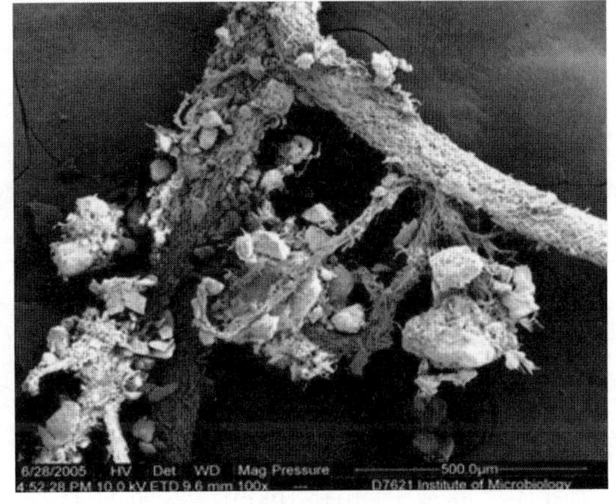

图3-7-7 石果衣假根对沙粒的固定

五、石果衣共生菌抗旱存活能力分析

根据前人关于地衣型和非地衣型真菌和藻类抗旱存活能力研究中，地衣共生菌在干燥条件下可存活5周（3个种）或6周（7个种）；而共生藻可存活6周（13个种）或41周（2个种）（Ahmadjian and Hale，1973）。在对自由生活的陆生真菌研究中，石生真菌被认为是抗逆性极强的自由生活真核生物（Gorbushina et al.，2003），其中，外生瓶梗菌（*Exophiala* sp.）和暗巩膜菌（*Phaeosclera* sp.）在干燥脱水32天后仍然存活，而束孢菌（*Sarcinomyces* sp.）和忍冬枝孢（*Cladosporium* sp.）在干燥脱水32天后死亡（Sterflinger和Krumbein，1995）。此外，Gorbushina等（2008）研究发现石生真菌石生束孢菌（*Sarcinomyces petricola*）在空气干燥8个星期（56天）后仍然存活。

然而，根据对荒漠地衣石果衣共生菌藻抗旱存活能力分析结果表明，其共生菌可以在干燥和饥饿胁迫下存活 7 个月，在饥饿胁迫下 8 个月后，不仅仍能存活，而且其长势与对照相当。由此可见，共生菌在饥饿单胁迫下的存活能力远比在干燥和饥饿双胁迫下更强（Zhang and Wei，2011）。

石果衣的共生藻已被鉴定为柯氏复球藻（*Diplosphaera chodatii* Bialosuknia）。据前人研究发现，柯氏复球藻也可自由生活并广泛分布于世界干旱荒漠地区（Flechtner et al.，2008）。自由生活的陆生单细胞藻类小球藻（*Chlorella vulgaris*）在干旱胁迫 1 个月后仅有 1% 的存活率（Agrwal and Singh，2001），而丝状藻类，如原型微鞘藻（*Microcoleus chthonoplastes*）、鲍氏席藻（*Phormidium bohneri*）、厚壁根枝藻（*Rhizoclonium crassipellitum*）、毛鞘丝藻（*Lyngbya mesotricha*）和米氏伪枝藻（*Scytonema millei*）在干燥胁迫下可存活半个月至 2 个月（Gupta and Agrawal，2006），也是上述实验中最长的耐旱存活期。

作者团队对石果衣共生藻柯氏复球藻所进行的实验表明，在干燥胁迫下能够存活 2 个月，从而也显示出比较强的抗旱性能（Zhang and Wei，2011）。

实验表明，分离自石果衣的共生菌在饥饿胁迫下的抗旱存活期最长可达 7 个月，这就为石果衣共生菌在干旱荒漠条件下与其共生藻柯氏复球藻进入共生状态提供了长达 7 个月的机会，从而为该地衣在干旱荒漠中形成优势种提供了条件。

至于石果衣共生菌在饥饿单胁迫下 8 个月以后的存活期的临界点，以及每隔半年或更短期的不断降雨间隔后，该种共生菌的存活期的临界点有待进一步研究。

尽管其共生藻柯氏复球藻的抗旱临界点仅为 2 个月，但是，在间歇式地零星降雨后导致菌藻同步释放，以及自然界存在的能够抗旱长达 7 个月的共生菌，都为石果衣的共生菌、藻在干旱荒漠中实现共生联合，不断形成地衣及其繁衍提供了广泛机会。

该实验结果为对分离自石果衣的共生菌、藻进行人工培养及其野外接种以形成地衣结皮固沙、固碳的干旱沙漠微型生物地毯工程的可行性提供了科学依据；也为从该种地衣共生菌中筛选抗旱基因时需要的干旱胁迫临界点提供了数据参考（Zhang and Wei，2011）。

六、石果衣菌藻人工繁殖及其野外接种技术的研究

通过液体培养途径，是沙漠生物地毯工程所需要的地衣型菌、藻大规模繁殖必由之途。在不同摇床转速条件下，石果衣共生菌的生长量存在显著性差异。当摇床转速为 150 r/min 时，共生菌生长量最高，与其他水平具有显著性差异，其次分别为 120 r/min 和 180 r/min。它们两者之间共生菌的生长量没有显著差异。由此表明，共生菌液体培养中摇床的最适转速为 150 r/min。

在优化复合液体培养基中，石果衣共生菌的生长延迟期为 30 天，然后进入对数生长期，约 45 天，期间生长量迅速增加，在培养 75 天后（干重 1.474 g）进入稳定期，90 天后（干重 1.536 g）其生长量达到最大值。在 LB 培养基中，共生菌的延迟期为 45 天，对数生长期约 30 天，生长 75 天后（干重 0.196 g）进入稳定期，90 天后（干重 0.197 g）达到最大值。通过对比分析显示，在培养 90 天时，优化培养基中生长量是基础培养基生长量的近 8 倍。

在不同温度条件下，共生藻柯氏复球藻的生长量存在显著性差异。共生藻在 22℃ 时的生长量最高，其次分别为 16℃、28℃、34℃和 10℃。由此表明，共生藻的最适生长温度为 22℃。

在不同温度条件下，共生藻柯氏复球藻的生长量存在显著性差异。共生藻在 22℃ 时的生长量最高，其次分别为 16℃、28℃、34℃和 10℃。由此表明，共生藻的最适生长温度为 22℃。

在不同 pH 条件下，共生藻柯氏复球藻的生长量存在显著性差异。共生藻在 pH 为 5 时的生长量最高，其次分别为 pH6、pH4、pH7、pH8 和 pH9。由此表明，共生藻的最适生长 pH 为 5（张涛，2011）。

不同地衣种类在一定程度上影响着微型生物结皮表面的微型地貌。在制订沙漠生物地毯工程方案设计时，应该结合不同地区条件，选用能够导致微型生物结皮表面呈现凹凸不平的地衣种类组成，以利于降水下渗。

当前本实验室正在进行石果衣共生菌、藻的液体繁殖和野外沙地接种试验阶段。

七、转基因技术在沙漠生物地毯工程中应用的可行性研究

荒漠中降水极少，多为阵雨性质，微管植物得不到正常水分供应，几乎是寸草不生！然而，地衣却是荒漠中“物竞天择，适者生存”的最大优胜者！因此，地衣作为荒漠微型生物结皮中的优势类群，是沙漠生物地毯工程的主要组成部分之一，是抗逆境基因资源的宝库。如果能够利用荒漠中优胜者地衣的相关基因改造草皮植物，也是科学家值得探索的途径之一。

转基因工程技术是人造优良品种的革命性技术突破。然而，转基因粮食作物品种的食用安全问题，一直是人们所关心的重大民生问题之一，更是科学家应该破解的重大科学问题之一。

关于转基因品种与传统杂交品种都是基因的转移，因而都是安全的论点，有将复杂问题简单化之嫌。传统杂交品种的基因转移是通过生物系统本身自我调节和适应以后而接受的结果。至于转基因品种的基因转移，则省去了生物系统本身自我调节和适应的过程，而是人为的强制行为。人工强制装配的外源基因，除了起到了抗病、抗虫等作用之外，在基因组表达过程中究竟还扮演了何等角色？人们一无所知。作为基因型和表型之间连接桥梁的代谢物组学的任务就在于定量描述生物内源性代谢物质的整体及其对内因和外因变化的应答规律。看来，探明这一规律，可能是破解转基因粮食作物品种是否安全这一重大科学问题的关键所在。不过，在尚未破解转基因粮食品种安全问题之前，非粮食的转基因品种还是可行的。为了用于固沙防护体系的转基因草皮植物，虽然生态安全问题是不可排除的，但是，与食品安全和固沙防护体系的意义相比，毕竟还是次要问题。

荒漠生物竞争中的优胜者地衣是抗逆境基因资源的宝库。为了从荒漠地衣中筛选抗逆境基因资源，本实验室已率先构建了地衣型真菌，即荒漠地衣石果衣真菌的全长 cDNA 文库；从中鉴定了 110 个基因。其中与已知基因同源性在 75% 以上者 42 个，占总数中的 38. 18%；与已知基因同源性在 74%~55% 者 29 个，占总数的 26. 86%；与已知基因同源性在 54% 以下者 34 个，占总数的 30. 91%；与已知基因毫无同源性者 5 个，占总数的 4. 55%（Wang 等，2011），为用于沙漠生物地毯工程的耐旱转基因草皮植物提供基因资源储备。

八、展望

无论是以微型生物结皮为主体，混有稀疏人工植被的为一体的沙漠生物地毯工程，还是以耐旱转基因草皮为主体，混有稀疏人工植被的为一体的沙漠生物地毯工程的成功实施之后，流沙移动和沙尘暴灾害将被遏制或缓解，实现人和大自然的和谐，为将大面积沙漠变为发展节水日光温室养殖产业基地成为可能。

参考文献

慈龙骏等. 2005.《中国的荒漠化及其防治》. 北京：高等教育出版社

邸醒民 . 2005. 沙坡头沙漠科学研究 50 年回顾.《中国沙漠研究与治理 50 年》. 北京：海洋出版社

冯金朝，陈荷生，康跃虎，等. 1995. 腾格里沙漠. 沙坡头地区人工植被蒸散耗水与水量平衡的研究. 植物学报，37（10）：815-821

李守中. 2005.《中国沙漠研究与治理 50 年》. 北京：海洋出版社

李新荣，石庆辉，张景光，等 . 1998. 沙坡头地区人工植被演变过程种植物多样性变化的研究. 中国沙漠，18（增）：23-29

李新荣，肖洪浪，刘立超，等. 2005. 腾格里沙漠沙坡头地区固沙植被对生物多样性恢复的长期影响. 中国沙漠，25（2）：173-181

刘萌，魏江春. 2012. 腾格里沙漠沙坡头地区地衣物种多样性研究（待发表）

魏江春. 2005. 沙漠生物地毯工程—干旱沙漠治理的新途径. 干旱区研究，22（3）：287-288

杨喜林. 2005. 沙坡头铁路防沙研究工作回忆.《中国沙漠研究与治理 50 年》. 北京：海洋出版社

张涛. 2011. 荒漠地衣共生菌藻营养繁殖及其抗旱生物学初步研究（博士论文）

张新时. 1994. 毛乌素沙地的生态背景及其草地建设的原则与优化模式. 植物生态学报，18（1）：1-16

赵兴梁. 1988. 腾格里沙漠沙坡头地区流沙治理研究. 银川：宁夏人民出版社

Agrwal S. C, Singh V. 2001. Viability of dried cells, and suvivability and reproduction under water stress, low light, heat and UV exposure in Chlorella vulgaris. Isr J Plant Sci, 2001, 49: 27-32

Ahmadjian V, & Hale M. E, eds. 1973. The Lichens. New York: Academic Press

Belnap, J. , B. Büdel, O. L. Lange . 2001. Biological Soil Crusts: Characteristics and Distribution in J. Belnap O. L. Lange (Eds.) Biological Soil Crusts: Structure, Function, and Management China. Geomorphology, 59: 237-246

Duan Zhenghu et al. 2004. Evolution of soil properties on stabilized sand in the Tengger Desert, China. Geomorphology, 237-246

FlechtnerV. R. , Johansen J. R. , Belnap J. 2008. The biological soil crusts of the San Nicolas Islands: enigmatic algae from a geographically isolated ecosystem. Western North American Naturalist, 68, 405-436

Galloway D. J. 1985. Flora of New Zealand Lichens P. D. Hasselberg, Wellington, New Zealand: Government Printer, 1-662

Gorbushina A. A. , Kotlova E R, Sherstneva O. A. 2008. Cellular responses of microcolonial rock fungi to long-term desiccation and subsequent rehydration. Stud Mycol, 61: 91-97

Gorbushina A. A. 2003. Microcolonial fungi: survival potential of terrestrial vegetative structures. Astrobiology, 3: 543-554

Gupta S. , Agrawal S. C. 2006 . Survival of blue-green and green algae under stress conditions. Folia Microbiol, 51: 121-128

Li Xinrong. 2005. Influence of variation of soil spatial heterogeneity on vegetation restoration Science in China Ser. D Earth Science, 48 (11): 2020-2031

Long Li-Qun, Li Xin-Rong . 2003. Effects of soil microbiotic crusts on seedling survival and seedling growth of two annual olants . Journal of Desert Research, 23 (6): 659 (2003) (全文: 656-660) (中国沙漠, 全文: 656-660)

Moore W. G. 1973. A Dictionary of Geography. 刘伉等译. 1984. 地理学词典. 北京: 商务印书馆

Shi Sha et al. 2004. The effect on different vegetation structure to soil water contents in Shapoto region. Journal of the CUN (Natural Sciences Edition) 13 (2): 139

Sterflinger K. , Krumbein W. E. 1995. Multiple stress factors affecting growth of rock-inhabiting black fungi. Bot Acta, 108: 490-496

Wang Yan-Yan, Tao Zhang, Qi-Ming Zhou et al. 2011. Construction and characterization of a full-length cDNA library from mycobiont of Endocarpon pusillum (lichen-forming Ascomycota) World Journal of Microbiology and Biotechnology 27 (12): 2879-2884. DOI 10. 1007/s 11274-011-0768-5

Xiao Hong-Lang et al. 2003a. Succession of plant-soil system in the process of mobile dunes stabilization. Journal of Desert Research, 23 (6): 608

Xiao Hong-Lang et al. 2003b. Impact of evolution of plant-soil system on the water environment during the mobile dunes stabilization. Acta Pedologica Sinica 40 (6): 810-814

Yang Jun, Wei Jiang-Chun. 2012. Lichen species diversity and their ecological significance in the Tengger Desert, China (To be published)

Zhang Tao, Wei Jiang-Chun. 2011. Survival analyses of symbionts isolated from Endocarpon pusillum Hedwig to desiccation and starvation stress. SCIENCE CHINA Life Sciences, 54 (5): 480-489

Desert lichens in Shapotou region of Tengger Desert and bio-carpet engineering

YANG Jun[1, 2] WEI Jiang-Chun[1*]

[1]*The State Key Laboratory of Mycology, Institute of Microbiology, Chinese Academy of Sciences, Beijing 100101, China*

[2]*Department of Biology, Beijing Polytechnic, Beijing 100029, China*

Abstract: This paper is dealt with the desert lichens and bio-carpet engineering. Twenty three lichens in the crust microbiota of Shapotou Desert Research Station (SDRS) of CAS are reported in this paper. Among them 2 species new to science and have been published earlier, 1 genus and 6 species are new to China. The ecological succession after the establishment of the artificial vegetation protection system in the Shapotou region of the southeast fringe of the Tengger Desert was analyzed. The artificial vegetation provided a habitat suitable to developing the microbiotic crusts formed by crust microbiota including cyanobacteria, lichens, mosses etc. The artificial vegetation with water pump effect expended the deep soil water during the long process of ecological succession by the action of the water balance, and led itself to decline. On the contrary, the crust microbiota without water pump effect and with the function of drought resistance, sand and carbon fixation developed well. Such a result provided a scientific basis for the feasibility of constructing "bio-carpet engineering" on the arid desert. By means of the technique of isolation and inoculation of the crust microbiota, the lichen *Endocarpon pusillum* as one of the species resources for the "bio-carpet engineering" is selected. A study of the drought resistant transgenic sward plants using the drought resistant genes from the desert lichens is carried out to improve the "bio-carpet engineering".

Key words: artificial vegetation, crust microbiota, *Endocarpon pusillum*

腾格里沙漠沙坡头地区荒漠地衣与生物地毯工程

杨军[1,2] 魏江春[1*]

[1] 中国科学院微生物研究所真菌学国家重点实验室 北京 100101

[2] 北京电子科技职业学院生物工程学院 北京 100029

摘　要：文章论述了荒漠地衣与“沙漠生物地毯工程”。在沙坡头结皮微型生物中发现了 23 种地衣，其中两个新

Supported by the National Natural Science Foundation of China (No. 30670004) and Main Direction Program of Knowledge Innovation of Chinese Academy of Sciences (No. KSCXZ-YW-Z-1020 and KSCXZ-EW-J-6).

*Corresponding author. E-mail: weijc2004@126.com

Received: 28-06-2013, accepted: 03-09-2013

This paper was originally published in *Mycosystema*, 33（5）: 1025-1035, 2014

种已发表，一属 6 种为中国新记录。对于在腾格里沙漠东南角沙坡头地区人工植被固沙防护体系建成后的生态演替进行了分析。由于人工植被为形成结皮的微型生物提供了适宜的生长环境而导致微型生物结皮的形成和发育。在水分平衡规律的作用下漫长生态演替过程中，具有抽水机效应的人工植被使沙土深层水分消耗殆尽，从而导致人工植被自身逐年衰退。然而，与此相反的是无抽水机效应而具有固沙、固碳和抗旱功能的结皮微型生物却逐年形成并发育。这一结果为借助于结皮微型生物的接种技术在干旱沙漠构建“沙漠生物地毯工程”的可行性提供了科学依据。为了优化“沙漠生物地毯工程”利用荒漠地衣耐旱基因以构建转基因草地植物的研究也正在进行中。该研究是“沙漠生物地毯工程”基础研究的组成部分。

关键词：人工植被，结皮微型生物，石果衣

INTRODUCTION

Storm is a physical process of the atmospheric flow (Moore 1975). It cannot be controlled by the human beings. Sand-dust storm depends upon whether the sand and dust on the surface of the earth are covered and controlled or not. This can be controlled by the human beings.

The Shapotou region is located in the southeast fringe of the Tengger Desert at 37°32′N and 105°02′E with an elevation of 1 339m, belonging to steppified desert zone, also a transitional zone between desert and oasis (Li *et al.* 1998; Li *et al.* 2000). The mean annual precipitation in Shapotou region is 180.2mm, and mean annual evaporation 3 000mm. Mean annual air temperature is 10.0°C, with minimum of -25.1°C and maximum of 38.1°C; annual sunshine duration is 3 264h; mean annual wind velocity 2.9 ms-1, and annual number of dust-storm days is 59 (Li 2005b). The classification of the arid region used here follows McGinnies *et al.* (1968) and Noy-Meir (1973). Extreme arid (E): less than 60–100mm mean annual precipitation; Arid (A): from 60–100mm to 150–250mm; Semiarid (S): from 150–250mm to 250–500mm. The Shapotou region is considered as arid region.

The artificial vegetation protection system without irrigation in the Shapotou section of the Baotou-Lanzhou railway was established in 1956. The coverage of the artificial vegetation was 15%–25% during the first decade since the artificial vegetation had been cultivated, and from the second decades to 2005, the coverage of the artificial vegetation was reduced to 6%–9%. At the same time, the microbiotic crusts developed well, and during the fourth decades the coverage of the microbiotic crusts developed into more than 90%, and their thicknesses reach about 11mm. So, the artificial vegetation protection system was replaced by the new protection system of the dominant microbiotic crusts with some annual herbs and scattered survived artificial shrubs (Li 2005a).

The microbiotic crusts result from an intimate association between soil particles and cyanobacteria, algae, microfungi, lichens, and bryophytes. Soil particles are aggregated through gathered by the presence and activity of these biota, and the resultant living crust covers the surface of the ground as a coherent layer (Belnap & Lange 2001). These biota are called crust microbiota in this paper.

The water balance principle is an essential prerequisite of the ecological engineering by artificial vegetation in the arid and semi-arid regions (Zhang 1994). Only 1%–5% of the water absorbed by the root system of the terrestrial plant from soil is utilized for metabolism, and 95%–99% of it evaporated through the transpiration (Zhou 2004). Such a terrestrial plant looks like a water pump in the evaporation of the soil moisture in the arid region. The artificial vegetation in Shapotou region looks like a large number of water pumps which were installed in the arid desert.

The artificial vegetation protection system with effect of water pump expended the deep soil water in the Shapotou region during the 40 years ecological succession under the water balance principle, and led itself to decline year by year. On the contrary, the crust microbiota without effect of water pump and with the function of drought resistance, sand and carbon fixation developed well year after year. Then, the declined artificial vegetation protection system was replaced by the new protection system of the dominant carpet like crust microbiota, and mixed up with some annual herbs and scattered survived artificial shrubs. This is a living example of the "natural selection, or the survival of the fittest" (Darwin 1859).

The desert lichens in China, excluding Shapotou region, were studied earlier by the Swedish lichenologist Magnusson (Magnusson 1940, 1944). Recently, some authors mentioned: "We did not find lichens as a component of crusts in the Shapotou region" (Li *et al.* 2002, 2004). As the matter of fact, during the investigation of the crust microbiota in the Shapotou region of the Tengger Desert we have found that lichens are quite common and even dominant in the crust microbiota.

A lichen is a symbiotic association of a fungus (mycobiont) and an alga (phycobiont) or a cyanobacterium (cyanobiont). In the association the fungus produces a thallus or body, within which the phycobiont or cyanobiont is housed (Ahmadjian 1993). The mycobiont in the symbiotic association is a specialized or so called lichen-forming fungus belonging mainly to the Ascomycota and a few to the Basidiomycota in the kingdom Fungi. The scientific name of the lichen-forming fungus is also considered as that of the lichen. The alga and the cyanobacterium have their own scientific names in the kingdom Plantae and Bacteria respectively.

The "desert bio-carpet engineering" was suggested (Wei 2005) to construct the carpet-like crust microbiota on the surface of the arid desert by the biotech.

Biological diversity is diversity of species containing diversity of genes in the diversity of ecological systems of the biosphere. Desert lichen biodiversity is diversity of lichen species containing diversity of genes in different desert regions and the pool of lichen species resources for "desert bio-carpet engineering". This study is one of the part of the basic research for the "desert bio-carpet engineering".

1 MATERIALS AND METHODS

The materials examined were collected from the desert area in the Shapotou region of the Tengger Desert. The identification of the materials was carried out by the methods of morphology, anatomy, and chemistry. The dissecting microscope Motic SMZ-168 and the compound microscope Zeiss Axioscope 2 plus were available for studies of the morphology and anatomy. A standardized thin-layer chromatographic (TLC) method for identification of lichen products was used (Culberson 1972; Culberson & Kristinsson 1970). All the specimens examined are preserved in Herbarium Mycologicum Academiae Sinicae-Lichenes (HMAS-L).

The quadrat sampling method was available for the investigation of the coverage of crust microbiota, artificial vegetation, and naked sand soil and their growth and decline during the succession of the artificial vegetation. The four artificial vegetation zones cultivated from 1956, 1964, 1981, and 1990 respectively in the Shapotou Desert Research Station (SDRS) of the Chinese Academy of Sciences (CAS) were selected. The different vegetation zones represent the different ages of the microbiotic crusts. The eldest crust is in the 1956 artificial vegetation zone, and the youngest one in 1990 zone.

2 RESULTS AND DISCUSSIONS

2.1 The growth and decline of the artificial vegetation and crust microbiota

The coverages of the artificial vegetation, crust microbiota, and naked sand soil in the different ages of the artificial vegetation zones in SDRS of CAS were investigated in 2007. Fig. 1A & B clearly illustrates that the coverages of the artificial vegetation were maintained at the stable level of 19.42%–21.79% in the different ages from 17 to 51 years. The coverage of the

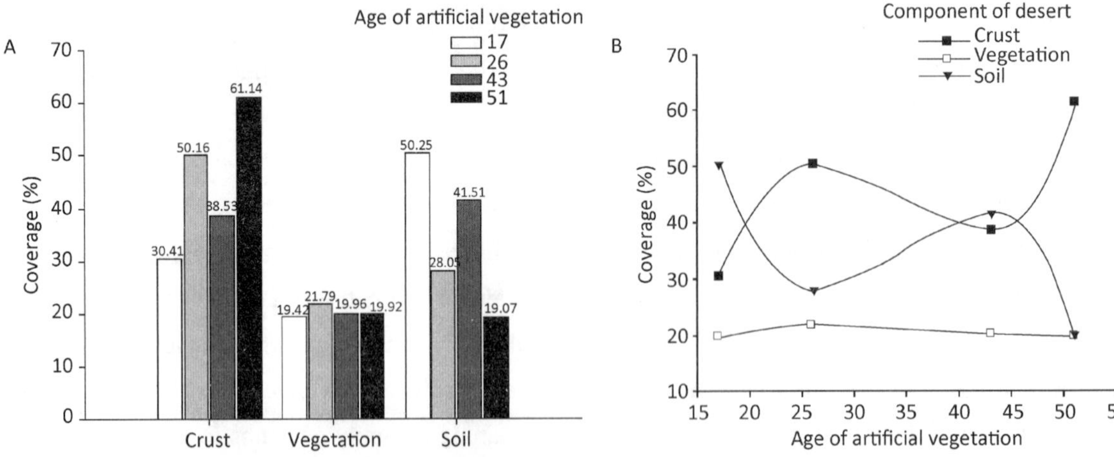

Fig. 1 A & B, the coverages of the artificial vegetation, crust microbiota and naked sand soil were changed in the different years.

naked sand soil was decreased from 50.25% to 19.07% from the age of 17 to 51 years. The coverage of the crust microbiota, however, was increased from 0% to 30.41% during the first 17 years, and from 30.41% to 61.14% during the 17th to 51th years.

It is clear that the coverage of the artificial vegetation was reduced to 19.42% by the action of the water balance during 17 years and still maintained at the same level (19.92%) during 51 years. On the contrary, the crust microbiota were formed and developed, and their coverage was increased from 0% to 61.14% during 51 years when the artificial vegetation was cultivated. The coverage of the crust microbiota in this case should be increased by 1.2% per year. So, it is estimated that the coverage of the crust microbiota on the all sand soil surface, including that under the artificial vegetation will require 83 years after the cultivation of artificial vegetation in the arid desert. In this case, the execution of the "desert bio-carpet engineering" would shorten such a protracted process of succession by a big margin.

2.2 The coverage, growth and decline of the cyanobacteria, lichens, mosses, and naked sand soil in the crust microbiota

As to the lichens in the crust microbiota, some authors indicated that they did not find lichen as a component of crust in the SDRS of CAS (Li *et al.* 2002, 2004). It's a fact that lichens are very common, and even dominant in the crust microbiota in the region. Twenty three lichens belonging to 17 genera, 14 families, and 8 orders in the Ascomycota in the crust microbiota of Shapotou region are reported in this paper. Among them 2 species including one as a subspecies with the mark "♣" were described by the authors earlier and 1 genus with the mark "#" and 6 species with the mark "♥" are new to China. These lichens are included in the checklist of the lichens from the region studied.

Appendix

The check list of the lichens from shapotou region Tengger Desert

Candelariales

Candelariaceae

Candelariella rosulans (Müll.Arg.) Zahlbr.

Shapotou, Cuiliugou: 1 684m, Augt. 10, 2008. Zhang T. & Cao S.N. SPT3; 1 619m, April. 13, 2011. Liu M. SPT11032; Xiangshan, 1 377m, April 17, 2011. Liu M. SPT11045, SPT11046, SPT11047; Shapotou, on the way to 1 410m, April 19, 2011. Zhou Q.M. & Cao S.N. ZW11016 (Liu & Wei 2013).

Lecanorales

Lecanoraceae

Lecania mongolica H. Magn.

Shapotou: Aug. 7, 2003. Wei J.C. & Yang J. SPT021, 022, 024; May 26, 2004. Yang J. & Wang H.Y. SPT065 & 066; Aug. 26, 2006. Yang J. & Zhang T. SPT285.

Lecanora argopholis (Ach.) Ach.

Shapotou, Cuiliugou: 1 669m, July 3, 2010. Cao S.N. & Liu W.J. SPT10011 (Liu & Wei 2013).

Parmeliaceae

Xanthoparmelia camtschadalis (Ach.) Hale

Shapotou, on the way to Qingshan: 1

410m, April 19, 2011. Zhou Q.M. & Cao S.N. ZW11016.

Psoraceae

Psora decipiens (Hedw.) Hoffm.

Shapotou, Cuiliugou: 1 600–1 669m, July 3, 2010. Cao S.N. & Liu W.J. SPT10007, SPT10016, SPT10033; April 9–16, 2010. Zhang T. SPT10071, SPT10104; Qingshan, 1 535–1 569m, April 19, 2011. Zhou Q.M. & Cao S.N. ZW11005, ZW11023, ZW11030, ZW11044, ZW11058, ZW11071 (Liu & Wei 2013).

Stereocaulaceae

Squamarina lentigera (G.H. Weber) Poelt

Shapotou, Cuiliugou: 1 600m, Aug. 28, 2008. Zhang T. Z08062, Z08063, Z08064; May 28, 2009. Zhang T. NX09011, NX09013, NX09023, NX09024 & SPT10096 (2010.4.9); On the north of railway: 1 400m, April 17, 2007. Yang J. & Zhang T. SPT231, & April 10, 2010. Zhang T. SPT10037 (Liu & Wei 2013).

Ostropales

Asterothyriaceae

Gyalidea asteriscus (Anzi) Aptroot & Lücking #

Subsp. ***gracilispora*** Yang & Wei ♣

Shapoyou: Aug. 6, 2003. Wei J.C. & Yang J. SPT372.

Thelotremataceae

Diploschistes muscorum (Scop.) R. Sant.

Shapotou, Cuiliugou: 1 604m, April 13, 2011. Liu M. SPT11038 (Liu & Wei 2013).

Teloschistales

Caliciaceae

Pyxine cocoës (Sw.) Nyl.

Shapotou: on the north of railway, 1 400m, Aug. 25, 2007. Yang J. & Zahng T. SPT368, SPT230; Qingshan, 1 420m, April 19, 2011. Liu M. SPT11079; Cuiliugou, 1 684m, Aug. 29, 2008. Zhang T. & Cao S.N. SPT1-4 & May 29, 2009. Zhang T. Z08071, NX09027 & April 9, 2010. SPT10069; Qingshan, 1 535m, April 19, 2011. Zhou Q.M. & Cao S.N. ZW11037 (Liu & Wei 2013).

Physciaceae

Rinodina terrestris Tomin

Shapotou: on the north of railway, 1 400m, Aug. 13, 2004. Yang J. SPT140 & Aug. 26, 2006. Yang J. & Zhang T. SPT285 (Liu & Wei 2013).

Teloschistaceae

Fulgensia bracteata (Hoffm.) Räsänen

Shapotou, Cuiliugou: 1 600m, April 9, 2010. Zhang T. SPT10088 (Liu & Wei 2013).

Fulgensia desertorum (Tomin) Poelt

Shapotou: Cuiliugou, 1 600–1 669m, April 9–16, 2010. Zhang T. SPT10033, SPT10076, SPT10094, SPT10095 & July 5, 2010. Cao S.N. & Liu W.J. SPT10015, SPT10027; Qingshan, 1 360m, April 19, 2011. Liu M. SPT11058, SPT11060 (Liu & Wei 2013).

Fulgensia fulgens (Sw.) Elenkin ♥

Shapotou: May 26, 2004. Yang J. & Wang H.Y. SPT057.

Seirophora lacunosa (Rupr.) Frödén

≡*Teloschistes lacunosa* (Rupr.) Savicz

Shapotou: Cuiliugou, 1 600m, April 17, 2010. Zhang T. SPT10110 (Liu & Wei 2013).

Seirophora villosa (Ach.) Frödén

=*Teloschistes brevior* (Nyl.) Vain. [14] (Magnusson 1940).

Shapotou: Cuiliughou, 1 669m, July 3, 2010. Cao S.N. & Liu W.J. SPT10019 (Liu & Wei 2013).

Lichinales

Lichinaceae

Heppia lutosa (Ach.)Nyl. ♥

Shapotou: Aug. 7, 2003. Wei J.C. & Yang J. SPT004, 032; May 26, 2004. Yang J. Wang H.Y. SPT078, 097; Aug. 26, 2006. Yang J. & Zhang T. SPT 343, 352, 353, 357, 371.

Peltigerales

Collemataceae

Collema coccophorum Tuck. ♥

Shapotou: Aug. 6–9, 2003. Wei J.C. *et al.* SPT041, 042, 046, 049, 050, 053, 054, 055, 056, 057, 058, 059, 060, 061; July 13, 2004. Yang J. & Wang H.Y. SPT069, 070, 071, 072, 073, 074, 075, 076, 077, 078, 079, 080, 081, 082, 083, 084, 085, 086, 087; Nov. 4, 2006. Yang J. & Zhang E.R. SPT176, 177, 178, 179, 180, 181, 182, 183, 184, 185, 186, 187, 188, 189, 190.

Rhizocarpales

Catillariaceae

Toninia alutacea (Anzi) Jatta ♥

Shapotou: Aug. 6, 2003. Wang H.Y. & Yang J. SPT012.

Verrucariales

Verrucariaceae

Endocarpon aridum P.M. McCarthy ♥

Shapotou: Aug. 6–9, 2003. Yang J. *et al.* SPT006, 007, 008, 009, 010, 013, 014, 017, 018 & 026.

Endocarpon *crystallinum* J.C. Wei & J. Yang ♣

Shapotou: Xiangshan, April 17, 2007. Yang J. & Zhang T. SPT-202.

Endocarpon pallidum Ach.

Shapotou: Aug. 6–9, 2003. Wei J.C. SPT035, 044, 046, 051, 052, 055 & 056; Aug. 12–13, 2003. Yang J. & Wang H.Y. SP081, 083, 086, 087, 096, 114, 115 & 117.

Endocarpon pusillum Hedw.

Shapotou: Aug. 6–9, 2003. Wei J.C. *et al.* SPT067, 068, 070, 071, 076, 077, 097, 098, 099, 100, 101, 102, 104, 105, 106, 107, 108 & 109; July 13, 2004. Yang J. & Wang H.Y. SPT111, 112, 123, 125, 126, 130, 132, 134, 141, 142, 143, 144, 151, 169 & 175; Nov. 4, 2006. Yang J. & Zhang E.R. SPT201, 211, 212, 213.

Endocarpon rogersii P.M. McCarthy ♥

Shapotou: Aug. 6–9, 2003. Yang J. *et al.* SPT031, 032, 062.

The result of investigation based on 155 quadrats showed that the crust microbiota in Shapotou region mainly consist of cyanobacteria, lichens, and mosses. The alteration of the coverage of cyanobacteria, lichens, mosses and also the naked sand soil is showed that the cyanobacteria and the naked sand soil were decreased, and the lichens and mosses increased in the different ages of the artificial vegetation zones (Fig. 2). among them the lichens are predominant component.

2.3 The lichen communities and their ecological significance

There are 3 lichen communities in the crust microbiota in the Shapotou region, i.e., the *Collema coccophorum* community, the *C. coccophorum-Endocarpon pusillum* community, and the *E. pusillum* community. The *C. coccophorum* community is the pioneer one,

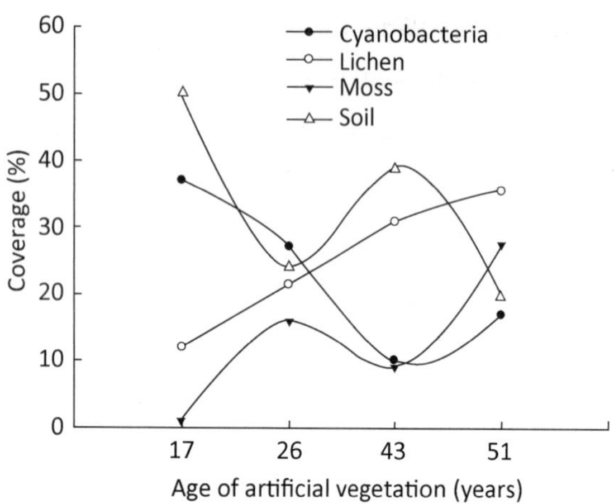

Fig. 2 The growth and decline of the different organisms in the microbiotic crusts.

which is the first to be formed on the surface of the sand during the ecological succession, and both the pioneer and dominant species is the same *C. coccophorum*. The *C. coccophorum-E. pusillum* community is the secondary one. The constructive species is both *C. coccophorum* and *E. pusillum*. The third one is *E. pusillum* community, and the constructive species is *E. pusillum* its self. The lichens of the other genera in the crust microbiota are scattered. Among them the coverage of *Collema* spp. in crust microbiota is 71.9%, and that of *E. pusillum* is 27.6%, and others are only 0.5% (Fig. 3).

The surface of the sand soil are covered by the desert lichens with their squamose thalli (Fig. 4A), and the sand crystals are wrapped with the branched lichen rhizines, hyphae, and also fishnet like mycelia (Fig. 4B–F).

The data above mentioned show that desert lichens are the dominant organisms in the crust microbiota. The data show that the mycobiont isolated from the desert lichen *Endocarpon pusillum* can survive for seven months under desiccation stress in combination with starvation stress. The phycobiont of it *Diplosphaera chodatii* Bialosuknia can survive for two months under desiccation stress. The result provides a scientific basis for further research of the reproduction biology of the desert lichens and arid "desert bio-carpet engineering" to fix sand and carbon (Zhang & Wei 2011; Ding *et al*. 2013). Based on small-scale sequencing results from the full-length cDNA library of the mycobiont of the same *E. pusillum*, 111 genes were identified. Among them 11 genes shared no homology with any known fungal genes (Wang *et al*. 2011). So, the desert lichens are not only the treasure-house of species resources, but also that of the extrinsic genes for the transgenic drought-resistant grasses for the "desert bio-carpet engineering".

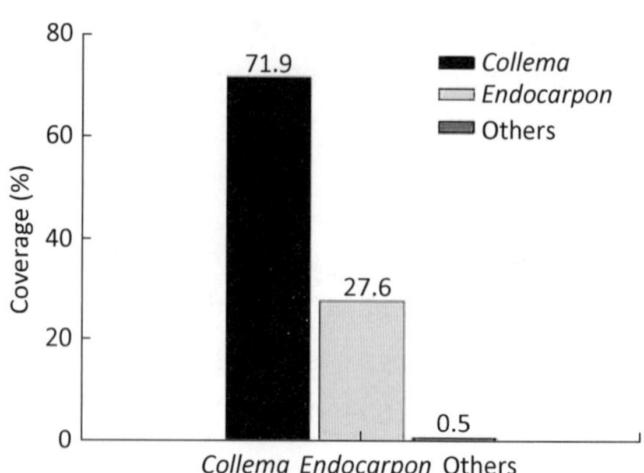

Fig. 3 The coverage of the lichens in the crust microbiota.

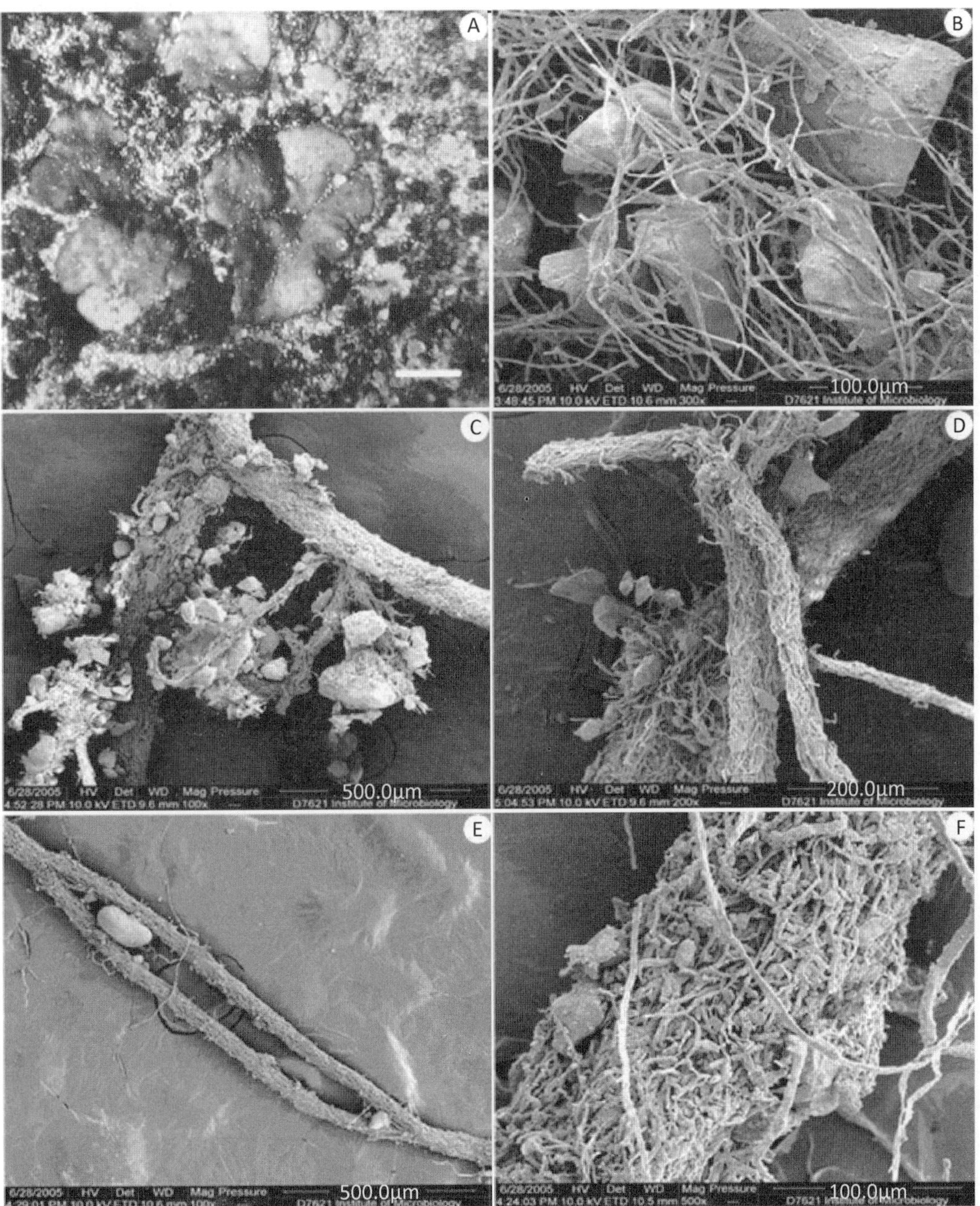

Fig. 4 A: *Endocarpon pusillum* covers the surface of sand soil with its squamose thalli (bar=1mm); B: *Collema* sp. wraps the sand crystals with its fishnet like mycelia (bar=1mm) (SEM); C: *E. pusillum* fixes the sand crystals with its branched rhizines (bar=500μm) (SEM); D: *E. pusillum* fixes the sand crystals with its branched rhizines (bar=200μm) (SEM); E: A reniform sand crystal is clamped by a branched rhizine of a lichen (bar=500μm) (SEM); F: The sand crystals are knitted in lichen mycelia (bar=100μm) (SEM).

3 CONCLUSIONS

The artificial vegetation protection system was replaced by the dominant crust microbiota during a long process of ecological succession under the water balance. Following the natural law to implement the “desert bio-carpet engineering” by the isolation and culture of the crust microbiota, and their artificial inoculation on the arid desert to form the crust microbiota and using the drought resistant genes from the desert lichens to construct the drought resistant transgenic sward plants is a new way to choose for arid desert control. The desert lichen biodiversity for “desert bio-carpet engineering” is a treasure-house of species resources containing diversity of the adversity-resistant genes. The lichen *Endocarpon pusillum* with the features of drought resistance, sand and carbon fixation is selected as a species resource for the bio-carpet engineering to develop in the fields. The “desert bio-carpet engineering” will provide a fixed and stable surface of the sandy desert suitable for the cultivated industry in the green house.

Acknowledgement: The authors are very grateful to Ms. W. Guo for making technique treatments of some figures in the paper.

[REFERENCES]

Ahmadjian V, 1993. The lichen symbiosis. John Wiley, New York. 255

Belnap BB, Lange OL, 2001. Biological soil crusts: charateristics and distribution. In: Belnap J, Lange OL (eds.) Biological soil crusts: structure, function, and management. Springer, New York. 3-30

Culberson CF, 1972. Improved conditions and new data for identification of lichen products by standardized thin-layer chromatographic method. *Journal of Chromatography A*, 72(1): 113-125

Culberson CF, Kristinsson HD, 1970. A standardized method for the identification of lichen products. *Journal of Chromatography A*, 46: 85-93

Darwin C, 1859. The origin of species (with a special introduction by Julian Huxlev). 1958 by the New American Library of World Literature, INC. Times Miror, New York and Toronto, The New English Library Limited, London. 1-479

Ding LP, Zhou QM, Wei JC, 2013. Estimation of *Endocarpon pusillum* Hedwig carbon budget in the Tengger Desert based on its photosynthetic rate. *Science China Life Sciences*, 56(9): 1-8

Li SZ, 2005a. Effect of mechanical disturbance to microbiotic crusts on soil hydrological process in revegetated sanddune regions. The Research on the Sandy Desert and Control for 50 Years. Ocean Press, Beijing. 252-259 (in Chinese)

Li XR, 2005b. Influence of variation of soil spatial heterogeneity on vegetation restoration. *Science in China Series D: Earth Sciences*, 48(11): 2020-2031

Li XR, Shi QH, Zhang JG, Liu LC, Chen HS, 1998. Study of plant diversity changes during the artificial vegetation evolution processes in the Shapotou region. *Journal of Desert Research*, 18(Suppl. 4): 23-29

Li XR, Wang XP, Zhang JG, 2002. Microbiotic soil crust and its effect on vegetation and habitat on artificially stabilized desert dunes in Tengger Desert, North China. *Biology and Fertility of Soils*, 35:

147-154

Li XR, Xiao HL, Zhang JG, Wang XP, 2004. Long-term ecosystem effects of sand-binding vegetation in the Tengger Desert, Northern China. *Restoration Ecology*, 12(3): 376-390

Li XR, Zhang JG, Wang XP, Liu LC, Xiao HL, 2000. Study on soil microbiotic crust and its influences on sand-fixing vegetation in arid desert region. *Acta Botanica Sinica*, 42(9): 965-970

Liu M, Wei JC, 2013. Lichen diversity in Shapotou region of Tengger Desert, China. *Mycosystema*, 32(1): 42-50

Magnusson AH, 1940. Lichens from Central Asia I. Rep. Sci. Exped. N.W. China S. Hedin – The Sino-Swedish expedition – (Publ.13). XI. Bot.I: 1-168. Pl.1-12, 1 folded map. f. 1-3

Magnusson AH, 1944. Lichens from Central Asia II. Rep. Sci. Exped. N.W. China S. Hedin – The Sino-Swedish expedition – (Publ.13). XI. Bot.II: 1-68. Pl.1-8, 1 text map

McGinnies WG, Goldman BJ, Paylore P (eds.), 1968. Deserts of the world. Univ. Arizona, Tucson. 188

Moore WG, 1975. A dictionary of geography – definitions and explanations of terms used in physical geography. Adam & Charles Black, London. 1-439

Noy-Meir I, 1973. Desert ecosystems: environment and producers. *Annual Review of Ecology and Systematics*, 4: 25-51

Wang YY, Zhang T, Zhou QM, Wei JC, 2011. Construction and characterization of a full-length cDNA library from mycobiont of *Endocarpon pusillum* (lichen-forming Ascomycota). *World Journal of Microbiology and Biotechnology*, 27(12): 2879-2884

Wei JC, 2005. The desert biocarpet engineering – a new way to control the arid desert. *Arid Zone Research*, 22(3): 287-288 (in Chinese)

Zhang T, Wei JC, 2011. Survival analyses of symbionts isolated from *Endocarpon pusillum* Hedwig to desiccation and starvation stress. *Science in China Life Sciences*, 54(5): 480-489

Zhang XS, 1994. Principles and optimal models for development of Maowusu sandy glassland. *Acta Phytoecologica Sinica*, 18(1): 1-16

Zhou YL (ed.), 2004. Plant biology. 2nd edition. Higher Education Press, Beijing. 1-555 (in Chinese)

三种荒漠地衣共生菌藻的耐热性研究

陈凯[1,2]　魏江春[1*]

[1]中国科学院微生物研究所真菌学国家重点实验室 北京 100101

[2]中国科学院大学 北京 100049

摘　要：作为沙漠生物地毯工程研究的组成部分，本文对荒漠地衣石果衣 *Endocarpon pusillum*、节瘤微孢衣 *Acarospora nodulosa* 以及荒漠微孢衣 *A. schleicheri* 的共生菌藻进行了耐热性研究。结果表明，三种荒漠地衣的共生菌、藻在湿润条件下对温度的最大忍受力仅为 50℃，而在干燥条件下石果衣的共生菌藻的最大忍受力均为 75℃，而其他两种地衣共生菌藻的最大忍受力均为 80℃。三种荒漠地衣共生菌、藻分别对高温胁迫的耐受力，在干燥状态下均比在湿润状态下明显增高。

关键词：生物地毯工程，湿润，干燥，温度

Heat tolerance of the mycobionts and phycobionts from three desert lichens

CHEN Kai[1,2]　WEI Jiang-Chun[1*]

[1]State Key Laboratory of Mycology, Institute of Microbiology, Chinese Academy of Sciences, Beijing, 100101, China.

[2]University of Chinese Academy of Sciences, Beijing, 100049, China.

Abstract: This paper deals with the heat tolerance of mycobionts and phycobionts isolated from desert lichens *Endocarpon pusillum*, *Acarospora nodulosa*, and *A. schleicheri*. This study is a part of the research program of the desert biological carpet engineering. The results showed that the mycobionts and phycobionts of the three desert lichens can only tolerate the temperature of 50℃ under the moist conditions. However, under the arid conditions both the mycobiont and phycobiont of *E. pusillum* can tolerate the temperature of 75℃, and those of the other two lichens can tolerate the temperature of 80℃. It is concluded that the high-temperature tolerance of mycobionts and phycobionts isolated from desert lichens is much stronger under arid conditions than that under moist conditions.

[*]Corresponding author. E-mail: weijc2004@126.com

收稿日期: 2014-03-10, 接受日期: 2014-04-17

本文原载于《菌物学报》, 5: 1007-1014, 2015.

Key words: biological carpet engineering, moist, arid, temperature

地衣是由一种地衣型真菌和一种藻类或蓝细菌形成的稳定的共生生态系统。由于其结构的特异性与功能的适应性，使其广泛分布于从两极到热带雨林的多样生态系统中（Lawrey 1984）。

地衣是世界上最耐高温的生物之一。已有的研究结果表明，在干燥状态下大部分地衣具有很强的耐热性，例如，条斑鳞茶渍 *Squamarina lentigera* 置于 90℃高温 30 分钟，其共生藻仍能保持活性，而地卷 *Peltigera rufescens*、莲座石蕊 *Cladonia pocillum* 甚至在 100℃高温下 30 分钟后，其共生藻仍能存活（Lange 1953）。但是当完全复水时，地衣的耐热性与干燥状态下相比则较低，例如浅绿绣球衣 *Chrondropsis semiviridis* 在干燥状态下直到 105℃才完全失活；而完全复水后，在 55℃即完全失活（Rogers 1971）。虽然对地衣体的耐高温极限已经做了大量的研究，但在纯培养条件下的地衣共生菌藻的耐热性表现迄今尚无相关报道。

对非地衣型真菌与藻类的耐热性研究表明，土壤真菌灰绿曲霉 *Aspergillus glaucus* 在 70℃下能存活 60min；赭曲霉 *Aspergillus ochraceus* 在 90℃下能存活 10min（Jesenská *et al.* 1993）；在干燥状态下，绿绵菌 *Spongiochloris typical* 在 100℃下至少能存活 60min（McLean 1967）；而在湿润状态下，原型微鞘藻 *Microcoleus chthonoplastes* 和中胞鞘丝藻 *Lyngbya mesotricha* 在 45℃下 20min 便会死亡（Gupta & Agrawal 2006）。

荒漠化是当今世界重大的生态难题之一。在干旱荒漠地区，地衣能够固定沙土以减少沙尘暴的发生，并能通过光合作用固定二氧化碳以发挥其碳汇作用（Ding *et al.* 2013）。因此，对荒漠地衣及其共生菌藻的抗逆性研究具有重要的的生态学意义。本研究的目的在于通过实验获得纯培养条件下的地衣共生菌和共生藻的抗高温能力的数据，为“沙漠地毯工程”提供科学依据。

1 材料与方法

1.1 研究材料产地

用于菌藻分离的石果衣 *Endocarpon pusillum*、节瘤微孢衣 *Acarospora nodulosa* 和荒漠微孢衣 *A. schleicheri* 的新鲜地衣体采自宁夏中卫沙坡头地区（北纬 37°32′，东经 105°02′；海拔 1 720m）。该地区位于腾格里沙漠南缘，年均降雨量 186mm，年均蒸发量 2 800mm，夏季沙表温度最高可达 74℃（Chen *et al.* 1991）。

1.2 地衣共生菌藻的分离培养

用于菌藻分离培养的三种地衣的共生菌均由孢子释放法获得，并在 PDA 培养基中培养。石果衣的共生藻从其孢子与子实层藻同步释放的菌藻群体中分离而获得，经分子系统学方法鉴定该共生藻为柯氏复球藻 *Diplosphaera chodatii*（Zhang & Wei 2011）。节瘤微孢衣和荒漠微孢衣的共生藻由组织培养法获得，经分子系统学方法鉴定该两种地衣的共生藻均为褪色共球藻 *Trebouxia decolorans*。上述三种地衣的共生藻均在液体 BBM 培养基中培养。

1.3 地衣共生菌藻的耐热性

地衣共生菌藻的耐热性实验分别在干燥状态和湿润状态下进行。高温处理方法参考 Sterflinger（1998）。

共生菌在 PDA 培养基中培养 2 个月后，利用无菌不锈钢滤网过滤培养液，收集共生菌的菌丝球，并用无菌滤纸吸干表面水分。检测共生菌在湿润状态下的耐热性时，取适量菌体于1.5mL 无菌离心管中进行高温处理。高温处理在水浴锅内进行，温度范围从 40–60℃，温度梯度为 5℃，误差范围为±1℃，每个温度下分别处理 30、60、90、120、150、180min。检测共生菌在干燥状态下的耐热性时，取适量菌体于无菌培养皿内，将培养皿放入含有吸水变色硅胶的玻璃干燥器内（相对湿度≤10%，室温 20–27℃，室内光强）进行干燥处理，一周之后取出，并将干燥的菌体转到 1.5mL 无菌离心管中进行高温处理。温度范围从 50℃到 80℃，温度梯度为 5℃，每个温度梯度下处理时间为 30、60、90、120、150、180min。高温处理后，取小块菌体（直径≈1mm）接种到 PDA 培养基并在生化培养箱（18℃）内培养 4 周，观察生长情况，并利用扫描电镜和透射电镜对共生菌的形态进行观察。

共生藻在 BBM 培养基内培养 2 个月后，取 1mL 藻悬浮液于 1.5mL 无菌离心管中，低速离心使藻细胞沉淀，转速为 6 000r/min，用移液枪吸干上层液体。检测湿润状态下的藻细胞耐热性时，直接对吸干上层液体的藻细胞进行高温处理。高温处理同样在水浴锅内进行，温度范围从 40℃到 60℃，温度梯度为 5℃，误差范围为±1℃，每个温度下分别处理 30、60、90、120、150、180min。检测共生藻在干燥状态下的耐热性时，离心并吸干上层液体之后，将离心管口打开，横放在无菌培养皿内，然后将培养皿放入含有吸水变色硅胶的玻璃干燥器内（相对湿度≤10%，室温 20–27℃，室内光强）进行干燥处理，一周之后取出进行高温处理。温度范围从 50℃到 80℃，温度梯度为 5℃，每个温度梯度下处理时间为 30、60、90、120、150、180min。高温处理后，在每个离心管内添加 1mL 无菌水，使藻细胞重新悬浮，然后将悬浮液涂布于固体 BBM 培养基平板，置于光照培养箱（18℃，9 600Lux）内培养 4 周，观察生长情况，并利用扫描电镜和透射电镜对共生藻的形态进行观察。

以上每个实验处理都设 3 个重复，并用一组未经高温处理的正常活体共生菌藻（湿润和干燥）作为对照。

1.4 显微镜观察地衣共生菌藻形态

1.4.1 扫描电子显微镜：取适量菌丝块或藻细胞，加适量 2.5%戊二醛 4℃固定过夜；然后用 0.1mol/L 磷酸盐缓冲液(pH7.2)漂洗 3 次，10min / 次；50%-70%-85%-95%乙醇梯度脱水，各一次，15min / 次 100%乙醇脱水 3 次，15min / 次；然后进行二氧化碳临界点干燥(BAL-TEC CPD030)；最后进行喷金——离子溅射（BAL-TEC SCD005），并在扫描电镜（scanning electron microscopy，SEM，Quanta 200，FEI 公司）下观察。

1.4.2 透射电子显微镜：取适量菌丝块或藻细胞，在 2.5%戊二醛溶液中前固定，并用 0.1mol/L 磷酸钠缓冲液（pH7.2）漂洗；然后用 1%锇酸（OsO_4）固定 2h，0.1mol/L 磷酸钠缓冲液（pH 7.2）漂洗。然后进行乙醇梯度脱水，50%-70%-85%-95%乙醇各一次，100%乙醇脱水 3 次。依次向样品渗透 1：1 比例的乙醇和树脂 LR White 的混合液，2h；然后纯树脂过夜。之后样品放入包埋板包埋，60℃聚合 24h。最后用 Lecia uc7 切片机切成 60nm 的切片醋酸双氧铀一柠檬酸铅染色，并在透射电镜（transmission electron microscopy，TEM，JEM-1400，JEOL 公司）下观察。

2 结果与分析

2.1 菌体存活力比较

在湿润条件下，供试三种荒漠地衣的共生菌在 40℃时能够存活 180min，在 45℃时仅能存活 60min，在 50℃时便不能存活（表 1）。

表 1 湿润状态下共生菌的耐热性

Table 1 Heat tolerance of moist mycobionts

共生菌 Mycobiont	温度 Temperature (℃)	时间 Duration (min)					
		30	60	90	120	150	180
Endocarpon pusillum	40	+	+	+	+	+	+
	45	+	+	—	—	—	—
	50	—	—	—	—	—	—
	55	—	—	—	—	—	—
	60	—	—	—	—	—	—
Acarospora nodulosa	40	+	+	+	+	+	+
	45	+	—	—	—	—	—
	50	—	—	—	—	—	—
	55	—	—	—	—	—	—
	60	—	—	—	—	—	—
Acarospora schleicheri	40	+	+	+	+	+	+
	45	+	—	—	—	—	—
	50	—	—	—	—	—	—
	55	—	—	—	—	—	—
	60	—	—	—	—	—	—

注：“＋”表示存活，“－”表示死亡

Note: "+", alive; "—", dead.

在干燥状态下，石果衣的共生菌 60℃时能够存活 180min，在 65℃时能存活 120min，在 70℃时能存活 60min，在 75℃时便不能存活。其他两种供试地衣的共生菌则在 65℃时均能存活 180min，而节瘤微孢衣的共生菌在 70℃时能存活 120min，荒漠微孢衣的共生菌则能存活 90min，而在 75℃时，前者可存活 60min，后者仅能存活 30min，二者在 80℃时则均无法存活（表 2）。

2.2 藻细胞存活力比较

在湿润条件下，石果衣的共生藻柯氏复球藻 *Diplosphaera chodatii* 以及两种微孢衣的同种共生藻褪色共球藻 *Trebouxia decolorans* 在 40℃时均能存活 180min，在 45℃时均能存活 150 分钟，在 50℃时均不能存活（表 3）。在干燥状态下，前者能在 65℃时能存活 180min，在 70℃时能存活 150min，在 75℃时则不能存活；后者在 60℃时存活 180min，65℃时存活 150min，70℃时存活 90min，75℃时存活 30min，80℃时不能存活（表 4）。

3 种荒漠地衣的共生菌及共生藻在湿润条件下对温度的最大忍受力均为 50℃，而在干燥条件下的最大忍受力石果衣的共生菌藻均为

75℃，而其他两种地衣共生菌藻的最大忍受力均为 80℃。3 种荒漠地衣共生菌、藻分别对高温胁迫的耐受力，在干燥状态下均比在湿润状态下明显增高。

表 2 干燥状态下共生菌的耐热性

Table 2 Heat tolerance of dehydrated mycobionts

共生菌 Mycobiont	温度 Temperature (℃)	时间 Duration (min)					
		30	60	90	120	150	180
Endocarpon pusillum	50	+	+	+	+	+	+
	55	+	+	+	+	+	+
	60	+	+	+	+	+	+
	65	+	+	+	+	—	—
	70	+	+	—	—	—	—
	75	—	—	—	—	—	—
	80	—	—	—	—	—	—
Acarospora nodulosa	50	+	+	+	+	+	+
	55	+	+	+	+	+	+
	60	+	+	+	+	+	+
	65	+	+	+	+	+	+
	70	+	+	+	+	—	—
	75	+	+	—	—	—	—
	80	—	—	—	—	—	—
Acarospora schleicheri	50	+	+	+	+	+	+
	55	+	+	+	+	+	+
	60	+	+	+	+	+	+
	65	+	+	+	+	+	+
	70	+	+	+	—	—	—
	75	+	—	—	—	—	—
	80	—	—	—	—	—	—

注："+"表示存活，"—"表示死亡

Note: "+", alive; "—", dead.

表 3 湿润状态下共生藻的耐热性

Table 3 Heat tolerance of moist phycobionts

共生藻 Phycobiont	温度 Temperature (℃)	时间 Duration (min)					
		30	60	90	120	150	180
Diplosphaera chodatii	40	+	+	+	+	+	+
	45	+	+	+	+	+	—
	50	—	—	—	—	—	—
	55	—	—	—	—	—	—
	60	—	—	—	—	—	—
Trebouxia decolorans	40	+	+	+	+	+	+
	45	+	+	+	+	+	—
	50	—	—	—	—	—	—

	55	—	—	—	—	—	—
	60	—	—	—	—	—	—

注：“+”表示存活，“—”表示死亡

Note: “+”, alive; “—”, dead.

表 4 干燥状态下共生藻的耐热性

Table 4 Heat tolerance of dehydrated phycobionts

共生藻 Phycobiont	温度 Temperature (℃)	时间 Duration (min)					
		30	60	90	120	150	180
Diplosphaera chodatii	50	+	+	+	+	+	+
	55	+	+	+	+	+	+
	60	+	+	+	+	+	+
	65	+	+	+	+	+	+
	70	+	+	+	+	+	—
	75	—	—	—	—	—	—
	80	—	—	—	—	—	—
Trebouxia decolorans	50	+	+	+	+	+	+
	55	+	+	+	+	+	+
	60	+	+	+	+	+	+
	65	+	+	+	+	+	—
	70	+	+	+	—	—	—
	75	+	—	—	—	—	—
	80	—	—	—	—	—	—

注：“+”表示存活，“—”表示死亡

Note: “+”, alive; “—”, dead.

2.3 菌体存活力形态学观察

经过高温处理之后，在扫描电镜下，3 种菌种的对照组菌丝表面光滑完整，无破损情况，生长状态良好（图 1A、B、C）。而经过 80℃高温处理 30min 之后，3 种菌丝都出现了不同程度的破损（图 1D、E、F）。

在透射电子显微镜下，*E. pusillum* 正常菌丝内原生质分布均匀且有层次感，少数细胞器轮廓清晰可辨（图 1G）；经过 80℃高温处理 30min 之后，原生质固缩，层次感消失，细胞器已无法辨认（图 1J）。*A. nodulosa* 正常菌丝内原生质分布均匀，细胞器轮廓清晰可辨，且有数量较多的小液泡均匀分布在整个原生质内（图 1H）；经过 80℃高温处理 30min 之后，均匀分布的小液泡合为一个大液泡，其他细胞器已无法辨认，原生质层次感消失（图 1K）。*A. schleicheri* 正常菌丝内，与 *A. nodulosa* 类似，原生质分布均匀有层次感，细胞器轮廓清晰可辨（图 1I）；经过 80℃高温处理 30min 之后，原生质层次感消失，出现大液泡，细胞器轮廓无法辨认（图 1L）。

2.4 藻细胞存活力形态学观察

经过高温处理之后，藻细胞的形态学观察扫描电镜结果显示，经过高温处理之后，两种共生藻细胞的外部形态与对照相比无明显差异，均呈现均匀圆球状且未发生破裂。而透射电镜结果显示，正常藻细胞内部原生质分布均匀有层次感，且能观察到呈片层状细胞器轮廓；经过50℃高温处理30min之后，细胞器不可见，原生质虽均匀分布但层次感消失。

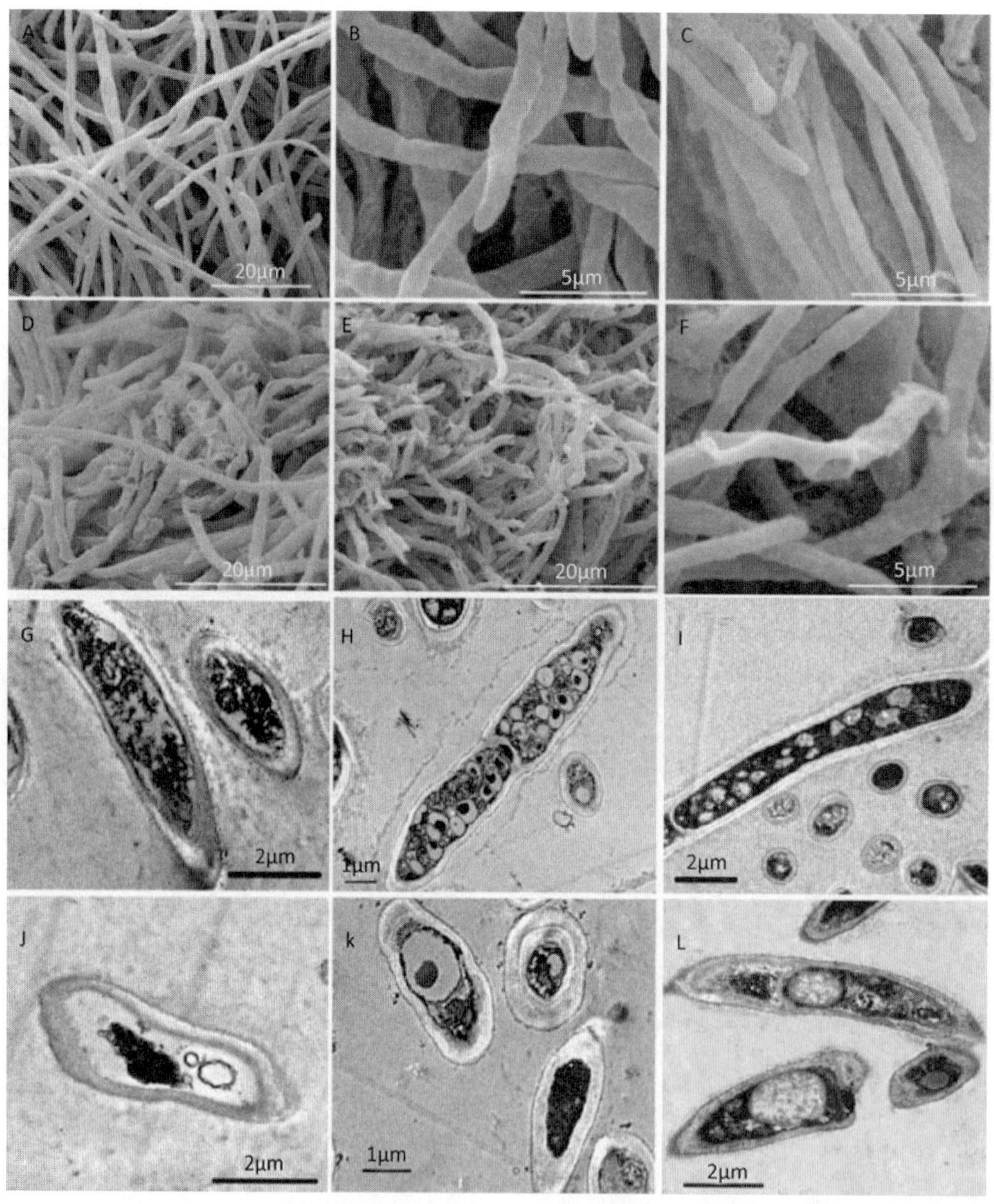

图 1 共生菌高温处理之后形态图 A—F：扫描电镜；G—L：透射电镜. A、G：*E. pusillum* 菌丝，对照；B、H：*A. nodulosa* 菌丝，对照；C、I：*A. schleicheri* 菌丝，对照；D、J：干燥状态 80℃处理 30min，*E. pusillum* 菌丝；E、K：

干燥状态 80℃处理 30min，*A. nodulosa* 菌丝；F、L：干燥状态 80℃处理 30min，*A. schleicheri* 菌丝.

Fig. 1 Morphology of mycobionts after high-temperature processing. A—F: SEM; G—L: TEM. A, G: Hyphae of *E. pusillum*, control group; B, H: Hyphae of *A. nodulosa*, control group; C, I: Hyphae of *A. schleicheri*, control group; D, J: Hyphae of *E. pusillum*, dehydrated, treated at 80℃ for 30min; E, K: Hyphae of *A. nodulosa*, dehydrated, treated at 80℃ for 30min ; F, L: Hyphae of *A. schleicheri*, dehydrated, treated at 80℃ for 30min.

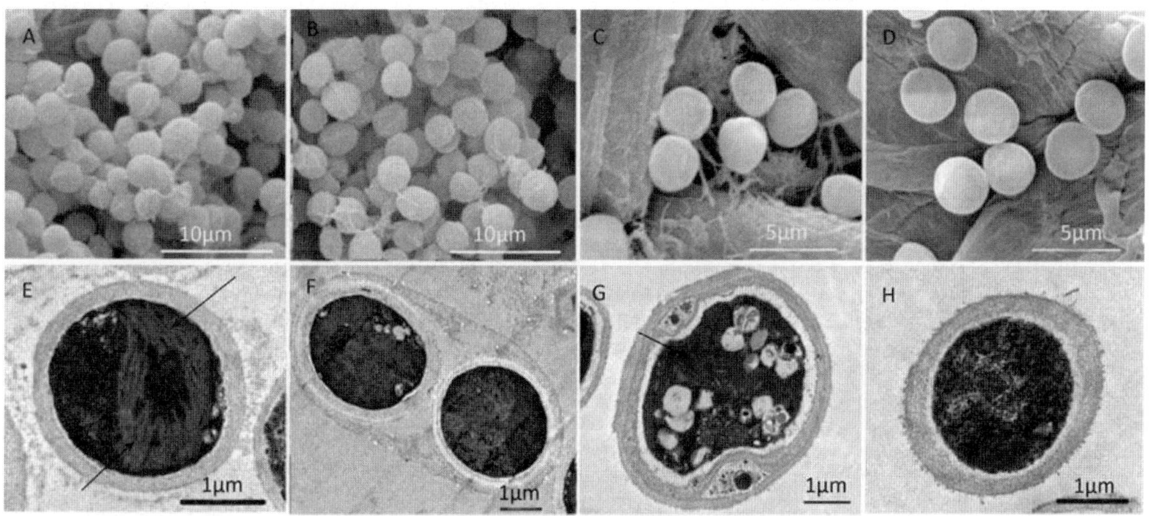

图 2 共生藻高温处理之后形态图 A—D：扫描电镜；E—H：透射电镜. A、E：*E. pusillum* 共生藻，对照；B、F：湿润状态 50℃处理 30min，*E. pusillum* 共生藻细胞；C、G：*A. nodulosa* 和 *A. schleicheri* 共生藻，对照；D、H：湿润状态 50℃处理 30min，*A. nodulosa* 和 *A. schleicheri* 共生藻细胞.

Fig. 2 Morphology of phycobionts after high-temperature processing. A—D: SEM; E—H: TEM. A, E: Algae cells of *E. pusillum*, control group; B, F: Algae cells of *E. pusillum*, moist, treated at 50℃ for 30min; C, G: Algae cells of *A. nodulosa* and *A. schleicheri*, control group; D, H: Algae cells of *A. nodulosa* and *A. schleicheri*, treated at 50℃ for 30min.

3 讨论

以上实验结果表明，本研究所选用的材料中，无论是地衣共生菌或共生藻，在干燥状态下对高温胁迫的耐受能力要比在湿润状态下高，这一结果与地衣对高温胁迫的耐受规律——耐干热而不耐湿热（Kappen & Smith 1980；MacFarlane & Kershaw 1980）结果相一致。同一属的地衣 *A. nodulosa* 与 *A. schleicheri* 的共生菌耐高温能力相似，且都比 *E. pusillum* 的耐高温能力强。地衣及其纯培养菌藻对干燥下的高温耐受力强于对湿润下的高温耐受力的原因，可能因为它们在干燥条件下处于休眠状态，因而耐受力较强；在湿润条件下则处于生活状态，因而耐受力较弱。休眠状态往往是真菌度过极端环境的方式。

从以上供试的 3 种荒漠地衣中分离培养的共生菌藻，在湿润和干燥条件下对温度的耐受力数据是沙漠生物地毯工程前期研究中进行菌藻野外接种时选择适宜环境条件和时段的重要依据。

[REFERENCES]

Chen H, Kang Y, Feng J, 1991. Preliminary study on the plant growth and water balance in Shapotou area, Tengger Desert. *Journal of Desert Research*, 11(2): 1-10

Ding LP, Zhou QM, Wei JC, 2013. Estimation of *Endocarpon pusillum* Hedwig carbon budget in the Tengger Desert based on its photosynthetic rate. *Science China Life Sciences*, 56(9): 848-855

Gupta S, Agrawal SC, 2006. Survival of blue-green and green algae under stress conditions. *Folia microbiologica*, 51(2): 121-128

Jesenská Z, Piecková E, Bernat D, 1993. Heat resistance of fungi from soil. *International Journal of Food Microbiology*, 19(3): 187-192

Kappen L, Smith CW, 1980. Heat tolerance of two *Cladonia* species and *Campylopus praemorsus* in a hot steam vent area of Hawaii. *Oecologia*, 47(2): 184-189

Lange OL, 1953. Hitze-und Trockenresistenz der Flechten in Beziehung zu ihrer Verbreitung. *Flora*, 140: 39-97

Lawrey JD, 1984. Biology of lichenized fungi. Praeger Publishers, New York.

MacFarlane JD, Kershaw KA, 1980. Physilogical-environmental in lichens. *New Phytologist*, 84(4): 669-685

McLean RJ, 1967. Desiccation and heat resistance of the green alga *Spongiochloris typica*. *Canadian Journal of Botany*, 45(11): 1933-1938

Rogers RW, 1971. Distribution of the lichen *Chondropsis semiviridis* in relation to its heat and drought resistance. *New Phytologist*, 70(6): 1069-1077

Sterflinger K, 1998. Temperature and NaCl-tolerance of rock-inhabiting meristematic fungi. *Antonie van Leeuwenhoek*, 74(4): 271-281

Zhang T, Wei JC, 2011. Survival analyses of symbionts isolated from *Endocarpon pusillum* Hedwig to desiccation and starvation stress. *Science China Life Sciences*, 54(5): 480-489

第四篇　地衣型真菌基因组学
Genomics of Lichenized Fungi

Construction and characterization of a full-length cDNA library from mycobiont of *Endocarpon pusillum* (lichen-forming Ascomycota)

Yan-Yan Wang · Tao Zhang · Qi-Ming Zhou · Jiang-Chun Wei

Received: 24 January 2011 / Accepted: 23 April 2011 / Published online: 5 May 2011

Abstract The first full-length cDNA library for lichenized fungi was constructed from cultured mycobiont of the arid desert lichen *Endocarpon pusillum*. Based on small-scale sequencing results, 111 genes of the lichenized fungi were identified for the first time, among which 11 genes shared no homology with any known fungal genes. Real-time PCR showed that the size of the mycobiont genome is 39.13 Mb and the copy number of ribosomal RNA gene repeat units is 43. The results of this study will be valuable for the ongoing lichen genome-sequencing project and the large-scale identification of functional genes from lichenized fungi.

Keywords Lichen · Lichenized fungus · Lichen-forming fungus · RNA isolation · Real time PCR

Introduction

Lichen is a symbiotic association comprising a specific fungus (referred to as lichenized fungus, lichen-forming fungus or mycobiont) together with a photobiont partner, which may be an alga (referred to as phycobiont) or a cyanobacterium (referred to as cyanobiont) or even both. Lichen is distributed with rather wide ecological amplitude, including Polar Regions, deserts and other extreme habitats. As terrestrial organism, lichen is pioneer organism in the rock-weathering and pedogenic processes. Lichen is able to regulate its water content and physiological activity rapidly in response to environmental stresses, and can revive after being desiccated for many months (Ilse et al. 2008).

With the passing of years, many sequences or genes of lichen-forming fungi were identified. However, most of the identified sequences or genes were used to resolve the phylogenetic positions of a variety of lichen species. The most widely used sequences are from the region of the nuclear ribosomal RNA gene repeat units, including SSU (small subunit) rDNA, LSU (large subunit) rDNA, ITS (internal transcribed spacer) rDNA, IGS (intergenic spacer) and group I intron (Palice et al. 2005; Lindblom and Ekman 2006; Staiger et al. 2006; Zhou et al. 2006; Gutierrez et al. 2007). Moreover, β-tubulin, RNA polymerase II and microsatellite markers were also used in some phylogenetic studies (Reeb et al. 2004; Walser et al. 2005; Buschbom 2007). Meanwhile, only a few studies have been published on functional genes from lichenized fungi, and these identified functional genes include *pyrG* in *Solorina crocea* (Sinnemann et al. 2000), *XPH1* in *Xanthoria parietina* (Scherrer et al. 2002), *pks* genes in *Xanthoparmelia semiviridis* (Chooi et al. 2008), *Solorina crocea* (Gagunashvili et al. 2009), *Dirinaria applanata* (Valarmathi et al. 2009), and *lip*3 in *Cladonia grayi* (Joneson et al. 2011). Construction of genomic or cDNA library is a general way for gene isolation. However, up till now only three genomic or cDNA libraries from mycobiont or lichen thalli have been described, which are genomic library from the mycobiont of *Xanthoria parietina* (Scherrer et al. 2002), genomic library from the whole lichens *Solorina crocea* (Sinnemann et al. 2000), and cDNA library from *Cladonia rangiferina* (Junttila et al. 2009), and very few

Y.-Y. Wang · J.-C. Wei
College of Life Sciences, Shandong Agriculture University, Taian 271018, China

Y.-Y. Wang · T. Zhang · Q.-M. Zhou (✉) · J.-C. Wei (✉)
Key Laboratory of Systematic Mycology and Lichenology, Institute of Microbiology, Chinese Academy of Sciences, Beijing 100101, China
e-mail: zhqm@im.ac.cn

J.-C. Wei
e-mail: weijc2004@126.com

This paper was originally published in *World Journal of Microbiology and Biotechnology*, 27（12）: 2879-2884, 2011.

functional genes were identified from these libraries. In the present study, we constructed a full-length cDNA library from the mycobiont of the desert lichen *E. pusillum* and some functional genes were identified from this library. Our work will be valuable for the ongoing lichen genome-sequencing project and the large-scale identification of functional genes from lichenized fungi.

Materials and methods

Materials and culture

The lichen materials of *E. pusillum* used in this study were collected from the Shapotou Region of Ningxia, south-eastern edge of the Tengger desert, China (latitude 37.40°N, longitude 105.00°E) and were stored at −80°C.

The spore discharge from a perithecium (Yoshimura et al. 2002) was used to isolate the mycobiont from *E. pusillum*. The apothecia of the mycobiont were attached to the underside of the lid of a plastic Petri dish, and the ascospores were synchronously discharged together with the algal cells of the phycobiont onto the surface of agar medium. A single colony of the mycobiont formed from a discharged ascospore was transferred to potato dextrose liquid medium. Cultures were incubated with shaking (120 rpm) at 20°C in the dark for 3 weeks.

The identity of the mycobiont culture from *E. pusillum* was verified by ITS rDNA sequence comparison using primers ITS1 and ITS4 (White et al. 1990). The voucher specimen of *E. pusillum* is deposited in the Herbarium Mycologicum Academiae Sinicae-Lichenes (HMAS-L).

RNA extraction

Three methods were used for the total RNA extraction from mycobiont. Fresh tissue samples of 300 mg were powdered in liquid nitrogen using mortar and pestle, and RNA was extracted with TRIzol reagent (Invitrogen, USA) (Michael et al. 2006) or by using the NucloSpin® RNA Plant Kit (Macherey–Nagel, Germany) according to the manufacturer's instruction. In the third method, the total RNA was crudely extracted using TRIzol reagent, and then filtered through NucleoSpin® Filter units after adjusting RNA binding conditions with Solution RA1 from the NucloSpin® RNA Plant Kit. Concentration and purity of RNA were assessed by measuring A_{230}, A_{260} and A_{280} in a NanoDrop instrument (NanoDrop Technologies, Wilmington, USA).

Construction of cDNA library

The construction of the full-length cDNA library was carried out using the Creator™ SMART™ cDNA library Kit (Clontech, USA) using standard protocols as outlined by the supplier. First-strand synthesis was initiated with an Oligo (dT) primer. The double strand (ds) cDNAs were synthesized and amplified by Long-Distance PCR (LD-PCR), and then were digested with restriction endonuclease *Sfi* I to generate cohesive ends. After fractionated by column chromatography, the cDNAs were ligated into the vector pDNR-LIB (*Sfi* I digested). The ratio of cDNAs to vector was 3:2. The ligation products were electroporated into electrocompetent *E. coli* strain DH5α cells. The electroporated cells were cultured in 1 mL of Luria–Bertani medium (LB) for 45 min at 37°C. One μL of cultured cells were diluted 1:10,000 with LB medium, and 100 μL of diluted solution was spread on a LB/chloramphenicol plate. The plates were incubated at 37°C overnight. The number of colonies in each plate was counted to determine the titer (cfu/ml). The titer was calculated according to the following formula: titer = (colony number/plating volume) × dilution factor. The original library was amplified by spreading on 100 plates overnight. Colonies were collected and stored in 30% glycerol at −80°C.

cDNA sequence analysis

Two hundred clones were randomly selected from the library to estimate the rate of insertion and the average size of inserted fragments by PCR using primers M13R and M13F (according to the manufacturer's protocol) which annealed to the upstream and downstream regions of the insertion site of vector pDNR-LIB. One hundred and eleven cDNA inserts whose lengths were more than 500 bp were sequenced using primers M13R and M13F, and walking reactions were carried out to complete insert sequences if necessary. The trimmed data were assembled using SeqMan program (DNASTAR package, DNASTAR Inc., USA). The DNAMAN software (Lynnon BioSoft, USA) was used to identify coding sequences (CDS) in the sequences and to translate them into protein sequences. Homology analysis of the obtained sequences against all known fungal protein sequences was performed using BLASTP program and GenBank database (http://www.ncbi.nlm.nih.gov). The results were marked with "high" if the homology scores over run 75%, or marked with "medium" when the homology scores were between 55 and 74%, or with "low" for rate lower than 54%, or with "N" for no hits.

Genomic DNA isolation

The genomic DNA was extracted from three-week cultured mycobiont using the cetyl trimethylammonium bromide (CTAB) method (Clarke 2009). The Fungal DNA Mini Kit (Omega Bio-Tek, USA) was used to purify DNA. The concentrations of DNA samples were determined in a Qubit fluorometer (Invitrogen, USA).

Real time quantitative PCR for genome sizing

Thirty-two primer pairs designed from 15 new genes, which were screened from the full-length cDNA library, were used to prepare the PCR fragment starters. Each 50 μL of PCR mixture contained 2 μL of mycobiont genomic DNA, 25 μL of 2× Taq Mix (CwBio, China), 1 μL of each primer (10 μM) and 21 μL of ddH_2O. PCR was performed with denaturation for 10 min at 95°C, followed by 40 cycles of 30 s at 95°C, 30 s at 60°C and 30 s at 72°C. The PCR fragment starters were observed on 1% agarose gel and those with clear band and without smear were purified with PCR purification kit (Omega Bio-Tek, USA). Concentrations of the purified products were determined in a Qubit fluorometer (Invitrogen, USA).

Four different concentrations of genomic DNA and six different concentrations of PCR fragment starters were used for each gene to reduce the variations. All sample analyses were repeated at least three times. Each 20 μL of quantitative PCR (qPCR) mixture contained 0.4 μL of each primer (10 μM), 0.4 μL of ROX Reference, 2 μL of template with 6.8 μL ddH_2O and 10 μL of SYBR Premix Ex Taq (2×) (TaKaRa). The qPCR was performed with the beginning for 2 min at 50°C; 10 min at 95°C followed by 40 cycles of 15 s at 95°C, 30 s at 60°C and 45 s at 72°C. Fluorescence intensity data were collected at the end of the extension step (72°C, 45 s). The qPCR was performed in an ABI 7300 real time PCR instrument (Applied Biosystems Inc., USA).

The amplification efficiency (E) was calculated using a series of concentrations (six gradients) of PCR fragment starters with each primer pair. Ct is the cycle number at which the fluorescence generated within a reaction crosses the fluorescence threshold. In this study, the threshold was automatically analysed by the ABI 7300 system software (Applied Biosystems Inc., USA). The size of genome is determined by the analysis of samples with known concentrations of genomic DNA (G) and PCR fragment starters (P), and the formula of this method is: Length of genome = length of P × E^($Ct_G - Ct_P$) × quantity (G) / quantity (P), in which E = 10^(−1/slope) (Armaleo and May 2009).

Real time qPCR for determination of the copy number of rRNA gene repeat units

The copy number of rRNA gene repeat units was determined by relative quantification PCR (Corradi et al. 2007; Herrera et al. 2009). Three genes (*SCD*, *ATA* and *RPB11a*), whose amplification efficiency are nearly same and close to the theoretical calculated value ($E_{theoretic} = 2$), were chosen as the single-copy genes according to the size of genome and the BLAST analysis in other fungal genomes. These three genes were considered as reference gene and their Ct were defined as Ct_{ref}. ITS1, which locates in each ribosomal RNA gene repeat unit, was considered as test gene and its Ct was defined as Ct_{test}. The following primers were used to amplify ITS1: forward primer, 5′-CCGTCTCCAACCCTTGTCT AT-3′; reverse primer, 5′-TTCGCTGCGTTCTTCATCG-3′.

Two serial dilutions of genome DNA were used as template in each reference reaction and test reaction. Each experiment was performed for three separate times. The qPCR was performed in an ABI 7300 real time PCR instrument (Applied Biosystems Inc., USA), under the same conditions used for the above absolute qPCR. The copy number of rRNA gene repeat units was calculated by the formula of 2^($Ct_{ref} - Ct_{test}$).

Results

Sequencing results showed that the ITS rDNA sequence from mycobiont was the same as that from lichen, indicating that the culture derived from the thallus of *E. pusillum*. These two sequences were deposited in GenBank with the accession number HM237334 for mycobiont and HM237333 for lichen.

Three methods (TRIzol reagent, NucloSpin® RNA Plant Kit and the combination of above two methods) were used to extract RNA from mycobiont. The yield and quality of total RNA isolated by these methods are presented in Table 1. The results of agarose gel electrophoresis showed that the bands of 28S rRNA and 18S rRNA were clear, indicating no RNA degradation occurred during extraction. However, the A_{260}/A_{280} and A_{260}/A_{230} ratios indicated that total RNAs extracted by combination of TRIzol reagent and NucloSpin® RNA Plant Kit were most suitable for cDNA library construction. It was found that the optimal ratio of cDNA and vector is 3:2 for cDNA library construction. And the titer of the full-length cDNA library was 1.5×10^6 cfu/ml. In order to determine the cloning efficiency and size distribution of inserts, 200 cDNA clones were randomly isolated and the inserts were amplified by PCR using primers M13R and M13F. The PCR products of 191 clones were larger than 250 bp, while the PCR product of empty vector is 250 bp. Thus, the rate of insertion is

Table 1 Quali-quantitative evaluation of different RNA extraction methods

Method	A_{260}/A_{280}	A_{260}/A_{230}	Concentration (ng/μL)
Trizol reagent	1.70	1.91	313
NucloSpin® RNA plant kit	2.01	2.13	83
Combination of both	2.06	2.36	284

95.5%. It was found that PCR products from 111 clones were larger than 750 bp, which implied that the cDNA inserts in these clones were larger than 500 bp. BLAST analysis revealed that 42 CDSes showed high homology to known fungal genes (identities >75%), 58 CDSes had some homology, and the others did not produce any hits (Table 2). Based on the sequencing results, 32 primer pairs for 15 identified genes were designed for qPCR.

Six serially diluted PCR products (P) were used to generate the standard curve, and four concentrations of test DNA sample (G) were used to diminish variation. The result from *Glu* demonstrated the calculating process (Table 3). A stand curve [$y = 1.0272 - 3.2839$ x ($R^2 = 0.9996$)] was generated using the data. Ten primer pairs for 10 genes were chosen to determine genome size (Table 4) due to their similar and appropriate amplification efficiency ($E \approx 2$).

The data of ten genes for sizing genome are shown in Fig. 1. The average size of the genome of the mycobiont is about 39.13 Mb.

The formula of $2^\wedge(Ct_{ref} - Ct_{test})$ was used to calculated the copy number of rRNA gene repeat units, in which each average Ct value of reference or test gene was calculated by three replications. For each gene, the average copy number (±standard error) was generated from templates with two different concentrations. The results indicate that the copy number of rRNA gene repeat units is 43 (Fig. 2).

Discussion

Although the cDNA or genomic DNA libraries have been constructed in many fungi species (Hayes et al. 1994; Zhang et al. 2007), only very few reports are there related to the construction of cDNA library from lichens or lichenized fungi. Junttila et al. (2009) developed four methods for RNA isolation from the reindeer lichen *Cladonia rangiferina* and constructed a mixed cDNA library including the genes from both mycobiont and photobiont to demonstrate the suitability of the RNA extraction methods. Joneson et al. (2011) created cDNA libraries for genes with different expression profiles during the development of lichen from the mycobiont of *Cladonia grayi* and its phycobiont partner *Asterochloris* sp. with SSH (suppression subtractive hybridization) method. However, these libraries are not full-length library, and most genes isolated from these libraries are incomplete.

The full-length cDNA library is a powerful tool for isolating and identifying new genes from lichen, and high quality messenger RNA (mRNA) is the key in construction of full-length cDNA library. Since the yield of total RNA from lichen or lichen mycobiont varies very much using different RNA isolation methods or different culture time (Brink et al. 1995; Junttila et al. 2009), it is difficult to determine whether the amount of total RNA from lichenized fungi is significantly lower than other non-lichen forming fungi. It is undoubted that natural lichens contain a significant amount of RNA, though they may grow very slowly. However, the content of mRNAs would be less, because mRNAs make up only 1–6% of total RNA in eukaryotic cells and the majority of RNAs are ribosomal RNAs (Lucas et al. 1977; Sturani et al. 1979). It is generally considered that the total RNA content in lichen mycobiont is not much lower than that in fast growth free living fungus, but the concentrations of mRNAs may be very low because the lichen thalli are dormant after field collection (Honegger 2003). Thus fungal mRNAs need to be extracted from fresh culture of mycobiont. Though a lot of mycobionts from different lichen species have been isolated in the laboratory, it is difficult to obtain enough materials for RNA extraction because lichen mycobionts grow very slowly compared to those of free living fungi (Brink et al. 1995). We improved the culture conditions so that the liquid cultures of mycobiont from *E. pusillum* can be harvested within 3 weeks and the biomass is sufficient for RNA manipulation.

Polysaccharides and secondary metabolites are prominent in the mycelia of the mycobiont. These substances may coprecipitate with RNA during extraction, decreasing RNA yield and quality (McNeil et al. 1984; Logemann et al. 1987). We employed three different methods for extracting total RNA from the mycobiont of *E. pusillum*. The polysaccharides and secondary metabolites present in mycelia of the mycobiont were not removed by conventional Trizol method, yielding total RNA with $A_{260}/A_{280} < 1.8$. The yield of RNA with NucloSpin® RNA Plant Kit, a commercial RNA extraction kit, is much lower than that with the conventional Trizol method, but the RNA purity is greater. The combination method can overcome these limitations and yield total RNA extract with good quality and quantity. This RNA extract is sufficient and suitable for cDNA library construction.

The titer of the primary library was 1.5×10^6 cfu/ml, and the rate of insertion was 95.5%. These results indicated that the library was constructed successfully. Using this cDNA library, 111 genes were identified from lichenized fungi for the first time in this study. However, the functions of the proteins encoded by these genes were speculated from BLASTP results and will be studied in the further experiments. Generally, the genes that had significant homology to those in other fungi could be regarded as possessing the same function. Though there were hits to other fungi in the protein BLAST analysis, the genes marked with "medium" or "low" should be treated as encoding proteins of uncertain functions. Noticeably, there are 11 genes producing no hits, indicating that these genes share no homology to known fungal genes. It suggests that

Table 2 Genes from cDNA clones of *Endocarpon pusillum* based on their CDS comparison with other fungal genes through BLASTP search against the protein databases at NCBI

Acronym	Clone no.	Size of nucleic acid (bp)	Size of CDS (bp)	GeneBank accession	Best hit in the NCBI fungal databases	
					Accession and description	Homology scores
FCD	L1	1432	582	HM193176	XP_003070080.1, delta (12) fatty acid desaturase [*Coccidioides posadasii* C735 delta SOWgp]	High
H4	L2	712	302[a]	HM193177	XP_385667.1, H4 NEUCR Histone H4 [*Gibberella zeae* PH-1]	High
GST	L3	804	549	HM193178	XP_001264814.1, GPI anchored serine-threonine rich protein [*Neosartorya fischeri* NRRL 181]	Low
TRO-1	L4	724	486	HM193179	XP_002789119.1, tropomyosin-1 [*Paracoccidioides brasiliensis* Pb01]	High
C3/C4	L5	1603	1218	HM193180	XP_002375707.1, C-3 sterol dehydrogenase/C-4 decarboxylase [*Aspergillus flavus* NRRL3357]	Low
Cap20	L10	1267	678	HM193181	XP_002143598.1, pathogenesis associated protein Cap20, putative [*Penicillium marneffei* ATCC 18224]	Low
L16	L12	1029	795	HM193182	XP_002148776.1, mitochondrial large ribosomal subunit protein L16, putative [*Penicillium marneffei* ATCC 18224]	High
TMI	L14	1100	822	HM193183	EEH21729.1, transmembrane BAX inhibitor motif-containing protein [*Paracoccidioides brasiliensis* Pb03]	High
L10-B	L15	773	528	HM193184	XP_002797193.1, 60S ribosomal protein L10-B [*Paracoccidioides brasiliensis* Pb01]	High
PDH-E1	09-168	1448	1170	HM193194	EFR03435.1, pyruvate dehydrogenase E1 component subunit alpha [*Arthroderma gypseum* CBS 118893]	High
ZBC	09-175	1342	1113	HM193195	XP_003069029.1, Fungal Zn binuclear cluster domain containing protein [*Coccidioides posadasii* C735 delta SOWgp]	Low
HP1	09-179	850	270	HM193196	XP_001225850.1, hypothetical protein CHGG_08194 [*Chaetomium globosum* CBS 148.51]	Low
SP	09-232	912	651	HM193193	XP_001262077.1, signal peptidase complex component, putative [*Neosartorya fischeri* NRRL 181]	Low
L17	09-299	789	564	HM193197	XP_001586516.1, 60S ribosomal protein L17 [*Sclerotinia sclerotiorum* 1980]	High
HS	09-1622	1014	711	HM193198	XP_001933310.1, 30 kDa heat shock protein [*Pyrenophora tritici-repentis* Pt-1C-BFP]	Low
HP2	09-1711	854	561	HM193199	XP_002796243.1, conserved hypothetical protein [*Paracoccidioides brasiliensis* Pb01]	Low
SCD	09-1712	1016	801	HM193200	XP_001935870.1, short chain dehydrogenase [*Pyrenophora tritici-repentis* Pt-1C-BFP]	Low
Ubc	09-2911	815	441	HM193201	EFQ34182.1, ubiquitin-conjugating enzyme [*Glomerella graminicola* M1.001]	High
L21-B	09-2920	751	477	HM193202	XP_003066603.1, 60S ribosomal protein L21-B, putative [*Coccidioides posadasii* C735 delta SOWgp]	High
DP1	09-2921	815	639	HM193203	CBF75423.1, TPA: DUF814 domain protein, putative [*Aspergillus nidulans* FGSC A4]	Medium
AC	09-11314	1349	960	HM193204	XP_001244490.1, ADP, ATP carrier protein [*Coccidioides immitis* RS]	High
AF	09-11318	766	534	HM193205	XP_003071126.1, ARD/ARD' family protein [*Coccidioides posadasii* C735 delta SOWgp]	Medium

Table 2 continued

Acronym	Clone no.	Size of nucleic acid (bp)	Size of CDS (bp)	GeneBank accession	Best hit in the NCBI fungal databases	
					Accession and description	Homology scores
PRE2	13	1172	849	HM193134	EEH47596.1, proteasome component PRE2 [*Paracoccidioides brasiliensis* Pb18]	High
Ran	24	1118	651	HM193135	XP_002145774.1, GTP-binding nuclear protein Ran, putative [*Penicillium marneffei* ATCC 18224]	High
Prp	29	782	249	HM193136	XP_001597498.1, predicted protein [*Sclerotinia sclerotiorum* 1980]	Medium
RER1	33	981	582	HM193137	EEH04498.1, RER1 protein [*Ajellomyces capsulatus* G186AR]	Medium
Gns1	113	1496	1026	HM193138	XP_001266255.1, fatty acid elongase (Gns1), putative [*Neosartorya fischeri* NRRL 181]	Medium
Cyc	121	722	339	HM193139	XP_370188.2, cytochrome c [*Magnaporthe oryzae* 70-15]	High
MBF-1	123	610	393	HM193140	XP_002550109.1, multiprotein bridging factor 1 [*Candida tropicalis* MYA-3404]	Low
HP3	131	928	429	HM193141	XP_001801683.1, hypothetical protein SNOG_11441 [*Phaeosphaeria nodorum* SN15]	Low
Vid24	143	1783	1284	HM193142	XP_001264673.1, vesicle-mediated transport protein Vid24, putative [*Neosartorya fischeri* NRRL 181]	Medium
HP4	144	832	549	HM193143	XP_001932060.1, conserved hypothetical protein [*Pyrenophora tritici-repentis* Pt-1C-BFP]	Low
Rad24	147	1267	795	HM193144	EEH11002.1, DNA damage checkpoint protein rad24 [*Ajellomyces capsulatus* G186AR]	High
ssp1	152	1308	636	HM193145	EEQ87856.1, serine/threonine-protein kinase ssp1 [*Ajellomyces dermatitidis* ER-3]	Low
L12	179	717	150	HM193146	XP_002620177.1, 60S ribosomal protein L12 [*Ajellomyces dermatitidis* SLH14081]	High
HP5	182	1354	1101	HM193147	XP_002150683.1, conserved hypothetical protein [*Penicillium marneffei* ATCC 18224]	Low
RPB11a	205	821	372	HM193148	XP_002379141.1, DNA-directed RNA polymerase II subunit RPB11a [*Aspergillus flavus* NRRL3357]	Medium
TFP	206	834	522	HM193149	XP_002479299.1, TCTP family protein [*Talaromyces stipitatus* ATCC 10500]	Medium
HP6	207	562	251[a]	HM193150	XP_001826093.1, hypothetical protein [*Aspergillus oryzae* RIB40]	Low
S26	213	736	363	HM193151	XP_002790158.1, 40S ribosomal protein S26 [*Paracoccidioides brasiliensis* Pb01]	High
Tsa	215	1182	975	HM193152	XP_002793894.1, transaldolase [*Paracoccidioides brasiliensis* Pb01]	High
Pup3	228	861	618	HM193153	XP_002148948.1 proteasome component Pup3, putative [*Penicillium marneffei* ATCC 18224]	High
HP7	236	856	642	HM193154	XP_001220985.1, hypothetical protein CHGG_01764 [*Chaetomium globosum* CBS 148.51]	Low
L14-A	238	663	441	HM193155	XP_003072088.1, 60S ribosomal protein L14-A, putative [*Coccidioides posadasii* C735 delta SOWgp]	Medium
Gtd	239	664	399	HM193156	XP_961585.1, glutaredoxin [*Neurospora crassa* OR74A]	Low
H2A	244	740	417	HM193157	XP_002341942.1, histone H2A [*Talaromyces stipitatus ATCC* 10500]	High
Ubq	256	1152	915	HM193158	XP_001210780.1, ubiquitin [*Aspergillus terreus* NIH2624]	High
Ata	259	1123	855	HM193159	XP_003065057.1, GNAT family acetyltransferase, putative [*Coccidioides posadasii* C735 delta SOWgp]	Medium
Utp6	269	1276	1203	HM193160	XP_002627802.1, rRNA processing protein Utp6 [*Ajellomyces dermatitidis* SLH14081]	Low
IFH	272	829	645	HM193161	XP_001261612.1, isochorismatase family hydrolase [*Neosartorya fischeri* NRRL 181]	Medium
GSR	273	1244	813	HM193162	EEH49791.1, glucan synthesis regulatory protein [*Paracoccidioides brasiliensis* Pb18]	Medium

Table 2 continued

Acronym	Clone no.	Size of nucleic acid (bp)	Size of CDS (bp)	GeneBank accession	Best hit in the NCBI fungal databases	
					Accession and description	Homology scores
HP8	284	1069	693	HM193163	XP_002382903.1, conserved hypothetical protein [*Aspergillus flavus* NRRL3357]	Low
ASS-9	315	919	234	HM193210	YP_001648752.1, ATP synthase subunit 9 [*Mycosphaerella graminicola*]	High
S8-A	301	1032	603	HM193209	XP_002583604.1, 40S ribosomal protein S8-A [*Uncinocarpus reesii* 1704]	High
HP9	300	1883	831	HM193164	XP_001588060.1, hypothetical protein SS1G_11303[*Sclerotinia sclerotiorum* 1980]	Low
SSC1	323	1334	900	HM193211	XP_001537067.1, heat shock protein SSC1, mitochondrial precursor [*Ajellomyces capsulatus* NAm1]	High
Ubc-H	325	815	276	HM193212	XP_001268055.1, ubiquitin-conjugating enzyme h [*Aspergillus clavatus* NRRL 1]	Medium
S14	334	698	453	HM193214	XP_002796425.1, 40S ribosomal protein S14 [*Paracoccidioides brasiliensis* Pb01]	High
TRA	342	1160	936	HM193215	CBF74996.1, TPA: thioredoxin reductase GliT [*Aspergillus nidulans* FGSC A4]	Medium
HP10	350	1150	528	HM193216	XP_001593081.1, hypothetical protein SS1G_06003 [*Sclerotinia sclerotiorum* 1980]	Low
HP11	354	1424	648	HM193217	XP_002340233.1, conserved hypothetical protein [*Talaromyces stipitatus* ATCC 10500]	Low
L33-A	358	684	327	HM193218	XP_002582873.1, 60S ribosomal protein L33-A [*Uncinocarpus reesii* 1704]	High
Tom20	361	897	510	HM193219	XP_001258255.1, mitochondrial import receptor subunit (Tom20), putative [*Neosartorya fischeri* NRRL 181]	Medium
L13	367	921	657	HM193220	XP_002482338.1, 60S ribosomal protein L13 [*Talaromyces stipitatus* ATCC 10500]	Medium
PQR	372	972	816	HM193222	XP_002381633.1, PQ loop repeat protein [*Aspergillus flavus* NRRL3357]	Low
DP2	376	921	546	HM193223	XP_002379190.1, UPF0041 domain protein [*Aspergillus flavus* NRRL3357]	Medium
L126	386	688	429	HM193207	EEH48947.1, 60S ribosomal protein L126 [*Paracoccidioides brasiliensis* Pb18]	High
L15	394	845	561	HM193224	XP_002797110.1, 60S ribosomal protein L15 [*Paracoccidioides brasiliensis* Pb01]	High
SMNR	397	716	399	HM193225	XP_003069554.1, sulphur metabolism negative regulator, putative [*Coccidioides posadasii* C735 delta SOWgp]	High
NPA	412	855	477	HM193226	EEH09382.1, nascent polypeptide-associated complex subunit beta [*Ajellomyces capsulatus* G186AR]	High
DP3	416	856	378	HM193227	EEH09286.1, PAP2 domain-containing protein [*Ajellomyces capsulatus* G186AR]	Medium
NTF-2	419	688	375	HM193228	XP_001558550.1, nuclear transport factor 2 [*Botryotinia fuckeliana* B05.10]	Medium
Dcp1	428	967	231	HM193230	EER36623.1, decapping enzyme Dcp1 [*Ajellomyces capsulatus* H143]	Medium
SDC	434	1046	642	HM193231	XP_003067984.1, SH3 domain containing protein [*Coccidioides posadasii* C735 delta SOWgp]	Medium
CCT	438	836	528	HM193208	XP_001257473.1, Ctr copper transporter, putative [*Neosartorya fischeri* NRRL 181]	Medium
HP12	456	908	567	HM193232	XP_002479933.1, hypothetical protein TSTA_027900 [*Talaromyces stipitatus* ATCC 10500]	Low
TktA	460	947	570	HM193233	XP_003069218.1, transketolase TktA, putative [*Coccidioides posadasii* C735 delta SOWgp]	High
HP13	489	1060	816	HM193234	XP_002495848.1, ZYRO0C04378p [*Zygosaccharomyces rouxii*]	Low
CDC-10	492	1348	1017	HM193235	XP_002620551.1, cell division control protein 10 [*Ajellomyces dermatitidis* SLH14081]	High
L34	495	584	351	HM193236	XP_001265868.1, ribosomal protein L34 protein, putative [*Neosartorya fischeri* NRRL 181]	Medium
Aut7/ IDI7	506	828	354	HM193237	XP_750493.1, autophagic death protein Aut7/IDI-7 [*Aspergillus fumigatus* Af293]	High

Table 2 continued

Acronym	Clone no.	Size of nucleic acid (bp)	Size of CDS (bp)	GeneBank accession	Best hit in the NCBI fungal databases	
					Accession and description	Homology scores
HP14	513	781	351	HM193238	XP_002375538.1, conserved hypothetical protein [*Aspergillus flavus* NRRL3357]	Low
L37	518	513	276	HM193239	XP_001547251.1, 60S ribosomal protein L37 [*Botryotinia fuckeliana* B05.10]	High
FBP	527	1936	1563	HM193241	EEH47631.1, FK506-binding protein [*Paracoccidioides brasiliensis* Pb18]	Medium
Coa	548	1025	741	HM193242	XP_001264163.1, Coatomer subunit delta, putative [*Neosartorya fischeri* NRRL 181]	High
HP14	A11	869	669	HM193165	EEH09407.1, conserved hypothetical protein [*Ajellomyces capsulatus* G186AR]	Low
Gap1	A14	891	675	HM193166	XP_002478659.1, amino acid permease (Gap1), putative [*Talaromyces stipitatus* ATCC 10500]	Medium
Pro II	A20	665	390	HM193167	XP_003070680.1, Profilin II, putative [*Coccidioides posadasii* C735 delta SOWgp]	Medium
AKA	A42	1287	834	HM193168	XP_002376267.1, adenylate kinase, putative [*Aspergillus flavus* NRRL3357]	Medium
S3	A53	979	792	HM193169	XP_002480514.1, 40S ribosomal protein S3, putative [*Talaromyces stipitatus* ATCC 10500]	High
OTS	A54	1456	1407	HM193170	XP_002481545.1, oligosaccharyl transferase subunit (beta), putative [*Talaromyces stipitatus* ATCC 10500]	Medium
ETIF-1A	A93	902	453	HM193172	XP_001241882.1, eukaryotic translation initiation factor 1A [*Coccidioides immitis* RS]	High
DP4	A113	675	531	HM193174	XP_001257683.1, UPF0220 domain protein [*Neosartorya fischeri* NRRL 181]	Medium
L7a	A112	952	765	HM193173	EEQ85142.1, ribosomal protein L7a [*Ajellomyces dermatitidis* ER-3]	High
NQO	A75	806	675	HM193171	XP_002144283.1, NADH-quinone oxidoreductase, 23 kDa subunit, putative [*Penicillium marneffei* ATCC 18224]	High
LaeA	A117	2246	1008	HM193175	XP_001934837.1, methyltransferase LaeA [*Pyrenophora tritici-repentis* Pt-1C-BFP]	High
L5	09-2913	806	762[a]	HM193185	XP_960578.1, 60S ribosomal protein L5 [*Neurospora crassa* OR74A]	High
LNDP	09-11324	1162	1095[a]	HM193186	XP_002374624.1, LipA and NB-ARC domain protein [*Aspergillus flavus* NRRL3357]	Low
DDP	306	695	582[a]	HM193187	XP_001268204.1, DnaJ domain protein [*Aspergillus clavatus* NRRL 1]	High
Glu	466	719	495	HQ335009	XP_961585.1, glutaredoxin [*Neurospora crassa* OR74A]	Medium
HP15	252	1021	699	HM193191	No hit	N
HP16	368	757	429	HM193221	No hit	N
HP17	331	678	300	HM193213	No hit	N
HP18	341	672	339	HM193206	No hit	N
HP19	550	796	108	HM193192	No hit	N
HP20	A4	764	201	HM193188	No hit	N
HP21	A5	575	159	HM193189	No hit	N
HP22	A37	1056	240	HM193190	No hit	N
HP23	421	720	174	HM193229	No hit	N
HP24	525	643	222	HM193240	No hit	N
HP25	557	847	201	HM193243	No hit	N

[a] incomplete 3′ end

Table 3 The data of *Glu* used to generate a standard curve and to calculate the genome size

P_{Glu}	Well	ng DNA/rxn	Actual Ct	Trendline Ct	G_{Glu}	Well	ng DNA/rxn	Actual Ct	Average Ct
Conc. 1	A1	0.00676	8.26721	8.1534	Conc. 1	H1	31.2	13.8429	13.6200
	A2	0.00676	8.06356			H2	31.2	13.1965	
	A3	0.00676	8.16939			H3	31.2	13.8205	
Conc.2	B1	0.000676	11.5847	11.4370	Conc. 2	I1	3.10	17.0289	16.8994
	B2	0.000676	11.378			I2	3.10	17	
	B3	0.000676	11.4118			I3	3.10	16.6692	
Conc.3	C1	0.000169	13.421	13.4139	Conc. 3	J1	0.62	19.2766	19.1390
	C2	0.000169	13.3738			J2	0.62	19.0261	
	C3	0.000169	13.3202			J3	0.62	19.1144	
Conc.4	D1	0.00004225	15.497	15.3891	Conc. 4	K1	0.31	20.2435	20.2893
	D2	0.00004225	15.3671			K2	0.31	20.2468	
	D3	0.00004225	15.2732			K3	0.31	20.3777	
Conc.5	E1	0.00001056	17.3983	17.3626					
	E2	0.00001056	17.3622						
	E3	0.00001056	17.3132						
Conc.6	F1	0.00000264	19.4074	19.3449					
	F2	0.00000264	19.3886						
	F3	0.00000264	19.3269						

Table 4 The primers designed for identified genes from the cDNA library to determine the genome size

Target gene	Primer name	Sequence 5′-3′	Size of amplicon (bp)
SCD	scd-F	CACAGATAGCGGCACGACGAAA	126
(09-1712)	scd-R	GCAGCAAGACGAACCAAACCG	
L10-B	L10b-F	CCAAATCCTTCTCTCGGTCCGCA	188
(L15)	L10b-R	GCACCATCAATCAACACTCTACCCTC	
Glu	glu-F	GACCTGGTGTTCTCACTGTGC	183
(466)	glu-R	ACGAAGATACGAGGGACCGTT	
HP11	hp11-F	AACAATGTCATCCAGCACCGT	185
(354)	hp11-R	TGAGGAGAGATTTGCGGGTTTG	
Utp6	utp6-F	ACTGGTCAACGCAGGGTGT	158
(269)	utp6-R	TGATGGGTGTAGGCGAAGAGC	
Ata	ata-F	CTCAGCCTTCATCGTCATCCG	182
(259)	ata-R	CGTCTCGTCCTCTTCGTCGT	
RPB11a	rpb-F	ACCCTTGCTGCTCTTACCGA	199
(205)	rpb-R	CGGGCTTGTTCGTCATCTTGA	
HS	hs-F	GCAACCATACCGAGCAACTTT	180
(09-1622)	hs-R	TCGTCCAGGAGAGAGAACAGT	
RER1	rer1-F	TGAACCTCTTCCTCGCTTTCC	217
(33)	rer1-R	TCAAACCAACTCGCCACGAA	
CDC-10	cdc10-F	TTGAGCCAGGGAGTATCGGT	193
(492)	cdc10-R	ACGGACTGAAACCAACCTCG	

these 11 genes may be unique genes in lichenized fungi, which should play important roles in lichen symbiosis. These genes isolated from the cDNA library not only provide information about the expression of genes in mycobiont, but would also be helpful for the further study on functional genomics in lichenized fungus, as the functional characterization of these genes is one of the main challenges in modern lichenology.

Fig. 1 The estimated genome size for *E. pusillum*. The height of the *bar* represents the genome size (mega base on the ordinate) calculated using the primer pairs for the genes listed below the abscissa (*SCD, L10-B, Glu, HP11, Utp6, Ata, RPB11a, HS, RER1, CDC-10*). Four different concentrations (31.2, 3.1, 0.62 and 0.31 ng/μL) of genomic DNA (G) were applied to calculate the genome size

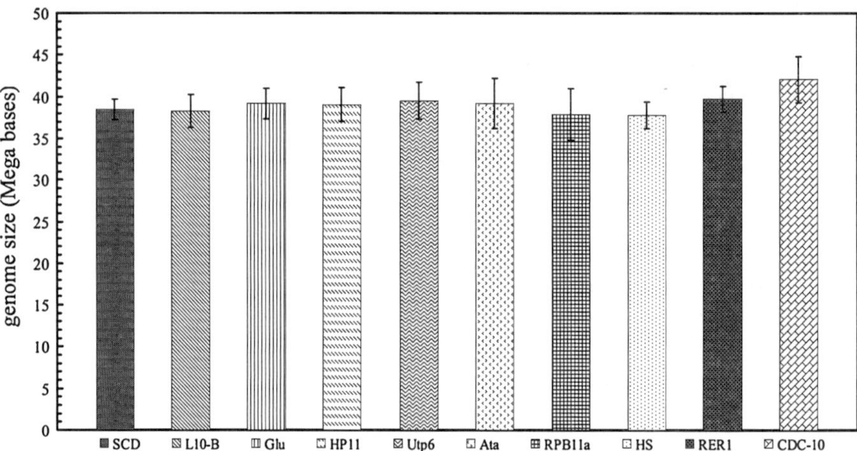

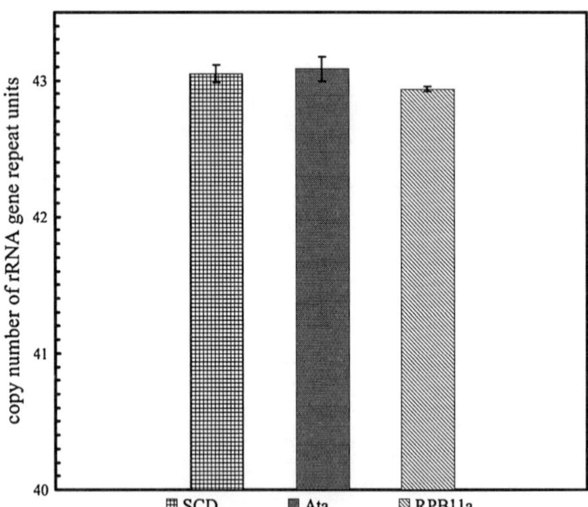

Fig. 2 The estimated copy number of rRNA gene repeat units. The height of the *bar* represents the copy number calculated using the primer pairs of three genes (*SCD, Ata and RPB11a*) listed below the abscissa. Each result was calculated with two concentrations of genomic DNA

The accurate measurement of the concentration of DNA sample is very important for absolute quantification using real time PCR. We found that irreproducible results would be generated if the samples were measured with UV absorbance method, which is due to the possible residual UV absorbing impurities in DNA sample. Thus a Qubit fluorometer using fluorescent dyes, instead of UV-absorbance method, was used to determine the concentration of DNA in a sample and the results were replicable and consistent (Armaleo and May 2009).

The absolute quantitative PCR method for genome sizing was verified with genomes of known size (Wilhelm et al. 2003). We applied this method to size the genome of the mycobiont from *E. pusillum*, which has not been explored before. Armaleo and May (2009) measured genome sizes of the mycobiont and photobiont from *Cladonia grayi* using three genes which were *Pks*, *Tub* and *Cut*. However, we found that their primers had low match to our mycobiont because no band was observed after gel electrophoresis. Therefore, 32 primer pairs from 15 identified genes were designed to be applied in qPCR. Finally, ten primer pairs with the similar and reasonable amplification efficiency were screened to determine the genome size of the mycobiont from *E. pusillum*. Statistical analysis showed that there was no significant difference between the results calculated by these ten genes. Generally, the lichen-forming fungus is classified into Pezizomycota (James et al. 2006; Hibbett et al. 2007), and the size of genomes of most species in Pezizomycota are between the 15 and 40 Mb (http://www.zbi.ee/fungal-genomesize), which is consistent with our results.

As multi-copy sequence, ribosomal RNA (rRNA) gene repeat unit is easy to be obtained by PCR (DePriest 1994; Maleszka and Clark-Walker 1993). Each unit consists of the parts of SSU rDNA, LSU rDNA, ITS rDNA, IGS and group I intron, and these parts can be used to resolve the phylogenetic relationship on different classification levels because their evolutionary rates vary remarkably. Thus, rRNA is the most common gene used in fungal molecular systematics analysis. Moreover, the copy number polymorphism of rRNA gene repeat units has been observed in other fungi (Belkhiri and Klassen 1996; Fournier et al. 1986; Garber et al. 1988; Howlett et al. 1997; Maleszka and Clark-Walker 1990), which will be valuable for population study. Here, we determined the copy number of rRNA gene repeat unit of the lichenized fungi for the first time, which will be helpful to investigate the relationship among the copy number polymorphism, sexual reproduction and morphological differentiation in various lichen species or community.

In this study, a full-length cDNA library from the mycobiont of an arid desert lichen *E. pusillum* has been successfully constructed and more than one hundred functional genes from lichenized fungi were obtained for the first time. This work will facilitate research on genes involved in stress-resistance in lichenized fungi, and enlarge our knowledge on genes in lichens. The result of the size of genome and the copy number of rRNA gene will facilitate the study on sequencing the genome of this mycobiont in assembly and annotation. This work will span the gap in our knowledge of molecular understanding of lichens.

Acknowledgments We are indebted to Dr. Bin Liu and Miss Dan Guo (College of life science, Nankai University) for proving us the accurate concentration of nucleic acids using Qubit system. The authors would like to thank Professor Chuan-Peng Liu and an anonymous reviewer for their critical reading of this paper and valuable suggestions. The present study was supported in part by the Ministry of Science and Technology of the People's Republic of China (MOST, No. 2007AA021405) and the Knowledge Innovation Program of Chinese Academy of Sciences (KSCX2-YW-Z-1020, 2010-Biols-CAS-0104).

References

Armaleo D, May S (2009) Sizing the fungal and algal genomes of the lichen *Cladonia grayi* through quantitative PCR. Symbiosis 49: 43–51

Belkhiri A, Klassen GR (1996) Diverged 5 s rRNA sequences adjacent to 5 s rRNA genes in the rDNA of *Pythium pachycaule*. Curr Genet 29:287–292

Brink JJ, Ahmadjian V, Fröberg L, Goldsmith S (1995) Time course of RNA accumulation in cultures of lichen fungi. Cryptogam Bot 5:55–59

Buschbom J (2007) Migration between continents: geographical structure and long-distance gene flow in *Porpidia flavicunda* (lichen-forming Ascomycota). Mol Ecol 16:1835–1846

Chooi YH, Stalker DM, Davis MA, Fujii I, Elix JA, Louwhoff SH, Lawrie AC (2008) Cloning and sequence characterization of a non-reducing polyketide synthase gene from the lichen *Xanthoparmelia semiviridis*. Mycol Res 112:147–161

Clarke JD (2009) Cetyltrimenthyl ammonium bromide (CTAB) DNA miniprep for plant DNA isolation. Cold Spring Harb Protoc doi: 10.1101/pdb.prot5177

Corradi N, Croll D, Colard A, Kuhn G, Ehinger M, Sanders IR (2007) Gene copy number polymorphisms in an arbuscular mycorrhizal fungal population. Appl Environ Microb 73:366–369

DePriest PT (1994) Variation in the *Cladonia chlorophaea* complex II: ribosomal DNA variation in a Southern Appalachian population. Bryologist 97:117–126

Fournier P, Gaillardin C, Persuy MA, Klootwijk J, Heerikhuizen H (1986) Heterogeneity in the ribosomal family of the yeast *Yarrowia lipolytica*: genomic organization and segregation studies. Gene 42:273–282

Gagunashvili AN, Davidsson SP, Jonsson ZO, Andresson OS (2009) Cloning and heterologous transcription of a polyketide synthase gene from the lichen *Solorina crocea*. Mycol Res 113:354–363

Garber RC, Turgeon BG, Selker EU, Yoder OC (1988) Organization of ribosomal RNA genes in the fungus *Cochliobolus heterostrophus*. Curr Genet 14:573–582

Gutierrez G, Blanco O, Divakar PK, Lumbsch HT, Crespo A (2007) Patterns of group I intron presence in nuclear SSU rDNA of the lichen family Parmeliaceae. J Mol Evol 64:181–195

Hayes CK, Klemsdal S, Lorito M, Pietro AD, Peterbauer C, Nakas JP, Tronsmo A, Harman GE (1994) Isolation and sequence of an endochitinase-encoding gene from a cDNA library of *Trichoderma harzianum*. Gene 138:143–148

Herrera ML, Vallor AC, Gelfond JA, Patterson TF, Wickes BL (2009) Strain-dependent variation in 18S ribosomal DNA copy numbers in *Aspergillus fumigatus*. J Clin Microbiol 47:1325–1332

Hibbett DS, Binder M, Bischoff JF et al (2007) A higher-level phylogenetic classification of the fungi. Mycol Res 111:509–547

Honegger R (2003) The impact of different long-term storage conditions on the viability of lichen-forming Ascomycees and their green algal photobiont, *Trebouxia* spp. Plant Biol 5:324–330

Howlett BJ, Rolls BD, Cozijnsen AJ (1997) Organisation of ribosomal DNA in the ascomycete *Leptosphaeria maculans*. Microbiol Res 152:261–267

Ilse K, Richard B, Ayala H, Thomas H, Nash III (2008) Desiccation-tolerance in lichens: a review. Bryologist 111:576–593

James TY, Kauff F, Conrad L et al (2006) Reconstructing the early evolution of fungi using a six-gene phylogeny. Nature 443:818–822

Joneson S, Armaleo D, Lutzoni F (2011) Fungal and algal gene expression in early developmental stages of lichen-symbiosis. Mycologia 103:291–306

Junttila S, Lim KJ, Rudd S (2009) Optimization and comparison of different methods for RNA isolation for cDNA library construction from the reindeer lichen *Cladonia rangiferina*. BMC Res Notes 2:204–208

Lindblom L, Ekman S (2006) Genetic variation and population differentiation in the lichen-forming ascomycete *Xanthoria parietina* on the island Storfosna, central Norway. Mol Ecol 15:1545–1559

Logemann J, Schell J, Willmitzer L (1987) Improved method for the isolation of RNA from plant tissues. Anal Biochem 163:16–20

Lucas MC, Jacobson JW, Giles NH (1977) Characterization and in vitro translation of polyadenylated messenger ribonucleic acid from *Neurospora crassa*. J Bacteriol 130:1192–1198

Maleszka R, Clark-Walker GD (1990) Magnification of the rDNA cluster in *Kluyveromyces lactis*. Mol Gen Genet 223:342–344

Maleszka R, Clark-Walker GD (1993) Yeasts have a four-fold variation in ribosomal DNA copy number. Yeast 9:53–58

McNeil M, Darvill AG, Fry SC, Albersheim P (1984) Structure and function of the primary cell walls of plants. Annu Rev Biochemistry 53:625–663

Michael A, Connolly, Peter A, Clausen, James G, Lazar (2006) Preparation of RNA from plant tissue using trizol. Cold Spring Harb Protoc doi:10.1101/pdb.prot4105

Palice Z, Schmitt I, Lumbsch HT (2005) Molecular data confirm that *Omphalina foliacea* is a lichen-forming basidiomycete. Mycol Res 109:447–451

Reeb V, Lutzoni F, Roux C (2004) Contribution of RPB2 to multilocus phylogenetic studies of the euascomycetes (Pezizomycotina, Fungi) with special emphasis on the lichen-forming Acarosporaceae and evolution of polyspory. Mol Phylogenet Evol 32:1036–1060

Scherrer S, Haisch A, Honegger R (2002) Characterization and expression of XPH1, the hydrophobin gene of the lichen-forming ascomycete *Xanthoria parietina*. New Phytol 154:175–184

Sinnemann SJ, Andrésson óS, Brown DW, Miao VPW (2000) Cloning and heterologous expression of *Solorina crocea pyrG*. Curr Genet 37:333–338

Staiger B, Kalb K, Grube M (2006) Phylogeny and phenotypic variation in the lichen family Graphidaceae (Ostropomycetidae, Ascomycota). Mycol Res 110:765–772

Sturani E, Costantini MG, Martegani E, Alberghina L (1979) Level and turnover of polyadenylate-containing ribonucleic acid in *Neurospora crassa* in different steady states of growth. Eur J Biochem 99:1–7

Valarmathi R, Hariharan GN, Venkataraman G, Parida A (2009) Characterization of a non-reducing polyketide synthase gene from lichen *Dirinaria applanata*. Phytochemistry 70:721–729

Walser JC, Holderegger R, Gugerli F, Hoebee SE, Scheidegger C (2005) Microsatellites reveal regional population differentiation and isolation in *Lobaria pulmonaria*, an epiphytic lichen. Mol Ecol 14:457–467

White TJ, Bruns T, Lee S, Taylor J (1990) Amplification and direct sequencing of fungal ribosomal RNA genes for phylogenetics. In: Innis M, Gelfand D, Sninsky J, White T (eds) PCR protocols: a guide to methods and applications. Academic Press, San Diego, CA, pp 315–322

Wilhelm J, Pingoud A, Hahn M (2003) Real-time PCR-based method for the estimation of genome sizes. Nucleic Acids Res 31:e56

Yoshimura I, Yamamoto Y, Nakano T, Finnie J (2002) Isolation and culture of lichen photobionts and mycobionts. In: Kranner I, Beckett RP, Varma AK (eds) Protocols in lichenology. Culturing, biochemistry, ecophysiology and use in biomonitoring. Springer, Berlin, pp 3–33

Zhang CS, Cao YQ, Wang ZK, Yin YQ, Peng GX, Xia YX (2007) A method to construct cDNA library of the entomopathogenic fungus, *Metarhizium anisopliae*, in the hemolymph of the infected locust. Mol Biotechnol 36:23–31

Zhou QM, Guo SY, Huang MR, Wei JC (2006) A study of the genetic variability of *Rhizoplaca chrysoleuca* using DNA sequences and secondary metabolic substances. Mycologia 98:57–67

Genome characteristics reveal the impact of lichenization on lichen-forming fungus *Endocarpon pusillum* Hedwig (Verrucariales, Ascomycota)

Yan-Yan Wang[1,2†], Bin Liu[3†], Xin-Yu Zhang[1†], Qi-Ming Zhou[1†], Tao Zhang[1,2], Hui Li[1,2], Yu-Fei Yu[1], Xiao-Ling Zhang[1], Xi-Yan Hao[3], Meng Wang[3], Lei Wang[3*] and Jiang-Chun Wei[1*]

Abstract

Background: Lichen is a classic mutualistic organism and the lichenization is one of the fungal symbioses. The lichen-forming fungus *Endocarpon pusillum* is living in symbiosis with the green alga *Diplosphaera chodatii* Bialsuknia as a lichen in the arid regions.

Results: 454 and Illumina technologies were used to sequence the genome of *E. pusillum*. A total of 9,285 genes were annotated in the 37.5 Mb genome of *E. pusillum*. Analyses of the genes provided direct molecular evidence for certain natural characteristics, such as homothallic reproduction and drought-tolerance. Comparative genomics analysis indicated that the expansion and contraction of some protein families in the *E. pusillum* genome reflect the specific relationship with its photosynthetic partner (*D. chodatii*). Co-culture experiments using the lichen-forming fungus *E. pusillum* and its algal partner allowed the functional identification of genes involved in the nitrogen and carbon transfer between both symbionts, and three lectins without signal peptide domains were found to be essential for the symbiotic recognition in the lichen; interestingly, the ratio of the biomass of both lichen-forming fungus and its photosynthetic partner and their contact time were found to be important for the interaction between these two symbionts.

Conclusions: The present study lays a genomic analysis of the lichen-forming fungus *E. pusillum* for demonstrating its general biological features and the traits of the interaction between this fungus and its photosynthetic partner *D. chodatii*, and will provide research basis for investigating the nature of its drought resistance and symbiosis.

Keyword: Mycobiont, Phycobiont, Lichenization, Symbiosis, Symbiosis-related gene, Photosynthetic products

Background

A lichen is a symbiotic association of a fungus (*mycobiont*) and a photosynthetic partner (*photobiont*), which may be an alga (*phycobiont*) or a cyanobacterium (*cyanobiont*). In the association the fungus produces a *thallus*, or body, within which the photobionts are housed [1]. Around 20% of all Fungi and 40% of all Ascomycota are lichen-forming. Recent estimates of global diversity suggest that there are between 17,500 and 20,000 species [2].

Most lichens and isolated lichen-forming fungi grow extremely slow, but the lichen-symbiosis is a very successful association as lichens can survive in almost all adverse terrestrial conditions [3]. They are also famous for their particular secondary products, which are frequently used as antibacterial and antiviral compounds [4,5]. The lichen-forming fungi differ from non-lichenized fungi by their adaptations to symbiosis with photobiont [6]. This mutualistic association, as called lichenization, is one of the most important fungal lifestyles and the lichenization, considered by some researchers, has evolved many times in the phylogeny of fungi [7,8], and also some major fungal lineages may have derived from lichen symbiotic ancestors [9].

The principal problem about lichenization is the necessity of fungal propagules meeting a suitable photosynthetic

* Correspondence: wanglei@nankai.edu.cn; weijc2004@126.com
†Equal contributors
[3]TEDA School of Biological Sciences and Biotechnology, Nankai University, Tianjin 300457, China
[1]State Key Laboratory of Mycology, Institute of Microbiology, Chinese Academy of Sciences, Beijing 100101, China
Full list of author information is available at the end of the article

This paper was originally published in *BMC genomics*, 15: 34, 2014.

partner for the resynthesis of the symbiosis [3,10]. The recognition step is complicated, involving many morphological and molecular changes. Scanning electron microscopy (SEM) has been used to investigate the changes in morphology during the early resynthesis of the lichen thallus [11-13]. However, few studies have explored the resynthesis events in lichen using molecular tools. Previous studies suggest that the mycobiont-derived lectins (sugar-specific, cell agglutinating proteins) may play a key role in recognition [14-17] between both symbionts.

The interdependent relationship between mycobiont and photobiont is the foundation of lichenization, which required for both symbionts to maintain each other. In the lichen thallus, the photobiont provides its mycobiont with photosynthetic products [18,19], previous reports showed that lichen-forming fungi absorb polyol (ribitol, sorbitol, and erythritol) or glucose from algae or cyanobacteria, respectively [20,21], and in most green algae lichens, the hyphae of the mycobionts wrap tightly around photobiont cells, thereby protecting the photobiont cells from a range of biotic and abiotic stress, including drought, high light and mechanical damage. The protection from light-injury is associated with secondary metabolic substances, such as melanins, produced by lichen-forming fungi [22]. However, little is known about the signals and mechanisms that lead to symbiotic recognition and maintenance in lichen, although it can be predicted that some metabolic products and macromolecules are essential.

The whole life cycle of lichen is rarely observed in the laboratory or in nature [11,13]. The detail of lichenization, whereby a lichen-forming fungus contacts with a compatible photosynthetic partner, recognizes it and captures it, has become a hot topic in recent studies [23,24]. *Endocarpon pusillum* is the most successful cases on artificial resynthesis of the fertile (perithecia-bearing) thalli from isolated mycobiont and phycobiont [12], and it will be very useful to reveal the origin of symbiosis between fungi and photosynthetic organisms.

Endocarpon is a special genus, which have hymenial algal cells in their perithecia and the ascospores are discharged together with the hymenial algal cells [25]. The systematics and physiology of *E. pusillum* (Figure 1) have been studied well. The mycobiont *E. pusillum* exhibits much stronger desiccation-tolerant than other non-lichenized fungi as it can survive for 7 months under desiccation stress in combination with starvation stress [26]. Although a large number of fungal genomes have been published, no lichen-forming is included, and only two mitochondrial genomes from lichen-forming fungi (*Peltigera membranacea* and *Peltigera malacea*) and a transcriptome from the lichen *Cladonia rangiferina* have been reported until now [27,28]. Therefore, the genome of *E. pusillum* was sequenced and analyzed to ascertain the biological features of this lichen-forming fungus and the traits of lichenization.

Results and discussion

General features of the genome

The genome of the lichen-forming fungus *E. pusillum* was sequenced to about 78-fold coverage using both 454 and Illumina technologies (Additional file 1: Table S1). All sequences were assembled into 908 scaffolds (> 2 kb; N50, 178 kb) containing 1,731 contigs, with a genome size of 37.5 Mb (Table 1), which was almost identical to the result calculated by real-time polymerase chain

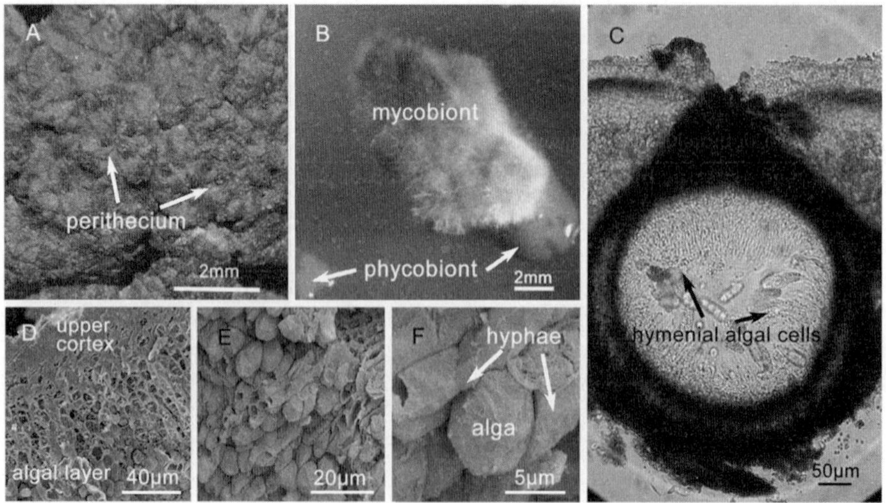

Figure 1 ***Endocarpon pusillum*. A.** The lichen *E. pusillum*. **B.** The isolated mycobiont and phycobiont [26]. **C.** Cross section of a perithecium with hymenial algal cells inside. **D.** Cross section of a thallus under scanning electron microscopy (SEM). **E.** The algal layer (SEM). **F.** An algal cell is clasped and surrounded by some hyphae (SEM).

Table 1 Main features of the *Endocarpon pusillum* genome

Features	*Endocarpon pusillum*
Assembly size/Mb	37.5
Scaffold N50/kb	178
Coverage/fold	78
G + C content	46.01%
GC Exonic	51.73%
GC Intronic	47.05%
Repeat rate	1.68%
Protein-coding genes	9,285
Gene density	250.8 per Mbp
Exons per genes	2.53
tRNAs	72
rRNAs	19
SM (Secondary Metabolism) genes	28
TE	15%

reaction (PCR) [29]. The average GC content of the genome is 46.1%, and exonic region has a 4% higher GC content than the intronic region. Repetitive sequences represent 15% of the genome. A circular map was generated for the 30 largest scaffolds to illustrate the genome features more clearly (Figure 2).

The whole project has been deposited at DDBJ/EMBL/GenBank under accession number APWS00000000. A total of 9,285 protein-coding genes were predicted, and 1,479 (15.7%) of these genes have no significant matches to known proteins from public databases. A total of 2,754, 3,787 and 7,589 proteins were assigned to Gene Ontology (GO) terms, the eukaryotic orthologous groups (KOG) and functional catalogue (FunCat) databases, respectively. The distributions of the top 10 GO, KOG, and FunCat terms of the sequences are presented in Figure 3.

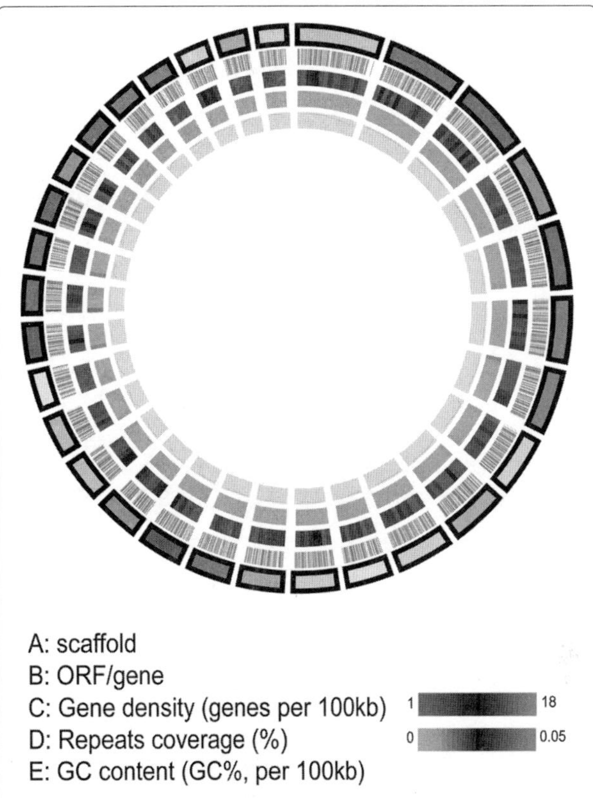

Figure 2 Genome of *Endocarpon pusillum*. A-E represent the circles from the outside to inside. **A:** Scaffolds of the genome, filtered by size (≥ 280 Kb). **B:** ORFs/genes. **C:** Gene density represented as number of genes per 100 kb (non-overlapping, window size 100 kb). **D:** Percentage coverage of repetitive sequences (non-overlapping windows, window size 100 kb). **E:** GC content was estimated by the percent GC in 100 kb non-overlapping windows.

Phylogenetic analysis of *E. pusillum*

The orthologous genes from the lichen-forming fungus *E. pusillum* and 14 other non lichen-forming fungi whose genomes were available were identified using Inparanoid [30] with default parameters (Figure 4). Phylogenetic analysis was performed using 1,893 single-copy orthologous genes identified among the genomes of above mentioned 15 fungi from the subkingdom Dikarya, and a linearized phylogenetic tree was constructed with estimates of the divergence times among these taxa (Figure 4). The phylogenomic analysis shows that the *E. pusillum* lineage is more closely related to the human pathogen *Exophiala dermatitidis*, which is the anamorph species of *Capronia* belonging to the Chaetothyriales, and the *E. pusillum* belongs to the Verrucariales. Both the orders belong to the same subclass Chaetothyriomycetidae, and the divergence between the lichenized Verrucariales and nonlichenized Chaetothyriales occurred approximately 131 million years (Myr) ago (Figure 4). This result is consistence with the phylogenetic analysis for Ascomycota using ribosomal RNA [9].

Repeat-induced point mutation

Repeat-induced point mutation (RIP) is a gene silencing mechanism that can cause C-G to T-A mutations on repetitive DNA sequences, and the mutations from C to T mostly occur at CpA dinucleotides [31]. According to the method proposed by Margolin *et al.* [32], sequences with a high TA/AT ratio (>0.89) and a low (CA + TG)/(AC + GT) ratio (< 1.03) are thought to indicate RIP [33].

A quantitative alignment-based method, RIPCAL [34], was used to search for evidence for RIP in *E. pusillum* genome. The RIP indices of the repetitive sequence are 1.46927 and 0.697168, and those of non-repetitive sequence are 0.565627 and 1.42277, which indicates that

A

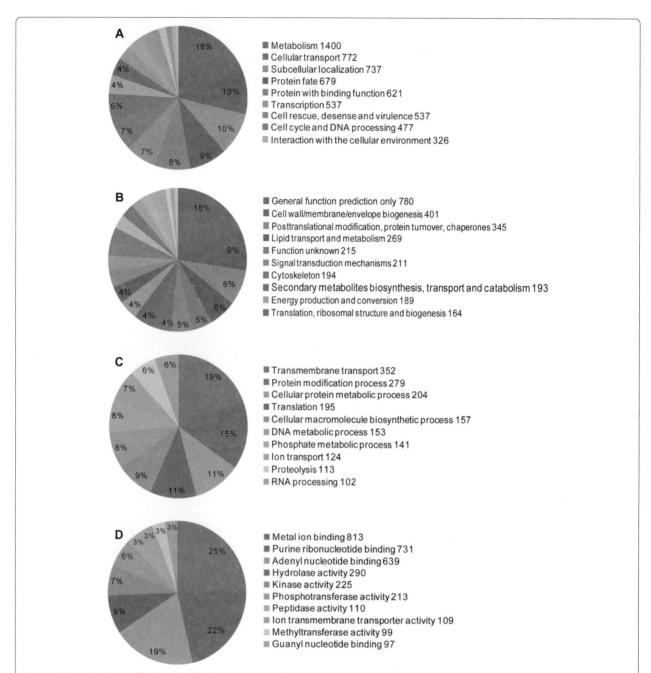

■ Metabolism 1400
■ Cellular transport 772
■ Subcellular localization 737
■ Protein fate 679
■ Protein with binding function 621
■ Transcription 537
■ Cell rescue, desense and virulence 537
■ Cell cycle and DNA processing 477
■ Interaction with the cellular environment 326

B

■ General function prediction only 780
■ Cell wall/membrane/envelope biogenesis 401
■ Posttranslational modification, protein turnover, chaperones 345
■ Lipid transport and metabolism 269
■ Function unknown 215
■ Signal transduction mechanisms 211
■ Cytoskeleton 194
■ Secondary metabolites biosynthesis, transport and catabolism 193
■ Energy production and conversion 189
■ Translation, ribosomal structure and biogenesis 164

C

■ Transmembrane transport 352
■ Protein modification process 279
■ Cellular protein metabolic process 204
■ Translation 195
■ Cellular macromolecule biosynthetic process 157
■ DNA metabolic process 153
■ Phosphate metabolic process 141
■ Ion transport 124
■ Proteolysis 113
■ RNA processing 102

D

■ Metal ion binding 813
■ Purine ribonucleotide binding 731
■ Adenyl nucleotide binding 639
■ Hydrolase activity 290
■ Kinase activity 225
■ Phosphotransferase activity 213
■ Peptidase activity 110
■ Ion transmembrane transporter activity 109
■ Methyltransferase activity 99
■ Guanyl nucleotide binding 97

Figure 3 Functional classification for the *Endocarpon pusillum* genome. A. FunCat database classification in the second level. **B.** KOG database classification. **C.** The top 10 Gene Ontology biological process. **D.** The top 10 Gene Ontology molecular functions.

the regions of repetitive sequence in *E. pusillum* have undergone RIP. In addition, the percentage of genes in multigene families shows that *E. pusillum* has a low proportion of genes in multigene families (Figure 5A). The analysis of amino acid similarities in multigene families among eight fungi indicates that there are fewer best-matching (≥ 80%) genes in the *E. pusillum* genome (Figure 5B), which implies that some genes in the multigene families in *E. pusillum* are RIP mutated. However, the function of RIP in lichen-forming fungi still remains unclear.

The mating systems and reproduction

The genes involved in the mating process were identified in *E. pusillum* genome. The result shows that there is a single mating type (*MAT*) locus containing both a *MATα* and HMG domain in the genome, which provides the molecular evidence for homothallic lifestyle of *E. pusillum* (Figure 6). Genes for conidiophore development and asexual reproduction were also identified, together with those involved in fruit body development and sexual reproduction (Additional file 1: Table S2), and genes with

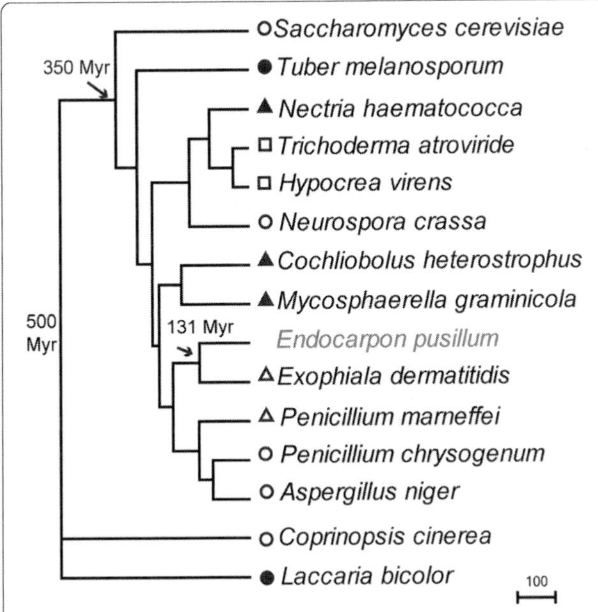

Figure 4 Phylogenetic analysis of 15 fungi based on 1,893 orthologous genes. The times of divergence of clades are indicated in millions of years by arrows. The symbols correspond to certain model fungi (open circles), mycorrhizal fungi (filled circles), plant pathogen (filled triangles), animal pathogen (open triangles) and fungi symbiosis with fungi (open squares). *C. cinerea* and *L. bicolor* belonging to the Basidiomycota were used as an outgroup.

the function of pheromone precursor (alpha-factor like), pheromone receptor (for alpha-factor like pheromone), ascus development (rhamnogalacturonase B), and receptors preventing improper sexual development have been lost in *E. pusillum*. However, it need more experimental evidence to determine whether these missing genes contribute to the special reproduction mode of *E. pusillum*.

Secreted proteins

Secreted proteins are important effectors that modulate the interaction between pathogenic microbes and hosts [35-37], and many small secreted proteins (SSPs) are virulence factors [38]. However, *E. pusillum* has fewer secreted proteins than other phytopathogenic fungi (e.g. *Fusarium graminearum*, *Magnaporthe oryzae* and *Metarhizium anisopliae*) according to the blast results (Additional file 1: Table S3), and the function of 66% of SSPs is unknown. Lichen-forming fungus and pathogenic fungi are different, even though they both live with another organism and absorb nutrients from their partner. The analysis of the secreted proteins suggests that the SSP-mediated interaction between symbionts in *E. pusillum* is weaker than the virulence in parasitism (pathogenic fungi).

Secondary metabolism

To date, several polyketide synthase (PKS) genes have been identified in different lichen-forming fungi [39-42], but no non-ribosomal peptide synthetase (NRPS) genes have been found. Analysis of the lichen-forming fungus *E. pusillum* genome revealed 2 NRPS genes and 14 PKS genes (Additional file 1: Table S4). The information of genes for secondary metabolism is helpful to the research of the secondary metabolites of *E. pusillum* and the synthesis conditions.

Candidate genes for drought tolerance

At least 93 genes involving in these drought resistant mechanisms were identified in the genome of *E. pusillum* (Additional file 1: Table S5), which will provide us many important clues to resolve the question about why *E. pusillum* can survive in extreme drought environment. Recently, most studies on mechanisms about desiccation-

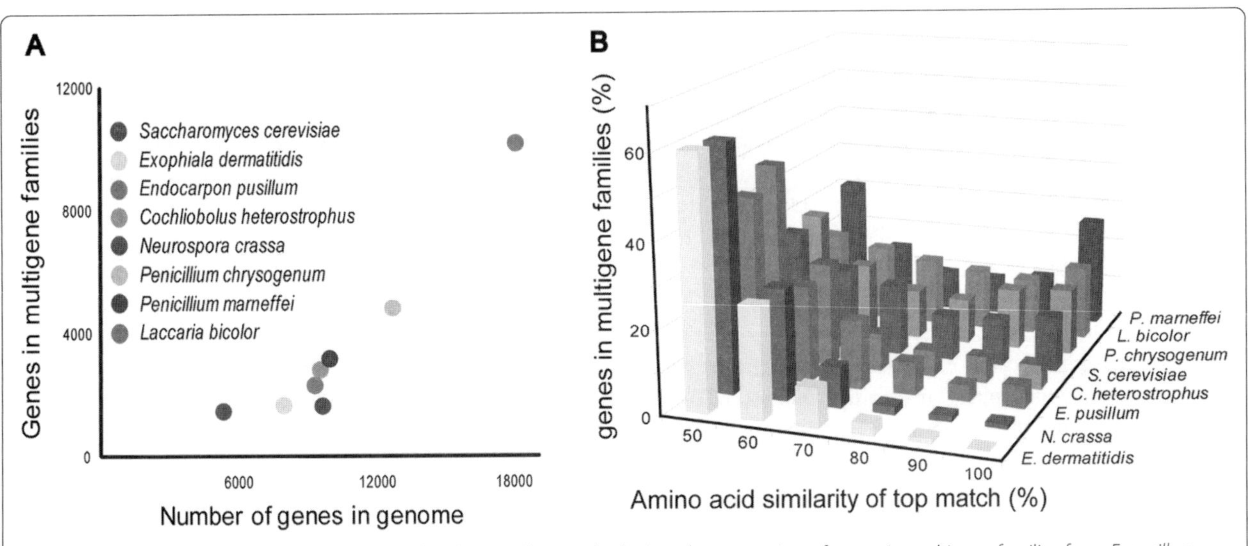

Figure 5 Proportion of genes in multi-gene families. A. The graph displays the proportion of genes in multigene families from *E. pusillum* and other fungi. **B.** Histogram of paralogous gene numbers with different levels of amino acid similarity in *E. pusillum* and other fungi.

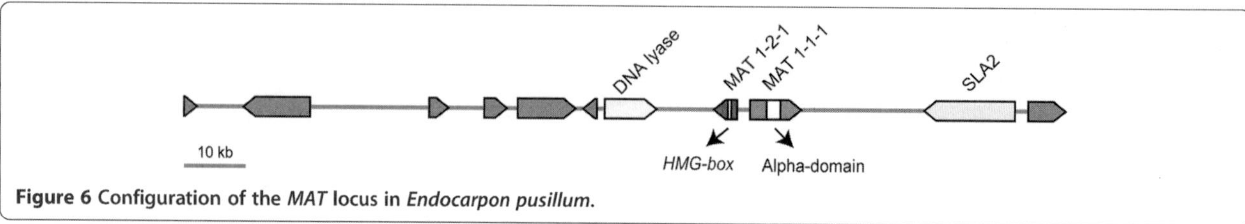

Figure 6 Configuration of the *MAT* locus in *Endocarpon pusillum*.

tolerance in lichens have focused on the scavenging reactive oxygen species [43]. However, genes involved in regulating osmotic pressure, correcting protein misfolding, and scavenging the reactive oxygen species were identified in *E. pusillum* genome, which were worthy of further investigation.

Multigene families expansion

Up to 1,889 protein families (containing at least two genes in all selected species) were identified from *E. pusillum* and 14 other fungi using Markov cluster (MCL) method. Based on BadiRate analysis, we found that 1,129 protein families were expanded and 760 protein families had undergone contraction in *E. pusillum*. The outlier results generated by BadiRate analysis are presented in Table 2. The expansion of these protein families might reflect specific characteristics of lichen-forming fungus *E. pusillum*.

Protein families involved in signal transduction

Compared with other fungi, the lichen-forming fungi are special because they can form mutualistic symbiosis with photosynthetic organisms and this specialty is reflected in its genome. Little is known about the functional genes in lichens, and most expanded protein families are function-unknown in *E. pusillum* (Table 2); however, it is likely that many of these expansion families are related to symbiosis. For example, the expressions of WD-repeat domain-containing proteins are up-regulated in the early developmental stages of lichen-symbiosis in *Cladonia grayi* [23].

Similarly, G protein-coupled receptors (GPCRs), which sense molecules outside the cell and activate signal transduction pathways inside the cell [44], are also likely to provide lichen-forming fungi with a strong ability to response to such signals, and all GPCR families have undergone obvious expansion (Table 3).

Heterokaryon incompatibility protein families

There are 261 genes in the *E. pusillum* genome that were annotated to have heterokaryon incompatibility (HET)-related functions, and 182 of them are homologous to those in *N. crassa* and *Podospora anserina* (Additional file 1: Table S6). HI (Heterokaryon incompatibility) is a common characteristic among filamentous fungi; it can prevent the formation of heterokaryotic cells in which two different genomes coexist. Once the fusion between two individuals with incompatible *het* loci occurs, the HET genes would trigger the HI reaction, which is characterized as growth inhibition, repression of asexual sporulation, hyphal compartmentation and death of the heterokaryotic cell [45-47]. However, the biological significance of HI is still unknown. The expansion of HET protein families suggests that *E. pusillum* is likely to have a strict regulation of its vegetation [48], and may represent a strategy for stable genotype by defending against the transfer of exogenous genes.

Lineage-specific protein families involved in self-splicing of insertions

The genes encoding ribonuclease H and transposase, which are predicted to be able to target some insertions, such as group I introns that can be stably integrated into the genome following the reverse splicing reaction [49], belong to lineage-specific multigene families in *E. pusillum* (Table 2).

The genomic sequencing data reveal that there are three group I introns in the large-subunit ribosomal DNA (LSU rDNA) of *E. pusillum* Z07020 at position 856, 2169, and 2721 (corresponding to the sequence of *Pichia methylivora* with NCBI accession number EU011611), and one group I intron in the small-subunit ribosomal DNA (SSU rDNA) at position 1769 (corresponding to the sequence of *Saccharomyces cerevisiae* with NCBI accession number Z75578).

Group I introns can self-splice themselves from an RNA transcript, requiring protein factors to facilitate the correct folding of the ribozyme core [49]. There are multiple group I introns present in the nuclear rDNA of many lichen-forming fungi [50-53], and they are considered to decrease the growth rate of lichen-forming fungi by interfering with rRNA maturation [54]. It is considered an adaption for lichenization: most lichens grow very slowly because the symbiosis would be disrupted if the mycobionts too many nutrients from the photobionts. Therefore, it is reasonable that the protein families involved in RNA reverse splicing have undergone expansion in a lichen-forming fungal genome that harbors abundant self-splicing introns.

Protein families involved in transport

The exchange of ions and metabolites between the mycobiont and photobiont is the foundation of the symbiotic

Table 2 Expanded and lineage-specific protein families in *Endocarpon pusillum*

Protein family ID	Number of proteins in *E. pusillum*	p-value	HMMPfam description
group10696	20	0	Hypothetical protein CIMG_07368
group10024	15	0	Ankyrin and HET domain-containing protein
group14580	14	0	Ankyrin repeat
group10771	13	0	P-loop containing nucleoside triphosphate hydrolases
group11690	13	0	MFS general substrate transporter
group10071	11	0	Putative transposase
group10011	10	0	Hypothetical protein NECHADRAFT_87861
group14903	8	0	Heterokaryon incompatibility protein
group17019	7	1E-07	DNA/RNA polymerases
group10674	10	3E-07	Magnesium transport protein CorA, transmembrane region
group10674	10	3E-07	Hypothetical protein AN9301.2
group10038	8	5E-07	Hypothetical protein SS1G_06795
group10340	8	5E-07	Conserved hypothetical protein
group15154	6	5E-07	Hypothetical protein CHGG_10108
group16626	6	5E-07	Hypothetical protein
group18347	6	5E-07	Hypothetical protein
group10002	9	2.4E-06	Ribonuclease H-like
group10015	9	2.4E-06	Ankyrin repeat
group10097	7	4.6E-06	Ankyrin repeat
group10201	7	4.6E-06	ARM repeat
group10224	7	4.6E-06	Heterokaryon incompatibility protein
group10843	7	4.6E-06	Predicted protein
group11854	7	4.6E-06	Cytochrome P450
group10000	11	0.000001	P-loop containing nucleoside triphosphate hydrolases
group10705	5	0.000005	Hypothetical protein
group11683	5	0.000005	Dimeric alpha + beta barrel
group11909	5	0.000005	Unnamed protein product
group13886	5	0.000005	Ankyrin repeat
group15448	5	0.000005	FabD/lysophospholipase-like
group16624	5	0.000005	Protein kinase-like (PK-like)
group16962	5	0.000005	Hypothetical protein NFIA_004500
group17139	5	0.000005	Cytochrome P450
group18335	5	0.000005	Predicted protein
group18344	5	0.000005	Conserved hypothetical protein
group19016	5	0.000005	Hypothetical protein
group19022	5	0.000005	Hypothetical protein
group10704	8	1.78E-05	Conserved hypothetical protein
group10239	6	3.81E-05	P-loop containing nucleoside triphosphate hydrolases
group10495	6	3.81E-05	Unnamed protein product
group10677	6	3.81E-05	Unnamed protein product
group11482	6	3.81E-05	Predicted protein
group11606	6	3.81E-05	Hypothetical protein AOR_1_1448144
group15933	6	3.81E-05	Purine and uridine phosphorylases
group16361	6	3.81E-05	Hypothetical protein SS1G_13793

Table 2 Expanded and lineage-specific protein families in *Endocarpon pusillum* *(Continued)*

group16513	6	3.81E-05	Conserved hypothetical protein
group17376	6	3.81E-05	Hypothetical protein
group10014	9	4.84E-05	POZ domain
group10175	9	4.84E-05	Hypothetical protein CIMG_05613
group10238	4	4.94E-05	WD40 repeat-like
group11337	4	4.94E-05	TPR-like
group12140	4	4.94E-05	WD40 repeat-like
group16268	4	4.94E-05	Hypothetical protein MGYG_01629
group17166	4	4.94E-05	Predicted protein
group17167	4	4.94E-05	Alpha/beta-Hydrolases
group17374	4	4.94E-05	Hypothetical protein FRAAL5044
group17553	4	4.94E-05	Protein kinase-like (PK-like)
group17955	4	4.94E-05	Conserved hypothetical protein
group18350	4	4.94E-05	Hypothetical protein CHGG_08502
group19010	4	4.94E-05	NTF2-like
Lineage-specific protein families			
Protein family ID	Number of proteins in *E. pusillum*	p-value	Pfam description
group19669	4	4.94E-05	Ribonuclease H-like
group19844	4	4.94E-05	Transposase
group19847	4	4.94E-05	Hypothetical protein
group19848	4	4.94E-05	Hypothetical protein
group19850	4	4.94E-05	Protein kinase-like (PK-like)
group19853	4	4.94E-05	Ankyrin repeat

association. For this reason, genes encoding transporter proteins are expected to exhibit some traits in the lichen-forming fungus genome. There are 10 genes belonging to the magnesium transport protein CorA family, which showed significant expansion in the *E. pusillum* genome (Table 2). The CorA proteins were the first family to be identified that could transport Mg^{2+} in bacteria [55,56]. The mechanism of action of the CorA family has been well characterized in yeast and members of this family can transfer Mg^{2+} both into and out of the cell [57]. Magnesium is the most important divalent cation in cells, and is particularly important for photosynthesis [58]. Magnesium deficiency has been reported to affect plant photosynthesis and growth [59,60]. In the lichen thallus, the special structure of the symbionts (Figure 1D–F) makes it difficult for algal cells to absorb magnesium from the environment. However, genes for magnesium transport protein are expanded in *E. pusillum*, which led us to suspect that during symbiosis the mycobiont could provide magnesium to the phycobiont to meet the needs of their life cycles.

Table 3 Evolutionary changes of GPCR protein families in *Endocarpon pusillum*

GPCRs family	Number of protein in *E. pusillum*	Evolutionary changes in *E. pusillum*	Annotation
Class A	6	Gain	Rhodopsin-like family
Class B	2	Gain	Secretin receptor
Class D	5	Gain	Fungal mating pheromone receptor
Class E	1	Gain	cAMP receptor

A striking finding is that most nitrogen transporter families are expanded but most sugar transporter families have been lost from the *E. pusillum* genome (Table 4). Though the mechanisms of substance transfer in lichen are not completely understood, it has been reported that the mycobiont absorbs certain carbohydrates generated by photosynthesis of the photobiont [61]. The reduction of sugar transporter genes, especially those for common sugars, such as glucose and fructose, which can be utilized by many organisms, suggests that the mycobiont simply absorbs certain uncommon carbohydrates from their photosynthetic partner. Therefore, the genes for many unnecessary sugar transporters have been lost during evolution. The lichen-forming fungi do not rely on common sugars because these carbon sources can be used by numerous microorganisms and lichen cannot compete with them for its slow growth and metabolism.

Table 4 Evolutionary changes of transporter protein families in *Endocarpon pusillum* (partial results)

Transporter family	Number of protein in *E. pusillum*	Evolutionary changes in *E. pusillum*	Annotation
1.A.11.3.3	3	Gain	Ammonium transporter-Hebeloma cylindrosporum
2.A.3.10.20	1	Gain	Lysine/arginine permease-*Candida albicans*
2.A.3.4.2	16	Gain	Gaba permease-Emericella nidulans-*Aspergillus nidulans*
2.A.3.8.4	6	Gain	High affinity methionine permease-*Saccharomyces cerevisiae*
2.A.39.3.3	1	Gain	Uridine permease-Saccharomyces cerevisiae
8.A.9.1.1	3	Gain	Amino acid transport related protein (RBAT)-*Oryctolagus cuniculus*
2.A.1.1.10	0	Loss	Maltose permease MAL6T-*Saccharomyces cerevisiae*
2.A.1.1.11	1	Loss	General alpha-glucoside permease-*Saccharomyces cerevisiae*
2.A.1.1.33	0	Loss	Hexose transporter-*Kluyveromyces lactis*
2.A.1.1.38	2	Loss	Sugar transporter STL1-*Saccharomyces cerevisiae*
2.A.1.1.39	3	Loss	High-affinity glucose transporter-*Kluyveromyces lactis*
2.A.1.1.57	1	Loss	Monosaccharide transporter-*Aspergillus niger*
2.A.1.1.58	0	Loss	Monosaccharide transporter-*Aspergillus niger*
2.A.1.1.68	0	Loss	Glucose transporter/Sensor OS-*Pichia stipitis*
2.A.1.1.9	0	Loss	Lactose permease-Kluyvero myces lactis
2.A.1.2.23	1	Loss	Fructose facilitator-Zygosaccharomyces bailii

By contrast, the expansion of nitrogen transporters suggests that the lichen-forming fungi need to transfer (export or import) various nitrogen sources. Nitrogen is an indispensable substance for the growth of organisms, and lichens whose photosynthetic partners are green algae must obtain inorganic and organic nitrogen [62-64]. There is no evidence for nitrogen transfer between symbionts in lichen. However, owing to the particular structure of lichen thallus in which the photobiont cells are tightly entwined by mycelia (Figure 1D–F) and the photobiont cells cannot absorb substances from their surroundings directly, it can be deduced that the necessary components for their growth, such as ions and nitrogen sources, must be transferred from the mycobiont to the photobiont. The phycobiont may be have a preference for certain forms of nitrogen; for example, in the alga-bacteria association, the algae prefer NH_4^+ and NO_3^-, which are release by bacteria when cultured in organic nitrogen conditions [65]. Thus, the mycobiont may absorb and transform the nitrogen forms that are preferred by its photosynthetic partner. Therefore, the expansion of nitrogen transporters should be related to this physiological process. Recently, it was reported that the ammonium transporter, MEP-α, in lichen-forming fungi was obtained though horizontal gene transfer [66], which can partly explain the mechanism of the expansion of the nitrogen transporter superfamily. This gene appears to have been lost in some groups of lichens, such as in those that are symbiotic with nitrogen-fixing cyanobacteria or those that inhabit high-nitrogen niches [67]. However, for the lichens whose photobiont is a green alga that cannot fix nitrogen, the expansion of nitrogen transporter families is of great significance in promoting the growth of the lichen thallus, as seen in *E. pusillum*.

Expression analysis of symbiosis-related genes by quantitative real-time PCR (qRT-PCR)

A qRT-PCR experiment was performed to determine the changes of some symbiosis-related genes in the lichen-forming fungus at transcription level when *E. pusillum* pre-contacted with its photosynthetic partner, *Diplosphaera chodatii.*

Symbiosis-related genes selection

Thirty-two genes encoding lectins were identified in the *E. pusillum* genome. They belong to four superfamilies: concanavalin A-like lectin, fucose-specific lectin, mannose-binding lectin and ricin B-related lectin. For the qRT-PCR assay, six lectin genes were selected as representatives of the four superfamilies, according to the difference in the number of transmembrane helices and signal peptides (Additional file 1: Table S7).

The expansions of nitrogen transporters and the contractions of sugar transporters imply that they are closely related to maintenance of both symbionts. Hence, certain genes encoding these transporters were chosen for the qRT-PCR assay. Among them, two ammonium transporters and one nitrate transporter that displays high homology to those in other fungi (Additional file 1: Table S8) were included. It has been reported that eight sugar transporters are able to transfer photosynthesis products in other fungi, and they are predicted to be functional at the interface of lichen association [68-75]. Thus, the homologous proteins were identified in the *E. pusillum* genome (Additional file 1: Table S9). The genes sharing homology with known transporters were obtained using BlastP and each of them appeared in several BlastP results; their expressions were determined by the qRT-PCR assay (Additional file 1: Table S10). Nitrogen metabolism and sugar metabolism in fungal cells would be active after nitrogen and carbohydrates were assimilated or transferred. Therefore, genes encoding glutamate synthase and nitrite reductase (Additional file 1: Table S8), and that encoding the Golgi GDP-mannose transporter, which is functional in the glycosylation of secreted proteins [76] and five enzymes involved in galactose and nucleotide sugar metabolism (Additional file 1: Table S10) were also included. The gene encoding the tetracycline resistance protein was chosen as a control, because it recognizes and exports tetracycline from the cell [77,78] and is expected to show no expression change under the experimental conditions in the present study.

Differences in expression levels of genes involved in symbiosis under co-culture conditions

From the results presented in Figure 7, the expression levels of most genes increased only under the condition for experimental group IV (Table 5) (weight ratio of lichen-forming fungus and phycobiont is 10:3, and culture time is 72 hours), which indicated that the contact time and biomass ratio of both symbionts affects their recognition and nutrient transfer significantly.

Three lectin genes (F481_01961, F481_04092, F481_02882) showed significant increases ($\log_2$ fold change > 2) in their expression level in experiment group IV compared with the control group B (only the mycobiont was cultured in BBM for 72 h) (Table 5; Figure 7). Interestingly, the common aspect of these lectins is that there are no predicted signal peptides in their amino acid sequences (Additional file 1: Table S7) (http://elm.eu.org) [79]. Lectins are sugar-binding proteins that play an important roles in cell recognition [80]. The first lectin discovered in lichens was that found in *Peltigera canina* and *Peltigera polydactyla* [17]. Thereafter, some studies showed that lectins secreted by lichen-forming fungi may be involved in recognizing their compatible algae [15,16,81]. Lectins, as surface proteins, can directly contact compatible phycobionts, and the contact requires receptor sites on the surface of the phycobiont to which the lectins binds [82,83]. Recent studies on lectins in *P. membranacea* indicated that the expression of galectin *lec-1* was influenced by the presence of the phycobiont [84], and the evolution of galectin *lec-2* was driven by interaction with different strains of the phycobiont *Nostoc* [85]. Galectins, together with cellular slime mold lectin discoidin I, do not have signal peptides in eukaryotic cells and both of them have intracellular and extracellular localizations [86,87]. Our study showed that lectins without signal peptides could play a major role in symbiotic recognition and provided a practical guide for screening for lectins participating in the interaction between lichen-forming fungi and their photosynthetic partners.

An ammonium transporter (F481_01640) and a nitrate transporter (F481_07695) were up-regulated ($\log_2$ fold change > 2) in experimental group IV (Figure 7). BBM contains nitrate ions; therefore, the up-regulation of the nitrate transporter (F481_07695) is expected. However, the up-regulation of the ammonium transporter (F481_01640) suggests that ammonium could be produced from nitrate in *E. pusillum* and transferred to its photosynthetic partner, because there is no ammonium ion in BBM. The differences in the expressions of glutamate synthase (F481_06302) and nitrite reductase (F481_07075) indicate that nitrogen metabolism is altered in the lichen-forming fungus when it meets its compatible alga. It has been reported that algae from different lichens can utilize organic and inorganic nitrogen [62-64]. Thus, we can conclude that the pattern of nitrogen transfer in *E. pusillum* is that the mycobiont absorbs various nitrogen sources from the environment and can convert them to different forms *in vivo* to meet the algae's demand for nitrogen, based on the expansion of nitrogen transporter families in *E. pusillum* genome and the result of co-culture experiments.

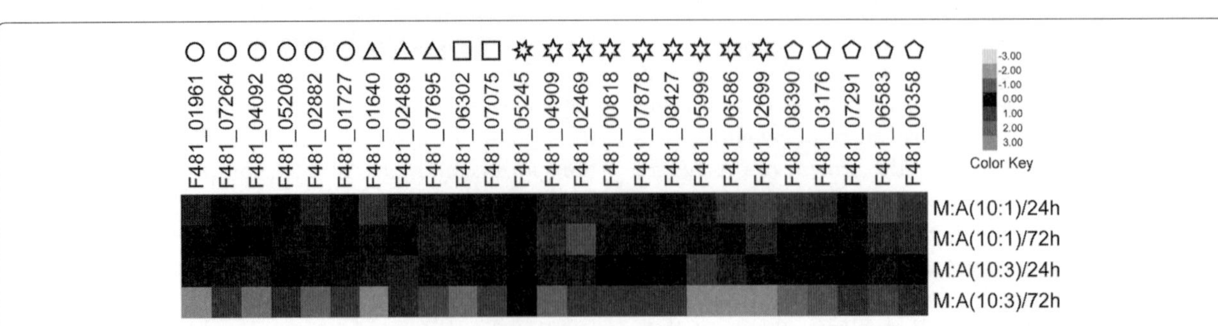

Figure 7 Relative transcription levels of certain symbiosis-related genes in *Endocarpon pusillum* when cultured with *Diplosphaera chodatii*. The names of the genes are listed on the top of the picture. Circles represent genes encoding lectins, triangles represent N transporter genes, squares represent gene encoding enzymes of nitrogen metabolism, a heptagon represents the tetracycline resistance protein gene, hexagons represent sugar transporter genes, and pentagons represent genes encoding enzymes of sugar metabolism. The culture conditions are displayed at the right of the picture, in which the ratios of mycobiont (M) and phycobiont (A) and incubation time are given. The $\log_2$ relative expression with respect to the control group is illustrated in the heat map.

Table 5 Sample treatments to identify differentially expressed genes

Experiment group	Sample	Weight ratio (mycobiont: Phycobiont)	Incubation time (h)
Control A	Mycobiont	-	24
Control B	Mycobiont	-	72
Control C	Phycobiont	-	72
I	Mycobiont + phycobiont	10:1	24
II	Mycobiont + phycobiont	10:3	24
III	Mycobiont + phycobiont	10:1	72
IV	Mycobiont + phycobiont	10:3	72

The lichen-forming fungus cannot obtain any organic nutrition from BBM. However, six of the symbiosis-related sugar transporter genes (F481_00818, F481_07878, F481_08427, F481_05999, F481_06586, and F481_02699) show significant differences in expression in experiment group IV (Figure 7). This suggests that some carbohydrates are released into the BBM by *D. chodatii*, and the transporters encoded by these genes may be responsible for transferring these carbohydrates into the fungal cells. The upregulation of the expression of the Golgi GDP-mannose transporter (F481_08390) under experimental group IV conditions, bearing in mind that a previous study proved that this transporter participates in glycosylation [76], indicated that the sugar metabolic pathway is active in *E. pusillum* after obtaining carbohydrates. Furthermore, the enzymes (F481_03176, F481_07291, F481_06583, and F481_00358) involved in galactose and nucleotide sugar metabolism also showed high expressions levels, which demonstrated that the carbohydrates produced by the alga and absorbed by the fungus are indeed used for fungal cell life activities.

Native and non-native sugars utilized by *E. pusillum*

The carbohydrates used for growth and metabolism of lichen-forming fungus originate from the photosynthetic products of its photosynthetic partner, however, the forms of the photosynthetic products absorbed by different lichen-forming fungi vary depending on their different photobionts. It has been report that lichen-forming fungi absorb many polyols or glucose from algae or cyanobacteria, respectively [20,21], and the photosynthetic product transferred from D. chodatii to E. pusillum was sorbitol [20]. Additionally, previous studies showed that some monosaccharides, such as glucose, and disaccharides, such as trehalose and sucrose, were transferred between non-lichenized symbionts [68-75]. Because some homology structures are frequently found in different sugar transporters from diverse fungi, the specificities of these transporters may be low. Therefore, 11 different carbohydrates were used to confirm the function of sugar transporters in E. pusillum and the potential carbohydrates transferred into this fungus.

The most significant up-regulation in expression of the examined genes was in the sample with trehalose for 24 hours. However, the high expression levels were not maintained over the next 48 hours (Figure 8). This suggests that trehalose cannot be absorbed by the lichen-forming fungus as an energy source, but is likely to be a signal molecule for the activation of the gene expression in E. pusillum. Trehalose is involved in many functions besides osmoprotection [88], and acts as a signal molecule that is possibly exported from bacteria to plants to regulate the carbon and nitrogen metabolism of plants in plant-bacteria interactions [89]. Trehalose can be

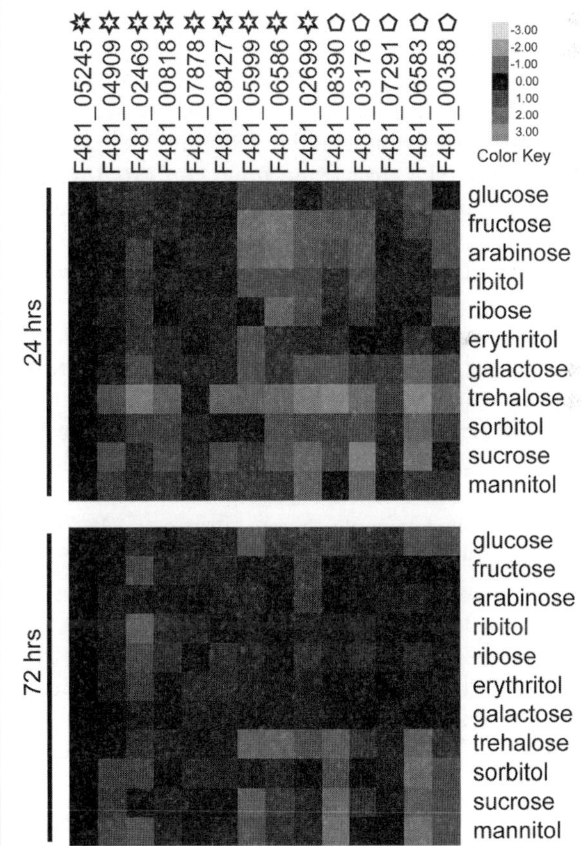

Figure 8 Relative transcription levels of sugar transfer-related genes in *Endocarpon pusillum* after supplementation with different carbohydrates. The names of the genes are listed on the top of the picture. Heptagons represent the gene encoding the tetracycline transporter, hexagons represent sugar transporter genes, and pentagons represent genes encoding enzymes of sugar metabolism. The culture time is displayed on the left and sugars added into BBM are listed on the right. The $\log_2$ relative expression with respect to the control group is illustrated in the heat map.

converted into trehalose-6-phosphate (T6P), which inhibits hexokinase activity and regulates glycolysis. Over expression of the trehalose 6-phospate synthase (TPS) gene induces the expression of several genes involved in stress tolerance, and carbon and nitrogen metabolism [90,91]. In the alga-invertebrate association, trehalose produced by the alga was shown to be present in certain symbiotic interfaces [92]. Although no research has shown a link between trehalose and lichen symbiosis interaction, our results suggest that trehalose may act as a signal molecule and would play an important role in the symbiotic interaction in E. pusillum.

Four carbohydrates (glucose, sucrose, sorbitol, and mannitol) increased the transcription levels of sugar transporters or enzymes involved in sugar metabolism in *E. pusillum* at 24 and 72 hours (Figure 8). These results suggest that these carbohydrates maintain the metabolism of the lichen-forming fungus over a long period, which implies that *E. pusillum* may absorb them as its carbon sources through some common sugar transporters. Although the transcription levels of some genes slightly fluctuated at 24 and 72 hours in the presence other sugars (arabinose, erythritol, fructose, galactose, ribitol and ribose) in the BBM, no genes exhibited a clear rising trend compared with the control group, especially those involved in sugar metabolism (Figure 8). This indicates that these sugars cannot be utilized as carbon sources by *E. pusillum*.

The transcription levels of some sugar transporters, such as F481_05999, F481_06586, and F481_2699, were up-regulated when various sugars were added to the BBM, indicating that they can transfer more than one carbohydrate into *E. pusillum*. Some genes showed different changes in their transcription levels, either in trend or extent, compared with those in the co-culture experiment. For example, sugar transporter F481_02469 showed no differential expression in the co-culture experiment, but was up-regulated slightly under some sugars treatments. In addition, the transcription levels of sugar transporters F481_05999, F481_06586, and F481_02699 were up-regulated more obviously in the co-culture experiment (Figures 7 and 8). These phenomena imply that there is some induction mechanism between the symbionts; i.e., the expressions of certain genes in a lichen-forming fungus are influenced by its photosynthetic partner to absorb organic carbon effectively.

Ion chromatography was used to detect glucose, sucrose, sorbitol and mannitol in the filtrates of experiment group IV, and controls B and C, to determine whether they could be released by *D. chodatii* and to determine the conditions under which this alga secretes organic carbon into the BBM. Sorbitol (12.6 mg/100 mL), glucose (0.8 mg/100 mL) and sucrose (0.8 mg/100 mL) have been detected in group IV (Figure 9). Thus, sorbitol showed the highest accumulation; however, there were other, unidentified, carbohydrates, which suggested that the lichen-forming fungus *E. pusillum* preferentially absorbs sorbitol. By contrast, no mannitol was found in this sample. There were no detectable carbohydrates in the control group B and C (the phycobiont and the lichen-forming fungus cultured respectively in BBM), which demonstrated that the pure cultured *D. chodatii* does not release carbon. Previous experiments for the phycobionts of *Ramalina crassa* and *Ramalina subbreviuscula* indicated that the pure cultured phycobiont released ribitol to the medium [19]. However, Hawksworth (1984) suggested that the photobiont had lost its ability to release carbohydrate after isolation from the lichen thallus, and that the lichen-forming fungus exerted some specific control on the photobiont cells that lead them to secrete carbohydrates [21]. Our result implies that *E. pusillum* can control the sugar export of *D. chodatii*, and that the biomass ratio and contact time are crucial for the interaction between the two symbionts.

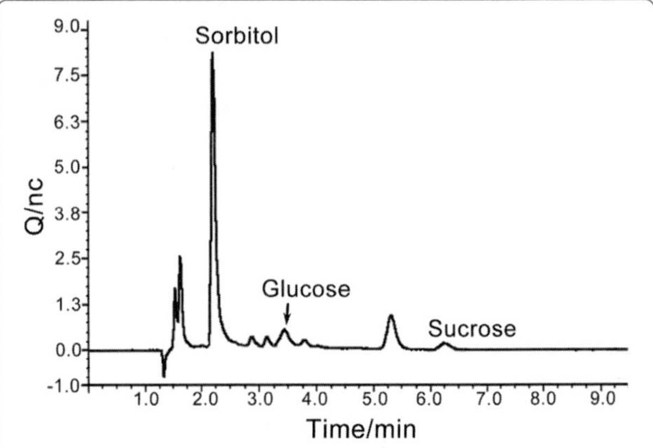

Figure 9 Ion chromatogram of co-cultured sample from experimental group IV. The sample from experiment group IV (mycobiont: phycobiont = 10:3; culture for 72 hrs) was diluted 25 times before four carbohydrates (glucose, mannitol, sorbitol and sucrose) were measured using ion chromatography.

Although mannitol is not released by D. chodatii, it can still be utilized by E. pusillum; it is likely to be because mannitol and sorbitol are isomers. This suggests that E. pusillum has the potential to use non-natural carbon sources. This phenomenon is likely to occur in other lichen-forming fungi, which would contribute to discovering the mechanism of the algal switch [93-95] and could provide clues for the evolution of lichenization.

Conclusions

Approximately 40% of all Ascomycota are lichen-forming; thus, lichenization is regarded as one of the most important fungal lifestyles [96]. Hence, genomic information for lichen-forming fungi would expand the knowledge of fungi. In the present study, we report, for the first time, the characteristic of the lichen-forming fungal genome, which displays many features that are different to other fungi.

This is the first study to report that the lichen-forming fungal genome have undergone RIP. Genes for mating system, secondary metabolism, and the drought-related mechanisms were indentified in *E. pusillum* genome, which are worth being investigated in the future. The evolution analysis of multigene families indicated the expansion and contraction in *E. pusillum* genome reveal the effect of lichenization on lichen-forming fungi.

Co-culture experiments suggest that the lectins without signal peptides would be likely to play an essential role in the recognition of lichen symbiosis, and one of the most striking findings in these experiments is that an appropriate weigh ratio of lichen-forming fungus and its photosynthetic partner and sufficient contact time are vital for their recognition and mutual influence. We also confirmed that the most important natural carbon source for *E. pusillum* is sorbitol transferred from *D. chodatii*; however, this lichen-forming fungus can also use other non-natural carbohydrates under the pure culture condition.

A mycobiont-phycobiont interplay model is shown as Figure 10. The model reflects aspects of the recognition and interaction of the lichen thallus in *E. pusillum* and is likely to be applicable to other lichens, especially those whose photobionts are algae. This study provides a valuable genomic resource for future research in screening functional genes including drought-tolerance genes from lichens and would be useful for investigating the formation and divergence on the functional biology between lichenized and nonlichenized fungi.

Methods

Fungal strains

The lichen-forming fungal strain of *E. pusillum* Z07020 (HMAS-L-300199) was isolated by a single-spore discharge from the perithecium of lichen *E. pusillum* collected from Shapotou Desert Research Station (SDRS) of the Chinese Academy of Sciences (CAS) in the Tengger Desert of northern China [26]. The isolates were grown on 1.5% water agar for 1–2 weeks, and then cultured at room temperature after transfer to potato dextrose liquid medium.

Genome sequencing and assembly

The genome of the lichen-forming fungus E. pusillum was sequenced using high-throughput next-generation sequencing technology and the sequencing platforms were Roche 454 and Illumina Solexa systems. Genomic libraries containing 8-kb inserts were constructed and 1,394,086 paired-end reads (281.9 Mb) were generated using the 454 Roche GS FLX system. The Illumina adaptors were ligated onto the genomic DNA fragments, and DNA fragments with estimated sizes of 0.5 kb to 3 kb were selected using gel-electrophoresis. Libraries were

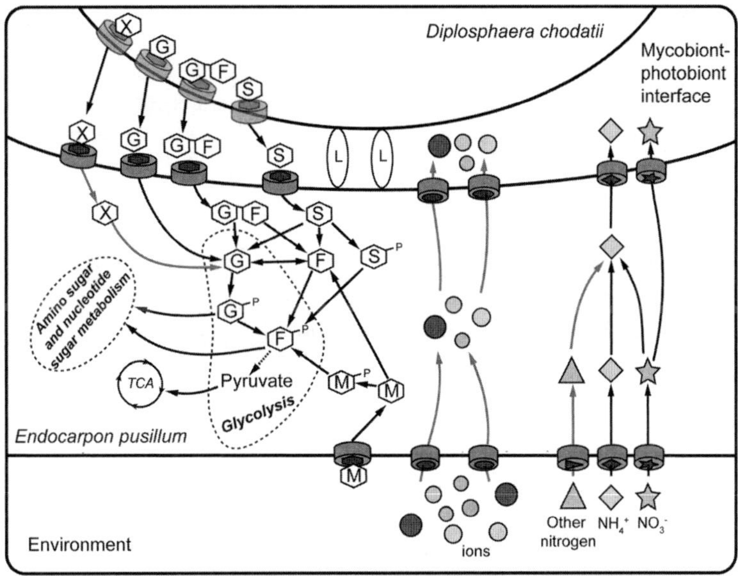

Figure 10 Summary of the interactions between symbionts in lichen *Endocarpon pusillum*. The lectins lacking signal peptides, which are shown as ellipse marked with L, act as recognition factors in the direct contact between *E. pusillum* and *D. chodatii*. After it is captured by the compatible fungal partner, the photosynthetic products sorbitol (S), glucose (G) and sucrose (G-F) are released from the phycobiont *D. chodatii*. These sugars are then absorbed through sugar transporters of *E. pusillum* and converted into glucose or fructose to support fungal metabolism. In addition, there are a small amount of other uncertainty carbon sources (X) released by the phycobiont, and the mycobiont can also utilize mannitol (M) from environment. At the same time, organic and inorganic nitrogen, together with various ions in the environment, are transferred into *E. pusillum*. In fungal cells, these substances may be converted into the forms preferred by the phycobiont and are then delivered to *D. chodatii*.

PCR-amplified using Phusion polymerase. Sequencing libraries were denatured with sodium hydroxide and diluted in hybridization buffer for loading onto a single lane of an Illumina GA flow cell. A Solexa sequencer generated the mate-paired reads (7,155,072 reads, 716 Mb) and paired-end reads (18,176,986 reads, 1818 Mb). Solexa sequencing paired-end reads and mate-paired reads were assembled by SOAPdenovo [97], which adopts the de Bruijn graph data structure to construct contigs.

Gene prediction and annotation

We used Augustus [98], GeneID [99], and GeneMark-ES [100] programs to predict the gene models for E. pusillum. A final set of gene models was selected by EvidenceModeler [101]. An ab initio prediction was carried on using the annotated information of A. fumigatus as a reference. All predicted gene models were subjected to GO [102], KOG [103], FunCat [104] and Kyoto Encyclopedia of Genes and Genomes (KEGG) database analysis [105]. Protein domains were predicted using InterProScan [106] against various domain libraries (HMMPfam, superfamily, HMMTigr, and HMMSmart). Repetitive elements were identified by blasting against the RepeatMasker library (http://www.repeatmasker.org/). Non-coding RNAs were predicted according to the Rfam database [107], and tRNAs were predicted using tRNAscan-SE [108]. Pseudogenes and rRNAs were designated using PseudoGene [109] and RNAmmer [110] respectively.

Orthologous gene and phylogeny analysis of *E. pusillum*

The sequences of corresponding orthologous genes from 15 fungi were aligned using ClustalW [111]. A maximum parsimony (MP) phylogenomic tree was created using the concatenated amino acid sequences in phylogenetic analysis using parsimony (PAUP) [112], and a bootstrap analysis with 1000 replications was performed to evaluate the reliability of the phylogenetic tree. The divergence time between species was estimated using the r8s method [113]. The time of divergence between Ascomycota and Basidiomycota is set as 500 Myr, and the time between Pezizomycotina and Saccharomycotina was 350 Myr [114].

Multigene families and evolution analysis

The families that were absent from the most recent common ancestor were chosen to analyze the evolution of protein families [115]. Multigene families were identified using the MCL method [116]. Whole genome blast analyses against Transporter Classification Database (http://www.tcdb.org/tcdb/) and GPCRDB (http://www.gpcr.org/7tm/) database were performed to identify genes exhibiting difference between E. pusillum and other fungi. All proteins in the genome were blast searched against the database with an e-value cutoff of $\leq e^{-10}$ and at least 40% identity over 60% coverage. A new software, BadiRate [117], was used in the evolutionary analysis of multigene families to estimate rates of gene gain and loss. Families with p-values less than 0.01 were considered to have experienced significant expansion or contraction.

Mating type

To detect the sexual cycle of E. pusillum, genes involved in the mating process, incompatibility, ascomata and conidiophore development, and HET were identified using BlastP against related genes from A. niger, A. nidulans [118], N. crassa [119] and P. anserine [120,121].

Analysis of genes involved in secondary metabolism

Genes encoding PKS, NRPS, and NRPS-PKS hybrid genes in the genome of E. pusillum were analyzed with the program SMURF (http://www.jcvi.org/smurf/index.php). Modulation analysis and domain extraction of different NRPS or PKS proteins were conducted by Blast searching against the SBSPKS database [122].

Secreted proteins

The potential secreted proteins of *E. pusillum* and other fungi, including *F. graminearum*, *M. oryzae*, and *M. anisopliae*, were predicted by SignalP 3.0 analysis using a hidden Markov model [http://www.cbs.dtu.dk/services/SignalP/].

Quantitative RT-PCR

RNA was extracted from cultured lichen-forming fungus *E. pusillum*. cDNA synthesis and relative quantitative RT-PCR were carried out as described previously [28]. For each treatment, qRT-PCR was performed in an Applied Biosystems 7500 Real-Time PCR system (Applied Biosystems, USA). The data were analyzed using the $2^{-\Delta Ct}$ method.

Gene-stability measure of reference genes using geNORM

Vandesomepele et al. (2002) developed an algorithm named geNORM that determines the expression stability of reference genes [123]. The calculated gene-stability measure (M) relies on the principle that the expression ratio of two ideal internal reference genes is identical in all samples, regardless of the experimental condition or cell type. Ten genes whose expressions showed no difference in comparative transcriptome data between control and drought-stress conditions (unpublished data) were chosen to determine the best reference genes (Additional file 1: Table S11). Ultimately, we chose the tetracycline resistance protein (F481_ 05245) gene as the reference gene.

Treatment of samples for co-culture experiments

Four treatments were carried out, comprising two weight ratios (10:1 and 10:3) for the lichen-forming fungus and phycobiont and two culture times (24 h and 72 h) (Table 5), to investigate whether the ratio and contact time between the lichen-forming fungus and phycobiont has an effect on gene expression in *E. pusillum* when both symbionts are incubated together.

The experiment was performed in Bold's basal medium (BBM) (does not contain carbohydrate), and under 12 h illuminations per 24 h, which allow the algal cells to produce carbohydrates by photosynthesis.

Treatment of samples for sugar transfer experiments

To confirm the capability of *E. pusillum* to utilize carbohydrates, polyols (mannitol, sorbitol, ribitol, and erythritol), monosaccharides (glucose, fructose, arabinose, ribose, and galactose) and disaccharides (trehalose and sucrose), respectively, were added to BBM containing only the lichen-forming fungus, and the transcript levels of genes involved in sugar transport and metabolism were determined after culturing for 24 and 72 hrs.

Determination of photosynthetic products by ion chromatography

Ion chromatography was used to determine glucose, sucrose, sorbitol and mannitol in co-cultured samples. The analytes were separated on a CarboPac™ PA1 (4 mm × 250 mm) anion exchange column using 200 mmol/L NaOH as mobile phase at flow rate of 1.0 mL/min and detected with a pulsed amperometric detector.

Availability of supporting data

The data sets supporting the results of this article are included within the article (and its additional files).

Additional file

Additional file 1: Genomic analysis of *Endocarpon pusillum*. The file contains additional information on genomic properties and qRT-PCR assays, comprising 10 tables provided in separate excel sheets. **Table S1** summarizes the main features of the primary sequence data. **Table S2** provides information on genes encoding proteins involved in sexual and asexual reproduction in *E. pusillum*. **Table S3** is a comparison of the number of secreted proteins between *E. pusillum* and other phytopathogenic fungi. **Table S4** lists the domain structures of predicted *E. pusillum* PKS and NRPS genes. **Table S5** lists the genes involving in drought resistant mechanisms in *E. pusillum*. **Table S6** provides information on genes involved in heterokaryon incompatibility in *E. pusillum*. **Table S7** lists the lectins measured by qRT-PCR analysis. **Table S8** lists genes that encode nitrogen transporters or proteins involved in nitrogen metabolism in *E. pusillum*, whose transcriptions were determined by qRT-PCR analysis. **Table S9** lists the homologous genes involved in symbiotic fungal sucrose and monosaccharide transporters in *E. pusillum*. **Table S10** lists genes that encode sugar transporters or proteins involved in sugar metabolism in *E. pusillum*, whose transcriptions were determined by qRT-PCR analysis. **Table S11** lists the candidate reference genes for qRT-PCR analysis in *E. pusillum*.

Abbreviations

BBM: Bold's basal medium; bp: Base pair; FunCat: Functionally annotation; GO: Gene ontology; GPCR: G-protein coupled receptor; HET: Heterokaryon incompatibility; HI: Heterokaryon incompatibility; HMAS: Mycological herbarium of microbiology institute,the chinese academy of sciences; KEGG: Kyoto encyclopedia of genes and genomes database; KOG: Eukaryotic clusters of orthologous groups; LSUrDNA: Large-subunit ribosomal DNA; NCBI: National center for biotechnology information; MAT: Mating-type; MCL: Markov cluster; MFS: Major facilitator superfamily; Myr: million years; NRPS: Non-ribosomal peptide synthetase; PCR: Polymerase chain reaction; PKS: Polyketide synthase; qRT-PCR: Quantitative real-time PCR; RIP: Repeat-induced point mutation; SEM: Scanning electron microscopy; SSUrDNA: Small-subunit ribosomal DNA; T6P: Trehalose-6-phosphate; TPS: Trehalose 6-phospate synthase; SSP: Small secret protein.

Competing interests

There are no competing financial interests associated with this work.

Authors' contributions

YYW performed the biological assays, analyzed the genomic and qRT-PCR data and wrote the draft manuscript. BL was in charge of the genome sequencing and gene annotation. XYZ provided the analysis of genome data. QMZ took part in writing the draft manuscript, designed the biological experiments and analyzed the data. TZ isolated and determined the strain of lichen-forming fungus *E. pusillum* and its phycobiont *D. chodatii*, and optimized their culture for the study. HL provided the genes concerning sexual cycle of fungi and constructed two color figures (Figures 1 and 10). YFY and XLZ participated in the analysis of genome data. XYH and MW assisted with the genome sequencing. LW initiated the study and finalized the manuscript. JCW initiated and designed the study, wrote parts of the manuscript, and revised and finalized the manuscript. All authors read and approved the final version of the manuscript.

Acknowledgments

This research was supported by the National Natural Science Foundation of China (No. 31070018, 81171524, 31270003, 31000010 and 31070360), the National 973 Program of China Grant (2013CB733904) and the Knowledge Innovation Program of the Chinese Academy of Sciences (KSCX2-EW-J-6).

Accession number

The whole project has been deposited at DDBJ/EMBL/GenBank under accession number APWS00000000.

Author details

[1]State Key Laboratory of Mycology, Institute of Microbiology, Chinese Academy of Sciences, Beijing 100101, China. [2]University of Chinese Academy of Sciences, Beijing 100049, China. [3]TEDA School of Biological Sciences and Biotechnology, Nankai University, Tianjin 300457, China.

Received: 7 October 2013 Accepted: 14 January 2014
Published: 17 January 2014

References

1. Ahmadjian V: *The lichen symbiosis*. New York: John Wiley; 1993.
2. Kirk PMCP, Minter DW, Stalpers JA: *Dictionary of the fungi*. 10th edition. Wallingford: CABI; 2008.
3. Nash TH: *Lichen biology. 2nd edn. Cambridge*. New York: Cambridge University Press; 2008.
4. Ingolfsdottir K: *Usnic acid. Phytochemistry* 2002, **61**:729–736.
5. Burkholder PR, Evans AW, McVeigh I, Thornton HK: **Antibiotic activity of lichens.** *P Natl Acad Sci USA* 1944, **30**:250.
6. Honegger R: **The lichen symbiosis—what is so spectacular about it?** *The Lichenologist* 1998, **30**:193–212.
7. Wedin M, Döring H, Gilenstam G: **Saprotrophy and lichenization as options for the same fungal species on different substrata: environmental plasticity and fungal lifestyles in the Stictis-Conotrema complex.** *New Phytol* 2004, **164**:459–465.
8. Gargas A, DePriest PT, Grube M, Tehler A: **Multiple origins of lichen symbioses in fungi suggested by SSU rDNA phylogeny.** *Science-New York then Washington* 1995:1492–1492.
9. Lutzoni F, Pagel M, Reeb V: **Major fungal lineages are derived from lichen symbiotic ancestors.** *Nature* 2001, **411**:937–940.
10. Rikkinen J: **Molecular studies on cyanobacterial diversity in lichen symbioses.** *Lichens: from genome to ecosystems in a changing world MycoKeys* 2013, **6**:3–32.

11. Ahmadjian V, Jacobs JB, Russell LA: **Scanning electron-microscope study of early lichen synthesis.** *Science* 1978, **200**:1062–1064.
12. Ahmadjian VHH: **The culture and synthesis of *Endocarpon pusillum* and Staurothele clopima.** *Lichenologist* 1970, **4**:13.
13. Trembley ML, Ringli C, Honegger R: **Morphological and molecular analysis of early stages in the resynthesis of the lichen Baeomyces rufus.** *Mycol Res* 2002, **106**:768–776.
14. Guillot J, Konska G: **Lectins in higher fungi.** *Biochem Syst Ecol* 1997, **25**:203–230.
15. Kardish NSL, Flemminger N, Galun M: **Lectin from the lichen Nephroma laevigatum. Localization and function.** *Symbiosis* 1991, **11**:47–62.
16. Andrews GA, Chavey PS, Smith JE: **Reactivity of lichen lectins with blood typed canine erythrocytes.** *Res Vet Sci* 1992, **53**:315–319.
17. Lockhart C, Rowell P, Stewart W: **Phytohaemagglutinins from the nitrogen-fixing lichens Peltigera canina and P. polydactyla.** *FEMS Microbiology Letters* 1978, **3**:127–130.
18. Richardson DHS, Hill DJ, Smith DC: **Lichen physiology.** *New Phytol* 1968, **67**:469–486.
19. Komiya T, Shibata S: **Polyols produced by the cultured phyco-and mycobionts of some Ramalina species.** *Phytochemistry* 1971, **10**:695–699.
20. Hill DJ, Ahmadjia V: **Relationship between carbohydrate movement and symbiosis in lichens with Green-Algae.** *Planta* 1972, **103**:267.
21. Hawksworth DL, Hill DJ: **The lichen-forming fungi.** *Blackie* 1984, **5**:62–69.
22. Gauslaa Y, Solhaug KA: **Fungal melanins as a sun screen for symbiotic green algae in the lichen Lobaria pulmonaria.** *Oecologia* 2001, **126**:462–471.
23. Joneson S, Armaleo D, Lutzoni F: **Fungal and algal gene expression in early developmental stages of lichen-symbiosis.** *Mycologia* 2011, **103**:291–306.
24. Suzanne Joneson FL: **Compatibility and thigmotropism in the lichen symbiosis: a reappraisal.** *Symbiosis* 2009, **47**:109–115.
25. Awasthi DD: *A hand book of lichens.* Bishen Singh Mahendra Pal Singh: Dehradun; 2000.
26. Zhang T, Wei J: **Survival analyses of symbionts isolated from *Endocarpon pusillum* Hedwig to desiccation and starvation stress.** *Sci China Life Sci* 2011, **54**:480–489.
27. Xavier BB, Miao VPW, Jonsson ZO, Andresson OS: **Mitochondrial genomes from the lichenized fungi Peltigera membranacea and Peltigera malacea: Features and phylogeny.** *Fungal Biol-Uk* 2012, **116**:802–814.
28. Junttila SM, Rudd S: **Characterization of a transcriptome from a non-model organism, Cladonia rangiferina, the grey reindeer lichen, using high-throughput next generation sequencing and EST sequence data.** *BMC Genomics* 2012, **13**:575.
29. Wang YY, Zhang T, Zhou QM, Wei JC: **Construction and characterization of a full-length cDNA library from mycobiont of *Endocarpon pusillum* (lichen-forming Ascomycota).** *World J Microb Biot* 2011, **27**:2873–2884.
30. Ostlund G, Schmitt T, Forslund K, Kostler T, Messina DN, Roopra S, Frings O, Sonnhammer ELL: **InParanoid 7: new algorithms and tools for eukaryotic orthology analysis.** *Nucleic Acids Res* 2010, **38**:D196–D203.
31. Cambareri EB, Jensen BC, Schabtach E, Selker EU: **Repeat-Induced G-C to a-T Mutations in Neurospora.** *Science* 1989, **244**:1571–1575.
32. Margolin BS, Garrett-Engele PW, Stevens JN, Fritz DY, Garrett-Engele C, Metzenberg RL, Selker EU: **A methylated Neurospora 5S rRNA pseudogene contains a transposable element inactivated by repeat-induced point mutation.** *Genetics* 1998, **149**:1787–1797.
33. Selker EU, Tountas NA, Cross SH, Margolin BS, Murphy JG, Bird AP, Freitag M: **The methylated component of the Neurospora crassa genome.** *Nature* 2003, **422**:893–897.
34. Hane JK, Oliver RP: **RIPCAL: a tool for alignment-based analysis of repeat-induced point mutations in fungal genomic sequences.** *Bmc Bioinformatics* 2008, **9**:478–490.
35. Hogenhout SA, Van der Hoorn RAL, Terauchi R, Kamoun S: **Emerging concepts in effector biology of plant-associated organisms.** *Mol Plant Microbe In* 2009, **22**:115–122.
36. Plett JM, Kemppainen M, Kale SD, Kohler A, Legue V, Brun A, Tyler BM, Pardo AG, Martin F: **A secreted effector protein of laccaria bicolor is required for symbiosis development.** *Curr Biol* 2011, **21**:1197–1203.
37. Kloppholz S, Kuhn H, Requena N: **A secreted fungal effector of glomus intraradices promotes symbiotic biotrophy.** *Curr Biol* 2011, **21**:1204–1209.
38. Deller S, Hammond-Kosack KE, Rudd JJ: **The complex interactions between host immunity and non-biotrophic fungal pathogens of wheat leaves.** *J plant physiol* 2011, **168**:63–71.
39. Armaleo D, Sun XM, Culberson C: **Insights from the first putative biosynthetic gene cluster for a lichen depside and depsidone.** *Mycologia* 2011, **103**:741–754.
40. Wang Y, Kim JA, Cheong YH, Joshi Y, Koh YJ, Hur JS: **Isolation and characterization of a reducing polyketide synthase gene from the lichen-forming fungus usnea longissima.** *J Microbiol* 2011, **49**:473–480.
41. Kim JA, Hong SG, Cheong YH, Koh YJ, Hur JS: **A new reducing polyketide synthase gene from the lichen-forming fungus Cladonia metacorallifera.** *Mycologia* 2012, **104**:362–370.
42. Wang Y, Kim JA, Cheong YH, Koh YJ, Hur JS: **Isolation and characterization of a non-reducing polyketide synthase gene from the lichen-forming fungus Usnea longissima.** *Mycol Prog* 2012, **11**:75–83.
43. Kranner I, Beckett R, Hochman A, Nash TH: **Desiccation-tolerance in lichens: a review.** *Bryologist* 2008, **111**:576–593.
44. Gilman AG: **G-Proteins - Transducers of receptor-generated signals.** *Annu Rev Biochem* 1987, **56**:615–649.
45. Glass NL, Jacobson DJ, Shiu PK: **The genetics of hyphal fusion and vegetative incompatibility in filamentous ascomycete fungi.** *Annu Rev Genet* 2000, **34**:165–186.
46. Saupe SJ, Clave C, Begueret J: **Vegetative incompatibility in filamentous fungi: Podospora and Neurospora provide some clues.** *Curr Opin Microbiol* 2000, **3**:608–612.
47. Fedorova ND, Badger JH, Robson GD, Wortman JR, Nierman WC: **Comparative analysis of programmed cell death pathways in filamentous fungi.** *BMC Genomics* 2005, **6**:177–191.
48. Saupe SJ: **Molecular genetics of heterokaryon incompatibility in filamentous ascomycetes.** *Microbiol Mol Biol R* 2000, **64**:489.
49. Haugen P, Simon DM, Bhattacharya D: **The natural history of group I introns.** *Trends Genet* 2005, **21**:111–119.
50. Gargas A, Taylor JW: **Phylogeny of discomycetes and early radiations of the apothecial ascomycotina inferred from Ssu Rdna sequence data.** *Exp Mycol* 1995, **19**:7–15.
51. Bhattacharya D, Friedl T, Helms G: **Vertical evolution and intragenic spread of lichen-fungal group I introns.** *J Mol Evol* 2002, **55**:74–84.
52. Gargas A, DePriest PT, Taylor JW: **Positions of multiple insertions in SSU rDNA of lichen-forming fungi.** *Mol Biol Evol* 1995, **12**:208–218.
53. Hibbett DS: **Phylogenetic evidence for horizontal transmission of group I introns in the nuclear ribosomal DNA of mushroom-forming fungi.** *Mol Biol Evol* 1996, **13**:903–917.
54. Depriest PT: **Molecular innovations in lichen systematics - the use of ribosomal and intron nucleotide-sequences in the cladonia-chlorophaea complex.** *Bryologist* 1993, **96**:314–325.
55. Smith RL, Banks JL, Snavely MD, Maguire ME: **Sequence and topology of the CorA magnesium transport systems of Salmonella typhimurium and Escherichia coli. Identification of a new class of transport protein.** *J Biol Chem* 1993, **268**:14071–14080.
56. Kehres DG, Maguire ME: **Structure, properties and regulation of magnesium transport proteins.** *Biometals* 2002, **15**:261–270.
57. Liu GJ, Martin DK, Gardner RC, Ryan PR: **Large Mg2+–dependent currents are associated with the increased expression of ALR1 in Saccharomyces cerevisiae.** *FEMS Microbiology Letters* 2002, **213**:231–237.
58. Gardner RC: **Genes for magnesium transport.** *Curr Opin Plant Biol* 2003, **6**:263–267.
59. Zhao H, Zhou Q, Zhou M, Li C, Gong X, Liu C, Qu C, Si W, Hong F: **Magnesium deficiency results in damage of nitrogen and carbon cross-talk of maize and improvement by cerium addition.** *Biol Trace Elem Res* 2012, **148**:102–109.
60. Mcswain BD, Tsujimoto HY, Arnon DI: **Photochemical activity and components of membrane preparations from blue-green-algae.4. Effects of magnesium and chloride-Ions on light-induced electron-transport in membrane fragments from a blue-green-alga.** *Biochimica Et Biophysica Acta* 1976, **423**:313–322.
61. Smith D, Muscatine L, Lewis D: **Carbohydrate movement from autotrophs to heterotrophs in parasitic and mutualistic symbiosis.** *Biol Rev Camb Philos Soc* 1969, **44**:17–90.
62. Crittenden PD: **The effect of oxygen deprivation on inorganic nitrogen uptake in an Antarctic macrolichen.** *Lichenologist* 1996, **28**:347–354.
63. Crittenden PD: **Nutrient exchange in an Antarctic macrolichen during summer snowfall snow melt events.** *New Phytol* 1998, **139**:697–707.
64. Dahlman L, Persson J, Palmqvist K, Nasholm T: **Organic and inorganic nitrogen uptake in lichens.** *Planta* 2004, **219**:459–467.

65. Berman T, Chava S: **Algal growth on organic compounds as nitrogen sources.** *J Plankton Res* 1999, **21**:1423–1437.
66. McDonald TR, Dietrich FS, Lutzoni F: **Multiple horizontal gene transfers of ammonium transporters/ammonia permeases from prokaryotes to eukaryotes: toward a new functional and evolutionary classification.** *Mol Biol Evol* 2012, **29**:51–60.
67. McDonald TR, Mueller O, Dietrich FS, Lutzoni F: **High-throughput genome sequencing of lichenizing fungi to assess gene loss in the ammonium transporter/ammonia permease gene family.** *Bmc Genomics* 2013, **14**:225.
68. Wahl R, Wippel K, Goos S, Kämper J, Sauer N: **A novel high-affinity sucrose transporter is required for virulence of the plant pathogen Ustilago maydis.** *Plos Biol* 2010, **8**:e1000303.
69. Fang WG: **St Leger RJ: Mrt, a gene unique to fungi, encodes an oligosaccharide transporter and facilitates rhizosphere competency in metarhizium robertsii.** *Plant Physiology* 2010, **154**:1549–1557.
70. Helber N, Wippel K, Sauer N, Schaarschmidt S, Hause B, Requena N: **A versatile monosaccharide transporter that operates in the arbuscular mycorrhizal fungus glomus sp is crucial for the symbiotic relationship with plants.** *Plant Cell* 2011, **23**:3812–3823.
71. Voegele RT, Struck C, Hahn M, Mendgen K: **The role of haustoria in sugar supply during infection of broad bean by the rust fungus Uromyces fabae.** *P Natl Acad Sci USA* 2001, **98**:8133–8138.
72. Doehlemann G, Molitor F, Hahn M: **Molecular and functional characterization of a fructose specific transporter from the gray mold fungus Botrytis cinerea.** *Fungal Genet Biol* 2005, **42**:601–610.
73. Wiese J, Kleber R, Hampp R, Nehls U: **Functional characterization of the Amanita muscaria monosaccharide transporter, Am Mst1.** *Plant Biology* 2000, **2**:278–282.
74. Polidori E, Ceccaroli P, Saltarelli R, Guescini M, Menotta M, Agostini D, Palma F, Stocchi V: **Hexose uptake in the plant symbiotic ascomycete Tuber borchii Vittadini: biochemical features and expression pattern of the transporter TBHXT1.** *Fungal Genet Biol* 2007, **44**:187–198.
75. Schussler A, Martin H, Cohen D, Fitz M, Wipf D: **Characterization of a carbohydrate transporter from symbiotic glomeromycotan fungi.** *Nature* 2006, **444**:933–936.
76. Abe M, Noda Y, Adachi H, Yoda K: **Localization of GDP-mannose transporter in the Golgi requires retrieval to the endoplasmic reticulum depending on its cytoplasmic tail and coatomer.** *J Cell Sci* 2004, **117**:5687–5696.
77. Schnappinger D, Hillen W: **Tetracyclines: antibiotic action, uptake, and resistance mechanisms.** *Arch Microbiol* 1996, **165**:359–369.
78. Walsh C: **Molecular mechanisms that confer antibacterial drug resistance.** *Nature* 2000, **406**:775–781.
79. Gould CM, Diella F, Via A, Puntervoll P, Gemund C, Chabanis-Davidson S, Michael S, Sayadi A, Bryne JC, Chica C, *et al*: **ELM: the status of the 2010 eukaryotic linear motif resource.** *Nucleic Acids Res* 2010, **38**:D167–D180.
80. Sharon N, Lis H: **Lectins as cell recognition molecules.** *Science* 1989, **246**:227–234.
81. Sacristan M, Millanes AM, Legaz ME, Vicente C: **A lichen lectin specifically binds to the alpha-1,4-Polygalactoside moiety of urease located in the cell wall of homologous algae.** *Plant Signal Behav* 2006, **1**:23–27.
82. Galun MKN: **Lectins as determinants of symbiotic specificity in lichens.** *Cryptog Bot* 1995, **5**:144–148.
83. Vivas M, Sacristan M, Legaz ME, Vicente C: **The cell recognition model in chlorolichens involving a fungal lectin binding to an algal ligand can be extended to cyanolichens.** *Plant Biol (Stuttg)* 2010, **12**:615–621.
84. Miao VPW, Manoharan SS, Snaebjarnarson V, Andresson OS: **Expression of lec-1, a mycobiont gene encoding a galectin-like protein in the lichen Peltigera membranacea.** *Symbiosis* 2012, **57**:23–31.
85. Manoharan SS, Miao VP, Andresson OS: **LEC-2, a highly variable lectin in the lichen.** *Syrnbiosis* 2012, **58**:91–98.
86. Liener IE, Sharon N, Goldstein IJ: *The Lectins: properties, functions, and applications in biology and medicine.* Orlando: Academic; 1986.
87. Barondes SH, Cooper DN, Gitt MA, Leffler H: **Galectins. Structure and function of a large family of animal lectins.** *J Biol Chem* 1994, **269**:20807–20810.
88. Iturriaga G, Suarez R, Nova-Franco B: **Trehalose metabolism: from osmoprotection to signaling.** *Int J Mol Sci* 2009, **10**:3793–3810.
89. Suarez R, Wong A, Ramirez M, Barraza A, Orozco Mdel C, Cevallos MA, Lara M, Hernandez G, Iturriaga G: **Improvement of drought tolerance and grain yield in common bean by overexpressing trehalose-6-phosphate synthase in rhizobia.** *Mol Plant Microbe Interact* 2008, **21**:958–966.
90. Blazquez MA, Lagunas R, Gancedo C, Gancedo JM: **Trehalose-6-Phosphate, a new regulator of yeast glycolysis that inhibits hexokinases.** *Febs Letters* 1993, **329**:51–54.
91. Muller J, Boller T, Wiemken A: **Trehalose and trehalase in plants: recent developments.** *Plant Science* 1995, **112**:1–9.
92. Pardy RL, Spargo B, Crowe JH: **Release of trehalose by symbiotic algae.** *Symbiosis* 1989, **7**:149–158.
93. Nelsen MP, Gargas A: **Dissociation and horizontal transmission of codispersing lichen symbionts in the genus Lepraria (Lecanorales: Stereocaulaceae).** *New Phytol* 2008, **177**:264–275.
94. Piercey-Normore MD, DePriest PT: **Algal switching among lichen symbioses.** *American Journal of Botany* 2001, **88**:1490–1498.
95. Piercey-Normore MD: **The lichen-forming ascomycete Evernia mesomorpha associates with multiple genotypes of Trebouxia jamesii.** *New Phytol* 2006, **169**:331–344.
96. Mats Wedin HDGG: **Saprotrophy and lichenization as options for the same fungal species on different substrata: environmental plasticity and fungal lifestyles in the Stictis–*Conotrema complex*.** *New Phytol* 2004, **164**:7.
97. Luo R, Liu B, Xie Y, Li Z, Huang W, Yuan J, He G, Chen Y, Pan Q, Liu Y: **SOAPdenovo2: an empirically improved memory-efficient short-read de novo assembler.** *GigaScience* 2012, **1**:1–6.
98. Stanke M, Morgenstern B: **AUGUSTUS: a web server for gene prediction in eukaryotes that allows user-defined constraints.** *Nucleic Acids Res* 2005, **33**:W465–W467.
99. Parra G, Blanco E, Guigó R: **Geneid in drosophila.** *Genome research* 2000, **10**:511–515.
100. Ter-Hovhannisyan V, Lomsadze A, Chernoff YO, Borodovsky M: **Gene prediction in novel fungal genomes using an ab initio algorithm with unsupervised training.** *Genome research* 2008, **18**:1979–1990.
101. Haas BJ, Salzberg SL, Zhu W, Pertea M, Allen JE, Orvis J, White O, Buell CR, Wortman JR: **Automated eukaryotic gene structure annotation using EVidenceModeler and the program to assemble spliced alignments.** *Genome Biol* 2008, **9**:R7.
102. Ashburner M, Ball CA, Blake JA, Botstein D, Butler H, Cherry JM, Davis AP, Dolinski K, Dwight SS, Eppig JT, *et al*: **Gene ontology: tool for the unification of biology.** *Nat Genet* 2000, **25**:25–29.
103. Sonnhammer ELL, Koonin EV: **Orthology, paralogy and proposed classification for paralog subtypes.** *Trends Genet* 2002, **18**:619–620.
104. Ruepp A, Zollner A, Maier D, Albermann K, Hani J, Mokrejs M, Tetko I, Guldener U, Mannhaupt G, Munsterkotter M, Mewes HW: **The FunCat, a functional annotation scheme for systematic classification of proteins from whole genomes.** *Nucleic Acids Res* 2004, **32**:5539–5545.
105. Kanehisa M, Goto S, Kawashima S, Okuno Y, Hattori M: **The KEGG resource for deciphering the genome.** *Nucleic Acids Res* 2004, **32**:D277–D280.
106. Zdobnov EM, Apweiler R: **InterProScan–an integration platform for the signature-recognition methods in InterPro.** *Bioinformatics* 2001, **17**:847–848.
107. Griffiths-Jones S, Moxon S, Marshall M, Khanna A, Eddy SR, Bateman A: **Rfam: annotating non-coding RNAs in complete genomes.** *Nucleic Acids Res* 2005, **33**:D121–D124.
108. Lowe TM, Eddy SR: **tRNAscan-SE: A program for improved detection of transfer RNA genes in genomic sequence.** *Nucleic Acids Res* 1997, **25**:955–964.
109. Karro JE, Yan Y, Zheng D, Zhang Z, Carriero N, Cayting P, Harrrison P, Gerstein M: **Pseudogene. org: a comprehensive database and comparison platform for pseudogene annotation.** *Nucleic Acids Res* 2007, **35**:D55–D60.
110. Lagesen K, Hallin P, Rødland EA, Stærfeldt H-H, Rognes T, Ussery DW: **RNAmmer: consistent and rapid annotation of ribosomal RNA genes.** *Nucleic Acids Res* 2007, **35**:3100–3108.
111. Thompson JD, Higgins DG, Gibson TJ: **Clustal-W - Improving the sensitivity of progressive multiple sequence alignment through sequence weighting, position-specific gap penalties and weight matrix choice.** *Nucleic Acids Res* 1994, **22**:4673–4680.
112. Baxevanis AD: *Current protocols in bioinformatics. In Book Current protocols in bioinformatics (Editor ed.^eds.).* City: Wiley; 2003.
113. Sanderson MJ: **r8s: inferring absolute rates of molecular evolution and divergence times in the absence of a molecular clock.** *Bioinformatics* 2003, **19**:301–302.
114. Padovan ACB, Sanson GFO, Brunstein A, Briones MRS: **Fungi evolution revisited: application of the penalized likelihood method to a bayesian fungal phylogeny provides a new perspective on phylogenetic

relationships and divergence dates of ascomycota groups. *J Mol Evol* 2005, **60**:726–735.
115. Li H, Coghlan A, Ruan J, Coin LJ, Heriche JK, Osmotherly L, Li RQ, Liu T, Zhang Z, Bolund L, *et al*: **TreeFam: a curated database of phylogenetic trees of animal gene families.** *Nucleic Acids Res* 2006, **34**:D572–D580.
116. Enright AJ, Van Dongen S, Ouzounis CA: **An efficient algorithm for large-scale detection of protein families.** *Nucleic Acids Res* 2002, **30**:1575–1584.
117. Librado P, Vieira FG, Rozas J: **BadiRate: estimating family turnover rates by likelihood-based methods.** *Bioinformatics* 2012, **28**:279–281.
118. Galagan JE, Calvo SE, Cuomo C, Ma LJ, Wortman JR, Batzoglou S, Lee SI, Basturkmen M, Spevak CC, Clutterbuck J, *et al*: **Sequencing of Aspergillus nidulans and comparative analysis with A-fumigatus and A-oryzae.** *Nature* 2005, **438**:1105–1115.
119. Glass NL, Kaneko I: **Fatal attraction: Nonself recognition and heterokaryon incompatibility in filamentous fungi.** *Eukaryot Cell* 2003, **2**:1–8.
120. Dementhon K, Saupe SJ, Clave C: **Characterization of IDI-4, a bZIP transcription factor inducing autophagy and cell death in the fungus Podospora anserina.** *Mol Microbiol* 2004, **53**:1625–1640.
121. Saupe S, Descamps C, Turcq B, Begueret J: **Inactivation of the podospora-anserina vegetative incompatibility locus Het-C, whose product resembles a glycolipid transfer protein, drastically impairs ascospore production.** *P Natl Acad Sci USA* 1994, **91**:5927–5931.
122. Anand S, Prasad MVR, Yadav G, Kumar N, Shehara J, Ansari MZ, Mohanty D: **SBSPKS: structure based sequence analysis of polyketide synthases.** *Nucleic Acids Res* 2010, **38**:W487–W496.
123. Vandesompele J, De Preter K, Pattyn F, Poppe B, Van Roy N, De Paepe A, Speleman F: **Accurate normalization of real-time quantitative RT-PCR data by geometric averaging of multiple internal control genes.** *Genome Biol* 2002, **3**:research0034.

doi:10.1186/1471-2164-15-34
Cite this article as: Wang *et al*.: Genome characteristics reveal the impact of lichenization on lichen-forming fungus *Endocarpon pusillum* Hedwig (Verrucariales, Ascomycota). *BMC Genomics* 2014 **15**:34.

Comparative transcriptome analysis of the lichen-forming fungus *Endocarpon pusillum* elucidates its drought adaptation mechanisms

WANG YanYan[1,2], ZHANG XinYu[1], ZHOU QiMing[1], ZHANG XiaoLing[1] & WEI JiangChun[1,3*]

[1]*State Key Laboratory of Mycology, Institute of Microbiology, Chinese Academy of Sciences, Beijing 100101, China;*
[2]*University of Chinese Academy of Sciences, Beijing 100049, China;*
[3]*State Key Laboratory of Crop Stress Biology for Arid Areas, NWAFU, Yangling 712100, China*

Received March 21, 2014; accepted June 6, 2014

The lichen-forming fungus was isolated from the desert lichen *Endocarpon pusillum* that is extremely drought resistant. To understand the molecular mechanisms of drought resistance in the fungus, we employed RNA-seq and quantitative real-time PCR to compare and characterize the differentially expressed genes in pure culture at two different water levels and with that in desiccated lichen. The comparative transcriptome analysis indicated that a total of 1781 genes were differentially expressed between samples cultured under normal and PEG-induced drought stress conditions. Similar to those in drought resistance plants and non-lichenized fungi, the common drought-resistant mechanisms were differentially expressed in *E. pusillum*. However, the expression change of genes involved in osmotic regulation in *E. pusillum* is different, which might be the evidence for the feature of drought adaptation. Interestingly, different from other organisms, some genes involved in drought adaption mechanisms showed significantly different expression patterns between the presence and absence of drought stress in *E. pusillum*. The expression of 23 candidate stress responsive genes was further confirmed by quantitative real-time PCR using dehydrated *E. pusillum* lichen thalli. This study provides a valuable resource for future research on lichen-forming fungi and shall facilitate future functional studies of the specific genes related to drought resistance.

lichen, mycobiont, dehydration, drought adaption, drought resistant

Citation: Wang YY, Zhang XY, Zhou QM, Zhang XL, Wei JC. Comparative transcriptome analysis of the lichen-forming fungus *Endocarpon pusillum* elucidates its drought adaptation mechanisms. Sci China Life Sci, 2014, 57: 1–12, doi: 10.1007/s11427-014-4760-9

Lichens are symbiotic products between fungi (the mycobionts) and photosynthetic partners (the photobionts). Each lichen species is made up of one mycobiont and at least one photobiont, and the photosynthetic partner can be either an alga or a cyanobacterium. Lichens are widely distributed across many types of ecological niches, ranging from cold to hot deserts and other extreme habitats. Unlike vascular plants, lichens lack active mechanisms for controlling water content. As a result, their water content tends to fluctuate widely based on water availability in the environment [1]. Indeed, lichens can lose water rapidly when the environment is dry and they can also rapidly recover to normal water content when water becomes available [2]. Specifically, different from most plants, a low water content is typically nonlethal to lichens and most lichens can withstand drying to water contents of 5% or less for a long period of time [3,4]. Furthermore, in the presence of water, not only does the water content recover, lichens can rapidly return to normal physiological state to carry out photosynthesis and respiration during rehydration [5,6]. As a previous study

*Corresponding author (email: weijc2004@126.com)

This paper was originally published in *Science China Life Sciences*, 58（1）: 1-12, 2015.

indicated, even in a low water content state, some lichens could maintain active metabolism [7].

A number of studies have examined the effects of water availability on the morphology, physiology, and survival of lichens [1,6,8–13]. However, little is known about the molecular mechanisms of drought resistance in lichen. Many recent studies have examined the molecular mechanisms of drought resistance in plants and non-lichenized fungi. Whether similar mechanisms are involved in lichens remains unknown. In plants and non-lichenized fungi, desiccation can induce several cellular stresses, such as hyperosmolarity, hyperoxidation, hyper-ionicity and protein misfolding and aggregation [14]. To survive the desiccation, the organisms need to deal with these cellular damages. Three major stress-response pathways are known to be involved in protection from these stresses, including osmoregulation (to modulate intracellular ion concentration); DNA and protein damage repair (to prevent DNA damage and protein misfolding and degradation); and antioxidation (to scavenge reactive oxygen species (ROS)) [15,16]. Our limited understanding of the molecular mechanisms of drought resistance in lichens has been mainly due to the lack of genetic tools for analyzing lichens. However, recent advances in genomics technologies and the information of whole-genomes and transcriptomes from lichen-forming fungi and their photosynthetic partners are making the studies of drought resistance in lichens feasible [17,18].

Since lichen is the symbiotic product of two interacting partners, most of its biological characteristics are the result of interaction between the symbiotic partners. However, in drought resistance, the mycobiont partner seems to be the main contributor in lichens [19–21], probably due to the fact that within lichens, the photobionts are housed by the fungal tissues, which effectively protect their photosynthetic partners from desert climate. For example, a previous study indicated that, separately, the lichen-forming fungus *Endocarpon pusillum* has a much stronger drought resistant ability than its algal partner *Diplosphaera chodatii* [22]. Therefore, to understand the mechanisms of drought resistance in lichens, it is reasonable to focus on the fungal partner.

In the present study, the transcriptomes of the lichen-forming fungus *E. pusillum*, whose genome has been sequenced [18], are analyzed in order to investigate the genetic mechanisms underlying drought resistance. In addition, representative genes that showed differential expressions in response to desiccation were further investigated in fresh lichen thallus *E. pusillum* during natural dehydration, to elucidate whether the pure-culture isolated mycobiont and the symbiotic mycobiont possess the same drought resistance mechanisms.

1 Materials and methods

1.1 Materials and stress treatments

The mycobiont of *E. pusillum* was isolated from the specimen Z07020, originally collected from Shapotou Region of Ningxia, south-eastern edge of the Tengger desert, China (latitude 37.40°N, longitude 105.00°E). The fungal isolate was grown on 1.5% water agar for 1–2 weeks, and then transferred to Potato Dextrose Broth (PDB) medium cultured at room temperature. The artificial desiccation treatment was performed on the isolated lichen-forming fungus using polyethylene glycol (PEG), which is a non-permeable osmolyte and can create a severe water deficit [23]. The PEG-induced drought stress was used to investigate the drought-resistant of plants and fungi [24,25]. The mycobiont was desiccated under the condition of cultured in the medium containing 20% PEG for 3 weeks.

The natural dehydration treatment was performed on the lichen thalli *E. pusillum* collected from Shapotou Desert in the same location as specimen Z07020. Under laboratory conditions, the fresh samples were completely rehydrated in deionized water for 5 min and excess water on the surface of thallus was removed using filter paper. The samples were then incubated at room temperature to naturally dehydrate for 0, 40, 60, 150 and 240 min. At the end of each treatment, the samples were frozen immediately in liquid nitrogen and RNAs were extracted, then reverse-transcribed into cDNA as the template of quantitative real-time PCR (qRT-PCR).

1.2 cDNA library construction and DNA sequencing

Total RNA was extracted using the Trizol (Invitrogen, USA) extraction method according to the manufacturer's protocol. Poly-A mRNA was isolated with oligo-dT-coupled beads from 40 μg total RNA of each sample and then sheared, and the isolated RNA samples were used for first strand cDNA synthesis which was random hexamers and Superscript II reverse transcriptase. After end repair and addition of a 3′-dA overhang, the cDNA was ligated to Illumina paired-end adapter oligo mix, and size selected to about 200 bp fragments by gel purification. After 16 PCR cycles, the libraries were sequenced using Illumina GAIIx (Illumina, USA) and the paired-end sequencing module.

1.3 Mapping assembly of the transcriptome and differential expression analysis

After removing the adapter sequences, the reads were mapped to the *E. pusillum* genome using PASA [26] and ORFs were found from the PASA assembly. Augustus [27], GeneID [28], and GeneMark-ES [29] programs were used to predict the gene models for the genome, and then the gene models generated by EVM [30] were updated by PASA. All predicted gene models were subjected to Gene Ontology (GO) [31], EuKaryotic Orthologous Groups (KOG) [32], FunCat [33] and Kyoto Encyclopedia of Genes and Genomes (KEGG) database analysis [34].

Raw array data were normalized using the ARRAYSTAR software (Dnastar, USA), and we used

NOISeq [35] statistical method to identify differentially expressed gene between normal and drought stress cultivations. As suggested by NOISeq authors, a gene is declared as differentially expressed one if the probability (*P* value) is higher than 0.8.

1.4 Quantitative RT-PCR analysis

Total RNA of the lichen-forming fungus *E. pusillum* was extracted from cultured fungal mycelium and lichen thalli respectively. cDNA syntheses were carried out according to the protocol described in the manual of Reverse Transcriptome System (Promega, USA). Real-time PCR was carried out using ABI 7500 real time PCR system (Applied Biosystems, USA). The data were analyzed using the $2^{-\Delta C_t}$ method [36].

2 Results

2.1 Functional classification of the transcriptome of *E. pusillum* upon water stress

In the present study, we found that 1781 genes (probability>0.8) were differentially expressed in the desiccation condition (20% PEG) in *E. pusillum* when compared with no water stressed condition (0% PEG). The differentially expressed genes are 19.18% of 9285 genes annotated in the 37.5 Mb genome of *E. pusillum* [18]. Among the 1781 genes, 1004 were up-regulated and 777 were down-regulated. Among them, 620 genes (35%) were annotated as having unknown functions.

Using the Blast2GO platform [37], we classified the differentially expressed genes with GO terms according to their functions, and we found that most abundant GO terms were distributed in biological process, molecular functions and cellular components (Figure 1). Genes classified in the categories of "intracellular component", "cellular metabolic process", "transferase activity", "primary metabolic process" and "macromolecule metabolic process" were significantly up-regulated. As expected, the GO term "membrane component" was significantly overrepresented among the down-regulated genes, consistent with the strong effect of desiccation stress on the integrity of membranes. Surprisingly, very few genes belonging to the category of "stress response" were differentially expressed, with only three genes up-regulated and four down-regulated in the total of 51 genes. This result suggests that genes involved in PEG-induced stress response in *E. pusillum* are different from other organisms; therefore, there are likely new mechanisms of drought resistance in lichens.

When the differentially expressed genes were mapped to the KEGG using the Blast2GO platform (Figure 2), we found that genes whose expression level changed most markedly ($P \leqslant 0.0001$) were related to ribosome and oxidative phosphorylation. Genes in several pathways were induced, including mRNA translation, vitamin B6 metabolism, synthesis and degradation of ketone bodies, terpenoid backbone biosynthesis, oxidative phosphorylation, steroid biosynthesis, pyrimidine metabolism, and purine metabolism. Among these pathways, over 30% of all genes in the first five pathways mentioned above were up-regulated. However, only a few genes were significantly repressed and no genes were significantly up-regulated in nitrogen and galactose metabolisms ($P \geqslant 0.1$). Most of these metabolism pathways exhibited active responses under the desiccation treatment, especially the pathways involved in essential function of the cell, such as ribosome, oxidative phosphorylation, pyrimidine metabolism, and purine metabolism.

2.2 The expression changes of gene commonly involved in stress responses in *E. pusillum*

Continuous exposure to drought leads to oxidative stress and induces defense mechanisms to scavenge the ROS that may result in significant damage to cell structure, it also induces osmotic pressure of cells and cells need to regulate this pressure by accumulating osmolytes [38]. Moreover, the osmolyte accumulation under drought can lead to misfolding and aggregation of proteins [39], and these changes can potentially limit cells' responses to desiccation [40]. Therefore, to better understand the drought resistant mechanism in *E. pusillum*, we investigated the expression changes of genes commonly involved in the response to stresses in other organisms, such as oxidative stress, osmotic regulation and post-translational processing under our drought treatment.

Previous research has revealed that oxidative defense mechanism in other organisms involved a number of specific enzymes, such as superoxide dismutase (SOD), catalase, peroxidases and auxiliary enzymes; and low-molecular-weight antioxidants, such as tripeptide glutathione (GSH) and ascorbate [16,41,42]. In this study, three genes encoding antioxidation enzymes were induced in the *E. pusillum* genome under PEG-induced drought stress, including one for SOD and two for peroxidase (Table S1 in Supporting Information). By comparison, more genes involved in low-molecular-weight antioxidants were induced in *E. pusillum*, including those coding for thioredoxin, glutathionc, and vitamin B6 (Table S1 in Supporting Information). We identified seven highly expressed thioredoxin-like genes, whose expressions were significantly up-regulated in *E. pusillum* under desiccation. In addition, five genes related to glutathione S-transferase (GST), which were involved in glutathione metabolism, were up-regulated under desiccation in *E. pusillum*. This result indicates that *E. pusillum* regulates the glutathione metabolism in response to drought-induced oxidative stress. To our surprise, among the three genes for trehalose-6-phosphate synthase in *E. pusillum* (Table S2 in Supporting Information), none of them is differentially expressed (probability<0.8). Similarly,

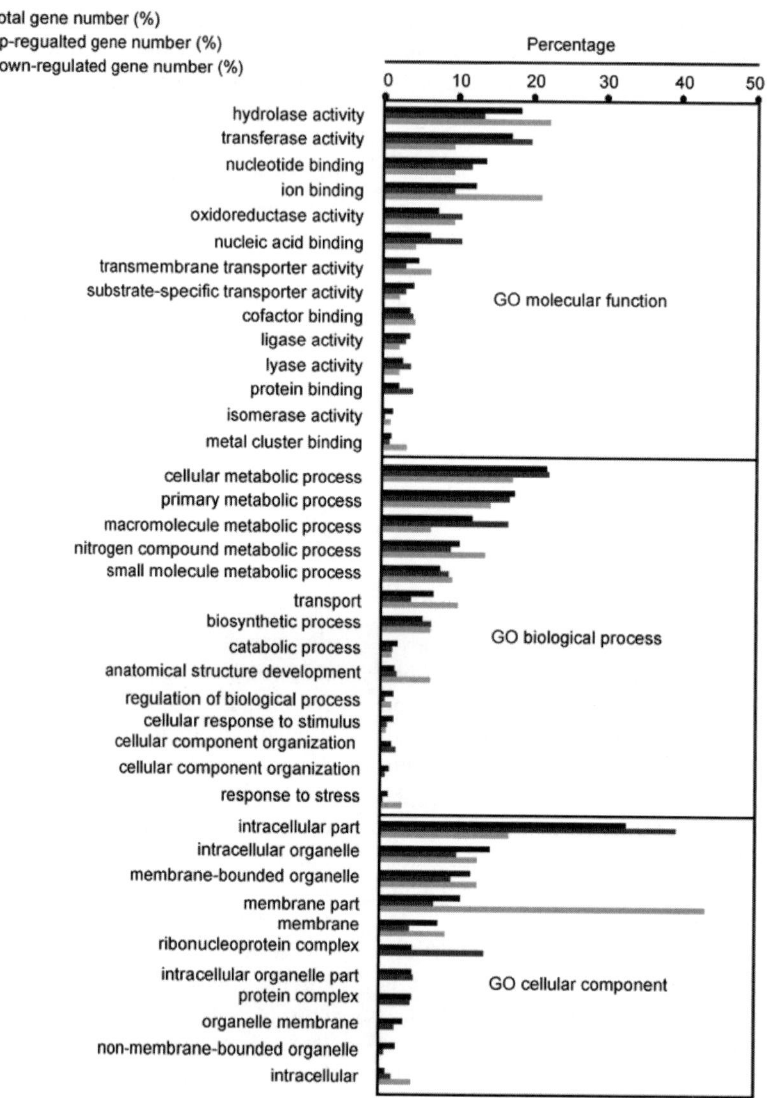

Figure 1 Gene Ontology classification of differentially expressed genes upon water stress in *E. pusillum* according to GO groups based on molecular function, biological process, and cellular component. The percentages of the total matched genes, up-and down-regulated genes contained in a particular GO group, which were labeled as black, dark grey, and light grey bars, respectively.

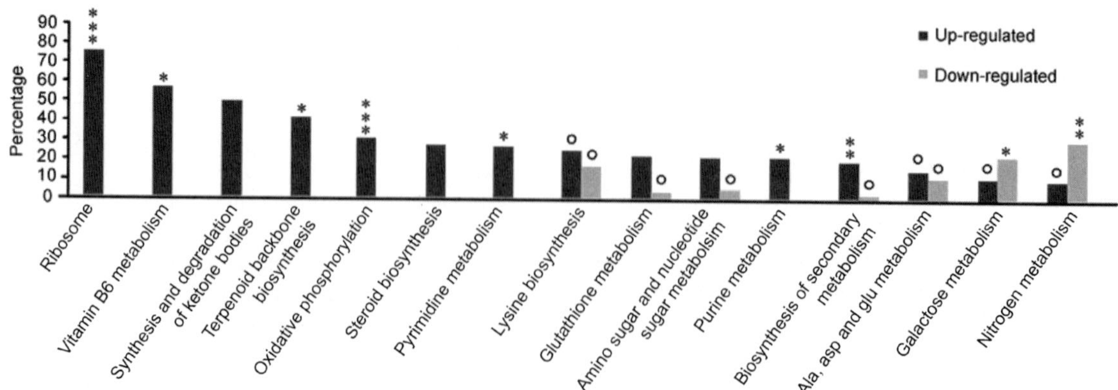

Figure 2 Functional classification of differentially expressed genes upon water stress in *E. pusillum* according to KEGG analysis. The percentage of up- and down-regulated genes in a particular KEGG pathway was labeled as dark grey and light grey bars, respectively. Asterisks indicate the significance differences of differentially expressed genes. *, $P \leqslant 0.01$; **, $P \leqslant 0.001$; ***, $P \leqslant 0.0001$; °, $P \geqslant 0.1$.

no other osmotic regulators were differentially expressed in *E. pusillum* under drought. However, we did identify 11 chaperone genes (including the heat shock protein (HSP), GroEL, and chaperonins) and 10 genes in proteasomes in *E. pusillum* (Table S3 in Supporting Information) to be significantly induced under PEG treatment.

Under stressed conditions, effective signal transduction is needed to communicate information from the external environment to the inside of a cell. Therefore, the expression of genes involved in signal transduction is expected to vary widely between normal and stress conditions. In this study, a total of 40 genes are annotated as related to signaling pathway in the transcriptome of *E. pusillum*, including calcium-mediated signaling, MAPK signaling pathway, TOR signaling pathway, mTOR signaling pathway, and G protein coupled receptor protein signaling pathway. However, only four genes (one for calcium-mediated signaling and three for MAPK signaling pathway) were differentially expressed (Table S4 in Supporting Information), and all four were down-regulated in *E. pusillum* under PEG-induced stress condition.

Transcription factors (TFs) function as key regulators of gene expression in response to environmental stresses. A total of 25 genes encoding putative TFs in *E. pusillum* genome were differentially expressed under PEG-induced stress. These 25 genes belonged to15 groups, i.e., zinc finger, TFIIA, heat shock, MADS, P53, NAC, DIP2, MYB, bZIP, Nus A, CBF, FOX, MEIS1, GATA, and Cmr1. Some of these TFs had been shown to function under abiotic stresses in plants and fungi [43–47]. For *E. pusillum* cultured under PEG-induced water limitations, seven of the 25 were up-regulated and 18 were down-regulated (Table S5 in Supporting Information). This result might be useful for the study of gene transcription regulation in lichen-forming fungi.

2.3 Other differentially expressed genes revealed drought-adaptive mechanism of *E. pusillum*

Except for genes commonly involved in the response to stresses in other organisms mentioned above, we identified some other genes that were differentially expressed in PEG-stressed *E. pusillum* from the transcriptome analysis (Figure 2). These differentially expressed genes can be classified into several categories: (i) Genes encode ribosomal proteins. The most significantly up-regulated transcripts are ribosomal proteins, and over 70% of ribosomal proteins were induced under PEG-induced drought. (ii) Genes involved in pyrimidine and purine metabolisms. The high transcriptional levels of these genes in stressed transcriptome (Table S6 in Supporting Information) indicate higher synthetic rates of DNA and RNA in *E. pusillum* under desiccation. (iii) Genes involved in oxidative phosphorylation. Oxidative phosphorylation is one of the pathways that showed the most significant differential expression of genes. Specifically, genes for ATPase, NADH-ubiquinone oxidoreductase, Fe-S protein, and cytochrome C oxidase were all up-regulated under drought stress, and no gene in this pathway was down-regulated (Table S7 in Supporting Information). (iv) Genes involved in the nitrogen metabolisms. These genes were significantly repressed under desiccation (Table S8 in Supporting Information), which were marked in Figure 3. As a key compound in cellular metabolism, glutamate is synthesized using ammonium as a substrate through the functions of glutamate synthase and asparagine synthase. In this pathway, the generation of ammonium from formamide and nitroalkane is catalyzed by acetamidase/formamidase and 2-nitropropane dioxygenase. These enzymes were all repressed (Table S8 in Supporting Information), suggesting that the *E. pusillum* cells reduced the synthesis of glutamate through nitrogen metabolism under PEG-induced desiccation. However, the induced glutathione metabolism provides compensatory glutamate by the catalysis of glutathione S-transferase (Figure 3; Table S8 in Supporting Information). It has been reported that nitrogen metabolism was affected by some stressed conditions in other organisms [48,49] although the mechanism is not clear. In this study, the expression changes of genes involved in nitrogen metabolism were likely the results of feedback inhibition of induced glutathione metabolism because the latter were known to play an important role under drought stress. (v) Genes involved in carbon metabolism, including nucleotide sugar metabolism and galactose metabolism (Table S8 in Supporting Information). Before galactose can be metabolized through the glycolysis pathway, it must be converted into glyceraldehydes-3P by galactokinase, galactose oxidase, and galactonate dehydratase (Figure 4). Among the differentially expressed genes in *E. pusillum*, those coding for galactose oxidase and galactonate dehydratase, which are indispensable enzymes in the galactose metabolism, were significantly down-regulated (Table

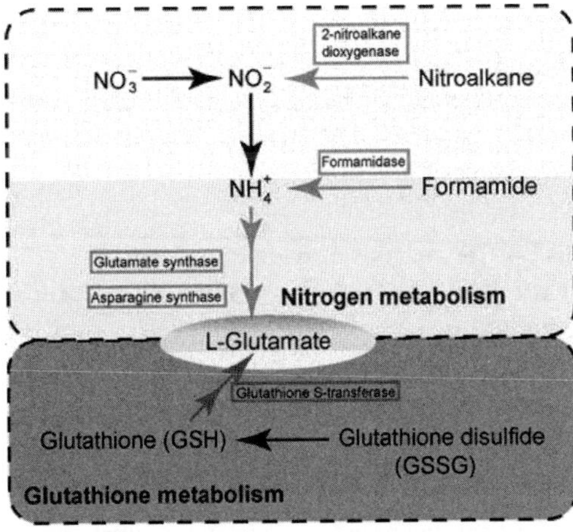

Figure 3 A diagram of nitrogen and glutathione metabolism, in which enzymes are encoded by genes either up- (labeled in red) or down-regulated (labeled in green) by water stress in *E. pusillum*.

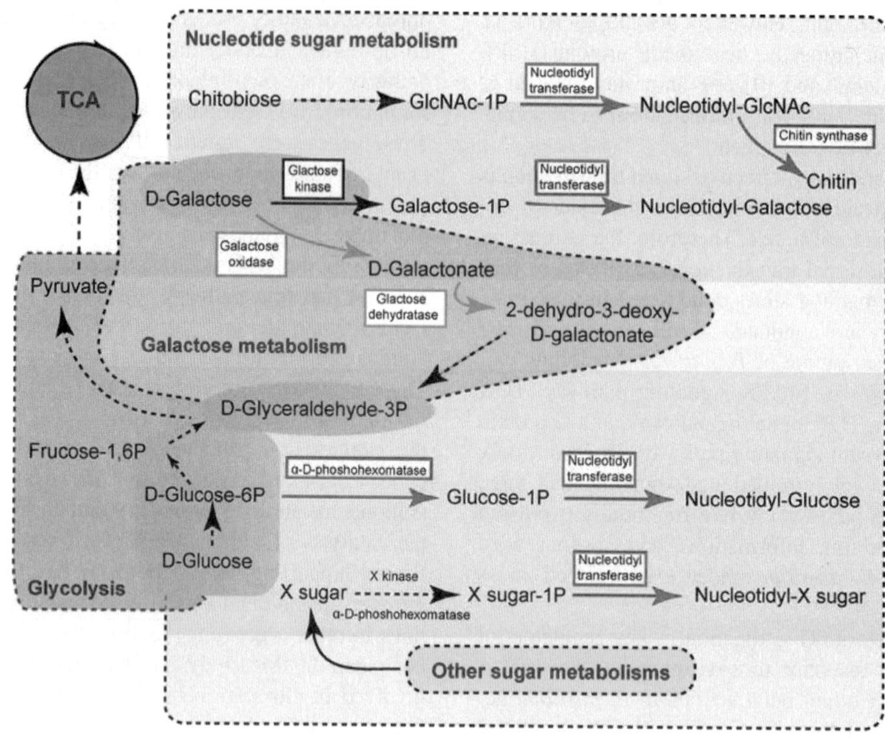

Figure 4 A diagram of carbon metabolism, including nucleotide sugar metabolism, galactose metabolism, and glycolysis, in which enzymes are encoded by genes either up- (labeled in red) or down-regulated (labeled in green) by water stress in *E. pusillum*.

S8 in Supporting Information). However, the accumulated galactose might be converted into nucleotide galactose through amino sugar and nucleotide metabolism. The common reaction in this metabolism is that sugars can be converted into nucleotide sugars by kinases, α-D-phosphohexomutase, and nucleotidyltransferase. The expressions of the last two enzymes were up-regulated under drought stress (Table S8 in Supporting Information), which suggested that more nucleotide sugars were produced from monosaccharides and disaccharides in *E. pusillum* under desiccation. In addition, the up-regulation of the chitin synthase gene increases the amount of chitin, potentially making the cell wall thicker and preventing water loss and the damages caused by desiccation.

2.4 qRT-PCR analysis of transcript levels for differentially expressed genes during dehydration process of lichen thallus

Using qRT-PCR, a set of 23 representative drought response-related genes (Table 1) were selected to further monitor their transcript levels in *E. pusillum* thallus under natural dehydration stress. These genes are involved in posttranslational modification, osmotic regulation, energy providing, damage repairing (including ribosome protein, pyrimidine and purine metabolism), and carbon metabolism (including nucleotide sugar metabolism and galactose metabolism).

Because the degree of the stress for lichen thalli grown under natural conditions is difficult to determine, a dehydration process was performed for fresh lichen thalli of *E. pusillum* in laboratory to monitor the expression of lichen-forming fungal genes undergoing different degrees of drought stress. The lichen thalli were totally rehydrated, and the relative water content was detected over the dehydration process (Figure 5). Within the first 120 min, the dehydration rate is fast and it did not exhibit significant change after 150 min. The relative water content was close to zero after dehydration for 150 min, which means that cells should have lost all the free water and are under severe drought stress at this time point. Here, the expressions of the selected genes

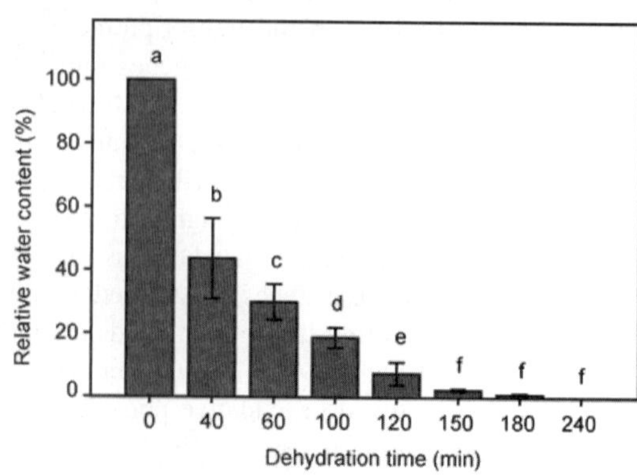

Figure 5 Relative water content in dehydrating lichen thalli of *E. pusillum* at different dehydration time period (n=3).

at this time point were set to 1. A series of time points (0, 40, 60, 150 and 240 min) were chosen to investigate the transcript levels of candidate genes, and the samples dehydrated for 150 and 240 min were regarded as samples under drought stress. For comparison of the relative expression at different time points, a heat-map was generated using the $\log_2$ fold change.

From the heat-map representation (Figure 6), the test genes can be classified into four classes according to the cluster analysis. In the first cluster (marked as blue), the

Table 1 The selected genes for qRT-PCR to verify the expression levels

Term	Gene	Annotation	Expression level (stress)	Expression level (control)
Ribosome	F481_05528	40S ribosomal protein S3Ae	739.977	193.079
	F481_05538	Ribosomal protein S10	1380.97	257.81
Purine and pyrimidine metabolism	F481_00996	Nucleoside diphosphate kinase	4654.23	209.721
	F481_02100	dUTP pyrophosphatase	108.252	17.3186
Peroxidase	F481_03410	Dyp-type peroxidase	111.807	16.5827
Glutathione metabolism	F481_02789	Glutathione S-transferase (GST), C-terminal domain	227.393	69.5137
	F481_06421	Thioredoxin-like	1736	241.976
Vitamin B6 metabolism	F481_00322	Pyridoxal 5'-phosphate (PLP) synthase	1206.83	91.5978
	F481_02092	Pyridoxal kinase	14.9954	3.74325
Oxidative phosphorylation	F481_04003	Complex 1 LYR protein	242.525	49.5297
	F481_02166	ATPase, F0 complex, subunit B, mitochondrial	815.057	121.74
	F481_01557	Cytochrome c oxidase, subunit VIb	1274.81	329.341
Proteasome	F481_04061	Proteasome, subunit alpha/beta	399.412	62.1101
	F481_03276	Proteasome, subunit beta type 3	552.606	69.6354
Molecular chaperone	F481_05773	GroES-like	1284.79	178.943
	F481_07753	Heat shock protein 70kD	314.865	63.8575
Trehalose-6-phosohate synthase	F481_01316	Trehalose-6-phosohate synthase	61.1314	37.7043
Nitrogen metabolism	F481_06302	Glutamate synthase subunit alpha	14.3953	124.632
	F481_07075	Nitritereductase	3.40505	158.605
Galactose metabolism	F481_03176	D-galactonate dehydratase	10.0886	159.008
	F481_05966	Galactose oxidase, central domain	7.7875	31.8316
Nucleotide sugar metabolism	F481_07291	Nucleotidyltransferases	145.625	29.9006
	F481_00358	N-acetyltransferase	238.396	64.8396

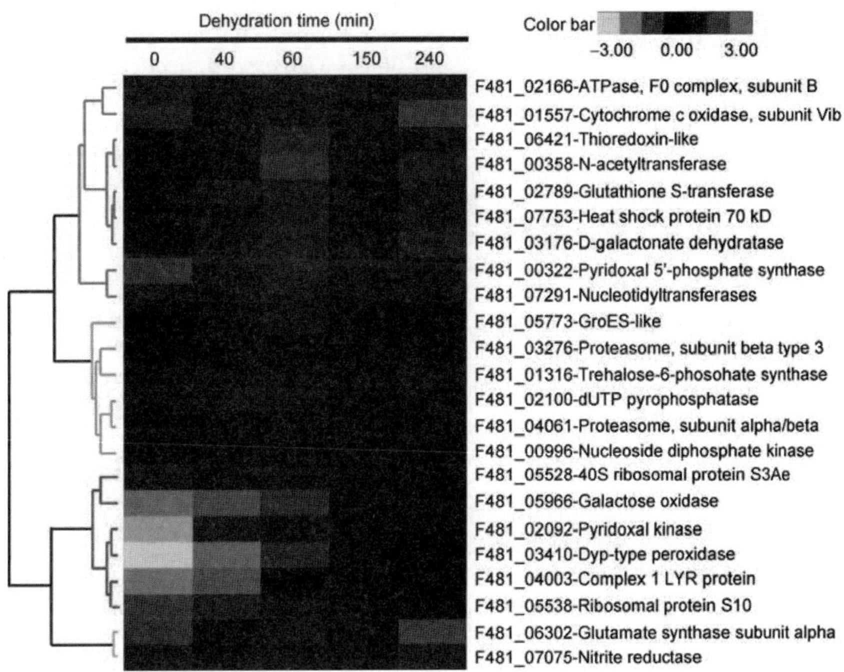

Figure 6 Relative transcript levels of stress-related genes in *E. pusillum* under natural dehydration process.

transcript levels of these genes increased during the first 60 min of dehydration, and were then significantly down-regulated from 60 to 150 min. A striking feature of these genes was that they were up-regulated at low water content from 150 to 240 min, and the relative expressions at 240 min were higher than those at the beginning. The result indicated that the expressions of these genes were correlated to drought stress so they were likely to be related to drought-resistant mechanisms. Genes involved in antioxidation response mechanisms (vitamin B6 metabolism and glutathione), HSP, oxidative phosphorylation, and sugar metabolisms (nucleotide sugar metabolism and galactose metabolism) belonged to this class. In the second cluster (marked as yellow), the transcript levels of these genes do not exhibit significant fluctuations during the whole dehydration process, although the relative expressions of these genes at 150 min were slightly lower than those at other time points. Genes coding for proteins involved in purine and pyrimidine metabolism, proteasome, the trehalose-6-phosohate synthase (TPS), and GroES belonged to this class. At the first time point, the cells were undergoing long period drought and the expressions of these genes are not yet decreased dramatically, which suggests that these genes are highly expressed under both the wet and drought periods, therefore they seem to be house-keeping genes. In the third cluster (marked as red), the trend of gene expression during the first three time points is the same as that in the first cluster, but the up-regulation trend is more significant. The relative expressions of genes in this cluster at the last two time points were obviously higher than those at the first two or three time points, suggesting that these genes are highly expressed under drought; therefore, they can be regarded as drought-resistant related genes. Genes coding for ribosome, enzymes of galactose metabolism, and antioxidant mechanisms (vitamin B6 metabolism and peroxidase) belonged to this class. In the fourth cluster (marked as green), the relative expression during the five time points was present as a bell-shaped curve, with a significant declining at the low water level. The genes in this cluster are not the typical drought-resistant genes, such as those involved in nitrogen metabolism.

Generally, after rehydration for 60 min, the relative expressions of these 23 selected genes reach a maximum, which implies that it needs to take some time to reflect non-drought state at transcript level in lichen-forming fungi. The results above indicate that the trends of expression of these drought-related genes vary widely. Genes in the first three clusters can be considered as drought-resistant genes, because they exhibited increased expression level when the water content was low after 150 min of dehydration. While the genes in the second cluster did not exhibit significant variation in their expressions, they nonetheless play very important roles under stressful conditions, because they are important housekeeping genes and their expressions are apt to be constant even under extreme desiccation.

3 Discussion

The model lichen-forming fungus *E. pusillum* was used to explore the drought stress response mechanisms in the present study. Our analyses identified 1781 differentially expressed genes, which are potential targets for further investigating the stress response mechanisms and characterizing the function of stress-related genes.

A large number of differentially expressed genes were classified into metabolism process (Figure 1), which suggests that *E. pusillum* had active metabolism under PEG-induced drought stress. This phenomenon is different from other drought-resistant organisms [50–52], where metabolic processes were largely suppressed during desiccation including PEG-induced stress. This result suggests that different from most other organisms, *E. pusillum* is a drought-adapted organism.

3.1 The common drought-resistant mechanisms in *E. pusillum*

The drought-resistant mechanisms, including reduce oxidative stress, osmotic regulation, and post-translational processing, in *E. pusillum* were compared to those in other drought-resistant plants and fungi. Our results show that *E. pusillum* has multiple mechanisms to deal with oxidative stress. Vitamin B6 has three forms (pyridoxine, pyridoxal, and pyridoxamine) and is an enzymatic cofactor participating in many biochemical reactions [53]. Recent research showed that vitamin B6 played a critical role in resistance to oxidative stress. For example, vitamin B6 is able to protect cells against cell death induced by ROS in the filamentous fungus *Cercospora nicotianae* [54]. Similarly, the loss of *pdx1* gene which encodes a pyridoxine synthase in *Arabidopsis thaliana* can result in hypersensitive to osmotic stress and oxidative stress [55]. Therefore, it is not surprising that the expressions of genes involved in vitamin B6 biosynthesis in *E. pusillum* are significantly induced under desiccation (Table S1 in Supporting Information, Figure 2). Several antioxidant mechanisms, such as SOD and glutathione, have been reported in lichens [56]. However, this is the first report demonstrating that the low-molecular-weight antioxidant vitamin B6 is related to oxidative stress response and drought resistance in lichen-forming fungi.

Molecular chaperones direct protein folding and assembly, including refolding denatured proteins induced by stresses. There are many different families of chaperones, such as HSP, GroEL, and chaperonins. Each family acts to aid protein folding in different ways. For example, HSPs are induced mainly in response to high temperatures and other cellular stresses damages. In contrast, GroEL generally participates in response to the reduction in folding efficiency induced by macromolecule aggregation. The roles of proteasome are to degrade unnecessary or damaged proteins

[57,58], and it has been found that genes encoding proteasomes are up-regulated under stress [59,60]. The up-regulated expression of chaperone genes demonstrates that *E. pusillum* can likely cope with problems associated with protein misfolding and aggregation induced by desiccation through up-regulating its main chaperones. In addition, the increasing expression of proteasome implies that the misfolded proteins were likely quickly degraded [61].

At present, our knowledge of osmotic regulation is mainly established in plants [62], and a variety of osmolytes have been identified, including sugar, proline, malondialdehyde and potassium [63,64]. Among them, the sugar trehalose is the most important osmotic regulator in fungi [65,66]. No osmotic regulators were differentially expressed indicated that these ubiquitous osmotic regulators found in other organisms were not induced in *E. pusillum* under PEG-induced drought stress. Such a result would imply either that there are other osmotic regulators in *E. pusillum*, or that these known regulators are constitutively expressed in this lichen-forming fungus under both the normal and water-stressed conditions because of its drought adaptive character.

Recent studies have shown that an abiotic stress can trigger a set of common stress response signaling pathways in fungi, such as the cyclic AMP (cAMP) signaling pathway, Ca^{2+}-dependent protein kinases (CDPKs), mitogen-activated protein kinases (MAPKs), protein kinase C (PKC), and Hog1 MAPK pathway [67–70]. One genes encoding for calcium-mediated signaling and three genes for MAPK signaling pathway were differentially expressed indicates that at least two signaling pathways participate in the drought response in lichen-forming fungus *E. pusillum*, and the functions of the differentially expressed genes are worthy of further investigation.

3.2 Drought-adaptive mechanisms in *E. pusillum*

In consideration of the strong drought tolerance characteristics of this lichen, we think that some different drought response mechanisms exist in *E. pusillum*. The ribosome is an intricate ribonucleo protein complex responsible for polypeptide synthesis in all living cells, so the expression changes of ribosomal proteins affect the translation process. Generally, the expressions of ribosomal proteins are regulated in response to environmental changes and cellular needs, such as nutrition increase, heat stress, and starvation (amino acid/nitrogen deprivation), with reduced expression of these proteins under stressful and starvation conditions [71–73]. The expressions of these proteins also change through the growth phases in a stationary culture, with increased expressions during the exponential growth phase and decreased expression during stationary phase. Recent research discovered that ribosomal subunits could act as the regulatory elements or filters which mediate interactions between particular mRNAs and components of the translation machinery, thus providing a mechanism for translational control [74]. Therefore, it has been suggested that the increased expression of ribosomal proteins in *E. pusillum* could meet the demand of the expression of specific mRNAs required for stress response. Similarly, this phenomenon has also been found in other desiccation tolerant organisms [75–77]. The PEG-induced stress affects and accelerates the transcription and translation process, which can account for the active drought response mechanisms that more materials, such as nucleotides and amino acids, are needed to repair the DNA, RNA and protein damages caused by desiccation.

Almost all aerobic organisms need ATP to supply energy for metabolism, and oxidative phosphorylation is the dominating ATP synthesis pathway in mitochondrion. In other plants and fungi, the ATPase and ATP synthesis pathway have shown to be suppressed under different stress treatments [78,79]. However, KEGG assignment illustrates that most metabolic pathways were active in *E. pusillum* under PEG-induced stress (Figure 2). Our results suggest that *E. pusillum*, unlike other drought-sensitive organism, is more adapted to drought stress so its oxidative phosphorylation is also active under PEG-induced drought stress.

The differential expression of genes involved in galactose and nucleotide sugar metabolisms suggests that the changing carbohydrate metabolism of the mycobiont under desiccation likely leads to reduced monosaccharides and increased polysaccharides. Many previous studies have shown that carbohydrate metabolism is highly susceptible to desiccation [80–82], but most changes were linked to adjusting osmotic pressure. However, few studies have related nucleotide sugar metabolism to drought resistance. Nucleotide sugars are the universal sugar donors for the formation of polysaccharides, glycoproteins, proteoglycans, glycolipids, and glycosylated secondary metabolites [83], and enzymes involved in nucleotide sugar production are important because of the potential to manipulate the composition of cell walls through substrate level control [84]. It is known that glycosylated hormones and secondary metabolites play important roles in plant resistance against biotic and abiotic stresses [85,86]. In lichens, several light-absorbing secondary metabolites produced by fungi have also shown to provide protection for their photosynthetic partners against photo damage [87]. Therefore, it can be argued that the increased expression of genes related to nucleotide sugar synthesis is related to drought-resistant in *E. pusillum*. However, further research is needed in order to identify if glycosylated secondary metabolites indeed provide protection for its algae.

In conclusion, these differentially expressed genes in *E. pusillum* can be summarized in damage repair, energy provision, and carbon metabolisms-provide protection according to their function. Together, these results suggest that *E. pusillum* is adapted to water limitations and drought.

3.3 The comparison of transcript levels for differentially expressed genes between PEG-stressed and dehydrated *E. pusillum*

The transcript levels of drought response-related genes were verified based on the lichen thalli dehydration experiment, together with data from comparative transcriptome analysis. The qRT-PCR result is consistent with that from RNA-seq data in which the expressions of these drought-resistant genes were up-regulated when lichen-forming fungus *E. pusillum* was cultured after adding 20% PEG. However, the expressions of genes involved in galactose metabolism were different between the RNA-seq data and the result from RT-PCR. This was likely due to the specific environmental treatment effect. During the dehydration performed in the laboratory, the photosynthetic partner provided very limited carbohydrates to lichen-forming fungus. In contrast, the mycobiont under PEG-induced drought stress was grown in PDB medium containing plenty of carbohydrates and other nutrients. Thus, the dehydration experiment actually included both drought and starvation stresses, but the sample used in transcriptome sequencing only experienced drought stress. Hence, the difference for genes involved in galactose metabolism reflects the variations in stresses. In summary, the drought-resistant and drought-adaptive mechanisms for the lichen-forming fungus *E. pusillum* under PEG-induced stress and for its lichen thalli under natural dehydration stress were essentially the same.

Our transcriptome analyses have identified many differentially expressed genes with unknown functions. These genes may be involved in novel drought response pathways or mechanisms to contribute to the strong drought resistance in *E. pusillum*. The detailed roles of these genes await further investigation.

This work was supported by the National Natural Science Foundation of China (31270071), the Knowledge Innovation Program of the Chinese Academy of Sciences (KSCX2-EW-J-6), and the State Key Laboratory of Crop Stress Biology for Arid Areas, NWAFU.

1 Brown D, Rapsch S, Beckett A, Ascaso C. The effect of desiccation on cell shape in the lichen *Parmelia sulcata* taylor. New Phytol, 1987, 105: 295–299

2 Bewley JD. Physiological-aspects of desiccation tolerance. Annu Rev Plant Phys, 1979, 30: 195–238

3 Oliver MJ, Derek Bewley J. Desiccation—tolerance of plant tissues: a mechanistic overview. Hortic Rev, 1997, 18: 171–213

4 Kappen L. Response to Extreme Environments. New York: Academic Press, 1973. 311–380

5 Aubert S, Juge C, Boisson AM, Gout E, Bligny R. Metabolic processes sustaining the reviviscence of lichen *Xanthoria elegans* (link) in high mountain environments. Planta, 2007, 226: 1287–1297

6 Kranner I. Glutathione status correlates with different degrees of desiccation tolerance in three lichens. New Phytol, 2002, 154: 451–460

7 Lange OL. Experimentell-Ökologische Untersuchungen an Flechten der Negev-Wüste. I. CO_2-Gaswechsel von Ramalina maciformis (Del.) Bory unter kontrotlierten Bedingungen im Laboratorium. Flora Abt B, 1969, 158: 324–359

8 Ascaso C, Brown D, Rapsch S. The ultrastructure of the phycobiont of desiccated and hydrated lichens. Lichenologist, 1986, 18: 37–46

9 Kranner I, Cram WJ, Zorn M, Wornik S, Yoshimura I, Stabentheiner E, Pfeifhofer HW. Antioxidants and photoprotection in a lichen as compared with its isolated symbiotic partners. Proc Natl Acad Sci USA, 2005, 102: 3141–3146

10 Nash T, Reiner A, Demmig-Adams B, Kilian E, Kaiser W, Lange O. The effect of atmospheric desiccation and osmotic water stress on photosynthesis and dark respiration of lichens. New Phytol, 1990, 116: 269–276

11 Tuba Z, Csintalan Z, Proctor MCF. Photosynthetic responses of a moss, *Tortula ruralis*, ssp. *ruralis*, and the lichens *Cladonia convoluta* and *C. Furcata* to water deficit and short periods of desiccation, and their ecophysiological significance: a baseline study at present-day CO_2 concentration. New Phytol, 1996, 133: 353–361

12 Calatayud A, Deltoro VI, Barreno E, Tascon SD. Changes in *in vivo* chlorophyll fluorescence quenching in lichen thalli as a function of water content and suggestion of zeaxanthin-associated photoprotection. Physiol Plantarum, 1997, 101: 93–102

13 Weissman L, Garty J, Hochman A. Characterization of enzymatic antioxidants in the lichen *Ramalina lacera* and their response to rehydration. Appl Environ Microb, 2005, 71: 6508–6514

14 França MB, Panek AD, Eleutherio ECA. The role of cytoplasmic catalase in dehydration tolerance of *Saccharomyces cerevisiae*. Cell Stress Chaperon, 2005, 10: 167

15 Jiang MY, Zhang JH. Water stress-induced abscisic acid accumulation triggers the increased generation of reactive oxygen species and up-regulates the activities of antioxidant enzymes in maize leaves. J Exp Bot, 2002, 53: 2401–2410

16 Kranner I, Birtic S. A modulating role for antioxidants in desiccation tolerance. Integr Comp Biol, 2005, 45: 734–740

17 Junttila S, Laiho A, Gyenesei A, Rudd S. Whole transcriptome characterization of the effects of dehydration and rehydration on *Cladonia rangiferina*, the grey reindeer lichen. BMC Genomics, 2013, 14: 870

18 Wang YY, Liu B, Zhang XY, Zhou QM, Zhang T, Li H, Yu YF, Zhang XL, Hao XY, Wang M, Wang L, Wei JC. Genome characteristics reveal the impact of lichenization on lichen-forming fungus *Endocarpon pusillum* Hedwig (Verrucariales, Ascomycota). BMC Genomics, 2014, 15: 34

19 Veerman J, Vasil'ev S, Paton GD, Ramanauskas J, Bruce D. Photoprotection in the lichen *Parmelia sulcata*: the origins of desiccation-induced fluorescence quenching. Plant Physiol, 2007, 145: 997–1005

20 Scheidegger C, Schroeter B, Frey B. Structural and functional processes during water vapour uptake and desiccation in selected lichens with green algal photobionts. Planta, 1995, 197: 399–409

21 Holder J, Wynn-Williams D, Rull Perez F, Edwards H. Raman spectroscopy of pigments and oxalates in situ within epilithic lichens: *Acarospora* from the Antarctic and Mediterranean. New Phytol, 2000, 145: 271–280

22 Zhang T, Wei J. Survival analyses of symbionts isolated from *Endocarpon pusillum* Hedwig to desiccation and starvation stress. Sci China Life Sci, 2011, 54: 480–489

23 Lagerwer J, Eagle HE, Ogata G. Control of osmotic pressure of culture solutions with polyethylene glycol. Science, 1961, 133: 1486–1487

24 Caruso A, Chefdor F, Carpin S, Depierreux C, Delmotte FM, Kahlem G, Morabito D. Physiological characterization and identification of genes differentially expressed in response to drought induced by PEG 6000 in *Populus canadensis* leaves. J Plant Physiol, 2008, 165: 932–941

25 Vivas A, Marulanda A, Ruiz-Lozano JM, Barea JM, Azcon R. Influence of a *Bacillus* sp on physiological activities of two arbuscular mycorrhizal fungi and on plant responses to PEG-induced drought stress. Mycorrhiza, 2003, 13: 249–256

26 Haas BJ, Delcher AL, Mount SM, Wortman JR, Smith Jr RK, Hannick LI, Maiti R, Ronning CM, Rusch DB, Town CD. Improving the *Arabidopsis* genome annotation using maximal transcript alignment

assemblies. Nucleic Acids Res, 2003, 31: 5654–5666

27 Stanke M, Morgenstern B. Augustus: a web server for gene prediction in eukaryotes that allows user-defined constraints. Nucleic Acids Res, 2005, 33: W465–467

28 Parra G, Blanco E, Guigó R. Geneid in *Drosophila*. Genome Res, 2000, 10: 511–515

29 Ter-Hovhannisyan V, Lomsadze A, Chernoff YO, Borodovsky M. Gene prediction in novel fungal genomes using an *ab initio* algorithm with unsupervised training. Genome Res, 2008, 18: 1979–1990

30 Haas BJ, Salzberg SL, Zhu W, Pertea M, Allen JE, Orvis J, White O, Buell CR, Wortman JR. Automated eukaryotic gene structure annotation using EVidenceModeler and the program to assemble spliced alignments. Genome Biol, 2008, 9: R7

31 Ashburner M, Ball CA, Blake JA, Botstein D, Butler H, Cherry JM, Davis AP, Dolinski K, Dwight SS, Eppig JT, Harris MA, Hill DP, Issel-Tarver L, Kasarskis A, Lewis S, Matese JC, Richardson JE, Ringwald M, Rubin GM, Sherlock G, Consortium GO. Gene ontology: tool for the unification of biology. Nat Genet, 2000, 25: 25–29

32 Sonnhammer ELL, Koonin EV. Orthology, paralogy and proposed classification for paralog subtypes. Trends Genet, 2002, 18: 619–620

33 Ruepp A, Zollner A, Maier D, Albermann K, Hani J, Mokrejs M, Tetko I, Guldener U, Mannhaupt G, Munsterkotter M, Mewes HW. The funcat, a functional annotation scheme for systematic classification of proteins from whole genomes. Nucleic Acids Res, 2004, 32: 5539–5545

34 Kanehisa M, Goto S, Kawashima S, Okuno Y, Hattori M. The KEGG resource for deciphering the genome. Nucleic Acids Res, 2004, 32: D277–280

35 Tarazona S, García-Alcalde F, Dopazo J, Ferrer A, Conesa A. Differential expression in RNA-Seq: a matter of depth. Genome Res, 2011, 21: 2213–2223

36 Schmittgen TD, Livak KJ. Analyzing real-time PCR data by the comparative C_t method. Nat Protoc, 2008, 3: 1101–1108

37 Gotz S, Garcia-Gomez JM, Terol J, Williams TD, Nagaraj SH, Nueda MJ, Robles M, Talon M, Dopazo J, Conesa A. High-throughput functional annotation and data mining with the Blast2GO suite. Nucleic Acids Res, 2008, 36: 3420–3435

38 Bray EA. Molecular responses to water deficit. Plant Physiol, 1993, 103: 1035–1040

39 Ben-Zvi AP, Goloubinoff P. Review: mechanisms of disaggregation and refolding of stable protein aggregates by molecular chaperones. J Struct Biol, 2001, 135: 84–93

40 Welch AZ, Gibney PA, Botstein D, Koshland DE. TOR and RAS pathways regulate desiccation tolerance in *Saccharomyces cerevisiae*. Mol Biol Cell, 2013, 24: 115–128

41 Nyska A, Kohen R. Oxidation of biological systems: oxidative stress phenomena, antioxidants, redox reactions, and methods for their quantification. Toxicol Pathol, 2002, 30: 620–650

42 Navrot N, Rouhier N, Gelhaye E, Jacquot JP. Reactive oxygen species generation and antioxidant systems in plant mitochondria. Physiol Plantarum, 2007, 129: 185–195

43 Tamai KT, Liu XD, Silar P, Sosinowski T, Thiele DJ. Heat-shock transcription factor activates yeast metallothionein gene-expression in response to heat and glucose starvation via distinct signaling pathways. Mol Cell Biol, 1994, 14: 8155–8165

44 Damveld RA, Arentshorst M, Franken A, vanKuyk PA, Klis FM, van den Hondel CAMJJ, Ram AFJ. The *Aspergillus niger* MADS-box transcription factor RlmA is required for cell wall reinforcement in response to cell wall stress. Mol Microbiol, 2005, 58: 305–319

45 Guo M, Chen Y, Du Y, Dong YH, Guo W, Zhai S, Zhang HF, Dong SM, Zhang ZG, Wang YC, Wang P, Zheng XB. The bZIP transcription factor MoAP1 mediates the oxidative stress response and is critical for pathogenicity of the rice blast fungus *Magnaporthe oryzae*. Plos Pathog, 2011, 7: e1001302

46 Voitsik AM, Muench S, Deising HB, Voll LM. Two recently duplicated maize NAC transcription factor paralogs are induced in response to *Colletotrichum graminicola* infection. BMC Plant Biol, 2013, 13: 85

47 Nicholls S, Straffon M, Enjalbert B, Nantel A, Macaskill S, Whiteway M, Brown AJP. Msn2-and Msn4-like transcription factors play no obvious roles in the stress responses of the fungal pathogen *Candida albicans*. Eukaryot Cell, 2004, 3: 1111–1123

48 Debouba M, Maâroufi-Dghimi H, Suzuki A, Ghorbel MH, Gouia H. Changes in growth and activity of enzymes involved in nitrate reduction and ammonium assimilation in tomato seedlings in response to NaCl stress. Ann Bot, 2007, 99: 1143–1151

49 Planchet E, Rannou O, Ricoult C, Boutet-Mercey S, Maia-Grondard A, Limami AM. Nitrogen metabolism responses to water deficit act through both abscisic acid (ABA)-dependent and independent pathways in *Medicago truncatula* during post-germination. J Exp Bot, 2011, 62: 605–615

50 Dittami SM, Scornet D, Petit JL, Segurens B, Da Silva C, Corre E, Dondrup M, Glatting KH, Konig R, Sterck L, Rouze P, Van de Peer Y, Cock JM, Boyen C, Tonon T. Global expression analysis of the brown alga *Ectocarpus siliculosus* (Phaeophyceae) reveals large-scale reprogramming of the transcriptome in response to abiotic stress. Genome Biol, 2009, 10: R66

51 Foito A, Byrne SL, Shepherd T, Stewart D, Barth S. Transcriptional and metabolic profiles of *Lolium perenne* L. Genotypes in response to a PEG-induced water stress. Plant Biotechnol J, 2009, 7: 719–732

52 Gulez G, Dechesne A, Workman CT, Smets BF. Transcriptome dynamics of *Pseudomonas putida* KT2440 under water stress. Appl Environ Microbiol, 2012, 78: 676–683

53 Mooney S, Leuendorf JE, Hendrickson C, Hellmann H. Vitamin B6: a long known compound of surprising complexity. Molecules, 2009, 14: 329–351

54 Bilski P, Li M, Ehrenshaft M, Daub M, Chignell C. Vitamin B6 (pyridoxine) and its derivatives are efficient singlet oxygen quenchers and potential fungal antioxidants. Photochem Photobiol, 2000, 71: 129–134

55 Chen H, Xiong L. Pyridoxine is required for post-embryonic root development and tolerance to osmotic and oxidative stresses. Plant J, 2005, 44: 396–408

56 Kranner I, Beckett R, Hochman A, Nash TH. Desiccation-tolerance in lichens: a review. Bryologist, 2008, 111: 576–593

57 Grune T, Reinheckel T, Davies K. Degradation of oxidized proteins in mammalian cells. FASEB J, 1997, 11: 526–534

58 Sassa H, Oguchi S, Inoue T, Hirano H. Primary structural features of the 20S proteasome subunits of rice *Oryza sativa*. Gene, 2000, 250: 61–66

59 Wu P, Wang X, Qin GX, Liu T, Jiang YF, Li MW, Guo XJ. Microarray analysis of the gene expression profile in the midgut of silkworm infected with cytoplasmic polyhedrosis virus. Mol Biol Rep, 2011, 38: 333–341

60 Rajjou L, Belghazi M, Huguet R, Robin C, Moreau A, Job C, Job D. Proteomic investigation of the effect of salicylic acid on *Arabidopsis* seed germination and establishment of early defense mechanisms. Plant Physiol, 2006, 141: 910–923

61 Huang H, Moller IM, Song SQ. Proteomics of desiccation tolerance during development and germination of maize embryos. J Proteomics, 2012, 75: 1247–1262

62 Chaves MM, Maroco JP, Pereira JS. Understanding plant responses to drought-from genes to the whole plant. Funct Plant Biol, 2003, 30: 239–264

63 Shao HB, Liang ZS, Shao MA. Osmotic regulation of 10 wheat (*Triticum aestivum* L.) genotypes at soil water deficits. Colloid Surface B, 2006, 47: 132–139

64 Ashraf M, Foolad M. Roles of glycine betaine and proline in improving plant abiotic stress resistance. Environ Exp Bot, 2007, 59: 206–216

65 Hounsa CG, Brandt EV, Thevelein J, Hohmann S, Prior BA. Role of trehalose in survival of *Saccharomyces cerevisiae* under osmotic stress. Microbiology, 1998, 144: 671–680

66 Ashraf M. Inducing drought tolerance in plants: recent advances. Bi-

otechnol Adv, 2010, 28: 169–183
67 Bahn YS, Xue C, Idnurm A, Rutherford JC, Heitman J, Cardenas ME. Sensing the environment: lessons from fungi. Nat Rev Microbiol, 2007, 5: 57–69
68 Alonso-Monge R, Roman E, Arana DM, Pla J, Nombela C. Fungi sensing environmental stress. Clin Microbiol Infec, 2009, 15: 17–19
69 Fernandes L, Araujo MA, Amaral A, Reis VC, Martins NF, Felipe MS. Cell signaling pathways in *Paracoccidioides brasiliensis*— inferred from comparisons with other fungi. Genet Mol Res, 2005, 4: 216–231
70 Brown SM, Campbell LT, Lodge JK. *Cryptococcus neoformans*, a fungus under stress. Curr Opin Microbiol, 2007, 10: 320–325
71 Li B, Nierras CR, Warner JR. Transcriptional elements involved in the repression of ribosomal protein synthesis. Mol Cell Biol, 1999, 19: 5393–5404
72 Warner JR. Synthesis of ribosomes in *Saccharomyces cerevisiae*. Microbiol Mol Biol Rev, 1989, 53: 256–271
73 Warner JR. The economics of ribosome biosynthesis in yeast. Trends Biochem Sci, 1999, 24: 437–440
74 Mauro VP, Edelman GM. The ribosome filter hypothesis. Proc Natl Acad Sci USA, 2002, 99: 12031–12036
75 Adhikari BN, Wall DH, Adams BJ. Desiccation survival in an Antarctic nematode: molecular analysis using expressed sequenced tags. BMC Genomics, 2009, 10: 69
76 Angelovici R, Galili G, Fernie AR, Fait A. Seed desiccation: a bridge between maturation and germination. Trends Plant Sci, 2010, 15: 211–218
77 Pearson GA, Hoarau G, Lago-Leston A, Coyer JA, Kube M, Reinhardt R, Henckel K, Serrao ET, Corre E, Olsen JL. An expressed sequence tag analysis of the intertidal brown seaweeds *Fucus serratus* (L.) and *F. vesiculosus* (L.) (Heterokontophyta, Phaeophyceae) in response to abiotic stressors. Mar Biotechnol, 2010, 12: 195–213
78 Piper PW. The heat-shock and ethanol stress responses of yeast exhibit extensive similarity and functional overlap. FEMS Microbiol Lett, 1995, 134: 121–127
79 Tezara W, Mitchell VJ, Driscoll SD, Lawlor DW. Water stress inhibits plant photosynthesis by decreasing coupling factor and ATP. Nature, 1999, 401: 914–917
80 Toldi O, Tuba Z, Scott P. Vegetative desiccation tolerance: is it a goldmine for bioengineering crops? Plant Sci, 2009, 176: 187–199
81 Valliyodan B, Nguyen HT. Understanding regulatory networks and engineering for enhanced drought tolerance in plants. Curr Opin Plant Biol, 2006, 9: 189–195
82 Abdel Latef AAH. Influence of arbuscular mycorrhizal fungi and copper on growth, accumulation of osmolyte, mineral nutrition and antioxidant enzyme activity of pepper (*Capsicum annuum* L.). Mycorrhiza, 2011, 21: 495–503
83 Strominger JL. Nucleotide intermediates in the biosynthesis of heteropolymeric polysaccharides. Biophys J, 1964, 4: 139–153
84 Gibeaut DM. Nucleotide sugars and glycosyltransferases for synthesis of cell wall matrix polysaccharides. Plant Physiol Bioch, 2000, 38: 69–80
85 Roberts MR, Warner SA, Darby R, Lim EK, Draper J, Bowles DJ. Differential regulation of a glucosyl transferase gene homologue during defence responses in tobacco. J Exp Bot, 1999, 50: 407–410
86 O'Donnell PJ, Truesdale MR, Calvert CM, Dorans A, Roberts MR, Bowles DJ. A novel tomato gene that rapidly responds to wound- and pathogen-related signals. Plant J, 1998, 14: 137–142
87 Gauslaa Y, Solhaug KA. Fungal melanins as a sun screen for symbiotic green algae in the lichen *Lobaria pulmonaria*. Oecologia, 2001, 126: 462–471

Supporting Information

Table S1 The differential expression of genes putatively involved in antioxidant mechanisms

Table S2 The differential expression of genes putatively involved in osmotic regulation

Table S3 The differential expression of genes putatively involved in post-translational processing

Table S4 The differential expression of genes putatively involved in signal transduction pathway

Table S5 The differential expression of genes putatively involved in transcriptional factor

Table S6 The differential expression of genes putatively involved in pyrimidine and purine metabolism

Table S7 The differential expression of genes putatively involved in oxidative phosphorylation

Table S8 The differential expression of genes putatively involved in nitrogen and carbon metabolism

The supporting information is available online at life.scichina.com and link.springer.com. The supporting materials are published as submitted, without typesetting or editing. The responsibility for scientific accuracy and content remains entirely with the authors.

第五篇　部分药用真菌
Some Medicinal Fungi

第一章　地衣内生菌
Endolichenic Fungus

Ambuic acid and torreyanic acid derivatives from the endolichenic fungus *Pestalotiopsis* sp.

Gang Ding, Yan Li, Shaobin Fu, Shuchun Liu, Jiangchun Wei, and Yongsheng Che

J. Nat. Prod., **2009**, 72 (1), 182-186• DOI: 10.1021/np800733y • Publication Date (Web): 31 December 2008

This paper was originally published in *J. Nat. Prod.*, 72（1）：182-186，2009.

Ambuic Acid and Torreyanic Acid Derivatives from the Endolichenic Fungus *Pestalotiopsis* sp.

Gang Ding,[†,‡] Yan Li,[†,‡] Shaobin Fu,[†] Shuchun Liu,[†] Jiangchun Wei,[†] and Yongsheng Che*[,†]

Key Laboratory of Systematic Mycology & Lichenology, Institute of Microbiology, Chinese Academy of Sciences, Beijing 100080, People's Republic of China, and Graduate School of Chinese Academy of Sciences, Beijing 100039, People's Republic of China

Received November 16, 2008

Six new ambuic acid (**1**) derivatives (**2**−**7**) and a new torreyanic acid analogue (**8**) have been isolated from the crude extract of endophytic fungus *Pestalotiopsis* sp. inhabiting the lichen *Clavaroids* sp. The structures of these compounds were elucidated primarily by NMR and MS methods, and their absolute configurations were assigned by application of the CD excitation chirality method. Compounds **1** and **2** displayed antimicrobial activity against the Gram-positive bacterium *Staphylococcus aureus*.

Endolichenic fungi are microorganisms living in the thalli of lichens that are analogous to the plant endophytic fungi inhabiting the intercellular spaces of the hosts.[1] Although several secondary metabolites have been isolated from lichens and their fungal mycobionts,[2−7] only three heptaketides, corynesporol, herbarin, and 1-hydroxydehydroherbarin, have been previously reported from an endolichenic fungus *Corynespora* sp. inhabiting the cavern beard lichen *Usnea cavernosa*.[8] In a search for new bioactive natural products from fungal species of unique niches, chemical investigations of fungi living with lichens were recently initiated in our laboratory. In this study, the fungus *Pestalotiopsis* sp. was isolated from the lichen *Clavaroids* sp. collected from Bawang Mountain, Hainan Province, People's Republic of China. The fungus was grown in a solid-substrate fermentation culture. Its organic solvent extract showed antimicrobial activity against the Gram-positive bacterium *Staphylococcus aureus* (ATCC 6538). Bioassay-directed fractionation of the extract led to the isolation of ambuic acid (**1**)[9] and seven new metabolites (**2**−**8**). Compounds **2**−**7** are ambuic acid derivatives, and **8** is closely related to the known dimeric quinone torreyanic acid (**9**).[10] Details of the isolation, structure elucidation, and biological activity of these compounds are presented herein.

Ambuic acid (**1**) was a major component of the crude extract, and its structure was identified by comparison of the NMR and MS data with those reported.[9] Ambuic acid is a highly functionalized cyclohexenone and was initially isolated from the rainforest plant endophytic fungi *Pestalotiopsis* spp. and *Monochaetia* sp. as an antifungal agent.

The molecular formula of compound **2** was established as $C_{21}H_{28}O_7$ (eight degrees of unsaturation) on the basis of its HRESIMS (*m/z* 415.1723 $[M + Na]^+$; Δ −0.4 mmu). The extra 42 mass units, compared to that of **1**, suggested the presence of an acetyl group. Analysis of the ^1H and ^{13}C NMR spectroscopic data of **2** revealed structural similarity to those of **1**, except that the oxygenated methylene protons (H_2-18) were shifted downfield to δ_H 4.86 and 4.92, respectively in **2**. In addition, NMR resonances corresponding to an acetyl group (δ_H 2.00; δ_C 20.7 and 172.3) were observed, indicating that the C-18 oxygen of **2** was acylated. It was confirmed by an HMBC correlation from H_2-18 to the carboxyl carbon at δ_C 172.3. The gross structure of compound **2** was established as shown.

Compound **3** gave a molecular formula of $C_{19}H_{24}O_7$ (eight degrees of unsaturation) by analysis of its HRESIMS (*m/z* 387.1415 $[M + Na]^+$; Δ −0.1 mmu). Comparison of the ^1H and ^{13}C NMR spectroscopic data of **3** with those of **1** revealed a resonance for one more ketone functionality (δ_C 213.4) and the absence of signals for a methylene unit (δ_H 1.27; δ_C 32.5), suggesting that the C-15 methylene carbon in **1** was oxidized to a ketone in **3**. HMBC correlations from H_2-13, H_2-14, H_2-16, and H_3-17 to C-15 were observed to support this structure as shown.

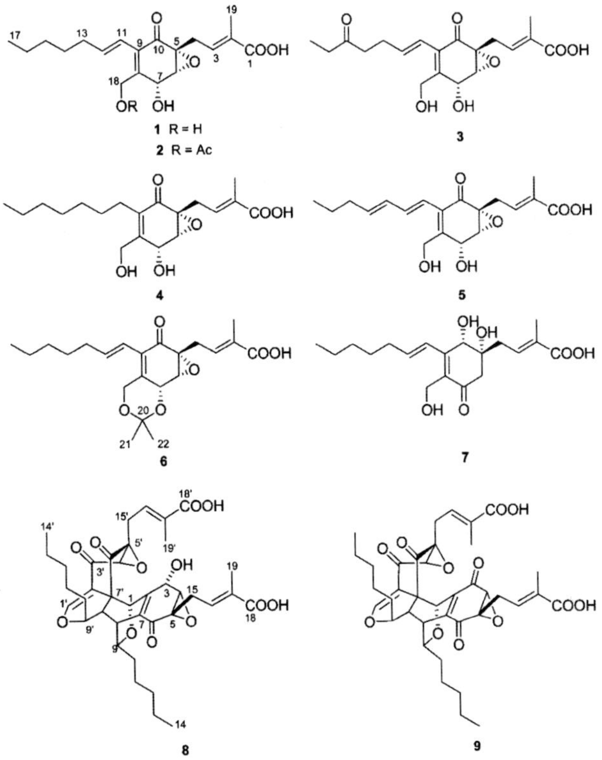

The molecular formula of compound **4** was determined to be $C_{19}H_{28}O_6$ (six degrees of unsaturation) on the basis of HRESIMS analysis (*m/z* 375.1788 $[M + Na]^+$; Δ −1.0 mmu). The ^1H and ^{13}C NMR spectra of **4** displayed the resonances for structural fragments similar to those presented in the spectra of **1**, except that the C-11/C-12 olefin was reduced to two methylene units. This observation was confirmed by analysis of relevant $^1H-^1H$ COSY and HMBC data. Interpretation of the HRESIMS and NMR data (Table 1) of **5** enabled assignment of its molecular formula as $C_{19}H_{24}O_6$ (*m/z* 371.1465 $[M + Na]^+$; Δ −0.2 mmu), with two less hydrogens than that of **1**. Detailed comparison of the NMR spectroscopic data between **5** and **1** revealed that the two mutually coupled methylene units (C-13−C-14) in **1** were oxidized to an olefin moiety in **5**. Further analysis of the COSY spectral data of

* To whom correspondence should be addressed. Tel: +86 10 82618785. Fax: +86 10 82618785. E-mail: cheys@im.ac.cn.

† Institute of Microbiology.

‡ Graduate School of Chinese Academy of Sciences.

Table 1. NMR Spectroscopic Data for **2** and **3** (in CD_3OD) and **4** and **5** (in acetone-d_6)

	compound **2**		compound **3**		compound **4**		compound **5**	
position	δ_H[a] (J in Hz)	δ_C[b], mult.	δ_H[a] (J in Hz)	δ_C[c], mult.	δ_H[a] (J in Hz)	δ_C[b] mult.	δ_H[a] (J in Hz)	δ_C[d], mult.
1		171.2, qC		167.8, qC		168.5, qC		168.7, qC
2		132.0, qC		131.6, qC		131.0, qC		130.1, qC
3	6.62, t (7.0)	136.4, CH	6.61, t (7.0)	136.8, CH	6.71, t (7.0)	136.2, CH	6.72, t (7.5)	136.2, CH
4	2.77, dd (16, 7.0)	28.6, CH_2	2.76, dd (16, 7.0)	28.7, CH_2	2.86, m	28.7, CH_2	2.84, dd (16, 7.5)	28.5, CH_2
	2.72, dd (16, 7.0)		2.69, dd (16, 7.0)				2.75, dd (16, 7.5)	
5		61.4, qC		61.3, qC		59.9, qC		60.7, qC
6	3.69, d (2.5)	60.9, CH	3.70, d (2.5)	61.1, CH	3.80, d (2.5)	60.2, CH	3.83, d (2.5)	60.5, CH
7	4.66, br s	65.8, CH	4.76, br s	65.9, CH	4.89, br s	65.8, CH	4.96, br s	66.4, CH
8		145.4, qC		151.2, qC		151.6, qC		151.0, qC
9		134.3, qC		131.6, qC		133.0, qC		131.1, qC
10		195.7, qC		195.9, qC		195.2, qC		195.3, qC
11	6.10, d (16)	122.8 CH	6.12, d (16)	123.5, CH	2.46, m; 2.17, m	26.3, CH_2	6.27, d (16)	122.8, CH
12	5.80, td (16, 6.5)	140.8, CH	5.79, td (16, 6.5)	138.4, CH	1.26, m	30.5, CH_2	6.57, dd (16, 11)	137.2, CH
13	2.13, m	34.4, CH_2	2.34, m	28.8, CH_2	1.26, m	30.5, CH_2	6.15, dd (16, 11)	132.1, CH
14	1.38, m	29.8, CH_2	2.54, t (7.5)	42.2, CH_2	1.26, m	29.8, CH_2	5.80, td (16, 7.0)	137.4, CH
15	1.27, m	32.5, CH_2		213.4, qC	1.26, m	32.4, CH_2	2.08, m	35.5, CH_2
16	1.27, m	23.6, CH_2	2.44, m	36.7, CH_2	1.26, m	23.0, CH_2	1.42, m	23.0, CH_2
17	0.85, t (7.0)	14.4, CH_3	0.96, t (7.5)	8.0, CH_3	0.86, t (6.5)	14.2, CH_3	0.89, t (7.5)	13.9, CH_3
18	4.92, d (13),	62.4, CH_2	4.44, d (13)	60.3, CH_2	4.47, d (14)	60.2, CH_2	4.53, d (13)	60.3, CH_3
	4.86, d (13)		4.34, d (13)		4.44, d (14)		4.90, d (13)	
19	1.81, s	12.9, CH_3	1.86, s	12.8, CH_3	1.86, s	12.7, CH_3	1.86, s	12.7, CH_3
AcO		172.3, qC						
	2.00, s	20.7, CH_3						

[a] Recorded at 500 MHz. [b] Recorded at 150 MHz. [c] Recorded at 125 MHz. [d] Recorded at 100 MHz.

Table 2. NMR Spectroscopic Data for **6** (in $CDCl_3$) and **7** (in CD_3OD)

	compound **6**		compound **7**		
position	δ_H[a] (J in Hz)	δ_C[b], mult.	δ_H[a] (J in Hz)	δ_C[c], mult.	key HMBC (H → C#)
1		169.8, qC		172.0, qC	
2		130.2, qC		133.0, qC	
3	6.77, t (7.5)	136.7, CH	6.92, t (7.5)	136.8, CH	1, 2, 4, 5, 19
4	3.00, dd (16, 7.5)	28.8, CH_2	2.56, d (7.5)	38.6, CH_2	2, 3, 5, 6, 10
	2.79, dd (16, 7.5)				
5		59.4, qC		75.5, qC	
6	3.67, d (2.0)	60.8, CH	2.72, d (17)	48.5, CH_2	4, 5, 7, 8
			2.38, d (17)		
7	4.79, br s	65.0, CH		199.3, qC	
8		146.7, qC		133.0, qC	
9		125.5, qC		153.0, qC	
10		193.6, qC	4.31, s	69.9, CH	4, 5, 6, 8, 9, 11
11	5.94, m	119.7, CH	6.73, d (16)	128.0, CH	8, 9, 10, 13
12	5.94, m	139.2, CH	6.51, td (16, 7.0)	142.9, CH	9, 13, 14
13	2.13, m	33.8, CH_2	2.25, m	35.0, CH_2	11, 12, 14, 15
14	1.41, m	28.8, CH_2	1.46, m	29.8, CH_2	12, 13, 15, 16
15	1.29, m	31.4, CH_2	1.31, m	32.6, CH_2	13, 14, 16, 17
16	1.29, m	22.5, CH	1.31, m	23.5, CH_2	14, 15, 17
17	0.89, t (7.0)	14.0, CH_3	0.87, t (7.0)	13.1, CH_3	15, 16
18	4.55, s	55.5, CH_2	4.40, d (12)	54.5, CH_2	7, 8, 9
			4.36, d (12)		
19	1.91, s	12.5, CH_3	1.82, s	14.3, CH_3	1, 2, 3
20		101.0, qC			
21	1.49, s	24.1, CH_3			
22	1.47, s	24.0, CH_3			

[a] Recorded at 500 MHz. [b] Recorded at 150 MHz. [c] Recorded at 125 MHz.

5 confirmed this observation, leading to the assignment of its gross structure as shown.

The molecular formula of compound **6** was established as $C_{22}H_{30}O_6$ (eight degrees of unsaturation) on the basis of HRESIMS analysis (m/z 413.1970 $[M + Na]^+$; Δ −3.5 mmu), which was consistent with its NMR data (Table 2). Interpretation of the ^1H, ^{13}C, and HMQC spectroscopic data of **6** revealed the presence of the same highly functionalized cyclohexenone core structure as that found in **1**, but additional resonances corresponding to two methyl singlets (δ_C 1.47 and 1.49, respectively) and one quaternary carbon at δ_C 101.0 were observed in the NMR spectra of **6**, implying that **6** could be an acetonide of **1** originating from the addition of an acetone unit with the OH groups attached to C-7 and C-18. HMBC correlations from H-7, H_2-18 and from H_3-21/22 to C-20 were observed, and therefore, the gross structure of **6** was depicted. Since acetone was never used as a solvent in the isolation and purification of these compounds, compound **6** should be a naturally occurring metabolite rather than an artifact resulting from the isolation process.

The relative configurations of **2**−**6** were determined by analysis of the $^1H-^1H$ coupling constants and NOESY data. The C-11/C-12 double bond in compounds **2**, **3**, **5**, and **6** and the C-13/C-14 olefin in **5** were all assigned E-geometry on the basis of the large coupling constant observed for corresponding olefinic protons (16 Hz), and the same assignment was made for the C-2/C-3 olefin in **2**−**6** by NOESY correlation of H_2-4 with H_3-19. The small vicinal coupling constant of 2.0 Hz between H-6 and H-7 in compounds **2**−**6** suggested a *cis* relationship between these two protons, and the NOESY correlation of H_2-4 with H-6 indicated that these protons have the same orientation with respect to the cyclohexenone ring. The absolute configurations of **2**−**6** were assigned by application of the CD excitation chirality method. The CD spectra of **2**−**6** all showed positive and negative Cotton effects at near 340 and 240 nm, respectively, which closely resembled those of macrophorin A[11] and (+)-epoxydon.[12] The coupling constants observed between H-6 and H-7 in **2**−**6** (2.5 Hz in **2**−**5** and 2.0 Hz in **6**) were also similar to those of the corresponding protons in macrophorin A and (+)-epoxydon, suggesting the 5*R*, 6*R*, and 7*R* absolute configuration for **2**−**6**.

Compound **7** was assigned the molecular formula $C_{19}H_{28}O_6$ (six degrees of unsaturation) by analysis of its HRESIMS (*m/z* 375.1779 $[M + Na]^+$; Δ −0.1 mmu). The ^1H, ^{13}C, and HMQC NMR spectroscopic data for **7** revealed the presence of the same side chains as those attached to C-5, C-8, and C-9 in **1**, but significant structural changes were observed for the remaining portion of the molecule. Therefore, the HMQC and HMBC for those unaccounted resonances in **7** were analyzed in order to establish its gross structure. HMBC correlations from H-3 to C-5 and from H_2-4 to C-5, C-6, and C-10 led to the connection of C-5 to C-4, C-6, and C-10, whereas those from H_2-6 to C-7 and C-8 indicated that both C-6 and C-8 were connected to the ketone carbon C-7. Additional HMBC cross-peaks from H-10 to C-8 and C-9 permitted completion of the cyclohexenone moiety of **7**. The side chains that were identical to those in compounds **1**−**6** were located at C-5, C-8, and C-9 by relevant HMBC correlations. Collectively, these data allowed assignment of structure **7** as shown. The relative configurations of the C-2/C-3 and C-11/C-12 olefins in **7** were determined on the basis of the $^1H-^1H$ *J*-values and NOED data and by analogy to those in **1**. In NOED experiments, upon irradiation of H-10, enhancement was observed for H_2-4, indicating that those protons are on the same face of the cyclohexenone ring. Although the absolute configuration of **7** could not be directly assigned by analysis of its CD spectrum, C-5 was presumed to have the *R*-configuration based on biogenetic considerations, whereas the absolute configuration of the stereogenic center C-10 was deduced to be *S* by NOED results.

The molecular formula of compound **8** was determined to be $C_{38}H_{46}O_{12}$ (16 degrees of unsaturation) on the basis of HRESIMS analysis (*m/z* 693.2910 $[M - H]^-$; Δ +0.7 mmu). Analysis of the ^1H, ^{13}C, and HMQC NMR data of **8** revealed four methyl groups, 10 methylene units, eight methines (six oxygenated), three quaternary carbons (two oxygenated), eight aromatic/olefinic carbons (three of which were protonated), and five carbonyl carbons. These data, together with three unobserved exchangeable protons, accounted for all the ^1H and ^{13}C NMR resonances for **1**. Interpretation of the COSY NMR data of **8** identified four isolated proton spin systems, which were C-3−C-4, C-15−C-16, C-15′−C-16′, and C-14−C-14′. Further analysis of the 2D NMR data, especially HMBC, revealed the presence of two units of 2-methyl-2-butenoic acid. In addition, the correlations in the HMBC spectrum from H-3 to C-2, C-5, and C-7; from H-4 to C-2, C-6, and C-15; and from H_2-15 to C-4 and C-6 established the cyclohexenone ring, with one 2-methyl-2-butenoic acid moiety attached to C-5. Those from H-1 to C-3 and from H-8 to C-6 led to the connection of C-1 to C-2 and C-7 to C-8, respectively. In turn, correlations from H-8 to C-7′ and from H-8′ to C-1 and C-7′ indicated that C-7′ was connected to both C-1 and C-8′, completing the cyclohexane ring fused to the cyclohexenone ring at C2/C7. Key HMBC correlations from H-1 to C-9 and from H-9 to C-1 established the ether linkage between these two carbons. Further correlations from H-1 to C-2′ and C-6′; from H-1′ to C-2′ and C-3′; from H-1′ to C-9′; and from H-9′ to C-1′ allowed assignment of the 3,4-dihydro-2*H*-pyran moiety fused to the cyclohexane ring at C7′/C8′, with one ketone carbon at δ_C 189.8 (C-3′) attached to C-2′, and the remaining carbon at δ_C 201.7 (C-6′) attached to C-7′. HMBC from H-4′ to C-2′, C-3′, C-5′, and C-15′ and from H_2-15′ to C-4′ and C-6′ completed the cyclohexane-1,4-dione substructure with C-15′ attached to C-5′. Since two exchangeable protons in **8** were already accounted for by the presence of two carboxylic acid units, the remaining exchangeable proton was assigned to OH-3 to complete the gross structure of **8**.

The identified structure was similar to the dimeric quinone torreyanic acid (**9**), isolated from the plant endophytic fungus *Pestalotiopsis macrospore* with selective cytotoxicity against human cancer cell lines.[10] Therefore, the relative configuration for the majority of structure **8** was deduced by analogy to torreyanic acid (**9**), except for H-3 and H-4, which were assigned on the basis of their coupling constant value (2.0 Hz) and NOESY data. From the CD spectrum of **8**, the inverse octant rule could not be applied to assign the absolute configuration of C-3 due to the unclear data seen in the range 330−350 nm. Considering the absolute stereochemistry established for **9** by synthesis,[13] the absolute configuration of C-3 was assigned as *S*.

Compounds **1**−**8** were evaluated for antimicrobial activity against a panel of bacteria and fungi, including the bacteria *Staphylococcus aureus* (ATCC 6538), *Streptococcus pneumoniae* (CGMCC 1.1692), and *Escherichia coli* (CGMCC 1.2340), the yeasts *Candida albicans* (ATCC 10231) and *Geotrichum candidum* (AS2.498), and the fungus *Aspergillus fumigatus* (ATCC 10894). Only compounds **1** and **2** showed activity against the Gram-positive bacterium *S. aureus* (ATCC 6538), with IC_{50} values of 43.9 and 27.8 μM, respectively (the positive control AMP showed an IC_{50} value of 1.40 μM), whereas compounds **3**−**8** did not show noticeable *in vitro* antibacterial or antifungal activities against the above-mentioned organisms (IC_{50} > 50 μM). Even though ambuic acid (**1**) was initially isolated as an antifungal agent with moderate activities against several plant pathogenic fungi,[9] compounds **1**−**8** did not display activity against *Aspergillus fumigatus* (ATCC 10894) in our assays due to the selection of different fungal species as targets.

Compounds **2**−**7** are new analogues of the known compound ambuic acid (**1**), but differ from **1** by the presence of different aliphatic side chains at C-9 (in **2**−**5**) and substitution pattern (in **7**), as well as the presence of an unit of acetonide (in **6**). Compound **8** is closely related to the known dimeric quinone torreyanic acid (**9**),[10] with the only difference being an OH group at C-3 instead of the ketone functionality. The known compounds ambuic acid (**1**) and torreyanic acid (**9**) have been synthesized, and the possible biosynthetic pathways, as well as their biogenetic relationships, have been described.[13] The biosynthesis of **2**−**7** probably proceeds in a manner similar to that of **1**, whereas **8** might be a reduction product (ketone functionality C-3) of **9**.

The isolation of ambuic acid (**1**) and its heterodimer **8** further corroborates the biogenetic hypothesis that terroyanic acid (**9**) could be generated through the oxidation, cyclization, and Diels−Alder dimerization of **1**.[13] Compounds **2**−**8** are the first secondary metabolites to be reported from the endolichenic fungus *Pestalotiopsis* sp. and the second examples of natural products discovered from endolichenic fungal sources, implying that endolichenic fungi could prove to be valuable sources of new bioactive natural products.

Experimental Section

General Experimental Procedures. Optical rotations were measured on a Perkin-Elmer 241 polarimeter, and UV data were recorded on a Shimadzu Biospec-1601 spectrophotometer. CD spectra were recorded on a JASCO J-815 spectropolarimeter, using CH_3OH as solvent. IR data were recorded using a Nicolet Magna-IR 750 spectrophotometer. ^1H and ^{13}C NMR data were acquired with Varian Mercury-400, -500, and -600 spectrometers using solvent signals ($CDCl_3$; δ_H 7.26/δ_C 77.7; CD_3OD; δ_H 3.35/δ_C 49.9; acetone-d_6; δ_H 2.05/δ_C 29.8, 206.0) as references. The HMQC and HMBC experiments were optimized for 145.0 and 8.0 Hz, respectively. ESIMS data were recorded on a Bruker Esquire 3000^{plus} spectrometer, and HRESIMS data were obtained using a Bruker APEX III 7.0 T spectrometer.

Table 3. NMR Spectroscopic Data for **8** in CD_3OD

position	δ_H[a] (J in Hz)	δ_C[b], mult.	key HMBC (H → C#)
1	5.34, s	73.4, CH	2, 3, 7, 9, 2′, 7′, 8′
2		153.9, qC	
3	4.29, d (2.0)	64.8, CH	1, 2, 5, 7
4	3.61, d (2.0)	60.8, CH	2, 6, 15
5		61.2, qC	
6		191.1, qC	
7		133.1, qC	
8	3.28, m	38.9, CH	2, 6, 7, 7′, 9′
9	3.98, t (6.0)	73.1, CH	1, 7, 8′, 11
10	1.20, m; 1.40, m	35.3, CH_2	8, 9, 11, 12
11	1.20, m	25.8, CH_2	9, 10, 12, 13
12	1.20, m	32.8, CH_2	10, 11, 13, 14
13	1.12, m; 0.98, m	23.5, CH_2	11, 12, 14
14	0.80, t (6.0)	14.3, CH_3	12, 13
15	2.82, dd (13, 6.5); 2.70, dd (13, 6.0)	28.6, CH_2	4, 6, 16, 17
16	6.62, dd (6.5, 6.0)	135.9, CH	5, 15, 18, 19
17		133.1, qC	
18		171.0, qC	
19	1.83, s	8.0, CH_3	16, 17, 18
1′	7.79, s	159.6, CH	2′, 3′, 7′, 9′
2′		114.7, qC	
3′		189.8, qC	
4′	3.75, s	67.4, CH	2′, 3′, 6′, 15′
5′		65.3, qC	
6′		201.7, qC	
7′		51.7, qC	
8′	2.67, d (5.0)	37.5, qC	1, 7, 9, 2′, 6′, 7′
9′	4.39, dd (8.5, 4.5)	83.1, CH	1′, 7′, 8
10′	1.20, m	34.3, CH_2	
11′	1.20, m	26.6, CH_2	
12′	1.20, m	32.3, CH_2	
13′	1.20, m	23.5, CH_2	
14′	0.84, t (6.0)	14.2, CH_3	12′, 13′
15′	3.08, dd (14, 6.5); 2.39, dd (14, 6.0)	29.6, CH_2	4′, 6′, 16,′ 17′
16′	6.62, dd (6.5, 6.0)	135.1, CH	15′, 17′, 18′, 19′,
17′		133.2, qC	
18′		171.0, qC	
19′	1.82, s	12.8, CH_3	16′, 17′, 18′

[a] Recorded at 500 MHz. [b] Recorded at 125 MHz.

Fungal Material. The culture of *Pestalotiopsis* sp. was isolated by one of authors (S.F.) from samples of the lichen *Clavaroids* sp. collected from Bawang Mountain, Hainan Province, in May 2007. The isolate was characterized as an unidentified species of *Pestalotiopsis* by one of authors (J.W.) based on sequence analysis of the ITS region of the rDNA and assigned the accession number WN1 in J.W.'s culture collection at the Institute of Microbiology, Chinese Academy of Sciences, Beijing. The fungal strain was cultured on slants of potato dextrose agar (PDA) at 25 °C for 10 days. The agar plugs were used to inoculate 250 mL Erlenmeyer flasks, each containing 50 mL of media (0.4% glucose, 1% malt extract, and 0.4% yeast extract), and the final pH of the media was adjusted to 6.5 before sterilization. Flask cultures were incubated at 25 °C on a rotary shaker at 170 rpm for five days. Fermentation was carried out in four 500 mL Fernbach flasks each containing 75 g of rice. Spore inoculum was prepared by suspension in sterile, distilled H_2O to give a final spore/cell suspension of 1 × 10^6/mL. Distilled H_2O (100 mL) was added to each flask, and the contents were soaked overnight before autoclaving at 15 lb/in.^{2} for 30 min.[14] After cooling to room temperature, each flask was inoculated with 5.0 mL of the spore inoculum and incubated at 25 °C for 40 days.

Extraction and Isolation. The fermented rice substrate was freeze-dried and extracted with EtOAc (3 × 500 mL), and the organic solvent was evaporated to dryness under vacuum to afford a crude extract (10.0 g), which was fractionated by silica column chromatography (CC) (5 × 25 cm) using $CH_2Cl_2-CH_3OH$ gradient elution. The fraction (50 mg) eluted with 99:1 $CH_2Cl_2-CH_3OH$ was separated by semipreparative reversed-phase HPLC (Agilent Zorbax SB-C_{18} column; 5 μm; 9.4 × 250 mm; 2 mL/min; 60% MeCN in H_2O over 5 min, 60−90% MeCN over 85 min) to afford **2** (2.5 mg; t_R 21.5 min) and **8** (1.8 mg; t_R 30.3 min). Another fraction eluted with 99:1 $CH_2Cl_2-CH_3OH$ was separated by Sephadex LH-20 CC using MeOH as eluent, and one subfraction (30 mg) was further purified by reversed-phase HPLC (60% MeOH in H_2O for 5 min, followed by 60−85% MeOH for 40 min) to afford **4** (2.0 mg; t_R 18.0 min), **5** (2.5 mg; t_R 25.6 min), and **1** (8.0 mg; t_R 28.2 min). The fraction (380 mg) eluted with 98:2 $CH_2Cl_2-CH_3OH$ was chromatographed on a Sephadex LH-20 column using CH_3OH as solvent, and one subfraction (30 mg) was purified by HPLC (29% MeCN in H_2O as eluent) to afford **3** (2.3 mg, t_R 28.0 min) and **6** (1.9 mg, t_R 23.0 min). The fraction (200 mg) eluted with 97:3 $CH_2Cl_2-CH_3OH$ was again separated by Sephadex LH-20 CC eluted with CH_3OH, and one subfraction (35 mg) was purified by reversed-phase HPLC (25% MeOH in H_2O for 2 min, followed by 25−40% for 42 min) to afford **7** (2.0 mg, t_R 27.1 min).

Ambuic acid (1): ^1H NMR, ^{13}C NMR, and the ESIMS data were consistent with the literature.[9]

Compound 2: colorless oil; $[\alpha]_D$ +106 (c 0.1, MeOH); UV (MeOH) λ_{max} (ϵ) 213 (19 800), 239 (20 300), 294 (17 900) nm; IR (neat) ν_{max} 3395 (br), 2932, 1739, 1685, 1650, 1379, 1233 cm^{-1}; ^1H and ^{13}C NMR data see Table 1; HRESIMS m/z 415.1723 (calcd for $C_{21}H_{28}O_7Na$, 415.1719).

Compound 3: colorless oil; $[\alpha]_D$ +124 (c 0.1, MeOH); UV (MeOH) λ_{max} (ϵ) 214 (23 500), 234 (23 700), 294 (20 400) nm; IR (neat) ν_{max} 3426 (br), 2933, 1739, 1686, 1650, 1378, 1232 cm^{-1}; ^1H and ^{13}C NMR data see Table 1; HRESIMS m/z 387.1415 (calcd for $C_{19}H_{24}O_7Na$, 387.1414).

Compound 4: colorless oil; $[\alpha]_D$ +105 (c 0.1, MeOH); UV (MeOH) λ_{max} (ϵ) 213 (14 700), 231 (18 200), 291 (7600) nm; IR (neat) ν_{max} 3400 (br), 2929, 1686, 1649, 1379, 1235 cm^{-1}; ^1H and ^{13}C NMR data see Table 1; HRESIMS m/z 375.1788 (calcd for $C_{19}H_{28}O_6Na$, 375.1778).

Compound 5: colorless oil; $[\alpha]_D$ +154 (c 0.1, MeOH); UV (MeOH) λ_{max} (ϵ) 213 (20 800), 232 (21 400), 298 (17 400) nm; IR (neat) ν_{max} 3390 (br), 3187 (br), 2920, 2851, 1685, 1642, 1464, 1418 cm^{-1}; ^1H and ^{13}C NMR data see Table 1; HRESIMS m/z 371.1465 (calcd for $C_{19}H_{24}O_6Na$, 371.1463).

Compound 6: colorless oil; $[\alpha]_D$ +116 (c 0.1, MeOH); UV (MeOH) λ_{max} (ϵ) 213 (14 400), 227 (15 800), 285 (5100) nm; IR (neat) ν_{max} 3379 (br), 2929, 1685, 1649, 1379, 1237 cm^{-1}; ^1H and ^{13}C NMR data see Table 2; HRESIMS m/z 413.1970 (calcd for $C_{22}H_{30}O_6Na$, 413.1935).

Compound 7: colorless oil; $[\alpha]_D$ +135 (*c* 0.1, MeOH); UV (MeOH) λ_{max} (ϵ) 213 (14 100), 231 (18 300), 295 (6600) nm; IR (neat) ν_{max} 3352 (br), 2928, 1685, 1650, 1382, 1266 cm^{-1}; ^1H NMR, ^{13}C NMR, and HMBC data see Table 2; HRESIMS *m/z* 375.1779 (calcd for $C_{19}H_{28}O_6Na$, 375.1778).

Compound 8: white, amorphous powder; $[\alpha]_D$ +154 (*c* 0.1, MeOH); UV (MeOH) λ_{max} (ϵ) 213 (20 800), 232 (21 400), 298 (17 400) nm; IR (neat) ν_{max} 3370 (br), 2932, 1738, 1685, 1651, 1378, 1232 cm^{-1}; ^1H NMR, ^{13}C NMR, and HMBC data see Table 3; HRESIMS *m/z* 693.2910 (calcd for $C_{38}H_{45}O_{12}$, 693.2917).

Antimicrobial and Antifungal Bioassays. Antimicrobial and antifungal bioassays were conducted in triplicate by following the National Center for Clinical Laboratory Standards (NCCLS) recommendations.[15] The bacterial strains *Staphylococcus aureus* (ATCC 6538), *Streptococcus pneumoniae* (CGMCC 1.1692), and *Escherichia coli* (CGMCC 1.2340) were grown on Mueller-Hinton agar, the yeasts, *Candida albicans* (ATCC 10231) and *Geotrichum candidum* (AS2.498), were grown on Sabouraud dextrose agar, and the fungus, *Aspergillus fumigatus* (ATCC 10894), was grown on potato dextrose agar. Targeted microbes (3−4 colonies) were prepared from broth culture (bacteria: 37 °C for 24 h; fungus: 28 °C for 48 h), and the final spore suspensions of bacteria (in MHB medium), yeasts (in SDB medium), and *Aspergillus fumigatus* (in PDB medium) were 10^6 and 10^5 cells/mL and 10^4 mycelial fragments/mL, respectively. Test samples (10 mg/mL as stock solution in DMSO and serial dilutions) were transferred to a 96-well clear plate in triplicate, and the suspension of the test organisms was added to each well, achieving a final volume of 200 μL (antimicrobial peptide AMP, streptomycin, and fluconazole were used as positive controls). After incubation, the absorbance at 595 nm was measured with a microplate reader (TECAN), and the inhibition rate was calculated and plotted versus test concentrations to afford the IC_{50}.

Acknowledgment. We gratefully acknowledge financial support from the National Basic Research Project of China (Grant 2009CB522302), the National Project of Science and Technology of China (Grant 2008BAI63B01), and the Key Project of Hi-Tech Research and Development of China (Grant 2007AA021506).

Supporting Information Available: ^1H NMR, ^{13}C NMR, and CD spectra of compounds **2−8**. This material is available free of charge via the Internet at http://pubs.acs.org.

References and Notes

(1) Arnold, A. E. *Fungal Biol. Rev.* **2007**, *21*, 51–66.
(2) Rezanka, T.; Guschina, I. A. *J. Nat. Prod.* **1999**, *62*, 1675–1677.
(3) Kumar, K. C. S.; Müller, K. *J. Nat. Prod.* **1999**, *62*, 821–823.
(4) Lohézic-Le Dévéhat, F.; Tomasi, S.; Elix, J. A.; Bernard, A.; Rouaud, I.; Uriac, P.; Boustie, J. *J. Nat. Prod.* **2007**, *70*, 1218–1220.
(5) Rezanka, T.; Sigler, K. *J. Nat. Prod.* **2007**, *70*, 1487–1491.
(6) Davies, J.; Wang, H.; Taylor, T.; Warabi, K.; Huang, X. H.; Andersen, R. J. *Org. Lett.* **2005**, *7*, 5233–5236.
(7) Kinoshita, K.; Yamamoto, Y.; Takatori, K.; Koyama, K.; Takahashi, K.; Kawai, K.; Yoshimura, I. *J. Nat. Prod.* **2005**, *68*, 1723–1727.
(8) Paranagama, P. A.; Wijeratne, E. M. K.; Burns, A. M.; Marron, M. T.; Gunatilaka, M. K.; Arnold, A. E.; Gunatilaka, A. A. L. *J. Nat. Prod.* **2007**, *70*, 1700–1705.
(9) Li, J. Y.; Harper, J. K.; Grant, D. M.; Tombe, B. O.; Bharal, B.; Hess, W. M.; Strobel, G. A *Phytochemistry* **2001**, *56*, 463–468.
(10) Lee, J.; Stroble, G.; Lobkovsky, E.; Clardy, J. *J. Org. Chem.* **1996**, *61*, 3232–3233.
(11) Sassa, T.; Yoshikoshi, H. *Agric. Biol. Chem.* **1983**, *47*, 187–189.
(12) Sekiguchi, J.; Gaucher, G. M. *Biochem. J.* **1979**, *182*, 445–453.
(13) Li, C.; Johnson, R. P.; Porco, J. A., Jr. *J. Am. Chem. Soc.* **2003**, *125*, 5095–5106.
(14) Che, Y.; Gloer, J. B.; Wicklow, D. T. *J. Nat. Prod.* **2002**, *65*, 399–402.
(15) Li, E.; Jiang, L.; Guo, L.; Zhang, H.; Che, Y. *Bioorg. Med. Chem.* **2008**, *16*, 7894–7899.

NP800733Y

第二章　冬虫夏草

Ophiocordyceps sinensis

冬虫夏草及其相关类群的分子系统学分析

魏鑫丽[1,2]　印象初[3]　郭英兰[1]　沈南英[4]　魏江春[1]

[1]中国科学院微生物研究所真菌地衣系统学重点实验室，北京 100080; [2]中国科学院研究生院，北京 100039; [3]河北大学生命科学学院，保定 071002; [4]青海畜牧兽医科学院，西宁 810003

摘　要：为了探明冬虫夏草及其相关类群的亲缘关系，以冬虫夏草、中国被毛孢及中华束丝孢 (= 冬虫夏草头孢 = 蝙蝠蛾多毛孢) 共 6 株菌种作为内群和一株蛹草拟青霉作为外群进行了 DNA 随机多态型 (RAPD) 分析。此外，基于上述供试材料又在内群中增加了一株蝙蝠蛾拟青霉，并对内群和外群样品的 nrDNA 间隔区（ITS）碱基序列进行了测定；对于测定的 8 条序列连同来自 GenBank 中的 4 条相关序列进行了分子系统学分析。结果表明：中华束丝孢和中国被毛孢均系冬虫夏草菌的无性型。按照国际植物命名法规，中国被毛孢应为冬虫夏草菌无性型的正确名称。而蝙蝠蛾拟青霉为不同于冬虫夏草菌的另一种真菌；该名称由于不合格发表而不被国际植物命名法规所承认。

关键词：DNA 随机扩增多态型，间隔区，中华束丝孢，中国被毛孢，无性型，蝙蝠蛾拟青霉

中图分类号：Q939.5　**文献标识码**：A　**文章编号**：1672-6472（2006）02-0192-0202

Analyses of molecular systematics on *Cordyceps sinensis* and its related taxa

WEI Xin-Li[1,2]　YIN Xiang-Chu[3]　GUO Ying-Lan[1]　SHEN Nan-Ying[4]　WEI Jiang-Chun[1]

(*[1]Systematic Mycology & Lichenology Laboratory, Institute of Microbiology, Chinese Academy of Sciences, Beijing, China 100080; [2]Graduate School of Chinese Academy of Sciences, Beijing, China 100039; [3]Hebei University, Baoding 071002 [4]Qinghai Animal Husbandry and Veterinary Medicine Academy, Xining 810003*)

ABSTRACT: Six samples of *Cordyceps sinensis*, *Hirsutella sinensis* and *Synnematium sinense*（= *Cephalosporium dongchongxiacao* nom.nud. = *Hirsutella hepiali*, nom.nud.）as ingroup and a sample of *Paecilomyces militaris* as outgroup were available for the systematic analysis based on RAPD method. Besides, a sample of *Paecilomyces hepiali* was added to the above mentioned ingroup. The 7 samples including the sample of *Paecilomyces hepiali* of the ingroup and a sample of its outgroup were used for sequencing of the nrDNA ITS region. Both the sequences from the samples of ingroup and outgroup together with the 4 corresponding sequences from GenBank were furnished for phylogenetic analysis.

The result of the analyses is that *Synnematium sinense* (nom. inval.) was proved to be the anamorph of *C. sinensis*, but the *Hirsutella sinensis* is accepted as a correct name according to the International Code of Botanical Nomenclature.

As to *Paecilomyces hepiali* (nom. inval.), it was proved to be a distinct species outside *Cordyceps sinensis* and its anamorph. This species name remains to be validly published in the future.

KEYWORDS: RAPD, ITS region, *Synnematium sinense*, *Hirsutella sinensis*, anamorph, *Paecilomyces hepiali*,

---

该项目得到中国科学院微生物研究所创新经费的资助（编号 0110）
作者联系邮址：weijc2004@126.com
收原稿日期：2005-10-20，收修改稿日期：2005-12-08

本文原载于《菌物学报》, 25（2）：192-202，2006.

1 前言

冬虫夏草是我国的珍贵中药材之一。所谓冬虫夏草，并非虫与草，而是虫与菌，即生长在我国高寒山地特有的昆虫，蝙蝠蛾 (*Hepialus* spp.) 幼虫躯体上的一种子囊真菌。这种真菌属于真菌界（*Fungi*）的子囊菌门 (Ascomycota)、子囊菌纲 (Ascomycetes)、粪壳菌亚纲（Sordariomycetidae)、肉座菌目 (Hypocreales)、麦角菌科 (Clavicipitaceae)、虫草菌属 (*Cordyceps* (Frey) Link.) 之一种，其拉丁学名为 *Cordyceps sinensis* (Berkeley) Saccardo。

冬虫夏草具有独特的医疗保健作用，而其药材价值也不断提升，因而，引起有关药用真菌资源研发科学工作者的日益关注。多年来，为了对冬虫夏草菌进行人工培植和液体深层培养，科学家们从冬虫夏草上分离培养出多种丝状真菌菌种；其中有的已被国家有关部门批准作为中药一类新药投入生产。同为国家中药一类新药的“金水宝胶囊”(1987)和“百令胶囊”(1988）的包装盒和说明书中均标明其成分为“发酵虫草菌粉”。其中前者的菌种为蝙蝠蛾拟青霉 *Paecilomyces hepiali* W.H. Chen et R.Q. Dai（戴如琴等，1989）是分离自云南迪庆藏族自治州白马雪山产的冬虫夏草。中国医学科学院药物研究所从青海省化隆县采集的冬虫夏草上也分离获得了该菌，菌种编号为 Cs-4，作为金水宝胶囊的生产菌在江西国药厂投入生产，其培养温度为 25-35°C，培养周期为 3-7 天。后者的菌种是青海畜牧兽医科学研究院沈南英等在 1983 年从青海产的新鲜冬虫夏草子囊孢子、虫体组织、子座和虫体外菌丝中分离出的一种相同的真菌。其中有的菌落上多次形成了类似于天然冬虫夏草的子座，从而证明该分离物正是冬虫夏草菌。该菌在中国专利(China Patent)No.85101971 (1985) 中起用的名称为冬虫夏草头孢 *Cephalosporium dongchongxiacao* N.Y. Shen et al. 原词尾误为"-cae")，菌种号为 Cs-C-Q80，培养温度为 15-18°C，培养周期为 40-45 天。

不过，无论是在中国专利（China Patent）No.85101971 (1985) 中或食用菌刊物中（梁佩琼和陆大京，1990）所出现的冬虫夏草头孢的拉丁学名，均因无拉丁文描述或拉丁文特征提要而不符合国际植物命名法规要求，系不合格发表（ICBN-Art.41.3)；何况头孢属的拉丁学名“*Cephalosporium* Corda ” (1839), 已被归入顶孢属 “*Acremonium* Link ” (1809) 而成为后者的同物异名 (Gams, 1971; Kirk et al., 2001)。

蝙蝠蛾多毛孢 *Hirsutella hepiali* C.T. Chen et N.Y. Shen 曾被作为冬虫夏草头孢的替代名称而向专利局做了补报（1985)，属于未发表的无效名称。后来，该名称虽在食用菌刊物中发表（张显耻和何道珍，1995)，但是，同样由于无拉丁文描述或拉丁文特征提要而系不合格发表（ICBN- Art.41.3)。

陈庆涛等（1984）从四川康定产冬虫夏草上分离出一种被命名为“中国拟青霉” *Paecilomyces sinensis* C.T. Chen 的真菌，并认为它可能是冬虫夏草的无性型。

刘锡进等将分离自四川康定产冬虫夏草菌核及子实体上的，形态类似于冬虫夏草头孢的菌种作为冬虫夏草菌的无性型进行了描述，并命名为中国被毛孢 *Hirsutella sinensis* X.J. Liu et al.）予以正式发表（刘锡进等，1989)。

中华束丝孢 *Synnematium sinense* X.C. Yin et N.Y. Shen（1990）作为冬虫夏草头孢和蝙蝠蛾多毛孢两个名称的替代者而以新种予以正式发表。在发表时虽然配有拉丁文描述，但是，由于未指明模式标本是否处于非代谢状态及其存放地而成为不合格发表（ICBN-Art.8.2 & 9.14)。此外，束丝孢属的拉丁学名“*Synnematium* Speare” (1920) 已被证实系被毛孢属

"*Hirsutella* Pat." (1892) 的晚出同物异名而被归入后者（Evans & Samson, 1982; Kirk et al., 2001）。

冬虫夏草菌无性型菌种的科学名称方面存在的混乱现象引起了我国系统生物学家的关注。根据沈南英等（1983）及刘锡进等（1989）的工作，梁宗琦（1994）首先承认了中国被毛孢为冬虫夏草菌的无性型。近年来，通过个体发育生物学（莫明和等，2001）和分子系统生物学（赵锦等，1999；刘作易，1999；李增智等，2000；王宁等，2000；黄勃，2001；Chen et al., 2001; Liu et al., 2001；Chen et al., 2002; 章卫民等，2002）等实验数据及其比较分析，进一步确认了中国被毛孢是冬虫夏草菌的无性型。而通过分子系统学的分析结果还表明，中国拟青霉并非冬虫夏草菌的无性型（赵锦等，1999；李增智等，2001；黄勃，2001；Chen et al., 2002）。

由于以蝙蝠蛾拟青霉为生产菌种的金水宝胶囊和以中华束丝孢（=冬虫夏草头孢=蝙蝠蛾多毛孢）为生产菌种的百令胶囊各自的包装盒及说明书中均标明其成分为"发酵虫草菌粉"。因此，查明这两种真菌之间的亲缘关系实属必要。虽然，上述有关菌种名称为不合格发表 (ICBN-Art.8.2 & 41.3)，但是，迄今为止，它们与冬虫夏草菌之间的亲缘关系尚缺乏分子系统学的实验证据。因此，本文是针对蝙蝠蛾拟青霉和中华束丝孢（=冬虫夏草头孢=蝙蝠蛾多毛孢）与冬虫夏草菌及其无性型菌种之间的亲缘关系所进行的分子系统学分析。

关于至灵胶囊、宁心宝胶囊和心肝宝胶囊的生产菌与冬虫夏草菌之间亲缘关系的研究虽然未被纳入本实验过程，但是，根据国家中药品种保护及国家食品药品监督管理部门的有关资料，至灵胶囊的生产菌是被孢霉属 (*Mortierella* sp.)的一种真菌，属于接合菌门，而冬虫夏草菌则属于子囊菌门，二者亲缘关系相距甚远，因而，与冬虫夏草无关。宁心宝胶囊的生产菌虫草头孢 *Cephalosporium sinense* C.T. Chen (nom. nud.)与冬虫夏草之间的关系，根据其发酵生产周期为 3-5 天以及温度为 25-35°C 便可推知，它并非冬虫夏草菌。至于心肝宝胶囊的生产菌粉红胶霉 *Gliocladium roseum* (Link) Bain，则是一种真菌寄生菌。它所在的胶霉属 *Gliocladium* Corda 是肉座菌科 (Hypocreaceae) 小灿球赤壳菌属 *Spahaerostibella* 的无性型。因此，无论是至灵胶囊、宁心宝胶囊或心肝宝胶囊的生产菌似均与冬虫夏草菌无关。不过，由于这些菌种并未被纳入本实验过程，因而在本文中将不做进一步讨论。

至于分离自冬虫夏草而又未涉及冬虫夏草产品的其他丝状真菌（Jiang & Yao, 2002; 蒋毅和姚一建，2003）则未被包括在本文范围之内。

2 材料、方法与结果

通过 DNA 随机扩增多型性 (RAPD) 分析以探明中华束丝孢，中国被毛孢与冬虫夏草菌之间亲缘关系的研究。

2.1 材料

供试材料的编号、名称、产地及提供者见表 1。

2.2 方法

总 DNA 提取：采用改良 CTAB 法（Rogers & Bendich, 1988）。

随机引物的选用：从华美公司试剂盒中选择了 5 个随机引物：S23（AGTCAGCCAC），S33（CAGAACCCAC），S61（TTCGAGCCAG），OPV-14（AGATCCCGCC）和 OPW-06（AGGCCCGATG）。

表 1 供试材料编号、名称、产地及提供者

Table1. Examined materials number, name, locality, and supplier

编 号 No.	菌株号 Isolate no.	名 称 Name	产地（提供者） Locality (Supplier)
内群 (Ingroup)			
无性型 (Anamorph)			
*S5=Ph	无	蝙蝠蛾拟青霉 *Paecilomyces hepiali* W.H. Chen et R.Q. Dai, 1989	青海(Qinghai) (沈南英)
G2=Hs-S	Guo-2	中国被毛孢 *Hirsutella sinensis* X.J. Liu et al.，1989	四川 (Sichuan)（郭英兰）
S1=Ss-Q	0207	中华束丝孢 *Synnematium sinense* X.C. Yin & N.Y. Shen，1990	青海 (Qinghai)（沈南英）
S2=Ss-Q	上 020	中华束丝孢 *Synnematium sinense* X.C. Yin & N.Y. Shen	青海 (Qinghai)（沈南英）
S3=Ss	KD0205 F9	中华束丝孢 *Synnematium sinense* X.C. Yin & N.Y. Shen	青海 (Qinghai)（沈 沧）
有性型 (Teleomorph)			
S4=Cs-Q	无	冬虫夏草 *Cordyceps sinensis* (Berkeley) Saccardo	青海 (Qinghai)（沈南英）
G3=Cs-S	无	冬虫夏草 *Cordyceps sinensis* (Berkeley) Saccardo	四川 (Sichuan)（郭英兰）
外群 (Outgroup)			
G4=Pm-Guo	无	蛹草拟青霉 *Paecilomyces militaris* Z.Q. Liang	吉林 (Jilin) （郭英兰）

注：*标记的样品未用于 RAPD 扩增和分析. (The sample marked with * was not included in RAPD analysis.)

RAPD 反应体系(25μL)含有：25mmol/L $MgCl_2$ 1.5μL，2.5mmol/L dNTP 1.5μL，10μmol/L 引物 2μL，10×Buffer 2.5μL，3U/ μL Taq 酶 0.5μL，模板（稀释 50 倍）0.5μL，无菌去离子水 16.5μL。扩增反应：95℃预变性 5min 一次；94℃1min，36℃1min，72℃2min，共 40 个循环；最后 72℃2min。扩增产物在 1.5%琼脂糖凝胶中电泳，80V，2h。

数据分析：RAPD 扩增谱带用 BandScan 软件处理，将每条谱带视为一个性状，有谱带的赋值为 1，该位置无带的赋值为 0；利用 PAUP*4.08b (Swofford, 2002) 对其进行分析（Distance method；bootstrap=1000；UPGMA），最终得到反映菌株之间系统关系的树型图。

2.3 结果

中华束丝孢，中国被毛孢与冬虫夏草菌之间亲缘关系的 RAPD 分析（对 5 个 RAPD-PCR 随机引物所扩增）结果图谱及其树型图分别见图 1~6。

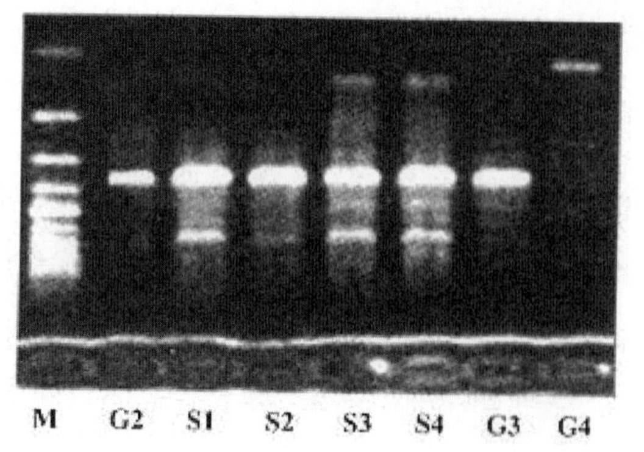

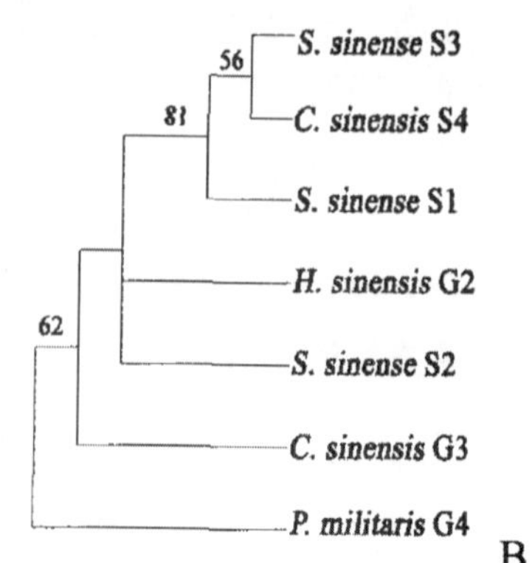

图 1 引物 S33 RAPD 分析结果

A-电泳图谱（左）；B-树型图（bootstrap=1000），节间数字代表 bootstrap 值(50%以下未显示)（右）

Fig.1 The RAPD result of primer S33: A-electrophoresis photo (left); B-tree (bootstrap=1000), the numbers in each node represents bootstrap support value, and the numbers lower than 50 were not shown (right)

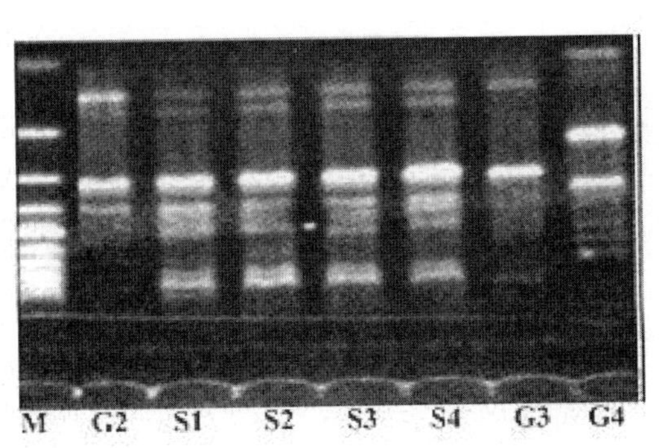

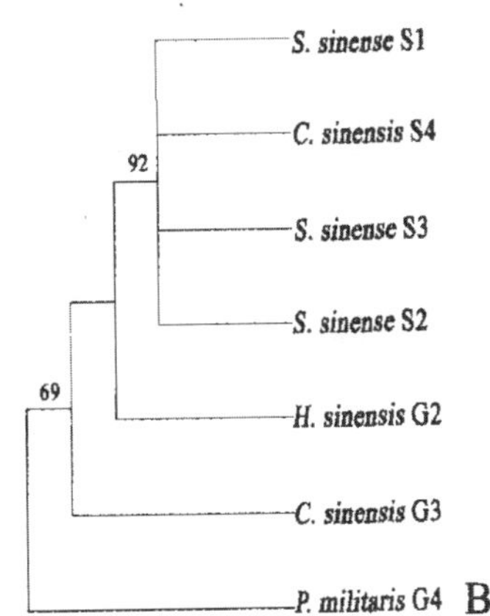

图 2 引物 S23 RAPD 分析结果

A-电泳图谱（左）；B-树型图（bootstrap=1000），节间数字代表 bootstrap 值(50%以下未显示)（右）

Fig.2 The RAPD result of primer S23: A-electrophoresis photo (left); B-tree (bootstrap=1000), the numbers in each node represents bootstrap support value, and the numbers lower than 50 were not shown (right)

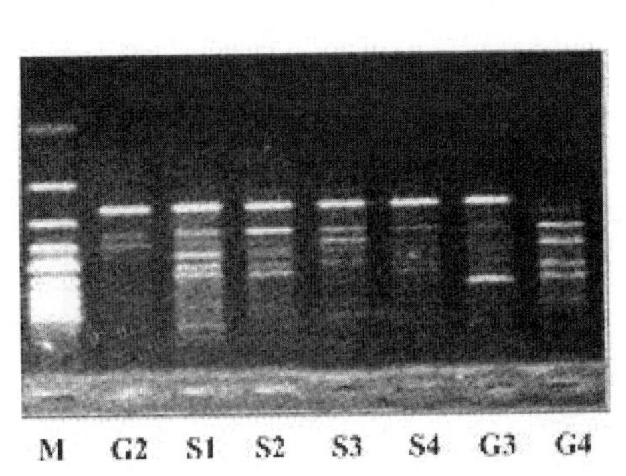

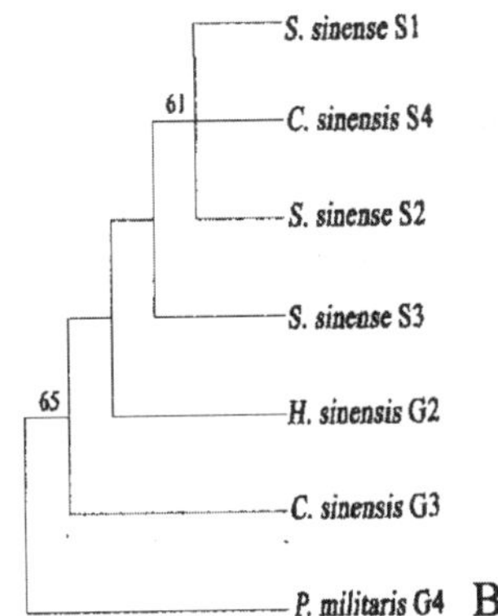

图 3 引物 S61 RAPD 分析结果

A. 电泳图谱（左）；B. 树型图（bootstrap=1000），节间数字代表 bootstrap 值(50%以下未显示)（右）

Fig.3 The RAPD result of primer S61: A-electrophoresis photo (left); B-tree (bootstrap=1000), the numbers in each node represents bootstrap support value, and the numbers lower than 50 were not shown (right)

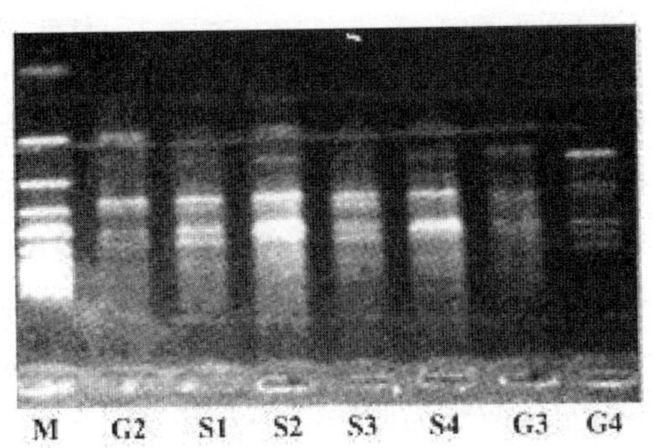

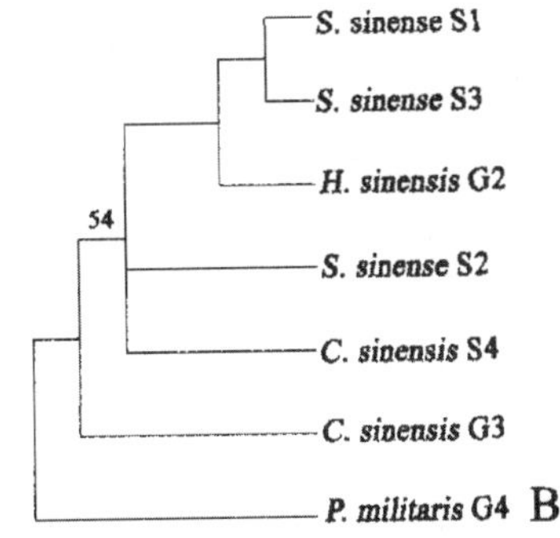

图 4 引物 OPV-14 RAPD 分析结果

A. 电泳图谱（左）；B-树型图（bootstrap=1000），节间数字代表 bootstrap 值(50%以下未显示)（右）

Fig.4 The RAPD result of primer OPV-14: A-electrophoresis photo (left); B-tree (bootstrap=1000), the numbers in each node represents bootstrap support value, and the numbers lower than 50 were not shown (right)

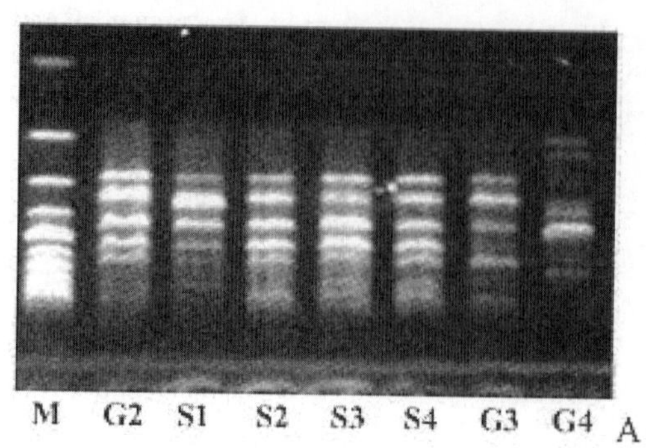

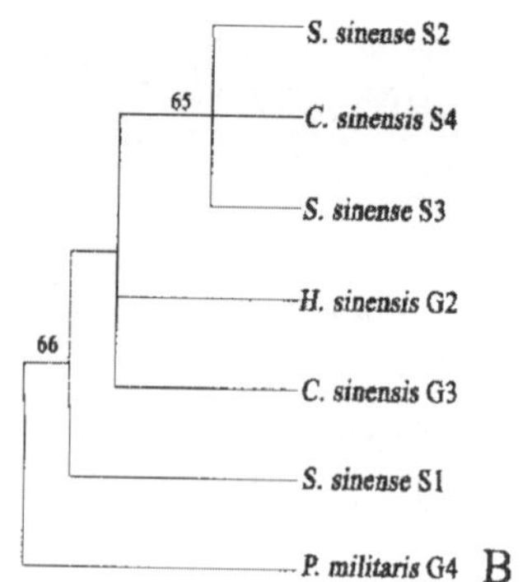

图 5 引物 OPW-06 RAPD 分析结果

A. 电泳图谱（左）；B. 树型图（bootstrap=1000），节间数字代表 bootstrap 值(50%以下未显示)（右）

Fig.5 The RAPD result of primer OPW-06: A-electrophoresis photo (left); B-tree (bootstrap=1000), the numbers in each node represents bootstrap support value, and the numbers lower than 50 were not shown (right)

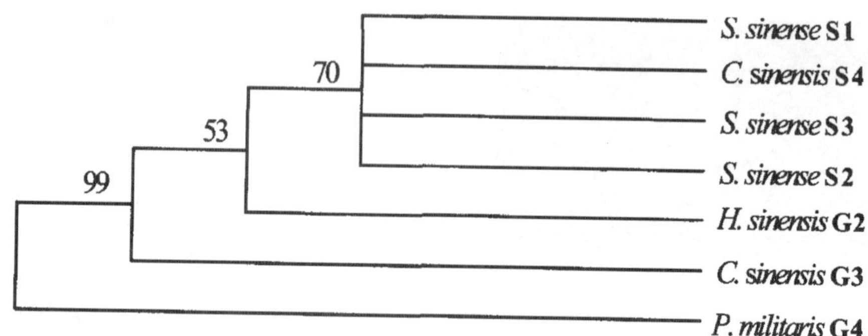

图6 利用PAUP分析5条引物的RAPD图谱所得的树型图 (Distance method; bootstrap=1000; UPGMA)，节间数字代表 bootstrap 值 (50%以下未显示).

Fig.6 The tree based on analyses of six RAPD primers' electrophoresis results (Distance method; bootstrap=1000; UPGMA), the numbers in each node represents bootstrap support value, and the numbers lower than 50 were not shown

上图 1~5 电泳图及其树型图图 6 表明，中国被毛孢、中华束丝孢和冬虫夏草菌各菌株一起形成了一个独立的类群。虽然由于仅一条随机引物所扩增的产物多型性不够丰富而导致树型图中的 bootstrap 值不高，但是，该类群与作为外群的蛹草拟青霉（G4）明显分开。图 6 中 bootstrap 值为 70%的三株中华束丝孢及一株冬虫夏草菌所形成的类群 1 和 bootstrap 值为 53%的中国被毛孢独自为一类群 2；而类群 1 和 2 与 G3 为编号的冬虫夏草菌独自作为类群 3 共同组成的类群，其 bootstrap 值为 99%。这一结果既表明冬虫夏草菌 *C. sinensis*，中国被毛孢 *H. sinensis* 和中华束丝孢 *S. sinense* 的各菌株为同一物种，又表明其种内的遗传多样性。

3 核糖体 RNA 基因 (nr DNA) 间隔区(ITS) 碱基序列分析

通过 nrDNA ITS 区(ITS1+5.8S+ITS2)碱基序列的系统生物学分析以探明蝙蝠蛾拟青霉与冬虫夏草菌有性型、无性型各菌株之间亲缘关系的研究。

3.1 材料

供试材料的编号，名称、产地及提供者见表 1 及 2。

表 2 来自 GenBank 的 nrDNA-ITS 区碱基序列

Table2. The nrDNA-ITS region sequences from GenBank

名 称 (Name)	GenBank 编号 (GenBank accession number)
冬虫夏草 *Cordyceps sinensis* (Berk.) Saccardo	AJ488272
冬虫夏草 *Cordyceps sinensis* (Berk.) Saccardo	AJ309357
中国被毛孢 *Hirsutella sinensis* Liu et al.	AJ309353
蛹草拟青霉 *Paecilomyces militaris* Liang	AY725790

3.2 方法

总 DNA 提取：同上。

PCR 扩增、测序及分析：通过 PCR 从供试菌株总 DNA 中进行核 rDNA ITS 区 (ITS1-5.8S-ITS2) 碱基序列的扩增；引物选用 ITS4 和 ITS5 (White et al., 1990)；PCR 产物由上海博亚和基康公司测序。测序结果利用软件 DNAMAN 4.0 进行比对；利用 Mega 2 进行系统学分析（Kumar et al., 2001）；采用 Kimura 2-parameter 模式，存在缺失时成对删除；构建系统关系树采用 NJ 法，重复运算 1000 次（bootstrap=1000）。

3.3 结果

利用 Mega2 对得到的蝙蝠蛾拟青霉，冬虫夏草有性型和无性型各菌株的 nrDNA ITS 区碱基序列数据进行分析，利用 Kimura 2-parameter 模式所得的树型图见图 7。包括 4 个来自 GenBank 数据在内的 12 个样品的 rDNA ITS 区碱基序列比对结果表明，其中 9 个样品，包括 3 株冬虫夏草菌（G3 及 L95D05，AJ488272），4 株中华束丝孢（S1, S2, S3, S4）和 2 株中国被毛孢（G2，L95DRZ-7）的碱基序列基本一致，属于同一物种。作为外群的 2 株蛹草拟青霉（G4, AY725790) 的碱基序列则明显地有别于上述 9 株菌。至于蝙蝠蛾拟青霉 (S5)的碱基序列也和外群蛹草拟青霉一样，明显地有别于上述 9 株菌，但又因为存在 30 多个碱基突变位点而与蛹草拟青霉分开。

以上述序列数据为基础而形成的树型图（图 7）表明，由中国被毛孢 (G2) 和中华束丝孢（S3）聚成的小类群连同由冬虫夏草菌 (AJ488272) 和中华束丝孢（S1, S2, S4) 聚成的另一类群共同组成了一大类群，其 bootstrap 值为 97%；而由冬虫夏草菌 (G3, L95D05) 和中国被毛孢（L95DRZ-7）聚成的另一类群，其 bootstrap 值为 98%。这两大类群所形成的单系类群的 bootstrap 值则为 100%。这一结果令人信服地表明，作为冬虫夏草菌有性型和无性型的 9 株菌同属一个物种。作为外群蛹草拟青霉姐妹群的蝙蝠蛾拟青霉与外群一起则处于冬虫夏草菌的姐妹群地位。

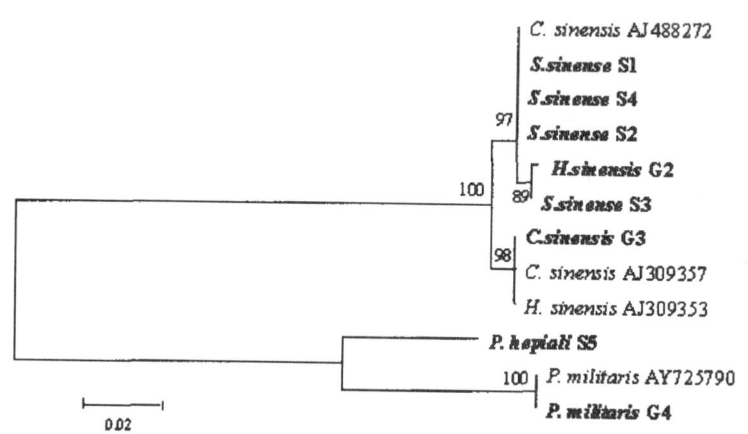

图 7 基于 rDNA ITS 区碱基序列所得的 NJ 树型图

Nucleotide: Kimura 2-parameter, pairwise deletion, bootstrap=1000; 节间数字代表 bootstrap 支持值 (50%以下未显示); bar 代表遗传距离 (黑体标记的样品序列为本文作者所测，白体标记的样品序列来自 GenBank)

Fig. 7 The NJ tree based on nrDNA ITS region sequences.

Nucleotide: Kimura 2-parameter, pairwise deletion, bootstrap=1000. The numbers in each node represents bootstrap support value, and the numbers lower than 50 were not shown. Genetic distance scale=0.02. The samples marked with boldface were examined by the authors, and others obtained from GenBank

4 讨论

由于中华束丝孢 *Synnematium sinense* X.C. Yin et N.Y. Shen (1990)，冬虫夏草头孢 (*Cephalosporium dongchongxiacaonis* N.Y. Shen et al. (中国专利，1985；梁佩琼和陆大京，1990) 及蝙蝠蛾多毛孢 *Hirsutella hepiali* C.T. Chen et N.Y. Shen (张显耻和何道珍，1995）是基于同一菌种的不同名称，因此，本实验所用青海化隆的菌种名称为中华束丝孢（=冬虫夏草头孢=蝙蝠蛾多毛孢)。通过其 DNA 随机多态型（RAPD）及 nrDNA 间隔区（ITS）碱基序列的分子系统学分析结果表明，中华束丝孢及中国被毛孢和冬虫夏草 *Cordyceps sinensis* (Berkeley) Saccardo 同属一个物种；换言之，前三个名称以及中国被毛孢所代表的菌种，都是冬虫夏草菌的无性型。符合国际植物命名法规的无性型名称则为中国被毛孢，其拉丁学名为 *Hirsutella sinensis* X.J. Liu et al.（1989)。

虽然在冬虫夏草菌的无性型名称使用上由于受国际植物命名法规的约束而有所取舍，但是，冬虫夏草菌无型性的发现、命名、描述和发表终究是由我国科学家在对不同产地的冬虫夏草菌的研究中分别实现的。

实验结果还表明，蝙蝠蛾拟青霉 *Paecilomyces hepiali* W.H. Chen et R.Q. Dai Cs-4 并非冬虫夏草菌的无性型，而是不同于冬虫夏草菌无性型的另一种真菌。此外，由于作者在发表新种时的模式标本为活的培养物而成为不合格发表（ICBN- Art 8.2)。

基于上述实验结果，作为研究对象的内群中的供试菌种（表 1)，实际上分别隶属于两种：(1) 冬虫夏草菌及其无性型；(2) 蝙蝠蛾拟青霉，其有性型迄今尚不明确。

现将本实验内群中的冬虫夏草菌及其无性型和蝙蝠蛾拟青霉以及作为外群的蛹草拟青霉共三种真菌的全型、有性型及无性型名称的现状分列于下：

内群 (ingroup):
1 冬虫夏草 *Cordyceps sinensis* (Berkeley) Saccardo
全型名称 (nomen holomorphum):
冬虫夏草
Cordyceps sinensis (Berkeley) Saccardo, *Michelia* 1: 320 (1878).
≡*Sphaeria sinensis* Berkeley in Hooker, *Lond. Journ. Bot.* 2: 207 (1843).
有性型名称 (nomen teleomorphum):
冬虫夏草
Cordyceps sinensis (Berkeley) Saccardo (1878).
无性型名称 (nomen anamorphum):
中国被毛孢
Hirsutella sinensis X.J. Liu, Y.L. Guo, Y.X. Yu et W. Zeng, *Acta Mycologica Sinica* 8:37 (1989).
同种异名：冬虫夏草头孢
Syn.: *Cephalosporium dongchongxiacao*（原文词尾误为-cae）见中国专利(China Patent), No.8510197 (1985), nom.nud. (ICBN Art.41.3); [in Liang & Lu, *Edible Fungi* 3：2-3 (1990)]见梁佩琼和陆大京，食用菌 3：2-3 (1990), nom.nud. (ICBN Art.41.3)。
同种异名：中华束丝孢
Syn.: *Synnematium sinense* X.C. Yin et N.Y. Shen, *Acta Biologica Plateau Sinica* 9: 1 (1990), nom.nud. (ICBN Art.41.3).
同种异名：蝙蝠蛾多毛孢
Syn.: *Hirsutella hepiali* C.T. Chen et N.Y. Shen in Zhang & He, *Edible Fungi* 6：6 (1995) 见张显耻和何道珍，食用菌 6：6 (1995), nom. nud. (ICBN Art.41.3)。
2 蝙蝠蛾拟青霉：
全型名称不明 (nomen holomorphum ignotum)
有性型名称不明 (nomen teleomorphum ignotum):
无性型名称 (nomen anamorphum):
蝙蝠蛾拟青霉
Paecilomyces hepialid W.H. Chen et R.Q. Dai, *Acta Agricuturae Universitatis Pekinensis* 15 (2):223 (1989), nom. inval. (ICBN- Art 8.2).
外群：
3 蛹虫草(*Cordyceps militaris*):
全型名称(nomen holomorphum):
蛹虫草
Cordyceps militaris (L.) Link, Handb. III, p.347 (1833) (ut Cordiceps).
≡*Clavaria militaris* L., Sp. Pl. ed. 1. II:1182 (1753).
有性型名称 (nomen teleomorphum):
蛹虫草
Cordyceps militaris (L.) Link
无性型名称 (nomen anamorphum):

蛹草拟青霉

Paecilomyces militaris Z.Q. Liang, *Acta Edulis Fungi* 8 (4):29-30 (2001), nom. inval. (ICBN-Art.8.2 & 9.14).

致 谢 本文作者感谢青海沈沧先生赠送冬虫夏草菌种；诚挚感谢郭守玉博士在 DNA 数据分析软件使用中的帮助和建设性的意见。

[REFERENCES]

Chen QT, Xiao S R, Shi ZY, 1984. *Paecilomyces sinensis* sp. nov. and its connection with *Cordyceps sinensis*. *Acta Mycologica Sinica*, **3** (1): 24~28 (in Chinese)

Chen YQ, Wang N, Qu LH, Li TH, Zhang WM, 2001. Determination of the anamorph of *Cordyceps sinensis* inferred from the analysis of the ribosomal DNA internal transcribed spacers and 5.8S rDNA. *Biochemical Systematics and Ecology*, **29**: 597~607

Chen, YQ., Wang N., Zhou H, Qu LH, 2002. Differentiation of medicinal *Cordyceps* species by rDNA ITS sequence analysis. Planta Med **68**: 635~639

Dai RQ, Lan JL, Chen WH, Li XM, Chen QT, Shen CY, 1989. Research on *Paecilomyces hepiali* Chen et Dai, sp. nov.. *Acta Agriculturae Universitatis Pekinensis*. **15** (2): 221~225 (in Chinese)

Evans HC, Samson RA, 1982. Cordyceps species and their anamorphs pathogenic on ants in tropical forest ecosystems 1. The Cephalotes (Myrmicinae) complex. *Transactions of the British Mycological Society* (TBMS), **79**: 431~453

Gams W, 1971. Cephalosporium-artige Schimmelpilze. G. Fischer, Stuttgart: 1~262

Huang B, 2001. A Study of the molecular systematics of some important groups of entomogenous fungi. A dissertation of the requirements for a Ph. D. degree from University of Science and Technology of China. Hefei. 1~148 (in Chinese)

Jiang Y, Yao YJ, 2002. Names related to *Cordyceps sinensis* anamorph. *Mycotaxon* **84**: 245~254

Jiang Y, Yao YJ, 2003.Anamorphic fungi related to *Cordyceps sinensis*. *Mycosystema*, **22** (1): 161~176 (in Chinese)

Kirk PM, Cannon PF, David JC, Stalpers JA, 2001. Answorth & Bisby's Dictionary of the Fungi (Ninth Edition). CAB International. 1~655

Kumar S, Tamura K, Jakobsen IB, Nei M, 2001. MEGA2: Molecular evolutionary genetics analysis software. Arizona State University, Tempe, Arizona, USA.

Li ZZ, Huang B, Li CR, Fan MZ, 2000. Molecular evidence for anamorph determination of *Cordyceps sinensis* (Berk.) Sacc. *Mycosystema*, **19** (1): 60~64 (in Chinese)

Liang PQ, Lu DJ, 1990. New progress of biological studies on *Cordyceps sinensis*. *Edible Fungi*, **12** (3): 2~3 (in Chinese)

Liang ZQ, 1994. Cordyceps and its artificial culture. *Collection of Guizhou Agric Coll* (26): 1~21 (in Chinese)

Liu XJ, Guo YL, Yu YX, Zeng W, 1989. Isolation and identification of the anamorphic state of *Cordyceps sinensis* (Berk.) Sacc. *Acta Mycologica Sinica*, **8** (1): 35~40 (in Chinese)

Liu Z Y, 1999. Studies on relationship between *Cordyceps* spp. and their anamorphs. A dissertation of the requirements for a Ph. D. degree from Huazhong Agricultural University. Wuhan 1~179 (in Chinese)

Liu ZY, Yao YJ, Liang ZQ, Liu AY, Pegler DN, Chase MW, 2001. Molecular evidence for the anamorph-teleomorph connection in *Cordyceps sinensis*. *Mycological Research* **105** (7): 827~832

Mo MH, Chi SQ, Zhang KQ, 2001. Microcycle conidiation of *Cordyceps sinensis* and anamorph isolation. *Mycosystema*, **20** (4): 482~485 (in Chinese)

Rogers SO, Bendich AJ, 1988. Extraction of DNA from plant tissues. Plant Molecular Biology Manual A6:1-10. Kluwer Academic Publishers, Dordrecht

Shen NY,Zeng I, Zhang XC, 1983. Isolation of *Cordyceps sinensis*. *Edible Fungi*, (5): 1~3 (in Chinese)

Swofford DL, 2002. PAUP: phylogenetic analysis using parsimony (and other methods). Sinauer Associates, Sunderland, Mass.

Wang N, Chen YQ, Zhang WM, Li TH, Qu LH, 2000. Molecular evidences indicating multiple origins in the entomogenous *Cordyceps*. *Acta Scientiarum Naturalium Universitatis Sunyatseni*, **39** (4): 70~73 (in Chinese)

Werner G, Barrie FR, Burdet HM, Chaloner WG, Demoulin V, Hawksworth DL, Jørgensen PM, Nicolson DH, Silva PC, Trehane P, McNeill J, 1994. International Code of Botanical Nomenclature. Koeltz Scientific Books, D-61453 Königstein, Germany. 1~389

White TJ, Bruns T, Lee S, Taylor JW, 1990. Amplication and direct sequencing of fungal ribosomal RNA genes for phylogenetics. In: Innis MA, Gelfand DH, Sninsky JJ, White TJ (eds). PCR protocols: a guide to methods and applications. Academic Press, San Diego, California, 315~322

Yin XC, Shen NY, 1990. The conidial state of *Cordyceps sinensis* (Berk.) Sacc.-*Synnematium sinense* Yin et Shen, sp. nov.. Acta Biologica Plateau Sinica, 9: 1~5 (in Chinese)

Zhang XC, He DZ, 1995. Primary studies on ferment technics of liquid cultures of *Cordyceps sinensis*. *Edible Fungi*, **17** (6): 5~6 (in Chinese)

Zhang WM, Li TH, Chen YQ, Qu LH, Zhong H, Xu XP, 2002. Molecular study on anamorph of *Cordyceps sinensis* from Tibet. *Microbiology*, **29** (3): 54~57 (in Chinese)

Zhao J, Wang N, Chen YQ, Li TH, Qu LH, 1999. Molecular identification for the asexual stage of *Cordyceps sinensis*. *Acta Scientiarum Naturalium Universitatis Sunyatseni*, **38** (1): 121~123 (in Chinese)

[附中文参考文献]

陈庆涛, 肖生荣, 施至用, 1984. 中国拟青霉新种及其与虫草的关系 真菌学报, **3** (1): 24~28

戴如琴, 兰江丽, 陈伟华, 李晓明, 陈庆涛, 沈崇尧, 1989. 蝙蝠蛾拟青霉新种的研究 北京农业大学学报, **15** (2):221~225

黄勃, 2001.一些重要虫生真菌的分子系统学研究。中国科学技术大学博士学位论文. 合肥. 1~148

蒋毅, 姚一建, 2003. 冬虫夏草无性型研究概况. 菌物系统, **22** (1): 161~176

李增智, 黄勃, 李春如, 樊美珍, 2000. 确证冬虫夏草无性型的分子生物学证据. 菌物系统, **19** (1): 60~64

梁佩琼, 陆大京, 1990. 冬虫夏草生物学研究新进展. 食用菌, (3): 2~3

梁宗琦, 1994. 虫草及其人工培养. 贵州农学院丛刊 增刊: 1-21。

刘锡进, 郭英兰, 愈永信, 曾纬, 1989. 冬虫夏草菌无性阶段的分离和鉴定. 真菌学报, **8** (1): 35~40

刘作易, 1999. 虫草属及其无性型关系的研究. 华中农业大学博士学位论文. 武汉. 1~179

莫明和, 迟胜起, 张克勤, 2001. 冬虫夏草的微循环产孢及其无性型的分离. 菌物系统, **20** (4):482~485

沈南英，曾璐，张显耻 1983 冬虫夏草菌得分离 食用菌 **1983** (5): 1-3。

王宁, 陈月琴, 章卫民, 李泰辉, 屈良鹄, 2000. 虫草属多源起源的分子生物学证据. 中山大学学报 (自然科学版), **39** (4): 70~73

印象初, 沈南英, 1990. 冬虫夏草菌 *Cordyceps sinensis* (Berk.) Sacc.的无性世代–中华束丝孢 *Synnematium sinense* Yin et Shen sp.nov. 高原生物学集刊, 第九集: 1~5

张显耻, 何道珍, 1995. 中国冬虫夏草菌液体发酵工艺初探. 食用菌, (6): 5~6

章卫民, 李泰辉, 陈月琴, 屈良鹄, 钟韩, 徐学平, 2002.西藏冬虫夏草无性型的分子生物学研究. 微生物学通报, **29** (3): 54~57

赵锦, 王宁, 陈月琴, 李泰辉, 屈良鹄, 1999. 冬虫夏草无性型的分子鉴定. 中山大学学报 (自然科学版), **38** (1): 121~123

现代工业化培植的冬虫夏草物种鉴定与成分检测

魏江春[1*]，魏鑫丽[1]，郑维发[2]，郭威[1]，柳润东[1]
[1]中国科学院微生物研究所真菌学国家重点实验室，北京，100101
[2]江苏师范大学，江苏省药用植物生物技术重点实验室，徐州，221116

摘要：本文就东阳光药业集团公司工厂化培植的冬虫夏草有性型通过分子系统学分析进行物种鉴定以及对杭州中美华东制药有限公司工厂化培养的冬虫夏草无性型菌丝粉通过 HPLC 法和 ^{1}H NMR 指纹图谱法进行的检测予以报道。最后，对于现代工业化培植的冬虫夏草在人类健康与生物经济时代的意义予以展望。
关键词：冬虫夏草，工业化培植，物种鉴定，成分检测

Species identification and component detection of the fungus *Ophiocordyceps sinensis* cultivated by modern industry

WEI Jiang-Chun[1*], WEI Xin-Li[1], Zheng Wei-Fa[2], GUO Wei[1], LIU Run-Dong[1]
[1] *State Key Laboratory of Mycology, Institute of Microbiology, Chinese Academy of Sciences, Beijing 100101, China*
[2]*The Key Laboratory of Biotechnology for Medicinal Plant of Jiangsu Province，Jiangsu Normal University，Xuzhou,Jiangsu, China, 221116*

Abstract: This paper deals with the molecular systematic analysis of the sexual fungal species *Ophicordyceps sinensis* cultivated by Dongyangguang Pharmaceutical Co. Ltd and component detection of its asexual fungus cultivated by Hangzhou Zhongmeihuadong Pharmaceutical Co. Ltd. by HPLC. and ^{1}H-NMR （^{1}H-nuclear magnetic resonance） finger print. Finally, a short prospect of significance of the fungus *O. sinensis* cultivated by modern industry in human health and bio-economic era is given as well.
Key words: *Ophiocordyceps sinensis*, cultivated by industry, species identification, component detection.

冬虫夏草是生长在青藏高原海拔 3000 米以上草甸土壤草丛中的一种名贵药材。它主要产于中国青海、西藏、四川、云南、甘肃和贵州等喜马拉雅地区的高寒地带。冬虫夏草是我国民间应用历史较早的高级滋补保健珍品，被誉为中华医药瑰宝。“冬虫夏草”顾名思义系冬季为虫，夏季为草。实际上它既非动物界的虫，又非植物界的草，而系真菌的真菌。

在以形态解剖学为基础的旧分类系统中，冬虫夏草属于原麦角菌科（Clavicipitaceae, sensu lato）中广义的虫草菌属（*Cordyceps* sensu lato）；在以分子系统学分析为基础的新分类系统中，广义的虫草菌属被分归为三科， 即麦角菌科[Clavicipitaceae （Lindau） Earle ex Rogerson, *sensu stricto*）]，虫草菌科（Cordycipitaceae Kreisel 1969 ex G. H. Sung, J. M. Sung, Hywel-Jones & Spatafora），及线虫草菌科（Ophiocordycipitaceae *G. H. Sung*, *J. M.* Sung, Hywel-Jones & Spatafora）。冬虫夏草划归线虫草属（Ophiocordyceps Petch）。

冬虫夏草的虫菌关系在于真菌侵染蝙蝠蛾幼虫以后，将幼虫躯体逐步转变为寄生菌的菌体，最终致死幼虫，只残留其干涸的薄层虫皮。此种生物学现象类似于昆虫中常见的拟寄生（parasitoidism）（H. C. J.

* Correspongding author. Email:weijc2004@126.com

Godfray1994），即寄生物通过寄生最终导致寄主生物死亡的一种生物学现象；而寄生现象中的寄主生物最终不被寄生物所致死。这是拟寄生与寄生之间的重要区别。

因此，冬虫夏草是真菌拟寄生现象的典型代表，它并非共生现象中的任何一种。至于从冬虫夏草上分离得到的其它微生物，是生物多样性在自然界无处不在的反映。在任何一种生物体上，包括在冬虫夏草上，均可分离到多种不同的其他微生物。因此，在进行生物学实验中务必在严格的无菌条件下操作才能保证实验结果的正确无误。

迄今为止，人们概念中的冬虫夏草是从青藏高原草甸人工采挖的冬虫夏草。全国中药市场的冬虫夏草也均来自人工采挖（图 4）。曾据新华社西宁消息，每年春夏季节，地处中国「三江源」地区的青海玉树藏族自治州曲玛莱县的草场上，就汇聚大批人群挖掘虫草，导致当地的生态系统遭到破坏。据玉树州曲玛莱县农牧局官员称，因虫草采挖每年破坏的草皮面积就高达 10.8 万至 18 万平方米。而青海省社科院多年研究生态问题的马生林副研究员说，年年如此采挖虫草，会导致「三江源」的源头断流。因此，采挖冬虫夏草对于号称为“中国水塔”的三江源地区生态正在造成严重威胁。

世界著名未来学家阿尔文·托夫勒（A.·Toffler）在他 1980 年出版的“第三次浪潮”中预言，农业经济和工业经济将要进入信息经济和生物经济时代。实际上，当前已经是信息经济的成熟和繁荣阶段，而生物经济尚处于起步和成长阶段。面对全球正在成长起步阶段的生物经济时代以及我国冬虫夏草工厂化产业在科学技术支撑下的快速发展，以狂挖冬虫夏草，破坏生态为特征的冬虫夏草原始生产方式有待改变!

工厂化种植冬虫夏草是利用冬虫夏草无性型在现代化工厂无菌条件下对蝙蝠蛾幼虫进行人工接种后，培养出完全等同于野生冬虫夏草（图 4）的工业化冬虫夏草产品（图 1,2,3），年产量在 10 吨以上。

本文将对现代工厂化培植的冬虫夏草有性型通过分子系统学分析进行物种鉴定，对其无性型工业培养物通过 HPLC 法和 ^{1}H NMR 指纹图谱法进行成分分析和比较，为冬虫夏草无性型工业化培养产物质量控制奠定实验基础。

1. 材料与方法

1.1. 工业化培植的冬虫夏草有性型物种鉴定材料与方法

材料：用于内群分析的工厂化种植冬虫夏草（标记为“●”：子实体2，菌丝体3，）来自东阳光冬虫夏草产业基地（图1,2,3）；野生冬虫夏草（图4）以及冬虫夏草无性型菌种（中国被毛孢）（标记为“○”）来自中国科学院微生物研究所魏江春实验室；其余三种未标记的野生冬虫夏草（AB067749，EU570957，KC305892）来自GenBank。用于外群分析的蝙蝠蛾拟青霉[（*Paecilomyces hepiali*）及蛹虫草（标记为“○”）来自中国科学院微生物研究所魏实验室。其他样品（EF555097, AY725790, EU825996）来自GenBank（表1）。

图 1 东阳光药业有限公司人工培植冬虫夏草幼小阶段（李文佳）

Fig. 1 Young stage of artificial cultivation of *Ophiocordyceps sinensis* from Dongyangguang Pharmaceutical Co. Ltd （photo taken by Li W-J）

图 2 东阳光药业有限公司工厂化人工培植的冬虫夏草产品

Fig. 2 The Fungus *Ophiocordyceps sinensis* from Dongyangguang Pharmaceutical Co. Ltd （photo taken by JCW）

表 1　用于分子系统分析的样品信息

Table 1　Information of the samples for molecular systematic analyses

菌株号 No. of strain	样品名称 Name of samples	样品来源 Sources of sample
2 & 3	工厂培植冬虫夏草有性型（标记为“●”:子实体 2，菌核中菌丝体 3，图 3） *Ophiocordyceps sinensis* sexual （marked with“●”:fruiting body 2,mycelia in sclerotium 3）	东阳光药业有限公司（图 1,2,3） *Dongyangguang Pharmaceutical Co. Ltd*（Fig.1,2）
4	冬虫夏草无性型（标记为“●”） *O. sinensis* asexual isolate 130508-KD-2B（marked with“●”）	东阳光药业有限公司 Dongyangguang Pharmaceutical Co. Ltd
5	冬虫夏草无性型（标记为“●”） *O. sinensis* asexual isolate 20110514 （marked with“●”）	东阳光药业有限公司 Dongyangguang Pharmaceutical Co. Ltd
6	冬虫夏草无性型（标记为“●”） *O. sinensis* asexual isolate H01-20140924-03 （marked with“●”）	东阳光药业有限公司 Dongyangguang Pharmaceutical Co. Ltd
G3	野生冬虫夏草有性型（标记为“○”） *Cordyceps sinensis* ≡*Ophiocordyceps sinensis* sexual （marked with“○”）	魏实验室样品（图 3） Samples from J.C. Wei's Lab（Fig. 3）
	冬虫夏草有性型 *Cordyceps sinensis* ≡*Ophiocordyceps sinensis* sexual	基因序列资料库 AB067749 from GenBank
	冬虫夏草有性型 *Ophiocordyceps sinensis* sexual	基因序列资料库 EU570957 from GenBank
	冬虫夏草有性型 *Ophiocordyceps sinensis* sexual	基因序列资料库 KC305892 from GenBank
G2	冬虫夏草无性型（标记为“○”） *Hirsutella sinensis* = *O. sinensis* asexual （marked with“○”）	魏实验室样品 Samples from J.C. Wei's Lab
S5	蝙蝠蛾拟青霉（标记为“○”） *Paecilomyces hepiali* （2008）（marked with“○”）	魏实验室样品 Samples from J.C. Wei's Lab
	蝙蝠蛾拟青霉 *Paecilomyces hepiali*（2008）	基因序列资料库 EF555097 from GenBank
G4	蛹虫草（标记为“○”） *Cordyceps militaris*（marked with“○”）	魏实验室样品 Samples from J.C. Wei's Lab
	蛹虫草 *Cordyceps militaris*	基因序列资料库 AY725790 from GenBank
	蛹虫草 *Cordyceps militaris*	基因序列资料库 EU825996 from GenBank

图 3　东阳光药业有限公司工厂化人工培植的冬虫夏草产品菌核中白色菌丝体

Fig. 3　White fungal mycelia in the sclerotium of *Ophiocordyceps sinensis* （photo taken by WJC）

图 4　从青藏高原采挖的野生冬虫夏草

Fig. 4　The Fugus *Ophiocordyceps sinensis* collected　from the field of Qing-Zang Plateau （photo taken by WJC

方法：利用 Mega5 对所获得的 ITS 序列进行分析，采用 K2+G 模式以获得 ML 树形图

总 DNA 提取：采用改良 CTAB 法（Rogers & Bendich, 1988）。核 rDNA ITS 扩增及系统学分析：引物选用 ITS4 和 ITS5 （White *et al.*, 1990），PCR 产物由铂尚公司测序。测序结果利用软件 DNAMAN 4.0 进行比对，利用 Mega 5 进行系统学分析（Tamura *et al.*, 2011），采用 K2+G 模式，构建系统关系树采用 ML 法，重复运算 1000 次（bootstrap=1000）

1.2　工业化液体培养的冬虫夏草无性型有关成分分析

冬虫夏草无性型菌体分别为：百令胶囊（杭州中美华东制药有限公司提供，图 5）、冬虫夏草无性型 7 天培养菌丝体（中国科学院微生物研究所姚一建实验室提供）、冬虫夏草无性型菌丝体提取物（杭州中美华东制药有限公司提供）。以野生冬虫夏草（中国科学院微生物研究所魏实验室提供）为参考指标，分别检测腺苷、肌苷、甘露醇以及多糖含量。

图 5　杭州中美华东制药有限公司冬虫夏草无性型菌体百令胶囊产品
（照片由王松和王瑞提供）

Fig. 5　The capsules of anamorphic *Ophiocordyceps sinensis* from
Hangzhou Huadong Zhongmeihuadong Pharmaceutical Co. Ltd
（photo taken by Wang S. & Wang R.）

样品常温干燥后经初步粉碎，再经液氮研磨。以 20%乙醇水溶液超声 10 分钟提取核苷类化合物。提取核苷后的粉末以 100°C 热水提取 3 小时，减压浓缩后冻干，提取多糖。肌苷、腺苷以 HPLC 法检测（标准品自 Sigma 公司订购），甘露醇以检测试剂盒测定（法国-伯桑特，北京泰乐祺科技有限公司经销）；多糖以硫酸蒽酮法测定（20%乙醇提取物与热水提取物之和）。

1.3　工业化液体培养的冬虫夏草无性型 1H NMR 指纹图谱测定

取 50 mg 野生冬虫夏草或其冬虫夏草培养物的提取物溶于 600 μl 的 DMSO-d6（550 μl）和 100 mM 磷酸钠（50 μl）混合溶液（pH7.4）。样品完全溶解后转移到一核磁共振管中，以核磁共振波谱仪（400 MHz，Bruker BioSpin, Billarica, USA）按 Zheng et al. （2009）所述的方法测定样品的 ^{1}H NMR 图谱。质子信号的大致归属按照 Zheng et al.（2011）所述方法进行。

2. 结果与讨论

2.1　工业化培植的冬虫夏草有性型物种鉴定结果与讨论

分析结果见图 6 系统树。

内群分枝 1（右上）：东阳光人工培养冬虫夏草的子实体（fruiting body 2）菌丝和菌核中菌丝体（mycelia 3）与 GenBank 中序列号为 AB067749 的冬虫夏草（*Cordyceps sinensis* AB067749）三者完全等同；该三者与 GenBank 中序列号为 KC305892 的冬虫夏草（*Ophiocordyceps sinensis*KC305892）具微弱差异，自展值

为 90%；上述四个样品均取自冬虫夏草菌的全型，它们之间完全等同，其自展值为 100%。

内群分枝 2（左中）：东阳光用于人工培养冬虫夏草菌的三株无性型菌株 H01-20140924-03，130508-KD-2B 和 20110514 和与魏实验室根据可靠材料自测的冬虫夏草菌（*Cordyceps sinensis* G3），GenBank 中序列号为 EU570957 的冬虫夏草菌（*Ophiocordyceps sinensis*）完全相同，其自展值为 100%；与魏实验室根据可靠材料自测的冬虫夏草菌无性型中国被毛孢（*Hirsutella sinensis* G2）相差较小，其自展值为 89%。

系统树结果表明，所有内群（in-group）（即分枝 1+分枝 2）中的样品均为冬虫夏草菌，其自展值为 100%。至于内群中个别样品之间的差异系冬虫夏草菌种内个体之间的差异所致。

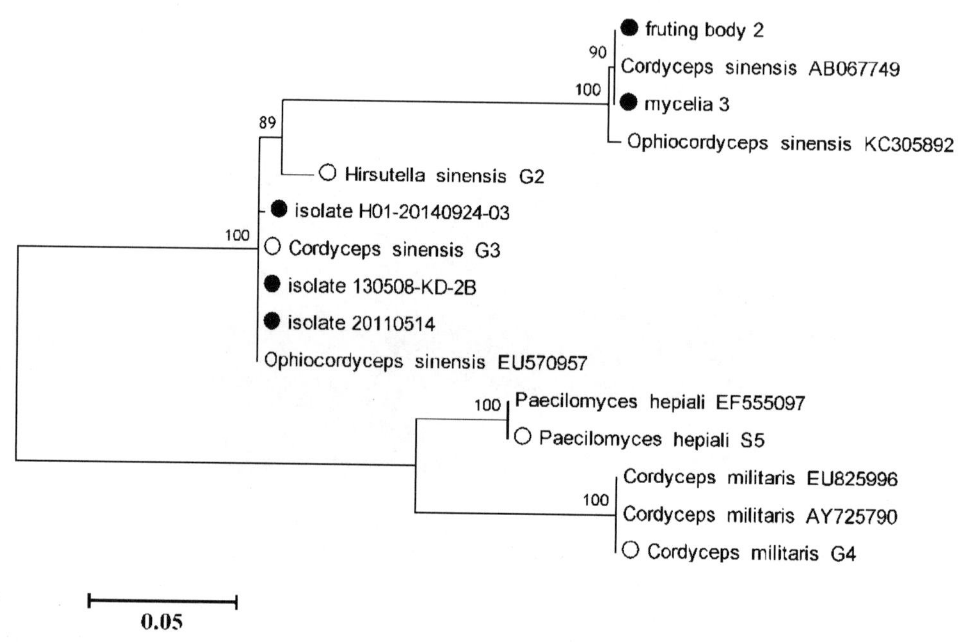

图 6 基于核 rDNA ITS 序列所得的 ML 系统树：内群（右上+左中），外群（右下）。节间数字代表自展值或支持值（bootstrap）.

Fig. 6 The ML tree based on nrDNA ITS sequences, bootstrap =1000. The numbers in each node represents bootstrap support value, and the numbers lower than 50 were not shown. In-group is on the left middle and right above, and out-group on the right below.

考虑到不断地从冬虫夏草原产地取回土壤及蓼科（Polygonaceae）植物根茎，可能会导致原产地生态破坏以及能否持续供应土壤及蓼科植物根茎的问题。因此，建议一方面对于冬虫夏草原产地土壤进行仔细而全面的分析研究，以便将当地土壤改造为适合于蝙蝠蛾幼虫完成其生活史和栽培冬虫夏草菌的土壤。另方面，在蝙蝠蛾幼虫饲养中，可以通过植物组织培养对蝙蝠蛾幼虫喜食的蓼科植物，如头花蓼（*Polygonum capitatum* Buch.-Ham. ex D. Don）和珠芽蓼（*Polygonum viviparum* L.） 植物进行工厂化人工栽培方法的研究；此外，还可进行蓼科植物替代品，如胡萝卜等，作为蝙蝠蛾幼虫的食物（沈南英等，1983）。

2.2 工业化液体培养的冬虫夏草无性型有关成分比较

检测结果表明，冬虫夏草菌丝粉相关产品中的肌苷的含量均远远高于野生冬虫夏草，腺苷的含量均低于野生冬虫夏草。而甘露醇和多糖的含量与野生冬虫夏草相近。七天短期培养的菌丝粉未能检出肌苷含量（表 2）。

表 2 野生冬虫夏草与其培养物的有效成分分析（n=3）

Table 2 Analysis of active constituents from field-grown Ophiocordyceps sinenesis and its cultured mycelia（n=3）

Test items 检测项目	Wild *Ophiocordyceps sinensis* 野生冬虫夏草	Mycelia grown for 7 days 7 天培养菌丝粉	Mycelia extracts 菌丝粉提取物	Mycelia from Corbrin Capsules 百令胶囊菌丝粉
Adenosine 腺苷	359.8 ±14.5μg/g sclerotia 359.8±14.5μg/g 菌核	640 ±15.9μg/g Mycelia 640 ±15.9μg/g 菌丝体	4.4 ±0.23 mg/g extracts 4.4±0.23g/g 提取物	939.2 ±31.2 μg/g Mycelia 939.2 ±31.2μg/g 菌丝粉
Inosine 肌苷	1123.36±55.2μg/g sclerotia 1123.36±55.2μg/g 菌核	Not detected 未检测出	2.39±0.13 μg/g extracts 2.39 ±0.13μg/g 提取物	84.71±9.11 μg/g Mycelia 84.71±9.11 μg/g 菌丝粉
Mannitol 甘露醇	120.34 ±3.15mg/g sclerotia 120.34±3.15 mg/g 菌核	181.76 ±3.19mg/g Mycelia 181.76±3.19 mg/g 菌丝体	270 ±7.88mg/g extracts 270 ±7.88mg/g 提取物	165.19±15.8 mg/g mycelia 165.19 ±15.8mg/g 菌丝粉
Polysaccharides 多糖	90.73 ±5.47mg/g sclerotia 90.73±5.47 mg/g 菌核	114.55 ±8.97mg/g mycelia 114.55 ±8.97mg/g 菌丝体	168.55±12.4 mg/g extracts 168.55 ±12.4mg/g 提取物	80.087±6.36 mg/g Mycelia 80.087 ±6.36mg/g 菌丝粉

2.3 工业化液体培养的冬虫夏草无性型有关成分 1H NMR 图谱信号分析

野生冬虫夏草的 ^{1}H NMR 图谱中，在δ 5.21，4.85 以及 3~4 ppm 之间有明显的多糖类质子信号。显示多糖是野生冬虫夏草主要免疫调节成分。在多糖的含量与组成方面，7 天液体培养的菌丝粉与野生冬虫夏草极为相似；而菌丝体提取物和百令胶囊则与野生冬虫夏草存在一定的差异（图 6）。此外，参试样品的质子信号在δ 1~3 ppm 的高场区存在明显差异。野生冬虫夏草和百令胶囊在δ 1.16 ppm 出有一明显的三重峰，表明百令胶囊与野生冬虫夏草有一定的相似度。冬虫夏草的 7 天发酵粉和提取物在δ 1~3 ppm 有较为复杂的质子共振信号，表现出一定的相似性，但提取物 ^{1}H NMR 在δ 0.89 ppm 出有一宽的质子工作信号（图 7）。由此可见，野生冬虫夏草及其不同培养物的 ^{1}H NMR 图谱有着不同的质子共振信号。冬虫夏草菌丝粉相关产品中首检成分含量差别可能与菌丝生长阶段与培养条件有关。这一提示意味着培养条件对于菌种有效成分含量具有某种调控作用。不过本实验检测的只是菌丝体代谢产物质子的总信号，虽然受检成分不等于有效成分，但可以区分或辨别野生冬虫夏草与人工培养产品的有效手段。

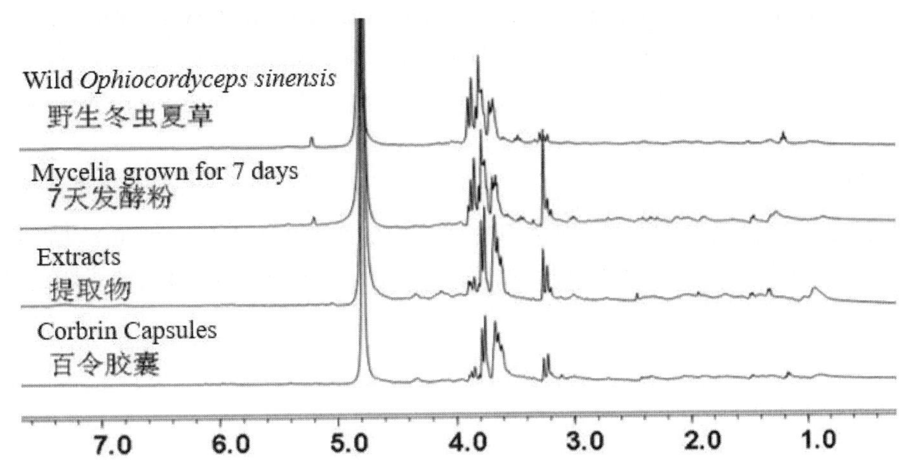

图 7 冬虫夏草及其相关菌体提取物质 ^{1}H NMR 指纹图谱

Fig. 7 The ^{1}H NMR fingerprint of extracts from *Ophiocordyceps sinensis* and its allies.

2.4 从生物经济时代看现代工业化冬虫夏草产业前景

现代医学研究认为冬虫夏草在调节免疫功能，抗肾间质纤维化，保护肾小管上皮细胞，以及治疗糖尿病肾病等方面的作用均不可低估。此外，冬虫夏草及其制剂在治疗肾脏病上的应用,决不仅限上述 4 方面（谌贻璞 2009）。不过抑制肾间质纤维化的活性组分迄今尚未探明，有待深入研究。因此，在探索冬虫夏草对人类健康贡献方面尚有极大空间，如对文献记载的各种相关疾病疗效及其活性组分的破解，对人类健康将具有更大贡献。由于冬虫夏草两种工业化产品的问世，克服了野生冬虫夏草资源匮乏的难题。

世界著名未来学家阿尔文·托夫勒预言中的信息经济已经进入相当成熟和繁荣阶段，而预言中的生物经济尚处于成长起步阶段。难于人工培养的冬虫夏草两种工业产品的问世，堪称为生物经济时代虫草产业现代工业化开发之先河，为保持人类健康提供丰富医疗保健资源开辟了现代工业化的新途径。

展望未来，结合冬虫夏草基因组数据（Hu X. *et al.* 2013），在破解冬虫夏草以及有关虫草对各种相关疾病疗效及其活性组分与代谢组研究的基础上，沿着合成生物学的方向，通过新生命与新生物药制造工厂途径，冬虫夏草产业在生物经济时代为全人类健康做出更大贡献的梦想必将成为现实。

[REFERENCES]

Chen YP. 2009. The application of Cordyceps sinensis and its preparations in renal disease is worthy of attention *CJITWN* （Chinese Journal of Integrated Traditional and Western Nephrology）, 10（10）:847-848（in Chinese）.

Godfray HCJ. 1994. Parasitoids: Behavioral and Evolutionary Ecology. *In*: H. C. J. Godfray. *Environmental Entomology*, Princeton Universit Press, **24（2）**:483-484.

Hu X, Zhang YJ, Xiao GH, Zheng P, Xia YL, Zhang XY, Stleger RJ, Liu XZ, & Wang CS*. 2013. Genome Survey uncovers the secrets of sex and lifestyle in caterpillar fungus. *Chinese Science Bulletin*, **58（23）**:2846-2854（2013）.

Rogers SO, Bendich AJ. 1988. Extraction of DNA from plant tissues. *In*: Gelvin SB, Schilperoort RA （eds）. Plant Molecular Biology ManualA6. Boston: Kluwer Academic Publishers: 1-10

Shen NY, Zeng L, Zhang XC, Zhou ZR, Shen Z, Wei SL. 1983. Feeding patterns of *Hepialus armoricanus* Oberthur larvae. Journal of Specialty Science, 03: 19-20

Tamura K, Peterson D, Peterson N, Stecher G, Nei M, Kumar S. 2011. MEGA5: molecular evolutionary genetics analysis using maximum likelihood, evolutionary distance, and maximum parsimony methods. *Molecular Biology and Evolution*, **28:** 2731–2739.

White TJ, Bruns TD, Lee SB, Taylor JW. 1990. Amplification and direct sequencing of fungal ribosomal RNA genes for phylogenetics. *In*: Innis MA, Gelfand DH, Sninsky JJ, White TJ, editors. PCR Protocols. San Diego: Academic Press: 315-322.

Zheng W, Miao K, Zhao Y, Pan S, Zhang M, Jiang H. 2009. Nitric oxide mediates the fungal-elicitor enhanced biosynthesis of antioxidant polyphenols in submerged cultures of *Inonotus obliquus*. Microbiology, 155:3440–3448

Zheng W, Zhao Y, Zheng X, Liu Y, Pan S, Dai Y, Liu F. 2011. Production of antioxidant and antitumor metabolites by submerged cultures of *Inonotus obliquus* cocultured with *Phellinus punctatus*. Applied Microbiology and Biotechnology, 89:157–167

[附中文参考文献]

谌贻璞. 2009. 冬虫夏草及其制剂在肾脏病中的应用值得重视. 中国中西医结合肾病杂志，10（10）：847-848。

沈南英，曾璐，张显耻，周中蓉，叁智，魏世录. 1983. 虫草蝙蝠蛾幼虫的食性研究. 特产科学杂志，03：19-20.

第三章　斜纤孔菌

Inonotus obliquus

野生桦褐孔菌及其深层发酵产物的弱极性成分研究

项小燕[1,2],顾　琪[2],郑维发[2*],张小平[1],魏江春[3*]

(1. 安徽师范大学生命科学学院,安徽 芜湖　241000; 2. 江苏省药用植物生物技术重点实验室,江苏 徐州　221116; 3. 中国科学院微生物研究所,北京　100080)

桦褐孔菌 *Inonotus obliquus* 属担子菌亚门、层菌纲,非褐菌目、多孔菌科、褐卧孔菌属。分布于北美、芬兰、波兰、俄罗斯及我国黑龙江、吉林等地区,寄生于桦树、榆树、赤杨等的树皮下或活立木的树皮下或砍代后树木的枯干上[1]。该菌在俄罗斯已有一百多年的应用历史,是具有显著的药理活性而对人体没有任何副作用的一种药用真菌[2]。研究表明,桦褐孔菌对糖尿病、传染性病毒及多种癌症有显著的疗效[3]。化学成分研究显示,桦褐孔菌含有甾体类[4]、黄酮类[5]和黑色素[6]类等化合物。现代药理学表明,三萜类化合物有改善血液循环、调整血压、降胆固醇的作用[7],并可在体外抑制 HIV的增殖[2]。因此,桦褐孔菌具有重大的药学价值。然而,桦褐孔菌在野生环境下生长极为缓慢,远远满足不了日益增长的市场需求。为此,有很多学者开始尝试人工固体栽培研究。然而,大量实践证明,固体栽培生长周期长,极易被杂菌污染,并且受场地和季节的限制[8],目前国内外尚无培养成功的先例。而深层发酵培养则生长周期短,产量高,不受场地和季节的限制,是解决桦褐孔菌资源危机的重要途径。本研究对两种培养基培养的桦褐孔菌深层发酵产物和野生子实体的弱极性物质作了比较,以期为桦褐孔菌深层培养体系甾体和芳香类化合物代谢研究奠定基础。

1　实验部分

1.1　实验材料:桦褐孔菌采自牡丹江镜泊湖火山口,标本由俄罗斯真菌学家 Margarita A B教授鉴定。菌种经 PDA斜面纯化培养后接入液体培养基,在 26℃、140 r/min的黑暗条件下振荡培养 10 d形成一级种子,分别接入装有 300 mL基本培养基和桦树汁培养基的三角瓶中,接种量为 10%,培养条件同上。基本培养基(A)的配方为(%):葡萄糖 2.5,蛋白胨 0.3,酵母膏 0.1, KH_2PO_4 0.1, $MgSO_4 \cdot 7H_2O$ 0.05, $CaCl_2$ 0.01, pH自然。桦树汁培养基(B)的配方为在基本培养基的基础上加 5% 桦树木屑。培养结束后,离心分离菌丝体,并用蒸馏水洗涤数次,60 ℃烘干至恒重。

1.2　样品的提取:精确称取基本培养基(A)和桦树汁培养基(B)的发酵产物以及干燥的野生子实体(C)各 5.0 g,充分粉碎后,用 60~ 90℃石油醚 50 mL在索式提取器中回流提取 2次,每次 2 h。合并提取液,离心取上清。低温减压回收溶剂,得浅黄色粉末,提取率分别为 0.34%(A)和 0.26%(B)和 1.02%(C)。精确称取各提取物 0.1 mg,溶于 1 mL丙酮(色谱纯),作为进样样品。

1.3　GC-MS分析:气相色谱-质谱联用仪(美国安捷伦公司 HP6890/5973N型)。气相色谱条件:色谱柱,为弹性石英毛细管柱(30 m× 0.25 mm× 0.25 μm);气化室温度:280 ℃;载气:高纯 He;流量:1 mL/min;程序升温:初温 70℃,以 10℃ /min的速度升至 270 ℃,等热 5 min;分流比 50∶1;进样量:1 μL。质谱条件:EI源;离子源温度:230 ℃;电子能量:70 eV;质量范围:30~ 550 amu;扫描周期,1 s;扫描范围,30~ 550 *m/z*。各组分成分鉴定:将分离的各化合物质谱图在质谱图库中电脑检索比较,选择与质谱图库中匹配度在 95% 以上的化合物作为被鉴定化合物[9]。各成分的质量分数用峰面积归一化法分析。

2　结果

从桦褐孔菌基本培养基的发酵产物中检测出 30个化合物,包括 7种甾体类衍生物(质量分数和为 40.1%),4种芳香类化合物(质量分数和为 2.33%)和 18种脂肪族化合物;从桦树汁培养基发酵产物中检测出 29个化合物,其中甾体类化合物 3种(质量分数和为 18.11%),芳香族化合物 2种(质量分数和为 2.7%),脂肪族化合物 24种;从野生子实体中检测出 25种化合物,其中甾体类化合物 7种(质量分数和为 69.04%),芳香族化合物 2种(质量分数和为 2.96%),脂肪族化合物 14种。从甾体类化合物组成

* 收稿日期:2005-08-25
基金项目:江苏省高校自然科学研究计划重大项目(2005KJA36012)
* 通讯作者　郑维发　Tel/Fax:(0516)83403179　E-mail:yyzw@xznu.edu.cn

本文原载于《中草药》, 37(5): 670-672, 2006.

来看，野生桦褐孔菌子实体弱极性成分以羊毛甾醇为主，占61. 38%，远远高于两种培养基发酵产物甾体类化合物质量分数和。而基本培养基深层发酵产物中出现质量分数较高的麦角甾醇及其衍生物，桦树汁培养基深层发酵产物中则出现质量分数较高的桦褐孔菌醇和麦角甾醇。就芳香族化合物而言，深层发酵产物和野生子实体在种类和质量分数基本一致。两种培养基的深层发酵产物中都含有邻苯二甲酸单-2-乙基己酯，质量分数分别为1. 49%（A）和2. 35%（B），而野生子实体则含有总量仅为3. 96%的邻苯二甲酸酯衍生物，见表1

3　讨论

表1　野生桦褐孔菌与其深层发酵产物的弱极性成分及相对质量分数

Table 1　Non-polar constituents and relative contents from wild *I. obliquus* and its submerged zymotic products

编号	化合物	质量分数%		
		A	B	C
甾体类化合物				
1	羊毛甾醇	3. 47	1. 72	61. 38
2	桦褐孔菌醇	8. 71	25. 04	-
3	24-亚甲基(3β)-羊毛甾-8-烯-3-醇	0. 38	3. 48 -	-
4	9(11)-二氢麦角甾基苯甲酸酯	0. 24	-	-
5	苯甲酸蒽麦角四烯酯	-	-	0. 27
6	麦角甾醇	21. 53	27. 35	3. 51
7	c-麦角甾醇	0. 23	-	0. 65
8	(3α, 5α, 22*E*)-麦角甾-7, 22-二烯-3-醇	-	-	0. 47
9	羽扇豆醇	-	-	0. 52
10	4, 4, 6a, 6b, 8a, 11, 12, 14b-八甲基-1, 2, 3, 4, 4a, 5, 6, 6a, 6b, 7, 8, 8a, 9, 10, 11, 12, 12a, 14b-十八氢苉-3-醇	-	-	2. 24
11	(5*R*, 6*R*)-(5, 6-二氢)-5, 6-二羟基-10′-脱辅基-α-阿洛胡萝卜酸甲酯	5. 95		
芳香族化合物				
12	邻苯二甲酸单-2-乙基己酯	1. 49	2. 35	-
13	2-甲基-2-苯基十三烷	0. 13	-	-
14	邻苯二甲酸二-2-甲基庚酯	0. 53	-	1. 94
15	邻苯二甲酸二异丁酯	0. 18	0. 35	1. 02
脂肪族化合物				
16	十六烷	0. 08	-	0. 51
17	十七烷	0. 11	0. 43	-
18	十八烷	-	0. 49	0. 93
19	十九烷	-	0. 57	1. 88
20	二十烷	0. 42	0. 65	-
21	二十一烷	0. 34	0. 48	0. 95
22	二十二烷	-	0. 71	-
23	二十三烷	-	-	1. 57
24	二十四烷	-	0. 31	1. 23
25	二十五烷	-	0. 21	-
26	二十六烷	0. 09	0. 14	0. 91
27	二十七烷	-	0. 29	0. 45
28	二十八烷	-	-	2. 04
29	2, 6, 10-三甲基十四烷	-	-	0. 10
30	2, 6, 10-三甲基十五烷	-	0. 20	0. 83
31	2, 6, 10-三甲基十六烷	-	0. 21	1. 92
32	2, 6, 10, 14-四甲基十六烷	-	0. 25	1. 31
33	2, 6, 11, 15-四甲基十六烷	0. 09		-
34	9-己基十七烷	-	0. 75	-
35	5, 5-二甲基-4-(3-甲基-1, 3-丁间二烯基)1-氧螺[2. 5]辛烷	0. 99	-	-
36	(9, 12)-十八二烯酸乙酯	0. 83	0. 73	-
37	油酸	-	0. 87	-
38	(6*E*)-2, 6, 10, 15, 19, 23-六甲基-2, 6, 10, 14, 18, 22-二十四碳六烯	0. 93	1. 48	-
39	2-己基-1-癸醇	-	-	0. 73
40	戊二酸二丁酯	-	0. 14	-
41	甲基丁二酸二丁酯	0. 10	-	-
42	十六烷酸乙酯	0. 29	-	-
43	己二酸二异丁酯	0. 04	-	-
44	十八酸乙酯	1. 40	-	-
45	十六甲基十七酸甲酯	-	0. 29	-
46	十六烷酸-2-羟基-1, 3-丙二酯	0. 59	-	-
47	(*Z*)-9-十八烯酸-2-羟基-1-羟甲基乙酯	-	0. 67	-
48	(*Z*, *Z*)-9, 12-十八二烯酸-2-羟基-1-羟甲基乙酯	3. 19	13. 72	-
50	(*Z*, *Z*)-9, 12-十八二烯酸-2, 3-二羟基丙酯	0. 76	0. 50	-
51	(*Z*)-9-十八酰胺	1. 02	0. 60	-
其他类化合物				
52	6α, 7, 10α-三甲基-十二氢-1*H*-苯并[f]苯并吡喃	0. 65	0. 22	-
53	2, 4α, 8, 8-四甲基十氢环丙[d]萘	1. 06	-	-

A：基本培养基发酵产物；B：桦树汁培养基发酵产物；C：野生子实体

A：zymotic products on basic medium；B：zymotic products on birch juice medium；C：wild sporophore

自然条件下，桦褐孔菌受到多种病菌的感染和昆虫吞噬的压力，导致桦褐孔菌体内合成大量的抗病毒、抗细菌和真菌以及昆虫拒食物质。研究表明，羊毛甾类化合物具有抗病毒[10]、杀虫[11]活性。这可能是野生桦褐孔菌子实体积累大量羊毛甾醇的原因。深层发酵产物中只含有少量的羊毛甾类化合物而积累大量的麦角甾醇。从真菌的甾体类化合物生物合成途径来看，麦角甾醇是羊毛甾醇经过多个酶促反应后形成的产物。深层发酵条件下由于没有病菌感染和昆虫吞噬的压力，导致羊毛甾类化合物进一步转化生成麦角甾醇。这可能是发酵产物中羊毛甾醇的含量低于野生子实体的主要原因。从生物合成途径来看，芳香类化合物的生物合成或源自莽草酸途径，或源自多酮的环化[13]，但无论是哪种途径其最初的前体都是与甾体类化合物生物合成的最初前体相同，即都是乙酰辅酶A。因此，生长环境的改变导致乙酰辅酶A参与不同类别化合物合成的比例发生改变。

生长条件的变化是桦褐孔菌深层发酵产物与野

生子实体的弱极性成分组成差异的主要原因。从发酵产物中笔者获得了高于野生子实体的活性物质——麦角甾醇。它是维持真菌生物膜结构和功能的必须成分[14,15],也是合成维生素 D_2 的前体[16]。因此,以基本培养基培养的桦褐孔菌深层发酵产物也具有重要的药学价值。至于桦树汁培养基和基本培养基在弱极性成分组成差异的原因以及如何提高发酵产物甾体类化合物的量有待于进一步研究。

References

[1] Huang L N. Mysterious folk medical fungi in Russia- *Inonotus obliquus* [J]. *Edib Fungi China* (中国食用菌), 2002, 21 (4): 7-8.

[2] Lin B X. Medical fungi- *Inonotus obliquus* [J]. *Strait Pharm J* (海峡药学), 2004, 16(6): 74-76.

[3] Wass S P, Weis A L. Therapeutic effect of substances occurring in higher basidionmycetes mushrooms: a modern perspective [J]. *Crit Rev Innun*, 1999, 19: 65-69.

[4] He J, Feng X Z. Studies on chemical constituents of *Fuscoporia oblique* [J]. *Chin Tradit Herb Drugs* (中草药), 2001, 32 (1): 4-6.

[5] Cui Y, Kim D S, Park K C. Antioxidant efect of *Inonotus obliquus* [J]. *J Ethnopharm*, 2005, 96: 79-85.

[6] Babitskaya V G, Scherba V V, Lionnikova N V, *et al*. Melanin complex from medicinal mushroom *Inonotus obliquus* Pilat (Chaga) (Aphyllophoromycetideae) [J]. *Inter J Med Mush*, 2001, 4: 139-145.

[7] Rzymowska J. The effect of aqueous extracts from *Inonotus obliquus* on the mitotic index and enzyme activities [J]. *Boll Chim Garm*, 1998, 137(1): 13-16.

[8] Jiang Y J. Studies on cultivated character of *Inonotus obliquus* [J]. *Fujian J Agric Sci* (福建农业学报), 2004, 19(2): 92-95.

[9] Zheng W F, Shi F, Wang L, *et al*. Chemical components of non-polar fractions from ethanol extract of radixes of Daphne genkwa and their inhibition to acute inflammation [J]. *Pharm J Chin PLA* (解放军药学学报), 2004, 20(1): 18-21.

[10] Ge X C, Wu J Y. Tanshinone production and isoprenoid pathways in Salvi miltiorrhiza hairy roots induced by Ag+ and yeast elicitor [J]. *Plant Sci*, 2005, 168: 487-491.

[11] Liu Z R, Hu Y C. Extraction, separation and bio-assay of *Daphne genkwa* active components [J]. *J Central South Forest Univ* (中南林学院学报), 2000, 20(4): 15-19.

[12] Veen M, Stahl U, Lang C. Combined overexpression of genes of the ergosterol biosynthetic pathway leads to accumulation of sterols in *Saccharomyces cerevisiae* [J]. *FEMS Yeast Res*, 2003, 4: 87-95.

[13] Dewick P M. *Medicinal Natural Products* [M]. West Sussex: John Wiley & Sons Ltd, 2002.

[14] Parks L W and Casey W M. Physiological implications of sterol biosynthesis in yeast [J]. *Annu Rev Microbiol*, 1995, 49: 95-116.

[15] Arnezeder C, Hample W A. Influence of growth rate on the accumulation of ergosterol in yeast-cells [J]. *Biotechnol Lett*, 1990, 12: 277-282.

[16] Mattila P, Maija A. Sterol and vitamin D2 contents in some wild and cultivated mushrooms [J]. *Food Chem*, 2002, 76: 293-298.

Aminophenols and mold-water-extracts affect the accumulation of flavonoids and their antioxidant activity in cultured mycelia of *Inonotus obliquus*

ZHENG Wei-Fa[1*] GU Qi[1**] CHEN Cai-Fa [1] YANG Shi-Zhao [1] WEI Jiang-Chun [2] CHU Cheng-Cai[3]

([1]Key Laboratory for Biotechnology on Medicinal Plants of Jiangsu Province, Xuzhou Normal University, Xuzhou 221116; [2]Institute of Microbiology, Chinese Academy of Sciences, Beijing 100080; [3]Institute of Genetics and Developmental Biology, Chinese Academy of Sciences, Beijing 100101)

Abstract: Flavonoids are one of the primary polyphenols in field-grown basidiomycetous fungus *Inonotus obliquus* responsible for the therapy of many diseases. In cultured *I. obliquus,* however, flavonoids are less accumulated, resulting in a significant reduction in pharmacological activities. In this study, three aminophenols and water extracts of four molds were evaluated for their effects on the accumulation of flavonoids and antioxidant activity of cultured *I. obliquus*. L-Tyr, and the water extracts of *Aspergillus flavus* and *Mucor racemosus* evidently increased the accumulation of flavonoids, and suggested a greater advance towards those found in field-grown mycelia. Flavonoids in cultured *I. obliquus* consisted of more than four glycosides of quercetin, naringenin, kaempferol and Isorhamnetin. The antioxidant activity of cultured *I. obliquus* was positively associated with the accumulation of flavonoids. And the capacity for scavenging superoxide anion, hydroxyl and DPHH radicals was more effective in cultured *I. obliquus* with regulations of L-Tyr, *A. flavus* and *M. racemosus* water extracts.

Key words: submerged culture, total flavonoids, phenylalanine ammonia-lyase, chalcone synthase, free radical scavenging activity

氨基酸和霉菌水提物对深层发酵桦褐孔菌菌丝体黄酮积累及其抗氧化活性的影响

郑维发[1*] 顾琪[1**] 陈才法[1] 杨士钊[1] 魏江春[2] 储成才[3]

([1]徐州师范大学江苏省药用植物生物技术重点实验室 徐州 221116；[2]中国科学院微生物研究所 北京 100080；[3]中国科学院遗传与发育生物学研究所 北京 100101)

Supported by key project of grant from natural science foundations of education department of Jiangsu province (No.05KJA36012), and a grant of natural science foundations from the government of Jiangsu province (BK2006034).
*Corresponding author. E-mail: yyzw@xznu.edu.cn;**Contribute equally to the paper.
Received:2006-05-12, Accepted:2006-12-06

This paper was originally published in *Mycosystema*, 26（3）: 414-426, 2007.

摘 要：黄酮类化合物是桦褐孔菌菌丝体中多酚类化合物的重要组成部分，也是该菌治疗众多疾病的有效成分之一。然而人工培养桦褐孔菌黄酮等酚类化合物积累甚少，导致药理活性的明显下降。为此，我们研究了3种氨基酸和4种霉菌水提物对深层发酵桦褐孔菌黄酮的积累及其抗氧化能力的影响。在所试验的3种氨基酸和4种霉菌水提物中，L-酪氨酸黄曲霉和毛霉水提物能有效地增加该菌黄酮的积累。人工培养菌体中的黄酮至少由4种黄酮苷组成，苷元分别是槲皮素、柚皮素、山奈酚和异鼠李素。深层发酵菌丝体具有一定的抗氧化能力，并与总黄酮的含量呈正相关。由L-酪氨酸，黄曲霉和毛霉水提物调控生长的桦褐孔菌菌丝体，能有效地清除超氧阴离子、羟基自由基和DPPH自由基。

关键词：深层发酵，总黄酮，苯丙氨酸解氨酶，查尔酮合酶，自由基清除活性

中图分类号：Q939.5 **文献标识码**：A **文章编号**：1672-6472（2007）03-0414-426

INTRODUCTION

The basidiomycetous fungus *Inonotus obliquus* (Fr.) Pilat (Hymenochaetaceae) generally grows on trunks of living birch in Far East of Russia, Northeast China and other adjacent countries at latitudes of 45°N~50°N. This fungus has been used for literally more than four centuries (Molitoris, 1994) as an effective agent to treat breast cancer, hepatoma, gastrointestinal cancers and tuberculosis without incurring any unacceptable toxicity (Saar, 1991; Huang, 2002). It has been reported that *I. obliquus* produces large sums of polyphenols (Cui *et al.*, 2005) including melanins (Babitskaia *et al.*, 2000), flavonoids (He & Feng, 2001) in field environment, and demonstrated strong activities for scavenging free radicals (Cui *et al.*, 2005), inhibiting tumor cell proliferation (Burczyk *et al.*, 1996),. A number of studies also reveled that flavonoids possessed various biological activities including inhibiting platelet aggregation, quenching free radicals, preventing tumor cell proliferation (Formica & Regelson, 1995; Dixon & Steele, 1999; Kanadaswami *et al.*, 2005). Therefore, flavonoids are the active principles in *I. obliquus* responsible for the effective therapy of many diseases.

Geographically, *I. obliquus* distributes in frigid region and grows in an exceptionally slow rate. In addition, this fungus shows highly host specificity and extremely low inhabitation possibility. Consequently, natural reserve of *I. obliquus* is being exhausted. Previous works on the culture of *I. obliquus* have for the most part focused on mycelia biomass, polysaccharides (Kim *et al.*, 2005) and melanins (Kukulyanskaia *et al.*, 2002), with little attention paid to the factors affecting the biosynthesis of flavonoids. The metabolites of the cultured mycelia mainly consist of immunomodulatory polysaccharides (Kim *et al.*, 2005). In field environment, fungi growth is frequently affected by environmental factors. These factors act as pro-oxidative agents (Romero Alvira *et al.*, 1995; Abrahamsson *et al.*, 2003; Valencia & Kochevar, 2006), producing oxidative stress and hereby resulting in the production of reactive oxygen species (ROS) within the cells of mycelia. In the past decades, however, relatively little is known about the stress-induced accumulation of polyphenols, particularly flavonoids in cultured mycelia of *I. obliquus*.

It has been revealed that biosynthesis of flavonoids begins with the conversion of phenylalanine to cinnamic acid, the conversion of tyrosine to 4-hydroxyl cinnamic acid and subsequently the synthesis of naringenin chalcone (Dewick, 2002) (Fig.1), during which phenylalanine ammonia-lyase (PAL) and chalcone synthase (CHS) represent the key enzymes driving for the rate-limiting chemical reactions in biosynthesis of flavonoids (Moore *et al.*, 2002). In this study, we report the effects of aminophenols and four molds water extracts on the activities of PAL and CHS, accumulation of flavonoids and the capacity for scavenging free radicals in cultured mycelia of *I. obliquus*.

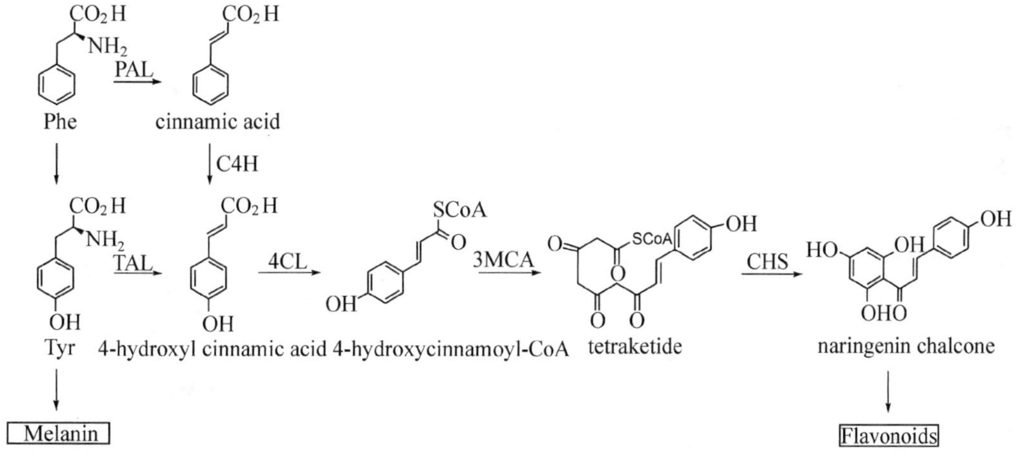

Fig.1 Biosynthetic pathway of flavonoids. PAL: phenylalanine ammonia-lyase; TAL: tyrosine ammonia-lyase; C4H: cinnamate 4-hydroxylase; 4CL: 4-coumarate coenzyme A ligase; CHS: chalcone synthase; MCA: malonyl-CoA.

MATERIALS AND METHODS

Microorganism

I. obliquus was collected in Changbai Mountains of Mudanjiang Region, Northeast China. The vouch specimen (KLBMP04005) identified by Russian mycologist Prof. Margarita A Bondartseva was preserved in the Herbarium of Key Laboratory for Biotechnology on Medicinal Plants of Jiangsu Province, P.R. China. This fungus was successfully isolated from mycelia tissue using potato dextrose agar (PDA) slant and subcultured each three months. The slants were incubated at 25℃ for seven days and then stored in refrigerator (4℃).

Seed culture

I. obliquus was initially grown on PDA medium in a Petri dish for seven days and then transferred to the seed culture medium by punching out 15mm^2 of the agar plate culture for 10 peaces with a sterilized self-designed cutter. Seed culture was grown in a 500mL-conical flask with 200mL medium containing glucose (2%), peptone (0.35%), yeast extract (2%), KH_2PO_4 (0.01%), $MgSO_4 \cdot 7H_2O$

(0.05%) and distilled water without modification of pH level (5.5). The cultivation was performed on a rotary shaker incubator (Shen Neng Bao Cai, Shanghai China) at 26℃ and 140r/min for seven days.

Mycelia culture

The mycelia culture for the regulations of the tested aminophenols and molds were performed, unless otherwise stated, in triplicate for all experiments by submerged culture on a rotary incubator using the medium and culture conditions identical to that of seed culture. The medium for submerged culture was inoculated with 10% (v/v) seed culture (about 1.5g/L dry mycelia), and incubated with a 1000mL-conical flask containing 400mL medium.

Aminophenol regulation

L-Phe, L-Tyr and L-Trp (Sigma, St Louis MO, USA) were aseptically supplemented to the culture at a concentration of 0.5mmol/L on the third day after incubation. The activity of PAL and CHS was determined on day 10, and the flavonoids determined on day 15.

Simulated pathogen attack

The molds including *Aspergillus flavus*, *Rhizopus stolonifer*, *Penicillium citrinum* and *Mucor racemosus* were all obtained from the fungal collection center of Key Lab for Biotech on Med Plants of Jiangsu Province. For preparation of water extracts of test molds, 1g lyophilized mycelia from submerged culture were homogenized with 20mL of 50mmol/L PBS buffer (pH 7.4) in an ice bath. The supernatant was filtered aseptically using 0.22μm syringe filters. For simulated pathogen attack, 800μL supernatant was supplemented into the culture media on the third day after incubation. The activities of PAL and CHS were determined on day 10, and the total flavonoids determined on day 15.

Determination of PAL and CHS activity

Crude enzymes were prepared according to the method previously described (Mori *et al.*, 2001). For PAL activity assay, the reaction mixture contained 200μL of 25mmol/L L-Phe, 700μL crude enzyme, and 2100μL of 50mmol/L Tris-HCl buffer (pH 8.8) was incubated at 37℃ for 30min, then ceased by supplementing 5mL EtOAc. The extraction was performed three times and the organic layer was evaporated to dryness (D'Cunha, 2005). The residue was dissolved in 150μL MeOH for HPLC assay (Waters Corp., Milford, MA). Column (Inertsil®, Japan): 250mm × 4.6mm, 7μm C_{18} silica; mobile phase: HAc/H_2O/MeOH: 1/39/60 (v/v/v); flow rate: 1mL/min; wavelength: 280nm; column temperature: 30℃; injection volume: 20μL.

For CHS activity, 20μL of 50mmol/L 4-hydroxycinnamoyl-CoA, 20μL of 100mmol/L malonyl-CoA and 700μL crude enzymes were mixed with 2100μL 50mmol/L Tris-HCl Buffer (pH 7.0). The reaction was initialed by incubation at 37℃ for 1h and ceased by supplementing 5mL EtOAc (Zuurbier *et al.*, 1998). The reaction product was isolated as the procedure described in PAL activity assay. Mobile phase: HAc/H_2O/MeOH: 1/24/75 (v/v/v); wavelength: 310nm. The other conditions were identical to that of PAL activity assay.

Total soluble protein in the mycelia was determined using Coomassie blue-dye binding assay as described previously (Bradford, 1976). Bovine serum albumin (Sigma, St. Louis MO, USA) was taken as the standard protein. The levels of PAL and CHS were expressed as pkat/mg prot.

Estimation, preparation and composition analysis of total flavonoids

The total flavonoids were estimated according to the method previously described (Meda *et al.*, 2005), and quantified using absorbance of rutin at 415nm against its concentration (mg/L). The mean of three readings was used as the level of mg of rutin equivalents (RE)/g of mycelia.

Composition analysis was performed by hydrolysis. Briefly, 100mg total flavonoids were dissolved in a solution consisting of 40mL MeOH and 10mL 25% HCl, and hydrolyzed at 80℃ for 2h. The hydrolysate was filtered for HPLC analysis. Mobile phase: MeOH/H_2O (0.4% H_3PO_4): 60/40 (v/v); wavelength: 360nm. The other conditions were identical to that of PAL activity assay.

Free radical scavenging activity

Well-powdered lyophilized mycelia (15-day-old) were extracted with petroleum (b.p. 30℃~60℃) at 40℃ for three times with 2h each for elimination of non-polar fractions. The petroleum-treated mycelia were extracted with 85% EtOH by ultrasonication at 50℃ for three times (1h each). The extract was then placed in 0℃~4℃ refrigerator overnight followed by centrifugation to precipitate possible lipids and sterols. The supernatant was evaporated to dryness for antioxidant assay. Pyrogallol (3mmol/L, Sigma, St. Louis, MO) was used for assaying the capacity of scavenging superoxide anion (Xiao *et al.*, 1999), 1,10-Phenanthroline (5mmol/L, Sigma, St. Louis, MO) for hydroxyl radical (Jing *et al.*, 1996), and 2,2-Dipenyl-1-picrylhydrazyl (DPPH) (0.1mmol/L, Sigma, St. Louis, MO) for DPPH radical (Wang *et al.*, 2004). The capacity for scavenging free radicals was indicated as 50% effective concentration (EC_{50}), which was calculated as previously described (Mortensen *et al.*, 1998). Vitamin C (Biodee, Beijing China) was used as positive control.

Statistical analysis

The data acquired in the experiment were processed using SPSS 10.0 software. The assumptions of analysis of variance were confirmed to be statistically significant when $p < 0.05$. The results were expressed as mean ± S.E.

RESULTS

PAL and CHS activity

As shown in Fig.2, PAL activity was increased from about 69pkat/mg protein (control level) to about 123pkat/mg protein following the supplement of L-Phe. In comparison, the supplement of L-Trp and L-Tyr into the culture of *I. obliquus* also resulted in an evident increment in PAL activity, reaching 104.00pkat/mg and 110.55pkat/mg protein, respectively (Fig.2). The water extracts of *A.*

flavus and *M. racemosus* evidently increased the PAL activity when compared to that of control, reaching 114.07pkat/mg and 90.88pkat/mg protein, respectively. Supplementation of water extracts from other two tested molds, however, resulted in a distinctive reduction in PAL activity (Fig.3). CHS activity of *I. obliquus* was less affected by the four tested molds. Among the molds tested, only *P. citrinum* showed a slight enhancement on CHS activity, reaching 49.50pkat/mg protein. The other three made no difference in CHS activity in cultured mycelia of *I. obliquus* (Fig.3).

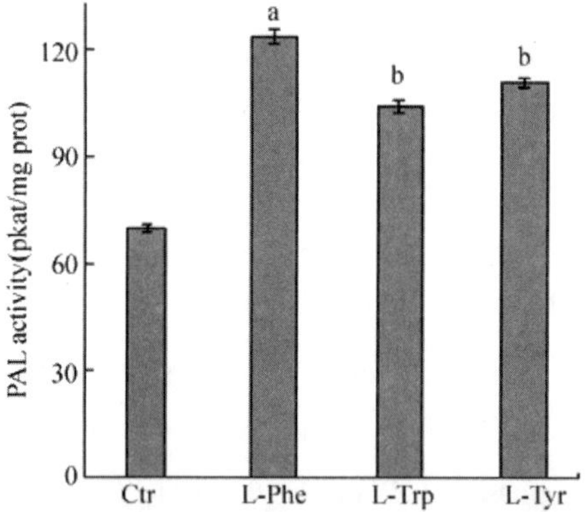

Fig.2 PAL activity in cultured mycelia of *I. obliquus* with regulations of three aminophenols. $^{a}P<0.01$, $^{b}P<0.05$ vs control. Data were expressed as mean ± S.E. (SPSS).

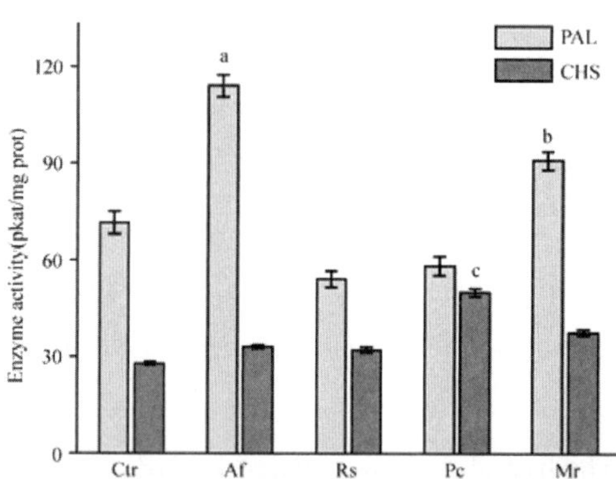

Fig.3 The activities of PAL and CHS from cultured mycelia with regulations of the water extracts of the four tested molds. Ctr: control; Af: *Aspergillus flavus*; Rs: *Rhizopus stolonifer*; Pc: *Penicillium citrinum*; Mr: *Mucor racemosus*. $^{a}P<0.001$,$^{b}P<0.01$ vs PAL control; $^{c}P<0.05$ vs CHS control. Data were expressed as mean ± S.E. (SPSS).

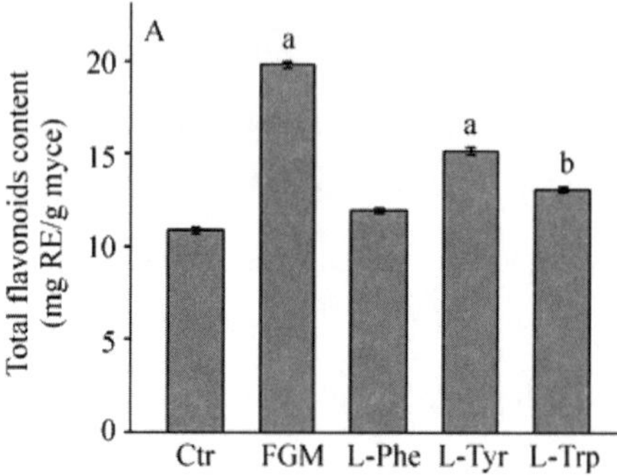

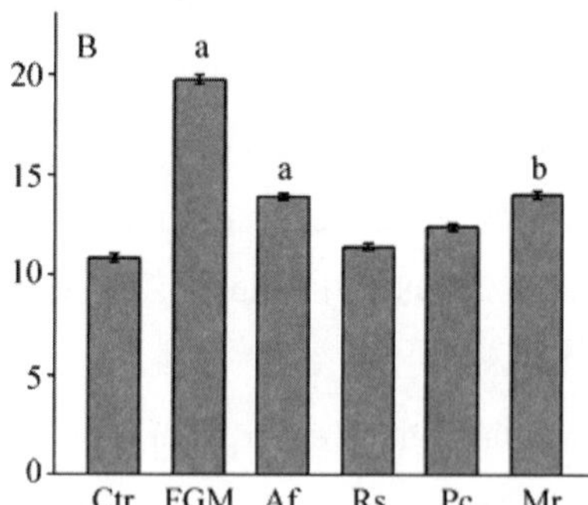

Fig.4 The content of total flavonoids in mycelia of *I. obliquus* from submerged culture with regulations of tested amino acids and water extract of the four tested molds. Ctr: control; FGM: field-grown mycelia; Af: *Aspergillus flavus*; Rs: *Rhizopus stolonifer*; Pc: *Penicillium citrinum*; Mr: *Mucor racemosus*. $^{a}P<0.01$, $^{b}P<0.05$ vs control. Data were expressed as mean ± S.E. (SPSS).

Accumulation of flavonoids

Addition of aminophenols into the culture of *I. obliquus* also affected the accumulation of total flavonoids. As shown in Fig.4, in the mycelia from the culture with L-Tyr and L-Trp, the total flavonoids reached 15.17mg/g and 13.08mg/g respectively, evidently higher than 10.85mg/g found in control and reduced the difference with 19.81mg/g found in the field-grown mycelia. In the culture supplementing L-Phe, however, no significant enhancement in flavonoids was observed (Fig.4-A). The accumulation of flavonoids was also influenced by the tested molds. In the culture supplementing water extracts of *A. flavus* and *M. racemosus* the level of total flavonoids reached 13.98mg/g and 13.90mg/g, respectively (Fig.4-B). In comparison, the water extracts of *R. stolonifer* and *P. citrinum* showed no remarkable enhancement on the accumulation of flavonoids (Fig.4-B).

HPLC analysis showed that the flavonoids in cultured mycelia of *I. obliquus* consisted of flavone glycosides. In HPLC spectrum, the retention time of the main peaks was found to be less than 5min in the mobile phase of MeOH/H_2O (0.4%H_3PO_4):60/40 (v/v). Following hydrolysis, however, at least seven recognizable peaks appeared in HPLC spectrum, and four of them were determined to be quercetin, naringenin, kaempferol and isorhamnetin by comparing with authentic standards (Fig.5).

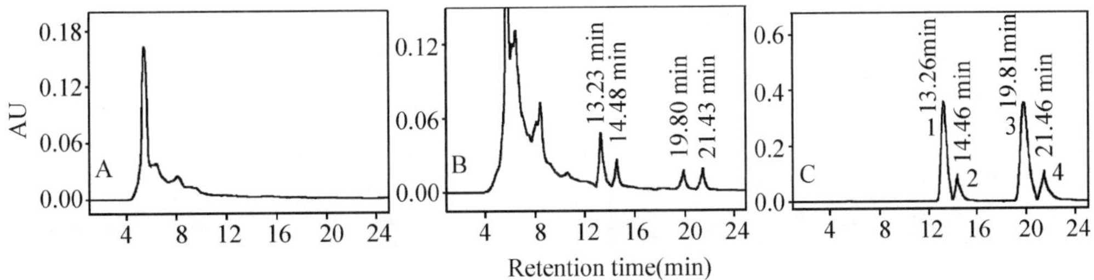

Fig.5 Composition analysis of total flavonoids. A: Unhydrolyzed total flavonoids; B:Hydrolyzed total flavonoids; C: Standards. 1.Quercetin; 2.Naringenin; 3. Kaempferol; 4. Isorhamnetin.

Free radical scavenging activity

The antioxidant activity of cultured *I. obliquus* was affected by tested aminophenols and water extracts of tested molds. For scavenging DPPH radical, the lowest EC_{50} was found at 58.13μg/mL EtOH extract of mycelia with regulation of L-Tyr, and at 62.11μg/mL and 60.59μg/mL in the mycelia with regulations of *A. flavus* and *M. racemosus*, respectively (Fig.6-A). The capacity for scavenging superoxide anion was also effected by some tested molds. As shown in Fig 6-B, significant scavenging efficacy was observed in EtOH extracts of mycelia with regulations of L-Tyr and *A. flavus* and *M. racemosus* in contrast to that of control. The evident increase in capacity for scavenging hydroxyl radical was also observed in the mycelia from culture supplemented with L-Tyr and water extracts of *A. flavus* and *M. racemosus*, and EC_{50} of the mycelia with these regulations was almost equal to that of Vc (Fig.6-C). The EtOH extracts of mycelia with regulations of other two amino acids and other two tested molds, however, showed no enhanced capacity of scavenging free radicals (Fig.6).

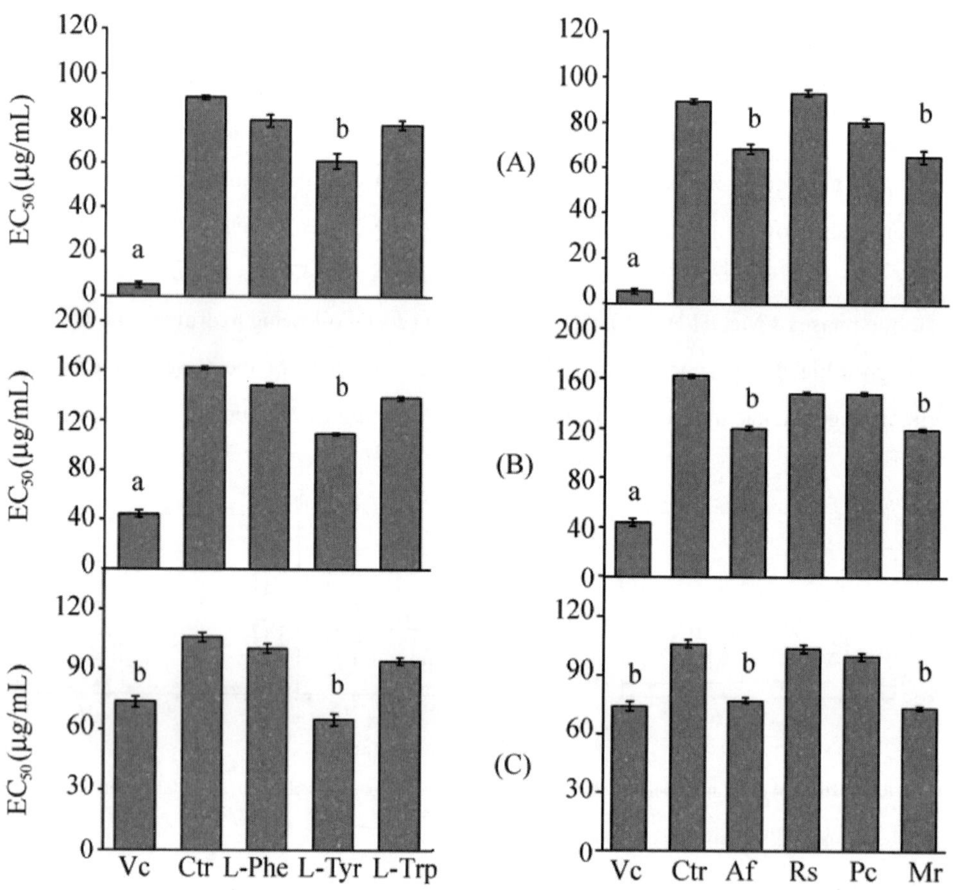

Fig.6 The scavenging capacity of EtOH extracts of cultured mycelia of *I. obliquus*. A: DPPH; B: Superoxide anion; C: Hydroxyl radical. The scavenging capacity was expressed as the half effective concentration (EC_{50}). The concentration of DPPH, pyrogallol and 1,10-phenanthroline used for the assay was 3mmol/L, 0.4mmol/L and 0.2mmol/L, respectively. Vc: vitamin C; Ctr: control; Af: *Aspergillus flavus*; Rs: *Rhizopus stolonifer*; Pc: *Penicillium citrinum*; Mr: *Mucor racemosus*. $^{a}P<0.001$, $^{b}P<0.05$ vs control. Data were expressed as mean ± S.E. (SPSS).

DISCUSSION

In the present study, we found that aminophenols (L-Trp and L-Tyr) and water extracts of *A. flavus* and *M. racemosus* significantly increased the accumulation of flavonoids in cultured mycelia of *I. obliquus* and their antioxidant activities. The biomass of the cultured mycelia, however, all reached around 9g/L in all treatments without being affected by these manipulations (Fig.7). It has been reported that the antioxidant activity of plants is responsible for their therapeutic effect against cancer, cardiovascular disease and diabetes (Anderson *et al.*, 2004; Stanner *et al.*, 2004). Therefore, following the increase of flavonoids, the therapeutic effects of cultured mycelia of *I. obliquus* will be enhanced.

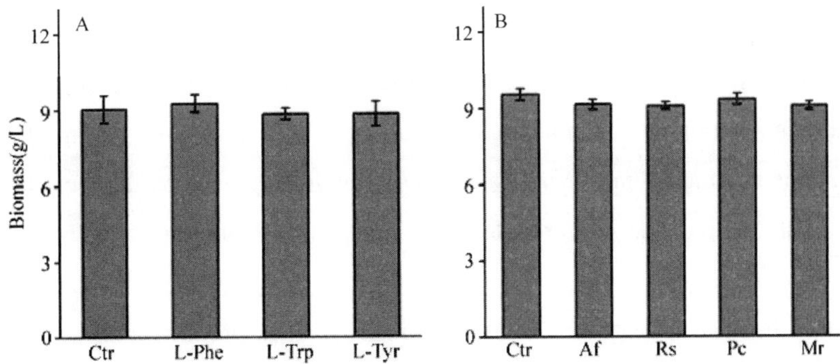

Fig.7 The biomass of mycelia from the culture supplemented with tested aminophenols and mold water extracts. A: Effects of aminophenols; B: Effects of mold water extracts. Note: ctr: control; Af: *Aspergillus flavus*; Rs: *Rhizopus stolonifer*; Pc: *Penicillium citrinum*; Mr: *Mucor racemosus*. Data were expressed as mean ± S.E. (SPSS).

In our study, we noticed that the crude enzymes in *I. obliquus* indicated a higher capacity for transforming L-Phe to cinnamic acid, and a higher capacity for transforming 4-hydroxylcinnamoyl-CoA to naringenin chalcone in the culture with addition of aminophenols and water extract of *A. flavus* and *M. racemosus*. The increase in transforming capacities triggered by aminophenols and the two tested molds was, at least in part, responsible for the enhancement on the activity of PAL. These findings were evidenced by an enhanced accumulation of flavonoids in cultured mycelia of *I. obliquus* with regulations of L-Tyr, *A. flavus* and *M. racemosus*. HPLC analysis disclosed that flavonoids consisted of at least four antioxidant flavone glycosides including quercetin, naringenin, kaempferol and isorhamnetin (Chen *et al.*, 2005; Chen *et al.*, 2006; Dai *et al.*, 2006; Park *et al.*, 2006). Therefore, the capacity for scavenging free radicals was significantly enhanced following an increased accumulation of flavonoids in cultured mycelia of *I. obliquus*. In phenylpropanoid metabolism, PAL and CHS represent the rate limiting reactions for the biosynthesis of flavonoids (Moore *et al.*, 2002). Our results showed that tryptophan induced a significant increase in PAL activity in *I. obliquus*, which is similar with those found in plant pathogen fungi *Ustilago maydis* (Kim *et al.*, 2001). In *Rhizoctonia solani*, the supplementation of L-Tyr and L-Phe into the culture also induced an evident enhancement of PAL activity (Kalghatgi & Subba Rao, 1976), which is in well accordance with our observations in *I. obliquus*. Theoretically, L-Phe is the direct substrate of PAL for synthesizing cinnamic acid. However, supplementing L-Phe into the culture was not responded by a significant increase in total flavonoids. Instead, a distinctive enhancement on accumulation of total flavonoids was observed in the culture with addition of L-Tyr. It has generally been accepted that all plants appear to have the ability to deaminate phenylalanine via the PAL, but the corresponding transformation of tyrosine by tyrosine ammonia-lyase (TAL) is more restricted, being mainly limited to members of the grass family (Dewick, 2002). In some plant pathogen fungi, TAL shows far less activity than that of PAL, and has very limited capacity for transforming tyrosine (Kalghatgi & Subba Rao, 1976). In our study, supplementing L-Tyr also resulted in maximum increase in the accumulation of flavonoids in cultured mycelia of *I. obliquus* among the three tested aminophenols. This finding seems to indicate that *I. obliquus* has a broad specificity PAL, also deaminating tyrosine to produce 4-hydroxylcinamic acid, the key intermediate for biosynthesis of chalcone (Dewick, 2002).

Pathogen attack has been recognized as a pro-oxidative agent, producing oxidative stress to the living organisms (Romero Alvira *et al.*, 1995). In higher plants, accumulation of flavonoids in response to pathogen attack is seen in many species. And their importance as antibiotic phytoalexins is well established (Dixon & Paiva, 1995). In the terms of molecular biology, plants usually initiate an enhanced gene expression of PAL and CHS in response to pathogen attack (Christensen *et al.*, 1998a; Christensen *et al.*, 1998b). In our study, water extracts of *A. flavus* and *M. racemosus* also triggered an enhancement on PAL activity, and subsequent increase in the accumulation of total flavonoids and antioxidant activity in cultured mycelia of *I. obliquus*. However, CHS activity was found to be less affected by the presence of water extracts of tested molds. Therefore, the exposure to tested molds predominantly caused an enhanced gene transcription for PAL, and the enhanced gene transcription of PAL underlines the defense mechanism against pathogen attack, in a way different from that in higher plants. In our study, the simulated pathogen attack was performed by supplementing filtered water extracts of tested molds into the culture of *I. obliquus*, implicating the existence of signaling substances in water extracts of some tested molds. It should be noted that supplementing water extracts of *A. flavus* and *M. racemosus* was responded by an enhanced accumulation of flavonoids and an enhanced the capacity for scavenging free radicals. This should be the results of a stress-induced signal transmission. Indeed, a single stimulation by water extracts of some tested molds only resulted in a limited increase (though significant) in accumulating flavonoids, and still showed less capacity for scavenging free radicals in cultured mycelia when compared to those of the total polyphenols found in field-grown mycelia of *I. obliquus* (Kim *et al.*, 2005). However, the antioxidant activity of total flavonoids in cultured mycelia of *I. obliquus* was 5-10 times of those found in *Ganoderma ludidum* and *Ganoderma tsugae* (Mau *et al.*, 2002).

I. obliquus has wide pharmaceutical application, particularly in the therapy of various tumors, cardiovascular diseases and tuberculosis (Saar, 1991; Huang, 2002). The pharmaceutical significance of *I. obliquus* has made it gradually exhausted in natural reserve. Scale-up culture of the fungus should be a possible solution for the crisis if performed under manipulations of adequate fed-batch of L-Tyr and continuous stimulation by some pathogen molds or other pro-oxidative agents.

[REFERENCES]

Abrahamsson K, Choo KS, Pedersen M, Johansson G, Snoeijs P, 2003. Effects of temperature on the production of hydrogen peroxide and volatile halocarbons by brackish-water algae. *Phytochemistry*, **64**: 725~734

Anderson RA, Broadhurst CL, Polansky MM, Schmidt WF, Khan A, Flanagan VP, Schoene NW, Graves DJ, 2004. Isolation and characterization of polyphenol type-a polymers from cinnamon with insulin-like biological activity. *J Agric Food Chem*, **52**: 65~70

Babitskaia VG, Shcherba VV, Ikonnikova NV, 2000. Melanin complex of the fungus *Inonotus obliquus*. *Prikl Biokhim Mikrobiol*, **36**: 439~444

Bradford E, 1976. A rapid and sensitive method for the quantification of microgram quantities of protein utilizing the principle of protein-dye binding. *Anal Biochem*, **72**: 248~254

Burczyk J, Gawron A, Slotwinska M, Smietana B, Terminska K, 1996. Antimitotic activity of aqueous extracts of *Inonotus obliquus*. *Boll Chim Farm*, **135**: 306~309

Chen CY, Milbury PE, Lapsley K, Blumberg JB, 2005. Flavonoids from almond skins are bioavailable and act synergistically with vitamins C and E to enhance hamster and human LDL resistance to oxidation. *J Nutr*, **135**: 1366~1373

Chen TJ, Jeng JY, Lin CW, Wu CY, Chen YC, 2006. Quercetin inhibition of ROS-dependent and–independent apoptosis in rat glioma C6 cells. *Toxicology*, **223**: 113~126

Christensen AB, Gregersen PL, Olsen CE, Collinge DB, 1998a. A flavonoid 7-*O*-methyltransferase is expressed in barley leaves in response to pathogen attack. *Plant Mol Biol*, **36**: 219~227

Christensen AB, Gregersen PL, Schröder J, Collinge DB, 1998b. A chalcone synthase with an unusual substrate preference is expressed in barley leaves in response to UV light and pathogen attack. *Plant Mol Biol*, **37**: 849~857

Cui Y, Kim DS, Parka KC, 2005. Antioxidant effect of *Inonotus obliquus*. *J Ethnopharmacol*, **96**: 79~85

Dai F, Miao Q, Zhou B, Yang L, Liu ZL, 2006. Protective effects of flavonols and their glycosides against free radical-induced oxidative hemolysis of red blood cells. *Life Sci*, **78**: 2488~2493

D'Cunha GB, 2005. Enrichment of phenylalanine ammonia lyase activity of *Rhodotorula yeast*. *Enzyme Microb Tech*, **36**: 498~502

Dewick PM, 2002. Medicinal natural products: a biosynthetic approach. John Wiley & Sons, Inc. Press, New York. 130-132

Dixon RA, Paiva NL, 1995. Stress-induced phenylpropanoid metabolism. *Plant Cell*, **7**: 1085~1097

Dixon RA, Steele CL, 1999. Flavonoids and isoflavonoids - a gold mine for metabolic engineering. *Trends Plant Sci*, **4**: 394~400

Formica JV, Regelson W, 1995. Review of the biology of Quercetin and related bioflavonoids. *Food Chem Toxicol*, **33**: 1061~1080

He J, Feng XZ, 2001. Studies on chemical constituents of *Fuscoparia obliqua*. *Chin Tradit Herbal Drugs*, **32**: 4~6 (in Chinese)

Huang NL, 2002. *Inonotus obliquus*. *Edible Fungi of China*, **21**: 7~8 (in Chinese)

Jing M, Cai YX, Li JR, Zhao H, 1996. 1,10-Phenanthroline-Fe^{2+} oxidative assay of hydroxyl radical produced by H_2O_2/ Fe^{2+}. *Prog Biochem Biophys*, **23**:553~555 (in Chinese)

Kalghatgi KK, Subba Rao PV, 1976. Regulation of L-phenylalanine ammonia-lyase from *Rhizoctonia solani*. *J Bacteriol*, **126**: 568~578

Kanadaswami C, Lee LT, Lee PP, Hwang JJ, Ke FC, Huang YT, Lee MT, 2005. The antitumor activities of flavonoids. *In Vivo*, **19**: 895~909

Kim SH, Kronstad JW, Ellis BE, 2001. Induction of phenylalanine ammonia-lyase activity by tryptophan in *Ustilago maydis*. *Phytochemistry*, **58**: 849~857

Kim YO, Han SB, Lee HW, Ahn HJ, Yoon YD, Jung JK, Kim HM, Shin CS, 2005. Immuno-stimulating effect of the endo polysaccharide produced by submerged culture of *Inonotus obliquus*. *Life Sci*, **77**: 2438~2456

Kukulyanskaia TA, Kurchenko NV, Kurchenko VP, Babitskaya VG, 2002.Physicochemical properties of melanins produced by the sterile form of *Inonotus obliquus* ("Chagi") in natural and cultivated fungus. *Prikl Biokhim Mikrobiol*, **38**: 68~72

Mau JL, Lin HC, Chen CC, 2002. Antioxidant properties of several medicinal mushrooms. *J Agric Food Chem*, **50**: 6072~6077

Meda A, Lamien CE, Romito M, Millogo J, Nacoulma OG, 2005. Determination of the total phenolic, flavonoids and praline contents in Burkina Fasan honey, as well as their radical scavenging activity. *Food Chem*, **91**: 571~577

Molitoris HP, 1994. Mushrooms in medicine. *Folia Microbiol*, **39**: 91~98

Moore BS, Hertweck C, Hopke JN, Izumikawa M, Kalaitzis JA, Nilsen G, O'Hare T, Piel J, Shipley PR, Xiang L, Austin MB, Noel JP, 2002. Plant-like biosynthetic pathways in bacteria: from benzoic acid to chalcone. *J Nat Prod*, **65**: 1956~1962

Mori T, Sakurai M, Sakuta M, 2001. Effects of conditioned medium on activities of PAL, CHS, DAHP synthase (DS-Co and DS-Mn) and anthocyanin production in suspension cultures of *Fragaria ananassa*. *Plant Sci*, **160**: 355~360

Mortensen SR, Brimijoin S, Hooper MJ, Padilla S, 1998. Comparison of the *in vitro* sensitivity of rat acetylcholinesterase to chlorpyrifos-oxon: what do tissue IC50 values represent? *Toxicol Appl Pharmacol*, **148**: 46~49

Park JS, Rho HS, Kim DH, Chang IS, 2006. Enzymatic preparation of Kaempferol from green tea seed and its antioxidant activity. *J Agric Food Chem*, **54**: 2951~2956

Romero Alvira D, Guerrero Navarro L, Gotor Lazaro MA, Roche Collado E, 1995. Oxidative stress and infectious pathology. *An Med Interna*, **12**: 139~149

Saar M, 1991. Fungi in Khanty folk medicine. *J Ethnopharmacol*, **31**: 175~179

Stanner SA, Hughes J, Kelly CN, Buttriss J, 2004. A review of the epidemiological evidence for the 'antioxidant hypothesis'. *Public Health Nutr*, **7**: 407~422

Valencia A, Kochevar IE, 2006. Ultraviolet A induces apoptosis via reactive oxygen species in a model for Smith-Lemli-Opitz syndrome. *Free Radic Biol Med*, **40**:641~650

Wang SY, Wu JH, Cheng SS, Lo CP, Chang HN, Shyur LF, Chang ST, 2004. Antioxidant activity of extracts from *Calocedrus formosana* leaf, bark, and heartwood. *J Wood Sci*, **50**: 422~426

Xiao HS, He WJ, Fu WQ, Cao HY, Fan ZN, 1999. A spectrophotometer method testing oxygen radicals. *Prog Biochem Biophys*, **26**: 180~182 (in Chinese)

Zuurbier KWM, Leser J, Berger T, Hofte AJP, Schröder G, Verpoorte R, Schröder J, 1998. 4-hydroxy-2-pyrone formation by chalcone and stilbene synthase with non-physiological substrates. *Phytochemistry*, **49**: 1945~1951

[附中文参考文献]

何坚，冯孝章，2001. 桦褐孔菌化学成分的研究. 中草药，**32**: 4~6

黄年来，2002. 俄罗斯神秘的民间药用真菌——桦褐孔菌. 中国食用菌，**21**: 7~8

金鸣，蔡亚欣，李金荣，赵辉，1996.邻二氮菲-Fe^{2+}氧化法检测H_2O_2/Fe^{2+}产生的羟自由基. 生物化学与生物物理进展，**23**: 553~555

萧华山，何文锦，傅文庆，曹红云，范子南，1999. 一种用分光光度计检测氧自由基的新方法. 生物化学与生物物理进展，**26**: 180~182

Involvements of *S*-nitrosylation and denitrosylation in the production of polyphenols by *Inonotus obliquus*

Weifa Zheng · Yubing Liu · Shenyuan Pan · Weihua Yuan · Yucheng Dai · Jiangchun Wei

Received: 17 December 2010 / Revised: 3 March 2011 / Accepted: 6 March 2011 / Published online: 6 April 2011

Abstract Nitric oxide (NO) has been evidenced to mediate biosynthesis of polyphenols in *Inonotus obliquus*. However, it remains unknown how NO regulates their biosynthesis. Here we show that higher cellular NO levels coincided with higher accumulation of *S*-nitrosothiols (SNO; the products of NO combined with a specific residue in glutathione or proteins) and polyphenols, and higher activity of denitrosylated *S*-nitrosoglutathione reductase (GSNOR) and thioredoxin reductase (TrxR). This homeostasis was breached by GSNOR or TrxR inhibitors. Inhibiting GSNOR boosted TrxR activity, but reduced SNO formation, coinciding with an enhanced production of polyphenols. Likewise, inhibiting TrxR increased GSNOR activity and SNO production, but downregulated accumulation of polyphenols. Inhibiting GSNOR or TrxR also modified the polyphenolic profiles of *I. obliquus*. Suppressing GSNOR-enhanced biosynthesis of phelligridins C and H, inoscavin C and methyl inoscavin B, but reduced that of phelligridin D, methyl inoscavin A, davallialactone and methyl davallialactone, the typical polyphenols in *I. obliquus*. Similarly, downregulating TrxR increased production of phelligridin D, methyl inoscavin A, davallialactone, and methyl davallialactone, but shrinking that of phelligridins C and H, methyl inoscavin B and inoscavin C. Thus, in *I. obliquus*, the state of *S*-nitrosylation and denitrosylation affects not only the accumulation of polyphenols, but also their metabolic profiles.

Keywords *Inonotus obliquus* · Nitric oxide · *S*-nitrosylation · Denitrosylation · Polyphenols

W. Zheng (✉) · Y. Liu · W. Yuan
Key Laboratory for Biotechnology on Medicinal Plants of Jiangsu Province, Xuzhou Normal University, Xuzhou 221116, China
e-mail: yyzw@xznu.edu.cn

S. Pan
School of Life Sciences, Xuzhou Normal University, Xuzhou, China

Y. Dai (✉)
Institute of Applied Ecology, Chinese Academy of Sciences, Shenyang 110116, China
e-mail: yuchengd@yahoo.com

J. Wei
Institute of Microbiology, Chinese Academy of Sciences, Beijing 110101, China

Introduction

The medicinal fungus *Inonotus obliquus* (Fr.) Pilat (*Hymenochaetaceae*) has been used as a folk remedy in Russia and Eastern Europe for more than four centuries, where its beneficial effects on treating a wide variety of human diseases, in the absence of any unacceptable toxic side effects, has become established (Zheng et al. 2009c). In nature, this fungus inhabits primarily the trunks of *Betula* trees in form of sclerotia and is termed “Chaga”. It produces a diverse range of metabolites including lanostane-type triterpenoids (Shin et al. 2002), styrylpyrone derivatives (also termed hispidin analogs) (Lee and Yun 2007) and melanins (Babitskaia et al. 2000). Among these are biologically active compounds possessing hypoglycaemic, hepato-protective, anti-fungal, antitumor, and antiviral activities (Zheng et al. 2009c). Recently, evidences have also been provided that styrylpyrone derivatives are potent antioxidant agents and an effective remedy for protecting

This paper was originally published in *Applied Microbiology and Biotechnology*, 90（5）: 1763-1772, 2011.

against oxidative stress-induced human diseases including cardiovascular, neurodegenerative, and autoimmune diseases (Zheng et al. 2009b).

In nature, this fungus is restricted to cold habitats (45° N–50° N latitude) and grows very slowly, suggesting that the naturally occuring sclerotial form is not a reliable source for obtaining these bioactive polyphenols (Zheng et al. 2009c). Previous attempts to grow *I. obliquus* under submerged culture conditions resulted in low production of polyphenols together with low pharmacological activities of the cultured products (Zheng et al. 2008). Recently, Zheng et al. (2009a) found that an exposure of *I. obliquus* to a fungal elicitor, the debris of cell walls of *Alternaria alternata*, led to an enhanced production of polyphenols and evidenced the involvement of nitric oxide (NO) in modulating metabolism of polyphenols by the fungus under submerged culture conditions (Zheng et al. 2009a). It has been made clear that NO acts as a key signaling molecule able to drive the expression of a battery of redox-regulated defense genes in higher plants (Durner et al. 1998; Durner and Klessig 1999), and hence the production of antimicrobial metabolites (Romero-Puertas and Delledonne 2003). It is also known that *S*-nitrosylation, the addition of a NO moiety to a specific cysteine residue in glutathione or proteins, to form *S*-nitrosothiols (SNO) or *S*-nitrosylated proteins emerged as a principal mechanism by which NO orchestrates cellular functions in *Arabidopsis* (Wang et al. 2008). Emerging data suggest that SNO and its turnover constitute a key mechanism to control cellular SNO levels. The addition of a NO moiety to the antioxidant tripetide glutathione forms *S*-nitrosoglutathione (GSNO), which may function as a mobile reservoir of NO bioactivity (Wang et al. 2008). GSNO reductase (GSNOR), an enzyme highly specific for GSNO in bacteria and mice, metabolizes this molecule and thereby reduces the cellular levels of SNO (Liu et al. 2001). The reduction in GSNOR activity in higher plants results in an increased accumulation of SNO and a reduced expression of defense-related genes (Wang et al. 2008). Recently, thioredoxin reductase (TrxR) was also discovered to mediate SNO denitrosylation, and to protect nitrosative stress in microorganisms, mammals, and plants (Benhar et al. 2009; Forrester et al. 2009). Yet, by far, relatively little is known about the involvements of these two denitrosylated enzymes in the biosynthesis of polyphenols by *I. obliquus*.

In natural habitat, the major class of polyphenols produced by *I. obliquus* comprises styrylpyrone derivates. Styrylpyrone is thought to be synthesized either from phenylalanine, via cinnamyl derivates (phenylpropanoid pathway) combined with acetate, malonate through the polyketide pathway, or from the condensation of 4-hydroxy-6-methyl-2-pyrone, which is formed by the reaction of three molecules of acetyl-SCoA and one molecule of 3,4-dihydroxybenzol-SCoA (Mo et al. 2004). In our previous study, we found that the exposure of *I. obliquus* mycelia to fungal elicitor enhanced the production of styrylpyrone derivatives and increased their capacity for scavenging free radicals (Zheng et al. 2009a). But it remains unknown how the changes in denitrosylation capacity affect the metabolic profiles of styrylpyrone derivatives.

In this study, we investigated the involvements of *S*-nitrosylation and denitrosylation in the biosynthesis of styrylpyrone derivates in *I. obliquus* in order to elucidate the impact of denitrosylated enzymes on the biosynthesis of polyphenols and thereby to gain insights into understanding of NO-mediated mechanisms for biosynthetic metabolism of polyphenols in submerged cultures of *I. obliquus*.

Materials and methods

Fungal materials, inoculum preparation, and conditions for submerged culture

I. obliquus (Persoon: Fries) Pila ATCC 28281 was obtained from the fungal collection center of the Institute of Microbiology, Chinese Academy of Sciences, Beijing China. The fungus was maintained on malt extract agar slants containing 3% malt extract, 0.3% peptone and 2% agar at pH 5.6. The slants were cultivated at 25 °C for 2 weeks. When the mycelia overgrew the slants, they were stored at 4 °C and subcultured every 3 months. The preparation of standardized inoculum was conducted as described (Zheng et al. 2009b). The inocula (0.1 g/l dry mycelia) were incubated for 7 days in 500-ml conical flask containing 200 ml medium consisting of glucose (2%), peptone (0.3%), KH_2PO_4 (0.01%), and $MgSO_4{\cdot}7H_2O$ (0.05%) with an initial pH value of 5.5 (not modified after autoclave; Zheng et al. 2009c).

Dynamic changes in production of NO, SNO and polyphenols, and the activity of GSNOR and TrxR in response to the addition of fungal elicitor

Fungal elicitor (cell walls of *A. alternata*), prepared as previously described (Zheng et al. 2009a), was added simultaneously with the inoculation of *I. obliquus* at the final concentrations of 20, 40, and 60 μg/l, respectively. These inoculated media were incubated by shaking on an orbital platform shaker (Shen Neng Bao Cai, Shanghai China) at 26 °C and 140 rpm for 15 days. Samples (20 ml) were taken 24 h (day 1) after inoculation followed by every other day up to the end of incubation for determining production of NO, SNO, and polyphenols, as well as the activity of GSNOR and TrxR. Cultures without supplementing fungal elicitor were used as controls.

NO-dependent biosynthesis of polyphenols

Sodium nitroprusside (NO donor; Sigma, MO, USA) and aminoguanidine (NO synthase inhibitor; Sigma, MO, USA) were added immediately after inoculation of *I. obliquus* at a final concentration of 0.1 mM for sodium nitroprusside and 40 mM for aminognanidine The inoculated media were incubated in the conditions identical to those quoted above. Samples were withdrawn 24 h (day 1) after inoculation followed by every other day up to the end of incubation for the assays of dynamic accumulation of mycelial biomass, total polyphenols and SNO, and time–course changes of GSNOR and TrxR activity. Controls consisted of the cultures with or without supplementing fungal elicitor.

Effects of GSNOR and TrxR activity on production of polyphenols

The GSNOR inhibitor mithramycin A (Sigma, MO, USA) and TrxR inhibitor auranofin (Sigma, MO, USA) was supplemented respectively into the cultures of *I. obliquus* 7 days after inoculation and fungal elicitor was added at a final concentration of 40 μg/l 2 h after adding the inhibitor of GSNOR and TrxR. The cultures were harvested 2 h after the addition of fungal elicitor for assaying production of polyphenols and SNO, and the activity of GSNOR, and TrxR.

Analytical protocols

Mycelial biomass was estimated as previously detailed (Zheng et al. 2009c). For measuring total mycelial polyphenols, mycelial samples were washed three times with pure water and then extracted following the methods as described (Zheng et al. 2011b). The levels of total mycelial polyphenols in the extracts were determined using the Folin–Ciocalteu reagent (Singleton and Rossi 1965). For determining cellular levels of NO and SNO, and the activity of GSNOR and TrxR, washed mycelial pellets were homogenized by ultrasonication with 50 mM Tris/HCl buffer (pH 8.0) for three times with 30 s for each. The homogenates were centrifuged at 10,000×*g* at 4 °C for 10 min, and the supernatants were used for the assays. NO was determined using commercial reagent kit (Jiangcheng Biotech Inc, Nanjing, China) following specifications. Cellular levels of SNO and GSNOR activity were detected as described, respectively (Feechan et al. 2005; Rusterucci et al. 2007). TrxR activity was estimated using DTNB method (Hill et al. 1997). Protein content was measured by the Coomassie blue-binding method (Bradford 1976).

Effects of GSNOR and TrxR activity on metabolic profiles of polyphenols

Effects of GSNOR and TrxR activity on metabolic profiles of polyphenols were determined by Nuclear Magnetic Resonance (NMR)-based metabolomic analysis. These include the measurements of ^{1}H NMR spectroscopy and proton assignments, and data reduction and pattern recognition. For measuring ^{1}H NMR spectroscopy, an aliquot of 50 ml culture liquid was centrifuged at 1,000×*g* for 10 min. The sediment was washed three times with pure water, which resulted in the isolation of mycelia from culture filtrates. The mycelia were disrupted by motar and pestle under liquid nitrogen and extracted with ethanol/acetone (1:1, *v*/*v*), and the supernatant was lyophilized. The culture filtrates were also lyophilized. Mycelial extracts and lyophilized powder of culture filtrates were integrated. For NMR spectral measurements, a total of 50 mg of each sample was dissolved in a mixed solution of 600 μl deuterated methyl disulfide (DMSO-d_6) and 50 μl of 100 mM sodium phosphate (prepared with D_2O; pH 7.4) in a NMR tube for ^{1}H-NMR (1D and 2D) measurements following the parameters previously described (Zheng et al. 2011a) on a 400-MHz NMR Spectrometer (Bruker). Proton signals were assigned by referencing the standards as detailed in our previous study (Zheng et al. 2011a). For differing changes of metabolic profiles in *I. obliquus* grown in the cultures with or without the addition of auranofin and mithramycin A, ^{1}H NMR spectra were integrated for data reduction and pattern recognition. The proton resonances of polyphenols in ^{1}H NMR spectra are present predominantly within the range between δ 10.0 and δ 6.0 ppm (Lee and Yun 2006), and thereby were automatically data-reduced to 400 integral segments of equal length (δ 0.01 ppm) using Mestrec 4.86 software (Mestrelab Research, Alicante, Spain), with each segment consisting of the integral of the NMR region to which it was associated. The data were normalized to total spectral area and centered scaling was applied before pattern recognition analysis. Principal component analysis (PCA) was performed using a mean-centered approach with SIMCA P-11 (Umetrics, Umea, Sweden) software following the procedures as those described previously (Zheng et al. 2011a).

Statistics

All the experiments were performed by ten independent experiments. Results from representative experiments are expressed as means±standard deviation. Data of all experiments were analyzed by *t* test using software Statistic Products and Social Solutions 11 (SPSS 11.0, SSPS Company, USA). The assumptions of analysis of variance were considered to be statistically significant at $p<0.05$.

Results

Effects of cellular NO levels on production of polyphenols and SNO, and the activity of GSNOR and TrxR

Figure 1 describes dynamic changes in endogenous production of NO, the accumulation of SNO and polyphenols, and the activity of GSNOR and TrxR in response to the addition of different levels of fungal elicitor. Control cultures reached maximum levels of all measured compounds and enzymatic activities at day 13 of cultivation (see Fig. 1). Additions of fungal elicitor at concentrations of 20, 40, and 60 μg/l increased amounts of compounds and enzymatic activities over the whole cultivation period with highest levels also at day 13 of cultivation (20 and 60 μg/l fungal elicitor) and at day 11 or 13 of cultivation (40 μg/l fungal elicitor). Addition of 40 μg/l fungal elicitor was best with increasing production of endogenous NO to 1,228.0± 24.2 nM compared to 876.0±29.8 nM in the controls, levels of total extracellular polyphenols to 367.6±15.6 mg/l compared to 185.5±7.8 mg/l in the controls, total mycelial polyphenols to 40.0±4.0 mg/g compared to 36.8±6.1 mg/g in the controls, and cellular levels of SNO to 31.5± 1.6 nM compared to 28.9±1.3 nM in the controls. Highest

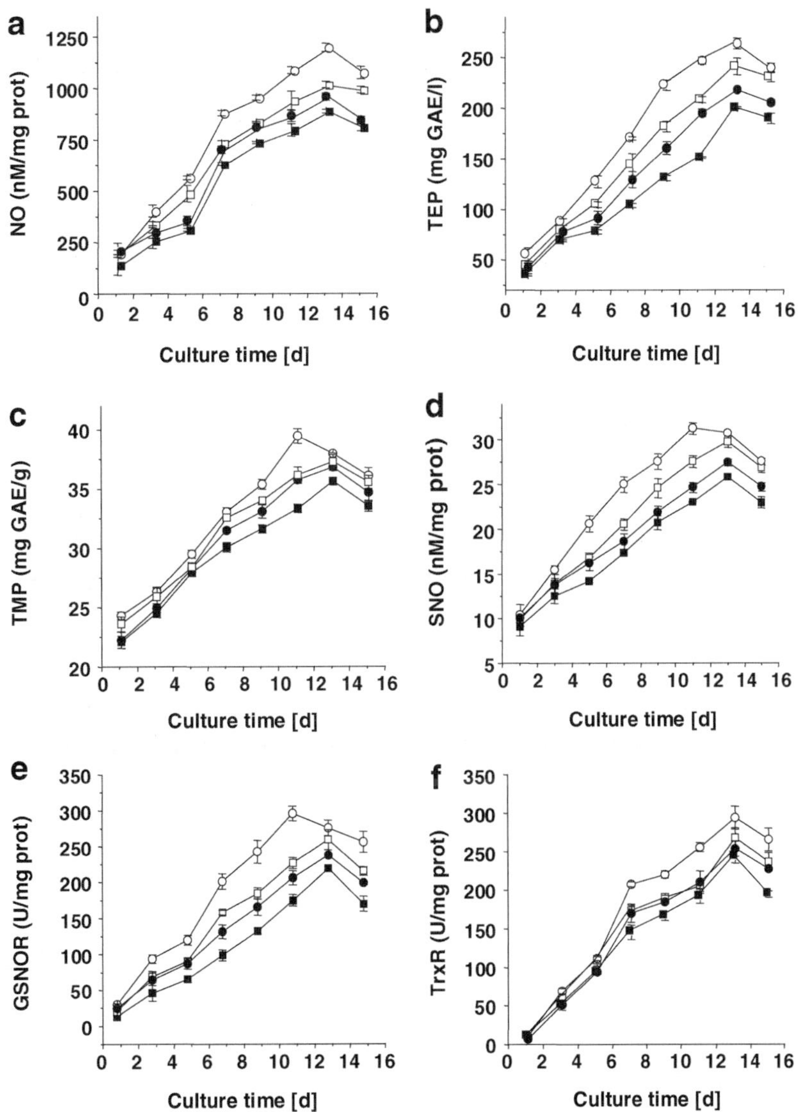

Fig. 1 Effects of fungal elicitor (FE) on endogenous formation of nitric oxide (NO) (**a**), accumulation of total extracellular polyphenols (TEP) (**b**), total mycelial polyphenols (TMP) (**c**), the homeostasis in production of *S*-nitrosothiols (SNO) (**d**) and the activity of SNOR (**e**) and TrxR (**f**) in submerged cultures of *I. obliquus*. Results are the mean of ten individual cultures and the error bars the standard deviations. *Filled square* control cultures, *empty square* fungal elicitor at a final concentration of 20 μg/l, *empty circle* fungal elicitor at a final concentration of 40 μg/l, *filled circle* fungal elicitor at a final concentration of 60 μg/l. The content of polyphenols is indicated as the gallic acid equivalents (GAE)

GSNOR activities in cultures with 40 μg/l were 329.9± 23.4 U/mg prot/min and highest TrxR activities 309.9± 21.2 U/mg prot/min compared to GSNOR activities of 234.9±3.7 U/mg prot/min and TrxR activities 254.2± 11.4 U/mg prot/min in control cultures.

Dependence of biosynthesis of polyphenols on NO production

In order to further evidence the effects of NO production on biosynthesis of polyphenols, we incubated the cultures of *I. obliquus* in the presence of fungal elicitor at 40 μg/l (stimulus for endogenous production of NO), sodium nitrupresside at 0.1 mM, and aminoguanidine at 0.4 mM. Figure 2 illustrates the dynamic changes in the accumulation of mycelial biomass, polyphenols, SNO, and the activity of GSNOR and TrxR following the addition of these modulators of NO production. Control cultures reached maximum levels of mycelial biomass and all measured compounds at day 13 of cultivation and enzymatic activities at day 11 of cultivation (see Fig. 2). Addition of fungal elicitor at 40 μg/l increased the amounts of mycelial

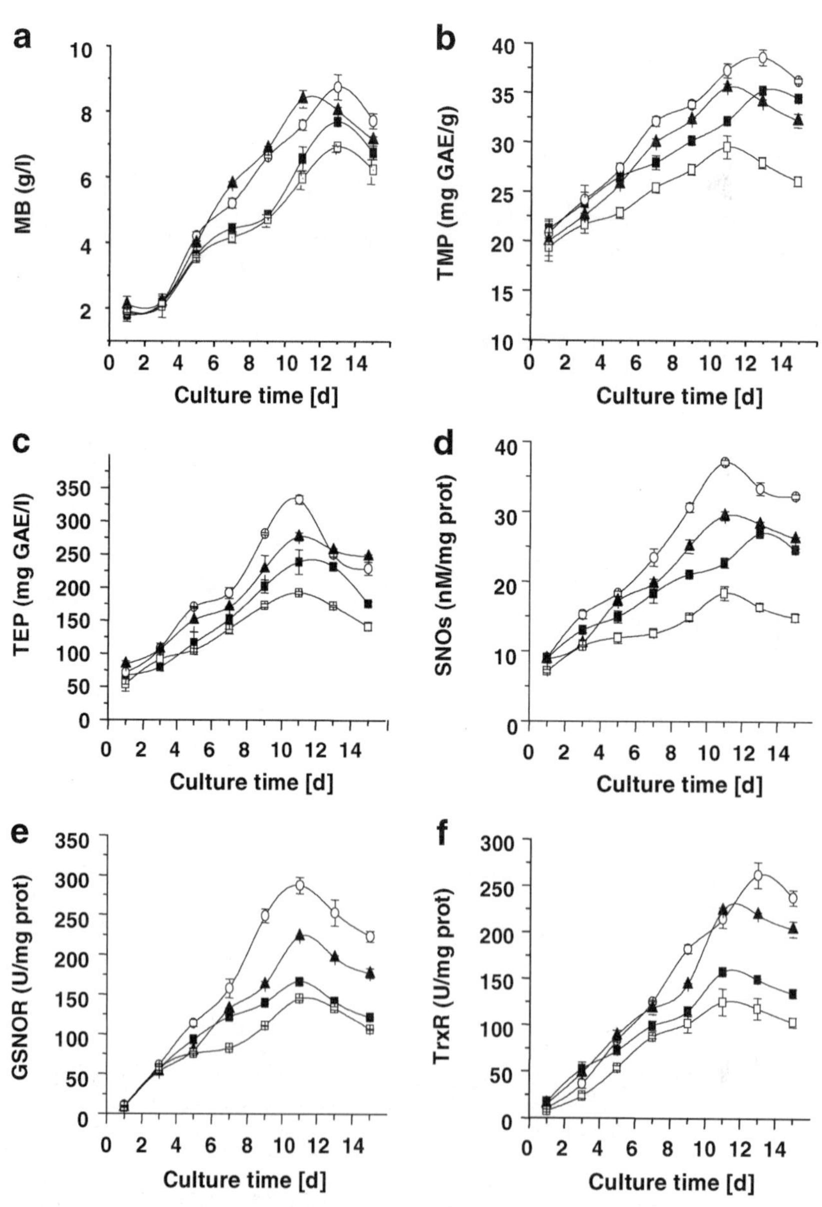

Fig. 2 Time-course of production of mycelial biomass (**a**), total mycelial polyphenols (TMP) (**b**), total extracellular polyphenols (TEP) (**c**), and SNO (**d**), the dynamic changes in the activity of GSNOR (**e**), and TrxR (**f**) in cultures of *I. obliquus* grown in non-supplemented medium, or exposed to (1) fungal elicitor (FE), (2) the NO donor sodium nitroprusside (SNP), and (3) NO synthase (NOS) inhibitor aminoguanidine (AG), indicating the dependence of biosynthesis of polyphenols on NO formation. Results are the mean of 10 independent experiments and error bars are standard deviations. *Filled square* control cultures, *filled triangle* fungal elicitor-added cultures (40 μg/l), *empty circle* sodium nitroprusside-added cultures (0.1 mM), *empty square* aminoguanidine-added cultures (40 mM)

biomass, total extracellular polyphenols and SNO, and enzymatic activities over the whole cultivation period with the levels at day 11 or 13 of cultivation. Addition of aminoguanidine at 0.4 mM, however, evidently reduced the amounts of mycelial biomass, total mycelial and extracellular polyphenols and SNO, and enzymatic activities over the whole cultivation period compared to those found in the control cultures. Addition of sodium nitrupresside at 0.1 mM was best with increasing the production of mycelial biomass to 8.9±0.9 against 7.4±0.4 g/l in the controls, levels of total mycelial polyphenols to 38.8±3.1 against 35.1±2.3 mg/g in the controls, levels of total extracellular polyphenols to 327.6±3.2 against 238.7±27.2 mg/l in the controls, and cellular levels of SNO to 37.6±0.3 against 27.2±0.3 in the controls. Highest GSNOR activities in cultures with sodium nitroprusside at 0.1 mM were 280.4±20.1 U/mg prot/min and TrxR activities 269.9±24.8 U/mg prot/min compared to GSNOR activities of 171.4±1.4 U/mg prot/min and TrxR activities of 158.0±2.5 U/mg prot/min in the control cultures.

Dependence of biosynthesis of polyphenols on GSNOR and TrxR activity

For determining the involvements of GSNOR and TrxR in the biosynthesis of polyphenols, GSNOR inhibitor mithramycin A and TrxR inhibitor auranofin was supplemented to the cultures of *I. obliquus* 2 h before adding fungal elicitor at 40 μg/l. Figure 3 depicts the changes in the activity of GSNOR and TrxR, and the accumulation of SNO and polyphenols in response to the addition of mithramycin A and auranofin followed by adding fungal elicitor. GSNOR activity was observed at 197.8±18.5 U/mg prot/min in control cultures, and these were enhanced to 265.8± 14.1 U/mg prot/min in fungal elicitor-added cultures. Following the addition of mithramycin A and auranofin,

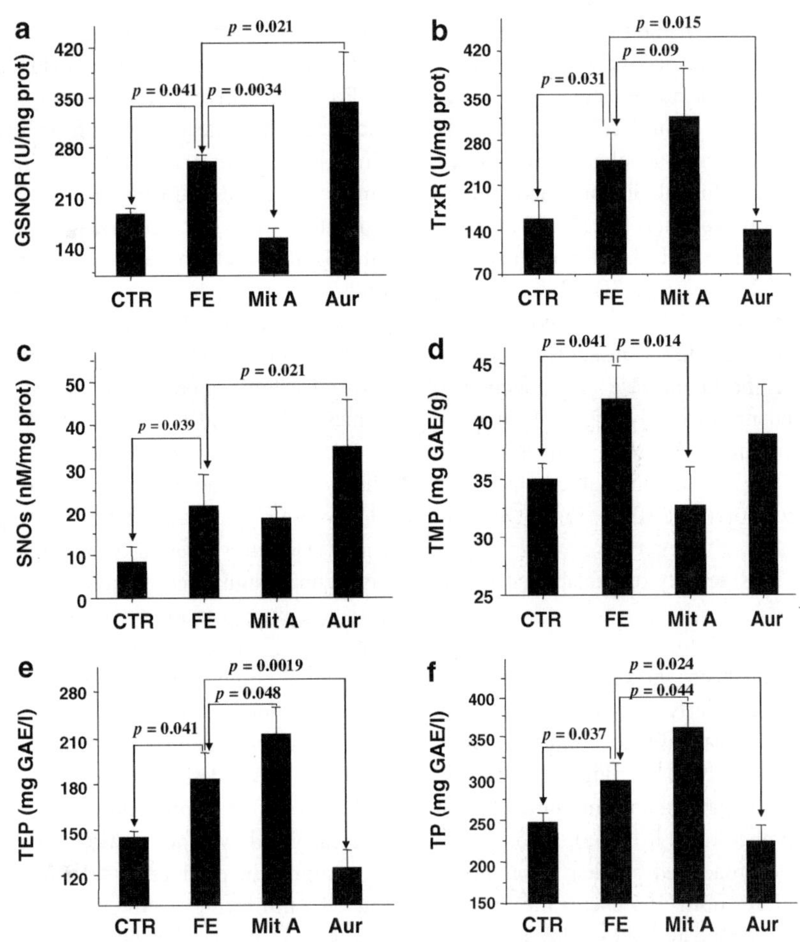

Fig. 3 Dependence of biosynthesis of polyphenols on GSNOR and TrxR activity. The activity of GSNOR (**a**) and TrxR (**b**) and the accumulation of S-nitrosothiols (SNOs) (**c**), total mycelial polyphenols (TMP) (**d**), total extracellular polyphenols (**e**), and total polyphenols (**f**) in *I. obliquus* grown in non-supplemented medium, or in the cultures supplemented by fungal elicitor (FE) (40 μg/l), by fungal elicitor at 40 μg/l and mithramycin A (Mit A) at 2.58 μM, or by fungal elicitor at 40 μg/l and auranofin (Aur) at 3.24 μM. Results are the mean of 10 individual cultures and error bars standard deviations

levels of GSNOR activity were modified to 154.8±14.0 and 347.3±32.6 U/mg prot/min, respectively (Fig. 3a). In control cultures, levels of TrxR activity were found at 157.9±33.4 U/mg prot/min, and these were upregulated to 253.5±61.3 U/mg/min prot in fungal elicitor-added cultures. Supplement of mithramycin A further increased TrxR activity to 315.0±67.4 U/mg prot/min, whereas addition of auranofin reduced TrxR activity (Fig. 3b). Cellular levels of SNO reached 7.7±3.1 nM/mg prot/min in control cultures and increased to 22.4±6.8 nM/mg prot/min in fungal elicitor-added cultures. Inhibiting GSNOR did not increase SNO levels, but suppressing TrxR triggered an upregulation of SNO to the levels of 35.7±9.9 nM/mg prot/min (Fig. 3c).

Accumulation of polyphenols was evidently affected by the activity of GSNOR or TrxR. As illustrated in Fig. 3, the levels of total mycelial polyphenol were seen at 35.8±3.0 mg/g in the control cultures and increased to 43.8±3.5 mg/g in fungal elicitor-added cultures. These were reduced to 37.8±5.3 and 38.4±6.1 mg/g in mithramycin A- and auranofin-added cultures, respectively (Fig. 3d). Simultaneously, total extracellular polyphenols were determined to be 147.0±9.5 mg/l in control cultures, and to be 187.5±26.9 mg/l in fungal elicitor-supplemented cultures. Addition of mithramycin A in fungal elicitor-added cultures enhanced total extracellular polyphenols to 218.6±32.2 mg/l, while addition of auranofin reduced total extracellular polyphenols to 121.5±23.2 mg/l (Fig. 3f). Based on the measurement of mycelial biomass (data not shown), total yields of polyphenols was calculated. Consequently, inhibiting GSNOR led to a significant increase in production of total polyphenols at amounts evidently higher than those seen in the cultures supplemented with fungal elicitor. In contrast, inhibiting TrxR triggered a remarkable reduction in total polyphenols accumulation (Fig. 3f).

Effects of GSNOR and TrxR activity on metabolic profiles of polyphenols

^{1}H NMR spectroscopic analysis showed that metabolic profiles of polyphenols, particularly styrylpyrone derivates differed in the control cultures and the cultures supplemented with fungal elicitor only, both fungal elicitor and mithramycin A, and both fungal elicitor and auranofin. In ^{1}H NMR spectra of control cultures (Fig. 4a), according to Lee and Yun (2006), the broadened singlets overlapping between δ 7.04 and 7.10 ppm represent the typical resonances of H-9 in styrylpyrone derivates. Also according to Lee and Yun (2006), typical downfield singlets between δ 8.10 and 8.36 ppm in the ^{1}H NMR spectra of the control cultures suggest the presence of the most common styrylpyrone derivates phelligridins C, D, and H. In addition, typical doublets with a coupling constant of 13.2 Hz at δ 5.76 and 5.66 ppm coincide with those of davallialactone and methyl davallialactone (Lee and Yun 2006). Combined with other typical resonances, doublets at δ 6.67 (J=16.0 Hz) and 7.40 (J=16.0 Hz), and 6.68 (J=16.0 Hz) and 7.33 ppm (J=16.0 Hz) represent the presence of inoscavins B and C, respectively (Zheng et al. 2011a). Moreover, the singlet at δ 6.44 ppm, and the doublets at δ 6.65 (J=16.0), 6.70 (J=8.4), and 6.84 ppm (J=8.4) suggested the existence of methyl inoscavin A (Zheng et al. 2010). In fungal elicitor-added cultures (see Fig. 4a), the signals representing inoscavins B and C, and methyl inoscavin A were all enhanced, conforming the earlier work by Zheng et al. (2009a). Following the inhibition of GSNOR (in mithramycin A-added cultures), the signals representing phelligridins C and H, inoscavin B and methyl inoscavin B [typically resonating at δ 6.39 s and δ 7.22 (d, J=2.0 Hz), 6.84 (d, J=8.4), 6.98 ppm (dd, J=8.0, 2.0 Hz)] were enhanced (Fig. 4a). Similarly, the inhibition of TrxR (in auranofin-added cultures) resulted in the reduction of the signals incarnating phelligridins C and H, inoscavins B and C, and methyl inoscavin B (Fig. 4a).

For quantitative differentiation of these signals among these styrylpyrone derivates from different culture conditions, chemometric methods including PCA were employed. Figure 4b describes the PCA score plot derived from the ^{1}H NMR spectra, indicating the discrepancies among the metabolites from control, fungal elicitor-, mithramycin A-, and auranofin-added cultures. For comparisons from the PCA model, the statistics for differentiating metabolic profiles by different culture conditions showed a high goodness of fit and predictability with R^2 values of 0.87 and Q^2 values of 0.79.

In order to identify the constituents responsible for the changes by the addition of mithramycin A and auranofin in PCA score plots as a reflection to the inhibition of GSNOR and TrxR, the datasets were standardized followed by orthogonal signal corrections to remove any unrelated variation in X variables (^{1}HNMR spectra) to Y variables (intensity of proton resonance) or variability in X orthogonal to Y. Figure 4c–f show the comparisons between fungal elicitor- and mithramycin A-added cultures in score (Fig. 4c, e) and loading plots (Fig. 4d, f). In PCA loading plots, the upward sections represent metabolites that were relatively higher in production than the downward sections, whereas the downward sections reveal the metabolites that were higher in production than the upward. As shown in Fig. 4, inhibiting GSNOR resulted in an increased production of phelligridins C and H, inoscavin C and methyl inoscavin B, and a reduced accumulation of phelligridin D, methyl inoscavin A, davallialactone, and methyl davallialactone (Fig. 4d). In comparison, inhibition of TrxR triggered a reduction in accumulating phelligridins C and H, inoscavin C and methyl inoscavin C, and a slight

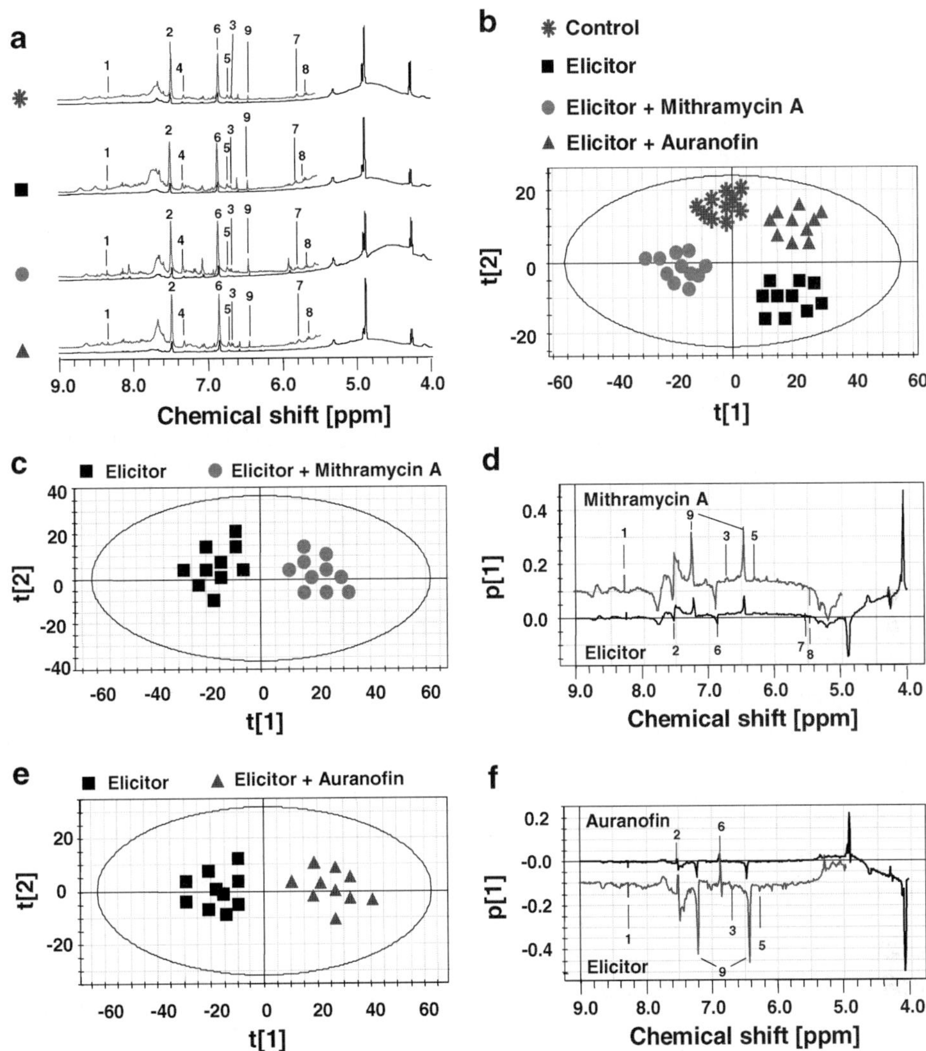

Fig. 4 Phenolic profiles of metabolites produced by *I. obliquus* grown in non-supplemented medium, in the cultures added fungal elicitor at 40 μg/l, or added fungal elicitor at 40 μg/l and mithramycin A (Mit A) at 2.58 μM, or added fungal elicitor at 40 μg/l and auranofin (Aur) at 3.24 μM. **a** ^{1}H NMR spectra, **b** PCA score plot derived from the ^{1}H NMR spectra, showing the discrepancies of phenolic profiles, **c–d** score and loading plots derived from ^{1}H NMR spectra of the metabolites by the fungus from cultures supplemented by fungal elicitor only and by both fungal elicitor and mithramycin A; **e–f** score and loading plots derived from ^{1}H NMR spectra of the metabolites by the fungus from cultures supplemented by fungal elicitor only and by both fungal elicitor and auranofin. *t[1]* and *t[2]* Scores for the first and second principal component, respectively; *p[1]* loading of the first principal component. *1* Phelligridin C, *2* phelligridin D, *3* phelligridin H, *4* inoscavin B, *5* inoscavin C, *6* inoscavin C, *7* methyl inoscavin A, *8* davallialactone, *9* methyl davallialactone (Zheng et al. 2011a)

increase in phelligridin D, methyl inoscavin A, davallialactone, and methyl davallialactone (Fig. 4f). For comparisons between the metabolites from mithramycin A- and auranofin-added cultures from the PCA model, the differences between the two comparisons showed high goodness of fit and predictability with R^2 values of 0.84 and 0.77 and Q^2 values of 0.89 and 0.91, respectively (Fig. 4c, e).

Discussion

It has been evidenced that NO mediates the fungal elicitor-induced biosynthesis of polyphenols in cell suspension cultures of higher plant *Hypericum perforatum* (Xu et al. 2005) and in submerged cultures of *I. obliquus* (Zheng et al. 2009a). In higher plants, SNO formation and its turnover regulate multiple modes of plant disease resistance and GSNOR positively correlates with the production of secondary metabolites and thereby enhances plant disease resistance (Feechan et al. 2005). In *Cryptococcus neoformans*, TrxR is believed to protect cells from oxidative and nitrosative stress (Missall and Lodge 2005). The data presented here show that NO-mediated biosynthesis of polyphenols in *I. obliquus* also involves the formation of SNO. Within a limited range of NO concentrations (<1,250 nM/mg prot), higher cellular levels of NO coincided with higher accumulation of SNO and polyphenols, as well as higher GSNOR and TrxR activity, and GSNOR and TrxR did not show an obvious denitrosylation to the *S*-nitrosylated enzymes responsible for biosynthesis of polyphenols under the tested NO concentrations. However, this homeostasis was disturbed following the inhibition of GSNOR or TrxR. Inhibiting GSNOR resulted in a significant rise in TrxR activity, and a reduced formation of SNO, which coincided with an enhanced production of polyphenols. On the other hand, inhibiting TrxR triggered a remarkable increase in GSNOR activity and in production of SNO, but a reduced accumulation of polyphenols. This suggests that TrxR is responsible for denitrosylation of the *S*-nitrosylated enzymes associated with polyphenol biosynthesis, whereas GSNOR inhibits the denitrosylation of SNO by TrxR and thereby hinders over-biosynthesis of polyphenols that are not necessary for fungal cell normal physiological activities in submerged cultures.

The thioredoxin system, which consists of thioredoxin (Trx), TrxR, and NADPH (Arners and Holmgren 2000), is thought to mediate both basal and stimulus-coupled protein denitrosylation (Forrester et al. 2009). Trx denitrosylates *S*-nitrosylated proteins through its dithiol moiety, thereby forming a reduced protein thiol (−SH) and oxidized Trx; oxidized Trx is reduced (and therefore reactivated) by the seleno-flavoprotein TrxR and NADPH (Benhar et al. 2009). In mammals, *S*-nitrosylation of Trx at cysteine 69 is required for scavenging reactive oxygen species and for preserving the redox regulatory activity of Trx (Haendeler et al. 2002). Owing to the expected existence of highly conserved active sequence of Cys-Gly-Pro-Cys (Collins and Messens 2010), it is reasonable to hypothesize that *S*-nitrosylation of Cys at certain point of Trx in *I. obliquus* is also required for denitrosylation of *S*-nitrosylated proteins contributing to the biosynthesis of polyphenols. The inability to denitrosylation by Trx/TrxR under basal conditions might be contributed by the inhibition from GSNOR via the removal of NO necessary for Trx denitrosylation.

Our results also showed that modification of GSNOR and TrxR activity led to the changes of metabolic profiles of polyphenols. It has been proposed that styrylpyrone derivates like davallialactone and phelligridins in *I. obliquus* are the products of enzymatic rearrangements of interfungin A (Lee and Yun 2007). Thus, changes of GSNOR and TrxR activity may affect these rearrangements leading to the changes in metabolic profiles of polyphenols.

In summary, GSNOR and TrxR are the two important denitrosylation enzymes that work together to control cellular levels of SNO, protect against nitrosative and oxidative stress and govern the biosynthesis of polyphenols not only in the accumulation but the metabolic profiles as well.

This study has examined the involvements of *S*-nitrosylation and denitrosylation in the biosynthesis of polyphenols. Further experiments are needed to determine the mechanisms how GSNOR control the denitrosylation of the enzymes closely associated with biosynthesis of polyphenols, which have already been undertaken in our lab. Regardless, this study provides undeniable evidences that GSNOR and TrxR are, in one way or another, involved in the biosynthesis of polyphenols in *I. obliquus*. Moreover, higher accumulation of polyphenols in GSNOR-inhibited cultures also gains insights into upregulation of bioactive metabolites by the fungus in submerged cultures. This finding is undoubtedly one of the important ways forward to better understand NO-mediated mechanisms in governing biosynthesis of polyphenols by *I. obliquus*.

Acknowledgments Financial support was provided by grants Natural Science Foundation of China (31070052, 30910103907), Natural Science Foundations of Jiangsu Province, China (BK2009084) and the Key Project of Natural Science Foundations of Jiangsu Province, China (BK2010027).

References

Arners ESJ, Holmgren A (2000) Physiological functions of thioredoxin and thioredoxin reductase. Eur J Biochem 267:6102–6109

Babitskaia VG, Shcherba VV, Ikonnikova NV (2000) Melanin complex of the fungus *Inonotus obliquus*. Prikl Biokhim Mikrobiol 36:439–444

Benhar M, Forrester MT, Stamler JS (2009) Protein denitrosylation: enzymatic mechanisms and cellular functions. Nat Rev Mol Cell Biol 10:721–732

Bradford MM (1976) A rapid and sensitive method for the quantitation of microgram quantities of protein utilizing the principle of protein-dye binding. Anal Biochem 72:248–254

Collins F-J, Messens J (2010) Structure, function, and mechanism of thioredoxin proteins. Antioxid Redox Signal 13:1205–1216

Durner J, Klessig DF (1999) Nitric oxide as a signal in plants. Curr Opin Biol 2:369–374

Durner J, Wendelienne D, Klessig DF (1998) Defense gene induction in tobacco by nitric oxide, cyclic GMP, and cyclic ADP ribose. Proc Natl Acad Sci USA 95:10328–10333

Feechan A, Kwon E, Yun B, Wang Y, Pallas J, Lake G (2005) A central role for *S*-nitrosothiols in plant disease resistance. Proc Natl Acad Sci USA 102:8045–8059

Forrester MT, Seth D, Hausladen A, Eyler CE, Foster MW, Mastumoto A, Benhar M, Marshall HE, Stamler JS (2009) Thioredoxin-interacting protein (Txnip) is a feedback regulator of *S*-nitrosylation. J Biol Chem 248:36160–36168

Haendeler J, Hoffmann J, Tischler V, Berk B, Zeiher A, Dimmeler S (2002) Redox regulatory and anti-apototic functions of thioredoxin depend on *S*-nitrosylation at cysteine 69. Nat Cell Biol 4:743–763

Hill KE, McCollum GW, Burk RF (1997) Determination of thioredoxin reductase activity in rat liver supernatant. Anal Biochem 253:123–125

Lee IK, Yun BS (2006) Hispidin analogs from the mushroom *Inonotus xeranticus* and their free radical scavenging activity. Bioorg Med Chem Lett 16:2376–2379

Lee IK, Yun BS (2007) Highly oxygenated and unsaturated metabolites providing a diversity of hispidin class antioxidants in the medicinal mushrooms *Inonotus* and *Phellinus*. Bioorg Med Chem 15:3309–3314

Liu L, Hausladen A, Zeng M, Que L, Heitman J, Stamler JSA (2001) A metabolic enzyme for *S*-nitrosothiol conserved from bacteria to humans. Nature 410:490–494

Missall TA, Lodge JK (2005) Thioredoxin reductase is essential for viability in the fungal pathogen *Cryptococcus neoformans*. Eukaryot Cell 4:487–489

Mo S, Wang S, Zhou G, Yang Y, Li Y, Chen X, Shi J (2004) Phelligridins C-F: cytotoxic pyrano[4-3-c][2]benzopyran-1,6-dione and furo[3,2-c]pyran-4-one derivatives from the fungus *Phellinus igniarius*. J Nat Prod 67:823–828

Romero-Puertas MC, Delledonne M (2003) Nitric oxide signaling in plant-pathogen interactions. IUBMB Life 55:579–583

Rusterucci C, Espunya MC, Diaz M, Chabannes M, Martinez MC (2007) *S*-Nitosoglutathione reductase affords protection against pathogens in *Abraidopsis*, both locally and systemically. Plant Physiol 143:1282–1292

Shin Y, Tamai Y, Terazawa M (2002) Triterpenoids, steroids, and a new sesquiterpene from *Inonotus obliquus* (Pers.: Fr.) Pilat. Int J Med Mushrooms 4:77–84

Singleton VL, Rossi JAJ (1965) Colorimetry of total phenolics with phosphomolybdic phosphotunstic acid reagents. Am J Enol Vitic 16:144–158

Wang Y-Q, Feechan A, Yun B-W, Shafiei R, Hofmann A, Taylor P, Xue P, Yang F-Q, Xie Z-S, Pallas JA, Chu C-C, Loake GJ (2008) *S*-Nitrosylation of AtSABP3 antagonizes the expression of plant immunity. J Biol Chem 284:2131–2137

Xu MJ, Dong JF, Zhu MY (2005) Nitric oxide mediates the fungal elicitor-induced hypericin production of *Hypericum perforatum* cell suspension cultures through a jasmonic-acid-dependent signal pathway. Plant Physiol 139:991–998

Zheng W, Zhao Y, Zhang M, Yin Z, Chen C, Wei Z (2008) Phenolic compounds from *Inonotus obliquus* and their immune stimulating effects. Mycosystema 27:39–47

Zheng W, Miao K, Zhao Y, Zhang M (2009a) Nitric oxide mediates fungal elicitor-enhanced biosynthesis of antioxidant polyphenols in *Inonotus obliquus* in submerged cultures. Microbiol 155:3340–3348

Zheng W, Zhang M, Zhao Y, Miao K, Jiang H (2009b) NMR-based metabonomic analysis on effect of light on production of antioxidant phenolic compounds in submerged cultures of *Inonotus obliquus*. Bioresour Technol 100:4481–4487

Zheng W, Zhang M, Zhao Y, Wang Y (2009c) Accumulation of antioxidant phenolic constituents in submerged cultures of *Inonotus obliquus*. Bioresour Technol 100:1327–1337

Zheng W, Zhang M, Zhao Y, Miao K, Pan S, Cao F, Dai Y (2011a) Analysis of antioxidant metabolites by solvent extraction from sclerotia of *Inonotus obliquus* (Chaga). Phytochem Anal 22:95–102

Zheng W, Zhao Y, Zheng X, Liu Y, Pan S, Dai Y, Liu F (2011b) Production of antioxidant and antitumor metabolites by submerged cultures of *Inonotus obliquus* co-cultured with *Phellinus punctatus*. Appl Microbiol Biotechnol 89:157–167

第六篇　真菌学科发展

Development of the Fungal Sciences

Introducing the MYCOSYSTEMA

MYCOSYSTEMA, an Annual Report of the Systematic Mycology and Lichenology Laboratory of the Institute of Microbiology, Academia Sinica, is an annual publication, carrying papers on the systematics and evolution of fungi, both lichenized and non–lichenized.

The rise and rapid development of molecular biology is undoubtedly a great breakthrough in biology. The techniques of molecular biology have spurred an overall development of biology and will continue doing so. But it must be remembered that molecular biology, in the final analysis, depends on systematics. In his "Study of Plant Communities", Henry J. Oosting (1956, p.30, second ed.) gave a brilliant exposition of the dependence of plant physiology on plant systematics: "The fallacy of doing detailed physiological studies with an unnamed plant is obvious. If the physiologist does not know the species, and sometimes the variety or strain, with which he is working, his conclusions will be limited to the particular group of plants he is using in his experiments. The studies of taxonomists, floristic geographers, and geneticists represent an accumulation of information and data upon which the physiologist can draw and which he can use to make generalizations and comparisons. All this information is connoted by the scientific name of the plant being studied." No recognition of the species of living things is possible until systematic studies have been carried out.

Therefore, it is necessary to carry out comprehensive analyses and research in the systematics and evolution of fungi according to the principles and methods of systematics from all angles (e.g. morphology, anatomy, ecology, physiology, biochemistry, genetics, chorology,) and at all levels (e.g. population biology, ontogeny, cytology, molecular biology). The scientific name of a fungus obtained after comprehensive analyses and research not only shows the position of the species in the natural system, but also contains information on its biological characteristics. Such a study provides an information storage and retrieval system with a high degree of predictivity for the development of the life sciences, for the exploitation and utilization of fungal including lichen resources and for the control of harmful fungi. Without such a system it would be impossible to select material for physiological or biochemical studies, including molecular biology. Even if an unnamed living organism had been chosen for study, its conclusions would be meaningless either in theory or in practice without the reference point provided by the name. Thus it is easy to see the dependence of experimental biology,including molecular biology, on systematics as a support system of biotechnology or bioengineering. That is why systematic and evolutionary studies of fungi are of considerable importance.

In view of the fact that fungi and lichens are scarcely known in many parts of China, as well as the requirements for exploiting and utilizing fungal and lichen resources, for controlling harmful fungi, and for developing life science, research in the systematics and evolution of fungi is indispensable and has been chosen as the long–term goal for the Laboratory. I

This paper was originally published in *Mycosystema*, 1: 1-2, 1988.

believe that the research work in the Laboratory will, in the course of time, be able to approach this goal with the discoveries and the extensive studies of more and more fungi and lichens from China and in a certain group from the world according to the principles and methods of systematics under the guidance of farreaching academic strategies. The MYCOSYSTEMA will mirror the research achievements of the Laboratory towards this end.

WEI Jiang-chun
Beijing 20 May, 1987

《真菌系统》发刊词

《真菌系统》为中国科学院微生物研究所真菌地衣系统学开放研究实验室年报，是报导关于地衣型与非地衣型真菌系统与演化研究论文的定期出版物。

分子生物学的兴起及迅速发展无疑是生物学中的重大突破。分子生物学技术已经并将继续推动生物学的全面发展。然而，不容忽视的是分子生物学，归根结底，是依赖于生物系统学的。欧斯汀(Oosting HJ)在他的“植物群落的研究”(1956, p.30; 中译本 1962, p.24)一书中给我们提供了一个关于植物生理学对植物系统分类学依存关系的精辟说明：“对于一种没有命名的植物进行详细的生理学研究显然是荒谬的。如果，植物生理学家对于他所研究的植物不知道是什么种，而且有时不知道是什么变种或生理小种，他的结论势必只局限于他试验中所用到的那些植物。分类学家、植物区系地理学家和遗传学家通过研究而积累资料和数据；生理学家能够吸取这些资料和数据，并且加以利用而作出概括和对比。所有这些资料都是通过被研究的植物的科学名称（学名）来联系的”。而这些科学名称只能是通过对生物进行系统分类的研究而获得。

因此，在真菌的系统分类与进化方面应该按照生物系统学的原理与方法从生物学的各个角度（如：形态学、解剖学、生态学、生理学、生物化学、遗传学以及分布学等）和不同层次（如：群体生物学、个体发育生物学、细胞学以及分子生物学等）进行综合性的研究与分析。经过这样的研究和分析而获得的真菌学名，不仅告诉我们这种真菌在生物系统中的位置，而且包含着一系列生物学性状的信息和资料。这样的研究将为现代生命科学的发展，为真菌和地衣资源的开发利用以及有害真菌的控制提供具有高度预期价值的信息存取系统。没有这样的系统，将不可能选用适宜的材料以进行生理学或生物化学，包括分子生物学的实验。即使利用无学名的生物进行了这样的实验，其结论无论是在理论上或实践上都是毫无意义的。因此，实验生物学，包括分子生物学，对于作为生物技术或生物工程支持系统的生物系统学的依存关系是显而易见的。

可以相信，由于我国的真菌、地衣还缺乏足够的研究，为了真菌、地衣资源的开发利用、有害真菌的控制以及现代生命科学的发展需要，真菌系统和演化的研究是必不可少的，因此，作为本实验室的长远目标。在敏锐的学术思想指导下，按照生物系统学的原理与方法，结合对国内或者有时在个别类群中对世界范围真菌、地衣的广泛研究及其属种的不断发现，本实验室的研究工作，必将日益接近这一目标。而《真菌系统》将反映本实验室在这一进程中的研究成果。

魏江春　　北京

一九八七年五月二十日

Current progress of systematics of lichenized fungi

WEI, JIANG - CHUN

Systematic Mycology & Lichenology Laboratory

Institute of Microbiology, Academia Sinica

Beijing 100080, China

Abstract. This paper in the first place presents the systematic position of the lichenized fungi in the organisms, and then summaries the morpho - geographical approach in combination with chemistry of lichen substances. Finally, the paper focuses on the molecular systematics of the lichenized fungi. Some techniques concerning the isolation of mycobiont DNA from the algal - fungal thalli, and the latest research results, including the author's own work are reviewed. The importance of the systematics to the exploitation of lichen resources is also concerned.

Keywords. Lichenized fungi; lichens; molecular systematics

There are about 20, 000 lichen species all over the world. Lichens produce characteristic lichen secondary metabolities, such as aliphatics (including lactonic acid, triterpenoids, alcohols and polysacharides) and aromatics (including benzoic and vulpinic acids, depsides, depsidones, quinones, xanthones and diphenylenoxides). Depside, depsidones, and diphenylenoxides are only found in lichens. Lichens and lichen substances can be useful as antibiotics, UV absorbers, antioxidants, anticancer agents, and dyes. As is well known, tetronic acid, depsides, and diphenylenoxide show antibiotic activity against gram negative bacteria, and a polysaccharide, GE - 3 from *Umbilicaria* inhibits the growth of sarcoma 180 in mice (Shibata et al., 1968). Recently, some authors reported that a GE - 3 sulfate derivative had activity against the AIDS virus (Hirabayashi et al., 1989). The norsolorinic acid from *Solorina crocea* has stronger inhibition effect than the known positive control (Yamazaki et al., 1988). It seems that lichens have great potentialities in resources. The systematic study of lichenized fungi provide an information storage and retrieval system with a high degree of predictivity for the exploitation and utilization of lichenized fungal resources. The present paper focuses on the current situation of the systematics of lichenized fungi.

Systematic Position of Lichenized Fungi

Before 1866 lichens were misunderstood as a part of the kingdom "Plantae" in a phylum and placed alongside Fungi, Algae, and Bryophytae. Since De Bary in 1866, followed by Schwendener in 1867, discovered the dual nature of lichens, a lichen was understood as an association of a fungus and an alga. Now. actually, lichens are treated as the specialized fungi that live in symbiosis with algal and / or cyanobacterial photobionts. They are extracellularly located

This paper was originally published in *Proceedings of the First Korea-China Joint Seminar for Mycology*. Seoul, Korea: 5-12, Dec. 2-5, 1993.

endosymbionts of lichen thalli, and they themselves are fungal－algal symbiosis ecosystem. This ecosystem is a brilliant example among all the ecosystems.

About 8%of terrestrial ecosystems are lichen dominated (Larson, 1987) . The dominace is defined not only by a high "lichen quotient", but also by a quantitative abundance that leads to high percentage (up to 100%) of ground cover by lichens (Honegger, 1991).

Since lichens, actually, are specially lichenized fungi, and lichen names refer only to the fungal component of the symbiosis. The characters, derived from a fungus, are regarded as most important for the classification of lichens, and both lichenized and non－lichenized fungi have to be classified in one fungal entity depending the fungal similarities only.

As is well known, lichens are lichenized fungi, the question is "what are fungi? "

In the first half of this century, "Fungi" were misunderstood as a part of the kingdom "Plantae" in the subdivision "Thallophyta" and placed alongside Bacteria, Lichens, and Algae (Fitzpatrick, 1930). The five kingdom system of Whittaker (1969) comprises Myxomycetes, Oomycetes and true fungi, but the Plasmodiophoromycetes and Hyphochytriomycetes are included in Prostista. At present, the "Fungi" on the basis of the latest knowledge in a combination with the ultrastructural data, cell wall chemistry and molecular biology in narrow sense could be called "true fungi" (Barr, 1992) or "Eumycetes" (Tai, 1979) and placed in the kingdom Eumycota, and include the Basidiomycota, Ascomycota, Zygomycota and Chytridiomycota. The "Fungi" inbroad sense

Table 1. The panomycetes belonging to three different kingdoms

Kingdoms	Organisms studied by mycologists (Hawksworth, 1991) Union of Fungi (Barr, 1992) Panomycetes (Wei, 1993)
Eumycota (Truefungi, Barr, 1992; Fungi, Hawks- worth, 1991)	Phyla: Basidiomycota Ascomycota Zygomycota Chytridiomycota
Chromista	Phylum Heterokonta subphylum: Pseudomycotina (=Phylum Pseudofungi in Cavalier-Smith, 1989) Classes: Omycetes Hyphochytriomycetes subphylum Labyrinthista Class Labyrinthulea
Protozoa	Phylum Myxomycota Phylum Plasmodiophoromycota

usually comprise Fungi and the fungal －like groups , such as the Oomycetes and Hyphochytriomycetes, called "Pseudofungi" by Cavalier－Smith (1989), belong with the kingdom Chromista, and the slime molds, protostelids, and Plasmodiophoromycota, belong in the kingdom Protozoa. All the "Fungi" in broad sense of Whittaker's five kingdom systems are polyphyletic and placed now in three different kingdoms: the kingdom Eumycota or Fungi, the kingdom Chromista, and the kingdom Protozoa (table 1). However, classification based on phylogeny at the kingdom level is inconsistent with applied biologists, such as phytopathologists. Attempts to include fungal - like groups in the "Kingdom Fungi" is also inconsistent with natural classification. The solution lies in recognizing polyphyletic assemblages that include kingdoms, or parts of kingdoms, that make up logical groups for the benefit of the applied biologists. Barr suggests these assemblages be called "Union of Fungi" (Barr, 1992), or "Organisms studied by mycologists" (Hawksworth, 1991), but I should prefer "Panomycetes", as a science "Panomycology" (Wei, 1993).

Because of a long historical tradition, the name "fungi" should be retained for the polyphyletic union, i. e. fungi in the popular or colloquial sense, not in the taxonomic sense for the monophyletic kingdom. To have two meanings for fungi is confusing. The kingdom of true fungi could be called the kingdom Eumycota (Barr, 1992).

The lichenized fungi are the true fungi and belong to the kingdom Eumycota, and mostly to Ascomycota. Lichenized true fungi are about 21% of all true fungi, and about 46% of Ascomycota, but about 98% of lichenized true fungi are Ascomycetes (Hawksworth et al., 1983; Hawksworth, 1988; Honegger, 1991). The systematic position of the lichens or lichenized true fungi in the evolution of organisms is widely distributed in the classification systems of true fungi, especially in those of Ascomycota (Hafellner, 1988).

Morpho - geographical approach in combination with chemistry of lichen substances

Use of chemistry in lichen systematics has a long history, and is now utilized by the most of lichenologists, and in no other plant group has chemistry been used so successfully as aids in identification as in lichens (Hale, 1983). Therefore, morpho - geographical approach in combination with chemistry of lichen substances has been being a major method in the study on modern systematics of lichens or lichenized true fungi so far. Another advance in the systematics of lichenized true fungi came with the use of TLC method and discovery of ascal characters of their apical apparatus.

Chemistry may be useful in interpreting families and genera, and taxa of lower taxonomic categories, if the results are seen in connection with morphological features. Pigments especially should not be overemphasized. Almost all discocarp lichens containing anthraquinones have been thought to belong to the same family *Teloschistaceae* until recent times. In reality, pigments of the anthraquinone type have been formed by several not closely related groups. Not all lichen substances are of the same value for taxonomy (Hafellner, 1988).

Lichenized groups are widely distributed in ascomycetes and are much rarer in basidiomycetes. Several of the lichenized groups have no clear relatives in ascomycetes, and therefore are classified in separate orders, e. g., Lecanorales and Verrucariales. Other groups not so rich in species can easily be attached to well - known orders of true fungi, e. g., Dothideales and Tricholomatales. Several families, e. g., *Peltigeraceae*, *Teloschistaceae*, and

Pertusariaceae, have been raised recently to order level, but this probably does not reflect real relationship. In Lecanorales, many isolated groups are included. This is a highly heterogeneous order. (Hafellner, 1988).

Gene flow in lichens was demonstrated by the analysis of secondary products in the progeny of individuals from natural populations of mixed chemotypes of the *Cladonia chlorophaea* complex. All of the chemotypes had been interpreted as distinct sibling species (C. and W. Culberson et al., 1988).

Molecular Systematics

The rapidly developed techniques of the molecular biology in combination with the morpho-geographical and chemical approach may assist in working out some of the evolutionary pathways by which groups have arisen.

In order to give a more precise definition to systematic position of some species of lichen genera *Umbilicaria* and *Lasallia* the DNA hybridization techniques were used in the Institute of Botany of the Academy of Sciences of the formerUkrainian SSR for the first time in the world. The DNA hybridization data of *Umbilicaria hirsuta*, *U. grisea* and *Lasallia pustulata* proved that the genetic distances between these species are considerable in both the homologies and the quantity of nucleotide substitutions. The 67% homologies and 3% nucleotide substitutions in hybrid duplexes show *U. hirsuta* and *U. grisea* to be two distinct species, and at the level of DNA divergence of *L. pustulata* considerably excel the differences at the species level which points to a solitary position of *L. pustulata* among those species studied (Blum and Kashevarov, , 1986).

As is well known, a lichen is a specialized true fungus that lives in symbiosis with algal and/or cyanobacterial photobiont, and sometimes both an algal and a cyanobacterial ones. Therefore, the purified DNA from a lichen may contain up to six different ribosomal DNA (rDNA) genomes. Such as the rDNA genomes of the true fungal nuclei and mitochondria; the green algal nuclei, mitochondria and chloroplasts, and that of a cyanobacterium. For sequence analysis of the lichenized fungal nuclear small subunit (18s) rDNA, the complicating sequence is the algal 18s rDNA. In order to sequence the 18S rDNA of the mycobiorts (i. e. the fungal component) from a lichen association there is a large choice of the following ways:

1) The non-algal primer pairs for the PCR were designed which amplify the fungal 18S rDNA but not the green algal 18S rDNA (Gargas and Taylor, 1992).

2) During the DNA extractton Novozym buffer is added. Novozym mostly damages hyphae, leaving algal cells sufficiently intact so that they do not release their DNA readily (Armaleo and Clere, 1991).

3) The DNA of the lichenized true fungi can be extracted from cultured mycobionts started from spores (Armaleo and Clere, 1991). The cultured mycobionts started from vegetative thalli of lichens (Yamamoto et al., 1985) are also be able to use for extraction of DNA.

4) Lecideine and sublecanoriene apothecia or rhizines can be used for extraction of DNA from mycobionts (Niu and Wei, 1993).

5) PCR amplification of rDNA sequences from single spores (Leeand Taylor, 1990) can be used for systematic analysis of lichenized true fungi.

In the case of the lichens from the genus *Sticta*, the thalli containing cyanobacterium are classified as *S. dufourii*, those containing green alga as *S. canariensis*. In the composite thallus, the two morphotypes are distinguished by their shape and colour, and can be separated under a dissecting microscope. The same rDNA segment (i. e. 25S rDNA) was amplified by PCR from each species studied, restricted, and digests were compared after electrophoresis. Of the 5 enzymes tested (Hpa II, Hae III, Nci I, Ava II, Taq I), none detected a difference between the members of the *S. dufourii/ canariensis* pair, where as two (Hpa II and Hae III) revealed differences between *S. weigelii* and the paired morphotypes. The method indicates that also in this case the paired mycobionts have very similar if not identical genomes, and correctly identifies the outgroup, even within the same genus (Armaleo and Clere, 1991).

A report of a comparison of rDNA nucleotide sequences from lichenized and non - lichenized true fungi was given by Gargas and Taylor during the Fourth International Mycological Congress in Germany. They have used the PCR to amplify specifically the fungal nuclear 18s rDNA from both lichenized and non - lichenized fungi. Their 18s genes amplified vary from 1. 8 to 3. 2Kb. Arrangements of orders within the Ascomycota remain controversial, and DNA sequences of this molecule may be used as the basis for a phylogeny which integrates both the lichenized and non - lichenized fungi. This study focuses on the relationships between the ascomycetous orders Pezizales, Helotiales, Caliciales and Lecanorales (Gargas and Taylor, 1990).

The genomic DNA of the mycobionts from nine species distributed in five families, i. e. *Lecanoracea e*, *Peltigeraceae*, *Teloschistaceae*, *Umbilicariaceae*, and as an outgroup *Dermatocarpaceae*, was extracted from herbarium collections. The 880 bp segment of rDNA between 5. 8S and 25S from nuclear DNA were amplified by PCR using the primer pair 5. 8SR (5' - TCGATGAAGAACGCAGCG - 3') and LR3 (5' - CCGTGTTTCAAGACGGG - 3') (Vilgalys' Lab manual). The amplified rDNA was restricted with enzyme AvaI, and the fragment patterns obtained from the nine species were compared after agarose electrophoresis. None restriction fragment having the same length is there among those of the five families, whereas more than one restriction fragment having the same length among the five species within the *Umbilicariaceae*. The PCR products from the four species within the family were restricted with HinfI, none detected a difference among them, where as with NciI, one restriction fragment having the same length occurs between *U. mammulata* and *U. esculenta*, and between *L. mayebare* and *L. rossica* respectively. From the data of the restriction analysis we could draw the conclusion that the two - genus system of *Lasallia* and *Umbilicaria* in the family *Umbilicariaceae* is consistent with the detailed analysis on the classification of the family (Wei, 1966; Wei and Niu, 1993).

Lasallia rossica "appears to be an intermediate taxon between *L. papulosa* and *L. pensylvanica* because of its dark brown lower surface of thalli. The rest of its Characteristics basically correspond to *L. papulora*. In addition, many collections of *L. papulosa* examined from North America also have a dark brown lower surfce of thalli as well. Thus, whether this species is in reality a seperate one remains an open question " (Wei & Jiang, 1993) In order to carry out a systematic analysis based on the combination of their molecular and morphological characters with the chemical data and distribution patterns, the internal transcribed spacer 2 (ITS2) sequences of the nuclear rDNA from *L. papulosa* and *L. rossica* were obtained using primer pair ITS3 and ITS4, and determined by ABI automatic DNA sequencing system. Because the ITS2 sequences of the nuclear rDNA evolve faster then others, and may vary among species within a genus or among papulations (White, Bruns, Lee, and Taylor, 1990).

The ITS3 (5' - GCATCGATGAAGAACGCAGC - 3') and ITS4 (5'-TCCTCCGCTTATTGATATAGC-3') oligonucleotide primer pair was used for the PCR to amplify the ITS2 fragment between 5.8s and 25s rDNA. The PCR products were purified and their sequencing was carried out using ABI automatic DNA sequencing system.

A total of 309 bp of the sequences including ITS2 between 5.8s and 25s nuclear rDNA from

Table 2. The sequences from ITS2 of *L. papulosa* and *L. rossica*.

L. papulosa	CGGCATTCCG	GGGGGCATGC	CTGTCCGAGC	GTCATTGCAC	40
L. rossica	T..T......	....	....T.....	......A...	
L. papulosa	CCCTCAAGCT	CCGCTTGGTG	TTGGGCCCCC	GTCCCCCGGG	80
L. rossica	.T.......C	.T.......A	......TGT.	....	
L. papulosa	ACGCGCCCGA	AAGCGATTGG	CGGCGCGGTC	CGACTTCGAG	120
L. rossica	...G...T..	...TT.G...	...T..A.C.	T......A..	
L. papulosa	CGTAGTAGTG	AC-TCCAAAC	CCGCTCCGGA	AGCCGGCAGG	160
L. rossica	......---A	..T.AA....	.....T....	....TC....	
L. papulosa	TCCGCCCCGG	TCAGACAACC	CGGTTGCACA	CTTCG-AC	198
L. rossica	CTG.G..-A.	C...TT....	.A..C.T-A.	....TT.T	

L. papulosa and *L. rossica* were determined. Partial sequences of them (198bp) are used for systematic analysis (table 2). Fourty eight (c. 24%) variable positions exist between *L. papulosa* and *L. rossica*.

A total of 187 bp were aligned from ITS2 of some *Lentinuss* species by Hibbett and Vilgalys (1993). From these data it is clear that between the two strains of *Lentinus lepideus* there are 5 (2.67%) variable positions in ITS2 sequence, and among the three strains of *L. ponderosus* 2 (c. 1%), whereas between *L. strigosus* and *L. trigrinus* 42 (22.46%) variable positions. These show that within species sequence variability in ITS2 is lower (< 3%), but interspecific sequence much higher (c. 22%). A similar results exist in comparison with *Lentinus* spp. and *Lasallia* spp. The molecular evidences support that *Lasallia rossica* probably be a distinct species from *L. papulosa*.

A combined analysis of the molecular and the morphological charaeters with their chemical data and distribution patterns support *L. rossica* to be a separate species.

It seems that comparative analysis of the phenotype in combination with genotype is important for systematic and evolutionary study not only on lichenized fungi but also on the other organisms.

Literature cited

Armaleo, Daniele and Clerc, Philippe (1991) Lichen Chimeras: DNA Analysis suggests that one Fungus forms two Morphotypes. *Experimental Mycology*, **15**: 1 – 10.

Barr, Donald J. (1992) Evolution and Kingdoms of organisms from the perspective of a mycologist. *Mycologia* **84** (1): 1 – 11.

Blum, O. B., and Kashevarov, G. P. (1986). DNA homologies as proof of the lichen genus *Lasallia* Merat (*Umbilicariaceae*). *Dokladai Akad. Nauk Ukr.* S. S. R. Ser. B1986 (**12**): 61 – 64 (in Russian).

Bubrick, P., Frensdorff, A., and Galun, M. (1985). Selectivity in thelichen symbiosis. in Brown, D. H., ed. Lichen Physiology and cell Biology. pp. 319 – 334. Plenum Press, New York and London.

Cavalier – Smith, T. (1981) Eukaryote kingdoms: seven or nine? *BioSystems* **14**: 461 – 481.

Culberson, C. F., Culberson, W. L., and Johnson, A. (1988) Gene flow in lichens. *Amer. J. Bot.* 75 (8): 1135 – 1139.

Gargas, Andrea and Taylor, John W. (1992) Polymerase Chain Reaction (PCR) Primers for Amplifying and Sequencing Nuclear 185 rDNA from Lichenized Fungi. *Mycologia* **84** (4): 589 – 592.

Hafellner, Josef (1988) Principles of Classification and Main Taxonomic Groups. in CRC Handbook of Lichenology vol. 111, Chapter X: 41 – 52.

Hale, Mason E. (1983) The Biology of Lichens. Edward Arnold.

Hawksworth, D. L. (1988) The Fungal Partner in Galun, M., ed. 1988, Handbook of Lichenology, vol. I, pp. 35 – 38.

Hawksworth, D. L., Sutton, B. C., and Ainsworth, D. C. (1983) Ainsworth and Bisby's Dictionary of the Fungi. Kew: Common wealth Mycol. Inst. 7th ed.

Hirabayashi, K., Iwata, S., Ito, M., Shigeta, S., Narui, T., and Shibata, S. (1989) in *Chem. Pharm. Bull.* 37: 2410.

Honegger, Rosmarie (1991) Functional Aspects of the Lichen Symbiosis. *Ann. Rev. Plant Physiol. Plant Mol. Biol.* **42**: 553 – 578.

Larson, D. W. (1987) The absorption and release of waterby lichens. *Bibl. Lichenol* **25**: 251 – 360.

Lee, steven B., and Taylor, John W. (1990) Isolation of DNA From fungal mycelia and single spores. pp. 282 – 287. in PCR protocols – A guide to methods and applications. Eds., Innis, M. A., Gelfand, D. H., Sninsky, J. J., and White, T. J., Academic Press, New York.

面向21世纪的菌物学

（代序言）

中国菌物学会理事长
中国科学院院士　　魏江春
中国科学院微生物所研究员

克里克与华生(Crick & Watson,1953)关于DNA双螺旋结构的发现是20世纪下半叶科学上最激动人心的重大突破。这一突破为后来生命科学中一系列重大进展(表1)奠定了基础,并直接促进了分子生物学的兴起。

表1　分子生物学进展中的重要事件

重要事件	时间和创始人
DNA双螺旋结构学说的提出	1953, Watson & Crick
中心法则的提出	1958, Crick
操纵子学说的提出	1961, Jacob & Monal
遗传密码的破译	1966, Khorana & Nirenberg
反转录酶的发现	1970, Temin & Baltimore
转基因在E. coli中的表达(遗传工程的产生)	1973, Cuken
DNA测序的实现	1977, Maxan Gibert
核酶的发现	1981, Cech et al.
朊病毒的发现	1982, Prusiner
PCR技术的发明	1985, Mullis

分子生物学的兴起为生命科学的各个分支学科注入了活力,为数学、物理学、化学以及技术科学提出了一系列新的课题,使生命科学处于当代自然科学的焦点,成为自然科学中一颗光彩夺目的明星。由于生命科学所取得的惊人成就,促使人们不得不提出21世纪将是生物学世纪的预言。面对21世纪的来临,世界三大科技强国——美国、德国及日本为实现国家目标和本国利益,都已确定了科技发展的重要领域和战略目标。美国的科技发展重要领域主要涉及基础研究、生物技术、农业科学、环境科技、空间科技等领域。白宫致国会的《塑造21世纪的科学和技术》的报告中,确定了基础研究五大"推动领域",涉及生命科学的有三条,其中生命起源被列人首条;生态系统被列人第二条;基因组研究列为第三条。此外,美国国家科技委员会《21世纪生物技术新的方向》蓝皮报告指出,生物技术研究已进人"第二次浪潮"。在生物技术的发展上,美国将维护他在全球生物技术发展中的领先地位。德国在面向21世纪的长期研究中强调抓住机遇,加速信息化社会进程的同时,大力发展生物科学与技术,力争在2000年成为欧洲生物科技第一强国。德国科研体制的结构性调整重点是促进生物技术和信息技术的发展。日本的科研重点转向生命科学,1997年的科研预算中,生命科学研究经费有了大幅度的增加(江凌勇等,1997)。很明显,关于21世纪是生物学世纪的预言正在变为现实。

关于如何面对生物学世纪的任务方面有各种不同的议论。由于分子生物学突飞猛进的发展成为生物学中的热点。生态系统的研究由于直接关系到自然与人类生存

本文原载于《吉林农业大学学报》, 20（20）: 1-4, 1998.

环境以及全球变化问题而受到多方重视。直接进行物种研究的生物分类学长期以来却得不到较强的支持。于是形成这样一种局面，一个微观——分子生物学，一个宏观——生态系统研究，多年来均得到较强的支持，而处于二者之间的物种研究——分类学却得不到应有的重视。有人认为这种“抓两头，带中间”的形势将继续存在于生物学世纪。然而，有人却提出，物种是基因性状即遗传性状的载体，是分子生物学及生态系统的基础。因而，在生物学世纪，生物分类学应该得到更强的支持，这就是“抓中间，带两头”。也有人主张，生物学世纪将以综合为主，分析(即微观研究，如分子生物学等)为次。而反对这一主张的观点是，分析任务尚未完成，综合时期尚未到来，因而生物学世纪将仍以分析为主。

实际上，由于分子生物学及其技术对生物学各分支领域的渗透，21 世纪的生物学必将在微观与宏观、基因型与表型相结合的基础上，从整体生物学(Integrative Biology)水平上开展综合分析；它将在分析中综合，在综合中分析的过程中不断向前发展。在发展过程中，将体现开发促进科学，科学指导开发的辨证关系。

在 21 世纪即将来临的时候，在此，我们将着重讨论生物学世纪中的菌物学问题。

通过 rRNA 碱基序列的系统发育学分析，整个生物被划分成细菌(Bacteria)、古菌(Archaca)及真核生物(Eukaryotes)三大超界(Domain)(Woese and Fox, 1977)。我们所讨论的“菌物”，是指真核生物超界中的真菌、卵菌及粘菌。

菌物学在系统演化研究中，两个世纪以来经历了植物学—真菌学—菌物学三个时期。自 19 世纪至 20 世纪 70 年代(1801～1969)，菌物学一直被当作植物学的组成部分进行研究。自 20 世纪 70 年代至 90 年代(1970～1990)，惠特克(Whittaker, 1969)将菌物从植物中分出，建立了真菌界。此后，菌物一直被作为真菌学进行研究。自 20 世纪 90 年代以来，由于分子系统学及超微结构的研究结果，惠特克的“真菌界”被证明在亲缘关系上并非一元的单系类群(Monophyletic group)，而是多元的复系类群(Polyphyletic group)。其中的卵菌门(Oonmycota)、丝壶菌门(Hyphochytriomycota)及网粘菌门(Labyrinthulomycota)在亲缘关系上远离真菌而更接近于硅藻与褐藻等生物类群。它们被从真菌中分出而划归管毛生物界(Stramenopila)。至于粘菌，在亲缘关系上则更远离于真菌这一事实早已被人们所熟知。不过，近年来已被证实，统称为粘菌(Myxomycetes)或裸菌(Gymnomycetes)的根肿菌门(Plasmodiophoromycota)、网柄粘菌门(Dictyosteliomycota)、集胞粘菌门(Acrasiomycota)及粘菌门(Myxomycota)在亲缘关系上也彼此远离而处于多元的复系类群状态(Alexopoulos 等, 1996)。这样以来，惠特克的“真菌界”所剩者只有担子菌门(Basidiomycota)、子囊菌门(Ascomycota)、接合菌门(Zygomycota)及壶菌门(Chytridiomycota)，其中也包括地衣或地衣型真菌，在亲缘关系上，它们被认为是一元的单系类群。这就是现在关于真菌界的实际范围。由于包括管毛生物界、真菌界以及复系类群的粘菌在内的这个多元的复系类群，习惯上一直是由真菌学家进行研究的，因而，它们往往连同真菌界的四大门类一起同时被置于同一专业教程或专著中进行论述。所以，为了方便，对它们需要有个普通称谓。这三大生物类群在国外，尚无统一的普通称谓，有的称它们为“由真菌学进行研究的生物”(Organisms studied by mycologists, Hawksworh, 1991)，有的则称为“真菌联合体”(Union of fungi, Barr, 1992)，也有以第一字母

"F"大写的"Fungi"表示真正的真菌，而以"f"小写的"fungi"表示菌物者(Alexopoulos 等，1996)等等。而在国内，它们曾被统称为"真菌"；自 1969 年以后被称为"真菌界"；为了表达粘菌并非真菌，也被统称为"广义的真菌"；近年来，被通称为菌物。因此，所谓"菌物"，实际上是真菌界、管毛生物界以及多元的粘菌这三大生物类群的总称。至于微生物，它并不是生物演化中的自然类群，完全不顾及其亲缘关系，而只是按照生物体的大小人为地划定一个范围，即微小的(＜10 μm)的、用肉眼难以辨认的生物，如病毒、立克次氏体、细菌、酵母菌、霉菌以及一些微观的藻类和原生动物等。

现就面向 21 世纪的菌物学提出下列看法：

1 菌物基因资源产业的崛起

在菌物开发资源利用中，存在着第一次开发、第二次开发和第三次开发等三个不同层次。所谓第一次开发，是指直接利用菌物有机体及其水溶性混合物以利于人类生活。这种只知其果而不知其因的第一次开发无异于李时珍时代的本草植物学时期。而第二次开发是指从菌物有机体中提取对人类生活有用的特定化学组分。这些化学组分的分子结构已经研究清楚。至于第三次开发是指在第二次开发的基础上，明确人类所需要的目标，有针对性地进行生物特定组分及其基因的开发，完全不同于"只知其果，不知其因"的盲目开发路线，从而直接利用基因或通过转基因工程菌实现工业化生产。在资源开发利用中，菌物为工农业生产和医药卫生以及人类活动的诸多领域作出了重要贡献。尤其在保障人类健康方面。如果说，在病毒方面由于发明了牛痘使人类平均寿命由 20 岁提高至 40 岁，那么，20 世纪 40 年代由于从真菌中发现了青霉素而促进了抗生素工业的崛起，从而使人类的平均寿命从 40 岁提高到 65 岁。抗生素工业是以容易培养，生长周期短而易于实现工业化生产的微生物资源为基础的，其中原核的放线菌发挥了重要作用。然而，易于工业化生产的微生物则为数甚少，经过人类半个世纪的开发和利用，这些微生物中可被开发利用的资源已基本枯竭。因此，美、英、日、加等发达国家一些大公司，在生物资源开发中已将注意力集中投向难于培养的菌物，如地衣、真菌、粘菌等的基因上，以实现第三个层次的开发利用。每一个物种就是一个独特的基因库。据估计，自然界实际存在的菌物为 100～150 万种(Hawksworth，1991；Systematics Agenda 2000，1994)，而已被人类认识的还不到 7 万种。这就意味着还有 93～143 万种菌物物种或基因尚待人们去发现、描述和开发与利用。因此，在三个层次上进行开发的同时，菌物第三次开发以及菌物基因资源产业的崛起可能是生物学世纪的特色之一。

2 菌物生物多样性的综合研究

由于菌物基因资源的开发和利用，必将进一步促进菌物生物多样性的研究。而发现和描述新的菌物物种将是菌物生物多样性研究的主要任务。标本馆和菌种库则是这一任务的重要组成部分，也是菌物基因资源产业的重要支撑系统。菌物物种编目工作只能在不断发现和描述新种的基础上才有意义。菌物物种是其基因或遗传性的载体，是生态系统的基础，因此，在物种、基因和生态系统三结合中进行菌物生物多样性的综合研究，才有利于探索菌物多样性的形成原因及其保护途径，从而成为 21 世纪的又一特色。

3 菌物系统与演化生物学

菌物系统与演化生物学实际上是关于

菌物在生物演化过程中所形成的结果及其历史进行研究的科学。对于它们在演化过程中所形成的结果进行研究远比对其形成历史进行研究容易得多。然而，无论是系统学或演化生物学，它们的任务往往是追求生物演化历史的重建。在漫长的地质历史中，大量生物类群惨遇灭绝，因而给重建演化历史带来了困难。在这一研究中，除了化石证据之外，主要依赖于不同角度的间接证据进行分析。而对菌物来说，已发现的化石是极其缺乏的。因此，在进行菌物系统与演化生物学研究中，主要还是依赖于间接证据进行分析。所得数据的分析方法将不会局限于三大派，即进化学派、表型学派和分支学派中的任何一种，基于分子生物学及其技术的迅速发展，从分子、细胞、组织、个体、群体与生态系统等各个层次和不同角度在整体生物学水平上进行综合分析，将会成为21世纪系统与演化生物学的主要方向。

4 人 才

整体生物学的兴起，势必导致多学科的相互交叉与综合分析，进而导致新学科和新领域的产生。缺乏思路，缺乏想象力，将不可能在科学领域中做出创新成果。而有无思路和想象力是和一个人有无丰富实践及广泛科学知识密切相关的。因此，21世纪的菌物学家将是具有广泛基础科学知识和现代化信息技术与丰富科学实践的科学家，是一专多能的科学家，是专心致志地与自己所研究的对象进行顽强斗争的科学家，是具有科学道德、宽阔胸怀、与周围同事合作共事、团结一致、为探求真理而奋斗的科学家。在老菌物学家的帮助和提携下，年轻菌物学家将会不断涌现，而且，必将青出于蓝而胜于蓝。

参 考 文 献

1 Alexopoulos C J, Mims C W and Blackcoell M. Introductory Mycology. John Wiley and Sons, InsNew York, Chichester, Brisbane, Toronto, Singapore. 1996

2 Barr D J. Evolution and kingdoms of organisms from the perstective of a mycologist - Mycologia 1992, 84(1): 1~11

3 Hawksworth D L. The fungal dimension of biodiversity: magnitude signifance and conservation. - Mycol. Res. 1991, 95(6): 641~655

4 Systematies Agenda 2000; Charting the Biosphere, Technical Report, 1994: 16

5 江凌勇，黄群，张民主等. 美、德、日三大科技强国面向21世纪的科技发展重点与战略目标. 中国科学院—科学发展报告. 1997

6 Watson J D and Crick F H C. Nature. 1953, 171: 737, 964

7 魏江春. 菌物多样性、系统性及其对人类发展的意义. 生物多样性, 1993, 1(1): 23~25

8 Whithaker R H. New concepts of Kingdoms of organisms - Science, 1969, 163: 150~160

9 Woese C R and George Fox E. Phylogenetic structure of the prokaryotic domain: The primary kingdoms. Proc. Narl. Acad. Sci. USA. 1977, 74(11): 5088~5090

中国科学院院士　魏江春

缅怀著名海洋藻类学家曾呈奎院士

我和曾呈奎先生的第一次相见是1971年7月在中国科学院计划工作会议上。当时中国科学院停止一切科学研究,全国大中小学自停课到忽视基础科学文化课程教育已经整整5年了。为了整顿科教事业,在周恩来总理建议下,当时的中国科学院副院长、北京大学校长周培元教授写了一篇强调基础科学与基础文化教育方面的文章发表在《光明日报》上。

中国科学院计划工作会议正是在这一背景下召开的。出席会议的有近百人,都是中国科学院各研究所的科技负责人。在生物领域著名植物分类学家林镕先生代表植物所,曾呈奎先生代表海洋所,朱弘复先生代表动物所出席会议。我当时是代表微生物所出席会议的。当时植物所科技处的负责人将一份关于恢复《中国植物志编委会》工作的申请报告交林镕先生过目。当时我正好坐在林镕先生旁边,所以顺便看到了那份报告的内容。在这一报告的启发下,我和曾先生商量,提出筹建《中国孢子植物志编委会》和恢复《中国动物志编委会》工作,连同《中国植物志编委会》一起召开一次"三志"工作会议的建议。当时得到曾先生的强力支持。随后我们先后与朱弘复先生以及各有关研究所出席会议的代表进行了接触,得到大家的一致支持。于是我们一致推举曾呈奎先生代表大家向大会正式提出了上述建议。当时得到刚刚恢复工作的生物学部副主任、生物局局长过兴先先生的支持。会后便得到当时主持中国科学院工作的武衡同志的批准。这就是1973年广州"三志"工作会议的背景。因此,曾呈奎先生不仅是《中国孢子植物志编委会》的发起人之一,而且在促成《中国植物志编委会》、《中国动物志编委会》和《中国孢子植物志编委会》(简称"三志")广州工作会议的召开做出了重要贡献。

后来作为《中国孢子植物志编委会》主编,曾老在主持中国孢子植物志编写工作中做了大量工作,从不因年事已高而放松工作,为中国孢子植物志编写工作付出了后半生的心血。中国孢子植物志编研中所取得的一系列重要成绩都渗透着曾老的心血。

本文原载于《蔚蓝色的丰碑: 人民科学家曾呈奎同志纪念文集》(李乃胜主编). 青岛: 青岛出版社: 40-41, 2006.

此外，曾老还将自己于1996年所获得的“香港求实科技基金会基础科技成就奖”12.5万元奖金全部捐出，在《中国孢子植物志编委会》下设立了《曾呈奎孢子植物分类学奖金》。《中国孢子植物志编委会》正副主编为奖金管理委员会成员。

曾老不仅在做学问方面是一位科学大家，而且他平易近人，热情和蔼，关心青年人的成长。当我回忆起和曾老一起工作时，他那和蔼可亲，实事求是，鼓励年轻人向上的高尚人格，使我永生不忘。因此，曾老无论是在做学问方面，还是做人方面，都是我们的楷模。在曾老逝世一周年之际，以此短文作为缅怀和纪念，愿曾老永垂不朽！

2006年1月8日　于北京中关村

新种发现的科学与实践意义

地球上丰富的生物物种多样性是物种长期演化的产物，也是人类可持续发展的基础。自从地球上生命诞生和物种形成以来，在38亿年漫长的生物演化过程中，作为基因载体的物种通过基因突变和重组引起基因组结构的变化，从而导致蛋白质中氨基酸序列的变化，进而表达为表型性状的差异，通过自然选择逐步分化进而形成了丰富多彩的生物物种多样性。作为生物繁殖和系统生物学，即分类学研究的基本单位，每一种生物都具有各自独特的基因库、结构与功能。这些物种资源及其基因库中的基因资源，是漫长的生物演化过程和自然选择的产物。据专家估计，地球上实际存在的生物物种约为1亿种，但只有140万种(占估计种数的1.4%)已被人类发现、认识、描述、定名和发表，还有98.6%的物种(即新种或新的基因库)等待人类去发现。新种的发现对于填补生物演化长链中的缺失环节具有重要的科学意义，对于开发新的物种资源和基因资源以支撑人类可持续发展，具有重要的战略意义。

自从20世纪70年代基因重组技术问世以来，它的发展从根本上改变了生物技术的本质，从而推动了分子生物学技术的产生，进而大大提高了基因资源在生物产业中的地位。通过基因重组技术，人类已经能够利用已知的140万个物种资源及其基因库中的基因资源，将单细胞生物改造成生产胰岛素、干扰素、生长素、病毒抗原等多种蛋白质的细胞工厂，今后还会出现更多生产新型产品的细胞工厂。不过，为了人类的可持续发展，不能将希望只寄托于1.4%的已知物种及其基因资源上，也不能寄托于偶然碰到的新物种或不可培养物种的基因资源上，更不能寄托于未经自然选择并可能具有巨大安全风险的人造基因和人造生命上，人类首先应探索未知的98.6%的新种及其基因库。

不过，物种是基因的载体，只研究基因而不研究物种，是无法全面发现、认识、掌握和利用新的物种资源及其基因资源的，犹如只取得人的指纹而不利用指纹库去查对其主人的姓名、照片及其档案资料，也无法认识并接近指纹的主人。因此，人类必须在表型与基因型相结合中，既研究基因，也研究物种及其表型性状；有组织、有计划地加强对于作为基因载体的新种的发现、认识、描述、定名和发表，不断补充和丰富人类的物种资源及其基因资源库，并发挥其在实践中的作用。这正是系统生物学，即生物系统分类学在人类可持续发展中所具有的重要战略意义。

魏江春简介：

中国科学院院士，地衣真菌学家，我国该学科的主要奠基人。现任中国科学院微生物研究所研究员、博士生导师、中国菌物学会名誉理事长、中国科学院《中国孢子植物志》编委会主编、中华人民共和国濒危物种科学委员会委员。魏江春1931年生于陕西咸阳，1955年毕业于西北农学院，1962年毕业于苏联科学院研究生院，先后获苏联科学副博士和科学博士学位。历任中国科学院微生物研究所副所长及学位委员会主任、美国芝加哥菲尔德自然博物馆植物系客座研究员、《国际孢子植物学报》编委等职务。

魏江春从事地衣真菌系统生物学研究50年来，在中国地衣生物多样性，尤其在西藏地衣、南极地衣和石蕊科地衣型真菌系统与演化生物学研究，以及《中国地衣志》不同科属的编研中做出了显著成绩。他通过表型与基因型相结合的综合分析方法，对世界范围内的石耳科系统与演化生物学进行了系统而深入的研究；在对石耳科地衣四属、五属系统进行修订的基础上，论述了石耳科地衣的新二属系统，并获得国内外同行的认同。他描述并发表了一系列新种、1新属和1新目——石耳目；为真菌的共同祖先起源理论提供了分子证据。目前，魏江春在地衣生物多样性及系统生物学研究基础上，继续进行荒漠地衣及其固沙生物学与抗逆基因的筛选与功能分析的研究。迄今，他已发表学术论文约90篇，专著4部，即《中国药用地衣》《西藏地衣》《中国地衣综览》(英文版)以及《亚洲石耳科地衣》(英文版)。他的研究成果曾获中国科学院科技进步特等奖(1986年)、国家海洋局科技进步特等奖(1996年)及二等奖(1997年)。

本文原载于《大自然》(*China Nature*), 3: 1, 2008.

试论我国真菌学科的布局与发展

魏江春

On the arrangement and development of mycology in China

Wei Jiang-Chun

中国科学院成立于 1949 年 11 月。为了发展新中国的真菌学科，中国科学院于 1952 年在本院植物研究所真菌组的基础上成立了真菌与植物病理研究室，由著名真菌学家戴芳澜教授任主任。1956 年真菌与植物病理研究室从植物所分出后成立中国科学院应用真菌学研究所，由著名真菌学家戴芳澜教授任所长，著名真菌学家邓叔群教授任副所长兼任真菌研究室主任，著名真菌学家王云章教授任真菌研究室副主任兼标本室主任。同年真菌学科被列入国家十二年科学规划。

应用真菌学研究所筹建期间，根据上级要填补空白学科的精神，植物学家们按照惯例认为，当时的空白学科苔藓和地衣属于植物，应由植物所负责填补；而戴芳澜教授则认为苔藓是植物，应由植物所负责填补；地衣并非植物，而是真菌，应由真菌所负责填补。这就是后来由应用真菌学研究所向苏联派出留学生学习地衣专业的历史背景。随着科学技术的飞速发展，地衣并非植物，而是真菌这一事实已被国内外科学界所普遍接受。而这一理念早在半个世纪以前就由我国真菌学的奠基人戴芳澜所长所坚持。

自从 1958 年底成立微生物所以后，在戴芳澜所长和邓叔群副所长兼真菌研究室主任的领导和主持下，真菌室的科研工作覆盖了包括地衣型真菌在内的真菌界（Fungi）有关门类的分类学研究。此外，还进行类似于真菌的管毛生物界（Chromista）的菌类和原生动物界（Protozoa）的黏菌的分类学研究。

十年动乱期间，科学研究与文化教育曾基本停止。为了拨乱反正，周恩来总理曾授意中国科学院副院长，中国科学院学部委员，北京大学副校长，著名物理学家周培元教授于上世纪七十年代初在光明日报发表文章，说明文化教育和基础科学的重要性。文中提到苔藓植物虽无应用价值，却在植物进化中具有重要的科学意义。在此背景下，全国中小学加强了基础科学文化教学工作；中国科学院于也于 1972 年夏季在北京召开了计划工作会议。我是出席计划会议的本所代表。当我得知出席会议的植物所代表林鎔先生拟向会议提出恢复《中国科学院中国植物志编辑委员会》工作的信息时，便去和出席会议的海洋所代表曾呈奎先生商议，以便借此机会提出成立《中国科学院中国孢子植物志编辑委员会》（简称“孢编会”）的倡议。随后又去和出席会议的动物所代表朱弘复先生商议，向他报告了上述情况，并试探恢复《中国科学院动物志编辑委员会》工作的可能性。在他们二人的大力支持下，又和其他有关研究所出席会议的代表们进行了充分协商并取得了共识，大家一致同意将成立孢编会和恢复原有动植物志编委会（简称“三志”）作为同一个方案，由海洋所代表曾呈奎代表大家在会上发言，正式提出。这一倡议被当时主持中国科学院工作的武衡批准后，由微生物所作为孢编会的依托单位进行筹备工作，由戴芳澜教授任主编。当时刚恢复工作并参与主持计划会议的过兴先在受理并促成批准这一倡议的过程中起到了积极作用。这就是 1973 年在广州召开的“三志工作会议”的历史背景。《中国科学院中国孢子植物志编辑委员会》的成立进一步带动和推进了我国真菌分类学的全面发展。戴芳澜教授是我国真菌学科的奠基人，是我国真菌学科发展和前进中的旗帜。由于戴芳澜教授在“三志工作会议”之前不幸逝世，正式成立的《中国科学院中国孢子植物志编辑委员会》主编由王云章先生代理。

Whittaker 早在 1959 年提出生物四界系统时就将真菌从植物界中分出并独立成界，从而除原生生物界之外，使真菌界与动物界和植物界处于鼎立之位。后来，受到“三志工作会议”及孢编会成立的鼓舞，在

本文原载于《忆五十载奋斗历程, 谱新世纪壮丽篇章（1958-2008）》（中国科学院微生物研究所编）: 89-92, 2008 年 12 月 3 日.本文部分字段有所调整。

中国植物学会的支持下，由所内外真菌学家的积极支持和筹备，中国植物学会《真菌学分会》于 1980 年正式成立。在所领导的支持下，《真菌学报》也随之于 1981 创刊。从此，我国真菌学科有了专门的全国性学术刊物。

由邓小平同志开创的改革开放时代之初的 1985 年，中国科学院为了保护和促进一批有基础，有积累，有骨干的学科，决定成立了全国首批开放研究实验室和两个开放研究所；本着面向全国，面向世界，面向未来的办科学方针予以资助和支持。在中科院生物局的推荐和专家论证与评比下，本所《真菌地衣系统学开放研究实验室》被列入科学院首批开放研究实验室行列。从那时起，本实验室遵循面向全国，面向世界，面向未来的方针走向全国，走向世界；与全国同行科学家研讨学科发展规划；与欧、美、日等国的同行科学家开展了广泛的学术和人员交流。开放实验室于 1991 年举办了一个月的“真菌分子系统学讲习班”，由美国杜克大学的 R.Vilgalys 主讲，学员来自国内外。

由本实验室创办的英文版科学年报《MYCOSYSTEMA》（图 1）作为世界公认的真菌学刊物曾被 DICTIONARY OF THE FUNGI （1995） 第八版所收录；隶属于 MYCOSYSTEMA 的系列专著《MYCOSYSTEMA MONOGRAPHICUM》也随之陆续问世。

由于分子生物学技术的飞速发展，貌似真菌的卵菌（Oomycetes）或称假菌（pseudofungi）和黏菌被从 Whittaker 的真菌界中分出并分别划归管毛生物界（Chromista）和原生动物界（Protozoa）。但是，由于这些并非真菌而貌似真菌的真核生物习惯上一直是由真菌学家研究的，因此，对它们连同真菌一起通常被通称为“由真菌学家研究的生物”（Organisms studied by mycologists，Hawksworth, 1991）；或“真菌联合体”（Union of fungi，Barr,1992）；或“菌物”（裘维藩，1991）；或以第一字母为小写的“fungi”（Bruns et al., 1991），而将真菌界的真菌则看作第一字母大写的“Fungi”。在由从事真菌学、假菌学以及黏菌学科研与教学工作者组成的中国植物学会真菌学分会基础上，于 1993 年成立了全国性学会：中国菌物学会。

建国近 60 年来，我国真菌学科，在党和政府的关心和支持下，在以戴芳澜所长和邓叔群副所长为代表的老一辈科学家的带领和影响下，团结全国大专院校和科研院所的真菌学家、假菌学家以及黏菌学家相互支持，团结奋进，走过了 56 年的光辉历程，为国家做出了一系列成绩和贡献。

在科技成果评价体系的引导下，随着时间的推移和科技人员的退休，本所的卵菌分类学和黏菌分类学已人去楼空。真菌分类学的研究力量也在逐年减少，从而使本学科面临后继无人的严峻形势。如不采取有效措施，以本所首任正副所长，我国真菌学科奠基人戴芳澜、邓叔群为代表的先辈科学家开创的中国真菌分类学科必将遭受自生自灭的后果。果真如此，对科学和人民生活有无影响呢？这直接涉及到对生物分类学包括菌物分类学或生物系统分类学（systematic biology）的认识问题。

认识一：发现新种有何意义？自然界的生物经历了 30 多亿年的进化过程已经形成了那麽多新种，何必还要你们去发现它们呢？

这显然是个误会！每一种生物都具有各自独特的基因库。地球上丰富的生物物种多样性也是基因库的多样性。它们是在生存斗争和自然选择的长期演化过程中，新种不断产生和灭绝的结果。据专家估计，地球上已经形成并存在的生物物种约为一亿种，其中只有约 140 万种已被人类所发现，仅占估计种数的 1.4%；尚有 9,860 万种，占估计种数的 98.6%，尚未被人类所发现。所谓新种，是指在地球生物圈中生存的，尚未被发现的物种首次被生物学家所发现。通过对于新发现物种的认识、描述、命名、发表并归入“物种知识库”，或称“物种资源信息存取系统”，作为人类可持续发展的物种与基因资源储备；也为不断填补人类关于生物演化长链中的缺失环节做出贡献。生物在它们的生存斗争中，新的物种在不断地产生，原有的物种也在不断地灭绝。因此，地球生物圈是一个相对稳定而不断变化着的生物多样性世界。只要地球存在，地球生物圈中的生物多样性就存在，生物系统分类学便不可缺少，将永远是人类可持续发展中无法回避的科学研究领域之一。

认识二：不用那麽多物种资源，也无须甚麽种名，随便从一把土样中照样可以筛选出需要的活性物质。

生物分类学研究的第一步就是从全国各种不同生态条件下获取物种并将其归入物种资源信息存取系统，为人类开发利用生物物种资源和基因资源多提供更多的选择机会。与只含有少数常见物种的随机土样相比，其筛选成功率之差别是显而易见的。

认识三：分子系统学是生物分类学或生物系统分类学的最佳选择。

分子生物学技术的飞速发展，为两个半世纪以来基于表型的生物分类学打开了进入基因型的大门。人类只有在表现型与基因型相结合的综合分析中才能正确认识生物物种及其相互之间的亲缘关系本质。只相信传统的表型分类学，或只是简单地倚赖基因型的分子系统学手段都是片面而不可取的。生物物种是基因的载体，只研究基因而不研究物种，是无法发现、认识、掌握和利用新的物种资源及其基因资源的，犹如只取得人的指纹而不知，也不去认识指纹的主人，再精确的指纹也毫无意义。发现和认识主人，进而查明其指纹，以及根据指纹查明其主人的过程，在某种意义上，犹如生物系统分类学研究中由表现型到基因型，再由基因型回到表现型，亦即表型与基因型相结合的综合分析过程。因此，人类必须在表型与基因型相结合中，既研究物种的表型性状，也研究物种的基因型结构，有组织、有计划地加强对于作为基因载体的新种进行发现、认识、描述、命名、发表并归入物种资源信息存取系统，探明它们之间的亲缘关系，不断补充和丰富人类关于物种资源及其基因资源库，为可再生的生物资源，包括菌物资源的开发利用提供信息存取系统，为资源开发利用提供更多的选择机会。因此，生物系统分类学，包括菌物分类学，是发现有用物种的强大后盾，在人类可持续发展中具有重要的战略意义。

尽管真菌分类学是本实验室的强项，但是，我们不能只沿着一条路走道底。我们必须按照国家目标和国际前沿拓展我们菌物科学的研究领域。

如果将生命系统知识看作为一部巨著，那么，经过自然长期演化所形成的生物多样性可被称之为“生命原版”(Life Original Edition: LOE)；经过传统的杂交育种而获得的优良新品种可被称之为生命再版（Life Secondary Edition: LSE)；通过遗传工程获得的转基因生物可被称之为生命三版（Life Third Edition: LTE)；而合成生物学则可被称之为生命四版（Life Fourth Edition: LFE)。

在拓展我们菌物科学的研究领域时，按照国家目标和国际前沿，无论是“生命原版”、“生命再版”、“生命三版”或“生命四版”均可被列入我们探索研究的计划名单。生命再版至四版均以生命原版为基础进行创新设计。因此，对生命原版的结构与功能的各有关研究领域仍是我们面临的主战场。

所谓“有所为，有所不为”的后半句不可用于生物分类学。作为具有 13 亿人口的发展中大国，在生命科学领域，我们可以“有所大为，有所中为，有所小为，绝不可有所不为”!

我想引用美国科学院院士，中国科学院外籍院士 Peter Raven 的一段话：“我们甚至不能将地球上的物种估计到一个确定的数量级，从我们影响人类前景的知识能力的角度来看，这是一个多么令人震惊的现实！很明显没有几个科学领域中，我们的知识还如此贫乏，然而没有哪一个科学领域与人类有如此直接的相关”(Peter Raven 美国密苏里植物园)。

最后，我愿以“浪淘沙”诗词一首结束此文：

菌物多样性，这边独丰，先辈为之献终生。
踏遍青山辨物种，积存备用。
日月行如风，家底未清。
生物分类谁继承？
表基结合创未来，还看后生！

图 1. 真菌地衣系统学开放研究实验室学术年报

参考文献

Barr DJ 1992. Evolution and kingdoms of organisms from the perspective of a mycologist. ***Mycologia* 84**（1）:1-11.

Bruns TD, White TJ and Taylor JW 1991. Fungal Molecular Systematics. ***Annu. Rev. Ecil. Syst.*22:**525-564.

Hawksworth DL 1991. The fungal dimension of biodiversity: magnitude, significance, and conservation. ***Mycol. Res.*95**（6）:641-655.

Hawksworth DL, Kirk PM, Sutton BC et al. 1995. Ainsworth & Biby's Dictionary of the fungi （Eighth Edition） CAB International, Loundon 1-616.

Whittaker, R. H.**1959** On the broad classification of organisms. ***Quart.Rev. Boil.* 34:**210-226.

裘维蕃 1991 对菌物学进展的前瞻 ***真菌学报* 10**（2）：81-84.

菌物生物多样性与人类可持续发展*

魏江春

（中国科学院微生物研究所真菌地衣系统学重点实验室 北京 100101）

摘要 文章对菌物生物多样性从下列三个角度进行了论述。首先，在自然界生物圈中的生物多样性包括物种、基因和生态系统三个层次；而物种多样性则是最基础和关键的。菌物是自然界物种多样性最丰富的两大类生物之一。其次，菌物分类学是研究菌物物种多样性的重要工具。第三，生物信息、菌种以及菌物原型参证的三个存取系统，作为菌物分类学的研究成果，是人类可持续发展所需要的可再生的生物资源的重要支撑系统。最后，为了加强对菌物物种多样性和基因多样性的研究、保护和开发，作者提出了4点建议。

关键词 物种多样性，分类学，演化系统生物学，存取系统，生物信息学，菌种库，标本馆，生物资源

DOI：10.3969/j.issn.1000-3045.2010.06.010

魏江春院士

1 引言

生物多样性通常被理解为物种多样性、基因多样性和生态系统多样性3个层次。物种是基因的载体；一个物种一个基因库；基因本身在生物个体之外是没有生存价值的。因此，没有物种便没有基因；没有物种多样性便没有基因多样性；没有物种，就无法进行结构基因组学和功能基因组学的研究、开发和利用。因此，生物多样性实际上是指生存于地球生物圈多种多样生态系统中的，含有多种多样基因的物种多样性；简言之，在多样性的生态系统中生存着丰富的物种多样性。

* 修改稿收到日期：2010年10月8日

由于人类关于生物物种多样性，尤其是微生物以及菌物物种多样性的知识还非常贫乏，大量微生物及菌物物种尚无法从生态系统中分离培养，因而无法对它们的物种进行全面认识、命名、研究、开发和利用，无法对其基因组进行结构与功能的研究、开发和利用。虽然可以借助于分子生物学技术直接从生态系统中获取部分基因片段加以研究、开发和利用，但并不知道这些基因是哪些物种所拥有；从而无法了解这些物种的基因组全貌及其结构与功能。因此，直接从生态系统中获取基因片段的方法并非人类利用生物资源的理想之途，只是在人类关于物种多样性知识非常贫乏时的无奈之举。

不断地认识自然界生物物种多样性，尤其是菌物物种多样性，给它们命名，可为研究它们的结构基因组学和功能基因组学、亲

本文原载于《中国科学院院刊》，25（6）：645-650, 2010.

缘关系与演化系统，为菌物物种资源和基因资源的开发利用提供综合信息存取系统，即所谓菌物分类学或菌物演化系统生物学，是菌物资源研究与开发中必不可少的上游环节。

以表型与基因型相结合的菌物分类学或菌物演化系统生物学的发展，经历了宏观形态学、微观解剖学、超微形态学与结构学以及部分基因的分子系统学的不同历史阶段。所谓的经典分类学或传统分类学只是菌物分类学发展历程中的不同历史阶段。今后的菌物分类学，或菌物演化系统生物学必将朝着表型组学与基因组学相结合的研究方向发展。

2 菌物物种多样性

所谓菌物(*pan fungi*)，是指由真菌学家研究的真核生物域原生动物界(*Protozoa*)中的黏菌(*slime moulds*)；菌藻界(*Chromista*)中的假菌(*pseudofungi*)等以及真菌界(*Fungi*)中的全部成员(图 1)。

至于微生物(*microorganisms*)，顾名思义是指一切体型微小的生物，包括原核生物域细菌界真细菌亚界和古细菌亚界(图 1)，以及真核生物域中真菌界的微观真菌，甚至无独立生存能力的非细胞大分子病毒等等。无论是菌物或微生物，从演化系统角度看，各自均非单系类群(*monophyletic group*)(同一祖先的全部后代)，而是由部分并系类群(*paraphyletic group*)(同一祖先的部分后代)和部分单系类群以及复系类群(*polyphyletic group*)(不同祖先的后代)组成的混合群。

地球生物圈中自然存在的表型与基因型彼此相似的生物个体的集合构成居群。表基彼此相似的居群构成物种。一个物种便是一个基因库，是生物圈中生物演化的基本单位。

我们以菌物中 3 大菌类生物之主角真菌为例，对其物种多样性作进一步分析。真菌主要是寄生于维管束植物，据专家的保守估计，全世界有 25 万种维管束植物，因此，按每种维管束植物寄生有 6 种真菌估计，真菌至少有 150 万种[2]。除了植物寄生真菌以外，植物内生真菌已被大量发现。按每种维管束植物有 4 种内生真菌估计，全世界就有 100 万种内生真菌[3]。按照这一估计，全世界有寄生和内生真菌至少有 250 万种。如果将腐生真菌、附生在植物茎叶和生长在岩石上的地衣型和非地衣型真菌以及地衣内生菌包括在内，必将远远超过 250 万种。然而，已被人类所认识和命名的真菌还不到 97 861 种[5]，仅占估计种数的 3.9%；尚有 96.1%的真菌有待人类去发现、认识、命名、描述、研究和开发利用。

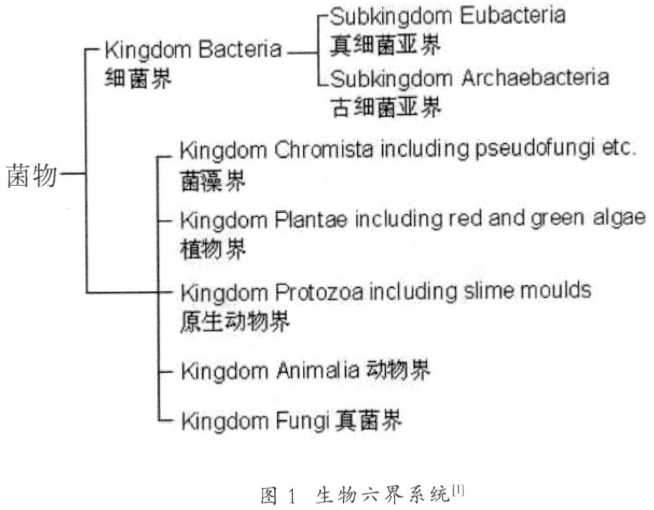

按照这一保守估计，作为世界生物多样性最丰富的国家之一，我国已知维管束植物按 3 万种计，寄生和内生真菌至少有 30 万种。如果将土壤腐生真菌以及附生在植物茎叶上的附生真菌(*phorophytic fungi*)和生长在岩石上的地衣型和非地衣型真菌以及地衣内生菌估计在内，必将远远

菌物
- Kingdom Bacteria 细菌界
  - Subkingdom Eubacteria 真细菌亚界
  - Subkingdom Archaebacteria 古细菌亚界
- Kingdom Chromista including pseudofungi etc. 菌藻界
- Kingdom Plantae including red and green algae 植物界
- Kingdom Protozoa including slime moulds 原生动物界
- Kingdom Animalia 动物界
- Kingdom Fungi 真菌界

图 1 生物六界系统[1]

超过 30 万种。然而，已被命名的真菌仅 1.3 万种，占估计种数的 0.23%；尚有 99.77%的真菌物种有待发现、认识、命名、描述、研究和开发利用。

3 菌物物种多样性与资源

世界上第一个抗菌素——青霉素是从真菌中发现的。它的发现和应用揭开了人类利用抗菌素的历史，使人类的平均寿命从 40 岁提高至 60 岁。然而，后来的抗菌素则以放线菌为主要来源。由于放线菌的物种数量有限，作为抗菌素资源正在趋于枯竭；而作为新药资源的真菌却日益显示其明显的上升优势。

以中华医药瑰宝之一的灵芝为例，在古代文献记载中不可能精确到现代的物种概念。所谓灵芝，实际上为灵芝科中 4 个属，即灵芝属、假芝属、网孢芝属和鸡冠孢芝属中的不同种类。它们在外形上虽有一定区别，其孢子特征却差异更加明显(图 2)。因此，我们的任务在于探明灵芝科的 4 个属中的哪一种或哪几种、甚至全部种类均对人类健康有益。

我国已知的冬虫夏草、桑黄、樟芝、云芝、槐耳等 300 多种药用真菌是保障人类健康的宝贵资源。

此外，我们于 2007 年采自海南岛的担子菌地衣 (*Multiclavula sinensis*) 的内生真菌 (*Pestalotiopsis sp.*) 中发现了 23 种次生代谢产物，其中 21 种在结构上是全新的，其中有的还具有抗癌和抗艾滋病毒活性[6]。

我们还从荒漠地衣石果衣 (*Endocarpon pusillum*) 真菌中发现 100 多种新的基因[7]；目前正在进行功能分析。而且，此类地衣物种在我国干旱荒漠中还起着重要的固沙和固碳作用。类似的荒漠地衣物种多样性广泛地分布于我国干旱和半干旱荒漠地带，是荒漠微型生物群落的优势成分，在缓解沙尘肆虐和全球变暖中具有不可忽视的积极意义；又是抗干旱、抗辐射、抗盐碱、抗重金属基因的资源宝库。

4 菌物物种多样性研究保护及其存取系统

地球生物圈中多样性的生态系统中生存着含有多样性基因的多样性生物物种。生物多样性是人类可持续发展的资源宝库。对于生物多样性的研究、保护和开发利用对于人类可持续发展具有重要的战略意义。

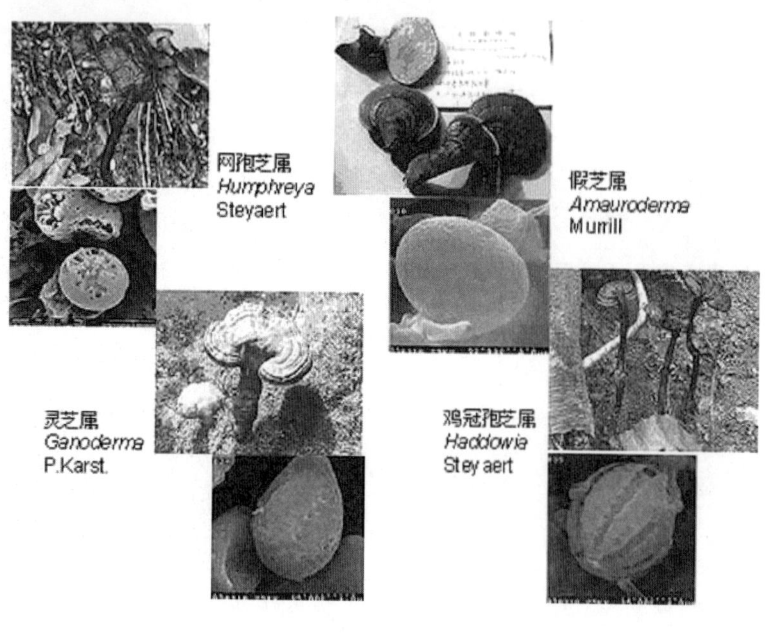

图 2　灵芝科 4 个属的代表种原型特征及其孢子特征[8]

对于菌物物种多样性来说，通过 3 大存取系统，即作为专著的物种资源信息存取系统，作为菌种库的物种资源和基因资源存取系统，以及作为标本馆的物种原型参证存取系统进行研究、保护、开发和利用，为生命科学研究和人类可持续发展提供支撑系统，在新药发现、可再生能源筛选以及环境治理与人类健康创新方面具有重要意义。

专著等信息存取系统：对自然界多种多样的菌物物种进行采集、分析、研究、分类、命名、描述；按照菌物物种各自的演化关系，以物种为基本单位，将它们排列成不同等级的演化系统，存入菌物种系综合信息存取系统，并向世界公开发表；为生命科学研究和生物资源开发利用提供菌物物种资源信息存取系统，如地区和世界范围的物种分类和演化系统专著(图 3)。

标本馆：菌物的活体菌种和生长在自然界多种多样生态系统中的菌物物种原型在表型特征上是截然不同的。自然界多种多样的菌物物种具有极为多样而复杂的外部形态和内部结构。保存于菌种库中的活体菌种则只呈现出肉眼可见的菌落和显微镜下的菌丝和孢子。仅仅根据这些特征，包括由ATGC的不同排列组合构成的基因，根本无法辨认生长在自然界多种多样的菌物物种原型。为了辨认和对接菌种库中的活体菌种和自然界的物种原型，将采自自然界中多种多样的菌物物种原型制作成可供长期保存的菌物标本，保存于国家菌物标本馆，作为菌物物种原型的证据，为生命科学研究和物种原型提供参考证据的存取系统(图 3)。

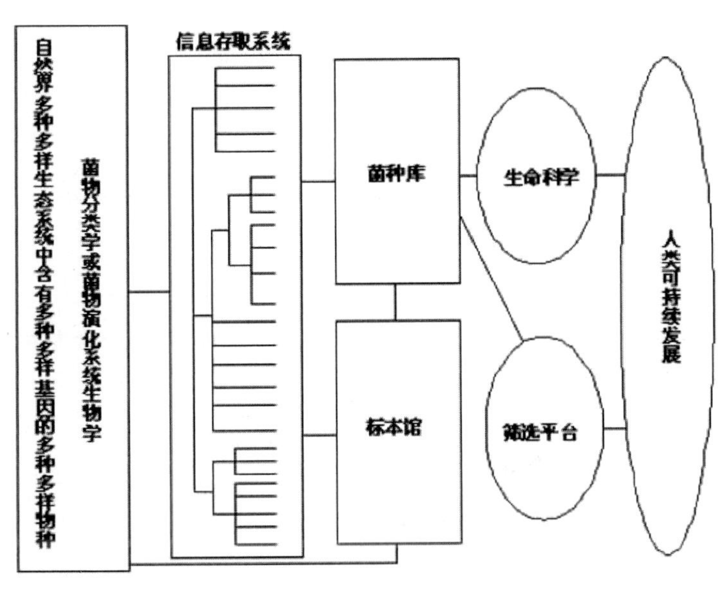

图 3 菌物物种多样性研究保护开发及其 3 大存取系统示意图

菌物分类学家或菌物演化系统生物学家，在对地球生物圈多样性生态系统中的菌物物种多样性进行采集、分析、研究、识别、归类、命名、描述等一系列研究的基础上，使其研究结果不断丰富上述 3 个存取系统，是新世纪生命科学研究和菌物资源研究开发和利用中不可缺少的上游环节和重要支撑系统。这 3 个存取系统越丰富，则资源筛选的基数就越大，筛选效率就越高，对生命科学的深入研究的支撑力度就越强。如果能将我国菌物分类学或菌物演化系统生物学及其 3 个存取系统作为一个整体予以重视并给予强力资助，必将在可再生资源创新与人

菌种库：对于菌物物种多样性的有效保护在于迁地保护，即从自然界多种多样的菌物物种中采集并分离培养出尽可能多的活体菌种。将多样性菌物的活体菌种保藏于作为物种资源和基因资源存取系统的菌种库；为生命科学研究和生物资源开发利用提供菌物物种资源和基因资源的存取系统（图 3）。

类可持续发展中起到重要作用。

5 建议

菌物物种多样性是人类可持续发展中与人类健康与环境保障密切相关的可再生资源宝库。保护生态系统多样性是物种多样性，因而也是基因多样性保护的关键。在此基础上，进行采集、分析、识别、归类、命名、描述等一系列研究；将其研究结果储备于上述三个存取系统，是菌物物种资源和基因资源保护、研究、开发和利用的最佳选择。而菌物分类学，或菌物演化系统生物学是这一最佳选择的基础。然而，近年来，菌物分类学的科技队伍处于后继乏人的危机处境。

“当你失去它时才真正感到它的重要性时”将为时太晚。对于一个正在和平崛起的发展中大国来说，我国在科技发展中的该领域反而倒退至上一世纪 20 年代的水平将是难以想象的。因此，谨提出以下建议：

（1）鉴于管理生物多样性的国家部级和局级部门不少于 4 个，因而，建议国家在国务院设立生物多样性及生物资源办公室，负责统一规划和协调有关部局的管理工作。

（2）鉴于菌物物种多样性的独特性和丰富性，建议成立国家菌物生物多样性与基因组学研究所。

（3）鉴于从事菌物物种多样性及其分类学研究的科技队伍后继无人，建议给他（她）们以稳定的经费支持。

（4）鉴于对菌物物种多样性的知识十分贫乏，建议组织菌物分类学家有计划、有步骤地进行菌物物种世纪大普查，不断丰富三大存取系统，为国家战略需求提供丰富的菌物物种资源和基因资源储备。菌物分类学及其三大存取系统与新药筛选等国家需求相结合，必将发挥菌物物种资源和基因资源的巨大潜力，为国家的科技创新做出重要贡献。

通过上述国家战略需求任务的实施，必将不断地培养和锻炼出成批的年轻菌物分类学家等科技人才和创新科技团队。

主要参考文献

1 Cavalier-Smith T. Only six kingdoms of life, Proc. R. Soc. Lond. B., 2004,271 (1 545): 1 251-62. doi: 10.1098/rspb.2004.2705, PMID 15306349, PMC 1691724, http://www.cladocera.de/protozoa/cavalier-smith_2004_prs.pdf, retrieved 2010-04-29.

2 Hawksworth D L. The fungal demention of biodiversity:magnitude, significance and conservation. Mycol. Res., 1991, 95(6):641-655.

3 Petrini O, Sieber T N, Toti L et al. Ecology, Metabolit Production and Substrate Utilization in Endophytic Fungi. Natural Toxins, 1992,1:185-196.

4 Strobel G, Daisy B. Bioprospecting for Microbial Endophytes and Their Natural Products. Microbiology and Molecular Biology Reviews, 2003, 67(4): 491-502.

5 Kirk P M, Cannon P F, Minter D W et al. Dictionary of the Fungi 10th edition CABI Europe-UK, 2008.

6 Ding G, Li Y, Fu S et al. Ambuic Acid and Torreyanic Acid Derivatives from the Endolichenic Fungus Pestalotiopsis sp. J. Nat. Prod., 2009, 72 (1):182-186, DOI: 10.1021/np800733y, Publication Date (Web): 31 December 2008. Downloaded from http://pubs.acs.org on February 23, 2009(SCI).

7 Wang Y Y, Zhou Q M, Wei J C. Construction of a full-length cDNA library from mycobiont of Endocarpon pusillum (in press).

8 吴兴亮，戴玉成. 中国灵芝图鉴. 北京: 科学出版社, 2005.

The Biodiversity of Pan-fungi and the Sustainable Development of Human Beings

Wei Jiangchun

(Key Laboratory of Systematic Mycology & Lichenology, IM, CAS 100101 Beijing)

Abstract The present paper deals with the biodiversity of pan-fungi from the following three angles: (1)The biodiversity consists of species, genetic and ecosystem diversities in the biosphere of the nature. The species diversity is the most important of the three. The pan-fungi are one of the two large groups of the organisms which have the most rich species diversity in the nature.(2)The pan-fungal taxonomy is the most important tool to study the species diversity of the pan-fungi. (3)The three storage and retrieval systems of bioinformatics, culture collections and herbaria of pan-fungal prototype references, as the results of pan-fungal taxonomy are the most important support system to biological resources as renewable one for the sustainable development of human beings. Finally, to strengthen the study, conservation and development of the pan-fungal species and genetic diversity, four suggestions are given by the author.

Keywords species diversity, taxonomy, evolutionary systematic biology, storage and retrieval systems, bioinformatics, culture collections, herbarium, biological resources

魏江春 中国科学院院士，中科院微生物研究所研究员。1931 年 11 月出生，陕西咸阳人。1955 年毕业于西北农学院，1962 年毕业于苏联科学院研究生院，先后获生物科学副博士和博士学位。中国地衣学主要奠基者。历任中科院微生物所副所长、学位委员会主任、院属真菌地衣系统学重点实验室主任及学委会主任、微生物资源前期开发国家重点实验室学委会主任、中国菌物学会理事长；现任中国菌物学会名誉理事长、《中国孢子植物志》编委会主编、国际生物多样性中国委员会顾问委员、国家濒危物种科委会委员等。曾论述石耳科新二属系统；发表地衣新属、新科、新目各 1 个，论文 100 余篇，专著 4 册。先后分别获中科院及国家海洋局科技进步奖特等奖。E-mail:weijc2004@126.com

《菌物学报》三十年回眸与展望

魏江春*

中国科学院微生物研究所真菌地衣系统学重点实验室 北京 100101

摘　要：就下列三方面进行论述：首先就《菌物学报》的简史进行了回顾。对《真菌学报》及《Mycosystema》的创刊及其背景进行了简短叙述。当《Mycosystema》已被《Dictionary of the Fungi》第八版作为世界25种菌物科学定期刊物之一列入国际刊物名录后，学会和开放实验室决定将《真菌学报》并入《Mycosystema》；就中文刊名启用《菌物学报》的背景进行了回顾。其次，就菌物多样性及其分类学与演化系统生物学、《菌物学报》所扮演的三大存取系统之一——菌物信息存取系统的重要性进行了论述。再次，对《菌物学报》的过去和现在的历史性贡献予以肯定；对其今后的作用抱有期望。

关键词：简史，菌物期刊，三大存取系统，原型参证，学术交流

A glance back at the thirty years' Mycosystema and prospect

WEI Jiang-Chun*

The Key Laboratory of Systematic Mycology & Lichenology, Institute of Microbiology, Chinese Academy of Sciences, Beijing 100101, China

Abstract: The present paper deals with the following three points: 1. A brief history of starting publication of the mycological journal in China. The first issue of the *Acta Mycologica Sinica* in Chinese was published in 1982. The first issue of the *Mycosystema* in English was published in 1988, which as one of the 25 periodicals in mycology of the world was listed by the *Dictionary of the Fungi* (8th Edition). In this case, the *Acta Mycologica Sinica* was merged into the *Mycosystema* in 1997. And the volume ordinal numbers of the *Acta Mycologica Sinica* was continued to use for that of the merged *Mycosystema*. 2. The three storage and retrieval systems of bioinformatics (publications), culture collections and herbaria of pan-fungal specimens as prototype references of the pan-fungal taxonomy are the most important support system to biological resources. 3. The *Mycosystema* as one of the three storage and retrieval systems played the leading role in the academic exchange of the mycology at home and abroad. The author hopes to see the *Mycosystema* will play more important role in the academic exchange of the mycology.

Key words: a brief history, mycological journal, three storage and retrieval systems, prototype references, academic exchange

*Corresponding author. E-mail: weijc2004@126.com

本文原载于《菌物学报》, 30（1）: 1-4, 2011.

《菌物学报》自创刊以来已经走过了整整三十年。三十年来，《菌物学报》在我国菌物科学发展与人才成长中，在与国内外菌物科学家进行学术交流中，做出了重要的历史性贡献，也是改革开放带给我国菌物科学的重要成果之一，值得庆贺。

此刻，对于《菌物学报》的创刊简史及其前身与后来的发展与演变过程进行回顾；对于《菌物学报》未来发展的希望和期待进行展望，是对《菌物学报》三十年来所做重要贡献的最佳庆贺。

三十年回眸

自从我国真菌学科的奠基人和领军人，杰出的真菌学大师戴芳澜和邓叔群开创我国真菌学科，并于1956年成立中国科学院应用真菌学研究所以来，我国真菌学科不断发展繁荣，真菌学家队伍不断成长壮大，学科积累雄厚扎实。在此基础上，在出席1972年中国科学院计划工作会议的微生物研究所和海洋研究所代表倡议下，经中国科学院批准，于1973年正式成立了中国科学院中国孢子植物志编辑委员会，以组织全国孢子植物分类学家开展中国真菌志、地衣志、藻类志和苔藓志的编前研究和在研究基础上的编写工作。中国科学院微生物研究所为编委会的依托单位。

此后，在编委会依托单位中国科学院微生物研究所创办《孢子植物学报》的努力失败后，我们发起并团结和组织全国真菌学家和地衣学家，于1980年在中国植物学会内成立了真菌学分会。学会是科学家之家；学报是学术交流的窗口。在真菌学分会名义下，经微生物研究所批准，启动了《真菌学报》的创办工作。在排除各种干扰后，《真菌学报》(Acta Mycologica Sinica)中文版季刊于1982年正式问世；其封面展示了由Whittaker（1969）提出的，包括真菌、假菌和黏菌的真菌界在内的生物五界系统树，以表明《真菌学报》所涵盖的生物类群及其系统学地位。

中国科学院微生物研究所真菌地衣系统学开放实验室作为中国科学院第一批开放实验室于1985年成立后，我们创办了实验室英文版年报《Mycosystema》，以J.C.Wei和D.L.Hawksworth为合作主编；以国内外著名真菌学家和地衣学家为编委会成员。《Mycosystema》作为世界25种真菌学科刊物之一，被《Dictionary of the Fungi》第八版（Hawksworth *et al.* 1995，p.254）所收录。因此，在植物学会真菌分会基础上，经国家批准，于1993年成立的中国菌物学会与开放实验室研究决定，将中文版《真菌学报》与英文版实验室年报《Mycosystema》合并；采用《Mycosystema》为合并后的刊名；以《菌物学报》(《菌物系统》曾被短暂作为中文刊名）为合并后的中文刊名；沿用了中文版《真菌学报》的卷序；同时发表中英文稿件，尽量增加英文稿件的比例，以利于国际学术交流。

随着分子生物学技术的飞速发展及其在演化系统生物学中的应用，进一步揭示了Whittaker生物五界系统中的真菌界并非同一祖先后代的单系类群，而是除了单系类群的真菌界（Fungi）以外，还有原生动物界（Protozoa）中的并系类群黏菌以及藻菌界(Chromista)中的并系类群假菌等。因此，Whittaker的真菌界实为一复系生物类群。该复系类群曾被称为“Organisms studied by mycologists”(Hawksworth 1991)，即“由真菌学家研究的生物”，或“Union of fungi”(Barr 1992)，即“真菌联合体”。既然如此，涵盖这一复系类群生物的学报继续延用单系生物类群“真菌”为刊名，显然是不合时宜的。因而，菌物学会与开放实验室决定，合并后的中文刊名，改用由裘维蕃提出的“菌物”为涵盖该复系生物类群刊物的刊名，以反映演化系统生物学的最新进展。这是中文刊名《真菌学报》改名为《菌物学报》的历史背景。

当然，作为复系类群的菌物所涵盖的生物类群包括单系类群的真菌界（Fungi），藻菌界(Chromista)中的并系类群假菌等，以及原生动物界（Protozoa）中的并系类群黏菌（Cavalier-Smith 2004)。因而，“菌物”只是代表真核生物三个生物界中这一复系菌类生物的普通名称，而非单一生物界的学名；正如“微生物”是代表多个原核及真核生物界中微小生物的普通名称一样，亦非单一生物界的学名。

《真菌学报》创刊后的第一任主编为王云章，第二任主编为余永年，第三任主编为白金铠，第四届主编以及两刊合并后的《菌物学报》主编为庄剑云，现任主编为戴玉成。学报是学会和学科的窗口，为我国菌物科学的发展做出了重大的历史性贡献。学报主编在这一历史性贡献中付出了他们的智慧和辛劳。

我国菌物学科的开创、发展、壮大和不断繁荣，是在戴芳澜、邓叔群作为奠基人和领军人的基础上，全国菌物学家团结奋进的结果；其中中国科学院微生物研究所真菌研究室，真菌与地衣系统学开放实验室、重点实验室的科学家们发挥了应有的引擎作用。而拥有决策权的中国科学院微生物研究所党政领导班子中具有本实验室代表的话语权，以及院所各级领导的支持至关重要。

展望未来

《菌物学报》是由中国科学院微生物研究所和中国菌物学会主办，是中国科学院主管的学术期刊之一；是以菌物生物多样性及其系统分类学与演化系统生物学为主，涵盖菌物科学各领域的国家核心刊物，是与国内外进行菌物学术交流的重要窗口。

物种是基因的载体，基因本身在生物个体之外是没有生存价值的，一个物种一个基因库。因此，没有物种便没有基因，便无法进行基因组学的研究、开发和利用。保护物种，便是保护基因；保护物种多样性，便是保护基因多样性。因此，所谓生物多样性，是指在多样性的生态系统中生存着含有基因多样性的物种多样性。

在我国多气候带的生态系统多样性中生存的极为丰富的菌物物种多样性，是我国取之不尽，用之不竭的重要生物资源宝库。自然界中的物种多样性绝非一成不变。生物在其漫长的演化过程中，新的物种在不断产生，濒危物种在不断灭绝。不断地认识自然界生物物种多样性，尤其是菌物物种多样性，给它们命名，为研究它们的基因组学、亲缘关系与演化系统，为菌物物种资源和基因资源的开发利用提供三大存取系统，即作为论文和专著的物种资源信息存取系统；作为菌种库的物种资源和基因资源存取系统；以及作为标本馆的物种原型参证存取系统，是菌物系统分类学和菌物演化系统生物学研究的重要内容，是生命科学研究和人类可持续发展的重要支撑体系，是菌物资源研发中必不可少的上游环节；在农业发展、环境治理、新药发现与人类健康方面具有重要的战略意义（魏江春 2010）。

《菌物学报》正是三大存取系统中物种资源信息存取系统的重要平台，是我国菌物科学与国际进行学术交流的重要窗口。只要地球还存在，菌物物种多样性就会存在和演化。发现、认识、掌握菌物物种多样性及其分类学与演化系统生物学就是不可缺少的。作为菌物科学三大存取系统之一——物种资源信息存取系统的《菌物学报》就是不可取代的。

奖励作者在 SCI 刊物上发表文章；重奖发表在影响因子更高的 SCI 刊物上的文章作者，并以此作为博士论文答辩、成果鉴定、职称评定的衡量标准，是我国学术论文大量外流的动力；是国内学术刊物难以进入国际一流学术刊物行列的主要障碍。

我相信，如果上述动力和障碍一旦被排除，在全国菌物学家的团结奋进中；在主编和编委们的努力下，为了便于国际学术交流，不断增加英文稿件比例，增大中文稿件中的英文摘要篇幅，在我国菌物科技工作者的优秀论文支撑下，《菌物学报》（Mycosystema）必将挤入国际菌物科学一流学术刊物行列，成为国际菌物学家投稿的选择之一。届时，我国菌物科学才能称得上与我国作为日益和平崛起的发展中大国地位相适应。

无论是孢子植物志编委会、学会、学报或实验室，都是本学科发展、繁荣的见证，来之不易，理应珍惜。以继承和创新为前提，在全国菌物学家团结奋进中，我国菌物科学必将更加发展、繁荣；我国菌物科技队伍必将更加壮大、优化，为生命科学以及我国直至全人类可持续发展做出更大贡献。

2010 年 12 月

菌物学报

[REFERENCES]

Barr DJ, 1992. Evolution and kingdoms of organisms from the perspective of a mycologist. *Mycologia*, 84(1): 1-11

Cavalier-Smith T, 2004. Only six kingdoms of life. *Proceedings of the Royal Society of London, Series B*, 271(1545): 1251-1262

Hawksworth DL, 1991. The fungal dimension of biodivdersity: magnitude, signifance, and conservatrion. *Mycological Research*, 95(6): 641-655

Hawksworth DL, Kirk PM, Sutton BC, Pegler DN, 1995. Ainsworth & Bisby's dictionary of the Fungi. 8th edition. CAB International, Oxon, UK. 1-616

Wei JC, 2010. The Biodiversity of Pan-fungi and the sustainable development of human beings. *Bulletin of Chinese Academy of Sciences*, 25(6): 645-650 (in Chinese)

Whittaker RH, 1969. New concepts of Kingdoms of organisms. *Science*, 163: 150-160

[附中文参考文献]

魏江春，2010. 菌物生物多样性与人类可持续发展. 中国科学院院刊，25(6): 645-650

On biodiversity of fungi and its systematic biology with three storage and retrieval systems

—Celebrate the 88th* anniversary of the distinguished mycologist Prof. Korf's birth

WEI Jiang-Chun**

The State Key Laboratory of Mycology, Institute of Microbiology, Chinese Academy of Sciences, Beijing 100101, China

Abstract: The concepts of pan-fungi, fungi, and lichens, their biodiversity, and systematic biology of lichen-forming fungi with three storage and retrieval systems are discussed in this paper. A short prospect of the lichen resources is also given.

Key words: biodiversity, systematic biology, storage and retrieval systems

真菌生物多样性及其系统生物学与三大存取系统

—祝贺杰出的真菌学家 Korf 教授 88 华诞

魏江春**

中国科学院微生物研究所真菌学国家重点实验室 北京 100101

摘　要： 讨论了菌物、真菌和地衣的概念，它们的生物多样性，地衣型真菌的系统生物学及其三大存取系统。同时，对地衣资源研究进行了展望。

关键词： 生物多样性，系统生物学存取系统

INTRODUCTION

The organisms belonging to the Kingdom Fungi together with the fungal analogues belonging to the Kingdoms Chromista and Protozoa can be called pan-fungi, or "Junwu" in the Chinese pronunciation, also the organisms studied by mycologists (Hawksworth & Greuter 1989), and the union of fungi (Barr 1992). Members of the Kingdom Fungi

*The 88th anniversary of somebody's birth is called "Rice Birthday" in China, because the number "88" is equivalent to "八十八" in Chinese character, and when arranged vertically it becomes the Chinese character "米" which means "rice". The "Tea Birthday" is the 108th anniversary of somebody's birth in China, because the word "tea" is "茶" in Chinese character. The upper part of this character means two tens (20), and the lower part of it is also "88" as arranged vertically. So, 20 + 88 = 108. A proverb in China says that the Rice Birthday is coming and the Tea Birthday is in sight.

**Corresponding author. E-mail: weijc2004@126.com

Received: 18-02-2013, accepted: 14-03-2013

This paper was originally published in *Mycosystema*, 32（3）: 316-320, 2013.

are generally saprobic and symbiotic organisms. The symbiotic fungi including mutualistic (lichens, endophytes, mycorrhizae and rumen ones), commensalistic and parasitic or antagonistic are frequent, abundant and widespread. The mycorrhizae in association with roots of a plant are found in most, perhaps 85%, of plant species.

A lichen is an ecological association or community of a lichen-forming fungus (*mycobiont*) in symbiosis with a photosynthetic partner (*photobiont*), which may be an alga (phycobiont) or a cyanobacterium (cyanobiont). "The association is not a simple mixture, however, but one in which the fungus produces a thallus, or body, within which the photobionts are housed" (Ahmadjian 1993).

The scientific name of a lichen is that of a lichen-forming fungus in the community. An alga or a cyanobacterium in the community has its own scientific name in the evolutionary system of algae and bacteria respectively.

Cephalodia are small gall-like structures found in some species of lichens that contain cyanobacterial symbionts like the insect galls formed by some herbivorous insects as their own microhabitats in some plants.

Some lichenicolous fungi live on some lichens as a parasite, parasymbiont, or saprobe, and there even exist lichenicolous lichens. In addition, some endolichenic fungi live inside some lichen tissues without showing external signs of its presence.

Biodiversity of fungi

Biodiversity is usually understood as diversity of species, genes, and ecological systems. The gene itself has no living value *in vitro* (Zhang 2000). Each individual of the organisms is the sector of genes. The living value of the genes is dependent upon the activity of the sector *in vivo*. A group of conspecific individuals living in same place at the same time is a population. A species composed of one to several populations. One species is one gene pool. So, there is no gene and genetic diversity without species and species diversity in the nature, and it is then impossible to do the research and development of the structural and functional genomics without species. Therefore biodiversity is the species diversity containing genetic diversity living in the diversity of ecological systems of the biosphere. In short, biodiversity by definition is the species diversity in the diverse ecological systems of the biosphere.

About 97,330 known species belonging to 75,337 genera, 560 families, 140 orders, 36 classes, and 6 phyla in the Kingdom Fungi (Kirk *et al.* 2008).

Currently the estimate of 1.5 million species of fungi (Hawksworth 1991) and that of 1 million species of endophytic fungi (Petrini *et al.* 1992; Strobel & Daisy 2003) worldwide remains to be widely accepted.

The so-called lichen biodiversity, in fact, is the species biodiversity of the lichen-forming fungi, algae, and cyanobacteria in the lichen associations, or ecological systems in symbiosis. Around 20% of all fungi and 40% of all Ascomycota are lichen-forming. Most recent estimates of global diversity suggest that between 17,500 and 20,000 species are known, with a further 1,500 lichennicolous fungi (Kirk *et al.* 2008) and a further numerous endolichen fungi. Unfortunately, such a rich lichen biodiversity, *i.e.* rich lichen species diversity containing abundant anti-adversity gene pools, is a virgin group of biological resources.

The knowledge of biodiversity of the photobionts in lichens is still poor in comparison with that of lichen-forming fungi. About 44 genera of algae and cyanobacteria have been reported as lichen photobionts. Due to the uncertain taxonomic positions of many of these photobionts, the number of genera is considered as approximations only (Gasulla *et al.* 2012). It is estimated that only 25 genera were typical lichen photobionts (Ahmadjian 1993). The most common cyanobionts are *Nostoc*, *Scytonema*, *Stigonema*, *Gloeocapsa*, and *Calothrix*, in order of frequency (Büdel 1992). Green algal photobionts include *Asterochloris*, *Trebouxia*, *Trentepohlia*, *Coccomyxa*, and *Dictyochloropsis* (Gaertner 1992). These authors assessed that more than 50% of all lichen species are associated with *Trebouxia* and *Asterochloris*. However, this is just an estimation since the photobiont genera are reported in only 2% of the described species (Tschermak-Woess 1989), mostly due to the difficulties of isolating and characterizing the algae from lichen thalli (Gasulla *et al.* 2012).

Existence and development of the humankinds have been dependent upon the natural resources together with the intelligence of human beings. The natural resources can be divided into two kinds, reproducible and non-reproducible. Sustainable development of the humankinds has to relay on the reproducible resources. Fungi in the nature are the most important reproducible resources.

Systematic biology of lichen-forming fungi with three storage and retrieval systems

The systematic biology of fungi with three storage and retrieval systems is the key to recognize the biodiversity of fungi and to open the treasure-house of fungal resources in the nature.

As may be inferred from this, the collection, recognition, classification, nomenclature, description, and research of species, *i.e.* the systematic biology is the advanced, indispensable, and basic research for both life sciences and development of biological resources. Lichens are the virgin group of the biological resources. The research on biodiversity and systematic biology of the lichen-forming fungi is indispensable for investigation of lichen resources.

Recognition of lichens, *i.e.* the biodiversity and systematic biology of lichens based on both phenotypic and genotypic data with three storage and retrieval systems, are the first step for exploration and research of the lichens with great potentialities of nature resources.

1. The storage & retrieval system of lichen bio-information is the publications & databases of systematic biology of lichen-forming and non lichen-forming fungi.

2. The storage & retrieval system of culture collection is the feeding center for both exploring fungal resources and research in life sciences.

3. The storage & retrieval system of the fungal and lichen prototype is fungal and lichen herbaria. One cannot recognize and find the lichen-forming fungi and most of the non lichen-forming fungi in the nature based on culture collections only, unless with the help of herbaria. So, herbaria are the bridges between fungal culture collections and the lichen-forming fungi and most of the non lichen-forming fungi in nature. More than 400 thousand collections of prototype fungi are preserved in the Herbarium of Mycology, Chinese Academy of Sciences (HMAS) including more than 120 thousand prototype lichens.

The above mentioned three systems are the bridges between systematic & functional biology of lichens. The recognition, classification, and nomenclature of the lichen-forming fungi are indispensable for both exploration of lichen resources and study of life sciences.

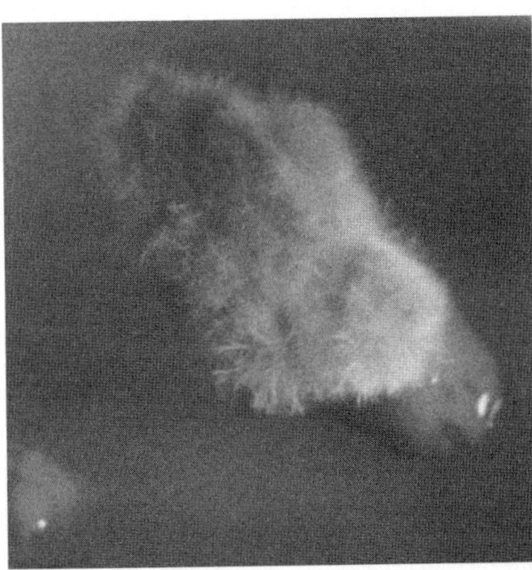

Fig. 1 The cultured mycobiont *Endocarpon pusillum* Hedwig (whitish to flavo-ochraceous) and phycobiont *Diplosphaera chodatii* Bialosuknia (green) isolated from the lichen *Endocarpon pusillum* (Zhang & Wei 2011).

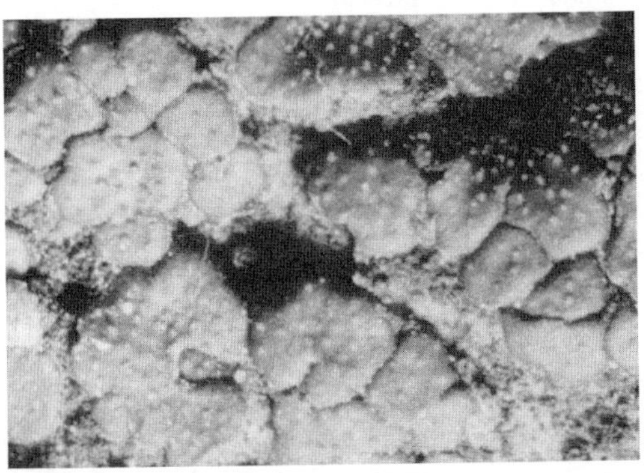

Fig. 2 Prototype of the lichen *Endocarpon pusillum* in nature and preserved in the herbarium HMAS-L.

From 1753 to 1977 when the "Three-domain system" introduced by Carl Woese (Woese & Fox 1977), the biological classification or systematic biology were carried out mainly based on phenotypes. Such a phenotypic era of biological classification proceeded for 224 years. The era of molecular systematics in systematic biology started from 1977. Many classification systems based on phenotype have been modified by the analysis of molecular systematics, but all the taxa of the modified classification systems are still described by phenotypic characteristics only. According to the Darwin's common descent theory, the research of systematic biology should be based on both phenotype and genotype, and all taxa of organisms with diagnoses and descriptions of both phenotype and genotype are indispensable for the future.

The systematic biology should be carried out by phenomics in the combination of genomics and environmics.

Biology of lichen resources

According to the more recent investigations, the biological activities of lichens and lichen substances can be divided into antibiotic, antitumour and antimutagenic, allergenic, enzyme and plant growth inhibitory activities, and also activity against HIV. Research in the field of cultivating lichens and their symbionts open a door to the mass production of lichen substances and their pharmaceutical and technical application (Huneck & Yoshimura 1996).

In the Shapotou region located in the southeast fringe of the Tengger Desert of China, the artificial vegetation with water pump effect expended the deep soil water during the long process of ecological succession under the role of water balance declined year by year. On the contrary, the microbiotic crusts dominated by lichens without water pump effect and with sand-fixating function developed well year by year, and then replaced the artificial vegetation.

The experiments showed that the lichen-forming fungus *Endocarpon pusillum* can survive for seven months under desiccation stress in combination with starvation stress (Zhang & Wei 2011). The results provided a scientific basis for the feasibility of constructing "Bio-carpet Engineering" by means of isolation and inoculation of the microbiotic organisms, including the mycobionts and photobionts to form lichens on the arid desert.

In order to improve the "Bio-carpet

Engineering" it is necessary also to study the drought resistant transgenic sward plants using the drought resistant genes from the desert lichens.

To recognize, transform, and utilize the nature to support the sustainable development of the human beings is the responsibility of scientists. For lichenologists, combining the systematic biology with three storage and retrieval systems, symbiotic biology, and R. and D. of lichen resources for proving the health and environment are our obligatory task.

[REFERENCES]

Ahmadjian V, 1993. The lichen symbiosis. John Wiley & Sons, Inc., New York. 238

Barr DJ, 1992. Evolution and kingdoms of organisms from the perspective of a mycologist. *Mycologia*, 84(1): 1-11

Büdel B, 1992. Taxonomy of lichenized procarytic blue-green algae. In: Reisser W (ed.) Algae and symbioses: plants, animals, fungi, viruses. Interactions explored. Biopress Ltd., Bristol, United Kingdom. 301-324

Gasulla F, Herrero J, Esteban-Carrasco A, Ros-Barceló A, Barreno F, Zapata JM, Guéra A, 2012. Photosynthesis in lichen: light reactions and protective mechanisms, advances in photosynthesis - fundamental aspects. Najafpour M (ed.) ISBN: 978- 953-307-928-8, InTech, Available from: http://www.intechopen.com/books/advances-in-photosynthesisfundamental-aspects/photosynthesis-in-lichen-light-reactions-and-protective-mechanisms

Gaetner G, 1992. Taxonomy of symbiotic eukaryotic algae. In: Reisser W (ed.) Algae and symbioses: plants, animals, fungi, viruses. Interactions explored. Biopress Ltd., Bristol, UK. 325-338

Hawksworth DL, 1991. The fungal demention of biodiversity: magnitude, significance, and conservation. *Mycology Research*, 95(6): 641-655

Hawksworth LD, Greuter W, 1989. Report of the first meeting of a working group on lists of names in current use. *Taxon*, 38: 142-148

Huneck S, Yoshimura I, 1996. Identification of lichen substances. Springer. 1-493

Kirk PM, Cannon PF, Minter DW, Stalpers JA, 2008. Dictionary of the fungi. CABI, Wallingford. 771

Petrini O, Sieber TN, Toti L, Viret O, 1992. Ecology, metabolit production, and substrate utilization in endophytic fungi. *Natural Toxins*, 1: 185-196

Strobel G, Daisy B, 2003. Bioprospecting for microbial endophytes and their natural products. *Microbiology and Molecular Biology Reviews*, 67(4): 491-502

Tschermak-Woess E, 1989. Developmental studies in trebouxioid algae and taxonomical consequences. *Plant Systematics and Evolution*, 164(1): 161-195

Woese C, Fox G, 1977. Phylogenetic structure of the prokaryotic domain: the primary kingdoms. *Proceedings of the National Academy Sciences USA*, 74(11): 5088-5090

Zhang T, Wei JC, 2011 Survuval analyses of symbionts isolated from *Endocarpon pusillum* Hedwig to desiccation and starvation stress. *Science China Life Sciences*, 54(5): 480-489

Zhang YJ (ed.), 2000. Molecular genetics. Science Press, Beijing. 465

魏江春著作目录
（按发表年代顺序排列）

1. **ВЭЙ Цзян-Чунь. 1961**. К флоре лишайников северо-восточной части Карельского Перешейка Ленигpадской Области. [Vej T. C.（**Wei Jiang-Chun**）. **1961**. Ad floram Lichenum partis septenrionali-oreintalis Isthmi Karelici regionis Leningradensis]（in Russian & Latin）. Bot. Materialy，Notulae System. eSect. Cryptogam. Inst. Bot. nomine V. L. Komarovii Acad. Sci. URSS，14：6-14.
2. **ВЭЙ Цзян-Чунь**. **1962**. Дополнение к флоре лишайников северо-восточной части Карельского Перешейка Ленигpадской Области. [Vej T. C.（**Wei Jiang-Chun**）. **1962**. Oddenda Ad floram Lichenum partis septentrionati-orientalis Isthmi Karelici regionibus Leningradensis]（in Russian & Latin）. Bot. Materialy，Notulae System. eSect. Cryptog. Inst. Bot. nomine V. L. Komarovii Acad. Sci. URSS，15（1）：8-12.
3. **ВЭЙ Цзян-Чунь. 1962**. Лихенофлора северо-восточной части Карельского Перешейка（Ленигpадской Область）Бот. Журн. АН СССР，47（6）：830-837. [Vej T. C.（**Wei Jiang-Chun**）. **1962**. Lichenoflora of the North-Eastern part of the Karelian isthmus（Leningrad Region）（in Russian & English）. Botanicheskii Zurnal（Botanical Journal），47（6）：830-837.]
4. **魏江春. 1966**. 疱脐衣属的一新亚属. 植物分类学报，11（1）：1-10. [Wei Jiang-Chun（Vej T. C.）. 1966. A new subgenus of *Lasallia* Merat emend. Wei（Vej）. Acta Phytotaxonomica Sinica，11（1）：1-8 with 2 plates.]
5. **魏江春**，陈健斌. **1974**. 珠穆朗玛峰地区地衣区系资料. 珠峰地区科考报告，生物与高山生理）. 北京：科学出版社：173-18.（Materials for the lichen flora of the Mount Qomolangma region in Southern Xizang，China）
6. **魏江春. 1977**. 奇妙的地衣. 植物杂志，3：37; 4：41.（Marvellous lichens）
7. **魏江春**，姜玉梅. **1980**. 西藏梅衣科地衣新种. 植物分类学报，18（3）：386-388.（Species novae lichenum e Parmeliaceis in regione xizangensi）
8. **魏江春. 1981**. 中国地衣标本集（第一辑）. 植物研究，1（3）：81-91.（Lichenes sinenses exsiccatae）
9. **Wei Jiang-Chun**，Jiang Yu-Mei. **1981**. A biogeographical analysis of the lichen flora of Mr. Qomolangma region in Xizang. Proceedings of Symposium on Qinghai-Xizang（Tibet）Plateau（Beijing，China）Vol. 2：Environment and Ecology of Qinghai-Xizang Plateau：1145-1151.（珠穆朗玛峰地区地衣区系的生物地理学分析）
10. **魏江春. 1982**. 中国疱脐衣属的新种与新资料. 真菌学报，1（1）：19-26.（Some new species and materials of lichen genus *Lasallia* from China）
11. **魏江春**，姜玉梅. **1982**. 西藏地衣新类群与新资料. 植物分类学报，20（4）：496-501.（New materials for lichen flora from Xizang）
12. **魏江春. 1983**. 地衣的物种概念与进化论，进化论选集：北京：科学出版社：169-170.（The species concept in lichens and the theory of evolution）
13. **魏江春**. **1983**. 中国黄梅衣属地衣的初步订正. 真菌学报，2（4）：221-227. [A taxonomic revision of lichen genus *Xanthoparmelia*（Vain.）Hale from China]
14. **魏江春. 1984**. 中国脐鳞属地衣的初步研究. 真菌学报，3（4）：207-213.（A preliminary study of the lichen genus *Rhizoplaca* from China）
15. **Wei Jiang-Chun**. **1984**. A new isidiate species of *Hypogymnia* in China. Acta Mycologica Sinica，3（3）：214-216.（袋衣属一裂芽新种）
16. **魏江春. 1984**. 关于西北地区低等植物资源的开发利用问题. 中国干旱半干旱地区农业通讯，63（1984）.（Problem on the development and utilization of lower plant resources in the region of northwest China）
17. **魏江春**，陈健斌，姜玉梅，等. **1985**. 中国石蕊科地衣研究之一：筛蕊属的订正研究. 真菌学报，4（1）：55-59. [**Wei J. C.**，Chen J. B.，Jiang Y. M. **1985**. Studies on lichen family Cladoniaceae in China Ⅰ：a revision of *Cladia* Nyl. Acta Mycologica Sinica，4（1）：55-59.]
18. **Wei Jiang-Chun**，Chen Jian-Bin，Jiang Yu-Mei. **1986**. Studies on lichen family Cladoniaceae in China II：the lichen genus *Cladina*

Nyl. Acta Mycologica Sinica，5（4）：240-250.（中国石蕊科地衣研究之二——鹿蕊属）

19. **Wei Jiang-Chun**. **1986**. Notes on isidiate species of *Hypogymnia* in Asia. Acta Mycologica Sinica，Supplement I：379-385.（亚洲袋衣属裂芽种类札记）

20. **Wei Jiang-Chun**，Chen Jian-Bin，Jiang Yu-Mei. **1986**. Notes on lichen genus *Lobaria* in China. Acta Mycologica Sinica，Supplement I：363-378.（中国肺衣属简志）

21. **魏江春**. **1986**. 关于子囊菌分类的全系统问题，第二届全国真菌地衣学术讨论会，学术报告及论文摘要汇编：31-36.（On the integrative system of ascomycetous classification）

22. **Wei Jiang-Chun**，Jiang Yu-Mei. **1988**. A conspectus of the lichenized Ascomycetes. Umbilicariaceae in China. **Mycosystema**，**1**：73-106.（中国地衣型子囊菌石耳科纲要）

23. **Wei Jiang-Chun**. **1988**. Introducing the *Mycosystema*. Mycosystema，1：1-2.（《真菌系统》发刊词）

24. **Wei Jiang-Chun**，Jiang Yu-Mei. **1989**. The status of the genus *Llanoa* Dodge and the delimitation of the genera in the Umbilicariaceae（Ascomycotina）. Mycosystema，2：135-150.（子囊菌亚门石耳科属级分类及拉诺属地位的研究）

25. 姜玉梅，**魏江春**. **1989**. 中国条衣属的初步研究. Acta Byrolichenologica Asiatica，1（1-2）：43-52.（A preliminary study on *Everniastrum* from China）

26. 姜玉梅，**魏江春**. **1990**. 袋衣属一新种. 真菌学报，9（4）：293-295.（A new species of *Hypogymnia*）

27. **魏江春**. **1990**. 某些地衣的分布类型及种系分化的初步分析，第三届全国真菌地衣学术讨论会论文及论文摘要汇编：236-237.（Preliminary study of distribution pattern and species differentiation of some lichens）

28. **魏江春**. **1990**. 英国、美国、苏联的真菌地衣分类研究简况，第三届全国真菌地衣学术讨论会及论文摘要汇编：87-89.（The research situation of fungal and lichen systematics in the United Kingdom，United States and Soviet Union）

29. **Wei Jiang-Chun**，Jiang Yu-Mei. **1991**. Some foliiclous lichens in Xishuangbaana，China. // D. J. Galloway（ed.）. 1991. Tropical lichens：their systematics，conservation and ecology. Oxford：Clarendon Press：Systematics Association Special Volume No. 43：201-216.

30. **魏江春**. **1991**. 真菌的系统学现状与展望. 微生物科技信息，6：1-9.（The present situation and prospect of fungal systematics）

31. **Wei Jiang-Chun**，Biazrov. **1991**. Some disjunctions and vicarisms in the Umbilicariaceae（Ascomycotina）. Mycosystema，4：65-72.

32. 姜玉梅，**魏江春**. **1991**. 网脊石耳及网脊平盘石耳分类中微形态学新证据. 真菌学报，10：326-328.（New micromorphologic evidences for classification of *Umbilicaria decussate* and *U. lyngel*）

33. **Wei Jiang-Chun**. **1992**. Some species new to science and distribution in Umbilicariaceae（Ascomycota）. Mycosystema，5：73-88. [石耳科（子囊菌门）的新种和新分布]

34. **魏江春**. **1992**. 菌物系统学进展. 微生物科技信息，2：60-62.（The progress of pan fungal systematics）

35. **Wei Jiang-Chun**. **1993**. The lectotypification of Some species in the *Umbilicariaceae described* by Linnaeus or Hoffmann. Mycosystema，Supplement 5：1-17.

36. **魏江春**. **1993**. 菌物多样性、系统性及其对人类发展中的意义. 生物多样性，1（1）：23-25.（Biological diversity and systematicness of panomycetes，and their significance to the development of human Beijing.）

37. **Wei Jiang-Chun**. **1993**. Current progress of systematics of lichenized fungi. Proceedings of the first Korea-China Joint Seminar for Mycology，Seoul：5-12.

38. Jiang，Yu-Mei，**Wei**，**Jiang-Chun**. **1993**. A new species of *Everniastrum* containing diffractaic acid. Lichenologist，25（1）：57-60.

39. Niu Yong-Chun，**Wei Jiang-Chun**. **1993**. Variations in ITS2 sequences of nuclear rDNA from two Lasallia species and their systematic significance. Mycosystema，6：25-29.

40. **Wei Jiang-Chun**，Jiang Yu-Mei，Guo Shou-Yu. **1994**. Studies on the lichen family Cladoniaceae in China III：a new genus to China：*Thysanothecium*. Mycosystema，7：23-27.

41. Guo Shou-Yu，**Wei Jiang-Chun**，et al. **1994**. Studies on the lichen family Cladoniaceae in China IV：species of *Cladonia* new to China Ascomycota）. Mycosystema，7：29-35.

42. **魏江春**，胡玉琛，姜玉梅，陈健斌. **1994**. 粗皮松萝群体分化性状的研究. 真菌学报，13（3）：199-207.（A study on divergent characters of populations in *Usnea montis-fuji* Mot.）

43. **魏江春**，牛永春. **1994**. 石耳科rDNA多型性分析及其系统学意义. 中国科学院真菌地衣系统学开放实验室年报：8-12.

（RFLP-analysis of DNA in the Umbilicariaceae and its significance in systematics）

44. **魏江春**. **1995**. 亚洲东部石耳科地衣的分布与分化. 中国科学院系统与进化生物学术讨论会论文摘要汇编：92-93.（The distribution and differentiation of lichen species in Umbilicariaceae from E-Asia）

45. **Wei Jiang-Chun**，Jiang Yu-Mei. **1996**. A new species of *Umbilicaria*. Mycosystema，8-9：65-70.

46. **魏江春**. **1996**. 中国生物多样性及其所受威胁情况——地衣. 中国生物多样性国情研究报告：29-30.（The biological diversity of China and their endangered situation-Lichens）

47. 时向阳，**魏江春**，姜玉梅，等. **1997**. 提取地衣真菌总DNA的简便方法. 北京林业大学学报，19（4）：46-50.

48. **魏江春**. **1998**. 地衣 // 裘维蕃主编. 菌物学大全 第六编. 北京：科学出版社：445-548.

49. **魏江春**. **1998**. 面向21世纪的菌物学. 吉林农业大学学报，20（20）：1-4.（Extensive mycology facing 21 century）

50. **Wei Jiang-Chun**，Jiang Yu-Mei. **1998**. Chemical revision of *Hypogymnia hengduanensis*. The Bryologist，101（4）：556-557.（SCI）

51. **魏江春**. **1998**. 石耳科地衣系统学的综合研究分析. 中国生物系统学研究回顾与展望，1998：131-137

52. **Wei J. C**. and Jiang Y. M. **1999**. A lichen genus *Brodoa*，new to China in Parmeliaceae. Mycosystema，18（4）：445-448.

53. **Wei J. C.**，Titov A. N. **2001**. De caliciales（s.l.）sinicae. Novitates Systematice Plantarum non Vascularium，34：102-108.（in Russian）

54. **Wei J. C.**，T. Ahti. **2002**. *Cetradonia*，a new genus in the new family Cetradoniaceae（Lecanorales，Ascomycota）. Lichenologist，34（1）：19-31.（SCI）（One genus new to science）

55. **Wei J. C.**，Abdulla A. **2003**. The lichen genus *Pseudevernia* Zopf. in China. Mycosystema，22（1）：26-29.

56. Liu H. J.，**Wei J. C**. **2003**. Taxonomic revision of six taxa in the lichen genus *Collema* from China. Mycotaxon，86：349-358.（SCI）

57. Ren Q.，Zhao Z. T.，**Wei J. C. 2003**. A lichen genus *Varicellaria* Nylander from China. Mycosystema，22（2）：216-218.

58. Li H. M.，Wang H. Y.，**Wei J. C. 2003**. Diversity of mycobiont substances from three *Xanthoria* species and its significance in systematic biology of lichens. Mycosystema，22（3）：364-368.

59. 王海英，李红梅，石楠，**魏江春**. **2003**. 五种石黄衣中次生代谢产物的高压液相色谱分析. 菌物系统，22（4）：536-541.

60. Liu H. J.，**Wei J. C. 2003**. Two new taxa of the lichen genus *Collema* from China. Mycosystema，22（4）：531-533.

61. 魏鑫丽，**魏江春**. **2003**. 用于PCR扩增的DNA简易制备法现状. 菌物研究，1（1）：52-54.

62. 周启明，黄满荣，**魏江春**. **2004**. 蜈蚣衣科SSU rDNA中I型内含子的初步研究. 菌物学报，23（1）：56-62.

63. Huang M. R.，**Wei J. C. 2004**. Three new taxa of *Stereocaulon* from China. Mycotaxon，90（2）：469-472.（SCI）

64. Wei X. L.，**Wei J. C. 2005**. A study on delimitation of *Rhizoplaca chrysoleuca* group on comprehensive data. Mycosystema，24（1）：24-28.

65. Wei X. L.，**Wei J. C. 2005**. Two new species of *Hypogymnia*（Lecanorales，Ascomycota）with pruinose lobe tips from China. Mycotaxon，94（1）：155-158.（SCI）

66. **魏江春**. **2005**. 沙漠生物地毯工程——干旱沙漠治理的新途径. 干旱区研究，22（3）：287-288. [**Wei J. C. 2005**. Desert biological carpet engineering – a new way to control the arid desert. Arid Zone Research，22（3）：287-288].（in Chinese）

67. 张恩然，黄满荣，**魏江春**，赵遵田. **2005**. 中国珊瑚枝属地衣之新记录和稀有种. 菌物学报，24（3）：356-359.

68. Zhang T.，**Wei J. C. 2006**. The lichens of Mts. Fanjingshan in Guizhou province. Journal of Fungal Research，4（1）：1-13.

69. 魏鑫丽，印象初，郭英兰，沈南英，**魏江春**. **2006**. 冬虫夏草及其相关真菌的分子系统学分析. 菌物学报，25（2）：192-202.

70. Zhou Q. M.，**Wei J. C**. Ahti T.，Stenroos S. **2006**. The systematic position of *Gymnoderma* and *Cetradonia* based on SSU nrDNA sequences. J. Hattori Bot. Lab.，No. 100：871-880.（SCI）

71. Zhou Q. M.，Guo S. Y.，Huang M. R.，**Wei J. C. 2006**. A study of the genetic variability of *Rhizoplaca* using DNA sequences and secondary metabolic substances. Mycologia，98（1）：57-67.（SCI）

72. Zhou Q. M.，**Wei J. C. 2006**. A new genus and species *Rhizoplacopsis weichingii* in a new family Rhizoplacopsidaceae. Mycosystema，25（3）：376-385.

73. Manrong Huang，**Jiangchun Wei**. **2006**. Overlooked taxa of *Stereocaulon*（Stereocaulaceae，Lecanorales） in China. Nova Hedwigia，82（3-4）：437-445.（SCI）

74. 项小燕，顾琪，郑维发，张小平，**魏江春**. **2006**. 野生桦褐孔菌及其深层发酵产物的弱极性成分研究. 中草药，37（5）：670-672.（Chinese Traditional and Herbal Drugs）

75. Zhou Q. M.，**Wei J. C. 2007**. A new order Umbilicariales J. C. Wei & Q. M. Zhou（Ascomycota）. Mycosystema，26（1）：40-45.

76. 赵艳霞，郑维发，**魏江春. 2007**. 脱落酸对盐藻SZ205药理活性成分积累的影响. 解放军药学学报，23（1）：13-16.

77. Wei X. L. & **Wei J. C. 2007**. Taxonomic Study on Lichen Genus *Hypogymnia* in China（1）. Lichenology，6（2）：160.

78. Zheng Wei-Fa，Gu Qi，Chen Cai-Fa，Yang Ahi-Zhao，**Wei Jiang-Chun**，Chu Cheng-Cai. **2007**. Aminophenols and mold-water-extracts affect the accumulation of flavonoids and their antioxidant activity in cultured mycelia of Inonotus obliquus. Mycosystema，26（3）：414-426.

79. 杜彩华，郑维发，赵艳霞，**魏江春**，储成才. **2007**. 短小芽孢杆菌对盐藻SZ205的生物量和β2胡萝卜素产量的影响. 植物研究，27（4）：469-472（下接477）.（BULLETIN OF BOTAN ICAL RESEARCH）

80. Miao X. L.，Jia Z. F.，Meng Q. F.，**Wei J. C. 2007**. Some species of Graphidaceae（Ostropales，Ascomycota） rare and new to China. Mycosystema，26（4）：493-506.

81. **魏江春. 2008**. 新种发现的科学与实践意义（刊首）. 大自然（China Nature），141：1（2008：3：1）.

82. Meng Q. F. and **Wei J. C. 2008**. A lichen genus *Diorygma*（Graphidaceae，Ascomycota） in China. Mycosystema，27（4）：525-531

83. Jia Z. F.，**Wei J. C. 2008**. *Graphis fujianensis*，a new species of Graphidaceae from China. Mycotaxon，104：107-109.（SCI）

84. Wang H. Y.，Chen J. B.，**Wei J. C. 2008**. A new species of *Melanelixia*（Parmeliaceae） from China. Mycotaxon，104：185-188.（SCI）

85. **Wei Jiang-Chun. 2008**. The present status and prospect of systematic biology of Fungi and their allies. Proceeding of Japan Pan Asia Pacific Mycology Forum Symposium，2008：1-13.

86. 付少彬，**魏江春. 2008**. 真菌共祖起源的分子证据. 菌物学报，27(2)：217-223. [Fu S. B.，**Wei J. C. 2008**. Molecular evidence for common descent of true fungi. Mycosystema，27（2）：217-223.]

87. Su M. and **Wei J. C. 2008**. Research on liquid culture of mycobiont and photobiont isolated from *Cladonia pyxidata*. Journal of Fungal Research，6（1）：57-62.

88. Yang Jun，**Jiang-Chun Wei. 2008**. *Endocarpon crystallinum*，the new lichen species from semiarid desert in China. Mycotaxon，106：445-448.（SCI）

89. **魏江春. 2008**. 真菌学科的布局与发展. 中国科学院微生物研究所（编辑）. 忆五十载奋斗历程，谱新世纪壮丽篇章（1958-2008）：89-92.

90. Wang H. Y.，Chen J. B. & **Wei J. C. 2009**. A phylogenetic analysis of *Melanelia tomini* and four new records of brown parmelioid lichens from China. Mycotaxon，107：163-173.（SCI）

91. Gang Ding，Yan Li，Shaobin Fu，Shuchun Liu，**Jiangchun Wei**，and Yongsheng Che. **2009**. Ambuic acid and torreyanic acid derivatives from the endolichenic fungus *Pestalotiopsis* sp. J. Nat. Prod.，72（1），182-186.（SCI）

92. Huajie Liu & **Jiangchun Wei. 2009**. A brief overview of and key to species of *Collema* from China. Mycotaxon，108：9-29.（SCI）

93. Davydov E. A.，**J. C. Wei. 2009**. *Boreoplaca ultrafrigida*（Umbilicariales），the correct name for Rhizoplacopsis weichingii. Mycotaxon，108：301-305.（SCI）

94. Jia Ze-Feng & **Jiang-Chun Wei. 2009**. A new isidiate species of *Graphis*（lichenized *Ascomycotina*） from China. Mycotaxon，110：27-30.（SCI）

95. Jia Z F & **Wei J C. 2009**. A new species *Thalloloma microsporum*（Graphidaceae，Ostropales，Ascomycota）. Mycotaxon，107：197-199.（SCI）

96. Yang Jun & **Jiang-Chun Wei. 2009**. A new subspecies of *Gyalidea asteriscus* from China. Mycotaxon，109：373-377.（SCI）

97. 曹叔楠，**魏江春. 2009**. 荒漠地衣糙聚盘衣共生菌耐旱生物学研究及液体优化培养. 菌物学报，28（6）：790-796.

98. 韩乐琳，**魏江春. 2009**. 南极地衣提取物抗氧化能力的初步研究. 菌物学报，28（6）：846-849.

99. Wang H. Y.，Guo S. Y.，Huang M. R.，Lumbsch H. T.，**Wei J. C. 2010**. Ascomycota has faster evolutionary rate and higher species diversity than Basidiomycota（Fungi）. Sci China Life Sci，Volume 53，Issue 10：1163-1169. [王海英，郭守玉，黄满荣，Lumbsch H. T.，**魏江春**. 2010. 子囊菌较担子菌具有更快的进化速率和更高的物种多样性. 中国科学，40（8）：731-737.]

100. **魏江春. 2010**. 菌物生物多样性与人类可持续发展. 中国科学院院刊，25（6）：645-650.

101. 郭英兰，肖培根，**魏江春. 2010**. 论冬虫夏草生物学与可持续利用. 中国现代中药，12（11）：3-8.

102. Xahidin Hurnisa，Abdulla Abbas & **Jiang-Chun Wei**，**2010**. *Caloplaca tianshanensis*（lichen-forming Ascomycota），a new

species of subgenus *Pyrenodesmia* from China. Mycotaxon，114：1-6.

103. **魏江春. 2011**. 《菌物学报》三十年回眸与展望. 菌物学报，30（1）：1-4.

104. Zhang Tao & **WEI JiangChun. 2011**. Survival analyses of symbionts isolated from *Endocarpon pusillum* Hedwig to desiccation and starvation stress. SCIENCE CHINA Life Sciences，54（5）：480-489.

105. Yan-Yan WANG，Tao ZHANG，Qi-Ming ZHOU and **Jiang-Chun WEI. 2011**. Construction and characterization of a full-length cDNA library from mycobiont of *Endocarpon pusillum*（lichen-forming Ascomycota）. World Journal of Microbiology and Biotechnology，27（12）：2879-2884.

106. Weifa Zheng，Yubing Liu，Shenyuan Pan，Weihua Yuan，Yucheng Dai，**Jiangchun Wei. 2011**. Involvements of S-nitrosylation and denitrosylation in the production of polyphenols by *Inonotus obliquus*. Applied Microbiology and Biotechnology. Volume 90，Issue 5：1763-1772.

107. Xin_Li Wei，**Jiang-Chun Wei. 2012**. A study of the pruinose species of *Hypogymnia*（Parmeliaceae，AQscomycota）from China. The Lichenologist，44（6）：783-793.

108. 刘萌，**魏江春. 2013**. 腾格里沙漠沙坡头地区地衣物种多样性研究. 菌物学报，32(1):42-50. [LIU Meng，**WEI Jiang-Chun. 2013**. Lichen diversity in Shapotou region of Tengger Desert，China. **Mycosystema，32（1）**：42-50]

109. **Wei Jiang-Chun. 2013**. On biodiversity of fungi and its systematic biology with three storage and retrieval systems. Mycosystema，32（3）：316-320.

110. Ding LiPing，Zhou QiMing，**WEI JiangChun. 2013**. Estimation of *Endocarpon pusillum* Hedwig carbon budget in the Tengger Desert based on its photosynthetic rate. SCIENCE CHINA Life Sciences，56（9）：848-855.

111. Shunan Cao，Xinli Wei，Qiming Zhou，**Jiangchun Wei. 2013**. *Phyllobaeis crustacea* sp. nov. from China. Mycotaxon，126：31-36.

112. **魏江春. 2013**. 荒漠地衣结皮固沙研究进展//李文华（主编）. 中国当代生态需研究，生态系统恢复卷，第七节. 北京：科学出版社：337-346.

113. Yan-Yan Wang，Bin Liu，Xin-Yu Zhang，Qi-Ming Zhou，Tao Zhang，Hui Li，Yu-Fei Yu，Xiao-Ling Zhang，Xi-Yan Hao，Meng Wang，Lei Wang，**Jiang-Chun Wei. 2014**. Genome characteristics reveal the impact of lichenization on lichen-forming fungus *Endocarpon pusillum* Hedwig（Verrucariales，Ascomycota）. BMC genomics，15：34.

114. 王瑞芳，王立松，**魏江春. 2014**. 同岛衣属一新种——粉头同岛衣（梅衣科，子囊菌门）. 菌物学报. 33（1）：19-22. [*Allocetraria capitata* sp. nov.（Parmeliaceae，Ascomycota）from China]

115. YANG Jun，**WEI Jiang-Chun. 2014**. Desert lichens in Shapotou region of Tengger Desert and bio-carpet engineering. Mycosystema，33（5）：1025-1035.

116. Yan-Yan Wang，Xin-Yu Zhang，Qi-Ming Zhou，Xiao-Ling Zhang，**Jiang-Chun Wei. 2015**. Comparative transcriptome analysis of the lichen-forming fungus *Endocarpon pusillum* elucidates its drought adaptation mechanisms. SCIENCE CHINA Life Sciences，58（1）：89-100.

117. **魏江春**，魏鑫丽，郑维发，郭威，柳润东. **2016**. 现代工业化培植的冬虫夏草物种鉴定与成分检测. 菌物学报，35（4）.

专著：

1. **魏江春**主编. **1982**. 中国药用地衣. 北京：科学出版社：1-65，图版1-15.

2. **魏江春**，姜玉梅. **1986**. 西藏地衣. 北京：科学出版社：1-130，图版1-24.

3. **Wei Jiang-Chun**. **1991**. An enumeration of lichens in China international academic publishers. Beijing：Science Press：1-278. [中国地衣综览（修订版） 见中国生物物种名录-科学出版社. 2016]

4. **Wei Jiang-Chun**，Jiang Yu-Mei. **1993**. The Umbilicariaceae in Asia（Ascomycota）. International Academic Publishers：1-217.

5. **ВЭЙ Цзян-Чунь. 1995.** Анализ систематики и географии лишайников Сем. Umlilicariaceae Восточной Азии [стр.1-140（автореферат1-33）] -Защита состоится 24 мая 1995 г. в 15.00 час. на Заседании Специадизированного Совета Д CO_2. 46.01 по защите диссертаций на соискание ученой степени доктора биологических наук при Ботаническом институте им. В. Л. Комарова РАН по адресу：197376 С. Петербург，ул. Проф. Попова 2，зал Ученого Совета.

6. 贾泽峰，**魏江春**. **2016**. 中国地衣志 文字衣科（Ⅰ）. 北京：科学出版社（印刷中）

7. **魏江春**. **2016**. 中国地衣志 石耳科. 北京：科学出版社（待出版）

8. **魏江春**. **2016**. 中国地衣志 石蕊科. 北京：科学出版社（待出版）

9. **魏江春**. **2016**. 中国地衣属志. 北京：科学出版社（待出版）

编著：

10. **魏江春**. **1980**. 地衣名词及名称. 北京：科学出版社：1-73.

11. 郑儒永，**魏江春**，胡鸿钧，余永年，等. **1990**. 孢子植物名词及名称. 北京：科学出版社：1-961.

12. **魏江春**主编. **2005**. 中国经济真菌企事业大全. 北京：中国农业大学出版社：1-275.

培养的研究生名录

中国科学院:

硕士生：卢效德，马承华，王娇红，张永利，柳润东

硕士生及博士生：高向群，郭守玉，周启明，黄满荣，杨军，李慧，蒋淑华

博士生：刘华杰，陈林海，郭林，魏鑫丽，王海英，张涛，丁利平，曹叔楠，王延延

博士后：黄亦存，牛永春，张克勤

河北大学:

硕士生：魏鑫丽，王海英，李红梅，张 涛，石 妍，苏 敏，付少彬，韩乐琳，曹叔楠

山东农业大学:

硕士生：高 斌，李 静，刘萌，王延延，孟庆峰，苗晓琳，孙俊杰，王瑞芳，张恩然

博士生：贾泽峰

代培研究生:

北京首都师范大学：

硕士生：李学东

山东师范大学：

硕士生：刘华杰，贾泽峰，任 强，赵俊祯

齐齐哈尔大学：

硕士生：刘文婧，杨博宇

吉林农业大学：

硕士生：李丽，张杰

哈尔滨工业大学：

硕士生：郑方圆

短期进修研究生:

新疆大学：

博士生：吾尔妮莎（Xahidin，Hurnisa）

后　　记

在此谨向各位领导和国内外朋友以及家乡代表对我生日的贺词与祝福表示衷心感谢！对于曾专程来自京内外出席我生日聚会的朋友们，特别是专程来自韩国顺天国立大学（Sunchon National University）的许宰铣教授（Prof.Jae-Seoun Hur）深表谢意！

在本选集编辑过程中，得到室内外，所内外同事和朋友的支持，尤其是在资料收集和整理，部分纸质资料的电子化处理，以及与出版社的联络等自始至终得到郭威的帮助，对此表示感谢！最后对科学出版社王静及其编辑团队编辑人员的辛勤劳作表示感谢！

希望本书的问世，能为本学科的后来者在我国地衣型真菌学科的发展中，结合国家经济建设需要，面向本学科的世界前沿做出更大成绩中有所借鉴。

魏江春
2016年3月1日
于北京中关村